IBC®

INTERNATIONAL BUILDING CODE®

A Member of the International Code Family®

2021

ICC
INTERNATIONAL CODE COUNCIL®

2021 International Building Code®

Date of First Publication: October 23, 2020

First Printing: October 2020

ISBN: 978-1-60983-955-0 (soft-cover edition)
ISBN: 978-1-60983-956-7 (loose-leaf edition)

T025567

PRINTED IN THE USA

PREFACE

Introduction

The *International Building Code®* (IBC®) establishes minimum requirements for building systems using prescriptive and performance-related provisions. It is founded on broad-based principles that make possible the use of new materials and new building designs. This 2021 edition is fully compatible with all of the International Codes® (I-Codes®) published by the International Code Council® (ICC®), including the *International Energy Conservation Code®* (IECC®), *International Existing Building Code®* (IEBC®), *International Fire Code®* (IFC®), *International Fuel Gas Code®* (IFGC®), *International Green Construction Code®* (IgCC®), *International Mechanical Code®* (IMC®), *International Plumbing Code®* (IPC®), *International Private Sewage Disposal Code®* (IPSDC®), *International Property Maintenance Code®* (IPMC®), *International Residential Code®* (IRC®), *International Swimming Pool and Spa Code®* (ISPSC®), *International Wildland-Urban Interface Code®* (IWUIC®), *International Zoning Code®* (IZC®) and *International Code Council Performance Code®* (ICCPC®).

In addition to the codes themselves, the code development process brings together building professionals on a regular basis. It provides an international forum for discussion and deliberation about building design, construction methods, safety, performance requirements, technological advances and innovative products.

The I-Codes, including the IBC, are used in a variety of ways in both the public and private sectors. Most industry professionals are familiar with the I-Codes as the basis of laws and regulations in communities across the US and in other countries. However, the impact of the codes extends well beyond the regulatory arena, as they are used in a variety of nonregulatory settings, including:

- Voluntary compliance programs such as those promoting sustainability, energy efficiency and disaster resistance.
- The insurance industry, to estimate and manage risk, and as a tool in underwriting and rate decisions.
- Certification and credentialing of individuals involved in the fields of building design, construction and safety.
- Certification of building and construction-related products.
- US federal agencies, to guide construction in an array of government-owned properties.
- Facilities management.
- "Best practices" benchmarks for designers and builders, including those who are engaged in projects in jurisdictions that do not have a formal regulatory system or a governmental enforcement mechanism.
- College, university and professional school textbooks and curricula.
- Reference works related to building design and construction.

In addition to the codes themselves, the code development process brings together building professionals on a regular basis. It provides an international forum for discussion and deliberation about building design, construction methods, safety, performance requirements, technological advances and innovative products.

Development

This 2021 edition presents the code as originally issued, with changes reflected in the 2003 through 2018 editions and further changes approved by the ICC Code Development Process through 2019. A new edition such as this is promulgated every 3 years.

This code is intended to establish provisions that adequately protect public health, safety and welfare; that do not unnecessarily increase construction costs; that do not restrict the use of new materials, products or methods of construction; and that do not give preferential treatment to particular types or classes of materials, products or methods of construction.

Maintenance

The IBC is kept up to date through the review of proposed changes submitted by code enforcement officials, industry representatives, design professionals and other interested parties. Proposed changes are carefully considered through an open code development process in which all interested and affected parties may participate.

The ICC Code Development Process reflects principles of openness, transparency, balance, due process and consensus, the principles embodied in OMB Circular A-119, which governs the federal government's use of private-sector standards. The ICC process is open to anyone; there is no cost to participate, and people can participate without travel cost through the ICC's cloud-based app, cdpAccess®. A broad cross section of interests are represented in the ICC Code Development Process. The codes, which are updated regularly, include safeguards that allow for emergency action when required for health and safety reasons.

In order to ensure that organizations with a direct and material interest in the codes have a voice in the process, the ICC has developed partnerships with key industry segments that support the ICC's important public safety mission. Some code development committee members were nominated by the following industry partners and approved by the ICC Board:

- American Institute of Architects (AIA)
- National Association of Home Builders (NAHB)
- National Association of State Fire Marshals (NASFM)

The code development committees evaluate and make recommendations regarding proposed changes to the codes. Their recommendations are then subject to public comment and council-wide votes. The ICC's governmental members—public safety officials who have no financial or business interest in the outcome—cast the final votes on proposed changes.

The contents of this work are subject to change through the code development cycles and by any governmental entity that enacts the code into law. For more information regarding the code development process, contact the Codes and Standards Development Department of the ICC.

While the I-Code development procedure is thorough and comprehensive, the ICC, its members and those participating in the development of the codes disclaim any liability resulting from the publication or use of the I-Codes, or from compliance or noncompliance with their provisions. The ICC does not have the power or authority to police or enforce compliance with the contents of this code.

IBC Primary Code Development Committees

In each code development cycle, code change proposals to this code are considered at the Committee Action Hearings by 11 different code development committees. Four of these committees have primary responsibility for designated chapters and appendices as shown in the following table:

IBC CODE COMMITTEES

IBC—Egress Code Development Committee [BE]:	Chapters 10, 11, Appendix E
IBC—Fire Safety Code Development Committee [BF]:	Chapters 7, 8, 9, 14, 26
IBC—General Code Development Committee [BG]:	Chapters 2, 3, 4, 5, 6, 12, 27, 28, 29, 30, 31, 32, 33, Appendices A, B, C, D, K, N, O
IBC—Structural Code Development Committee [BS]:	Chapters 15, 16, 17, 18, 19, 20, 21, 22, 23, 24, 25, Appendices F, G, H, I, J, L, M

Code Development Committee Responsibilities (Letter Designations in Front of Section Numbers)

As mentioned in the preceding material, in each code development cycle, code change proposals to this code are considered at the Committee Action Hearings by 11 different code development committees.

Code change proposals to sections of the code that are preceded by a bracketed letter designation, such as [A], will be considered by a committee other than the building code committee listed for the chapter or appendix on the preceding page. For example, proposed code changes to Section [F] 307.1.1 will be considered by the International Fire Code Development Committee during the Committee Action Hearing in the 2021 (Group A) code development cycle.

The bracketed letter designations for committees responsible for portions of this code are as follows:

[A] = Administrative Code Development Committee

[BE] = IBC—Egress Code Development Committee

[BF] = IBC—Fire Safety Code Development Committee

[BG] = IBC—General Code Development Committee

[BS] = IBC—Structural Code Development Committee

[E] = International Commercial Energy Conservation Code Development Committee or International Residential Energy Conservation Code Development Committee

[EB] = International Existing Building Code Development Committee

[F] = International Fire Code Development Committee

[FG] = International Fuel Gas Code Development Committee

[M] = International Mechanical Code Development Committee

[P] = International Plumbing Code Development Committee

For the development of the 2024 edition of the I-Codes, there will be two groups of code development committees and they will meet in separate years, as shown in the following Code Development Hearings Table.

Code change proposals submitted for code sections that have a letter designation in front of them will be heard by the respective committee responsible for such code sections. Because different committees hold Committee Action Hearings in different years, proposals for the IBC will be heard by committees in both the 2021 (Group A) and the 2022 (Group B) code development cycles.

For instance, every section of Chapter 16 is the responsibility of the IBC—Structural Code Development Committee. As noted in the preceding table, that committee will hold its Committee Action Hearings in 2022 to consider code change proposals for the chapters for which it is responsible. Therefore any proposals received for Chapter 16 of this code will be assigned to the IBC—Structural Code Development Committee and will be considered in 2022, during the Group B code change cycle.

It is very important that anyone submitting code change proposals understands which code development committee is responsible for the section of the code that is the subject of the code change proposal. For further information on the Code Development Committee responsibilities, please visit the ICC website at www.iccsafe.org/current-code-development-cycle.

CODE DEVELOPMENT HEARINGS

Group A Codes (Heard in 2021, Code Change Proposals Deadline: January 11, 2021)	Group B Codes (Heard in 2022, Code Change Proposals Deadline: January 10, 2022)
International Building Code – Egress (Chapters 10, 11, Appendix E) – Fire Safety (Chapters 7, 8, 9, 14, 26) – General (Chapters 2–6, 12, 27–33, Appendices A, B, C, D, K, N)	Administrative Provisions (Chapter 1 of all codes except IECC, IRC and IgCC; IBC Appendix O; the appendices titled "Board of Appeals" for all codes except IECC, IRC, IgCC, ICCPC and IZC; administrative updates to currently referenced standards; and designated definitions)
International Fire Code	**International Building Code** – Structural (Chapters 15–25, Appendices F, G, H, I, J, L, M)
International Fuel Gas Code	**International Existing Building Code**
International Mechanical Code	**International Energy Conservation Code—Commercial**
International Plumbing Code	**International Energy Conservation Code—Residential** – IECC—Residential – IRC—Energy (Chapter 11)
International Property Maintenance Code	**International Green Construction Code (Chapter 1)**
International Private Sewage Disposal Code	**International Residential Code** – IRC—Building (Chapters 1–10, Appendices AE, AF, AH, AJ, AK, AL, AM, AO, AQ, AR, AS, AT, AU, AV, AW)
International Residential Code – IRC—Mechanical (Chapters 12–23) – IRC—Plumbing (Chapters 25–33, Appendices AG, AI , AN, AP)	
International Swimming Pool and Spa Code	
International Wildland-Urban Interface Code	
International Zoning Code	

Note: Proposed changes to the ICCPC will be heard by the code development committee noted in brackets [ ] in the text of the ICCPC.

Marginal Markings

Solid vertical lines in the margins within the body of the code indicate a technical change from the requirements of the 2018 edition. Deletion indicators in the form of an arrow (➡) are provided in the margin where an entire section, paragraph, exception or table has been deleted or an item in a list of items or a row of a table has been deleted.

A single asterisk [*] placed in the margin indicates that text or a table has been relocated within the code. A double asterisk [**] placed in the margin indicates that the text or table immediately following it has been relocated there from elsewhere in the code. The following table indicates such relocations in the 2021 edition of the IBC.

RELOCATIONS

2021 LOCATION	2018 LOCATION
508.5–508.5.11	419.1–419.9
904.12	904.14
904.13	904.12
904.14	904.13
1010.2	1010.1.9
1010.2.1	1010.1.9.6
1010.2.2	1010.1.9.1
1010.2.3	1010.1.9.2
1010.2.4	1010.1.9.4
1010.2.5	1010.1.9.5
1010.2.6	1010.1.9.6.1
1010.2.7	1010.1.9.12
1010.2.8	1010.1.4.4
1010.2.9	1010.1.10
1010.2.9.3	1010.1.10.1
1010.2.9.4	1010.1.10.2
1010.2.10	1010.1.9.3
1010.2.11	1010.1.9.10
1010.2.12	1010.1.9.9
1010.2.13	1010.1.9.8
1010.2.13.1	1010.1.9.8.1
1010.2.14	1010.1.9.7
1010.2.15	1010.1.9.11
1010.3	1010.1.4
1010.3.1	1010.1.4.1
Table 1010.3.1(1)	Table 1010.1.4.1(1)
Table 1010.3.1(2)	Table 1010.1.4.1(2)
1010.3.1.1	1010.1.4.1.1
1010.3.1.2	1010.1.4.1.2
1010.3.2	1010.1.4.2
1010.3.3	1010.1.4.3
1029.1	1028.4
1029.2	1028.4.2
1029.3	1028.4.2

(continued)

RELOCATIONS—continued

2021 LOCATION	2018 LOCATION
1107.3	1109.14
1110.14	1109.12.1
1605.2	605.3.2
1607.14.2.2	1607.13.3
1607.14.4.3	1607.13.5.2.1
2304.12.2.6	2304.12.3
2304.12.2.6.1	2304.12.3.1
2304.12.2.7	2304.12.4
2304.12.2.8	2304.12.5
3301.2.1	1511.2

Coordination of the International Codes

The coordination of technical provisions is one of the strengths of the ICC family of model codes. The codes can be used as a complete set of complementary documents, which will provide users with full integration and coordination of technical provisions. Individual codes can also be used in subsets or as stand-alone documents. To make sure that each individual code is as complete as possible, some technical provisions that are relevant to more than one subject area are duplicated in some of the model codes. This allows users maximum flexibility in their application of the I-Codes.

Italicized Terms

Terms italicized in code text, other than document titles, are defined in Chapter 2. The terms selected to be italicized have definitions that the user should read carefully to better understand the code. Where italicized, the Chapter 2 definition applies. If not italicized, common-use definitions apply.

Note: In Sections 1903 through 1905, italics indicate provisions that differ from ACI 318.

Adoption

The International Code Council maintains a copyright in all of its codes and standards. Maintaining copyright allows the ICC to fund its mission through sales of books, in both print and electronic formats. The ICC welcomes adoption of its codes by jurisdictions that recognize and acknowledge the ICC's copyright in the code, and further acknowledge the substantial shared value of the public/private partnership for code development between jurisdictions and the ICC.

The ICC also recognizes the need for jurisdictions to make laws available to the public. All I-Codes and I-Standards, along with the laws of many jurisdictions, are available for free in a nondownloadable form on the ICC's website. Jurisdictions should contact the ICC at adoptions@iccsafe.org to learn how to adopt and distribute laws based on the IBC in a manner that provides necessary access, while maintaining the ICC's copyright.

To facilitate adoption, several sections of this code contain blanks for fill-in information that needs to be supplied by the adopting jurisdiction as part of the adoption legislation. For this code, please see:

Section 101.1. Insert: [NAME OF JURISDICTION]

Section 103.1. Insert: [NAME OF DEPARTMENT]

Section 1612.3. Insert: [NAME OF JURISDICTION]

Section 1612.3. Insert: [DATE OF ISSUANCE]

Effective Use of the International Building Code

The IBC is a model code that provides minimum requirements to safeguard the public health, safety and general welfare of the occupants of new and existing buildings and structures. The IBC is fully compatible with the ICC family of codes, including: the IECC, IEBC, IFC, IFGC, IgCC, IMC, IPC, IPSDC, IPMC, IRC, ISPSC, IWUIC, IZC and ICCPC.

The IBC addresses structural strength, means of egress, sanitation, adequate lighting and ventilation, accessibility, energy conservation and life safety in regard to new and existing buildings, facilities and systems. The codes are promulgated on a 3-year cycle to allow for new construction methods and technologies to be incorporated into the codes. Alternative materials, designs and methods not specifically addressed in the code can be approved by the building official where the proposed materials, designs or methods comply with the intent of the provisions of the code (see Section 104.11).

The IBC applies to all occupancies, including one- and two-family dwellings and townhouses that are not within the scope of the IRC. The IRC is referenced for coverage of detached one- and two-family dwellings and townhouses as defined in the exception to Section 101.2 and the definition for "Townhouse" in Chapter 2. The IRC can also be used for the construction of live/work units (as defined in Section 508.5) and small bed and breakfast-style hotels where there are five or fewer guest rooms and the hotel is owner occupied. The IBC applies to all types of buildings and structures unless exempted. Work exempted from permits is listed in Section 105.2.

ARRANGEMENT AND FORMAT OF THE 2021 IBC

Before applying the requirements of the IBC, it is beneficial to understand its arrangement and format. The IBC, like other codes published by ICC, is arranged and organized to follow sequential steps that generally occur during a plan review or inspection.

The following table shows how the IBC is divided. The three tables following that show IBC requirements that are correlated with other I-Codes. Lastly, the ensuing chapter-by-chapter synopsis details the scope and intent of the provisions of the IBC.

CHAPTER TOPICS

Chapters	Subjects
1–2	Administration and definitions
3	Use and occupancy classifications
4, 31	Special requirements for specific occupancies or elements
5–6	Height and area limitations based on type of construction
7–9	Fire resistance and protection requirements
10	Requirements for evacuation
11	Specific requirements to allow use and access to a building for persons with disabilities
12–13, 27–30	Building systems, such as lighting, HVAC, lumbing fixtures, elevators
14–26	Structural components—performance and stability
32	Encroachment outside of property lines
33	Safeguards during construction
35	Referenced standards
Appendices A–O	Appendices

IFC Correlated Topics

The IBC requirements for hazardous materials, fire-resistance-rated construction, interior finish, fire protection systems, means of egress, emergency and standby power, and temporary structures are directly correlated with the requirements of the IFC. The following table shows chapters/sections of the IBC that are correlated with the IFC:

IBC/IFC CORRELATED TOPICS

IBC Chapter/Section	IFC Chapter/Section	Subject
Sections 307, 414, 415	Chapters 50–67	Hazardous materials and Group H requirements
Chapter 7	Chapter 7	Fire-resistance-rated construction (fire and smoke protection features in the IFC)
Chapter 8	Chapter 8	Interior finish, decorative materials and furnishings
Chapter 9	Chapter 9	Fire protection systems
Chapter 10	Chapter 10	Means of egress
Chapter 27	Section 604	Standby and emergency power
Section 3103	Chapter 31	Temporary structures

IMC Correlated Topics

The IBC requirements for smoke control systems, and smoke and fire dampers are directly correlated to the requirements of the IMC. IBC Chapter 28 is a reference to the IMC and the IFGC for chimneys, fireplaces and barbecues, and all aspects of mechanical systems. The following table shows chapters/sections of the IBC that are correlated with the IMC:

IBC/IMC CORRELATED TOPICS

IBC Chapter/Section	IMC Chapter/Section	Subject
Section 717	Section 607	Smoke and fire dampers
Section 909	Section 513	Smoke control

IPC Correlated Topics

The IBC requirements for plumbing fixtures and toilet rooms are directly correlated to the requirements of the IPC. The following table shows chapters/sections of the IBC that are correlated with the IPC:

IBC/IPC CORRELATED TOPICS

IBC Chapter/Section	IPC Chapter/Section	Subject
Chapter 29	Chapters 3 & 4	Plumbing fixtures and facilities

Chapter 1 Scope and Administration

Chapter 1 establishes the limits of applicability of the code and describes how the code is to be applied and enforced. Chapter 1 is in two parts, Part 1— Scope and Application (Sections 101-102) and Part 2—Administration and Enforcement (Sections 103-116). Section 101 identifies which buildings and structures come under its purview and references other I-Codes as applicable. Standards and codes are scoped to the extent referenced (see Section 102.4).

The building code is intended to be adopted as a legally enforceable document and it cannot be effective without adequate provisions for its administration and enforcement. The provisions of Chapter 1 establish the authority and duties of the building official appointed by the authority having jurisdiction and also establish the rights and privileges of the design professional, contractor and property owner.

Chapter 2 Definitions

All terms that are defined in the code are listed alphabetically in Chapter 2. While a defined term may be used in one chapter or another, the meaning provided in Chapter 2 is applicable throughout the code.

Where understanding a term's definition is especially key to or necessary for understanding a particular code provision, the term is shown in italics. This is true only for those terms that have a meaning that is unique to the code. In other words, the generally understood meaning of a term or phrase might not be sufficient or consistent with the meaning prescribed by the code; therefore, it is essential that the code-defined meaning be known.

Guidance regarding tense, gender and plurality of defined terms as well as guidance regarding terms not defined in this code is provided.

Chapter 3 Occupancy Classification and Use

Chapter 3 provides for the classification of buildings, structures and parts thereof based on the purpose or purposes for which they are used. Section 302 identifies the groups into which all buildings, structures and parts thereof must be classified. Sections 303 through 312 identify the occupancy characteristics of each group classification. In some sections, specific group classifications having requirements in common are collectively organized such that one term applies to all. For example, Groups A-1, A-2, A-3, A-4 and A-5 are individual groups for assembly-type buildings. The general term "Group A," however, includes each of these individual groups. Other groups include Business (B), Educational (E), Factory (F-1, F-2), High Hazard (H-1, H-2, H-3, H-4, H-5), Institutional (I-1, I-2, I-3, I-4), Mercantile (M), Residential (R-1, R-2, R-3, R-4), Storage (S-1, S-2) and Utility (U). In some occupancies, the smaller number means a higher hazard, but that is not always the case.

Defining the use of the buildings is very important as it sets the tone for the remaining chapters of the code. Occupancy works with the height, area and construction type requirements in Chapters 5 and 6, as well as the special provisions in Chapter 4, to determine "equivalent risk," or providing a reasonable level of protection or life safety for building occupants. The determination of equivalent risk involves three interdependent considerations: (1) the level of fire hazard associated with the specific occupancy of the facility; (2) the reduction of fire hazard by limiting the floor area and the height of the building based on the fuel load (combustible contents and burnable building components); and (3) the level of overall fire resistance provided by the type of construction used for the building. The greater the potential fire hazards indicated as a function of the group, the lesser the height and area allowances for a particular construction type.

Occupancy classification also plays a key part in organizing and prescribing the appropriate protection measures. As such, threshold requirements for fire protection and means of egress systems are based on occupancy classification (see Chapters 9 and 10). Other sections of the code also contain requirements respective to the classification of building groups. For example, Section 706 specifies requirements for fire wall fire-resistance ratings that are tied to the occupancy classification of a building and Section 803.11 contains interior finish requirements that are dependent upon the occupancy classification. The use of the space, rather than the occupancy of the building, is utilized for determining occupant loading (Section 1004) and live loading (Section 1607).

Over the useful life of a building, the activities in the building will evolve and change. Where the provisions of the code address uses differently, moving from one activity to another or from one level of activity to another is, by definition, a change of occupancy. The new occupancy must be in compliance with the applicable provisions.

Chapter 4 Special Detailed Requirements Based on Occupancy and Use

Chapter 4 contains the requirements for protecting special uses and occupancies, which are supplemental to the remainder of the code. Chapter 4 contains provisions that may alter requirements found elsewhere in the code; however, the general requirements of the code still apply unless modified within the chapter. For example, the height and area limitations established in Chapter 5 apply to all special occupancies unless Chapter 4 contains height and area limitations. In this case, the limitations in Chapter 4 supersede those in other sections. An example of this is the height and area limitations for open parking garages given in Section 406.5.4, which supersede the limitations given in Sections 504 and 506.

In some instances, it may not be necessary to apply the provisions of Chapter 4. For example, if a covered mall building complies with the provisions of the code for Group M, Section 402 does not apply; however, other sections that address a use, process or operation must be applied to that specific occupancy, such as stages and platforms, special amusement buildings and hazardous materials (Sections 410, 411 and 414).

The chapter includes requirements for buildings and conditions that apply to one or more groups, such as high-rise buildings, underground buildings or atriums. Special uses may also imply specific occupancies and operations, such as for Group H, hazardous materials, application of flammable finishes, drying rooms, organic coatings and combustible storage or hydrogen fuel gas rooms, all of which are coordinated with the IFC. Unique consideration is taken for special use areas, such as covered mall buildings, motor-vehicle-related occupancies, special amusement buildings and aircraft-related occupancies. Special facilities within other occupancies are considered, such as stages and platforms, motion picture projection rooms, children's play structures and storm shelters. Finally, in order that the overall package of protection features can be easily understood, unique considerations for specific occupancies are addressed: Groups I-1, I-2, I-3, R-1, R-2, R-3 and R-4; and ambulatory care facilities and live/work units.

Chapter 5 General Building Heights and Areas

Chapter 5 contains the provisions that regulate the minimum type of construction for area limits and height limits based on the occupancy of the building. Height and area increases (including allowances for basements, mezzanines and equipment platforms) are permitted based on open frontage for fire department access, separation and the type of sprinkler protection provided (Sections 503-506, 510). These thresholds are reduced for buildings over three stories in height in accordance with Sections 506.2.1 and 506.2.2. Provisions include the protection and/or separation of incidental uses (Table 509.1), accessory occupancies (Section 508.2) and mixed uses in the same building (Sections 506.2.2, 508.3, 508.4 and 510). Unlimited area buildings are permitted in certain occupancies when they meet special provisions (Section 507). Live/work units are provided for in Section 508.5.

Tables 504.3, 504.4 and 506.2 are the keystones in setting thresholds for building size based on the building's use and the materials with which it is constructed. If one then looks at Tables 504.3, 504.4 and 506.2, the relationship among group classification, allowable heights and areas and types of construction becomes apparent. Respective to each group classification, the greater the fire-resistance rating of structural elements, as represented by the type of construction, the greater the floor area and height allowances. The greater the potential fire hazards indicated as a function of the group, the lesser the height and area allowances for a particular construction type. Starting in the 2015 edition, the table that once contained both height and area has been separated and these three new tables address the topics individually. In addition, the tables list criteria for buildings with and without automatic sprinkler systems.

Chapter 6 Types of Construction

The interdependence of these fire safety considerations can be seen by first looking at Tables 601 and 705.5, which show the fire-resistance ratings of the principal structural elements comprising a building in relation to the five classifications for types of construction. Type I construction is the classification that generally requires the highest fire-resistance ratings for structural elements, whereas Type V construction, which is designated as a combustible type of construction, generally requires the least amount of fire-resistance-rated structural elements. The greater the potential fire hazards indicated as a function of the group, the lesser the height and area allowances for a particular construction type. Section 603 includes a list of combustible elements that can be part of a noncombustible building (Types I and II construction).

Chapter 7 Fire and Smoke Protection Features

The provisions of Chapter 7 present the fundamental concepts of fire performance that all buildings are expected to achieve in some form. This chapter identifies the acceptable materials, techniques and methods by which proposed construction can be designed and evaluated against to determine a building's ability to limit the impact of fire. The fire-resistance-rated construction requirements within Chapter 7 provide passive resistance to the spread and effects of fire. Types of separations addressed include fire walls, fire barriers, fire partitions, horizontal assemblies, smoke barriers and smoke partitions. A fire produces heat that can weaken structural components and smoke products that cause property damage and place occupants at risk. The requirements of Chapter 7 work in unison with height and area requirements (Chapter 5), active fire detection and suppression systems (Chapter 9) and occupant egress requirements (Chapter 10) to contain a fire should it occur while helping ensure occupants are able to safely exit.

Chapter 8 Interior Finishes

This chapter contains the performance requirements for controlling fire growth within buildings by restricting interior finish and decorative materials. Past fire experience has shown that interior finish and decorative materials are key elements in the development and spread of fire. The provisions of Chapter 8 require materials used as interior finishes and decorations to meet certain flame-spread index or flame-propagation criteria based on the relative fire hazard associated with the occupancy. As smoke is also a hazard associated with fire, this chapter contains limits on the smoke development characteristics of interior finishes. The performance of the material is evaluated based on test standards.

Chapter 9 Fire Protection and Life Safety Systems

Chapter 9 prescribes the minimum requirements for active systems of fire protection equipment to perform the following functions: detect a fire; alert the occupants or fire department of a fire emergency; and control smoke and control or extinguish the fire. Generally, the requirements are based on the occupancy, the height and the area of the building, because these are the factors that most affect fire-fighting capabilities and the relative hazard of a specific building or portion thereof. This chapter parallels and is substantially duplicated in Chapter 9 of the IFC; however, the IFC Chapter 9 also contains periodic testing criteria that are not contained in the IBC. In addition, the special fire protection system requirements based on use and occupancy found in IBC Chapter 4 are duplicated in IFC Chapter 9 as a user convenience.

Chapter 10 Means of Egress

The general criteria set forth in Chapter 10 regulating the design of the means of egress are established as the primary method for protection of people in buildings by allowing timely relocation or evacuation of building occupants. Both prescriptive and performance language is utilized in this chapter to provide for a basic approach in the determination of a safe exiting system for all occupancies. It addresses all portions of the egress system (i.e., exit access, exits and exit discharge) and includes design requirements as well as provisions regulating individual components. The requirements detail the size, arrangement, number and protection of means of egress components. Functional and operational characteristics also are specified for the components that will permit their safe use without special knowledge or effort. The means of egress protection requirements work in coordination with other sections of the code, such as protection of vertical openings (see Chapter 7), interior finish (see Chapter 8), fire suppression and detection systems (see Chapter 9) and numerous others, all having an impact on life safety. Chapter 10 of the IBC is duplicated in Chapter 10 of the IFC; however, the IFC contains one additional section on the means of egress system in existing buildings.

Chapter 11 Accessibility

Chapter 11 contains provisions that set forth requirements for accessibility of buildings and their associated sites and facilities for people with physical disabilities. The fundamental philosophy of the code on the subject of accessibility is that everything is required to be accessible. This is reflected in the basic applicability requirement (see Section 1103.1). The code's scoping requirements then address the conditions under which accessibility is not required in terms of exceptions to this general mandate. While the IBC contains scoping provisions for accessibility (for example, what, where and how many), ICC A117.1, *Accessible and Usable Buildings and Facilities*, is the referenced standard for the technical provisions (in other words, how).

There are many accessibility issues that not only benefit people with disabilities, but also provide a tangible benefit to people without disabilities. This type of requirement can be set forth in the code as generally applicable without necessarily identifying it specifically as an accessibility-related issue. Such a requirement would then be considered as having been "mainstreamed." For example, visible alarms are located in Chapter 9 and accessible means of egress and ramp requirements are addressed in Chapter 10.

Accessibility criteria for existing buildings are addressed in the IEBC.

Appendix E is supplemental information included in the code to address accessibility for items in the 2010 *ADA Standards for Accessible Design* that were not typically enforceable through the standard traditional building code enforcement approach system (for example, beds, room signage). The IRC references Chapter 11 for accessibility provisions; therefore, this chapter may be applicable to housing covered under the IRC.

Chapter 12 Interior Environment

Chapter 12 provides minimum standards for the interior environment of a building. The standards address the minimum sizes of spaces, minimum temperature levels, and minimum light and ventilation levels. The collection of requirements addresses limiting sound transmission through walls, ventilation of attic spaces and under floor spaces (crawl spaces). Finally, the chapter provides minimum standards for toilet and bathroom construction, including privacy shielding and standards for walls, partitions and floors to resist water intrusion and damage.

Chapter 13 Energy Efficiency

The purpose of Chapter 13 is to provide minimum design requirements that will promote efficient utilization of energy in buildings. The requirements are directed toward the design of building envelopes with adequate thermal resistance and low air leakage, and toward the design and selection of mechanical, water heating, electrical and illumination systems that promote effective use of depletable energy resources. For the specifics of these criteria, Chapter 13 requires design and construction in compliance with the IECC.

Chapter 14 Exterior Walls

This chapter addresses requirements for exterior walls of buildings. Minimum standards for wall covering materials, installation of wall coverings and the ability of the wall to provide weather protection are provided. This chapter also requires exterior walls that are close to lot lines, or that are bearing walls for certain types of construction, to comply with the minimum fire-resistance ratings specified in Chapters 6 and 7. The installation of each type of wall covering, be it wood, masonry, vinyl, metal composite material or an exterior insulation and finish system, is critical to its long-term performance in protecting the interior of the building from the elements and the spread of fire. Limitations on the use of combustible materials on exterior building elements such as balconies, eaves, decks and architectural trim are also addressed in this chapter.

Chapter 15 Roof Assemblies and Rooftop Structures

Chapter 15 provides standards for both roof assemblies and structures that sit on top of the roofs of buildings. The criteria address roof construction and covering, including the weather-protective barrier at the roof and, in most circumstances, a fire-resistant barrier. The chapter is prescriptive in nature and is based on decades of experience with various traditional materials, but it also addresses newer products such as photovoltaic shingles. These prescriptive rules are very important for satisfying performance of one type of roof covering or another. Section 1511 addresses rooftop structures, including penthouses, tanks, towers and spires. Rooftop penthouses larger than prescribed in this chapter must be treated as a story under Chapter 5.

Chapter 16 Structural Design

Chapter 16 prescribes minimum structural loading requirements for use in the design and construction of buildings and structural components. It includes minimum design loads, assignment of risk categories and permitted design methodologies. Standards are provided for minimum design loads (live, dead, snow, wind, rain, flood, ice and earthquake as well as the required load combinations). The application of these loads and adherence to the serviceability criteria will enhance the protection of life and property. The chapter references and relies on many nationally recognized design standards. A key standard is the American Society of Civil Engineers' *Minimum Design Loads for Buildings and Other Structures* (ASCE 7). Structural design must address the conditions of the site and location. Therefore, maps are provided of rainfall, seismic, snow and wind criteria in different regions.

Chapter 17 Special Inspections and Tests

Chapter 17 provides a variety of procedures and criteria for testing materials and assemblies, labeling materials and assemblies and special inspection of structural assemblies. This chapter expands on the inspections of Chapter 1 by requiring special inspection where indicated and, in some cases, structural observation. It also spells out additional responsibilities for the owner, contractor, design professionals and special inspectors. Proper assembly of structural components, proper quality of materials used and proper application of materials are essential to ensuring that a building, once constructed, complies with the structural and fire-resistance minimums of the code and the approved design. To determine this compliance often requires continuous or frequent inspection and testing. Chapter 17 establishes standards for special inspection, testing and reporting of the work to the building official.

Chapter 18 Soils and Foundations

Chapter 18 provides criteria for geotechnical and structural considerations in the selection, design and installation of foundation systems to support the loads from the structure above. This chapter includes requirements for soils investigation and site preparation for receiving a foundation, including the allowed load-bearing values for soils and for protecting the foundation from water intrusion. Section 1808 addresses the basic requirements for all foundation types. Later sections address foundation requirements that are specific to shallow foundations and deep foundations. Due care must be exercised in the planning and design of foundation systems based on obtaining sufficient soils information, the use of accepted engineering procedures, experience and good technical judgment.

Chapter 19 Concrete

This chapter provides minimum accepted practices for the design and construction of buildings and structural components using concrete—both plain and reinforced. Chapter 19 relies primarily on the reference to American Concrete Institute (ACI) 318, *Building Code Requirements for Structural Concrete*. This chapter also includes references to additional standards. Structural concrete must be designed and constructed to comply with this code and all listed standards. There are specific sections of the chapter addressing concrete slabs, anchorage to concrete and shotcrete. Because of the variable properties of material and numerous design and construction options available in the uses of concrete, due care and control throughout the construction process is necessary.

Chapter 20 Aluminum

Chapter 20 contains standards for the use of aluminum in building construction. Only the structural applications of aluminum are addressed. This chapter does not address the use of aluminum in specialty products such as storefront or window framing or architectural hardware. The use of aluminum in heating, ventilating or air-conditioning systems is addressed in the IMC. This chapter references national standards from the Aluminum Association for use of aluminum in building construction, AA ASM 35, *Aluminum Sheet Metal Work in Building Construction*, and AA ADM, *Aluminum Design Manual*. By utilizing the standards set forth, a proper application of this material can be obtained.

Chapter 21 Masonry

This chapter provides comprehensive and practical requirements for masonry construction. The provisions of Chapter 21 require minimum accepted practices and the use of standards for the design and construction of masonry structures. The provisions address: material specifications and test methods; types of wall construction; criteria for engineered and empirical designs; and required details of construction, including the execution of construction. Masonry design methodologies including allowable stress design, strength design and empirical design are covered by provisions of this chapter. Also addressed are masonry fireplaces and chimneys, masonry heaters and glass unit masonry. Fire-resistant construction using masonry is also required to comply with Chapter 7. Masonry foundations are also subject to the requirements of Chapter 18.

Chapter 22 Steel

Chapter 22 provides the requirements necessary for the design and construction of structural steel (including composite construction), cold-formed steel, steel joists, steel cable structures and steel storage racks. This chapter specifies appropriate design and construction standards for these types of structures. It also provides a road map of the applicable technical requirements for steel structures. Because steel is a noncombustible building material, it is commonly associated with Types I and II construction; however, it is permitted to be used in all types of construction. Chapter 22 requires that the design and use of steel materials be in accordance with the specifications and standards of the American Institute of Steel Construction, the American Iron and Steel Institute, the Steel Joist Institute and the American Society of Civil Engineers.

Chapter 23 Wood

This chapter provides minimum requirements for the design of buildings and structures that use wood and wood-based products. The chapter is organized around three design methodologies: allowable stress design (ASD), load and resistance factor design (LRFD) and conventional light-frame construction. Included in this chapter are references to design and manufacturing standards for various wood and wood-based products; general construction requirements; design criteria for lateral force-resisting systems and specific requirements for the application of the three design methods. In general, only Type III, IV or V buildings may be constructed of wood.

Chapter 24 Glass and Glazing

This chapter establishes regulations for glass and glazing that, when installed in buildings and structures, are subjected to wind, snow and dead loads. Engineering and design requirements are included in the chapter. Additional structural requirements are found in Chapter 16. Another concern of this chapter is glass and glazing used in areas where it is likely to be impacted by the occupants. Section 2406 identifies hazardous locations where glazing installed must either be safety glazing or blocked to prevent human impact. Safety glazing must meet stringent standards and be appropriately marked or identified. Additional requirements are provided for glass and glazing in guards, handrails, elevator hoistways and elevator cars, as well as in athletic facilities.

Chapter 25 Gypsum Board, Gypsum Panel Products and Plaster

Chapter 25 contains the provisions and referenced standards that regulate the design, construction and quality of gypsum board, gypsum panel products and plaster. It also addresses reinforced gypsum concrete. These represent the most common interior and exterior finish materials in the building industry. This chapter primarily addresses quality-control-related issues with regard to material specifications and installation requirements. Most products are manufactured under the control of industry standards. The building official or inspector primarily needs to verify that the appropriate product is used and properly installed for the intended use and location. While often simply used as wall and ceiling coverings, proper design and application are necessary to provide weather resistance and required fire protection for both structural and nonstructural building components.

Chapter 26 Plastic

The use of plastics in building construction and components is addressed in Chapter 26. This chapter provides standards addressing foam plastic insulation, foam plastics used as interior finish and trim, and other plastic veneers used on the inside or outside of a building. Plastic siding is regulated by Chapter 14. Sections 2606 through 2611 address the use of light-transmitting plastics in various configurations such as walls, roof panels, skylights, signs and as glazing. Requirements for the use of fiber-reinforced polymers, fiberglass-reinforced polymers and reflective plastic core insulation are also contained in this chapter. Additionally, requirements specific to the use of wood-plastic composites and plastic lumber are contained in this chapter. Some plastics exhibit rapid flame spread and heavy smoke density characteristics when exposed to fire. Exposure to the heat generated by a fire can cause some plastics to deform, which can affect their performance. The requirements and limitations of this chapter are necessary to control the use of plastic and foam plastic products such that they do not compromise the safety of building occupants.

Chapter 27 Electrical

Since electrical systems and components are an integral part of almost all structures, it is necessary for the code to address the installation of such systems. For this purpose, Chapter 27 references the National Electrical Code (NEC). In addition, Section 2702 addresses emergency and standby power requirements. Such systems must comply with the *IFC* and referenced standards. This section also provides references to the various code sections requiring emergency and standby power, such as high-rise buildings and buildings containing hazardous materials.

Chapter 28 Mechanical Systems

Nearly all buildings will include mechanical systems. This chapter provides references to the IMC and the IFGC for the design and installation of mechanical systems. In addition, Chapter 21 of this code is referenced for masonry chimneys, fireplaces and barbecues.

Chapter 29 Plumbing Systems

Chapter 29 regulates the minimum number of plumbing fixtures that must be provided for every type of building. This chapter also regulates the location of the required fixtures in various types of buildings. This section requires separate facilities for males and females except for certain types of small occupancies. The regulations in this chapter come directly from Chapters 3 and 4 of the IPC.

Chapter 30 Elevators and Conveying Systems

Chapter 30 provides standards for the installation of elevators into buildings. Referenced standards provide the requirements for the elevator system and mechanisms. Detailed standards are provided in the chapter for hoistway enclosures, machine rooms and requirements for sizing of elevators. Beginning in the 2015 edition of this code, the elevator lobby requirements were moved from Chapter 7 to Chapter 30 to pull all the elevator-related construction requirements together. New provisions were added in the 2009 edition for fire service access elevators required in high-rise buildings and for the optional choice of occupant evacuation elevators (see Section 403).

Chapter 31 Special Construction

Chapter 31 contains a collection of regulations for a variety of unique structures and architectural features. Pedestrian walkways and tunnels connecting two buildings are addressed in Section 3104. Membrane and air-supported structures are addressed by Section 3102. Safeguards for swimming pool safety are addressed by way of reference to the ISPSC in Section 3109. Standards for temporary structures, including permit requirements, are provided in Section 3103. Structures as varied as awnings, marquees, signs, telecommunication and broadcast towers and automatic vehicular gates are also addressed (see Sections 3105 through 3108 and 3110).

Chapter 32 Encroachments into the Public Right-of-way

Buildings and structures from time to time are designed to extend over a property line and into the public right-of-way. Local regulations outside of the building code usually set limits to such encroachments, and such regulations take precedence over the provisions of this chapter. Standards are provided for encroachments below grade for structural support, vaults and areaways. Encroachments above grade are divided into below 8 feet, 8 feet to 15 feet, and above 15 feet, because of headroom and vehicular height issues. This includes steps, columns, awnings, canopies, marquees, signs, windows and balconies. Similar architectural features above grade are also addressed. Pedestrian walkways must also comply with Chapter 31.

Chapter 33 Safeguards During Construction

Chapter 33 provides safety requirements during construction and demolition of buildings and structures. These requirements are intended to protect the public from injury and adjoining property from damage. In addition the chapter provides for the progressive installation and operation of exit stairways and standpipe systems during construction.

Chapter 34 Reserved

During the 2015 code change cycle the membership voted to delete Chapter 34, Existing Structures, from this code and reference the IEBC. The provisions that were in Chapter 34 appear in the IEBC. Former Sections 3402 through 3411 appear as IEBC Chapter 4 and Section 3412 as Chapter 14.

Chapter 35 Referenced Standards

The code contains numerous references to standards that are used to regulate materials and methods of construction. Chapter 35 contains a comprehensive list of all standards that are referenced in the code, including the appendices. The standards are part of the code to the extent of the reference to the standard (see Section 102.4). Compliance with the referenced standard is necessary for compliance with this code. By providing specifically adopted standards, the construction and installation requirements necessary for compliance with the code can be readily determined. The basis for code compliance is, therefore, established and available on an equal basis to the building official, contractor, designer and owner.

Chapter 35 is organized in a manner that makes it easy to locate specific standards. It lists all of the referenced standards, alphabetically, by acronym of the promulgating agency of the standard. Each agency's standards are then listed in either alphabetical or numeric order based upon the standard identification. The list also contains the title of the standard; the edition (date) of the standard referenced; any addenda included as part of the ICC adoption; and the section or sections of this code that reference the standard.

Appendices

Appendices are provided in the IBC to offer optional or supplemental criteria to the provisions in the main chapters of the code. Appendices provide additional information for administration of the Department of Building Safety as well as standards not typically administered by all building departments. Appendices have the same force and effect as the first 35 chapters of the IBC only when explicitly adopted by the jurisdiction.

Appendix A Employee Qualifications

Effective administration and enforcement of the family of International Codes depends on the training and expertise of the personnel employed by the jurisdiction and their knowledge of the codes. Section 103 of the code establishes the Department of Building Safety and calls for the appointment of a building official and deputies such as plans examiners and inspectors. Appendix A provides standards for experience, training and certification for the building official and the other staff mentioned in Chapter 1.

Appendix B Board of Appeals

Section 113 requires the establishment of a board of appeals to hear appeals regarding determinations made by the building official. Appendix B provides qualification standards for members of the board as well as operational procedures of such board.

Appendix C Group U—Agricultural Buildings

Appendix C provides a more liberal set of standards for the construction of agricultural buildings, rather than strictly following the utility building provision, reflective of their specific usage and limited occupant load. The provisions of this appendix, when adopted, allow reasonable heights and areas commensurate with the risk of agricultural buildings.

Appendix D Fire Districts

Fire districts have been a tool used to limit conflagration hazards in areas of a city with intense and concentrated development. More frequently used under the model codes that preceded the IBC, this appendix is provided to allow jurisdictions to continue the designation and use of fire districts. Fire district standards restrict certain occupancies within the district, as well as setting higher minimum construction standards.

Appendix E Supplementary Accessibility Requirements

The Architectural and Transportation Barriers Compliance Board (U.S. Access Board) has revised and updated its accessibility guidelines for buildings and facilities covered by the Americans with Disabilities Act (ADA) and the Architectural Barriers Act (ABA). Appendix E includes scoping requirements contained in the *2010 ADA Standards for Accessible Design* that are not in Chapter 11 and not otherwise mentioned or mainstreamed throughout the code. Items in the appendix address subjects not typically addressed in building codes (for example, beds, room signage, transportation facilities).

Appendix F Rodentproofing

The provisions of this appendix are minimum mechanical methods to prevent the entry of rodents into a building. These standards, when used in conjunction with cleanliness and maintenance programs, can significantly reduce the potential of rodents invading a building.

Appendix G Flood-resistant Construction

Appendix G is intended to fulfill the flood-plain management and administrative requirements of the National Flood Insurance Program (NFIP) that are not included in the code. Communities that adopt the IBC and Appendix G will meet the minimum requirements of NFIP as set forth in Title 44 of the Code of Federal Regulations.

Appendix H Signs

Appendix H gathers in one place the various code standards that regulate the construction and protection of outdoor signs. Whenever possible, this appendix provides standards in performance language, thus allowing the widest possible application.

Appendix I Patio Covers

Appendix I provides standards applicable to the construction and use of patio covers. It is limited in application to patio covers accessory to dwelling units. Covers of patios and other outdoor areas associated with restaurants, mercantile buildings, offices, nursing homes or other nondwelling occupancies would be subject to standards in the main code and not this appendix.

Appendix J Grading

Appendix J provides standards for the grading of properties. This appendix also provides standards for administration and enforcement of a grading program including permit and inspection requirements. Appendix J was originally developed in the 1960s and used for many years in jurisdictions throughout the western United States. It is intended to provide consistent and uniform code requirements anywhere grading is considered an issue.

Appendix K Administrative Provisions

Appendix K primarily provides administrative provisions for jurisdictions adopting and enforcing NFPA 70—the *National Electrical Code* (NEC). The provisions contained in this appendix are compatible with administrative and enforcement provisions contained in Chapter 1 of the IBC and the other I-Codes. Annex H of NFPA 70 also contains administrative provisions for the NEC; however, some of its provisions are not compatible with IBC Chapter 1. Section K110 also contains technical provisions that are unique to this appendix and are in addition to technical standards of NFPA 70.

Appendix L Earthquake Recording Instrumentation

The purpose of this appendix is to foster the collection of ground motion data, particularly from strong-motion earthquakes. When this ground motion data is synthesized, it may be useful in developing future improvements to the earthquake provisions of the IBC.

Appendix M Tsunami-Generated Flood Hazards

Addressing a tsunami risk for all types of construction in a tsunami hazard zone through building code requirements would typically not be cost effective, making tsunami-resistant construction impractical at an individual building level. However, this appendix does allow the adoption and enforcement of requirements for tsunami hazard zones that regulate the presence of high-risk or high-hazard structures.

Appendix N Replicable Buildings

Many jurisdictions have recognized the need for some form of expedited review process for replicable buildings. By codifying the approach contained in the ICC G1-2010 *Guideline for Replicable Buildings*, this appendix provides jurisdictions with a means of incorporating replicable building requirements into their building code adoption process. The intent is to streamline the plan review process at the local level by removing redundant reviews.

Appendix O Performance-based Application

Appendix O provides an optional design, review and approval framework for use by the building official. It simply extracts the relevant administrative provisions from the ICCPC into a more concise, usable appendix format for a jurisdiction confronted with such a need. Typical uses would include cases of alternate methods in Chapter 1 and select areas of the code that require a rational analysis, such as Section 909.

TABLE OF CONTENTS

CHAPTER 1
SCOPE AND ADMINISTRATION

User notes:

About this chapter: *Chapter 1 establishes the limits of applicability of the code and describes how the code is to be applied and enforced. Chapter 1 is in two parts: Part 1—Scope and Application (Sections 101–102) and Part 2—Administration and Enforcement (Sections 103–116). Section 101 identifies which buildings and structures come under its purview and references other I-Codes as applicable. Standards and codes are scoped to the extent referenced (see Section 102.4).*

This code is intended to be adopted as a legally enforceable document and it cannot be effective without adequate provisions for its administration and enforcement. The provisions of Chapter 1 establish the authority and duties of the code official appointed by the authority having jurisdiction and also establish the rights and privileges of the design professional, contractor and property owner. Chapter 1 is largely concerned with maintaining "due process of law" in enforcing the building performance criteria contained in the body of the code.

Code development reminder: *Code change proposals to this chapter will be considered by the Administrative Code Development Committee during the 2022 (Group B) Code Development Cycle.*

PART 1—SCOPE AND APPLICATION

SECTION 101 SCOPE AND GENERAL REQUIREMENTS

[A] 101.1 Title. These regulations shall be known as the *Building Code* of **[NAME OF JURISDICTION]**, hereinafter referred to as "this code."

[A] 101.2 Scope. The provisions of this code shall apply to the construction, *alteration*, relocation, enlargement, replacement, *repair*, equipment, use and occupancy, location, maintenance, removal and demolition of every building or structure or any appurtenances connected or attached to such buildings or structures.

Exception: Detached one- and two-family *dwellings* and *townhouses* not more than three *stories above grade plane* in height with a separate *means of egress*, and their accessory structures not more than three *stories above grade plane* in height, shall comply with this code or the *International Residential Code*.

[A] 101.2.1 Appendices. Provisions in the appendices shall not apply unless specifically adopted.

[A] 101.3 Purpose. The purpose of this code is to establish the minimum requirements to provide a reasonable level of safety, health and general welfare through structural strength, *means of egress*, stability, sanitation, light and *ventilation*, energy conservation, and for providing a reasonable level of life safety and property protection from the hazards of fire, *explosion* or *dangerous* conditions, and to provide a reasonable level of safety to fire fighters and emergency responders during emergency operations.

[A] 101.4 Referenced codes. The other codes specified in Sections 101.4.1 through 101.4.7 and referenced elsewhere in this code shall be considered to be part of the requirements of this code to the prescribed extent of each such reference.

[A] 101.4.1 Gas. The provisions of the *International Fuel Gas Code* shall apply to the installation of gas piping from the point of delivery, gas appliances and related accessories as covered in this code. These requirements apply to gas piping systems extending from the point of delivery to the inlet connections of appliances and the installation and operation of residential and commercial gas appliances and related accessories.

[A] 101.4.2 Mechanical. The provisions of the *International Mechanical Code* shall apply to the installation, *alterations*, *repairs* and replacement of mechanical systems, including equipment, appliances, fixtures, fittings and appurtenances, including ventilating, heating, cooling, air-conditioning and refrigeration systems, incinerators and other energy-related systems.

[A] 101.4.3 Plumbing. The provisions of the *International Plumbing Code* shall apply to the installation, *alteration*, *repair* and replacement of plumbing systems, including equipment, appliances, fixtures, fittings and appurtenances, and where connected to a water or sewage system and all aspects of a medical gas system. The provisions of the *International Private Sewage Disposal Code* shall apply to private sewage disposal systems.

[A] 101.4.4 Property maintenance. The provisions of the *International Property Maintenance Code* shall apply to *existing structures* and premises; equipment and facilities; light, *ventilation,* space heating, sanitation, life and fire safety hazards; responsibilities of *owners*, operators and occupants; and occupancy of existing premises and structures.

[A] 101.4.5 Fire prevention. The provisions of the *International Fire Code* shall apply to matters affecting or relating to structures, processes and premises from the hazard of fire and *explosion* arising from the storage, handling or use of structures, materials or devices; from conditions hazardous to life, property or public welfare in the occupancy of structures or premises; and from the construction, extension, *repair*, *alteration* or removal of fire suppression, *automatic sprinkler systems* and alarm systems or fire hazards in the structure or on the premises from occupancy or operation.

[A] 101.4.6 Energy. The provisions of the *International Energy Conservation Code* shall apply to all matters governing the design and construction of buildings for energy efficiency.

[A] 101.4.7 Existing buildings. The provisions of the *International Existing Building Code* shall apply to matters governing the *repair*, *alteration*, *change of occupancy, addition* to and relocation of *existing buildings*.

SECTION 102 APPLICABILITY

[A] 102.1 General. Where there is a conflict between a general requirement and a specific requirement, the specific requirement shall be applicable. Where, in any specific case, different sections of this code specify different materials, methods of construction or other requirements, the most restrictive shall govern.

[A] 102.2 Other laws. The provisions of this code shall not be deemed to nullify any provisions of local, state or federal law.

[A] 102.3 Application of references. References to chapter or section numbers, or to provisions not specifically identified by number, shall be construed to refer to such chapter, section or provision of this code.

[A] 102.4 Referenced codes and standards. The codes and standards referenced in this code shall be considered to be part of the requirements of this code to the prescribed extent of each such reference and as further regulated in Sections 102.4.1 and 102.4.2.

[A] 102.4.1 Conflicts. Where conflicts occur between provisions of this code and referenced codes and standards, the provisions of this code shall apply.

[A] 102.4.2 Provisions in referenced codes and standards. Where the extent of the reference to a referenced code or standard includes subject matter that is within the scope of this code or the International Codes specified in Section 101.4, the provisions of this code or the International Codes specified in Section 101.4, as applicable, shall take precedence over the provisions in the referenced code or standard.

[A] 102.5 Partial invalidity. In the event that any part or provision of this code is held to be illegal or void, this shall not have the effect of making void or illegal any of the other parts or provisions.

[A] 102.6 Existing structures. The legal occupancy of any structure existing on the date of adoption of this code shall be permitted to continue without change, except as otherwise specifically provided in this code, the *International Existing Building Code*, the *International Property Maintenance Code* or the *International Fire Code*.

[A] 102.6.1 Buildings not previously occupied. A building or portion of a building that has not been previously occupied or used for its intended purpose in accordance with the laws in existence at the time of its completion shall comply with the provisions of this code or the *International Residential Code*, as applicable, for new construction or with any current permit for such occupancy.

[A] 102.6.2 Buildings previously occupied. The legal occupancy of any building existing on the date of adoption of this code shall be permitted to continue without change, except as otherwise specifically provided in this code, the *International Fire Code* or *International Property Maintenance Code*, or as is deemed necessary by the *building official* for the general safety and welfare of the occupants and the public.

PART 2—ADMINISTRATION AND ENFORCEMENT

SECTION 103 CODE COMPLIANCE AGENCY

[A] 103.1 Creation of enforcement agency. The [INSERT NAME OF DEPARTMENT] is hereby created and the official in charge thereof shall be known as the *building official*. The function of the agency shall be the implementation, administration and enforcement of the provisions of this code.

[A] 103.2 Appointment. The *building official* shall be appointed by the chief appointing authority of the jurisdiction.

[A] 103.3 Deputies. In accordance with the prescribed procedures of this jurisdiction and with the concurrence of the appointing authority, the *building official* shall have the authority to appoint a deputy building official, other related technical officers, inspectors and other employees. Such employees shall have powers as delegated by the *building official*.

SECTION 104 DUTIES AND POWERS OF BUILDING OFFICIAL

[A] 104.1 General. The *building official* is hereby authorized and directed to enforce the provisions of this code. The *building official* shall have the authority to render interpretations of this code and to adopt policies and procedures in order to clarify the application of its provisions. Such interpretations, policies and procedures shall be in compliance with the intent and purpose of this code. Such policies and procedures shall not have the effect of waiving requirements specifically provided for in this code.

[A] 104.2 Applications and permits. The *building official* shall receive applications, review *construction documents* and issue *permits* for the erection, and *alteration*, demolition and moving of buildings and structures, inspect the premises for which such *permits* have been issued and enforce compliance with the provisions of this code.

[A] 104.2.1 Determination of substantially improved or substantially damaged existing buildings and structures in flood hazard areas. For applications for reconstruction, rehabilitation, *repair*, *alteration*, *addition* or other improvement of *existing buildings or structures* located in *flood hazard areas*, the *building official* shall determine if the proposed work constitutes *substantial improvement* or *repair* of *substantial damage*. Where the *building official* determines that the proposed work constitutes *substantial improvement* or *repair* of *substan-*

tial damage, and where required by this code, the *building official* shall require the building to meet the requirements of Section 1612 or Section R322 of the *International Residential Code*, as applicable.

[A] 104.3 Notices and orders. The *building official* shall issue necessary notices or orders to ensure compliance with this code.

[A] 104.4 Inspections. The *building official* shall make the required inspections, or the *building official* shall have the authority to accept reports of inspection by *approved agencies* or individuals. Reports of such inspections shall be in writing and be certified by a responsible officer of such *approved agency* or by the responsible individual. The *building official* is authorized to engage such expert opinion as deemed necessary to report on unusual technical issues that arise, subject to the approval of the appointing authority.

[A] 104.5 Identification. The *building official* shall carry proper identification when inspecting structures or premises in the performance of duties under this code.

[A] 104.6 Right of entry. Where it is necessary to make an inspection to enforce the provisions of this code, or where the *building official* has reasonable cause to believe that there exists in a structure or on a premises a condition that is contrary to or in violation of this code that makes the structure or premises unsafe, *dangerous* or hazardous, the *building official* is authorized to enter the structure or premises at reasonable times to inspect or to perform the duties imposed by this code, provided that if such structure or premises be occupied that credentials be presented to the occupant and entry requested. If such structure or premises is unoccupied, the *building official* shall first make a reasonable effort to locate the *owner* or other person having charge or control of the structure or premises and request entry. If entry is refused, the *building official* shall have recourse to the remedies provided by law to secure entry.

[A] 104.7 Department records. The *building official* shall keep official records of applications received, *permits* and certificates issued, fees collected, reports of inspections, and notices and orders issued. Such records shall be retained in the official records for the period required for retention of public records.

[A] 104.8 Liability. The *building official*, member of the board of appeals or employee charged with the enforcement of this code, while acting for the jurisdiction in good faith and without malice in the discharge of the duties required by this code or other pertinent law or ordinance, shall not thereby be civilly or criminally rendered liable personally and is hereby relieved from personal liability for any damage accruing to persons or property as a result of any act or by reason of an act or omission in the discharge of official duties.

[A] 104.8.1 Legal defense. Any suit or criminal complaint instituted against an officer or employee because of an act performed by that officer or employee in the lawful discharge of duties and under the provisions of this code shall be defended by legal representatives of the jurisdiction until the final termination of the proceedings. The *building official* or any subordinate shall not be liable for cost in any action, suit or proceeding that is instituted in pursuance of the provisions of this code.

[A] 104.9 Approved materials and equipment. Materials, equipment and devices *approved* by the *building official* shall be constructed and installed in accordance with such approval.

[A] 104.9.1 Used materials and equipment. Materials that are reused shall comply with the requirements of this code for new materials. Used equipment and devices shall not be reused unless *approved* by the *building official*.

[A] 104.10 Modifications. Where there are practical difficulties involved in carrying out the provisions of this code, the *building official* shall have the authority to grant modifications for individual cases, upon application of the *owner* or the owner's authorized agent, provided that the *building official* shall first find that special individual reason makes the strict letter of this code impractical, the modification is in compliance with the intent and purpose of this code and that such modification does not lessen health, *accessibility*, life and fire safety or structural requirements. The details of action granting modifications shall be recorded and entered in the files of the department of building safety.

[A] 104.10.1 Flood hazard areas. The *building official* shall not grant modifications to any provision required in *flood hazard areas* as established by Section 1612.3 unless a determination has been made that:

1. A showing of good and sufficient cause that the unique characteristics of the size, configuration or topography of the site render the elevation standards of Section 1612 inappropriate.
2. A determination that failure to grant the variance would result in exceptional hardship by rendering the lot undevelopable.
3. A determination that the granting of a variance will not result in increased flood heights, additional threats to public safety, extraordinary public expense, cause fraud on or victimization of the public, or conflict with existing laws or ordinances.
4. A determination that the variance is the minimum necessary to afford relief, considering the *flood* hazard.
5. Submission to the applicant of written notice specifying the difference between the *design flood elevation* and the elevation to which the building is to be built, stating that the cost of flood insurance will be commensurate with the increased risk resulting from the reduced floor elevation, and stating that construction below the *design flood elevation* increases risks to life and property.

[A] 104.11 Alternative materials, design and methods of construction and equipment. The provisions of this code are not intended to prevent the installation of any material or to prohibit any design or method of construction not specifically prescribed by this code, provided that any such alternative has been *approved*. An alternative material, design or method of construction shall be *approved* where the *building official* finds that the proposed alternative meets all of the following:

1. The alternative material, design or method of construction is satisfactory and complies with the intent of the provisions of this code,
2. The material, method or work offered is, for the purpose intended, not less than the equivalent of that prescribed in this code as it pertains to the following:
   2.1. Quality.
   2.2. Strength.
   2.3. Effectiveness.
   2.4. *Fire resistance.*
   2.5. Durability.
   2.6. Safety.

Where the alternative material, design or method of construction is not approved, the *building official* shall respond in writing, stating the reasons why the alternative was not approved.

[A] 104.11.1 Research reports. Supporting data, where necessary to assist in the approval of materials or assemblies not specifically provided for in this code, shall consist of valid research reports from *approved* sources.

[A] 104.11.2 Tests. Whenever there is insufficient evidence of compliance with the provisions of this code, or evidence that a material or method does not conform to the requirements of this code, or in order to substantiate claims for alternative materials or methods, the *building official* shall have the authority to require tests as evidence of compliance to be made without expense to the jurisdiction. Test methods shall be as specified in this code or by other recognized test standards. In the absence of recognized and accepted test methods, the *building official* shall approve the testing procedures. Tests shall be performed by an *approved agency*. Reports of such tests shall be retained by the *building official* for the period required for retention of public records.

SECTION 105
PERMITS

[A] 105.1 Required. Any *owner* or owner's authorized agent who intends to construct, enlarge, alter, *repair*, move, demolish or change the occupancy of a building or structure, or to erect, install, enlarge, alter, *repair*, remove, convert or replace any electrical, gas, mechanical or plumbing system, the installation of which is regulated by this code, or to cause any such work to be performed, shall first make application to the *building official* and obtain the required *permit*.

[A] 105.1.1 Annual permit. Instead of an individual *permit* for each *alteration* to an already *approved* electrical, gas, mechanical or plumbing installation, the *building official* is authorized to issue an annual *permit* upon application therefor to any person, firm or corporation regularly employing one or more qualified tradespersons in the building, structure or on the premises owned or operated by the applicant for the *permit*.

[A] 105.1.2 Annual permit records. The person to whom an annual *permit* is issued shall keep a detailed record of *alterations* made under such annual *permit*. The *building official* shall have access to such records at all times or such records shall be filed with the *building official* as designated.

[A] 105.2 Work exempt from permit. Exemptions from *permit* requirements of this code shall not be deemed to grant authorization for any work to be done in any manner in violation of the provisions of this code or any other laws or ordinances of this jurisdiction. *Permits* shall not be required for the following:

Building:

1. One-story detached accessory structures used as tool and storage sheds, playhouses and similar uses, provided that the floor area is not greater than 120 square feet (11 m^2).
2. Fences not over 7 feet (2134 mm) high.
3. Oil derricks.
4. Retaining walls that are not over 4 feet (1219 mm) in height measured from the bottom of the footing to the top of the wall, unless supporting a surcharge or impounding Class I, II or IIIA liquids.
5. Water tanks supported directly on grade if the capacity is not greater than 5,000 gallons (18 925 L) and the ratio of height to diameter or width is not greater than 2:1.
6. Sidewalks and driveways not more than 30 inches (762 mm) above adjacent grade, and not over any *basement* or *story* below and are not part of an *accessible route*.
7. Painting, papering, tiling, carpeting, cabinets, counter tops and similar finish work.
8. Temporary motion picture, television and theater stage sets and scenery.
9. Prefabricated *swimming pools* accessory to a Group R-3 occupancy that are less than 24 inches (610 mm) deep, are not greater than 5,000 gallons (18 925 L) and are installed entirely above ground.
10. Shade cloth structures constructed for nursery or agricultural purposes, not including service systems.
11. Swings and other playground equipment accessory to detached one- and two-family *dwellings*.
12. Window awnings in Group R-3 and U occupancies, supported by an *exterior wall* that do not project more than 54 inches (1372 mm) from the *exterior wall* and do not require additional support.
13. Nonfixed and movable fixtures, cases, racks, counters and partitions not over 5 feet 9 inches (1753 mm) in height.

Electrical:

1. **Repairs and maintenance:** Minor repair work, including the replacement of lamps or the connection of *approved* portable electrical equipment to *approved* permanently installed receptacles.
2. **Radio and television transmitting stations:** The provisions of this code shall not apply to electrical equipment used for radio and television transmissions, but do apply to equipment and wiring for a power supply and the installations of towers and antennas.
3. **Temporary testing systems:** A *permit* shall not be required for the installation of any temporary system required for the testing or servicing of electrical equipment or apparatus.

Gas:

1. Portable heating appliance.
2. Replacement of any minor part that does not alter approval of equipment or make such equipment unsafe.

Mechanical:

1. Portable heating appliance.
2. Portable ventilation equipment.
3. Portable cooling unit.
4. Steam, hot or chilled water piping within any heating or cooling equipment regulated by this code.
5. Replacement of any part that does not alter its approval or make it unsafe.
6. Portable evaporative cooler.
7. Self-contained refrigeration system containing 10 pounds (4.54 kg) or less of refrigerant and actuated by motors of 1 horsepower (0.75 kW) or less.

Plumbing:

1. The stopping of leaks in drains, water, soil, waste or vent pipe, provided, however, that if any concealed trap, drain pipe, water, soil, waste or vent pipe becomes defective and it becomes necessary to remove and replace the same with new material, such work shall be considered as new work and a *permit* shall be obtained and inspection made as provided in this code.
2. The clearing of stoppages or the repairing of leaks in pipes, valves or fixtures and the removal and reinstallation of water closets, provided that such repairs do not involve or require the replacement or rearrangement of valves, pipes or fixtures.

[A] 105.2.1 Emergency repairs. Where equipment replacements and repairs must be performed in an emergency situation, the *permit* application shall be submitted within the next working business day to the *building official.*

[A] 105.2.2 Public service agencies. A *permit* shall not be required for the installation, *alteration* or repair of generation, transmission, distribution or metering or other related equipment that is under the ownership and control of public service agencies by established right.

[A] 105.3 Application for permit. To obtain a *permit,* the applicant shall first file an application therefor in writing on a form furnished by the department of building safety for that purpose. Such application shall:

1. Identify and describe the work to be covered by the *permit* for which application is made.
2. Describe the land on which the proposed work is to be done by legal description, street address or similar description that will readily identify and definitely locate the proposed building or work.
3. Indicate the use and occupancy for which the proposed work is intended.
4. Be accompanied by *construction documents* and other information as required in Section 107.
5. State the valuation of the proposed work.
6. Be signed by the applicant, or the applicant's authorized agent.
7. Give such other data and information as required by the *building official.*

[A] 105.3.1 Action on application. The *building official* shall examine or cause to be examined applications for *permits* and amendments thereto within a reasonable time after filing. If the application or the *construction documents* do not conform to the requirements of pertinent laws, the *building official* shall reject such application in writing, stating the reasons therefor. If the *building official* is satisfied that the proposed work conforms to the requirements of this code and laws and ordinances applicable thereto, the *building official* shall issue a *permit* therefor as soon as practicable.

[A] 105.3.2 Time limitation of application. An application for a *permit* for any proposed work shall be deemed to have been abandoned 180 days after the date of filing, unless such application has been pursued in good faith or a *permit* has been issued; except that the *building official* is authorized to grant one or more extensions of time for additional periods not exceeding 90 days each. The extension shall be requested in writing and justifiable cause demonstrated.

[A] 105.4 Validity of permit. The issuance or granting of a *permit* shall not be construed to be a *permit* for, or an approval of, any violation of any of the provisions of this code or of any other ordinance of the jurisdiction. *Permits* presuming to give authority to violate or cancel the provisions of this code or other ordinances of the jurisdiction shall not be valid. The issuance of a *permit* based on *construction documents* and other data shall not prevent the *building official* from requiring the correction of errors in the *construction documents* and other data. The *building official* is authorized to prevent occupancy or use of a structure where in violation of this code or of any other ordinances of this jurisdiction.

[A] 105.5 Expiration. Every *permit* issued shall become invalid unless the work on the site authorized by such *permit* is commenced within 180 days after its issuance, or if the work authorized on the site by such *permit* is suspended or abandoned for a period of 180 days after the time the work is commenced. The *building official* is authorized to grant, in writing, one or more extensions of time, for periods not more than 180 days each. The extension shall be requested in writing and justifiable cause demonstrated.

[A] 105.6 Suspension or revocation. The *building official* is authorized to suspend or revoke a *permit* issued under the provisions of this code wherever the *permit* is issued in error or on the basis of incorrect, inaccurate or incomplete information, or in violation of any ordinance or regulation or any of the provisions of this code.

[A] 105.7 Placement of permit. The building *permit* or copy shall be kept on the site of the work until the completion of the project.

SECTION 106
FLOOR AND ROOF DESIGN LOADS

[A] 106.1 Live loads posted. In commercial or industrial buildings, for each floor or portion thereof designed for *live loads* exceeding 50 psf (2.40 kN/m^2), such design *live loads* shall be conspicuously posted by the *owner* or the owner's authorized agent in that part of each *story* in which they apply, using durable signs. It shall be unlawful to remove or deface such notices.

[A] 106.2 Issuance of certificate of occupancy. A certificate of occupancy required by Section 111 shall not be issued until the floor load signs, required by Section 106.1, have been installed.

[A] 106.3 Restrictions on loading. It shall be unlawful to place, or cause or permit to be placed, on any floor or roof of a building, structure or portion thereof, a *load* greater than is permitted by this code.

SECTION 107
CONSTRUCTION DOCUMENTS

[A] 107.1 General. Submittal documents consisting of *construction documents*, statement of *special inspections*, geotechnical report and other data shall be submitted in two or more sets, or in a digital format where allowed by the building official, with each *permit* application. The *construction documents* shall be prepared by a *registered design professional* where required by the statutes of the jurisdiction in which the project is to be constructed. Where special conditions exist, the *building official* is authorized to require additional *construction documents* to be prepared by a *registered design professional.*

> **Exception:** The *building official* is authorized to waive the submission of *construction documents* and other data not required to be prepared by a *registered design professional* if it is found that the nature of the work applied for is such that review of *construction documents* is not necessary to obtain compliance with this code.

[A] 107.2 Construction documents. *Construction documents* shall be in accordance with Sections 107.2.1 through 107.2.8.

[A] 107.2.1 Information on construction documents. *Construction documents* shall be dimensioned and drawn on suitable material. Electronic media documents are permitted to be submitted where *approved* by the *building official. Construction documents* shall be of sufficient clarity to indicate the location, nature and extent of the work proposed and show in detail that it will conform to the provisions of this code and relevant laws, ordinances, rules and regulations, as determined by the *building official.*

[A] 107.2.2 Fire protection system shop drawings. Shop drawings for the *fire protection systems* shall be submitted to indicate conformance to this code and the *construction documents* and shall be *approved* prior to the start of system installation. Shop drawings shall contain all information as required by the referenced installation standards in Chapter 9.

[A] 107.2.3 Means of egress. The *construction documents* shall show in sufficient detail the location, construction, size and character of all portions of the *means of egress* including the path of the exit discharge to the *public way* in compliance with the provisions of this code. In other than occupancies in Groups R-2, R-3, and I-1, the *construction documents* shall designate the number of occupants to be accommodated on every floor, and in all rooms and spaces.

[A] 107.2.4 Exterior wall envelope. *Construction documents* for all buildings shall describe the *exterior wall envelope* in sufficient detail to determine compliance with this code. The *construction documents* shall provide details of the *exterior wall envelope* as required, including flashing, intersections with dissimilar materials, corners, end details, control joints, intersections at roof, eaves or parapets, means of drainage, water-resistive barrier and details around openings.

The *construction documents* shall include manufacturer's installation instructions that provide supporting documentation that the proposed penetration and opening details described in the *construction documents* maintain the weather resistance of the *exterior wall envelope.* The supporting documentation shall fully describe the *exterior wall* system that was tested, where applicable, as well as the test procedure used.

[A] 107.2.5 Exterior balconies and elevated walking surfaces. Where balconies or other elevated walking surfaces have *weather-exposed surfaces*, and the structural framing is protected by an impervious moisture barrier, the *construction documents* shall include details for all elements of the impervious moisture barrier system. The *construction documents* shall include manufacturer's installation instructions.

[A] 107.2.6 Site plan. The *construction documents* submitted with the application for *permit* shall be accompanied by a site plan showing to scale the size and location of new construction and *existing structures* on the *site*, distances from *lot lines*, the established street grades and the proposed finished grades and, as applicable, *flood hazard areas*, *floodways*, and *design flood elevations*; and it shall be drawn in accordance with an accurate boundary line survey. In the case of demolition, the *site* plan shall show construction to be demolished and the location and size of *existing structures* and construction that are to remain on the *site* or plot. The *building official* is authorized to waive or modify the requirement for a site plan where the application for *permit* is for *alteration* or *repair* or where otherwise warranted.

[A] 107.2.6.1 Design flood elevations. Where *design flood elevations* are not specified, they shall be established in accordance with Section 1612.3.1.

[A] 107.2.7 Structural information. The *construction documents* shall provide the information specified in Section 1603.

107.2.8 Relocatable buildings. *Construction documents* for *relocatable buildings* shall comply with Section 3112.

[A] 107.3 Examination of documents. The *building official* shall examine or cause to be examined the accompanying submittal documents and shall ascertain by such examinations whether the construction indicated and described is in accordance with the requirements of this code and other pertinent laws or ordinances.

[A] 107.3.1 Approval of construction documents. When the *building official* issues a *permit*, the *construction documents* shall be *approved*, in writing or by stamp, as "Reviewed for Code Compliance." One set of *construction documents* so reviewed shall be retained by the *building official*. The other set shall be returned to the applicant, shall be kept at the site of work and shall be open to inspection by the *building official* or a duly authorized representative.

[A] 107.3.2 Previous approvals. This code shall not require changes in the *construction documents*, construction or designated occupancy of a structure for which a lawful *permit* has been heretofore issued or otherwise lawfully authorized, and the construction of which has been pursued in good faith within 180 days after the effective date of this code and has not been abandoned.

[A] 107.3.3 Phased approval. The *building official* is authorized to issue a *permit* for the construction of foundations or any other part of a building or structure before the *construction documents* for the whole building or structure have been submitted, provided that adequate information and detailed statements have been filed complying with pertinent requirements of this code. The holder of such *permit* for the foundation or other parts of a building or structure shall proceed at the holder's own risk with the building operation and without assurance that a *permit* for the entire structure will be granted.

[A] 107.3.4 Design professional in responsible charge. Where it is required that documents be prepared by a *registered design professional*, the *building official* shall be authorized to require the *owner* or the owner's authorized agent to engage and designate on the building *permit* application a *registered design professional* who shall act as the *registered design professional in responsible charge*. If the circumstances require, the *owner* or the owner's authorized agent shall designate a substitute *registered design professional in responsible charge* who shall perform the duties required of the original *registered design professional in responsible charge*. The *building official* shall be notified in writing by the *owner* or the owner's authorized agent if the *registered design professional in responsible charge* is changed or is unable to continue to perform the duties.

The *registered design professional in responsible charge* shall be responsible for reviewing and coordinating submittal documents prepared by others, including phased and deferred submittal items, for compatibility with the design of the building.

[A] 107.3.4.1 Deferred submittals. Deferral of any submittal items shall have the prior approval of the *building official*. The *registered design professional in responsible charge* shall list the *deferred submittals* on the *construction documents* for review by the *building official*.

Documents for *deferred submittal* items shall be submitted to the *registered design professional in responsible charge* who shall review them and forward them to the *building official* with a notation indicating that the *deferred submittal* documents have been reviewed and found to be in general conformance to the design of the building. The *deferred submittal* items shall not be installed until the *deferred submittal* documents have been *approved* by the *building official*.

[A] 107.4 Amended construction documents. Work shall be installed in accordance with the *approved construction documents*, and any changes made during construction that are not in compliance with the *approved construction documents* shall be resubmitted for approval as an amended set of *construction documents*.

[A] 107.5 Retention of construction documents. One set of *approved construction documents* shall be retained by the *building official* for a period of not less than 180 days from date of completion of the permitted work, or as required by state or local laws.

SECTION 108
TEMPORARY STRUCTURES AND USES

[A] 108.1 General. The *building official* is authorized to issue a *permit* for temporary structures and temporary uses. Such *permits* shall be limited as to time of service, but shall not be permitted for more than 180 days. The *building official* is authorized to grant extensions for demonstrated cause.

[A] 108.2 Conformance. Temporary structures and uses shall comply with the requirements in Section 3103.

[A] 108.3 Temporary power. The *building official* is authorized to give permission to temporarily supply and use power in part of an electric installation before such installation has been fully completed and the final certificate of completion has been issued. The part covered by the temporary certificate shall comply with the requirements specified for temporary lighting, heat or power in NFPA 70.

[A] 108.4 Termination of approval. The *building official* is authorized to terminate such *permit* for a temporary structure or use and to order the temporary structure or use to be discontinued.

SECTION 109
FEES

[A] 109.1 Payment of fees. A *permit* shall not be valid until the fees prescribed by law have been paid, nor shall an amendment to a *permit* be released until the additional fee, if any, has been paid.

[A] 109.2 Schedule of permit fees. Where a *permit* is required, a fee for each *permit* shall be paid as required, in accordance with the schedule as established by the applicable governing authority.

[A] 109.3 Permit valuations. The applicant for a *permit* shall provide an estimated *permit* value at time of application. *Permit* valuations shall reflect the total value of work, including materials and labor, for which the *permit* is being issued, such as electrical, gas, mechanical, plumbing equipment and permanent systems. If, in the opinion of the *building official*, the valuation is underestimated on the application, the *permit* shall be denied, unless the applicant can show detailed estimates to meet the approval of the *building official*. Final building *permit* valuation shall be set by the *building official*.

[A] 109.4 Work commencing before permit issuance. Any person who commences any work before obtaining the necessary *permits* shall be subject to a fee established by the *building official* that shall be in addition to the required *permit* fees.

[A] 109.5 Related fees. The payment of the fee for the construction, *alteration*, removal or demolition for work done in connection to or concurrently with the work authorized by a building *permit* shall not relieve the applicant or holder of the *permit* from the payment of other fees that are prescribed by law.

[A] 109.6 Refunds. The *building official* is authorized to establish a refund policy.

SECTION 110
INSPECTIONS

[A] 110.1 General. Construction or work for which a *permit* is required shall be subject to inspection by the *building official* and such construction or work shall remain visible and able to be accessed for inspection purposes until *approved*. Approval as a result of an inspection shall not be construed to be an approval of a violation of the provisions of this code or of other ordinances of the jurisdiction. Inspections presuming to give authority to violate or cancel the provisions of this code or of other ordinances of the jurisdiction shall not be valid. It shall be the duty of the *owner* or the owner's authorized agent to cause the work to remain visible and able to be accessed for inspection purposes. Neither the *building official* nor the jurisdiction shall be liable for expense entailed in the removal or replacement of any material required to allow inspection.

[A] 110.2 Preliminary inspection. Before issuing a *permit*, the *building official* is authorized to examine or cause to be examined buildings, structures and *sites* for which an application has been filed.

[A] 110.3 Required inspections. The *building official*, upon notification, shall make the inspections set forth in Sections 110.3.1 through 110.3.12.

[A] 110.3.1 Footing and foundation inspection. Footing and foundation inspections shall be made after excavations for footings are complete and any required reinforcing steel is in place. For concrete foundations, any required forms shall be in place prior to inspection. Materials for the foundation shall be on the job, except where concrete is ready mixed in accordance with ASTM C94, the concrete need not be on the job.

[A] 110.3.2 Concrete slab and under-floor inspection. Concrete slab and under-floor inspections shall be made after in-slab or under-floor reinforcing steel and building service equipment, conduit, piping accessories and other ancillary equipment items are in place, but before any concrete is placed or floor sheathing installed, including the subfloor.

[A] 110.3.3 Lowest floor elevation. In *flood hazard areas*, upon placement of the *lowest floor*, including the *basement*, and prior to further vertical construction, the elevation certification required in Section 1612.4 or the *International Residential Code*, as applicable, shall be submitted to the *building official*.

[A] 110.3.4 Frame inspection. Framing inspections shall be made after the roof deck or sheathing, all framing, *fireblocking* and bracing are in place and pipes, chimneys and vents to be concealed are complete and the rough electrical, plumbing, heating wires, pipes and ducts are *approved*.

[A] 110.3.5 Types IV-A, IV-B and IV-C connection protection inspection. In buildings of Types IV-A, IV-B and IV-C construction, where connection *fire-resistance ratings* are provided by wood cover calculated to meet the requirements of Section 2304.10.1, inspection of the wood cover shall be made after the cover is installed, but before any other coverings or finishes are installed.

[A] 110.3.6 Lath, gypsum board and gypsum panel product inspection. Lath, *gypsum board* and *gypsum panel product* inspections shall be made after lathing, *gypsum board* and *gypsum panel products*, interior and exterior, are in place, but before any plastering is applied or *gypsum board* and *gypsum panel product* joints and fasteners are taped and finished.

Exception: *Gypsum board* and *gypsum panel products* that are not part of a fire-resistance-rated assembly or a shear assembly.

[A] 110.3.7 Weather-exposed balcony and walking surface waterproofing. Where balconies or other elevated walking surfaces have *weather-exposed surfaces*, and the structural framing is protected by an impervious moisture barrier, all elements of the impervious moisture barrier system shall not be concealed until inspected and *approved*.

Exception: Where *special inspections* are provided in accordance with Section 1705.1.1, Item 3.

[A] 110.3.8 Fire- and smoke-resistant penetrations. Protection of joints and penetrations in fire-resistance-rated assemblies, *smoke barriers* and *smoke partitions* shall not be concealed from view until inspected and *approved*.

[A] 110.3.9 Energy efficiency inspections. Inspections shall be made to determine compliance with Chapter 13 and shall include, but not be limited to, inspections for: envelope insulation *R*- and *U*-values, *fenestration U*-value, duct system *R*-value, and HVAC and water-heating equipment efficiency.

[A] 110.3.10 Other inspections. In addition to the inspections specified in Sections 110.3.1 through 110.3.9, the *building official* is authorized to make or require other inspections of any construction work to ascertain compliance with the provisions of this code and other laws that are enforced by the department of building safety.

[A] 110.3.11 Special inspections. For *special inspections*, see Chapter 17.

[A] 110.3.12 Final inspection. The final inspection shall be made after all work required by the building *permit* is completed.

[A] 110.3.12.1 Flood hazard documentation. If located in a *flood hazard area*, documentation of the elevation of the *lowest floor* as required in Section 1612.4 shall be submitted to the *building official* prior to the final inspection.

[A] 110.4 Inspection agencies. The *building official* is authorized to accept reports of *approved* inspection agencies, provided that such agencies satisfy the requirements as to qualifications and reliability.

[A] 110.5 Inspection requests. It shall be the duty of the holder of the building *permit* or their duly authorized agent to notify the *building official* when work is ready for inspection. It shall be the duty of the *permit* holder to provide access to and means for inspections of such work that are required by this code.

[A] 110.6 Approval required. Work shall not be done beyond the point indicated in each successive inspection without first obtaining the approval of the *building official*. The *building official*, upon notification, shall make the requested inspections and shall either indicate the portion of the construction that is satisfactory as completed, or notify the *permit* holder or the *permit* holder's agent wherein the same fails to comply with this code. Any portions that do not comply shall be corrected and such portion shall not be covered or concealed until authorized by the *building official*.

SECTION 111
CERTIFICATE OF OCCUPANCY

[A] 111.1 Change of occupancy. A building or structure shall not be used or occupied in whole or in part, and a *change of occupancy* of a building or structure or portion thereof shall not be made, until the *building official* has issued a certificate of occupancy therefor as provided herein. Issuance of a certificate of occupancy shall not be construed as an approval of a violation of the provisions of this code or of other ordinances of the *jurisdiction*. Certificates presuming to give authority to violate or cancel the provisions of this code or other ordinances of the *jurisdiction* shall not be valid.

Exception: Certificates of occupancy are not required for work exempt from *permits* in accordance with Section 105.2.

[A] 111.2 Certificate issued. After the *building official* inspects the building or structure and does not find violations of the provisions of this code or other laws that are enforced by the department, the *building official* shall issue a certificate of occupancy that contains the following:

1. The *permit* number.
2. The address of the structure.
3. The name and address of the *owner* or the owner's authorized agent.
4. A description of that portion of the structure for which the certificate is issued.
5. A statement that the described portion of the structure has been inspected for compliance with the requirements of this code.
6. The name of the *building official*.
7. The edition of the code under which the *permit* was issued.
8. The use and occupancy, in accordance with the provisions of Chapter 3.
9. The type of construction as defined in Chapter 6.
10. The design *occupant load*.
11. Where an *automatic sprinkler system* is provided, whether the sprinkler system is required.
12. Any special stipulations and conditions of the building *permit*.

[A] 111.3 Temporary occupancy. The *building official* is authorized to issue a temporary certificate of occupancy before the completion of the entire work covered by the *permit*, provided that such portion or portions shall be occupied safely. The *building official* shall set a time period during which the temporary certificate of occupancy is valid.

[A] 111.4 Revocation. The *building official* is authorized to suspend or revoke a certificate of occupancy or completion issued under the provisions of this code, in writing, wherever the certificate is issued in error, or on the basis of incorrect information supplied, or where it is determined that the building or structure or portion thereof is in violation of the provisions of this code or other ordinance of the *jurisdiction*.

SECTION 112
SERVICE UTILITIES

[A] 112.1 Connection of service utilities. A person shall not make connections from a utility, a source of energy, fuel, or power, or a water system or sewer system to any building or system that is regulated by this code for which a *permit* is required, until approved by the *building official*.

[A] 112.2 Temporary connection. The *building official* shall have the authority to authorize the temporary connection of the building or system to the utility, the source of energy, fuel, or power, or the water system or sewer system for the purpose of testing systems or for use under a temporary approval.

[A] 112.3 Authority to disconnect service utilities. The *building official* shall have the authority to authorize disconnection of utility service to the building, structure or system regulated by this code and the referenced codes and standards in case of emergency where necessary to eliminate an immediate hazard to life or property or where such utility connection has been made without the approval required by Section 112.1 or 112.2. The *building official* shall notify the serving utility, and wherever possible the *owner* or the owner's authorized agent and occupant of the building, structure or service system of the decision to disconnect prior to taking such action. If not notified prior to disconnecting, the *owner* or the owner's authorized agent or occupant of the building, structure or service system shall be notified in writing, as soon as practical thereafter.

SECTION 113
MEANS OF APPEALS

[A] 113.1 General. In order to hear and decide appeals of orders, decisions or determinations made by the *building official* relative to the application and interpretation of this code, there shall be and is hereby created a board of appeals. The board of appeals shall be appointed by the applicable governing authority and shall hold office at its pleasure. The board shall adopt rules of procedure for conducting its business and shall render all decisions and findings in writing to the appellant with a duplicate copy to the *building official*.

[A] 113.2 Limitations on authority. An application for appeal shall be based on a claim that the true intent of this code or the rules legally adopted thereunder have been incorrectly interpreted, the provisions of this code do not fully apply or an equivalent or better form of construction is proposed. The board shall not have authority to waive requirements of this code or interpret the administration of this code.

[A] 113.3 Qualifications. The board of appeals shall consist of members who are qualified by experience and training to pass on matters pertaining to building construction and are not employees of the jurisdiction.

[A] 113.4 Administration. The *building official* shall take immediate action in accordance with the decision of the board.

SECTION 114
VIOLATIONS

[A] 114.1 Unlawful acts. It shall be unlawful for any person, firm or corporation to erect, construct, alter, extend, *repair*, move, remove, demolish or occupy any building, structure or equipment regulated by this code, or cause same to be done, in conflict with or in violation of any of the provisions of this code.

[A] 114.2 Notice of violation. The *building official* is authorized to serve a notice of violation or order on the person responsible for the erection, construction, *alteration*, extension, *repair*, moving, removal, demolition or occupancy of a building or structure in violation of the provisions of this code, or in violation of a *permit* or certificate issued under the provisions of this code. Such order shall direct the discontinuance of the illegal action or condition and the abatement of the violation.

[A] 114.3 Prosecution of violation. If the notice of violation is not complied with promptly, the *building official* is authorized to request the legal counsel of the jurisdiction to institute the appropriate proceeding at law or in equity to restrain, correct or abate such violation, or to require the removal or termination of the unlawful occupancy of the building or structure in violation of the provisions of this code or of the order or direction made pursuant thereto.

[A] 114.4 Violation penalties. Any person who violates a provision of this code or fails to comply with any of the requirements thereof or who erects, constructs, alters or repairs a building or structure in violation of the *approved construction documents* or directive of the *building official*, or of a *permit* or certificate issued under the provisions of this code, shall be subject to penalties as prescribed by law.

SECTION 115
STOP WORK ORDER

[A] 115.1 Authority. Where the *building official* finds any work regulated by this code being performed in a manner contrary to the provisions of this code or in a *dangerous* or unsafe manner, the *building official* is authorized to issue a stop work order.

[A] 115.2 Issuance. The stop work order shall be in writing and shall be given to the *owner* of the property, the owner's authorized agent or the person performing the work. Upon issuance of a stop work order, the cited work shall immediately cease. The stop work order shall state the reason for the order and the conditions under which the cited work is authorized to resume.

[A] 115.3 Emergencies. Where an emergency exists, the *building official* shall not be required to give a written notice prior to stopping the work.

[A] 115.4 Failure to comply. Any person who shall continue any work after having been served with a stop work order, except such work as that person is directed to perform to remove a violation or unsafe condition, shall be subject to fines established by the authority having *jurisdiction*.

SECTION 116
UNSAFE STRUCTURES AND EQUIPMENT

[A] 116.1 Unsafe conditions. Structures or existing equipment that are or hereafter become unsafe, insanitary or deficient because of inadequate *means of egress* facilities, inadequate light and *ventilation*, or that constitute a fire hazard, or are otherwise *dangerous* to human life or the public welfare, or that involve illegal or improper occupancy or inadequate maintenance, shall be deemed an unsafe condition. Unsafe structures shall be taken down and removed or made safe, as the *building official* deems necessary and as provided for in this section. A vacant structure that is not secured against unauthorized entry shall be deemed unsafe.

[A] 116.2 Record. The *building official* shall cause a report to be filed on an unsafe condition. The report shall state the occupancy of the structure and the nature of the unsafe condition.

[A] 116.3 Notice. If an unsafe condition is found, the *building official* shall serve on the *owner* of the structure, or the owner's authorized agent, a written notice that describes the condition deemed unsafe and specifies the required repairs or improvements to be made to abate the unsafe condition, or that requires the unsafe structure to be demolished within a stipulated time. Such notice shall require the person thus notified to declare immediately to the *building official* acceptance or rejection of the terms of the order.

[A] 116.4 Method of service. Such notice shall be deemed properly served where a copy thereof is served in accordance with one of the following methods:

1. A copy is delivered to the *owner* personally.
2. A copy is sent by certified or registered mail addressed to the *owner* at the last known address with the return receipt requested.
3. A copy is delivered in any other manner as prescribed by local law.

If the certified or registered letter is returned showing that the letter was not delivered, a copy thereof shall be posted in a conspicuous place in or about the structure affected by such notice. Service of such notice in the foregoing manner on the owner's authorized agent shall constitute service of notice on the *owner*.

[A] 116.5 Restoration or abatement. Where the structure or equipment determined to be unsafe by the *building official* is restored to a safe condition, the owner, the owner's authorized agent, operator or occupant of a structure, premises or equipment deemed unsafe by the *building official* shall abate or cause to be abated or corrected such unsafe conditions either by repair, rehabilitation, demolition or other approved corrective action. To the extent that repairs, *alterations* or *additions* are made or a *change of occupancy* occurs during the restoration of the structure, such *repairs*, *alterations*, *additions* and *change of occupancy* shall comply with the requirements of the *International Existing Building Code*.

CHAPTER 2

DEFINITIONS

User notes:

About this chapter: *Codes, by their very nature, are technical documents. Every word, term and punctuation mark can add to or change the meaning of a technical requirement. It is necessary to maintain a consensus on the specific meaning of each term contained in the code. Chapter 2 performs this function by stating clearly what specific terms mean for the purposes of the code.*

Code development reminder: *Code change proposals to sections preceded by the designation [A] or [BS] will be considered by one of the code development committees meeting during the 2022 (Group B) Code Development Cycle.*

SECTION 201 GENERAL

201.1 Scope. Unless otherwise expressly stated, the following words and terms shall, for the purposes of this code, have the meanings shown in this chapter.

201.2 Interchangeability. Words used in the present tense include the future; words stated in the masculine gender include the feminine and neuter; the singular number includes the plural and the plural, the singular.

201.3 Terms defined in other codes. Where terms are not defined in this code and are defined in the *International Energy Conservation Code*, *International Fuel Gas Code*, *International Fire Code*, *International Mechanical Code* or *International Plumbing Code*, such terms shall have the meanings ascribed to them as in those codes.

201.4 Terms not defined. Where terms are not defined through the methods authorized by this section, such terms shall have ordinarily accepted meanings such as the context implies.

SECTION 202 DEFINITIONS

[BG] 24-HOUR BASIS. The actual time that a person is an occupant within a facility for the purpose of receiving care. It shall not include a facility that is open for 24 hours and is capable of providing care to someone visiting the facility during any segment of the 24 hours.

[BS] AAC MASONRY. *Masonry* made of autoclaved aerated concrete (AAC) units, manufactured without internal reinforcement and bonded together using thin- or thick-bed *mortar*.

[BE] ACCESSIBLE. A site, *building*, *facility* or portion thereof that complies with Chapter 11.

[BE] ACCESSIBLE MEANS OF EGRESS. A continuous and unobstructed way of egress travel from any accessible point in a *building* or *facility* to a *public way*.

[BE] ACCESSIBLE ROUTE. A continuous, unobstructed path that complies with Chapter 11.

[BE] ACCESSIBLE UNIT. A *dwelling unit* or *sleeping unit* that complies with this code and the provisions for *Accessible units* in ICC A117.1.

[BS] ACCREDITATION BODY. An *approved*, third-party organization that is independent of the grading and inspection agencies, and the lumber mills, and that initially accredits and subsequently monitors, on a continuing basis, the competency and performance of a grading or inspection agency related to carrying out specific tasks.

[A] ADDITION. An extension or increase in floor area, number of *stories* or height of a building or structure.

[BS] ADHERED MASONRY VENEER. *Veneer* secured and supported through the adhesion of an *approved* bonding material applied to an *approved* backing.

[BS] ADOBE CONSTRUCTION. Construction in which the exterior *load-bearing* and *nonload-bearing walls* and partitions are of unfired clay *masonry* units, and floors, roofs and interior framing are wholly or partly of wood or other *approved* materials.

[F] AEROSOL CONTAINER. A metal can or plastic container up to a maximum size of 33.8 fluid ounces (1000 ml), or a glass bottle up to a maximum size of 4 fluid ounces (118 ml), designed and intended to dispense an aerosol.

[F] AEROSOL PRODUCT. A combination of a container, a propellant and a material that is dispensed. Aerosol products shall be classified by means of the calculation of their chemical heats of combustion and shall be designated Level 1, Level 2 or Level 3.

Level 1 aerosol products. Those with a total chemical heat of combustion that is less than or equal to 8,600 British thermal units per pound (Btu/lb) (20 kJ/g).

Level 2 aerosol products. Those with a total chemical heat of combustion that is greater than 8,600 Btu/lb (20 kJ/g), but less than or equal to 13,000 Btu/lb (30 kJ/g).

Level 3 aerosol products. Those with a total chemical heat of combustion that is greater than 13,000 Btu/lb (30 kJ/g).

[BS] AGGREGATE. In roofing, crushed stone, crushed slag or water-worn gravel used for surfacing for *roof coverings*.

[BG] AGRICULTURAL BUILDING. A structure designed and constructed to house farm implements, hay, grain, poultry, livestock or other horticultural products. This structure shall not be a place of human habitation or a place of employment where agricultural products are processed, treated or packaged, nor shall it be a place used by the public.

[BF] AIR-IMPERMEABLE INSULATION. An insulation having an air permeance equal to or less than 0.02 l/s × m^2 at 75 pa pressure differential tested in accordance with ASTM E283 or ASTM E2178.

[BG] AIR-INFLATED STRUCTURE. A structure that uses air-pressurized membrane beams, arches or other elements to enclose space. Occupants of such a structure do not occupy the pressurized area used to support the structure.

[BG] AIR-SUPPORTED STRUCTURE. A structure wherein the shape of the structure is attained by air pressure and occupants of the structure are within the elevated pressure area. *Air-supported structures* are of two basic types:

[BE] AISLE. An unenclosed *exit access* component that defines and provides a path of egress travel.

[BE] AISLE ACCESSWAY. That portion of an *exit access* that leads to an *aisle*.

[F] ALARM NOTIFICATION APPLIANCE. A *fire alarm system* component such as a bell, horn, speaker, light or text display that provides audible, tactile or visible outputs, or any combination thereof.

[F] ALARM SIGNAL. A signal indicating an emergency requiring immediate action, such as a signal indicative of fire.

[F] ALARM VERIFICATION FEATURE. A feature of *automatic* fire detection and alarm systems to reduce unwanted alarms wherein *smoke detectors* report alarm conditions for a minimum period of time, or confirm alarm conditions within a given time period, after being automatically reset, in order to be accepted as a valid alarm-initiation signal.

[BS] ALLOWABLE STRESS DESIGN. A method of proportioning structural members, such that elastically computed stresses produced in the members by *nominal loads* do not exceed *specified* allowable stresses (also called "working stress design").

[A] ALTERATION. Any construction or renovation to an *existing structure* other than *repair* or *addition*.

[BE] ALTERNATING TREAD DEVICE. A device that has a series of steps between 50 and 70 degrees (0.87 and 1.22 rad) from horizontal, usually attached to a center support rail in an alternating manner so that the user does not have both feet on the same level at the same time.

[BG] AMBULATORY CARE FACILITY. Buildings or portions thereof used to provide medical, surgical, psychiatric, nursing or similar care on a less than *24-hour basis* to persons who are rendered *incapable of self-preservation* by the services provided or staff has accepted responsibility for care recipients already incapable.

[BG] ANCHOR BUILDING. An exterior perimeter building of a group other than H having direct access to a *covered or open mall building* but having required *means of egress* independent of the mall.

[BS] ANCHORED MASONRY VENEER. *Veneer* secured with *approved* mechanical fasteners to an *approved* backing.

[BF] ANNULAR SPACE. The opening around the penetrating item.

[F] ANNUNCIATOR. A unit containing one or more indicator lamps, alphanumeric displays or other equivalent means in which each indication provides status information about a circuit, condition or location.

[A] APPROVED. Acceptable to the *building official*.

[A] APPROVED AGENCY. An established and recognized agency that is regularly engaged in conducting tests, furnishing inspection services or furnishing product certification where such agency has been *approved* by the *building official*.

[BS] APPROVED FABRICATOR. An established and qualified person, firm or corporation *approved* by the *building official* pursuant to Chapter 17 of this code.

[A] APPROVED SOURCE. An independent person, firm or corporation, *approved* by the *building official*, who is competent and experienced in the application of engineering principles to materials, methods or systems analyses.

[BS] AREA (for masonry).

Gross cross-sectional. The *area* delineated by the out-to-out *specified dimensions* of *masonry* in the plane under consideration.

Net cross-sectional. The *area* of *masonry units*, grout and *mortar* crossed by the plane under consideration based on out-to-out *specified dimensions*.

[BG] AREA, BUILDING. The area included within surrounding *exterior walls*, or *exterior walls* and *fire walls*, exclusive of vent *shafts* and *courts*. Areas of the building not provided with surrounding walls shall be included in the *building area* if such areas are included within the horizontal projection of the roof or floor above.

[BE] AREA OF REFUGE. An area where persons unable to use *stairways* can remain temporarily to await instructions or assistance during emergency evacuation.

[BE] AREA OF SPORT ACTIVITY. That portion of an indoor or outdoor space where the play or practice of a sport occurs.

[BG] AREAWAY. A subsurface space adjacent to a building open at the top or protected at the top by a grating or guard.

ASSEMBLY SEATING, MULTILEVEL. See "*Multilevel assembly seating*."

[BG] ATRIUM. A vertical space that is closed at the top, connecting two or more *stories* in Group I-2 and I-3 occupancies or three or more *stories* in all other occupancies.

[BG] ATTIC. The space between the ceiling framing of the top *story* and the underside of the roof.

[F] AUDIBLE ALARM NOTIFICATION APPLIANCE. A notification appliance that alerts by the sense of hearing.

[F] AUTOMATIC. As applied to fire protection devices, a device or system providing an emergency function without the necessity for human intervention and activated as a result of a predetermined temperature rise, rate of temperature rise or combustion products.

[F] AUTOMATIC FIRE-EXTINGUISHING SYSTEM. An *approved* system of devices and equipment which automatically detects a fire and discharges an *approved* fire-extinguishing agent onto or in the area of a fire.

[F] AUTOMATIC SMOKE DETECTION SYSTEM. A *fire alarm system* that has initiation devices that utilize *smoke detectors* for protection of an area such as a room or space with detectors to provide early warning of fire.

[F] AUTOMATIC SPRINKLER SYSTEM. An *automatic sprinkler system*, for fire protection purposes, is an integrated system of underground and overhead piping designed in accordance with fire protection engineering standards. The system includes a suitable water supply. The portion of the system above the ground is a network of specially sized or hydraulically designed piping installed in a structure or area, generally overhead, and to which *automatic* sprinklers are connected in a systematic pattern. The system is usually activated by heat from a fire and discharges water over the *fire area*.

[F] AUTOMATIC WATER MIST SYSTEM. A system consisting of a water supply, a pressure source and a distribution piping system with attached nozzles, which, at or above a minimum operating pressure defined by its listing, discharges water in fine droplets meeting the requirements of NFPA 750 for the purpose of the control, suppression or extinguishment of a fire. Such systems include wet-pipe, dry-pipe and preaction types. The systems are designed as engineered, preengineered, local-application or total-flooding systems.

[F] AVERAGE AMBIENT SOUND LEVEL. The root mean square, A-weighted sound pressure level measured over a 24-hour period, or the time any person is present, whichever time period is less.

[BG] AWNING. An architectural projection that provides weather protection, identity or decoration and is partially or wholly supported by the building to which it is attached. An awning is composed of a lightweight *frame structure* over which a covering is attached.

[BF] BACKING. The wall or surface to which the *veneer* is secured.

[BE] BALANCED DOOR. A door equipped with double-pivoted hardware so designed as to cause a semicounterbalanced swing action when opening.

[F] BALED COTTON. A natural seed fiber wrapped in and secured with industry accepted materials, usually consisting of burlap, woven polypropylene, polyethylene or cotton or sheet polyethylene, and secured with steel, synthetic or wire bands or wire; also includes linters (lint removed from the cottonseed) and motes (residual materials from the ginning process).

[F] BALED COTTON, DENSELY PACKED. Cotton made into banded bales with a packing density of not less than 22 pounds per cubic foot (360 kg/m^3), and dimensions complying with the following: a length of 55 inches (1397 mm), a width of 21 inches (533.4 mm) and a height of 27.6 to 35.4 inches (701 to 899 mm).

[BS] BALLAST. In roofing, *ballast* comes in the form of large stones or paver systems or light-weight interlocking paver systems and is used to provide uplift resistance for roofing systems that are not adhered or mechanically attached to the roof deck.

[F] BARRICADE. A structure that consists of a combination of walls, floor and roof, which is designed to withstand the rapid release of energy in an *explosion* and which is fully confined, partially vented or fully vented; or other effective method of shielding from *explosive* materials by a natural or artificial barrier.

Artificial barricade. An artificial mound or revetment a minimum thickness of 3 feet (914 mm).

Natural barricade. Natural features of the ground, such as hills, or timber of sufficient density that the surrounding exposures that require protection cannot be seen from the magazine or building containing *explosives* when the trees are bare of leaves.

[BS] BASE FLOOD. The *flood* having a 1-percent chance of being equaled or exceeded in any given year.

[BS] BASE FLOOD ELEVATION. The elevation of the *base flood*, including wave height, relative to the National Geodetic Vertical Datum (NGVD), North American Vertical Datum (NAVD) or other datum specified on the *Flood Insurance Rate Map* (FIRM).

[BG] BASEMENT. A *story* that is not a *story above grade plane* (see "*Story above grade plane*"). This definition of "Basement" does not apply to the provisions of Section 1612 for *flood loads*.

[BS] BASEMENT (for flood loads). The portion of a building having its floor subgrade (below ground level) on all sides. This definition of "Basement" is limited in application to the provisions of Section 1612.

[BS] BEARING WALL STRUCTURE. A building or other structure in which vertical *loads* from floors and roofs are primarily supported by walls.

[BS] BED JOINT. The horizontal layer of *mortar* on which a *masonry unit* is laid.

[BE] BLEACHERS. Tiered seating supported on a dedicated structural system and two or more rows high and is not a *building element* (see "*Grandstand*").

[BG] BOARDING HOUSE. A building arranged or used for lodging for compensation, with or without meals, and not occupied as a single-family unit.

[F] BOILING POINT. The temperature at which the vapor pressure of a *liquid* equals the atmospheric pressure of 14.7 pounds per square inch (psia) (101 kPa) or 760 mm of mercury. Where an accurate boiling point is unavailable for the material in question, or for mixtures which do not have a constant boiling point, for the purposes of this classification, the 20-percent evaporated point of a distillation performed in accordance with ASTM D86 shall be used as the boiling point of the *liquid*.

[BS] BRACED WALL LINE. A straight line through the building plan that represents the location of the lateral resistance provided by the wall bracing.

[BS] BRACED WALL PANEL. A full-height section of wall constructed to resist in-plane shear *loads* through interaction of framing members, sheathing material and anchors. The panel's length meets the requirements of its particular bracing method and contributes toward the total amount of bracing required along its *braced wall line*.

[BE] BREAKOUT. For revolving doors, a process whereby wings or door panels can be pushed open manually for *means of egress* travel.

[BS] BRICK.

Calcium silicate (sand lime brick). A pressed and subsequently autoclaved unit that consists of sand and lime, with or without the inclusion of other materials.

Clay or shale. A solid or hollow *masonry unit* of *clay or shale*, usually formed into a rectangular *prism*, then burned or fired in a kiln; *brick* is a ceramic product.

Concrete. A concrete *masonry unit* made from Portland cement, water, and suitable aggregates, with or without the inclusion of other materials.

[A] BUILDING. Any structure utilized or intended for supporting or sheltering any occupancy.

BUILDING AREA. See "*Area, building*."

[BG] BUILDING ELEMENT. A fundamental component of building construction, specified in TABLE 601, which may or may not be of fire-resistance-rated construction and is constructed of materials based on the building type of construction.

BUILDING HEIGHT. See "*Height, building*."

[BG] BUILDING LINE. The line established by law, beyond which a building shall not extend, except as specifically provided by law.

[A] BUILDING OFFICIAL. The officer or other designated authority charged with the administration and enforcement of this code, or a duly authorized representative.

[BS] BUILDING-INTEGRATED PHOTOVOLTAIC (BIPV) PRODUCT. A building product that incorporates *photovoltaic modules* and functions as a component of the building envelope.

[BS] BUILDING-INTEGRATED PHOTOVOLTAIC ROOF PANEL (BIPV ROOF PANEL). A *photovoltaic panel* that functions as a component of the building envelope.

[BS] BUILT-UP ROOF COVERING. Two or more layers of felt cemented together and surfaced with a cap sheet, mineral *aggregate*, smooth coating or similar surfacing material.

➡ **[BG] CANOPY.** A permanent structure or architectural projection of rigid construction over which a covering is attached that provides weather protection, identity or decoration. A *canopy* is permitted to be structurally independent or supported by attachment to a building on one or more sides.

[F] CAPACITOR ENERGY STORAGE SYSTEM. A stationary, rechargeable energy storage system consisting of capacitors, chargers, controls and associated electrical equipment designed to provide electrical power to a building or facility. The system is typically used to provide standby or emergency power, an uninterruptable power supply, load shedding, load sharing or similar capabilities.

[F] CARBON DIOXIDE EXTINGUISHING SYSTEMS. A system supplying carbon dioxide (CO_2) from a pressurized vessel through fixed pipes and nozzles. The system includes a manual- or *automatic*-actuating mechanism.

[F] CARBON MONOXIDE ALARM. A single- or multiple-station alarm intended to detect carbon monoxide gas and alert occupants by a distinct audible signal. It incorporates a sensor, control components and an alarm notification appliance in a single unit.

[F] CARBON MONOXIDE DETECTOR. A device with an integral sensor to detect carbon monoxide gas and transmit an alarm signal to a connected alarm control unit.

[BG] CARE SUITE. In Group I-2 occupancies, a group of treatment rooms, care recipient sleeping rooms and the support rooms or spaces and circulation space within the suite where staff are in attendance for supervision of all care recipients within the suite, and the suite is in compliance with the requirements of Section 407.4.4.

[BS] CAST STONE. A building stone manufactured from Portland cement concrete precast and used as a *trim*, *veneer* or facing on or in buildings or structures.

[F] CEILING LIMIT. The maximum concentration of an airborne contaminant to which one may be exposed. The ceiling limits utilized are those published in DOL 29 CFR Part 1910.1000. The ceiling Recommended Exposure Limit (REL-C) concentrations published by the US National Institute for Occupational Safety and Health (NIOSH), Threshold Limit Value—Ceiling (TLV-C) concentrations published by the American Conference of Governmental Industrial Hygienists (ACGIH), Ceiling Workplace Environmental Exposure Level (WEEL-Ceiling) Guides published by the American Industrial Hygiene Association (AIHA), and other *approved*, consistent measures are allowed as surrogates for hazardous substances not listed in DOL 29 CFR Part 1910.1000.

[BF] CEILING RADIATION DAMPER. A *listed* device installed in a ceiling membrane of a fire-resistance-rated floor/ceiling or roof/ceiling assembly to limit automatically the radiative heat transfer through an air inlet/outlet opening. Ceiling radiation dampers include air terminal units, ceiling dampers and ceiling air diffusers. Ceiling radiation dampers are classified for use in either static systems that will automatically shut down in the event of a fire, or in dynamic systems that continue to operate during a fire. A dynamic ceiling radiation damper is tested and rated for closure under elevated temperature airflow.

[BG] CELL (Group I-3 occupancy). A room within a *housing unit* in a detention or correctional facility used to confine inmates or prisoners.

[BS] CELL (masonry). A void space having a gross cross-sectional *area* greater than $1^{1}/_{2}$ square inches (967 mm^2).

[BG] CELL TIER. Levels of *cells* vertically stacked above one another within a *housing unit.*

[BS] CEMENT PLASTER. A mixture of Portland or blended cement, Portland cement or blended cement and hydrated lime, *masonry* cement or plastic cement and aggregate and other *approved* materials as specified in this code.

[BF] CERAMIC FIBER BLANKET. A high-temperature mineral wool insulation material made of alumina-silica ceramic or calcium magnesium silicate soluble fibers and weighing 4 to 10 pounds per cubic foot (pcf) (64 to 160 kg/m^3).

[BS] CERTIFICATE OF COMPLIANCE. A certificate stating that materials and products meet specified standards or that work was done in compliance with *approved construction documents*.

[A] CHANGE OF OCCUPANCY. Either of the following shall be considered as a change of occupancy where this code requires a greater degree of safety, accessibility, structural strength, fire protection, *means of egress*, ventilation or sanitation than is existing in the current building or structure:

1. Any change in the occupancy classification of a building or structure.
2. Any change in the purpose of, or a change in the level of activity within, a building or structure.

[M] CHIMNEY. A primarily vertical structure containing one or more flues, for the purpose of carrying gaseous products of combustion and air from a fuel-burning appliance to the outdoor atmosphere.

Factory-built chimney. A *listed* and *labeled chimney* composed of factory-made components, assembled in the field in accordance with manufacturer's instructions and the conditions of the listing.

Masonry chimney. A field-constructed chimney composed of solid *masonry units*, bricks, stones, or concrete.

Metal chimney. A field-constructed chimney of metal.

[M] CHIMNEY TYPES.

High-heat appliance type. An *approved* chimney for removing the products of combustion from fuel-burning, high-heat appliances producing combustion gases in excess of 2,000°F (1093°C) measured at the appliance flue outlet (see Section 2113.11.3).

Low-heat appliance type. An *approved* chimney for removing the products of combustion from fuel-burning, low-heat appliances producing combustion gases not in excess of 1,000°F (538°C) under normal operating conditions, but capable of producing combustion gases of 1,400°F (760°C) during intermittent forces firing for periods up to 1 hour. Temperatures shall be measured at the appliance flue outlet.

Masonry type. A field-constructed chimney of solid *masonry units* or stones.

Medium-heat appliance type. An *approved* chimney for removing the products of combustion from fuel-burning, medium-heat appliances producing combustion gases not exceeding 2,000°F (1093°C) measured at the appliance flue outlet (see Section 2113.11.2).

[BE] CIRCULATION PATH. An exterior or interior way of passage from one place to another for pedestrians.

[F] CLEAN AGENT. Electrically nonconducting, volatile or gaseous fire extinguishant that does not leave a residue upon vaporation.

[BF] CLIMATE ZONE. A geographical region that has been assigned climatic criteria as specified in Chapters 3 [CE] and 3 [RE] of the *International Energy Conservation Code.*

[BG] CLINIC, OUTPATIENT. Buildings or portions thereof used to provide *medical care* on less than a *24-hour basis* to persons who are not rendered *incapable of self-preservation* by the services provided.

[F] CLOSED SYSTEM. The *use* of a *solid* or *liquid hazardous material* involving a closed vessel or system that remains closed during normal operations where vapors emitted by the product are not liberated outside of the vessel or system and the product is not exposed to the atmosphere during normal operations; and all *uses* of *compressed gases*. Examples of closed systems for *solids* and *liquids* include product conveyed through a piping system into a closed vessel, system or piece of equipment.

[BS] COASTAL A ZONE. Area within a *special flood hazard area*, landward of a V zone or landward of an open coast without mapped *coastal high-hazard areas*. In a *coastal A zone*, the principal source of *flooding* must be astronomical tides, storm surges, seiches or tsunamis, not riverine *flooding*. During the *base flood* conditions, the potential for breaking wave height shall be greater than or equal to $1^{1}/_{2}$ feet (457 mm). The inland limit of the *coastal A zone* is (a) the *Limit of Moderate Wave Action* if delineated on a FIRM, or (b) designated by the authority having jurisdiction.

[BS] COASTAL HIGH-HAZARD AREA. Area within the *special flood hazard area* extending from offshore to the inland limit of a primary dune along an open coast and any other area that is subject to high-velocity wave action from storms or seismic sources, and shown on a *Flood Insurance Rate Map* (FIRM) or other flood hazard map as velocity Zone V, VO, VE or V1-30.

[BS] COLLAR JOINT. Vertical longitudinal space between *wythes* of *masonry* or between *masonry wythe* and backup construction that is permitted to be filled with *mortar* or grout.

[BS] COLLECTOR. A horizontal *diaphragm* element parallel and in line with the applied force that collects and transfers *diaphragm* shear forces to the vertical elements of the lateral force-resisting system or distributes forces within the *diaphragm*, or both.

[BF] COMBINATION FIRE/SMOKE DAMPER. A *listed* device installed in ducts and air transfer openings designed to close *automatically* upon the detection of heat and resist the passage of flame and smoke. The device is installed to operate automatically, controlled by a smoke detection system, and where required, is capable of being positioned from a *fire command center*.

[BS] COMBINED PILE RAFT. A geotechnical composite construction that combines the bearing effect of both foundation elements, raft and piles, by taking into account interactions between the foundation elements and the subsoil.

[F] COMBUSTIBLE DUST. Finely divided *solid* material that is 420 microns or less in diameter and which, when dispersed in air in the proper proportions, could be ignited by a flame, spark or other source of ignition. Combustible dust will pass through a US No. 40 standard sieve.

[F] COMBUSTIBLE FIBERS. Readily ignitable and free-burning materials in a fibrous or shredded form, such as cocoa fiber, cloth, cotton, excelsior, hay, hemp, henequen, istle, jute, kapok, oakum, rags, sisal, Spanish moss, straw, tow, wastepaper, certain synthetic fibers or other like materials. This definition does not include densely packed *baled cotton*.

[F] COMBUSTIBLE LIQUID. A *liquid* having a closed cup *flash point* at or above 100°F (38°C). Combustible liquids shall be subdivided as follows:

The category of combustible liquids does not include *compressed gases* or *cryogenic fluids* or *liquids* that do not have a fire point when tested in accordance with ASTM D92.

Class II. *Liquids* having a closed cup *flash point* at or above 100°F (38°C) and below 140°F (60°C).

Class IIIA. *Liquids* having a closed cup *flash point* at or above 140°F (60°C) and below 200°F (93°C).

Class IIIB. *Liquids* having a closed cup *flash point* at or above 200°F (93°C).

[F] COMMERCIAL MOTOR VEHICLE. A motor vehicle used to transport passengers or property where the motor vehicle meets one of the following:

1. Has a gross vehicle weight rating of 10,000 pounds (4540 kg) or more.
2. Is designed to transport 16 or more passengers, including the driver.

[BE] COMMON PATH OF EGRESS TRAVEL. That portion of *exit access* travel distance measured from the most remote point of each room, area or space to that point where the occupants have separate and distinct access to two *exits* or *exit access* doorways.

[BE] COMMON USE. Interior or exterior *circulation paths*, rooms, spaces or elements that are not for public use and are made available for the shared use of two or more people.

[F] COMPRESSED GAS. A material or mixture of materials that meets both of the following:

1. Is a gas at 68°F (20°C) or less at 14.7 pounds per square inch atmosphere (psia) (101 kPa) of pressure.
2. Has a *boiling point* of 68°F (20°C) or less at 14.7 psia (101 kPa) which is either liquefied, nonliquefied or in solution, except those gases which have no other health- or physical-hazard properties are not considered to be compressed until the pressure in the packaging exceeds 41 psia (282 kPa) at 68°F (20°C).

The states of a compressed gas are categorized as follows:

1. Nonliquefied compressed gases are gases, other than those in solution, which are in a packaging under the charged pressure and are entirely gaseous at a temperature of 68°F (20°C).
2. Liquefied compressed gases are gases that, in a packaging under the charged pressure, are partially *liquid* at a temperature of 68°F (20°C).
3. Compressed gases in solution are nonliquefied gases that are dissolved in a solvent.
4. Compressed gas mixtures consist of a mixture of two or more compressed gases contained in a packaging, the hazard properties of which are represented by the properties of the mixture as a whole.

[BS] CONCRETE.

Carbonate aggregate. Concrete made with aggregates consisting mainly of calcium or magnesium carbonate, such as limestone or dolomite, and containing 40 percent or less quartz, chert or flint.

Cellular. A lightweight insulating concrete made by mixing a preformed foam with Portland cement slurry and having a dry unit weight of approximately 30 pcf (480 kg/m^3).

Lightweight aggregate. Concrete made with aggregates of expanded clay, shale, slag or slate or sintered fly ash or any natural lightweight aggregate meeting ASTM C330 and possessing equivalent fire-resistance properties and weighing 85 to 115 pcf (1360 to 1840 kg/m^3).

Perlite. A lightweight insulating concrete having a dry unit weight of approximately 30 pcf (480 kg/m^3) made with perlite concrete aggregate. Perlite aggregate is produced from a volcanic rock which, when heated, expands to form a glass-like material of cellular structure.

Sand-lightweight. Concrete made with a combination of expanded clay, shale, slag, slate, sintered fly ash, or any natural lightweight aggregate meeting ASTM C330 and possessing equivalent fire-resistance properties and natural sand. Its unit weight is generally between 105 and 120 pcf (1680 and 1920 kg/m^3).

Siliceous aggregate. Concrete made with normal-weight aggregates consisting mainly of silica or compounds other than calcium or magnesium carbonate, which contains more than 40-percent quartz, chert or flint.

Vermiculite. A light weight insulating concrete made with *vermiculite* concrete aggregate which is laminated micaceous material produced by expanding the ore at high temperatures. When added to a Portland cement slurry the resulting concrete has a dry unit weight of approximately 30 pcf (480 kg/m^3).

[BG] CONGREGATE LIVING FACILITIES. A building or part thereof that contains *sleeping units* where residents share bathroom or kitchen facilities, or both.

[F] CONSTANTLY ATTENDED LOCATION. A designated location at a facility staffed by trained personnel on a continuous basis where alarm or supervisory signals are monitored and facilities are provided for notification of the fire department or other emergency services.

[A] CONSTRUCTION DOCUMENTS. Written, graphic and pictorial documents prepared or assembled for describing the design, location and physical characteristics of the elements of a project necessary for obtaining a building *permit*.

[BG] CONSTRUCTION TYPES. See Section 602.

Type I. See Section 602.2.

Type II. See Section 602.2.

Type III. See Section 602.3.

Type IV. See Section 602.4.

Type V. See Section 602.5.

[BF] CONTINUOUS INSULATION (ci). Insulating material that is continuous across all structural members without thermal bridges other than fasteners and service openings. It is installed on the interior or exterior, or is integral to any opaque surface of the building envelope.

[F] CONTROL AREA. Spaces within a building where quantities of *hazardous materials* not exceeding the maximum allowable quantities per control area are stored, dispensed, *used* or handled. See the definition of "*Outdoor control area*" in the *International Fire Code*.

[BS] CONTROLLED LOW-STRENGTH MATERIAL. A self-compacted, cementitious material used primarily as a backfill in place of compacted fill.

[BS] CONVENTIONAL LIGHT-FRAME CONSTRUCTION. Construction whose primary structural elements are formed by a system of repetitive wood-framing members. See Section 2308 for conventional *light-frame construction* provisions.

[BG] CORNICE. A projecting horizontal molded element located at or near the top of an architectural feature.

[BE] CORRIDOR. An enclosed *exit access* component that defines and provides a path of egress travel.

CORRIDOR, OPEN-ENDED. See "*Open-ended corridor*."

[BF] CORRIDOR DAMPER. A *listed* device intended for use where air ducts penetrate or terminate at horizontal openings in the ceilings of fire-resistance-rated corridors, where the *corridor* ceiling is permitted to be constructed as required for the *corridor* walls.

[BS] CORROSION RESISTANCE. The ability of a material to withstand deterioration of its surface or its properties when exposed to its environment.

[F] CORROSIVE. A chemical that causes visible destruction of, or irreversible alterations in, living tissue by chemical action at the point of contact. A chemical shall be considered corrosive if, when tested on the intact skin of albino rabbits by the method described in DOTn 49 CFR, Part 173.137, such chemical destroys or changes irreversibly the structure of the tissue at the point of contact following an exposure period of 4 hours. This term does not refer to action on inanimate surfaces.

[BG] COURT. An open, uncovered space, unobstructed to the sky, bounded on three or more sides by exterior building walls or other enclosing devices.

[BG] COVERED MALL BUILDING. A single building enclosing a number of tenants and occupants, such as retail stores, drinking and dining establishments, entertainment and amusement facilities, passenger transportation terminals, offices and other similar uses wherein two or more tenants have a main entrance into one or more malls. *Anchor buildings* shall not be considered as a part of the covered mall building. The term "covered mall building" shall include *open mall buildings* as defined below.

Mall. A roofed or covered common pedestrian area within a *covered mall building* that serves as access for two or more tenants and not to exceed three levels that are open to each other. The term "mall" shall include open malls as defined below.

Open mall. An unroofed common pedestrian way serving a number of tenants not exceeding three levels. Circulation at levels above grade shall be permitted to include open exterior balconies leading to exits discharging at grade.

Open mall building. Several structures housing a number of tenants, such as retail stores, drinking and dining establishments, entertainment and amusement facilities, offices, and other similar uses, wherein two or more tenants have a main entrance into one or more open malls. *Anchor buildings* are not considered as a part of the open mall building.

[BS] CRIPPLE WALL. A framed stud wall extending from the top of the foundation to the underside of floor framing for the lowest occupied floor level.

[F] CRITICAL CIRCUIT. A circuit that requires continuous operation to ensure safety of the structure and occupants.

[BS] CROSS-LAMINATED TIMBER. A prefabricated engineered wood product consisting of not less than three layers of solid-sawn lumber or *structural composite lumber* where the adjacent layers are cross oriented and bonded with structural adhesive to form a solid wood element.

[F] CRYOGENIC FLUID. A *liquid* having a *boiling point* lower than -150°F (-101°C) at 14.7 pounds per square inch atmosphere (psia) (an absolute pressure of 101 kPa).

[BG] CUSTODIAL CARE. Assistance with day-to-day living tasks; such as assistance with cooking, taking medication, bathing, using toilet facilities and other tasks of daily living. *Custodial care* includes persons receiving care who have the ability to respond to emergency situations and evacuate at a slower rate and/or who have mental and psychiatric complications.

[BS] DALLE GLASS. A decorative composite glazing material made of individual pieces of glass that are embedded in a cast matrix of concrete or epoxy.

DAMPER. See "*Ceiling radiation damper,*" "*Combination fire/smoke damper,*" "*Corridor damper,*" "*Fire damper*" and "*Smoke damper.*"

[BS] DANGEROUS. Any building, structure or portion thereof that meets any of the conditions described below shall be deemed *dangerous*:

1. The building or structure has collapsed, has partially collapsed, has moved off its foundation or lacks the necessary support of the ground.
2. There exists a significant risk of collapse, detachment or dislodgment of any portion, member, appurtenance or ornamentation of the building or structure under permanent, routine, or frequent *loads;* under actual loads already in effect; or under snow, wind, rain, *flood,* earthquake, or other environmental loads when such *loads* are imminent.

[F] DAY BOX. A portable magazine designed to hold *explosive* materials constructed in accordance with the requirements for a Type 3 magazine as defined and classified in Chapter 56 of the *International Fire Code*.

[BS] DEAD LOAD. The weight of materials of construction incorporated into the building, including but not limited to walls, floors, roofs, ceilings, *stairways*, built-in partitions, finishes, cladding and other similarly incorporated architectural and structural items, and the weight of fixed service equipment, including cranes and material handling systems.

[BS] DECORATIVE GLASS. A carved, leaded or *Dalle glass* or glazing material whose purpose is decorative or artistic, not functional; whose coloring, texture or other design qualities or components cannot be removed without destroying the glazing material and whose surface, or assembly into which it is incorporated, is divided into segments.

[F] DECORATIVE MATERIALS. All materials applied over the building *interior finish* for decorative, acoustical or other effect including, but not limited to, curtains, draperies, fabrics and streamers; and all other materials utilized for decorative effect including, but not limited to, bulletin boards, artwork, posters, photographs, batting, cloth, cotton, hay, stalks, straw, vines, leaves, trees, moss and similar items, foam plastics and materials containing foam plastics. Decorative materials do not include wall coverings, ceiling coverings, floor coverings, ordinary window shades, *interior finish* and materials 0.025 inch (0.64 mm) or less in thickness applied directly to and adhering tightly to a substrate.

[BS] DEEP FOUNDATION. A deep foundation is a foundation element that does not satisfy the definition of a *shallow foundation*.

[BE] DEFEND-IN-PLACE. A method of emergency response that engages building components and trained staff to provide occupant safety during an emergency. Emergency response involves remaining in place, relocating within the building, or both, without evacuating the building.

[A] DEFERRED SUBMITTAL. Those portions of the design that are not submitted at the time of the application and that are to be submitted to the *building official* within a specified period.

[F] DEFLAGRATION. An exothermic reaction, such as the extremely rapid oxidation of a flammable dust or vapor in air, in which the reaction progresses through the unburned material at a rate less than the velocity of sound. A deflagration can have an explosive effect.

[BF] DELAYED-ACTION CLOSER. A *self-closing* device that incorporates a delay prior to the initiation of closing. Delayed-action closers are mechanical devices with an adjustable delay.

[F] DELUGE SYSTEM. A sprinkler system employing open sprinklers attached to a piping system connected to a water supply through a valve that is opened by the operation of a detection system installed in the same areas as the sprinklers. When this valve opens, water flows into the piping system and discharges from all sprinklers attached thereto.

[BS] DESIGN EARTHQUAKE GROUND MOTION. The earthquake ground motion that buildings and structures are specifically proportioned to resist in Section 1613.

[BS] DESIGN FLOOD. The *flood* associated with the greater of the following two areas:

1. Area with a flood plain subject to a 1-percent or greater chance of *flooding* in any year.
2. Area designated as a *flood hazard area* on a community's flood hazard map, or otherwise legally designated.

[BS] DESIGN FLOOD ELEVATION. The elevation of the "*design flood,*" including wave height, relative to the datum specified on the community's legally designated flood hazard map. In areas designated as Zone AO, the *design flood elevation* shall be the elevation of the highest existing grade of the building's perimeter plus the depth number (in feet) specified on the flood hazard map. In areas designated as Zone AO where a depth number is not specified on the map, the depth number shall be taken as being equal to 2 feet (610 mm).

DESIGN PROFESSIONAL, REGISTERED. See "*Registered design professional.*"

DESIGN PROFESSIONAL IN RESPONSIBLE CHARGE, REGISTERED. See "*Registered design professional in responsible charge.*"

[BS] DESIGN STRENGTH. The product of the nominal strength and a *resistance factor* (or strength reduction factor).

[BS] DESIGNATED SEISMIC SYSTEM. Those nonstructural components that require design in accordance with Chapter 13 of ASCE 7 and for which the component importance factor, I_p, is greater than 1 in accordance with Section 13.1.3 of ASCE 7.

[F] DETACHED BUILDING. A separate single-story building, without a *basement* or crawl space, used for the storage or *use* of *hazardous materials* and located an *approved* distance from all structures.

[BS] DETAILED PLAIN CONCRETE STRUCTURAL WALL. See Section 1905.1.1.

[BE] DETECTABLE WARNING. A standardized surface feature built in or applied to walking surfaces or other elements to warn visually impaired persons of hazards on a *circulation path*.

[F] DETECTOR, HEAT. A fire detector that senses heat—either abnormally high temperature or rate of rise, or both.

[F] DETONATION. An exothermic reaction characterized by the presence of a shock wave in the material which establishes and maintains the reaction. The reaction zone progresses through the material at a rate greater than the velocity of sound. The principal heating mechanism is one of shock compression. Detonations have an explosive effect.

[BG] DETOXIFICATION FACILITIES. Facilities that provide treatment for substance abuse, serving care recipients who are *incapable of self-preservation* or who are harmful to themselves or others.

[BS] DIAPHRAGM. A horizontal or sloped system acting to transmit lateral forces to vertical elements of the lateral force-resisting system. When the term "*diaphragm*" is used, it shall include horizontal bracing systems.

[BS] DIMENSIONS. This definition applies only to Chapter 21.

Nominal. The *specified dimension* plus an allowance for the *joints* with which the units are to be laid. Nominal *dimensions* are usually stated in whole numbers. Thickness is given first, followed by height and then length.

Specified. *Dimensions* specified for the manufacture or construction of a unit, *joint* or element.

[BE] DIRECT ACCESS. A path of travel from a space to an immediately adjacent space through an opening in the common wall between the two spaces.

[F] DISPENSING. The pouring or transferring of any material from a container, tank or similar vessel, whereby vapors, dusts, fumes, mists or gases are liberated to the atmosphere.

DOOR, BALANCED. See "*Balanced door.*"

DOOR, LOW-ENERGY POWER-OPERATED. See "*Low-energy power-operated door.*"

DOOR, POWER-ASSISTED. See "*Power-assisted door.*"

DOOR, POWER-OPERATED. See "*Power-operated door.*"

DOORWAY, EXIT ACCESS. See "*Exit access doorway.*"

[BG] DORMITORY. A space in a building where group sleeping accommodations are provided in one room, or in a series of closely associated rooms, for persons not members of the same family group, under joint occupancy and single management, as in college *dormitories* or fraternity houses.

[BF] DRAFTSTOP. A material, device or construction installed to restrict the movement of air within open spaces of concealed areas of building components such as crawl spaces, floor/ceiling assemblies, roof/ceiling assemblies and *attics*.

DRAG STRUT. See "*Collector.*"

[BS] DRILLED SHAFT. A cast-in-place *deep foundation* element, also referred to as a caisson, drilled pier or bored pile, constructed by drilling a hole (with or without permanent casing or drilling fluid) into soil or rock and filling it with fluid concrete after the drilling equipment is removed.

[BS] DRY FLOODPROOFING. A combination of design modifications that results in a building or structure, including the attendant utilities and equipment and sanitary facilities, being water tight with walls substantially impermeable to the passage of water and with structural components having the capacity to resist *loads* as identified in ASCE 7.

[F] DRY-CHEMICAL EXTINGUISHING AGENT. A powder composed of small particles, usually of sodium bicarbonate, potassium bicarbonate, urea-potassium-based bicarbonate, potassium chloride or monoammonium phosphate, with added particulate material supplemented by special treatment to provide resistance to packing, resistance to moisture absorption (caking) and the proper flow capabilities.

[A] DWELLING. A building that contains one or two *dwelling units* used, intended or designed to be used, rented, leased, let or hired out to be occupied for living purposes.

[A] DWELLING UNIT. A single unit providing complete, independent living facilities for one or more persons, including permanent provisions for living, sleeping, eating, cooking and sanitation.

[BG] DWELLING UNIT, EFFICIENCY. A *dwelling unit* where all permanent provisions for living, sleeping, eating and cooking are contained in a single room.

DWELLING UNIT OR SLEEPING UNIT, MULTISTORY. See "*Multistory unit.*"

[BE] EGRESS COURT. A *court* or *yard* which provides access to a *public way* for one or more *exits*.

[BG] ELECTRIC VEHICLE CHARGING STATION. One or more vehicle spaces served by an electric vehicle charging system.

[BF] ELECTRICAL CIRCUIT PROTECTIVE SYSTEM. A specific construction of devices, materials, or coatings installed as a fire-resistive barrier system applied to electrical system components, such as cable trays, conduits and other raceways, open run cables and conductors, cables, and conductors.

[F] ELEVATOR GROUP. A grouping of elevators in a building located adjacent or directly across from one another that responds to common hall call buttons.

[F] EMERGENCY ALARM SYSTEM. A system to provide indication and warning of emergency situations involving *hazardous materials*.

[F] EMERGENCY CONTROL STATION. An *approved* location on the premises where signals from emergency equipment are received and which is staffed by trained personnel.

[BE] EMERGENCY ESCAPE AND RESCUE OPENING. An operable exterior window, door or other similar device that provides for a means of escape and access for rescue in the event of an emergency.

[F] EMERGENCY POWER SYSTEM. A source of *automatic* electric power of a required capacity and duration to operate required life safety, fire alarm, detection and ventilation systems in the event of a failure of the primary power. Emergency power systems are required for electrical loads where interruption of the primary power could result in loss of human life or serious injuries.

[F] EMERGENCY VOICE/ALARM COMMUNICATIONS. Dedicated manual or *automatic* facilities for originating and distributing voice instructions, as well as alert and evacuation signals pertaining to a fire emergency, to the occupants of a building.

[BF] EMITTANCE. The ratio of radiant heat flux emitted by a specimen to that emitted by a blackbody at the same temperature and under the same conditions.

[BE] EMPLOYEE WORK AREA. All or any portion of a space used only by employees and only for work. *Corridors*, toilet rooms, kitchenettes and break rooms are not employee work areas.

[F] ENERGY STORAGE SYSTEM, ELECTROCHEMICAL. An energy storage system that stores energy and produces electricity using chemical reactions. It includes, among others, battery ESS and capacitor ESS.

[BS] ENGINEERED WOOD RIM BOARD. A full-depth *structural composite lumber*, *wood structural panel*, structural glued laminated timber or *prefabricated wood I-joist* member designed to transfer horizontal (shear) and vertical (compression) *loads*, provide attachment for *diaphragm* sheathing, siding and exterior deck ledgers, and provide lateral support at the ends of floor or roof joists or rafters.

ENTRANCE, PUBLIC. See "*Public entrance.*"

ENTRANCE, RESTRICTED. See "*Restricted entrance.*"

ENTRANCE, SERVICE. See "*Service entrance.*"

[BG] EQUIPMENT PLATFORM. An unoccupied, elevated platform used exclusively for mechanical systems or industrial process equipment, including the associated elevated walkways, *stairways, alternating tread devices* and ladders necessary to access the platform (see Section 505.3).

[BS] ESSENTIAL FACILITIES. Buildings and other structures that are intended to remain operational in the event of extreme environmental loading from *flood*, wind, snow or earthquakes.

[F] EXHAUSTED ENCLOSURE. An appliance or piece of equipment that consists of a top, a back and two sides providing a means of local exhaust for capturing gases, fumes, vapors and mists. Such enclosures include laboratory hoods, exhaust fume hoods and similar appliances and equipment used to locally retain and exhaust the gases, fumes, vapors and mists that could be released. Rooms or areas provided with general *ventilation*, in themselves, are not exhausted enclosures.

[A] EXISTING BUILDING. A building erected prior to the date of adoption of the appropriate code, or one for which a legal building *permit* has been issued.

[BS] EXISTING STRUCTURE. A structure erected prior to the date of adoption of the appropriate code, or one for which a legal building *permit* has been issued.

[BE] EXIT. That portion of a *means of egress* system between the *exit access* and the *exit discharge* or *public way*. Exit components include exterior exit doors at the *level of exit discharge*, *interior exit stairways* and *ramps*, *exit passageways*, *exterior exit stairways* and *ramps* and *horizontal exits*.

EXIT, HORIZONTAL. See "*Horizontal exit.*"

[BE] EXIT ACCESS. That portion of a *means of egress* system that leads from any occupied portion of a building or structure to an *exit*.

[BE] EXIT ACCESS DOORWAY. A door or access point along the path of egress travel from an occupied room, area or space where the path of egress enters an intervening room, *corridor*, *exit access stairway* or *ramp*.

[BE] EXIT ACCESS RAMP. A *ramp* within the *exit access* portion of the *means of egress* system.

[BE] EXIT ACCESS STAIRWAY. A *stairway* within the *exit access* portion of the *means of egress* system.

[BE] EXIT DISCHARGE. That portion of a *means of egress* system between the termination of an *exit* and a *public way*.

[BE] EXIT DISCHARGE, LEVEL OF. The *story* at the point at which an *exit* terminates and an *exit discharge* begins.

[BE] EXIT PASSAGEWAY. An *exit* component that is separated from other interior spaces of a building or structure by fire-resistance-rated construction and opening protectives, and provides for a protected path of egress travel in a horizontal direction to an *exit* or to the *exit discharge*.

[BF] EXPANDED VINYL WALL COVERING. Wall covering consisting of a woven textile backing, an expanded vinyl base coat layer and a nonexpanded vinyl skin coat. The expanded base coat layer is a homogeneous vinyl layer that contains a blowing agent. During processing, the blowing agent decomposes, causing this layer to expand by forming closed cells. The total thickness of the wall covering is approximately 0.055 inch to 0.070 inch (1.4 mm to 1.78 mm).

[F] EXPLOSION. An effect produced by the sudden violent expansion of gases, which may be accompanied by a shock wave or disruption, or both, of enclosing materials or structures. An explosion could result from any of the following:

1. Chemical changes such as rapid oxidation, *deflagration* or *detonation*, decomposition of molecules and runaway polymerization (usually *detonations*).
2. Physical changes such as pressure tank ruptures.
3. Atomic changes (nuclear fission or fusion).

[F] EXPLOSIVE. A chemical compound, mixture or device, the primary or common purpose of which is to function by explosion. The term includes, but is not limited to: dynamite, black powder, pellet powder, initiating explosives, detonators, safety fuses, squibs, detonating cord, igniter cord, and igniters. The term "explosive" includes any material determined to be within the scope of USC Title 18: Chapter 40 and also includes any material classified as an explosive other than consumer fireworks, 1.4G by the *hazardous materials* regulations of DOTn 49 CFR Parts 100-185.

High explosive. Explosive material, such as dynamite, which can be caused to detonate by means of a No. 8 test blasting cap when unconfined.

Low explosive. Explosive material that will burn or deflagrate when ignited. It is characterized by a rate of reaction that is less than the speed of sound. Examples of low explosives include, but are not limited to: black powder; safety fuse; igniters; igniter cord; fuse lighters; fireworks; and propellants, 1.3C.

Mass-detonating explosives. Division 1.1, 1.2 and 1.5 explosives alone or in combination, or loaded into various types of ammunition or containers, most of which can be expected to explode virtually instantaneously when a small portion is subjected to fire, severe concussion, impact, the impulse of an initiating agent or the effect of a considerable discharge of energy from without. Materials that react in this manner represent a mass explosion hazard. Such an explosive will normally cause severe structural damage to adjacent objects. Explosive propagation could occur immediately to other items of ammunition and explosives stored sufficiently close to and not adequately protected from the initially exploding pile with a time interval short enough so that two or more quantities must be considered as one for quantity-distance purposes.

UN/DOTn Class 1 explosives. The former classification system used by DOTn included the terms "high" and "low" explosives as defined herein. The following terms further define explosives under the current system applied by DOTn for all explosive materials defined as hazard Class 1 materials. Compatibility group letters are used in concert with the division to specify further limitations on each division noted (i.e., the letter G identifies the material as a pyrotechnic substance or article containing a pyrotechnic substance and similar materials).

Division 1.1. Explosives that have a mass explosion hazard. A mass explosion is one which affects almost the entire *load* instantaneously.

Division 1.2. Explosives that have a projection hazard but not a mass explosion hazard.

Division 1.3. Explosives that have a fire hazard and either a minor blast hazard or a minor projection hazard or both, but not a mass explosion hazard.

Division 1.4. Explosives that pose a minor explosion hazard. The explosive effects are largely confined to the package and no projection of fragments of appreciable size or range is to be expected. An external fire must not cause virtually instantaneous explosion of almost the entire contents of the package.

Division 1.5. Very insensitive explosives. This division is comprised of substances that have a mass explosion hazard, but that are so insensitive there is very little probability of initiation or of transition from burning to *detonation* under normal conditions of transport.

Division 1.6. Extremely insensitive articles which do not have a mass explosion hazard. This division is comprised of articles that contain only extremely insensitive detonating substances and which demonstrate a negligible probability of accidental initiation or propagation.

[BE] EXTERIOR EXIT RAMP. An *exit* component that serves to meet one or more *means of egress* design requirements, such as required number of *exits* or *exit access* travel distance, and is open to *yards*, *courts* or *public ways*.

[BE] EXTERIOR EXIT STAIRWAY. An *exit* component that serves to meet one or more *means of egress* design requirements, such as required number of *exits* or *exit access* travel distance, and is open to *yards*, *courts* or *public ways*.

[BF] EXTERIOR INSULATION AND FINISH SYSTEMS (EIFS). EIFS are nonstructural, nonload-bearing, *exterior wall* cladding systems that consist of an insulation board attached either adhesively or mechanically, or both, to the substrate; an integrally reinforced base coat and a textured protective finish coat.

[BF] EXTERIOR INSULATION AND FINISH SYSTEMS (EIFS) WITH DRAINAGE. An EIFS that incorporates a means of drainage applied over a *water-resistive barrier*.

[BF] EXTERIOR SURFACES. *Weather-exposed surfaces.*

[BF] EXTERIOR WALL. A wall, bearing or nonbearing, that is used as an enclosing wall for a building, other than a *fire wall*, and that has a slope of 60 degrees (1.05 rad) or greater with the horizontal plane.

[BF] EXTERIOR WALL COVERING. A material or assembly of materials applied on the exterior side of *exterior walls* for the purpose of providing a weather-resisting barrier, insulation or for aesthetics, including but not limited to, *veneers*, siding, *exterior insulation and finish systems*, architectural trim and embellishments such as *cornices*, soffits, facias, gutters and leaders.

[BF] EXTERIOR WALL ENVELOPE. A system or assembly of *exterior wall* components, including *exterior wall* covering materials, that provides protection of the building structural members, including framing and sheathing materials, and conditioned interior space, from the detrimental effects of the exterior environment.

[BF] F RATING. The time period that the *through-penetration firestop system or perimeter fire containment system* limits the spread of fire through the penetration or void.

[BF] FABRIC PARTITION. A partition consisting of a finished surface made of fabric, without a continuous rigid backing, that is directly attached to a framing system in which the vertical framing members are spaced greater than 4 feet (1219 mm) on center.

[BS] FABRICATED ITEM. Structural, *load*-bearing or lateral *load*-resisting members or assemblies consisting of materials assembled prior to installation in a building or structure, or subjected to operations such as heat treatment, thermal cutting, cold working or reforming after manufacture and prior to installation in a building or structure. Materials produced in accordance with standards referenced by this code, such as rolled structural steel shapes, steel reinforcing bars, *masonry units* and *wood structural panels*, or in accordance with a referenced standard that provides requirements for quality control done under the supervision of a third-party quality control agency, are not "*fabricated items*."

[F] FABRICATION AREA. An area within a semiconductor fabrication facility and related research and development areas in which there are processes using hazardous production materials. Such areas are allowed to include ancillary rooms or areas such as dressing rooms and offices that are directly related to the fabrication area processes.

[A] FACILITY. All or any portion of buildings, structures, site improvements, elements and pedestrian or vehicular routes located on a *site*.

[BS] FACTORED LOAD. The product of a *nominal load* and a load factor.

[BS] FENESTRATION. Products classified as either *vertical fenestration* or *skylights and sloped glazing*, installed in such a manner as to preserve the weather-resistant barrier of the wall or roof in which they are installed. *Fenestration* includes products with glass or other transparent or translucent materials.

[BS] FENESTRATION, VERTICAL. Windows that are fixed or movable, opaque doors, glazed doors, glazed block and combination opaque and glazed doors installed in a wall at less than 15 degrees from the vertical.

[BS] FIBERBOARD. A fibrous, homogeneous panel made from lignocellulosic fibers (usually wood or cane) and having a density of less than 31 pounds per cubic foot (pcf) (497 kg/m^3) but more than 10 pcf (160 kg/m^3).

[BS] FIBER-CEMENT (BACKER BOARD, SIDING, SOFFIT, TRIM AND UNDERLAYMENT) PRODUCTS. Manufactured thin section composites of hydraulic cementitious matrices and discrete nonasbestos fibers.

[BF] FIBER-REINFORCED POLYMER. A polymeric composite material consisting of reinforcement fibers, such as glass, impregnated with a fiber-binding polymer which is then molded and hardened. Fiber-reinforced polymers are permitted to contain cores laminated between fiber-reinforced polymer facings.

FIELD NAILING. See "*Nailing, field*."

FIRE ALARM BOX, MANUAL. See "*Manual fire alarm box*."

[F] FIRE ALARM CONTROL UNIT. A system component that receives inputs from *automatic* and manual *fire alarm* devices and may be capable of supplying power to detection devices and transponders or off-premises transmitters. The control unit may be capable of providing a transfer of power to the notification appliances and transfer of condition to relays or devices.

[F] FIRE ALARM SIGNAL. A signal initiated by a *fire alarm-initiating device* such as a *manual fire alarm box*, *automatic fire detector*, waterflow switch or other device whose activation is indicative of the presence of a fire or fire signature.

[F] FIRE ALARM SYSTEM. A system or portion of a combination system consisting of components and circuits arranged to monitor and annunciate the status of *fire alarm* or *supervisory signal-initiating devices* and to initiate the appropriate response to those signals.

[BF] FIRE AREA. The aggregate floor area enclosed and bounded by *fire walls*, *fire barriers*, *exterior walls* or *horizontal assemblies* of a building. Areas of the building not provided with surrounding walls shall be included in the fire area if such areas are included within the horizontal projection of the roof or floor next above.

[BF] FIRE BARRIER. A fire-resistance-rated wall assembly of materials designed to restrict the spread of fire in which continuity is maintained.

[F] FIRE COMMAND CENTER. The principal attended or unattended location where the status of detection, alarm communications and control systems is displayed, and from which the systems can be manually controlled.

[BF] FIRE DAMPER. A *listed* device installed in ducts and air transfer openings designed to close automatically upon detection of heat and resist the passage of flame. Fire dampers are classified for use in either static systems that will automatically shut down in the event of a fire, or in dynamic systems that continue to operate during a fire. A dynamic fire damper is tested and rated for closure under elevated temperature airflow.

[F] FIRE DETECTOR, AUTOMATIC. A device designed to detect the presence of a fire signature and to initiate action.

[BF] FIRE DOOR. The door component of a *fire door assembly*.

[BF] FIRE DOOR ASSEMBLY. Any combination of a *fire door*, frame, hardware and other accessories that together provide a specific degree of fire protection to the opening.

FIRE DOOR ASSEMBLY, FLOOR. See "*Floor fire door assembly*."

[BF] FIRE EXIT HARDWARE. *Panic hardware* that is *listed* for use on *fire door assemblies*.

[F] FIRE LANE. A road or other passageway developed to allow the passage of fire apparatus. A fire lane is not necessarily intended for vehicular traffic other than fire apparatus.

[BF] FIRE PARTITION. A vertical assembly of materials designed to restrict the spread of fire in which openings are protected.

[BF] FIRE PROTECTION RATING. The period of time that an opening protective will maintain the ability to confine a fire as determined by tests specified in Section 716. Ratings are stated in hours or minutes.

[F] FIRE PROTECTION SYSTEM. *Approved* devices, equipment and systems or combinations of systems used to detect a fire, activate an alarm, extinguish or control a fire, control or manage smoke and products of a fire or any combination thereof.

[BF] FIRE PROTECTIVE CURTAIN ASSEMBLY. An assembly consisting of a fabric curtain, a bottom bar, guides, a coil, and an operating and closing system.

[BF] FIRE RESISTANCE. That property of materials or their assemblies that prevents or retards the passage of excessive heat, hot gases or flames under conditions of use.

[F] FIRE SAFETY FUNCTIONS. Building and fire control functions that are intended to increase the level of life safety for occupants or to control the spread of harmful effects of fire.

[BF] FIRE SEPARATION DISTANCE. The distance measured from the building face to one of the following:

1. The closest interior *lot line*.
2. To the centerline of a street, an alley or *public way*.
3. To an imaginary line between two buildings on the lot.

The distance shall be measured at right angles from the face of the wall.

[BF] FIRE WALL. A fire-resistance-rated wall having protected openings, which restricts the spread of fire and extends continuously from the foundation to or through the roof, with sufficient structural stability under fire conditions to allow collapse of construction on either side without collapse of the wall.

[BF] FIRE WINDOW ASSEMBLY. A window constructed and glazed to give protection against the passage of fire.

[BF] FIREBLOCKING. Building materials, or materials *approved* for use as fireblocking, installed to resist the free passage of flame to other areas of the building through concealed spaces.

[M] FIREPLACE. A hearth and fire chamber or similar prepared place in which a fire may be made and which is built in conjunction with a chimney.

[BS] FIREPLACE THROAT. The opening between the top of the firebox and the smoke chamber.

[BF] FIRE-RATED GLAZING. Glazing with either a *fire protection rating* or a *fire-resistance rating*.

[BF] FIRE-RESISTANCE RATING. The period of time a *building element*, component or assembly maintains the ability to confine a fire, continues to perform a given structural function, or both, as determined by the tests, or the methods based on tests, prescribed in Section 703.

[BF] FIRE-RESISTANT JOINT SYSTEM. An assemblage of specific materials or products that are designed, tested and fire-resistance rated in accordance with either ASTM E1966 or UL 2079 to resist for a prescribed period of time the passage of fire through *joints* made in or between fire-resistance-rated assemblies.

[BS] FIRE-RETARDANT-TREATED WOOD. Wood products that, when impregnated with chemicals by a pressure process or other means during manufacture, exhibit reduced surface-building characteristics and resist propagation of fire.

FIRESTOP, MEMBRANE-PENETRATION. See "*Membrane-penetration firestop*."

FIRESTOP, PENETRATION. See "*Penetration firestop*."

FIRESTOP SYSTEM, THROUGH-PENETRATION. See "*Through-penetration firestop system*."

[F] FIREWORKS. Any composition or device for the purpose of producing a visible or audible effect for entertainment purposes by combustion, *deflagration* or *detonation* that meets the definition of 1.4G fireworks or 1.3G fireworks.

Fireworks, 1.3G. Large fireworks devices, which are explosive materials, intended for use in fireworks displays and designed to produce audible or visible effects by combustion, *deflagration* or *detonation*. Such 1.3G fireworks include, but are not limited to, firecrackers containing more than 130 milligrams (2 grains) of explosive composition, aerial shells containing more than 40 grams of pyrotechnic composition, and other display pieces which exceed the limits for classification as 1.4G fireworks. Such 1.3G fireworks are also described as fireworks, UN0335 by the DOTn.

Fireworks, 1.4G. Small fireworks devices containing restricted amounts of pyrotechnic composition designed primarily to produce visible or audible effects by combustion or deflagration that complies with the construction, chemical composition and labeling regulations of the DOTn for fireworks, UN0336, and the US Consumer Product Safety Commission (CPSC) as set forth in CPSC 16 CFR: Parts 1500 and 1507.

[BG] FIXED BASE OPERATOR (FBO). A commercial business granted the right by the airport sponsor to operate on an airport and provide aeronautical services, such as fueling, hangaring, *tie-down* and parking, aircraft rental, aircraft maintenance and flight instruction.

[BE] FIXED SEATING. Furniture or fixture designed and installed for the use of sitting and secured in place including bench-type seats and seats with or without backs or armrests.

[BF] FLAME SPREAD. The propagation of flame over a surface.

[BF] FLAME SPREAD INDEX. A comparative measure, expressed as a dimensionless number, derived from visual measurements of the spread of flame versus time for a material tested in accordance with ASTM E84 or UL 723.

[F] FLAMMABLE GAS. A material that is a gas at 68°F (20°C) or less at 14.7 pounds per square inch atmosphere (psia) (101 kPa) of pressure [a material that has a *boiling point* of 68°F (20°C) or less at 14.7 psia (101 kPa)], which also meets one of the following:

1. Is ignitable at 14.7 psia (101 kPa) when in a mixture of 13 percent or less by volume with air.

2. Has a flammable range at 14.7 psia (101 kPa) with air of at least 12 percent, regardless of the lower limit.

The limits specified shall be determined at 14.7 psi (101 kPa) of pressure and a temperature of 68°F (20°C) in accordance with ASTM E681.

[F] FLAMMABLE LIQUEFIED GAS. A liquefied compressed gas which, under a charged pressure, is partially liquid at a temperature of 68°F (20°C) and which is flammable.

[F] FLAMMABLE LIQUID. A *liquid* having a closed cup *flash point* below 100°F (38°C). Flammable liquids are further categorized into a group known as Class I liquids. The Class I category is subdivided as follows:

Class IA. *Liquids* having a *flash point* below 73°F (23°C) and a *boiling point* below 100°F (38°C).

Class IB. *Liquids* having a *flash point* below 73°F (23°C) and a *boiling point* at or above 100°F (38°C).

Class IC. *Liquids* having a *flash point* at or above 73°F (23°C) and below 100°F (38°C). The category of flammable liquids does not include *compressed gases* or *cryogenic fluids,* or liquids that do not have a fire point when tested in accordance with ASTM D92.

[F] FLAMMABLE MATERIAL. A material capable of being readily ignited from common sources of heat or at a temperature of 600°F (316°C) or less.

[F] FLAMMABLE SOLID. A *solid*, other than a blasting agent or *explosive*, that is capable of causing fire through friction, absorption or moisture, spontaneous chemical change, or retained heat from manufacturing or processing, or which has an ignition temperature below 212°F (100°C) or which burns so vigorously and persistently when ignited as to create a serious hazard. A chemical shall be considered a flammable *solid* as determined in accordance with the test method of CPSC 16 CFR; Part 1500.44, if it ignites and burns with a self-sustained flame at a rate greater than 0.1 inch (2.5 mm) per second along its major axis.

[F] FLAMMABLE VAPORS OR FUMES. The concentration of flammable constituents in air that exceeds 25 percent of their *lower flammable limit (LFL)*.

[F] FLASH POINT. The minimum temperature in degrees Fahrenheit at which a *liquid* will give off sufficient vapors to form an ignitable mixture with air near the surface or in the container, but will not sustain combustion. The flash point of a *liquid* shall be determined by appropriate test procedure and apparatus as specified in ASTM D56, ASTM D93 or ASTM D3278.

[BE] FLIGHT. A continuous run of rectangular treads, *winders* or combination thereof from one landing to another.

FLOOD, DESIGN. See "*Design flood*."

[BS] FLOOD DAMAGE-RESISTANT MATERIALS. Any construction material capable of withstanding direct and prolonged contact with floodwaters without sustaining any damage that requires more than cosmetic *repair*.

FLOOD ELEVATION, DESIGN. See "*Design flood elevation*."

[BS] FLOOD HAZARD AREA. The greater of the following two areas:

1. The area within a flood plain subject to a 1-percent or greater chance of *flooding* in any year.
2. The area designated as a flood hazard area on a community's flood hazard map, or otherwise legally designated.

FLOOD HAZARD AREAS, SPECIAL. See "*Special flood hazard area*."

[BS] FLOOD INSURANCE RATE MAP (FIRM). An official map of a community on which the Federal Emergency Management Agency (FEMA) has delineated both the *special flood hazard areas* and the risk premium zones applicable to the community.

[BS] FLOOD INSURANCE STUDY. The official report provided by the Federal Emergency Management Agency containing the *Flood Insurance Rate Map* (FIRM), the Flood Boundary and *Floodway* Map (FBFM), the water surface elevation of the *base flood* and supporting technical data.

[BS] FLOOD or FLOODING. A general and temporary condition of partial or complete inundation of normally dry land from:

1. The overflow of inland or tidal waters.
2. The unusual and rapid accumulation or runoff of surface waters from any source.

[BS] FLOODWAY. The channel of the river, creek or other watercourse and the adjacent land areas that must be reserved in order to discharge the *base flood* without cumulatively increasing the water surface elevation more than a designated height.

[BE] FLOOR AREA, GROSS. The floor area within the inside perimeter of the *exterior walls* of the building under consideration, exclusive of vent *shafts* and *courts*, without deduction for *corridors*, *stairways*, *ramps*, closets, the thickness of interior walls, columns or other features. The floor area of a building, or portion thereof, not provided with surrounding *exterior walls* shall be the usable area under the horizontal projection of the roof or floor above. The *gross floor area* shall not include *shafts* with no openings or interior *courts*.

[BE] FLOOR AREA, NET. The actual occupied area not including unoccupied accessory areas such as *corridors*, *stairways*, *ramps*, toilet rooms, mechanical rooms and closets.

[BF] FLOOR FIRE DOOR ASSEMBLY. A combination of a *fire door*, a frame, hardware and other accessories installed in a horizontal plane, which together provide a specific degree of fire protection to a through-opening in a fire-resistance-rated floor (see Section 712.1.13.1).

[BF] FOAM PLASTIC INSULATION. A plastic that is intentionally expanded by the use of a foaming agent to produce a reduced-density plastic containing voids consisting of open or closed cells distributed throughout the plastic for thermal insulating or acoustical purposes and that has a density less than 20 pounds per cubic foot (pcf) (320 kg/m^3).

[F] FOAM-EXTINGUISHING SYSTEM. A special system discharging a foam made from concentrates, either mechanically or chemically, over the area to be protected.

[BE] FOLDING AND TELESCOPIC SEATING. Tiered seating having an overall shape and size that is capable of being reduced for purposes of moving or storing and is not a *building element.*

[BG] FOOD COURT. A public seating area located in the *mall* that serves adjacent food preparation tenant spaces.

[BG] FOSTER CARE FACILITIES. Facilities that provide care to more than five children, $2^1/_2$ years of age or less.

[BS] FOUNDATION PIER. This definition applies only to Chapter 21.

An isolated vertical foundation member whose horizontal dimension measured at right angles to its thickness does not exceed three times its thickness and whose height is equal to or less than four times its thickness.

[BS] FRAME STRUCTURE. A building or other structure in which vertical *loads* from floors and roofs are primarily supported by columns.

[F] FUEL CELL POWER SYSTEM, STATIONARY. A stationary energy-generation system that converts the chemical energy of a fuel and oxidant to electric energy (DC or AC electricity) by an electrochemical process.

Field-fabricated fuel cell power system. A *stationary fuel cell power system* that is assembled at the job site and is not a preengineered or prepackaged factory-assembled fuel cell power system.

Preengineered fuel cell power system. A *stationary fuel cell power system* consisting of components and modules that are produced in a factory and shipped to the job site for assembly.

Prepackaged fuel cell power system. A *stationary fuel cell power system* that is factory assembled as a single, complete unit and shipped as a complete unit for installation at the job site.

[BS] GABLE. The triangular portion of a wall beneath the end of a dual-slope, pitched, or mono-slope roof or portion thereof and above the top plates of the story or level of the ceiling below.

[BE] GAMING. To deal, operate, carry on, conduct, maintain or expose for play any game played with cards, dice, equipment or any mechanical, electromechanical or electronic device or machine for money, property, checks, credit or any representative of value except where occurring at private home or operated by a charitable or educational organization.

[BE] GAMING AREA. Single or multiple areas of a building or facility where gaming machines or tables are present and *gaming* occurs, including but not limited to, primary casino gaming areas, VIP gaming areas, high-roller gaming areas, bar tops, lobbies, dedicated rooms or spaces such as in retail or restaurant establishments, sports books and tournament areas.

[BE] GAMING MACHINE TYPE. Categorization of gaming machines per type of game played on them, including, but not limited to, slot machines, video poker and video keno.

[BE] GAMING TABLE TYPE. Categorization of gaming tables per the type of game played on them, including, but not limited to, baccarat, bingo, blackjack/21, craps, pai gow, poker, roulette.

[F] GAS CABINET. A fully enclosed, ventilated noncombustible enclosure used to provide an isolated environment for *compressed gas* cylinders in storage or *use.* Doors and access ports for exchanging cylinders and accessing pressure-regulating controls are allowed to be included.

[F] GAS DETECTION SYSTEM. A system or portion of a combination system that utilizes one or more stationary sensors to detect the presence of a specified gas at a specified concentration and initiate one or more responses required by this code, such as notifying a responsible person, activating an alarm signal, or activating or deactivating equipment. A self-contained gas detection and alarm device is not classified as a gas detection system.

[F] GAS ROOM. A separately ventilated, fully enclosed room in which only *compressed gases* and associated equipment and supplies are stored or *used.*

[F] GASEOUS HYDROGEN SYSTEM. An assembly of piping, devices and apparatus designed to generate, store, contain, distribute or transport a nontoxic, gaseous hydrogen-containing mixture having not less than 95-percent hydrogen gas by volume and not more than 1-percent oxygen by volume. Gaseous hydrogen systems consist of items such as *compressed gas* containers, reactors and appurtenances, including pressure regulators, pressure relief devices, manifolds, pumps, compressors and interconnecting piping and tubing and controls.

[BF] GLASS FIBERBOARD. Fibrous glass roof insulation consisting of inorganic glass fibers formed into rigid boards using a binder. The board has a top surface faced with asphalt and kraft reinforced with glass fiber.

[BS] GLASS MAT GYPSUM PANEL. A *gypsum panel* consisting of a noncombustible core primarily of gypsum, surfaced with glass mat partially or completely embedded in the core.

[BS] GRADE (LUMBER). The classification of lumber in regard to strength and utility in accordance with American Softwood Lumber Standard DOC PS 20 and the grading rules of an *approved* lumber rules-writing agency.

[BE] GRADE FLOOR EMERGENCY ESCAPE AND RESCUE OPENING. An *emergency escape and rescue opening* located such that the bottom of the clear opening is not more than 44 inches (1118 mm) above or below the finished ground level adjacent to the opening.

[BG] GRADE PLANE. A reference plane representing the average of finished ground level adjoining the building at *exterior walls*. Where the finished ground level slopes away from the *exterior walls*, the reference plane shall be established by the lowest points within the area between the building and the *lot line* or, where the *lot line* is more than 6 feet (1829 mm) from the building, between the building and a point 6 feet (1829 mm) from the building.

GRADE PLANE, STORY ABOVE. See "*Story above grade plane.*"

[BE] GRANDSTAND. Tiered seating supported on a dedicated structural system and two or more rows high and is not a *building element* (see "*Bleachers*").

[BG] GREENHOUSE. A structure or thermally isolated area of a building that maintains a specialized sunlit environment used for and essential to the cultivation, protection or maintenance of plants.

[BG] GROSS LEASABLE AREA. The total floor area designed for tenant occupancy and exclusive use. The area of tenant occupancy is measured from the centerlines of joint partitions to the outside of the tenant walls. All tenant areas, including areas used for storage, shall be included in calculating *gross leasable area*.

[BG] GROUP HOME. A facility for social rehabilitation, substance abuse or mental health problems that contains a group housing arrangement that provides *custodial care* but does not provide *medical care*.

[BE] GUARD. A building component or a system of building components located at or near the open sides of elevated walking surfaces that minimizes the possibility of a fall from the walking surface to a lower level.

[BG] GUESTROOM. A room used or intended to be used by one or more guests for living or sleeping purposes.

[BS] GYPSUM BOARD. The generic name for a family of sheet products consisting of a noncombustible core primarily of gypsum with paper surfacing.

[BS] GYPSUM PANEL PRODUCT. The general name for a family of sheet products consisting essentially of gypsum complying with the standards specified in Table 2506.2 and Table 2507.2, and Chapter 35. *Gypsum board* and *glass mat gypsum panels* are examples of *gypsum panel products*.

[BS] GYPSUM PLASTER. A mixture of calcined gypsum or calcined gypsum and lime and aggregate and other *approved* materials as specified in this code.

[BS] GYPSUM SHEATHING. *Gypsum panel products* specifically manufactured with enhanced water resistance for use as a substrate for exterior surface materials.

[BS] GYPSUM VENEER PLASTER. *Gypsum plaster* applied to an *approved* base in one or more coats normally not exceeding $^1/_4$ inch (6.4 mm) in total thickness.

[BS] GYPSUM WALLBOARD. A *gypsum board* used primarily as an interior surfacing for building structures.

[BG] HABITABLE SPACE. A space in a building for living, sleeping, eating or cooking. Bathrooms, toilet rooms, closets, halls, storage or utility spaces and similar areas are not considered *habitable spaces*.

[F] HALOGENATED EXTINGUISHING SYSTEM. A fire-extinguishing system using one or more atoms of an element from the halogen chemical series: fluorine, chlorine, bromine and iodine.

[F] HANDLING. The deliberate transport by any means to a point of storage or *use*.

[BE] HANDRAIL. A horizontal or sloping rail intended for grasping by the hand for guidance or support.

[BS] HARDBOARD. A fibrous-felted, homogeneous panel made from lignocellulosic fibers consolidated under heat and pressure in a hot press to a density not less than 31 pcf (497 kg/m^3).

HARDWARE. See "*Fire exit hardware*" and "*Panic hardware.*"

[F] HAZARDOUS MATERIALS. Those chemicals or substances that are *physical hazards* or *health hazards* as classified in Section 307 and the *International Fire Code*, whether the materials are in usable or waste condition.

[F] HAZARDOUS PRODUCTION MATERIAL (HPM). A *solid*, *liquid* or gas associated with semiconductor manufacturing that has a degree-of-hazard rating in health, flammability or instability of Class 3 or 4 as ranked by NFPA 704 and which is *used* directly in research, laboratory or production processes which have as their end product materials that are not hazardous.

[BS] HEAD JOINT. Vertical *mortar joint* placed between *masonry units* within the *wythe* at the time the *masonry units* are laid.

[F] HEALTH HAZARD. A classification of a chemical for which there is statistically significant evidence that acute or chronic health effects are capable of occurring in exposed persons. The term "health hazard" includes chemicals that are *toxic* or *highly toxic*, and *corrosive*.

HEAT DETECTOR. See "*Detector, heat.*"

[BG] HEIGHT, BUILDING. The vertical distance from *grade plane* to the average height of the highest roof surface.

[BS] HELICAL PILE. Manufactured steel *deep foundation* element consisting of a central shaft and one or more helical bearing plates. A *helical pile* is installed by rotating it into the ground. Each helical bearing plate is formed into a screw thread with a uniform defined pitch.

[F] HELIPAD. A structural surface that is used for the landing, taking off, taxiing and parking of helicopters.

[F] HELIPORT. An area of land or water or a structural surface that is used, or intended for use, for the landing and taking off of helicopters, and any appurtenant areas that are used, or intended for use, for heliport buildings or other heliport facilities.

[F] HELISTOP. The same as "heliport," except that no fueling, defueling, maintenance, repairs or storage of helicopters is permitted.

[F] HIGHER EDUCATION LABORATORY. Laboratories in Group B occupancies used for educational purposes above the 12th grade. Storage, use and handling of chemicals in such laboratories shall be limited to purposes related to testing, analysis, teaching, research or developmental activities on a nonproduction basis.

[F] HIGHLY TOXIC. A material which produces a lethal dose or lethal concentration that falls within any of the following categories:

1. A chemical that has a median lethal dose (LD_{50}) of 50 milligrams or less per kilogram of body weight when administered orally to albino rats weighing between 200 and 300 grams each.
2. A chemical that has a median lethal dose (LD_{50}) of 200 milligrams or less per kilogram of body weight when administered by continuous contact for 24 hours (or less if death occurs within 24 hours) with the bare skin of albino rabbits weighing between 2 and 3 kilograms each.
3. A chemical that has a median lethal concentration (LC_{50}) in air of 200 parts per million by volume or less of gas or vapor, or 2 milligrams per liter or less of mist, fume or dust, when administered by continuous inhalation for 1 hour (or less if death occurs within 1 hour) to albino rats weighing between 200 and 300 grams each.

Mixtures of these materials with ordinary materials, such as water, might not warrant classification as *highly toxic*. While this system is basically simple in application, any hazard evaluation that is required for the precise categorization of this type of material shall be performed by experienced, technically competent persons.

[BF] HIGH-PRESSURE DECORATIVE EXTERIOR-GRADE COMPACT LAMINATE (HPL). Panels consisting of layers of cellulose fibrous material impregnated with thermosetting resins and bonded together by a high-pressure process to form a homogeneous nonporous core suitable for exterior use.

[BF] HIGH-PRESSURE DECORATIVE EXTERIOR-GRADE COMPACT LAMINATE (HPL) SYSTEM. An *exterior wall covering* fabricated using HPL in a specific assembly including *joints*, seams, attachments, substrate, framing and other details as appropriate to a particular design.

[BG] HIGH-RISE BUILDING. A building with an occupied floor located more than 75 feet (22 860 mm) above the lowest level of fire department vehicle access.

[A] HISTORIC BUILDINGS. Any building or structure that is one or more of the following:

1. Listed or certified as eligible for listing by the State Historic Preservation Officer or the Keeper of the National Register of Historic Places, in the National Register of Historic Places.
2. Designated as historic under an applicable state or local law.
3. Certified as a contributing resource within a National Register, state designated or locally designated historic district.

[BF] HORIZONTAL ASSEMBLY. A fire-resistance-rated floor or *roof assembly* of materials designed to restrict the spread of fire in which continuity is maintained.

[BE] HORIZONTAL EXIT. An *exit* component consisting of fire-resistance-rated construction and opening protectives intended to compartmentalize portions of a building thereby creating refuge areas that afford safety from the fire and smoke from the area of fire origin.

[BG] HOSPITALS AND PSYCHIATRIC HOSPITALS. Facilities that provide care or treatment for the medical, psychiatric, obstetrical, or surgical treatment of care recipients who are *incapable of self-preservation*.

[BG] HOUSING UNIT. A *dormitory* or a group of *cells* with a common dayroom in Group I-3.

HPM. See "*Hazardous Production Material*."

[F] HPM ROOM. A room used in conjunction with or serving a Group H-5 occupancy, where *HPM* is stored or *used* and which is classified as a Group H-2, H-3 or H-4 occupancy.

[BS] HURRICANE-PRONE REGIONS. Areas vulnerable to hurricanes defined as:

1. The US Atlantic Ocean and Gulf of Mexico coasts where the basic design wind speed, *V*, for Risk Category II buildings is greater than 115 mph (51.4 m/s);
2. Hawaii, Puerto Rico, Guam, Virgin Islands and American Samoa.

[F] HYDROGEN FUEL GAS ROOM. A room or space that is intended exclusively to house a *gaseous hydrogen system*.

[BS] ICE-SENSITIVE STRUCTURE. A structure for which the effect of an atmospheric ice *load* governs the design of a structure or portion thereof. This includes, but is not limited to, lattice structures, guyed masts, overhead lines, light suspension and cable-stayed bridges, aerial cable systems (e.g., for ski lifts or logging operations), amusement rides, open catwalks and platforms, flagpoles and signs.

[F] IMMEDIATELY DANGEROUS TO LIFE AND HEALTH (IDLH). The concentration of airborne contaminants which poses a threat of death, immediate or delayed permanent adverse health effects, or effects that could prevent escape from such an environment. This contaminant concentration level is established by the National Institute of Occupational Safety and Health (NIOSH) based on both toxicity and flammability. It generally is expressed in parts per million by volume (ppmv/v) or milligrams per cubic meter (mg/m^3). If adequate data do not exist for precise establishment of IDLH concentrations, an independent certified industrial hygienist, industrial toxicologist, appropriate regulatory agency or other source *approved* by the *building official* shall make such determination.

[BS] IMPACT LOAD. The *load* resulting from moving machinery, elevators, craneways, vehicles and other similar forces and kinetic *loads*, pressure and possible surcharge from fixed or moving *loads*.

[BS] IMPACT PROTECTIVE SYSTEM. Construction that has been shown by testing to withstand the impact of test missiles and that is applied, attached or locked over exterior glazing.

[BG] INCAPABLE OF SELF-PRESERVATION. Persons who, because of age, physical limitations, mental limitations, chemical dependency or medical treatment, cannot respond as an individual to an emergency situation.

[F] INCOMPATIBLE MATERIALS. Materials that, when mixed, have the potential to react in a manner that generates heat, fumes, gases or byproducts which are hazardous to life or property.

[BS] INDIVIDUAL TRUSS MEMBER. A truss chord or truss web.

[F] INERT GAS. A gas that is capable of reacting with other materials only under abnormal conditions such as high temperatures, pressures and similar extrinsic physical forces. Within the context of the code, inert gases do not exhibit either physical or health hazard properties as defined (other than acting as a simple asphyxiant) or hazard properties other than those of a *compressed gas*. Some of the more common inert gases include argon, helium, krypton, neon, nitrogen and xenon.

[F] INITIATING DEVICE. A system component that originates transmission of a change-of-state condition, such as in a *smoke detector*, *manual fire alarm box* or supervisory switch.

[BF] INSULATING SHEATHING. A rigid panel or board insulation material having a thermal resistance of not less than R-2 of the core material with properties suitable for use on walls, floors, roofs or foundations.

[BE] INTENDED TO BE OCCUPIED AS A RESIDENCE. This refers to a *dwelling unit* or *sleeping unit* that can or will be used all or part of the time as the occupant's place of abode.

[BE] INTERIOR EXIT RAMP. An *exit* component that serves to meet one or more *means of egress* design requirements, such as required number of *exits* or *exit access* travel distance, and provides for a protected path of egress travel to the *exit discharge* or *public way*.

[BE] INTERIOR EXIT STAIRWAY. An *exit* component that serves to meet one or more *means of egress* design requirements, such as required number of *exits* or *exit access* travel distance, and provides for a protected path of egress travel to the *exit discharge* or *public way*.

[BF] INTERIOR FINISH. Interior finish includes *interior wall* and *ceiling finish* and *interior floor finish*.

[BF] INTERIOR FLOOR FINISH. The exposed floor surfaces of buildings including coverings applied over a finished floor or *stair*, including risers.

[BF] INTERIOR FLOOR-WALL BASE. *Interior floor finish trim* used to provide a functional or decorative border at the intersection of walls and floors.

[BF] INTERIOR SURFACES. Surfaces other than weather exposed surfaces.

[BF] INTERIOR WALL AND CEILING FINISH. The exposed *interior surfaces* of buildings, including but not limited to: fixed or movable walls and partitions; toilet room privacy partitions; columns; ceilings; and interior wainscoting, paneling or other finish applied structurally or for decoration, acoustical correction, surface insulation, structural fire resistance or similar purposes, but not including *trim*.

[BS] INTERLAYMENT. A layer of felt or nonbituminous saturated felt not less than 18 inches (457 mm) wide, shingled between each course of a wood-shake *roof covering*.

[BS] INTERMODAL SHIPPING CONTAINER. A six-sided steel unit originally constructed as a general cargo container used for the transport of goods and materials.

[BF] INTUMESCENT FIRE-RESISTANT COATINGS. Thin film liquid mixture applied to substrates by brush, roller, spray or trowel which expands into a protective foamed layer to provide fire-resistant protection of the substrates when exposed to flame or intense heat.

[BS] JOINT. The opening in or between adjacent assemblies that is created due to building tolerances, or is designed to allow independent movement of the building in any plane caused by thermal, seismic, wind or any other loading.

[A] JURISDICTION. The governmental unit that has adopted this code.

[BF] L RATING. The air leakage rating of a *through penetration firestop system* or a fire-resistant *joint* system when tested in accordance with UL 1479 or UL 2079, respectively.

[A] LABEL. An identification applied on a product by the manufacturer that contains the name of the manufacturer, the function and performance characteristics of the product or material and the name and identification of an *approved agency*, and that indicates that the representative sample of the product or material has been tested and evaluated by an *approved agency* (see Section 1703.5, "*Manufacturer's designation*" and "*Mark*").

[A] LABELED. Equipment, materials or products to which has been affixed a *label*, seal, symbol or other identifying *mark* of a nationally recognized testing laboratory, *approved* agency or other organization concerned with product evaluation that maintains periodic inspection of the production of the above-labeled items and whose labeling indicates either that the equipment, material or product meets identified standards or has been tested and found suitable for a specified purpose.

[F] LABORATORY SUITE. A fire-rated, enclosed laboratory area providing one or more laboratory spaces within a Group B educational occupancy that includes ancillary uses such as offices, bathrooms and corridors that are contiguous with the laboratory area, and are constructed in accordance with Section 428.

LEVEL OF EXIT DISCHARGE. See "*Exit discharge, level of.*"

[F] LIFE SAFETY SYSTEMS. Systems, devices and equipment that enhance or facilitate evacuation, smoke control, compartmentation and isolation.

[BF] LIGHT-DIFFUSING SYSTEM. Construction consisting in whole or in part of lenses, panels, grids or baffles made with light-transmitting plastics positioned below independently mounted electrical light sources, skylights or light-transmitting plastic roof panels. Lenses, panels, grids and baffles that are part of an electrical fixture shall not be considered as a light-diffusing system.

[BS] LIGHT-FRAME CONSTRUCTION. Construction whose vertical and horizontal structural elements are primarily formed by a system of repetitive wood or cold-formed steel framing members.

[BF] LIGHT-TRANSMITTING PLASTIC ROOF PANELS. Structural plastic panels other than *skylights* that are fastened to structural members, or panels or sheathing and that are used as light-transmitting media in the plane of the roof.

[BF] LIGHT-TRANSMITTING PLASTIC WALL PANELS. Plastic materials that are fastened to structural members, or to structural panels or sheathing, and that are used as light-transmitting media in *exterior walls*.

[BS] LIMIT OF MODERATE WAVE ACTION. Line shown on FIRMs to indicate the inland limit of the $1^1/_2$-foot (457 mm) breaking wave height during the *base flood*.

[BS] LIMIT STATE. A condition beyond which a structure or member becomes unfit for service and is judged to be no longer useful for its intended function (serviceability *limit state*) or to be unsafe (strength *limit state*).

[F] LIQUID. A material that has a melting point that is equal to or less than 68°F (20°C) and a *boiling point* that is greater than 68°F (20°C) at 14.7 pounds per square inch absolute (psia) (101 kPa). When not otherwise identified, the term "liquid" includes both *flammable* and *combustible liquids*.

[F] LIQUID STORAGE ROOM. A room classified as a Group H-3 occupancy used for the storage of *flammable* or *combustible liquids* in a closed condition.

[F] LIQUID USE, DISPENSING AND MIXING ROOM. A room in which Class I, II and IIIA *flammable* or *combustible liquids* are *used*, dispensed or mixed in open containers.

[A] LISTED. Equipment, materials, products or services included in a list published by an organization acceptable to the *building official* and concerned with evaluation of products or services that maintains periodic inspection of production of listed equipment or materials or periodic evaluation of services and whose listing states either that the equipment, material, product or service meets identified standards or has been tested and found suitable for a specified purpose.

[BS] LIVE LOAD. A *load* produced by the use and occupancy of the building or other structure that does not include construction or environmental *loads* such as wind load, snow *load*, rain *load*, earthquake *load*, *flood load* or *dead load*.

[BS] LIVE LOAD, ROOF. A *load* on a roof produced:

1. During maintenance by workers, equipment and materials; or
2. During the life of the structure by movable objects such as planters or other similar small decorative appurtenances that are not occupancy related.

[BG] LIVE/WORK UNIT. A *dwelling unit* or *sleeping unit* in which a significant portion of the space includes a nonresidential use that is operated by the tenant.

[BS] LOAD AND RESISTANCE FACTOR DESIGN (LRFD). A method of proportioning structural members and their connections using load and *resistance factors* such that no applicable *limit state* is reached when the structure is subjected to appropriate load combinations. The term "LRFD" is used in the design of steel and wood structures.

[BS] LOAD EFFECTS. Forces and deformations produced in structural members by the applied *loads*.

[BS] LOAD FACTOR. A factor that accounts for deviations of the actual *load* from the *nominal load*, for uncertainties in the analysis that transforms the *load* into a *load effect*, and for the probability that more than one extreme *load* will occur simultaneously.

[BS] LOADS. Forces or other actions that result from the weight of building materials, occupants and their possessions, environmental effects, differential movement and restrained dimensional changes. Permanent *loads* are those *loads* in which variations over time are rare or of small magnitude, such as *dead loads*. All other *loads* are variable loads (see "*Nominal loads*").

[BG] LODGING HOUSE. A one-family dwelling where one or more occupants are primarily permanent in nature and rent is paid for guest rooms.

[A] LOT. A portion or parcel of land considered as a unit.

[A] LOT LINE. A line dividing one lot from another, or from a street or any public place.

[BE] LOW-ENERGY POWER-OPERATED DOOR. A swinging, sliding or folding door that opens automatically upon an action by a pedestrian such as pressing a push plate or waving a hand in front of a sensor. The door closes automatically, and operates with decreased forces and decreased speeds (see "*Power-assisted door*" and "*Power-operated door*").

[F] LOWER FLAMMABLE LIMIT (LFL). The minimum concentration of vapor in air at which propagation of flame will occur in the presence of an ignition source. The LFL is sometimes referred to as "LEL" or "lower explosive limit."

[BS] LOWEST FLOOR. The *lowest floor* of the lowest enclosed area, including *basement*, but excluding any unfinished or flood-resistant enclosure, usable solely for vehicle parking, building access or limited storage provided that such enclosure is not built so as to render the structure in violation of Section 1612.

[BS] MAIN WINDFORCE-RESISTING SYSTEM. An assemblage of structural elements assigned to provide support and stability for the overall structure. The system generally receives wind loading from more than one surface

MALL BUILDING, COVERED and MALL BUILDING, OPEN. See "*Covered mall building*."

[F] MANUAL FIRE ALARM BOX. A manually operated device used to initiate an *alarm signal*.

[A] MANUFACTURER'S DESIGNATION. An identification applied on a product by the manufacturer indicating that a product or material complies with a specified standard or set of rules (see "*Label*" and "*Mark*").

[A] MARK. An identification applied on a product by the manufacturer indicating the name of the manufacturer and the function of a product or material (see "*Label*" and "*Manufacturer's designation*").

[BG] MARQUEE. A *canopy* that has a top surface which is sloped less than 25 degrees from the horizontal and is located less than 10 feet (3048 mm) from operable openings above or adjacent to the level of the marquee.

[BS] MASONRY. A built-up construction or combination of building units or materials of clay, shale, concrete, glass, gypsum, stone or other *approved* units bonded together with or without *mortar* or grout or other accepted methods of joining.

Glass unit masonry. Masonry composed of glass units bonded by *mortar*.

Plain masonry. Masonry in which the tensile resistance of the masonry is taken into consideration and the effects of stresses in reinforcement are neglected.

Reinforced masonry. Masonry construction in which reinforcement acting in conjunction with the masonry is used to resist forces.

Solid masonry. Masonry consisting of solid *masonry units* laid contiguously with the *joints* between the units filled with *mortar*.

Unreinforced (plain) masonry. Masonry in which the tensile resistance of masonry is taken into consideration and the resistance of the reinforcing steel, if present, is neglected.

[BS] MASONRY UNIT. *Brick*, tile, stone, glass block or concrete block conforming to the requirements specified in Section 2103.

Hollow. A *masonry unit* whose net cross-sectional *area* in any plane parallel to the load-bearing surface is less than 75 percent of its gross cross-sectional *area* measured in the same plane.

Solid. A *masonry unit* whose net cross-sectional *area* in every plane parallel to the load-bearing surface is 75 percent or more of its gross cross-sectional *area* measured in the same plane.

[BG] MASS TIMBER. Structural elements of Type IV construction primarily of solid, built-up, panelized or engineered wood products that meet minimum cross-section dimensions of Type IV construction.

[BF] MASTIC FIRE-RESISTANT COATINGS. Liquid mixture applied to a substrate by brush, roller, spray or trowel that provides fire-resistant protection of a substrate when exposed to flame or intense heat.

[BE] MEANS OF EGRESS. A continuous and unobstructed path of vertical and horizontal egress travel from any occupied portion of a building or structure to a *public way*. A *means of egress* consists of three separate and distinct parts: the *exit access*, the *exit* and the *exit discharge*.

[BF] MECHANICAL EQUIPMENT SCREEN. A rooftop structure, not covered by a roof, used to aesthetically conceal plumbing, electrical or mechanical equipment from view.

[BG] MECHANICAL-ACCESS ENCLOSED PARKING GARAGE. An enclosed parking garage that employs parking machines, lifts, elevators or other mechanical devices for vehicle moving from and to street level and in which public occupancy in the garage is prohibited in all areas except the vehicle access bay.

[BG] MECHANICAL-ACCESS OPEN PARKING GARAGES. *Open parking garages* employing parking machines, lifts, elevators or other mechanical devices for vehicles moving from and to street level and in which public occupancy is prohibited above the street level.

[BG] MEDICAL CARE. Care involving medical or surgical procedures, nursing or for psychiatric purposes.

[BF] MEMBRANE PENETRATION. A breach in one side of a floor-ceiling, roof-ceiling or wall assembly to accommodate an item installed into or passing through the breach.

[BG] MEMBRANE-COVERED CABLE STRUCTURE. A nonpressurized structure in which a mast and cable system provides support and tension to the membrane weather barrier and the membrane imparts stability to the structure.

[BG] MEMBRANE-COVERED FRAME STRUCTURE. A nonpressurized building wherein the structure is composed of a rigid framework to support a tensioned membrane which provides the weather barrier.

[BF] MEMBRANE-PENETRATION FIRESTOP. A material, device or construction installed to resist for a prescribed time period the passage of flame and heat through openings in a protective membrane in order to accommodate cables, cable trays, conduit, tubing, pipes or similar items.

[BF] MEMBRANE-PENETRATION FIRESTOP SYSTEM. An assemblage consisting of a fire-resistance-rated floor-ceiling, roof-ceiling or wall assembly, one or more penetrating items installed into or passing through the breach in one side of the assembly and the materials or devices, or both, installed to resist the spread of fire into the assembly for a prescribed period of time.

[BE] MERCHANDISE PAD. A merchandise pad is an area for display of merchandise surrounded by *aisles*, permanent fixtures or walls. Merchandise pads contain elements such as nonfixed and movable fixtures, cases, racks, counters and partitions as indicated in Section 105.2 from which customers browse or shop.

[BF] METAL COMPOSITE MATERIAL (MCM). A factory-manufactured panel consisting of metal skins bonded to both faces of a solid plastic core.

[BF] METAL COMPOSITE MATERIAL (MCM) SYSTEM. An *exterior wall covering* fabricated using MCM in a specific assembly including *joints*, seams, attachments, substrate, framing and other details as appropriate to a particular design.

[BS] METAL ROOF PANEL. An interlocking metal sheet having a minimum installed weather exposure of 3 square feet (0.279 m^2) per sheet.

[BS] METAL ROOF SHINGLE. An interlocking metal sheet having an installed weather exposure less than 3 square feet (0.279 m^2) per sheet.

[BG] MEZZANINE. An intermediate level or levels between the floor and ceiling of any *story* and in accordance with Section 505.

[BS] MICROPILE. A *micropile* is a bored, grouted-in-place *deep foundation* element that develops its load-carrying capacity by means of a bond zone in soil, bedrock or a combination of soil and bedrock.

[BF] MINERAL BOARD. A rigid felted thermal insulation board consisting of either felted mineral fiber or cellular beads of expanded aggregate formed into flat rectangular units.

[BF] MINERAL FIBER. Insulation composed principally of fibers manufactured from rock, slag or glass, with or without binders.

[BF] MINERAL WOOL. Synthetic vitreous fiber insulation made by melting predominately igneous rock or furnace slag, and other inorganic materials, and then physically forming the melt into fibers.

[BS] MODIFIED BITUMEN ROOF COVERING. One or more layers of polymer-modified asphalt sheets. The sheet materials shall be fully adhered or mechanically attached to the substrate or held in place with an *approved ballast* layer.

[BS] MORTAR. A mixture consisting of cementitious materials, fine aggregates, water, with or without admixtures, that is used to construct unit masonry assemblies.

[BS] MORTAR, SURFACE-BONDING. A mixture to bond concrete *masonry units* that contains hydraulic cement, glass fiber reinforcement with or without inorganic fillers or organic modifiers and water.

[BE] MULTILEVEL ASSEMBLY SEATING. Seating that is arranged in distinct levels where each level is composed of either multiple rows, or a single row of box seats accessed from a separate level.

[F] MULTIPLE-STATION ALARM DEVICE. Two or more single-station alarm devices that can be interconnected such that actuation of one causes all integral or separate audible alarms to operate. A multiple-station alarm device can consist of one single-station alarm device having connections to other detectors or to a *manual fire alarm box*.

[F] MULTIPLE-STATION SMOKE ALARM. Two or more single-station alarm devices that are capable of interconnection such that actuation of one causes the appropriate *alarm signal* to operate in all interconnected alarms.

[BE] MULTISTORY UNIT. A *dwelling unit* or *sleeping unit* with *habitable space* located on more than one story.

[BF] NAILABLE SUBSTRATE. A product or material such as framing, sheathing or furring, composed of wood, wood-based materials or other materials providing equivalent fastener withdrawal resistance.

[BS] NAILING, BOUNDARY. A special nailing pattern required by design at the boundaries of *diaphragms*.

[BS] NAILING, EDGE. A special nailing pattern required by design at the edges of each panel within the assembly of a *diaphragm* or *shear wall*.

[BS] NAILING, FIELD. Nailing required between the sheathing panels and framing members at locations other than *boundary nailing* and *edge nailing*.

[BS] NATURALLY DURABLE WOOD. The heartwood of the following species except for the occasional piece with corner sapwood, provided 90 percent or more of the width of each side on which it occurs is heartwood.

Decay resistant. Redwood, cedar, black locust and black walnut.

Termite resistant. Redwood, Alaska yellow cedar, Eastern red cedar and Western red cedar.

[BS] NOMINAL LOADS. The magnitudes of the *loads* specified in Chapter 16 (dead, live, soil, wind, snow, rain, *flood* and earthquake).

[BS] NOMINAL SIZE (LUMBER). The commercial size designation of width and depth, in standard sawn lumber and glued-laminated lumber *grades*; somewhat larger than the standard net size of dressed lumber, in accordance with DOCPS 20 for sawn lumber and with the ANSI/AWC NDS for glued-laminated lumber.

[BG] NONCOMBUSTIBLE MEMBRANE STRUCTURE. A membrane structure in which the membrane and all component parts of the structure are noncombustible.

[BF] NONCOMBUSTIBLE PROTECTION (FOR MASS TIMBER). Noncombustible material, in accordance with Section 703.6, designed to increase the *fire-resistance rating* and delay the combustion of *mass timber*.

[BS] NONSTRUCTURAL CONCRETE. Any element made of plain or reinforced concrete that is not part of a structural system required to transfer either gravity or lateral *loads* to the ground.

[F] NORMAL TEMPERATURE AND PRESSURE (NTP). A temperature of 70°F (21°C) and a pressure of 1 atmosphere [14.7 psia (101 kPa)].

[BE] NOSING. The leading edge of treads of *stairs* and of landings at the top of *stairway flights*.

NOTIFICATION ZONE. See "*Zone, notification*."

[F] NUISANCE ALARM. An alarm caused by mechanical failure, malfunction, improper installation or lack of proper maintenance, or an alarm activated by a cause that cannot be determined.

[BG] NURSING HOMES. Facilities that provide care, including both intermediate care facilities and skilled nursing facilities where any of the persons are *incapable of self-preservation*.

[BE] OCCUPANT LOAD. The number of persons for which the *means of egress* of a building or portion thereof is designed.

[BG] OCCUPIABLE SPACE. A room or enclosed space designed for human occupancy in which individuals congregate for amusement, educational or similar purposes or in which occupants are engaged at labor, and which is equipped with *means of egress* and light and *ventilation* facilities meeting the requirements of this code.

[BG] OPEN PARKING GARAGE. A structure or portion of a structure with the openings as described in Section 406.5.2 on two or more sides that is used for the parking or storage of private motor vehicles as described in Section 406.5.3.

[F] OPEN SYSTEM. The *use* of a *solid* or *liquid hazardous material* involving a vessel or system that is continuously open to the atmosphere during normal operations and where vapors are liberated, or the product is exposed to the atmosphere during normal operations. Examples of open systems for *solids* and *liquids* include dispensing from or into open beakers or containers, dip tank and plating tank operations.

[BE] OPEN-AIR ASSEMBLY SEATING. Seating served by *means of egress* that is not subject to smoke accumulation within or under a structure and is open to the atmosphere.

[BE] OPEN-ENDED CORRIDOR. An interior *corridor* that is open on each end and connects to an exterior *stairway* or *ramp* at each end with no intervening doors or separation from the *corridor*.

[BF] OPENING PROTECTIVE. A *fire door* assembly, fire shutter assembly, *fire window assembly* or glass-block assembly in a fire-resistance-rated wall or partition.

[F] OPERATING BUILDING. A building occupied in conjunction with the manufacture, transportation or *use* of explosive materials. Operating buildings are separated from one another with the use of intraplant or intraline distances.

[BS] ORDINARY PRECAST STRUCTURAL WALL. See Section 1905.1.1.

[BS] ORDINARY REINFORCED CONCRETE STRUCTURAL WALL. See Section 1905.1.1.

[BS] ORDINARY STRUCTURAL PLAIN CONCRETE WALL. See Section 1905.1.1.

[F] ORGANIC PEROXIDE. An organic compound that contains the bivalent -O-O- structure and which may be considered to be a structural derivative of hydrogen peroxide where one or both of the hydrogen atoms have been replaced by an organic radical. Organic peroxides can pose an *explosion* hazard (*detonation* or *deflagration*) or they can be shock sensitive. They can also decompose into various unstable compounds over an extended period of time.

Class I. Those formulations that are capable of *deflagration* but not *detonation*.

Class II. Those formulations that burn very rapidly and that pose a moderate reactivity hazard.

Class III. Those formulations that burn rapidly and that pose a moderate reactivity hazard.

Class IV. Those formulations that burn in the same manner as ordinary combustibles and that pose a minimal reactivity hazard.

Class V. Those formulations that burn with less intensity than ordinary combustibles or do not sustain combustion and that pose no reactivity hazard.

Unclassified detonable. Organic peroxides that are capable of *detonation*. These peroxides pose an extremely high *explosion* hazard through rapid explosive decomposition.

[BS] ORTHOGONAL. To be in two horizontal directions, at 90 degrees (1.57 rad) to each other.

[BS] OTHER STRUCTURES. This definition applies only to Chapters 16 through 23.

Structures, other than buildings, for which loads are specified in Chapter 16.

OUTPATIENT CLINIC. See "*Clinic, outpatient*."

[A] OWNER. Any person, agent, operator, entity, firm or corporation having any legal or equitable interest in the property; or recorded in the official records of the state, county or municipality as holding an interest or title to the property; or otherwise having possession or control of the property, including the guardian of the estate of any such person, and the executor or administrator of the estate of such person if ordered to take possession of real property by a court.

[F] OXIDIZER. A material that readily yields oxygen or other *oxidizing gas*, or that readily reacts to promote or initiate combustion of combustible materials and, if heated or contaminated, can result in vigorous self-sustained decomposition.

Class 4. An oxidizer that can undergo an explosive reaction due to contamination or exposure to thermal or physical shock and that causes a severe increase in the burning rate of combustible materials with which it comes into contact. Additionally, the oxidizer causes a severe increase in the burning rate and can cause spontaneous ignition of combustibles.

Class 3. An oxidizer that causes a severe increase in the burning rate of combustible materials with which it comes in contact.

Class 2. An oxidizer that will cause a moderate increase in the burning rate of combustible materials with which it comes in contact.

Class 1. An oxidizer that does not moderately increase the burning rate of combustible materials.

[F] OXIDIZING GAS. A gas that can support and accelerate combustion of other materials more than air does.

[BS] PANEL (PART OF A STRUCTURE). The section of a floor, wall or roof comprised between the supporting frame of two adjacent rows of columns and girders or column bands of floor or roof construction.

[BE] PANIC HARDWARE. A door-latching assembly incorporating a device that releases the latch upon the application of a force in the direction of egress travel. See "*Fire exit hardware*."

[BS] PARTICLEBOARD. A generic term for a panel primarily composed of cellulosic materials (usually wood), generally in the form of discrete pieces or particles, as distinguished from fibers. The cellulosic material is combined with synthetic resin or other suitable bonding system by a process in which the interparticle bond is created by the bonding system under heat and pressure.

[BF] PENETRATION FIRESTOP. A *through-penetration* firestop or a *membrane-penetration firestop*.

[BG] PENTHOUSE. An enclosed, unoccupied rooftop structure used for sheltering mechanical and electrical equipment, tanks, elevators and related machinery, *stairways,* and vertical *shaft* openings.

[BS] PERFORMANCE CATEGORY. A designation of *wood structural panels* as related to the panel performance used in Chapter 23.

[BF] PERIMETER FIRE CONTAINMENT SYSTEM. An assemblage of specific materials or products that is designed to resist for a prescribed period of time the passage of fire through voids created at the intersection of exterior curtain wall assemblies and fire-resistance-rated floor or floor/ceiling assemblies.

[BS] PERMANENT INDIVIDUAL TRUSS MEMBER DIAGONAL BRACING (PITMDB). Structural member or assembly intended to permanently stabilize the *PITMRs*.

[BS] PERMANENT INDIVIDUAL TRUSS MEMBER RESTRAINT (PITMR). Restraint that is used to prevent local buckling of an individual truss chord or web member because of the axial forces in the *individual truss member*.

[A] PERMIT. An official document or certificate issued by the *building official* that authorizes performance of a specified activity.

[A] PERSON. An individual, heirs, executors, administrators or assigns, and also includes a firm, partnership or corporation, its or their successors or assigns, or the agent of any of the aforesaid.

[BG] PERSONAL CARE SERVICE. The care of persons who do not require *medical care*. Personal care involves responsibility for the safety of the persons while inside the building

[BE] PHOTOLUMINESCENT. Having the property of emitting light that continues for a length of time after excitation by visible or invisible light has been removed.

[BS] PHOTOVOLTAIC MODULE. A complete, environmentally protected unit consisting of solar cells, optics and other components, exclusive of tracker, designed to generate DC power when exposed to sunlight.

[BS] PHOTOVOLTAIC PANEL. A collection of modules mechanically fastened together, wired and designed to provide a field-installable unit.

[BS] PHOTOVOLTAIC PANEL SYSTEM. A system that incorporates discrete *photovoltaic panels*, that converts solar radiation into electricity, including rack support systems.

[BS] PHOTOVOLTAIC SHINGLES. A *roof covering* resembling shingles that incorporates *photovoltaic modules*.

[F] PHYSICAL HAZARD. A chemical for which there is evidence that it is a *combustible liquid*, *cryogenic fluid*, *explosive*, flammable (*solid*, *liquid* or gas), *organic peroxide* (*solid* or *liquid*), *oxidizer* (*solid* or *liquid*), *oxidizing gas*, *pyrophoric* (*solid*, *liquid* or gas), *unstable (reactive) material* (*solid*, *liquid* or gas) or *water-reactive material* (*solid* or *liquid*).

[F] PHYSIOLOGICAL WARNING THRESHOLD LEVEL. A concentration of airborne contaminants, normally expressed in parts per million (ppm) or milligrams per cubic meter (mg/m^3), that represents the concentration at which persons can sense the presence of the contaminant due to odor, irritation or other quick-acting physiological response. When used in conjunction with the permissible exposure limit (PEL) the physiological warning threshold levels are those consistent with the classification system used to establish the PEL. See the definition of "*Permissible exposure limit (PEL)*" in the *International Fire Code*.

PLACE OF RELIGIOUS WORSHIP. See "*Religious worship, place of*."

[BF] PLASTIC COMPOSITE. A generic designation that refers to wood/plastic composites, plastic lumber and similar materials.

[BF] PLASTIC GLAZING. Plastic materials that are glazed or set in a frame or sash or are otherwise supported.

[BF] PLASTIC LUMBER. A manufactured product made primarily of plastic materials (filled or unfilled) which is generally rectangular in cross section.

[BG] PLATFORM. A raised area within a building used for worship, the presentation of music, plays or other entertainment; the head table for special guests; the raised area for lecturers and speakers; boxing and wrestling rings; theater-in-the-round *stages*; and similar purposes wherein, other than horizontal sliding curtains, there are no overhead hanging curtains, drops, scenery or stage effects other than lighting and sound. A temporary platform is one installed for not more than 30 days.

[BG] PLAY STRUCTURE. A structure composed of one or more components, where the user enters a play environment.

[BF] POLYPROPYLENE SIDING. A shaped material, made principally from polypropylene homopolymer, or copolymer, which in some cases contains fillers or reinforcements, that is used to clad *exterior walls* of buildings.

[BS] PORCELAIN TILE. Ceramic tile having an absorption of 0.5 percent or less in accordance with Table 10 of ANSI A137.1, or Tables 4 or 5 of ANSI A137.3.

[BS] POSITIVE ROOF DRAINAGE. A design that accounts for deflections from all *design loads* and has sufficient additional slope to ensure that drainage of the roof occurs within 48 hours of precipitation.

[BE] POWER-ASSISTED DOOR. Swinging door which opens by reduced pushing or pulling force on the door-operating hardware. The door closes automatically after the pushing or pulling force is released and functions with decreased forces. See "*Low-energy power-operated door*" and "*Power-operated door.*"

[BE] POWER-OPERATED DOOR. Swinging, sliding, or folding door which opens automatically when approached by a pedestrian or opens automatically upon an action by a pedestrian. The door closes automatically and includes provisions such as presence sensors to prevent entrapment. See "*Low energy power-operated door*" and "*Power-assisted door.*"

[BS] PREFABRICATED WOOD I-JOIST. Structural member manufactured using sawn or *structural composite lumber* flanges and *wood structural panel* webs bonded together with exterior exposure adhesives, which forms an "I" cross-sectional shape.

[BS] PRESERVATIVE-TREATED WOOD. Wood products that, when impregnated with chemicals by a pressure process or other means during manufacture, exhibit reduced susceptibility to damage by fungi, insects or marine borers.

[BS] PRESTRESSED MASONRY. *Masonry* in which internal stresses have been introduced to counteract potential tensile stresses in *masonry* resulting from applied *loads.*

[BG] PRIMARY STRUCTURAL FRAME. The primary structural frame shall include all of the following structural members:

1. The columns.
2. Structural members having direct connections to the columns, including girders, beams, trusses and spandrels.
3. Members of the floor construction and roof construction having direct connections to the columns.
4. Members that are essential to the vertical stability of the *primary structural frame* under gravity loading.

[BG] PRIVATE GARAGE. A building or portion of a building in which motor vehicles used by the *owner* or tenants of the building or buildings on the premises are stored or kept, without provisions for repairing or servicing such vehicles for profit.

[BG] PROSCENIUM WALL. The wall that separates the *stage* from the auditorium or assembly seating area.

PSYCHIATRIC HOSPITALS. See "*Hospitals and psychiatric hospitals.*"

[BE] PUBLIC ENTRANCE. An entrance that is not a *service entrance* or a *restricted entrance.*

[A] PUBLIC WAY. A street, alley or other parcel of land open to the outside air leading to a street, that has been deeded, dedicated or otherwise permanently appropriated to the public for public use and which has a clear width and height of not less than 10 feet (3048 mm).

[BE] PUBLIC-USE AREAS. Interior or exterior rooms or spaces that are made available to the general public.

[BG] PUZZLE ROOM. A puzzle room is a type of *special amusement area* in which occupants are encouraged to solve a challenge to escape from a room or series of rooms.

[F] PYROPHORIC. A chemical with an auto-ignition temperature in air, at or below a temperature of 130°F (54.4°C).

[F] PYROTECHNIC COMPOSITION. A chemical mixture that produces visible light displays or sounds through a self-propagating, heat-releasing chemical reaction which is initiated by ignition.

[BF] RADIANT BARRIER. A material having a low-*emittance* surface of 0.1 or less installed in building assemblies.

[BE] RAMP. A walking surface that has a running slope steeper than one unit vertical in 20 units horizontal (5-percent slope).

RAMP, EXIT ACCESS. See "*Exit access ramp.*"

RAMP, EXTERIOR EXIT. See "*Exterior exit ramp.*"

RAMP, INTERIOR EXIT. See "*Interior exit ramp.*"

[BG] RAMP-ACCESS OPEN PARKING GARAGES. *Open parking garages* employing a series of continuously rising floors or a series of interconnecting ramps between floors permitting the movement of vehicles under their own power from and to the street level.

[A] RECORD DRAWINGS. Drawings ("as builts") that document the location of all devices, appliances, wiring sequences, wiring methods and connections of the components of a *fire alarm system* as installed.

[BF] REFLECTIVE PLASTIC CORE INSULATION. An insulation material packaged in rolls, that is less than $^1/_2$ inch (12.7 mm) thick, with not less than one exterior low-*emittance* surface (0.1 or less) and a core material containing voids or cells.

[A] REGISTERED DESIGN PROFESSIONAL. An individual who is registered or licensed to practice their respective design profession as defined by the statutory requirements of the professional registration laws of the state or *jurisdiction* in which the project is to be constructed.

[A] REGISTERED DESIGN PROFESSIONAL IN RESPONSIBLE CHARGE. A *registered design professional* engaged by the owner or the owner's authorized agent to review and coordinate certain aspects of the project, as determined by the *building official*, for compatibility with the design of the building or structure, including submittal documents prepared by others, deferred submittal documents and phased submittal documents.

[BG] RELIGIOUS WORSHIP, PLACE OF. A building or portion thereof intended for the performance of religious services.

[A] RELOCATABLE BUILDING. A partially or completely assembled building constructed and designed to be reused multiple times and transported to different building sites.

[A] REPAIR. The reconstruction, replacement or renewal of any part of an existing building for the purpose of its maintenance or to correct damage.

[BG] REPAIR GARAGE. A building, structure or portion thereof used for servicing or repairing motor vehicles.

[BS] REROOFING. The process of recovering or replacing an existing *roof covering*. See "*Roof recover*" and "*Roof replacement*."

[BG] RESIDENTIAL AIRCRAFT HANGAR. An accessory building less than 2,000 square feet (186 m^2) and 20 feet (6096 mm) in *building height* constructed on a one- or two-family property where aircraft are stored. Such use will be considered as a residential accessory use incidental to the dwelling.

[BS] RESISTANCE FACTOR. A factor that accounts for deviations of the actual strength from the *nominal strength* and the manner and consequences of failure (also called "strength reduction factor").

[BE] RESTRICTED ENTRANCE. An entrance that is made available for common use on a controlled basis, but not public use, and that is not a *service entrance*.

[BG] RETRACTABLE AWNING. A retractable *awning* is a cover with a frame that retracts against a building or other structure to which it is entirely supported.

[BS] RISK CATEGORY. A categorization of buildings and *other structures* for determination of *flood*, wind, snow, ice and earthquake *loads* based on the risk associated with unacceptable performance.

[BS] RISK-TARGETED MAXIMUM CONSIDERED EARTHQUAKE (MCE_R) GROUND MOTION RESPONSE ACCELERATIONS. The most severe earthquake effects considered by this code, determined for the orientation that results in the largest maximum response to horizontal ground motions and with adjustment for targeted risk.

[BS] ROOF ASSEMBLY (For application to Chapter 15 only). A system designed to provide weather protection and resistance to design *loads*. The system consists of a *roof covering* and roof deck or a single component serving as both the *roof covering* and the roof deck. A roof assembly can include an *underlayment*, a thermal barrier, insulation or a *vapor retarder*.

[BS] ROOF COATING. A fluid-applied, adhered coating used for roof maintenance or *roof repair*, or as a component of a *roof covering* system or *roof assembly*.

[BS] ROOF COVERING. The covering applied to the roof deck for weather resistance, fire classification or appearance.

ROOF COVERING SYSTEM. See "*Roof assembly*."

[BS] ROOF DECK. The flat or sloped surface constructed on top of the *exterior walls* of a building or other supports for the purpose of enclosing the *story* below, or sheltering an area, to protect it from the elements, not including its supporting members or vertical supports.

ROOF DRAINAGE, POSITIVE. See "*Positive roof drainage*."

[BS] ROOF RECOVER. The process of installing an additional *roof covering* over a prepared existing *roof covering* without removing the existing *roof covering*.

[BS] ROOF REPAIR. Reconstruction or renewal of any part of an existing roof for the purposes of correcting damage or restoring pre-damage condition.

[BS] ROOF REPLACEMENT. The process of removing the existing *roof covering*, repairing any damaged substrate and installing a new *roof covering*.

[BG] ROOF VENTILATION. The natural or mechanical process of supplying conditioned or unconditioned air to, or removing such air from, *attics*, cathedral ceilings or other enclosed spaces over which a roof assembly is installed.

[BG] ROOFTOP STRUCTURE. A structure erected on top of the roof deck or on top of any part of a building.

[BS] RUNNING BOND. The placement of *masonry units* such that *head joints* in successive courses are horizontally offset at least one-quarter the unit length.

[BG] SALLYPORT. A security vestibule with two or more doors or gates where the intended purpose is to prevent continuous and unobstructed passage by allowing the release of only one door or gate at a time.

[BE] SCISSOR STAIRWAY. Two interlocking *stairways* providing two separate paths of egress located within one *exit* enclosure.

[BS] SCUPPER. An opening in a wall or parapet that allows water to drain from a roof.

[BG] SECONDARY STRUCTURAL MEMBERS. The following structural members shall be considered secondary members and not part of the *primary structural frame*:

1. Structural members not having direct connections to the columns.
2. Members of the floor construction and roof construction not having direct connections to the columns.
3. Bracing members that are not designated as part of a *primary structural frame* or bearing wall.

[BS] SEISMIC DESIGN CATEGORY. A classification assigned to a structure based on its *risk category* and the severity of the *design earthquake ground motion* at the site.

[BS] SEISMIC FORCE-RESISTING SYSTEM. That part of the structural system that has been considered in the design to provide the required resistance to the prescribed seismic forces.

[BF] SELF-CLOSING. As applied to a *fire door* or other opening protective, means equipped with an device that will ensure closing after having been opened.

[BE] SELF-LUMINOUS. Illuminated by a self-contained power source, other than batteries, and operated independently of external power sources.

SELF-PRESERVATION, INCAPABLE OF. See "*Incapable of self-preservation.*"

[BG] SELF-SERVICE STORAGE FACILITY. Real property designed and used for the purpose of renting or leasing individual storage spaces to customers for the purpose of storing and removing personal property on a self-service basis.

[F] SERVICE CORRIDOR. A fully enclosed passage used for transporting *HPM* and purposes other than required *means of egress*.

[BE] SERVICE ENTRANCE. An entrance intended primarily for delivery of goods or services.

[BF] SHAFT. An enclosed space extending through one or more *stories* of a building, connecting vertical openings in successive floors, or floors and roof.

[BF] SHAFT ENCLOSURE. The walls or construction forming the boundaries of a *shaft*.

[BS] SHALLOW FOUNDATION. A *shallow foundation* is an individual or strip footing, a mat foundation, a slab-on-grade foundation or a similar foundation element.

[BS] SHEAR WALL. This definition applies only to Chapter 23.

A wall designed to resist lateral forces parallel to the plane of a wall.

Shear wall, perforated. A *wood structural panel* sheathed wall with openings, that has not been specifically designed and detailed for force transfer around openings.

Shear wall segment, perforated. A section of shear wall with full-height sheathing that meets the height-to-width ratio limits of Section 4.3.4 of AWC SDPWS.

[BS] SHINGLE FASHION. A method of installing roof or wall coverings, *water-resistive barriers,* flashing or other building components such that upper layers of material are placed overlapping lower layers of material to provide for drainage via gravity and moisture control.

[BS] SINGLE-PLY MEMBRANE. A roofing membrane that is field applied using one layer of membrane material (either homogeneous or composite) rather than multiple layers.

[F] SINGLE-STATION SMOKE ALARM. An assembly incorporating the detector, the control equipment and the alarm-sounding device in one unit, operated from a power supply either in the unit or obtained at the point of installation.

[BG] SITE. A parcel of land bounded by a *lot line* or a designated portion of a public right-of-way.

[BS] SITE CLASS. A classification assigned to a site based on the types of soils present and their engineering properties as defined in Section 1613.2.2.

[BS] SITE COEFFICIENTS. The values of F_a and F_v indicated in Table 1613.2.3(1) and Table 1613.2.3(2), respectively.

[BG] SITE-FABRICATED STRETCH SYSTEM. A system, fabricated on site and intended for acoustical, tackable or aesthetic purposes, that is composed of three elements:

1. A frame (constructed of plastic, wood, metal or other material) used to hold fabric in place.
2. A core material (infill, with the correct properties for the application).
3. An outside layer, composed of a textile, fabric or vinyl, that is stretched taut and held in place by tension or mechanical fasteners via the frame.

[BS] SKYLIGHT, UNIT. A factory-assembled, glazed *fenestration* unit, containing one panel of glazing material that allows for natural lighting through an opening in the *roof assembly* while preserving the weather-resistant barrier of the roof.

[BS] SKYLIGHTS AND SLOPED GLAZING. Glass or other transparent or translucent glazing material installed at a slope of 15 degrees (0.26 rad) or more from vertical. *Unit skylights*, *tubular daylighting devices*, glazing materials, solariums, *sunrooms*, roofs and sloped walls are included in this definition.

[A] SLEEPING UNIT. A single unit that provides rooms or spaces for one or more persons, includes permanent provisions for sleeping and can include provisions for living, eating and either sanitation or kitchen facilities but not both. Such rooms and spaces that are also part of a dwelling unit are not sleeping units.

[F] SMOKE ALARM. A single- or multiple-station alarm responsive to smoke. See "*Multiple-station smoke alarm*" and "*Single-station smoke alarm.*"

[BF] SMOKE BARRIER. A continuous membrane, either vertical or horizontal, such as a wall, floor or ceiling assembly, that is designed and constructed to restrict the movement of smoke.

[BG] SMOKE COMPARTMENT. A space within a building separated from other interior areas of the building by *smoke barriers*, including interior walls and *horizontal assemblies.*

[BF] SMOKE DAMPER. A *listed* device installed in ducts and air transfer openings designed to resist the passage of smoke. The device is installed to operate automatically, controlled by a smoke detection system, and where required, is capable of being positioned from a *fire command center.*

[F] SMOKE DETECTOR. A *listed* device that senses visible or invisible particles of combustion.

[BF] SMOKE PARTITION. A wall assembly that extends from the top of the foundation or floor below to the underside of the floor or roof sheathing, deck or slab above or to the underside of the ceiling above where the ceiling membrane is constructed to limit the transfer of smoke.

[BF] SMOKE-DEVELOPED INDEX. A comparative measure, expressed as a dimensionless number, derived from measurements of smoke obscuration versus time for a material tested in accordance with ASTM E84.

[BF] SMOKEPROOF ENCLOSURE. An *exit stairway* or *ramp* designed and constructed so that the movement of the products of combustion produced by a fire occurring in any part of the building into the enclosure is limited.

[BE] SMOKE-PROTECTED ASSEMBLY SEATING. Seating served by *means of egress* that is not subject to smoke accumulation within or under a structure for a specified design time by means of passive design or by mechanical ventilation.

[BG] SOFT CONTAINED PLAY EQUIPMENT STRUCTURE. A play structure containing one or more components where the user enters a play environment that utilizes pliable materials.

[F] SOLID. A material that has a melting point, decomposes or sublimes at a temperature greater than 68°F (20°C).

[BG] SPECIAL AMUSEMENT AREA. A special amusement area is any temporary or permanent building or portion thereof that is occupied for amusement, entertainment or educational purposes and is arranged in a manner that:

1. Makes the means of egress path not readily apparent due to visual or audio distractions.
2. Intentionally confounds identification of the means of egress path.
3. Otherwise makes the means of egress path not readily available because of the nature of the attraction or mode of conveyance through the building or structure.

[BG] SPECIAL EVENT STRUCTURE. Any ground-supported structure, *platform, stage, stage* scaffolding or rigging, canopy, tower or similar structure supporting entertainment-related equipment or signage.

[BS] SPECIAL FLOOD HAZARD AREA. The land area subject to flood hazards and shown on a *Flood Insurance Rate Map* or other flood hazard map as Zone A, AE, A1-30, A99, AR, AO, AH, V, VO, VE or V1-30.

[BS] SPECIAL INSPECTION. Inspection of construction requiring the expertise of an *approved special inspector* in order to ensure compliance with this code and the approved *construction documents*.

Continuous special inspection. *Special inspection* by the *special inspector* who is present continuously when and where the work to be inspected is being performed.

Periodic special inspection. *Special inspection* by the *special inspector* who is intermittently present where the work to be inspected has been or is being performed.

[BS] SPECIAL INSPECTOR. A qualified person employed or retained by an *approved* agency and *approved* by the *building official* as having the competence necessary to inspect a particular type of construction requiring *special inspection*.

[BS] SPECIFIED COMPRESSIVE STRENGTH OF MASONRY, f'_m. Minimum compressive strength, expressed as force per unit of net cross-sectional area, required of the *masonry* used in construction by the *approved construction documents*, and upon which the project design is based. Whenever the quantity f'_m is under the radical sign, the square root of numerical value only is intended and the result has units of pounds per square inch (psi) (MPa).

[BF] SPLICE. The result of a factory and/or field method of joining or connecting two or more lengths of a *fire-resistant joint system* into a continuous entity.

SPORT ACTIVITY, AREA OF. See "*Area of sport activity.*"

[F] SPRAY ROOM. A room designed to accommodate spraying operations.

[BF] SPRAY-APPLIED FOAM PLASTIC. Single- and multiple-component, spray-applied foam plastic insulation used in nonstructural applications that are installed at locations wherein the material is applied in a liquid or frothed state, permitted to free rise and cure in situ.

[BF] SPRAYED FIRE-RESISTANT MATERIALS. Cementitious or fibrous materials that are sprayed to provide fire-resistant protection of the substrates.

[BG] STAGE. A space within a building utilized for entertainment or presentations, which includes overhead hanging curtains, drops, scenery or stage effects other than lighting and sound.

[BE] STAIR. A change in elevation, consisting of one or more risers.

[BE] STAIRWAY. One or more *flights* of *stairs*, either exterior or interior, with the necessary landings and platforms connecting them, to form a continuous and uninterrupted passage from one level to another.

STAIRWAY, EXIT ACCESS. See "*Exit access stairway.*"

STAIRWAY, EXTERIOR EXIT. See "*Exterior exit stairway.*"

STAIRWAY, INTERIOR EXIT. See "*Interior exit stairway.*"

STAIRWAY, SCISSOR. See "*Scissor stairway.*"

[BE] STAIRWAY, SPIRAL. A *stairway* having a closed circular form in its plan view with uniform section-shaped treads attached to and radiating from a minimum-diameter supporting column.

[F] STANDBY POWER SYSTEM. A source of *automatic* electric power of a required capacity and duration to operate required building, *hazardous materials* or *ventilation* systems in the event of a failure of the primary power. Standby power systems are required for electrical loads where interruption of the primary power could create hazards or hamper rescue or fire-fighting operations.

[F] STANDPIPE, TYPES OF. Standpipe types are as follows:

Automatic dry. A dry standpipe system, normally filled with pressurized air, that is arranged through the use of a device, such as dry pipe valve, to admit water into the system piping *automatically* upon the opening of a hose valve. The water supply for an *automatic* dry standpipe system shall be capable of supplying the system demand.

Automatic wet. A wet standpipe system that has a water supply that is capable of supplying the system demand *automatically*.

Manual dry. A dry standpipe system that does not have a permanent water supply attached to the system. Manual dry standpipe systems require water from a fire department pumper to be pumped into the system through the fire department connection in order to meet the system demand.

Manual wet. A wet standpipe system connected to a water supply for the purpose of maintaining water within the system but does not have a water supply capable of delivering the system demand attached to the system. Manual-wet standpipe systems require water from a fire department pumper (or the like) to be pumped into the system in order to meet the system demand.

Semiautomatic dry. A dry standpipe system that is arranged through the use of a device, such as a deluge valve, to admit water into the system piping upon activation of a remote control device located at a hose connection. A remote control activation device shall be provided at each hose connection. The water supply for a semiautomatic dry standpipe system shall be capable of supplying the system demand.

[F] STANDPIPE SYSTEM, CLASSES OF. Standpipe classes are as follows:

Class I system. A system providing $2^1/_2$-inch (64 mm) hose connections to supply water for use by fire departments and those trained in handling heavy fire streams.

Class II system. A system providing $1^1/_2$-inch (38 mm) hose stations to supply water for use primarily by the building occupants or by the fire department during initial response.

Class III system. A system providing $1^1/_2$-inch (38 mm) hose stations to supply water for use by building occupants and $2^1/_2$-inch (64 mm) hose connections to supply a larger volume of water for use by fire departments and those trained in handling heavy fire streams.

[BS] STEEL CONSTRUCTION, COLD-FORMED. That type of construction made up entirely or in part of *steel structural members* cold formed to shape from sheet or strip steel such as roof deck, floor and wall panels, studs, floor joists, roof joists and other structural elements.

[BS] STEEL ELEMENT, STRUCTURAL. Any *steel structural member* of a building or structure consisting of rolled shapes, pipe, hollow structural sections, plates, bars, sheets, rods or steel castings other than cold-formed steel or *steel joist* members.

[BS] STEEL JOIST. Any *steel structural member* of a building or structure made of hot-rolled or cold-formed solid or open-web sections, or riveted or welded bars, strip or sheet steel members, or slotted and expanded, or otherwise deformed rolled sections.

[BF] STEEP SLOPE. A roof slope 2 units vertical in 12 units horizontal (17-percent slope) or greater.

[BS] STONE MASONRY. *Masonry* composed of field, quarried or *cast stone* units bonded by *mortar*.

[F] STORAGE, HAZARDOUS MATERIALS. The keeping, retention or leaving of hazardous materials in closed containers, tanks, cylinders, or similar vessels; or vessels supplying operations through closed connections to the vessel.

[BS] STORAGE RACKS, STEEL. Cold-formed or hot-rolled steel structural members which are formed into *steel storage racks*, including pallet storage racks, movable-shelf racks, rack-supported systems, automated storage and retrieval systems (stacker racks), push-back racks, pallet-flow racks, case-flow racks, pick modules and rack-supported platforms. Other types of racks, such as drive-in or drive-through racks, cantilever racks, portable racks or racks made of materials other than steel, are not considered storage racks for the purpose of this code.

[BS] STORAGE RACKS, STEEL CANTILEVERED. A framework or assemblage composed of cold-formed or hot-rolled steel structural members, primarily in the form of vertical columns, extended bases, horizontal arms projecting from the faces of the columns, and longitudinal (down-aisle) bracing between columns. There may be shelf beams between the arms, depending on the products being stored; this definition does not include other types of racks such as pallet storage racks, drive-in racks, drive-through racks, or racks made of materials other than steel.

[BG] STORM SHELTER. A building, structure or portions thereof, constructed in accordance with ICC 500 and designated for use during a severe wind storm event, such as a hurricane or tornado.

Community storm shelter. A storm shelter not defined as a "Residential storm shelter."

Residential storm shelter. A storm shelter serving occupants of *dwelling units* and having an *occupant load* not exceeding 16 persons.

[BG] STORY. That portion of a building included between the upper surface of a floor and the upper surface of the floor or roof next above (see "*Basement*," "*Building height*," "*Grade plane*" and "*Mezzanine*"). A story is measured as the vertical distance from top to top of two successive tiers of beams or finished floor surfaces and, for the topmost story, from the top of the floor finish to the top of the ceiling joists or, where there is not a ceiling, to the top of the roof rafters.

[BG] STORY ABOVE GRADE PLANE. Any *story* having its finished floor surface entirely above *grade plane*, or in which the finished surface of the floor next above is:

1. More than 6 feet (1829 mm) above *grade plane*; or
2. More than 12 feet (3658 mm) above the finished ground level at any point.

[BS] STRENGTH. This term is defined two ways, the first for use in Chapter 16 and the second for use in Chapter 21.

For Chapter 16:

Nominal strength. The capacity of a structure or member to resist the effects of loads, as determined by computations using specified material strengths and dimensions and equations derived from accepted principles of structural mechanics or by field tests or laboratory tests of scaled models, allowing for modeling effects and differences between laboratory and field conditions.

Required strength. Strength of a member, cross section or connection required to resist factored loads or related internal moments and forces in such combinations as stipulated by these provisions.

Strength design. A method of proportioning structural members such that the computed forces produced in the members by factored loads do not exceed the member design strength [also called "load and resistance factor design" (LRFD)]. The term "strength design" is used in the design of concrete and masonry structural elements.

For Chapter 21:

Design strength. Nominal strength multiplied by a strength reduction factor.

Nominal strength. Strength of a member or cross section calculated in accordance with these provisions before application of any strength-reduction factors.

Required strength. Strength of a member or cross section required to resist factored loads.

[BS] STRUCTURAL COMPOSITE LUMBER. Structural member manufactured using wood elements bonded together with exterior adhesives. Examples of *structural composite lumber* are:

[BS] STRUCTURAL GLUED-LAMINATED TIMBER. An engineered, stress-rated product of a timber laminating plant, composed of assemblies of specially selected and prepared wood laminations in which the grain of all laminations is approximately parallel longitudinally and the laminations are bonded with adhesives.

[BS] STRUCTURAL OBSERVATION. The visual observation of the structural system by a *registered design professional* for general conformance to the *approved construction documents*.

[A] STRUCTURE. That which is built or constructed.

[BS] SUBSTANTIAL DAMAGE. Damage of any origin sustained by a structure whereby the cost of restoring the structure to its before-damaged condition would equal or exceed 50 percent of the market value of the structure before the damage occurred.

[BS] SUBSTANTIAL IMPROVEMENT. Any *repair*, reconstruction, rehabilitation, *alteration*, *addition* or other improvement of a building or structure, the cost of which equals or exceeds 50 percent of the market value of the structure before the improvement or repair is started. If the structure has sustained *substantial damage*, any repairs are considered *substantial improvement* regardless of the actual *repair* work performed. The term does not, however, include either:

1. Any project for improvement of a building required to correct existing health, sanitary or safety code violations identified by the *building official* and that are the minimum necessary to assure safe living conditions.
2. Any *alteration* of a historic structure provided that the *alteration* will not preclude the structure's continued designation as a historic structure.

[BG] SUNROOM. A one-*story* structure attached to a building with a glazing area in excess of 40 percent of the gross area of the structure's *exterior walls* and roof.

[F] SUPERVISING STATION. A facility that receives signals and at which personnel are in attendance at all times to respond to these signals.

[F] SUPERVISORY SERVICE. The service required to monitor performance of guard tours and the operative condition of fixed suppression systems or other systems for the protection of life and property.

[F] SUPERVISORY SIGNAL. A signal indicating the need of action in connection with the supervision of guard tours, the fire suppression systems or equipment or the maintenance features of related systems.

[F] SUPERVISORY SIGNAL-INITIATING DEVICE. An initiation device, such as a valve supervisory switch, water-level indicator or low-air pressure switch on a dry-pipe sprinkler system, whose change of state signals an off-normal condition and its restoration to normal of a fire protection or life safety system, or a need for action in connection with guard tours, fire suppression systems or equipment or maintenance features of related systems.

[BS] SUSCEPTIBLE BAY. A roof or portion thereof with either of the following:

1. A slope less than $^1/_4$-inch per foot (0.0208 rad).
2. On which water is impounded, in whole or in part, and the secondary drainage system is functional but the primary drainage system is blocked.

A roof surface with a slope of $^1/_4$-inch per foot (0.0208 rad) or greater towards points of free drainage is not a susceptible bay.

[BG] SWIMMING POOL. Any structure intended for swimming, recreational bathing or wading that contains water over 24 inches (610 mm) deep. This includes in-ground, above-ground and on-ground pools; hot tubs; spas and fixed-in-place wading pools.

[BF] T RATING. The time period that the *penetration firestop system*, including the penetrating item, limits the maximum temperature rise to 325°F (163°C) above its initial temperature through the penetration on the nonfire side when tested in accordance with ASTM E814 or UL 1479.

[BG] TECHNICAL PRODUCTION AREA. Open elevated areas or spaces intended for entertainment technicians to walk on and occupy for servicing and operating entertainment technology systems and equipment. Galleries, including fly and lighting galleries, gridirons, catwalks, and similar areas are designed for these purposes.

[BG] TENSILE MEMBRANE STRUCTURE. A membrane structure having a shape that is determined by tension in the membrane and the geometry of the support structure. Typically, the structure consists of both flexible elements (e.g., membrane and cables), nonflexible elements (e.g., struts, masts, beams and arches) and the anchorage (e.g., supports and foundations). This includes frame-supported *tensile membrane structures*.

[F] TENT. A structure, enclosure, *umbrella structure* or shelter, with or without sidewalls or drops, constructed of fabric or pliable material supported in any manner except by air or the contents it protects (see "*Umbrella structure*").

[BF] TERMINATED STOPS. Factory feature of a door frame where the stops of the door frame are terminated not more than 6 inches (152 mm) from the bottom of the door frame. Terminated stops are also known as "hospital stops" or "sanitary stops."

[BG] THERMAL ISOLATION. A separation of conditioned spaces, between a *sunroom* and a *dwelling unit*, consisting of existing or new walls, doors or windows.

[BF] THERMOPLASTIC MATERIAL. A plastic material that is capable of being repeatedly softened by increase of temperature and hardened by decrease of temperature.

[BF] THERMOSETTING MATERIAL. A plastic material that is capable of being changed into a substantially nonreformable product when cured.

[BF] THROUGH PENETRATION. A breach in both sides of a floor, floor-ceiling or wall assembly to accommodate an item passing through the breaches.

[BF] THROUGH-PENETRATION FIRESTOP SYSTEM. An assemblage consisting of a fire-resistance-rated floor, floor-ceiling, or wall assembly, one or more penetrating items passing through the breaches in both sides of the assembly and the materials or devices, or both, installed to resist the spread of fire through the assembly for a prescribed period of time.

[BS] TIE, WALL. Metal connector that connects *wythes* of *masonry* walls together.

[BS] TIE-DOWN (HOLD-DOWN). A device used to resist uplift of the chords of *shear walls*.

[BS] TILE, STRUCTURAL CLAY. A hollow *masonry unit* composed of burned clay, shale, fire clay or mixture thereof, and having parallel cells.

[F] TIRES, BULK STORAGE OF. Storage of tires where the area available for storage exceeds 20,000 cubic feet (566 m^3).

[A] TOWNHOUSE. A single-family *dwelling unit* constructed in a group of three or more attached units in which each unit extends from the foundation to roof and with open space on at least two sides.

[F] TOXIC. A chemical falling within any of the following categories:

1. A chemical that has a median lethal dose (LD_{50}) of more than 50 milligrams per kilogram, but not more than 500 milligrams per kilogram of body weight when administered orally to albino rats weighing between 200 and 300 grams each.
2. A chemical that has a median lethal dose (LD_{50}) of more than 200 milligrams per kilogram, but not more than 1,000 milligrams per kilogram of body weight when administered by continuous contact for 24 hours (or less if death occurs within 24 hours) with the bare skin of albino rabbits weighing between 2 and 3 kilograms each.
3. A chemical that has a median lethal concentration (LC_{50}) in air of more than 200 parts per million, but not more than 2,000 parts per million by volume of gas or vapor, or more than 2 milligrams per liter but not more than 20 milligrams per liter of mist, fume or dust, when administered by continuous inhalation for 1 hour (or less if death occurs within 1 hour) to albino rats weighing between 200 and 300 grams each.

[BG] TRANSIENT. Occupancy of a *dwelling unit* or *sleeping unit* for not more than 30 days.

[BG] TRANSIENT AIRCRAFT. Aircraft based at another location and that is at the transient location for not more than 90 days.

TREATED WOOD. See *"Fire-retardant-treated wood"* and *"Preservative-treated wood."*

[BF] TRIM. Picture molds, chair rails, baseboards, *handrails*, door and window frames and similar decorative or protective materials used in fixed applications.

[F] TROUBLE SIGNAL. A signal initiated by the *fire alarm system* or device indicative of a fault in a monitored circuit or component.

[BS] TSUNAMI DESIGN GEODATABASE. The ASCE database (version 2016-1.0) of *Tsunami Design Zone* maps and associated design data for the states of Alaska, California, Hawaii, Oregon and Washington.

[BS] TSUNAMI DESIGN ZONE. An area identified on the *Tsunami Design Zone* map between the shoreline and the inundation limit, within which certain structures designated in Chapter 16 are designed for or protected from inundation.

[BS] TUBULAR DAYLIGHTING DEVICE (TDD). A non-operable *fenestration* unit primarily designed to transmit daylight from a roof surface to an interior ceiling via a tubular conduit. The basic unit consists of an exterior glazed weathering surface, a light-transmitting tube with a reflective interior surface, and an interior-sealing device such as a translucent ceiling panel. The unit can be factory assembled, or field-assembled from a manufactured kit.

[BE] TYPE A UNIT. A *dwelling unit* or *sleeping unit* designed and constructed for accessibility in accordance with this code and the provisions for *Type A units* in ICC A117.1.

[BE] TYPE B UNIT. A *dwelling unit* or *sleeping unit* designed and constructed for accessibility in accordance with this code and the provisions for *Type B units* in ICC A117.1, consistent with the design and construction requirements of the federal Fair Housing Act.

[F] UMBRELLA STRUCTURE. A structure, enclosure or shelter with or without sidewalls or drops, constructed of fabric or pliable material supported by a central pole or poles (see "*Tent*").

[BS] UNDERLAYMENT. One or more layers of a material that is applied to a steep-slope *roof covering* deck under the *roof covering* and resists liquid water that penetrates the *roof covering*.

[BS] UNDERPINNING. The alteration of an existing foundation to transfer *loads* to a lower elevation using new piers, piles or other permanent structural support elements installed below the existing foundation.

UNIT SKYLIGHT. See "*Skylight, unit.*"

[F] UNSTABLE (REACTIVE) MATERIAL. A material, other than an *explosive*, which in the pure state or as commercially produced, will vigorously polymerize, decompose, condense or become self-reactive and undergo other violent chemical changes, including *explosion*, when exposed to heat, friction or shock, or in the absence of an inhibitor, or in the presence of contaminants, or in contact with *incompatible materials*. Unstable (reactive) materials are subdivided as follows:

Class 1. Materials that in themselves are normally stable but which can become unstable at elevated temperatures and pressure.

Class 2. Materials that in themselves are normally unstable and readily undergo violent chemical change but do not detonate. This class includes materials that can undergo chemical change with rapid release of energy at *normal temperatures and pressures*, and that can undergo violent chemical change at elevated temperatures and pressures.

Class 3. Materials that in themselves are capable of *detonation* or of explosive decomposition or explosive reaction but which require a strong initiating source or which must be heated under confinement before initiation. This class includes materials that are sensitive to thermal or mechanical shock at elevated temperatures and pressures.

Class 4. Materials that in themselves are readily capable of *detonation* or *explosive* decomposition or explosive reaction at *normal temperatures and pressures*. This class includes materials that are sensitive to mechanical or localized thermal shock at *normal temperatures and pressures*.

[F] USE (MATERIAL). Placing a material into action, including *solids*, *liquids* and gases.

[BG] VAPOR DIFFUSION PORT. An assembly constructed or installed within a roof assembly at an opening in the roof deck to convey water vapor from an unvented *attic* to the outside atmosphere.

[BF] VAPOR PERMEABLE. The property of having a moisture vapor permeance rating of 5 perms (2.9×10^{-10} kg/Pa $\times$ s $\times$ m^2) or greater, when tested in accordance with Procedure A or Procedure B of ASTM E96. A vapor permeable material permits the passage of moisture vapor.

[BF] VAPOR RETARDER CLASS. A measure of a material or assembly's ability to limit the amount of moisture that passes through that material or assembly. Vapor retarder class shall be defined using the desiccant method with Procedure A of ASTM E96 as follows:

Class I: 0.1 perm or less.

Class II: $0.1 < \text{perm} \leq 1.0$ perm.

Class III: $1.0 < \text{perm} \leq 10$ perm.

[BS] VEGETATIVE ROOF. An assembly of interacting components designed to waterproof a building's top surface that includes, by design, vegetation and related landscape elements.

[BS] VEHICLE BARRIER. A component or a system of components, near open sides or walls of garage floors or ramps that act as a restraint for vehicles.

[BG] VEHICULAR GATE. A gate that is intended for use at a vehicular entrance or exit to a facility, building or portion thereof, and that is not intended for use by pedestrian traffic.

[BF] VENEER. A facing attached to a wall for the purpose of providing ornamentation, protection or insulation, but not counted as adding strength to the wall.

[M] VENTILATION. The natural or mechanical process of supplying conditioned or unconditioned air to, or removing such air from, any space.

[BF] VINYL SIDING. A shaped material, made principally from rigid polyvinyl chloride (PVC), that is used as an *exterior wall covering*.

[F] VISIBLE ALARM NOTIFICATION APPLIANCE. A notification appliance that alerts by the sense of sight.

[BG] WALKWAY, PEDESTRIAN. A walkway used exclusively as a pedestrian trafficway.

[BS] WALL. This definition applies only to Chapter 21.

A vertical element with a horizontal length-to-thickness ratio greater than three, used to enclose space.

Cavity wall. A wall built of *masonry units* or of concrete, or a combination of these materials, arranged to provide an airspace within the wall, and in which the inner and outer parts of the wall are tied together with metal ties.

Dry-stacked, surface-bonded wall. A wall built of concrete *masonry units* where the units are stacked dry, without *mortar* on the bed or *head joints*, and where both sides of the wall are coated with a surface-bonding *mortar*.

Parapet wall. The part of any wall entirely above the roof line.

[BS] WALL, LOAD-BEARING. Any wall meeting either of the following classifications:

1. Any metal or wood stud wall that supports more than 100 pounds per linear foot (1459 N/m) of vertical load in addition to its own weight.
2. Any *masonry,* concrete or *mass timber* wall that supports more than 200 pounds per linear foot (2919 N/m) of vertical *load* in addition to its own weight.

[BS] WALL, NONLOAD-BEARING. Any wall that is not a *load-bearing wall*.

[F] WATER-REACTIVE MATERIAL. A material that explodes; violently reacts; produces *flammable*, *toxic* or other hazardous gases; or evolves enough heat to cause autoignition or ignition of combustibles upon exposure to water or moisture. Water-reactive materials are subdivided as follows:

Class 3. Materials that react explosively with water without requiring heat or confinement.

Class 2. Materials that react violently with water or have the ability to boil water. Materials that produce *flammable*, *toxic* or other hazardous gases or evolve enough heat to cause autoignition or ignition of combustibles upon exposure to water or moisture.

Class 1. Materials that react with water with some release of energy, but not violently.

[BF] WATER-RESISTIVE BARRIER. A material behind an *exterior wall covering* that is intended to resist liquid water that has penetrated behind the exterior covering from further intruding into the *exterior wall* assembly.

[BF] WEATHER-EXPOSED SURFACES. Surfaces of walls, ceilings, floors, roofs, soffits and similar surfaces exposed to the weather except the following:

1. Ceilings and roof soffits enclosed by walls, fascia, bulkheads or beams that extend not less than 12 inches (305 mm) below such ceiling or roof soffits.
2. Walls or portions of walls beneath an unenclosed roof area, where located a horizontal distance from an open exterior opening equal to not less than twice the height of the opening.
3. Ceiling and roof soffits located a minimum horizontal distance of 10 feet (3048 mm) from the outer edges of the ceiling or roof soffits.

[F] WET-CHEMICAL EXTINGUISHING SYSTEM. A solution of water and potassium-carbonate-based chemical, potassium-acetate-based chemical or a combination thereof, forming an extinguishing agent.

[BE] WHEELCHAIR SPACE. A space for a single wheelchair and its occupant.

[BS] WIND SPEED, V. Basic design wind speeds.

[BS] WIND SPEED, V_{asd}. *Allowable stress design* wind speeds.

[BS] WINDBORNE DEBRIS REGION. Areas within *hurricane-prone regions* located:

1. Within 1 mile (1.61 km) of the mean high-water line where an Exposure D condition exists upwind at the waterline and the basic design wind speed, V, is 130 mph (58 m/s) or greater; or
2. In areas where the basic design wind speed is 140 mph (63 m/s) or greater.

For *Risk Category* II buildings and structures and Risk Category III buildings and structures, except health care facilities, the windborne debris region shall be based on Figure 1609.3.(1). For *Risk Category* IV buildings and structures and *Risk Category* III health care facilities, the windborne debris region shall be based on Figure 1609.3(2).

[BE] WINDER. A tread with nonparallel edges.

WINDFORCE-RESISTING SYSTEM, MAIN. See "*Main windforce-resisting system*."

[BS] WIRE BACKING. Horizontal strands of tautened wire attached to surfaces of vertical supports which, when covered with the building paper, provide a backing for *cement plaster*.

[F] WIRELESS PROTECTION SYSTEM. A system or a part of a system that can transmit and receive signals without the aid of wire.

[BS] WOOD SHEAR PANEL. A wood floor, roof or wall component sheathed to act as a *shear wall* or *diaphragm*.

[BS] WOOD STRUCTURAL PANEL. A panel manufactured from *veneers*, wood strands or wafers or a combination of *veneer* and wood strands or wafers bonded together with waterproof synthetic resins or other suitable bonding systems. Examples of *wood structural panels* are:

[BS] WOOD/PLASTIC COMPOSITE. A composite material made primarily from wood or cellulose-based materials and plastic.

[F] WORKSTATION. A defined space or an independent principal piece of equipment using *HPM* within a *fabrication area* where a specific function, laboratory procedure or research activity occurs. *Approved* or *listed hazardous materials storage cabinets*, *flammable liquid* storage cabinets or gas cabinets serving a workstation are included as part of the workstation. A workstation is allowed to contain *ventilation* equipment, fire protection devices, detection devices, electrical devices and other processing and scientific equipment.

[BS] WYTHE. Each continuous, vertical section of a wall, one *masonry unit* in thickness.

[BG] YARD. An open space, other than a *court*, unobstructed from the ground to the sky, except where specifically provided by this code, on the lot on which a building is situated.

[F] ZONE. A defined area within the protected premises. A zone can define an area from which a signal can be received, an area to which a signal can be sent or an area in which a form of control can be executed.

[F] ZONE, NOTIFICATION. An area within a building or facility covered by notification appliances which are activated simultaneously.

CHAPTER 3
OCCUPANCY CLASSIFICATION AND USE

User note:

About this chapter: *Chapter 3 provides the criteria by which buildings and structures are classified into use groups and occupancies. Through the balance of the code, occupancy classification is fundamental in the setting of features of construction; occupant safety requirements, especially building limitations; means of egress; fire protection systems; and interior finishes.*

SECTION 301
SCOPE

301.1 General. The provisions of this chapter shall control the classification of all buildings and structures as to occupancy and use. Different classifications of occupancy and use represent varying levels of hazard and risk to building occupants and adjacent properties.

SECTION 302
OCCUPANCY CLASSIFICATION AND USE DESIGNATION

302.1 Occupancy classification. Occupancy classification is the formal designation of the primary purpose of the building, structure or portion thereof. Structures shall be classified into one or more of the occupancy groups specified in this section based on the nature of the hazards and risks to building occupants generally associated with the intended purpose of the building or structure. An area, room or space that is intended to be occupied at different times for different purposes shall comply with all applicable requirements associated with such potential multipurpose. Structures containing multiple occupancy groups shall comply with Section 508. Where a structure is proposed for a purpose that is not specified in this section, such structure shall be classified in the occupancy it most nearly resembles based on the fire safety and relative hazard. Occupied roofs shall be classified in the group that the occupancy most nearly resembles, according to the fire safety and relative hazard, and shall comply with Section 503.1.4 .

1. Assembly (see Section 303): Groups A-1, A-2, A-3, A-4 and A-5.
2. Business (see Section 304): Group B.
3. Educational (see Section 305): Group E.
4. Factory and Industrial (see Section 306): Groups F-1 and F-2.
5. High Hazard (see Section 307): Groups H-1, H-2, H-3, H-4 and H-5.
6. Institutional (see Section 308): Groups I-1, I-2, I-3 and I-4.
7. Mercantile (see Section 309): Group M.
8. Residential (see Section 310): Groups R-1, R-2, R-3 and R-4.
9. Storage (see Section 311): Groups S-1 and S-2.
10. Utility and Miscellaneous (see Section 312): Group U.

302.2 Use designation. Occupancy groups contain subordinate uses having similar hazards and risks to building occupants. Uses include, but are not limited to, those functional designations specified within the occupancy group descriptions in Section 302.1. Certain uses require specific limitations and controls in accordance with the provisions of Chapter 4 and elsewhere in this code.

SECTION 303
ASSEMBLY GROUP A

303.1 Assembly Group A. Assembly Group A occupancy includes, among others, the use of a building or structure, or a portion thereof, for the gathering of persons for purposes such as civic, social or religious functions; recreation, food or drink consumption or awaiting transportation.

303.1.1 Small buildings and tenant spaces. A building or tenant space used for assembly purposes with an *occupant load* of less than 50 persons shall be classified as a Group B occupancy.

303.1.2 Small assembly spaces. The following rooms and spaces shall not be classified as Assembly occupancies:

1. A room or space used for assembly purposes with an *occupant load* of less than 50 persons and accessory to another occupancy shall be classified as a Group B occupancy or as part of that occupancy.
2. A room or space used for assembly purposes that is less than 750 square feet (70 m^2) in area and accessory to another occupancy shall be classified as a Group B occupancy or as part of that occupancy.

303.1.3 Associated with Group E occupancies. A room or space used for assembly purposes that is associated with a Group E occupancy is not considered a separate occupancy.

303.1.4 Accessory to places of religious worship. Accessory religious educational rooms and religious auditoriums with *occupant loads* of less than 100 per room or space are not considered separate occupancies.

303.1.5 Special amusement areas. Special *amusement areas* shall comply with Section 411.

303.2 Assembly Group A-1. Group A-1 occupancy includes assembly uses, usually with fixed seating, intended for the production and viewing of the performing arts or motion pictures including, but not limited to:

Motion picture theaters

Symphony and concert halls

Television and radio studios admitting an audience

Theaters

303.3 Assembly Group A-2. Group A-2 occupancy includes assembly uses intended for food and/or drink consumption including, but not limited to:

Banquet halls

Casinos (*gaming areas*)

Nightclubs

Restaurants, cafeterias and similar dining facilities (including associated commercial kitchens)

Taverns and bars

303.4 Assembly Group A-3. Group A-3 occupancy includes assembly uses intended for worship, recreation or amusement and other assembly uses not classified elsewhere in Group A including, but not limited to:

Amusement arcades

Art galleries

Bowling alleys

Community halls

Courtrooms

Dance halls (not including food or drink consumption)

Exhibition halls

Funeral parlors

Greenhouses for the conservation and exhibition of plants that provide public access

Gymnasiums (without spectator seating)

Indoor *swimming pools* (without spectator seating)

Indoor tennis courts (without spectator seating)

Lecture halls

Libraries

Museums

Places of religious worship

Pool and billiard parlors

Waiting areas in transportation terminals

303.5 Assembly Group A-4. Group A-4 occupancy includes assembly uses intended for viewing of indoor sporting events and activities with spectator seating including, but not limited to:

Arenas

Skating rinks

Swimming pools

Tennis courts

303.6 Assembly Group A-5. Group A-5 occupancy includes assembly uses intended for participation in or viewing outdoor activities including, but not limited to:

Amusement park structures

Bleachers

Grandstands

Stadiums

SECTION 304
BUSINESS GROUP B

304.1 Business Group B. Business Group B occupancy includes, among others, the use of a building or structure, or a portion thereof, for office, professional or service-type transactions, including storage of records and accounts. Business occupancies shall include, but not be limited to, the following:

Airport traffic control towers

Ambulatory care facilities

Animal hospitals, kennels and pounds

Banks

Barber and beauty shops

Car wash

Civic administration

Clinic, outpatient

Dry cleaning and laundries: pick-up and delivery stations and self-service

Educational occupancies for students above the 12th grade including *higher education laboratories*

Electronic data processing

Food processing establishments and commercial kitchens not associated with restaurants, cafeterias and similar dining facilities not more than 2,500 square feet (232 m^2) in area

Laboratories: testing and research

Motor vehicle showrooms

Post offices

Print shops

Professional services (architects, attorneys, dentists, physicians, engineers, etc.)

Radio and television stations

Telephone exchanges

Training and skill development not in a school or academic program (this shall include, but not be limited to, tutoring centers, martial arts studios, gymnastics and similar uses regardless of the ages served, and where not classified as a Group A occupancy)

304.2 Airport traffic control towers. Airport traffic control towers shall comply with Section 412.2.

304.3 Ambulatory care facilities. *Ambulatory care facilities* shall comply with Section 422.

304.4 Higher education laboratories. *Higher education laboratories* shall comply with Section 428.

SECTION 305
EDUCATIONAL GROUP E

305.1 Educational Group E. Educational Group E occupancy includes, among others, the use of a building or structure, or a portion thereof, by six or more persons at any one time for educational purposes through the 12th grade.

305.1.1 Accessory to places of religious worship. Religious educational rooms and religious auditoriums, which are accessory to *places of religious worship* in accordance with Section 303.1.4 and have *occupant loads* of less than 100 per room or space, shall be classified as Group A-3 occupancies.

305.2 Group E, day care facilities. This group includes buildings and structures or portions thereof occupied by more than five children older than $2^1/_2$ years of age who receive educational, supervision or *personal care services* for fewer than 24 hours per day.

305.2.1 Within places of religious worship. Rooms and spaces within *places of religious worship* providing such day care during religious functions shall be classified as part of the primary occupancy.

305.2.2 Five or fewer children. A facility having five or fewer children receiving such day care shall be classified as part of the primary occupancy.

305.2.3 Five or fewer children in a dwelling unit. A facility such as the above within a *dwelling unit* and having five or fewer children receiving such day care shall be classified as a Group R-3 occupancy or shall comply with the *International Residential Code*.

305.3 Storm shelters in Group E occupancies. *Storm shelters* shall be provided for Group E occupancies where required by Section 423.5.

SECTION 306
FACTORY GROUP F

306.1 Factory Industrial Group F. Factory Industrial Group F occupancy includes, among others, the use of a building or structure, or a portion thereof, for assembling, disassembling, fabricating, finishing, manufacturing, packaging, repair or processing operations that are not classified as a Group H hazardous or Group S storage occupancy.

306.2 Moderate-hazard factory industrial, Group F-1. Factory industrial uses that are not classified as Factory Industrial F-2 Low Hazard shall be classified as F-1 Moderate Hazard and shall include, but not be limited to, the following:

Aircraft (manufacturing, not to include repair)
Appliances
Athletic equipment
Automobiles and other motor vehicles
Bakeries
Beverages: over 16-percent alcohol content
Bicycles
Boats
Brooms or brushes
Business machines
Cameras and photo equipment
Canvas or similar fabric
Carpets and rugs (includes cleaning)
Clothing
Construction and agricultural machinery
Disinfectants
Dry cleaning and dyeing
Electric generation plants
Electronics
Energy storage systems (ESS) in dedicated use buildings
Engines (including rebuilding)
Food processing establishments and commercial kitchens not associated with restaurants, cafeterias and similar dining facilities more than 2,500 square feet (232 m^2) in area
Furniture
Hemp products
Jute products
Laundries
Leather products
Machinery
Metals
Millwork (sash and door)
Motion pictures and television filming (without spectators)
Musical instruments
Optical goods
Paper mills or products
Photographic film
Plastic products
Printing or publishing
Recreational vehicles
Refuse incineration
Shoes
Soaps and detergents
Textiles
Tobacco
Trailers
Upholstering
Water/sewer treatment facilities
Wood; distillation
Woodworking (cabinet)

306.2.1 Aircraft manufacturing facilities. Aircraft manufacturing facilities shall comply with Section 412.6.

306.3 Low-hazard factory industrial, Group F-2. Factory industrial uses that involve the fabrication or manufacturing of noncombustible materials that during finishing, packing or processing do not involve a significant fire hazard shall be classified as F-2 occupancies and shall include, but not be limited to, the following:

Beverages: up to and including 16-percent alcohol content

Brick and masonry

Ceramic products

Foundries

Glass products

Gypsum

Ice

Metal products (fabrication and assembly)

SECTION 307
HIGH-HAZARD GROUP H

[F] 307.1 High-hazard Group H. High-hazard Group H occupancy includes, among others, the use of a building or structure, or a portion thereof, that involves the manufacturing, processing, generation or storage of materials that constitute a physical or *health hazard* in quantities in excess of those allowed in *control areas* complying with Section 414, based on the maximum allowable quantity limits for *control areas* set forth in Tables 307.1(1) and 307.1(2). Hazardous occupancies are classified in Groups H-1, H-2, H-3, H-4 and H-5 and shall be in accordance with this section, the requirements of Section 415 and the *International Fire Code*. *Hazardous materials* stored, or used on top of roofs or canopies, shall be classified as outdoor storage or use and shall comply with the *International Fire Code*.

[F] 307.1.1 Uses other than Group H. An occupancy that stores, uses or handles *hazardous materials* as described in one or more of the following items shall not be classified as Group H, but shall be classified as the occupancy that it most nearly resembles.

1. Buildings and structures occupied for the application of flammable finishes, provided that such buildings or areas conform to the requirements of Section 416 and the *International Fire Code*.
2. Wholesale and retail sales and storage of flammable and combustible liquids in mercantile occupancies conforming to the *International Fire Code*.
3. Closed piping system containing *flammable or combustible liquids* or gases utilized for the operation of machinery or equipment.
4. Cleaning establishments that utilize *combustible liquid* solvents having a *flash point* of 140°F (60°C) or higher in closed systems employing equipment *listed* by an *approved* testing agency, provided that this occupancy is separated from all other areas of the building by 1-hour *fire barriers* constructed in accordance with Section 707 or 1-hour *horizontal assemblies* constructed in accordance with Section 711, or both.
5. Cleaning establishments that utilize a liquid solvent having a *flash point* at or above 200°F (93°C).
6. Liquor stores and distributors without bulk storage.
7. Refrigeration systems.
8. The storage or utilization of materials for agricultural purposes on the premises.
9. Stationary storage battery systems installed in accordance with the *International Fire Code*.
10. *Corrosive* personal or household products in their original packaging used in retail display.
11. Commonly used *corrosive* building materials.
12. Buildings and structures occupied for *aerosol product* storage, aerosol cooking spray products or plastic aerosol 3 products shall be classified as Group S-1, provided that such buildings conform to the requirements of the *International Fire Code*.
13. Display and storage of nonflammable solid and nonflammable or noncombustible liquid *hazardous materials* in quantities not exceeding the maximum allowable quantity per *control area* in Group M or S occupancies complying with Section 414.2.5.
14. The storage of black powder, smokeless propellant and small arms primers in Groups M and R-3 and special industrial *explosive* devices in Groups B, F, M and S, provided such storage conforms to the quantity limits and requirements prescribed in the *International Fire Code*.
15. Stationary fuel cell power systems installed in accordance with the *International Fire Code*.
16. Capacitor energy storage systems in accordance with the *International Fire Code*.
17. Group B *higher education laboratory* occupancies complying with Section 428 and Chapter 38 of the *International Fire Code*.
18. Distilling or brewing of beverages conforming to the requirements of the *International Fire Code*.
19. The storage of beer, distilled spirits and wines in barrels and casks conforming to the requirements of the *International Fire Code*.

TABLE 307.1(1)
MAXIMUM ALLOWABLE QUANTITY PER CONTROL AREA OF HAZARDOUS MATERIALS POSING A PHYSICAL HAZARD[a, j, m, n, p]

MATERIAL	CLASS	GROUP WHEN THE MAXIMUM ALLOWABLE QUANTITY IS EXCEEDED	STORAGE[b]			USE-CLOSED SYSTEMS[b]			USE-OPEN SYSTEMS[b]	
			Solid pounds (cubic feet)	Liquid gallons (pounds)	Gas (cubic feet at NTP)	Solid pounds (cubic feet)	Liquid gallons (pounds)	Gas (cubic feet at NTP)	Solid pounds (cubic feet)	Liquid gallons (pounds)
Combustible dust	NA	H-2	See Note q	NA	NA	See Note q	NA	NA	See Note q	NA
Combustible fiber[q]	Loose	H-3	(100)	NA	NA	(100)	NA	NA	(20)	NA
	Baled[o]		(1,000)			(1,000)			(200)	
Combustible liquid[c, i]	II	H-2 or H-3	NA	120[d, e]	NA	NA	120[d]	NA	NA	30[d]
	IIIA	H-2 or H-3		330[d, e]			330[d]			80[d]
	IIIB	NA		13,200[e, f]			13,200[f]			3,300[f]
Cryogenic flammable	NA	H-2	NA	45[d]	NA	NA	45[d]	NA	NA	10[d]
Cryogenic inert	NA	NA	NA	NA	NL	NA	NA	NL	NA	NA
Cryogenic oxidizing	NA	H-3	NA	45[d]	NA	NA	45[d]	NA	NA	10[d]
Explosives	Division 1.1	H-1	1[e, g]	(1)[e, g]	NA	0.25[g]	(0.25)[g]	NA	0.25[g]	(0.25)[g]
	Division 1.2	H-1	1[e, g]	(1)[e, g]		0.25[g]	(0.25)[g]		0.25[g]	(0.25)[g]
	Division 1.3	H-1 or H-2	5[e, g]	(5)[e, g]		1[g]	(1)[g]		1[g]	(1)[g]
	Division 1.4	H-3	50[e, g]	(50)[e, g]		50[g]	(50)[g]		NA	NA
	Division 1.4G	H-3	125[e, l]	NA		NA	NA		NA	NA
	Division 1.5	H-1	1[e, g]	(1)[e, g]		0.25[g]	(0.25)[g]		0.25[g]	(0.25)[g]
	Division 1.6	H-1	1[e, g]	NA		NA	NA		NA	NA
Flammable gas	Gaseous	H-2	NA	NA	1,000[d,e]	NA	NA	1,000[d,e]	NA	NA
	Liquefied			(150)[d,e]	NA		(150)[d,e]	NA		
Flammable liquid[c]	IA	H-2 or H-3	NA	30[d, e]	NA	NA	30[d]	NA	NA	10[d]
	IB and IC			120[d, e]			120[d]			30[d]
Flammable liquid, combination (IA, IB, IC)	NA	H-2 or H-3	NA	120[d, e, h]	NA	NA	120[d, h]	NA	NA	30[d, h]
Flammable solid	NA	H-3	125[d, e]	NA	NA	125[d]	NA	NA	25[d]	NA
Inert gas	Gaseous	NA	NA	NA	NL	NA	NA	NL	NA	NA
	Liquefied	NA	NA	NA	NL	NA	NA	NL	NA	NA
Organic peroxide	UD	H-1	1[e, g]	(1)[e, g]	NA	0.25[g]	(0.25)[g]	NA	0.25[g]	(0.25)[g]
	I	H-2	5[d, e]	(5)[d, e]		1[d]	(1)[d]		1[d]	(1)[d]
	II	H-3	50[d, e]	(50)[d, e]		50[d]	(50)[d]		10[d]	(10)[d]
	III	H-3	125[d, e]	(125)[d, e]		125[d]	(125)[d]		25[d]	(25)[d]
	IV	NA	NL	NL		NL	NL		NL	NL
	V	NA	NL	NL		NL	NL		NL	NL

(continued)

TABLE 307.1(1)—continued
MAXIMUM ALLOWABLE QUANTITY PER CONTROL AREA OF HAZARDOUS MATERIALS POSING A PHYSICAL HAZARD[a, j, m, n, p]

MATERIAL	CLASS	GROUP WHEN THE MAXIMUM ALLOWABLE QUANTITY IS EXCEEDED	STORAGE[b]			USE-CLOSED SYSTEMS[b]			USE-OPEN SYSTEMS[b]	
			Solid pounds (cubic feet)	Liquid gallons (pounds)	Gas (cubic feet at NTP)	Solid pounds (cubic feet)	Liquid gallons (pounds)	Gas (cubic feet at NTP)	Solid pounds (cubic feet)	Liquid gallons (pounds)
Oxidizer	4	H-1	1[g]	(1)[e, g]	NA	0.25[g]	(0.25)[g]	NA	0.25[g]	(0.25)[g]
	3[k]	H-2 or H-3	10[d, e]	(10)[d, e]		2[d]	(2)[d]		2[d]	(2)[d]
	2	H-3	250[d, e]	(250)[d, e]		250[d]	(250)[d]		50[d]	(50)[d]
	1	NA	4,000[e, f]	(4,000)[e, f]		4,000[f]	(4,000)[f]		1,000[f]	(1,000)[f]
Oxidizing gas	Gaseous	H-3	NA	NA	1,500[d,e]	NA	NA	1,500[d,e]	NA	NA
	Liquefied			(150)[d, e]	NA		(150)[d,e]	NA		
Pyrophoric	NA	H-2	4[e, g]	(4)[e, g]	50[e, g]	1[g]	(1)[g]	10[e, g]	0	0
Unstable (reactive)	4	H-1	1[e, g]	(1)[e, g]	10[e, g]	0.25[g]	(0.25)[g]	2[e, g]	0.25[g]	(0.25)[g]
	3	H-1 or H-2	5[d, e]	(5)[d, e]	50[d, e]	1[d]	(1)[d]	10[d, e]	1[d]	(1)[d]
	2	H-3	50[d, e]	(50)[d, e]	750[d, e]	50[d]	(50)[d]	750[d, e]	10[d]	(10)[d]
	1	NA	NL	NL	NL	NL	NL	NL	NL	NL
Water reactive	3	H-2	5[d, e]	(5)[d, e]	NA	5[d]	(5)[d]	NA	1[d]	(1)[d]
	2	H-3	50[d, e]	(50)[d, e]		50[d]	(50)[d]		10[d]	(10)[d]
	1	NA	NL	NL		NL	NL		NL	NL

For SI: 1 cubic foot = 0.028 m^3, 1 pound = 0.454 kg, 1 gallon = 3.785 L.
NL = Not Limited; NA = Not Applicable; UD = Unclassified Detonable.

a. For use of control areas, see Section 414.2.

b. The aggregate quantity in use and storage shall not exceed the quantity specified for storage.

c. The quantities of alcoholic beverages in retail and wholesale sales occupancies shall not be limited provided the liquids are packaged in individual containers not exceeding 1.3 gallons. In retail and wholesale sales occupancies, the quantities of medicines, foodstuffs or consumer products, and cosmetics containing not more than 50 percent by volume of water-miscible liquids with the remainder of the solutions not being flammable, shall not be limited, provided that such materials are packaged in individual containers not exceeding 1.3 gallons.

d. Maximum allowable quantities shall be increased 100 percent in buildings equipped throughout with an automatic sprinkler system in accordance with Section 903.3.1.1. Where Note e also applies, the increase for both notes shall be applied accumulatively.

e. Maximum allowable quantities shall be increased 100 percent when stored in approved storage cabinets, day boxes, gas cabinets, gas rooms or exhausted enclosures or in listed safety cans in accordance with Section 5003.9.10 of the *International Fire Code.* Where Note d also applies, the increase for both notes shall be applied accumulatively.

f. Quantities shall not be limited in a building equipped throughout with an automatic sprinkler system in accordance with Section 903.3.1.1.

g. Allowed only in buildings equipped throughout with an automatic sprinkler system in accordance with Section 903.3.1.1.

h. Containing not more than the maximum allowable quantity per control area of Class IA, IB or IC flammable liquids.

i. The maximum allowable quantity shall not apply to fuel oil storage complying with Section 605.4.2 of the *International Fire Code.*

j. Quantities in parentheses indicate quantity units in parentheses at the head of each column.

k. A maximum quantity of 220 pounds of solid or 22 gallons of liquid Class 3 oxidizers is allowed when such materials are necessary for maintenance purposes, operation or sanitation of equipment when the storage containers and the manner of storage are approved.

l. Net weight of the pyrotechnic composition of the fireworks. Where the net weight of the pyrotechnic composition of the fireworks is not known, 25 percent of the gross weight of the fireworks, including packaging, shall be used.

m. For gallons of liquids, divide the amount in pounds by 10 in accordance with Section 5003.1.2 of the *International Fire Code.*

n. For storage and display quantities in Group M and storage quantities in Group S occupancies complying with Section 414.2.5, see Tables 414.2.5(1) and 414.2.5(2).

o. Densely packed baled cotton that complies with the packing requirements of ISO 8115 shall not be included in this material class.

p. The following shall not be included in determining the maximum allowable quantities:
   1. Liquid or gaseous fuel in fuel tanks on vehicles.
   2. Liquid or gaseous fuel in fuel tanks on motorized equipment operated in accordance with the *International Fire Code.*
   3. Gaseous fuels in piping systems and fixed appliances regulated by the *International Fuel Gas Code.*
   4. Liquid fuels in piping systems and fixed appliances regulated by the *International Mechanical Code.*
   5. Alcohol-based hand rubs classified as Class I or II liquids in dispensers that are installed in accordance with Sections 5705.5 and 5705.5.1 of the *International Fire Code.* The location of the alcohol-based hand rub (ABHR) dispensers shall be provided in the construction documents.

q. Where manufactured, generated or used in such a manner that the concentration and conditions create a fire or explosion hazard based on information prepared in accordance with Section 414.1.3.

TABLE 307.1(2)
[F] TABLE 307.1(2)
MAXIMUM ALLOWABLE QUANTITY PER CONTROL AREA OF HAZARDOUS MATERIALS POSING A HEALTH HAZARD[a, c, f, h, i]

MATERIAL	STORAGE[b]			USE-CLOSED SYSTEMS[b]			USE-OPEN SYSTEMS[b]	
	Solid pounds[d, e]	Liquid gallons (pounds)[d, e]	Gas cubic feet at NTP (pounds)[d]	Solid pounds[d]	Liquid gallons (pounds)[d]	Gas cubic feet at NTP (pounds)[d]	Solid pounds[d]	Liquid gallons (pounds)[d]
Corrosives	5,000	500	Gaseous 810[e] Liquefied (150)	5,000	500	Gaseous 810[e] Liquefied (150)	1,000	100
Highly Toxic	10	(10)	Gaseous 20[g] Liquefied (4)[g]	10	(10)	Gaseous 20[g] Liquefied (4)[g]	3	(3)
Toxic	500	(500)	Gaseous 810[e] Liquefied (150)[e]	500	(500)	Gaseous 810[e] Liquefied (150)[e]	125	(125)

For SI: 1 cubic foot = 0.028 m^3, 1 pound = 0.454 kg, 1 gallon = 3.785 L.
a. For use of control areas, see Section 414.2.
b. The aggregate quantity in use and storage shall not exceed the quantity specified for storage.
c. In retail and wholesale sales occupancies, the quantities of medicines, foodstuffs or consumer products, and cosmetics containing not more than 50 percent by volume of water-miscible liquids and with the remainder of the solutions not being flammable, shall not be limited, provided that such materials are packaged in individual containers not exceeding 1.3 gallons.
d. Maximum allowable quantities shall be increased 100 percent in buildings equipped throughout with an approved automatic sprinkler system in accordance with Section 903.3.1.1. Where Note e also applies, the increase for both notes shall be applied accumulatively.
e. Maximum allowable quantities shall be increased 100 percent where stored in approved storage cabinets, gas cabinets or exhausted enclosures as specified in the *International Fire Code*. Where Note d also applies, the increase for both notes shall be applied accumulatively.
f. For storage and display quantities in Group M and storage quantities in Group S occupancies complying with Section 414.2.5, see Tables 414.2.5(1) and 414.2.5(2).
g. Allowed only where stored in approved exhausted gas cabinets or exhausted enclosures as specified in the *International Fire Code*.
h. Quantities in parentheses indicate quantity units in parentheses at the head of each column.
i. For gallons of liquids, divide the amount in pounds by 10 in accordance with Section 5003.1.2 of the *International Fire Code*.

[F] 307.2 Hazardous materials. *Hazardous materials* in any quantity shall conform to the requirements of this code, including Section 414 , and the *International Fire Code*.

[F] 307.3 High-hazard Group H-1. Buildings and structures containing materials that pose a detonation hazard shall be classified as Group H-1. Such materials shall include, but not be limited to, the following:

Detonable *pyrophoric* materials

Explosives:

Division 1.1

Division 1.2

Division 1.3

Division 1.4

Division 1.5

Division 1.6

Organic peroxides, unclassified detonable

Oxidizers, Class 4

Unstable (reactive) materials, Class 3 detonable and Class 4

[F] 307.3.1 Occupancies containing explosives not classified as H-1. The following occupancies containing *explosive* materials shall be classified as follows:

1. Division 1.3 *explosive* materials that are used and maintained in a form where either confinement or configuration will not elevate the hazard from a mass fire to mass *explosion* hazard shall be allowed in H-2 occupancies.
2. Articles, including articles packaged for shipment, that are not regulated as a Division 1.4 *explosive* under Bureau of Alcohol, Tobacco, Firearms and Explosives regulations, or unpackaged articles used in process operations that do not propagate a detonation or deflagration between articles shall be allowed in H-3 occupancies.

[F] 307.4 High-hazard Group H-2. Buildings and structures containing materials that pose a deflagration hazard or a hazard from accelerated burning shall be classified as Group H-2. Such materials shall include, but not be limited to, the following:

Class I, II or IIIA *flammable or combustible liquids* that are used or stored in normally open containers or systems, or in closed containers or systems pressurized at more than 15 pounds per square inch gauge (103.4 kPa).

Combustible dusts where manufactured, generated or used in such a manner that the concentration and conditions create a fire or explosion hazard based on information prepared in accordance with Section 414.1.3.

Cryogenic fluids, flammable.

Flammable gases.

Organic peroxides, Class I.

Oxidizers, Class 3, that are used or stored in normally open containers or systems, or in closed containers or systems pressurized at more than 15 pounds per square inch gauge (103 kPa).

Pyrophoric liquids, solids and gases, nondetonable.

Unstable (reactive) materials, Class 3, nondetonable.

Water-reactive materials, Class 3.

[F] 307.5 High-hazard Group H-3. Buildings and structures containing materials that readily support combustion or that pose a *physical hazard* shall be classified as Group H-3. Such materials shall include, but not be limited to, the following:

Class I, II or IIIA *flammable or combustible liquids* that are used or stored in normally closed containers or systems pressurized at 15 pounds per square inch gauge (103.4 kPa) or less

Combustible fibers, other than densely packed *baled cotton*, where manufactured, generated or used in such a manner that the concentration and conditions create a fire or *explosion* hazard based on information prepared in accordance with Section 414.1.3

Consumer *fireworks*, 1.4G (Class C, Common)

Cryogenic fluids, oxidizing

Flammable solids

Organic peroxides, Class II and III

Oxidizers, Class 2

Oxidizers, Class 3, that are used or stored in normally closed containers or systems pressurized at 15 pounds per square inch gauge (103 kPa) or less

Oxidizing gases

Unstable (reactive) materials, Class 2

Water-reactive materials, Class 2

[F] 307.6 High-hazard Group H-4. Buildings and structures containing materials that are *health hazards* shall be classified as Group H-4. Such materials shall include, but not be limited to, the following:

Corrosives

Highly toxic materials

Toxic materials

[F] 307.7 High-hazard Group H-5. Semiconductor fabrication facilities and comparable research and development areas in which hazardous production materials (HPM) are used and the aggregate quantity of materials is in excess of those specified in Table 307.1(1) and [F] Table 307.1(2) shall be classified as Group H-5. Such facilities and areas shall be designed and constructed in accordance with Section 415.11.

[F] 307.8 Multiple hazards. Buildings and structures containing a material or materials representing hazards that are classified in one or more of Groups H-1, H-2, H-3 and H-4 shall conform to the code requirements for each of the occupancies so classified.

SECTION 308
INSTITUTIONAL GROUP I

308.1 Institutional Group I. Institutional Group I occupancy includes, among others, the use of a building or structure, or a portion thereof, in which care or supervision is provided to persons who are or are *incapable of self-preservation* without physical assistance or in which persons are detained for penal or correctional purposes or in which the liberty of the occupants is restricted. Institutional occupancies shall be classified as Group I-1, I-2, I-3 or I-4.

308.2 Institutional Group I-1. Institutional Group I-1 occupancy shall include buildings, structures or portions thereof for more than 16 persons, excluding staff, who reside on a *24-hour basis* in a supervised environment and receive *custodial care*. Buildings of Group I-1 shall be classified as one of the occupancy conditions specified in Section 308.2.1 or 308.2.2 and shall comply with Section 420. This group shall include, but not be limited to, the following:

Alcohol and drug centers

Assisted living facilities

Congregate care facilities

Group homes

Halfway houses

Residential board and care facilities

Social rehabilitation facilities

308.2.1 Condition 1. This occupancy condition shall include buildings in which all persons receiving *custodial care* who, without any assistance, are capable of responding to an emergency situation to complete building evacuation.

308.2.2 Condition 2. This occupancy condition shall include buildings in which there are any persons receiving *custodial care* who require limited verbal or physical assistance while responding to an emergency situation to complete building evacuation.

308.2.3 Six to 16 persons receiving custodial care. A facility housing not fewer than six and not more than 16 persons receiving *custodial care* shall be classified as Group R-4.

308.2.4 Five or fewer persons receiving custodial care. A facility with five or fewer persons receiving *custodial care* shall be classified as Group R-3 or shall comply with the *International Residential Code* provided an *automatic sprinkler system* is installed in accordance with Section 903.3.1.3 or Section P2904 of the *International Residential Code*.

308.3 Institutional Group I-2. Institutional Group I-2 occupancy shall include buildings and structures used for *medical care* on a *24-hour basis* for more than five persons who are *incapable of self-preservation*. This group shall include, but not be limited to, the following:

Foster care facilities

Detoxification facilities

Hospitals

Nursing homes

Psychiatric hospitals

308.3.1 Occupancy conditions. Buildings of Group I-2 shall be classified as one of the occupancy conditions specified in Section 308.3.1.1 or 308.3.1.2 and shall comply with Section 407.

308.3.1.1 Condition 1. This occupancy condition shall include facilities that provide nursing and *medical care* but do not provide emergency care, surgery, obstetrics or in-patient stabilization units for psychiatric or detoxification, including but not limited to *nursing homes* and *foster care facilities*.

308.3.1.2 Condition 2. This occupancy condition shall include facilities that provide nursing and *medical care* and could provide emergency care, surgery, obstetrics or in-patient stabilization units for psychiatric or detoxification, including but not limited to *hospitals*.

308.3.2 Five or fewer persons receiving *medical care*. A facility with five or fewer persons receiving *medical care* shall be classified as Group R-3 or shall comply with the *International Residential Code* provided an *automatic sprinkler system* is installed in accordance with Section 903.3.1.3 or Section P2904 of the *International Residential Code*.

308.4 Institutional Group I-3. Institutional Group I-3 occupancy shall include buildings and structures that are inhabited by more than five persons who are under restraint or security. A Group I-3 facility is occupied by persons who are generally *incapable of self-preservation* due to security measures not under the occupants' control. This group shall include, but not be limited to, the following:

Correctional centers

Detention centers

Jails

Prerelease centers

Prisons

Reformatories

Buildings of Group I-3 shall be classified as one of the occupancy conditions specified in Sections 308.4.1 through 308.4.5 and shall comply with Section 408.

308.4.1 Condition 1. This occupancy condition shall include buildings in which free movement is allowed from sleeping areas, and other spaces where access or occupancy is permitted, to the exterior via *means of egress* without restraint. A Condition 1 facility is permitted to be constructed as Group R.

308.4.2 Condition 2. This occupancy condition shall include buildings in which free movement is allowed from sleeping areas and any other occupied *smoke compartment* to one or more other *smoke compartments*. Egress to the exterior is impeded by locked exits.

308.4.3 Condition 3. This occupancy condition shall include buildings in which free movement is allowed within individual *smoke compartments*, such as within a residential unit composed of individual *sleeping units* and group activity spaces, where egress is impeded by remote-controlled release of *means of egress* from such a *smoke compartment* to another *smoke compartment*.

308.4.4 Condition 4. This occupancy condition shall include buildings in which free movement is restricted from an occupied space. Remote-controlled release is provided to permit movement from *sleeping units*, activity spaces and other occupied areas within the *smoke compartment* to other *smoke compartments*.

308.4.5 Condition 5. This occupancy condition shall include buildings in which free movement is restricted from an occupied space. Staff-controlled manual release is provided to permit movement from *sleeping units*, activity spaces and other occupied areas within the *smoke compartment* to other *smoke compartments*.

308.5 Institutional Group I-4, day care facilities. Institutional Group I-4 occupancy shall include buildings and structures occupied by more than five persons of any age who receive *custodial care* for fewer than 24 hours per day by persons other than parents or guardians; relatives by blood, marriage or adoption; and in a place other than the home of the person cared for. This group shall include, but not be limited to, the following:

Adult day care

Child day care

308.5.1 Classification as Group E. A child day care facility that provides care for more than five but not more than 100 children $2^1/_2$ years or less of age, where the rooms in which the children are cared for are located on a *level of exit discharge* serving such rooms and each of these child care rooms has an *exit* door directly to the exterior, shall be classified as Group E.

308.5.2 Within a place of religious worship. Rooms and spaces within *places of religious worship* providing such care during religious functions shall be classified as part of the primary occupancy.

308.5.3 Five or fewer persons receiving care. A facility having five or fewer persons receiving *custodial care* shall be classified as part of the primary occupancy.

308.5.4 Five or fewer persons receiving care in a dwelling unit. A facility such as the above within a *dwelling unit* and having five or fewer persons receiving *custodial care* shall be classified as a Group R-3 occupancy or shall comply with the *International Residential Code*.

SECTION 309
MERCANTILE GROUP M

309.1 Mercantile Group M. Mercantile Group M occupancy includes, among others, the use of a building or structure or a portion thereof for the display and sale of merchandise, and involves stocks of goods, wares or merchandise incidental to such purposes and where the public has access. Mercantile occupancies shall include, but not be limited to, the following:

Department stores

Drug stores

Markets

Greenhouses for display and sale of plants that provide public access.

Motor fuel-dispensing facilities

Retail or wholesale stores

Sales rooms

309.2 Quantity of hazardous materials. The aggregate quantity of nonflammable solid and nonflammable or noncombustible liquid *hazardous materials* stored or displayed in a single *control area* of a Group M occupancy shall not exceed the quantities in Table 414.2.5(1).

309.3 Motor fuel-dispensing facilities. Motor fuel-dispensing facilities shall comply with Section 406.7.

SECTION 310 RESIDENTIAL GROUP R

310.1 Residential Group R. Residential Group R includes, among others, the use of a building or structure, or a portion thereof, for sleeping purposes when not classified as an Institutional Group I or when not regulated by the *International Residential Code*. Group R occupancies not constructed in accordance with the *International Residential Code* as permitted by Sections 310.4.1 and 310.4.2 shall comply with Section 420.

310.2 Residential Group R-1. Residential Group R-1 occupancies containing *sleeping units* where the occupants are primarily *transient* in nature, including:

Boarding houses (*transient*) with more than 10 occupants

Congregate living facilities (*transient*) with more than 10 occupants

Hotels (*transient*)

Motels (*transient*)

310.3 Residential Group R-2. Residential Group R-2 occupancies containing *sleeping units* or more than two *dwelling units* where the occupants are primarily permanent in nature, including:

Apartment houses

Congregate living facilities (nontransient) with more than 16 occupants

Boarding houses (*nontransient*)

Convents

Dormitories

Fraternities and sororities

Monasteries

Hotels (nontransient)

Live/work units

Motels (nontransient)

Vacation timeshare properties

310.4 Residential Group R-3. Residential Group R-3 occupancies where the occupants are primarily permanent in nature and not classified as Group R-1, R-2, R-4 or I, including:

Buildings that do not contain more than two *dwelling units*

Care facilities that provide accommodations for five or fewer persons receiving care

Congregate living facilities (nontransient) with 16 or fewer occupants

Boarding houses (nontransient)

Convents

Dormitories

Fraternities and sororities

Monasteries

Congregate living facilities (*transient*) with 10 or fewer occupants

Boarding houses (*transient*)

Lodging houses (*transient*) with five or fewer *guest rooms* and 10 or fewer occupants

310.4.1 Care facilities within a dwelling. Care facilities for five or fewer persons receiving care that are within a single-family dwelling are permitted to comply with the *International Residential Code* provided an *automatic sprinkler system* is installed in accordance with Section 903.3.1.3 or Section P2904 of the *International Residential Code*.

310.4.2 Lodging houses. Owner-occupied *lodging houses* with five or fewer *guest rooms* and 10 or fewer total occupants shall be permitted to be constructed in accordance with the *International Residential Code*, provided that an *automatic sprinkler system* is installed in accordance with Section 903.3.1.3 or Section P2904 of the *International Residential Code*.

310.5 Residential Group R-4. Residential Group R-4 occupancy shall include buildings, structures or portions thereof for more than five but not more than 16 persons, excluding staff, who reside on a *24-hour basis* in a supervised residential environment and receive *custodial care*. Buildings of Group R-4 shall be classified as one of the occupancy conditions specified in Section 310.5.1 or 310.5.2. This group shall include, but not be limited to, the following:

Alcohol and drug centers

Assisted living facilities

Congregate care facilities

Group homes

Halfway houses

Residential board and care facilities

Social rehabilitation facilities

Group R-4 occupancies shall meet the requirements for construction as defined for Group R-3, except as otherwise provided for in this code.

310.5.1 Condition 1. This occupancy condition shall include buildings in which all persons receiving *custodial*

care, without any assistance, are capable of responding to an emergency situation to complete building evacuation.

310.5.2 Condition 2. This occupancy condition shall include buildings in which there are any persons receiving *custodial care* who require limited verbal or physical assistance while responding to an emergency situation to complete building evacuation.

SECTION 311
STORAGE GROUP S

311.1 Storage Group S. Storage Group S occupancy includes, among others, the use of a building or structure, or a portion thereof, for storage that is not classified as a hazardous occupancy.

311.1.1 Accessory storage spaces. A room or space used for storage purposes that is accessory to another occupancy shall be classified as part of that occupancy.

311.1.2 Combustible storage. High-piled stock or rack storage, or *attic*, under-floor and concealed spaces used for storage of combustible materials, shall be in accordance with Section 413.

311.2 Moderate-hazard storage, Group S-1. Storage Group S-1 occupancies are buildings occupied for storage uses that are not classified as Group S-2, including, but not limited to, storage of the following:

Aerosol products, Levels 2 and 3

Aircraft hangar (storage and repair)

Bags: cloth, burlap and paper

Bamboos and rattan

Baskets

Belting: canvas and leather

Beverages over 16-percent alcohol content

Books and paper in rolls or packs

Boots and shoes

Buttons, including cloth covered, pearl or bone

Cardboard and cardboard boxes

Clothing, woolen wearing apparel

Cordage

Dry boat storage (indoor)

Furniture

Furs

Glues, mucilage, pastes and size

Grains

Horns and combs, other than celluloid

Leather

Linoleum

Lumber

Motor vehicle *repair garages* complying with the maximum allowable quantities of *hazardous materials* specified in Table 307.1(1) (see Section 406.8)

Photo engravings

Resilient flooring

Self-service storage facility (mini-storage)

Silks

Soaps

Sugar

Tires, bulk storage of

Tobacco, cigars, cigarettes and snuff

Upholstery and mattresses

Wax candles

311.2.1 Aircraft hangers. Aircraft hangars used for storage or repair shall comply with Section 412.3.

311.2.2 Motor vehicle repair garages Motor vehicle *repair garages* shall comply with Section 406.8.

311.3 Low-hazard storage, Group S-2. Storage Group S-2 occupancies include, among others, buildings used for the storage of noncombustible materials such as products on wood pallets or in paper cartons with or without single thickness divisions; or in paper wrappings. Such products are permitted to have a negligible amount of plastic *trim*, such as knobs, handles or film wrapping. Group S-2 storage uses shall include, but not be limited to, storage of the following:

Asbestos

Beverages up to and including 16-percent alcohol

Cement in bags

Chalk and crayons

Dairy products in nonwaxed coated paper containers

Dry cell batteries

Electrical coils

Electrical motors

Empty cans

Food products

Foods in noncombustible containers

Fresh fruits and vegetables in nonplastic trays or containers

Frozen foods

Glass

Glass bottles, empty or filled with noncombustible liquids

Gypsum board

Inert pigments

Ivory

Meats

Metal cabinets

Metal desks with plastic tops and *trim*

Metal parts

Metals

Mirrors

Oil-filled and other types of distribution transformers

Public parking garages, open or enclosed

Porcelain and pottery

Stoves

Talc and soapstones

Washers and dryers

311.3.1 Public parking garages. Public parking garages shall comply with Section 406.4 and the additional requirements of Section 406.5 for *open parking garages* or Section 406.6 for enclosed parking garages.

SECTION 312
UTILITY AND MISCELLANEOUS GROUP U

312.1 General. Buildings and structures of an accessory character and miscellaneous structures not classified in any specific occupancy shall be constructed, equipped and maintained to conform to the requirements of this code commensurate with the fire and life hazard incidental to their occupancy. Group U shall include, but not be limited to, the following:

Agricultural buildings

Aircraft hangars, accessory to a one- or two-family residence (see Section 412.4)

Barns

Carports

Communication equipment structures with a *gross floor area* of less than 1,500 square feet (139 m^2)

Fences more than 7 feet (2134 mm) in height

Grain silos, accessory to a residential occupancy

Livestock shelters

Private garages

Retaining walls

Sheds

Stables

Tanks

Towers

312.1.1 Greenhouses. *Greenhouses* not classified as another occupancy shall be classified as Use Group U.

312.2 Private garages and carports. *Private garages* and carports shall comply with Section 406.3.

312.3 Residential aircraft hangars. Aircraft hangars accessory to a one- or two-family residence shall comply with Section 412.4.

CHAPTER 4
SPECIAL DETAILED REQUIREMENTS BASED ON OCCUPANCY AND USE

User note:

About this chapter: *Chapter 4 provides detailed criteria for special uses and occupancies. The unique characteristics of a live/work unit as opposed to a 30-story high-rise building call for specific standards for each. Twenty-seven sections address covered and open mall buildings, atriums, hospitals, stages, buildings where hazardous materials are used and stored, jails and prisons, ambulatory care facilities and storm shelters, among other special occupancy issues.*

SECTION 401
SCOPE

401.1 Detailed occupancy and use requirements. In addition to the occupancy and construction requirements in this code, the provisions of this chapter apply to the occupancies and use described herein.

SECTION 402
COVERED MALL AND OPEN MALL BUILDINGS

402.1 Applicability. The provisions of this section shall apply to buildings or structures defined herein as *covered or open mall buildings* not exceeding three floor levels at any point nor more than three *stories above grade plane*. Except as specifically required by this section, *covered and open mall buildings* shall meet applicable provisions of this code.

Exceptions:

1. Foyers and lobbies of Group B, R-1 and R-2 occupancies are not required to comply with this section.
2. Buildings need not comply with the provisions of this section where they totally comply with other applicable provisions of this code.

402.1.1 Open mall building perimeter line. For the purpose of this code, a perimeter line shall be established. The perimeter line shall encircle all buildings and structures that comprise the *open mall building* and shall encompass any open-air interior walkways, open-air courtyards or similar open-air spaces. The perimeter line shall define the extent of the *open mall building. Anchor buildings* and parking structures shall be outside of the perimeter line and are not considered as part of the *open mall building*.

402.2 Open space. A *covered mall building* and attached *anchor buildings* and parking garages shall be surrounded on all sides by a permanent open space or not less than 60 feet (18 288 mm). An *open mall building* and *anchor buildings* and parking garages adjoining the perimeter line shall be surrounded on all sides by a permanent open space of not less than 60 feet (18 288 mm).

Exception: The permanent open space of 60 feet (18 288 mm) shall be permitted to be reduced to not less than 40 feet (12 192 mm), provided that the following requirements are met:

1. The reduced open space shall not be allowed for more than 75 percent of the perimeter of the *covered or open mall building* and *anchor buildings*.
2. The *exterior wall* facing the reduced open space shall have a *fire-resistance rating* of not less than 3 hours.
3. Openings in the *exterior wall* facing the reduced open space shall have opening protectives with a *fire protection rating* of not less than 3 hours.
4. Group E, H, I or R occupancies are not located within the *covered or open mall building* or *anchor buildings*.

402.3 Lease plan. Each *owner* of a *covered mall building* or of an *open mall building* shall provide both the building and fire departments with a lease plan showing the location of each occupancy and its *exits* after the certificate of occupancy has been issued. Modifications or changes in occupancy or use from that shown on the lease plan shall not be made without prior approval of the *building official*.

402.4 Construction. The construction of *covered and open mall buildings*, *anchor buildings* and parking garages associated with a mall building shall comply with Sections 402.4.1 through 402.4.3.

402.4.1 Area and types of construction. The *building area* and type of construction of *covered mall* or *open mall buildings*, *anchor buildings* and parking garages shall comply with this section.

402.4.1.1 Covered and open mall buildings. The *building area* of any *covered mall* or *open mall building* shall not be limited provided that the *covered mall* or *open mall building* does not exceed three floor levels at any point nor three *stories above grade plane*, and is of Type I, II, III or IV construction.

402.4.1.2 Anchor buildings. The *building area* and *building height* of any *anchor building* shall be based on the type of construction as required by Section 503 as modified by Sections 504 and 506.

Exception: The *building area* of any *anchor building* shall not be limited provided that the *anchor building* is not more than three *stories above grade plane*, and is of Type I, II, III or IV construction.

402.4.1.3 Parking garage. The *building area* and *building height* of any parking garage shall be based on the type of construction as required by Sections 406.5 and 406.6, respectively.

402.4.2 Fire-resistance-rated separation. Fire-resistance-rated separation is not required between tenant spaces and the *mall*. Fire-resistance-rated separation is not required between a *food court* and adjacent tenant spaces or the *mall*.

402.4.2.1 Tenant separations. Each tenant space shall be separated from other tenant spaces by a *fire partition* complying with Section 708. A tenant separation wall is not required between any tenant space and the *mall*.

402.4.2.2 Anchor building separation. An *anchor building* shall be separated from the *covered or open mall building* by *fire walls* complying with Section 706.

Exceptions:

1. *Anchor buildings* of not more than three stories *above grade plane* that have an occupancy classification the same as that permitted for tenants of the *mall building* shall be separated by 2-hour fire-resistance-rated *fire barriers* complying with Section 707.
2. The *exterior walls* of *anchor buildings* separated from an *open mall building* by an *open mall* shall comply with Table 705.5.

402.4.2.2.1 Openings between anchor building and mall. Except for the separation between Group R-1 *sleeping units* and the *mall*, openings between *anchor buildings* of Type IA, IB, IIA or IIB construction and the *mall* need not be protected.

402.4.2.3 Parking garages. An attached garage for the storage of passenger vehicles having a capacity of not more than nine persons and *open parking garages* shall be considered as a separate building where it is separated from the *covered or open mall building* or *anchor building* by not less than 2-hour *fire barriers* constructed in accordance with Section 707 or *horizontal assemblies* constructed in accordance with Section 711, or both.

Parking garages, which are separated from *covered mall buildings*, *open mall buildings* or *anchor buildings*, shall comply with the provisions of Table 705.5.

Pedestrian walkways and tunnels that connect garages to *mall buildings* or *anchor buildings* shall be constructed in accordance with Section 3104.

402.4.3 Open mall construction. Floor assemblies in, and roof assemblies over, the *open mall* of an *open mall building* shall be open to the atmosphere for not less than 20 feet (9096 mm), measured perpendicular from the face of the tenant spaces on the lowest level, from edge of balcony to edge of balcony on upper floors and from edge of roof line to edge of roof line. The openings within, or the unroofed area of, an *open mall* shall extend from the lowest/grade level of the open mall through the entire roof assembly. Balconies on upper levels of the *mall* shall not project into the required width of the opening.

402.4.3.1 Pedestrian walkways. *Pedestrian walkways* connecting balconies in an *open mall* shall be located not less than 20 feet (9096 mm) from any other *pedestrian walkway*.

[F] 402.5 Automatic sprinkler system. *Covered* and *open mall buildings* and buildings connected shall be equipped throughout with an *automatic sprinkler system* in accordance with Section 903.3.1.1, which shall comply with all of the following:

1. The *automatic sprinkler system* shall be complete and operative throughout occupied space in the *mall building* prior to occupancy of any of the tenant spaces. Unoccupied tenant spaces shall be similarly protected unless provided with *approved* alternative protection.
2. Sprinkler protection for the *mall* of a *covered mall building* shall be independent from that provided for tenant spaces or *anchor buildings*.
3. Sprinkler protection for the tenant spaces of an *open mall building* shall be independent from that provided for *anchor buildings*.
4. Sprinkler protection shall be provided beneath exterior circulation balconies located adjacent to an *open mall*.
5. Where tenant spaces are supplied by the same system, they shall be independently controlled.

Exception: An *automatic sprinkler system* shall not be required in spaces or areas of *open parking garages* separated from the covered or *open mall building* in accordance with Section 402.4.2.3 and constructed in accordance with Section 406.5.

402.6 Interior finishes and features. *Interior finishes* within the *mall* and installations within the *mall* shall comply with Sections 402.6.1 through 402.6.4.

402.6.1 Interior finish *Interior wall* and *ceiling finishes* within the *mall* of a *covered mall building* and within the *exits* of *covered or open mall buildings* shall have a minimum *flame spread index* and *smoke-developed index* of Class B in accordance with Chapter 8. *Interior floor finishes* shall meet the requirements of Section 804.

402.6.2 Kiosks. Kiosks and similar structures (temporary or permanent) located within the *mall* of a *covered mall building* or within the perimeter line of an *open mall building* shall meet the following requirements:

1. Combustible kiosks or other structures shall not be located within a *covered* or *open mall* unless constructed of any of the following materials:
   1.1. *Fire-retardant-treated* wood complying with Section 2303.2.
   1.2. Foam plastics having a maximum heat release rate not greater than 100 kW (105 Btu/h) when tested in accordance with the exhibit booth protocol in UL 1975 or when tested in accordance with NFPA 289 using the 20 kW ignition source.
   1.3. Aluminum composite material (ACM) meeting the requirements of Class A *interior finish* in accordance with Chapter 8 when tested as an assembly in the maximum thickness intended.
2. Kiosks or similar structures located within the *mall* shall be provided with *approved automatic sprinkler system* and detection devices.
3. The horizontal separation between kiosks or groupings thereof and other structures within the *mall* shall be not less than 20 feet (6096 mm).
4. Each kiosk or similar structure or groupings thereof shall have an area not greater than 300 square feet (28 m^2).

402.6.3 Play structures. Play structures located within a *mall building* or within the perimeter line of an *open mall building* shall comply with Section 424. The horizontal separation between *play structures*, kiosks and similar structures within the *mall* shall be not less than 20 feet (6096 mm).

402.6.4 Plastic signs. Plastic signs affixed to the storefront of any tenant space facing a *mall* or *open mall* shall be limited as specified in Sections 402.6.4.1 through 402.6.4.5.

402.6.4.1 Area. Plastic signs shall be not more than 20 percent of the wall area facing the *mall.*

402.6.4.2 Height and width. Plastic signs shall be not greater than 36 inches (914 mm) in height, except that where the sign is vertical, the height shall be not greater than 96 inches (2438 mm) and the width shall be not greater than 36 inches (914 mm).

402.6.4.3 Location. Plastic signs shall be located not less than 18 inches (457 mm) from adjacent tenants.

402.6.4.4 Plastics other than foam plastics. Plastics other than foam plastics used in signs shall be light-transmitting plastics complying with Section 2606.4 or shall have a self-ignition temperature of 650°F (343°C) or greater when tested in accordance with ASTM D1929, and a *flame spread index* not greater than 75 and *smoke-developed index* not greater than 450 when tested in the manner intended for use in accordance with ASTM E84 or UL 723 or meet the acceptance criteria of Section 803.1.1.1 when tested in accordance with NFPA 286.

402.6.4.4.1 Encasement. Edges and backs of plastic signs in the *mall* shall be fully encased in metal.

402.6.4.5 Foam plastics. Foam plastics used in signs shall have flame-retardant characteristics such that the sign has a maximum heat-release rate of 150 kilowatts when tested in accordance with UL 1975 or when tested in accordance with NFPA 289 using the 20 kW ignition source, and the foam plastics shall have the physical characteristics specified in this section. Foam plastics used in signs installed in accordance with Section 402.6.4 shall not be required to comply with the *flame spread* and *smoke-developed indices* specified in Section 2603.3.

402.6.4.5.1 Density. The density of foam plastics used in signs shall be not less than 20 pounds per cubic foot (pcf) (320 kg/ m^3).

402.6.4.5.2 Thickness. The thickness of foam plastic signs shall not be greater than $^1/_2$ inch (12.7 mm).

[F] 402.7 Emergency systems. *Covered and open mall buildings*, *anchor buildings* and associated parking garages shall be provided with emergency systems complying with Sections 402.7.1 through 402.7.5.

[F] 402.7.1 Standpipe system. *Covered and open mall buildings* shall be equipped throughout with a *standpipe system* as required by Section 905.3.3.

[F] 402.7.2 Smoke control. *Atriums* connecting three or more *stories* in a *covered mall building* shall be provided with a smoke control system in accordance with Section 909.

[F] 402.7.3 Emergency power. *Covered mall buildings* greater than 50,000 square feet (4645 m^2) in area and *open mall buildings* greater than 50,000 square feet (4645 m^2) within the established perimeter line shall be provided with emergency power that is capable of operating the *emergency voice/alarm communication system* in accordance with Section 2702.

[F] 402.7.4 Emergency voice/alarm communication system. Where the total floor area is greater than 50,000 square feet (4645 m^2) within either a *covered mall building* or within the perimeter line of an *open mall building*, an *emergency voice/alarm communication system* shall be provided.

The fire department shall have access to any *emergency voice/alarm communication systems* serving a *mall*, required or otherwise. The systems shall be provided in accordance with Section 907.5.2.2.

[F] 402.7.5 Fire department access to equipment. Rooms or areas containing controls for air-conditioning systems or *fire protection systems* shall be identified for use by the fire department.

402.8 Means of egress. *Covered mall buildings*, *open mall buildings* and each tenant space within a *mall building* shall be provided with *means of egress* as required by this section and this code. Where there is a conflict between the requirements of this code and the requirements of Sections 402.8.1 through 402.8.8, the requirements of Sections 402.8.1 through 402.8.8 shall apply.

402.8.1 Mall width. For the purpose of providing required egress, *malls* are permitted to be considered as corridors but need not comply with the requirements of Section 1005.1 of this code where the width of the *mall* is as specified in this section.

402.8.1.1 Minimum width. The aggregate clear egress width of the *mall* in either a *covered or open mall building* shall be not less than 20 feet (6096 mm). The *mall* width shall be sufficient to accommodate the *occupant load* served. Any portion of the minimum required aggregate egress width shall be not less than 10 feet (3048 mm) measured to a height of 8 feet (2438 mm) between any projection of a tenant space bordering the *mall* and the nearest kiosk, vending machine, bench, display opening, *food court* or other obstruction to *means of egress* travel.

402.8.2 Determination of occupant load. The *occupant load* permitted in any individual tenant space in a *covered or open mall building* shall be determined as required by this code. *Means of egress* requirements for individual tenant spaces shall be based on the *occupant load* thus determined.

402.8.2.1 Occupant formula. In determining required *means of egress* of the *mall*, the number of occupants for whom *means of egress* are to be provided shall be based on *gross leasable area* of the *covered or open mall building* (excluding *anchor buildings*) and the *occupant load factor* as determined by Equation 4-1.

$$OLF = (0.00007)(GLA) + 25 \quad \textbf{(Equation 4-1)}$$

where:

OLF = The *occupant load* factor (square feet per person).

GLA = The *gross leasable area* (square feet).

Exception: Tenant spaces attached to a *covered or open mall building* but with a *means of egress* system that is totally independent of the open mall of an *open mall building* or of a *covered mall building* shall not be considered as *gross leasable area* for determining the required *means of egress* for the *mall building*.

402.8.2.2 OLF range. The *occupant load* factor (*OLF*) is not required to be less than 30 and shall not exceed 50.

402.8.2.3 Anchor buildings. The *occupant load* of *anchor buildings* opening into the *mall* shall not be included in computing the total number of occupants for the *mall*.

402.8.2.4 Food courts. The *occupant load* of a *food court* shall be determined in accordance with Section 1004. For the purposes of determining the *means of egress* requirements for the *mall*, the *food court occupant load* shall be added to the *occupant load* of the covered or *open mall building* as calculated in Section 402.8.2.1.

402.8.3 Number of means of egress. Wherever the distance of travel to the *mall* from any location within a tenant space used by persons other than employees is greater than 75 feet (22 860 mm) or the tenant space has an *occupant load* of 50 or more, not fewer than two *means of egress* shall be provided.

402.8.4 Arrangements of means of egress. Assembly occupancies with an *occupant load* of 500 or more located within a *covered mall building* shall be so located such that their entrance will be immediately adjacent to a principal entrance to the mall and shall have not less than one-half of their required *means of egress* opening directly to the exterior of the *covered mall building*. Assembly occupancies located within the perimeter line of an *open mall building* shall be permitted to have their main *exit* open to the *open mall*.

402.8.4.1 Anchor building means of egress. Required *means of egress* for *anchor buildings* shall be provided independently from the *mall means of egress* system. The *occupant load* of *anchor buildings* opening into the *mall* shall not be included in determining *means of egress* requirements for the *mall*. The path of egress travel of *malls* shall not exit through *anchor buildings*. Malls terminating at an *anchor building* where other *means of egress* has not been provided shall be considered as a dead-end *mall*.

402.8.5 Distance to exits. Within each individual tenant space in a *covered* or *open mall building*, the distance of travel from any point to an *exit* or entrance to the *mall* shall be not greater than 200 feet (60 960 mm).

The distance of travel from any point within a *mall* of a *covered mall building* to an *exit* shall be not greater than 200 feet (60 960 mm). The maximum distance of travel from any point within an *open mall* to the perimeter line of the *open mall building* shall be not greater than 200 feet (60 960 mm).

402.8.6 Access to exits. Where more than one *exit* is required, they shall be so arranged that it is possible to travel in either direction from any point in a *mall* of a *covered mall building* to separate *exits* or from any point in an *open mall* of an *open mall building* to two separate locations on the perimeter line, provided that neither location is an *exterior wall* of an *anchor building* or parking garage. The width of an *exit passageway* or *corridor* from a *mall* shall be not less than 66 inches (1676 mm).

Exception: Access to *exits* is permitted by way of a dead-end *mall* that does not exceed a length equal to twice the width of the *mall* measured at the narrowest location within the dead-end portion of the *mall*.

402.8.6.1 Exit passageways. Where *exit passageways* provide a secondary *means of egress* from a tenant space, the *exit passageways* shall be constructed in accordance with Section 1024.

402.8.7 Service areas fronting on exit passageways. Mechanical rooms, electrical rooms, building service areas and service elevators are permitted to open directly into *exit passageways*, provided that the *exit passageway* is separated from such rooms with not less than 1-hour *fire barriers* constructed in accordance with Section 707 or *horizontal assemblies* constructed in accordance with Section 711, or both. The *fire protection rating* of openings in the *fire barriers* shall be not less than 1 hour.

402.8.8 Security grilles and doors. Horizontal sliding or vertical security grilles or doors that are a part of a required *means of egress* shall conform to the following:

1. Doors and grilles shall remain in the full open position during the period of occupancy by the general public.
2. Doors or grilles shall not be brought to the closed position when there are 10 or more persons occupying spaces served by a single *exit* or 50 or more persons occupying spaces served by more than one *exit*.
3. The doors or grilles shall be openable from within without the use of any special knowledge or effort where the space is occupied.
4. Where two or more *exits* are required, not more than one-half of the *exits* shall be permitted to include either a horizontal sliding or vertical rolling grille or door.

SECTION 403 HIGH-RISE BUILDINGS

403.1 Applicability. *High-rise buildings* shall comply with Sections 403.2 through 403.6.

Exceptions: The provisions of Sections 403.2 through 403.6 shall not apply to the following buildings and structures:

1. Airport traffic control towers in accordance with Section 412.2.
2. *Open parking garages* in accordance with Section 406.5.
3. The portion of a building containing a Group A-5 occupancy in accordance with Section 303.6.
4. Special industrial occupancies in accordance with Section 503.1.1.
5. Buildings containing any one of the following:
   5.1. A Group H-1 occupancy.
   5.2. A Group H-2 occupancy in accordance with Section 415.8, 415.9.2, 415.9.3 or 426.1.
   5.3. A Group H-3 occupancy in accordance with Section 415.8.

403.2 Construction. The construction of *high-rise buildings* shall comply with the provisions of Sections 403.2.1 through 403.2.3.

403.2.1 Reduction in fire-resistance rating. The *fire-resistance rating* reductions specified in Sections 403.2.1.1 and 403.2.1.2 shall be allowed in buildings that have sprinkler control valves equipped with supervisory initiating devices and water-flow initiating devices for each floor.

403.2.1.1 Type of construction. The following reductions in the minimum *fire-resistance rating* of the *building elements* in TABLE 601 shall be permitted as follows:

1. For buildings not greater than 420 feet (128 m) in *building height*, the *fire-resistance rating* of the *building elements* in Type IA construction shall be permitted to be reduced to the minimum *fire-resistance ratings* for the *building elements* in Type IB.

   Exception: The required *fire-resistance rating* of columns supporting floors shall not be reduced.

2. In other than Group F-1, H-2, H-3, H-5, M and S-1 occupancies, the *fire-resistance rating* of the *building elements* in Type IB construction shall be permitted to be reduced to the *fire-resistance ratings* in Type IIA.
3. The *building height* and *building area* limitations of a building containing *building elements* with reduced *fire-resistance ratings* shall be permitted to be the same as the building without such reductions.

403.2.1.2 Shaft enclosures. For buildings not greater than 420 feet (128 m) in *building height*, the required *fire-resistance rating* of the *fire barriers* enclosing vertical *shafts*, other than *interior exit stairway* and elevator hoistway enclosures, is permitted to be reduced to 1 hour where *automatic* sprinklers are installed within the *shafts* at the top and at alternate floor levels.

[BS] 403.2.2 Structural integrity of interior exit stairways and elevator hoistway enclosures. For *high-rise buildings of Risk Category* III or IV in accordance with Section 1604.5, and for all buildings that are more than 420 feet (128 m) in *building height*, enclosures for *interior exit stairways* and elevator hoistway enclosures shall comply with Sections 403.2.2.1 through 403.2.2.4.

[BS] 403.2.2.1 Wall assembly materials—soft body impact. The panels making up the enclosures for *interior exit stairways* and elevator hoistway enclosures shall meet or exceed Soft Body Impact Classification Level 2 as measured by the test method described in ASTM C1629/C1629M.

[BS] 403.2.2.2 Wall assembly materials—hard body impact. The panels making up the enclosures for *inte-*

rior exit stairways and elevator hoistway enclosures that are not exposed to the interior of the enclosures for *interior exit stairways* or elevator hoistway enclosure shall be in accordance with one of the following:

1. The wall assembly shall incorporate not fewer than two layers of impact-resistant panels, each of which meets or exceeds Hard Body Impact Classification Level 2 as measured by the test method described in ASTM C1629/C1629M.
2. The wall assembly shall incorporate not fewer than one layer of impact-resistant panels that meet or exceed Hard Body Impact Classification Level 3 as measured by the test method described in ASTM C1629/C1629M.
3. The wall assembly incorporates multiple layers of any material, tested in tandem, that meets or exceeds Hard Body Impact Classification Level 3 as measured by the test method described in ASTM C1629/C1629M.

[BS] 403.2.2.3 Concrete and masonry walls. Concrete or masonry walls shall be deemed to satisfy the requirements of Sections 403.2.2.1 and 403.2.2.2.

[BS] 403.2.2.4 Other wall assemblies. Any other wall assembly that provides impact resistance equivalent to that required by Sections 403.2.2.1 and 403.2.2.2 for Hard Body Impact Classification Level 3, as measured by the test method described in ASTM C1629/C1629M, shall be permitted.

403.2.3 Sprayed fire-resistant materials (SFRM). The bond strength of the SFRM installed throughout the building shall be in accordance with Table 403.2.3.

TABLE 403.2.3
MINIMUM BOND STRENGTH

HEIGHT OF BUILDING[a]	SFRM MINIMUM BOND STRENGTH
Up to 420 feet	430 psf
Greater than 420 feet	1,000 psf

For SI: 1 foot = 304.8 mm, 1 pound per square foot (psf) = 0.0479 kW/m^2.
a. Above the lowest level of fire department vehicle access.

[F] 403.3 Automatic sprinkler system. Buildings and structures shall be equipped throughout with an *automatic sprinkler system* in accordance with Section 903.3.1.1 and a secondary water supply where required by Section 403.3.3.

Exception: An *automatic sprinkler system* shall not be required in spaces or areas of telecommunications equipment buildings used exclusively for telecommunications equipment, associated electrical power distribution equipment, batteries and standby engines, provided that those spaces or areas are equipped throughout with an *automatic* fire detection system in accordance with Section 907.2 and are separated from the remainder of the building by not less than 1-hour *fire barriers* constructed in accordance with Section 707 or not less than 2-hour *horizontal assemblies* constructed in accordance with Section 711, or both.

[F] 403.3.1 Number of sprinkler risers and system design. Each sprinkler system zone in buildings that are more than 420 feet (128 m) in *building height* shall be supplied by not fewer than two risers. Each riser shall supply sprinklers on alternate floors. If more than two risers are provided for a zone, sprinklers on adjacent floors shall not be supplied from the same riser.

[F] 403.3.1.1 Riser location. Sprinkler risers shall be placed in *interior exit stairways* and *ramps* that are remotely located in accordance with Section 1007.1.

[F] 403.3.2 Water supply to required fire pumps. In all buildings that are more than 420 feet (128 m) in *building height* and buildings of Type IVA and IVB construction that are more than 120 feet (36 576 mm) in *building height*, required fire pumps shall be supplied by connections to not fewer than two water mains located in different streets. Separate supply piping shall be provided between each connection to the water main and the pumps. Each connection and the supply piping between the connection and the pumps shall be sized to supply the flow and pressure required for the pumps to operate.

Exception: Two connections to the same main shall be permitted provided that the main is valved such that an interruption can be isolated so that the water supply will continue without interruption through not fewer than one of the connections.

[F] 403.3.3 Secondary water supply. An *automatic* secondary on-site water supply having a capacity not less than the hydraulically calculated sprinkler demand, including the hose stream requirement, shall be provided for *high-rise buildings* assigned to *Seismic Design Category* C, D, E or F as determined by Section 1613. An additional fire pump shall not be required for the secondary water supply unless needed to provide the minimum design intake pressure at the suction side of the fire pump supplying the *automatic sprinkler system*. The secondary water supply shall have a duration of not less than 30 minutes as determined by the occupancy hazard classification in accordance with NFPA 13.

[F] 403.3.4 Fire pump room. Fire pumps shall be located in rooms protected in accordance with Section 913.2.1.

[F] 403.4 Emergency systems. The detection, alarm and emergency systems of *high-rise buildings* shall comply with Sections 403.4.1 through 403.4.8.

[F] 403.4.1 Smoke detection. Smoke detection shall be provided in accordance with Section 907.2.13.1.

[F] 403.4.2 Fire alarm system. A *fire alarm* system shall be provided in accordance with Section 907.2.13.

[F] 403.4.3 Standpipe system. A *high-rise building* shall be equipped with a *standpipe system* as required by Section 905.3.

[F] 403.4.4 Emergency voice/alarm communication system. An *emergency voice/alarm communication system* shall be provided in accordance with Section 907.5.2.2.

[F] 403.4.5 Emergency communication coverage. In-building, two-way emergency responder communication coverage shall be provided in accordance with Section 510 of the *International Fire Code*.

[F] 403.4.6 Fire command. A *fire command center* complying with Section 911 shall be provided in a location *approved* by the fire code official.

[F] 403.4.7 Smoke removal. To facilitate smoke removal in post-fire salvage and overhaul operations, buildings and structures shall be equipped with natural or mechanical *ventilation* for removal of products of combustion in accordance with one of the following:

1. Easily identifiable, manually operable windows or panels shall be distributed around the perimeter of each floor at not more than 50-foot (15 240 mm) intervals. The area of operable windows or panels shall be not less than 40 square feet (3.7 m^2) per 50 linear feet (15 240 mm) of perimeter.

   Exceptions:

   1. In Group R-1 occupancies, each *sleeping unit* or suite having an *exterior wall* shall be permitted to be provided with 2 square feet (0.19 m^2) of venting area in lieu of the area specified in Item 1.
   2. Windows shall be permitted to be fixed provided that glazing can be cleared by fire fighters.
2. Mechanical air-handling equipment providing one exhaust air change every 15 minutes for the area involved. Return and exhaust air shall be moved directly to the outside without recirculation to other portions of the building.
3. Any other *approved* design that will produce equivalent results.

[F] 403.4.8 Standby and emergency power. A *standby power system* complying with Section 2702 and Section 3003 shall be provided for the standby power loads specified in Section 403.4.8.3. An *emergency power system* complying with Section 2702 shall be provided for the emergency power loads specified in Section 403.4.8.4.

[F] 403.4.8.1 Equipment room. If the *standby or emergency power system* includes a generator set inside a building, the system shall be located in a separate room enclosed with 2-hour *fire barriers* constructed in accordance with Section 707 or *horizontal assemblies* constructed in accordance with Section 711, or both. System supervision with manual start and transfer features shall be provided at the *fire command center*.

Exception: In Group I-2, Condition 2, manual start and transfer features for the critical branch of the emergency power are not required to be provided at the *fire command center*.

[F] 403.4.8.2 Fuel line piping protection. Fuel lines supplying a generator set inside a building shall be separated from areas of the building other than the room the generator is located in by one of the following methods:

1. A fire-resistant pipe-protection system that has been tested in accordance with UL 1489. The system shall be installed as tested and in accordance with the manufacturer's installation instructions, and shall have a rating of not less than 2 hours. Where the building is protected throughout with an *automatic sprinkler system* installed in accordance with Section 903.3.1.1, the required rating shall be reduced to 1 hour.
2. An assembly that has a *fire-resistance rating* of not less than 2 hours. Where the building is protected throughout with an *automatic sprinkler system* installed in accordance with Section 903.3.1.1 or 903.3.1.2, the required fire-resistance rating shall be reduced to 1 hour.
3. Other approved methods.

[F] 403.4.8.3 Standby power loads. The following are classified as standby power loads:

1. *Ventilation* and *automatic* fire detection equipment for *smokeproof* enclosures.
2. Elevators..
3. Where elevators are provided in a *high-rise building* for *accessible* means of egress, fire service access or occupant self-evacuation, the *standby power system* shall also comply with Sections 1009.4, 3007 or 3008, as applicable.

[F] 403.4.8.4 Emergency power loads. The following are classified as emergency power loads:

1. Exit signs and *means of egress* illumination required by Chapter 10.
2. Elevator car lighting.
3. *Emergency voice/alarm communications systems*.
4. *Automatic* fire detection systems.
5. *Fire alarm* systems.
6. Electrically powered fire pumps.
7. Power and lighting for the *fire command center* required by Section 403.4.6.

403.5 Means of· egress and evacuation. The *means of egress* in *high-rise buildings* shall comply with Sections 403.5.1 through 403.5.5.

403.5.1 Remoteness of interior exit stairways. Required *interior exit stairways* shall be separated by a distance not less than 30 feet (9144 mm) or not less than one-fourth of the length of the maximum overall diagonal dimension of the building or area to be served, whichever is less. The distance shall be measured in a straight line between the nearest points of the enclosure surrounding the *interior exit stairways*. In buildings with three or more *interior exit stairways*, not fewer than two of the *interior exit* *stairways* shall comply with this section. Interlocking or *scissor stairways* shall be counted as one *interior exit stairway*.

403.5.2 Additional interior exit stairway. For buildings other than Group R-2 and their ancillary spaces that are more than 420 feet (128 m) in *building height*, one additional *interior exit stairway* meeting the requirements of Sections 1011 and 1023 shall be provided in addition to the minimum number of *exits* required by Section 1006.3. The total capacity of any combination of remaining *interior exit stairways* with one *interior exit stairway* removed shall be not less than the total capacity required by Section 1005.1. *Scissor stairways* shall not be considered the additional *interior exit stairway* required by this section.

Exceptions:

1. An additional *interior exit stairway* shall not be required to be installed in buildings having elevators used for occupant self-evacuation in accordance with Section 3008.
2. An additional *interior exit stairway* shall not be required for other portions of the building where the highest occupiable floor level in those areas is less than 420 feet (128 m) in *building height*.

403.5.3 Stairway door operation. *Stairway* doors other than the exit discharge doors shall be permitted to be locked from the *stairway* side. *Stairway* doors that are locked from the *stairway* side shall be capable of being unlocked simultaneously without unlatching upon a signal from the *fire command center*.

403.5.3.1 Stairway communication system. A telephone or other two-way communications system connected to an *approved constantly attended station* shall be provided at not less than every fifth floor in each *stairway* where the doors to the *stairway* are locked.

403.5.4 Smokeproof enclosures. Every required *interior exit stairway* serving floors more than 75 feet (22 860 mm) above the lowest level of fire department vehicle access shall be a *smokeproof enclosure* in accordance with Sections 909.20 and 1023.12.

403.5.5 Luminous egress path markings. Luminous egress path markings shall be provided in accordance with Section 1025.

➡ **403.6 Elevators.** Elevator installation and operation in *high-rise buildings* shall comply with Chapter 30 and Sections 403.6.1 and 403.6.2.

403.6.1 Fire service access elevator. In buildings with an occupied floor more than 120 feet (36 576 mm) above the lowest level of fire department vehicle access, not fewer than two fire service access elevators, or all elevators, whichever is less, shall be provided in accordance with Section 3007. Each fire service access elevator shall have a capacity of not less than 3,500 pounds (1588 kg) and shall comply with Section 3002.4.

403.6.2 Occupant evacuation elevators. Where installed in accordance with Section 3008, passenger elevators for general public use shall be permitted to be used for occupant self-evacuation.

SECTION 404
ATRIUMS

404.1 General. The provisions of Sections 404.1 through 404.11 shall apply to buildings containing *atriums*. *Atriums* are not permitted in buildings or structures classified as Group H.

Exception: Vertical openings that comply with Sections 712.1.1 through 712.1.3, and Sections 712.1.9 through 712.1.14.

404.2 Use. The floor of the *atrium* shall not be used for other than low fire hazard uses and only *approved* materials and decorations in accordance with the *International Fire Code* shall be used in the *atrium* space.

Exception: The *atrium* floor area is permitted to be used for any *approved* use where the individual space is provided with an *automatic sprinkler system* in accordance with Section 903.3.1.1.

[F] 404.3 Automatic sprinkler protection. An *approved automatic sprinkler system* shall be installed throughout the entire building.

Exceptions:

1. That area of a building adjacent to or above the *atrium* need not be sprinklered provided that portion of the building is separated from the *atrium* portion by not less than 2-hour *fire barriers* constructed in accordance with Section 707 or *horizontal assemblies* constructed in accordance with Section 711, or both.
2. Where the ceiling of the *atrium* is more than 55 feet (16 764 mm) above the floor, sprinkler protection at the ceiling of the *atrium* is not required.

[F] 404.4 Fire alarm system. A *fire alarm* system shall be provided in accordance with Section 907.2.14.

404.5 Smoke control. A smoke control system shall be installed in accordance with Section 909.

Exceptions:

1. In other than Group I-2, and Group I-1, Condition 2, smoke control is not required for *atriums* that connect only two *stories*.
2. A smoke control system is not required for *atriums* connecting more than two *stories* when all of the following are met:
   - 2.1. Only the two lowest *stories* shall be permitted to be open to the *atrium*.
   - 2.2. All *stories* above the lowest two *stories* shall be separated from the *atrium* in accordance with the provisions for a *shaft* in Section 713.4.

404.6 Enclosure of atriums. *Atrium* spaces shall be separated from adjacent spaces by a 1-hour *fire barrier* constructed in accordance with Section 707 or a *horizontal assembly* constructed in accordance with Section 711, or both.

Exceptions:

1. A *fire barrier* is not required where a glass wall forming a *smoke partition* is provided. The glass wall shall comply with all of the following:
   - 1.1. *Automatic* sprinklers are provided along both sides of the separation wall and doors, or on the room side only if there is not a walkway on the *atrium* side. The sprinklers shall be located between 4 inches and 12 inches (102 mm and 305 mm) away from the glass and at intervals along the glass not greater than 6 feet (1829 mm). The sprinkler system shall be designed so that the entire surface of the glass is wet upon activation of the sprinkler system without obstruction;
   - 1.2. The glass wall shall be installed in a gasketed frame in a manner that the framing system deflects without breaking (loading) the glass before the sprinkler system operates; and
   - 1.3. Where glass doors are provided in the glass wall, they shall be either *self-closing* or automatic-closing.
2. A *fire barrier* is not required where a glass-block wall assembly complying with Section 2110 and having a $^3/_4$-hour *fire protection rating* is provided.
3. A *fire barrier* is not required between the *atrium* and the adjoining spaces of up to three floors of the *atrium* provided that such spaces are accounted for in the design of the smoke control system.
4. A *fire barrier* is not required between the *atrium* and the adjoining spaces where the *atrium* is not required to be provided with a smoke control system.
5. A *horizontal assembly* is not required between the *atrium* and openings for escalators complying with Section 712.1.3.
6. A *horizontal assembly* is not required between the *atrium* and openings for *exit access stairways* and *ramps* complying with Item 4 of Section 1019.3.

[F] 404.7 Standby power. Equipment required to provide smoke control shall be provided with standby power in accordance with Section 909.11.

404.8 Interior finish. The *interior finish* of walls and ceilings of the *atrium* shall be not less than Class B. Sprinkler protection shall not result in a reduction in class.

404.9 Exit access travel distance. *Exit access* travel distance for areas open to an *atrium* shall comply with the requirements of Section 1017.

404.10 Exit stairways in an atrium. Where an *atrium* contains an *interior exit stairway* all the following shall be met:

1. The entry to the exit stairway is the edge of the closest riser of the exit stairway.
2. The entry of the exit stairway shall have access from a minimum of two directions.
3. The distance between the entry to an exit stairway in an atrium and the entrance to a minimum of one exit stairway enclosed in accordance with Section 1023.2 shall comply with the separation required by Section 1007.1.1.
4. Exit access travel distance shall be measured to the closest riser of the exit stairway.
5. Not more than 50 percent of the exit stairways shall be located in the same atrium.

404.11 Interior exit stairway discharge. Discharge of *interior exit stairways* through an *atrium* shall be in accordance with Section 1028.

SECTION 405
UNDERGROUND BUILDINGS

405.1 General. The provisions of Sections 405.2 through 405.9 apply to building spaces having a floor level used for human occupancy more than 30 feet (9144 mm) below the finished floor of the lowest *level of exit discharge.*

Exceptions: The provisions of Section 405 are not applicable to the following buildings or portions of buildings:

1. One- and two-family *dwellings*, sprinklered in accordance with Section 903.3.1.3.
2. Parking garages provided with *automatic sprinkler systems* in compliance with Section 405.3.
3. Fixed guideway transit systems.
4. *Grandstands, bleachers*, stadiums, arenas and similar facilities.
5. Where the lowest *story* is the only *story* that would qualify the building as an underground building and has an area not greater than 1,500 square feet (139 m^2) and has an *occupant load* less than 10.
6. Pumping stations and other similar mechanical spaces intended only for limited periodic use by service or maintenance personnel.

405.2 Construction requirements. The underground portion of the building shall be of Type I construction.

[F] 405.3 Automatic sprinkler system. The highest *level of exit discharge* serving the underground portions of the building and all levels below shall be equipped with an *automatic* *sprinkler system* installed in accordance with Section 903.3.1.1. Water-flow switches and control valves shall be supervised in accordance with Section 903.4.

405.4 Compartmentation. Compartmentation shall be in accordance with Sections 405.4.1 through 405.4.3.

405.4.1 Number of compartments. A building having a floor level more than 60 feet (18 288 mm) below the finished floor of the lowest *level of exit discharge* shall be divided into not fewer than two compartments of approximately equal size. Such compartmentation shall extend through the highest *level of exit discharge* serving the underground portions of the building and all levels below.

Exception: The lowest *story* need not be compartmented where the area is not greater than 1,500 square feet (139 m^2) and has an *occupant load* of less than 10.

405.4.2 Smoke barrier penetration. The compartments shall be separated from each other by a *smoke barrier* in accordance with Section 709. Penetrations between the two compartments shall be limited to plumbing and electrical piping and conduit that are firestopped in accordance with Section 714. Doorways shall be protected by *fire door assemblies* that comply with Section 716, automatic-closing by smoke detection in accordance with Section 716.2.6.6 and installed in accordance with NFPA 105 and Section 716.2.2.1. Where provided, each compartment shall have an air supply and an exhaust system independent of the other compartments.

405.4.3 Elevators. Where elevators are provided, each compartment shall have direct access to an elevator. Where an elevator serves more than one compartment, an enclosed elevator lobby shall be provided and shall be separated from each compartment by a *smoke barrier* in accordance with Section 709. Doorways in the *smoke barrier* shall be protected by *fire door assemblies* that comply with Section 716, shall comply with the smoke and draft control assembly requirements of Section 716.2.2.1 with the UL 1784 test conducted without an artificial bottom seal, and shall be automatic-closing by smoke detection in accordance with Section 716.2.6.6.

405.5 Smoke control system. A smoke control system shall be provided in accordance with Sections 405.5.1 and 405.5.2.

405.5.1 Control system. A smoke control system is required to control the migration of products of combustion in accordance with Section 909 and the provisions of this section. Smoke control shall restrict movement of smoke to the general area of fire origin and maintain *means of egress* in a usable condition.

405.5.2 Compartment smoke control system. Where compartmentation is required, each compartment shall have an independent smoke control system. The system shall be automatically activated and capable of manual operation in accordance with Sections 907.2.18 and 907.2.19.

[F] 405.6 Fire alarm systems. A *fire alarm* system shall be provided where required by Sections 907.2.18 and 907.2.19.

405.7 Means of egress. *Means of egress* shall be in accordance with Sections 405.7.1 and 405.7.2.

405.7.1 Number of exits. Each floor level shall be provided with not fewer than two *exits*. Where compartmentation is required by Section 405.4, each compartment shall have not fewer than one *exit* and not fewer than one *exit access* doorway into the adjoining compartment.

405.7.2 Smokeproof enclosure. Every required *stairway* serving floor levels more than 30 feet (9144 mm) below the finished floor of its *level of exit discharge* shall comply with the requirements for a *smokeproof enclosure* as provided in Section 1023.12.

[F] 405.8 Standby and emergency power. A *standby power system* complying with Section 2702 shall be provided for the standby power loads specified in Section 405.8.1. An *emergency power system* complying with Section 2702 shall be provided for the emergency power loads specified in Section 405.8.2.

[F] 405.8.1 Standby power loads. The following are classified as standby power loads:

1. Smoke control system.
2. *Ventilation* and *automatic* fire detection equipment for *smokeproof* enclosures.
3. Elevators, as required in Section 3003.

[F] 405.8.2 Emergency power loads. The following are classified as emergency power loads:

1. *Emergency voice/alarm communications systems.*
2. *Fire alarm systems.*
3. *Automatic* fire detection systems.
4. Elevator car lighting.
5. *Means of egress* and *exit* sign illumination as required by Chapter 10.
6. Fire pumps.

[F] 405.9 Standpipe system. The underground building shall be equipped throughout with a *standpipe system* in accordance with Section 905.

SECTION 406
MOTOR-VEHICLE-RELATED OCCUPANCIES

406.1 General. All motor-vehicle-related occupancies shall comply with Section 406.2. *Private garages* and carports shall also comply with Section 406.3. Open public parking garages shall also comply with Sections 406.4 and 406.5. Enclosed public parking garages shall also comply with Sections 406.4 and 406.6. Motor fuel-dispensing facilities shall also comply with Section 406.7. *Repair garages* shall also comply with Section 406.8.

406.2 Design. *Private garages* and carports, open and enclosed public parking garages, motor fuel-dispensing facilities and *repair garages* shall comply with Sections 406.2.1 through 406.2.9.

406.2.1 Automatic garage door openers and *vehicular gates*. Automatic garage door openers shall be listed and labeled in accordance with UL 325. Where provided, *automatic vehicular gates* shall comply with Section 3110.

406.2.2 Clear height. The clear height of each floor level in vehicle and pedestrian traffic areas shall be not less than 7 feet (2134 mm). Canopies under which fuels are dispensed shall have a clear height in accordance with Section 406.7.2.

Exception: A lower clear height is permitted for a parking tier in *mechanical-access open parking garages* where approved by the *building official.*

406.2.3 Accessible parking spaces. Where parking is provided, accessible parking spaces, access aisles and vehicular routes serving accessible parking shall be provided in accordance with Section 1106.

406.2.4 Floor surfaces. Floor surfaces shall be of concrete or similar approved noncombustible and nonabsorbent materials. The area of floor used for the parking of automobiles or other vehicles shall be sloped to facilitate the movement of liquids to a drain or toward the main vehicle entry doorway. The surface of vehicle fueling pads in motor fuel-dispensing facilities shall be in accordance with Section 406.7.1.

Exceptions:

1. Asphalt parking surfaces shall be permitted at ground level for public parking garages and private carports.
2. Slip-resistant, nonabsorbent, *interior floor finishes* having a critical radiant flux not more than 0.45 W/cm^2, as determined by ASTM E648 or NFPA 253, shall be permitted in *repair garages*.

406.2.5 Sleeping rooms. Openings between a motor vehicle-related occupancy and a room used for sleeping purposes shall not be permitted.

406.2.6 Fuel dispensing. The dispensing of fuel shall only be permitted in motor fuel-dispensing facilities in accordance with Section 406.7.

406.2.7 Electric vehicle charging stations and systems. Where provided, electric vehicle charging systems shall be installed in accordance with NFPA 70. Electric vehicle charging system equipment shall be *listed* and labeled in accordance with UL 2202. Electric vehicle supply equipment shall be *listed* and labeled in accordance with UL 2594. Accessibility to *electric vehicle charging stations* shall be provided in accordance with Section 1108.

406.2.8 Mixed occupancies and uses. Mixed uses shall be allowed in the same building as public parking garages and *repair garages* in accordance with Section 508.1. Mixed uses in the same building as an *open parking garage* are subject to Sections 402.4.2.3, 406.5.11, 508.1, 510.3, 510.4 and 510.7.

406.2.9 Equipment and appliances. Equipment and appliances shall be installed in accordance with Sections 406.2.9.1 through 406.2.9.3 and the *International Mechanical Code*, *International Fuel Gas Code* and NFPA 70.

406.2.9.1 Elevation of ignition sources. Equipment and appliances having an ignition source and located in hazardous locations and public garages, *private garages, repair garages*, automotive motor fuel-dispensing facilities and parking garages shall be elevated such that the source of ignition is not less than 18 inches (457 mm) above the floor surface on which the equipment or appliance rests. For the purpose of this section, rooms or spaces that are not part of the living space of a *dwelling unit* and that communicate directly with a *private garage* through openings shall be considered to be part of the *private garage.*

Exception: Elevation of the ignition source is not required for appliances that are listed as flammable vapor ignition resistant.

406.2.9.1.1 Parking garages. Connection of a parking garage with any room in which there is a fuel-fired appliance shall be by means of a vestibule providing a two-doorway separation, except that a single door is permitted where the sources of ignition in the appliance are elevated in accordance with Section 406.2.9.

Exception: This section shall not apply to appliance installations complying with Section 406.2.9.2 or 406.2.9.3.

406.2.9.2 Public garages. Appliances located in public garages, motor fuel-dispensing facilities, *repair garages* or other areas frequented by motor vehicles shall be installed not less than 8 feet (2438 mm) above the floor. Where motor vehicles are capable of passing under an appliance, the appliance shall be installed at the clearances required by the appliance manufacturer and not less than 1 foot (305 mm) higher than the tallest vehicle garage door opening.

Exception: The requirements of this section shall not apply where the appliances are protected from motor vehicle impact and installed in accordance with Section 406.2.9.1 and NFPA 30A.

406.2.9.3 Private garages. Appliances located in *private garages* and carports shall be installed with a minimum clearance of 6 feet (1829 mm) above the floor.

Exception: The requirements of this section shall not apply where the appliances are protected from motor vehicle impact and are installed in accordance with Section 406.2.9.1.

406.3 Private garages and carports. *Private garages* and carports shall comply with Sections 406.2 and 406.3, or they shall comply with Sections 406.2 and 406.4.

406.3.1 Classification. *Private garages* and carports shall be classified as Group U occupancies. Each *private garage* shall be not greater than 1,000 square feet (93 m^2) in area. Multiple *private garages* are permitted in a building where each *private garage* is separated from the other *private garages* by 1-hour *fire barriers* in accordance with Section 707, or 1-hour *horizontal assemblies* in accordance with Section 711, or both.

406.3.2 Separation. For other than *private garages* adjacent to dwelling units, the separation of *private garages* from other occupancies shall comply with Section 508. Separation of *private garages* from *dwelling units* shall comply with Sections 406.3.2.1 and 406.3.2.2.

406.3.2.1 Dwelling unit separation. The *private garage* shall be separated from the *dwelling unit* and its *attic* area by means of *gypsum board,* not less than $^1/_2$ inch (12.7 mm) in thickness, applied to the garage side. Garages beneath habitable rooms shall be separated from all habitable rooms above by not less than a $^5/_8$-inch (15.9 mm) Type X *gypsum board* or equivalent and $^1/_2$-inch (12.7 mm) *gypsum board* applied to structures supporting the separation from habitable rooms above the garage. Door openings between a *private garage* and the *dwelling unit* shall be equipped with either solid wood doors or solid or honeycomb core steel doors not less than $1^3/_8$ inches (34.9 mm) in thickness, or doors in compliance with Section 716.2.2.1 with a *fire protection rating* of not less than 20 minutes. Doors shall be *self-closing* and self-latching.

406.3.2.2 Ducts. Ducts in a *private garage* and ducts penetrating the walls or ceilings separating the *dwelling unit* from the garage, including its *attic* area, shall be constructed of sheet steel of not less than 0.019 inch (0.48 mm) in thickness and shall not have openings into the garage.

406.3.3 Carports. Carports shall be open on not fewer than two sides. Carports open on fewer than two sides shall be considered to be a garage and shall comply with the requirements for *private garages.*

406.3.3.1 Carport separation. A separation is not required between a Group R-3 and U carport, provided that the carport is entirely open on two or more sides and there are not enclosed areas above.

406.4 Public parking garages. Parking garages, other than *private garages*, shall be classified as public parking garages and shall comply with the provisions of Sections 406.2 and 406.4 and shall be classified as either an *open parking garage* or an enclosed parking garage. *Open parking garages* shall also comply with Section 406.5. Enclosed parking garages shall also comply with Section 406.6. See Section 510 for special provisions for parking garages.

406.4.1 Guards. *Guards* shall be provided in accordance with Section 1015. *Guards* serving as *vehicle barriers* shall comply with Sections 406.4.2 and 1015.

406.4.2 Vehicle barriers. *Vehicle barriers* not less than 2 feet 9 inches (835 mm) in height shall be placed where the vertical distance from the floor of a drive lane or parking space to the ground or surface directly below is greater than 1 foot (305 mm). *Vehicle barriers* shall comply with the loading requirements of Section 1607.10.

Exception: *Vehicle barriers* are not required in vehicle storage compartments in a mechanical access parking garage.

406.4.3 Ramps. Vehicle ramps shall not be considered as required *exits* unless pedestrian facilities are provided. Vehicle ramps that are utilized for vertical circulation as well as for parking shall not exceed a slope of 1 unit vertical in 15 units horizontal (6.67-percent slope).

406.5 Open parking garages. *Open parking garages* shall comply with Sections 406.2, 406.4 and 406.5.

406.5.1 Construction. *Open parking garages* shall be of Type I, II or IV construction. *Open parking garages* shall meet the design requirements of Chapter 16. For *vehicle barriers*, see Section 406.4.2.

406.5.2 Openings. For natural *ventilation* purposes, the exterior side of the structure shall have uniformly distributed openings on two or more sides. The area of such openings in *exterior walls* on a tier shall be not less than 20 percent of the total perimeter wall area of each tier. The aggregate length of the openings considered to be providing natural *ventilation* shall be not less than 40 percent of the perimeter of the tier. Interior walls shall be not less than 20 percent open with uniformly distributed openings.

Exception: Openings are not required to be distributed over 40 percent of the building perimeter where the required openings are uniformly distributed over two opposing sides of the building.

406.5.2.1 Openings below grade. Where openings below grade provide required natural *ventilation*, the outside horizontal clear space shall be one and one-half times the depth of the opening. The width of the horizontal clear space shall be maintained from grade down to the bottom of the lowest required opening.

406.5.3 Mixed occupancies and uses. Mixed uses shall be allowed in the same building as an *open parking garage* subject to the provisions of Sections 402.4.2.3, 406.5.11, 508.1, 510.3, 510.4 and 510.7.

406.5.4 Area and height. Area and height of *open parking garages* shall be limited as set forth in Chapter 5 for Group S-2 occupancies and as further provided for in Section 508.1.

406.5.4.1 Single use. Where the *open parking garage* is used exclusively for the parking or storage of private motor vehicles, and the building is without other uses, the area and height shall be permitted to comply with Table 406.5.4, along with increases allowed by Section 406.5.5.

Exception: The grade-level tier is permitted to contain an office, waiting and toilet rooms having a total combined area of not more than 1,000 square feet (93 m^2). Such area need not be separated from the *open parking garage.*

In *open parking garages* having a spiral or sloping floor, the horizontal projection of the structure at any cross section shall not exceed the allowable area per parking tier. In the case of an *open parking garage* having a continuous spiral floor, each 9 feet 6 inches (2896 mm) of height, or portion thereof, shall be considered under these provisions to be a tier.

406.5.5 Area and height increases. The allowable area and height of *open parking garages* shall be increased in accordance with the provisions of this section. Garages with sides open on three-fourths of the building's perimeter are permitted to be increased by 25 percent in area and one tier in height. Garages with sides open around the entire building's perimeter are permitted to be increased by 50 percent in area and one tier in height. For a side to be considered open under these provisions, the total area of openings along the side shall be not less than 50 percent of the interior area of the side at each tier and such openings shall be equally distributed along the length of the tier. For purposes of calculating the interior area of the side, the height shall not exceed 7 feet (2134 mm).

Allowable tier areas in Table 406.5.4 shall be increased for *open parking garages* constructed to heights less than the table maximum. The gross tier area of the garage shall not exceed that permitted for the higher structure. Not fewer than three sides of each such larger tier shall have continuous horizontal openings not less than 30 inches (762 mm) in clear height extending for not less than 80 percent of the length of the sides. All parts of such larger tier shall be not more than 200 feet (60 960 mm) horizontally from such an opening. In addition, each such opening shall face a street or *yard* with access to a street with a width of not less than 30 feet (9144 mm) for the full length of the opening, and *standpipes* shall be provided in each such tier.

Open parking garages of Type II construction, with all sides open, shall be unlimited in allowable area where the *building height* does not exceed 75 feet (22 860 mm). For a side to be considered open, the total area of openings along the side shall be not less than 50 percent of the interior area of the side at each tier and such openings shall be equally distributed along the length of the tier. For purposes of calculating the interior area of the side, the height shall not exceed 7 feet (2134 mm). All portions of tiers shall be within 200 feet (60 960 mm) horizontally from such openings or other natural *ventilation* openings as defined in Section 406.5.2. These openings shall be permitted to be provided in *courts* with a minimum dimension of 20 feet (6096 mm) for the full width of the openings.

406.5.6 Fire separation distance. *Exterior walls* and openings in *exterior walls* shall comply with Table 601 and Table 705.5. The distance to an adjacent *lot line* shall be determined in accordance with Section 705 and Table 705.5.

406.5.7 Means of egress. Where persons other than parking attendants are permitted, *open parking garages* shall meet the *means of egress* requirements of Chapter 10. Where persons other than parking attendants are not permitted, there shall be not fewer than two exit stairways. Each *exit stairway* shall be not less than 36 inches (914 mm) in width. Lifts shall be permitted to be installed for use of employees only, provided that they are completely enclosed by noncombustible materials.

[F] 406.5.8 Standpipe system. An *open parking garage* shall be equipped with a *standpipe system* as required by Section 905.3.

406.5.9 Enclosure of vertical openings. Enclosure shall not be required for vertical openings except as specified in Section 406.5.7.

406.5.10 Ventilation. *Ventilation*, other than the percentage of openings specified in Section 406.5.2, shall not be required.

406.5.11 Prohibitions. The following uses and alterations are not permitted:

1. Vehicle repair work.
2. Parking of buses, trucks and similar vehicles.
3. Partial or complete closing of required openings in *exterior walls* by tarpaulins or any other means.
4. Dispensing of fuel.

406.6 Enclosed parking garages. Enclosed parking garages shall comply with Sections 406.2, 406.4 and 406.6.

406.6.1 Heights and areas. Enclosed vehicle parking garages and portions thereof that do not meet the definition of *open parking garages* shall be limited to the allowable heights and areas specified in Sections 504 and

TABLE 406.5.4
OPEN PARKING GARAGES AREA AND HEIGHT

TYPE OF CONSTRUCTION	AREA PER TIER (square feet)	HEIGHT (in tiers)		
		Ramp access	Mechanical access — Automatic sprinkler system: No	Mechanical access — Automatic sprinkler system: Yes
IA	Unlimited	Unlimited	Unlimited	Unlimited
IB	Unlimited	12 tiers	12 tiers	18 tiers
IIA	50,000	10 tiers	10 tiers	15 tiers
IIB	50,000	8 tiers	8 tiers	12 tiers
IV	50,000	4 tiers	4 tiers	4 tiers

For SI: 1 square foot = 0.0929 m^2.

506 as modified by Section 507. Roof parking is permitted.

406.6.2 Ventilation. A mechanical *ventilation* system and exhaust system shall be provided in accordance with *Chapters 4* and 5 of the *International Mechanical Code*.

Exception: Mechanical *ventilation* shall not be required for enclosed parking garages that are accessory to one- and two-family *dwellings*.

[F] 406.6.3 Automatic sprinkler system. An enclosed parking garage shall be equipped with an *automatic sprinkler system* in accordance with Section 903.2.10.

406.6.4 Mechanical-access enclosed parking garages. *Mechanical-access enclosed parking garages* shall be in accordance with Sections 406.6.4.1 through 406.6.4.4.

406.6.4.1 Separation. *Mechanical-access enclosed parking garages* shall be separated from other occupancies and accessory uses by not less than 2-hour *fire barriers* constructed in accordance with Section 707 or by not less than 2-hour *horizontal assemblies* constructed in accordance with Section 711, or both.

406.6.4.2 Smoke removal. A mechanical smoke removal system, installed in accordance with Section 910.4, shall be provided for all areas containing a *mechanical-access enclosed parking garage*.

406.6.4.3 Fire control equipment room. Fire control equipment, consisting of the *fire alarm* control unit, mechanical *ventilation* controls and an emergency shutdown switch, shall be provided in a room located where the equipment is able to be accessed by the fire service from a secured exterior door of the building. The room shall be not less than 50 square feet (4.65 m^2) in area and shall be in a location that is approved by the fire code official.

406.6.4.3.1 Emergency shutdown switch. The mechanical parking system shall be provided with a manually activated emergency shutdown switch for use by emergency personnel. The switch shall be clearly identified and shall be in a location approved by the fire code official.

406.6.4.4 Fire department access doors. Access doors shall be provided in accordance with Section 3206.7 of the *International Fire Code*.

406.7 Motor fuel-dispensing facilities. Motor fuel-dispensing facilities shall comply with the *International Fire Code* and Sections 406.2 and 406.7.

406.7.1 Vehicle fueling pad. The vehicle shall be fueled on noncoated concrete or other *approved* paving material having a resistance not exceeding 1 megohm as determined by the methodology in CEN EN 1081.

406.7.2 Canopies. Canopies under which fuels are dispensed shall have a clear, unobstructed height of not less than 13 feet 6 inches (4115 mm) to the lowest projecting element in the vehicle drive-through area. Canopies and their supports over pumps shall be of noncombustible materials, *fire-retardant-treated wood* complying with Chapter 23, heavy timber complying with Section 2304.11 or construction providing 1-hour *fire resistance*. Combustible materials used in or on a *canopy* shall comply with one of the following:

1. Shielded from the pumps by a noncombustible element of the *canopy*, or heavy timber complying with Section 2304.11.
2. Plastics covered by aluminum facing having a thickness of not less than 0.010 inch (0.30 mm) or corrosion-resistant steel having a base metal thickness of not less than 0.016 inch (0.41 mm). The plastic shall have a *flame spread index* of 25 or less and a *smoke-developed index* of 450 or less when tested in the form intended for use in accordance with ASTM E84 or UL 723 and a self-ignition temperature of 650°F (343°C) or greater when tested in accordance with ASTM D1929.
3. Panels constructed of light-transmitting plastic materials shall be permitted to be installed in *canopies* erected over motor vehicle fuel-dispensing station fuel dispensers, provided that the panels are located not less than 10 feet (3048 mm) from any building on the same *lot* and face *yards* or streets not less than 40 feet (12 192 mm) in width on the other sides. The aggregate areas of plastics shall be not greater than 1,000 square feet (93 m^2). The maximum area of any individual panel shall be not greater than 100 square feet (9.3 m^2).

406.7.2.1 Canopies used to support gaseous hydrogen systems. *Canopies* that are used to shelter dispensing operations where flammable compressed gases are located on the roof of the *canopy* shall be in accordance with the following:

1. The *canopy* shall meet or exceed Type I construction requirements.
2. Operations located under *canopies* shall be limited to refueling only.
3. The *canopy* shall be constructed in a manner that prevents the accumulation of hydrogen gas.

406.8 Repair garages. *Repair garages* shall be constructed in accordance with the *International Fire Code* and Sections 406.2 and 406.8. This occupancy shall not include motor fuel-dispensing facilities, as regulated in Section 406.7.

406.8.1 Ventilation. *Repair garages* shall be mechanically ventilated in accordance with the *International Mechanical Code*. The *ventilation* system shall be controlled at the entrance to the garage.

[F] 406.8.2 Gas detection system. *Repair garages* used for repair of vehicles fueled by nonodorized gases including but not limited to hydrogen and nonodorized LNG, shall be provided with a *gas detection system* that complies with Section 916. The *gas detection system* shall be designed to detect leakage of nonodorized gaseous fuel. Where lubrication or chassis service pits are provided in garages used for repairing nonodorized LNG-fueled vehicles, gas sensors shall be provided in such pits.

[F] 406.8.2.1 System activation. Activation of a gas detection alarm shall result in all of the following:

1. Initiation of distinct audible and visual alarm signals in the *repair garage,* where the *ventilation* system is interlocked with gas detection.
2. Deactivation of all heating systems located in the *repair garage.*
3. Activation of the mechanical *ventilation* system, where the system is interlocked with gas detection.

[F] 406.8.2.2 Failure of the gas detection system. Failure of the *gas detection system* shall automatically deactivate the heating system, activate the mechanical ventilation system where the system is interlocked with the *gas detection system*, and cause a trouble signal to sound at an *approved* location.

[F] 406.8.3 Automatic sprinkler system. A *repair garage* shall be equipped with an *automatic sprinkler system* in accordance with Section 903.2.9.1.

SECTION 407
GROUP I-2

407.1 General. Occupancies in Group I-2 shall comply with the provisions of Sections 407.1 through 407.11 and other applicable provisions of this code.

407.2 Corridors continuity and separation. *Corridors* in occupancies in Group I-2 shall be continuous to the *exits* and shall be separated from other areas in accordance with Section 407.3 except spaces conforming to Sections 407.2.1 through 407.2.6.

407.2.1 Waiting and similar areas. Waiting areas, public-use areas or group meeting spaces constructed as required for *corridors* shall be permitted to be open to a *corridor*, only where all of the following criteria are met:

1. The spaces are not occupied as care recipient's sleeping rooms, treatment rooms, incidental uses in accordance with Section 509, or hazardous uses.
2. The open space is protected by an *automatic* fire detection system installed in accordance with Section 907.
3. The *corridors* onto which the spaces open, in the same *smoke compartment*, are protected by an *automatic* fire detection system installed in accordance with Section 907, or the *smoke compartment* in which the spaces are located is equipped throughout with quick-response sprinklers in accordance with Section 903.3.2.
4. The space is arranged so as not to obstruct access to the required *exits.*

407.2.2 Care providers' stations. Spaces for care providers', supervisory staff, doctors' and nurses' charting, communications and related clerical areas shall be permitted to be open to the *corridor*, where such spaces are constructed as required for *corridors*.

407.2.3 Psychiatric treatment areas. Areas wherein psychiatric care recipients who are *incapable of self-preservation* are housed, or group meeting or multipurpose therapeutic spaces other than incidental uses in accordance with Section 509, under continuous supervision by facility staff, shall be permitted to be open to the *corridor*, where the following criteria are met:

1. Each area does not exceed 1,500 square feet (140 m^2).
2. The area is located to permit supervision by the facility staff.
3. The area is arranged so as not to obstruct any access to the required *exits*.
4. The area is equipped with an *automatic* fire detection system installed in accordance with Section 907.2.
5. Not more than one such space is permitted in any one *smoke compartment*.
6. The walls and ceilings of the space are constructed as required for *corridors*.

407.2.4 Gift shops. Gift shops and associated storage that are less than 500 square feet (455 m^2) in area shall be permitted to be open to the *corridor* where such spaces are constructed as required for *corridors*.

407.2.5 Nursing home housing units. In Group I-2, Condition 1 occupancies, in areas where *nursing home* residents are housed, shared living spaces, group meeting or multipurpose therapeutic spaces shall be permitted to be open to the *corridor*, where all of the following criteria are met:

1. The walls and ceilings of the space are constructed as required for *corridors*.
2. The spaces are not occupied as resident sleeping rooms, treatment rooms, incidental uses in accordance with Section 509, or hazardous uses.
3. The open space is protected by an *automatic* fire detection system installed in accordance with Section 907.
4. The *corridors* onto which the spaces open, in the same *smoke compartment*, are protected by an *automatic* fire detection system installed in accordance with Section 907, or the *smoke compartment* in which the spaces are located is equipped throughout with quick-response sprinklers in accordance with Section 903.3.2.
5. The space is arranged so as not to obstruct access to the required *exits*.

407.2.6 Nursing home cooking facilities. In Group I-2, Condition 1 occupancies, rooms or spaces that contain a cooking facility with domestic cooking appliances shall be permitted to be open to the *corridor* where all of the following criteria are met:

1. The number of care recipients housed in the *smoke compartment* shall not be greater than 30.

2. The number of care recipients served by the cooking facility shall not be greater than 30.
3. Not more than one cooking facility area shall be permitted in a *smoke compartment*.
4. The *corridor* shall be a clearly identified space delineated by construction or floor pattern, material or color.
5. The space containing the domestic cooking facility shall be arranged so as not to obstruct access to the required *exit*.
6. The cooking appliance shall comply with Section 407.2.7.

407.2.7 Domestic cooking appliances. In Group I-2 occupancies, installation of cooking appliances used in domestic cooking facilities shall comply with all of the following:

1. The types of cooking appliances permitted shall be limited to ovens, cooktops, ranges, warmers and microwaves.
2. Domestic cooking hoods installed and constructed in accordance with Section 505 of the International Mechanical Code shall be provided over cooktops and ranges.
3. Cooktops and ranges shall be protected in accordance with Section 904.14.
4. A shut-off for the fuel and electrical power supply to the cooking equipment shall be provided in a location to which only staff has access.
5. A timer shall be provided that automatically deactivates the cooking appliances within a period of not more than 120 minutes.
6. A portable fire extinguisher shall be provided. Installation shall be in accordance with Section 906, and the extinguisher shall be located within a 30-foot (9144 mm) distance of travel from each domestic cooking appliance.

Exceptions:

1. Cooktops and ranges located within smoke compartments with no patient sleeping or patient care areas are not required to comply with this section.
2. Cooktops and ranges used for care recipient training or nutritional counseling are not required to comply with Item 3 of this section.

407.3 Corridor wall construction. *Corridor* walls shall be constructed as *smoke partitions* in accordance with Section 710.

407.3.1 Corridor doors. *Corridor* doors, other than those in a wall required to be rated by Section 509.4 or for the enclosure of a vertical opening or an *exit*, shall not have a required *fire protection rating* and shall not be required to be equipped with *self-closing* or automatic-closing devices, but shall provide an effective barrier to limit the transfer of smoke and shall be equipped with positive latching. Roller latches are not permitted. Other doors shall conform to Section 716.

407.3.1.1 Door construction. Doors in *corridors* not required to have a *fire protection rating* shall comply with the following:

1. Solid doors shall have close-fitting operational tolerances, head and jamb stops.
2. Dutch-style doors shall have an astragal, rabbet or bevel at the meeting edges of the upper and lower door sections. Both the upper and lower door sections shall have latching hardware. Dutch-style doors shall have hardware that connects the upper and lower sections to function as a single leaf.
3. To provide makeup air for exhaust systems in accordance with Section 1020.7, Exception 1, doors are permitted to have louvers or to have a clearance between the bottom of the door and the floor surface that is $^{2}/_{3}$ inch (19.1 mm) maximum.

407.4 Means of egress. Group I-2 occupancies shall be provided with *means of egress* complying with Chapter 10 and Sections 407.4.1 through 407.4.4. The fire safety and evacuation plans provided in accordance with Section 1002.2 shall identify the building components necessary to support a *defend-in-place* emergency response in accordance with Sections 403 and 404 of the *International Fire Code*.

407.4.1 Direct access to a corridor. Habitable rooms in Group I-2 occupancies shall have an exit access door leading directly to a *corridor*.

Exceptions:

1. Rooms with *exit* doors opening directly to the outside at ground level.
2. Rooms arranged as *care suites* complying with Section 407.4.4.

407.4.1.1 Locking devices. Locking devices that restrict access to a care recipient's room from the *corridor* and that are operable only by staff from the corridor side shall not restrict the *means of egress* from the care recipient's room.

Exceptions:

1. This section shall not apply to rooms in psychiatric treatment and similar care areas.
2. Locking arrangements in accordance with Section 1010.2.14.

407.4.2 Distance of travel. The distance of travel between any point in a Group I-2 occupancy sleeping room, not located in a *care suite*, and an *exit access* door in that room shall be not greater than 50 feet (15 240 mm).

407.4.3 Projections in nursing home corridors. In Group I-2, Condition 1 occupancies, where the *corridor* width is not less than 96 inches (2440 mm), projections shall be permitted for furniture where all of the following criteria are met:

1. The furniture is attached to the floor or to the wall.
2. The furniture does not reduce the clear width of the *corridor* to less than 72 inches (1830 mm) except where other encroachments are permitted in accordance with Section 1005.7.
3. The furniture is positioned on only one side of the *corridor.*
4. Each arrangement of furniture is 50 square feet (4.6 m^2) maximum in area.
5. Furniture arrangements are separated by 10 feet (3048 mm) minimum.
6. Placement of furniture is considered as part of the fire and safety plans in accordance with Section 1002.2.

407.4.4 Group I-2 care suites. *Care suites* in Group I-2 shall comply with Sections 407.4.4.1 through 407.4.4.4 and either Section 407.4.4.5 or 407.4.4.6.

407.4.4.1 Exit access through care suites. *Exit* access from all other portions of a building not classified as a *care suite* shall not pass through a *care suite.*

407.4.4.2 Separation. *Care suites* shall be separated from other portions of the building, including other *care suites,* by a *smoke partition* complying with Section 710.

407.4.4.3 Access to corridor. Every *care suite* shall have a door leading directly to an *exit access corridor* or *horizontal exit.* Movement from habitable rooms within a *care suite* shall not require more than 100 feet (30 480 mm) of travel within the *care suite* to a door leading to the *exit access corridor* or *horizontal exit.* Where a *care suite* is required to have more than one *exit access* door by Section 407.4.4.5.2 or 407.4.4.6.2, the additional door shall lead directly to an *exit access corridor, exit* or an adjacent suite.

407.4.4.4 Doors within care suites. Doors in *care suites* serving habitable rooms shall be permitted to comply with one of the following:

1. Manually operated horizontal sliding doors permitted in accordance with Exception 9 to Section 1010.1.2.
2. *Power-operated doors* permitted in accordance with Section 1010.1.2, Exception 7.
3. *Means of egress* doors complying with Section 1010.

407.4.4.5 Care suites containing sleeping room areas. Sleeping rooms shall be permitted to be grouped into *care suites* where one of the following criteria is met:

1. The *care suite* is not used as an *exit access* for more than eight care recipient beds.
2. The arrangement of the *care suite* allows for direct and constant visual supervision into the sleeping rooms by care providers.
3. An *automatic* smoke detection system is provided in the sleeping rooms and installed in accordance with NFPA 72.

407.4.4.5.1 Area. *Care suites* containing sleeping rooms shall be not greater than 7,500 square feet (696 m^2) in area.

Exception: *Care suites* containing sleeping rooms shall be permitted to be not greater than 10,000 square feet (929 m^2) in area where an *automatic* smoke detection system is provided throughout the *care suite* and installed in accordance with NFPA 72.

407.4.4.5.2 Exit access. Any sleeping room, or any *care suite* that contains sleeping rooms, of more than 1,000 square feet (93 m^2) shall have not fewer than two *exit access* doors from the *care suite* located in accordance with Section 1007.

407.4.4.6 Care suites not containing sleeping rooms. Areas not containing sleeping rooms, but only treatment areas and the associated rooms, spaces or circulation space, shall be permitted to be grouped into *care suites* and shall conform to the limitations in Sections 407.4.4.6.1 and 407.4.4.6.2.

407.4.4.6.1 Area. *Care suites* of rooms, other than sleeping rooms, shall have an area not greater than 12,500 square feet (1161 m^2).

Exception: *Care suites* not containing sleeping rooms shall be permitted to be not greater than 15,000 square feet (1394 m^2) in area where an *automatic* smoke detection system is provided throughout the *care suite* in accordance with Section 907.

407.4.4.6.2 Exit access. *Care suites,* other than sleeping rooms, with an area of more than 2,500 square feet (232 m^2) shall have not fewer than two *exit access* doors from the *care suite* located in accordance with Section 1007.

407.5 Smoke barriers. *Smoke barriers* shall be provided to subdivide every *story* used by persons receiving care, treatment or sleeping into not fewer than two *smoke compartments. Smoke barriers* shall be provided to subdivide other *stories* with an *occupant load* of 50 or more persons, into not fewer than two *smoke compartments.* The *smoke barrier* shall be in accordance with Section 709.

407.5.1 Smoke compartment size. Stories shall be divided into *smoke compartments* with an area of not more than 22,500 square feet (2092 m^2) in Group I-2 occupancies.

Exceptions:

1. A *smoke compartment* in Group I-2, Condition 2 is permitted to have an area of not more than 40,000 square feet (3716 m^2) provided that all patient sleeping rooms within that *smoke compartment* are configured for single patient occupancy and any suite within the *smoke compartment* complies with Section 407.4.4.

2. A *smoke compartment* in Group I-2, Condition 2 without patient sleeping rooms is permitted to have an area of not more than 40,000 square feet (3716 m^2).

407.5.2 Exit access travel distance. The distance of travel from any point in a *smoke compartment* to a *smoke barrier* door shall be not greater than 200 feet (60 960 mm).

407.5.3 Refuge area. Refuge areas shall be provided within each *smoke compartment.* The size of the refuge area shall accommodate the occupants and care recipients from the adjoining *smoke compartment.* Where a *smoke compartment* is adjoined by two or more *smoke compartments*, the minimum area of the refuge area shall accommodate the largest *occupant load* of the adjoining compartments. The size of the refuge area shall provide the following:

1. Not less than 30 net square feet (2.8 m^2) for each care recipient confined to bed or stretcher.
2. Not less than 6 square feet (0.56 m^2) for each ambulatory care recipient not confined to bed or stretcher and for other occupants.

Areas or spaces permitted to be included in the calculation of refuge area are corridors, sleeping areas, treatment rooms, lounge or dining areas and other low-hazard areas.

407.5.4 Independent egress. A *means of egress* shall be provided from each *smoke compartment* created by *smoke barriers* without having to return through the *smoke compartment* from which *means of egress* originated. *Smoke compartments* that do not contain an *exit* shall be provided with *direct access* to not less than two adjacent *smoke compartments*.

407.5.5 Horizontal assemblies. *Horizontal assemblies* supporting *smoke barriers* required by this section shall be designed to resist the movement of smoke. Elevator lobbies shall be in accordance with Section 3006.2.

407.6 Automatic-closing doors. Automatic-closing doors with hold-open devices shall comply with Sections 709.5 and 716.2.

407.6.1 Activation of automatic-closing doors. Automatic-closing doors on hold-open devices in accordance with Section 716.2.6.6 shall also close upon activation of a *fire alarm* system, an *automatic sprinkler system*, or both. The *automatic* release of the hold-open device on one door shall release all such doors within the same *smoke compartment.*

[F] 407.7 Automatic sprinkler system. *Smoke compartments* containing sleeping rooms shall be equipped throughout with an *automatic sprinkler* system in accordance with Sections 903.3.1.1 and 903.3.2.

[F] 407.8 Fire alarm system. A *fire alarm* system shall be provided in accordance with Section 907.2.6.

[F] 407.9 Automatic fire detection. *Corridors* in Group I-2, Condition 1 occupancies and spaces permitted to be open to the *corridors* by Section 407.2 shall be equipped with an *automatic* fire detection system.

Group I-2, Condition 2 occupancies shall be equipped with smoke detection as required in Section 407.2.

Exceptions:

1. Corridor smoke detection is not required where sleeping rooms are provided with *smoke detectors* that comply with UL 268. Such detectors shall provide a visual display on the corridor side of each sleeping room and an audible and visual alarm at the care provider's station attending each unit.
2. *Corridor* smoke detection is not required where sleeping room doors are equipped with automatic door-closing devices with integral *smoke detectors* on the unit sides installed in accordance with their listing, provided that the integral detectors perform the required alerting function.

407.10 Secured yards. Grounds are permitted to be fenced and gates therein are permitted to be equipped with locks, provided that safe dispersal areas having 30 net square feet (2.8 m^2) for bed and stretcher care recipients and 6 net square feet (0.56 m^2) for ambulatory care recipients and other occupants are located between the building and the fence. Such provided safe dispersal areas shall be located not less than 50 feet (15 240 mm) from the building they serve.

[F] 407.11 Electrical systems. In Group I-2 occupancies, the essential electrical system for electrical components, equipment and systems shall be designed and constructed in accordance with the provisions of Chapter 27 and NFPA 99.

SECTION 408
GROUP I-3

408.1 General. Occupancies in Group I-3 shall comply with the provisions of Sections 408.1 through 408.11 and other applicable provisions of this code (see Section 308.5).

408.2 Other occupancies. Buildings or portions of buildings in Group I-3 occupancies where security operations necessitate the locking of required *means of egress* shall be permitted to be classified as a different occupancy. Occupancies classified as other than Group I-3 shall meet the applicable requirements of this code for that occupancy where provisions are made for the release of occupants at all times.

Means of egress from detention and correctional occupancies that traverse other use areas shall, as a minimum, conform to requirements for detention and correctional occupancies.

Exception: It is permissible to exit through a *horizontal exit* into other contiguous occupancies that do not conform to detention and correctional occupancy egress provisions but that do comply with requirements set forth in the appropriate occupancy, as long as the occupancy is not a Group H use.

408.3 Means of egress. Except as modified or as provided for in this section, the *means of egress* provisions of Chapter 10 shall apply.

408.3.1 Door width. Doors to resident *sleeping units* shall have a clear width of not less than 28 inches (711 mm).

408.3.2 Sliding doors. Where doors in a *means of egress* are of the horizontal-sliding type, the force to slide the door to its fully open position shall be not greater than 50 pounds (220 N) with a perpendicular force against the door of 50 pounds (220 N).

408.3.3 Guard tower doors. A hatch or trap door not less than 16 square feet (610 m^2) in area through the floor and having dimensions of not less than 2 feet (610 mm) in any direction shall be permitted to be used as a portion of the *means of egress* from guard towers.

408.3.4 Spiral stairways. *Spiral stairways* that conform to the requirements of Section 1011.10 are permitted for access to and between staff locations.

408.3.5 Ship's ladders. Ship's ladders shall be permitted for egress from control rooms or elevated facility observation rooms in accordance with Section 1011.15.

408.3.6 Exit discharge. *Exits* are permitted to discharge into a fenced or walled courtyard. Enclosed *yards* or *courts* shall be of a size to accommodate all occupants, be located not less than 50 feet (15 240 mm) from the building and have an area of not less than 15 square feet (1.4 m^2) per person.

408.3.7 Sallyports. A *sallyport* shall be permitted in a *means of egress* where there are provisions for continuous and unobstructed passage through the *sallyport* during an emergency egress condition.

408.3.8 Interior exit stairway and ramp construction. One *interior exit stairway* or *ramp* in each building shall be permitted to have glazing installed in doors and interior walls at each landing level providing access to the *interior exit stairway* or *ramp*, provided that the following conditions are met:

1. The *interior exit stairway or ramp* shall not serve more than four floor levels.
2. *Exit* doors shall be not less than $^3/_4$-hour *fire door assemblies* complying with Section 716.
3. The total area of glazing at each floor level shall not exceed 5,000 square inches (3.2 m^2) and individual panels of glazing shall not exceed 1,296 square inches (0.84 m^2).
4. The glazing shall be protected on both sides by an *automatic sprinkler system*. The sprinkler system shall be designed to wet completely the entire surface of any glazing affected by fire when actuated.
5. The glazing shall be in a gasketed frame and installed in such a manner that the framing system will deflect without breaking (loading) the glass before the sprinkler system operates.
6. Obstructions, such as curtain rods, drapery traverse rods, curtains, drapes or similar materials shall not be installed between the automatic sprinklers and the glazing.

408.4 Locks. Egress doors are permitted to be locked in accordance with the applicable use condition. Doors from a refuge area to the outside are permitted to be locked with a key in lieu of locking methods described in Section 408.4.1. The keys to unlock the exterior doors shall be available at all times and the locks shall be operable from both sides of the door.

408.4.1 Remote release. Remote release of locks on doors in a *means of egress* shall be provided with reliable means of operation, remote from the resident living areas, to release locks on all required doors. In Occupancy Condition 3 or 4, the arrangement, accessibility and security of the release mechanisms required for egress shall be such that with the minimum available staff at any time, the lock mechanisms are capable of being released within 2 minutes.

Exception: Provisions for remote locking and unlocking of occupied rooms in Occupancy Condition 4 are not required provided that not more than 10 locks are necessary to be unlocked in order to move occupants from one *smoke compartment* to a refuge area within 3 minutes. The opening of necessary locks shall be accomplished with not more than two separate keys.

[F] 408.4.2 Power-operated doors and locks. *Power-operated* sliding doors or *power-operated* locks for swinging doors shall be operable by a manual release mechanism at the door. Emergency power shall be provided for the doors and locks in accordance with Section 2702.

Exceptions:

1. Emergency power is not required in facilities with 10 or fewer locks complying with the exception to Section 408.4.1.
2. Emergency power is not required where remote mechanical operating releases are provided.

408.4.3 Redundant operation. Remote release, mechanically operated sliding doors or remote release, mechanically operated locks shall be provided with a mechanically operated release mechanism at each door, or shall be provided with a redundant remote release control.

408.4.4 Relock capability. Doors remotely unlocked under emergency conditions shall not automatically relock when closed unless specific action is taken at the remote location to enable doors to relock.

408.5 Protection of vertical openings. Any vertical opening shall be protected by a shaft enclosure in accordance with Section 713, or shall be in accordance with Section 408.5.1.

408.5.1 Floor openings. Openings in floors within a *housing unit* are permitted without a shaft enclosure, provided that all of the following conditions are met:

1. The entire normally occupied areas so interconnected are open and unobstructed so as to enable observation of the areas by supervisory personnel.
2. *Means of egress* capacity is sufficient for all occupants from all interconnected *cell tiers* and areas.
3. The height difference between the floor levels of the highest and lowest *cell tiers* shall not exceed 23 feet (7010 mm).
4. Egress from any portion of the *cell tier* to an *exit* or *exit access* door shall not require travel on more than one additional floor level within the *housing unit*.

408.5.2 Shaft openings in communicating floor levels. Where a floor opening is permitted between communicating floor levels of a *housing unit* in accordance with Section 408.5.1, plumbing chases serving vertically staked individual cells contained with the *housing unit* shall be permitted without a shaft enclosure.

408.6 Smoke barrier. Occupancies in Group I-3 shall have *smoke barriers* complying with Sections 408.6 and 709 to divide every *story* occupied by residents for sleeping, or any other *story* having an *occupant load* of 50 or more persons, into not fewer than two *smoke compartments*.

Exception: Spaces having a direct exit to one of the following, provided that the locking arrangement of the doors involved complies with the requirements for doors at the *smoke barrier* for the use condition involved:

1. A *public way*.
2. A building separated from the resident housing area by a 2-hour fire-resistance-rated assembly or 50 feet (15 240 mm) of open space.
3. A secured *yard* or *court* having a holding space 50 feet (15 240 mm) from the housing area that provides 6 square feet (0.56 m^2) or more of refuge area per occupant, including residents, staff and visitors.

408.6.1 Smoke compartments. The number of residents in any *smoke compartment* shall be not more than 200. The distance of travel to a door in a *smoke barrier* from any room door required as *exit access* shall be not greater than 150 feet (45 720 mm). The distance of travel to a door in a *smoke barrier* from any point in a room shall be not greater than 200 feet (60 960 mm).

408.6.2 Refuge area. Not less than 6 net square feet (0.56 m^2) per occupant shall be provided on each side of each *smoke barrier* for the total number of occupants in adjoining *smoke compartments*. This space shall be readily available wherever the occupants are moved across the *smoke barrier* in a fire emergency.

408.6.3 Independent egress. A *means of egress* shall be provided from each *smoke compartment* created by *smoke barriers* without having to return through the *smoke compartment* from which *means of egress* originates.

408.7 Security glazing. In occupancies in Group I-3, windows and doors in 1-hour *fire barriers* constructed in accordance with Section 707, *fire partitions* constructed in accordance with Section 708 and *smoke barriers* constructed in accordance with Section 709 shall be permitted to have security glazing installed provided that the following conditions are met.

1. Individual panels of glazing shall not exceed 1,296 square inches (0.84 m^2).
2. The glazing shall be protected on both sides by an *automatic sprinkler system*. The sprinkler system shall be designed to, when actuated, wet completely the entire surface of any glazing affected by fire.
3. The glazing shall be in a gasketed frame and installed in such a manner that the framing system will deflect without breaking (loading) the glass before the sprinkler system operates.
4. Obstructions, such as curtain rods, drapery traverse rods, curtains, drapes or similar materials shall not be installed between the automatic sprinklers and the glazing.

408.8 Subdivision of resident housing areas. Sleeping areas and any contiguous day room, group activity space or other common spaces where residents are housed shall be separated from other spaces in accordance with Sections 408.8.1 through 408.8.4.

408.8.1 Occupancy Conditions 3 and 4. Each sleeping area in Occupancy Conditions 3 and 4 shall be separated from the adjacent common spaces by a smoke-tight partition where the distance of travel from the sleeping area through the common space to the *corridor* exceeds 50 feet (15 240 mm).

408.8.2 Occupancy Condition 5. Each sleeping area in Occupancy Condition 5 shall be separated from adjacent sleeping areas, *corridors* and common spaces by a smoke-tight partition. Additionally, common spaces shall be separated from the *corridor* by a smoke-tight partition.

408.8.3 Openings in room face. The aggregate area of openings in a solid sleeping room face in Occupancy Conditions 2, 3, 4 and 5 shall not exceed 120 square inches (0.77 m^2). The aggregate area shall include all openings including door undercuts, food passes and grilles. Openings shall be not more than 36 inches (914 mm) above the floor. In Occupancy Condition 5, the openings shall be closeable from the room side.

408.8.4 Smoke-tight doors. Doors in openings in partitions required to be smoke tight by Section 408.8 shall be substantial doors, of construction that will resist the passage of smoke. Latches and door closures are not required on *cell* doors.

408.9 Windowless buildings. For the purposes of this section, a windowless building or portion of a building is one with nonopenable windows, windows not readily breakable or without windows. Windowless buildings shall be provided with an engineered smoke control system to provide a tenable environment for exiting from the *smoke* *compartment* in the area of fire origin in accordance with Section 909 for each windowless *smoke compartment*.

[F] 408.10 Fire alarm system. A *fire alarm* system shall be provided in accordance with Section 907.2.6.3.

[F] 408.11 Automatic sprinkler system. Group I-3 occupancies shall be equipped throughout with an *automatic sprinkler system* in accordance with Section 903.2.6.

SECTION 409 MOTION PICTURE PROJECTION ROOMS

409.1 General. The provisions of Sections 409.1 through 409.5 shall apply to rooms in which ribbon-type cellulose acetate or other safety film is utilized in conjunction with electric arc, xenon or other light-source projection equipment that develops hazardous gases, dust or radiation. Where cellulose nitrate film is utilized or stored, such rooms shall comply with NFPA 40.

409.1.1 Projection room required. Every motion picture machine projecting film as mentioned within the scope of this section shall be enclosed in a projection room. Appurtenant electrical equipment, such as rheostats, transformers and generators, shall be within the projection room or in an adjacent room of equivalent construction.

409.2 Construction of projection rooms. Every projection room shall be of permanent construction consistent with the construction requirements for the type of building in which the projection room is located. Openings are not required to be protected.

The room shall have a floor area of not less than 80 square feet (7.44 m^2) for a single machine and not less than 40 square feet (3.7 m^2) for each additional machine. Each motion picture projector, floodlight, spotlight or similar piece of equipment shall have a clear working space of not less than 30 inches by 30 inches (762 mm by 762 mm) on each side and at the rear thereof, but only one such space shall be required between two adjacent projectors. The projection room and the rooms appurtenant thereto shall have a ceiling height of not less than 7 feet 6 inches (2286 mm). The aggregate of openings for projection equipment shall not exceed 25 percent of the area of the wall between the projection room and the auditorium. Openings shall be provided with glass or other *approved* material, so as to close completely the opening.

409.3 Projection room and equipment ventilation. *Ventilation* shall be provided in accordance with the *International Mechanical Code*.

409.3.1 Supply air. Each projection room shall be provided with adequate air supply inlets so arranged as to provide well-distributed air throughout the room. Air inlet ducts shall provide an amount of air equivalent to the amount of air being exhausted by projection equipment. Air is permitted to be taken from the outside; from adjacent spaces within the building, provided that the volume and infiltration rate are sufficient; or from the building air-conditioning system, provided that it is so arranged as to provide sufficient air when other systems are not in operation.

409.3.2 Exhaust air. Projection rooms are permitted to be exhausted through the lamp exhaust system. The lamp exhaust system shall be positively interconnected with the lamp so that the lamp will not operate unless there is the required airflow. Exhaust air ducts shall terminate at the exterior of the building in such a location that the exhaust air cannot be readily recirculated into any air supply system. The projection room *ventilation* system is permitted to also serve appurtenant rooms, such as the generator and rewind rooms.

409.3.3 Projection machines. Each projection machine shall be provided with an exhaust duct that will draw air from each lamp and exhaust it directly to the outside of the building. The lamp exhaust is permitted to serve to exhaust air from the projection room to provide room air circulation. Such ducts shall be of rigid materials, except for a flexible connector *approved* for the purpose. The projection lamp or projection room exhaust system, or both, is permitted to be combined but shall not be interconnected with any other exhaust or return system, or both, within the building.

409.4 Lighting control. Provisions shall be made for control of the auditorium lighting and the *means of egress* lighting systems of theaters from inside the projection room and from not less than one other convenient point in the building.

409.5 Miscellaneous equipment. Each projection room shall be provided with rewind and film storage facilities.

SECTION 410 STAGES, PLATFORMS AND TECHNICAL PRODUCTION AREAS

410.1 Applicability. The provisions of Sections 410.1 through 410.7 shall apply to all parts of buildings and structures that contain *stages* or *platforms* and similar appurtenances as herein defined.

410.2 Stages. *Stage* construction shall comply with Sections 410.2.1 through 410.2.7.

410.2.1 Stage construction. *Stages* shall be constructed of materials as required for floors for the type of construction of the building in which such *stages* are located.

Exception: *Stages* need not be constructed of the same materials as required for the type of construction provided that the construction complies with one of the following:

1. *Stages* of Type IIB or IV construction with a nominal 2-inch (51 mm) wood deck, provided that the *stage* is separated from other areas in accordance with Section 410.2.4.
2. In buildings of Type IIA, IIIA and VA construction, a fire-resistance-rated floor is not required, provided that the space below the *stage* is equipped with an *automatic sprinkler system* or *fire-extinguishing system* in accordance with Section 903 or 904.

3. In all types of construction, the finished floor shall be constructed of wood or *approved* noncombustible materials. Openings through stage floors shall be equipped with tight-fitting, solid wood trap doors with *approved* safety locks.

410.2.1.1 Stage height and area. Stage areas shall be measured to include the entire performance area and adjacent backstage and support areas not separated from the performance area by fire-resistance-rated construction. *Stage* height shall be measured from the lowest point on the *stage* floor to the highest point of the roof or floor deck above the *stage.*

410.2.2 Technical production areas: galleries, gridirons and catwalks. Beams designed only for the attachment of portable or fixed theater equipment, gridirons, galleries and catwalks shall be constructed of *approved* materials consistent with the requirements for the type of construction of the building; and a *fire-resistance rating* shall not be required. These areas shall not be considered to be floors, *stories, mezzanines* or levels in applying this code.

Exception: Floors of fly galleries and catwalks shall be constructed of any *approved* material.

410.2.3 Exterior stage doors. Where protection of openings is required, exterior *exit* doors shall be protected with *fire door assemblies* that comply with Section 716. Exterior openings that are located on the *stage* for *means of egress* or loading and unloading purposes, and that are likely to be open during occupancy of the theater, shall be constructed with vestibules to prevent air drafts into the auditorium.

410.2.4 Proscenium wall. Where the *stage* height is greater than 50 feet (15 240 mm), all portions of the *stage* shall be completely separated from the seating area by a *proscenium wall* with not less than a 2-hour *fire-resistance rating* extending continuously from the foundation to the roof.

410.2.5 Proscenium curtain. Where a *proscenium wall* is required to have a *fire-resistance rating*, the *stage* opening shall be provided with a fire curtain complying with NFPA 80, horizontal sliding doors complying with Section 716 having a *fire protection rating* of not less than 1 hour, or an *approved* water curtain complying with Section 903.3.1.1 or, in facilities not utilizing the provisions of *smoke-protected assembly seating* in accordance with Section 1030.6.2, a smoke control system complying with Section 909 or natural *ventilation* designed to maintain the smoke level not less than 6 feet (1829 mm) above the floor of the *means of egress*.

410.2.6 Scenery. Combustible materials used in sets and scenery shall meet the fire propagation performance criteria of Test Method 1 or Test Method 2, as appropriate, of NFPA 701, in accordance with Section 806 and the *International Fire Code*. Foam plastics and materials containing foam plastics shall comply with Section 2603 and the *International Fire Code*.

410.2.7 Stage ventilation. Emergency *ventilation* shall be provided for *stages* larger than 1,000 square feet (93 m^2) in floor area, or with a *stage* height greater than 50 feet (15 240 mm). Such *ventilation* shall comply with Section 410.2.7.1 or 410.2.7.2.

410.2.7.1 Roof vents. Two or more vents constructed to open automatically by *approved* heat-activated devices and with an aggregate clear opening area of not less than 5 percent of the area of the *stage* shall be located near the center and above the highest part of the *stage area*. Supplemental means shall be provided for manual operation of the ventilator. Curbs shall be provided as required for skylights in Section 2610.2. Vents shall be *labeled.*

[F] 410.2.7.2 Smoke control. Smoke control in accordance with Section 909 shall be provided to maintain the smoke layer interface not less than 6 feet (1829 mm) above the highest level of the assembly seating or above the top of the proscenium opening where a *proscenium wall* is provided in compliance with Section 410.2.4.

410.3 Platform construction. Permanent *platforms* shall be constructed of materials as required for the type of construction of the building in which the permanent *platform* is located. Permanent *platforms* are permitted to be constructed of *fire-retardant-treated wood* for Types I, II and IV construction where the *platforms* are not more than 30 inches (762 mm) above the main floor, and not more than one-third of the room floor area and not more than 3,000 square feet (279 m^2) in area. Where the space beneath the permanent *platform* is used for storage or any purpose other than equipment, wiring or plumbing, the floor assembly shall be not less than 1-hour fire-resistance-rated construction. Where the space beneath the permanent *platform* is used only for equipment, wiring or plumbing, the underside of the permanent *platform* need not be protected.

410.3.1 Temporary platforms. *Platforms* installed for a period of not more than 30 days are permitted to be constructed of any materials permitted by this code. The space between the floor and the *platform* above shall only be used for plumbing and electrical wiring to *platform* equipment.

410.4 Dressing and appurtenant rooms. Dressing and appurtenant rooms shall comply with Sections 410.4.1 and 410.4.2.

410.4.1 Separation from stage. The *stage* shall be separated from dressing rooms, scene docks, property rooms, workshops, storerooms and compartments appurtenant to the *stage* and other parts of the building by *fire barriers* constructed in accordance with Section 707 or *horizontal assemblies* constructed in accordance with Section 711, or both. The *fire-resistance rating* shall be not less than 2 hours for *stage* heights greater than 50 feet (15 240 mm) and not less than 1 hour for *stage* heights of 50 feet (15 240 mm) or less.

410.4.2 Separation from each other. Dressing rooms, scene docks, property rooms, workshops, storerooms and compartments appurtenant to the *stage* shall be separated from each other by not less than 1-hour *fire barriers* constructed in accordance with Section 707 or *horizontal assemblies* constructed in accordance with Section 711, or both.

410.5 Means of egress. Except as modified or as provided for in this section, the provisions of Chapter 10 shall apply.

410.5.1 Arrangement. Where two or more *exits* or *exit access doorways* from the *stage* are required in accordance with Section 1006.2, not fewer than one *exit* or *exit access doorway* shall be provided on each side of a *stage*.

410.5.2 Stairway and ramp enclosure. *Exit access stairways* and *ramps* serving a *stage* or *platform* are not required to be enclosed. *Exit access stairways* and *ramps* serving *technical production areas* are not required to be enclosed.

410.5.3 Technical production areas. *Technical production areas* shall be provided with *means of egress* and means of escape in accordance with Sections 410.5.3.1 through 410.5.3.5.

410.5.3.1 Number of means of egress. Not fewer than one *means of egress* shall be provided from *technical production areas*.

410.5.3.2 Exit access travel distance. The *exit access* travel distance shall be not greater than 300 feet (91 440 mm) for buildings without a sprinkler system and 400 feet (122 mm) for buildings equipped throughout with an *automatic sprinkler system* in accordance with Section 903.3.1.1.

410.5.3.3 Two means of egress. Where two *means of egress* are required, the *common path of travel* shall be not greater than 100 feet (30 480 mm).

Exception: A means of escape to a roof in place of a second *means of egress* is permitted.

410.5.3.4 Path of egress travel. The following *exit access* components are permitted where serving *technical production areas*:

1. *Stairways.*
2. *Ramps.*
3. *Spiral stairways.*
4. Catwalks.
5. *Alternating tread devices.*
6. Permanent ladders.

410.5.3.5 Width. The path of egress travel within and from technical support areas shall be not less than 22 inches (559 mm).

[F] 410.6 Automatic sprinkler system. *Stages* shall be equipped with an *automatic sprinkler system* in accordance with Section 903.3.1.1. Sprinklers shall be installed under the roof and gridiron and under all catwalks and galleries over the *stage*. Sprinklers shall be installed in dressing rooms, performer lounges, shops and storerooms accessory to such *stages*.

Exceptions:

1. Sprinklers are not required under *stage* areas less than 4 feet (1219 mm) in clear height that are utilized exclusively for storage of tables and chairs, provided that the concealed space is separated from the adjacent spaces by Type X *gypsum board* not less than $^5/_8$-inch (15.9 mm) in thickness.
2. Sprinklers are not required for *stages* 1,000 square feet (93 m^2) or less in area and 50 feet (15 240 mm) or less in height where curtains, scenery or other combustible hangings are not retractable vertically. Combustible hangings shall be limited to a single main curtain, borders, legs and a single backdrop.
3. Sprinklers are not required within portable orchestra enclosures on *stages*.

[F] 410.7 Standpipes. *Standpipe systems* shall be provided in accordance with Section 905.

SECTION 411
SPECIAL AMUSEMENT AREAS

411.1 General. *Special amusement areas* having an *occupant load* of 50 or more shall comply with the requirements for the appropriate Group A occupancy and Sections 411.1 through 411.7. *Special amusement areas* having an *occupant load* of less than 50 shall comply with the requirements for a Group B occupancy and Sections 411.1 through 411.7.

Exception: *Special amusement areas* that are without walls or a roof and constructed to prevent the accumulation of smoke need not comply with this section.

[F] 411.2 Automatic sprinkler system. Buildings containing *special amusement areas* shall be equipped throughout with an *automatic sprinkler system* in accordance with Section 903.3.1.1. Where the *special amusement area* is temporary, the sprinkler water supply shall be of an *approved* temporary means.

Exception: *Automatic* sprinklers are not required where the total floor area of a temporary *special amusement area* is less than 1,000 square feet (93 m^2) and the *exit access* travel distance from any point in the *special amusement area* to an exit is less than 50 feet (15 240 mm).

411.3 Fire alarm system. Buildings containing *special amusement areas* shall be equipped with an *automatic smoke detection system* in accordance with Section 907.2.13.

[F] 411.4 Emergency voice/alarm communications system. An *emergency voice/alarm communications system* shall be provided in accordance with Section 907.2.12.

411.5 Puzzle room exiting. *Puzzle room* exiting shall comply with one of the following:

1. Exiting in accordance with Chapter 10.
2. An alternative design approved by the building official.
3. Exits shall be open and readily available upon activation by the automatic fire alarm system, automatic sprinkler system, and a manual control at a constantly attended location.

411.6 Exit marking. Exit signs shall be installed at the required *exit* or *exit access doorways* serving *special amusement areas* in accordance with this section and Section 1013. *Approved* directional exit markings shall be provided. Where mirrors, mazes or other designs are utilized that disguise the path of egress travel such that they are not apparent, *approved* and *listed* low-level exit signs that comply with Section 1013.5, and directional path markings *listed* in accordance with UL 1994, shall be provided and located not more than 8 inches (203 mm) above the walking surface and on or near the path of egress travel. Such markings shall become visible in an emergency. The directional exit marking shall be activated by the *automatic smoke detection system* and the *automatic sprinkler system* in accordance with Section 907.2.12.

411.6.1 Photoluminescent exit signs. Where *photoluminescent exit* signs are installed, activating light source and viewing distance shall be in accordance with the listing and markings of the signs.

411.7 Interior finish. The *interior finish* in *special amusement areas* shall be Class A in accordance with Section 803.1.

SECTION 412
AIRCRAFT-RELATED OCCUPANCIES

412.1 General. Aircraft-related occupancies shall comply with Sections 412.1 through 412.7 and the *International Fire Code.*

412.2 Airport traffic control towers. The provisions of Sections 412.2.1 through 412.2.6 shall apply to airport traffic control towers occupied only for the following uses:

1. Airport traffic control cab.
2. Electrical and mechanical equipment rooms.
3. Airport terminal radar and electronics rooms.
4. Office spaces incidental to the tower operation.
5. Lounges for employees, including sanitary facilities.

412.2.1 Construction. The construction of airport traffic control towers shall comply with the provisions of Sections 412.2.1.1 through 412.2.1.3.

412.2.1.1 Type of construction. Airport traffic control towers shall be constructed to comply with the height limitations of Table 412.2.1.1.

TABLE 412.2.1.1
HEIGHT LIMITATIONS FOR AIRPORT TRAFFIC CONTROL TOWERS

TYPE OF CONSTRUCTION	HEIGHT[a] (feet)
IA	Unlimited
IB	240
IIA	100
IIB	85
IIIA	65

For SI: 1 foot = 304.8 mm, 1 square foot = 0.0929 m^2.

a. Height to be measured from *grade plane* to cab floor.

[BS] 412.2.1.2 Structural integrity of interior exit stairways and elevator hoistway enclosures. Enclosures for *interior exit stairways* and elevator hoistway enclosures shall comply with Section 403.2.2 in airport traffic control towers where the control cab is located more than 75 feet (22 860 mm) above the lowest level of fire department vehicle access.

412.2.1.3 Sprayed fire-resistant materials (SFRM). The bond strength of the SFRM installed in airport traffic control towers shall be in accordance with Section 403.2.3 where the control cab is located more than 75 feet (22 860 mm) above the lowest level of fire department vehicle access.

412.2.2 Means of egress and evacuation. The *means of egress* in airport traffic control towers shall comply with Sections 412.2.2.1 through 412.2.2.3.

412.2.2.1 Stairways. *Stairways* in airport traffic control towers shall be in accordance with Section 1011. *Exit stairways* shall be *smokeproof enclosures* complying with one of the alternatives provided in Section 909.20.

Exception: *Stairways* in airport traffic control towers are not required to comply with Section 1011.12.

412.2.2.2 Exit access. From observation levels, airport traffic control towers shall be permitted to have a single means of *exit access* for a distance of travel not greater than 100 feet (30 480 mm). *Exit access stairways* from the observation level need not be enclosed.

412.2.2.3 Number of exits. Not less than one *exit stairway* shall be permitted for airport traffic control towers of any height provided that the *occupant load* per floor is not greater than 15 and the area per floor does not exceed 1,500 square feet (140 m^2).

412.2.2.3.1 Interior finish. Where an airport traffic control tower is provided with only one *exit stairway*, *interior wall and ceiling finishes* shall be either Class A or Class B.

412.2.2.3.2 Exit separation. Where an airport traffic control tower is equipped throughout with an *automatic sprinkler system* in accordance with

Section 903.3.1.1 and two *exits* are required, the exit separation distance required by Section 1007 shall be not less than one-fourth of the length of the maximum overall dimension of the area served.

[F] 412.2.3 Emergency systems. The detection, alarm and emergency systems of airport traffic control towers shall comply with Sections 412.2.3.1 through 412.2.3.3.

[F] 412.2.3.1 Automatic smoke detection systems. Airport traffic control towers shall be provided with an *automatic* smoke detection system installed in accordance with Section 907.2.22.

[F] 412.2.3.2 Fire command center. A *fire command center* shall be provided in airport traffic control towers where the control cab is located more than 75 feet (22 860 mm) above the lowest level of fire department vehicle access. The *fire command center* shall comply with Section 911.

Exceptions:

1. The *fire command center* shall be located in the airport control tower or an adjacent contiguous building where building functions are interdependent.
2. The room shall be not less than 150 square feet (14 m^2) in area with a minimum dimension of 10 feet (3048 mm).
3. The following features shall not be required in an airport traffic control tower *fire command center.*
   - 3.1. Emergency voice/alarm control unit.
   - 3.2. Public address system.
   - 3.3. Status indicators and controls for the air distributions centers.
   - 3.4. Generator supervision devices, manual start and transfer features.
   - 3.5. Elevator emergency or standby power switches where emergency or standby power is provided.

[F] 412.2.3.3 Smoke removal. Smoke removal in airport traffic control towers shall be provided in accordance with Section 403.4.7.

[F] 412.2.4 Automatic sprinkler system. Where an occupied floor is located more than 35 feet (10 668 mm) above the lowest level of fire department vehicle access, airport traffic control towers shall be equipped with an *automatic sprinkler system* in accordance with Section 903.3.1.1.

[F] 412.2.4.1 Fire pump room. Fire pumps shall be located in rooms that are separated from all other areas of the building by 2-hour *fire barriers* constructed in accordance with Section 707 or 2-hour *horizontal assemblies* constructed in accordance with Section 711, or both.

Exception: Separation is not required for fire pumps physically separated in accordance with NFPA 20.

[F] 412.2.5 Protection of elevator wiring and cables. Wiring and cables serving elevators in airport traffic control towers shall be protected in accordance with Section 3007.8.1.

412.2.5.1 Elevators for occupant evacuation. Where provided in addition to an exit stairway, occupant evacuation elevators shall be in accordance with Section 3008.

412.2.6 Accessibility. Airport traffic control towers shall be *accessible* except as specified in Section 1104.4.

412.3 Aircraft hangars. Aircraft hangars shall be in accordance with Sections 412.3.1 through 412.3.6.

412.3.1 Exterior walls. *Exterior walls* located less than 30 feet (9144 mm) from *lot lines* or a *public way* shall have a *fire-resistance rating* not less than 2 hours.

412.3.2 Basements. Where hangars have *basements*, floors over *basements* shall be of Type IA construction and shall be made tight against seepage of water, oil or vapors. There shall not be openings or communication between *basements* and the hangar. Access to *basements* shall be from outside only.

412.3.3 Floor surface. Floors shall be graded and drained to prevent water or fuel from remaining on the floor. Floor drains shall discharge through an oil separator to the sewer or to an outside vented sump.

Exception: Aircraft hangars with individual lease spaces not exceeding 2,000 square feet (186 m^2) each in which servicing, repairing or washing is not conducted and fuel is not dispensed shall have floors that are graded toward the door, but shall not require a separator.

412.3.4 Heating equipment. Heating equipment shall be placed in another room separated by 2-hour *fire barriers* constructed in accordance with Section 707 or *horizontal assemblies* constructed in accordance with Section 711, or both. Entrance shall be from the outside or by means of a vestibule providing a two-doorway separation.

Exceptions:

1. Unit heaters and vented infrared radiant heating equipment suspended not less than 10 feet (3048 mm) above the upper surface of wings or engine enclosures of the highest aircraft that are permitted to be housed in the hangar need not be located in a separate room provided that they are mounted not less than 8 feet (2438 mm) above the floor in shops, offices and other sections of the hangar communicating with storage or service areas.
2. Entrance to the separated room shall be permitted by a single interior door provided that the sources of ignition in the appliances are not less than 18 inches (457 mm) above the floor.

412.3.5 Finishing. The process of "doping," involving use of a volatile flammable solvent, or of painting, shall be carried on in a separate *detached building* equipped with *automatic fire-extinguishing equipment* in accordance with Section 903.

[F] 412.3.6 Fire suppression. Aircraft hangars shall be provided with a fire suppression system designed in accordance with NFPA 409, based on the classification for the hangar given in Table 412.3.6.

Exception: Where a *fixed base operator* has separate repair facilities on site, Group II hangars operated by a *fixed base operator* used for storage of *transient aircraft* only shall have a fire suppression system, but the system is exempt from foam requirements.

[F] 412.3.6.1 Hazardous operations. Any Group III aircraft hangar according to Table 412.3.6 that contains hazardous operations including, but not limited to, the following shall be provided with a Group I or II fire suppression system in accordance with NFPA 409 as applicable:

1. Doping.
2. Hot work including, but not limited to, welding, torch cutting and torch soldering.
3. Fuel transfer.
4. Fuel tank repair or maintenance not including defueled tanks in accordance with NFPA 409, inerted tanks or tanks that have never been fueled.
5. Spray finishing operations.
6. Total fuel capacity of all aircraft within the unsprinklered single *fire area* in excess of 1,600 gallons (6057 L).
7. Total fuel capacity of all aircraft within the maximum single *fire area* in excess of 7,500 gallons (28 390 L) for a hangar with an *automatic sprinkler system* in accordance with Section 903.3.1.1.

[F] 412.3.6.2 Separation of maximum single fire areas. Maximum single *fire areas* established in accordance with hangar classification and construction type in Table 412.3.6 shall be separated by 2-hour *fire walls* constructed in accordance with Section 706. In determining the maximum single *fire area* as set forth in Table 412.3.6, ancillary uses that are separated from aircraft servicing areas by a *fire barrier* of not less than 1 hour, constructed in accordance with Section 707, shall not be included in the area.

412.4 Residential aircraft hangars. *Residential aircraft hangars* shall comply with Sections 412.4.1 through 412.4.5.

412.4.1 Fire separation. A hangar shall not be attached to a *dwelling* unless separated by a *fire barrier* having a *fire-resistance rating* of not less than 1 hour. Such separation shall be continuous from the foundation to the underside of the roof and unpierced except for doors leading to the *dwelling unit*. Doors into the *dwelling unit* shall be equipped with *self-closing* devices and conform to the requirements of Section 716 with a noncombustible raised sill not less than 4 inches (102 mm) in height. Openings from a hangar directly into a room used for sleeping purposes shall not be permitted.

412.4.2 Egress. A hangar shall provide two *means of egress*. One of the doors into the dwelling shall be considered as meeting only one of the two *means of egress*.

[F] 412.4.3 Smoke alarms. *Smoke alarms* shall be provided within the hangar in accordance with Section 907.2.22.

412.4.4 Independent systems. Electrical, mechanical and plumbing drain, waste and vent (DWV) systems installed within the hangar shall be independent of the systems installed within the dwelling. Building sewer lines shall be permitted to be connected outside the structures.

Exception: *Smoke detector* wiring and feed for electrical subpanels in the hangar.

412.4.5 Height and area limits. *Residential aircraft hangars* shall be not greater than 2,000 square feet (186 m^2) in area and 20 feet (6096 mm) in *building height*.

[F] 412.5 Aircraft paint hangars. Aircraft painting operations shall be conducted in an aircraft paint hangar that complies with the provisions of Sections 412.5.1 through

[F] TABLE 412.3.6
HANGAR FIRE SUPPRESSION REQUIREMENTS[a, b, c]

MAXIMUM SINGLE FIRE AREA (square feet)	TYPE OF CONSTRUCTION								
	IA	IB	IIA	IIB	IIIA	IIIB	IV	VA	VB
≥ 40,001	Group I	Group I	Group I	Group I	Group I	Group I	Group I	Group I	Group I
40,000	Group II	Group II	Group II	Group II	Group II	Group II	Group II	Group II	Group II
30,000	Group III	Group II	Group II	Group II	Group II	Group II	Group II	Group II	Group II
20,000	Group III	Group III	Group II	Group II	Group II	Group II	Group II	Group II	Group II
15,000	Group III	Group III	Group III	Group II	Group III	Group II	Group III	Group II	Group II
12,000	Group III	Group III	Group III	Group III	Group III	Group III	Group III	Group II	Group II
8,000	Group III	Group III	Group III	Group III	Group III	Group III	Group III	Group III	Group II
5,000	Group III	Group III	Group III	Group III	Group III	Group III	Group III	Group III	Group III

For SI: 1 foot = 304.8 mm, 1 square foot = 0.0929 m^2.
a. Aircraft hangars with a door height greater than 28 feet shall be provided with fire suppression for a Group I hangar regardless of maximum fire area.
b. Groups shall be as classified in accordance with NFPA 409.
c. Membrane structures complying with Section 3102 shall be classified as a Group IV hangar.

412.5.8. Buildings and structures, or parts thereof, used for the application of flammable finishes shall comply with the applicable provisions of Section 416.

[F] 412.5.1 Occupancy classification. Aircraft paint hangars shall be classified in accordance with the provisions of Section 307.1. Aircraft paint hangars shall comply with the applicable requirements of this code and the *International Fire Code* for such occupancy.

412.5.2 Construction. Aircraft paint hangars shall be of Type I or II construction.

[F] 412.5.3 Spray equipment cleaning operations. Spray equipment cleaning operations shall be conducted in a liquid use, dispensing and mixing room.

[F] 412.5.4 Operations. Only those *flammable liquids* necessary for painting operations shall be permitted in quantities less than the maximum allowable quantities per *control area* in Table 307.1(1). Spray equipment cleaning operations exceeding the maximum allowable quantities per *control area* in Table 307.1(1) shall be conducted in a liquid use, dispensing and mixing room.

[F] 412.5.5 Storage. Storage of *flammable or combustible liquids* exceeding the maximum allowable quantities per *control area* in Table 307.1(1) shall be in a *liquid storage room.*

[F] 412.5.6 Fire suppression. Aircraft paint hangars shall be provided with fire suppression as required by NFPA 409.

[F] 412.5.7 Ventilation. Aircraft paint hangars shall be provided with *ventilation* as required in the *International Mechanical Code.*

[F] 412.5.8 Electrical. Electrical equipment and devices within the aircraft paint hangar shall comply with NFPA 70.

[F] 412.5.8.1 Class I, Division I hazardous locations. The area within 10 feet (3048 mm) horizontally from aircraft surfaces and from the floor to 10 feet (3048 mm) above the aircraft surface shall be classified as a Class I, Division I location.

[F] 412.5.8.2 Class I, Division 2 hazardous locations. The area horizontally from aircraft surfaces between 10 feet (3048 mm) and 30 feet (9144 mm) and from the floor to 30 feet (9144 mm) above the aircraft surface shall be classified as a Class I, Division 2 location.

412.6 Aircraft manufacturing facilities. In buildings used for the manufacturing of aircraft, *exit access* travel distances indicated in Section 1017.1 shall be increased in accordance with the following:

1. The building shall be of Type I or II construction.
2. *Exit access* travel distance shall not exceed the distances given in Table 412.6.

412.6.1 Ancillary areas. Rooms, areas and spaces ancillary to the primary manufacturing area shall be permitted to egress through such area having a minimum height as indicated in Table 412.6. *Exit access* travel distance within the ancillary room, area or space shall not exceed that indicated in Table 1017.2 based on the occupancy classification of that ancillary area. Total *exit access* travel distance shall not exceed that indicated in Table 412.6.

[F] 412.7 Heliports and helistops. *Heliports* and *helistops* shall be permitted to be erected on buildings or other locations where they are constructed in accordance with Sections 412.7.1 through 412.7.5.

[F] 412.7.1 Size. The landing area for helicopters less than 3,500 pounds (1588 kg) shall be not less than 20 feet (6096 mm) in length and width. The landing area shall be surrounded on all sides by a clear area having an average width at roof level of 15 feet (4572 mm), and all widths shall be not less than 5 feet (1524 mm).

[F] 412.7.2 Design. Helicopter landing areas and the supports thereof on the roof of a building shall be noncombustible construction. Landing areas shall be designed to confine any flammable liquid spillage to the landing area itself and provisions shall be made to drain such spillage away from any *exit* or *stairway* serving the helicopter landing area or from a structure housing such *exit* or *stairway.* For structural design requirements, see Section 1607.6.

412.7.3 Means of egress. The *means of egress* from *heliports* and *helistops* shall comply with the provisions of Chapter 10. Landing areas located on buildings or structures shall have two or more *exits or access to exits.* For landing areas less than 60 feet (18 288 mm) in length or less than 2,000 square feet (186 m^2) in area, the second *means of egress* is permitted to be a fire escape, *alternating tread device* or ladder leading to the floor below.

[F] 412.7.4 Rooftop heliports and helistops. Rooftop *heliports* and *helistops* shall comply with NFPA 418.

TABLE 412.6
AIRCRAFT MANUFACTURING EXIT ACCESS TRAVEL DISTANCE

HEIGHT (feet)[b]	MANUFACTURING AREA (square feet)[a]					
	≥ 150,000	≥ 200,000	≥ 250,000	≥ 500,000	≥ 750,000	≥ 1,000,000
≥ 25	400	450	500	500	500	500
≥ 50	400	500	600	700	700	700
≥ 75	400	500	700	850	1,000	1,000
≥ 100	400	500	750	1,000	1,250	1,500

For SI: 1 foot = 304.8 mm.
a. Contiguous floor area of the aircraft manufacturing facility having the indicated height.
b. Minimum height from finished floor to bottom of ceiling or roof slab or deck.

[F] 412.7.5 Standpipe system. In buildings equipped with a *standpipe system*, the standpipe shall extend to the roof level in accordance with Section 905.3.6.

SECTION 413
COMBUSTIBLE STORAGE

413.1 General. High-piled stock or rack storage in any occupancy group shall comply with the *International Fire Code.*

413.2 Attic, under-floor and concealed spaces. *Attic*, under-floor and concealed spaces used for storage of combustible materials shall be protected on the storage side as required for 1-hour fire-resistance-rated construction. Openings shall be protected by assemblies that are *self-closing* and are of noncombustible construction or solid wood core not less than $1^3/_4$ inches (45 mm) in thickness.

Exception: Neither fire-resistance-rated construction nor opening protectives are required in any of the following locations:

1. Areas protected by *approved automatic sprinkler systems*.
2. Group R-3 and U occupancies.

SECTION 414
HAZARDOUS MATERIALS

[F] 414.1 General. The provisions of Sections 414.1 through 414.6 shall apply to buildings and structures occupied for the manufacturing, processing, dispensing, use or storage of *hazardous materials*.

[F] 414.1.1 Other provisions. Buildings and structures with an occupancy in Group H shall comply with this section and the applicable provisions of Section 415 and the *International Fire Code*.

[F] 414.1.2 Materials. The safe design of hazardous material occupancies is material dependent. Individual material requirements are found in Sections 307 and 415, the *International Mechanical Code* and the *International Fire Code*.

[F] 414.1.2.1 Aerosol products, aerosol cooking spray products and plastic aerosol 3 products. Level 2 and 3 aerosol products, aerosol cooking spray products and plastic aerosol 3 products shall be stored and displayed in accordance with the *International Fire Code*. See Section 311.2 and the *International Fire Code* for occupancy group requirements.

[F] 414.1.3 Information required. A report shall be submitted to the *building official* identifying the maximum expected quantities of *hazardous materials* to be stored, used in a *closed system* and used in an *open system*, and subdivided to separately address *hazardous material* classification categories based on Tables 307.1(1) and 307.1(2). The methods of protection from such hazards, including but not limited to *control areas*, fire protection systems and Group H occupancies shall be indicated in the report and on the *construction documents*. The opinion and report shall be prepared by a qualified person, firm or corporation *approved* by the *building official* and provided without charge to the enforcing agency.

For buildings and structures with an occupancy in Group H, separate floor plans shall be submitted identifying the locations of anticipated contents and processes so as to reflect the nature of each occupied portion of every building and structure.

[F] 414.2 Control areas. *Control areas* shall comply with Sections 414.2.1 through 414.2.5 and the *International Fire Code*.

Exception: *Higher education laboratories* in accordance with Section 428 and Chapter 38 of the *International Fire Code*.

[F] 414.2.1 Construction requirements. *Control areas* shall be separated from each other by *fire barriers* constructed in accordance with Section 707 or *horizontal assemblies* constructed in accordance with Section 711, or both.

[F] 414.2.2 Percentage of maximum allowable quantities. The percentage of maximum allowable quantities of *hazardous materials* per *control area* permitted at each floor level within a building shall be in accordance with Table 414.2.2.

[F] 414.2.3 Number. The maximum number of *control areas* within a building shall be in accordance with Table 414.2.2. For the purposes of determining the number of *control areas* within a building, each portion of a building separated by one or more *fire walls* complying with Section 706 shall be considered a separate building.

[F] 414.2.4 Fire-resistance rating requirements. The required *fire-resistance rating* for *fire barriers* shall be in accordance with Table 414.2.2. The floor assembly of the *control area* and the construction supporting the floor of the *control area* shall have a *fire-resistance rating* of not less than 2 hours.

Exception: The floor assembly of the *control area* and the construction supporting the floor of the *control area* are allowed to be 1-hour fire-resistance-rated in buildings of Types IIA, IIIA, IV and VA construction, provided that both of the following conditions exist:

1. The building is equipped throughout with an *automatic sprinkler system* in accordance with Section 903.3.1.1.
2. The building is three or fewer *stories above grade plane*.

[F] 414.2.5 Hazardous material in Group M display and storage areas and in Group S storage areas. *Hazardous materials* located in Group M and Group S occupancies shall be in accordance with Sections 414.2.5.1 through 414.2.5.3.

[F] TABLE 414.2.2
DESIGN AND NUMBER OF CONTROL AREAS

STORY		PERCENTAGE OF THE MAXIMUM ALLOWABLE QUANTITY PER CONTROL AREA[a]	NUMBER OF CONTROL AREAS PER STORY	FIRE-RESISTANCE RATING FOR FIRE BARRIERS IN HOURS[b]
Above grade plane	Higher than 9	5	1	2
	7–9	5	2	2
	6	12.5	2	2
	5	12.5	2	2
	4	12.5	2	2
	3	50	2	1
	2	75	3	1
	1	100	4	1
Below grade plane	1	75	3	1
	2	50	2	1
	Lower than 2	Not Allowed	Not Allowed	Not Allowed

a. Percentages shall be of the maximum allowable quantity per control area shown in Tables 307.1(1) and 307.1(2), with all increases allowed in the notes to those tables.

b. Separation shall include fire barriers and horizontal assemblies as necessary to provide separation from other portions of the building.

[F] 414.2.5.1 Nonflammable solids and nonflammable and noncombustible liquids. The aggregate quantity of nonflammable solid and nonflammable or noncombustible liquid hazardous materials permitted within a single *control area* of a Group M display and storage area, a Group S storage area or an outdoor *control area* is permitted to exceed the maximum allowable quantities per *control area* specified in Tables 307.1(1) and 307.1(2) without classifying the building or use as a Group H occupancy, provided that the materials are displayed and stored in accordance with the *International Fire Code* and quantities do not exceed the maximum allowable specified in Table 414.2.5(1).

[F] 414.2.5.2 Flammable and combustible liquids. In Group M occupancy wholesale and retail sales uses, indoor storage of *flammable and combustible liquids* shall not exceed the maximum allowable quantities per *control area* as indicated in Table 414.2.5(2), provided that the materials are displayed and stored in accordance with the *International Fire Code*.

[F] 414.2.5.3 Aerosol products, aerosol cooking spray products or plastic aerosol 3 products. The maximum quantity of *aerosol products*, aerosol cooking spray products or plastic aerosol 3 products in Group M occupancy retail display areas, storage areas adjacent to retail display areas and retail storage areas shall be in accordance with the *International Fire Code*.

[F] 414.3 Ventilation. Rooms, areas or spaces in which *explosive, corrosive, combustible, flammable or highly toxic* dusts, mists, fumes, vapors or gases are or have the potential to be emitted due to the processing, *use*, handling or storage of materials shall be mechanically ventilated where required by this code, the *International Fire Code* or the *International Mechanical Code*.

Emissions generated at workstations shall be confined to the area in which they are generated as specified in the *International Fire Code* and the *International Mechanical Code*.

[F] 414.4 Hazardous material systems. Systems involving *hazardous materials* shall be suitable for the intended application. Controls shall be designed to prevent materials from entering or leaving process or reaction systems at other than the intended time, rate or path. *Automatic* controls, where provided, shall be designed to be fail safe.

[F] 414.5 Inside storage, dispensing and use. The inside storage, dispensing and *use* of *hazardous materials* shall be in accordance with Sections 414.5.1 through 414.5.3 of this code and the *International Fire Code*.

[F] 414.5.1 Explosion control. *Explosion* control shall be provided in accordance with the *International Fire Code* as required by Table 414.5.1 where quantities of *hazardous materials* specified in that table exceed the maximum allowable quantities in Table 307.1(1) or where a structure, room or space is occupied for purposes involving *explosion* hazards as required by Section 415 or the *International Fire Code*.

[F] 414.5.2 Emergency or standby power. Where required by the *International Fire Code* or this code, mechanical *ventilation*, treatment systems, temperature control, alarm, detection or other electrically operated systems shall be provided with emergency or standby power in accordance with Section 2702. For storage and use areas for *highly toxic or toxic* materials, see Sections 6004.2.2.8 and 6004.3.4.2 of the *International Fire Code*.

**[F] TABLE 414.2.5(1)
MAXIMUM ALLOWABLE QUANTITY PER INDOOR AND OUTDOOR CONTROL AREA IN GROUP M AND S OCCUPANCIES OF NONFLAMMABLE SOLIDS AND NONFLAMMABLE AND NONCOMBUSTIBLE LIQUIDS[d, e, f]**

CONDITION		MAXIMUM ALLOWABLE QUANTITY PER CONTROL AREA	
Material[a]	Class	Solids (pounds)	Liquids (gallons)
A. Health-hazard materials—nonflammable and noncombustible solids and liquids			
1. Corrosives[b, c]	Not Applicable	9,750	975
2. Highly toxics	Not Applicable	20[b, c]	2[b, c]
3. Toxics[b, c]	Not Applicable	1,000[k]	100
B. Physical-hazard materials—nonflammable and noncombustible solids and liquids			
1. Oxidizers[b, c]	4	Not Allowed	Not Allowed
	3	1,350[g]	115
	2	2,250[h]	225
	1	18,000[i, j]	1,800[i, j]
2. Unstable (reactives)[b, c]	4	Not Allowed	Not Allowed
	3	550	55
	2	1,150	115
	1	Not Limited	Not Limited
3. Water reactives	3[b, c]	550	55
	2[b, c]	1,150	115
	1	Not Limited	Not Limited

For SI: 1 pound = 0.454 kg, 1 gallon = 3.785 L.

a. Hazard categories are as specified in the *International Fire Code*.

b. Maximum allowable quantities shall be increased 100 percent in buildings that are sprinklered in accordance with Section 903.3.1.1. Where Note c also applies, the increase for both notes shall be applied accumulatively.

c. Maximum allowable quantities shall be increased 100 percent where stored in approved storage cabinets, in accordance with the *International Fire Code*. Where Note b also applies, the increase for both notes shall be applied accumulatively.

d. See Table 414.2.2 for design and number of control areas.

e. Allowable quantities for other hazardous material categories shall be in accordance with Section 307.

f. Maximum quantities shall be increased 100 percent in outdoor control areas.

g. Maximum amounts shall be increased to 2,250 pounds where individual packages are in the original sealed containers from the manufacturer or packager and do not exceed 10 pounds each.

h. Maximum amounts shall be increased to 4,500 pounds where individual packages are in the original sealed containers from the manufacturer or packager and do not exceed 10 pounds each.

i. The permitted quantities shall not be limited in a building equipped throughout with an automatic sprinkler system in accordance with Section 903.3.1.1.

j. Quantities are unlimited in an outdoor control area.

k. Maximum allowable quantities of consumer products shall be increased to 10,000 pounds where individual packages are in the original, sealed containers from the manufacturer and the toxic classification is exclusively based on the LC threshold and no other hazardous materials classifications apply.

**[F] TABLE 414.2.5(2)
MAXIMUM ALLOWABLE QUANTITY OF FLAMMABLE AND COMBUSTIBLE LIQUIDS IN WHOLESALE AND RETAIL SALES OCCUPANCIES PER CONTROL AREA[a]**

TYPE OF LIQUID	MAXIMUM ALLOWABLE QUANTITY PER CONTROL AREA (gallons)		
	Sprinklered in accordance with Note b densities and arrangements	Sprinklered in accordance with Tables 5704.3.6.3(4) through 5704.3.6.3(8) and 5704.3.7.5.1 of the *International Fire Code*	Nonsprinklered
Class IA	60	60	30
Class IB, IC, II and IIIA	7,500[c]	15,000[c]	1,600
Class IIIB	Unlimited	Unlimited	13,200

For SI: 1 foot = 304.8 mm, 1 square foot = 0.0929 m^2, 1 gallon = 3.785 L, 1 gallon per minute per square foot = 40.75 L/min/m^2.

a. Control areas shall be separated from each other by not less than a 1-hour *fire barrier* wall.

b. To be considered as sprinklered, a building shall be equipped throughout with an approved automatic sprinkler system with a design providing minimum densities as follows:

1. For uncartoned commodities on shelves 6 feet or less in height where the ceiling height does not exceed 18 feet, quantities are those permitted with a minimum sprinkler design density of Ordinary Hazard Group 2.
2. For cartoned, palletized or racked commodities where storage is 4 feet 6 inches or less in height and where the ceiling height does not exceed 18 feet, quantities are those permitted with a minimum sprinkler design density of 0.21 gallon per minute per square foot over the most remote 1,500-square-foot area.

c. Where wholesale and retail sales or storage areas exceed 50,000 square feet in area, the maximum allowable quantities are allowed to be increased by 2 percent for each 1,000 square feet of area in excess of 50,000 square feet, up to not more than 100 percent of the table amounts. A control area separation is not required. The cumulative amounts, including amounts attained by having an additional control area, shall not exceed 30,000 gallons.

[F] TABLE 414.5.1
EXPLOSION CONTROL REQUIREMENTS[a, h]

MATERIAL	CLASS	EXPLOSION CONTROL METHODS	
		Barricade construction	Explosion (deflagration) venting or explosion (deflagration) prevention systems[b]
HAZARD CATEGORY			
Combustible dusts[c]	—	Not Required	Required
Cryogenic flammables	—	Not Required	Required
Explosives	Division 1.1	Required	Not Required
	Division 1.2	Required	Not Required
	Division 1.3	Not Required	Required
	Division 1.4	Not Required	Required
	Division 1.5	Required	Not Required
	Division 1.6	Required	Not Required
Flammable gas	Gaseous	Not Required	Required
	Liquefied	Not Required	Required
Flammable liquid	IA[d]	Not Required	Required
	IB[e]	Not Required	Required
Organic peroxides	U	Required	Not Permitted
	I	Required	Not Permitted
Oxidizer liquids and solids	4	Required	Not Permitted
Pyrophoric gas	—	Not Required	Required
Unstable (reactive)	4	Required	Not Permitted
	3 Detonable	Required	Not Permitted
	3 Nondetonable	Not Required	Required
Water-reactive liquids and solids	3	Not Required	Required
	2[g]	Not Required	Required
SPECIAL USES			
Acetylene generator rooms	—	Not Required	Required
Electrochemical energy storage system[i]	—	Not Required	Required
Energy storage system[i]	—	Not Required	Required
Grain processing	—	Not Required	Required
Liquefied petroleum gas-distribution facilities	—	Not Required	Required
Where explosion hazards exist[f]	Detonation	Required	Not Permitted
	Deflagration	Not Required	Required

a. See Section 414.1.3.
b. See the International Fire Code.
c. Combustible dusts where manufactured, generated or used in such a manner that the concentration and conditions create a fire or explosion hazard based on information prepared in accordance with Section 104.8.2 of the International Fire Code. See definition of "Combustible dust" in Chapter 2.
d. Storage or use.
e. In open use or dispensing.
f. Rooms containing dispensing and use of hazardous materials where an explosive environment can occur because of the characteristics or nature of the hazardous materials or as a result of the dispensing or use process.
g. A method of explosion control shall be provided where Class 2 water-reactive materials can form potentially explosive mixtures.
h. Explosion venting is not required for Group H-5 fabrication areas complying with Section 415.11.1 and the International Fire Code.
i. Where explosion control is required in Section 1207 of the *International Fire Code.*

[F] 414.5.2.1 Exempt applications. Emergency or standby power is not required for the mechanical *ventilation* systems provided for any of the following:

1. Storage of Class IB and IC flammable and combustible liquids in closed containers not exceeding 6.5 gallons (25 L) capacity.
2. Storage of Class 1 and 2 oxidizers.
3. Storage of Class II, III, IV and V organic peroxides.
4. Storage of asphyxiant, irritant and radioactive gases.

[F] 414.5.2.2 Fail-safe engineered systems. Standby power for mechanical ventilation, treatment systems and temperature control systems shall not be required where an approved fail-safe engineered system is installed.

[F] 414.5.3 Spill control, drainage and containment. Rooms, buildings or areas occupied for the storage of solid and liquid *hazardous materials* shall be provided with a means to control spillage and to contain or drain off spillage and fire protection water discharged in the storage area where required in the *International Fire Code*. The methods of spill control shall be in accordance with the *International Fire Code*.

[F] 414.6 Outdoor storage, dispensing and use. The outdoor storage, dispensing and use of *hazardous materials* shall be in accordance with the *International Fire Code*.

[F] 414.6.1 Weather protection. Where weather protection is provided for sheltering outdoor *hazardous material* storage or use areas, such areas shall be considered outdoor storage or *use* where the weather protection structure complies with Sections 414.6.1.1 through 414.6.1.3.

[F] 414.6.1.1 Walls. Walls shall not obstruct more than one side of the structure.

Exception: Walls shall be permitted to obstruct portions of multiple sides of the structure, provided that the obstructed area is not greater than 25 percent of the structure's perimeter.

[F] 414.6.1.2 Separation distance. The distance from the structure to buildings, *lot lines*, *public ways* or *means of egress* to a *public way* shall be not less than the distance required for an outside *hazardous material* storage or use area without weather protection.

[F] 414.6.1.3 Noncombustible construction. The overhead structure shall be of *approved* noncombustible construction with a maximum area of 1,500 square feet (140 m^2).

Exception: The maximum area is permitted to be increased as provided by Section 506.

SECTION 415
GROUPS H-1, H-2, H-3, H-4 AND H-5

[F] 415.1 General. The provisions of Sections 415.1 through 415.11 shall apply to the storage and use of hazardous materials in excess of the maximum allowable quantities per *control area* listed in Section 307.1.

[F] 415.2 Compliance. Buildings and structures with an occupancy in Group H shall comply with the applicable provisions of Section 414 and the *International Fire Code*.

[F] 415.3 Automatic fire detection systems. Group H occupancies shall be provided with an automatic fire detection system in accordance with Section 907.2.

[F] 415.4 Automatic sprinkler system. Group H occupancies shall be equipped throughout with an *automatic sprinkler system* in accordance with Section 903.2.5.

[F] 415.5 Emergency alarms. Emergency alarms for the detection and notification of an emergency condition in Group H occupancies shall be provided as set forth herein.

[F] 415.5.1 Storage. An approved manual emergency alarm system shall be provided in buildings, rooms or areas used for storage of hazardous materials. Emergency alarm-initiating devices shall be installed outside of each interior *exit* or exit access door of storage buildings, rooms or areas. Activation of an emergency alarm-initiating device shall sound a local alarm to alert occupants of an emergency situation involving hazardous materials.

[F] 415.5.2 Dispensing, use and handling. Where hazardous materials having a hazard ranking of 3 or 4 in accordance with NFPA 704 are transported through *corridors*, *interior exit stairways* or *ramps*, or exit passageways, there shall be an emergency telephone system, a local manual alarm station or an approved alarm-initiating device at not more than 150-foot (45 720 mm) intervals and at each exit and *exit access doorway* throughout the transport route. The signal shall be relayed to an approved central, proprietary or remote station service or constantly attended on-site location and shall initiate a local audible alarm.

[F] 415.5.3 Supervision. Emergency alarm systems required by Section 415.5.1 or 415.5.2 shall be electrically supervised and monitored by an approved central, proprietary or remote station service or shall initiate an audible and visual signal at a constantly attended on-site location.

[F] 415.5.4 Emergency alarm systems. *Emergency alarm systems* required by Section 415.5.1 or 415.5.2 shall be provided with emergency or standby power in accordance with Section 2702.2.

[F] 415.6 Fire separation distance. Group H occupancies shall be located on property in accordance with the other provisions of this chapter. In Groups H-2 and H-3, not less than 25 percent of the perimeter wall of the occupancy shall be an *exterior wall*.

[F] 415.6.1 Rooms for flammable or combustible liquid use, dispensing or mixing in open systems. Rooms for *flammable or combustible liquid use*, dispensing or mixing in open systems having a floor area of not more than 500 square feet (46.5 m^2) need not be located on the outer perimeter of the building where they are in accordance with the *International Fire Code* and NFPA 30.

[F] 415.6.2 Liquid storage rooms and rooms for flammable or combustible liquid use in closed systems. *Liquid storage rooms* and rooms for *flammable or combustible liquid use* in closed systems, having a floor area of not more than 1,000 square feet (93 m^2) need not be located on the outer perimeter where they are in accordance with the *International Fire Code* and NFPA 30.

[F] 415.6.3 Spray paint booths. Spray paint booths that comply with the *International Fire Code* need not be located on the outer perimeter.

[F] 415.6.4 Group H occupancy minimum fire separation distance. Regardless of any other provisions, buildings containing Group H occupancies shall be set back to the *minimum fire separation distance* as set forth in Sections 415.6.4.1 through 415.6.4.4. Distances shall be measured from the walls enclosing the occupancy to *lot lines*, including those on a public way. Distances to assumed *lot lines* established for the purpose of determining *exterior wall* and opening protection are not to be used to establish the minimum *fire separation distance* for buildings on sites where explosives are manufactured or used where separation is provided in accordance with the quantity distance tables specified for *explosive* materials in the *International Fire Code*.

[F] 415.6.4.1 Group H-1. Group H-1 occupancies shall be set back not less than 75 feet (22 860 mm) and not less than required by the *International Fire Code*.

Exception: *Fireworks* manufacturing buildings separated in accordance with NFPA 1124.

[F] 415.6.4.2 Group H-2. Group H-2 occupancies shall be set back not less than 30 feet (9144 mm) where the area of the occupancy is greater than 1,000 square feet (93 m^2) and it is not required to be located in a *detached building*.

[F] 415.6.4.3 Groups H-2 and H-3. Group H-2 and H-3 occupancies shall be set back not less than 50 feet (15 240 mm) where a *detached building* is required (see Table 415.6.5).

[F] 415.6.4.4 Explosive materials. Group H-2 and H-3 occupancies containing materials with *explosive* characteristics shall be separated as required by the *International Fire Code*. Where separations are not specified, the distances required shall be determined by a technical report issued in accordance with Section 414.1.3.

[F] 415.6.5 Detached buildings for Group H-1, H-2 or H-3 occupancy. The storage or use of hazardous materials in excess of those amounts specified in Table 415.6.5 shall be in accordance with the applicable provisions of Sections 415.7 and 415.8.

[F] 415.6.5.1 Wall and opening protection. Where a *detached building* is required by Table 415.6.5, wall and opening protection based on *fire separation distance* is not required.

[F] 415.7 Special provisions for Group H-1 occupancies. Group H-1 occupancies shall be in detached buildings not used for other purposes. Roofs shall be of lightweight construction with suitable thermal insulation to prevent sensitive material from reaching its decomposition temperature. Group H-1 occupancies containing materials that are in themselves both physical and health hazards in quantities exceeding the maximum allowable quantities per *control area* in Table 307.1(2) shall comply with requirements for both Group H-1 and H-4 occupancies.

[F] 415.7.1 Floors in storage rooms. Floors in storage areas for organic peroxides, *pyrophoric* materials and unstable (reactive) materials shall be of liquid-tight, noncombustible construction.

[F] 415.8 Special provisions for Group H-2 and H-3 occupancies. Group H-2 and H-3 occupancies containing quantities of hazardous materials in excess of those set forth in Table 415.6.5 shall be in *detached buildings* used for manufacturing, processing, dispensing, use or storage of hazardous materials. Materials specified for Group H-1 occupancies in Section 307.3 are permitted to be located within Group H-2 or H-3 *detached buildings* provided that the amount of materials per *control area* do not exceed the maximum allowed quantity specified in Table 307.1(1).

[F] 415.8.1 Multiple hazards. Group H-2 or H-3 occupancies containing materials that are in themselves both physical and health hazards in quantities exceeding the maximum allowable quantities per *control area* in Table 307.1(2) shall comply with requirements for Group H-2, H-3 or H-4 occupancies as applicable.

[F] 415.8.2 Separation of incompatible materials. Hazardous materials other than those specified in Table 415.6.5 shall be allowed in manufacturing, processing, dispensing, use or storage areas when separated from *incompatible materials* in accordance with the provisions of the *International Fire Code*.

[F] 415.8.3 Water reactives. Group H-2 and H-3 occupancies containing water-reactive materials shall be resistant to water penetration. Piping for conveying liquids shall not be over or through areas containing water reactives, unless isolated by *approved* liquid-tight construction.

Exception: Fire protection piping shall be permitted over or through areas containing water reactives without isolating it with liquid-tight construction.

[F] 415.8.4 Floors in storage rooms. Floors in storage areas for organic peroxides, oxidizers, *pyrophoric* materials, unstable (reactive) materials and water-reactive solids and liquids shall be of liquid-tight, noncombustible construction.

[F] 415.8.5 Waterproof room. Rooms or areas used for the storage of water-reactive solids and liquids shall be constructed in a manner that resists the penetration of water through the use of waterproof materials. Piping carrying water for other than *approved automatic sprinkler systems* shall not be within such rooms or areas.

[F] 415.9 Group H-2. Occupancies in Group H-2 shall be constructed in accordance with Sections 415.9.1 through 415.9.3 and the *International Fire Code.*

[F] 415.9.1 Flammable and combustible liquids. The storage, handling, processing and transporting of flammable and combustible liquids in Group H-2 and H-3 occupancies shall be in accordance with Sections 415.9.1.1 through 415.9.1.9, the *International Mechanical Code* and the *International Fire Code.*

[F] 415.9.1.1 Mixed occupancies. Where the storage tank area is located in a building of two or more occupancies and the quantity of liquid exceeds the maximum allowable quantity for one *control area*, the use shall be completely separated from adjacent occupancies in accordance with the requirements of Section 508.4.

[F] 415.9.1.1.1 Height exception. Where storage tanks are located within a building not more than one *story above grade plane*, the height limitation of Section 504 shall not apply for Group H.

[F] 415.9.1.2 Tank protection. Storage tanks shall be noncombustible and protected from physical damage. *Fire barriers* or *horizontal assemblies* or both around the storage tanks shall be permitted as the method of protection from physical damage.

[F] 415.9.1.3 Tanks. Storage tanks shall be approved tanks conforming to the requirements of the *International Fire Code.*

[F] 415.9.1.4 Leakage containment. A liquid-tight containment area compatible with the stored liquid shall be provided. The method of spill control, drainage control and secondary containment shall be in accordance with the *International Fire Code.*

Exception: Rooms where only double-wall storage tanks conforming to Section 415.9.1.3 are used to store Class I, II and IIIA flammable and combustible liquids shall not be required to have a leakage containment area.

[F] TABLE 415.6.5
DETACHED BUILDING REQUIRED

A DETACHED BUILDING IS REQUIRED WHERE THE QUANTITY OF MATERIAL EXCEEDS THAT SPECIFIED HEREIN			
Material	**Class**	**Solids and Liquids (tons)[a, b]**	**Gases (cubic feet)[a, b]**
Explosives	Division 1.1	Maximum Allowable Quantity	Not Applicable
	Division 1.2	Maximum Allowable Quantity	
	Division 1.3	Maximum Allowable Quantity	
	Division 1.4	Maximum Allowable Quantity	
	Division 1.4[c]	1	
	Division 1.5	Maximum Allowable Quantity	
	Division 1.6	Maximum Allowable Quantity	
Oxidizers	Class 4	Maximum Allowable Quantity	Maximum Allowable Quantity
Unstable (reactives) detonable	Class 3 or 4	Maximum Allowable Quantity	Maximum Allowable Quantity
Oxidizer, liquids and solids	Class 3	1,200	Not Applicable
	Class 2	2,000	Not Applicable
Organic peroxides	Detonable	Maximum Allowable Quantity	Not Applicable
	Class I	Maximum Allowable Quantity	Not Applicable
	Class II	25	Not Applicable
	Class III	50	Not Applicable
Unstable (reactives) nondetonable	Class 3	1	2,000
	Class 2	25	10,000
Water reactives	Class 3	1	Not Applicable
	Class 2	25	Not Applicable
Pyrophoric gases[d]	Not Applicable	Not Applicable	2,000

For SI: 1 ton = 906 kg, 1 cubic foot = 0.02832 m^3, 1 pound = 0.454 kg.

a. For materials that are detonable, the distance to other buildings or lot lines shall be in accordance with Section 415.6 of this code or Chapter 56 of the *International Fire Code* based on trinitrotoluene (TNT) equivalence of the material, whichever is greater.

b. "Maximum Allowable Quantity" means the maximum allowable quantity per control area set forth in Table 307.1(1).

c. Limited to Division 1.4 materials and articles, including articles packaged for shipment, that are not regulated as an explosive under Bureau of Alcohol, Tobacco, Firearms and Explosives (BATF) regulations or unpackaged articles used in process operations that do not propagate a detonation or deflagration between articles, provided that the net explosive weight of individual articles does not exceed 1 pound.

d. Detached buildings are not required, for gases in gas rooms that support H-5 fabrication facilities where the gas room is separated from other areas by a fire barrier with a fire-resistance rating of not less than 2 hours and the gas is located in a gas cabinet that is internally sprinklered, equipped with continuous leak detection, automatic shutdown and is not manifolded upstream of pressure controls. Additionally, the gas supply is limited to cylinders that do not exceed 125 pounds (57 kg) water capacity in accordance with 49 CFR 173.192 for Hazard Zone A toxic gases.

[F] 415.9.1.5 Leakage alarm. An *approved* automatic alarm shall be provided to indicate a leak in a storage tank and room. The alarm shall sound an audible signal, 15 dBa above the ambient sound level, at every point of entry into the room in which the leaking storage tank is located. An *approved* sign shall be posted on every entry door to the tank storage room indicating the potential hazard of the interior room environment, or the sign shall state, "WARNING, WHEN ALARM SOUNDS, THE ENVIRONMENT WITHIN THE ROOM MAY BE HAZARDOUS." The leakage alarm shall be supervised in accordance with Chapter 9 to transmit a trouble signal.

[F] 415.9.1.6 Tank vent. Storage tank vents for Class I, II or IIIA liquids shall terminate to the outdoor air in accordance with the *International Fire Code.*

[F] 415.9.1.7 Room ventilation. Storage tank areas storing Class I, II or IIIA liquids shall be provided with mechanical *ventilation*. The mechanical *ventilation* system shall be in accordance with the *International Mechanical Code* and the *International Fire Code*.

[F] 415.9.1.8 Explosion venting. Where Class I liquids are being stored, explosion venting shall be provided in accordance with the *International Fire Code*.

[F] 415.9.1.9 Tank openings other than vents. Tank openings other than vents from tanks inside buildings shall be designed to ensure that liquids or vapor concentrations are not released inside the building.

[F] 415.9.2 Liquefied petroleum gas facilities. The construction and installation of liquefied petroleum gas facilities shall be in accordance with the requirements of this code, the *International Fire Code*, the *International Fuel Gas Code*, the *International Mechanical Code* and NFPA 58.

[F] 415.9.3 Dry cleaning plants. The construction and installation of dry cleaning plants shall be in accordance with the requirements of this code, the *International Mechanical Code*, the *International Plumbing Code* and NFPA 32. Dry cleaning solvents and systems shall be classified in accordance with the *International Fire Code*.

[F] 415.10 Groups H-3 and H-4. Groups H-3 and H-4 shall be constructed in accordance with the applicable provisions of this code and the *International Fire Code*.

[F] 415.10.1 Flammable and combustible liquids. The storage, handling, processing and transporting of flammable and combustible liquids in Group H-3 occupancies shall be in accordance with Section 415.9.1.

[F] 415.10.2 Gas rooms. Where gas rooms are provided, such rooms shall be separated from other areas by not less than 1-hour *fire barriers* constructed in accordance with Section 707 or *horizontal assemblies* constructed in accordance with Section 711, or both.

[F] 415.10.3 Floors in storage rooms. Floors in storage areas for *corrosive* liquids and *highly toxic* or toxic materials shall be of liquid-tight, noncombustible construction.

[F] 415.10.4 Separation of highly toxic solids and liquids. *Highly toxic* solids and liquids not stored in *approved* hazardous materials storage cabinets shall be isolated from other hazardous materials storage by not less than 1-hour *fire barriers* constructed in accordance with Section 707 or *horizontal assemblies* constructed in accordance with Section 711, or both.

[F] 415.11 Group H-5. In addition to the requirements set forth elsewhere in this code, Group H-5 shall comply with the provisions of Sections 415.11.1 through 415.11.12 and the *International Fire Code.*

[F] 415.11.1 Fabrication areas. *Fabrication areas* shall comply with Sections 415.11.1.1 through 415.11.1.8.

[F] 415.11.1.1 Hazardous materials. Hazardous materials and hazardous production materials (HPM) shall comply with Sections 415.11.1.1.1 and 415.11.1.1.2.

[F] 415.11.1.1.1 Aggregate quantities. The aggregate quantities of hazardous materials stored and used in a single *fabrication area* shall not exceed the quantities set forth in Table 415.11.1.1.1.

Exception: The quantity limitations for any hazard category in Table 415.11.1.1.1 shall not apply where the *fabrication area* contains quantities of hazardous materials not exceeding the maximum allowable quantities per *control area* established by Tables 307.1(1) and 307.1(2).

[F] 415.11.1.1.2 Hazardous production materials. The maximum quantities of hazardous production materials (HPM) stored in a single *fabrication area* shall not exceed the maximum allowable quantities per *control area* established by Table 307.1(1) and Table 307.1(2).

[F] 415.11.1.2 Separation. *Fabrication areas*, whose sizes are limited by the quantity of hazardous materials allowed by Table 415.11.1.1.1, shall be separated from each other, from *corridors* and from other parts of the building by not less than 1-hour *fire barriers* constructed in accordance with Section 707 or *horizontal assemblies* constructed in accordance with Section 711, or both.

Exceptions:

1. Doors within such *fire barrier* walls, including doors to *corridors*, shall be only *self-closing fire door assemblies* having a *fire protection rating* of not less than $^3/_4$ hour.
2. Windows between *fabrication areas* and *corridors* are permitted to be fixed glazing *listed* and labeled for a *fire protection rating* of not less than $^3/_4$ hour in accordance with Section 716.

[F] 415.11.1.3 Location of occupied levels. Occupied levels of *fabrication areas* shall be located at or above the first *story above grade plane*.

[F] TABLE 415.11.1.1.1
QUANTITY LIMITS FOR HAZARDOUS MATERIALS IN A SINGLE FABRICATION AREA IN GROUP H-5[a]

HAZARD CATEGORY		SOLIDS (pounds per square foot)	LIQUIDS (gallons per square foot)	GAS (cubic feet @ NTP/square foot)
PHYSICAL-HAZARD MATERIALS				
Combustible dust		Note b	Not Applicable	Not Applicable
Combustible fiber	Loose	Note b	Not Applicable	Not Applicable
	Baled	Notes b and c		
Combustible liquid	II	Not Applicable	0.01	Not Applicable
	IIIA		0.02	
	IIIB		Not Limited	
Combination Class	I, II and IIIA		0.04	
Cryogenic gas	Flammable	Not Applicable	Not Applicable	Note d
	Oxidizing			1.25
Explosives		Note b	Note b	Note b
Flammable gas	Gaseous	Not Applicable	Not Applicable	Note d
	Liquefied			Note d
Flammable liquid	IA	Not Applicable	0.0025	Not Applicable
	IB		0.025	
	IC		0.025	
Combination Class	IA, IB and IC		0.025	
Combination Class	I, II and IIIA		0.04	
Flammable solid		0.001	Not Applicable	Not Applicable
Organic peroxide	Unclassified detonable	Note b	Not Applicable	Not Applicable
	Class I	Note b		
	Class II	0.025		
	Class III	0.1		
	Class IV	Not Limited		
	Class V	Not Limited		
Oxidizing gas	Gaseous	Not Applicable	Not Applicable	1.25
	Liquefied			1.25
Combination of gaseous and liquefied				1.25
Oxidizer	Class 4	Note b	Note b	Not Applicable
	Class 3	0.003	0.03	
	Class 2	0.003	0.03	
	Class 1	0.003	0.03	
Combination Class	1, 2, 3	0.003	0.03	
Pyrophoric materials		0.01	0.00125	Notes d and e
Unstable (reactive)	Class 4	Note b	Note b	Note b
	Class 3	0.025	0.0025	Note b
	Class 2	0.1	0.01	Note b
	Class 1	Not Limited	Not Limited	Not Limited
Water reactive	Class 3	0.01[f]	0.00125	Not Applicable
	Class 2	0.25	0.025	
	Class 1	Not Limited	Not Limited	
HEALTH-HAZARD MATERIALS				
Corrosives		Not Limited	Not Limited	Not Limited
Highly toxic		Not Limited	Not Limited	Note d
Toxics		Not Limited	Not Limited	Note d

For SI: 1 pound = 0.454 kg, 1 pound per square foot = 4.882 kg/m^2, 1 gallon per square foot = 40.7 L/m^2, 1 cubic foot @ NTP/square foot = 0.305 m^3 @ NTP/m^2, 1 cubic foot = 0.02832 m^3.

a. Hazardous materials within piping shall not be included in the calculated quantities.
b. Quantity of hazardous materials in a single fabrication shall not exceed the maximum allowable quantities per control area in Tables 307.1(1) and 307.1(2).
c. Densely packed baled cotton that complies with the packing requirements of ISO 8115 shall not be included in this material class.
d. The aggregate quantity of flammable, pyrophoric, toxic and highly toxic gases shall not exceed the greater of 0.2 cubic feet at NTP/square foot or 9,000 cubic feet at NTP.
e. The aggregate quantity of pyrophoric gases in the building shall not exceed the amounts set forth in Table 415.6.5.
f. Quantity of Class 3 water-reactive solids in a single tool shall not exceed 1 pound.

[F] 415.11.1.4 Floors. Except for surfacing, floors within *fabrication areas* shall be of noncombustible construction.

Openings through floors of *fabrication areas* are permitted to be unprotected where the interconnected levels are used solely for mechanical equipment directly related to such *fabrication areas* (see Section 415.11.1.5).

Floors forming a part of an occupancy separation shall be liquid tight.

[F] 415.11.1.5 Shafts and openings through floors. Elevator hoistways, vent *shafts* and other openings through floors shall be enclosed where required by Sections 712 and 713. Mechanical, duct and piping penetrations within a *fabrication area* shall not extend through more than two floors. The *annular space* around penetrations for cables, cable trays, tubing, piping, conduit or ducts shall be sealed at the floor level to restrict the movement of air. The *fabrication area*, including the areas through which the ductwork and piping extend, shall be considered to be a single conditioned environment.

[F] 415.11.1.6 Ventilation. Mechanical exhaust *ventilation* at the rate of not less than 1 cubic foot per minute per square foot [0.0051 $m^3/(s \times m^2)$] of floor area shall be provided throughout the portions of the *fabrication area* where HPM are used or stored. The exhaust air duct system of one *fabrication area* shall not connect to another duct system outside that *fabrication area* within the building.

A *ventilation* system shall be provided to capture and exhaust gases, fumes and vapors at workstations.

Two or more operations at a *workstation* shall not be connected to the same exhaust system where either one or the combination of the substances removed could constitute a fire, explosion or hazardous chemical reaction within the exhaust duct system.

Exhaust ducts penetrating *fire barriers* constructed in accordance with Section 707 or *horizontal assemblies* constructed in accordance with Section 711 shall be contained in a *shaft* of equivalent fire-resistance-rated construction. Exhaust ducts shall not penetrate *fire walls*.

Fire dampers shall not be installed in exhaust ducts.

[F] 415.11.1.7 Transporting hazardous production materials to fabrication areas. HPM shall be transported to *fabrication areas* through enclosed piping or tubing systems that comply with Section 415.11.7, through *service corridors* complying with Section 415.11.3, or in *corridors* as permitted in the exception to Section 415.11.2. The handling or transporting of HPM within *service corridors* shall comply with the *International Fire Code*.

[F] 415.11.1.8 Electrical. Electrical equipment and devices within the *fabrication area* shall comply with NFPA 70. The requirements for hazardous locations need not be applied where the average air change is not less than four times that set forth in Section 415.11.1.6 and where the number of air changes at any location is not less than three times that required by Section 415.11.1.6. The use of recirculated air shall be permitted.

[F] 415.11.1.8.1 Workstations. Workstations shall not be energized without adequate exhaust *ventilation*. See Section 415.11.1.6 for workstation exhaust *ventilation requirements*.

[F] 415.11.2 Corridors. *Corridors* shall comply with Chapter 10 and shall be separated from *fabrication areas* as specified in Section 415.11.1.2. *Corridors* shall not contain HPM and shall not be used for transporting such materials except through closed piping systems as provided in Section 415.11.7.4.

Exception: Where existing *fabrication areas* are altered or modified, HPM is allowed to be transported in existing *corridors*, subject to the following conditions:

1. Nonproduction HPM is allowed to be transported in *corridors* if utilized for maintenance, lab work and testing.
2. Where existing *fabrication areas* are altered or modified, HPM is allowed to be transported in existing *corridors*, subject to the following conditions:
   - 2.1. Corridors. *Corridors* adjacent to the *fabrication area* where the alteration work is to be done shall comply with Section 1020 for a length determined as follows:
     - 2.1.1. The length of the common wall of the *corridor* and the *fabrication area*; and
     - 2.1.2. For the distance along the *corridor* to the point of entry of HPM into the *corridor* serving that *fabrication area*.
   - 2.2. *Emergency alarm system*. There shall be an emergency telephone system, a local manual alarm station or other *approved* alarm-initiating device within corridors at not more than 150-foot (45 720 mm) intervals and at each exit and doorway. The signal shall be relayed to an *approved* central, proprietary or remote station service or the *emergency control station* and shall initiate a local audible alarm.
   - 2.3. Pass-throughs. *Self-closing* doors having a *fire protection rating* of not less than 1 hour shall separate pass-throughs from existing *corridors*. Pass-throughs shall be constructed as required for the corridors and protected by an *approved automatic sprinkler system*.

[F] 415.11.3 Service corridors. *Service corridors* within a Group H-5 occupancy shall comply with Sections 415.11.3.1 through 415.11.3.4.

[F] 415.11.3.1 Use conditions. *Service corridors* shall be separated from *corridors* as required by Section 415.11.1.2. *Service corridors* shall not be used as a required *corridor*.

[F] 415.11.3.2 Mechanical ventilation. Service corridors shall be mechanically ventilated as required by Section 415.11.1.6 or at not less than six air changes per hour.

[F] 415.11.3.3 Means of egress. The distance of travel from any point in a *service corridor* to an *exit*, *exit access corridor* or door into a *fabrication area* shall be not greater than 75 feet (22 860 mm). Dead ends shall be not greater than 4 feet (1219 mm) in length. There shall be not less than two *exits*, and not more than one-half of the required *means of egress* shall require travel into a *fabrication area*. Doors from *service corridors* shall swing in the direction of egress travel and shall be *self-closing*.

[F] 415.11.3.4 Minimum width. The clear width of a *service corridor* shall be not less than 5 feet (1524 mm), or 33 inches (838 mm) wider than the widest cart or truck used in the *service corridor*, whichever is greater.

[F] 415.11.4 Emergency alarm system. *Emergency alarm systems* shall be provided in accordance with this section and Sections 415.5.1 and 415.5.2. The maximum allowable quantity per *control area* provisions shall not apply to *emergency alarm systems* required for HPM.

[F] 415.11.4.1 Service corridors. An *emergency alarm system* shall be provided in *service corridors*, with not fewer than one alarm device in each *service corridor*.

[F] 415.11.4.2 Corridors and interior exit stairways and ramps. Emergency alarms for *corridors, interior exit stairways* and *ramps* and *exit passageways* shall comply with Section 415.5.2.

[F] 415.11.4.3 Liquid storage rooms, HPM rooms and gas rooms. Emergency alarms for liquid storage rooms, HPM rooms and gas rooms shall comply with Section 415.5.1.

[F] 415.11.4.4 Alarm-initiating devices. An *approved* emergency telephone system, local alarm manual pull stations, or other *approved* alarm-initiating devices are allowed to be used as emergency alarm-initiating devices.

[F] 415.11.4.5 Alarm signals. Activation of the *emergency alarm system* shall sound a local alarm and transmit a signal to the *emergency control station*.

[F] 415.11.5 Storage of hazardous production materials. Storage of hazardous production materials (HPM) in *fabrication areas* shall be within *approved* or *listed* storage cabinets or gas cabinets or within a *workstation*. The storage of HPM in quantities greater than those specified in Section 5004.2 of the *International Fire Code* shall be in liquid storage rooms, HPM rooms or gas rooms as appropriate for the materials stored. The storage of other hazardous materials shall be in accordance with other applicable provisions of this code and the *International Fire Code*.

[F] 415.11.6 HPM rooms, gas rooms, *liquid storage room* construction. HPM rooms, gas rooms and liquid shall be constructed in accordance with Sections 415.11.6.1 through 415.11.6.9.

[F] 415.11.6.1 HPM rooms and gas rooms. HPM rooms and gas rooms shall be separated from other areas by *fire barriers* constructed in accordance with Section 707 or *horizontal assemblies* constructed in accordance with Section 711, or both. The *fire-resistance rating* shall be not less than 2 hours where the area is 300 square feet (27.9 m^2) or more and not less than 1 hour where the area is less than 300 square feet (27.9 m^2).

[F] 415.11.6.2 Liquid storage rooms. Liquid storage rooms shall be constructed in accordance with the following requirements:

1. Rooms greater than 500 square feet (46.5 m^2) in area, shall have not fewer than one exterior door *approved* for fire department access.
2. Rooms shall be separated from other areas by *fire barriers* constructed in accordance with Section 707 or *horizontal assemblies* constructed in accordance with Section 711, or both. The *fire-resistance rating* shall be not less than 1 hour for rooms up to 150 square feet (13.9 m^2) in area and not less than 2 hours where the room is more than 150 square feet (13.9 m^2) in area.
3. Shelving, racks and wainscotting in such areas shall be of noncombustible construction or wood of not less than 1-inch (25 mm) nominal thickness or *fire-retardant-treated wood* complying with Section 2303.2.
4. Rooms used for the storage of Class I flammable liquids shall not be located in a *basement*.

[F] 415.11.6.3 Floors. Except for surfacing, floors of HPM rooms and liquid storage rooms shall be of noncombustible liquid-tight construction. Raised grating over floors shall be of noncombustible materials.

[F] 415.11.6.4 Location. Where HPM rooms, liquid storage rooms and gas rooms are provided, they shall have not fewer than one *exterior wall* and such wall shall be not less than 30 feet (9144 mm) from *lot lines*, including *lot lines* adjacent to *public ways*.

[F] 415.11.6.5 Explosion control. Explosion control shall be provided where required by Section 414.5.1.

[F] 415.11.6.6 Exits. Where two *exits* are required from HPM rooms, liquid storage rooms and gas rooms, one shall be directly to the outside of the building.

[F] 415.11.6.7 Doors. Doors in a *fire barrier* wall, including doors to *corridors*, shall be *self-closing fire door assemblies* having a *fire protection rating* of not less than $^3/_4$ hour.

[F] 415.11.6.8 Ventilation. Mechanical exhaust ventilation shall be provided in liquid storage rooms, HPM rooms and gas rooms at the rate of not less than 1 cubic foot per minute per square foot (0.044 L/s/m^2) of floor area or six air changes per hour.

Exhaust ventilation for gas rooms shall be designed to operate at a negative pressure in relation to the surrounding areas and direct the exhaust ventilation to an exhaust system.

[F] 415.11.6.9 Emergency alarm system. An *approved emergency alarm system* shall be provided for HPM rooms, liquid storage rooms and gas rooms.

Emergency alarm-initiating devices shall be installed outside of each interior exit door of such rooms.

Activation of an emergency alarm-initiating device shall sound a local alarm and transmit a signal to the *emergency control station*.

An *approved* emergency telephone system, local alarm manual pull stations or other *approved* alarm-initiating devices are allowed to be used as emergency alarm-initiating devices.

[F] 415.11.7 Piping and tubing. Hazardous production materials piping and tubing shall comply with this section and ASME B31.3.

[F] 415.11.7.1 HPM having a health-hazard ranking of 3 or 4. Systems supplying HPM liquids or gases having a health-hazard ranking of 3 or 4 shall be welded throughout, except for connections, to the systems that are within a ventilated enclosure if the material is a gas, or an *approved* method of drainage or containment is provided for the connections if the material is a liquid.

[F] 415.11.7.2 Location in service corridors. Hazardous production materials supply piping or tubing in *service corridors* shall be exposed to view.

[F] 415.11.7.3 Excess flow control. Where HPM gases or liquids are carried in pressurized piping above 15 pounds per square inch gauge (psig) (103.4 kPa), excess flow control shall be provided. Where the piping originates from within a *liquid storage room*, *HPM room* or *gas room*, the excess flow control shall be located within the *liquid storage room*, *HPM room* or *gas room*. Where the piping originates from a bulk source, the excess flow control shall be located as close to the bulk source as practical.

[F] 415.11.7.4 Installations in corridors and above other occupancies. The installation of HPM piping and tubing within the space defined by the walls of corridors and the floor or roof above, or in concealed spaces above other occupancies, shall be in accordance with Sections 415.11.7.1 through 415.11.7.3 and the following conditions:

1. Automatic sprinklers shall be installed within the space unless the space is less than 6 inches (152 mm) in the least dimension.
2. *Ventilation* not less than six air changes per hour shall be provided. The space shall not be used to convey air from any other area.
3. Where the piping or tubing is used to transport HPM liquids, a receptor shall be installed below such piping or tubing. The receptor shall be designed to collect any discharge or leakage and drain it to an *approved* location. The 1-hour enclosure shall not be used as part of the receptor.
4. HPM supply piping and tubing and nonmetallic waste lines shall be separated from the corridor and from occupancies other than Group H-5 by *fire barriers* or by an approved method or assembly that has a *fire-resistance rating* of not less than 1 hour. Access openings into the enclosure shall be protected by approved fire-protection-rated assemblies.
5. Readily accessible manual or automatic remotely activated fail-safe emergency shutoff valves shall be installed on piping and tubing other than waste lines at the following locations:
    5.1. At branch connections into the *fabrication area*.
    5.2. At entries into *corridors*.

Exception: Transverse crossings of the *corridors* by supply piping that is enclosed within a ferrous pipe or tube for the width of the *corridor* need not comply with Items 1 through 5.

[F] 415.11.7.5 Identification. Piping, tubing and HPM waste lines shall be identified in accordance with ANSI A13.1 to indicate the material being transported.

[F] 415.11.8 Gas detection systems. A *gas detection system* complying with Section 916 shall be provided for HPM gases where the *physiological warning threshold level* of the gas is at a higher level than the accepted permissible exposure limit (PEL) for the gas and for flammable gases in accordance with Sections 415.11.8.1 through 415.11.8.2.

[F] 415.11.8.1 Where required. A *gas detection system* shall be provided in the areas identified in Sections 415.11.8.1.1 through 415.11.8.1.4.

[F] 415.11.8.1.1 Fabrication areas. A *gas detection system* shall be provided in fabrication areas where HPM gas is used in the *fabrication area*.

[F] 415.11.8.1.2 HPM rooms. A *continuous gas detection system* shall be provided in HPM rooms where HPM gas is used in the room.

[F] 415.11.8.1.3 Gas cabinets, exhausted enclosures and gas rooms. A *gas detection system* shall be provided in gas cabinets and exhausted enclosures for HPM gas. A *gas detection system* shall be provided in gas rooms where HPM gases are not located in gas cabinets or exhausted enclosures.

[F] 415.11.8.1.4 Corridors. Where HPM gases are transported in piping placed within the space defined by the walls of a corridor and the floor or roof above the corridor, a *gas detection system* shall be provided where piping is located and in the corridor.

> **Exception:** A *gas detection system* is not required for occasional transverse *crossings* of the corridors by supply piping that is enclosed in a ferrous pipe or tube for the width of the corridor.

[F] 415.11.8.2 Gas detection system operation. The *gas detection system* shall be capable of monitoring the room, area or equipment in which the HPM gas is located at or below all the following gas concentrations:

1. Immediately *dangerous* to life and health (IDLH) values where the monitoring point is within an *exhausted enclosure*, ventilated enclosure or *gas cabinet*.
2. Permissible exposure limit (PEL) levels where the monitoring point is in an area outside an *exhausted enclosure*, ventilated enclosure or *gas cabinet*.
3. For flammable gases, the monitoring detection threshold level shall be vapor concentrations in excess of 25 percent of the lower flammable limit (LFL) where the monitoring is within or outside an *exhausted enclosure*, ventilated enclosure or *gas cabinet*.
4. Except as noted in this section, monitoring for *highly toxic* and toxic gases shall also comply with Chapter 60 of the *International Fire Code*.

[F] 415.11.8.2.1 Alarms. The *gas detection system* shall initiate a local alarm and transmit a signal to the *emergency control station* when a short-term hazard condition is detected. The alarm shall be both visual and audible and shall provide warning both inside and outside the area where the gas is detected. The audible alarm shall be distinct from all other alarms.

[F] 415.11.8.2.2 Shutoff of gas supply. The gas detection system shall automatically close the shutoff valve at the source on gas supply piping and tubing related to the system being monitored for which gas is detected when a short-term hazard condition is detected. Automatic closure of shutoff valves shall comply with the following:

1. Where the gas detection sampling point initiating the gas detection system alarm is within a *gas cabinet* or *exhausted enclosure*, the shutoff valve in the *gas cabinet* or *exhausted enclosure* for the specific gas detected shall automatically close.
2. Where the gas detection sampling point initiating the gas detection system alarm is within a room and compressed gas containers are not in gas cabinets or an *exhausted enclosure*, the shutoff valves on all gas lines for the specific gas detected shall automatically close.
3. Where the gas detection sampling point initiating the gas detection system alarm is within a piping distribution manifold enclosure, the shutoff valve supplying the manifold for the compressed gas container of the specific gas detected shall automatically close.

> **Exception:** Where the gas detection sampling point initiating the gas detection system alarm is at the use location or within a gas valve enclosure of a branch line downstream of a piping distribution manifold, the shutoff valve for the branch line located in the piping distribution manifold enclosure shall automatically close.

[F] 415.11.9 Manual fire alarm system. An *approved* manual *fire alarm* system shall be provided throughout buildings containing Group H-5. Activation of the alarm system shall initiate a local alarm and transmit a signal to the *emergency control station*. The *fire alarm* system shall be designed and installed in accordance with Section 907.

[F] 415.11.10 Emergency control station. An *emergency control station* shall be provided in accordance with Sections 415.11.10.1 through 415.11.10.3.

[F] 415.11.10.1 Location. The *emergency control station* shall be located on the premises at an *approved* location outside the *fabrication area*.

[F] 415.11.10.2 Staffing. Trained personnel shall continuously staff the *emergency control station*.

[F] 415.11.10.3 Signals. The *emergency control station* shall receive signals from emergency equipment and alarm and detection systems. Such emergency equipment and alarm and detection systems shall include, but not be limited to, the following where such equipment or systems are required to be provided either in this chapter or elsewhere in this code:

1. *Automatic sprinkler system* alarm and monitoring systems.
2. Manual *fire alarm* systems.
3. *Emergency alarm systems*.
4. *Gas detection systems*.
5. Smoke detection systems.
6. Emergency power system.

7. Automatic detection and alarm systems for *pyrophoric* liquids and Class 3 water-reactive liquids required in Section 2705.2.3.4 of the *International Fire Code*.
8. Exhaust *ventilation* flow alarm devices for *pyrophoric* liquids and Class 3 water-reactive liquids cabinet exhaust *ventilation* systems required in Section 2705.2.3.4 of the *International Fire Code*.

[F] 415.11.11 Emergency power system. An emergency power system shall be provided in Group H-5 occupancies in accordance with Section 2702. The emergency power system shall supply power automatically to the electrical systems specified in Section 415.11.11.1 when the normal electrical supply system is interrupted.

[F] 415.11.11.1 Required electrical systems. Emergency power shall be provided for electrically operated equipment and connected control circuits for the following systems:

1. HPM exhaust *ventilation systems*.
2. HPM *gas cabinet ventilation systems*.
3. HPM *exhausted enclosure ventilation systems*.
4. HPM *gas room ventilation systems*.
5. HPM gas detection systems.
6. *Emergency alarm systems*.
7. Manual and automatic *fire alarm systems*.
8. *Automatic sprinkler system* monitoring and alarm systems.
9. Automatic alarm and detection systems for *pyrophoric* liquids and Class 3 water-reactive liquids required in Section 2705.2.3.4 of the *International Fire Code*.
10. Flow alarm switches for *pyrophoric* liquids and Class 3 water-reactive liquids cabinet exhaust *ventilation systems* required in Section 2705.2.3.4 of the *International Fire Code*.
11. Electrically operated systems required elsewhere in this code or in the *International Fire Code* applicable to the use, storage or handling of HPM.

[F] 415.11.11.2 Exhaust ventilation systems. Exhaust *ventilation* systems are allowed to be designed to operate at not less than one-half the normal fan speed on the emergency power system where it is demonstrated that the level of exhaust will maintain a safe atmosphere.

[F] 415.11.12 Automatic sprinkler system protection in exhaust ducts for HPM. An *approved automatic sprinkler system* shall be provided in exhaust ducts conveying gases, vapors, fumes, mists or dusts generated from HPM in accordance with Sections 415.11.12.1 through 415.11.12.3 and the *International Mechanical Code*.

[F] 415.11.12.1 Metallic and noncombustible nonmetallic exhaust ducts. An *approved automatic sprinkler system* shall be provided in metallic and noncombustible nonmetallic exhaust ducts where all of the following conditions apply:

1. Where the largest cross-sectional diameter is equal to or greater than 10 inches (254 mm).
2. The ducts are within the building.
3. The ducts are conveying flammable gases, vapors or fumes.

[F] 415.11.12.2 Combustible nonmetallic exhaust ducts. *Automatic sprinkler system* protection shall be provided in combustible nonmetallic exhaust ducts where the largest cross-sectional diameter of the duct is equal to or greater than 10 inches (254 mm).

Exception: Ducts need not be provided with automatic sprinkler protection as follows:

1. Ducts *listed* or *approved* for applications without *automatic sprinkler system* protection.
2. Ducts not more than 12 feet (3658 mm) in length installed below ceiling level.

[F] 415.11.12.3 Automatic sprinkler locations. Sprinkler systems shall be installed at 12-foot (3658 mm) intervals in horizontal ducts and at changes in direction. In vertical ducts, sprinklers shall be installed at the top and at alternate floor levels.

SECTION 416
SPRAY APPLICATION OF FLAMMABLE FINISHES

[F] 416.1 General. The provisions of this section shall apply to the construction, installation and use of buildings and structures, or parts thereof, for the spray application of flammable finishes. Operations and equipment shall comply with the *International Fire Code*.

[F] 416.2 Spray rooms. *Spray rooms* shall be enclosed with not less than 1-hour *fire barriers* constructed in accordance with Section 707 or *horizontal assemblies* constructed in accordance with Section 711, or both. Floors shall be waterproofed and drained in an *approved* manner.

[F] 416.2.1 Construction. Walls and ceilings of *spray rooms* shall be constructed of noncombustible materials or the interior surface shall be completely covered with noncombustible materials. Aluminum shall not be used.

[F] 416.2.2 Surfaces. The *interior surfaces* of *spray rooms* shall be smooth and shall be so constructed to permit the free passage of exhaust air from all parts of the interior and to facilitate washing and cleaning, and shall be so designed to confine residues within the room.

[F] 416.2.3 Ventilation. Mechanical *ventilation* and interlocks with the spraying operation shall be in accordance with the *International Fire Code* and *International Mechanical Code*.

[F] 416.3 Spraying spaces. Spraying spaces shall be ventilated with an exhaust system to prevent the accumulation of flammable mist or vapors in accordance with the *International Mechanical Code*. Where such spaces are not separately enclosed, noncombustible spray curtains shall be provided to restrict the spread of flammable vapors.

[F] 416.3.1 Surfaces. The *interior surfaces* of spraying spaces shall be smooth; shall be so constructed to permit the free passage of exhaust air from all parts of the interior and to facilitate washing and cleaning; and shall be so designed to confine residues within the spraying space. Aluminum shall not be used.

[F] 416.4 Spray booths. Spray booths shall be designed, constructed and operated in accordance with the *International Fire Code.*

[F] 416.5 Fire protection. An *automatic sprinkler system* or *fire-extinguishing system* shall be provided in all spray rooms and spray booths, and shall be installed in accordance with Chapter 9.

SECTION 417 DRYING ROOMS

[F] 417.1 General. A drying room or dry kiln installed within a building shall be constructed entirely of *approved* noncombustible materials or assemblies of such materials regulated by the *approved* rules or as required in the general and specific sections of this chapter for special occupancies and where applicable to the general requirements of the *International Mechanical Code.*

[F] 417.2 Piping clearance. Overhead heating pipes shall have a clearance of not less than 2 inches (51 mm) from combustible contents in the dryer.

[F] 417.3 Insulation. Where the operating temperature of the dryer is 175°F (79°C) or more, metal enclosures shall be insulated from adjacent combustible materials by not less than 12 inches (305 mm) of airspace, or the metal walls shall be lined with $^{1}/_{4}$-inch (6.4 mm) insulating mill board or other approved equivalent insulation.

[F] 417.4 Fire protection. Drying rooms designed for high-hazard materials and processes, including special occupancies as provided for in Chapter 4, shall be protected by an *approved automatic fire-extinguishing system* complying with the provisions of Chapter 9.

SECTION 418 ORGANIC COATINGS

[F] 418.1 Building features. Manufacturing of organic coatings shall be done only in buildings that do not have pits or *basements.*

[F] 418.2 Location. Organic coating manufacturing operations and operations incidental to or connected therewith shall not be located in buildings having other occupancies.

[F] 418.3 Process mills. Mills operating with close clearances and that process flammable and heat-sensitive materials, such as nitrocellulose, shall be located in a *detached building* or noncombustible structure.

[F] 418.4 Tank storage. Storage areas for *flammable and combustible liquid* tanks inside of structures shall be located at or above grade and shall be separated from the processing area by not less than 2-hour *fire barriers* constructed in accordance with Section 707 or *horizontal assemblies* constructed in accordance with Section 711, or both.

[F] 418.5 Nitrocellulose storage. Nitrocellulose storage shall be located on a detached pad or in a separate structure or a room enclosed with not less than 2-hour *fire barriers* constructed in accordance with Section 707 or *horizontal assemblies* constructed in accordance with Section 711, or both.

[F] 418.6 Finished products. Storage rooms for finished products that are *flammable or combustible liquids* shall be separated from the processing area by not less than 2-hour *fire barriers* constructed in accordance with Section 707 or *horizontal assemblies* constructed in accordance with Section 711, or both.

*

SECTION 419 ARTIFICIAL DECORATIVE VEGETATION

[F] 419.1 Artificial decorative vegetation. Artificial decorative vegetation exceeding 6 feet (1830 mm) in height and permanently installed outdoors within 5 feet (1524 mm) of a building, or on the roof of a building, shall comply with Section 321.1 of the *International Fire Code.*

Exception: Artificial decorative vegetation located more than 30 feet (9144 mm) from the exterior wall of a building.

SECTION 420 GROUPS I-1, R-1, R-2, R-3 AND R-4

420.1 General. Occupancies in Groups I-1, R-1, R-2, R-3 and R-4 shall comply with the provisions of Sections 420.1 through 420.11 and other applicable provisions of this code.

420.2 Separation walls. Walls separating *dwelling units* in the same building, walls separating *sleeping units* in the same building and walls separating *dwelling* or *sleeping units* from other occupancies contiguous to them in the same building shall be constructed as *fire partitions* in accordance with Section 708.

420.3 Horizontal separation. Floor assemblies separating *dwelling units* in the same buildings, floor assemblies separating *sleeping units* in the same building and floor assemblies separating dwelling or *sleeping units* from other occupancies contiguous to them in the same building shall be constructed as *horizontal assemblies* in accordance with Section 711.

[F] 420.4 Automatic sprinkler system. Group R occupancies shall be equipped throughout with an *automatic sprinkler system* in accordance with Section 903.2.8. Group I-1 occupancies shall be equipped throughout with an *automatic sprinkler system* in accordance with Section 903.2.6. Quick-response or residential *automatic* sprinklers shall be installed in accordance with Section 903.3.2.

[F] 420.5 Fire alarm systems and smoke alarms. *Fire alarm* systems and smoke alarms shall be provided in Group

I-1, R-1 and R-2 occupancies in accordance with Sections 907.2.6, 907.2.8 and 907.2.9, respectively. Single- or multiple-station smoke alarms shall be provided in Groups I-1, R-2, R-3 and R-4 in accordance with Section 907.2.11.

420.6 Smoke barriers in Group I-1, Condition 2. *Smoke barriers* shall be provided in Group I-1, Condition 2 to subdivide every *story* used by persons receiving care, treatment or sleeping and to provide other *stories* with an *occupant load* of 50 or more persons, into not fewer than two *smoke compartments*. Such *stories* shall be divided into *smoke compartments* with an area of not more than 22,500 square feet (2092 m^2) and the distance of travel from any point in a *smoke compartment* to a *smoke barrier* door shall not exceed 200 feet (60 960 mm). The *smoke barrier* shall be in accordance with Section 709.

420.6.1 Refuge area. Refuge areas shall be provided within each *smoke compartment*. The size of the refuge area shall accommodate the occupants and care recipients from the adjoining *smoke compartment*. Where a *smoke compartment* is adjoined by two or more *smoke compartments*, the minimum area of the refuge area shall accommodate the largest occupant load of the adjoining compartments. The size of the refuge area shall provide the following:

1. Not less than 15 net square feet (1.4 m^2) for each care recipient.
2. Not less than 6 net square feet (0.56 m^2) for other occupants.

Areas or spaces permitted to be included in the calculation of the refuge area are corridors, lounge or dining areas and other low-hazard areas.

420.7 Group I-1 assisted living housing units. In Group I-1 occupancies, where a *fire-resistance corridor* is provided in areas where assisted living residents are housed, shared living spaces, group meeting or multipurpose therapeutic spaces open to the *corridor* shall be in accordance with all of the following criteria:

1. The walls and ceilings of the space are constructed as required for *corridors*.
2. The spaces are not occupied as resident sleeping rooms, treatment rooms, incidental uses in accordance with Section 509, or hazardous uses.
3. The open space is protected by an *automatic* fire detection system installed in accordance with Section 907.
4. In Group I-1, Condition 1, the *corridors* onto which the spaces open are protected by an *automatic* fire detection system installed in accordance with Section 907, or the spaces are equipped throughout with quick-response sprinklers in accordance with Section 903.3.2.
5. In Group I-1, Condition 2, the *corridors* onto which the spaces open, in the same *smoke compartment*, are protected by an *automatic* fire detection system installed in accordance with Section 907, or the *smoke compartment* in which the spaces are located is equipped throughout with quick-response sprinklers in accordance with Section 903.3.2.
6. The space is arranged so as not to obstruct access to the required *exits*.

420.8 Group I-1 cooking facilities. In Group I-1 occupancies, rooms or spaces that contain a cooking facility with domestic cooking appliances shall be permitted to be open to the *corridor* where all of the following criteria are met:

1. In Group I-1, Condition 1 occupancies, the number of care recipients served by one cooking facility shall not be greater than 30.
2. In Group I-1, Condition 2 occupancies, the number of care recipients served by one cooking facility and within the same *smoke compartment* shall not be greater than 30.
3. The space containing the cooking facilities shall be arranged so as not to obstruct access to the required *exit*.
4. The cooking appliances shall comply with Section 420.9.

420.9 Domestic cooking appliances. In Group I-1 occupancies, installation of cooking appliance used in domestic cooking facilities shall comply with all of the following:

1. The types of cooking appliances permitted shall be limited to ovens, cooktops, ranges, warmers and microwaves.
2. Domestic cooking hoods installed and constructed in accordance with Section 505 of the International Mechanical Code shall be provided over cooktops or ranges.
3. Cooktops and ranges shall be protected in accordance with Section 904.14.
4. A shutoff for the fuel and electrical supply to the cooking equipment shall be provided in a location to which only staff has access.
5. A timer shall be provided that automatically deactivates the cooking appliances within a period of not more than 120 minutes.
6. A portable fire extinguisher shall be provided. Installation shall be in accordance with Section 906 and the extinguisher shall be located within a 30-foot (9144 mm) distance of travel from each domestic cooking appliance.

Exceptions:

1. Cooking facilities provided within care recipients' individual *dwelling units* are not required to comply with this section.
2. Cooktops and ranges used for care-recipient training or nutritional counseling are not required to comply with Item 3 of this section

420.10 Group R cooking facilities. In Group R occupancies, cooking appliances used for domestic cooking operations shall be in accordance with Section 917.2 of the *International Mechanical Code*.

420.11 Group R-2 dormitory cooking facilities. Domestic cooking appliances for use by residents of Group R-2 college *dormitories* shall be in accordance with Sections 420.11.1 and 420.11.2.

420.11.1 Cooking appliances. Where located in Group R-2 college *dormitories*, domestic cooking appliances for use by residents shall be in compliance with all of the following:

1. The types of domestic cooking appliances shall be limited to ovens, cooktops, ranges, warmers, coffee makers and microwaves.
2. Domestic cooking appliances shall be limited to approved locations.
3. Cooktops and ranges shall be protected in accordance with Section 904.14.
4. Cooktops and ranges shall be provided with a domestic cooking hood installed and constructed in accordance with Section 505 of the *International Mechanical Code*.

420.11.2 Cooking appliances in sleeping rooms. Cooktops, ranges and ovens shall not be installed or used in sleeping rooms.

SECTION 421
HYDROGEN FUEL GAS ROOMS

[F] 421.1 General. Where required by the *International Fire Code*, *hydrogen fuel gas rooms* shall be designed and constructed in accordance with Sections 421.1 through 421.7.

[F] 421.2 Location. *Hydrogen fuel gas rooms* shall not be located below grade.

[F] 421.3 Design and construction. *Hydrogen fuel gas rooms* not classified as Group H shall be separated from other areas of the building in accordance with Section 509.1.

[F] 421.3.1 Pressure control. *Hydrogen fuel gas rooms* shall be provided with a ventilation system designed to maintain the room at a negative pressure in relation to surrounding rooms and spaces.

[F] 421.3.2 Windows. Operable windows in interior walls shall not be permitted. Fixed windows shall be permitted where in accordance with Section 716.

[F] 421.4 Exhaust ventilation. *Hydrogen fuel gas rooms* shall be provided with mechanical exhaust *ventilation* in accordance with the applicable provisions of Section 502.16.1 of the *International Mechanical Code*.

[F] 421.5 Gas detection system. *Hydrogen fuel gas rooms* shall be provided with a *gas detection system* that complies with Sections 421.5.1, 421.5.2, and 916.

[F] 421.5.1 System activation. Activation of a gas detection alarm shall result in both of the following:

1. Initiation of distinct audible and visible alarm signals both inside and outside of the *hydrogen fuel gas room*.
2. *Automatic* activation of the mechanical exhaust ventilation system.

[F] 421.5.2 Failure of the gas detection system. Failure of the *gas detection system* shall automatically activate the mechanical exhaust *ventilation* system, stop hydrogen generation, and cause a trouble signal to sound at an *approved* location.

[F] 421.6 Explosion control. *Explosion* control shall be provided where required by Section 414.5.1.

[F] 421.7 Standby power. Mechanical *ventilation* and *gas detection systems* shall be provided with a *standby power system* in accordance with Section 2702.

SECTION 422
AMBULATORY CARE FACIILITIES

422.1 General. Occupancies classified as *ambulatory care facilities* shall comply with the provisions of Sections 422.1 through 422.6 and other applicable provisions of this code.

422.2 Separation. *Ambulatory care facilities* where the potential for four or more care recipients are to be *incapable of self-preservation* at any time shall be separated from adjacent spaces, *corridors* or tenants with a *fire partition* installed in accordance with Section 708.

422.3 Smoke compartments. Where the aggregate area of one or more *ambulatory care facilities* is greater than 10,000 square feet (929 m^2) on one *story*, the *story* shall be provided with a *smoke barrier* to subdivide the *story* into not fewer than two *smoke compartments*. The area of any one such *smoke compartment* shall be not greater than 22,500 square feet (2092 m^2). The distance of travel from any point in a *smoke compartment* to a *smoke barrier* door shall be not greater than 200 feet (60 960 mm). The *smoke barrier* shall be installed in accordance with Section 709 with the exception that *smoke barriers* shall be continuous from outside wall to an outside wall, a floor to a floor, or from a *smoke barrier* to a *smoke barrier* or a combination thereof.

422.3.1 Means of egress. Where *ambulatory care facilities* require smoke compartmentation in accordance with Section 422.3, the fire safety evacuation plans provided in accordance with Section 1002.2 shall identify the building components necessary to support a *defend-in-place* emergency response in accordance with Sections 403 and 404 of the *International Fire Code*.

422.3.2 Refuge area. Not less than 30 net square feet (2.8 m^2) for each nonambulatory care recipient shall be provided within the aggregate area of *corridors*, care recipient rooms, treatment rooms, lounge or dining areas and other low-hazard areas within each *smoke compartment*. Each occupant of an *ambulatory care facility* shall be provided with access to a refuge area without passing through or utilizing adjacent tenant spaces.

422.3.3 Independent egress. A *means of egress* shall be provided from each *smoke compartment* created by *smoke barriers* without having to return through the *smoke compartment* from which *means of egress* originated.

[F] 422.4 Automatic sprinkler systems. *Automatic sprinkler systems* shall be provided for *ambulatory care facilities* in accordance with Section 903.2.2.

[F] 422.5 Fire alarm systems. A *fire alarm* system shall be provided for *ambulatory care facilities* in accordance with Section 907.2.2.

[F] 422.6 Electrical systems. In *ambulatory care facilities*, the essential electrical system for electrical components, equipment and systems shall be designed and constructed in accordance with the provisions of Chapter 27 and NFPA 99.

422.7 Domestic cooking. Installation of cooking appliances used in domestic cooking facilities shall comply with all of the following:

1. The types of cooking appliances permitted are limited to ovens, cooktops, ranges, warmers and microwaves.
2. Domestic cooking hoods installed and constructed in accordance with Section 505 of the International Mechanical Code shall be provided over cooktops or ranges.
3. A shutoff for the fuel and electrical supply to the cooking equipment shall be provided in a location to which only staff has access.
4. A timer shall be provided that automatically deactivates the cooking appliances within a period of not more than 120 minutes.
5. A portable fire extinguisher shall be provided. Installation shall be in accordance with Section 906 and the extinguisher shall be located within a 30-foot (9144 mm) distance of travel from each domestic cooking appliance.

SECTION 423
STORM SHELTERS

423.1 General. This section applies to the construction of storm shelters constructed as separate detached buildings or constructed as rooms or spaces within buildings for the purpose of providing protection from storms that produce high winds, such as tornadoes and hurricanes, during the storm. This section specifies where *storm shelters* are required and provides requirements for the design and construction of *storm shelters*. Design of facilities for use as emergency shelters after the storm are outside the scope of ICC 500 and shall comply with Table 1604.5 as a *Risk Category* IV Structure.

423.2 Construction. *Storm shelters* shall be constructed in accordance with this code and ICC 500 and shall be designated as hurricane shelters, tornado shelters, or combined hurricane and tornado shelters. Buildings or structures that are also designated as emergency shelters shall also comply with Table 1604.5 as *Risk Category* IV structures.

Any *storm shelter* not required by this section shall be permitted to be constructed, provided that such structures meet the requirements of this code and ICC 500.

423.3 Occupancy classification. The occupancy classification for a *storm shelter* shall be determined in accordance with this section.

423.3.1 Dedicated storm shelters. A facility designed to be occupied solely as a *storm shelter* shall be classified as Group A-3 for the determination of requirements other than those covered in ICC 500.

Exceptions:

1. The occupancy category for dedicated storm shelters with an occupant load of fewer than 50 persons as determined in accordance with ICC 500 shall be in accordance with Section 303.
2. The occupancy category for a dedicated residential storm shelter shall be the Group R occupancy served.

423.3.2 Storm shelters within host buildings. Where designated *storm shelters* are constructed as a room or space within a host building that will normally be occupied for other purposes, the requirements of this code for the occupancy of the building, or the individual rooms or spaces thereof, shall apply unless otherwise required by ICC 500.

423.4 Critical emergency operations. In areas where the shelter design wind speed for tornados in accordance with Figure 304.2(1) of ICC 500 is 250 mph, 911 call stations, emergency operation centers and fire, rescue, ambulance and police stations shall comply with Table 1604.5 as a *Risk Category* IV structure and shall be provided with a *storm shelter* constructed in accordance with ICC 500.

423.5 Group E occupancies. In areas where the shelter design wind speed for tornados is 250 mph in accordance with Figure 304.2(1) of ICC 500, all Group E occupancies with an *occupant load* of 50 or more shall have a storm shelter constructed in accordance with ICC 500.

Exceptions:

1. Group E day care facilities.
2. Group E occupancies accessory to *places of religious worship*.
3. Buildings meeting the requirements for shelter design in ICC 500.

423.5.1 Required occupant capacity. The required occupant capacity of the *storm shelter* shall include all of the buildings on the site and shall be the greater of the following:

1. The total *occupant load* of the classrooms, vocational rooms and offices in the Group E occupancy.
2. The *occupant load* of the largest indoor assembly space that is associated with the Group E occupancy.

Exceptions:

1. Where a new building is being added on an existing Group E site, and where the new building is not of sufficient size to

accommodate the required occupant capacity of the *storm shelter* for all of the buildings on the site, the storm shelter shall at a minimum accommodate the required occupant capacity for the new building.

2. Where approved by the *building official,* the required occupant capacity of the shelter shall be permitted to be reduced by the occupant capacity of any existing *storm shelters* on the site.

423.5.2 Location. Storm shelters shall be located within the buildings they serve or shall be located where the maximum distance of travel from not fewer than one exterior door of each building to a door of the shelter serving that building does not exceed 1,000 feet (305 m).

SECTION 424
PLAY STRUCTURES

424.1 General. Play structures installed inside all occupancies covered by this code that exceed 10 feet (3048 mm) in height or 150 square feet (14 m^2) in area shall comply with Sections 424.2 through 424.5.

424.2 Materials. Play structures shall be constructed of noncombustible materials or of combustible materials that comply with the following:

1. *Fire-retardant-treated* wood complying with Section 2303.2.
2. Light-transmitting plastics complying with Section 2606.
3. Foam plastics (including the pipe foam used in *soft-contained play equipment structures*) having a maximum heat-release rate not greater than 100 kilowatts when tested in accordance with UL 1975 or when tested in accordance with NFPA 289, using the 20 kW ignition source.
4. Aluminum composite material (ACM) meeting the requirements of Class A *interior finish* in accordance with Chapter 8 when tested as an assembly in the maximum thickness intended for use.
5. Textiles and films complying with the fire propagation performance criteria contained in Test Method 1 or Test Method 2, as appropriate, of NFPA 701.
6. Plastic materials used to construct rigid components of *soft-contained play equipment structures* (such as tubes, windows, panels, junction boxes, pipes, slides and decks) exhibiting a peak rate of heat release not exceeding 400 kW/ m^2 when tested in accordance with ASTM E1354 at an incident heat flux of 50 kW/m^2 in the horizontal orientation at a thickness of 6 mm.
7. Ball pool balls, used in *soft-contained play equipment structures*, having a maximum heat-release rate not greater than 100 kilowatts when tested in accordance with UL 1975 or when tested in accordance with NFPA 289, using the 20 kW ignition source. The minimum specimen test size shall be 36 inches by 36 inches (914 mm by 914 mm) by an average of 21 inches (533 mm) deep, and the balls shall be held in a box constructed of galvanized steel poultry netting wire mesh.
8. Foam plastics shall be covered by a fabric, coating or film meeting the fire propagation performance criteria contained in Test Method 1 or Test Method 2, as appropriate, of NFPA 701.
9. The floor covering placed under the children's *play structure* shall exhibit a Class I *interior floor finish* classification, as described in Section 804, when tested in accordance with ASTM E648 or NFPA 253.
10. Interior finishes for structures exceeding 600 square feet (56 m^2) in area or 10 feet (3048 mm) in height shall have a flame spread index not greater than that specified in Table 803.13 for the occupancy group and location designated. Interior wall and ceiling finish materials tested in accordance with NFPA 286 and meeting the acceptance criteria of Section 803.1.1.1, shall be permitted to be used where a Class A classification in accordance with ASTM E84 or UL 723 is required.

[F] 424.3 Fire protection. Play structures shall be provided with the same level of *approved* fire suppression and detection devices required for other structures in the same occupancy.

424.4 Separation. Play structures shall have a horizontal separation from building walls, partitions and from elements of the *means of egress* of not less than 5 feet (1524 mm). Play structures shall have a horizontal separation from other *play structures* of not less than 20 feet (6090 mm).

424.5 Area limits. Play structures shall be not greater than 600 square feet (56 m^2) in area, unless a special investigation, acceptable to the *building official*, has demonstrated adequate fire safety.

424.5.1 Design. *Play structures* exceeding 600 square feet (56 m^2) in area or 10 feet (3048 mm) in height shall be designed in accordance with Chapter 16.

SECTION 425
HYPERBARIC FACILITIES

425.1 Hyperbaric facilities. Hyperbaric facilities shall meet the requirements contained in Chapter 14 of NFPA 99.

SECTION [F] 426
COMBUSTIBLE DUSTS, GRAIN PROCESSING AND STORAGE

[F] 426.1 General. The provisions of Sections 426.1.1 through 426.1.7 shall apply to buildings in which materials that produce *combustible dusts* are stored or handled. Buildings that store or handle *combustible dusts* shall comply with the applicable provisions of the *International Fire Code*. Where required by the fire code official, NFPA 652 and the applicable provisions of NFPA 61, NFPA 85, NFPA 120, NFPA 484, NFPA 654, NFPA 655 and NFPA 664 shall apply.

[F] 426.1.1 Type of construction and height exceptions. Buildings shall be constructed in compliance with the height, number of *stories* and area limitations specified in Sections 504 and 506; except that where erected of Type I or II construction, the heights and areas of grain elevators and similar structures shall be unlimited, and where of Type IV construction, the maximum *building height* shall be 65 feet (19 812 mm) and except further that, in isolated areas, the maximum *building height* of Type IV structures shall be increased to 85 feet (25 908 mm).

[F] 426.1.2 Grinding rooms. Every room or space occupied for grinding or other operations that produce *combustible dusts* in such a manner that the room or space is classified as a Group H-2 occupancy shall be enclosed with *fire barriers* constructed in accordance with Section 707 or *horizontal assemblies* constructed in accordance with Section 711, or both. The *fire-resistance rating* of the enclosure shall be not less than 2 hours where the area is not more than 3,000 square feet (279 m^2), and not less than 4 hours where the area is greater than 3,000 square feet (279 m^2).

[F] 426.1.3 Conveyors. Conveyors, chutes, piping and similar equipment passing through the enclosures of rooms or spaces shall be constructed dirt tight and vapor tight, and be of *approved* noncombustible materials complying with Chapter 30.

[F] 426.1.4 Explosion control. *Explosion* control shall be provided as specified in the *International Fire Code*, or spaces shall be equipped with the equivalent mechanical *ventilation* complying with the *International Mechanical Code*.

[F] 426.1.5 Grain elevators. Grain elevators, malt houses and buildings for similar occupancies shall not be located within 30 feet (9144 mm) of interior *lot lines* or structures on the same *lot*, except where erected along a railroad right-of-way.

[F] 426.1.6 Coal pockets. Coal pockets located less than 30 feet (9144 mm) from interior *lot lines* or from structures on the same *lot* shall be constructed of not less than Type IB construction. Where more than 30 feet (9144 mm) from interior *lot lines*, or where erected along a railroad right-of-way, the minimum type of construction of such structures not more than 65 feet (19 812 mm) in *building height* shall be Type IV.

[F] 426.1.7 Tire rebuilding. Buffing operations shall be located in a room separated from the remainder of the building housing the tire rebuilding or tire recapping operation by a 1-hour *fire barrier*.

Exception: Buffing operations are not required to be separated where all of the following conditions are met:

1. Buffing operations are equipped with an *approved* continuous automatic water-spray system directed at the point of cutting action.
2. Buffing machines are connected to particle-collecting systems providing a minimum air movement of 1,500 cubic feet per minute (cfm) (0.71 m^3/s) in volume and 4,500 feet per minute (fpm) (23 m/s) in-line velocity.
3. The collecting system shall discharge the rubber particles to an *approved* outdoor noncombustible or fire-resistant container, which is emptied at frequent intervals to prevent overflow.

SECTION 427
MEDICAL GAS SYSTEMS

[F] 427.1 General. Medical gases at health care-related facilities intended for patient or veterinary care shall comply with Sections 427.2 through 427.2.3 in addition to requirements of Chapter 53 of the *International Fire Code*.

[F] 427.2 Interior supply location. Medical gases shall be located in areas dedicated to the storage of such gases without other storage or uses. Where containers of medical gases in quantities greater than the permitted amount are located inside the buildings, they shall be located in a 1-hour exterior room, 1-hour interior room or a *gas cabinet* in accordance with Section 427.2.1, 427.2.2 or 427.2.3, respectively. Rooms or areas where medical gases are stored or used in quantities exceeding the maximum allowable quantity per control area as set forth in TABLE 307.1(1) and [F] TABLE 307.1(2) shall be in accordance with Group H occupancies.

[F] 427.2.1 One-hour exterior room. A 1-hour exterior room shall be a room or enclosure separated from the remainder of the building by *fire barriers* constructed in accordance with Section 707 or *horizontal assemblies* constructed in accordance with Section 711, or both, with a *fire-resistance rating* of not less than 1 hour. Openings between the room or enclosure and interior spaces shall be provided with *self-closing* smoke- and draft-control assemblies having a *fire protection rating* of not less than 1 hour. Rooms shall have not less than one *exterior wall* that is provided with not less than two vents. Each vent shall have a minimum free air opening of not less than 36 square inches (232 cm^2) for each 1,000 cubic feet (28 m^3) at normal temperature and pressure (NTP) of gas stored in the room and shall be not less than 72 square inches (465 cm^2) in aggregate free opening area. One vent shall be within 6 inches (152 mm) of the floor and one shall be within 6 inches (152 mm) of the ceiling. Rooms shall be provided with not fewer than one automatic fire sprinkler to provide container cooling in case of fire.

[F] 427.2.2 One-hour interior room. Where an *exterior wall* cannot be provided for the room, a 1-hour interior room shall be provided and shall be a room or enclosure separated from the remainder of the building by *fire barriers* constructed in accordance with Section 707 or *horizontal assemblies* constructed in accordance with Section 711, or both, with a *fire-resistance rating* of not less than 1 hour. Openings between the room or enclosure and interior spaces shall be provided with *self-closing* smoke- and draft-control assemblies having a *fire protection rating* of not less than 1 hour. An *automatic sprinkler* *system* shall be installed within the room. The room shall be exhausted through a duct to the exterior. Supply and exhaust ducts shall be enclosed in a 1-hour rated *shaft enclosure* from the room to the exterior. Approved mechanical *ventilation* shall comply with the *International Mechanical Code* and be provided with a minimum rate of 1 cubic foot per minute per square foot (0.00508 $m^3/s/m^2$) of the area of the room.

[F] 427.2.3 Gas cabinets. Gas cabinets shall be constructed in accordance with Section 5003.8.6 of the *International Fire Code* and shall comply with the following:

1. Cabinets shall be exhausted to the exterior through a dedicated exhaust duct system installed in accordance with Chapter 5 of the *International Mechanical Code.*
2. Supply and exhaust ducts shall be enclosed in a 1-hour rated *shaft enclosure* from the cabinet to the exterior. The average velocity of *ventilation* at the face of access ports or windows shall be not less than 200 feet per minute (1.02 m/s) with a minimum of 150 feet per minute (0.76 m/s) at any point of the access port or window.
3. Cabinets shall be provided with an *automatic sprinkler system* internal to the cabinet.

SECTION 428
HIGHER EDUCATION LABORATORIES

[F] 428.1 Scope. *Higher education laboratories* complying with the requirements of Sections 428.1 through 428.4 shall be permitted to exceed the maximum allowable quantities of *hazardous materials* in *control areas* set forth in Tables 307.1(1) and [F] 307.1(2) without requiring classification as a Group H occupancy. Except as specified in Section 428, such laboratories shall comply with all applicable provisions of this code and the *International Fire Code.*

[F] 428.2 Application. The provisions of Section 428 shall be applied as exceptions or additions to applicable requirements of this code. Unless specifically modified by Section 428, the storage, *use* and handling of *hazardous materials* shall comply with all other provisions in Chapters 38 and 50 through 67 of the *International Fire Code* and this code for quantities not exceeding the maximum allowable quantity.

[F] 428.3 Laboratory suite construction. Where *laboratory suites* are provided, they shall be constructed in accordance with this section and Chapter 38 of the *International Fire Code.* The number of *laboratory suites* and percentage of maximum allowable quantities of *hazardous materials* in *laboratory suites* shall be in accordance with Table 428.3.

[F] 428.3.1 Separation from other nonlaboratory areas. *Laboratory suites* shall be separated from other portions of the building in accordance with the most restrictive of the following:

1. *Fire barriers* and *horizontal assemblies* as required in Table 428.3. *Fire barriers* shall be constructed in accordance with Section 707 and *horizontal assemblies* constructed in accordance with Section 711.

   Exception: Where an individual *laboratory suite* occupies more than one *story*, the *fire-resistance rating* of intermediate floors contained within the *laboratory suite* shall comply with the requirements of this code.

2. Separations as required by Section 508.

[F] 428.3.2 Separation from other laboratory suites. *Laboratory suites* shall be separated from other *laboratory suites* in accordance with Table 428.3.

[F] TABLE 428.3
DESIGN AND NUMBER OF LABORATORY SUITES PER FLOOR

FLOOR LEVEL		PERCENTAGE OF THE MAXIMUM ALLOWABLE QUANTITY PER LAB SUITE[a]	NUMBER OF LAB SUITES PER FLOOR	FIRE-RESISTANCE RATING FOR FIRE BARRIERS IN HOURS[b]
Above Grade Plane	21+	Not allowed	Not Permitted	Not Permitted
	16-20	25	1	2[c]
	11-15	50	1	2[c]
	7-10	50	2	2[c]
	4-6	75	4	1
	3	100	4	1
	1-2	100	6	1
Below Grade Plane	1	75	4	1
	2	50	2	1
	Lower than 2	Not Allowed	Not Allowed	Not Allowed

a. Percentages shall be of the maximum allowable quantity per control area shown in Tables 307.1(1) and 307.1(2), with all increases allowed in the footnotes to those tables.

b. Fire barriers shall include walls, floors and ceilings necessary to provide separation from other portions of the building.

c. Vertical fire barriers separating laboratory suites from other spaces on the same floor shall be permitted to be 1-hour fire-resistance rated.

[F] 428.3.3 Floor assembly fire resistance. The floor assembly supporting *laboratory suites* and the construction supporting the floor of *laboratory suites* shall have a *fire-resistance rating* of not less than 2 hours.

> **Exception:** The floor assembly of the *laboratory suites* and the construction supporting the floor of the *laboratory suites* are allowed to be 1-hour *fire-resistance* rated in buildings of Types IIA, IIIA and VA construction, provided that the building is three or fewer *stories.*

[F] 428.3.4 Maximum number. The maximum number of *laboratory suites* shall be in accordance with Table 428.3. Where a building contains both *laboratory suites* and *control areas,* the total number of *laboratory suites* and *control areas* within a building shall not exceed the maximum number *of laboratory suites* in accordance with Table 428.3.

[F] 428.3.5 Means of egress. *Means of egress* shall be in accordance with Chapter 10.

[F] 428.3.6 Standby or emergency power. Standby or emergency power shall be provided in accordance with Section 414.5.2 where *laboratory suites* are located above the sixth *story above grade plane* or located in a *story* below *grade plane.*

[F] 428.3.7 Ventilation. *Ventilation* shall be in accordance with Chapter 7 of NFPA 45, and the *International Mechanical Code.*

[F] 428.3.8 Liquid-tight floor. Portions of *laboratory suites* where *hazardous materials* are present shall be provided with a liquid-tight floor.

[F] 428.3.9 Automatic sprinkler systems. Buildings containing *laboratory suites* shall be equipped throughout with an *automatic sprinkler system* in accordance with Section 903.3.1.1.

[F] 428.4 Percentage of maximum allowable quantity in each laboratory suite. The percentage of maximum allowable quantities of *hazardous materials* in each *laboratory suite* shall be in accordance with Table 428.3.

CHAPTER 5
GENERAL BUILDING HEIGHTS AND AREAS

User note:

About this chapter: *Chapter 5 establishes the limits to which a building can be built. Building height, number of stories and building area are specified in this chapter. Chapter 5 must be used in conjunction with the occupancies established in Chapter 3 and the types of construction established in Chapter 6. This chapter also specifies the impact that mezzanines, accessory occupancies and mixed occupancies have on the overall size of a building.*

SECTION 501
GENERAL

501.1 Scope. The provisions of this chapter control the height and area of structures hereafter erected and *additions* to *existing structures*.

SECTION 502
BUILDING ADDRESS

[F] 502.1 Address identification. New and existing buildings shall be provided with *approved* address identification. The address identification shall be legible and placed in a position that is visible from the street or road fronting the property. Address identification characters shall contrast with their background. Address numbers shall be Arabic numbers or alphabetical letters. Numbers shall not be spelled out. Each character shall be a minimum of 4 inches (102 mm) high with a minimum stroke width of $^1/_2$ inch (12.7 mm). Where required by the fire code official, address identification shall be provided in additional approved locations to facilitate emergency response. Where access is by means of a private road and the building address cannot be viewed from the public way, a monument, pole or other approved sign or means shall be used to identify the structure. Address identification shall be maintained.

SECTION 503
GENERAL BUILDING HEIGHT AND AREA LIMITATIONS

503.1 General. Unless otherwise specifically modified in Chapter 4 and this chapter, *building height*, number of *stories* and *building area* shall not exceed the limits specified in Sections 504 and 506 based on the type of construction as determined by Section 602 and the occupancies as determined by Section 302 except as modified hereafter. *Building height*, number of *stories* and *building area* provisions shall be applied independently. For the purposes of determining area limitations, height limitations and type of construction, each portion of a building separated by one or more *fire walls* complying with Section 706 shall be considered to be a separate building.

503.1.1 Special industrial occupancies. Buildings and structures designed to house special industrial processes that require large areas and unusual *building heights* to accommodate craneways or special machinery and equipment, including, among others, rolling mills; structural metal fabrication shops and foundries; or the production and distribution of electric, gas or steam power, shall be exempt from the *building height*, number of *stories* and *building area* limitations specified in Sections 504 and 506.

503.1.2 Buildings on same lot. Two or more buildings on the same lot shall be regulated as separate buildings or shall be considered as portions of one building where the *building height*, number of *stories* of each building and the aggregate *building area* of the buildings are within the limitations specified in Sections 504 and 506. The provisions of this code applicable to the aggregate building shall be applicable to each *building*.

503.1.3 Type I construction. Buildings of Type I construction permitted to be of unlimited tabular *building heights and areas* are not subject to the special requirements that allow unlimited area buildings in Section 507 or unlimited *building height* in Sections 503.1.1 and 504.3 or increased *building heights and areas* for other types of construction.

503.1.4 Occupied roofs. A roof level or portion thereof shall be permitted to be used as an occupied roof provided the occupancy of the roof is an occupancy that is permitted by Table 504.4 for the *story* immediately below the roof. The area of the occupied roofs shall not be included in the *building area* as regulated by Section 506. An occupied roof shall not be included in the *building height* or number of *stories* as regulated by Section 504, provided that the *penthouses* and other enclosed *rooftop structures* comply with Section 1511.

Exceptions:

1. The occupancy located on an occupied roof shall not be limited to the occupancies allowed on the *story* immediately below the roof where the building is equipped throughout with an *automatic sprinkler system* in accordance with Section 903.3.1.1 or 903.3.1.2 and occupant notification in accordance with Sections 907.5.2.1 and 907.5.2.3 is provided in the area of the occupied roof. *Emergency voice/alarm communication* system notification per Section 907.5.2.2 shall also be provided in the area of the occupied roof where such system is required elsewhere in the building.

2. Assembly occupancies shall be permitted on roofs of open parking spaces of Type I or Type II construction, in accordance with the exception to Section 903.2.1.6.

503.1.4.1 Enclosures over occupied roof areas. Elements or structures enclosing the occupied roof areas shall not extend more than 48 inches (1220 mm) above the surface of the occupied roof.

Exception: *Penthouses* constructed in accordance with Section 1511.2 and towers, domes, spires and cupolas constructed in accordance with Section 1511.5.

SECTION 504
BUILDING HEIGHT AND NUMBER OF STORIES

504.1 General. The height, in feet, and the number of *stories* of a building shall be determined based on the type of construction, occupancy classification and whether there is an *automatic sprinkler system* installed throughout the building.

Exception: The *building height* of one-*story* aircraft hangars, aircraft paint hangars and buildings used for the manufacturing of aircraft shall not be limited where the building is provided with an *automatic sprinkler system* or *automatic fire-extinguishing system* in accordance with Chapter 9 and is entirely surrounded by *public ways* or *yards* not less in width than one and one-half times the *building height.*

504.1.1 Unlimited area buildings. The height of unlimited area buildings shall be designed in accordance with Section 507.

504.1.2 Special provisions. The special provisions of Section 510 permit the use of special conditions that are exempt from, or modify, the specific requirements of this chapter regarding the allowable heights of buildings based on the occupancy classification and type of construction, provided the special condition complies with the provisions specified in Section 510.

504.2 Mixed occupancy. In a building containing mixed occupancies in accordance with Section 508, no individual occupancy shall exceed the height and number of *story* limits specified in this section for the applicable occupancies.

504.3 Height in feet. The maximum height, in feet, of a building shall not exceed the limits specified in Table 504.3.

Exception: Towers, spires, steeples and other *rooftop structures* shall be constructed of materials consistent with the required type of construction of the building except where other construction is permitted by Section 1511.2.4. Such structures shall not be used for habitation or storage. The structures shall be unlimited in height where of noncombustible materials and shall not extend more than 20 feet (6096 mm) above the allowable *building height* where of combustible materials (see Chapter 15 for additional requirements).

504.4 Number of stories. The maximum number of *stories above grade plane* of a building shall not exceed the limits specified in Table 504.4.

SECTION 505
MEZZANINES AND EQUIPMENT PLATFORMS

505.1 General. *Mezzanines* shall comply with Section 505.2. *Equipment platforms* shall comply with Section 505.3.

505.2 Mezzanines. A *mezzanine* or *mezzanines* in compliance with Section 505.2 shall be considered a portion of the *story* below. Such *mezzanines* shall not contribute to either the *building area* or number of *stories* as regulated by Section 503.1. The area of the *mezzanine* shall be included in determining the *fire area.* The clear height above and below the *mezzanine* floor construction shall be not less than 7 feet (2134 mm).

505.2.1 Area limitation. The aggregate area of a *mezzanine* or *mezzanines* within a room shall be not greater than one-third of the floor area of that room or space in which they are located. The enclosed portion of a room shall not be included in a determination of the floor area of the room in which the *mezzanine* is located. In determining the allowable *mezzanine* area, the area of the *mezzanine* shall not be included in the floor area of the room.

Exceptions:

1. The aggregate area of *mezzanines* in buildings and structures of Type I or II construction for special industrial occupancies in accordance with Section 503.1.1 shall be not greater than two-thirds of the floor area of the room.
2. The aggregate area of *mezzanines* in buildings and structures of Type I or II construction shall be not greater than one-half of the floor area of the room in buildings and structures equipped throughout with an *approved automatic sprinkler system* in accordance with Section 903.3.1.1 and an *approved emergency voice/alarm communication system* in accordance with Section 907.5.2.2.
3. The aggregate area of a *mezzanine* within a *dwelling unit* that is located in a building equipped throughout with an *approved automatic sprinkler system* in accordance with Section 903.3.1.1 or 903.3.1.2 shall not be greater than one-half of the floor area of the room, provided that:
    - 3.1. Except for enclosed closets and bathrooms, the *mezzanine* shall be open to the room in which such *mezzanine* is located;
    - 3.2. The opening to the room shall be unobstructed except for walls not more than 42 inches (1067 mm) in height, columns and posts; and
    - 3.3. Exceptions to Section 505.2.3 shall not be permitted.

505.2.1.1 Aggregate area of mezzanines and equipment platforms. Where a room contains both a *mezzanine* and an *equipment platform*, the aggregate area of the two raised floor levels shall be not greater than two-thirds of the floor area of that room or space in which they are located. The area of the *mezzanine* shall not exceed the area determined in accordance with Section 505.2.1.

505.2.2 Means of egress. The *means of egress* for *mezzanines* shall comply with the applicable provisions of Chapter 10.

505.2.3 Openness. A *mezzanine* shall be open and unobstructed to the room in which such *mezzanine* is located except for walls not more than 42 inches (1067 mm) in height, columns and posts.

Exceptions:

1. *Mezzanines* or portions thereof are not required to be open to the room in which the *mezzanines* are located, provided that the *occupant load* of the aggregate area of the enclosed space is not greater than 10.
2. A *mezzanine* having two or more exits or access to exits is not required to be open to the room in which the *mezzanine* is located.
3. *Mezzanines* or portions thereof are not required to be open to the room in which the *mezzanines* are located, provided that the aggregate floor area of the enclosed space is not greater than 10 percent of the *mezzanine* area.
4. In industrial facilities, *mezzanines* used for control equipment are permitted to be glazed on all sides.
5. In occupancies other than Groups H and I, which are no more than two *stories* above grade plane and equipped throughout with an *automatic sprinkler system* in accordance with Section 903.3.1.1, a *mezzanine* having two or more *exits* or *access to exits* shall not be required to be open to the room in which the *mezzanine* is located.

TABLE 504.3
ALLOWABLE BUILDING HEIGHT IN FEET ABOVE GRADE PLANE[a]

OCCUPANCY CLASSIFICATION	TYPE OF CONSTRUCTION												
	See Footnotes	Type I		Type II		Type III		Type IV				Type V	
		A	B	A	B	A	B	A	B	C	HT	A	B
A, B, E, F, M, S, U	NS[b]	UL	160	65	55	65	55	65	65	65	65	50	40
	S	UL	180	85	75	85	75	270	180	85	85	70	60
H-1, H-2, H-3, H-5	NS[c, d] S	UL	160	65	55	65	55	120	90	65	65	50	40
H-4	NS[c, d]	UL	160	65	55	65	55	65	65	65	65	50	40
	S	UL	180	85	75	85	75	140	100	85	85	70	60
I-1 Condition 1, I-3	NS[d, e]	UL	160	65	55	65	55	65	65	65	65	50	40
	S	UL	180	85	75	85	75	180	120	85	85	70	60
I-1 Condition 2, I-2	NS[d, e, f]	UL	160	65	55	65	55	65	65	65	65	50	40
	S	UL	180	85									
I-4	NS[d, g]	UL	160	65	55	65	55	65	65	65	65	50	40
	S	UL	180	85	75	85	75	180	120	85	85	70	60
R[h]	NS[d]	UL	160	65	55	65	55	65	65	65	65	50	40
	S13D	60	60	60	60	60	60	60	60	60	60	50	40
	S13R	60	60	60	60	60	60	60	60	60	60	60	60
	S	UL	180	85	75	85	75	270	180	85	85	70	60

For SI: 1 foot = 304.8 mm.

UL = Unlimited; NS = Buildings not equipped throughout with an automatic sprinkler system; S = Buildings equipped throughout with an automatic sprinkler system installed in accordance with Section 903.3.1.1; S13R = Buildings equipped throughout with an automatic sprinkler system installed in accordance with Section 903.3.1.2; S13D = Buildings equipped throughout with an automatic sprinkler system installed in accordance with Section 903.3.1.3.

a. See Chapters 4 and 5 for specific exceptions to the allowable height in this chapter.
b. See Section 903.2 for the minimum thresholds for protection by an automatic sprinkler system for specific occupancies.
c. New Group H occupancies are required to be protected by an automatic sprinkler system in accordance with Section 903.2.5.
d. The NS value is only for use in evaluation of existing building height in accordance with the *International Existing Building Code*.
e. New Group I-1 and I-3 occupancies are required to be protected by an automatic sprinkler system in accordance with Section 903.2.6. For new Group I-1 occupancies Condition 1, see Exception 1 of Section 903.2.6.
f. New and existing Group I-2 occupancies are required to be protected by an automatic sprinkler system in accordance with Section 903.2.6 and Section 1103.5 of the *International Fire Code*.
g. For new Group I-4 occupancies, see Exceptions 2 and 3 of Section 903.2.6.
h. New Group R occupancies are required to be protected by an automatic sprinkler system in accordance with Section 903.2.8.

TABLE 504.4
ALLOWABLE NUMBER OF STORIES ABOVE GRADE PLANE[a, b]

OCCUPANCY CLASSIFICATION	TYPE OF CONSTRUCTION												
	See Footnotes	Type I		Type II		Type III		Type IV				Type V	
		A	B	A	B	A	B	A	B	C	HT	A	B
A-1	NS	UL	5	3	2	3	2	3	3	3	3	2	1
	S	UL	6	4	3	4	3	9	6	4	4	3	2
A-2	NS	UL	11	3	2	3	2	3	3	3	3	2	1
	S	UL	12	4	3	4	3	18	12	6	4	3	2
A-3	NS	UL	11	3	2	3	2	3	3	3	3	2	1
	S	UL	12	4	3	4	3	18	12	6	4	3	2
A-4	NS	UL	11	3	2	3	2	3	3	3	3	2	1
	S	UL	12	4	3	4	3	18	12	6	4	3	2
A-5	NS	UL	UL	UL	UL	UL	UL	1	1	1	UL	UL	UL
	S	UL	UL	UL	UL	UL	UL	UL	UL	UL	UL	UL	UL
B	NS	UL	11	5	3	5	3	5	5	5	5	3	2
	S	UL	12	6	4	6	4	18	12	9	6	4	3
E	NS	UL	5	3	2	3	2	3	3	3	3	1	1
	S	UL	6	4	3	4	3	9	6	4	4	2	2
F-1	NS	UL	11	4	2	3	2	3	3	3	4	2	1
	S	UL	12	5	3	4	3	10	7	5	5	3	2
F-2	NS	UL	11	5	3	4	3	5	5	5	5	3	2
	S	UL	12	6	4	5	4	12	8	6	6	4	3
H-1	NS[c, d]	1	1	1	1	1	1	NP	NP	NP	1	1	NP
	S							1	1	1			
H-2	NS[c, d]	UL	3	2	1	2	1	1	1	1	2	1	1
	S							2	2	2			
H-3	NS[c, d]	UL	6	4	2	4	2	3	3	3	4	2	1
	S							4	4	4			
H-4	NS[c, d]	UL	7	5	3	5	3	5	5	5	5	3	2
	S	UL	8	6	4	6	4	8	7	6	6	4	3
H-5	NS[c, d]	4	4	3	3	3	3	2	2	2	3	3	2
	S							3	3	3			
I-1 Condition 1	NS[d, e]	UL	9	4	3	4	3	4	4	4	4	3	2
	S	UL	10	5	4	5	4	10	7	5	5	4	3
I-1 Condition 2	NS[d, e]	UL	9	4	3	4	3	3	3	3	4	3	2
	S	UL	10	5				10	6	4			
I-2	NS[d, f]	UL	4	2	1	1	NP	NP	NP	NP	1	1	NP
	S	UL	5	3				7	5	1			
I-3	NS[d, e]	UL	4	2	1	2	1	2	2	2	2	2	1
	S	UL	5	3	2	3	2	7	5	3	3	3	2
I-4	NS[d, g]	UL	5	3	2	3	2	3	3	3	3	1	1
	S	UL	6	4	3	4	3	9	6	4	4	2	2
M	NS	UL	11	4	2	4	2	4	4	4	4	3	1
	S	UL	12	5	3	5	3	12	8	6	5	4	2

(continued)

TABLE 504.4—continued
ALLOWABLE NUMBER OF STORIES ABOVE GRADE PLANE[a, b]

OCCUPANCY CLASSIFICATION	TYPE OF CONSTRUCTION												
	See Footnotes	Type I		Type II		Type III		Type IV				Type V	
		A	B	A	B	A	B	A	B	C	HT	A	B
R-1[h]	NS[d]	UL	11	4	4	4	4	4	4	4	4	3	2
	S13R	4	4									4	3
	S	UL	12	5	5	5	5	18	12	8	5	4	3
R-2[h]	NS[d]	UL	11	4	4	4	4	4	4	4	4	3	2
	S13R	4	4	4								4	3
	S	UL	12	5	5	5	5	18	12	8	5	4	3
R-3[h]	NS[d]	UL	11	4	4	4	4	4	4	4	4	3	3
	S13D	4	4									3	3
	S13R	4	4									4	4
	S	UL	12	5	5	5	5	18	12	5	5	4	4
R-4[h]	NS[d]	UL	11	4	4	4	4	4	4	4	4	3	2
	S13D	4	4									3	2
	S13R	4	4									4	3
	S	UL	12	5	5	5	5	18	12	5	5	4	3
S-1	NS	UL	11	4	2	3	2	4	4	4	4	3	1
	S	UL	12	5	4	4	4	10	7	5	5	4	2
S-2	NS	UL	11	5	3	4	3	4	4	4	5	4	2
	S	UL	12	6	4	5	4	12	8	5	6	5	3
U	NS	UL	5	4	2	3	2	4	4	4	4	2	1
	S	UL	6	5	3	4	3	9	6	5	5	3	2

UL = Unlimited; NP = Not Permitted; NS = Buildings not equipped throughout with an automatic sprinkler system; S = Buildings equipped throughout with an automatic sprinkler system installed in accordance with Section 903.3.1.1; S13R = Buildings equipped throughout with an automatic sprinkler system installed in accordance with Section 903.3.1.2; S13D = Buildings equipped throughout with an automatic sprinkler system installed in accordance with Section 903.3.1.3.

a. See Chapters 4 and 5 for specific exceptions to the allowable height in this chapter.
b. See Section 903.2 for the minimum thresholds for protection by an automatic sprinkler system for specific occupancies.
c. New Group H occupancies are required to be protected by an automatic sprinkler system in accordance with Section 903.2.5.
d. The NS value is only for use in evaluation of existing *building height* in accordance with the *International Existing Building Code*.
e. New Group I-1 and I-3 occupancies are required to be protected by an automatic sprinkler system in accordance with Section 903.2.6. For new Group I-1 occupancies, Condition 1, see Exception 1 of Section 903.2.6.
f. New and existing Group I-2 occupancies are required to be protected by an automatic sprinkler system in accordance with Section 903.2.6 and 1103.5 of the *International Fire Code*.
g. For new Group I-4 occupancies, see Exceptions 2 and 3 of Section 903.2.6.
h. New Group R occupancies are required to be protected by an automatic sprinkler system in accordance with Section 903.2.8.

505.3 Equipment platforms. *Equipment platforms* in buildings shall not be considered as a portion of the floor below. Such *equipment platforms* shall not contribute to either the *building area* or the number of *stories* as regulated by Section 503.1. The area of the *equipment platform* shall not be included in determining the *fire area* in accordance with Section 903. *Equipment platforms* shall not be a part of any *mezzanine* and such platforms and the walkways, *stairways, alternating tread devices* and ladders providing access to an *equipment platform* shall not serve as a part of the *means of egress* from the building.

505.3.1 Area limitation. The aggregate area of all *equipment platforms* within a room shall be not greater than two-thirds of the area of the room in which they are located. Where an *equipment platform* is located in the same room as a *mezzanine,* the area of the *mezzanine* shall be determined by Section 505.2.1 and the combined aggregate area of the *equipment platforms* and *mezzanines* shall be not greater than two-thirds of the room in which they are located. The area of the *mezzanine* shall not exceed the area determined in accordance with Section 505.2.1.

505.3.2 Automatic sprinkler system. Where located in a building that is required to be protected by an *automatic sprinkler system, equipment platforms* shall be fully protected by sprinklers above and below the platform, where required by the standards referenced in Section 903.3.

505.3.3 Guards. *Equipment platforms* shall have *guards* where required by Section 1015.2.

SECTION 506
BUILDING AREA

506.1 General. The floor area of a building shall be determined based on the type of construction, occupancy classification, whether there is an *automatic sprinkler system* installed throughout the building and the amount of building frontage on *public way* or open space.

506.1.1 Unlimited area buildings. Unlimited area buildings shall be designed in accordance with Section 507.

506.1.2 Special provisions. The special provisions of Section 510 permit the use of special conditions that are exempt from, or modify, the specific requirements of this chapter regarding the allowable areas of buildings based on the occupancy classification and type of construction, provided the special condition complies with the provisions specified in Section 510.

506.1.3 Basements. *Basements* need not be included in the total allowable floor area of a building provided the total area of such *basements* does not exceed the area permitted for a one-*story above grade plane* building.

506.2 Allowable area determination. The allowable area of a building shall be determined in accordance with the applicable provisions of Sections 506.2.1, 506.2.2 and 506.3.

506.2.1 Single-occupancy buildings. The allowable area of each *story* of a single-occupancy building shall be determined in accordance with Equation 5-1:

$$A_a = A_t + (NS \times I_f) \qquad \textbf{(Equation 5-1)}$$

where:

A_a = Allowable area (square feet).

A_t = Tabular allowable area factor (NS, S1, S13R or S13D value, as applicable) in accordance with Table 506.2.

NS = Tabular allowable area factor in accordance with Table 506.2 for nonsprinklered building (regardless of whether the building is sprinklered).

I_f = Area factor increase due to frontage (percent) as calculated in accordance with Section 506.3.

The allowable area per story of a single-occupancy building with a maximum of three stories above grade shall be determined by Equation 5-1. The total allowable area of a single-occupancy building more than three *stories above grade plane* shall be determined in accordance with Equation 5-2:

$$A_a = [A_t + (NS \times I_f)] \times S_a \qquad \textbf{(Equation 5-2)}$$

where:

A_a = Allowable area (square feet).

A_t = Tabular allowable area factor (NS, S13R, S13D or SM value, as applicable) in accordance with Table 506.2.

NS = Tabular allowable area factor in accordance with Table 506.2 for a nonsprinklered building (regardless of whether the building is sprinklered).

I_f = Area factor increase due to frontage (percent) as calculated in accordance with Section 506.3.

S_a = 3 where the actual number of stories above grade plane exceeds three, or

S_a = 4 where the building is equipped throughout with an automatic sprinkler system installed in accordance with Section 903.3.1.2.

The actual area of any individual floor shall not exceed the allowable area per Equation 5-1.

506.2.2 Mixed-occupancy buildings. The allowable area of each *story* of a mixed-occupancy building shall be determined in accordance with the applicable provisions of, Section 508.3.2 for nonseparated occupancies and Section 508.4.2 for separated occupancies.

For buildings with more than three *stories above grade plane,* the total *building area* shall be such that the aggregate sum of the ratios of the actual area of each *story* divided by the allowable area of such stories, determined in accordance with Equation 5-3 based on the applicable provisions of Section 508.1, shall not exceed three.

$$A_a = [A_t + (NS \times I_f)] \qquad \textbf{(Equation 5-3)}$$

A_a = Allowable area (square feet).

A_t = Tabular allowable area factor (*NS*, S13R, S13D or SM value, as applicable) in accordance with Table 506.2.

NS = Tabular allowable area factor in accordance with Table 506.2 for a nonsprinklered building, regardless of whether the building is sprinklered.

I_f = Area factor increase due to frontage (percent) as calculated in accordance with Section 506.3.

Exception: For buildings designed as separated occupancies under Section 508.4 and equipped throughout with an automatic sprinkler system installed in accordance with Section 903.3.1.2, the total building area shall be such that the aggregate sum of the ratios of the actual area of each story divided by the allowable area of such stories determined in accordance with Equation 5-3 based on the applicable provisions of Section 508.1, shall not exceed four.

506.2.2.1 Group H-2 or H-3 mixed occupancies. For a building containing Group H-2 or H-3 occupancies, the allowable area shall be determined in accordance with Section 508.4.2, with the sprinkler system increase applicable only to the portions of the building not classified as Group H-2 or H-3.

506.3 Frontage increase. Every building shall adjoin or have access to a *public way* to receive an area factor increase based on frontage. Area factor increase shall be determined in accordance with Sections 506.3.1 through 506.3.3.

506.3.1 Minimum percentage of perimeter. To qualify for an area factor increase based on frontage, a building shall have not less than 25 percent of its perimeter on a *public way* or open space. Such open space shall be either on the same lot or dedicated for public use and shall be accessed from a street or approved *fire lane.*

TABLE 506.2
ALLOWABLE AREA FACTOR (A_t = NS, S1, S13R, S13D or SM, as applicable) IN SQUARE FEET[a, b]

OCCUPANCY CLASSIFICATION	SEE FOOTNOTES	TYPE OF CONSTRUCTION											
		Type I		Type II		Type III		Type IV				Type V	
		A	B	A	B	A	B	A	B	C	HT	A	B
A-1	NS	UL	UL	15,500	8,500	14,000	8,500	45,000	30,000	18,750	15,000	11,500	5,500
	S1	UL	UL	62,000	34,000	56,000	34,000	180,000	120,000	75,000	60,000	46,000	22,000
	SM	UL	UL	46,500	25,500	42,000	25,500	135,000	90,000	56,250	45,000	34,500	16,500
A-2	NS	UL	UL	15,500	9,500	14,000	9,500	45,000	30,000	18,750	15,000	11,500	6,000
	S1	UL	UL	62,000	38,000	56,000	38,000	180,000	120,000	75,000	60,000	46,000	24,000
	SM	UL	UL	46,500	28,500	42,000	28,500	135,000	90,000	56,250	45,000	34,500	18,000
A-3	NS	UL	UL	15,500	9,500	14,000	9,500	45,000	30,000	18,750	15,000	11,500	6,000
	S1	UL	UL	62,000	38,000	56,000	38,000	180,000	120,000	75,000	60,000	46,000	24,000
	SM	UL	UL	46,500	28,500	42,000	28,500	135,000	90,000	56,250	45,000	34,500	18,000
A-4	NS	UL	UL	15,500	9,500	14,000	9,500	45,000	30,000	18,750	15,000	11,500	6,000
	S1	UL	UL	62,000	38,000	56,000	38,000	180,000	120,000	75,000	60,000	46,000	24,000
	SM	UL	UL	46,500	28,500	42,000	28,500	135,000	90,000	56,250	45,000	34,500	18,000
A-5	NS	UL	UL	UL	UL	UL	UL	UL	UL	UL	UL	UL	UL
	S1												
	SM												
B	NS	UL	UL	37,500	23,000	28,500	19,000	108,000	72,000	45,000	36,000	18,000	9,000
	S1	UL	UL	150,000	92,000	114,000	76,000	432,000	288,000	180,000	144,000	72,000	36,000
	SM	UL	UL	112,500	69,000	85,500	57,000	324,000	216,000	135,000	108,000	54,000	27,000
E	NS	UL	UL	26,500	14,500	23,500	14,500	76,500	51,000	31,875	25,500	18,500	9,500
	S1	UL	UL	106,000	58,000	94,000	58,000	306,000	204,000	127,500	102,000	74,000	38,000
	SM	UL	UL	79,500	43,500	70,500	43,500	229,500	153,000	95,625	76,500	55,500	28,500
F-1	NS	UL	UL	25,000	15,500	19,000	12,000	100,500	67,000	41,875	33,500	14,000	8,500
	S1	UL	UL	100,000	62,000	76,000	48,000	402,000	268,000	167,500	134,000	56,000	34,000
	SM	UL	UL	75,000	46,500	57,000	36,000	301,500	201,000	125,625	100,500	42,000	25,500
F-2	NS	UL	UL	37,500	23,000	28,500	18,000	151,500	101,000	63,125	50,500	21,000	13,000
	S1	UL	UL	150,000	92,000	114,000	72,000	606,000	404,000	252,500	202,000	84,000	52,000
	SM	UL	UL	112,500	69,000	85,500	54,000	454,500	303,000	189,375	151,500	63,000	39,000
H-1	NS[c]	21,000	16,500	11,000	7,000	9,500	7,000	10,500	10,500	10,500	10,500	7,500	NP
	S1												
H-2	NS[c]	21,000	16,500	11,000	7,000	9,500	7,000	10,500	10,500	10,500	10,500	7,500	3,000
	S1												
	SM												
H-3	NS[c]	UL	60,000	26,500	14,000	17,500	13,000	25,500	25,500	25,500	25,500	10,000	5,000
	S1												
	SM												
H-4	NS[c, d]	UL	UL	37,500	17,500	28,500	17,500	72,000	54,000	40,500	36,000	18,000	6,500
	S1	UL	UL	150,000	70,000	114,000	70,000	288,000	216,000	162,000	144,000	72,000	26,000
	SM	UL	UL	112,500	52,500	85,500	52,500	216,000	162,000	121,500	108,000	54,000	19,500
H-5	NS[c, d]	UL	UL	37,500	23,000	28,500	19,000	72,000	54,000	40,500	36,000	18,000	9,000
	S1	UL	UL	150,000	92,000	114,000	76,000	288,000	216,000	162,000	144,000	72,000	36,000
	SM	UL	UL	112,500	69,000	85,500	57,000	216,000	162,000	121,500	108000	54,000	27,000

(continued)

TABLE 506.2—continued
ALLOWABLE AREA FACTOR (A_t = NS, S1, S13R, S13D or SM, as applicable) IN SQUARE FEET[a, b]

OCCUPANCY CLASSIFICATION	SEE FOOTNOTES	TYPE OF CONSTRUCTION											
		Type I		Type II		Type III		Type IV				Type V	
		A	B	A	B	A	B	A	B	C	HT	A	B
I-1	NS[d, e]	UL	55,000	19,000	10,000	16,500	10,000	54,000	36,000	18,000	18,000	10,500	4,500
	S1	UL	220,000	76,000	40,000	66,000	40,000	216,000	144,000	72,000	72,000	42,000	18,000
	SM	UL	165,000	57,000	30,000	49,500	30,000	162,000	108,000	54,000	54,000	31,500	13,500
I-2	NS[d, f]	UL	UL	15,000	11,000	12,000	NP	36,000	24,000	12,000	12,000	9,500	NP
	S1	UL	UL	60,000	44,000	48,000	NP	144,000	96,000	48,000	48,000	38,000	NP
	SM	UL	UL	45,000	33,000	36,000	NP	108,000	72,000	36,000	36,000	28,500	NP
I-3	NS[d, e]	UL	UL	15,000	10,000	10,500	7,500	36,000	24,000	12,000	12,000	7,500	5,000
	S1	UL	UL	60,000	40,000	42,000	30,000	144,000	96,000	48,000	48,000	30,000	20,000
	SM	UL	UL	45,000	30,000	31,500	22,500	108,000	72,000	36,000	36,000	22,500	15,000
I-4	NS[d, g]	UL	60,500	26,500	13,000	23,500	13,000	76,500	51,000	25,500	25,500	18,500	9,000
	S1	UL	121,000	106,000	52,000	94,000	52,000	306,000	204,000	102,000	102,000	74,000	36,000
	SM	UL	181,500	79,500	39,000	70,500	39,000	229,500	153,000	76,500	76,500	55,500	27,000
M	NS	UL	UL	21,500	12,500	18,500	12,500	61,500	41,000	26,625	20,500	14,000	9,000
	S1	UL	UL	86,000	50,000	74,000	50,000	246,000	164,000	102,500	82,000	56,000	36,000
	SM	UL	UL	64,500	37,500	55,500	37,500	184,500	123,000	76,875	61,500	42,000	27,000
R-1[h]	NS[d] S13R	UL	UL	24,000	16,000	24,000	16,000	61,500	41,000	25,625	20,500	12,000	7,000
	S1	UL	UL	96,000	64,000	96,000	64,000	246,000	164,000	102,500	82,000	48,000	28,000
	SM	UL	UL	72,000	48,000	72,000	48,000	184,500	123,000	76,875	61,500	36,000	21,000
R-2[h]	NS[d] S13R	UL	UL	24,000	16,000	24,000	16,000	61,500	41,000	25,625	20,500	12,000	7,000
	S1	UL	UL	96,000	64,000	96,000	64,000	246,000	164,000	102,500	82,000	48,000	28,000
	SM	UL	UL	72,000	48,000	72,000	48,000	184,500	123,000	76,875	61,500	36,000	21,000
R-3[h]	NS[d] S13D S13R S1 SM	UL	UL	UL	UL	UL	UL	UL	UL	UL	UL	UL	UL
R-4[h]	NS[d] S13D S13R	UL	UL	24,000	16,000	24,000	16,000	61,500	41,000	25,625	20,500	12,000	7,000
	S1	UL	UL	96,000	64,000	96,000	64,000	246,000	164,000	102,500	82,000	48,000	28,000
	SM	UL	UL	72,000	48,000	72,000	48,000	184,500	123,000	76,875	61,500	36,000	21,000
S-1	NS	UL	48,000	26,000	17,500	26,000	17,500	76,500	51,000	31,875	25,500	14,000	9,000
	S1	UL	192,000	104,000	70,000	104,000	70,000	306,000	204,000	127,500	102,000	56,000	36,000
	SM	UL	144,000	78,000	52,500	78,000	52,500	229,500	153,000	95,625	76,500	42,000	27,000
S-2	NS	UL	79,000	39,000	26,000	39,000	26,000	115,500	77,000	48,125	38,500	21,000	13,500
	S1	UL	316,000	156,000	104,000	156,000	104,000	462,000	308,000	192,500	154,000	84,000	54,000
	SM	UL	237,000	117,000	78,000	117,000	78,000	346,500	231,000	144,375	115,500	63,000	40,500
U	NS[i]	UL	35,500	19,000	8,500	14,000	8,500	54,000	36,000	22,500	18,000	9,000	5,500
	S1	UL	142,000	76,000	34,000	56,000	34,000	216,000	144,000	90,000	72,000	36,000	22,000
	SM	UL	106,500	57,000	25,500	42,000	25,500	162,000	108,000	67,500	54,000	27,000	16,500

(continued)

TABLE 506.2—continued
ALLOWABLE AREA FACTOR (A_t = NS, S1, S13R, S13D or SM, as applicable) IN SQUARE FEET[a, b]

For SI: 1 square foot = 0.0929 m^2.

UL = Unlimited; NP = Not Permitted; NS = Buildings not equipped throughout with an automatic sprinkler system; S1 = Buildings a maximum of one story above grade plane equipped throughout with an automatic sprinkler system installed in accordance with Section 903.3.1.1; SM = Buildings two or more stories above grade plane equipped throughout with an automatic sprinkler system installed in accordance with Section 903.3.1.1; S13R = Buildings equipped throughout with an automatic sprinkler system installed in accordance with Section 903.3.1.2; S13D = Buildings equipped throughout with an automatic sprinkler system installed in accordance with Section 903.3.1.3.

a. See Chapters 4 and 5 for specific exceptions to the allowable area in this chapter.
b. See Section 903.2 for the minimum thresholds for protection by an automatic sprinkler system for specific occupancies.
c. New Group H occupancies are required to be protected by an automatic sprinkler system in accordance with Section 903.2.5.
d. The NS value is only for use in evaluation of existing building area in accordance with the *International Existing Building Code*.
e. New Group I-1 and I-3 occupancies are required to be protected by an automatic sprinkler system in accordance with Section 903.2.6. For new Group I-1 occupancies, Condition 1, see Exception 1 of Section 903.2.6.
f. New and existing Group I-2 occupancies are required to be protected by an automatic sprinkler system in accordance with Section 903.2.6 and Section 1103.5 of the *International Fire Code*.
g. New Group I-4 occupancies see Exceptions 2 and 3 of Section 903.2.6.
h. New Group R occupancies are required to be protected by an automatic sprinkler system in accordance with Section 903.2.8.
i. The maximum allowable area for a single-story nonsprinklered Group U greenhouse is permitted to be 9,000 square feet, or the allowable area shall be permitted to comply with Table C102.1 of Appendix C.

506.3.2 Minimum frontage distance. To qualify for an area factor increase based on frontage, the *public way* or open space adjacent to the building perimeter shall have a minimum distance *(W)* of 20 feet (6096 mm) measured at right angles from the building face to any of the following:

1. The closest interior lot line.
2. The entire width of a street, alley or *public way.*
3. The exterior face of an adjacent building on the same property.

The frontage increase shall be based on the smallest *public way* or open space that is 20 feet (6096 mm) or greater, and the percentage of building perimeter having a minimum 20 feet (6096 mm) *public way* or open space.

506.3.3 Amount of increase. The area factor increase based on frontage shall be determined in accordance with Table 506.3.3.

506.3.3.1 Section 507 buildings. Where a building meets the requirements of Section 507, as applicable, except for compliance with the minimum 60-foot (18 288 mm) *public way* or *yard* requirement, the area factor increase based on frontage shall be determined in accordance with Table 506.3.3.1.

SECTION 507 UNLIMITED AREA BUILDINGS

507.1 General. The area of buildings of the occupancies and configurations specified in Sections 507.1 through 507.13 shall not be limited. Basements not more than one *story* below *grade plane* shall be permitted.

507.1.1 Accessory occupancies. Accessory occupancies shall be permitted in unlimited area buildings in accordance with the provisions of Section 508.2, otherwise the requirements of Sections 507.3 through 507.13 shall be applied, where applicable.

TABLE 506.3.3
FRONTAGE INCREASE FACTOR[a]

PERCENTAGE OF BUILDING PERIMETER	OPEN SPACE (feet)			
	0 to less than 20	20 to less than 25	25 to less than 30	30 or greater
0 to less than 25	0	0	0	0
25 to less than 50	0	0.17	0.21	0.25
50 to less than 75	0	0.33	0.42	0.50
75 to 100	0	0.50	0.63	0.75

TABLE 506.3.3.1
SECTION 507 BUILDINGS[a]

PERCENTAGE OF BUILDING PERIMETER	OPEN SPACE (feet)					
	30 to less than 35	35 to less than 40	40 to less than 45	45 to less than 50	50 to less than 55	55 to less than 60
0 to less than 25	0	0	0	0	0	0
25 to less than 50	0.29	0.33	0.38	0.42	0.46	0.50
50 to less than 75	0.58	0.67	0.75	0.83	0.92	1.00
75 to 100	0.88	1.00	1.13	1.25	1.38	1.50

a. Interpolation is permitted.

507.2 Measurement of open spaces. Where Sections 507.3 through 507.13 require buildings to be surrounded and adjoined by *public ways* and *yards*, those open spaces shall be determined as follows:

1. *Yards* shall be measured from the building perimeter in all directions to the closest interior *lot lines* or to the exterior face of an opposing building located on the same *lot*, as applicable.
2. Where the building fronts on a *public way*, the entire width of the *public way* shall be used.

507.2.1 Reduced open space. The *public ways* or *yards* of 60 feet (18 288 mm) in width required in Sections 507.3, 507.4, 507.5, 507.6 and 507.12 shall be permitted to be reduced to not less than 40 feet (12 192 mm) in width, provided that the following requirements are met:

1. The reduced width shall not be allowed for more than 75 percent of the perimeter of the building.
2. The *exterior walls* facing the reduced width shall have a *fire-resistance rating* of not less than 3 hours.
3. Openings in the *exterior walls* facing the reduced width shall have opening protectives with a *fire protection rating* of not less than 3 hours.

507.3 Nonsprinklered, one-story buildings. The area of a Group F-2 or S-2 building not more than one *story* in height shall not be limited where the building is surrounded and adjoined by *public ways* or *yards* not less than 60 feet (18 288 mm) in width.

507.4 Sprinklered, one-story buildings. The area of a Group A-4 building not more than one *story above grade plane* of other than Type V construction, or the area of a Group B, F, M or S building no more than one *story above grade plane* of any construction type, shall not be limited where the building is provided with an *automatic sprinkler system* throughout in accordance with Section 903.3.1.1 and is surrounded and adjoined by *public ways* or *yards* not less than 60 feet (18 288 mm) in width.

Exceptions:

1. Buildings and structures of Type I or II construction for rack storage facilities that do not have access by the public shall not be limited in height, provided that such buildings conform to the requirements of Sections 507.4 and 903.3.1.1 and Chapter 32 of the *International Fire Code*.
2. The *automatic sprinkler system* shall not be required in areas occupied for indoor participant sports, such as tennis, skating, swimming and equestrian activities in occupancies in Group A-4, provided that the following criteria are met:
   - 2.1. *Exit* doors directly to the outside are provided for occupants of the participant sports areas.
   - 2.2. The building is equipped with a *fire alarm system* with *manual fire alarm boxes* installed in accordance with Section 907.
   - 2.3. An *automatic sprinkler system* is provided in storage rooms, press boxes, concession booths or other spaces ancillary to the sport activity space.

507.4.1 Mixed occupancy buildings with Groups A-1 and A-2. Group A-1 and A-2 occupancies of other than Type V construction shall be permitted within mixed occupancy buildings of unlimited area complying with Section 507.4, provided that the following criteria are met:

1. Group A-1 and A-2 occupancies are separated from other occupancies as required for separated occupancies in Section 508.4.4 with no reduction allowed in the *fire-resistance rating* of the separation based upon the installation of an *automatic sprinkler system*.
2. Each area of the portions of the building used for Group A-1 or A-2 occupancies shall not exceed the maximum allowable area permitted for such occupancies in Section 503.1.
3. *Exit* doors from Group A-1 and A-2 occupancies shall discharge directly to the exterior of the building.

507.5 Two-story buildings. The area of a Group B, F, M or S building not more than two *stories above grade plane* shall not be limited where the building is equipped throughout with an *automatic sprinkler system* in accordance with Section 903.3.1.1 and is surrounded and adjoined by *public ways* or *yards* not less than 60 feet (18 288 mm) in width.

507.6 Group A-3 buildings of Type II construction. The area of a Group A-3 building not more than one *story above grade plane*, used as a *place of religious worship*, community hall, dance hall, exhibition hall, gymnasium, lecture hall, indoor *swimming pool* or tennis court of Type II construction, shall not be limited provided that the following criteria are met:

1. The building shall not have a *stage* other than a *platform*.
2. The building shall be equipped throughout with an *automatic sprinkler system* in accordance with Section 903.3.1.1.
3. The building shall be surrounded and adjoined by *public ways* or *yards* not less than 60 feet (18 288 mm) in width.

507.7 Group A-3 buildings of Type III and IV construction. The area of a Group A-3 building of Type III or IV construction, with not more than one *story above grade plane* and used as a *place of religious worship*, community hall, dance hall, exhibition hall, gymnasium, lecture hall, indoor *swimming pool* or tennis court, shall not be limited provided that the following criteria are met:

1. The building shall not have a *stage* other than a *platform*.
2. The building shall be equipped throughout with an *automatic sprinkler system* in accordance with Section 903.3.1.1.

3. The assembly floor shall be located at or within 21 inches (533 mm) of street or grade level and all *exits* are provided with *ramps* complying with Section 1012 to the street or grade level.
4. The building shall be surrounded and adjoined by *public ways* or *yards* not less than 60 feet (18 288 mm) in width.

507.8 Group H-2, H-3 and H-4 occupancies. Group H-2, H-3 and H-4 occupancies shall be permitted in unlimited area buildings containing Group F or S occupancies in accordance with Sections 507.4 and 507.5 and the provisions of Sections 507.8.1 through 507.8.4.

507.8.1 Allowable area. The aggregate floor area of Group H occupancies located in an unlimited area building shall not exceed 10 percent of the area of the building or the area limitations for the Group H occupancies as specified in Section 506 based on the perimeter of each Group H floor area that fronts on a *public way* or open space.

507.8.1.1 Located within the building. The aggregate floor area of Group H occupancies not located at the perimeter of the building shall not exceed 25 percent of the area limitations for the Group H occupancies as specified in Section 506.

507.8.1.1.1 Rooms for flammable or combustible liquid use, dispensing or mixing in open systems. Rooms for *flammable* or *combustible liquid* use, dispensing or mixing in open systems having a floor area of not more than 500 square feet (46.5 m^2) need not be located on the outer perimeter of the building where they are in accordance with the *International Fire Code* and NFPA 30.

507.8.1.1.2 Liquid storage rooms and rooms for flammable or combustible liquid use in closed systems. Liquid storage rooms and rooms for *flammable or combustible liquid* use in closed systems having a floor area of not more than 1,000 square feet (93 m^2) need not be located on the outer perimeter where they are in accordance with the *International Fire Code* and NFPA 30.

507.8.1.1.3 Spray paint booths. Spray paint booths that comply with the *International Fire Code* need not be located on the outer perimeter.

507.8.2 Located on building perimeter. Except as provided for in Section 507.8.1.1, Group H occupancies shall be located on the perimeter of the building. In Group H-2 and H-3 occupancies, not less than 25 percent of the perimeter of such occupancies shall be an *exterior wall*.

507.8.3 Occupancy separations. Group H occupancies shall be separated from the remainder of the unlimited area building and from each other in accordance with Table 508.4.

507.8.4 Height limitations. For two-*story*, unlimited area buildings, Group H occupancies shall not be located more than one *story above grade plane* unless permitted based on the allowable height and number of *stories* and feet as specified in Section 504 based on the type of construction of the unlimited area building.

507.9 Unlimited mixed occupancy buildings with Group H-5. The area of a Group B, F, H-5, M or S building not more than two *stories above grade plane* shall not be limited where the building is equipped throughout with an *automatic sprinkler system* in accordance with Section 903.3.1.1, and is surrounded and adjoined by *public ways* or *yards* not less than 60 feet (18 288 mm) in width, provided that the following criteria are met:

1. Buildings containing Group H-5 occupancy shall be of Type I or II construction.
2. Each area used for Group H-5 occupancy shall be separated from other occupancies as required in Sections 415.11 and 508.4.
3. Each area used for Group H-5 occupancy shall not exceed the maximum allowable area permitted for such occupancies in Section 503.1 including modifications of Section 506.

   Exception: Where the Group H-5 occupancy exceeds the maximum allowable area, the Group H-5 shall be subdivided into areas that are separated by 2-hour *fire barriers*.

507.10 Aircraft paint hangar. The area of a Group H-2 aircraft paint hangar not more than one *story above grade plane* shall not be limited where such aircraft paint hangar complies with the provisions of Section 412.5 and is surrounded and adjoined by *public ways* or *yards* not less in width than one and one-half times the *building height*.

507.11 Group E buildings. The area of a Group E building not more than one *story above grade plane*, of Type II, IIIA or IV construction, shall not be limited provided that the following criteria are met:

1. Each classroom shall have not less than two *means of egress*, with one of the *means of egress* being a direct exit to the outside of the building complying with Section 1022.
2. The building is equipped throughout with an *automatic sprinkler system* in accordance with Section 903.3.1.1.
3. The building is surrounded and adjoined by *public ways* or *yards* not less than 60 feet (18 288 mm) in width.

507.12 Motion picture theaters. In buildings of Type II construction, the area of a motion picture theater located on the first *story above grade plane* shall not be limited where the building is provided with an *automatic sprinkler system* throughout in accordance with Section 903.3.1.1 and is surrounded and adjoined by *public ways* or *yards* not less than 60 feet (18 288 mm) in width.

507.13 Covered and open mall buildings and anchor buildings. The area of *covered and open mall buildings* and *anchor buildings* not exceeding three *stories* in height that comply with Section 402 shall not be limited.

SECTION 508
MIXED USE AND OCCUPANCY

508.1 General. Each portion of a building shall be individually classified in accordance with Section 302.1. Where a building contains more than one occupancy group, the building or portion thereof shall comply with the applicable provisions of Section 508.2, 508.3, 508.4 or 508.5, or a combination of these sections.

Exceptions:

1. Occupancies separated in accordance with Section 510.
2. Where required by Table 415.6.5, areas of Group H-1, H-2 and H-3 occupancies shall be located in a *detached building* or structure.

508.2 Accessory occupancies. Accessory occupancies are those occupancies that are ancillary to the main occupancy of the building or portion thereof. Accessory occupancies shall comply with the provisions of Sections 508.2.1 through 508.2.4.

508.2.1 Occupancy classification. Accessory occupancies shall be individually classified in accordance with Section 302.1. The requirements of this code shall apply to each portion of the building based on the occupancy classification of that space.

508.2.2 Allowable building height. The allowable height and number of *stories* of the building containing accessory occupancies shall be in accordance with Section 504 for the main occupancy of the building.

508.2.3 Allowable building area. The allowable area of the building shall be based on the applicable provisions of Section 506 for the main occupancy of the building. Aggregate accessory occupancies shall not occupy more than 10 percent of the floor area of the *story* in which they are located and shall not exceed the tabular values for nonsprinklered buildings in Table 506.2 for each such accessory occupancy.

508.2.4 Separation of occupancies. No separation is required between accessory occupancies and the main occupancy.

Exceptions:

1. Group H-2, H-3, H-4 and H-5 occupancies shall be separated from all other occupancies in accordance with Section 508.4.
2. Group I-1, R-1, R-2 and R-3 *dwelling units* and *sleeping units* shall be separated from other *dwelling* or *sleeping units* and from accessory occupancies contiguous to them in accordance with the requirements of Section 420.

508.3 Nonseparated occupancies. Buildings or portions of buildings that comply with the provisions of this section shall be considered as nonseparated occupancies.

508.3.1 Occupancy classification. Nonseparated occupancies shall be individually classified in accordance with Section 302.1. The requirements of this code shall apply to each portion of the building based on the occupancy classification of that space. In addition, the most restrictive provisions of Chapter 9 that apply to the nonseparated occupancies shall apply to the total nonseparated occupancy area.

508.3.1.1 High-rise buildings. Where nonseparated occupancies occur in a *high-rise building*, the most restrictive requirements of Section 403 that apply to the nonseparated occupancies shall apply throughout the *high-rise building*.

508.3.1.2 Group I-2, Condition 2 occupancies. Where one of the nonseparated occupancies is Group I-2, Condition 2, the most restrictive requirements of Sections 407, 509 and 712 shall apply throughout the *fire area* containing the Group I-2 occupancy. The most restrictive requirements of Chapter 10 shall apply to the path of egress from the Group I-2, Condition 2 occupancy up to and including the *exit discharge.*

508.3.2 Allowable building area, height and number of stories. The allowable *building area, height* and number of *stories* of the building or portion thereof shall be based on the most restrictive allowances for the occupancy groups under consideration for the type of construction of the building in accordance with Section 503.1.

508.3.3 Separation. No separation is required between nonseparated occupancies.

Exceptions:

1. Group H-2, H-3, H-4 and H-5 occupancies shall be separated from all other occupancies in accordance with Section 508.4.
2. Group I-1, R-1, R-2 and R-3 *dwelling units* and *sleeping units* shall be separated from other *dwelling* or *sleeping units* and from other occupancies contiguous to them in accordance with the requirements of Section 420.

508.4 Separated occupancies. Buildings or portions of buildings that comply with the provisions of this section shall be considered as separated occupancies.

508.4.1 Occupancy classification. Separated occupancies shall be individually classified in accordance with Section 302.1. Each separated space shall comply with this code based on the occupancy classification of that portion of the building. The most restrictive provisions of Chapter 9 that apply to the separate occupancies shall apply to the total nonfire-barrier-separated occupancy areas. Occupancy separations that serve to define *fire area* limits established in Chapter 9 for requiring a *fire protection system* shall also comply with Section 901.7.

508.4.2 Allowable building area. In each *story*, the *building area* shall be such that the sum of the ratios of the actual *building area* of each separated occupancy divided by the allowable *building* area of each separated occupancy shall not exceed 1.

508.4.3 Allowable building height and number of stories. Each separated occupancy shall comply with the *building height limitations* and *story* limitations based on the type of construction of the building in accordance with Section 503.1.

Exception: Special provisions of Section 510 shall permit occupancies at *building heights* and number of *stories* other than provided in Section 503.1.

508.4.4 Separation. Individual occupancies shall be separated from adjacent occupancies in accordance with Table 508.4.

508.4.4.1 Construction. Required separations shall be *fire barriers* constructed in accordance with Section 707 or *horizontal assemblies* constructed in accordance with Section 711, or both, so as to completely separate adjacent occupancies. *Mass timber* elements serving as *fire barriers* or *horizontal assemblies* to separate occupancies in Type IV-B or IV-C construction shall be separated from the interior of the building with an *approved* thermal barrier consisting of *gypsum board* that is not less than $^1/_2$ inch (12.7 mm) in thickness or a material that is tested in accordance with and meets the acceptance criteria of both the Temperature Transmission Fire Test and the Integrity Fire Test of NFPA 275.

508.5 Live/work units A *live/work unit* shall comply with Sections 508.5 through 508.5.11.

Exception: *Dwelling or sleeping units* that include an office that is less than 10 percent of the area of the *dwelling unit* are permitted to be classified as *dwelling units* with accessory occupancies in accordance with Section 508.2.

508.5.1 Limitations. The following shall apply to live/work areas:

1. The *live/work unit* is permitted to be not greater than 3,000 square feet (279 m^2) in area.
2. The nonresidential area is permitted to be not more than 50 percent of the area of each *live/work unit*.
3. The nonresidential area function shall be limited to the first or main floor only of the *live/work unit*.
4. Not more than five nonresidential workers or employees are allowed to occupy the nonresidential area at any one time.

508.5.2 Occupancies. *Live/work units* shall be classified as a Group R-2 occupancy. Separation requirements found in Sections 420 and 508 shall not apply within the *live/work unit* where the *live/work unit* is in compliance with Section 508.5. Nonresidential uses that would otherwise be classified as either a Group H or S occupancy shall not be permitted in a *live/work unit*.

Exception: Storage shall be permitted in the *live/work unit* provided that the aggregate area of storage in the nonresidential portion of the *live/work unit* shall be limited to 10 percent of the space dedicated to nonresidential activities.

508.5.3 Means of egress. Except as modified by this section, the *means of egress* components for a *live/work unit* shall be designed in accordance with Chapter 10 for the function served.

TABLE 508.4
REQUIRED SEPARATION OF OCCUPANCIES (HOURS)[f]

OCCUPANCY	A, E		I-1[a], I-3, I-4		I-2		R[a]		F-2, S-2[b], U		B[e], F-1, M, S-1		H-1		H-2		H-3, H-4		H-5	
	S	NS	S	NS	S	NS	S	NS	S	NS	S	NS	S	NS	S	NS	S	NS	S	NS
A, E	N	N	1	2	2	NP	1	2	N	1	1	2	NP	NP	3	4	2	3	2	NP
I-1[a], I-3, I-4	1	2	N	N	2	NP	1	NP	1	2	1	2	NP	NP	3	NP	2	NP	2	NP
I-2	2	NP	2	NP	N	N	2	NP	2	NP	2	NP	NP	NP	3	NP	2	NP	2	NP
R[a]	1	2	1	NP	2	NP	N	N	1[c]	2[c]	1	2	NP	NP	3	NP	2	NP	2	NP
F-2, S-2[b], U	N	1	1	2	2	NP	1[c]	2[c]	N	N	1	2	NP	NP	3	4	2	3	2	NP
B[e], F-1, M, S-1	1	2	1	2	2	NP	1	2	1	2	N	N	NP	NP	2	3	1	2	1	NP
H-1	NP	NP	NP	NP	NP	NP	NP	NP	NP	NP	NP	NP	N	NP	NP	NP	NP	NP	NP	NP
H-2	3	4	3	NP	3	NP	3	NP	3	4	2	3	NP	NP	N	NP	1	NP	1	NP
H-3, H-4	2	3	2	NP	2	NP	2	NP	2	3	1	2	NP	NP	1	NP	1[d]	NP	1	NP
H-5	2	NP	2	NP	2	NP	2	NP	2	NP	1	NP	NP	NP	1	NP	1	NP	N	NP

S = Buildings equipped throughout with an automatic sprinkler system installed in accordance with Section 903.3.1.1.
NS = Buildings not equipped throughout with an automatic sprinkler system installed in accordance with Section 903.3.1.1.
N = No separation requirement.
NP = Not Permitted.

a. See Section 420.
b. The required separation from areas used only for private or pleasure vehicles shall be reduced by 1 hour but not to less than 1 hour.
c. See Sections 406.3.2 and 406.6.4.
d. Separation is not required between occupancies of the same classification.
e. See Section 422.2 for *ambulatory care facilities*.
f. Occupancy separations that serve to define fire area limits established in Chapter 9 for requiring fire protection systems shall also comply with Section 707.3.10 and Table 707.3.10 in accordance with Section 901.7.

508.5.4 Egress capacity. The egress capacity for each element of the *live/work unit* shall be based on the *occupant load* for the function served in accordance with Table 1004.5.

508.5.5 Spiral stairways. *Spiral stairways* that conform to the requirements of Section 1011.10 shall be permitted.

508.5.6 Vertical openings. Floor openings between floor levels of a *live/work unit* are permitted without enclosure.

[F] 508.5.7 Fire protection. The *live/work unit* shall be provided with a monitored *fire alarm* system where required by Section 907.2.9 and an *automatic sprinkler system* in accordance with Section 903.2.8.

508.5.8 Structural. Floors within a *live/work unit* shall be designed for the *live loads* in Table 1607.1, based on the function within the space.

508.5.9 Accessibility. *Accessibility* shall be designed in accordance with Chapter 11 for the function served.

508.5.10 Ventilation. The applicable *ventilation* requirements of the *International Mechanical Code* shall apply to each area within the *live/work unit* for the function within that space.

508.5.11 Plumbing facilities. The nonresidential area of the *live/work unit* shall be provided with minimum plumbing facilities as specified by Chapter 29, based on the function of the nonresidential area. Where the nonresidential area of the *live/work unit* is required to be accessible by Section 1108.6.2.1, the plumbing fixtures specified by Chapter 29 shall be accessible.

SECTION 509
INCIDENTAL USES

509.1 General Incidental uses located within single occupancy or mixed occupancy buildings shall comply with the provisions of this section. Incidental uses are ancillary functions associated with a given occupancy that generally pose a greater level of risk to that occupancy and are limited to those uses specified in Table 509.1.

Exception: Incidental uses within and serving a *dwelling unit* are not required to comply with this section.

[F]TABLE 509.1
INCIDENTAL USES

ROOM OR AREA	SEPARATION AND/OR PROTECTION
Furnace room where any piece of equipment is over 400,000 Btu per hour input	1 hour or provide automatic sprinkler system
Rooms with boilers where the largest piece of equipment is over 15 psi and 10 horsepower	1 hour or provide automatic sprinkler system
Refrigerant machinery room	1 hour or provide automatic sprinkler system
Hydrogen fuel gas rooms, not classified as Group H	1 hour in Group B, F, M, S and U occupancies; 2 hours in Group A, E, I and R occupancies.
Incinerator rooms	2 hours and provide automatic sprinkler system
Paint shops, not classified as Group H, located in occupancies other than Group F	2 hours; or 1 hour and provide automatic sprinkler system
In Group E occupancies, laboratories and vocational shops not classified as Group H	1 hour or provide automatic sprinkler system
In Group I-2 occupancies, laboratories not classified as Group H	1 hour and provide automatic sprinkler system
In *ambulatory care facilities*, laboratories not classified as Group H	1 hour or provide automatic sprinkler system
Laundry rooms over 100 square feet	1 hour or provide automatic sprinkler system
In Group I-2, laundry rooms over 100 square feet	1 hour
Group I-3 cells and Group I-2 patient rooms equipped with padded surfaces	1 hour
In Group I-2, physical plant maintenance shops	1 hour
In ambulatory care facilities or Group I-2 occupancies, waste and linen collection rooms with containers that have an aggregate volume of 10 cubic feet or greater	1 hour
In other than ambulatory care facilities and Group I-2 occupancies, waste and linen collection rooms over 100 square feet	1 hour or provide automatic sprinkler system
In ambulatory care facilities or Group I-2 occupancies, storage rooms greater than 100 square feet	1 hour
➡ Electrical installations and transformers	See Sections 110.26 through 110.34 and Sections 450.8 through 450.48 of NFPA 70 for protection and separation requirements.

For SI: 1 square foot = 0.0929 m^2, 1 pound per square inch (psi) = 6.9 kPa, 1 British thermal unit (Btu) per hour = 0.293 watts, 1 horsepower = 746 watts, 1 gallon = 3.785 L, 1 cubic foot = 0.0283 m^3.

509.2 Occupancy classification. Incidental uses shall not be individually classified in accordance with Section 302.1. Incidental uses shall be included in the building occupancies within which they are located.

509.3 Area limitations. Incidental uses shall not occupy more than 10 percent of the *building area* of the *story* in which they are located.

509.4 Separation and protection. The incidental uses specified in Table 509.1 shall be separated from the remainder of the building or equipped with an *automatic sprinkler system*, or both, in accordance with the provisions of that table.

509.4.1 Separation. Where Table 509.1 specifies a fire-resistance-rated separation, the incidental uses shall be separated from the remainder of the *building* by a *fire barrier* constructed in accordance with Section 707 or a *horizontal assembly* constructed in accordance with Section 711, or both. Construction supporting 1-hour *fire barriers* or *horizontal assemblies* used for incidental use separations in buildings of Type IIB, IIIB and VB construction is not required to be fire-resistance rated unless required by other sections of this code.

509.4.1.1 Type IV-B and IV-C construction. Where Table 509.1 specifies a fire-resistance-rated separation, *mass timber* elements serving as *fire barriers* or *horizontal assemblies* in Type IV-B or IV-C construction shall be separated from the interior of the incidental use with an *approved* thermal barrier consisting of *gypsum board* that is not less than $^1/_2$ inch (12.7 mm) in thickness or a material that is tested in accordance with and meets the acceptance criteria of both the Temperature Transmission Fire Test and the Integrity Fire Test of NFPA 275.

509.4.2 Protection. Where Table 509.1 permits an *automatic sprinkler system* without a *fire barrier*, the incidental uses shall be separated from the remainder of the building by construction capable of resisting the passage of smoke. The walls shall extend from the top of the foundation or floor assembly below to the underside of the ceiling that is a component of a fire-resistance-rated floor assembly or roof assembly above or to the underside of the floor or roof sheathing, deck or slab above. Doors shall be self- or automatic-closing upon detection of smoke in accordance with Section 716.2.6.6. Doors shall not have air transfer openings and shall not be undercut in excess of the clearance permitted in accordance with NFPA 80. Walls surrounding the incidental use shall not have air transfer openings unless provided with *smoke dampers* in accordance with Section 710.8.

509.4.2.1 Protection limitation. Where an *automatic sprinkler system* is provided in accordance with Table 509.1, only the space occupied by the incidental use need be equipped with such a system.

SECTION 510
SPECIAL PROVISIONS

510.1 General. The provisions in Sections 510.2 through 510.9 shall permit the use of special conditions that are exempt from, or modify, the specific requirements of this chapter regarding the allowable *building heights and areas* of buildings based on the occupancy classification and type of construction, provided the special condition complies with the provisions specified in this section for such condition and other applicable requirements of this code. The provisions of Sections 510.2 through 510.8 are to be considered independent and separate from each other.

510.2 Horizontal building separation allowance. A building shall be considered as separate and distinct buildings for the purpose of determining area limitations, continuity of *fire walls*, limitation of number of *stories* and type of construction where the following conditions are met:

1. The buildings are separated with a *horizontal assembly* having a *fire-resistance rating* of not less than 3 hours. Where vertical offsets are provided as part of a *horizontal assembly*, the vertical offset and the structure supporting the vertical offset shall have a *fire-resistance rating* of not less than 3 hours.
2. The building below, including the *horizontal assembly*, is of Type IA construction.
3. *Shaft, stairway, ramp* and escalator enclosures through the *horizontal assembly* shall have not less than a 2-hour *fire-resistance rating* with opening protectives in accordance with Section 716.

   Exception: Where the enclosure walls below the *horizontal assembly* have not less than a 3-hour *fire-resistance rating* with opening protectives in accordance with Section 716, the enclosure walls extending above the *horizontal assembly* shall be permitted to have a 1-hour *fire-resistance rating*, provided that the following conditions are met:

   1. The building above the *horizontal assembly* is not required to be of Type I construction.
   2. The enclosure connects fewer than four stories.
   3. The enclosure opening protectives above the *horizontal assembly* have a *fire protection rating* of not less than 1 hour.

4. *Interior exit stairways* located within the Type IA building are permitted to be of combustible materials where the following requirements are met:

   4.1. The building above the Type IA building is of Type III, IV, or V construction.

   4.2. The *stairway* located in the Type IA building is enclosed by 3-hour fire-resistance-rated construction with opening protectives in accordance with Section 716.

5. The building or buildings above the *horizontal assembly* shall be permitted to have multiple Group A occupancy uses, each with an *occupant load* of less 300, or Group B, M, R or S occupancies.
6. The building below the *horizontal assembly* shall be protected throughout by an *approved automatic*

sprinkler system in accordance with Section 903.3.1.1, and shall be permitted to be any occupancy allowed by this code except Group H.

7, The maximum *building height* in feet (mm) shall not exceed the limits set forth in Section 504.3 for the building having the smaller allowable height as measured from the *grade plane*.

510.3 Group S-2 enclosed parking garage with Group S-2 open parking garage above. A Group S-2 enclosed parking garage with not more than one *story above grade plane* and located below a Group S-2 *open parking garage* shall be classified as a separate and distinct building for the purpose of determining the type of construction where the following conditions are met:

1. The allowable area of the building shall be such that the sum of the ratios of the actual area divided by the allowable area for each separate occupancy shall not exceed 1.
2. The Group S-2 enclosed parking garage is of Type I or II construction and is at least equal to the *fire-resistance* requirements of the Group S-2 *open parking garage*.
3. The height and the number of tiers of the Group S-2 *open parking garage* shall be limited as specified in Table 406.5.4.
4. The floor assembly separating the Group S-2 enclosed parking garage and Group S-2 *open parking garage* shall be protected as required for the floor assembly of the Group S-2 enclosed parking garage. Openings between the Group S-2 enclosed parking garage and Group S-2 *open parking garage*, except exit openings, shall not be required to be protected.
5. The Group S-2 enclosed parking garage is used exclusively for the parking or storage of private motor vehicles, but shall be permitted to contain an office, waiting room and toilet room having a total area of not more than 1,000 square feet (93 m^2) and mechanical equipment rooms associated with the operation of the building.

510.4 Parking beneath Group R. Where a maximum one *story above grade plane* Group S-2 parking garage, enclosed or open, or combination thereof, of Type I construction or open of Type IV construction, with grade entrance, is provided under a building of Group R, the number of *stories* to be used in determining the minimum type of construction shall be measured from the floor above such a parking area. The floor assembly between the parking garage and the Group R above shall comply with the type of construction required for the parking garage and shall also provide a *fire-resistance rating* not less than the mixed occupancy separation required in Section 508.4.

510.5 Group R-1 and R-2 buildings of Type IIIA construction. For buildings of Type IIIA construction in Groups R-1 and R-2, the maximum allowable height in Table 504.3 shall be increased by 10 feet (3048 mm) and the maximum allowable number of *stories* in Table 504.4 shall be increased by one where the first-floor assembly above the *basement* has a *fire-resistance rating* of not less than 3 hours and the floor area is subdivided by 2-hour fire-resistance-rated *fire walls* into areas of not more than 3,000 square feet (279 m^2).

510.6 Group R-1 and R-2 buildings of Type IIA construction. The height limitation for buildings of Type IIA construction in Groups R-1 and R-2 shall be increased to nine *stories* and 100 feet (30 480 mm) where the building is separated by not less than 50 feet (15 240 mm) from any other building on the *lot* and from *lot lines*, the *exits* are segregated in an area enclosed by a 2-hour fire-resistance-rated *fire wall* and the first floor assembly has a *fire-resistance rating* of not less than $1^1/_2$ hours.

510.7 Open parking garage beneath Groups A, I, B, M and R. *Open parking garages* constructed under Groups A, I, B, M and R shall not exceed the height and area limitations permitted under Section 406.5. The height and area of the portion of the building above the *open parking garage* shall not exceed the limitations in Section 503 for the upper occupancy. The height, in both feet and *stories*, of the portion of the building above the *open parking garage* shall be measured from *grade plane* and shall include both the *open parking garage* and the portion of the building above the parking garage.

510.7.1 Fire separation. *Fire barriers* constructed in accordance with Section 707 or *horizontal assemblies* constructed in accordance with Section 711 between the parking occupancy and the upper occupancy shall correspond to the required *fire-resistance rating* prescribed in Table 508.4 for the uses involved. The type of construction shall apply to each occupancy individually, except that structural members, including main bracing within the open parking structure, which is necessary to support the upper occupancy, shall be protected with the more restrictive fire-resistance-rated assemblies of the groups involved as shown in Table 601. *Means of egress* for the upper occupancy shall conform to Chapter 10 and shall be separated from the parking occupancy by *fire barriers* having not less than a 2-hour *fire-resistance rating* as required by Section 707 with *self-closing* doors complying with Section 716 or *horizontal assemblies* having not less than a 2-hour *fire-resistance rating* as required by Section 711, with *self-closing* doors complying with Section 716. *Means of egress* from the *open parking garage* shall comply with Section 406.5.

510.8 Group B or M buildings with Group S-2 open parking garage above. Group B or M occupancies located below a Group S-2 *open parking garage* of a lesser type of construction shall be considered as a separate and distinct building from the Group S-2 *open parking garage* for the purpose of determining the type of construction where the following conditions are met:

1. The buildings are separated with a *horizontal assembly* having a *fire-resistance rating* of not less than 2 hours.
2. The occupancies in the building below the *horizontal assembly* are limited to Groups B and M.

3. The occupancy above the *horizontal assembly* is limited to a Group S-2 *open parking garage.*
4. The building below the *horizontal assembly* is of Type IA construction.

   Exception: The building below the *horizontal assembly* shall be permitted to be of Type IB or II construction, but not less than the type of construction required for the Group S-2 *open parking garage* above, where the building below is not greater than *one story* in height above *grade plane.*
5. The height and area of the building below the *horizontal assembly* does not exceed the limits set forth in Section 503.
6. The height and area of the Group S-2 *open parking garage* does not exceed the limits set forth in Section 406.5. The height, in both feet and *stories*, of the Group S-2 *open parking garage* shall be measured from *grade plane* and shall include the building below the *horizontal assembly.*
7. *Exits* serving the Group S-2 *open parking garage* discharge at grade with direct and unobstructed access to a street or *public way* and are separated from the building below the *horizontal assembly* by 2-hour *fire barriers* constructed in accordance with Section 707 or 2-hour *horizontal assemblies* constructed in accordance with Section 711, or both.

510.9 Multiple buildings above a horizontal assembly. Where two or more buildings are provided above the *horizontal assembly* separating a Group S-2 parking garage or building below from the buildings above in accordance with the special provisions in Section 510.2, 510.3 or 510.8, the buildings above the *horizontal assembly* shall be regarded as separate and distinct buildings from each other and shall comply with all other provisions of this code as applicable to each separate and distinct building.

CHAPTER 6

TYPES OF CONSTRUCTION

User note:

About this chapter: *Chapter 6 establishes five types of construction in which each building must be categorized. This chapter looks at the materials used in the building (combustible or noncombustible) and the extent to which building elements such as the building frame, roof, wall and floor can resist fire. Depending on the type of construction and the specific building element, fire resistance of 1 to 3 hours is specified.*

SECTION 601
GENERAL

601.1 Scope. The provisions of this chapter shall control the classification of buildings as to type of construction.

SECTION 602
CONSTRUCTION CLASSIFICATION

602.1 General. Buildings and structures erected or to be erected, altered or extended in height or area shall be classified in one of the five *construction types* defined in Sections 602.2 through 602.5. The *building elements* shall have a *fire-resistance rating* not less than that specified in Table 601 and *exterior walls* shall have a *fire-resistance rating* not less than that specified in Table 705.5. Where required to have a *fire-resistance rating* by Table 601, *building elements* shall comply with the applicable provisions of Section 703.2. The protection of openings, ducts and air transfer openings in *building elements* shall not be required unless required by other provisions of this code.

602.1.1 Minimum requirements. A building or portion thereof shall not be required to conform to the details of a type of construction higher than that type which meets the minimum requirements based on occupancy even though certain features of such a building actually conform to a higher type of construction.

602.2 Types I and II. Types I and II construction are those types of construction in which the *building elements* specified in Table 601 are of noncombustible materials, except as permitted in Section 603 and elsewhere in this code.

602.3 Type III. Type III construction is that type of construction in which the *exterior walls* are of noncombustible materials and the interior *building elements* are of any material permitted by this code. *Fire-retardant-treated wood* framing and sheathing complying with Section 2303.2 shall be permitted within *exterior wall* assemblies of a 2-hour rating or less.

TABLE 601
FIRE-RESISTANCE RATING REQUIREMENTS FOR BUILDING ELEMENTS (HOURS)

BUILDING ELEMENT	TYPE I		TYPE II		TYPE III		TYPE IV				TYPE V	
	A	B	A	B	A	B	A	B	C	HT	A	B
Primary structural frame[f] (see Section 202)	3[a, b]	2[a, b, c]	1[b, c]	0[c]	1[b, c]	0	3[a]	2[a]	2[a]	HT	1[b, c]	0
Bearing walls												
Exterior[e, f]	3	2	1	0	2	2	3	2	2	2	1	0
Interior	3[a]	2[a]	1	0	1	0	3	2	2	1/HT[g]	1	0
Nonbearing walls and partitions Exterior	See Table 705.5											
Nonbearing walls and partitions Interior[d]	0	0	0	0	0	0	0	0	0	See Section 2304.11.2	0	0
Floor construction and associated secondary structural members (see Section 202)	2	2	1	0	1	0	2	2	2	HT	1	0
Roof construction and associated secondary structural members (see Section 202)	1 1/2[b]	1[b,c]	1[b,c]	0[c]	1[b,c]	0	1 1/2	1	1	HT	1[b,c]	0

For SI: 1 foot = 304.8 mm.

a. Roof supports: Fire-resistance ratings of primary structural frame and bearing walls are permitted to be reduced by 1 hour where supporting a roof only.

b. Except in Group F-1, H, M and S-1 occupancies, fire protection of structural members in roof construction shall not be required, including protection of primary structural frame members, roof framing and decking where every part of the roof construction is 20 feet or more above any floor immediately below. Fire-retardant-treated wood members shall be allowed to be used for such unprotected members.

c. In all occupancies, heavy timber complying with Section 2304.11 shall be allowed for roof construction, including primary structural frame members, where a 1-hour or less fire-resistance rating is required.

d. Not less than the fire-resistance rating required by other sections of this code.

e. Not less than the fire-resistance rating based on fire separation distance (see Table 705.5).

f. Not less than the fire-resistance rating as referenced in Section 704.10.

g. Heavy timber bearing walls supporting more than two floors or more than a floor and a roof shall have a fire resistance rating of not less than 1 hour.

602.4 Type IV. Type IV construction is that type of construction in which the *building elements* are *mass timber* or noncombustible materials and have *fire-resistance ratings* in accordance with Table 601. *Mass timber* elements shall meet the *fire-resistance-rating* requirements of this section based on either the *fire-resistance rating* of the *noncombustible protection*, the *mass timber*, or a combination of both and shall be determined in accordance with Section 703.2. The minimum dimensions and permitted materials for *building elements* shall comply with the provisions of this section and Section 2304.11. *Mass timber* elements of Types IV-A, IV-B and IV-C construction shall be protected with *noncombustible protection* applied directly to the *mass timber* in accordance with Sections 602.4.1 through 602.4.3. The time assigned to the *noncombustible protection* shall be determined in accordance with Section 703.6 and comply with Section 722.7.

Cross-laminated timber shall be labeled as conforming to ANSI/APA PRG 320 as referenced in Section 2303.1.4.

Exterior *load-bearing walls* and *nonload-bearing walls* shall be *mass timber* construction, or shall be of noncombustible construction.

Exception: Exterior load-bearing walls and nonload-bearing walls of Type IV-HT Construction in accordance with Section 602.4.4.

The interior *building elements* , including *nonload-bearing walls* and partitions, shall be of *mass timber* construction or of noncombustible construction.

Exception: Interior building elements and nonload-bearing walls and partitions of Type IV-HT construction in accordance with Section 602.4.4.

Combustible concealed spaces are not permitted except as otherwise indicated in Sections 602.4.1 through 602.4.4. Combustible stud spaces within light frame walls of Type IV-HT construction shall not be considered concealed spaces, but shall comply with Section 718.

In buildings of Type IV-A, IV-B, and IV-C construction with an occupied floor located more than 75 feet (22 860 mm) above the lowest level of fire department access, up to and including 12 *stories* or 180 feet (54 864 mm) above *grade plane*, *mass timber* interior exit and elevator hoistway enclosures shall be protected in accordance with Section 602.4.1.2. In buildings greater than 12 *stories* or 180 feet (54 864 mm) above *grade plane*, interior exit and elevator hoistway enclosures shall be constructed of noncombustible materials.

602.4.1 Type IV-A. *Building elements* in Type IV-A construction shall be protected in accordance with Sections 602.4.1.1 through 602.4.1.6. The required *fire-resistance rating* of noncombustible elements and protected *mass timber* elements shall be determined in accordance with Section 703.2.

602.4.1.1 Exterior protection. The outside face of *exterior walls* of *mass timber* construction shall be protected with *noncombustible protection* with a minimum assigned time of 40 minutes, as specified in Table 722.7.1(1). Components of the *exterior wall covering* shall be of noncombustible material except *water-resistive barriers* having a peak heat release rate of less than 150kW/m^2, a total heat release of less than 20 MJ/m^2 and an effective heat of combustion of less than 18MJ/kg as determined in accordance with ASTM E1354 and having a *flame spread index* of 25 or less and a *smoke-developed index* of 450 or less as determined in accordance with ASTM E84 or UL 723. The ASTM E1354 test shall be conducted on specimens at the thickness intended for use, in the horizontal orientation and at an incident radiant heat flux of 50 kW/m^2.

602.4.1.2 Interior protection. Interior faces of all *mass timber* elements, including the inside faces of exterior *mass timber* walls and *mass timber* roofs, shall be protected with materials complying with Section 703.3.

602.4.1.2.1 Protection time. *Noncombustible protection* shall contribute a time equal to or greater than times assigned in Table 722.7.1(1), but not less than 80 minutes. The use of materials and their respective protection contributions specified in Table 722.7.1(2) shall be permitted to be used for compliance with Section 722.7.1.

602.4.1.3 Floors. The floor assembly shall contain a noncombustible material not less than 1 inch (25 mm) in thickness above the *mass timber*. Floor finishes in accordance with Section 804 shall be permitted on top of the noncombustible material. The underside of floor assemblies shall be protected in accordance with Section 602.4.1.2.

602.4.1.4 Roofs. The *interior surfaces* of *roof assemblies* shall be protected in accordance with Section 602.4.1.2. *Roof coverings* in accordance with Chapter 15 shall be permitted on the outside surface of the *roof assembly.*

602.4.1.5 Concealed spaces. Concealed spaces shall not contain combustibles other than electrical, mechanical, fire protection, or plumbing materials and equipment permitted in plenums in accordance with Section 602 of the *International Mechanical Code*, and shall comply with all applicable provisions of Section 718. Combustible construction forming concealed spaces shall be protected in accordance with Section 602.4.1.2.

602.4.1.6 Shafts. *Shafts* shall be permitted in accordance with Sections 713 and 718. Both the *shaft* side and room side of *mass timber* elements shall be protected in accordance with Section 602.4.1.2.

602.4.2 Type IV-B. *Building elements* in Type IV-B construction shall be protected in accordance with Sections 602.4.2.1 through 602.4.2.6. The required *fire-resistance rating* of noncombustible elements or *mass timber* elements shall be determined in accordance with Section 703.2.

602.4.2.1 Exterior protection. The outside face of *exterior walls* of *mass timber* construction shall be protected with *noncombustible protection* with a minimum assigned time of 40 minutes, as specified in Table 722.7.1(1). Components of the *exterior wall covering* shall be of noncombustible material except *water-resistive barriers* having a peak heat release rate of less than 150kW/m^2, a total heat release of less than 20 MJ/m^2 and an effective heat of combustion of less than 18MJ/kg as determined in accordance with ASTM E1354, and having a *flame spread index* of 25 or less and a *smoke-developed index* of 450 or less as determined in accordance with ASTM E84 or UL 723. The ASTM E1354 test shall be conducted on specimens at the thickness intended for use, in the horizontal orientation and at an incident radiant heat flux of 50 kW/m^2.

602.4.2.2 Interior protection. Interior faces of all *mass timber* elements, including the inside face of exterior *mass timber* walls and *mass timber* roofs, shall be protected, as required by this section, with materials complying with Section 707.3.

602.4.2.2.1 Protection time. *Noncombustible protection* shall contribute a time equal to or greater than times assigned in Table 722.7.1(1), but not less than 80 minutes. The use of materials and their respective protection contributions specified in Table 722.7.1(2) shall be permitted to be used for compliance with Section 722.7.1.

602.4.2.2.2 Protected area. Interior faces of *mass timber* elements, including the inside face of exterior *mass timber walls* and *mass timber roofs*, shall be protected in accordance with Section 602.4.2.2.1.

Exceptions: Unprotected portions of mass timber ceilings and walls complying with Section 602.4.2.2.4 and the following:

1. Unprotected portions of mass timber ceilings and walls complying with one of the following:
   - 1.1. Unprotected portions of mass timber ceilings, including attached beams, shall be permitted and shall be limited to an area equal to 20 percent of the floor area in any dwelling unit or fire area.
   - 1.2. Unprotected portions of mass timber walls, including attached columns, shall be permitted and shall be limited to an area equal to 40 percent of the floor area in any dwelling unit or fire area.
   - 1.3. Unprotected portions of both walls and ceilings of mass timber, including attached columns and beams, in any dwelling unit or fire area shall be permitted in accordance with Section 602.4.2.2.3.
2. Mass timber columns and beams that are not an integral portion of walls or ceilings, respectively, shall be permitted to be unprotected without restriction of either aggregate area or separation from one another.

602.4.2.2.3 Mixed unprotected areas. In each *dwelling unit* or *fire area*, where both portions of ceilings and portions of walls are unprotected, the total allowable unprotected area shall be determined in accordance with Equation 6-1.

$$(U_{tc}/U_{ac}) + (U_{tw}/U_{aw}) \leq 1 \qquad \textbf{(Equation 6-1)}$$

where:

U_{tc} = Total unprotected mass timber ceiling areas.

U_{ac} = Allowable unprotected mass timber ceiling area conforming to Exception 1.1 of Section 602.4.2.2.2.

U_{tw} = Total unprotected mass timber wall areas.

U_{aw} = Allowable unprotected mass timber wall area conforming to Exception 1.2 of Section 602.4.2.2.2.

602.4.2.2.4 Separation distance between unprotected mass timber elements. In each *dwelling unit* or *fire area*, unprotected portions of *mass timber* walls and ceilings shall be not less than 15 feet (4572 mm) from unprotected portions of other walls and ceilings, measured horizontally along the ceiling and from other unprotected portions of walls measured horizontally along the floor.

602.4.2.3 Floors. The floor assembly shall contain a noncombustible material not less than 1 inch (25 mm) in thickness above the *mass timber*. Floor finishes in accordance with Section 804 shall be permitted on top of the noncombustible material. The underside of floor assemblies shall be protected in accordance with Section 602.4.1.2.

602.4.2.4 Roofs. The *interior surfaces* of roof assemblies shall be protected in accordance with Section 602.4.2.2 except, in nonoccupiable spaces, they shall be treated as a concealed space with no portion left unprotected. *Roof coverings* in accordance with Chapter 15 shall be permitted on the outside surface of the roof assembly.

602.4.2.5 Concealed spaces. Concealed spaces shall not contain combustibles other than electrical, mechanical, fire protection, or plumbing materials and equipment permitted in plenums in accordance with Section 602 of the *International Mechanical Code*, and shall comply with all applicable provisions of Section 718. Combustible construction forming concealed spaces shall be protected in accordance with Section 602.4.1.2.

602.4.2.6 Shafts. *Shafts* shall be permitted in accordance with Sections 713 and 718. Both the *shaft* side and room side of *mass timber* elements shall be protected in accordance with Section 602.4.1.2.

602.4.3 Type IV-C. *Building elements* in Type IV-C construction shall be protected in accordance with Sections 602.4.3.1 through 602.4.3.6. The required *fire-resistance rating* of *building elements* shall be determined in accordance with Section 703.2.

602.4.3.1 Exterior protection. The exterior side of walls of combustible construction shall be protected with *noncombustible protection* with a minimum assigned time of 40 minutes, as determined in Table 722.7.1(1). Components of the *exterior wall covering* shall be of noncombustible material except *water-resistive barriers* having a peak heat release rate of less than 150 kW/m^2, a total heat release of less than 20 MJ/m^2 and an effective heat of combustion of less than 18 MJ/kg as determined in accordance with ASTM E1354 and having a *flame spread index* of 25 or less and a *smoke-developed index* of 450 or less as determined in accordance with ASTM E84 or UL 723. The ASTM E1354 test shall be conducted on specimens at the thickness intended for use, in the horizontal orientation and at an incident radiant heat flux of 50 kW/m^2.

602.4.3.2 Interior protection. *Mass timber* elements are permitted to be unprotected.

602.4.3.3 Floors. Floor finishes in accordance with Section 804 shall be permitted on top of the floor construction.

602.4.3.4 Roof coverings. *Roof coverings* in accordance with Chapter 15 shall be permitted on the outside surface of the roof assembly.

602.4.3.5 Concealed spaces. Concealed spaces shall not contain combustibles other than electrical, mechanical, fire protection, or plumbing materials and equipment permitted in plenums in accordance with Section 602 of the *International Mechanical Code*, and shall comply with all applicable provisions of Section 718. Combustible construction forming concealed spaces shall be protected with *noncombustible protection* with a minimum assigned time of 40 minutes, as specified in Table 722.7.1(1).

602.4.3.6 Shafts. *Shafts* shall be permitted in accordance with Sections 713 and 718. *Shafts* and elevator hoistway and *interior exit stairway enclosures* shall be protected with *noncombustible protection* with a minimum assigned time of 40 minutes, as specified in Table 722.7.1(1), on both the inside of the *shaft* and the outside of the *shaft*.

602.4.4 Type IV-HT. Type IV-HT (Heavy Timber) construction is that type of construction in which the *exterior walls* are of noncombustible materials and the interior *building elements* are of solid wood, laminated heavy timber or *structural composite lumber* (SCL), without concealed spaces or with concealed spaces complying with Section 602.4.4.3. The minimum dimensions for permitted materials including solid timber, glued-laminated timber, SCL and *cross-laminated timber* (*CLT*) and the details of Type IV construction shall comply with the provisions of this section and Section 2304.11. *Exterior walls* complying with Section 602.4.4.1 or 602.4.4.2 shall be permitted. Interior walls and partitions not less than 1-hour fire-resistance rated or heavy timber conforming with Section 2304.11.2.2 shall be permitted.

602.4.4.1 Fire-retardant-treated wood in exterior walls. *Fire-retardant-treated wood* framing and sheathing complying with Section 2303.2 shall be permitted within *exterior wall* assemblies with a 2-hour rating or less.

602.4.4.2 Cross-laminated timber in exterior walls. *Cross-laminated timber* (*CLT*) not less than 4 inches (102 mm) in thickness complying with Section 2303.1.4 shall be permitted within *exterior wall* assemblies with a 2-hour rating or less. Heavy timber structural members appurtenant to the *CLT exterior wall* shall meet the requirements of Table 2304.11 and be fire-resistance rated as required for the *exterior wall*. The exterior surface of the cross-laminated timber and heavy timber elements shall be protected by one the following:

1. *Fire-retardant-treated wood* sheathing complying with Section 2303.2 and not less than $^{15}/_{32}$ inch (12 mm) thick.
2. *Gypsum board* not less than $^{1}/_{2}$ inch (12.7 mm) thick.
3. A noncombustible material.

602.4.4.3 Concealed spaces. Concealed spaces shall not contain combustible materials other than *building elements* and electrical, mechanical, fire protection, or plumbing materials and equipment permitted in plenums in accordance with Section 602 of the *International Mechanical Code*. Concealed spaces shall comply with applicable provisions of Section 718. Concealed spaces shall be protected in accordance with one or more of the following:

1. The building shall be sprinklered throughout in accordance with Section 903.3.1.1 and automatic sprinklers shall also be provided in the concealed space.
2. The concealed space shall be completely filled with noncombustible insulation.
3. Surfaces within the concealed space shall be fully sheathed with not less than $^{5}/_{8}$-inch Type X gypsum board.

Exception: Concealed spaces within interior walls and partitions with a 1-hour or greater fire-resistance rating complying with Section 2304.11.2.2 shall not require additional protection.

602.4.4.4 Exterior structural members. Where a horizontal separation of 20 feet (6096 mm) or more is provided, wood columns and arches conforming to heavy timber sizes complying with Section 2304.11 shall be permitted to be used externally.

602.5 Type V. Type V construction is that type of construction in which the structural elements, *exterior walls* and interior walls are of any materials permitted by this code.

SECTION 603 COMBUSTIBLE MATERIAL IN TYPES I AND II CONSTRUCTION

603.1 Allowable materials. Combustible materials shall be permitted in buildings of Type I or II construction in the following applications and in accordance with Sections 603.1.1 through 603.1.3:

1. *Fire-retardant-treated wood* shall be permitted in:
   1.1. Nonbearing partitions where the required *fire-resistance rating* is 2 hours or less except in *shaft enclosures* within Group I-2 occupancies and *ambulatory care facilities.*

   1.2. Nonbearing *exterior walls* where fire-resistance-rated construction is not required.

   1.3. Roof construction, including girders, trusses, framing and decking.

   Exceptions:

   1. In buildings of Type IA construction exceeding two *stories above grade plane*, *fire-retardant-treated wood* is not permitted in roof construction where the vertical distance from the upper floor to the roof is less than 20 feet (6096 mm).
   2. Group I-2, roof construction containing *fire-retardant-treated wood* shall be covered by not less than a Class A *roof covering* or roof assembly, and the roof assembly shall have a *fire-resistance rating* where required by the construction type.

   1.4. Balconies, porches, decks and exterior *stairways* not used as required exits on buildings three *stories* or less above grade plane.

2. Thermal and acoustical insulation, other than foam plastics, having a *flame spread index* of not more than 25.

   Exceptions:

   1. Insulation placed between two layers of noncombustible materials without an intervening airspace shall be allowed to have a *flame spread index* of not more than 100.
   2. Insulation installed between a finished floor and solid decking without intervening airspace shall be allowed to have a *flame spread index* of not more than 200.

3. Foam plastics in accordance with Chapter 26.
4. *Roof coverings* that have an A, B or C classification.
5. *Interior floor finish* and floor covering materials installed in accordance with Section 804.
6. Millwork such as doors, door frames, window sashes and frames.
7. *Interior wall and ceiling finishes* installed in accordance with Section 803.
8. *Trim* installed in accordance with Section 806.
9. Where not installed greater than 15 feet (4572 mm) above grade, show windows, nailing or furring strips and wooden bulkheads below show windows, including their frames, aprons and show cases.
10. Finish flooring installed in accordance with Section 805.
11. Partitions dividing portions of stores, offices or similar places occupied by one tenant only and that do not establish a *corridor* serving an *occupant load* of 30 or more shall be permitted to be constructed of *fire-retardant-treated* wood, 1-hour fire-resistance-rated construction or of wood panels or similar light construction up to 6 feet (1829 mm) in height.
12. *Stages* and *platforms* constructed in accordance with Sections 410.2 and 410.3, respectively.
13. Combustible *exterior wall coverings*, balconies and similar projections and bay or oriel windows in accordance with Chapter 14 and Section 705.2.3.1.
14. Blocking such as for handrails, millwork, cabinets and window and door frames.
15. Light-transmitting plastics as permitted by Chapter 26.
16. Mastics and caulking materials applied to provide flexible seals between components of *exterior wall* construction.
17. Exterior plastic *veneer* installed in accordance with Section 2605.2.
18. Nailing or furring strips as permitted by Section 803.15.
19. Heavy timber as permitted by Note c to Table 601 and Sections 602.4.4.4 and 705.2.3.1.
20. Aggregates, component materials and admixtures as permitted by Section 703.2.1.2.
21. Sprayed fire-resistant materials and intumescent and mastic fire-resistant coatings, determined on the basis of *fire resistance* tests in accordance with Section 703.2 and installed in accordance with Sections 1705.15 and 1705.16, respectively.
22. Materials used to protect penetrations in fire-resistance-rated assemblies in accordance with Section 714.
23. Materials used to protect *joints* in fire-resistance-rated assemblies in accordance with Section 715.
24. Materials allowed in the concealed spaces of buildings of Types I and II construction in accordance with Section 718.5.

25. Materials exposed within plenums complying with Section 602 of the *International Mechanical Code*.
26. Wall construction of freezers and coolers of less than 1,000 square feet (92.9 m^2), in size, lined on both sides with noncombustible materials and the building is protected throughout with an *automatic sprinkler system* in accordance with Section 903.3.1.1.
27. Wood nailers for parapet flashing and roof cants.

603.1.1 Ducts. The use of nonmetallic ducts shall be permitted where installed in accordance with the limitations of the *International Mechanical Code.*

603.1.2 Piping. The use of combustible piping materials shall be permitted where installed in accordance with the limitations of the *International Mechanical Code* and the *International Plumbing Code*.

603.1.3 Electrical. The use of electrical wiring methods with combustible insulation, tubing, raceways and related components shall be permitted where installed in accordance with the limitations of this code.

CHAPTER 7
FIRE AND SMOKE PROTECTION FEATURES

User note:

About this chapter: *Chapter 7 provides detailed requirements for fire-resistance-rated construction, including structural members, walls, partitions and horizontal assemblies. Other portions of the code describe where certain fire-resistance-rated elements are required. This chapter specifies how these elements are constructed, how openings in walls and partitions are protected and how penetrations of such elements are protected.*

SECTION 701
GENERAL

701.1 Scope. The provisions of this chapter shall govern the materials, systems and assemblies used for structural *fire resistance* and fire-resistance-rated construction separation of adjacent spaces to safeguard against the spread of fire and smoke within a building and the spread of fire to or from buildings.

SECTION 702
MULTIPLE-USE FIRE ASSEMBLIES

702.1 Multiple-use fire assemblies. Fire assemblies that serve multiple purposes in a building shall comply with all of the requirements that are applicable for each of the individual fire assemblies.

SECTION 703
FIRE-RESISTANCE RATINGS AND FIRE TESTS

703.1 Scope. Materials prescribed herein for *fire resistance* shall conform to the requirements of this chapter.

703.2 Fire resistance. The *fire-resistance rating* of *building elements*, components or assemblies shall be determined in accordance with Section 703.2.1 or 703.2.2 without the use of *automatic* sprinklers or any other fire suppression system being incorporated, or in accordance with Section 703.2.3.

703.2.1 Tested assemblies. A *fire-resistance rating* of *building elements,* components or assemblies shall be determined by the test procedures set forth in ASTM E119 or UL 263. The *fire-resistance rating* of penetrations and *fire-resistant joint systems* shall be determined in accordance with Sections 714 and 715, respectively.

703.2.1.1 Nonsymmetrical wall construction. Interior walls and partitions of nonsymmetrical construction shall be tested with both faces exposed to the furnace, and the assigned *fire-resistance rating* shall be the shortest duration obtained from the two tests conducted in compliance with ASTM E119 or UL 263. Where evidence is furnished to show that the wall was tested with the least fire-resistant side exposed to the furnace, subject to acceptance of the *building official*, the wall need not be subjected to tests from the opposite side (see Section 705.5 for *exterior walls*).

703.2.1.2 Combustible components. Combustible aggregates are permitted in gypsum and Portland cement concrete mixtures for fire-resistance-rated construction. Any component material or admixture is permitted in assemblies if the resulting tested assembly meets the *fire-resistance* test requirements of this code.

703.2.1.3 Restrained classification. Fire-resistance-rated assemblies tested under ASTM E119 or UL 263 shall not be considered to be restrained unless evidence satisfactory to the *building official* is furnished by the *registered design professional* showing that the construction qualifies for a restrained classification in accordance with ASTM E119 or UL 263. Restrained construction shall be identified on the *construction documents*.

703.2.1.4 Supplemental features. Where materials, systems or devices that have not been tested as part of a fire-resistance-rated assembly are incorporated into the *building element,* component or assembly, sufficient data shall be made available to the *building official* to show that the required *fire-resistance rating* is not reduced.

703.2.1.5 Exterior bearing walls. In determining the *fire-resistance rating* of exterior bearing walls, compliance with the ASTM E119 or UL 263 criteria for unexposed surface temperature rise and ignition of cotton waste due to passage of flame or gases is required only for a period of time corresponding to the required *fire-resistance rating* of an exterior nonbearing wall with the same *fire separation distance*, and in a building of the same group. Where the *fire-resistance rating* determined in accordance with this exception exceeds the *fire-resistance rating* determined in accordance with ASTM E119 or UL 263, the fire exposure time period, water pressure and application duration criteria for the hose stream test of ASTM E119 or UL 263 shall be based on the *fire-resistance rating* determined in accordance with this section.

703.2.2 Analytical methods. The fire resistance of *building elements* , components or assemblies established by an analytical method shall be by any of the methods listed in this section, based on the fire exposure and acceptance criteria specified in ASTM E119 or UL 263.

1. *Fire-resistance* designs documented in approved sources.
2. Prescriptive designs of fire-resistance-rated *building elements*, components or assemblies as prescribed in Section 721.
3. Calculations in accordance with Section 722.
4. Engineering analysis based on a comparison of *building element,* component or assemblies designs having *fire-resistance ratings* as determined by the test procedures set forth in ASTM E119 or UL 263.
5. *Fire-resistance* designs certified by an *approved* agency.

703.2.3 Approved alternate method. The *fire resistance* of *building elements*, components or assemblies not complying with Section 703.2.1 or 703.2.2 shall be permitted to be established by an alternative protection method in accordance with Section 104.11.

703.3 Noncombustibility tests. The tests indicated in Section 703.3.1 shall serve as criteria for acceptance of building materials as set forth in Sections 602.2, 602.3 and 602.4 in Types I, II, III and IV construction. The term "noncombustible" does not apply to the *flame spread* characteristics of *interior finish* or *trim* materials. A material shall not be classified as a noncombustible building construction material if it is subject to an increase in combustibility or *flame spread* beyond the limitations herein established through the effects of age, moisture or other atmospheric conditions.

703.3.1 Noncombustible materials. Materials required to be noncombustible shall be tested in accordance with ASTM E136. Alternately, materials required to be noncombustible shall be tested in accordance with ASTM E2652 using the acceptance criteria prescribed by ASTM E136.

Exception: Materials having a structural base of noncombustible material as determined in accordance with ASTM E136, or with ASTM E2652 using the acceptance criteria prescribed by ASTM E136, with a surfacing of not more than 0.125 inch (3.18 mm) in thickness having a flame spread index not greater than 50 when tested in accordance with ASTM E84 or UL 723 shall be acceptable as noncombustible.

703.4 Fire-resistance-rated glazing. Fire-resistance-rated glazing, when tested in accordance with ASTM E119 or UL 263 and complying with the requirements of Section 707, shall be permitted. Fire-resistance-rated glazing shall bear a *label* marked in accordance with Table 716.1(1) issued by an agency and shall be permanently identified on the glazing.

703.5 Marking and identification. Where there is an accessible concealed floor, floor-ceiling or *attic* space, *fire walls*, *fire barriers*, *fire partitions*, *smoke barriers* and *smoke partitions* or any other wall required to have protected openings or penetrations shall be effectively and permanently identified with signs or stenciling in the concealed space. Such identification shall:

1. Be located within 15 feet (4572 mm) of the end of each wall and at intervals not exceeding 30 feet (9144 mm) measured horizontally along the wall or partition.
2. Include lettering not less than 3 inches (76 mm) in height with a minimum $^{3}/_{8}$-inch (9.5 mm) stroke in a contrasting color incorporating the suggested wording, "FIRE AND/OR SMOKE BARRIER—PROTECT ALL OPENINGS," or other wording.

703.6 Determination of noncombustible protection time contribution. The time, in minutes, contributed to the *fire-resistance rating* by the *noncombustible protection* of *mass timber building elements*, components, or assemblies, shall be established through a comparison of assemblies tested using procedures set forth in ASTM E119 or UL 263. The test assemblies shall be identical in construction, loading and materials, other than the *noncombustible protection*. The two test assemblies shall be tested to the same criteria of structural failure with the following conditions:

1. Test Assembly 1 shall be without protection.
2. Test Assembly 2 shall include the representative noncombustible protection. The protection shall be fully defined in terms of configuration details, attachment details, joint sealing details, accessories and all other relevant details.

The *noncombustible protection* time contribution shall be determined by subtracting the *fire-resistance* time, in minutes, of Test Assembly 1 from the *fire-resistance* time, in minutes, of Test Assembly 2.

703.7 Sealing of adjacent mass timber elements. In buildings of Types IV-A, IV-B and IV-C construction, sealant or adhesive shall be provided to resist the passage of air in the following locations:

1. At abutting edges and intersections of mass timber building elements required to be fire-resistance rated.
2. At abutting intersections of mass timber building elements and building elements of other materials where both are required to be fire-resistance rated.

Sealants shall meet the requirements of ASTM C920. Adhesives shall meet the requirements of ASTM D3498.

Exception: Sealants or adhesives need not be provided where they are not a required component of a tested fire-resistance-rated assembly.

SECTION 704 FIRE-RESISTANCE RATING OF STRUCTURAL MEMBERS

704.1 Requirements. The *fire-resistance ratings* of structural members and assemblies shall comply with this section and the requirements for the type of construction as specified in Table 601. The *fire-resistance ratings* shall be not less than the ratings required for the fire-resistance-rated assemblies supported by the structural members.

Exception: *Fire barriers*, *fire partitions*, *smoke barriers* and *horizontal assemblies* as provided in Sections 707.5, 708.4, 709.4 and 711.2, respectively.

704.2 Column protection. Where columns are required to have protection to achieve a *fire-resistance rating*, the entire column shall be provided individual encasement protection by protecting it on all sides for the full column height, including connections to other structural members, with materials having the required *fire-resistance rating*. Where the column extends through a ceiling, the encasement protection shall be continuous from the top of the foundation or floor/ceiling assembly below through the ceiling space to the top of the column.

Exception: Columns that meet the limitations of Section 704.4.1.

704.3 Protection of the primary structural frame other than columns. Members of the *primary structural frame* other than columns that are required to have protection to achieve a *fire-resistance rating* and support more than two floors or one floor and roof, or support a *load-bearing wall* or a *nonload-bearing wall* more than two *stories* high, shall be provided individual encasement protection by protecting them on all sides for the full length, including connections to other structural members, with materials having the required *fire-resistance rating*.

Exception: Individual encasement protection on all sides shall be permitted on all exposed sides provided that the extent of protection is in accordance with the required *fire-resistance rating*, as determined in Section 703.

704.4 Protection of secondary structural members. *Secondary structural members* that are required to have protection to achieve a *fire-resistance rating* shall be protected by individual encasement protection.

704.4.1 Light-frame construction. Studs, columns and boundary elements that are integral elements in *walls* of *light-frame construction* and are located entirely between the top and bottom plates or tracks shall be permitted to have required *fire-resistance ratings* provided by the membrane protection provided for the *wall*.

704.4.2 Horizontal assemblies. *Horizontal assemblies* are permitted to be protected with a membrane or ceiling where the membrane or ceiling provides the required *fire-resistance rating* and is installed in accordance with Section 711.

704.5 Truss protection. The required thickness and construction of fire-resistance-rated assemblies enclosing trusses shall be based on the results of full-scale tests or combinations of tests on truss components or on *approved* calculations based on such tests that satisfactorily demonstrate that the assembly has the required *fire resistance*.

704.6 Attachments to structural members. The edges of lugs, brackets, rivets and bolt heads attached to structural members shall be permitted to extend to within 1 inch (25 mm) of the surface of the fire protection.

704.6.1 Secondary attachments to structural members. Where primary and secondary structural steel members require fire protection, secondary steel attachments to those structural members shall be protected with the same fire-resistive material and thickness as required for the structural member. The protection shall extend away from the structural member a distance of not less than 12 inches (305 mm), or shall be applied to the entire length where the attachment is less than 12 inches (305 mm) long. Where an attachment is hollow and the ends are open, the fire-resistive material and thickness shall be applied to both exterior and interior of the hollow steel attachment.

704.7 Reinforcing. Thickness of protection for concrete or masonry reinforcement shall be measured to the outside of the reinforcement except that stirrups and spiral reinforcement ties are permitted to project not more than 0.5 inch (12.7 mm) into the protection.

704.8 Embedments and enclosures. Pipes, wires, conduits, ducts or other service facilities shall not be embedded in the required fire protective covering of a structural member that is required to be individually encased.

704.9 Impact protection. Where the fire protective covering of a structural member is subject to impact damage from moving vehicles, the handling of merchandise or other activity, the fire protective covering shall be protected by corner guards or by a substantial jacket of metal or other noncombustible material to a height adequate to provide full protection, but not less than 5 feet (1524 mm) from the finished floor.

Exception: Corner protection is not required on concrete columns in parking garages.

704.10 Exterior structural members. Load-bearing structural members located within the *exterior walls* or on the outside of a building or structure shall be provided with the highest *fire-resistance rating* as determined in accordance with the following:

1. As required by Table 601 for the type of *building element* based on the type of construction of the building.
2. As required by Table 601 for exterior bearing walls based on the type of construction.
3. As required by Table 705.5 for *exterior walls* based on the *fire separation distance*.

704.11 Bottom flange protection. Fire protection is not required at the bottom flange of lintels, shelf angles and plates, spanning not more than 6 feet 4 inches (1931 mm) whether part of the *primary structural frame* or not, and from the bottom flange of lintels, shelf angles and plates not part of the structural frame, regardless of span.

704.12 Seismic isolation systems. *Fire-resistance ratings* for the isolation system shall meet the *fire-resistance rating* required for the columns, walls or other structural elements in which the isolation system is installed in accordance with Table 601. Isolation systems required to have a *fire-resistance rating* shall be protected with *approved* materials or construction assemblies designed to provide the same degree of *fire resistance* as the structural element in which the system is installed when tested in accordance with ASTM E119 or UL 263 (see Section 703.2).

Such isolation system protection applied to isolator units shall be capable of retarding the transfer of heat to the isolator unit in such a manner that the required gravity load-carrying capacity of the isolator unit will not be impaired after exposure to the standard time-temperature curve fire test prescribed in ASTM E119 or UL 263 for a duration not less than that required for the *fire-resistance rating* of the structure element in which the system is installed.

Such isolation system protection applied to isolator units shall be suitably designed and securely installed so as not to dislodge, loosen, sustain damage or otherwise impair its ability to accommodate the seismic movements for which the isolator unit is designed and to maintain its integrity for the purpose of providing the required fire-resistance protection.

704.13 Sprayed fire-resistant materials (SFRM). Sprayed fire-resistant materials (SFRM) shall comply with Sections 704.13.1 through 704.13.5.

704.13.1 Fire-resistance rating. The application of SFRM shall be consistent with the *fire-resistance rating* and the listing, including, but not limited to, minimum thickness and dry density of the applied SFRM, method of application, substrate surface conditions and the use of bonding adhesives, sealants, reinforcing or other materials.

704.13.2 Manufacturer's installation instructions. The application of SFRM shall be in accordance with the manufacturer's installation instructions. The instructions shall include, but are not limited to, substrate temperatures and surface conditions and SFRM handling, storage, mixing, conveyance, method of application, curing and *ventilation*.

704.13.3 Substrate condition. The SFRM shall be applied to a substrate in compliance with Sections 704.13.3.1 and 704.13.3.2.

704.13.3.1 Surface conditions. Substrates to receive SFRM shall be free of dirt, oil, grease, release agents, loose scale and any other condition that prevents adhesion. The substrates shall be free of primers, paints and encapsulants other than those fire tested and *listed* by a nationally recognized testing agency. Primed, painted or encapsulated steel shall be allowed, provided that testing has demonstrated that required adhesion is maintained.

704.13.3.2 Primers, paints and encapsulants. Where the SFRM is to be applied over primers, paints or encapsulants other than those specified in the listing, the material shall be field tested in accordance with ASTM E736. Where testing of the SFRM with primers, paints or encapsulants demonstrates that required adhesion is maintained, SFRM shall be permitted to be applied to primed, painted or encapsulated wide flange steel shapes in accordance with the following conditions:

1. The beam flange width does not exceed 12 inches (305 mm); or
2. The column flange width does not exceed 16 inches (400 mm); or
3. The beam or column web depth does not exceed 16 inches (400 mm).
4. The average and minimum bond strength values shall be determined based on not fewer than five bond tests conducted in accordance with ASTM E736. Bond tests conducted in accordance with ASTM E736 shall indicate an average bond strength of not less than 80 percent and an individual bond strength of not less than 50 percent, when compared to the bond strength of the SFRM as applied to clean, uncoated $^1/_8$-inch-thick (3.2 mm) steel plate.

704.13.4 Temperature. A minimum ambient and substrate temperature of 40°F (4.44°C) shall be maintained during and for not fewer than 24 hours after the application of the SFRM, unless the manufacturer's instructions allow otherwise.

704.13.5 Finished condition. The finished condition of SFRM applied to structural members or assemblies shall not, upon complete drying or curing, exhibit cracks, voids, spalls, delamination or any exposure of the substrate. Surface irregularities of SFRM shall be deemed acceptable.

SECTION 705
EXTERIOR WALLS

705.1 General. *Exterior walls* shall comply with this section.

705.2 Projections. *Cornices*, eave overhangs, exterior balconies and similar projections extending beyond the *exterior wall* shall conform to the requirements of this section and Section 1405. Exterior egress balconies and exterior *exit stairways* and *ramps* shall comply with Sections 1021 and 1027, respectively. Projections shall not extend any closer to the line used to determine the *fire separation distance* than shown in Table 705.2.

Exception: Buildings on the same lot and considered as portions of one building in accordance with Section 705.3 are not required to comply with this section for projections between the buildings.

TABLE 705.2
MINIMUM DISTANCE OF PROJECTION

FIRE SEPARATION DISTANCE (FSD) (feet)	MINIMUM DISTANCE FROM LINE USED TO DETERMINE FSD
0 to less than 2	Projections not permitted
2 to less than 3	24 inches
3 to less than 5	Two-thirds of FSD
5 or greater	40 inches

For SI: 1 foot = 304.8 mm; 1 inch = 25.4 mm.

705.2.1 Types I and II construction. Projections from walls of Type I or II construction shall be of noncombustible materials or combustible materials as allowed by Sections 705.2.3.1 and 705.2.4.

705.2.2 Type III, IV or V construction. Projections from walls of Type III, IV or V construction shall be of any *approved* material.

705.2.3 Projection protection. Projections extending to within 5 feet (1524 mm) of the line used to determine the *fire separation distance* shall be one of the following:

1. Noncombustible materials.
2. Combustible materials of not less than 1-hour fire-resistance-rated construction.
3. Heavy timber construction complying with Section 2304.11.
4. Fire-retardant-treated wood.
5. As permitted by Section 705.2.3.1.

Exception: Type VB construction shall be allowed for combustible projections in Group R-3 and U occupancies with a *fire separation distance* greater than or equal to 5 feet (1524 mm).

705.2.3.1 Balconies and similar projections. Balconies and similar projections of combustible construction other than *fire-retardant-treated wood* shall be *fire-resistance* rated where required by Table 601 for floor construction or shall be of heavy timber construction in accordance with Section 2304.11. The aggregate length of the projections shall not exceed 50 percent of the building's perimeter on each floor.

Exceptions:

1. On buildings of Types I and II construction, three *stories* or less above *grade plane*, *fire-retardant-treated wood* shall be permitted for balconies, porches, decks and exterior *stairways* not used as required exits.
2. Untreated wood and plastic composites that comply with ASTM D7032 and Section 2612 are permitted for pickets, rails and similar *guard* components that are limited to 42 inches (1067 mm) in height.
3. Balconies and similar projections on buildings of Types III, IV and V construction shall be permitted to be of Type V construction and shall not be required to have a *fire-resistance rating* where sprinkler protection is extended to these areas.
4. Where sprinkler protection is extended to the balcony areas, the aggregate length of the balcony on each floor shall not be limited.

705.2.4 Bay and oriel windows. Bay and oriel windows constructed of combustible materials shall conform to the type of construction required for the building to which they are attached.

Exception: *Fire-retardant-treated wood* shall be permitted on buildings three *stories* or less above *grade plane* of Type I, II, III or IV construction.

705.3 Buildings on the same lot. For the purposes of determining the required wall and opening protection, projections and *roof-covering* requirements, buildings on the same lot shall be assumed to have an imaginary line between them.

Where a new building is to be erected on the same lot as an *existing building*, the location of the assumed imaginary line with relation to the existing building shall be such that the *exterior wall* and opening protection of the *existing building* meet the criteria as set forth in Sections 705.5 and 705.8.

Exceptions:

1. Two or more buildings on the same lot shall be either regulated as separate buildings or shall be considered as portions of one building if the aggregate area of such buildings is within the limits specified in Chapter 5 for a single building. Where the buildings contain different occupancy groups or are of different types of construction, the area shall be that allowed for the most restrictive occupancy or construction.
2. Where an S-2 parking garage of Construction Type I or IIA is erected on the same lot as a Group R-2 building, and there is no *fire separation distance* between these buildings, then the adjoining *exterior walls* between the buildings are permitted to have occupant use openings in accordance with Section 706.8. However, opening protectives in such openings shall only be required in the *exterior wall* of the S-2 parking garage, not in the *exterior wall* openings in the R-2 building, and these opening protectives in the *exterior wall* of the S-2 parking garage shall be not less than $1^1/_2$-hour *fire protection rating*.

705.4 Materials. *Exterior walls* shall be of materials permitted by the building's type of construction.

705.5 Fire-resistance ratings. *Exterior walls* shall be *fire-resistance* rated in accordance with Table 601, based on the type of construction, and Table 705.5, based on the *fire separation distance*. The required *fire-resistance rating* of *exterior walls* with a *fire separation distance* of greater than 10 feet (3048 mm) shall be rated for exposure to fire from the inside. The required *fire-resistance rating* of *exterior walls* with a *fire separation distance* of less than or equal to 10 feet (3048 mm) shall be rated for exposure to fire from both sides.

705.6 Structural stability. *Exterior walls* shall extend to the height required by Section 705.11. Interior structural elements that brace the *exterior wall* but that are not located within the plane of the *exterior wall* shall have the minimum *fire-resistance rating* required in Table 601 for that structural element. Structural elements that brace the *exterior wall* but are located outside of the *exterior wall* or within the plane of the *exterior wall* shall have the minimum *fire-resistance rating* required in Table 601 and Table 705.5 for the *exterior wall.*

705.7 Unexposed surface temperature. Where protected openings are not limited by Section 705.8, the limitation on the rise of temperature on the unexposed surface of *exterior walls* as required by ASTM E119 or UL 263 shall not apply. Where protected openings are limited by Section 705.8, the limitation on the rise of temperature on the unexposed surface of *exterior walls* as required by ASTM E119 or UL 263 shall not apply provided that a correction is made for radiation from the unexposed *exterior wall* surface in accordance with the following formula:

$$A_e = A + (A_f \times F_{eo}) \quad \textbf{(Equation 7-1)}$$

where:

A_e = Equivalent area of protected openings.

A = Actual area of protected openings.

A_f = Area of exterior wall surface in the story under consideration exclusive of openings, on which the temperature limitations of ASTM E119 or UL 263 for walls are exceeded.

F_{eo} = An "equivalent opening factor" derived from Figure 705.7 based on the average temperature of the unexposed wall surface and the fire-resistance rating of the wall.

705.8 Openings. Openings in *exterior walls* shall comply with Sections 705.8.1 through 705.8.6.

705.8.1 Allowable area of openings. The maximum area of unprotected and protected openings permitted in an *exterior wall* in any *story* of a building shall not exceed the percentages specified in Table 705.8 based on the *fire separation distance* of each individual story.

Exceptions:

1. In other than Group H occupancies, unlimited unprotected openings are permitted in the first *story above grade plane* where the wall faces one of the following:
   1.1. A street and has a *fire separation distance* of more than 15 feet (4572 mm).
   1.2. An unoccupied space. The unoccupied space shall be on the same lot or dedicated for public use, shall be not less than 30 feet (9144 mm) in width and shall have access from a street by a posted fire lane in accordance with the *International Fire Code.*
2. Buildings whose exterior bearing walls, exterior nonbearing walls and exterior *primary structural frame* are not required to be fire-resistance rated shall be permitted to have unlimited unprotected openings.

705.8.2 Protected openings. Where openings are required to be protected, opening protectives shall comply with Section 716.

Exception: Opening protectives are not required where the building is equipped throughout with an *automatic sprinkler system* in accordance with Section 903.3.1.1 and the exterior openings are protected by a water curtain using automatic sprinklers *approved* for that use.

TABLE 705.5
FIRE-RESISTANCE RATING REQUIREMENTS FOR EXTERIOR WALLS BASED ON FIRE SEPARATION DISTANCE[a, d, g]

FIRE SEPARATION DISTANCE = X (feet)	TYPE OF CONSTRUCTION	OCCUPANCY GROUP H[e]	OCCUPANCY GROUP F-1, M, S-1[f]	OCCUPANCY GROUP A, B, E, F-2, I, R[i], S-2, U[h]
$X < 5$[b]	All	3	2	1
$5 \le X < 10$	IA, IVA	3	2	1
	Others	2	1	1
$10 \le X < 30$	IA, IB, IVA, IVB	2	1	1[c]
	IIB, VB	1	0	0
	Others	1	1	1[c]
$X \ge 30$	All	0	0	0

For SI: 1 foot = 304.8 mm.

a. Load-bearing exterior walls shall also comply with the fire-resistance rating requirements of Table 601.
b. See Section 706.1.1 for party walls.
c. Open parking garages complying with Section 406 shall not be required to have a fire-resistance rating.
d. The fire-resistance rating of an exterior wall is determined based upon the fire separation distance of the exterior wall and the story in which the wall is located.
e. For special requirements for Group H occupancies, see Section 415.6.
f. For special requirements for Group S aircraft hangars, see Section 412.3.1.
g. Where Table 705.8 permits nonbearing exterior walls with unlimited area of unprotected openings, the required fire-resistance rating for the exterior walls is 0 hours.
h. For a building containing only a Group U occupancy private garage or carport, the exterior wall shall not be required to have a fire-resistance rating where the fire separation distance is 5 feet (1523 mm) or greater.
i. For a Group R-3 building of Type II-B or Type V-B construction, the exterior wall shall not be required to have a fire-resistance rating where the fire separation distance is 5 feet (1523 mm) or greater.

705.8.3 Unprotected openings. Where unprotected openings are permitted, windows and doors shall be constructed of any *approved* materials. Glazing shall conform to the requirements of Chapters 24 and 26.

705.8.4 Mixed openings. Where both unprotected and protected openings are located in the *exterior wall* in any *story* of a building, the total area of openings shall be determined in accordance with the following:

$(A_p/a_p) + (A_u/a_u) \leq 1$ **(Equation 7-2)**

A_p = Actual area of protected openings, or the equivalent area of protected openings, A_e (see Section 705.7).

a_p = Allowable area of protected openings.

A_u = Actual area of unprotected openings.

a_u = allowable area of unprotected openings.

705.8.5 Vertical separation of openings. Openings in *exterior walls* in adjacent *stories* shall be separated vertically to protect against fire spread on the exterior of the buildings where the openings are within 5 feet (1524 mm) of each other horizontally and the opening in the lower *story* is not a protected opening with a *fire protection rating* of not less than $^3/_4$ hour. Such openings shall be separated vertically not less than 3 feet (914 mm) by spandrel girders, *exterior walls* or other similar assemblies that have a *fire-resistance rating* of not less than 1 hour, rated for exposure to fire from both sides, or by flame barriers that extend horizontally not less than 30 inches (762 mm) beyond the *exterior wall*. Flame barriers shall have a *fire-resistance rating* of not less than 1 hour. The unexposed surface temperature limitations specified in ASTM E119 or UL 263 shall not apply to the flame barriers unless otherwise required by the provisions of this code.

Exceptions:

1. This section shall not apply to buildings that are three *stories* or less above *grade plane*.
2. This section shall not apply to buildings equipped throughout with an *automatic sprinkler system* in accordance with Section 903.3.1.1 or 903.3.1.2.
3. *Open parking garages*.

705.8.6 Vertical exposure. For buildings on the same lot, opening protectives having a *fire protection rating* of not less than $^3/_4$ hour shall be provided in every opening that is less than 15 feet (4572 mm) vertically above the roof of an adjacent building or structure based on assuming an imaginary line between them. The opening protectives are required where the *fire separation distances* from the imaginary line to each building or structure are less than 15 feet (4572 mm).

Exceptions:

1. Opening protectives are not required where the *roof assembly* of the adjacent building or structure has a *fire-resistance rating* of not less than 1 hour for a minimum distance of 10 feet (3048

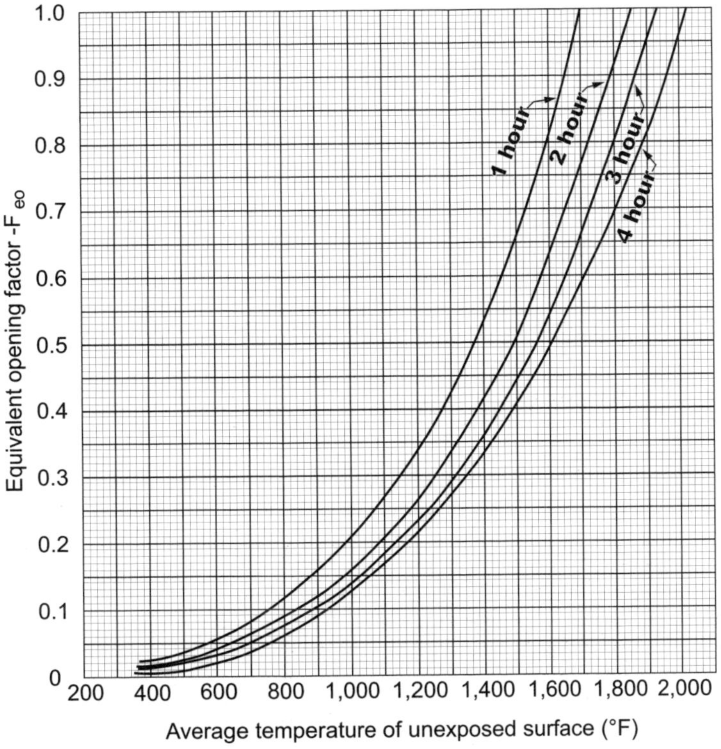

For SI: °C = [(°F) - 32] / 1.8.

FIGURE 705.7
EQUIVALENT OPENING FACTOR

mm) from the *exterior wall* facing the imaginary line and the entire length and span of the supporting elements for the fire-resistance-rated roof assembly has a *fire-resistance rating* of not less than 1 hour.

2. Buildings on the same lot and considered as portions of one building in accordance with Section 705.3 are not required to comply with Section 705.8.6.

705.9 Joints. *Joints* made in or between *exterior walls* required by this section to have a *fire-resistance rating* shall comply with Section 715.

Exception: *Joints* in *exterior walls* that are permitted to have unprotected openings.

705.9.1 Voids. The void created at the intersection of a floor/ceiling assembly and an exterior curtain wall assembly shall be protected in accordance with Section 715.4.

TABLE 705.8
MAXIMUM AREA OF EXTERIOR WALL OPENINGS BASED ON FIRE SEPARATION DISTANCE AND DEGREE OF OPENING PROTECTION

FIRE SEPARATION DISTANCE (feet)	DEGREE OF OPENING PROTECTION	ALLOWABLE AREA[a]
0 to less than 3[b, c, k]	Unprotected, Nonsprinklered (UP, NS)	Not Permitted[k]
	Unprotected, Sprinklered (UP, S)[i]	Not Permitted[k]
	Protected (P)	Not Permitted[k]
3 to less than 5[d, e]	Unprotected, Nonsprinklered (UP, NS)	Not Permitted
	Unprotected, Sprinklered (UP, S)[i]	15%
	Protected (P)	15%
5 to less than 10[e, f, j]	Unprotected, Nonsprinklered (UP, NS)	10%[h]
	Unprotected, Sprinklered (UP, S)[i]	25%
	Protected (P)	25%
10 to less than 15[e, f, g, j]	Unprotected, Nonsprinklered (UP, NS)	15%[h]
	Unprotected, Sprinklered (UP, S)[i]	45%
	Protected (P)	45%
15 to less than 20[f, g, j]	Unprotected, Nonsprinklered (UP, NS)	25%
	Unprotected, Sprinklered (UP, S)[i]	75%
	Protected (P)	75%
20 to less than 25[f, g, j]	Unprotected, Nonsprinklered (UP, NS)	45%
	Unprotected, Sprinklered (UP, S)[i]	No Limit
	Protected (P)	No Limit
25 to less than 30[f, g, j]	Unprotected, Nonsprinklered (UP, NS)	70%
	Unprotected, Sprinklered (UP, S)[i]	No Limit
	Protected (P)	No Limit
30 or greater	Unprotected, Nonsprinklered (UP, NS)	No Limit
	Unprotected, Sprinklered (UP, S)[i]	No Limit
	Protected (P)	No Limit

For SI: 1 foot = 304.8 mm.

UP, NS = Unprotected openings in buildings not equipped throughout with an automatic sprinkler system in accordance with Section 903.3.1.1.

UP, S = Unprotected openings in buildings equipped throughout with an automatic sprinkler system in accordance with Section 903.3.1.1.

P = Openings protected with an opening protective assembly in accordance with Section 705.8.2.

a. Values indicated are the percentage of the area of the exterior wall, per story.

b. For the requirements for fire walls of buildings with differing heights, see Section 706.6.1.

c. For openings in a fire wall for buildings on the same lot, see Section 706.8.

d. The maximum percentage of unprotected and protected openings shall be 25 percent for Group R-3 occupancies.

e. Unprotected openings shall not be permitted for openings with a fire separation distance of less than 15 feet for Group H-2 and H-3 occupancies.

f. The area of unprotected and protected openings shall not be limited for Group R-3 occupancies, with a fire separation distance of 5 feet or greater.

g. The area of openings in an open parking structure with a fire separation distance of 10 feet or greater shall not be limited.

h. Includes buildings accessory to Group R-3.

i. Not applicable to Group H-1, H-2 and H-3 occupancies.

j. The area of openings in a building containing only a Group U occupancy private garage or carport with a fire separation distance of 5 feet or greater shall not be limited.

k. For openings between S-2 parking garage and Group R-2 building, see Section 705.3, Exception 2.

705.10 Ducts and air transfer openings. Penetrations by air ducts and air transfer openings in fire-resistance-rated *exterior walls* required to have protected openings shall comply with Section 717.

Exception: Foundation vents installed in accordance with this code are permitted.

705.11 Parapets. Parapets shall be provided on *exterior walls* of buildings.

Exceptions: A parapet need not be provided on an *exterior wall* where any of the following conditions exist:

1. The wall is not required to be *fire-resistance rated* in accordance with Table 705.5 because of *fire separation distance*.
2. The building has an area of not more than 1,000 square feet (93 m^2) on any floor.
3. Walls that terminate at roofs of not less than 2-hour fire-resistance-rated construction or where the roof, including the deck or slab and supporting construction, is constructed entirely of noncombustible materials.
4. One-hour fire-resistance-rated *exterior walls* that terminate at the underside of the roof sheathing, deck or slab, provided that:
   4.1. Where the roof/ceiling framing elements are parallel to the walls, such framing and elements supporting such framing shall not be of less than 1-hour fire-resistance-rated construction for a width of 4 feet (1220 mm) for Groups R and U and 10 feet (3048 mm) for other occupancies, measured from the interior side of the wall.
   4.2. Where roof/ceiling framing elements are not parallel to the wall, the entire span of such framing and elements supporting such framing shall not be of less than 1-hour fire-resistance-rated construction.
   4.3. Openings in the roof shall not be located within 5 feet (1524 mm) of the 1-hour fire-resistance-rated *exterior wall* for Groups R and U and 10 feet (3048 mm) for other occupancies, measured from the interior side of the wall.
   4.4. The entire building shall be provided with not less than a Class B *roof covering*.
5. In Groups R-2 and R-3 where the entire building is provided with a Class C *roof covering*, the *exterior wall* shall be permitted to terminate at the underside of the roof sheathing or deck in Types III, IV and V construction, provided that one or both of the following criteria is met:
   5.1. The roof sheathing or deck is constructed of *approved* noncombustible materials or of *fire-retardant-treated wood* for a distance of 4 feet (1220 mm).
   5.2. The roof is protected with 0.625-inch (16 mm) Type X *gypsum board* directly beneath the underside of the roof sheathing or deck, supported by not less than nominal 2-inch (51 mm) ledgers attached to the sides of the roof framing members for a minimum distance of 4 feet (1220 mm).
6. Where the wall is permitted to have not less than 25 percent of the *exterior wall* areas containing unprotected openings based on *fire separation distance* as determined in accordance with Section 705.8.

705.11.1 Parapet construction. Parapets shall have the same *fire-resistance rating* as that required for the supporting wall, and on any side adjacent to a roof surface, shall have noncombustible faces for the uppermost 18 inches (457 mm), including counterflashing and coping materials. The height of the parapet shall be not less than 30 inches (762 mm) above the point where the roof surface and the wall intersect. Where the roof slopes toward a parapet at a slope greater than 2 units vertical in 12 units horizontal (16.7-percent slope), the parapet shall extend to the same height as any portion of the roof within a *fire separation distance* where protection of wall openings is required, but the height shall be not less than 30 inches (762 mm).

SECTION 706
FIRE WALLS

706.1 General. *Fire walls* shall be constructed in accordance with Sections 706.2 through 706.11. The extent and location of such *fire walls* shall provide a complete separation. Where a *fire wall* separates occupancies that are required to be separated by a *fire barrier* wall, the most restrictive requirements of each separation shall apply.

706.1.1 Party walls. Any wall located on a *lot line* between adjacent buildings, which is used or adapted for *joint* service between the two buildings, shall be constructed as a *fire wall* in accordance with Section 706. Party walls shall be constructed without openings and shall create separate buildings.

Exceptions:

1. Openings in a party wall separating an *anchor building* and a *mall* shall be in accordance with Section 402.4.2.2.1.
2. Party walls and *fire walls* are not required on *lot lines* dividing a building for ownership purposes where the aggregate height and area of the portions of the building located on both sides of the *lot line* do not exceed the maximum height and area requirements of this code. For the *building official's* review and approval, the official shall be provided with copies of dedicated access easements and contractual agreements that permit the *owners* of portions

of the building located on either side of the *lot line* access to the other side for purposes of maintaining fire and *life safety systems* necessary for the operation of the building.

706.2 Structural stability. *Fire walls* shall be designed and constructed to allow collapse of the structure on either side without collapse of the wall under fire conditions. *Fire walls* designed and constructed in accordance with NFPA 221 shall be deemed to comply with this section.

Exception: In *Seismic Design Categories* D through F, where double *fire walls* are used in accordance with NFPA 221, floor and roof sheathing not exceeding $^3/_4$ inch (19.05 mm) thickness shall be permitted to be continuous through the wall assemblies of *light frame construction*.

706.3 Materials. *Fire walls* shall be of any *approved* noncombustible materials.

Exception: Buildings of Type V construction.

706.4 Fire-resistance rating. *Fire walls* shall have a *fire-resistance rating* of not less than that required by Table 706.4.

TABLE 706.4
FIRE WALL FIRE-RESISTANCE RATINGS

GROUP	FIRE-RESISTANCE RATING (hours)
A, B, E, H-4, I, R-1, R-2, U	3[a]
F-1, H-3b, H-5, M, S-1	3
H-1, H-2	4[b]
F-2, S-2, R-3, R-4	2

a. In Type II or V construction, walls shall be permitted to have a 2-hour fire-resistance rating.
b. For Group H-1, H-2 or H-3 buildings, also see Sections 415.7 and 415.8.

706.5 Horizontal continuity. *Fire walls* shall be continuous from *exterior wall* to *exterior wall* and shall extend not less than 18 inches (457 mm) beyond the exterior surface of *exterior walls*.

Exceptions:

1. *Fire walls* shall be permitted to terminate at the interior surface of combustible exterior sheathing or siding provided that the *exterior wall* has a *fire-resistance rating* of not less than 1 hour for a horizontal distance of not less than 4 feet (1220 mm) on both sides of the *fire wall*. Openings within such *exterior walls* shall be protected by opening protectives having a *fire protection rating* of not less than $^3/_4$ hour.
2. *Fire walls* shall be permitted to terminate at the interior surface of noncombustible exterior sheathing, exterior siding or other noncombustible exterior finishes provided that the sheathing, siding or other exterior noncombustible finish extends a horizontal distance of not less than 4 feet (1220 mm) on both sides of the *fire wall*.
3. *Fire walls* shall be permitted to terminate at the interior surface of noncombustible exterior sheathing where the building on each side of the *fire wall* is protected by an *automatic sprinkler system* installed in accordance with Section 903.3.1.1 or 903.3.1.2.

706.5.1 Exterior walls. Where the *fire wall* intersects *exterior walls*, the *fire-resistance rating* and opening protection of the *exterior walls* shall comply with one of the following:

1. The *exterior walls* on both sides of the *fire wall* shall have a 1-hour *fire-resistance rating* with $^3/_4$-hour protection where opening protection is required by Section 705.8. The *fire-resistance rating* of the *exterior wall* shall extend not less than 4 feet (1220 mm) on each side of the intersection of the *fire wall* to *exterior wall*. *Exterior wall* intersections at *fire walls* that form an angle equal to or greater than 180 degrees (3.14 rad) do not need *exterior wall* protection.
2. Buildings or spaces on both sides of the intersecting *fire wall* shall assume to have an imaginary *lot line* at the *fire wall* and extending beyond the exterior of the *fire wall*. The location of the assumed line in relation to the *exterior walls* and the *fire wall* shall be such that the *exterior wall* and opening protection meet the requirements set forth in Sections 705.5 and 705.8. Such protection is not required for *exterior walls* terminating at *fire walls* that form an angle equal to or greater than 180 degrees (3.14 rad).

706.5.2 Horizontal projecting elements. *Fire walls* shall extend to the outer edge of horizontal projecting elements such as balconies, roof overhangs, canopies, *marquees* and similar projections that are within 4 feet (1220 mm) of the *fire wall*.

Exceptions:

1. Horizontal projecting elements without concealed spaces, provided that the *exterior wall* behind and below the projecting element has not less than 1-hour fire-resistance-rated construction for a distance not less than the depth of the projecting element on both sides of the *fire wall*. Openings within such *exterior walls* shall be protected by opening protectives having a *fire protection rating* of not less than $^3/_4$ hour.
2. Noncombustible horizontal projecting elements with concealed spaces, provided that a minimum 1-hour fire-resistance-rated wall extends through the concealed space. The projecting element shall be separated from the building by not less than 1-hour fire-resistance-rated construction for a distance on each side of the *fire wall* equal to the depth of the projecting element. The wall is not required to extend under the projecting element where the building *exterior wall* is not less than 1-hour *fire-resistance rated* for a distance on each side of the *fire wall* equal to the depth of the projecting element. Openings within such *exterior walls*

shall be protected by opening protectives having a *fire protection rating* of not less than $^3/_4$ hour.

3. For combustible horizontal projecting elements with concealed spaces, the *fire wall* need only extend through the concealed space to the outer edges of the projecting elements. The *exterior wall* behind and below the projecting element shall be of not less than 1-hour fire-resistance-rated construction for a distance not less than the depth of the projecting elements on both sides of the *fire wall*. Openings within such *exterior walls* shall be protected by opening protectives having a *fire protection rating* of not less than $^3/_4$ hour.

706.6 Vertical continuity. *Fire walls* shall extend from the foundation to a termination point not less than 30 inches (762 mm) above both adjacent roofs.

Exceptions:

1. Stepped buildings in accordance with Section 706.6.1.
2. Two-hour fire-resistance-rated walls shall be permitted to terminate at the underside of the roof sheathing, deck or slab, provided that:
   - 2.1. The lower *roof assembly* within 4 feet (1220 mm) of the wall has not less than a 1-hour *fire-resistance rating* and the entire length and span of supporting elements for the rated *roof assembly* has a *fire-resistance rating* of not less than 1 hour.
   - 2.2. Openings in the roof shall not be located within 4 feet (1220 mm) of the *fire wall*.
   - 2.3. Each building shall be provided with not less than a Class B *roof covering*.
3. Walls shall be permitted to terminate at the underside of noncombustible roof sheathing, deck or slabs where both buildings are provided with not less than a Class B *roof covering*. Openings in the roof shall not be located within 4 feet (1220 mm) of the *fire wall*.
4. In buildings of Types III, IV and V construction, walls shall be permitted to terminate at the underside of combustible roof sheathing or decks, provided that all of the following requirements are met:
   - 4.1. Roof openings are not less than 4 feet (1220 mm) from the *fire wall*.
   - 4.2. The roof is covered with a minimum Class B *roof covering*.
   - 4.3. The roof sheathing or deck is constructed of *fire-retardant-treated wood* for a distance of 4 feet (1220 mm) on both sides of the wall or the roof is protected with $^5/_8$-inch (15.9 mm) Type X *gypsum board* directly beneath the underside of the roof sheathing or deck, supported by not less than 2-inch (51 mm) nominal ledgers attached to the sides of the roof framing members for a distance of not less than 4 feet (1220 mm) on both sides of the *fire wall*.
5. In buildings designed in accordance with Section 510.2, *fire walls* located above the 3-hour *horizontal assembly* required by Section 510.2, Item 1 shall be permitted to extend from the top of this *horizontal assembly*.
6. Buildings with sloped roofs in accordance with Section 706.6.2.

706.6.1 Stepped buildings. Where a *fire wall* also serves as an *exterior wall* for a building and separates buildings having different roof levels, such wall shall terminate at a point not less than 30 inches (762 mm) above the lower roof level. *Exterior walls* above the *fire wall* extending more than 30 inches (762 mm) above the lower roof shall be of not less than 1-hour fire-resistance-rated construction from both sides with openings protected by fire assemblies having a *fire protection rating* of not less than $^3/_4$ hour. Portions of the *exterior walls* greater than 15 feet (4572 mm) above the lower roof shall be of nonfire-resistance-rated construction unless otherwise rated construction is required by other provisions of this code.

Exception: A *fire wall* serving as part of an *exterior wall* that separates buildings having different roof levels shall be permitted to terminate at the underside of the roof sheathing, deck or slab of the lower roof, provided that Items 1, 2 and 3 are met. The *exterior wall* above the *fire wall* is not required to be of fire-resistance-rated construction unless required by other provisions of this code.

1. The lower *roof assembly* within 10 feet (3048 mm) of the *fire wall* has not less than a 1-hour *fire-resistance rating*.
2. The entire length and span of supporting elements for the rated *roof assembly* shall have a *fire-resistance rating* of not less than 1 hour.
3. Openings in the lower roof shall not be located within 10 feet (3048 mm) of the *fire wall*.

706.6.2 Buildings with sloped roofs. Where a *fire wall* serves as an interior wall for a building, and the roof on one side or both sides of the *fire wall* slopes toward the *fire wall* at a slope greater than 2 units vertical in 12 units horizontal (2:12), the *fire wall* shall extend to a height equal to the height of the roof located 4 feet (1219 mm) from the *fire wall* plus 30 inches (762 mm). The extension of the *fire wall* shall be not less than 30 inches (762 mm).

706.7 Combustible framing in fire walls. Adjacent combustible members entering into a concrete or masonry *fire wall* from opposite sides shall not have less than a 4-inch (102 mm) distance between embedded ends. Where combustible members frame into hollow walls or walls of hollow units, hollow spaces shall be solidly filled for the full thickness of the wall and for a distance not less than 4 inches (102 mm) above, below and between the structural members, with noncombustible materials *approved* for *fireblocking*.

706.8 Openings. Each opening through a *fire wall* shall be protected in accordance with Section 716 and shall not exceed 156 square feet (15 m^2). The aggregate width of openings at any floor level shall not exceed 25 percent of the length of the wall.

Exceptions:

1. Openings are not permitted in party walls constructed in accordance with Section 706.1.1.
2. Openings shall not be limited to 156 square feet (15 m^2) where both buildings are equipped throughout with an *automatic sprinkler system* installed in accordance with Section 903.3.1.1.

706.9 Penetrations. Penetrations of *fire walls* shall comply with Section 714.

706.10 Joints. *Joints* made in or between *fire walls* shall comply with Section 715.

706.11 Ducts and air transfer openings. Ducts and air transfer openings shall not penetrate *fire walls*.

Exception: Penetrations by ducts and air transfer openings of *fire walls* that are not on a *lot line* shall be allowed provided that the penetrations comply with Section 717. The size and aggregate width of all openings shall not exceed the limitations of Section 706.8.

SECTION 707 FIRE BARRIERS

707.1 General. *Fire barriers* installed as required elsewhere in this code or the *International Fire Code* shall comply with this section.

707.2 Materials. *Fire barriers* shall be of materials permitted by the building type of construction.

707.3 Fire-resistance rating. The *fire-resistance rating* of *fire barriers* shall comply with this section.

707.3.1 Shaft enclosures. The *fire-resistance rating* of the *fire barrier* separating building areas from a *shaft* shall comply with Section 713.4.

707.3.2 Interior exit stairway and ramp construction. The *fire-resistance rating* of the *fire barrier* separating building areas from an *interior exit stairway* or *ramp* shall comply with Section 1023.1.

707.3.3 Enclosures for exit access stairways. The *fire-resistance rating* of the *fire barrier* separating building areas from an *exit access stairway* or *ramp* shall comply with Section 713.4.

707.3.4 Exit passageway. The *fire-resistance rating* of the *fire barrier* separating building areas from an *exit passageway* shall comply with Section 1024.3.

707.3.5 Horizontal exit. The *fire-resistance rating* of the separation between building areas connected by a *horizontal exit* shall comply with Section 1026.1.

707.3.6 Atriums. The *fire-resistance rating* of the *fire barrier* separating *atriums* shall comply with Section 404.6.

707.3.7 Incidental uses. The *fire barrier* separating incidental uses from other spaces in the building shall have a *fire-resistance rating* of not less than that indicated in Table 509.1.

707.3.8 Control areas. *Fire barriers* separating *control areas* shall have a *fire-resistance rating* of not less than that required in Section 414.2.4.

707.3.9 Separated occupancies. Where the provisions of Section 508.4 are applicable, the *fire barrier* separating mixed occupancies shall have a *fire-resistance rating* of not less than that indicated in Table 508.4 based on the occupancies being separated.

707.3.10 Fire areas. The *fire barriers*, *fire walls*, *horizontal assemblies* or combinations thereof separating a single occupancy into different *fire areas* shall have a *fire-resistance rating* of not less than that indicated in Table 707.3.10. The *fire barriers*, *fire walls*, *horizontal assemblies* or combinations thereof separating *fire areas* of mixed occupancies shall have a *fire-resistance rating* of not less than the highest value indicated in Table 707.3.10 for the occupancies under consideration.

TABLE 707.3.10
FIRE-RESISTANCE-RATING REQUIREMENTS FOR FIRE BARRIERS, FIRE WALLS OR HORIZONTAL ASSEMBLIES BETWEEN FIRE AREAS

OCCUPANCY GROUP	FIRE-RESISTANCE RATING (hours)
H-1, H-2	4
F-1, H-3, S-1	3
A, B, E, F-2, H-4, H-5, I, M, R, S-2	2
U	1

707.4 Exterior walls. Where *exterior walls* serve as a part of a required fire-resistance-rated *shaft,* or separation or enclosure for a *stairway, ramp* or *exit passageway*, such walls shall comply with the requirements of Section 705 for *exterior walls* and the fire-resistance-rated enclosure or separation requirements shall not apply.

Exceptions:

1. *Exterior walls* required to be *fire-resistance rated* in accordance with Section 1021 for exterior egress balconies, Section 1023.7 for *interior exit stairways and ramps,* Section 1024.8 for *exit passageways* and Section 1027.6 for *exterior exit stairways and ramps.*
2. *Exterior walls* required to be *fire-resistance rated* in accordance with Section 1207 of the *International Fire Code* for enclosure of energy storage systems.

707.5 Continuity. *Fire barriers* shall extend from the top of the foundation or floor/ceiling assembly below to the underside of the floor or roof sheathing, slab or deck above and shall be securely attached thereto. Such *fire barriers* shall be continuous through concealed space, such as the space above a suspended ceiling. *Joints* and voids at intersections shall comply with Sections 707.8 and 707.9

Exceptions:

1. *Shaft enclosures* shall be permitted to terminate at a top enclosure complying with Section 713.12.
2. *Interior exit stairway* and *ramp* enclosures required by Section 1023 and *exit access stairway* and *ramp* enclosures required by Section 1019 shall be permitted to terminate at a top enclosure complying with Section 713.12.
3. An *exit passageway* enclosure required by Section 1024.3 that does not extend to the underside of the roof sheathing, slab or deck above shall be enclosed at the top with construction of the same *fire-resistance rating* as required for the *exit passageway*.

707.5.1 Supporting construction. The supporting construction for a *fire barrier* shall be protected to afford the required *fire-resistance rating* of the *fire barrier* supported. Hollow vertical spaces within a *fire barrier* shall be fireblocked in accordance with Section 718.2 at every floor level.

Exceptions:

1. The maximum required *fire-resistance rating* for assemblies supporting *fire barriers* separating tank storage as provided for in Section 415.9.1.2 shall be 2 hours, but not less than required by Table 601 for the building construction type.
2. Supporting construction for 1-hour *fire barriers* required by Table 509.1 in buildings of Types IIB, IIIB and VB construction is not required to be *fire-resistance rated* unless required by other sections of this code.

707.6 Openings. Openings in a *fire barrier* shall be protected in accordance with Section 716. Openings shall be limited to a maximum aggregate width of 25 percent of the length of the wall, and the maximum area of any single opening shall not exceed 156 square feet (15 m^2). Openings in enclosures for *exit access stairways* and *ramps*, *interior exit stairways* and *ramps* and *exit passageways* shall also comply with Sections 1019, 1023.4 and 1024.5, respectively.

Exceptions:

1. Openings shall not be limited to 156 square feet (15 m^2) where adjoining floor areas are equipped throughout with an *automatic sprinkler system* in accordance with Section 903.3.1.1.
2. Openings shall not be limited to 156 square feet (15 m^2) or an aggregate width of 25 percent of the length of the wall where the opening protective is a *fire door* serving enclosures for *exit access stairways* and *ramps,* and *interior exit stairways* and *ramps.*
3. Openings shall not be limited to 156 square feet (15 m^2) or an aggregate width of 25 percent of the length of the wall where the opening protective has been tested in accordance with ASTM E119 or UL 263 and has a minimum *fire-resistance rating* not less than the *fire-resistance rating* of the wall.
4. *Fire window assemblies* permitted in *atrium* separation walls shall not be limited to a maximum aggregate width of 25 percent of the length of the wall.
5. Openings shall not be limited to 156 square feet (15 m^2) or an aggregate width of 25 percent of the length of the wall where the opening protective is a *fire door assembly* in a *fire barrier* separating an enclosure for *exit access stairways* and *ramps*, and *interior exit stairways* and *ramps* from an *exit passageway* in accordance with Section 1023.3.1.

707.7 Penetrations. Penetrations of *fire barriers* shall comply with Section 714.

707.7.1 Prohibited penetrations. Penetrations into enclosures for *exit access stairways* and *ramps*, *interior exit stairways* and *ramps*, and *exit passageways* shall be allowed only where permitted by Sections 1019, 1023.5 and 1024.6, respectively.

707.8 Joints. *Joints* made in or between *fire barriers*, and *joints* made at the intersection of *fire barriers* with underside of a fire-resistance-rated floor or roof sheathing, slab or deck above, and the exterior vertical wall intersection shall comply with Section 715.

707.9 Voids at intersections. The voids created at the intersection of a *fire barrier* and a nonfire-resistance-rated *roof assembly* or a nonfire-resistance-rated *exterior wall* assembly shall be filled. An *approved* material or system shall be used to fill the void, and shall be securely installed in or on the intersection for its entire length so as not to dislodge, loosen or otherwise impair its ability to accommodate expected building movements and to retard the passage of fire and hot gases.

707.10 Ducts and air transfer openings. Penetrations in a *fire barrier* by ducts and air transfer openings shall comply with Section 717.

SECTION 708
FIRE PARTITIONS

708.1 General. The following wall assemblies shall comply with this section:

1. Separation walls as required by Section 420.2 for Group I-1 and Group R occupancies.
2. Walls separating tenant spaces in *covered and open mall buildings* as required by Section 402.4.2.1.
3. *Corridor* walls as required by Section 1020.3.
4. Enclosed elevator lobby separation as required by Section 3006.3.
5. Egress balconies as required by Section 1021.2

6. Walls separating *ambulatory care facilities* from adjacent spaces, *corridors* or tenant as required by Section 422.2.
7. Walls separating *dwelling and sleeping units* in Groups R-1 and R-2 in accordance with Sections 907.2.8.1 and 907.2.9.1.
8. Vestibules in accordance with Section 1028.2.

708.2 Materials. The walls shall be of materials permitted by the building type of construction.

708.3 Fire-resistance rating. *Fire partitions* shall have a *fire-resistance rating* of not less than 1 hour.

Exceptions:

1. *Corridor* walls permitted to have a $^{1}/_{2}$-hour *fire-resistance rating* by Table 1020.2.
2. *Dwelling unit* and *sleeping unit* separations in buildings of Types IIB, IIIB and VB construction shall have *fire-resistance ratings* of not less than $^{1}/_{2}$ hour in buildings equipped throughout with an *automatic sprinkler system* in accordance with Section 903.3.1.1.

708.4 Continuity. *Fire partitions* shall extend from the top of the foundation or floor/ceiling assembly below and be securely attached to one of the following:

1. The underside of the floor or roof sheathing, deck or slab above.
2. The underside of a floor/ceiling or roof/ceiling assembly having a *fire-resistance rating* that is not less than the *fire-resistance rating* of the *fire partition*.

Exceptions:

1. *Fire partitions* shall not be required to extend into a crawl space below where the floor above the crawl space has a minimum 1-hour *fire-resistance rating*.
2. *Fire partitions* serving as a *corridor* wall shall not be required to extend above the lower membrane of a *corridor* ceiling provided that the *corridor* ceiling membrane is equivalent to *corridor* wall membrane, and either of the following conditions is met:
   2.1. The room-side membrane of the *corridor* wall extends to the underside of the floor or roof sheathing, deck or slab of a fire-resistance-rated floor or roof above.
   2.2. The building is equipped with an *automatic sprinkler system* installed throughout in accordance with Section 903.3.1.1 or 903.3.1.2, including *automatic* sprinklers installed in the space between the top of the *fire partition* and underside of the floor or roof sheathing, deck or slab above.
3. *Fire partitions* serving as a *corridor* wall shall be permitted to terminate at the upper membrane of the *corridor* ceiling assembly where the *corridor* ceiling is constructed as required for the *corridor* wall.
4. *Fire partitions* separating tenant spaces in a *covered or open mall building* complying with Section 402.4.2.1 shall not be required to extend above the underside of a ceiling. Such ceiling shall not be required to be part of a fire-resistance-rated assembly, and the *attic* or space above the ceiling at tenant separation walls shall not be required to be subdivided by *fire partitions*.

708.4.1 Supporting construction. The supporting construction for a *fire partition* shall have a *fire-resistance rating* that is equal to or greater than the required *fire-resistance rating* of the supported *fire partition*.

Exception: In buildings of Types IIB, IIIB and VB construction, the supporting construction requirement shall not apply to *fire partitions* separating tenant spaces in *covered and open mall buildings, fire partitions* separating *dwelling units*, *fire partitions* separating *sleeping units*, *fire partitions* serving as *corridor* walls, *fire partitions* separating *ambulatory care facilities* from adjacent spaces or *corridors, fire partitions* separating *dwelling and sleeping units* from Group R-1 and R-2 occupancies and *fire partitions* separating vestibules from the *level of exit discharge.*

708.4.2 Fireblocks and draftstops in combustible construction. In combustible construction where *fire partitions* do not extend to the underside of the floor or roof sheathing, deck or slab above, the space above and along the line of the *fire partition* shall be provided with one of the following:

1. *Fireblocking* up to the underside of the floor or roof sheathing, deck or slab above using materials complying with Section 718.2.1.
2. Draftstopping up to the underside of the floor or roof sheathing, deck or slab above using materials complying with Section 718.3.1 for floors or Section 718.4.1 for *attics*.

Exceptions:

1. Buildings equipped with an *automatic sprinkler system* installed throughout in accordance with Section 903.3.1.1, or in accordance with Section 903.3.1.2 provided that protection is provided in the space between the top of the *fire partition* and underside of the floor or roof sheathing, deck or slab above as required for systems complying with Section 903.3.1.1.
2. Where *corridor* walls provide a *sleeping unit* or *dwelling unit* separation, draftstopping shall only be required above one of the *corridor* walls.
3. In Group R-2 occupancies with fewer than four *dwelling units*, *fireblocking* and draftstopping shall not be required.
4. In Group R-2 occupancies up to and including four *stories* in height in buildings not exceeding 60 feet (18 288 mm) in height above *grade*

plane, the *attic* space shall be subdivided by *draftstops* into areas not exceeding 3,000 square feet (279 m^2) or above every two *dwelling units*, whichever is smaller.

5. In Group R-3 occupancies with fewer than three *dwelling units, fireblocking* and draftstopping shall not be required in floor assemblies.

708.5 Exterior walls. Where *exterior walls* serve as a part of a required fire-resistance-rated separation, such walls shall comply with the requirements of Section 705 for *exterior walls*, and the fire-resistance-rated separation requirements shall not apply.

Exception: *Exterior walls* required to be fire-resistance rated in accordance with Section 1021.2 for exterior egress balconies, Section 1023.7 for *interior exit stairways* and *ramps* and Section 1027.6 for exterior *exit stairways* and *ramps.*

708.6 Openings. Openings in a *fire partition* shall be protected in accordance with Section 716.

708.7 Penetrations. Penetrations of *fire partitions* shall comply with Section 714.

708.8 Joints. *Joints* made in or between *fire partitions* shall comply with Section 715.

708.9 Ducts and air transfer openings. Penetrations in a *fire partition* by ducts and air transfer openings shall comply with Section 717.

SECTION 709
SMOKE BARRIERS

709.1 General. Vertical and horizontal *smoke barriers* shall comply with this section.

709.2 Materials. *Smoke barriers* shall be of materials permitted by the building type of construction.

709.3 Fire-resistance rating. A 1-hour *fire-resistance rating* is required for *smoke barriers.*

Exception: *Smoke barriers* constructed of minimum 0.10-inch-thick (2.5 mm) steel in Group I-3 buildings.

709.4 Continuity. *Smoke barriers* shall form an effective membrane continuous from the top of the foundation or floor/ceiling assembly below to the underside of the floor or roof sheathing, deck or slab above, including continuity through concealed spaces, such as those found above suspended ceilings, and interstitial structural and mechanical spaces. The supporting construction shall be protected to afford the required *fire-resistance rating* of the wall or floor supported in buildings of other than Type IIB, IIIB or VB construction. *Smoke-barrier* walls used to separate *smoke compartments* shall comply with Section 709.4.1. *Smoke-barrier* walls used to enclose *areas of refuge* in accordance with Section 1009.6.4 or to enclose elevator lobbies in accordance with Section 405.4.3, 3007.6.2, or 3008.6.2 shall comply with Section 709.4.2.

Exception: *Smoke-barrier* walls are not required in interstitial spaces where such spaces are designed and constructed with ceilings or *exterior walls* that provide resistance to the passage of fire and smoke equivalent to that provided by the *smoke-barrier* walls.

709.4.1 Smoke-barrier assemblies separating smoke compartments. *Smoke-barrier* assemblies used to separate *smoke compartments* shall form an effective membrane enclosure that is continuous from an outside wall or *smoke barrier* wall to an outside wall or another *smoke barrier* wall and to the *horizontal assemblies.*

709.4.2 Smoke-barrier walls enclosing areas of refuge or elevator lobbies. *Smoke-barrier* walls used to enclose *areas of refuge* in accordance with Section 1009.6.4, or to enclose elevator lobbies in accordance with Section 405.4.3, 3007.6.2, or 3008.6.2, shall form an effective membrane enclosure that terminates at a *fire barrier* wall having a level of *fire protection rating* not less than 1 hour, another *smoke barrier* wall or an outside wall. A smoke and draft control door assembly as specified in Section 716.2.2.1.1 shall not be required at each elevator hoistway door opening or at each exit doorway between an *area of refuge* and the exit enclosure.

709.5 Openings. Openings in a *smoke barrier* shall be protected in accordance with Section 716.

Exceptions:

1. In Group I-1, Condition 2, Group I-2 and *ambulatory care facilities*, where a pair of opposite-swinging doors are installed across a corridor in accordance with Section 709.5.1, the doors shall not be required to be protected in accordance with Section 716. The doors shall be close fitting within operational tolerances, and shall not have a center mullion or undercuts in excess of $^3/_4$ inch (19.1 mm), louvers or grilles. The doors shall have head and jamb stops, and astragals or rabbets at meeting edges. Where permitted by the door manufacturer's listing, positive-latching devices are not required. Factory-applied or field-applied protective plates are not required to be labeled.
2. In Group I-1, Condition 2, Group I-2 and *ambulatory care facilities*, special purpose horizontal sliding, accordion or folding doors installed in accordance with Section 1010.3.3 and protected in accordance with Section 716.

709.5.1 Group I-2 and ambulatory care facilities. In Group I-2 and *ambulatory care facilities*, where doors protecting openings in *smoke barriers* are installed across a *corridor* and have hold-open devices, the doors shall be automatic-closing in accordance with Section 716.2.6.6. Such doors shall have a vision panel with fire-protection-rated glazing materials in fire-protection-rated frames, the area of which shall not exceed that tested.

709.6 Penetrations. Penetrations of *smoke barriers* shall comply with Section 714.

709.7 Joints. *Joints* made in or between *smoke barriers* shall comply with Section 715.

709.8 Ducts and air transfer openings. Penetrations in a *smoke barrier* by ducts and air transfer openings shall comply with Section 717.

SECTION 710
SMOKE PARTITIONS

710.1 General. *Smoke partitions* installed as required elsewhere in the code shall comply with this section.

710.2 Materials. The walls shall be of materials permitted by the building type of construction.

710.3 Fire-resistance rating. Unless required elsewhere in the code, *smoke partitions* are not required to have a *fire-resistance rating.*

710.4 Continuity. *Smoke partitions* shall extend from the top of the foundation or floor below to the underside of the floor or roof sheathing, deck or slab above or to the underside of the ceiling above where the ceiling membrane is constructed to limit the transfer of smoke.

710.5 Openings. Openings in *smoke partitions* shall comply with Sections 710.5.1 through 710.5.3.

710.5.1 Windows. Windows in *smoke partitions* shall be sealed to resist the free passage of smoke or be automatic-closing upon detection of smoke.

710.5.2 Doors. Doors in *smoke partitions* shall comply with Sections 710.5.2.1 through 710.5.2.3.

710.5.2.1 Louvers. Doors in *smoke partitions* shall not include louvers.

Exception: Where permitted in accordance with Section 407.3.1.1.

710.5.2.2 Smoke and draft control doors. Where required elsewhere in the code, doors in *smoke partitions* shall meet the requirements for a smoke and draft control door assembly tested in accordance with UL 1784. The air leakage rate of the door assembly shall not exceed 3.0 cubic feet per minute per square foot [$0.015424\ m^3/(s \times m^2)$] of door opening at 0.10 inch (24.9 Pa) of water for both the ambient temperature test and the elevated temperature exposure test. Installation of smoke doors shall be in accordance with NFPA 105.

710.5.2.2.1 Smoke and draft control door labeling. Smoke and draft control doors complying only with UL 1784 shall be permitted to show the letter "S" on the manufacturer's labeling.

710.5.2.3 Self- or automatic-closing doors. Where required elsewhere in the code, doors in *smoke partitions* shall be self- or automatic-closing by smoke detection in accordance with Section 716.2.6.6.

710.5.3 Pass-through openings in Group I-2, Condition 2. Where pass-through openings are provided in *smoke partitions* in Group I-2, Condition 2 occupancies, such openings shall comply with the following:

1. The smoke compartment in which the pass-through openings occur does not contain a patient care suite or sleeping room.
2. Pass-through openings are installed in a wall, door or vision panel that is not required to have a fire-resistance rating.
3. The top of the pass-through opening is located a maximum of 48 inches (1219 mm) above the floor.
4. The aggregate area of all such pass-through openings within a single room shall not exceed 80 square inches (0.05 m2).

710.6 Penetrations. The space around penetrating items shall be filled with an *approved* material to limit the free passage of smoke.

710.7 Joints. *Joints* shall be filled with an *approved* material to limit the free passage of smoke.

710.8 Ducts and air transfer openings. The space around a duct penetrating a *smoke partition* shall be filled with an *approved* material to limit the free passage of smoke. Air transfer openings in *smoke partitions* shall be provided with a *smoke damper* complying with Section 717.3.2.2.

Exception: Where the installation of a *smoke damper* will interfere with the operation of a required smoke control system in accordance with Section 909, *approved* alternative protection shall be utilized.

SECTION 711
FLOOR AND ROOF ASSEMBLIES

711.1 General. *Horizontal assemblies* shall comply with Section 711.2. Nonfire-resistance-rated floor and *roof assemblies* shall comply with Section 711.3.

711.2 Horizontal assemblies. *Horizontal assemblies* shall comply with Sections 711.2.1 through 711.2.6.

711.2.1 Materials. Assemblies shall be of materials permitted by the building type of construction.

711.2.2 Continuity. Assemblies shall be continuous without vertical openings, except as permitted by this section and Section 712.

711.2.3 Supporting construction. The supporting construction shall be protected to afford the required *fire-resistance rating* of the *horizontal assembly* supported.

Exception: In buildings of Type IIB, IIIB or VB construction, the construction supporting the *horizontal assembly* is not required to be fire-resistance rated at the following:

1. *Horizontal assemblies* at the separations of incidental uses as specified by Table 509.1

provided that the required *fire-resistance rating* does not exceed 1 hour.

2. *Horizontal assemblies* at the separations of *dwelling units* and *sleeping units* as required by Section 420.3.
3. *Horizontal assemblies* at *smoke barriers* constructed in accordance with Section 709.

711.2.4 Fire-resistance rating. The *fire-resistance rating* of *horizontal assemblies* shall comply with Sections 711.2.4.1 through 711.2.4.6 but shall be not less than that required by the building type of construction.

711.2.4.1 Separating mixed occupancies. Where the *horizontal assembly* separates mixed occupancies, the assembly shall have a *fire-resistance rating* of not less than that required by Section 508.4 based on the occupancies being separated.

711.2.4.2 Separating fire areas. Where the *horizontal assembly* separates a single occupancy into different *fire areas*, the assembly shall have a *fire-resistance rating* of not less than that required by Section 707.3.10.

711.2.4.3 Dwelling units and sleeping units. *Horizontal assemblies* serving as *dwelling or sleeping unit* separations in accordance with Section 420.3 shall be not less than 1-hour *fire-resistance-rated* construction.

Exception: *Horizontal assemblies* separating *dwelling units* and *sleeping units* shall be not less than $^1/_2$-hour fire-resistance-rated construction in a building of Types IIB, IIIB and VB construction, where the building is equipped throughout with an *automatic sprinkler system* in accordance with Section 903.3.1.1.

711.2.4.4 Separating smoke compartments. Where the *horizontal assembly* is required to be a *smoke barrier*, the assembly shall comply with Section 709.

711.2.4.5 Separating incidental uses. Where the *horizontal assembly* separates incidental uses from the remainder of the building, the assembly shall have a *fire-resistance rating* of not less than that required by Section 509.

711.2.4.6 Other separations. Where a *horizontal assembly* is required by other sections of this code, the assembly shall have a *fire-resistance rating* of not less than that required by that section.

711.2.5 Ceiling panels. Where the weight of lay-in ceiling panels, used as part of fire-resistance-rated floor/ceiling or roof/ceiling assemblies, is not adequate to resist an upward force of 1 pound per square foot (48 Pa), wire or other *approved* devices shall be installed above the panels to prevent vertical displacement under such upward force.

711.2.6 Unusable space. In 1-hour fire-resistance-rated floor/ceiling assemblies, the ceiling membrane is not required to be installed over unusable crawl spaces. In 1-hour fire-resistance-rated roof assemblies, the floor membrane is not required to be installed where unusable attic space occurs above.

711.3 Nonfire-resistance-rated floor and roof assemblies. Nonfire-resistance-rated floor, floor/ceiling, roof and roof/ceiling assemblies shall comply with Sections 711.3.1 and 711.3.2.

711.3.1 Materials. Assemblies shall be of materials permitted by the building type of construction.

711.3.2 Continuity. Assemblies shall be continuous without vertical openings, except as permitted by Section 712.

SECTION 712
VERTICAL OPENINGS

712.1 General. Each vertical opening shall comply in accordance with one of the protection methods in Sections 712.1.1 through 712.1.16.

712.1.1 Shaft enclosures. Vertical openings contained entirely within a *shaft enclosure* complying with Section 713 shall be permitted.

712.1.2 Individual dwelling unit. Unconcealed vertical openings totally within an individual residential *dwelling unit* and connecting four *stories* or less shall be permitted.

712.1.3 Escalator openings. Where a building is equipped throughout with an *automatic sprinkler system* in accordance with Section 903.3.1.1, vertical openings for escalators shall be permitted where protected in accordance with Section 712.1.3.1 or 712.1.3.2.

712.1.3.1 Opening size. Protection by a draft curtain and closely spaced sprinklers in accordance with NFPA 13 shall be permitted where the area of the vertical opening between *stories* does not exceed twice the horizontal projected area of the escalator. In other than Groups B and M, this application is limited to openings that do not connect more than four *stories*.

712.1.3.2 Automatic shutters. Protection of the vertical opening by approved shutters at every penetrated floor shall be permitted in accordance with this section. The shutters shall be of noncombustible construction and have a *fire-resistance rating* of not less than 1.5 hours. The shutter shall be so constructed as to close immediately upon the actuation of a smoke detector installed in accordance with Section 907.3.1 and shall completely shut off the well opening. Escalators shall cease operation when the shutter begins to close. The shutter shall operate at a speed of not more than 30 feet per minute (152.4 mm/s) and shall be equipped with a sensitive leading edge to arrest its progress where in contact with any obstacle, and to continue its progress on release therefrom.

712.1.4 Penetrations. Penetrations, concealed and unconcealed, shall be permitted where protected in accordance with Section 714.

712.1.5 Joints. *Joints* shall be permitted where complying with Section 712.1.5.1 or 712.1.5.2, as applicable.

712.1.5.1 Joints in or between horizontal assemblies. *Joints* made in or between *horizontal assemblies* shall comply with Section 715. The void created at the intersection of a floor/ceiling assembly and an exterior curtain wall assembly shall be permitted where protected in accordance with Section 715.4.

712.1.5.2 Joints in or between nonfire-resistance-rated floor assemblies. *Joints* in or between floor assemblies without a required *fire-resistance rating* shall be permitted where they comply with one of the following:

1. The *joint* shall be concealed within the cavity of a wall.
2. The *joint* shall be located above a ceiling.
3. The *joint* shall be sealed, treated or covered with an *approved* material or system to resist the free passage of flame and the products of combustion.

Exception: *Joints* meeting one of the exceptions specified in Section 715.3.

712.1.6 Ducts and air transfer openings. Penetrations by ducts and air transfer openings shall be protected in accordance with Section 717. Grease ducts shall be protected in accordance with the *International Mechanical Code*.

712.1.7 Atriums. *Atriums* complying with Section 404 that connect two or more *stories* in Group I-2 or I-3 occupancies or three or more *stories* in other occupancies shall be permitted.

Exceptions:

1. *Atriums* shall not be permitted within Group H occupancies.
2. Balconies or stories within Groups A-1, A-4 and A-5 and *mezzanines* that comply with Section 505 shall not be considered a story as it applies to this section

712.1.8 Masonry chimney. Approved vertical openings for masonry chimneys shall be permitted where the *annular space* is fireblocked at each floor level in accordance with Section 718.2.5.

712.1.9 Two-story openings. In other than Groups I-2 and I-3, a vertical opening that is not used as one of the applications specified in this section shall be permitted if the opening complies with all of the following items:

1. Does not connect more than two *stories*.
2. Does not penetrate a *horizontal assembly* that separates *fire areas* or *smoke barriers* that separate *smoke compartments*.
3. Is not concealed within the construction of a wall or a floor/ceiling assembly.
4. Is not open to a *corridor* in Group I and R occupancies.
5. Is not open to a *corridor* on nonsprinklered floors.
6. Is separated from floor openings and air transfer openings serving other floors by construction conforming to required *shaft enclosures*.

712.1.10 Parking garages. Vertical openings in parking garages for automobile ramps, elevators and duct systems shall comply with Section 712.1.10.1, 712.1.10.2 or 712.1.10.3, as applicable.

712.1.10.1 Automobile ramps. Vertical openings for automobile ramps in parking garages shall be permitted where constructed in accordance with Sections 406.5 and 406.6.

712.1.10.2 Elevators. Vertical openings for elevator hoistways in parking garages that serve only the parking garage, and complying with Sections 406.5 and 406.6, respectively, shall be permitted.

712.1.10.3 Duct systems. Vertical openings for mechanical exhaust or supply duct systems in parking garages complying with Sections 406.5 and 406.6, respectively, shall be permitted to be unenclosed where such duct system is contained within and serves only the parking garage.

712.1.11 Mezzanine. Vertical openings between a *mezzanine* complying with Section 505 and the floor below shall be permitted.

712.1.12 Exit access stairways and ramps. Vertical openings containing *exit access stairways* or *ramps* in accordance with Section 1019 shall be permitted.

712.1.13 Openings. Vertical openings for floor fire doors and access doors shall be permitted where protected by Section 712.1.13.1 or 712.1.13.2.

712.1.13.1 Horizontal fire door assemblies. Horizontal *fire door* assemblies used to protect openings in fire-resistance-rated *horizontal assemblies* shall be tested in accordance with NFPA 288, and shall achieve a *fire-resistance rating* not less than the assembly being penetrated. Horizontal *fire door* assemblies shall be labeled by an *approved* agency. The *label* shall be permanently affixed and shall specify the manufacturer, the test standard and the *fire-resistance rating*.

712.1.13.2 Access doors. Access doors shall be permitted in ceilings of fire-resistance-rated floor/ceiling and roof/ceiling assemblies, provided that such doors are tested in accordance with ASTM E119 or UL 263 as *horizontal assemblies* and labeled by an *approved agency* for such purpose.

712.1.14 Group I-3. In Group I-3 occupancies, vertical openings shall be permitted in accordance with Section 408.5.

712.1.15 Skylights. *Skylights* and other penetrations through a fire-resistance-rated roof deck or slab are permitted to be unprotected, provided that the structural integrity of the fire-resistance-rated *roof assembly* is maintained. Unprotected *skylights* shall not be permitted in *roof assemblies* required to be fire-resistance rated in accordance with Section 705.8.6. The supporting construction shall be protected to afford the required *fire-resistance rating* of the *horizontal assembly* supported.

712.1.16 Openings otherwise permitted. Vertical openings shall be permitted where allowed by other sections of this code.

SECTION 713
SHAFT ENCLOSURES

713.1 General. The provisions of this section shall apply to *shafts* required to protect openings and penetrations through floor/ceiling and roof/ceiling assemblies. *Interior exit stairways* and *ramps* shall be enclosed in accordance with Section 1023.

713.2 Construction. *Shaft enclosures* shall be constructed as *fire barriers* in accordance with Section 707 or *horizontal assemblies* in accordance with Section 711, or both.

713.3 Materials. *Shaft enclosures* shall be of materials permitted by the building type of construction.

713.4 Fire-resistance rating. *Shaft enclosures* shall have a *fire-resistance rating* of not less than 2 hours where connecting four *stories* or more, and not less than 1 hour where connecting less than four *stories*. The number of *stories* connected by the *shaft enclosure* shall include any *basements* but not any *mezzanines*. *Shaft enclosures* shall have a *fire-resistance rating* not less than the floor assembly penetrated, but need not exceed 2 hours. *Shaft enclosures* shall meet the requirements of Section 703.2.1.1.

713.5 Continuity. *Shaft enclosures* shall be constructed as *fire barriers* in accordance with Section 707 or *horizontal assemblies* constructed in accordance with Section 711, or both, and shall have continuity in accordance with Section 707.5 for *fire barriers* or Section 711.2.2 for *horizontal assemblies*, as applicable.

713.6 Exterior walls. Where *exterior walls* serve as a part of a required *shaft enclosure*, such walls shall comply with the requirements of Section 705 for *exterior walls* and the fire-resistance-rated enclosure requirements shall not apply.

Exception: *Exterior walls* required to be fire-resistance rated in accordance with Section 1021.2 for exterior egress balconies, Section 1023.7 for *interior exit stairways* and *ramps* and Section 1027.6 for exterior *exit stairways* and *ramps.*

713.7 Openings. Openings in a *shaft enclosure* shall be protected in accordance with Section 716 as required for *fire barriers*. Doors shall be self- or automatic-closing by smoke detection in accordance with Section 716.2.6.6.

713.7.1 Prohibited openings. Openings other than those necessary for the purpose of the *shaft* shall not be permitted in *shaft enclosures*.

713.8 Penetrations. Penetrations in a *shaft enclosure* shall be protected in accordance with Section 714 as required for *fire barriers*. Structural elements, such as beams or joists, where protected in accordance with Section 714 shall be permitted to penetrate a *shaft enclosure.*

713.8.1 Prohibited penetrations. Penetrations other than those necessary for the purpose of the *shaft* shall not be permitted in *shaft enclosures.*

Exception: *Membrane penetrations* shall be permitted on the outside of *shaft enclosures*. Such penetrations shall be protected in accordance with Section 714.4.2.

713.9 Joints. *Joints* in a *shaft enclosure* shall comply with Section 715.

713.10 Duct and air transfer openings. Penetrations of a *shaft enclosure* by ducts and air transfer openings shall comply with Section 717.

713.11 Enclosure at the bottom. *Shafts* that do not extend to the bottom of the building or structure shall comply with one of the following:

1. Be enclosed at the lowest level with construction of the same *fire-resistance rating* as the *lowest floor* through which the *shaft* passes, but not less than the rating required for the *shaft enclosure.*
2. Terminate in a room having a use related to the purpose of the *shaft*. The room shall be separated from the remainder of the building by *fire barriers* constructed in accordance with Section 707 or *horizontal assemblies* constructed in accordance with Section 711, or both. The *fire-resistance rating* and opening protectives shall be not less than the protection required for the *shaft enclosure.*
3. Be protected by *approved fire dampers* installed in accordance with their listing at the *lowest floor* level within the *shaft enclosure.*

Exceptions:

1. The fire-resistance-rated room separation is not required, provided that the only openings in or penetrations of the *shaft enclosure* to the interior of the building occur at the bottom. The bottom of the *shaft* shall be closed off around the penetrating items with materials permitted by Section 718.3.1 for draftstopping, or the room shall be provided with an *approved automatic sprinkler system.*
2. A *shaft enclosure* containing a waste or linen chute shall not be used for any other purpose and shall discharge in a room protected in accordance with Section 713.13.4.
3. The fire-resistance-rated room separation and the protection at the bottom of the *shaft* are not required provided that there are no combustibles in the *shaft* and there are no openings or other penetrations through the *shaft enclosure* to the interior of the building.

713.12 Enclosure at top. The top of *shaft enclosures* shall comply with one of the following:

1. Extend to the underside of the roof sheathing, deck or slab of the building, and the *roof assembly* shall comply with the requirements for the type of construction as specified in Table 601.

2. Terminate below the *roof assembly* and be enclosed at the top with construction of the same *fire-resistance rating* as the topmost floor penetrated by the shaft, but not less than the *fire-resistance rating* required for the shaft enclosure.
3. Extend past the *roof assembly* and comply with the requirements of Section 1511.

713.12.1 Penthouse mechanical rooms. A fire/*smoke damper* shall not be required at the penetration of the *rooftop structure* where *shaft enclosures* extend up through the *roof assembly* into a *rooftop structure* conforming to Section 1511. Ductwork in the *shaft* shall be connected directly to HVAC equipment.

713.13 Waste, recycling and linen chutes and incinerator rooms. Waste, recycling and linen chutes shall comply with the provisions of NFPA 82, Chapter 6 and shall meet the requirements of Sections 712 and 713.13.1 through 713.13.6. Incinerator rooms shall meet the provisions of Sections 713.13.4 and 713.13.5.

Exception: Chutes serving and contained within a single *dwelling unit.*

713.13.1 Waste, recycling and linen chute enclosures. A *shaft enclosure* containing a recycling, waste or linen chute shall not be used for any other purpose and shall be enclosed in accordance with Section 713.4. A *shaft enclosure* shall be permitted to contain recycling and waste chutes. Openings into the *shaft*, from access rooms and discharge rooms, shall be protected in accordance with this section and Section 716. Openings into chutes shall not be located in *corridors*. Doors into chutes shall be *self-closing*. Discharge doors shall be self-or automatic-closing upon the actuation of a smoke detector in accordance with Section 716.2.6.6, except that heat-activated closing devices shall be permitted between the *shaft* and the discharge room.

713.13.2 Materials. A *shaft enclosure* containing a waste, recycling, or linen chute shall be constructed of materials as permitted by the building type of construction.

713.13.3 Chute access rooms. Access openings for waste, recycling or linen chutes shall be located in rooms or compartments enclosed by not less than 1-hour *fire barriers* constructed in accordance with Section 707 or *horizontal assemblies* constructed in accordance with Section 711, or both. Openings into the access rooms shall be protected by opening protectives having a *fire protection rating* of not less than $^3/_4$ hour. Doors shall be self- or automatic-closing upon the detection of smoke in accordance with Section 716.2.6.6. The room or compartment shall be configured to allow the access door to the room or compartment to close and latch with the access panel to the chute in any position.

713.13.4 Chute discharge room. Table 509.1Waste, recycling or linen chutes shall discharge into an enclosed room separated by *fire barriers* with a *fire-resistance rating* not less than the required fire rating of the *shaft enclosure* and constructed in accordance with Section 707 or *horizontal assemblies* constructed in accordance with Section 711, or both. Openings into the discharge room from the remainder of the building shall be protected by opening protectives having a *fire protection rating* equal to the protection required for the *shaft enclosure.* Doors shall be self- or automatic-closing upon the detection of smoke in accordance with Section 716.2.6.6. Waste chutes shall not terminate in an incinerator room. Waste and linen rooms that are not provided with chutes need only comply with Table 509.1.

713.13.5 Incinerator room. Incinerator rooms shall comply with Table 509.1.

713.13.6 Automatic sprinkler system. An *approved automatic sprinkler system* shall be installed in accordance with Section 903.2.11.2.

713.14 Elevator, dumbwaiter and other hoistways. Elevator, dumbwaiter and other hoistway enclosures shall be constructed in accordance with Sections 712 and 713, and Chapter 30.

SECTION 714
PENETRATIONS

714.1 Scope. The provisions of this section shall govern the materials and methods of construction used to protect *through penetrations* and *membrane penetrations* of *horizontal assemblies* and fire-resistance-rated wall assemblies.

714.1.1 Ducts and air transfer openings. Penetrations of fire-resistance-rated walls by ducts that are not protected with *dampers* shall comply with Sections 714.3 through 714.4.3. Penetrations of *horizontal assemblies* not protected with a *shaft* as permitted by Section 717.6, and not required to be protected with *fire dampers* by other sections of this code, shall comply with Sections 714.5 through 714.6.2. Ducts and air transfer openings that are protected with *dampers* shall comply with Section 717.

714.2 Installation. A listed *penetration firestop* system shall be installed in accordance with the manufacturer's installation instructions and the listing criteria.

714.3 Installation details. Where sleeves are used, they shall be securely fastened to the assembly penetrated. The space between the item contained in the sleeve and the sleeve itself and any space between the sleeve and the assembly penetrated shall be protected in accordance with this section. Insulation and coverings on or in the penetrating item shall not penetrate the assembly unless the specific material used has been tested as part of the assembly in accordance with this section.

714.4 Fire-resistance-rated walls. Penetrations into or through *fire walls*, *fire barriers*, *smoke barrier* walls and *fire partitions* shall comply with Sections 714.4.1 through 714.4.3. Penetrations in *smoke barrier* walls shall also comply with Section 714.5.4.

714.4.1 Through penetrations. *Through penetrations* of fire-resistance-rated walls shall comply with Section 714.4.1.1 or 714.4.1.2.

Exception: Where the penetrating items are steel, ferrous or copper pipes, tubes or conduits, the *annular* *space* between the penetrating item and the fire-resistance-rated wall is permitted to be protected by either of the following measures:

1. In concrete or masonry walls where the penetrating item is a maximum 6-inch (152 mm) nominal diameter and the area of the opening through the wall does not exceed 144 square inches ($0.0929 m^2$), concrete, grout or *mortar* is permitted where installed the full thickness of the wall or the thickness required to maintain the *fire-resistance rating*.
2. The material used to fill the *annular space* shall prevent the passage of flame and hot gases sufficient to ignite cotton waste when subjected to ASTM E119 or UL 263 time-temperature fire conditions under a minimum positive pressure differential of 0.01 inch (2.49 Pa) of water at the location of the penetration for the time period equivalent to the *fire-resistance rating* of the construction penetrated.

714.4.1.1 Fire-resistance-rated assemblies. *Through penetrations* shall be protected using systems installed as tested in the *approved* fire-resistance-rated assembly.

714.4.1.2 Through-penetration firestop system. *Through penetrations* shall be protected by an *approved penetration firestop* system installed as tested in accordance with ASTM E814 or UL 1479, with a minimum positive pressure differential of 0.01 inch (2.49 Pa) of water and shall have an *F rating* of not less than the required *fire-resistance rating* of the wall penetrated.

714.4.2 Membrane penetrations. *Membrane penetrations* shall comply with Section 714.4.1. Where walls or partitions are required to have a *fire-resistance rating*, recessed fixtures shall be installed such that the required *fire resistance* will not be reduced.

Exceptions:

1. *Membrane penetrations* of maximum 2-hour fire-resistance-rated walls and partitions by steel electrical boxes that do not exceed 16 square inches ($0.0 103 m^2$) in area, provided that the aggregate area of the openings through the membrane does not exceed 100 square inches ($0.0645 m^2$) in any 100 square feet ($9.29 m^2$) of wall area. The *annular space* between the wall membrane and the box shall not exceed $^1/_8$ inch (3.2 mm). Such boxes on opposite sides of the wall or partition shall be separated by one of the following:
   - 1.1. By a horizontal distance of not less than 24 inches (610 mm) where the wall or partition is constructed with individual noncommunicating stud cavities.
   - 1.2. By a horizontal distance of not less than the depth of the wall cavity where the wall cavity is filled with cellulose loose-fill, rockwool or slag *mineral wool* insulation.
   - 1.3. By solid *fireblocking* in accordance with Section 718.2.1.
   - 1.4. By protecting both outlet boxes with *listed* putty pads.
   - 1.5. By other *listed* materials and methods.
2. *Membrane penetrations* by *listed* electrical boxes of any material, provided that such boxes have been tested for use in fire-resistance-rated assemblies and are installed in accordance with the instructions included in the listing. The *annular space* between the wall membrane and the box shall not exceed $^1/_8$ inch (3.2 mm) unless *listed* otherwise. Such boxes on opposite sides of the wall or partition shall be separated by one of the following:
   - 2.1. By the horizontal distance specified in the listing of the electrical boxes.
   - 2.2. By solid *fireblocking* in accordance with Section 718.2.1.
   - 2.3. By protecting both boxes with *listed* putty pads.
   - 2.4. By other *listed* materials and methods.
3. *Membrane penetrations* by electrical boxes of any size or type, that have been *listed* as part of a wall opening protective material system for use in fire-resistance-rated assemblies and are installed in accordance with the instructions included in the listing.
4. *Membrane penetrations* by boxes other than electrical boxes, provided that such penetrating items and the *annular space* between the wall membrane and the box, are protected by an *approved membrane penetration firestop system* installed as tested in accordance with ASTM E814 or UL 1479, with a minimum positive pressure differential of 0.01 inch (2.49 Pa) of water, and shall have an F and *T rating* of not less than the required *fire-resistance rating* of the wall penetrated and be installed in accordance with their listing.
5. The *annular space* created by the penetration of an automatic sprinkler, provided that it is covered by a metal escutcheon plate.
6. *Membrane penetrations* of maximum 2-hour fire-resistance-rated walls and partitions by steel electrical boxes that exceed 16 square inches ($0.0 103 m^2$) in area, or steel electrical boxes of any size having an aggregate area through the membrane exceeding 100 square inches ($0.0645 m^2$) in any 100 square feet ($9.29 m^2$) of wall area, provided that such penetrating items are protected by *listed* putty pads or other

listed materials and methods, and installed in accordance with the listing.

714.4.3 Dissimilar materials. Noncombustible penetrating items shall not connect to combustible items beyond the point of firestopping unless it can be demonstrated that the *fire-resistance* integrity of the wall is maintained.

714.5 Horizontal assemblies. Penetrations of a *fire-resistance-rated* floor, floor/ceiling assembly or the ceiling membrane of a roof/ceiling assembly not required to be enclosed in a *shaft* by Section 712.1 shall be protected in accordance with Sections 714.5.1 through 714.5.4.

714.5.1 Through penetrations. *Through penetrations* of *horizontal assemblies* shall comply with Section 714.5.1.1 or 714.5.1.2.

Exceptions:

1. Penetrations by steel, ferrous or copper conduits, pipes, tubes or vents or concrete or masonry items through a single fire-resistance-rated floor assembly where the *annular space* is protected with materials that prevent the passage of flame and hot gases sufficient to ignite cotton waste when subjected to ASTM E119 or UL 263 time-temperature fire conditions under a minimum positive pressure differential of 0.01 inch (2.49 Pa) of water at the location of the penetration for the time period equivalent to the *fire-resistance rating* of the construction penetrated. Penetrating items with a maximum 6-inch (152 mm) nominal diameter shall not be limited to the penetration of a single fire-resistance-rated floor assembly, provided that the aggregate area of the openings through the assembly does not exceed 144 square inches (92 900 mm^2) in any 100 square feet (9.3 m^2) of floor area.
2. Penetrations in a single concrete floor by steel, ferrous or copper conduits, pipes, tubes or vents with a maximum 6-inch (152 mm) nominal diameter, provided that the concrete, grout or *mortar* is installed the full thickness of the floor or the thickness required to maintain the *fire-resistance rating*. The penetrating items shall not be limited to the penetration of a single concrete floor, provided that the area of the opening through each floor does not exceed 144 square inches (92 900 mm^2).
3. Penetrations by *listed* electrical boxes of any material, provided that such boxes have been tested for use in fire-resistance-rated assemblies and installed in accordance with the instructions included in the listing.

714.5.1.1 Fire-resistance-rated assemblies. *Through penetrations* shall be protected using systems installed as tested in the *approved* fire-resistance-rated assembly.

714.5.1.2 Through-penetration firestop system. *Through penetrations* shall be protected by an *approved through-penetration firestop system* installed and tested in accordance with ASTM E814 or UL 1479, with a minimum positive pressure differential of 0.01 inch of water (2.49 Pa). The system shall have an *F rating/T rating* of not less than 1 hour but not less than the required rating of the floor penetrated.

Exceptions:

1. Floor penetrations contained and located within the cavity of a wall above the floor or below the floor do not require a *T rating*.
2. Floor penetrations by floor drains, tub drains or shower drains contained and located within the concealed space of a *horizontal assembly* do not require a *T rating*.
3. Floor penetrations of maximum 4-inch (102 mm) nominal diameter metal conduit or tubing penetrating directly into metal-enclosed electrical power switchgear do not require a *T rating*.

714.5.2 Membrane penetrations. Penetrations of membranes that are part of a *horizontal assembly* shall comply with Section 714.5.1.1 or 714.5.1.2. Where floor/ceiling assemblies are required to have a *fire-resistance rating*, recessed fixtures shall be installed such that the required *fire resistance* will not be reduced.

Exceptions:

1. *Membrane penetrations* by steel, ferrous or copper conduits, pipes, tubes or vents, or concrete or masonry items where the *annular space* is protected either in accordance with Section 714.5.1 or to prevent the free passage of flame and the products of combustion. The aggregate area of the openings through the membrane shall not exceed 100 square inches (64 500 mm^2) in any 100 square feet (9.3 m^2) of ceiling area in assemblies tested without penetrations.
2. Ceiling *membrane penetrations* of maximum 2-hour *horizontal assemblies* by steel electrical boxes that do not exceed 16 square inches (10 323 mm^2) in area, provided that the aggregate area of such penetrations does not exceed 100 square inches (44 500 mm^2) in any 100 square feet (9.29 m^2) of ceiling area, and the *annular space* between the ceiling membrane and the box does not exceed $^1/_8$ inch (3.2 mm).
3. *Membrane penetrations* by electrical boxes of any size or type, that have been *listed* as part of an opening protective material system for use in *horizontal assemblies* and are installed in accordance with the instructions included in the listing.
4. *Membrane penetrations* by *listed* electrical boxes of any material, provided that such boxes have been tested for use in fire-resis-

tance-rated assemblies and are installed in accordance with the instructions included in the listing. The *annular space* between the ceiling membrane and the box shall not exceed $^1/_8$ inch (3.2 mm) unless *listed* otherwise.

5. The *annular space* created by the penetration of a fire sprinkler, provided that it is covered by a metal escutcheon plate.
6. Noncombustible items that are cast into concrete *building elements* and that do not penetrate both top and bottom surfaces of the element.
7. The ceiling membrane of a maximum 2-hour fire-resistance-rated *horizontal assembly* is permitted to be interrupted with the double wood top plate of a wall assembly that is sheathed with Type X *gypsum wallboard*, provided that all penetrating items through the double top plates are protected in accordance with Section 714.5.1.1 or 714.5.1.2 and the ceiling membrane is tight to the top plates.
8. Ceiling *membrane penetrations* by listed luminaires (light fixtures) or by luminaires protected with *listed* materials, which have been tested for use in fire-resistance-rated assemblies and are installed in accordance with the instructions included in the listing.

714.5.3 Dissimilar materials. Noncombustible penetrating items shall not connect to combustible materials beyond the point of firestopping unless it can be demonstrated that the *fire-resistance* integrity of the *horizontal assembly* is maintained.

714.5.4 Penetrations in smoke barriers. Penetrations in *smoke barriers* shall be protected by an approved *through-penetration firestop system* installed and tested in accordance with the requirements of UL 1479 for air leakage. The *L rating* of the system measured at 0.30 inch (7.47 Pa) of water in both the ambient temperature and elevated temperature tests shall not exceed either of the following:

1. 5.0 cfm per square foot ($0.025 \text{ m}^3/\text{ s} \times \text{m}^2$) of penetration opening for each *through-penetration firestop system.*
2. A total cumulative leakage of 50 cfm ($0.024 \text{ m}^3/\text{s}$) for any 100 square feet (9.3 m^2) of wall area, or floor area.

714.6 Nonfire-resistance-rated assemblies. Penetrations of nonfire-resistance-rated floor or floor/ceiling assemblies or the ceiling membrane of a nonfire-resistance-rated roof/ceiling assembly shall meet the requirements of Section 713 or shall comply with Section 714.6.1 or 714.6.2.

714.6.1 Noncombustible penetrating items. Noncombustible penetrating items that connect not more than five *stories* are permitted, provided that the *annular space* is filled to resist the free passage of flame and the products of combustion with an *approved* noncombustible material or with a fill, void or cavity material that is tested and classified for use in *through-penetration firestop systems*.

714.6.2 Penetrating items. Penetrating items that connect not more than two *stories* are permitted, provided that the *annular space* is filled with an *approved* material to resist the free passage of flame and the products of combustion.

SECTION 715
JOINTS AND VOIDS

*

715.1 General. The provisions of this section shall govern the materials and methods of construction used to protect *joints* and voids in or between horizontal and vertical assemblies.

715.2 Installation. Systems or materials protecting *joints* and voids shall be securely installed in accordance with the manufacturer's installation instructions in or on the *joint* or void for its entire length so as not to dislodge, loosen or otherwise impair its ability to accommodate expected building movements and to resist the passage of fire and hot gases. *Fire-resistant joint systems* or systems used to protect voids at exterior curtain walls and fire-resistance-rated floor intersections shall also be installed in accordance with the listing criteria. **

715.3 Fire-resistance-rated assembly intersections. *Joints* installed in or between fire-resistance-rated walls, floor or floor/ceiling assemblies and roofs or roof/ceiling assemblies shall be protected by an approved *fire-resistant joint* system designed to resist the passage of fire for a time period not less than the required *fire-resistance rating* of the wall, floor or roof in or between which the system is installed.

Exception: *Fire-resistant joint systems* shall not be required for *joints* in the following locations:

1. Floors within a single *dwelling unit.*
2. Floors where the *joint* is protected by a shaft enclosure in accordance with Section 713.
3. Floors within *atriums* where the space adjacent to the *atrium* is included in the volume of the *atrium* for smoke control purposes.
4. Floors within *malls*.
5. Floors and ramps within parking garages or structures constructed in accordance with Sections 406.5 and 406.6.
6. *Mezzanine* floors.
7. Walls that are permitted to have unprotected openings.
8. Roofs where openings are permitted.
9. Control *joints* not exceeding a maximum width of 0.625 inch (15.9 mm) and tested in accordance with ASTM E119 or UL 263.
10. The intersection of exterior curtain wall assemblies and the roof slab or roof deck.

715.3.1 Fire test criteria. *Fire-resistant joint systems* shall be tested in accordance with the requirements of either ASTM E1966 or UL 2079. Nonsymmetrical wall *joint* systems shall be tested with both faces exposed to the furnace, and the assigned *fire-resistance rating* shall be the shortest duration obtained from the two tests. Where evidence is furnished to show that the wall was tested with the least fire-resistant side exposed to the furnace, subject to acceptance of the *building official*, the wall need not be subjected to tests from the opposite side.

Exception: For *exterior walls* with a horizontal *fire separation distance* greater than 10 feet (3048 mm), the joint system shall be required to be tested for interior fire exposure only.

715.4 Exterior curtain wall/fire-resistance-rated floor intersections. Voids created at the intersection of exterior curtain wall assemblies and fire-resistance-rated floor or floor/ceiling assemblies shall be protected with an *approved perimeter fire containment system* to prevent the interior spread of fire. Such systems shall provide an *F rating* for a time period not less than the *fire-resistance rating* of the floor or floor/ceiling assembly.

715.4.1 Fire test criteria. *Perimeter fire containment systems* shall be tested in accordance with the requirements of ASTM E2307.

Exception: Voids created at the intersection of the exterior curtain wall assemblies and floor assemblies where the vision glass extends to the finished floor level shall be permitted to be protected with an approved material to prevent the interior spread of fire. Such material shall be securely installed and capable of preventing the passage of flame and hot gases sufficient to ignite cotton waste where subjected to ASTM E119 time-temperature fire conditions under a minimum positive pressure differential of 0.01 inch (0.254 mm) of water column (2.5 Pa) for the time period not less than the fire-resistance rating of the floor assembly.

715.5 Exterior curtain wall/nonfire-resistance-rated floor assembly intersections. Voids created at the intersection of exterior curtain wall assemblies and nonfire-resistance-rated floor or floor/ceiling assemblies shall be filled with an *approved* material or system to retard the interior spread of fire and hot gases between *stories*.

715.6 Exterior curtain wall/vertical fire barrier intersections. Voids created at the intersection of nonfire-resistance-rated exterior curtain wall assemblies and vertical *fire barriers* shall be filled with an approved material or system to retard the interior spread of fire and hot gases.

715.7 Curtain wall spandrels. Height and *fire-resistance* requirements for curtain wall spandrels shall comply with Section 705.8.5. Where Section 705.8.5 does not require fire-resistance-rated curtain wall spandrels, the requirements of Sections 715.4 and 715.5 shall still apply to the intersection between the curtain wall spandrels and the floor.

715.8 Joints and voids in smoke barriers. *Fire-resistant joint systems* protecting *joints* in *smoke barriers*, and perimeter fire containment systems protecting voids at the intersection of a horizontal *smoke barrier* and an exterior curtain wall, shall be tested in accordance with the requirements of UL 2079 for air leakage. The L rating of the joint system shall not exceed 5 cubic feet per minute per linear foot ($0.00775\ m^3/s\ m$) of joint at 0.30 inch (74.7 Pa) of water for both the ambient temperature and elevated temperature tests.

SECTION 716
OPENING PROTECTIVES

716.1 General. Opening protectives required by other sections of this code shall comply with the provisions of this section and shall be installed in accordance with NFPA 80.

716.1.1 Alternative methods for determining fire protection ratings. The application of any of the alternative methods specified in this section shall be based on the fire exposure and acceptance criteria specified in NFPA 252, NFPA 257, UL 9, UL 10B or UL 10C. The required *fire resistance* of an opening protective shall be permitted to be established by any of the following methods or procedures:

1. Designs documented in *approved* sources.
2. Calculations performed in an *approved* manner.
3. Engineering analysis based on a comparison of opening protective designs having *fire protection ratings* as determined by the test procedures set forth in NFPA 252, NFPA 257, UL 9, UL 10B or UL 10C.
4. Alternative protection methods as allowed by Section 104.11.

716.1.2 Glazing. Glazing used in *fire door assemblies* and *fire window assemblies* shall comply with this section in addition to the requirements of Sections 716.2 and 716.3, respectively.

716.1.2.1 Safety glazing. *Fire-protection-rated* glazing and fire-resistance-rated glazing installed in *fire door* assemblies and *fire window assemblies* shall comply with the safety glazing requirements of Chapter 24 where applicable.

716.1.2.2 Marking fire-rated glazing assemblies. *Fire-rated glazing* assemblies shall be marked in accordance with Tables 716.1(1), 716.1(2) and 716.1(3).

716.1.2.2.1 Fire-rated glazing identification. For *fire-rated glazing*, the *label* shall bear the identification required in Tables 716.1(1) and 716.1(2). "D" indicates that the glazing is permitted to be used in *fire door assemblies* and meets the fire protection requirements of NFPA 252, UL 10B or UL 10C. "H" indicates that the glazing meets the hose stream requirements of NFPA 252, UL 10B or UL 10C. "T" indicates that the glazing meets the temperature requirements of Section 716.2.2.3.1. The placeholder "XXX" represents the fire-rating period, in minutes.

TABLE 716.1(1)
MARKING FIRE-RATED GLAZING ASSEMBLIES

FIRE TEST STANDARD	MARKING	DEFINITION OF MARKING
ASTM E119 or UL 263	W	Meets wall assembly criteria.
ASTM E119 or UL 263	FC	Meets floor/ceiling criteria[a]
NFPA 257 or UL 9	OH	Meets fire window assembly criteria including the hose stream test.
NFPA 252 or UL 10B or UL 10C	D	Meets fire door assembly criteria.
	H	Meets fire door assembly hose stream test.
	T	Meets 450°F temperature rise criteria for 30 minutes
—	XXX	The time in minutes of the fire resistance or fire protection rating of the glazing assembly.

For SI: °C = [(°F) – 32]/1.8.
a. See Section 2409.1

TABLE 716.1(2)
OPENING FIRE PROTECTION ASSEMBLIES, RATINGS AND MARKINGS

TYPE OF ASSEMBLY	REQUIRED WALL ASSEMBLY RATING (hours)		MINIMUM FIRE DOOR AND FIRE SHUTTER ASSEMBLY RATING (hours)	DOOR VISION PANEL SIZE[a]	FIRE-RATED GLAZING MARKING DOOR VISION PANEL[b,c]	MINIMUM SIDELIGHT/ TRANSOM ASSEMBLY RATING (hours)		FIRE-RATED GLAZING MARKING SIDE-LIGHT/TRANSOM PANEL	
						Fire protection	Fire resistance	Fire protection	Fire resistance
Fire walls and fire barriers having a required fire-resistance rating greater than 1 hour	4		3	See Note a	D-H-W-240	Not Permitted	4	Not Permitted	W-240
	3		3[d]	See Note a	D-H-W-180	Not Permitted	3	Not Permitted	W-180
	2		$1\frac{1}{2}$	100 sq. in.	≤100 sq. in. = D-H-90 >100 sq. in.=D-H-W-90	Not Permitted	2	Not Permitted	W-120
	$1\frac{1}{2}$		$1\frac{1}{2}$	100 sq. in.	≤100 sq. in. = D-H-90 >100 sq. in.= D-H-W-90	Not Permitted	$1\frac{1}{2}$	Not Permitted	W-90
Double fire walls constructed in accordance with NFPA 221	Single-wall assembly rating (hours)[e]	Each wall of the double-wall assembly (hours)[f]	—						
	4	3	3	See Note a	D-H-W-180	Not Permitted	3	Not Permitted	W-180
	3	2	$1\frac{1}{2}$	100 sq. in.	≤ 100 sq. in. = D-H-90 >100 sq. in.= D-H-W-90	Not Permitted	2	Not Permitted	W-120
	2	1	1	100 sq. in.	≤ 100 sq. in. = D-H-60 > 100 sq. in. = D-H-W-60	Not Permitted	1	Not Permitted	W-60
Enclosures for shafts, interior exit stairways and interior exit ramps.	2		$1\frac{1}{2}$	100 sq. in.[b]	≤100 sq. in. = D-H-90 > 100 sq. in.= D-H-T-W-90	Not Permitted	2	Not Permitted	W-120

(continued)

TABLE 716.1(2)—continued
OPENING FIRE PROTECTION ASSEMBLIES, RATINGS AND MARKINGS

TYPE OF ASSEMBLY	REQUIRED WALL ASSEMBLY RATING (hours)	MINIMUM FIRE DOOR AND FIRE SHUTTER ASSEMBLY RATING (hours)	DOOR VISION PANEL SIZE[a]	FIRE-RATED GLAZING MARKING DOOR VISION PANEL[b,c]	MINIMUM SIDELIGHT/TRANSOM ASSEMBLY RATING (hours)		FIRE-RATED GLAZING MARKING SIDE-LIGHT/TRANSOM PANEL	
					Fire protection	Fire resistance	Fire protection	Fire resistance
Horizontal exits in fire walls[g]	4	3	100 sq. in.	≤100 sq. in. = D-H-180 > 100 sq. in.=D-H-W-240	Not Permitted	4	Not Permitted	W-240
	3	3[d]	100 sq. in.	≤100 sq. in. = D-H-180 > 100 sq. in.=D-H-W-180	Not Permitted	3	Not Permitted	W-180
Fire barriers having a required fire-resistance rating of 1 hour: Enclosures for shafts, exit access stairways, exit access ramps, interior exit stairways and interior exit ramps; and exit passageway walls	1	1	100 sq. in.	≤100 sq. in. = D-H-60 >100 sq. in.=D-H-T-W-60	Not Permitted	1	Not Permitted	W-60
					Fire protection			
Other fire barriers	1	3/4	Maximum size tested	D-H	3/4[h]		D-H[h]	
Fire partitions: Corridor walls	1	1/3[a]	Maximum size tested	D-20	3/4[a]		D-H-OH-45	
	0.5	1/3[a]	Maximum size tested	D-20	1/3		D-H-OH-20	
Other fire partitions	1	3/4[i]	Maximum size tested	D-H-45	3/4		D-H-45	
	0.5	1/3	Maximum size tested	D-H-20	1/3		D-H-20	
Exterior walls	3	1 1/2	100 sq. in.[a]	≤100 sq. in. = D-H-90 > 100 sq. in = D-H-W-90	Not Permitted	3	Not Permitted	W-180
	2	1 1/2	Maximum size tested	D-H 90 or D-H-W-90	1 1/2[h]	2	D-H-OH-90[h]	W-120
					Fire protection			
	1	3/4	Maximum size tested	D-H-45	3/4[h]		D-H-45[h]	
					Fire protection			
Smoke barriers	1	1/3	Maximum size tested	D-20	3/4		D-H-OH-45	

(continued)

TABLE 716.1(2)—continued
OPENING FIRE PROTECTION ASSEMBLIES, RATINGS AND MARKINGS

For SI: 1 square inch = 645.2 mm.

a. Fire-resistance-rated glazing tested to ASTM E119 in accordance with Section 716.1.2.3 shall be permitted, in the maximum size tested.
b. Under the column heading "Fire-rated glazing marking door vision panel," W refers to the fire-resistance rating of the glazing, not the frame.
c. See Section 716.1.2.2.1 and Table 716.1(1) for additional permitted markings.
d. Two doors, each with a fire protection rating of $1^1/_2$ hours, installed on opposite sides of the same opening in a fire wall, shall be deemed equivalent in fire protection rating to one 3-hour fire door.
e. As required in Section 706.4.
f. As allowed in Section 4.6 of NFPA 221.
g. See Section 716.2.5.1.2.
h. Fire-protection-rated glazing is not permitted for fire barriers required by Section 1207 of the *International Fire Code* to enclose energy storage systems. Fire-resistance-rated glazing assemblies tested to ASTM E119 or UL 263, as specified in Section 716.1.2.3, shall be permitted.
i. Two doors, each with a fire rating of 20 minutes, installed on opposite sides of the same opening in a fire partition, shall be deemed equivalent in fire protection rating to one 45-minute fire door.

TABLE 716.1(3)
FIRE WINDOW ASSEMBLY FIRE PROTECTION RATINGS

TYPE OF WALL ASSEMBLY	REQUIRED WALL ASSEMBLY RATING (hours)	MINIMUM FIRE WINDOW ASSEMBLY RATING (hours)	FIRE-RATED GLAZING MARKING
Interior walls			
Fire walls	All	NP[a]	W-XXX[b]
Fire barriers	>1	NP[a]	W-XXX[b]
	1	NP[a]	W-XXX[b]
Atrium separations (Section 707.3.6), Incidental use areas (Section 707.3.7),[c] Mixed occupancy separations (Section 707.3.9)	1	$^3/_4$	OH-45 or W-60
Fire partitions	1	$^3/_4$	OH-45 or W-60
	0.5	$^1/_3$	OH-20 or W-30
Smoke barriers	1	$^3/_4$	OH-45 or W-60
Exterior walls	>1	$1^1/_2$	OH-90 or W-XXX[b]
	1	3/4	OH-45 or W-60
	0.5	1/3	OH-20 or W-30
Party wall	All	NP	Not Applicable

NP = Not Permitted.
a. Not permitted except fire-resistance-rated glazing assemblies tested to ASTM E119 or UL 263, as specified in Section 716.1.2.3.
b. XXX = The fire rating duration period in minutes, which shall be equal to the fire-resistance rating required for the wall assembly.
c. Fire-protection-rated glazing is not permitted for fire barriers required by Section 1207 of the *International Fire Code* to enclose energy storage systems. Fire-resistance-rated glazing assemblies tested to ASTM E119 or UL 263, as specified in Section 716.1.2.3, shall be permitted.

716.1.2.2.2 Fire-protection-rated glazing identification. For *fire-protection-rated* glazing, the *label* shall bear the following identification required in Tables 716.1(1) and 716.1(3): "OH – XXX." "OH" indicates that the glazing meets both the fire protection and the hose-stream requirements of NFPA 257 or UL 9 and is permitted to be used in fire window openings. The placeholder "XXX" represents the fire-rating period, in minutes.

716.1.2.2.3 Fire-resistance-rated glazing identification. For fire-resistance-rated glazing, the label shall bear the identification required in Section 703.4 and Table 716.1(1).

716.1.2.2.4 Fire-rated glazing that exceeds the code requirements. *Fire-rated glazing* assemblies marked as complying with hose stream requirements (H) shall be permitted in applications that do not require compliance with hose stream requirements. *Fire-rated glazing* assemblies marked as complying with temperature rise requirements (T) shall be permitted in applications that do not require compliance with temperature rise requirements. *Fire-rated glazing* assemblies marked with ratings (XXX) that exceed the ratings required by this code shall be permitted.

716.1.2.3 Fire-resistance-rated glazing. Fire-resistance-rated glazing tested as part of a fire-resistance-rated wall or floor/ceiling assembly in accordance with ASTM E119 or UL 263 and labeled in accordance with Section 703.4 shall not otherwise be required to comply with this section where used as part of a wall or floor/ceiling assembly.

716.1.2.3.1 Glazing in fire door and fire window assemblies. Fire-resistance-rated glazing shall be permitted in *fire door* and *fire window assemblies* where tested and installed in accordance with their listings and where in compliance with the requirements of this section.

716.2 Fire door assemblies. *Fire door assemblies* required by other sections of this code shall comply with the provisions of this section. *Fire door* frames with transom lights, sidelights or both shall be permitted in accordance with Section 716.2.5.4.

716.2.1 Testing requirements. Approved *fire door* and fire shutter assemblies shall be constructed of any material or assembly of component materials that conforms to the test requirements of Sections 716.2.1.1 through 716.2.1.4 and the *fire protection rating* indicated in Table 716.1(2).

Exceptions:

1. Labeled protective assemblies that conform to the requirements of this section or UL 10A, UL 14B and UL 14C for tin-clad *fire door assemblies*.
2. Floor *fire door assemblies* in accordance with Section 712.1.13.1.

716.2.1.1 Side-hinged or pivoted swinging doors. *Fire door* assemblies with side-hinged and pivoted swinging doors shall be tested in accordance with NFPA 252 or UL 10C. For tests conducted in accordance with NFPA 252, the fire test shall be conducted using the positive pressure method specified in the standard.

716.2.1.2 Other types of assemblies. *Fire door* assemblies with other types of doors, including swinging elevator doors, horizontal sliding *fire doors*, rolling steel *fire doors*, fire shutters, bottom- and side-hinged chute intake doors, and top-hinged chute discharge doors, shall be tested in accordance with NFPA 252 or UL 10B. For tests conducted in accordance with NFPA 252, the neutral pressure plane in the furnace shall be maintained as nearly equal to the atmospheric pressure as possible at the top of the door, as specified in the standard.

716.2.1.3 Glazing in transoms lights and sidelights in corridors and smoke barriers. Glazing material in any other part of the door assembly, including transom lights and sidelights, shall be tested in accordance with NFPA 257 or UL 9, including the hose stream test, in accordance with Section 716.3.1.1.

716.2.1.4 Smoke and draft control. *Fire door* assemblies that serve as smoke and draft control assemblies shall be tested in accordance with UL 1784.

716.2.2 Performance requirements. *Fire door assemblies* shall be installed in the assemblies specified in Table 716.1(2) and shall comply with the *fire protection rating* specified.

716.2.2.1 Door assemblies in corridors and smoke barriers. *Fire door* assemblies required to have a minimum *fire protection rating* of 20 minutes where located in *corridor* walls or *smoke barrier* walls having a *fire-resistance rating* in accordance with Table 716.1(2) shall be tested in accordance with NFPA 252 or UL 10C without the hose stream test.

Exceptions:

1. Viewports that require a hole not larger than 1 inch (25 mm) in diameter through the door, have not less than a 0.25-inch-thick (6.4 mm) glass disc and the holder is of metal that will not melt out where subject to temperatures of 1,700°F (927°C).
2. *Corridor* door assemblies in occupancies of Group I-2 shall be in accordance with Section 407.3.1.
3. Unprotected openings shall be permitted for *corridors* in multitheater complexes where each motion picture auditorium has not fewer than one-half of its required *exit* or *exit access doorways* opening directly to the exterior or into an *exit passageway*.
4. Horizontal sliding doors in *smoke barriers* that comply with Sections 408.6 and 408.8.4 in occupancies in Group I-3.

716.2.2.1.1 Smoke and draft control. The air leakage rate of the door assembly shall not exceed 3.0 cubic feet per minute per square foot (0.01524 $m^3/s \times m^2$) of door opening at 0.10 inch (24.9 Pa) of water for both the ambient temperature and elevated temperature tests. Louvers shall be prohibited. *Terminated stops* shall be prohibited on doors required by Section 405.4.3 to comply with Section 716.2.2.1 and prohibited on doors required by Item 3 of Section 3006.3, or Section 3007.6.3 or 3008.6.3 to comply with this section.

716.2.2.2 Door assemblies in other fire partitions. *Fire door* assemblies required to have a minimum *fire protection rating* of 20 minutes where located in other *fire partitions* having a *fire-resistance rating* of 0.5 hour in accordance with Table 716.1(2) shall be tested in accordance with NFPA 252, UL 10B or UL 10C with the hose stream test.

716.2.2.3 Doors in interior exit stairways and ramps and exit passageways. *Fire door* assemblies in *interior exit stairways* and *ramps* and *exit passageways* shall have a maximum transmitted temperature rise of not more than 450°F (250°C) above ambient at the end of 30 minutes of standard fire test exposure.

Exception: The maximum transmitted temperature rise is not required in buildings equipped throughout with an *automatic sprinkler system* installed in accordance with Section 903.3.1.1 or 903.3.1.2.

716.2.2.3.1 Glazing in doors. Fire-protection-rated glazing in excess of 100 square inches (0.065 m^2) is not permitted. Fire-resistance-rated glazing in excess of 100 square inches (0.065 m^2) shall be permitted in *fire doors*. Listed fire-resistance-rated glazing in a *fire door* shall have a maximum transmitted temperature rise in accordance with Section 716.2.2.3 when the *fire door* is tested in accordance with NFPA 252, UL 10B or UL 10C.

716.2.3 Fire doors *Fire doors* installed within a *fire door assembly* shall meet the fire rating indicated in Table 716.1(2).

716.2.4 Fire door frames. *Fire door* frames installed as part of a *fire door assembly* shall meet the fire rating indicated in Table 716.1(2).

716.2.5 Glazing in fire door assemblies. *Fire-rated glazing* conforming to the opening protection requirements in Section 716.2.1 shall be permitted in *fire door assemblies*.

716.2.5.1 Size limitations. Fire-resistance-rated glazing shall comply with the size limitations in Section 716.2.5.1.1. Fire-protection-rated glazing shall comply with the size limitations of NFPA 80, and as provided in Section 716.2.5.1.2.

716.2.5.1.1 Fire-resistance-rated glazing in door assemblies in fire walls and fire barriers rated greater than 1 hour. Fire-resistance-rated glazing tested to ASTM E119 or UL 263 and NFPA 252, UL 10B or UL 10C shall be permitted in *fire door assemblies* located in *fire walls* and in *fire barriers* in accordance with Table 716.1(2) to the maximum size tested and in accordance with their listings.

716.2.5.1.2 Fire-protection-rated glazing in door assemblies in fire walls and fire barriers rated greater than 1 hour. Fire-protection-rated glazing shall be prohibited in *fire walls* and *fire barriers* except as provided in Sections 716.2.5.1.2.1 and 716.2.5.1.2.2.

716.2.5.1.2.1 Horizontal exits. Fire-protection-rated glazing shall be permitted as vision panels in *self-closing* swinging *fire door assemblies* serving as horizontal exits in *fire walls* where limited to 100 square inches (0.065 m^2).

716.2.5.1.2.2 Fire barriers. Fire-protection-rated glazing shall be permitted in *fire doors* having a $1^1/_2$-hour *fire protection rating* intended for installation in *fire barriers*, where limited to 100 square inches (0.065 m^2).

716.2.5.2 Elevator, stairway and ramp protectives. Approved fire-protection-rated glazing used in *fire door assemblies* in elevator, *stairway and ramp enclosures* shall be so located as to furnish clear vision of the passageway or approach to the elevator, *stairway* or *ramp*.

716.2.5.3 Glazing in door assemblies in corridors and smoke barriers. In a 20-minute *fire door assembly*, the glazing material in the door itself shall have a minimum fire-protection-rated glazing of 20 minutes and shall be exempt from the hose stream test.

716.2.5.4 Fire door frames with transom lights and sidelights. Fire-protection-rated glazing shall be permitted in door frames with transom lights, sidelights or both, where a $^3/_4$-hour *fire protection rating* or less is required and in 2-hour fire-resistance-rated *exterior walls* in accordance with Table 716.1(2). *Fire door* frames with transom lights, sidelights or both, installed with fire-resistance-rated glazing tested as an assembly in accordance with ASTM E119 or UL 263 shall be permitted where a *fire protection rating* exceeding $^3/_4$ hour is required in accordance with Table 716.1(2).

716.2.5.4.1 Energy storage system separation. Fire-protection-rated glazing shall not be permitted in *fire door* frames with transom lights and sidelights in *fire barriers* required by Section 1207 of the *International Fire Code* to enclose energy storage systems.

716.2.6 Fire door hardware and closures. *Fire door* hardware and closures shall be installed on *fire door assemblies* in accordance with the requirements of this section.

716.2.6.1 Door closing. *Fire doors* shall be latching and self- or automatic-closing in accordance with this section.

Exceptions:

1. *Fire doors* located in common walls separating *sleeping units* in Group R-1 shall be permitted without automatic- or *self-closing* devices.
2. The elevator car doors and the associated hoistway enclosure doors at the floor level designated for recall in accordance with Section 3003.2 shall be permitted to remain open during Phase I emergency recall operation.

716.2.6.2 Latch required. Unless otherwise specifically permitted, single side-hinged *swinging fire doors* and both leaves of pairs of side-hinged swinging *fire doors* shall be provided with an active latch bolt that will secure the door when it is closed.

716.2.6.3 Chute intake door latching. Chute intake doors shall be positive latching, remaining latched and closed in the event of latch spring failure during a fire emergency.

716.2.6.4 Automatic-closing fire door assemblies. Automatic-closing *fire door assemblies* shall be *self-closing* in accordance with NFPA 80.

716.2.6.5 Delayed-action closers. Doors required to be *self-closing* and not required to be automatic closing shall be permitted to be equipped with *delayed-action closers*.

716.2.6.6 Smoke-activated doors. Automatic-closing doors installed in the following locations shall be permitted to have hold-open devices. Doors shall automatically close by the actuation of *smoke detectors* installed in accordance with Section 907.3 or by loss of power to the smoke *detector* or hold-open device.

Doors that are automatic-closing by smoke detection shall not have more than a 10-second delay before the door starts to close after the smoke detector is actuated. Automatic-closing doors that protect openings installed in the following locations shall comply with this section:

1. In walls that separate incidental uses in accordance with Section 509.4.
2. In *fire walls* in accordance with Section 706.8.
3. In *fire barriers* in accordance with Section 707.6.
4. In *fire partitions* in accordance with Section 708.6.
5. In *smoke barriers* in accordance with Section 709.5.
6. In *smoke partitions* in accordance with Section 710.5.2.3.
7. In *shaft enclosures* in accordance with Section 713.7.
8. In waste and linen chutes, discharge openings and access and discharge rooms in accordance with Section 713.13. Loading doors installed in waste and linen chutes shall meet the requirements of Sections 716.2.6.1 and 716.2.6.3.

716.2.6.7 Doors in pedestrian ways. Vertical sliding or vertical rolling steel *fire doors* in openings through which pedestrians travel shall be heat activated or activated by smoke detectors with alarm verification.

716.2.7 Swinging fire shutters. Where fire shutters of the swinging type are installed in exterior openings, not less than one row in every three vertical rows shall be arranged to be readily opened from the outside, and shall be identified by distinguishing marks or letters not less than 6 inches (152 mm) high.

716.2.8 Rolling fire shutters. Where fire shutters of the rolling type are installed, such shutters shall include *approved* automatic-closing devices.

716.2.9 Labeled protective assemblies. *Fire door* assemblies shall be labeled by an *approved agency*. The *labels* shall comply with NFPA 80, and shall be permanently affixed to the door or frame.

716.2.9.1 Fire door labeling requirements. *Fire doors* shall be *labeled* showing the name of the manufacturer or other identification readily traceable back to the manufacturer, the name or trademark of the third-party inspection agency, the *fire protection rating* and, where required for *fire doors* in *interior exit stairways and ramps* and *exit passageways* by Section 716.2.2.3, the maximum transmitted temperature end point. Smoke and draft control doors complying with UL 1784 shall be *labeled* as such and shall comply with Section 716.2.9.3. *Labels* shall be *approved* and permanently affixed. The *label* shall be applied at the factory or location where fabrication and assembly are performed.

716.2.9.1.1 Light kits, louvers and components. *Listed* light kits and louvers and their required preparations shall be considered as part of the *labeled* door where such installations are done under the listing program of the third-party agency. *Fire doors* and *fire door assemblies* shall be permitted to consist of components, including glazing, vision light kits and hardware that are *listed* or classified and *labeled* for such use by different third-party agencies.

716.2.9.2 Oversized doors. Oversized *fire doors* shall bear an oversized *fire door label* by an *approved agency* or shall be provided with a certificate of inspection furnished by an *approved* testing agency. Where a certificate of inspection is furnished by an *approved* testing agency, the certificate shall state that the door conforms to the requirements of design, materials and construction, but has not been subjected to the fire test.

716.2.9.3 Smoke and draft control door labeling requirements. Smoke and draft control doors complying with UL 1784 shall be *labeled* in accordance with Section 716.2.9.1 and shall show the letter "S" on the fire-rating *label* of the door. This marking shall indicate that the door and frame assembly are in compliance where *listed or labeled* gasketing is installed.

716.2.9.4 Fire door frame labeling requirements. *Fire door* frames shall be labeled showing the names of the manufacturer and the third-party inspection agency.

716.2.9.5 Labeling. *Fire-rated glazing* shall bear a *label* or other identification showing the name of the manufacturer, the test standard and information required in Table 716.1(1) that shall be issued by an *approved agency* and shall be permanently identified on the glazing.

716.2.9.6 Fire door operator labeling requirements. *Fire door* operators for horizontal sliding doors shall be *labeled* and *listed* for use with the assembly.

716.2.10 Installation of door assemblies in corridors and smoke barriers. Installation of smoke doors shall be in accordance with NFPA 105.

716.3 Fire window assemblies. *Fire window assemblies* required by other sections of this code shall comply with the provisions of this section.

716.3.1 Testing requirements. *Fire window assemblies* shall be constructed of any material or assembly of component materials that conforms to the test requirements of Sections 716.3.1.1 and 716.3.1.2 and the *fire protection rating* indicated in Table 716.1(3).

716.3.1.1 Testing under positive pressure. NFPA 257 or UL 9 shall evaluate fire-protection-rated glazing under positive pressure. Within the first 10 minutes of a test, the pressure in the furnace shall be adjusted so not less than two-thirds of the test specimen is above the neutral pressure plane, and the neutral pressure plane shall be maintained at that height for the balance of the test.

716.3.1.2 Nonsymmetrical glazing systems. Nonsymmetrical fire-protection-rated glazing systems in *fire partitions*, *fire barriers* or in *exterior walls* with a *fire separation distance* of 10 feet (3048 mm) or less pursuant to Section 705 shall be tested with both faces exposed to the furnace, and the assigned *fire protection rating* shall be the shortest duration obtained from the two tests conducted in compliance with NFPA 257 or UL 9.

716.3.2 Performance requirements. *Fire window assemblies* shall be installed in the assemblies and comply with the *fire protection rating* specified in Table 716.1(3).

716.3.2.1 Interior fire window assemblies. Fire-protection-rated glazing used in *fire window assemblies* located in *fire partitions* and *fire barriers* shall be limited to use in assemblies with a maximum *fire-resistance rating* of 1 hour in accordance with this section.

716.3.2.1.1 Where $^3/_4$-hour-fire-protection window assemblies permitted. Fire-protection-rated glazing requiring 45-minute opening protection in accordance with Table 716.1(3) shall be limited to *fire partitions* designed in accordance with Section 708 and *fire barriers* utilized in the applications set forth in Sections 707.3.6, 707.3.7 and 707.3.9 where the *fire-resistance rating* does not exceed 1 hour. Fire-resistance-rated glazing assemblies tested in accordance with ASTM E119 or UL 263 shall not be subject to the limitations of this section.

716.3.2.1.1.1 Energy storage system separation. *Fire-protection-rated glazing* is not permitted for use in *fire window assemblies* in *fire barriers* required by Section 1207 of the *International Fire Code* to enclose energy storage systems.

716.3.2.1.2 Area limitations. The total area of the glazing in fire-protection-rated window assemblies shall not exceed 25 percent of the area of a common wall with any room.

716.3.2.1.3 Where $^1/_3$-hour-fire-protection window assemblies permitted. Fire-protection-rated glazing shall be permitted in window assemblies tested to NFPA 257 or UL 9 in *fire partitions* requiring $^1/_3$-hour opening protection in accordance with Table 716.1(3).

716.3.3 Fire window frames. Fire window frames installed with a *fire window assembly* shall meet the fire-protection rating indicated in Table 716.1(3).

716.3.3.1 Window mullions. Metal mullions that exceed a nominal height of 12 feet (3658 mm) shall be protected with materials to afford the same *fire-resistance rating* as required for the wall construction in which the protective is located.

716.3.4 Fire-protection-rated glazing. Glazing in *fire window assemblies* shall be fire protection rated in accordance with this section and Table 716.1(3). Fire-protection-rated glazing in *fire window assemblies* shall be tested in accordance with and shall meet the acceptance criteria of NFPA 257 or UL 9. Openings in nonfire-resistance-rated *exterior wall* assemblies that require protection in accordance with Section 705.3, 705.8, 705.8.5 or 705.8.6 shall have a *fire protection rating* of not less than $^3/_4$ hour. Fire-protection-rated glazing in $^1/_2$-hour fire-resistance-rated partitions is permitted to have a 20-minute *fire protection rating*.

716.3.4.1 Glass and glazing. Glazing in *fire window assemblies* shall be fire-protection-rated glazing installed in accordance with and complying with the size limitations set forth in NFPA 80.

716.3.5 Labeled protective assemblies. Glazing in *fire window assemblies* shall be *labeled* by an *approved agency*. The *labels* shall comply with NFPA 80 and Section 716.3.5.2.

716.3.5.1 Fire window frames. Fire window frames shall be approved for the intended application.

716.3.5.2 Labeling requirements. Fire-protection-rated glazing shall bear a *label* or other identification showing the name of the manufacturer, the test standard and information required in Section 716.1.2.2.2 and Table 716.1(3) that shall be issued by an *approved agency* and permanently identified on the glazing.

716.3.6 Installation. *Fire window assemblies* shall be installed in accordance with the provisions of this section.

716.3.6.1 Closure. Fire-protection-rated glazing shall be in the fixed position or be automatic-closing and shall be installed in *labeled* frames.

716.4 Fire protective curtain assembly. *Approved fire protective curtain assemblies* shall be constructed of any materials or assembly of component materials tested without hose stream in accordance with UL 10D, and shall comply with the Sections 716.4.1 through 716.4.3

716.4.1 Label. *Fire protective curtain assemblies* used as opening protectives in fire-rated walls and *smoke partitions* shall be *labeled* in accordance with Section 716.2.9.

716.4.2 Smoke and draft control. *Fire protective curtain assemblies* used to protect openings where smoke and draft control assemblies are required shall comply with Section 716.2.1.4.

716.4.3 Installation. *Fire protective curtain assemblies* shall be installed in accordance with NFPA 80.

SECTION 717
DUCTS AND AIR TRANSFER OPENINGS

717.1 General. The provisions of this section shall govern the protection of duct penetrations and air transfer openings in assemblies required to be protected and duct penetrations in nonfire-resistance-rated floor assemblies.

717.1.1 Ducts and air transfer openings. Ducts transitioning horizontally between *shafts* shall not require a *shaft enclosure* provided that the duct penetration into each associated *shaft* is protected with *dampers* complying with this section.

717.1.2 Ducts that penetrate fire-resistance-rated assemblies without dampers. Ducts that penetrate fire-resistance-rated walls and are not required by this section to have *fire dampers* shall comply with the requirements of Sections 714.3 through 714.4.3. Ducts that penetrate *horizontal assemblies* not required to be contained within a *shaft* and not required by this section to have *fire dampers* shall comply with the requirements of Sections 714.5 through 714.6.2.

717.1.2.1 Ducts that penetrate nonfire-resistance-rated assemblies. The space around a duct penetrating a nonfire-resistance-rated floor assembly shall comply with Section 717.6.3.

717.2 Installation. *Fire dampers, smoke dampers, combination fire/smoke dampers* and *ceiling radiation dampers* located within air distribution and smoke control systems shall be installed in accordance with the manufacturer's instructions, the *dampers*' listing and Sections 717.2.1 through 717.2.3.

717.2.1 Smoke control system. Where the installation of a *fire damper* will interfere with the operation of a required smoke control system in accordance with Section 909, *approved* alternative protection shall be utilized. Where mechanical systems including ducts and *dampers* utilized for normal building ventilation serve as part of the smoke control system, the expected performance of these systems in smoke control mode shall be addressed in the rational analysis required by Section 909.4.

717.2.2 Hazardous exhaust ducts. *Fire dampers* for hazardous exhaust duct systems shall comply with the *International Mechanical Code*.

717.2.3 Static dampers. *Fire dampers* and *ceiling radiation dampers* that are *listed* for use in static systems shall only be installed in heating, *ventilation* and air-conditioning systems that are automatically shut down in the event of a fire.

717.3 Damper testing, ratings and actuation. Damper testing, ratings and actuation shall be in accordance with Sections 717.3.1 through 717.3.3.

717.3.1 Damper testing. *Dampers* shall be *listed* and *labeled* in accordance with the standards in this section.

1. *Fire dampers* shall comply with the requirements of UL 555.
2. *Smoke dampers* shall comply with the requirements of UL 555S.
3. *Combination fire/smoke dampers* shall comply with the requirements of both UL 555 and UL 555S.
4. *Ceiling radiation dampers* shall comply with the requirements of UL 555C or shall be tested as part of a fire-resistance-rated floor/ceiling or roof/ceiling assembly in accordance with ASTM E119 or UL 263.
5. *Corridor dampers* shall comply with requirements of both UL 555 and UL 555S. *Corridor dampers* shall demonstrate acceptable closure performance when subjected to 150 feet per minute (0.76 mps) velocity across the face of the *damper* during the UL 555 fire exposure test.

717.3.2 Damper rating. Damper ratings shall be in accordance with Sections 717.3.2.1 through 717.3.2.4.

717.3.2.1 Fire damper ratings. *Fire dampers* shall have the minimum rating specified in Table 717.3.2.1.

**TABLE 717.3.2.1
FIRE DAMPER RATING**

TYPE OF PENETRATION	MINIMUM DAMPER RATING (hours)
Less than 3-hour fire-resistance-rated assemblies	1.5
3-hour or greater fire-resistance-rated assemblies	3

717.3.2.2 Smoke damper ratings. *Smoke damper* leakage ratings shall be Class I or II. Elevated temperature ratings shall be not less than 250°F (121°C).

717.3.2.3 Combination fire/smoke damper ratings. *Combination fire/smoke dampers* shall have the minimum *rating* specified for *fire dampers* in Table 717.3.2.1 and shall have the minimum rating specified for *smoke dampers* in Section 717.3.2.2.

717.3.2.4 Corridor damper ratings. *Corridor dampers* shall have the following minimum ratings:

1. One-hour *fire-resistance rating.*
2. Class I or II leakage rating as specified in Section 717.3.2.2.

717.3.3 Damper actuation. *Damper* actuation shall be in accordance with Sections 717.3.3.1 through 717.3.3.5 as applicable.

717.3.3.1 Fire damper actuation. Primary heat responsive devices used to actuate *fire dampers* shall meet one of the following requirements:

1. The operating temperature shall be approximately 50°F (10°C) above the normal temperature within the duct system, but not less than 160°F (71°C).
2. The operating temperature shall be not more than 350°F (177°C) where located in a smoke control system complying with Section 909.

717.3.3.2 Smoke damper actuation. The *smoke damper* shall close upon actuation of a *listed smoke detector* or detectors installed in accordance with Section 907.3 and one of the following methods, as applicable:

1. Where a *smoke damper* is installed within a duct, a *smoke detector* shall be installed inside the duct or outside the duct with sampling tubes protruding into the duct. The detector or tubes within the duct shall be within 5 feet (1524 mm) of the *damper*. Air outlets and inlets shall not be located between the detector or tubes and the *damper*. The detector shall be *listed* for the air velocity, temperature and humidity anticipated at the point where it is installed. Other than in mechanical smoke control systems, *dampers* shall be closed upon fan shutdown where local smoke detectors require a minimum velocity to operate.
2. Where a *smoke damper* is installed above *smoke barrier* doors in a *smoke barrier*, a spot-type detector shall be installed on either side of the *smoke barrier* door opening. The detector shall be *listed* for releasing service if used for direct interface with the *damper*.
3. Where a *smoke damper* is installed within an air transfer opening in a wall, a spot-type detector shall be installed within 5 feet (1524 mm) horizontally of the *damper*. The detector shall be *listed* for releasing service if used for direct interface with the *damper*.
4. Where a *smoke damper* is installed in a corridor wall or ceiling, the *damper* shall be permitted to be controlled by a smoke detection system installed in the *corridor*.
5. Where a smoke detection system is installed in all areas served by the duct in which the *damper* will be located, the *smoke dampers* shall be permitted to be controlled by the smoke detection system.

717.3.3.3 Combination fire/smoke damper actuation. *Combination fire/smoke damper* actuation shall be in accordance with Sections 717.3.3.1 and 717.3.3.2. *Combination fire/smoke dampers* installed in smoke control system *shaft* penetrations shall not be activated by local area smoke detection unless it is secondary to the smoke management system controls.

717.3.3.4 Ceiling radiation damper actuation. The operating temperature of a *ceiling radiation damper* actuation device shall be 50°F (27.8°C) above the normal temperature within the duct system, but not less than 160°F (71°C).

717.3.3.5 Corridor damper actuation. *Corridor damper* actuation shall be in accordance with Sections 717.3.3.1 and 717.3.3.2.

717.4 Access and identification. Access and identification of fire and *smoke dampers* shall comply with Sections 717.4.1 through 717.4.2.

717.4.1 Access. *Fire* and *smoke dampers* shall be provided with an approved means of access that is large enough to permit inspection and maintenance of the *damper* and its operating parts. *Dampers* equipped with fusible links, internal operators, or both shall be provided with an access door that is not less than 12 inches (305 mm) square or provided with a removable duct section.

717.4.1.1 Access openings. The access shall not affect the integrity of *fire-resistance-rated* assemblies. The access openings shall not reduce the *fire-resistance rating* of the assembly. Access doors in ducts shall be tight fitting and suitable for the required duct construction.

717.4.1.2 Restricted access. Where space constraints or physical barriers restrict access to a damper for periodic inspection and testing, the *damper* shall be a single- or multi-blade type *damper* and shall comply with the remote inspection requirements of NFPA 80 or NFPA 105.

717.4.2 Identification. Access points shall be permanently identified on the exterior by a label having letters not less than $^1/_2$ inch (12.7 mm) in height reading: "FIRE/SMOKE DAMPER," "SMOKE DAMPER" or "FIRE DAMPER."

717.5 Where required. *Fire dampers, smoke dampers, combination fire/smoke dampers, ceiling radiation dampers* and *corridor dampers* shall be provided at the locations prescribed in Sections 717.5.1 through 717.5.7 and 717.6. Where an assembly is required to have both *fire dampers* and *smoke dampers, combination fire/smoke dampers* or a *fire damper* and a *smoke damper* shall be provided.

717.5.1 Fire walls. Ducts and air transfer openings permitted in *fire walls* in accordance with Section 706.11 shall be protected with *listed fire dampers* installed in accordance with their listing.

717.5.1.1 Horizontal exits. A *listed smoke damper* designed to resist the passage of smoke shall be provided at each point a duct or air transfer opening penetrates a *fire wall* that serves as a *horizontal exit.*

717.5.2 Fire barriers. Ducts and air transfer openings of *fire barriers* shall be protected with *listed fire dampers* installed in accordance with their listing. Ducts and air transfer openings shall not penetrate enclosures for *interior exit stairways* and *ramps* and *exit passageways*, except as permitted by Sections 1023.5 and 1024.6, respectively.

Exceptions: *Fire dampers* are not required at penetrations of *fire barriers* where any of the following apply:

1. Penetrations are tested in accordance with ASTM E119 or UL 263 as part of the fire-resistance-rated assembly.

2. Ducts are used as part of an *approved* smoke control system in accordance with Section 909 and where the use of a *fire damper* would interfere with the operation of a smoke control system.
3. Such walls are penetrated by fully ducted HVAC systems, have a required *fire-resistance rating* of 1 hour or less, are in areas of other than Group H and are in buildings equipped throughout with an *automatic sprinkler system* in accordance with Section 903.3.1.1 or 903.3.1.2. For the purposes of this exception, a fully ducted HVAC system shall be a duct system for conveying supply, return or exhaust air as part of the structure's HVAC system. Such a duct system shall be constructed of sheet steel not less than No. 26 gage thickness and shall be continuous from the air-handling appliance or equipment to the air outlet and inlet terminals. Nonmetal flexible air connectors shall be permitted in the following locations:
    - 3.1. At the duct connection to the air handling unit or equipment located within the mechanical room in accordance with Section 603.9 of the *International Mechanical Code.*
    - 3.2. From an overhead metal duct to a ceiling diffuser within the same room in accordance with Section 603.6.2 of the *International Mechanical Code.*

717.5.2.1 Horizontal exits. A *listed smoke damper* designed to resist the passage of smoke shall be provided at each point a duct or air transfer opening penetrates a *fire barrier* that serves as a *horizontal exit.*

717.5.3 Shaft enclosures. Shaft enclosures that are permitted to be penetrated by ducts and air transfer openings shall be protected with *listed* fire and *smoke dampers* installed in accordance with their listing.

Exceptions:

1. *Fire dampers* are not required at penetrations of *shafts* where any of the following criteria are met:
    - 1.1. Steel exhaust subducts having a wall thickness of not less than 0.0187 inch (0.4712 mm) are extended not less than 22 inches (559 mm) vertically in exhaust *shafts*, and an exhaust fan is installed at the upper terminus of the *shaft* that is powered continuously in accordance with Section 909.11, so as to maintain a continuous upward airflow to the outdoors.
    - 1.2. Penetrations are tested in accordance with ASTM E119 or UL 263 as part of the fire-resistance-rated assembly.
    - 1.3. Ducts are used as part of an *approved* smoke control system designed and installed in accordance with Section 909 and where the *fire damper* will interfere with the operation of the smoke control system.
    - 1.4. The penetrations are in parking garage exhaust or supply *shafts* that are separated from other building *shafts* by not less than 2-hour fire-resistance-rated construction.
2. In Group B and R occupancies equipped throughout with an *automatic sprinkler system* in accordance with Section 903.3.1.1, *smoke dampers* are not required at penetrations of *shafts* where all of the following criteria are met:
    - 2.1. Kitchen, clothes dryer, bathroom and toilet room exhaust openings are installed with steel exhaust subducts, having a wall thickness of not less than 0.0187 inch (0.4712 mm).
    - 2.2. The subducts extend not less than 22 inches (559 mm) vertically.
    - 2.3. An exhaust fan is installed at the upper terminus of the *shaft* that is powered continuously in accordance with the provisions of Section 909.11, so as to maintain a continuous upward airflow to the outdoors.
3. *Smoke dampers* are not required at penetration of exhaust or supply *shafts* in parking garages that are separated from other building *shafts* by not less than 2-hour fire-resistance-rated construction.
4. *Smoke dampers* are not required at penetrations of *shafts* where ducts are used as part of an *approved* mechanical smoke control system designed in accordance with Section 909 and where the *smoke damper* will interfere with the operation of the smoke control system.
5. *Fire dampers* and *combination fire/smoke dampers* are not required in kitchen and clothes dryer exhaust systems where *dampers* are prohibited by the *International Mechanical Code.*

717.5.3.1 Continuous upward airflow. *Fire dampers* and *smoke dampers* shall not be installed in *shafts* that are required to maintain a continuous upward airflow path where closure of the *damper* would result in the loss of the airflow.

717.5.4 Fire partitions. Ducts and air transfer openings that penetrate *fire partitions* shall be protected with *listed fire dampers* installed in accordance with their listing.

Exceptions: In occupancies other than Group H, *fire dampers* are not required where any of the following apply:

1. *Corridor* walls in buildings equipped throughout with an *automatic sprinkler system* in

accordance with Section 903.3.1.1 or 903.3.1.2 and the duct is protected as a *through penetration* in accordance with Section 714.

2. Tenant partitions in *covered and open mall buildings* where the walls are not required by provisions elsewhere in the code to extend to the underside of the floor or roof sheathing, slab or deck above.
3. The duct system is constructed of *approved* materials in accordance with the *International Mechanical Code* and the duct penetrating the wall complies with all of the following requirements:
    3.1. The duct shall not exceed 100 square inches (0.06 m^2).
    3.2. The duct shall be constructed of steel not less than 0.0217 inch (0.55 mm) in thickness.
    3.3. The duct shall not have openings that communicate the *corridor* with adjacent spaces or rooms.
    3.4. The duct shall be installed above a ceiling.
    3.5. The duct shall not terminate at a wall register in the fire-resistance-rated wall.
    3.6. A minimum 12-inch-long (305 mm) by 0.060-inch-thick (1.52 mm) steel sleeve shall be centered in each duct opening. The sleeve shall be secured to both sides of the wall and all four sides of the sleeve with minimum $1^1/_2$-inch by $1^1/_2$-inch by 0.060-inch (38 mm by 38 mm by 1.52 mm) steel retaining angles. The retaining angles shall be secured to the sleeve and the wall with No. 10 (M5) screws. The *annular space* between the steel sleeve and the wall opening shall be filled with mineral wool batting on all sides.
4. Such walls are penetrated by ducted HVAC systems, have a required *fire-resistance rating* of 1 hour or less, and are in buildings equipped throughout with an *automatic sprinkler system* in accordance with Section 903.3.1.1 or 903.3.1.2. For the purposes of this exception, a ducted HVAC system shall be a duct system for conveying supply, return or exhaust air as part of the structure's HVAC system. Such a duct system shall be constructed of sheet steel not less than No. 26 gage thickness and shall be continuous from the air-handling appliance or equipment to the air outlet and inlet terminals.

717.5.4.1 Corridors. Duct and air transfer openings that penetrate *corridors* shall be protected with *dampers* as follows:

1. A *corridor damper* shall be provided where *corridor* ceilings, constructed as required for the *corridor* walls as permitted in Section 708.4, Exception 3, are penetrated.
2. A *ceiling radiation damper* shall be provided where the ceiling membrane of a fire-resistance-rated floor-ceiling or roof-ceiling assembly, constructed as permitted in Section 708.4, Exception 2, is penetrated.
3. A *listed smoke damper* designed to resist the passage of smoke shall be provided at each point a duct or air transfer opening penetrates a *corridor* enclosure required to have smoke and draft control doors in accordance with Section 716.2.2.1.

Exceptions:

1. *Smoke dampers* are not required where the building is equipped throughout with an approved smoke control system in accordance with Section 909, and *smoke dampers* are not necessary for the operation and control of the system.
2. *Smoke dampers* are not required in *corridor* penetrations where the duct is constructed of steel not less than 0.019 inch (0.48 mm) in thickness and there are no openings serving the *corridor*.

717.5.5 Smoke barriers. A *listed smoke damper* designed to resist the passage of smoke shall be provided at each point a duct or air transfer opening penetrates a *smoke barrier*. *Smoke dampers* and smoke damper actuation methods shall comply with Section 717.3.3.2.

Exceptions:

1. *Smoke dampers* are not required where the openings in ducts are limited to a single *smoke compartment* and the ducts are constructed of steel.
2. *Smoke dampers* are not required in *smoke barriers* required by Section 407.5 for Group I-2, Condition 2—where the HVAC system is fully ducted in accordance with Section 603 of the *International Mechanical Code* and where buildings are equipped throughout with an *automatic sprinkler system* in accordance with Section 903.3.1.1 and equipped with quick-response sprinklers in accordance with Section 903.3.2.

717.5.6 Exterior walls. Ducts and air transfer openings in fire-resistance-rated *exterior walls* required to have protected openings in accordance with Section 705.10

shall be protected with *listed fire dampers* installed in accordance with their listing.

717.5.7 Smoke partitions. A *listed smoke damper* designed to resist the passage of smoke shall be provided at each point that an air transfer opening penetrates a *smoke partition*. *Smoke dampers* and *smoke damper* actuation methods shall comply with Section 717.3.3.2.

Exception: Where the installation of a *smoke damper* will interfere with the operation of a required smoke control system in accordance with Section 909, *approved* alternative protection shall be utilized.

717.6 Horizontal assemblies. Penetrations by ducts and air transfer openings of a floor, floor/ceiling assembly or the ceiling membrane of a roof/ceiling assembly shall be protected by a shaft enclosure that complies with Section 713 or shall comply with Sections 717.6.1 through 717.6.3.

717.6.1 Through penetrations. In occupancies other than Groups I-2 and I-3, a duct constructed of approved materials in accordance with the *International Mechanical Code* that penetrates a fire-resistance-rated floor/ceiling assembly that connects not more than two *stories* is permitted without *shaft enclosure* protection, provided that a *listed fire damper* is installed at the floor line or the duct is protected in accordance with Section 714.5. For air transfer openings, see Section 712.1.9.

Exception: A duct is permitted to penetrate three floors or less without a *fire damper* at each floor, provided that such duct meets all of the following requirements:

1. The duct shall be contained and located within the cavity of a wall and shall be constructed of steel having a minimum wall thickness of 0.0187 inches (0.4712 mm) (No. 26 gage).
2. The duct shall open into only one *dwelling unit* or *sleeping unit* and the duct system shall be continuous from the unit to the exterior of the building.
3. The duct shall not exceed 4-inch (102 mm) nominal diameter and the total area of such ducts shall not exceed 100 square inches (0.065 m^2) in any 100 square feet (9.3 m^2) of floor area.
4. The *annular space* around the duct is protected with materials that prevent the passage of flame and hot gases sufficient to ignite cotton waste where subjected to ASTM E119 or UL 263 time-temperature conditions under a minimum positive pressure differential of 0.01 inch (2.49 Pa) of water at the location of the penetration for the time period equivalent to the *fire-resistance rating* of the construction penetrated.
5. Grille openings located in a ceiling of a fire-resistance-rated floor/ceiling or roof/ceiling assembly shall be protected with a *listed ceiling radiation damper* installed in accordance with Section 717.6.2.1.

717.6.2 Membrane penetrations. Ducts and air transfer openings constructed of *approved* materials in accordance with the *International Mechanical Code* that penetrate the ceiling membrane of a fire-resistance-rated floor/ceiling or roof/ceiling assembly shall be protected with one of the following:

1. A *shaft enclosure* in accordance with Section 713.
2. A *listed ceiling radiation damper* installed at the ceiling line where a duct penetrates the ceiling of a fire-resistance-rated floor/ceiling or roof/ceiling assembly.

   Exceptions:

   1. A fire-resistance-rated assembly tested in accordance with ASTM E119 or UL 263 showing that *ceiling radiation dampers* are not required in order to maintain the *fire-resistance rating* of the assembly.
   2. Where exhaust duct or outdoor air duct penetrations protected in accordance with Section 714.5.2 are located within the cavity of a wall and do not pass through another dwelling unit or tenant space.
   3. Where duct and air transfer openings are protected with a duct outlet penetration system tested as part of a fire-resistance-rated assembly in accordance with ASTM E119 or UL 263.

3. A *listed ceiling radiation damper* installed at the ceiling line where a diffuser with no duct attached penetrates the ceiling of a fire-resistance-rated floor/ceiling or roof/ceiling assembly.

   Exceptions:

   1. A fire-resistance-rated assembly tested in accordance with ASTM E119 or UL 263 showing that *ceiling radiation dampers* are not required in order to maintain the *fire-resistance rating* of the assembly.
   2. Where duct and air transfer openings are protected with a duct outlet penetration system tested as part of a fire-resistance-rated assembly in accordance with ASTM E119 or UL 263.

717.6.2.1 Ceiling radiation dampers testing and installation. *Ceiling radiation dampers* shall be tested in accordance with Section 717.3.1. *Ceiling radiation dampers* shall be installed in accordance with the details specified in the fire-resistance-rated assembly and the manufacturer's instructions and the listing.

717.6.2.1.1 Dynamic systems. Only *ceiling radiation dampers labeled* for use in dynamic systems shall be installed in heating, *ventilation* and air-conditioning systems designed to operate with fans on during a fire.

717.6.2.1.2 Static systems. Static *ceiling radiation dampers* shall be provided with systems that are not designed to operate during a fire.

Exceptions:

1. Where a static ceiling radiation damper is installed at the opening of a duct, a smoke detector shall be installed inside the duct or outside the duct with sampling tubes protruding into the duct. The detector or tubes in the duct shall be within 5 feet (1524 mm) of the damper. Air outlets and inlets shall not be located between the detector or tubes and the damper. The detector shall be listed for the air velocity, temperature and humidity anticipated at the point where it is installed. Other than in mechanical smoke control systems, dampers shall be closed upon fan shutdown where local smoke detectors require a minimum velocity to operate.
2. Where a static ceiling radiation damper is installed in a ceiling, the ceiling radiation damper shall be permitted to be controlled by a smoke detection system installed in the same room or area as the ceiling radiation damper.
3. A static ceiling radiation damper shall be permitted to be installed in a room where an occupant sensor is provided within the room that will shut down the system.

717.6.3 Nonfire-resistance-rated floor assemblies. Duct systems constructed of *approved* materials in accordance with the *International Mechanical Code* that penetrate nonfire-resistance-rated floor assemblies shall be protected by any of the following methods:

1. A *shaft enclosure* in accordance with Section 713.
2. The duct connects not more than two *stories*, and the *annular space* around the penetrating duct is protected with an *approved* noncombustible material that resists the free passage of flame and the products of combustion.
3. In floor assemblies composed of noncombustible materials, a *shaft* shall not be required where the duct connects not more than three stories, the *annular space* around the penetrating duct is protected with an approved noncombustible material that resists the free passage of flame and the products of combustion and a *fire damper* is installed at each floor line.

   Exception: *Fire dampers* are not required in ducts within individual residential *dwelling units*.

717.7 Flexible ducts and air connectors. Flexible ducts and air connectors shall not pass through any fire-resistance-rated assembly. Flexible air connectors shall not pass through any wall, floor or ceiling.

SECTION 718
CONCEALED SPACES

718.1 General. *Fireblocking* and draftstopping shall be installed in combustible concealed locations in accordance with this section. *Fireblocking* shall comply with Section 718.2. Draftstopping in floor/ceiling spaces and attic spaces shall comply with Sections 718.3 and 718.4, respectively. The permitted use of combustible materials in concealed spaces of buildings of Type I or II construction shall be limited to the applications indicated in Section 718.5.

718.2 Fireblocking. In combustible construction, *fireblocking* shall be installed to cut off concealed draft openings (both vertical and horizontal) and shall form an effective barrier between floors, between a top story and a roof or attic space. *Fireblocking* shall be installed in the locations specified in Sections 718.2.2 through 718.2.7.

718.2.1 Fireblocking materials. *Fireblocking* shall consist of the following materials:

1. Two-inch (51 mm) nominal lumber.
2. Two thicknesses of 1-inch (25 mm) nominal lumber with broken lap joints.
3. One thickness of 0.719-inch (18.3 mm) *wood structural panels* with joints backed by 0.719-inch (18.3 mm) *wood structural panels.*
4. One thickness of 0.75-inch (19.1 mm) *particleboard* with joints backed by 0.75-inch (19 mm) *particleboard.*
5. One-half-inch (12.7 mm) *gypsum board.*
6. One-fourth-inch (6.4 mm) cement-based millboard.
7. Batts or blankets of *mineral wool*, *mineral fiber* or other *approved* materials installed in such a manner as to be securely retained in place.
8. Cellulose insulation tested in the form and manner intended for use to demonstrate its ability to remain in place and to retard the spread of fire and hot gases.
9, *Mass timber* complying with Section 2304.11.

718.2.1.1 Batts or blankets of mineral wool or mineral fiber. Batts or blankets of *mineral wool* or *mineral fiber* or other *approved* nonrigid materials shall be permitted for compliance with the 10-foot (3048 mm) horizontal *fireblocking* in walls constructed using parallel rows of studs or staggered studs.

718.2.1.2 Unfaced fiberglass. Unfaced fiberglass batt insulation used as *fireblocking* shall fill the entire cross section of the wall cavity to a minimum height of 16 inches (406 mm) measured vertically. Where piping, conduit or similar obstructions are encountered, the insulation shall be packed tightly around the obstruction.

718.2.1.3 Loose-fill insulation material. Loose-fill insulation material, insulating foam sealants and caulk materials shall not be used as a fireblock unless specifically tested in the form and manner intended for use to demonstrate its ability to remain in place and to retard the spread of fire and hot gases.

718.2.1.4 Fireblocking integrity. The integrity of fireblocks shall be maintained.

718.2.1.5 Double stud walls. Batts or blankets of mineral or glass fiber or other *approved* nonrigid materials shall be allowed as *fireblocking* in walls constructed using parallel rows of studs or staggered studs.

718.2.2 Concealed wall spaces. *Fireblocking* shall be provided in concealed spaces of stud walls and partitions, including furred spaces, and parallel rows of studs or staggered studs, as follows:

1. Vertically at the ceiling and floor levels.
2. Horizontally at intervals not exceeding 10 feet (3048 mm).

718.2.3 Connections between horizontal and vertical spaces. *Fireblocking* shall be provided at interconnections between concealed vertical stud wall or partition spaces and concealed horizontal spaces created by an assembly of floor joists or trusses, and between concealed vertical and horizontal spaces such as occur at soffits, drop ceilings, cove ceilings and similar locations.

718.2.4 Stairways. *Fireblocking* shall be provided in concealed spaces between *stair* stringers at the top and bottom of the run. Enclosed spaces under *stairways* shall comply with Section 1011.7.3.

718.2.5 Ceiling and floor openings. Where required by Section 712.1.8, Exception 1 of Section 714.5.1.2 or Section 714.6, *fireblocking* of the *annular space* around vents, pipes, ducts, chimneys and fireplaces at ceilings and floor levels shall be installed with a material specifically tested in the form and manner intended for use to demonstrate its ability to remain in place and resist the free passage of flame and the products of combustion.

718.2.5.1 Factory-built chimneys and fireplaces. Factory-built chimneys and fireplaces shall be fireblocked in accordance with UL 103 and UL 127.

718.2.6 Exterior wall coverings. *Fireblocking* shall be installed within concealed spaces of *exterior wall coverings* and other exterior architectural elements where permitted to be of combustible construction as specified in Section 1405 or where erected with combustible frames. *Fireblocking* shall be installed at maximum intervals of 20 feet (6096 mm) in either dimension so that there will be no concealed space exceeding 100 square feet (9.3 m^2) between *fireblocking*. Where wood furring strips are used, they shall be of *approved* wood of natural decay resistance or *preservative-treated wood*. If noncontinuous, such elements shall have closed ends, with not less than 4 inches (102 mm) of separation between sections.

Exceptions:

1. *Fireblocking* of *cornices* is not required in single-family *dwellings*. *Fireblocking* of *cornices* of a two-family *dwelling* is required only at the line of *dwelling unit* separation.
2. *Fireblocking* shall not be required where the *exterior wall covering* is installed on noncombustible framing and the face of the *exterior wall covering* exposed to the concealed space is covered by one of the following materials:
   2.1. Aluminum having a minimum thickness of 0.019 inch (0.5 mm).
   2.2. Corrosion-resistant steel having a base metal thickness not less than 0.016 inch (0.4 mm) at any point.
   2.3. Other *approved* noncombustible materials.
3. *Fireblocking* shall not be required where the *exterior wall covering* has been tested in accordance with, and complies with the acceptance criteria of, NFPA 285. The *exterior wall covering* shall be installed as tested in accordance with NFPA 285.

718.2.7 Concealed sleeper spaces. Where wood sleepers are used for laying wood flooring on masonry or concrete fire-resistance-rated floors, the space between the floor slab and the underside of the wood flooring shall be filled with an *approved* material to resist the free passage of flame and products of combustion or fireblocked in such a manner that open spaces under the flooring shall not exceed 100 square feet (9.3 m^2) in area and such space shall be filled solidly under permanent partitions so that communication under the flooring between adjoining rooms shall not occur.

Exceptions:

1. *Fireblocking* is not required for slab-on-grade floors in gymnasiums.
2. *Fireblocking* is required only at the juncture of each alternate lane and at the ends of each lane in a bowling facility.

718.3 Draftstopping in floors. Draftstopping shall be installed to subdivide floor/ceiling assemblies where required by Section 708.4.2. In other than Group R occupancies, draftstopping shall be installed to subdivide combustible floor/ceiling assemblies so that horizontal floor areas do not exceed 1,000 square feet (93 m^2).

Exception: Buildings equipped throughout with an *automatic sprinkler system* in accordance with Section 903.3.1.1.

718.3.1 Draftstopping materials. Draftstopping materials shall be not less than $^1/_2$-inch (12.7 mm) *gypsum board*, $^3/_8$-inch (9.5 mm) *wood structural panel*, $^3/_8$-inch (9.5 mm) *particleboard*, 1-inch (25-mm) nominal lumber, cement *fiberboard*, batts or blankets of mineral wool or glass fiber, or other *approved* materials adequately supported. The integrity of *draftstops* shall be maintained.

718.4 Draftstopping in attics. Draftstopping shall be installed to subdivide *attic* spaces where required by Section 708.4.2. In other than Group R, draftstopping shall be installed to subdivide combustible *attic* spaces and combustible concealed roof spaces such that any horizontal area does not exceed 3,000 square feet (279 m^2). *Ventilation* of concealed roof spaces shall be maintained in accordance with Section 1202.2.1.

Exception: Buildings equipped throughout with an *automatic sprinkler system* in accordance with Section 903.3.1.1.

718.4.1 Draftstopping materials. Materials utilized for draftstopping of *attic* spaces shall comply with Section 718.3.1.

718.4.1.1 Openings. Openings in the partitions shall be protected by *self-closing* doors with automatic latches constructed as required for the partitions.

718.5 Combustible materials in concealed spaces in Type I or II construction. Combustible materials shall not be permitted in concealed spaces of buildings of Type I or II construction.

Exceptions:

1. Combustible materials in accordance with Section 603.
2. Combustible materials exposed within plenums complying with Section 602 of the *International Mechanical Code*.
3. Class A *interior finish* materials classified in accordance with Section 803.
4. Combustible piping within partitions or shaft enclosures installed in accordance with the provisions of this code.
5. Combustible piping within concealed ceiling spaces installed in accordance with the *International Mechanical Code* and the *International Plumbing Code*.
6. Combustible insulation and covering on pipe and tubing, installed in concealed spaces other than plenums, complying with Section 720.7.

SECTION 719 FIRE-RESISTANCE REQUIREMENTS FOR PLASTER

719.1 Thickness of plaster. The minimum thickness of *gypsum plaster* or Portland *cement plaster* used in a fire-resistance-rated system shall be determined by the prescribed fire tests. The plaster thickness shall be measured from the face of the lath where applied to gypsum lath or metal lath.

719.2 Plaster equivalents. For *fire-resistance* purposes, $^1/_2$ inch (12.7 mm) of unsanded *gypsum plaster* shall be deemed equivalent to $^3/_4$ inch (19.1 mm) of one-to-three gypsum sand plaster or 1 inch (25 mm) of Portland cement sand plaster.

719.3 Noncombustible furring. In buildings of Types I and II construction, plaster shall be applied directly on concrete or masonry or on *approved* noncombustible plastering base and furring.

719.4 Double reinforcement. Plaster protection more than 1 inch (25 mm) in thickness shall be reinforced with an additional layer of *approved* lath embedded not less than 3/4 inch (19.1 mm) from the outer surface and fixed securely in place.

Exception: Solid plaster partitions or where otherwise determined by fire tests.

719.5 Plaster alternatives for concrete. In reinforced concrete construction, *gypsum plaster* or Portland *cement plaster* is permitted to be substituted for $^1/_2$ inch (12.7 mm) of the required poured concrete protection, except that a minimum thickness of $^3/_8$ inch (9.5 mm) of poured concrete shall be provided in reinforced concrete floors and 1 inch (25 mm) in reinforced concrete columns in addition to the plaster finish. The concrete base shall be prepared in accordance with Section 2510.7.

SECTION 720 THERMAL- AND SOUND-INSULATING MATERIALS

720.1 General. Insulating materials shall comply with the requirements of this section. Where a *flame spread index* or a *smoke-developed index* is specified in this section, such index shall be determined in accordance with ASTM E84 or UL 723. Any material that is subject to an increase in *flame spread index* or *smoke-developed index* beyond the limits herein established through the effects of age, moisture or other atmospheric conditions shall not be permitted. Insulating materials, when tested in accordance with the requirements of this section, shall include facings, when used, such as vapor retarders, vapor permeable membranes and similar coverings, and all layers of single and multilayer reflective foil insulation and similar materials.

Exceptions:

1. *Fiberboard* insulation shall comply with Chapter 23.
2. Foam plastic insulation shall comply with Chapter 26.
3. Duct and pipe insulation and duct and pipe coverings and linings in plenums shall comply with the *International Mechanical Code*.
4. All layers of single and multilayer reflective plastic core insulation shall comply with Section 2614.

720.2 Concealed installation. Insulating materials, where concealed as installed in buildings of any type of construction, shall have a *flame spread index* of not more than 25 and a *smoke-developed index* of not more than 450.

Exception: Cellulosic fiber loose-fill insulation complying with the requirements of Section 720.6 shall not be required to meet a *flame spread index* requirement but shall be required to meet a *smoke-developed index* of not more than 450 when tested in accordance with CAN/ULC S102.2.

720.2.1 Facings. Where such materials are installed in concealed spaces in buildings of Type III, IV or V construction, the *flame spread* and smoke-developed limitations do not apply to facings, coverings, and layers of reflective foil insulation that are installed behind and in substantial contact with the unexposed surface of the ceiling, wall or floor finish.

Exception: All layers of single and multilayer reflective plastic core insulation shall comply with Section 2614.

720.3 Exposed installation. Insulating materials, where exposed as installed in buildings of any type of construction, shall have a *flame spread index* of not more than 25 and a *smoke-developed index* of not more than 450.

Exception: Cellulosic fiber loose-fill insulation complying with the requirements of Section 720.6 shall not be required to meet a *flame spread index* requirement but shall be required to meet a *smoke-developed index* of not more than 450 when tested in accordance with CAN/ULC S102.2.

720.3.1 Attic floors. Exposed insulation materials installed on *attic* floors shall have a critical radiant flux of not less than 0.12 watt per square centimeter when tested in accordance with ASTM E970.

720.4 Loose-fill insulation. Loose-fill insulation materials that cannot be mounted in the ASTM E84 or UL 723 apparatus without a screen or artificial supports shall comply with the *flame spread* and smoke-developed limits of Sections 720.2 and 720.3 when tested in accordance with CAN/ULC S102.2.

Exception: Cellulosic fiber loose-fill insulation shall not be required to meet a *flame spread index* requirement when tested in accordance with CAN/ULC S102.2, provided that such insulation has a *smoke-developed index* of not more than 450 and complies with the requirements of Section 720.6.

720.5 Roof insulation. The use of combustible roof insulation not complying with Sections 720.2 and 720.3 shall be permitted in any type of construction provided that insulation is covered with *approved roof coverings* directly applied thereto.

720.6 Cellulosic fiber loose-fill insulation and self-supported spray-applied cellulosic insulation. Cellulosic fiber loose-fill insulation and self-supported spray-applied cellulosic insulation shall comply with CPSC 16 CFR Parts 1209 and 1404. Each package of such insulating material shall be clearly labeled in accordance with CPSC 16 CFR Parts 1209 and 1404.

720.7 Insulation and covering on pipe and tubing. Insulation and covering on pipe and tubing shall have a *flame spread index* of not more than 25 and a *smoke-developed index* of not more than 450.

Exception: Insulation and covering on pipe and tubing installed in plenums shall comply with the *International Mechanical Code*.

SECTION 721
PRESCRIPTIVE FIRE RESISTANCE

721.1 General. The provisions of this section contain prescriptive details of fire-resistance-rated *building elements*, components or assemblies. The materials of construction specified in Tables 721.1(1), 721.1(2) and 721.1(3) shall be assumed to have the *fire-resistance ratings* prescribed therein. Where materials that change the capacity for heat dissipation are incorporated into a fire-resistance-rated assembly, fire test results or other substantiating data shall be made available to the *building official* to show that the required *fire-resistance-rating* time period is not reduced.

721.1.1 Thickness of protective coverings. The thickness of fire-resistant materials required for protection of structural members shall be not less than set forth in Table 721.1(1), except as modified in this section. The figures shown shall be the net thickness of the protecting materials and shall not include any hollow space in back of the protection.

721.1.2 Unit masonry protection. Where required, metal ties shall be embedded in *bed joints* of unit masonry for protection of steel columns. Such ties shall be as set forth in Table 721.1(1) or be equivalent thereto.

721.1.3 Reinforcement for cast-in-place concrete column protection. Cast-in-place concrete protection for steel columns shall be reinforced at the edges of such members with wire ties of not less than 0.18 inch (4.6 mm) in diameter wound spirally around the columns on a pitch of not more than 8 inches (203 mm) or by equivalent reinforcement.

721.1.4 Plaster application. The finish coat is not required for plaster protective coatings where those coatings comply with the design mix and thickness requirements of Tables 721.1(1), 721.1(2) and 721.1(3).

721.1.5 Bonded prestressed concrete tendons. For members having a single tendon or more than one tendon installed with equal concrete cover measured from the nearest surface, the cover shall be not less than that set forth in Table 721.1(1). For members having multiple tendons installed with variable concrete cover, the average tendon cover shall be not less than that set forth in Table 721.1(1), provided that:

1. The clearance from each tendon to the nearest exposed surface is used to determine the average cover.
2. The clear cover for individual tendons shall not be less than one-half of that set forth in Table 721.1(1). A minimum cover of $^3/_4$ inch (19.1 mm) for slabs and 1 inch (25 mm) for beams is required for any aggregate concrete.
3. For the purpose of establishing a *fire-resistance rating*, tendons having a clear covering less than that set forth in Table 721.1(1) shall not contribute more than 50 percent of the required ultimate moment capacity for members less than 350 square inches (0.226 m^2) in cross-sectional area and 65 percent for larger members. For structural design purposes, however, tendons having a reduced cover are assumed to be fully effective.

TABLE 721.1(1)
MINIMUM PROTECTION OF STRUCTURAL PARTS BASED ON TIME PERIODS FOR VARIOUS NONCOMBUSTIBLE INSULATING MATERIALS[m]

STRUCTURAL PARTS TO BE PROTECTED	ITEM NUMBER	INSULATING MATERIAL USED	MINIMUM THICKNESS OF INSULATING MATERIAL FOR THE FOLLOWING FIRE-RESISTANCE PERIODS (inches)			
			4 hours	3 hours	2 hours	1 hour
1. Steel columns and all of primary trusses (continued)	1-1.1	Carbonate, lightweight and sand-lightweight aggregate concrete, members 6" × 6" or greater (not including sandstone, granite and siliceous gravel).[a]	$2^1/_2$	2	$1^1/_2$	1
	1-1.2	Carbonate, lightweight and sand-lightweight aggregate concrete, members 8" × 8" or greater (not including sandstone, granite and siliceous gravel).[a]	2	$1^1/_2$	1	1
	1-1.3	Carbonate, lightweight and sand-lightweight aggregate concrete, members 12" × 12" or greater (not including sandstone, granite and siliceous gravel).[a]	$1^1/_2$	1	1	1
	1-1.4	Siliceous aggregate concrete and concrete excluded in Item 1-1.1, members 6" × 6" or greater.[a]	3	2	$1^1/_2$	1
	1-1.5	Siliceous aggregate concrete and concrete excluded in Item 1-1.1, members 8" × 8" or greater.[a]	$2^1/_2$	2	1	1
	1-1.6	Siliceous aggregate concrete and concrete excluded in Item 1-1.1, members 12" × 12" or greater.[a]	2	1	1	1
	1-2.1	Clay or shale brick with brick and mortar fill.[a]	$3^3/_4$	—	—	$2^1/_4$
	1-3.1	4" hollow clay tile in two 2" layers; $^1/_2$" mortar between tile and column; $^3/_8$" metal mesh 0.046" wire diameter in horizontal joints; tile fill.[a]	4	—	—	—
	1-3.2	2" hollow clay tile; $^3/_4$" mortar between tile and column; $^3/_8$" metal mesh 0.046" wire diameter in horizontal joints; limestone concrete fill[a]; plastered with $^3/_4$" gypsum plaster.	3	—	—	—
	1-3.3	2" hollow clay tile with outside wire ties 0.08" diameter at each course of tile or $^3/_8$" metal mesh 0.046" diameter wire in horizontal joints; limestone or trap-rock concrete fill[a] extending 1" outside column on all sides.	—	—	3	—
	1-3.4	2" hollow clay tile with outside wire ties 0.08" diameter at each course of tile with or without concrete fill; $^3/_4$" mortar between tile and column.	—	—	—	2
	1-4.1	Cement plaster over metal lath wire tied to $^3/_4$" cold-rolled vertical channels with 0.049" (No. 18 B.W. gage) wire ties spaced 3" to 6" on center. Plaster mixed 1:$2^1/_2$ by volume, cement to sand.	—	—	$2^1/_2$[b]	$^7/_8$
	1-5.1	Vermiculite concrete, 1:4 mix by volume over paper-backed wire fabric lath wrapped directly around column with additional 2" × 2" 0.065"/0.065" (No. 16/16 B.W. gage) wire fabric placed $^3/_4$" from outer concrete surface. Wire fabric tied with 0.049" (No. 18B.W. gage) wire spaced 6" on center for inner layer and 2" on center for outer layer.	2	—	—	—
	1-6.1	Perlite or vermiculite gypsum plaster over metal lath wrapped around column and furred $1^1/_4$" from column flanges. Sheets lapped at ends and tied at 6" intervals with 0.049" (No. 18 B.W. gage) tie wire. Plaster pushed through to flanges.	$1^1/_2$	1	—	—
	1-6.2	Perlite or vermiculite gypsum plaster over self-furring metal lath wrapped directly around column, lapped 1" and tied at 6" intervals with 0.049" (No. 18 B.W. gage) wire.	$1^3/_4$	$1^3/_8$	1	—
	1-6.3	Perlite or vermiculite gypsum plaster on metal lath applied to $^3/_4$" cold-rolled channels spaced 24" apart vertically and wrapped flatwise around column.	$1^1/_2$	—	—	—
	1-6.4	Perlite or vermiculite gypsum plaster over two layers of $^1/_2$" plain full-length gypsum lath applied tight to column flanges. Lath wrapped with 1" hexagonal mesh of No. 20-gage wire and tied with doubled 0.035" diameter (No. 18 B.W. gage) wire ties spaced 23" on center. For three-coat work, the plaster mix for the second coat shall not exceed 100 pounds of gypsum to $2^1/_2$ cubic feet of aggregate for the 3-hour system.	$2^1/_2$	2	—	—

(continued)

**TABLE 721.1(1)—continued
MINIMUM PROTECTION OF STRUCTURAL PARTS BASED ON
TIME PERIODS FOR VARIOUS NONCOMBUSTIBLE INSULATING MATERIALS[m]**

STRUCTURAL PARTS TO BE PROTECTED	ITEM NUMBER	INSULATING MATERIAL USED	MINIMUM THICKNESS OF INSULATING MATERIAL FOR THE FOLLOWING FIRE-RESISTANCE PERIODS (inches)			
			4 hours	3 hours	2 hours	1 hour
1. Steel columns and all of primary trusses (continued)	1-6.5	Perlite or vermiculite gypsum plaster over one layer of $^1/_2$" plain full-length gypsum lath applied tight to column flanges. Lath tied with doubled 0.049" (No. 18 B.W. gage) wire ties spaced 23" on center and scratch coat wrapped with 1" hexagonal mesh 0.035" (No. 20 B.W. gage) wire fabric. For three-coat work, the plaster mix for the second coat shall not exceed 100 pounds of gypsum to $2^1/_2$ cubic feet of aggregate.	—	2	—	—
	1-7.1	Multiple layers of $^1/_2$" gypsum wallboard[c] adhesively[d] secured to column flanges and successive layers. Wallboard applied without horizontal joints. Corner edges of each layer staggered. Wallboard layer below outer layer secured to column with doubled 0.049" (No. 18 B.W. gage) steel wire ties spaced 15" on center. Exposed corners taped and treated.	—	—	2	1
	1-7.2	Three layers of $^5/_8$" Type X gypsum wallboard.[c] First and second layer held in place by $^1/_8$" diameter by $1^3/_8$" long ring shank nails with $^5/_{16}$" diameter heads spaced 24" on center at corners. Middle layer also secured with metal straps at mid-height and 18" from each end, and by metal corner bead at each corner held by the metal straps. Third layer attached to corner bead with 1" long gypsum wallboard screws spaced 12" on center.	—	—	$1^7/_8$	—
	1-7.3	Three layers of $^5/_8$" Type X gypsum wallboard,[c] each layer screw attached to $1^5/_8$" steel studs 0.018" thick (No. 25 carbon sheet steel gage) at each corner of column. Middle layer also secured with 0.049" (No. 18 B.W. gage) double-strand steel wire ties, 24" on center. Screws are No. 6 by 1" spaced 24" on center for inner layer, No. 6 by $1^5/_8$" spaced 12" on center for middle layer and No. 8 by $2^1/_4$" spaced 12" on center for outer layer.	—	$1^7/_8$	—	—
	1-8.1	Wood-fibered gypsum plaster mixed 1:1 by weight gypsum-to-sand aggregate applied over metal lath. Lath lapped 1" and tied 6" on center at all end, edges and spacers with 0.049" (No. 18 B.W. gage) steel tie wires. Lath applied over $^1/_2$" spacers made of $^3/_4$" furring channel with 2" legs bent around each corner. Spacers located 1" from top and bottom of member and not greater than 40" on center and wire tied with a single strand of 0.049" (No. 18 B.W. gage) steel tie wires. Corner bead tied to the lath at 6" on center along each corner to provide plaster thickness.	—	—	$1^5/_8$	—
	1-9.1	Minimum W8x35 wide flange steel column ($w/d \geq 0.75$) with each web cavity filled even with the flange tip with normal weight carbonate or siliceous aggregate concrete (3,000 psi minimum compressive strength with 145 pcf ± 3 pcf unit weight). Reinforce the concrete in each web cavity with a minimum No. 4 deformed reinforcing bar installed vertically and centered in the cavity, and secured to the column web with a minimum No. 2 horizontal deformed reinforcing bar welded to the web every 18" on center vertically. As an alternative to the No. 4 rebar, $^3/_4$" diameter by 3" long headed studs, spaced at 12" on center vertically, shall be welded on each side of the web mid-way between the column flanges.	—	—	—	See Note n
2. Webs or flanges of steel beams and girders (continued)	2-1.1	Carbonate, lightweight and sand-lightweight aggregate concrete (not including sandstone, granite and siliceous gravel) with 3" or finer metal mesh placed 1" from the finished surface anchored to the top flange and providing not less than 0.025 square inch of steel area per foot in each direction.	2	$1^1/_2$	1	1
	2-1.2	Siliceous aggregate concrete and concrete excluded in Item 2-1.1 with 3" or finer metal mesh placed 1" from the finished surface anchored to the top flange and providing not less than 0.025 square inch of steel area per foot in each direction.	$2^1/_2$	2	$1^1/_2$	1
	2-2.1	Cement plaster on metal lath attached to $^3/_4$" cold-rolled channels with 0.04" (No. 18 B.W. gage) wire ties spaced 3" to 6" on center. Plaster mixed $1:2^1/_2$ by volume, cement to sand.	—	—	$2^1/_2$[b]	$^7/_8$

(continued)

**TABLE 721.1(1)—continued
MINIMUM PROTECTION OF STRUCTURAL PARTS BASED ON
TIME PERIODS FOR VARIOUS NONCOMBUSTIBLE INSULATING MATERIALS[m]**

STRUCTURAL PARTS TO BE PROTECTED	ITEM NUMBER	INSULATING MATERIAL USED	MINIMUM THICKNESS OF INSULATING MATERIAL FOR THE FOLLOWING FIRE-RESISTANCE PERIODS (inches)			
			4 hours	3 hours	2 hours	1 hour
2. Webs or flanges of steel beams and girders (continued)	2-3.1	Vermiculite gypsum plaster on a metal lath cage, wire tied to 0.165" diameter (No. 8 B.W. gage) steel wire hangers wrapped around beam and spaced 16" on center. Metal lath ties spaced approximately 5" on center at cage sides and bottom.	—	$^{7}/_{8}$	—	—
	2-4.1	Two layers of $^{5}/_{8}$" Type X gypsum wallboard[c] are attached to U-shaped brackets spaced 24" on center. 0.018" thick (No. 25 carbon sheet steel gage) $1^{5}/_{8}$" deep by 1" galvanized steel runner channels are first installed parallel to and on each side of the top beam flange to provide a $^{1}/_{2}$" clearance to the flange. The channel runners are attached to steel deck or concrete floor construction with approved fasteners spaced 12" on center. U-shaped brackets are formed from members identical to the channel runners. At the bent portion of the U-shaped bracket, the flanges of the channel are cut out so that $1^{5}/_{8}$"-deep corner channels can be inserted without attachment parallel to each side of the lower flange. As an alternative, 0.021" thick (No. 24 carbon sheet steel gage) 1" × 2" runner and corner angles shall be used in lieu of channels, and the web cutouts in the U-shaped brackets shall not be required. Each angle is attached to the bracket with $^{1}/_{2}$"-long No. 8 self-drilling screws. The vertical legs of the U-shaped bracket are attached to the runners with one $^{1}/_{2}$"-long No. 8 self-drilling screw. The completed steel framing provides a $2^{1}/_{8}$" and $1^{1}/_{2}$" space between the inner layer of wallboard and the sides and bottom of the steel beam, respectively. The inner layer of wallboard is attached to the top runners and bottom corner channels or corner angles with $1^{1}/_{4}$"-long No. 6 self-drilling screws spaced 16" on center. The outer layer of wallboard is applied with $1^{3}/_{4}$"-long No. 6 self-drilling screws spaced 8" on center. The bottom corners are reinforced with metal corner beads.	—	—	$1^{1}/_{4}$	—
	2-4.2	Three layers of $^{5}/_{8}$" Type X gypsum wallboard[c] attached to a steel suspension system as described immediately above utilizing the 0.018" thick (No. 25 carbon sheet steel gage) 1" × 2" lower corner angles. The framing is located so that a $2^{1}/_{8}$" and 2" space is provided between the inner layer of wallboard and the sides and bottom of the beam, respectively. The first two layers of wallboard are attached as described immediately above. A layer of 0.035" thick (No. 20 B.W. gage) 1" hexagonal galvanized wire mesh is applied under the soffit of the middle layer and up the sides approximately 2". The mesh is held in position with the No. 6 $1^{5}/_{8}$"-long screws installed in the vertical leg of the bottom corner angles. The outer layer of wallboard is attached with No. 6 $2^{1}/_{4}$"-long screws spaced 8" on center. One screw is installed at the mid-depth of the bracket in each layer. Bottom corners are finished as described above.	—	$1^{7}/_{8}$	—	—
3. Bonded pretensioned reinforcement in prestressed concrete[e]	3-1.1	Carbonate, lightweight, sand-lightweight and siliceous[f] aggregate concrete				
		Beams or girders	4[g]	3[g]	$2^{1}/_{2}$	$1^{1}/_{2}$
		Solid[h]		2	$1^{1}/_{2}$	1

(continued)

**TABLE 721.1(1)—continued
MINIMUM PROTECTION OF STRUCTURAL PARTS BASED ON
TIME PERIODS FOR VARIOUS NONCOMBUSTIBLE INSULATING MATERIALS[m]**

STRUCTURAL PARTS TO BE PROTECTED	ITEM NUMBER	INSULATING MATERIAL USED	MINIMUM THICKNESS OF INSULATING MATERIAL FOR THE FOLLOWING FIRE-RESISTANCE PERIODS (inches)			
			4 hours	3 hours	2 hours	1 hour
4. Bonded or unbonded post-tensioned tendons in prestressed concrete[e, i]	4-1.1	Carbonate, lightweight, sand-lightweight and siliceous[f] aggregate concrete Unrestrained members:				
		Solid slabs[h]	—	2	$1^1/_2$	—
		Beams and girders[j]				
		8" wide		$4^1/_2$	$2^1/_2$	$1^3/_4$
		greater than 12" wide	3	$2^1/_2$	2	$1^1/_2$
	4-1.2	Carbonate, lightweight, sand-lightweight and siliceous aggregate Restrained members:[k]				
		Solid slabs[h]	$1^1/_4$	1	$^3/_4$	—
		Beams and girders[j]				
		8" wide	$2^1/_2$	2	$1^3/_4$	—
		greater than 12" wide	2	$1^3/_4$	$1^1/_2$	—
5. Reinforcing steel in reinforced concrete columns, beams girders and trusses	5-1.1	Carbonate, lightweight and sand-lightweight aggregate concrete, members 12" or larger, square or round. (Size limit does not apply to beams and girders monolithic with floors.)	$1^1/_2$	$1^1/_2$	$1^1/_2$	$1^1/_2$
		Siliceous aggregate concrete, members 12" or larger, square or round. (Size limit does not apply to beams and girders monolithic with floors.)	2	$1^1/_2$	$1^1/_2$	$1^1/_2$
6. Reinforcing steel in reinforced concrete joists[l]	6-1.1	Carbonate, lightweight and sand-lightweight aggregate concrete	$1^1/_4$	$1^1/_4$	1	$^3/_4$
	6-1.2	Siliceous aggregate concrete	$1^3/_4$	$1^1/_2$	1	$^3/_4$
7. Reinforcing and tie rods in floor and roof slabs[l]	7-1.1	Carbonate, lightweight and sand-lightweight aggregate concrete	1	1	$^3/_4$	$^3/_4$
	7-1.2	Siliceous aggregate concrete	$1^1/_4$	1	1	$^3/_4$

For SI: 1 inch = 25.4 mm, 1 square inch = 645.2 mm^2, 1 cubic foot = 0.0283 m^3, 1 pound per cubic foot = 16.02 kg/m^3.

a. Reentrant parts of protected members to be filled solidly.

b. Two layers of equal thickness with a $^3/_4$-inch airspace between.

c. For all of the construction with gypsum wallboard described in Table 721.1(1), gypsum base for veneer plaster of the same size, thickness and core type shall be permitted to be substituted for gypsum wallboard, provided that attachment is identical to that specified for the wallboard and the joints on the face layer are reinforced, and the entire surface is covered with not less than $^1/_{16}$-inch gypsum veneer plaster.

d. An approved adhesive qualified under ASTM E119 or UL 263.

e. Where lightweight or sand-lightweight concrete having an oven-dry weight of 110 pounds per cubic foot or less is used, the tabulated minimum cover shall be permitted to be reduced 25 percent, except that the reduced cover shall be not less than $^3/_4$ inch in slabs or $1^1/_2$ inches in beams or girders.

f. For solid slabs of siliceous aggregate concrete, increase tendon cover 20 percent.

g. Adequate provisions against spalling shall be provided by U-shaped or hooped stirrups spaced not to exceed the depth of the member with a clear cover of 1 inch.

h. Prestressed slabs shall have a thickness not less than that required in Table 721.1(3) for the respective fire-resistance time period.

i. Fire coverage and end anchorages shall be as follows: Cover to the prestressing steel at the anchor shall be $^1/_2$ inch greater than that required away from the anchor. Minimum cover to steel-bearing plate shall be 1 inch in beams and $^3/_4$ inch in slabs.

j. For beam widths between 8 inches and 12 inches, cover thickness shall be permitted to be determined by interpolation.

k. Interior spans of continuous slabs, beams and girders shall be permitted to be considered restrained.

l. For use with concrete slabs having a comparable fire endurance where members are framed into the structure in such a manner as to provide equivalent performance to that of monolithic concrete construction.

m. Generic fire-resistance ratings (those not designated as PROPRIETARY* in the listing) in GA 600 shall be accepted as if herein specified.

n. Additional insulating material is not required on the exposed outside face of the column flange to achieve a 1-hour fire-resistance rating.

TABLE 721.1(2)
RATED FIRE-RESISTANCE PERIODS FOR VARIOUS WALLS AND PARTITIONS [a, o, p]

MATERIAL	ITEM NUMBER	CONSTRUCTION	MINIMUM FINISHED THICKNESS FACE-TO-FACE[b] (inches)			
			4 hours	3 hours	2 hours	1 hour
1. Brick of clay or shale	1-1.1	Solid brick of clay or shale.[c]	6	4.9	3.8	2.7
	1-1.2	Hollow brick, not filled.	5.0	4.3	3.4	2.3
	1-1.3	Hollow brick unit wall, grout or filled with perlite vermiculite or expanded shale aggregate.	6.6	5.5	4.4	3.0
	1-2.1	4" nominal thick units not less than 75 percent solid backed with a hat-shaped metal furring channel $^{3}/_{4}$" thick formed from 0.021" sheet metal attached to the brick wall on 24" centers with approved fasteners, and $^{1}/_{2}$" Type X gypsum wallboard attached to the metal furring strips with 1"-long Type S screws spaced 8" on center.	—	—	5[d]	—
2. Combination of clay brick and load-bearing hollow clay tile	2-1.1	4" solid brick and 4" tile (not less than 40 percent solid).	—	8	—	—
	2-1.2	4" solid brick and 8" tile (not less than 40 percent solid).	12	—	—	—
3. Concrete masonry units	3-1.1[f, g]	Expanded slag or pumice.	4.7	4.0	3.2	2.1
	3-1.2[f, g]	Expanded clay, shale or slate.	5.1	4.4	3.6	2.6
	3-1.3[f]	Limestone, cinders or air-cooled slag.	5.9	5.0	4.0	2.7
	3-1.4[f, g]	Calcareous or siliceous gravel.	6.2	5.3	4.2	2.8
4. Solid concrete[h, i]	4-1.1	Siliceous aggregate concrete.	7.0	6.2	5.0	3.5
		Carbonate aggregate concrete.	6.6	5.7	4.6	3.2
		Sand-lightweight concrete.	5.4	4.6	3.8	2.7
		Lightweight concrete.	5.1	4.4	3.6	2.5
5. Glazed or unglazed facing tile, nonload-bearing	5-1.1	One 2" unit cored 15 percent maximum and one 4" unit cored 25 percent maximum with $^{3}/_{4}$" mortar-filled collar joint. Unit positions reversed in alternate courses.	—	$6^{3}/_{8}$	—	—
	5-1.2	One 2" unit cored 15 percent maximum and one 4" unit cored 40 percent maximum with $^{3}/_{4}$" mortar-filled collar joint. Unit positions side with $^{3}/_{4}$" gypsum plaster. Two wythes tied together every fourth course with No. 22 gage corrugated metal ties.	—	$6^{3}/_{4}$	—	—
	5-1.3	One unit with three cells in wall thickness, cored 29 percent maximum.	—	—	6	—
	5-1.4	One 2" unit cored 22 percent maximum and one 4" unit cored 41 percent maximum with $^{1}/_{4}$" mortar-filled collar joint. Two wythes tied together every third course with 0.030"(No. 22 galvanized sheet steel gage) corrugated metal ties.	—	—	6	—
	5-1.5	One 4" unit cored 25 percent maximum with $^{3}/_{4}$" gypsum plaster on one side.	—	—	$4^{3}/_{4}$	—
	5-1.6	One 4" unit with two cells in wall thickness, cored 22 percent maximum.	—	—	—	4
	5-1.7	One 4" unit cored 30 percent maximum with $^{3}/_{4}$" vermiculite gypsum plaster on one side.	—	—	$4^{1}/_{2}$	—
	5-1.8	One 4" unit cored 39 percent maximum with $^{3}/_{4}$" gypsum plaster on one side.	—	—	—	$4^{1}/_{2}$

(continued)

TABLE 721.1(2)—continued
RATED FIRE-RESISTANCE PERIODS FOR VARIOUS WALLS AND PARTITIONS [a, o, p]

MATERIAL	ITEM NUMBER	CONSTRUCTION	MINIMUM FINISHED THICKNESS FACE-TO-FACE[b] (inches)			
			4 hours	3 hours	2 hours	1 hour
6. Solid gypsum plaster	6-1.1	$^{3}/_{4}$" by 0.055" (No. 16 carbon sheet steel gage) vertical cold-rolled channels, 16" on center with 2.6-pound flat metal lath applied to one face and tied with 0.049" (No. 18 B.W. gage) wire at 6" spacing. Gypsum plaster each side mixed 1:2 by weight, gypsum to sand aggregate.	—	—	—	2[d]
	6-1.2	$^{3}/_{4}$" by 0.05" (No. 16 carbon sheet steel gage) cold-rolled channels 16" on center with metal lath applied to one face and tied with 0.049" (No. 18 B.W. gage) wire at 6" spacing. Perlite or vermiculite gypsum plaster each side. For three-coat work, the plaster mix for the second coat shall not exceed 100 pounds of gypsum to 2$^{1}/_{2}$ cubic feet of aggregate for the 1-hour system.	—	—	2$^{1}/_{2}$[d]	2[d]
	6-1.3	$^{3}/_{4}$" by 0.055" (No. 16 carbon sheet steel gage) vertical cold-rolled channels, 16" on center with $^{3}/_{8}$" gypsum lath applied to one face and attached with sheet metal clips. Gypsum plaster each side mixed 1:2 by weight, gypsum to sand aggregate.	—	—	—	2[d]
	6-2.1	Studless with $^{1}/_{2}$" full-length plain gypsum lath and gypsum plaster each side. Plaster mixed 1:1 for scratch coat and 1:2 for brown coat, by weight, gypsum to sand aggregate.	—	—	—	2[d]
	6-2.2	Studless with $^{1}/_{2}$" full-length plain gypsum lath and perlite or vermiculite gypsum plaster each side.	—	—	2$^{1}/_{2}$[d]	2[d]
	6-2.3	Studless partition with $^{3}/_{8}$" rib metal lath installed vertically adjacent edges tied 6" on center with No. 18 gage wire ties, gypsum plaster each side mixed 1:2 by weight, gypsum to sand aggregate.	—	—	—	2[d]
7. Solid perlite and Portland cement	7-1.1	Perlite mixed in the ratio of 3 cubic feet to 100 pounds of Portland cement and machine applied to stud side of 1$^{1}/_{2}$" mesh by 0.058-inch (No. 17 B.W. gage) paper-backed woven wire fabric lath wire-tied to 4"-deep steel trussed wire[j] studs 16" on center. Wire ties of 0.049" (No. 18 B.W. gage) galvanized steel wire 6" on center vertically.	—	—	3$^{1}/_{8}$[d]	—
8. Solid neat wood fibered gypsum plaster	8-1.1	$^{3}/_{4}$" by 0.055-inch (No. 16 carbon sheet steel gage) cold-rolled channels, 12" on center with 2.5-pound flat metal lath applied to one face and tied with 0.049" (No. 18 B.W.gage) wire at 6" spacing. Neat gypsum plaster applied each side.	—	—	2[d]	—
9. Solid wallboard partition	9-1.1	One full-length layer $^{1}/_{2}$" Type X gypsum wallboard[e] laminated to each side of 1" full-length V-edge gypsum coreboard with approved laminating compound. Vertical joints of face layer and coreboard staggered not less than 3".	—	—	2[d]	—
10. Hollow (studless) gypsum wallboard partition	10-1.1	One full-length layer of $^{5}/_{8}$" Type X gypsum wallboard[e] attached to both sides of wood or metal top and bottom runners laminated to each side of 1" × 6" full-length gypsum coreboard ribs spaced 2" on center with approved laminating compound. Ribs centered at vertical joints of face plies and joints staggered 24" in opposing faces. Ribs may be recessed 6" from the top and bottom.	—	—	—	2$^{1}/_{4}$[d]
	10-1.2	1" regular gypsum V-edge full-length backing board attached to both sides of wood or metal top and bottom runners with nails or 1$^{5}/_{8}$" drywall screws at 24" on center. Minimum width of runners 1$^{5}/_{8}$". Face layer of $^{1}/_{2}$" regular full-length gypsum wallboard laminated to outer faces of backing board with approved laminating compound.	—	—	4$^{5}/_{8}$[d]	—

(continued)

TABLE 721.1(2)—continued
RATED FIRE-RESISTANCE PERIODS FOR VARIOUS WALLS AND PARTITIONS [a, o, p]

MATERIAL	ITEM NUMBER	CONSTRUCTION	MINIMUM FINISHED THICKNESS FACE-TO-FACE[b] (inches)			
			4 hours	3 hours	2 hours	1 hour
11. Noncombustible studs—interior partition with plaster each side	11-1.1	$3^1/_4$" × 0.044" (No. 18 carbon sheet steel gage) steel studs spaced 24" on center. $^5/_8$" gypsum plaster on metal lath each side mixed 1:2 by weight, gypsum to sand aggregate.	—	—	—	$4^3/_4$[d]
	11-1.2	$3^3/_8$" × 0.055" (No. 16 carbon sheet steel gage) approved nailable[k] studs spaced 24" on center. $^5/_8$" neat gypsum wood-fibered plaster each side over $^3/_8$" rib metal lath nailed to studs with 6d common nails, 8" on center. Nails driven $1^1/_4$" and bent over.	—	—	$5^5/_8$	—
	11-1.3	4" × 0.044" (No. 18 carbon sheet steel gage) channel-shaped steel studs at 16" on center. On each side approved resilient clips pressed onto stud flange at 16" vertical spacing, $^1/_4$" pencil rods snapped into or wire tied onto outer loop of clips, metal lath wire-tied to pencil rods at 6" intervals, 1" perlite gypsum plaster, each side.	—	$7^5/_8$[d]	—	—
	11-1.4	$2^1/_2$" × 0.044" (No. 18 carbon sheet steel gage) steel studs spaced 16" on center. Wood fibered gypsum plaster mixed 1:1 by weight gypsum to sand aggregate applied on $^3/_4$-pound metal lath wire tied to studs, each side. $^3/_4$" plaster applied over each face, including finish coat.	—	—	$4^1/_4$[d]	—
12. Wood studs—interior partition with plaster each side	12-1.1[l, m]	2" × 4" wood studs 16" on center with $^5/_8$" gypsum plaster on metal lath. Lath attached by 4d common nails bent over or No. 14 gage by $1^1/_4$" by $^3/_4$" crown width staples spaced 6" on center. Plaster mixed $1{:}1^1/_2$ for scratch coat and 1:3 for brown coat, by weight, gypsum to sand aggregate.	—	—	—	$5^1/_8$
	12-1.2[l]	2" × 4" wood studs 16" on center with metal lath and $^7/_8$" neat wood-fibered gypsum plaster each side. Lath attached by 6d common nails, 7" on center. Nails driven $1^1/_4$" and bent over.	—	—	$5^1/_2$[d]	—
	12-1.3[l]	2" × 4" wood studs 16" on center with $^3/_8$" perforated or plain gypsum lath and $^1/_2$" gypsum plaster each side. Lath nailed with $1^1/_8$" by No. 13 gage by $^{19}/_{64}$" head plasterboard blued nails, 4" on center. Plaster mixed 1:2 by weight, gypsum to sand aggregate.	—	—	—	$5^1/_4$
	12-1.4[l]	2" × 4" wood studs 16" on center with $^3/_8$" Type X gypsum lath and $^1/_2$" gypsum plaster each side. Lath nailed with $1^1/_8$" by No. 13 gage by $^{19}/_{64}$" head plasterboard blued nails, 5" on center. Plaster mixed 1:2 by weight, gypsum to sand aggregate.	—	—	—	$5^1/_4$
13. Noncombustible studs—interior partition with gypsum wallboard each side	13-1.1	0.018" (No. 25 carbon sheet steel gage) channel-shaped studs 24" on center with one full-length layer of $^5/_8$" Type X gypsum wallboard[e] applied vertically attached with 1"-long No. 6 dry wall screws to each stud. Screws are 8" on center around the perimeter and 12" on center on the intermediate stud. Where applied horizontally, the Type X gypsum wallboard shall be attached to $3^5/_8$" studs and the horizontal joints shall be staggered with those on the opposite side. Screws for the horizontal application shall be 8" on center at vertical edges and 12" on center at intermediate studs.	—	—	—	$2^7/_8$[d]
	13-1.2	0.018" (No. 25 carbon sheet steel gage) channel-shaped studs 25" on center with two full-length layers of $^1/_2$" Type X gypsum wallboard[e] applied vertically each side. First layer attached with 1"-long, No. 6 drywall screws, 8" on center around the perimeter and 12" on center on the intermediate stud. Second layer applied with vertical joints offset one stud space from first layer using $1^5/_8$" long, No. 6 drywall screws spaced 9" on center along vertical joints, 12" on center at intermediate studs and 24" on center along top and bottom runners.	—	—	$3^5/_8$[d]	—
	13-1.3	0.055" (No. 16 carbon sheet steel gage) approved nailable metal studs[e] 24" on center with full-length $^5/_8$" Type X gypsum wallboard[e] applied vertically and nailed 7" on center with 6d cement-coated common nails. Approved metal fastener grips used with nails at vertical butt joints along studs.	—	—	—	$4^7/_8$

(continued)

TABLE 721.1(2)—continued
RATED FIRE-RESISTANCE PERIODS FOR VARIOUS WALLS AND PARTITIONS [a, o, p]

MATERIAL	ITEM NUMBER	CONSTRUCTION	MINIMUM FINISHED THICKNESS FACE-TO-FACE[b] (inches)			
			4 hours	3 hours	2 hours	1 hour
14. Wood studs—interior partition with gypsum wallboard each side	14-1.1[h, m]	2" × 4" wood studs 16" on center with two layers of $^{3}/_{8}$" regular gypsum wallboard[e] each side, 4d cooler[n] or wallboard[n] nails at 8" on center first layer, 5d cooler[n] or wallboard[n] nails at 8" on center second layer with laminating compound between layers, joints staggered. First layer applied full length vertically, second layer applied horizontally or vertically.	—	—	—	5
	14-1.2[l, m]	2" × 4" wood studs 16" on center with two layers $^{1}/_{2}$" regular gypsum wallboard[e] applied vertically or horizontally each side[k], joints staggered. Nail base layer with 5d cooler[n] or wallboard[n] nails at 8" on center face layer with 8d cooler[n] or wallboard[n] nails at 8" on center.	—	—	—	$5^{1}/_{2}$
	14-1.3[l, m]	2" × 4" wood studs 24" on center with $^{5}/_{8}$" Type X gypsum wallboard[e] applied vertically or horizontally nailed with 6d cooler[n] or wallboard[n] nails at 7" on center with end joints on nailing members. Stagger joints each side.	—	—	—	$4^{3}/_{4}$
	14-1.4[l]	2" × 4" fire-retardant-treated wood studs spaced 24" on center with one layer of $^{5}/_{8}$" Type X gypsum wallboard[e] applied with face paper grain (long dimension) parallel to studs. Wallboard attached with 6d cooler[n] or wallboard[n] nails at 7" on center.	—	—	—	$4^{3}/_{4}$[d]
	14-1.5[l, m]	2" × 4" wood studs 16" on center with two layers $^{5}/_{8}$" Type X gypsum wallboard[e] each side. Base layers applied vertically and nailed with 6d cooler[n] or wallboard[n] nails at 9" on center. Face layer applied vertically or horizontally and nailed with 8d cooler[n] or wallboard[n] nails at 7" on center. For nail-adhesive application, base layers are nailed 6" on center. Face layers applied with coating of approved wallboard adhesive and nailed 12" on center.	—	—	6	—
	14-1.6[l]	2" × 3" fire-retardant-treated wood studs spaced 24" on center with one layer of $^{5}/_{8}$" Type X gypsum wallboard[e] applied with face paper grain (long dimension) at right angles to studs. Wallboard attached with 6d cement-coated box nails spaced 7" on center.	—	—	—	$3^{5}/_{8}$[d]
15. Exterior or interior walls (continued)	15-1.1[l, m]	Exterior surface with $^{3}/_{4}$" drop siding over $^{1}/_{2}$" gypsum sheathing on 2" × 4" wood studs at 16" on center, interior surface treatment as required for 1-hour-rated exterior or interior 2" × 4" wood stud partitions. Gypsum sheathing nailed with $1^{3}/_{4}$" by No.11 gage by $^{7}/_{16}$" head galvanized nails at 8" on center. Siding nailed with 7d galvanized smooth box nails.	—	—	—	Varies
	15-1.2[l, m]	2" × 4" wood studs 16" on center with metal lath and $^{3}/_{4}$" cement plaster on each side. Lath attached with 6d common nails 7" on center driven to 1" minimum penetration and bent over. Plaster mix 1:4 for scratch coat and 1:5 for brown coat, by volume, cement to sand.	—	—	—	$5^{3}/_{8}$
	15-1.3[l, m]	2" × 4" wood studs 16" on center with $^{7}/_{8}$" cement plaster (measured from the face of studs) on the exterior surface with interior surface treatment as required for interior wood stud partitions in this table. Plaster mix 1:4 for scratch coat and 1:5 for brown coat, by volume, cement to sand.	—	—	—	Varies
	15-1.4	$3^{5}/_{8}$" No. 16 gage noncombustible studs 16" on center with $^{7}/_{8}$" cement plaster (measured from the face of the studs) on the exterior surface with interior surface treatment as required for interior, nonbearing, noncombustible stud partitions in this table. Plaster mix 1:4 for scratch coat and 1:5 for brown coat, by volume, cement to sand.	—	—	—	Varies[d]

(continued)

TABLE 721.1(2)—continued
RATED FIRE-RESISTANCE PERIODS FOR VARIOUS WALLS AND PARTITIONS [a, o, p]

MATERIAL	ITEM NUMBER	CONSTRUCTION	MINIMUM FINISHED THICKNESS FACE-TO-FACE[b] (inches)			
			4 hours	3 hours	2 hours	1 hour
15. Exterior or interior walls (continued)	15-1.5[m]	$2^1/_4$" × $3^3/_4$" clay face brick with cored holes over $^1/_2$" gypsum sheathing on exterior surface of 2" × 4" wood studs at 16" on center and two layers $^5/_8$" Type X gypsum wallboard[e] on interior surface. Sheathing placed horizontally or vertically with vertical joints over studs nailed 6" on center with $1^3/_4$" × No. 11 gage by $^7/_{16}$" head galvanized nails. Inner layer of wallboard placed horizontally or vertically and nailed 8" on center with 6d cooler[n] or wallboard[n] nails. Outer layer of wallboard placed horizontally or vertically and nailed 8" on center with 8d cooler[n] or wallboard[n] nails. Joints staggered with vertical joints over studs. Outer layer joints taped and finished with compound. Nail heads covered with joint compound. 0.035 inch (No. 20 galvanized sheet gage) corrugated galvanized steel wall ties $^3/_4$" by $6^5/_8$" attached to each stud with two 8d cooler[n] or wallboard[n] nails every sixth course of bricks.	—	—	10	—
	15-1.6[l, m]	2" × 6" fire-retardant-treated wood studs 16" on center. Interior face has two layers of $^5/_8$" Type X gypsum with the base layer placed vertically and attached with 6d box nails 12"on center. The face layer is placed horizontally and attached with 8d box nails 8" on center at joints and 12" on center elsewhere. The exterior face has a base layer of $^5/_8$" Type X gypsum sheathing placed vertically with 6d box nails 8" on center at joints and 12" on center elsewhere. An approved building paper is next applied, followed by self-furred exterior lath attached with $2^1/_2$", No. 12 gage galvanized roofing nails with a $^3/_8$" diameter head and spaced 6" on center along each stud. Cement plaster consisting of a $^1/_2$" brown coat is then applied. The scratch coat is mixed in the proportion of 1:3 by weight, cement to sand with 10 pounds of hydrated lime and 3 pounds of approved additives or admixtures per sack of cement. The brown coat is mixed in the proportion of 1:4 by weight, cement to sand with the same amounts of hydrated lime and approved additives or admixtures used in the scratch coat.	—	—	$8^1/_4$	—
	15-1.7[l, m]	2" × 6" wood studs 16" on center. The exterior face has a layer of $^5/_8$" Type X gypsum sheathing placed vertically with 6d box nails 8" on center at joints and 12" on center elsewhere. An approved building paper is next applied, followed by 1" by No. 18 gage self-furred exterior lath attached with 8d by $2^1/_2$"-long galvanized roofing nails spaced 6" on center along each stud. Cement plaster consisting of a $^1/_2$" scratch coat, a bonding agent and a $^1/_2$" brown coat and a finish coat is then applied. The scratch coat is mixed in the proportion of 1:3 by weight, cement to sand with 10 pounds of hydrated lime and 3 pounds of approved additives or admixtures per sack of cement. The brown coat is mixed in the proportion of 1:4 by weight, cement to sand with the same amounts of hydrated lime and approved additives or admixtures used in the scratch coat. The interior is covered with $^3/_8$" gypsum lath with 1" hexagonal mesh of 0.035 inch (No. 20 B.W. gage) woven wire lath furred out $^5/_{16}$" and 1" perlite or vermiculite gypsum plaster. Lath nailed with $1^1/_8$" by No. 13 gage by $^{19}/_{64}$" head plasterboard glued nails spaced 5" on center. Mesh attached by $1^3/_4$" by No. 12 gage by $^3/_8$" head nails with $^3/_8$" furrings, spaced 8" on center. The plaster mix shall not exceed 100 pounds of gypsum to $2^1/_2$ cubic feet of aggregate.	—	—	$8^3/_8$	—

(continued)

TABLE 721.1(2)—continued
RATED FIRE-RESISTANCE PERIODS FOR VARIOUS WALLS AND PARTITIONS [a, o, p]

MATERIAL	ITEM NUMBER	CONSTRUCTION	MINIMUM FINISHED THICKNESS FACE-TO-FACE[b] (inches) 4 hours	3 hours	2 hours	1 hour
15. Exterior or interior walls (continued)	15-1.8[l, m]	2" × 6" wood studs 16" on center. The exterior face has a layer of $^{5}/_{8}$" Type X gypsum sheathing placed vertically with 6d box nails 8" on center at joints and 12" on center elsewhere. An approved building paper is next applied, followed by $1^{1}/_{2}$" by No. 17 gage self-furred exterior lath attached with 8d by $2^{1}/_{2}$"-long galvanized roofing nails spaced 6" on center along each stud. Cement plaster consisting of a $^{1}/_{2}$" scratch coat and a $^{1}/_{2}$" brown coat is then applied. The plaster may be placed by machine. The scratch coat is mixed in the proportion of 1:4 by weight, plastic cement to sand. The brown coat is mixed in the proportion of 1:5 by weight, plastic cement to sand. The interior is covered with $^{3}/_{8}$" gypsum lath with 1" hexagonal mesh of No. 20-gage woven wire lath furred out $^{5}/_{16}$" and 1" perlite or vermiculite gypsum plaster. Lath nailed with $1^{1}/_{8}$" by No. 13 gage by $^{19}/_{64}$" head plasterboard glued nails spaced 5" on center. Mesh attached by $1^{3}/_{4}$" by No.12 gage by $^{3}/_{8}$" head nails with $^{3}/_{8}$" furrings, spaced 8" on center. The plaster mix shall not exceed 100 pounds of gypsum to $2^{1}/_{2}$ cubic feet of aggregate.	—	—	$8^{3}/_{8}$	—
	15-1.9	4" No. 18 gage, nonload-bearing metal studs, 16" on center, with 1" Portland cement lime plaster (measured from the back side of the $^{3}/_{4}$-pound expanded metal lath) on the exterior surface. Interior surface to be covered with 1" of gypsum plaster on $^{3}/_{4}$-pound expanded metal lath proportioned by weight—1:2 for scratch coat, 1:3 for brown, gypsum to sand. Lath on one side of the partition fastened to $^{1}/_{4}$" diameter pencil rods supported by No. 20 gage metal clips, located 16" on center vertically, on each stud. 3" thick mineral fiber insulating batts friction fitted between the studs.	—	—	$6^{1}/_{2}$[d]	—
	15-1.10	Steel studs 0.060" thick, 4" deep or 6" at 16" or 24" centers, with $^{1}/_{2}$" glass fiber-reinforced concrete (GFRC) on the exterior surface. GFRC is attached with flex anchors at 24" on center, with 5" leg welded to studs with two $^{1}/_{2}$"-long flare-bevel welds, and 4" foot attached to the GFRC skin with $^{5}/_{8}$"-thick GFRC bonding pads that extend $2^{1}/_{2}$" beyond the flex anchor foot on both sides. Interior surface to have two layers of $^{1}/_{2}$" Type X gypsum wallboard.[e] The first layer of wallboard to be attached with 1"-long Type S buglehead screws spaced 24" on center and the second layer is attached with $1^{5}/_{8}$"-long Type S screws spaced at 12" on center. Cavity is to be filled with 5" of 4 pcf (nominal) mineral fiber batts. GFRC has $1^{1}/_{2}$" returns packed with mineral fiber and caulked on the exterior.	—	—	$6^{1}/_{2}$	—
	15-1.11	Steel studs 0.060" thick, 4" deep or 6" at 16" or 24" centers, respectively, with $^{1}/_{2}$" glass fiber-reinforced concrete (GFRC) on the exterior surface. GFRC is attached with flex anchors at 24" on center, with 5" leg welded to studs with two $^{1}/_{2}$"-long flare-bevel welds, and 4" foot attached to the GFRC skin with $^{5}/_{8}$"-thick GFRC bonding pads that extend $2^{1}/_{2}$" beyond the flex anchor foot on both sides. Interior surface to have one layer of $^{5}/_{8}$" Type X gypsum wallboard[e], attached with $1^{1}/_{4}$"-long Type S buglehead screws spaced 12" on center. Cavity is to be filled with 5" of 4 pcf (nominal) mineral fiber batts. GFRC has $1^{1}/_{2}$" returns packed with mineral fiber and caulked on the exterior.	—	—	—	$6^{1}/_{8}$
	15-1.12[q]	2" × 6" wood studs at 16" with double top plates, single bottom plate; interior and exterior sides covered with $^{5}/_{8}$" Type X gypsum wallboard, 4′ wide, applied horizontally or vertically with vertical joints over studs, and fastened with $2^{1}/_{4}$" Type S drywall screws, spaced 12" on center. Cavity to be filled with $5^{1}/_{2}$" mineral wool insulation.	—	—	—	$6^{3}/_{4}$

(continued)

TABLE 721.1(2)—continued
RATED FIRE-RESISTANCE PERIODS FOR VARIOUS WALLS AND PARTITIONS [a, o, p]

MATERIAL	ITEM NUMBER	CONSTRUCTION	MINIMUM FINISHED THICKNESS FACE-TO-FACE[b] (inches)			
			4 hours	3 hours	2 hours	1 hour
15. Exterior or interior walls (continued)	15-1.13[q]	2" × 6" wood studs at 16" with double top plates, single bottom plate; interior and exterior sides covered with $^5/_8$" Type X gypsum wallboard, 4′ wide, applied vertically with all joints over framing or blocking and fastened with $2^1/_4$" Type S drywall screws, spaced 12" on center. R-19 mineral fiber insulation installed in stud cavity.	—	—	—	$6^3/_4$
	15-1.14[q]	2" × 6" wood studs at 16" with double top plates, single bottom plate; interior and exterior sides covered with $^5/_8$" Type X gypsum wallboard, 4′ wide, applied horizontally or vertically with vertical joints over studs, and fastened with $2^1/_4$" Type S drywall screws, spaced 7" on center.	—	—	—	$6^3/_4$
	15-1.15[q]	2" × 4" wood studs at 16" with double top plates, single bottom plate; interior and exterior sides covered with $^5/_8$" Type X gypsum wallboard and sheathing, respectively, 4′ wide, applied horizontally or vertically with vertical joints over studs, and fastened with $2^1/_4$" Type S drywall screws, spaced 12" on center. Cavity to be filled with $3^1/_2$" mineral wool insulation.	—	—	—	$4^3/_4$
	15-1.16[q]	2" × 6" wood studs at 24" centers with double top plates, single bottom plate; interior and exterior side covered with two layers of $^5/_8$" Type X gypsum wallboard, 4′ wide, applied horizontally with vertical joints over studs. Base layer fastened with $2^1/_4$" Type S drywall screws, spaced 24" on center and face layer fastened with Type S drywall screws, spaced 8" on center, wallboard joints covered with paper tape and joint compound, fastener heads covered with joint compound. Cavity to be filled with $5^1/_2$" mineral wool insulation.	—	—	8	—
	15-2.1[d]	$3^5/_8$" No. 16 gage steel studs at 24" on center or 2" × 4" wood studs at 24" on center. Metal lath attached to the exterior side of studs with minimum 1" long No. 6 drywall screws at 6" on center and covered with minimum $^3/_4$" thick Portland cement plaster. Thin veneer brick units of clay or shale complying with ASTM C1088, Grade TBS or better, installed in running bond in accordance with Section 1404.10. Combined total thickness of the Portland cement plaster, mortar and thin veneer brick units shall be not less than $1^3/_4$". Interior side covered with one layer of $^5/_8$"-thick Type X gypsum wallboard attached to studs with 1" long No. 6 drywall screws at 12" on center.	—	—	—	6
	15-2.2[d]	$3^5/_8$" No. 16 gage steel studs at 24" on center or 2" × 4" wood studs at 24" on center. Metal lath attached to the exterior side of studs with minimum 1" long No. 6 drywall screws at 6" on center and covered with minimum $^3/_4$" thick Portland cement plaster. Thin veneer brick units of clay or shale complying with ASTM C1088, Grade TBS or better, installed in running bond in accordance with Section 1404.10. Combined total thickness of the Portland cement plaster, mortar and thin veneer brick units shall be not less than 2". Interior side covered with two layers of $^5/_8$"-thick Type X gypsum wallboard. Bottom layer attached to studs with 1"-long No. 6 drywall screws at 24" on center. Top layer attached to studs with $1^5/_8$"-long No. 6 drywall screws at 12" on center.	—	—	$6^7/_8$	—
	15-2.3[d]	$3^5/_8$" No. 16 gage steel studs at 16" on center or 2" × 4" wood studs at 16" on center. Where metal lath is used, attach to the exterior side of studs with minimum 1"-long No. 6 drywall screws at 6" on center. Brick units of clay or shale not less than $2^5/_8$" thick complying with ASTM C216 installed in accordance with Section 1404.6 with a minimum 1" airspace. Interior side covered with one layer of $^5/_8$"-thick Type X gypsum wallboard attached to studs with 1"-long No. 6 drywall screws at 12" on center.	—	—	—	$7^7/_8$

(continued)

TABLE 721.1(2)—continued
RATED FIRE-RESISTANCE PERIODS FOR VARIOUS WALLS AND PARTITIONS [a, o, p]

MATERIAL	ITEM NUMBER	CONSTRUCTION	MINIMUM FINISHED THICKNESS FACE-TO-FACE[b] (inches)			
			4 hours	3 hours	2 hours	1 hour
15. Exterior or interior walls	15-2.4[d]	$3^5/_8$" No. 16 gage steel studs at 16" on center or 2" × 4" wood studs at 16" on center. Where metal lath is used, attach to the exterior side of studs with minimum 1"-long No. 6 drywall screws at 6" on center. Brick units of clay or shale not less than $2^5/_8$" thick complying with ASTM C216 installed in accordance with Section 1404.6 with a minimum 1" airspace. Interior side covered with two layers of $^5/_8$"-thick Type X gypsum wallboard. Bottom layer attached to studs with 1"-long No. 6 drywall screws at 24" on center. Top layer attached to studs with $1^5/_8$"-long No. 6 drywall screws at 12" on center.	—	—	$8^1/_2$	—
16. Exterior walls rated for fire resistance from the inside only in accordance with Section 705.5.	16-1.1[q]	2" × 4" wood studs at 16" centers with double top plates, single bottom plate; interior side covered with $^5/_8$" Type X gypsum wallboard, 4' wide, applied horizontally unblocked, and fastened with $2^1/_4$" Type S drywall screws, spaced 12" on center, wallboard joints covered with paper tape and joint compound, fastener heads covered with joint compound. Exterior covered with $^3/_8$" wood structural panels, applied vertically, horizontal joints blocked and fastened with 6d common nails (bright)—12" on center in the field, and 6" on center panel edges. Cavity to be filled with $3^1/_2$" mineral wool insulation. Rating established for exposure from interior side only.	—	—	—	$4^1/_2$
	16-1.2[q]	2" × 6" wood studs at 16" centers with double top plates, single bottom plate; interior side covered with $^5/_8$" Type X gypsum wallboard, 4' wide, applied horizontally or vertically with vertical joints over studs and fastened with $2^1/_4$" Type S drywall screws, spaced 12" on center, wallboard joints covered with paper tape and joint compound, fastener heads covered with joint compound, exterior side covered with $^7/_{16}$" wood structural panels fastened with 6d common nails (bright) spaced 12" on center in the field and 6" on center along the panel edges. Cavity to be filled with $5^1/_2$" mineral wool insulation. Rating established from the gypsum-covered side only.	—	—	—	$6^9/_{16}$
	16-1.3[q]	2" × 6" wood studs at 16" centers with double top plates, single bottom plates; interior side covered with $^5/_8$" Type X gypsum wallboard, 4'wide, applied vertically with all joints over framing or blocking and fastened with $2^1/_4$" Type S drywall screws spaced 7" on center. Joints to be covered with tape and joint compound. Exterior covered with $^3/_8$" wood structural panels, applied vertically with edges over framing or blocking and fastened with 6d common nails (bright) at 12" on center in the field and 6" on center on panel edges. R-19 mineral fiber insulation installed in stud cavity. Rating established from the gypsum-covered side only.	—	—	—	$6^1/_2$

For SI: 1 inch = 25.4 mm, 1 square inch = 645.2 mm^2, 1 cubic foot = 0.0283 m^3.

a. Staples with equivalent holding power and penetration shall be permitted to be used as alternate fasteners to nails for attachment to wood framing.

b. Thickness shown for brick and clay tile is nominal thicknesses unless plastered, in which case thicknesses are net. Thickness shown for concrete masonry and clay masonry is equivalent thickness defined in Section 722.3.1 for concrete masonry and Section 722.4.1.1 for clay masonry. Where all cells are solid grouted or filled with silicone-treated perlite loose-fill insulation; vermiculite loose-fill insulation; or expanded clay, shale or slate lightweight aggregate, the equivalent thickness shall be the thickness of the block or brick using specified dimensions as defined in Chapter 21. Equivalent thickness shall include the thickness of applied plaster and lath or gypsum wallboard, where specified.

c. For units in which the net cross-sectional area of cored brick in any plane parallel to the surface containing the cores is not less than 75 percent of the gross cross-sectional area measured in the same plane.

d. Shall be used for nonbearing purposes only.

e. For all of the construction with gypsum wallboard described in this table, gypsum base for veneer plaster of the same size, thickness and core type shall be permitted to be substituted for gypsum wallboard, provided that attachment is identical to that specified for the wallboard, and the joints on the face layer are reinforced and the entire surface is covered with not less than $^1/_{16}$-inch gypsum veneer plaster.

f. The fire-resistance time period for concrete masonry units meeting the equivalent thicknesses required for a 2-hour fire-resistance rating in Item 3, and having a thickness of not less than $7^5/_8$ inches is 4 hours where cores that are not grouted are filled with silicone-treated perlite loose-fill insulation; vermiculite loose-fill insulation; or expanded clay, shale or slate lightweight aggregate, sand or slag having a maximum particle size of $^3/_8$ inch.

g. The fire-resistance rating of concrete masonry units composed of a combination of aggregate types or where plaster is applied directly to the concrete masonry shall be determined in accordance with ACI 216.1/TMS 0216. Lightweight aggregates shall have a maximum combined density of 65 pounds per cubic foot.

h. See Note b. The equivalent thickness shall be permitted to include the thickness of cement plaster or 1.5 times the thickness of gypsum plaster applied in accordance with the requirements of Chapter 25.

(continued)

TABLE 721.1(2)—continued
RATED FIRE-RESISTANCE PERIODS FOR VARIOUS WALLS AND PARTITIONS [a, o, p]

i. Concrete walls shall be reinforced with horizontal and vertical temperature reinforcement as required by Chapter 19.
j. Studs are welded truss wire studs with 0.18 inch (No. 7 B.W. gage) flange wire and 0.18 inch (No. 7 B.W. gage) truss wires.
k. Nailable metal studs consist of two channel studs spot welded back to back with a crimped web forming a nailing groove.
l. Wood structural panels shall be permitted to be installed between the fire protection and the wood studs on either the interior or exterior side of the wood frame assemblies in this table, provided that the length of the fasteners used to attach the fire protection is increased by an amount not less than the thickness of the wood structural panel.
m. For studs with a slenderness ratio, l_e/d, greater than 33, the design stress shall be reduced to 78 percent of allowable F'_c. For studs with a slenderness ratio, l_e/d, not exceeding 33, the design stress shall be reduced to 78 percent of the adjusted stress F'_c calculated for studs having a slenderness ratio l_e/d of 33.
n. For properties of cooler or wallboard nails, see ASTM C514, ASTM C547 or ASTM F1667.
o. Generic fire-resistance ratings (those not designated as PROPRIETARY* in the listing) in the GA 600 shall be accepted as if herein specified.
p. NCMA TEK 5-8A shall be permitted for the design of fire walls.
q. The design stress of studs shall be equal to not more than 100 percent of the allowable F'_c calculated in accordance with Section 2306.

TABLE 721.1(3)
MINIMUM PROTECTION FOR FLOOR AND ROOF SYSTEMS[a, q]

FLOOR OR ROOF CONSTRUCTION	ITEM NUMBER	CEILING CONSTRUCTION	THICKNESS OF FLOOR OR ROOF SLAB (inches)				MINIMUM THICKNESS OF CEILING (inches)			
			4 hours	3 hours	2 hours	1 hour	4 hours	3 hours	2 hours	1 hour
1. Siliceous aggregate concrete	1-1.1	Slab (ceiling not required). Minimum cover over nonprestressed reinforcement shall be not less than $^3/_4$".[b]	7.0	6.2	5.0	3.5	—	—	—	—
2. Carbonate aggregate concrete	2-1.1		6.6	5.7	4.6	3.2	—	—	—	—
3. Sand-lightweight concrete	3-1.1		5.4	4.6	3.8	2.7	—	—	—	—
4. Lightweight concrete	4-1.1		5.1	4.4	3.6	2.5	—	—	—	—
5. Reinforced concrete	5-1.1	Slab with suspended ceiling of vermiculite gypsum plaster over metal lath attached to $^3/_4$" cold-rolled channels spaced 12" on center. Ceiling located 6" minimum below joists.	3	2	—	—	1	$^3/_4$	—	—
	5-2.1	$^5/_8$" Type X gypsum wallboard[c] attached to 0.018 inch (No.25 carbon sheet steel gage) by $^7/_8$" deep by $2^5/_8$" hat-shaped galvanized steel channels with 1"-long No. 6 screws. The channels are spaced 24" on center, span 35" and are supported along their length at 35" intervals by 0.033" (No. 21 galvanized sheet gage) galvanized steel flat strap hangers having formed edges that engage the lips of the channel. The strap hangers are attached to the side of the concrete joists with $^5/_{32}$" by $1^1/_4$"-long power-driven fasteners. The wallboard is installed with the long dimension perpendicular to the channels. End joints occur on channels and supplementary channels are installed parallel to the main channels, 12" each side, at end joint occurrences. The finished ceiling is located approximately 12" below the soffit of the floor slab.	—	—	$2^1/_2$	—	—	—	$^5/_8$	—

(continued)

TABLE 721.1(3)—continued
MINIMUM PROTECTION FOR FLOOR AND ROOF SYSTEMS[a, q]

FLOOR OR ROOF CONSTRUCTION	ITEM NUMBER	CEILING CONSTRUCTION	THICKNESS OF FLOOR OR ROOF SLAB (inches)				MINIMUM THICKNESS OF CEILING (inches)			
			4 hours	3 hours	2 hours	1 hour	4 hours	3 hours	2 hours	1 hour
6. Steel joists constructed with a poured reinforced concrete slab on metal lath forms or steel form units.[d, e]	6-1.1	Gypsum plaster on metal lath attached to the bottom cord with single No. 16 gage or doubled No. 18 gage wire ties spaced 6" on center. Plaster mixed 1:2 for scratch coat, 1:3 for brown coat, by weight, gypsum-to-sand aggregate for 2-hour system. For 3-hour system plaster is neat.	—	—	$2\frac{1}{2}$	$2\frac{1}{4}$	—	—	$\frac{3}{4}$	$\frac{5}{8}$
	6-2.1	Vermiculite gypsum plaster on metal lath attached to the bottom chord with single No. 16 gage or doubled 0.049-inch (No. 18 B.W. gage) wire ties 6" on center.	—	2	—	—	—	$\frac{5}{8}$	—	—
	6-3.1	Cement plaster over metal lath attached to the bottom chord of joists with single No. 16 gage or doubled 0.049" (No. 18 B.W. gage) wire ties spaced 6" on center. Plaster mixed 1:2 for scratch coat, 1:3 for brown coat for 1-hour system and 1:1 for scratch coat, $1:1\frac{1}{2}$ for brown coat for 2-hour system, by weight, cement to sand.	—	—	—	2	—	—	—	$\frac{5}{8}$[f]
	6-4.1	Ceiling of $\frac{5}{8}$" Type X wallboard[c] attached to $\frac{7}{8}$" deep by $2\frac{5}{8}$" by 0.021 inch (No. 25 carbon sheet steel gage) hat-shaped furring channels 12" on center with 1"-long No. 6 wallboard screws at 8" on center. Channels wire tied to bottom chord of joists with doubled 0.049 inch (No. 18 B.W. gage) wire or suspended below joists on wire hangers.[g]	—	—	$2\frac{1}{2}$	—	—	—	$\frac{5}{8}$	—
	6-5.1	Wood-fibered gypsum plaster mixed 1:1 by weight gypsum to sand aggregate applied over metal lath. Lath tied 6" on center to $\frac{3}{4}$" channels spaced $13\frac{1}{2}$" on center. Channels secured to joists at each intersection with two strands of 0.049 inch (No. 18 B.W. gage) galvanized wire.	—	—	$2\frac{1}{2}$	—	—	—	$\frac{3}{4}$	—
7. Reinforced concrete slabs and joists with hollow clay tile fillers laid end to end in rows $2\frac{1}{2}$" or more apart; reinforcement placed between rows and concrete cast around and over tile.	7-1.1	$\frac{5}{8}$" gypsum plaster on bottom of floor or roof construction.	—	—	8[h]	—	—	—	$\frac{5}{8}$	—
	7-1.2	None	—	—	—	$5\frac{1}{2}$[i]	—	—	—	—
8. Steel joists constructed with a reinforced concrete slab on top poured on a $\frac{1}{2}$"-deep steel deck.[e]	8-1.1	Vermiculite gypsum plaster on metal lath attached to $\frac{3}{4}$" cold-rolled channels with 0.049" (No. 18 B.W. gage) wire ties spaced 6" on center.	$2\frac{1}{2}$[j]	—	—	—	$\frac{3}{4}$	—	—	—

(continued)

TABLE 721.1(3)—continued
MINIMUM PROTECTION FOR FLOOR AND ROOF SYSTEMS[a, q]

FLOOR OR ROOF CONSTRUCTION	ITEM NUMBER	CEILING CONSTRUCTION	THICKNESS OF FLOOR OR ROOF SLAB (inches)				MINIMUM THICKNESS OF CEILING (inches)			
			4 hours	3 hours	2 hours	1 hour	4 hours	3 hours	2 hours	1 hour
9. 3"-deep cellular steel deck with concrete slab on top. Slab thickness measured to top.	9-1.1	Suspended ceiling of vermiculite gypsum plaster base coat and vermiculite acoustical plaster on metal lath attached at 6" intervals to $^3/_4$" cold-rolled channels spaced 12" on center and secured to $1^1/_2$" cold-rolled channels spaced 36" on center with 0.065" (No. 16 B.W. gage) wire. $1^1/_2$" channels supported by No. 8 gage wire hangers at 36" on center. Beams within envelope and with a $2^1/_2$" airspace between beam soffit and lath have a 4-hour rating.	$2^1/_2$	—	—	—	$1^1/_8$[k]	—	—	—
10. $1^1/_2$"-deep steel roof deck on steel framing. Insulation board, 30 pcf density, composed of wood fibers with cement binders of thickness shown bonded to deck with unified asphalt adhesive. Covered with a Class A or B roof covering.	10-1.1	Ceiling of gypsum plaster on metal lath. Lath attached to $^3/_4$" furring channels with 0.049" (No. 18 B.W. gage) wire ties spaced 6" on center. $^3/_4$" channel saddle tied to 2" channels with doubled 0.065" (No. 16 B.W. gage) wire ties. 2"channels spaced 36" on center suspended 2" below steel framing and saddle tied with 0.165" (No. 8 B.W. gage) wire. Plaster mixed 1:2 by weight, gypsum-to-sand aggregate.	—	—	$1^7/_8$	1	—	—	$^3/_4$[l]	$^3/_4$[l]
11. $1^1/_2$"-deep steel roof deck on steel-framing wood fiber insulation board, 17.5 pcf density on top applied over a 15-lb asphalt-saturated felt. Class A or B roof covering.	11-1.1	Ceiling of gypsum plaster on metal lath. Lath attached to $^3/_4$" furring channels with 0.049" (No. 18 B.W. gage) wire ties spaced 6" on center. $^3/_4$" channels saddle tied to 2" channels with doubled 0.065" (No. 16 B.W. gage) wire ties. 2" channels spaced 36" on center suspended 2" below steel framing and saddle tied with 0.165" (No. 8 B.W. gage) wire. Plaster mixed 1:2 for scratch coat and 1:3 for brown coat, by weight, gypsum-to-sand aggregate for 1-hour system. For 2-hour system, plaster mix is 1:2 by weight, gypsum-to-sand aggregate.	—	—	$1^1/_2$	1	—	—	$^7/_8$[g]	$^3/_4$[l]

(continued)

TABLE 721.1(3)—continued
MINIMUM PROTECTION FOR FLOOR AND ROOF SYSTEMS[a, q]

FLOOR OR ROOF CONSTRUCTION	ITEM NUMBER	CEILING CONSTRUCTION	THICKNESS OF FLOOR OR ROOF SLAB (inches)				MINIMUM THICKNESS OF CEILING (inches)			
			4 hours	3 hours	2 hours	1 hour	4 hours	3 hours	2 hours	1 hour
12. $1^1/_2$" deep steel roof deck on steel-framing insulation of rigid board consisting of expanded perlite and fibers impregnated with integral asphalt water-proofing; density 9 to 12 pcf secured to metal roof deck by $^1/_2$"-wide ribbons of waterproof, cold-process liquid adhesive spaced 6" apart. Steel joist or light steel construction with metal roof deck, insulation, and Class A or B built-up roof covering.[e]	12-1.1	Gypsum-vermiculite plaster on metal lath wire tied at 6" intervals to $^3/_4$" furring channels spaced 12" on center and wire tied to 2" runner channels spaced 32" on center. Runners wire tied to bottom chord of steel joists.	—	—	1	—	—	—	$^7/_8$	—
13. Double wood floor over wood joists spaced 16" on center.[m, n]	13-1.1	Gypsum plaster over $^3/_8$" Type X gypsum lath. Lath initially applied with not less than four $1^1/_8$" by No. 13 gage by $^{19}/_{64}$" head plasterboard blued nails per bearing. Continuous stripping over lath along all joist lines. Stripping consists of 3"-wide strips of metal lath attached by $1^1/_2$" by No. 11 gage by $^1/_2$" head roofing nails spaced 6" on center. Alternate stripping consists of 3"-wide 0.049" diameter wire stripping weighing 1 pound per square yard and attached by No. 16 gage by $1^1/_2$" by $^3/_4$" crown width staples, spaced 4" on center. Where alternate stripping is used, the lath nailing shall consist of two nails at each end and one nail at each intermediate bearing. Plaster mixed 1:2 by weight, gypsum-to-sand aggregate.	—	—	—	—	—	—	—	$^7/_8$
	13-1.2	Cement or gypsum plaster on metal lath. Lath fastened with $1^1/_2$" by No. 11 gage by $^7/_{16}$" head barbed shank roofing nails spaced 5" on center. Plaster mixed 1:2 for scratch coat and 1:3 for brown coat, by weight, cement to sand aggregate.	—	—	—	—	—	—	—	$^5/_8$
	13-1.3	Perlite or vermiculite gypsum plaster on metal lath secured to joists with $1^1/_2$" by No. 11 gage by $^7/_{16}$" head barbed shank roofing nails spaced 5" on center.	—	—	—	—	—	—	—	$^5/_8$
	13-1.4	$^1/_2$" Type X gypsum wallboard[c] nailed to joists with 5d cooler[o] or wallboard[o] nails at 6" on center. End joints of wallboard centered on joists.	—	—	—	—	—	—	—	$^1/_2$

(continued)

TABLE 721.1(3)—continued
MINIMUM PROTECTION FOR FLOOR AND ROOF SYSTEMS[a, q]

FLOOR OR ROOF CONSTRUCTION	ITEM NUMBER	CEILING CONSTRUCTION	THICKNESS OF FLOOR OR ROOF SLAB (inches)				MINIMUM THICKNESS OF CEILING (inches)			
			4 hours	3 hours	2 hours	1 hour	4 hours	3 hours	2 hours	1 hour
14. Plywood stressed skin panels consisting of $^5/_8$"-thick interior C-D (exterior glue) top stressed skin on 2" × 6" nominal (minimum) stringers. Adjacent panel edges joined with 8d common wire nails spaced 6" on center. Stringers spaced 12" maximum on center.	14-1.1	$^1/_2$"-thick wood fiberboard weighing 15 to 18 pounds per cubic foot installed with long dimension parallel to stringers or $^3/_8$" C-D (exterior glue) plywood glued and/or nailed to stringers. Nailing to be with 5d cooler[o] or wallboard[o] nails at 12" on center. Second layer of $^1/_2$" Type X gypsum wallboard[c] applied with long dimension perpendicular to joists and attached with 8d cooler[o] or wallboard[o] nails at 6" on center at end joints and 8" on center elsewhere. Wallboard joints staggered with respect to *fiberboard* joints.	—	—	—	—	—	—	—	1
15. Vermiculite concrete slab proportioned 1:4 (Portland cement to vermiculite aggregate) on a $1^1/_2$"-deep steel deck supported on individually protected steel framing. Maximum span of deck 6′-10" where deck is less than 0.019 inch (No. 26 carbon steel sheet gage) or greater. Slab reinforced with 4" × 8" 0.109/0.083" (No. $^{12}/_{14}$ B.W. gage) welded wire mesh.	15-1.1	None	—	—	—	3[j]	—	—	—	—

(continued)

TABLE 721.1(3)—continued
MINIMUM PROTECTION FOR FLOOR AND ROOF SYSTEMS[a, q]

FLOOR OR ROOF CONSTRUCTION	ITEM NUMBER	CEILING CONSTRUCTION	THICKNESS OF FLOOR OR ROOF SLAB (inches)				MINIMUM THICKNESS OF CEILING (inches)			
			4 hours	3 hours	2 hours	1 hour	4 hours	3 hours	2 hours	1 hour
16. Perlite concrete slab proportioned 1:6 (Portland cement to perlite aggregate) on a $1^1/_4$"-deep steel deck supported on individually protected steel framing. Slab reinforced with 4" × 8" 0.109/0.083" (No.$^{12}/_{14}$ B.W. gage) welded wire mesh.	16-1.1	None	—	—	—	$3^1/_2$[j]	—	—	—	—
17. Perlite concrete slab proportioned 1:6 (Portland cement to perlite aggregate) on a $^9/_{16}$"-deep steel deck supported by steel joists 4′ on center. Class A or B roof covering on top.	17-1.1	Perlite gypsum plaster on metal lath wire tied to $^3/_4$" furring channels attached with 0.065" (No. 16 B.W. gage) wire ties to lower chord of joists.	—	2[p]	2[p]	—	—	$^7/_8$	$^3/_4$	—

(continued)

TABLE 721.1(3)—continued
MINIMUM PROTECTION FOR FLOOR AND ROOF SYSTEMS[a, q]

FLOOR OR ROOF CONSTRUCTION	ITEM NUMBER	CEILING CONSTRUCTION	THICKNESS OF FLOOR OR ROOF SLAB (inches)				MINIMUM THICKNESS OF CEILING (inches)			
			4 hours	3 hours	2 hours	1 hour	4 hours	3 hours	2 hours	1 hour
18. Perlite concrete slab proportioned 1:6 (Portland cement to perlite aggregate) on $1^1/_4$"-deep steel deck supported on individually protected steel framing. Maximum span of deck 6′-10" where deck is less than 0.019" (No. 26 carbon sheet steel gage) and 8′-0" where deck is 0.019" (No. 26 carbon sheet steel gage) or greater. Slab reinforced with 0.042" (No. 19 B.W. gage) hexagonal wire mesh. Class A or B roof covering on top.	18-1.1	None	—	$2^1/_4$[p]	$2^1/_4$[p]	—	—	—	—	—
19. Floor and beam construction consisting of 3"-deep cellular steel floor unit mounted on steel members with 1:4 (proportion of Portland cement to perlite aggregate) perlite-concrete floor slab on top.	19-1.1	Suspended envelope ceiling of perlite gypsum plaster on metal lath attached to $^3/_4$" cold-rolled channels, secured to $1^1/_2$" cold-rolled channels spaced 42" on center supported by 0.203 inch (No. 6 B.W. gage) wire 36" on center. Beams in envelope with 3" minimum airspace between beam soffit and lath have a 4-hour rating.	2[p]	—	—	—	1[l]	—	—	—

(continued)

TABLE 721.1(3)—continued
MINIMUM PROTECTION FOR FLOOR AND ROOF SYSTEMS[a, q]

FLOOR OR ROOF CONSTRUCTION	ITEM NUMBER	CEILING CONSTRUCTION	THICKNESS OF FLOOR OR ROOF SLAB (inches)				MINIMUM THICKNESS OF CEILING (inches)			
			4 hours	3 hours	2 hours	1 hour	4 hours	3 hours	2 hours	1 hour
20. Perlite concrete proportioned 1:6 (Portland cement to perlite aggregate) poured to $^{1}/_{8}$" thickness above top of corrugations of $1^{5}/_{16}$"-deep galvanized steel deck maximum span 8′-0" for 0.024" (No. 24 galvanized sheet gage) or 6′-0" for 0.019" (No. 26 galvanized sheet gage) with deck supported by individually protected steel framing. Approved polystyrene foam plastic insulation board having a flame spread not exceeding 75 (1" to 4" thickness) with vent holes that approximate 3 percent of the board surface area placed on top of perlite slurry. A 2′ by 4′ insulation board contains six $2^{3}/_{4}$" diameter holes. Board covered with $2^{1}/_{4}$" minimum perlite concrete slab. Slab reinforced with mesh consisting of 0.042" (No.19 B.W. gage) galvanized steel wire twisted together to form 2" hexagons with straight 0.065" (No. 16 B.W. gage) galvanized steel wire woven into mesh and spaced 3". Alternate slab reinforcement shall be permitted to consist of 4" × 8", 0.109/0.238" (No. 12/4 B.W. gage), or 2" × 2", 0.083/0.083" (No. 14/14 B.W. gage) welded wire fabric. Class A or B roof covering on top.	20-1.1	None	—	—	Varies	—	—	—	—	—

(continued)

TABLE 721.1(3)—continued
MINIMUM PROTECTION FOR FLOOR AND ROOF SYSTEMS[a, q]

FLOOR OR ROOF CONSTRUCTION	ITEM NUMBER	CEILING CONSTRUCTION	THICKNESS OF FLOOR OR ROOF SLAB (inches)				MINIMUM THICKNESS OF CEILING (inches)			
			4 hours	3 hours	2 hours	1 hour	4 hours	3 hours	2 hours	1 hour
21. Wood joists, wood I-joists, floor trusses and flat or pitched roof trusses spaced a maximum 24" o.c. with $^1/_2$" wood structural panels with exterior glue applied at right angles to top of joist or top chord of trusses with 8 dnails. The wood structural panel thickness shall be not less than nominal $^1/_2$" nor less than required by Chapter 23.	21-1.1	Base layer $^5/_8$" Type X gypsum wallboard applied at right angles to joist or truss 24" o.c. with $1^1/_4$" Type S or Type W drywall screws 24" o.c. Face layer $^5/_8$" Type X gypsum wallboard or veneer base applied at right angles to joist or truss through base layer with $1^7/_8$" Type S or Type W drywall screws 12" o.c. at joints and intermediate joist or truss. Face layer Type G drywall screws placed 2" back on either side of face layer end joints, 12" o.c.	—	—	—	Varies	—	—	—	$1^1/_4$
22. Steel joists, floor trusses and flat or pitched roof trusses spaced a maximum 24" o.c. with $^1/_2$" wood structural panels with exterior glue applied at right angles to top of joist or top chord of trusses with No. 8 screws. The wood structural panel thickness shall be not less than nominal $^1/_2$" nor less than required by Chapter 23.	22-1.1	Base layer $^5/_8$" Type X gypsum board applied at right angles to steel framing 24" on center with 1" Type S dry wall screws spaced 24" on center. Face layer $^5/_8$" Type X gypsum board applied at right angles to steel framing attached through base layer with $1^5/_8$" Type S dry wall screws 12" on center at end joints and intermediate joints and $1^1/_2$" Type G dry wall screws 12 inches on center placed 2" back on either side of face layer end joints. Joints of the face layer are offset 24" from the joints of the base layer.	—	—	—	Varies	—	—	—	$1^1/_4$
23. Wood I-joist (minimum joist depth $9^1/_4$" with a minimum flange depth of $1^5/_{16}$" and a minimum flange cross-sectional area of 2.25 square inches) at 24" o.c. spacing with a minimum 1 × 4 ($^3/_4$" × 3.5"actual) ledger strip applied parallel to and covering the bottom of the bottom flange of each member, tacked in place. 2" mineral wool insulation, 3.5 pcf (nominal) installed adjacent to the bottom flange of the I-joist and supported by the 1 × 4 ledger strip.	23-1.1	$^1/_2$"-deep single-leg resilient channel 16" on center (channels doubled at wallboard end joints), placed perpendicular to the furring strip and joist and attached to each joist by $1^7/_8$" Type S drywall screws. $^5/_8$" Type C gypsum wallboard applied perpendicular to the channel with end joints staggered not less than 4′ and fastened with $1^1/_8$" Type S drywall screws spaced 7" on center. Wallboard joints to be taped and covered with joint compound.	—	—	—	Varies	—	—	—	$^5/_8$

(continued)

TABLE 721.1(3)—continued
MINIMUM PROTECTION FOR FLOOR AND ROOF SYSTEMS[a, q]

FLOOR OR ROOF CONSTRUCTION	ITEM NUMBER	CEILING CONSTRUCTION	THICKNESS OF FLOOR OR ROOF SLAB (inches)				MINIMUM THICKNESS OF CEILING (inches)			
			4 hours	3 hours	2 hours	1 hour	4 hours	3 hours	2 hours	1 hour
24. Wood I-joist (minimum I-joist depth $9^1/_4$" with a minimum flange depth of $1^1/_2$" and a minimum flange cross-sectional area of 5.25 square inches; minimum web thickness of $^3/_8$") @ 24" o.c., $1^1/_2$" mineral wool insulation (2.5 pcf-nominal) resting on hat-shaped furring channels.	24-1.1	Minimum 0.026" thick hat-shaped channel 16" o.c. (channels doubled at wallboard end joints), placed perpendicular to the joist and attached to each joist by $1^1/_4$" Type S drywall screws. $^5/_8$" Type C gypsum wallboard applied perpendicular to the channel with end joints staggered and fastened with $1^1/_8$" Type S drywall screws spaced 12" o.c. in the field and 8" o.c. at the wallboard ends. Wallboard joints to be taped and covered with joint compound.	—	—	—	Varies	—	—	—	$^5/_8$
25. Wood I-joist (minimum I-joist depth $9^1/_4$" with a minimum flange depth of $1^1/_2$" and a minimum flange cross-sectional area of 5.25 square inches; minimum web thickness of $^7/_{16}$") @ 24" o.c., $1^1/_2$" mineral wool insulation (2.5 pcf-nominal) resting on resilient channels.	25-1.1	Minimum 0.019"-thick resilient channel 16" o.c. (channels doubled at wallboard end joints), placed perpendicular to the joist and attached to each joist by $1^5/_8$" Type S drywall screws. $^5/_8$" Type C gypsum wallboard applied perpendicular to the channel with end joints staggered and fastened with 1" Type S drywall screws spaced 12" o.c. in the field and 8" o.c. at the wallboard ends. Wallboard joints to be taped and covered with joint compound.	—	—	—	Varies	—	—	—	$^5/_8$
26. Wood I-joist (minimum I-joist depth $9^1/_4$" with a minimum flange thickness of $1^1/_2$" and a minimum flange cross-sectional area of 2.25 square inches; minimum web thick-ness of $^3/_8$") @ 24" o.c.	26-1.1	Two layers of $^1/_2$" Type X gypsum wallboard applied with the long dimension perpendicular to the I-joists with end joints staggered. The base layer is fastened with $1^5/_8$" Type S dry wall screws spaced 12" o.c. and the face layer is fastened with 2" Type S drywall screws spaced 12" o.c. in the field and 8" o.c. on the edges. Face layer end joints shall not occur on the same I-joist as base layer end joints and edge joints shall be offset 24"from base layer joints. Face layer to also be attached to base layer with $1^1/_2$" Type G drywall screws spaced 8" o.c. placed 6" from face layer end joints. Face layer wallboard joints to be taped and covered with joint compound.	—	—	—	Varies	—	—	—	1

(continued)

TABLE 721.1(3)—continued
MINIMUM PROTECTION FOR FLOOR AND ROOF SYSTEMS[a, q]

FLOOR OR ROOF CONSTRUCTION	ITEM NUMBER	CEILING CONSTRUCTION	THICKNESS OF FLOOR OR ROOF SLAB (inches)				MINIMUM THICKNESS OF CEILING (inches)			
			4 hours	3 hours	2 hours	1 hour	4 hours	3 hours	2 hours	1 hour
27. Wood I-joist (minimum I-joist depth $9^1/_2$" with a minimum flange depth of $1^5/_{16}$" and a minimum flange cross-sectional area of 1.95 square inches; minimum web thickness of $^3/_8$") @ 24" o.c.	27-1.1	Minimum 0.019" thick resilient channel 16" o.c. (channels doubled at wallboard end joints), placed perpendicular to the joist and attached to each joist by $1^1/_4$" Type S drywall screws. Two layers of $^1/_2$" Type X gypsum wallboard applied with the long dimension perpendicular to the resilient channels with end joints staggered. The base layer is fastened with $1^1/_4$" Type S drywall screws spaced 12" o.c. and the face layer is fastened with $1^5/_8$" Type S drywall screws spaced 12" o.c. Face layer end joints shall not occur on the same I-joist as base layer end joints and edge joints shall be offset 24" from base layer joints. Face layer to also be attached to base layer with $1^1/_2$" Type G drywall screws spaced 8" o.c. placed 6" from face layer end joints. Face layer wallboard joints to be taped and covered with joint compound.	—	—	—	Varies	—	—	—	1
28. Wood I-joist (minimum I-joist depth $9^1/_4$" with a minimum flange depth of $1^1/_2$" and a minimum flange cross-sectional area of 2.25 square inches; minimum web thickness of $^3/_8$") @ 24" o.c. Unfaced fiberglass insulation or mineral wool insulation is installed between the I-joists supported on the upper surface of the flange by stay wires spaced 12" o.c.	28-1.1	Base layer of $^5/_8$" Type C gypsum wallboard attached directly to I-joists with $1^5/_8$" Type S drywall screws spaced 12" o.c. with ends staggered. Minimum 0.0179"-thick hat-shaped $^7/_8$-inch furring channel 16" o.c. (channels doubled at wallboard end joints), placed perpendicular to the joist and attached to each joist by $1^5/_8$" Type S drywall screws after the base layer of gypsum wallboard has been applied. The middle and face layers of $^5/_8$" Type C gypsum wallboard applied perpendicular to the channel with end joints staggered. The middle layer is fastened with 1" Type S drywall screws spaced 12" o.c. The face layer is applied parallel to the middle layer but with the edge joints offset 24" from those of the middle layer and fastened with $1^5/_8$" Type S drywall screws 8" o.c. The joints shall be taped and covered with joint compound.	—	—	—	Varies	—	—	$2^3/_4$	—
29. Channel-shaped 18 gage steel joists (minimum depth 8") spaced a maximum 24" o.c. supporting tongue-and-groove wood structural panels (nominal minimum $^3/_4$"-thick) applied perpendicular to framing members. Structural panels attached with $1^5/_8$" Type S-12 screws spaced 12" o.c.	29-1.1	Base layer $^5/_8$" Type X gypsum board applied perpendicular to bottom of framing members with $1^1/_8$" Type S-12 screws spaced 12" o.c. Second layer $^5/_8$" Type X gypsum board attached perpendicular to framing members with $1^5/_8$" Type S-12 screws spaced 12" o.c. Second layer joints offset 24" from base layer. Third layer $^5/_8$" Type X gypsum board attached perpendicular to framing members with $2^3/_8$" Type S-12 screws spaced 12" o.c. Third layer joints offset 12" from second layer joints. Hat-shaped $^7/_8$-inch rigid furring channels applied at right angles to framing members over third layer with two $2^3/_8$" TypeS-12 screws at each framing member. Face layer $^5/_8$" Type X gypsum board applied at right angles to furring channels with $1^1/_8$" Type S screws spaced 12" o.c.	—	—	Varies	—	—	—	$3^3/_8$	—

(continued)

TABLE 721.1(3)—continued
MINIMUM PROTECTION FOR FLOOR AND ROOF SYSTEMS[a, q]

FLOOR OR ROOF CONSTRUCTION	ITEM NUMBER	CEILING CONSTRUCTION	THICKNESS OF FLOOR OR ROOF SLAB (inches)				MINIMUM THICKNESS OF CEILING (inches)			
			4 hours	3 hours	2 hours	1 hour	4 hours	3 hours	2 hours	1 hour
30. Wood I-joist (minimum I-joist depth $9^1/_2$" with a minimum flange depth of $1^1/_2$" and a minimum flange cross-sectional area of 2.25 square inches; minimum web thickness of $^3/_8$") @ 24" o.c. Fiberglass insulation placed between I-joists supported by the resilient channels.	30-1.1	Minimum 0.019"-thick resilient channel 16" o.c. (channels doubled at wallboard end joints), placed perpendicular to the joists and attached to each joist by $1^1/_4$" Type S dry wall screws. Two layers of $^1/_2$" Type X gypsum wallboard applied with the long dimension perpendicular to the resilient channels with end joints staggered. The base layer is fastened with $1^1/_4$" Type S drywall screws spaced 12" o.c. and the face layer is fastened with $1^5/_8$" Type S drywall screws spaced 12" o.c. Face layer end joints shall not occur on the same I-joist as base layer end joints and edge joints shall be offset 24" from base layer joints. Face layer to be attached to base layer with $1^1/_2$" Type G drywall screws spaced 8" o.c. placed 6" from face layer end joints. Face layer wallboard joints to be taped and covered with joint compound.	—	—	—	Varies	—	—	—	1

For SI: 1 inch = 25.4 mm, 1 foot = 304.8 mm, 1 pound = 0.454 kg, 1 cubic foot = 0.0283 m^3, 1 pound per square inch = 6.895 kPa, 1 pound per linear foot = 1.4882 kg/m.

a. Staples with equivalent holding power and penetration shall be permitted to be used as alternate fasteners to nails for attachment to wood framing.

b. Where the slab is in an unrestrained condition, minimum reinforcement cover shall be not less than $1^5/_8$ inches for 4 hours (siliceous aggregate only); $1^1/_4$ inches for 4 and 3 hours; 1 inch for 2 hours (siliceous aggregate only); and $^3/_4$ inch for all other restrained and unrestrained conditions.

c. For all of the construction with gypsum wallboard described in this table, gypsum base for veneer plaster of the same size, thickness and core type shall be permitted to be substituted for gypsum wallboard, provided that attachment is identical to that specified for the wallboard, and the joints on the face layer are reinforced and the entire surface is covered with not less than $^1/_{16}$-inch gypsum veneer plaster.

d. Slab thickness over steel joists measured at the joists for metal lath form and at the top of the form for steel form units.

e. (a) The maximum allowable stress level for H-Series joists shall not exceed 22,000 psi.
   (b) The allowable stress for K-Series joists shall not exceed 26,000 psi, the nominal depth of such joist shall be not less than 10 inches and the nominal joist weight shall be not less than 5 pounds per linear foot.

f. Cement plaster with 15 pounds of hydrated lime and 3 pounds of approved additives or admixtures per bag of cement.

g. Gypsum wallboard ceilings attached to steel framing shall be permitted to be suspended with $1^1/_2$-inch cold-formed carrying channels spaced 48 inches on center, that are suspended with No. 8 SWG galvanized wire hangers spaced 48 inches on center. Cross-furring channels are tied to the carrying channels with No. 18 SWG galvanized wire hangers spaced 48 inches on center. Cross-furring channels are tied to the carrying channels with No. 18 SWG galvanized wire (double strand) and spaced as required for direct attachment to the framing. This alternative is applicable to those steel framing assemblies recognized under Note q.

h. Six-inch hollow clay tile with 2-inch concrete slab above.

i. Four-inch hollow clay tile with $1^1/_2$-inch concrete slab above.

j. Thickness measured to bottom of steel form units.

k. Five-eighths inch of vermiculite gypsum plaster plus $^1/_2$ inch of approved vermiculite acoustical plastic.

l. Furring channels spaced 12 inches on center.

m. Double wood floor shall be permitted to be either of the following:
   (a) Subfloor of 1-inch nominal boarding, a layer of asbestos paper weighing not less than 14 pounds per 100 square feet and a layer of 1-inch nominal tongue-and-groove finished flooring.
   (b) Subfloor of 1-inch nominal tongue-and-groove boarding or $^{15}/_{32}$-inch wood structural panels with exterior glue and a layer of 1-inch nominal tongue-and-groove finished flooring or $^{19}/_{32}$-inch wood structural panel finish flooring or a layer of Type I Grade M-1 particleboard not less than $^5/_8$-inch thick.

n. The ceiling shall be permitted to be omitted over unusable space, and flooring shall be permitted to be omitted where unusable space occurs above.

o. For properties of cooler or wallboard nails, see ASTM C514, ASTM C547 or ASTM F1667.

p. Thickness measured on top of steel deck unit.

q. Generic fire-resistance ratings (those not designated as PROPRIETARY* in the listing) in the GA 600 shall be accepted as if herein specified.

SECTION 722
CALCULATED FIRE RESISTANCE

722.1 General. The provisions of this section contain procedures by which the *fire resistance* of specific materials or combinations of materials is established by calculations. These procedures apply only to the information contained in this section and shall not be otherwise used. The calculated *fire resistance* of specific materials or combinations of materials shall be established by one of the following:

1. *Concrete*, concrete *masonry* and clay *masonry* assemblies shall be permitted in accordance with ACI 216.1/TMS 0216.
2. Precast and precast, prestressed *concrete* assemblies shall be permitted in accordance with PCI 124.
3. Steel assemblies shall be permitted in accordance with Chapter 5 of ASCE 29.
4. Exposed wood members and wood decking shall be permitted in accordance with Chapter 16 of ANSI/AWC NDS.

722.2 Concrete assemblies. The provisions of this section contain procedures by which the *fire-resistance ratings* of concrete assemblies are established by calculations.

722.2.1 Concrete walls. Cast-in-place and precast concrete walls shall comply with Section 722.2.1.1. Multiwythe concrete walls shall comply with Section 722.2.1.2. Joints between precast panels shall comply with Section 722.2.1.3. Concrete walls with *gypsum wallboard* or plaster finish shall comply with Section 722.2.1.4.

722.2.1.1 Cast-in-place or precast walls. The minimum equivalent thicknesses of cast-in-place or precast concrete walls for *fire-resistance ratings* of 1 hour to 4 hours are shown in Table 722.2.1.1. For solid walls with flat vertical surfaces, the equivalent thickness is the same as the actual thickness. The values in Table 722.2.1.1 apply to plain, reinforced or prestressed concrete walls.

TABLE 722.2.1.1
MINIMUM EQUIVALENT THICKNESS OF CAST-IN-PLACE OR PRECAST CONCRETE WALLS, LOAD-BEARING OR NONLOAD-BEARING

CONCRETE TYPE	MINIMUM SLAB THICKNESS (inches) FOR FIRE-RESISTANCE RATING OF				
	1 hour	$1^1/_2$ hours	2 hours	3 hours	4 hours
Siliceous	3.5	4.3	5.0	6.2	7.0
Carbonate	3.2	4.0	4.6	5.7	6.6
Sand-lightweight	2.7	3.3	3.8	4.6	5.4
Lightweight	2.5	3.1	3.6	4.4	5.1

For SI: 1 inch = 25.4 mm.

722.2.1.1.1 Hollow-core precast wall panels. For hollow-core precast concrete wall panels in which the cores are of constant cross section throughout the length, calculation of the equivalent thickness by dividing the net cross-sectional area (the gross cross section minus the area of the cores) of the panel by its width shall be permitted

722.2.1.1.2 Core spaces filled. Where all of the core spaces of hollow-core wall panels are filled with loose-fill material, such as expanded shale, clay or slag, or vermiculite or perlite, the *fire-resistance rating* of the wall is the same as that of a solid wall of the same concrete type and of the same overall thickness.

722.2.1.1.3 Tapered cross sections. The thickness of panels with tapered cross sections shall be that determined at a distance $2t$ or 6 inches (152 mm), whichever is less, from the point of minimum thickness, where t is the minimum thickness.

722.2.1.1.4 Ribbed or undulating surfaces. The equivalent thickness of panels with ribbed or undulating surfaces shall be determined by one of the following expressions:

For $s \geq 4t$, the thickness to be used shall be t

For $s \leq 2t$, the thickness to be used shall be t_e

For $4t > s > 2t$, the thickness to be used shall be

$$t + \left(\frac{4t}{s} - 1\right)(t_e - t) \qquad \textbf{(Equation 7-3)}$$

where:

s = Spacing of ribs or undulations.

t = Minimum thickness.

t_e = Equivalent thickness of the panel calculated as the net cross-sectional area of the panel divided by the width, in which the maximum thickness used in the calculation shall not exceed $2t$.

722.2.1.2 Multiwythe walls. For walls that consist of two *wythes* of different types of concrete, the *fire-resistance ratings* shall be permitted to be determined from Figure 722.2.1.2.

722.2.1.2.1 Two or more wythes. The *fire-resistance rating* for wall panels consisting of two or more *wythes* shall be permitted to be determined by the formula:

$$R = (R_1^{0.59} + R_2^{0.59} + ... + R_n^{0.59})^{1.7} \qquad \textbf{(Equation 7-4)}$$

where:

R = The fire endurance of the assembly, minutes.

R_1, R_2, *and* R_n = The fire endurances of the individual *wythes*, minutes. Values of $R_n^{0.59}$ for use in Equation 7-4 are given in Table 722.2.1.2(1). Calculated *fire-resistance ratings* are shown in Table 722.2.1.2(2).

722.2.1.2.2 Foam plastic insulation. The *fire-resistance ratings* of precast concrete wall panels consisting of a layer of foam plastic insulation sandwiched between two *wythes* of concrete shall be permitted to be determined by use of Equation 7-4. Foam plastic insulation with a total thickness of less than 1 inch (25 mm) shall be disregarded. The R_n value for thickness of foam plastic insulation of 1 inch (25 mm) or greater, for use in the calculation, is 5 minutes; therefore $R_n^{0.59} = 2.5$.

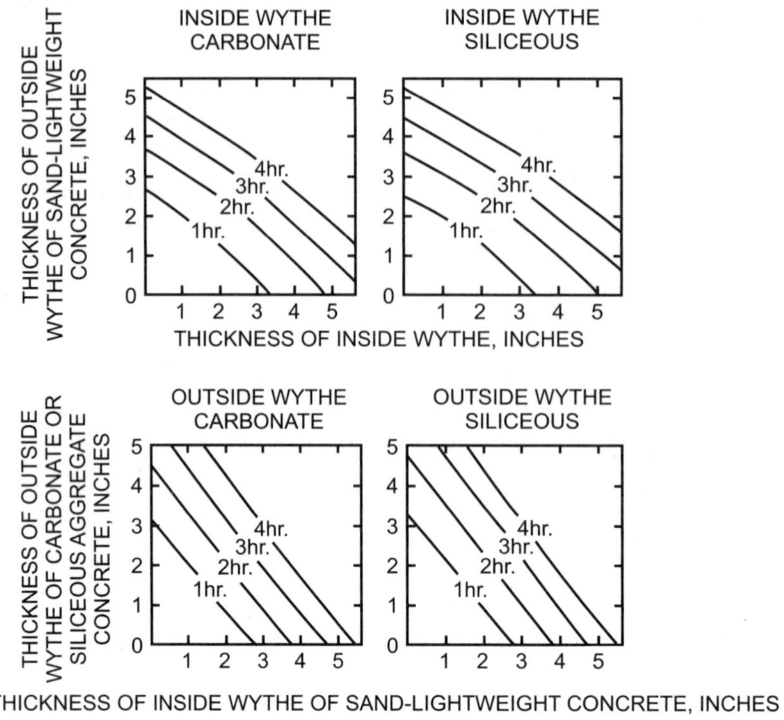

For SI: 1 inch = 25.4 mm.

FIGURE 722.2.1.2
FIRE-RESISTANCE RATINGS OF TWO-WYTHE CONCRETE WALLS

TABLE 722.2.1.2(1)
VALUES OF $R_n^{0.59}$ FOR USE IN EQUATION 7-4

TYPE OF MATERIAL	THICKNESS OF MATERIAL (inches)											
	$1^1/_2$	2	$2^1/_2$	3	$3^1/_2$	4	$4^1/_2$	5	$5^1/_2$	6	$6^1/_2$	7
Siliceous aggregate concrete	5.3	6.5	8.1	9.5	11.3	13.0	14.9	16.9	18.8	20.7	22.8	25.1
Carbonate aggregate concrete	5.5	7.1	8.9	10.4	12.0	14.0	16.2	18.1	20.3	21.9	24.7	27.2[c]
Sand-lightweight concrete	6.5	8.2	10.5	12.8	15.5	18.1	20.7	23.3	26.0[c]	Note c	Note c	Note c
Lightweight concrete	6.6	8.8	11.2	13.7	16.5	19.1	21.9	24.7	27.8[c]	Note c	Note c	Note c
Insulating concrete[a]	9.3	13.3	16.6	18.3	23.1	26.5[c]	Note c	Note c	Note c	Note c	Note c	Note c
Airspace[b]	—	—	—	—	—	—	—	—	—	—	—	—

For SI: 1 inch = 25.4 mm, 1 pound per cubic foot = 16.02 kg/m^3.

a. Dry unit weight of 35 pcf or less and consisting of cellular, perlite or vermiculite concrete.

b. The $R_n^{0.59}$ value for one $^1/_2$″ to $3^1/_2$″ airspace is 3.3. The $R_n^{0.59}$ value for two $^1/_2$″ to $3^1/_2$″ airspaces is 6.7.

c. The fire-resistance rating for this thickness exceeds 4 hours.

TABLE 722.2.1.2(2)
FIRE-RESISTANCE RATINGS BASED ON $R^{0.59}$

R[a], MINUTES	$R^{0.59}$
60	11.20
120	16.85
180	21.41
240	25.37

a. Based on Equation 7-4.

722.2.1.3 Joints between precast wall panels. Joints between precast concrete wall panels that are not insulated as required by this section shall be considered as openings in walls. Uninsulated joints shall be included in determining the percentage of openings permitted by Table 705.8. Where openings are not permitted or are required by this code to be protected, the provisions of this section shall be used to determine the amount of joint insulation required. Insulated joints shall not be considered openings for purposes of determining compliance with the allowable percentage of openings in Table 705.8.

722.2.1.3.1 Ceramic fiber joint protection. Figure 722.2.1.3.1 shows thicknesses of *ceramic fiber blankets* to be used to insulate joints between precast concrete wall panels for various panel thicknesses and for joint widths of $^3/_8$ inch (9.5 mm) and 1 inch (25 mm) for *fire-resistance ratings* of 1 hour to 4 hours. For joint widths between $^3/_8$ inch (9.5 mm) and 1 inch (25 mm), the thickness of *ceramic fiber blanket* is allowed to be determined by direct interpolation. Other tested and labeled materials are acceptable in place of *ceramic fiber blankets*.

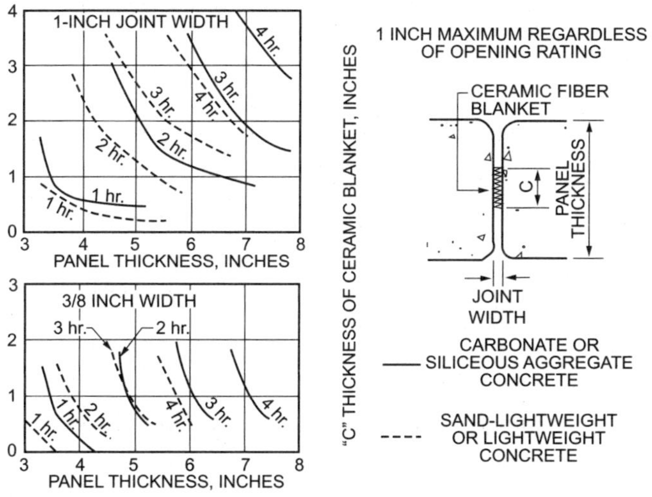

For SI: 1 inch = 25.4 mm.

FIGURE 722.2.1.3.1
CERAMIC FIBER JOINT PROTECTION

722.2.1.4 Walls with gypsum wallboard or plaster finishes. The *fire-resistance rating* of cast-in-place or precast concrete walls with finishes of *gypsum wallboard* or plaster applied to one or both sides shall be permitted to be calculated in accordance with the provisions of this section.

722.2.1.4.1 Nonfire-exposed side. Where the finish of *gypsum wallboard* or plaster is applied to the side of the wall not exposed to fire, the contribution of the finish to the total *fire-resistance rating* shall be determined as follows: The thickness of the finish shall first be corrected by multiplying the actual thickness of the finish by the applicable factor determined from Table 722.2.1.4(1) based on the type of aggregate in the concrete. The corrected thickness of finish shall then be added to the actual or equivalent thickness of concrete and *fire-resistance rating* of the concrete and finish determined from Tables 722.2.1.1 and 722.2.1.2(1) and Figure 722.2.1.2.

722.2.1.4.2 Fire-exposed side. Where *gypsum wallboard* or plaster is applied to the fire-exposed side of the wall, the contribution of the finish to the total *fire-resistance rating* shall be determined as follows: The time assigned to the finish as established by Table 722.2.1.4(2) shall be added to the *fire-resistance rating* determined from Tables 722.2.1.1 and 722.2.1.2(1) and Figure 722.2.1.2 for the concrete alone, or to the rating determined in Section 722.2.1.4.1 for the concrete and finish on the nonfire-exposed side.

TABLE 722.2.1.4(1)
MULTIPLYING FACTOR FOR FINISHES ON NONFIRE-EXPOSED SIDE OF WALL

TYPE OF FINISH APPLIED TO CONCRETE OR CONCRETE MASONRY WALL	TYPE OF AGGREGATE USED IN CONCRETE OR CONCRETE MASONRY			
	Concrete: siliceous or carbonate concrete masonry: siliceous or carbonate; solid clay brick	Concrete: sand-lightweight concrete masonry: clay tile; hollow clay brick; concrete masonry units of expanded shale and < 20% sand	Concrete: lightweight concrete masonry: concrete masonry units of expanded shale, expanded clay, expanded slag, or pumice < 20% sand	Concrete masonry: concrete masonry units of expanded slag, expanded clay, or pumice
Portland cement-sand plaster	1.00	0.75[a]	0.75[a]	0.50[a]
Gypsum-sand plaster	1.25	1.00	1.00	1.00
Gypsum-vermiculite or perlite plaster	1.75	1.50	1.25	1.25
Gypsum wallboard	3.00	2.25	2.25	2.25

For SI: 1 inch = 25.4 mm.

a. For Portland cement-sand plaster $^5/_8$ inch or less in thickness and applied directly to the concrete or concrete masonry on the nonfire-exposed side of the wall, the multiplying factor shall be 1.00.

TABLE 722.2.1.4(2)
TIME ASSIGNED TO FINISH MATERIALS ON FIRE-EXPOSED SIDE OF WALL

FINISH DESCRIPTION	TIME (minutes)
Gypsum wallboard	
$^3/_8$ inch	10
$^1/_2$ inch	15
$^5/_8$ inch	20
2 layers of $^3/_8$ inch	25
1 layer of $^3/_8$ inch, 1 layer of $^1/_2$ inch	35
2 layers of $^1/_2$ inch	40
Type X gypsum wallboard	
$^1/_2$ inch	25
$^5/_8$ inch	40
Portland cement-sand plaster applied directly to concrete masonry	See Note a
Portland cement-sand plaster on metal lath	
$^3/_4$ inch	20
$^7/_8$ inch	25
1 inch	30
Gypsum sand plaster on $^3/_8$-inch gypsum lath	
$^1/_2$ inch	35
$^5/_8$ inch	40
$^3/_4$ inch	50
Gypsum sand plaster on metal lath	
$^3/_4$ inch	50
$^7/_8$ inch	60
1 inch	80

For SI: 1 inch = 25.4 mm.

a. The actual thickness of Portland cement-sand plaster, provided that it is $^5/_8$ inch or less in thickness, shall be permitted to be included in determining the equivalent thickness of the masonry for use in Table 722.3.2.

722.2.1.4.3 Nonsymmetrical assemblies. For a wall without finish on one side or having different types or thicknesses of finish on each side, the calculation procedures of Sections 722.2.1.4.1 and 722.2.1.4.2 shall be performed twice, assuming either side of the wall to be the fire-exposed side. The *fire-resistance rating* of the wall shall not exceed the lower of the two values.

Exception: For an *exterior* wall with a *fire separation distance* greater than 5 feet (1524 mm) the fire shall be assumed to occur on the interior side only.

722.2.1.4.4 Minimum concrete fire-resistance rating. Where finishes applied to one or both sides of a concrete wall contribute to the *fire-resistance rating*, the concrete alone shall provide not less than one-half of the total required *fire-resistance rating*. Additionally, the contribution to the *fire resistance* of the finish on the nonfire-exposed side of a *load-bearing wall* shall not exceed one-half the contribution of the concrete alone.

722.2.1.4.5 Concrete finishes. Finishes on concrete walls that are assumed to contribute to the total *fire-resistance rating* of the wall shall comply with the installation requirements of Section 722.3.2.5.

722.2.2 Concrete floor and roof slabs. Reinforced and prestressed floors and roofs shall comply with Section 722.2.2.1. Multicourse floors and roofs shall comply with Sections 722.2.2.2 and 722.2.2.3, respectively.

722.2.2.1 Reinforced and prestressed floors and roofs. The minimum thicknesses of reinforced and prestressed concrete floor or roof slabs for *fire-resistance ratings* of 1 hour to 4 hours are shown in Table 722.2.2.1.

Exception: Minimum thickness shall not be required for floors and *ramps* within parking garages constructed in accordance with Sections 406.5 and 406.6.

TABLE 722.2.2.1
MINIMUM SLAB THICKNESS (inches)

CONCRETE TYPE	FIRE-RESISTANCE RATING (hours)				
	1	$1^1/_2$	2	3	4
Siliceous	3.5	4.3	5	6.2	7
Carbonate	3.2	4	4.6	5.7	6.6
Sand-lightweight	2.7	3.3	3.8	4.6	5.4
Lightweight	2.5	3.1	3.6	4.4	5.1

For SI: 1 inch = 25.4 mm.

722.2.2.1.1 Hollow-core prestressed slabs. For hollow-core prestressed concrete slabs in which the cores are of constant cross section throughout the length, the equivalent thickness shall be permitted to be obtained by dividing the net cross-sectional area of the slab including grout in the joints, by its width.

722.2.2.1.2 Slabs with sloping soffits. The thickness of slabs with sloping soffits (see Figure 722.2.2.1.2) shall be determined at a distance $2t$ or 6 inches (152 mm), whichever is less, from the point of minimum thickness, where t is the minimum thickness.

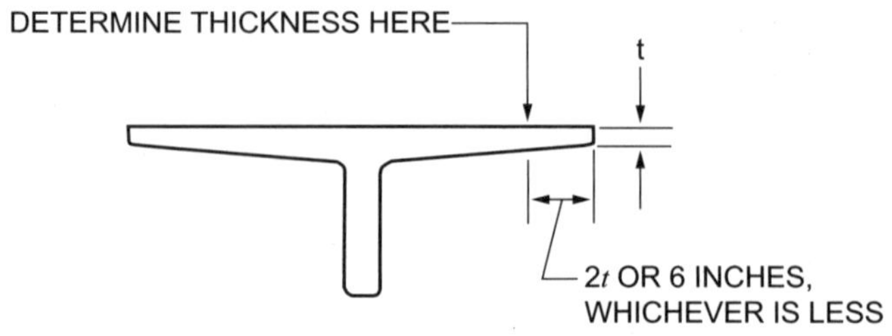

For SI: 1 inch = 25.4 mm.

FIGURE 722.2.2.1.2
DETERMINATION OF SLAB THICKNESS FOR SLOPING SOFFITS

722.2.2.1.3 Slabs with ribbed soffits. The thickness of slabs with ribbed or undulating soffits (see Figure 722.2.2.1.3) shall be determined by one of the following expressions, whichever is applicable:

For $s > 4t$, the thickness to be used shall be t

For $s \leq 2t$, the thickness to be used shall be t_e

For $4t > s > 2t$, the thickness to be used shall be

$$t + \left(\frac{4t}{s} - 1\right)(t_e - t) \qquad \textbf{(Equation 7-5)}$$

where:

s = Spacing of ribs or undulations.

t = Minimum thickness.

t_e = Equivalent thickness of the slab calculated as the net area of the slab divided by the width, in which the maximum thickness used in the calculation shall not exceed $2t$.

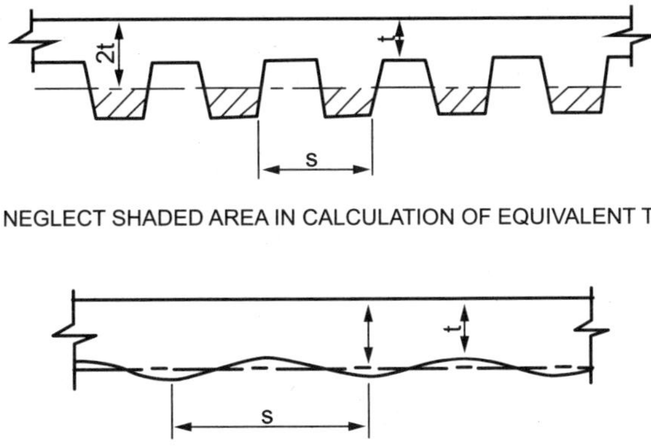

For SI: 1 inch = 25.4 mm.

FIGURE 722.2.2.1.3
SLABS WITH RIBBED OR UNDULATING SOFFITS

722.2.2.1.4 Flat plate concrete slabs with uniformly spaced hollow voids. Table 722.2.2.1 shall be used to determine the 1- and 2-hour fire-resistance ratings for flat plate concrete slabs with uniformly spaced hollow voids. The equivalent thickness of the slab shall be determined by dividing the net concrete volume of the slab by the floor area. The net concrete volume of the slab shall be equal to the volume of concrete of a solid slab minus the average concrete volume displaced by the hollow voids.

722.2.2.2 Multicourse floors. The *fire-resistance ratings* of floors that consist of a base slab of concrete with a topping (overlay) of a different type of concrete shall comply with Figure 722.2.2.2.

722.2.2.3 Multicourse roofs. The *fire-resistance ratings* of roofs that consist of a base slab of concrete with a topping (overlay) of an insulating concrete or with an insulating board and built-up roofing shall comply with Figures 722.2.2.3(1) and 722.2.2.3(2).

722.2.2.3.1 Heat transfer. For the transfer of heat, three-ply built-up roofing contributes 10 minutes to the *fire-resistance rating*. The *fire-resistance rating* for concrete assemblies such as those shown in Figure 722.2.2.3(1) shall be increased by 10 minutes. This increase is not applicable to those shown in Figure 722.2.2.3(2).

722.2.2.4 Joints in precast slabs. Joints between adjacent precast concrete slabs need not be considered in calculating the slab thickness provided that a concrete topping not less than 1 inch (25 mm) thick is used. Where concrete topping is not used, joints must be grouted to a depth of not less than one-third the slab thickness at the joint, but not less than 1 inch (25 mm), or the joints must be made fire resistant by other *approved* methods.

722.2.3 Concrete cover over reinforcement. The minimum thickness of concrete cover over reinforcement in concrete slabs, reinforced beams and prestressed beams shall comply with this section.

722.2.3.1 Slab cover. The minimum thickness of concrete cover to the positive moment reinforcement shall comply with Table 722.2.3(1) for reinforced concrete and Table 722.2.3(2) for prestressed concrete. These tables are applicable for solid or hollow-core one-way or two-way slabs with flat undersurfaces.

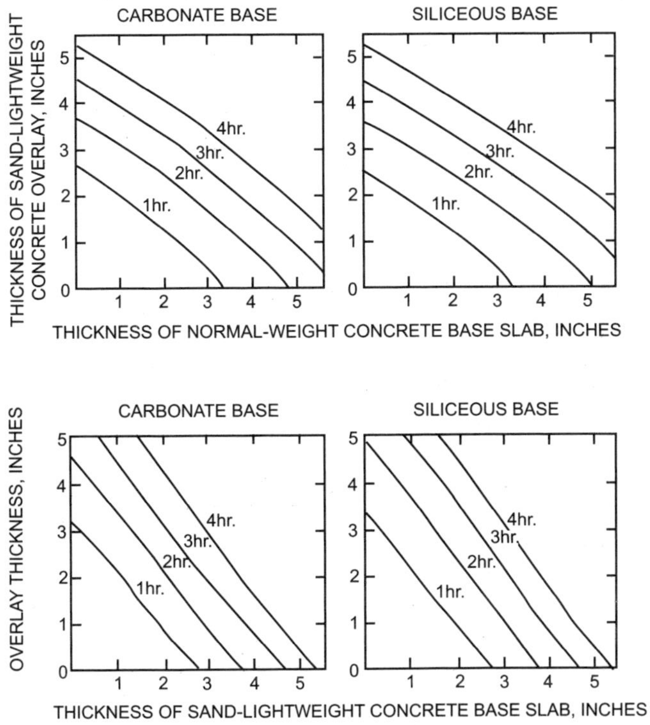

For SI: 1 inch = 25.4 mm.

FIGURE 722.2.2.2
FIRE-RESISTANCE RATINGS FOR TWO-COURSE CONCRETE FLOORS

These tables are applicable to slabs that are either cast in place or precast. For precast prestressed concrete not covered elsewhere, the procedures contained in PCI 124 shall be acceptable.

722.2.3.2 Reinforced beam cover. The minimum thickness of concrete cover to the positive moment reinforcement (bottom steel) for reinforced concrete beams is shown in Table 722.2.3(3) for *fire-resistance ratings* of 1 hour to 4 hours.

722.2.3.3 Prestressed beam cover. The minimum thickness of concrete cover to the positive moment prestressing tendons (bottom steel) for restrained and unrestrained prestressed concrete beams and stemmed units shall comply with the values shown in Tables 722.2.3(4) and 722.2.3(5) for *fire-resistance ratings* of 1 hour to 4 hours. Values in Table 722.2.3(4) apply to beams 8 inches (203 mm) or greater in width. Values in Table 722.2.3(5) apply to beams or stems of any width, provided that the cross-section area is not less than 40 square inches (25 806 mm^2). In case of differences between the values determined from Table 722.2.3(4) or 722.2.3(5), it is permitted to use the smaller value. The concrete cover shall be calculated in accordance with Section 722.2.3.3.1. The minimum concrete cover for nonprestressed reinforcement in prestressed concrete beams shall comply with Section 722.2.3.2.

722.2.3.3.1 Calculating concrete cover. The concrete cover for an individual tendon is the minimum thickness of concrete between the surface of the tendon and the fire-exposed surface of the beam, except that for ungrouted ducts, the assumed cover thickness is the minimum thickness of concrete between the surface of the duct and the fire-exposed surface of the beam. For beams in which two or more tendons are used, the cover is assumed to be the average of the minimum cover of the individual tendons. For corner tendons (tendons equal distance from the bottom and side), the minimum cover used in the calculation shall be one-half the actual value. For stemmed members with two or more prestressing tendons located along the vertical centerline of the stem, the average cover shall be the distance from the bottom of the member to the centroid of the tendons. The actual cover for any individual tendon shall be not less than one-half the smaller value shown in Tables 722.2.3(4) and 722.2.3(5), or 1 inch (25 mm), whichever is greater.

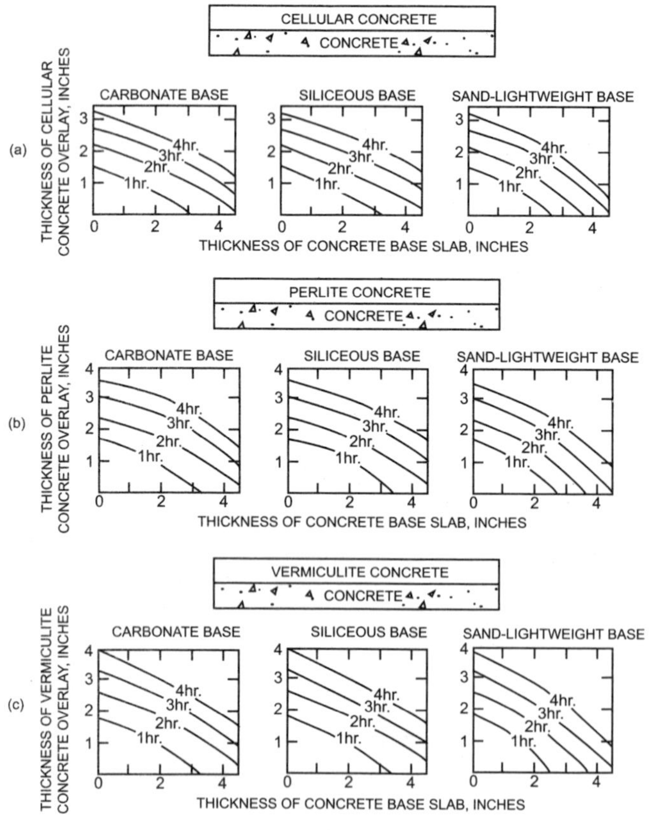

For SI: 1 inch = 25.4 mm.

FIGURE 722.2.2.3(1)
FIRE-RESISTANCE RATINGS FOR CONCRETE ROOF ASSEMBLIES

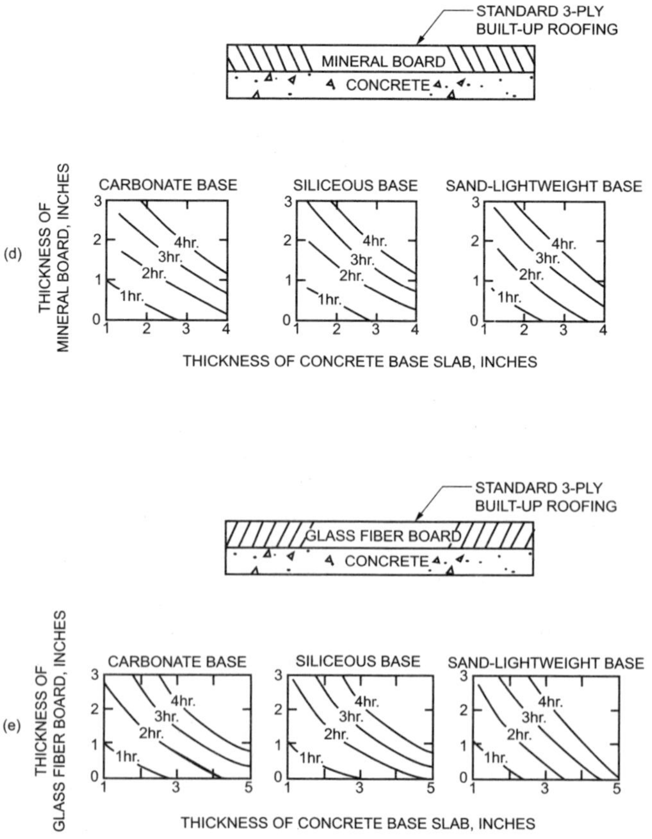

For SI: 1 inch = 25.4 mm.

FIGURE 722.2.2.3(2)
FIRE-RESISTANCE RATINGS FOR CONCRETE ROOF ASSEMBLIES

722.2.4 Concrete columns. Concrete columns shall comply with this section.

TABLE 722.2.4
MINIMUM DIMENSION OF CONCRETE COLUMNS (inches)

TYPES OF CONCRETE	FIRE-RESISTANCE RATING (hours)				
	1	$1^1/_2$	2[a]	3[a]	4[b]
Siliceous	8	9	10	12	14
Carbonate	8	9	10	11	12
Sand-lightweight	8	$8^1/_2$	9	$10^1/_2$	12

For SI: 1 inch = 25 mm.

a. The minimum dimension is permitted to be reduced to 8 inches for rectangular columns with two parallel sides not less than 36 inches in length.

b. The minimum dimension is permitted to be reduced to 10 inches for rectangular columns with two parallel sides not less than 36 inches in length.

722.2.4.1 Minimum size. The minimum overall dimensions of reinforced concrete columns for *fire-resistance ratings* of 1 hour to 4 hours for exposure to fire on all sides shall comply with this section.

722.2.4.1.1 Concrete strength less than or equal to 12,000 psi. For columns made with concrete having a specified compressive strength, f'_c, of less than or equal to 12,000 psi (82.7 MPa), the minimum dimension shall comply with Table 722.2.4.

722.2.4.1.2 Concrete strength greater than 12,000 psi. For columns made with concrete having a specified compressive strength, f'_c, greater than 12,000 psi (82.7 MPa), for *fire-resistance ratings* of 1 hour to 4 hours the minimum dimension shall be 24 inches (610 mm).

722.2.4.2 Minimum cover for R/C columns. The minimum thickness of concrete cover to the main longitudinal reinforcement in columns, regardless of the type of aggregate used in the concrete and the spec-

TABLE 722.2.3(1)
COVER THICKNESS FOR REINFORCED CONCRETE FLOOR OR ROOF SLABS (inches)

CONCRETE AGGREGATE TYPE	FIRE-RESISTANCE RATING (hours)									
	Restrained					Unrestrained				
	1	$1^1/_2$	2	3	4	1	$1^1/_2$	2	3	4
Siliceous	$^3/_4$	$^3/_4$	$^3/_4$	$^3/_4$	$^3/_4$	$^3/_4$	$^3/_4$	1	$1^1/_4$	$1^5/_8$
Carbonate	$^3/_4$	$^3/_4$	$^3/_4$	$^3/_4$	$^3/_4$	$^3/_4$	$^3/_4$	$^3/_4$	$1^1/_4$	$1^1/_4$
Sand-lightweight or lightweight	$^3/_4$	$^3/_4$	$^3/_4$	$^3/_4$	$^3/_4$	$^3/_4$	$^3/_4$	$^3/_4$	$1^1/_4$	$1^1/_4$

For SI: 1 inch = 25.4 mm.

TABLE 722.2.3(2)
COVER THICKNESS FOR PRESTRESSED CONCRETE FLOOR OR ROOF SLABS (inches)

CONCRETE AGGREGATE TYPE	FIRE-RESISTANCE RATING (hours)									
	Restrained					Unrestrained				
	1	$1^1/_2$	2	3	4	1	$1^1/_2$	2	3	4
Siliceous	$^3/_4$	$^3/_4$	$^3/_4$	$^3/_4$	$^3/_4$	$1^1/_8$	$1^1/_2$	$1^3/_4$	$2^3/_8$	$2^3/_4$
Carbonate	$^3/_4$	$^3/_4$	$^3/_4$	$^3/_4$	$^3/_4$	1	$1^3/_8$	$1^5/_8$	$2^1/_8$	$2^1/_4$
Sand-lightweight or lightweight	$^3/_4$	$^3/_4$	$^3/_4$	$^3/_4$	$^3/_4$	1	$1^3/_8$	$1^1/_2$	2	$2^1/_4$

For SI: 1 inch = 25.4 mm.

TABLE 722.2.3(3)
MINIMUM COVER FOR MAIN REINFORCING BARS OF REINFORCED CONCRETE BEAMS[c]
(APPLICABLE TO ALL TYPES OF STRUCTURAL CONCRETE)

RESTRAINED OR UNRESTRAINED[a]	BEAM WIDTH[b] (inches)	FIRE-RESISTANCE RATING (hours)				
		1	$1^1/_2$	2	3	4
Restrained	5	$^3/_4$	$^3/_4$	$^3/_4$	1[a]	$1^1/_4$[a]
	7	$^3/_4$	$^3/_4$	$^3/_4$	$^3/_4$	$^3/_4$
	≥10	$^3/_4$	$^3/_4$	$^3/_4$	$^3/_4$	$^3/_4$
Unrestrained	5	$^3/_4$	1	$1^1/_4$	—	—
	7	$^3/_4$	$^3/_4$	$^3/_4$	$1^3/_4$	3
	≥ 10	$^3/_4$	$^3/_4$	$^3/_4$	1	$1^3/_4$

For SI: 1 inch = 25.4 mm, 1 foot = 304.8 mm.

a. Tabulated values for restrained assemblies apply to beams spaced more than 4 feet on center. For restrained beams spaced 4 feet or less on center, minimum cover of $^3/_4$ inch is adequate for ratings of 4 hours or less.

b. For beam widths between the tabulated values, the minimum cover thickness can be determined by direct interpolation.

c. The cover for an individual reinforcing bar is the minimum thickness of concrete between the surface of the bar and the fire-exposed surface of the beam. For beams in which several bars are used, the cover for corner bars used in the calculation shall be reduced to one-half of the actual value. The cover for an individual bar must be not less than one-half of the value given in Table 722.2.3(3) nor less than $^3/_4$ inch.

ified compressive strength of concrete, f'_c, shall be not less than 1 inch (25 mm) times the number of hours of required *fire resistance* or 2 inches (51 mm), whichever is less.

722.2.4.3 Tie and spiral reinforcement. For concrete columns made with concrete having a specified compressive strength, f'_c, greater than 12,000 psi (82.7 MPa), tie and spiral reinforcement shall comply with the following:

1. The free ends of rectangular ties shall terminate with a 135-degree (2.4 rad) standard tie hook.
2. The free ends of circular ties shall terminate with a 90-degree (1.6 rad) standard tie hook.
3. The free ends of spirals, including at lap splices, shall terminate with a 90-degree (1.6 rad) standard tie hook.

The hook extension at the free end of ties and spirals shall be the larger of six bar diameters and the extension required by Section 25.3.2 of ACI 318. Hooks shall project into the core of the column.

722.2.4.4 Columns built into walls. The minimum dimensions of Table 722.2.4 do not apply to a reinforced concrete column that is built into a concrete or masonry wall provided that all of the following are met:

1. The *fire-resistance rating* for the wall is equal to or greater than the required rating of the column.
2. The main longitudinal reinforcing in the column has cover not less than that required by Section 722.2.4.2.
3. Openings in the wall are protected in accordance with Section 716.

Where openings in the wall are not protected as required by Section 716, the minimum dimension of columns required to have a *fire-resistance rating* of 3 hours or less shall be 8 inches (203 mm), and 10 inches

TABLE 722.2.3(4)
MINIMUM COVER FOR PRESTRESSED CONCRETE BEAMS 8 INCHES OR GREATER IN WIDTH[b]

RESTRAINED OR UNRESTRAINED[a]	CONCRETE AGGREGATE TYPE	BEAM WIDTH (inches)	FIRE-RESISTANCE RATING (hours)				
			1	$1^1/_2$	2	3	4
Restrained	Carbonate or siliceous	8	$1^1/_2$	$1^1/_2$	$1^1/_2$	$1^3/_4$[a]	$2^1/_2$[a]
	Carbonate or siliceous	≥ 12	$1^1/_2$	$1^1/_2$	$1^1/_2$	$1^1/_2$	$1^7/_8$[a]
	Sand lightweight	8	$1^1/_2$	$1^1/_2$	$1^1/_2$	$1^1/_2$	2[a]
	Sand lightweight	≥ 12	$1^1/_2$	$1^1/_2$	$1^1/_2$	$1^1/_2$	$1^5/_8$[a]
Unrestrained	Carbonate or siliceous	8	$1^1/_2$	$1^3/_4$	$2^1/_2$	5[c]	—
	Carbonate or siliceous	≥ 12	$1^1/_2$	$1^1/_2$	$1^7/_8$[a]	$2^1/_2$	3
	Sand lightweight	8	$1^1/_2$	$1^1/_2$	2	$3^1/_4$	—
	Sand lightweight	≥ 12	$1^1/_2$	$1^1/_2$	$1^5/_8$	2	$2^1/_2$

For SI: 1 inch = 25.4 mm, 1 foot = 304.8 mm.

a. Tabulated values for restrained assemblies apply to beams spaced more than 4 feet on center. For restrained beams spaced 4 feet or less on center, minimum cover of $^3/_4$ inch is adequate for 4-hour ratings or less.

b. For beam widths between 8 inches and 12 inches, minimum cover thickness can be determined by direct interpolation.

c. Not practical for 8-inch-wide beam but shown for purposes of interpolation.

TABLE 722.2.3(5)
MINIMUM COVER FOR PRESTRESSED CONCRETE BEAMS OF ALL WIDTHS

RESTRAINED OR UNRESTRAINED[a]	CONCRETE AGGREGATE TYPE	BEAM AREA[b] A (square inches)	FIRE-RESISTANCE RATING (hours)				
			1	$1^1/_2$	2	3	4
Restrained	All	$40 \le A \le 150$	$1^1/_2$	$1^1/_2$	2	$2^1/_2$	—
	Carbonate or siliceous	$150 < A \le 300$	$1^1/_2$	$1^1/_2$	$1^1/_2$	$1^3/_4$	$2^1/_2$
		$300 < A$	$1^1/_2$	$1^1/_2$	$1^1/_2$	$1^1/_2$	2
	Sand lightweight	$150 < A$	$1^1/_2$	$1^1/_2$	$1^1/_2$	$1^1/_2$	2
Unrestrained	All	$40 \le A \le 150$	2	$2^1/_2$	—	—	—
	Carbonate or siliceous	$150 < A \le 300$	$1^1/_2$	$1^3/_4$	$2^1/_2$	—	—
		$300 < A$	$1^1/_2$	$1^1/_2$	2	3[c]	4[c]
	Sand lightweight	$150 < A$	$1^1/_2$	$1^1/_2$	2	3[c]	4[c]

For SI: 1 inch = 25.4 mm, 1 foot = 304.8 mm, 1 square inch = 645.2 mm^2.

a. Tabulated values for restrained assemblies apply to beams spaced more than 4 feet on center. For restrained beams spaced 4 feet or less on center, minimum cover of $^3/_4$ inch is adequate for 4-hour ratings or less.

b. The cross-sectional area of a stem is permitted to include a portion of the area in the flange, provided that the width of the flange used in the calculation does not exceed three times the average width of the stem.

c. U-shaped or hooped stirrups spaced not to exceed the depth of the member and having a minimum cover of 1 inch shall be provided.

(254 mm) for columns required to have a *fire-resistance rating* of 4 hours, regardless of the type of aggregate used in the concrete.

722.2.4.5 Precast cover units for steel columns. See Section 722.5.1.4.

722.3 Concrete masonry. The provisions of this section contain procedures by which the *fire-resistance ratings* of concrete masonry are established by calculations.

722.3.1 Equivalent thickness. The equivalent thickness of concrete masonry construction shall be determined in accordance with the provisions of this section.

722.3.1.1 Concrete masonry unit plus finishes. The equivalent thickness of concrete masonry assemblies, T_{ea}, shall be computed as the sum of the equivalent thickness of the concrete *masonry unit*, T_e, as determined by Section 722.3.1.2, 722.3.1.3 or 722.3.1.4, plus the equivalent thickness of finishes, T_{ef}, determined in accordance with Section 722.3.2:

$$T_{ea} = T_e + T_{ef} \qquad \textbf{(Equation 7-6)}$$

722.3.1.2 Ungrouted or partially grouted construction. T_e shall be the value obtained for the concrete *masonry unit* determined in accordance with ASTM C140.

722.3.1.3 Solid grouted construction. The equivalent thickness, T_e, of solid grouted concrete *masonry units* is the actual thickness of the unit.

722.3.1.4 Airspaces and cells filled with loose-fill material. The equivalent thickness of completely filled hollow concrete masonry is the actual thickness of the unit where loose-fill materials are: sand, pea gravel, crushed stone, or slag that meet ASTM C33 requirements; pumice, scoria, expanded shale, expanded clay, expanded slate, expanded slag, expanded fly ash, or cinders that comply with ASTM C331; or perlite or vermiculite meeting the requirements of ASTM C549 and ASTM C516, respectively.

722.3.2 Concrete masonry walls. The *fire-resistance rating* of walls and partitions constructed of concrete *masonry units* shall be determined from Table 722.3.2. The rating shall be based on the equivalent thickness of the masonry and type of aggregate used.

722.3.2.1 Finish on nonfire-exposed side. Where plaster or *gypsum wallboard* is applied to the side of the wall not exposed to fire, the contribution of the finish to the total *fire-resistance rating* shall be determined as follows: The thickness of *gypsum wallboard* or plaster shall be corrected by multiplying the actual thickness of the finish by applicable factor determined from Table 722.2.1.4(1). This corrected thickness of finish shall be added to the equivalent thickness of masonry and the *fire-resistance rating* of the masonry and finish determined from Table 722.3.2.

722.3.2.2 Finish on fire-exposed side. Where plaster or *gypsum wallboard* is applied to the fire-exposed side of the wall, the contribution of the finish to the total *fire-resistance rating* shall be determined as follows: The time assigned to the finish as established by Table 722.2.1.4(2) shall be added to the *fire-resistance rating* determined in Section 722.3.2 for the masonry alone, or in Section 722.3.2.1 for the masonry and finish on the nonfire-exposed side.

722.3.2.3 Nonsymmetrical assemblies. For a wall without finish on one side or having different types or thicknesses of finish on each side, the calculation procedures of this section shall be performed twice, assuming either side of the wall to be the fire-exposed side. The *fire-resistance rating* of the wall shall not exceed the lower of the two values calculated.

Exception: For *exterior walls* with a *fire separation distance* greater than 5 feet (1524 mm), the fire shall be assumed to occur on the interior side only.

722.3.2.4 Minimum concrete masonry fire-resistance rating. Where the finish applied to a concrete masonry wall contributes to its *fire-resistance rating*, the masonry alone shall provide not less than one-half the total required *fire-resistance rating*.

722.3.2.5 Attachment of finishes. Installation of finishes shall be as follows:

1. *Gypsum wallboard* and gypsum lath applied to concrete masonry or concrete walls shall be secured to wood or steel furring members spaced not more than 16 inches (406 mm) on center (o.c.).

TABLE 722.3.2
MINIMUM EQUIVALENT THICKNESS (inches) OF BEARING OR NONBEARING CONCRETE MASONRY WALLS[a, b, c, d]

TYPE OF AGGREGATE	FIRE-RESISTANCE RATING (hours)														
	$^1/_2$	$^3/_4$	1	$1^1/_4$	$1^1/_2$	$1^3/_4$	2	$2^1/_4$	$2^1/_2$	$2^3/_4$	3	$3^1/_4$	$3^1/_2$	$3^3/_4$	4
Pumice or expanded slag	1.5	1.9	2.1	2.5	2.7	3.0	3.2	3.4	3.6	3.8	4.0	4.2	4.4	4.5	4.7
Expanded shale, clay or slate	1.8	2.2	2.6	2.9	3.3	3.4	3.6	3.8	4.0	4.2	4.4	4.6	4.8	4.9	5.1
Limestone, cinders or unexpanded slag	1.9	2.3	2.7	3.1	3.4	3.7	4.0	4.3	4.5	4.8	5.0	5.2	5.5	5.7	5.9
Calcareous or siliceous gravel	2.0	2.4	2.8	3.2	3.6	3.9	4.2	4.5	4.8	5.0	5.3	5.5	5.8	6.0	6.2

For SI: 1 inch = 25.4 mm.

a. Values between those shown in the table can be determined by direct interpolation.

b. Where combustible members are framed into the wall, the thickness of solid material between the end of each member and the opposite face of the wall, or between members set in from opposite sides, shall be not less than 93 percent of the thickness shown in the table.

c. Requirements of ASTM C55, ASTM C73, ASTM C90 or ASTM C744 shall apply.

d. Minimum required equivalent thickness corresponding to the hourly fire-resistance rating for units with a combination of aggregate shall be determined by linear interpolation based on the percent by volume of each aggregate used in manufacture.

2. *Gypsum wallboard* shall be installed with the long dimension parallel to the furring members and shall have all joints finished.
3. Other aspects of the installation of finishes shall comply with the applicable provisions of Chapters 7 and 25.

722.3.3 Multiwythe masonry walls. The *fire-resistance rating* of wall assemblies constructed of multiple *wythes* of masonry materials shall be permitted to be based on the *fire-resistance rating* period of each *wythe* and the continuous airspace between each *wythe* in accordance with the following formula:

$$R_A = (R_1^{0.59} + R_2^{0.59} + ... + R_n^{0.59} + A_1 + A_2 + ... + A_n)^{1.7}$$ **(Equation 7-7)**

where:

R_A = Fire-resistance rating of the assembly (hours).

$R_1, R_2, ..., R_n$ = *Fire-resistance rating* of wythes for 1, 2, *n* (hours), respectively.

$A_1, A_2, ..., A_n$ = 0.30, factor for each continuous airspace for 1, 2, ...*n*, respectively, having a depth of $^1/_2$ inch (12.7 mm) or more between wythes.

722.3.4 Concrete masonry lintels. *Fire-resistance ratings* for concrete masonry lintels shall be determined based on the nominal thickness of the lintel and the minimum thickness of concrete masonry or concrete, or any combination thereof, covering the main reinforcing bars, as determined in accordance with Table 722.3.4, or by *approved* alternate methods.

TABLE 722.3.4
MINIMUM COVER OF LONGITUDINAL REINFORCEMENT IN FIRE-RESISTANCE-RATED REINFORCED CONCRETE MASONRY LINTELS (inches)

NOMINAL WIDTH OF LINTEL (inches)	FIRE-RESISTANCE RATING (hours)			
	1	2	3	4
6	$1^1/_2$	2	—	—
8	$1^1/_2$	$1^1/_2$	$1^3/_4$	3
10 or greater	$1^1/_2$	$1^1/_2$	$1^1/_2$	$1^3/_4$

For SI:1 inch = 25.4 mm.

722.3.5 Concrete masonry columns. The *fire-resistance rating* of concrete masonry columns shall be determined based on the least plan dimension of the column in accordance with Table 722.3.5 or by *approved* alternate methods.

TABLE 722.3.5
MINIMUM DIMENSION OF CONCRETE MASONRY COLUMNS (inches)

FIRE-RESISTANCE RATING (hours)			
1	2	3	4
8 inches	10 inches	12 inches	14 inches

For SI: 1 inch = 25.4 mm.

722.4 Clay brick and tile masonry. The provisions of this section contain procedures by which the *fire-resistance ratings* of clay brick and tile masonry are established by calculations.

722.4.1 Masonry walls. The *fire-resistance rating* of masonry walls shall be based on the equivalent thickness as calculated in accordance with this section. The calculation shall take into account finishes applied to the wall and airspaces between *wythes* in multiwythe construction.

TABLE 722.4.1(3)
VALUES OF $R_n^{0.59}$

R_n 0.59	*R* (hours)
1	1.0
2	1.50
3	1.91
4	2.27

TABLE 722.4.1(4)
COEFFICIENTS FOR PLASTER, *pl* [a]

THICKNESS OF PLASTER (inch)	ONE SIDE	TWO SIDES
$^1/_2$	0.3	0.6
$^5/_8$	0.37	0.75
$^3/_4$	0.45	0.90

For SI: 1 inch = 25.4 mm.
a. Values specified in the table are for 1:3 sanded gypsum plaster.

TABLE 722.4.1(5)
REINFORCED MASONRY LINTELS

NOMINAL LINTEL WIDTH (inches)	MINIMUM LONGITUDINAL REINFORCEMENT COVER FOR FIRE RESISTANCE (inches)			
	1 hour	2 hours	3 hours	4 hours
6	$1^1/_2$	2	NP	NP
8	$1^1/_2$	$1^1/_2$	$1^3/_4$	3
10 or more	$1^1/_2$	$1^1/_2$	$1^1/_2$	$1^3/_4$

For SI: 1 inch = 25.4 mm.
NP = Not Permitted.

TABLE 722.4.1(6)
REINFORCED CLAY MASONRY COLUMNS

COLUMN SIZE	FIRE-RESISTANCE RATING (hours)			
	1	2	3	4
Minimum column dimension (inches)	8	10	12	14

For SI: 1 inch = 25.4 mm.

722.4.1.1 Equivalent thickness. The *fire-resistance ratings* of walls or partitions constructed of solid or hollow clay *masonry units* shall be determined from Table 722.4.1(1) or Table 722.4.1(2). The equivalent thickness of the clay *masonry unit* shall be determined by Equation 7-8 where using Table 722.4.1(1). The *fire-resistance rating* determined from Table 722.4.1(1) shall be permitted to be used in the calculated *fire-resistance rating* procedure in Section 722.4.2.

$T_e = V_n/LH$ **(Equation 7-8)**

where:

T_e = The equivalent thickness of the clay *masonry unit* (inches).

V_n = The net volume of the clay *masonry unit* (inch3).

L = The specified length of the clay *masonry unit* (inches).

H = The specified height of the clay *masonry unit* (inches).

722.4.1.1.1 Hollow clay units. The equivalent thickness, T_e, shall be the value obtained for hollow clay units as determined in accordance with Equation 7-8. The net volume, V_n, of the units shall be determined using the gross volume and percentage of void area determined in accordance with ASTM C67.

722.4.1.1.2 Solid grouted clay units. The equivalent thickness of solid grouted clay *masonry units* shall be taken as the actual thickness of the units.

722.4.1.1.3 Units with filled cores. The equivalent thickness of the hollow clay *masonry units* is the actual thickness of the unit where completely filled with loose-fill materials of: sand, pea gravel, crushed stone, or slag that meet ASTM C33 requirements; pumice, scoria, expanded shale, expanded clay, expanded slate, expanded slag, expanded fly ash, or cinders in compliance with ASTM C331; or perlite or vermiculite meeting the requirements of ASTM C549 and ASTM C516, respectively.

722.4.1.2 Plaster finishes. Where plaster is applied to the wall, the total *fire-resistance rating* shall be determined by the formula:

$R = (R_n^{0.59} + pl)^{1.7}$ **(Equation 7-9)**

where:

R = The *fire-resistance rating* of the assembly (hours).

R_n = The *fire-resistance rating* of the individual wall (hours).

pl = Coefficient for thickness of plaster.

TABLE 722.4.1(1)
FIRE-RESISTANCE PERIODS OF CLAY MASONRY WALLS

MATERIAL TYPE	MINIMUM REQUIRED EQUIVALENT THICKNESS FOR FIRE RESISTANCE[a, b, c] (inches)			
	1 hour	2 hours	3 hours	4 hours
Solid brick of clay or shale[d]	2.7	3.8	4.9	6.0
Hollow brick or tile of clay or shale, unfilled	2.3	3.4	4.3	5.0
Hollow brick or tile of clay or shale, grouted or filled with materials specified in Section 722.4.1.1.3	3.0	4.4	5.5	6.6

For SI: 1 inch = 25.4 mm.

a. Equivalent thickness as determined from Section 722.4.1.1.

b. Calculated fire resistance between the hourly increments specified shall be determined by linear interpolation.

c. Where combustible members are framed in the wall, the thickness of solid material between the end of each member and the opposite face of the wall, or between members set in from opposite sides, shall be not less than 93 percent of the thickness shown.

d. For units in which the net cross-sectional area of cored brick in any plane parallel to the surface containing the cores is not less than 75 percent of the gross cross-sectional area measured in the same plane.

TABLE 722.4.1(2)
FIRE-RESISTANCE RATINGS FOR BEARING STEEL FRAME BRICK VENEER WALLS OR PARTITIONS

WALL OR PARTITION ASSEMBLY	PLASTER SIDE EXPOSED (hours)	BRICK FACED SIDE EXPOSED (hours)
Outside facing of steel studs: $^1/_2$″ wood fiberboard sheathing next to studs, $^3/_4$″ airspace formed with $^3/_4$″ × $1^5/_8$″ wood strips placed over the fiberboard and secured to the studs; metal or wire lath nailed to such strips, $3^3/_4$″ brick *veneer* held in place by filling $^3/_4$″ airspace between the brick and lath with mortar. Inside facing of studs: $^3/_4$″ unsanded gypsum plaster on metal or wire lath attached to $^5/_{16}$″ wood strips secured to edges of the studs.	1.5	4
Outside facing of steel studs: 1″ insulation board sheathing attached to studs, 1″ airspace, and $3^3/_4$″ brick veneer attached to steel frame with metal ties every 5th course. Inside facing of studs: $^7/_8$″ sanded gypsum plaster (1:2 mix) applied on metal or wire lath attached directly to the studs.	1.5	4
Same as previous assembly except use $^7/_8$″ vermiculite-gypsum plaster or 1″ sanded gypsum plaster (1:2 mix) applied to metal or wire.	2	4
Outside facing of steel studs: $^1/_2$″ gypsum sheathing board, attached to studs, and $3^3/_4$″ brick veneer attached to steel frame with metal ties every 5th course. Inside facing of studs: $^1/_2$″ sanded gypsum plaster (1:2 mix) applied to $^1/_2$″ perforated gypsum lath securely attached to studs and having strips of metal lath 3 inches wide applied to all horizontal joints of gypsum lath.	2	4

For SI: 1 inch = 25.4 mm.

Values for $R_n^{0.59}$ for use in Equation 7-9 are given in Table 722.4.1(3). Coefficients for thickness of plaster shall be selected from Table 722.4.1(4) based on the actual thickness of plaster applied to the wall or partition and whether one or two sides of the wall are plastered.

722.4.1.3 Multiwythe walls with airspace. Where a continuous airspace separates multiple *wythes* of the wall or partition, the total *fire-resistance rating* shall be determined by the formula:

$$R = (R_1^{0.59} + R_2^{0.59} + ... + R_n^{0.59} + as)^{1.7} \quad \textbf{(Equation 7-10)}$$

where:

R = The fire-resistance rating of the assembly (hours).

R_1, R_2 and R_n = The fire-resistance rating of the individual *wythes* (hours).

as = Coefficient for continuous airspace.

Values for $R_n^{0.59}$ for use in Equation 7-10 are given in Table 722.4.1(3). The coefficient for each continuous airspace of $^1/_2$ inch to $3^1/_2$ inches (12.7 to 89 mm) separating two individual *wythes* shall be 0.3.

722.4.1.4 Nonsymmetrical assemblies. For a wall without finish on one side or having different types or thicknesses of finish on each side, the calculation procedures of this section shall be performed twice, assuming either side to be the fire-exposed side of the wall. The *fire resistance* of the wall shall not exceed the lower of the two values determined.

Exception: For *exterior walls* with a *fire separation distance* greater than 5 feet (1524 mm), the fire shall be assumed to occur on the interior side only.

722.4.2 Multiwythe walls. The *fire-resistance rating* for walls or partitions consisting of two or more dissimilar *wythes* shall be permitted to be determined by the formula:

$$R = (R_1^{0.59} + R_2^{0.59} + ... + R_n^{0.59})^{1.7} \quad \textbf{(Equation 7-11)}$$

where:

R = The *fire-resistance* rating of the assembly (hours).

R_1, R_2 and R_n = The *fire-resistance* rating of the individual *wythes* (hours).

Values for $R_n^{0.59}$ for use in Equation 7-11 are given in Table 722.4.1(3).

722.4.2.1 Multiwythe walls of different material. For walls that consist of two or more *wythes* of different materials (concrete or concrete *masonry units*) in combination with clay *masonry units*, the *fire-resistance rating* of the different materials shall be permitted to be determined from Table 722.2.1.1 for concrete; Table 722.3.2 for concrete *masonry units* or Table 722.4.1(1) or Table 722.4.1(2) for clay and tile masonry units.

722.4.3 Reinforced clay masonry lintels. *Fire-resistance ratings* for clay masonry lintels shall be determined based on the nominal width of the lintel and the minimum covering for the longitudinal reinforcement in accordance with Table 722.4.1(5).

722.4.4 Reinforced clay masonry columns. The *fire-resistance ratings* shall be determined based on the last plan dimension of the column in accordance with Table 722.4.1(6). The minimum cover for longitudinal reinforcement shall be 2 inches (51 mm).

722.5 Steel assemblies. The provisions of this section contain procedures by which the *fire-resistance ratings* of steel assemblies are established by calculations.

722.5.1 Structural steel columns. The *fire-resistance ratings* of structural steel columns shall be based on the size of the element and the type of protection provided in accordance with this section.

722.5.1.1 General. These procedures establish a basis for determining the fire resistance of column assemblies as a function of the thickness of fire-resistant material and, the weight, W, and heated perimeter, D, of structural steel columns. As used in these sections, W is the average weight of a structural steel column in pounds per linear foot. The heated perimeter, D, is the inside perimeter of the fire-resistant material in inches as illustrated in Figure 722.5.1(1).

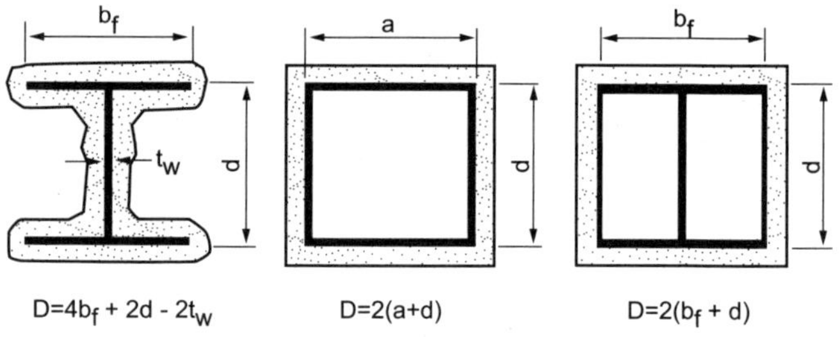

FIGURE 722.5.1(1)
DETERMINATION OF THE HEATED PERIMETER OF STRUCTURAL STEEL COLUMNS

722.5.1.1.1 Nonload-bearing protection. The application of these procedures shall be limited to column assemblies in which the fire-resistant material is not designed to carry any of the *load* acting on the column.

722.5.1.1.2 Embedments. In the absence of substantiating fire-endurance test results, ducts, conduit, piping, and similar mechanical, electrical, and plumbing installations shall not be embedded in any required fire-resistant materials.

722.5.1.1.3 Weight-to-perimeter ratio. Table 722.5.1(1) contains weight-to-heated-perimeter ratios (*W/D*) for both contour and box fire-resistant profiles, for the wide flange shapes most often used as columns. For different fire-resistant protection profiles or column cross sections, the weight-to-heated-perimeter ratios (*W/D*) shall be determined in accordance with the definitions given in this section.

TABLE 722.5.1(1)
W/D RATIOS FOR STEEL COLUMNS

STRUCTURAL SHAPE	CONTOUR PROFILE	BOX PROFILE
W14 × 233	2.55	3.65
× 211	2.32	3.35
× 193	2.14	3.09
× 176	1.96	2.85
× 159	1.78	2.60
× 145	1.64	2.39
× 132	1.56	2.25
× 120	1.42	2.06
× 109	1.29	1.88
× 99	1.18	1.72
× 90	1.08	1.58
× 82	1.23	1.68
× 74	1.12	1.53
× 68	1.04	1.41
× 61	0.928	1.28
× 53	0.915	1.21
× 48	0.835	1.10
× 43	0.752	0.99
W12 × 190	2.50	3.51
× 170	2.26	3.20
× 152	2.04	2.90
× 136	1.86	2.63
× 120	1.65	2.36
× 106	1.47	2.11
× 96	1.34	1.93
× 87	1.22	1.76
× 79	1.11	1.61
× 72	1.02	1.48
× 65	0.925	1.35
× 58	0.925	1.31
× 53	0.855	1.20
× 50	0.909	1.23
× 45	0.829	1.12
× 40	0.734	1.00
W10 × 112	1.81	2.57
× 100	1.64	2.33
× 88	1.45	2.08
× 77	1.28	1.85
× 68	1.15	1.66
× 60	1.01	1.48
× 54	0.922	1.34
× 49	0.84	1.23
× 45	0.888	1.24
× 39	0.78	1.09
× 33	0.661	0.93

(continued)

TABLE 722.5.1(1)—continued
W/D RATIOS FOR STEEL COLUMNS

STRUCTURAL SHAPE	CONTOUR PROFILE	BOX PROFILE
W8 × 67	1.37	1.94
× 58	1.20	1.71
× 48	1.00	1.44
× 40	0.849	1.23
× 35	0.749	1.08
× 31	0.665	0.97
× 28	0.688	0.96
× 24	0.591	0.83
× 21	0.577	0.77
× 18	0.499	0.67
W6 ×25	0.696	1.00
× 20	0.563	0.82
× 16	0.584	0.78
× 15	0.431	0.63
× 12	0.448	0.60
× 9	0.338	0.46
W5 ×19	0.644	0.93
× 16	0.55	0.80
W4 ×13	0.556	0.79

For SI: 1 pound per linear foot per inch = 0.059 kg/m/mm.

TABLE 722.5.1(2)
PROPERTIES OF CONCRETE

PROPERTY	NORMAL-WEIGHT CONCRETE	STRUCTURAL LIGHTWEIGHT CONCRETE
Thermal conductivity (k_c)	0.95 Btu/hr × ft × °F	0.35 Btu/hr × ft × °F
Specific heat (c_c)	0.20 Btu/lb °F	0.20 Btu/lb °F
Density (P_c)	145 lb/ft^3	110 lb/ft^3
Equilibrium (free) moisture content (m) by volume	4%	5%

For SI: 1 inch = 25.4 mm, 1 foot = 304.8 mm, 1 lb/ft^3 = 16.0185 kg/m^3, Btu/hr × ft × °F = 1.731 W/(m × K).

TABLE 722.5.1(3)
THERMAL CONDUCTIVITY OF CONCRETE OR CLAY MASONRY UNITS

DENSITY (d_m) OF UNITS (lb/ft^3)	THERMAL CONDUCTIVITY (*K*) OF UNITS (Btu/hr · ft · °F)
Concrete Masonry Units	
80	0.207
85	0.228
90	0.252
95	0.278
100	0.308
105	0.340
110	0.376
115	0.416
120	0.459
125	0.508
130	0.561
135	0.620
140	0.685
145	0.758
150	0.837
Clay Masonry Units	
120	1.25
130	2.25

For SI: 1 pound per cubic foot = 16.0185 kg/m^3, Btu/hr × ft × °F = 1.731 W/(m × K)

TABLE 722.5.1(4)
WEIGHT-TO-HEATED-PERIMETER RATIOS *(W/D)* FOR TYPICAL WIDE FLANGE BEAM AND GIRDER SHAPES

STRUCTURAL SHAPE	CONTOUR PROFILE	BOX PROFILE
W36 × 300	2.50	3.33
× 280	2.35	3.12
× 260	2.18	2.92
× 245	2.08	2.76
× 230	1.95	2.61
× 210	1.96	2.45
× 194	1.81	2.28
× 182	1.72	2.15
× 170	1.60	2.01
× 160	1.51	1.90
× 150	1.43	1.79
× 135	1.29	1.63
W33 × 241	2.13	2.86
× 221	1.97	2.64
× 201	1.79	2.42
× 152	1.53	1.94
× 141	1.43	1.80
× 130	1.32	1.67
× 118	1.21	1.53
W30 × 211	2.01	2.74
× 191	1.85	2.50
× 173	1.66	2.28
× 132	1.47	1.85
× 124	1.39	1.75
× 116	1.30	1.65
× 108	1.21	1.54
× 99	1.12	1.42
W27 × 178	1.87	2.55
× 161	1.70	2.33
× 146	1.55	2.12
× 114	1.39	1.76
× 102	1.24	1.59
× 94	1.15	1.47
× 84	1.03	1.33
W24 × 162	1.88	2.57
× 146	1.70	2.34
× 131	1.54	2.12
× 117	1.38	1.91
× 104	1.24	1.71
× 94	1.28	1.63
× 84	1.15	1.47
× 76	1.05	1.34

(continued)

TABLE 722.5.1(4)—continued
WEIGHT-TO-HEATED-PERIMETER RATIOS *(W/D)* FOR TYPICAL WIDE FLANGE BEAM AND GIRDER SHAPES

STRUCTURAL SHAPE	CONTOUR PROFILE	BOX PROFILE
W24 × 68	0.942	1.21
× 62	0.934	1.14
× 55	0.828	1.02
W21 × 147	1.87	2.60
× 132	1.68	2.35
× 122	1.57	2.19
× 111	1.43	2.01
× 101	1.30	1.84
× 93	1.40	1.80
× 83	1.26	1.62
× 73	1.11	1.44
× 68	1.04	1.35
W21 × 62	0.952	1.23
× 57	0.952	1.17
× 50	0.838	1.04
× 44	0.746	0.92
W18 × 119	1.72	2.42
× 106	1.55	2.18
× 97	1.42	2.01
× 86	1.27	1.80
× 76	1.13	1.60
× 71	1.22	1.59
× 65	1.13	1.47
× 60	1.04	1.36
× 55	0.963	1.26
× 50	0.88	1.15
× 46	0.878	1.09
× 40	0.768	0.96
× 35	0.672	0.85
W16 × 100	1.59	2.25
× 89	1.43	2.03
× 77	1.25	1.78
× 67	1.09	1.56
× 57	1.09	1.43
× 50	0.962	1.26
× 45	0.870	1.15
× 40	0.780	1.03
× 36	0.702	0.93
× 31	0.661	0.83
× 26	0.558	0.70
W14 × 132	1.89	3.00
× 120	1.71	2.75
× 109	1.57	2.52

(continued)

**TABLE 722.5.1(4)—continued
WEIGHT-TO-HEATED-PERIMETER RATIOS *(W/D)* FOR
TYPICAL WIDE FLANGE BEAM AND GIRDER SHAPES**

STRUCTURAL SHAPE	CONTOUR PROFILE	BOX PROFILE
W14 × 99	1.43	2.31
× 90	1.31	2.11
× 82	1.45	2.12
× 74	1.32	1.93
× 68	1.22	1.78
× 61	1.10	1.61
× 53	1.06	1.48
× 48	0.970	1.35
W14 × 43	0.874	1.22
× 38	0.809	1.09
× 34	0.725	0.98
× 30	0.644	0.87
× 26	0.628	0.79
× 22	0.534	0.68
W12 × 87	1.47	2.34
× 79	1.34	2.14
× 72	1.23	1.97
× 65	1.11	1.79
× 58	1.10	1.69
× 53	1.02	1.55
× 50	1.06	1.54
× 45	0.974	1.40
× 40	0.860	1.25
× 35	0.810	1.11
× 30	0.699	0.96
× 26	0.612	0.84
× 22	0.623	0.77
× 19	0.540	0.67
× 16	0.457	0.57
× 14	0.405	0.50
W10 × 112	2.17	3.38
× 100	1.97	3.07
× 88	1.74	2.75
× 77	1.54	2.45
× 68	1.38	2.20
× 60	1.22	1.97
× 54	1.11	1.79
× 49	1.01	1.64
× 45	1.06	1.59
× 39	0.94	1.40
× 33	0.77	1.20

(continued)

**TABLE 722.5.1(4)—continued
WEIGHT-TO-HEATED-PERIMETER RATIOS *(W/D)* FOR
TYPICAL WIDE FLANGE BEAM AND GIRDER SHAPES**

STRUCTURAL SHAPE	CONTOUR PROFILE	BOX PROFILE
W10 × 30	0.806	1.12
× 26	0.708	0.98
× 22	0.606	0.84
× 19	0.607	0.78
× 17	0.543	0.70
× 15	0.484	0.63
× 12	0.392	0.51
W8 × 67	1.65	2.55
× 58	1.44	2.26
× 48	1.21	1.91
× 40	1.03	1.63
× 35	0.907	1.44
× 31	0.803	1.29
× 28	0.819	1.24
× 24	0.704	1.07
× 21	0.675	0.96
× 18	0.583	0.84
× 15	0.551	0.74
× 13	0.483	0.65
× 10	0.375	0.51
W6 × 25	0.839	1.33
× 20	0.678	1.09
× 16	0.684	0.96
× 15	0.521	0.83
× 12	0.526	0.75
× 9	0.398	0.57
W5 × 19	0.776	1.24
× 16	0.664	1.07
W4 × 13	0.670	1.05

For SI: 1 pound per linear foot per inch = 0.059 kg/m/mm.

**TABLE 722.5.1(5)
FIRE RESISTANCE OF CONCRETE
MASONRY PROTECTED STEEL COLUMNS**

COLUMN SIZE	CONCRETE MASONRY DENSITY POUNDS PER CUBIC FOOT	MINIMUM REQUIRED EQUIVALENT THICKNESS FOR FIRE-RESISTANCE RATING OF CONCRETE MASONRY PROTECTION ASSEMBLY, T_e(inches)			
		1 hour	2 hours	3 hours	4 hours
W14 × 82	80	0.74	1.61	2.36	3.04
	100	0.89	1.85	2.67	3.40
	110	0.96	1.97	2.81	3.57
	120	1.03	2.08	2.95	3.73
W14 × 68	80	0.83	1.70	2.45	3.13
	100	0.99	1.95	2.76	3.49
	110	1.06	2.06	2.91	3.66
	120	1.14	2.18	3.05	3.82
W14 × 53	80	0.91	1.81	2.58	3.27
	100	1.07	2.05	2.88	3.62
	110	1.15	2.17	3.02	3.78
	120	1.22	2.28	3.16	3.94
W14 × 43	80	1.01	1.93	2.71	3.41
	100	1.17	2.17	3.00	3.74
	110	1.25	2.28	3.14	3.90
	120	1.32	2.38	3.27	4.05
W12 × 72	80	0.81	1.66	2.41	3.09
	100	0.91	1.88	2.70	3.43
	110	0.99	1.99	2.84	3.60
	120	1.06	2.10	2.98	3.76
W12 × 58	80	0.88	1.76	2.52	3.21
	100	1.04	2.01	2.83	3.56
	110	1.11	2.12	2.97	3.73
	120	1.19	2.23	3.11	3.89
W12 × 50	80	0.91	1.81	2.58	3.27
	100	1.07	2.05	2.88	3.62
	110	1.15	2.17	3.02	3.78
	120	1.22	2.28	3.16	3.94
W12 × 40	80	1.01	1.94	2.72	3.41
	100	1.17	2.17	3.01	3.75
	110	1.25	2.28	3.14	3.90
	120	1.32	2.39	3.27	4.06
W10 × 68	80	0.72	1.58	2.33	3.01
	100	0.87	1.83	2.65	3.38
	110	0.94	1.95	2.79	3.55
	120	1.01	2.06	2.94	3.72
W10 × 54	80	0.88	1.76	2.53	3.21
	100	1.04	2.01	2.83	3.57
	110	1.11	2.12	2.98	3.73
	120	1.19	2.24	3.12	3.90

(continued)

**TABLE 722.5.1(5)—continued
FIRE RESISTANCE OF CONCRETE
MASONRY PROTECTED STEEL COLUMNS**

COLUMN SIZE	CONCRETE MASONRY DENSITY POUNDS PER CUBIC FOOT	MINIMUM REQUIRED EQUIVALENT THICKNESS FOR FIRE-RESISTANCE RATING OF CONCRETE MASONRY PROTECTION ASSEMBLY, T_e(inches)			
		1 hour	2 hours	3 hours	4 hours
W10 × 45	80	0.92	1.83	2.60	3.30
	100	1.08	2.07	2.90	3.64
	110	1.16	2.18	3.04	3.80
	120	1.23	2.29	3.18	3.96
W10 × 33	80	1.06	2.00	2.79	3.49
	100	1.22	2.23	3.07	3.80
	110	1.30	2.34	3.20	3.96
	120	1.37	2.44	3.33	4.12
W8 × 40	80	0.94	1.85	2.63	3.33
	100	1.10	2.10	2.93	3.67
	110	1.18	2.21	3.07	3.83
	120	1.25	2.32	3.20	3.99
W8 × 31	80	1.06	2.00	2.78	3.49
	100	1.22	2.23	3.07	3.81
	110	1.29	2.33	3.20	3.97
	120	1.36	2.44	3.33	4.12
W8 × 24	80	1.14	2.09	2.89	3.59
	100	1.29	2.31	3.16	3.90
	110	1.36	2.42	3.28	4.05
	120	1.43	2.52	3.41	4.20
W8 × 18	80	1.22	2.20	3.01	3.72
	100	1.36	2.40	3.25	4.01
	110	1.42	2.50	3.37	4.14
	120	1.48	2.59	3.49	4.28
8 × 8 × $^1/_2$ wall thickness	80	0.77	1.66	2.44	3.13
	100	0.92	1.91	2.75	3.49
	110	1.00	2.02	2.89	3.66
	120	1.07	2.14	3.03	3.82
8 × 8 × $^3/_8$ wall thickness	80	0.91	1.84	2.63	3.33
	100	1.07	2.08	2.92	3.67
	110	1.14	2.19	3.06	3.83
	120	1.21	2.29	3.19	3.98
8 × 8 × $^1/_4$ wall thickness	80	1.10	2.06	2.86	3.57
	100	1.25	2.28	3.13	3.87
	110	1.32	2.38	3.25	4.02
	120	1.39	2.48	3.38	4.17
6 × 6 × $^1/_2$ wall thickness	80	0.82	1.75	2.54	3.25
	100	0.98	1.99	2.84	3.59
	110	1.05	2.10	2.98	3.75
	120	1.12	2.21	3.11	3.91

(continued)

**TABLE 722.5.1(5)—continued
FIRE RESISTANCE OF CONCRETE
MASONRY PROTECTED STEEL COLUMNS**

COLUMN SIZE	CONCRETE MASONRY DENSITY POUNDS PER CUBIC FOOT	MINIMUM REQUIRED EQUIVALENT THICKNESS FOR FIRE-RESISTANCE RATING OF CONCRETE MASONRY PROTECTION ASSEMBLY, T_e(inches)			
		1 hour	2 hours	3 hours	4 hours
6 × 6 × $^3/_8$ wall thickness	80	0.96	1.91	2.71	3.42
	100	1.12	2.14	3.00	3.75
	110	1.19	2.25	3.13	3.90
	120	1.26	2.35	3.26	4.05
6 × 6 × $^1/_4$ wall thickness	80	1.14	2.11	2.92	3.63
	100	1.29	2.32	3.18	3.93
	110	1.36	2.43	3.30	4.08
	120	1.42	2.52	3.43	4.22
4 × 4 × $^1/_2$ wall thickness	80	0.93	1.90	2.71	3.43
	100	1.08	2.13	2.99	3.76
	110	1.16	2.24	3.13	3.91
	120	1.22	2.34	3.26	4.06
4 × 4 × $^3/_8$ wall thickness	80	1.05	2.03	2.84	3.57
	100	1.20	2.25	3.11	3.88
	110	1.27	2.35	3.24	4.02
	120	1.34	2.45	3.37	4.17
4 × 4 × $^1/_4$ wall thickness	80	1.21	2.20	3.01	3.73
	100	1.35	2.40	3.26	4.02
	110	1.41	2.50	3.38	4.16
	120	1.48	2.59	3.50	4.30
6 double extra strong 0.864 wall thickness	80	0.59	1.46	2.23	2.92
	100	0.73	1.71	2.54	3.29
	110	0.80	1.82	2.69	3.47
	120	0.86	1.93	2.83	3.63
6 extra strong 0.432 wall thickness	80	0.94	1.90	2.70	3.42
	100	1.10	2.13	2.98	3.74
	110	1.17	2.22	3.11	3.89
	120	1.24	2.34	3.24	4.04
6 standard 0.280 wall thickness	80	1.14	2.12	2.93	3.64
	100	1.29	2.33	3.19	3.94
	110	1.36	2.43	3.31	4.08
	120	1.42	2.53	3.43	4.22
5 double extra strong 0.750 wall thickness	80	0.70	1.61	2.40	3.12
	100	0.85	1.86	2.71	3.47
	110	0.91	1.97	2.85	3.63
	120	0.98	2.02	2.99	3.79
5 extra strong 0.375 wall thickness	80	1.04	2.01	2.83	3.54
	100	1.19	2.23	3.09	3.85
	110	1.26	2.34	3.22	4.00
	20	1.32	2.44	3.34	4.14

(continued)

**TABLE 722.5.1(5)—continued
FIRE RESISTANCE OF CONCRETE
MASONRY PROTECTED STEEL COLUMNS**

COLUMN SIZE	CONCRETE MASONRY DENSITY POUNDS PER CUBIC FOOT	MINIMUM REQUIRED EQUIVALENT THICKNESS FOR FIRE-RESISTANCE RATING OF CONCRETE MASONRY PROTECTION ASSEMBLY, T_e(inches)			
		1 hour	2 hours	3 hours	4 hours
5 standard 0.258 wall thickness	80	1.20	2.19	3.00	3.72
	100	1.34	2.39	3.25	4.00
	110	1.41	2.49	3.37	4.14
	120	1.47	2.58	3.49	4.28
4 double extra strong 0.674 wall thickness	80	0.80	1.75	2.56	3.28
	100	0.95	1.99	2.85	3.62
	110	1.02	2.10	2.99	3.78
	120	1.09	2.20	3.12	3.93
4 extra strong 0.337 wall thickness	80	1.12	2.11	2.93	3.65
	100	1.26	2.32	3.19	3.95
	110	1.33	2.42	3.31	4.09
	120	1.40	2.52	3.43	4.23
4 standard 0.237 wall thickness	80	1.26	2.25	3.07	3.79
	100	1.40	2.45	3.31	4.07
	110	1.46	2.55	3.43	4.21
	120	1.53	2.64	3.54	4.34

For SI:1 inch = 25.4 mm, 1 pound per cubic feet = 16.02 kg/m^3.

Note: Tabulated values assume 1-inch air gap between masonry and steel section.

TABLE 722.5.1(6)
FIRE RESISTANCE OF CLAY MASONRY PROTECTED STEEL COLUMNS

COLUMN SIZE	CLAY MASONRY DENSITY, POUNDS PER CUBIC FOOT	MINIMUM REQUIRED EQUIVALENT THICKNESS FOR FIRE-RESISTANCE RATING OF CLAY MASONRY PROTECTION ASSEMBLY, T_e (inches)				COLUMN SIZE	CLAY MASONRY DENSITY, POUNDS PER CUBIC FOOT	MINIMUM REQUIRED EQUIVALENT THICKNESS FOR FIRE-RESISTANCE RATING OF CLAY MASONRY PROTECTION ASSEMBLY, T_e (inches)			
		1 hour	2 hours	3 hours	4 hours			1 hour	2 hours	3 hours	4 hours
W14 × 82	120	1.23	2.42	3.41	4.29	W10 × 68	120	1.27	2.46	3.26	4.35
	130	1.40	2.70	3.78	4.74		130	1.44	2.75	3.83	4.80
W14 × 68	120	1.34	2.54	3.54	4.43	W10 × 54	120	1.40	2.61	3.62	4.51
	130	1.51	2.82	3.91	4.87		130	1.58	2.89	3.98	4.95
W14 × 53	120	1.43	2.65	3.65	4.54	W10 × 45	120	1.44	2.66	3.67	4.57
	130	1.61	2.93	4.02	4.98		130	1.62	2.95	4.04	5.01
W14 × 43	120	1.54	2.76	3.77	4.66	W10 × 33	120	1.59	2.82	3.84	4.73
	130	1.72	3.04	4.13	5.09		130	1.77	3.10	4.20	5.13
W12 × 72	120	1.32	2.52	3.51	4.40	W8 × 40	120	1.47	2.70	3.71	4.61
	130	1.50	2.80	3.88	4.84		130	1.65	2.98	4.08	5.04
W12 × 58	120	1.40	2.61	3.61	4.50	W8 × 31	120	1.59	2.82	3.84	4.73
	130	1.57	2.89	3.98	4.94		130	1.77	3.10	4.20	5.17
W12 × 50	120	1.43	2.65	3.66	4.55	W8 × 24	120	1.66	2.90	3.92	4.82
	130	1.61	2.93	4.02	4.99		130	1.84	3.18	4.28	5.25
W12 × 40	120	1.54	2.77	3.78	4.67	W8 × 18	120	1.75	3.00	4.01	4.91
	130	1.72	3.05	4.14	5.10		130	1.93	3.27	4.37	5.34
STEEL TUBING						STEEL PIPE					
NOMINAL TUBE SIZE (inches)	CLAY MASONRY DENSITY, POUNDS PER CUBIC FOOT	MINIMUM REQUIRED EQUIVALENT THICKNESS FOR FIRE-RESISTANCE RATING OF CLAY MASONRY PROTECTION ASSEMBLY, T_e (inches)				NOMINAL PIPE SIZE (inches)	CLAY MASONRY DENSITY, POUNDS PER CUBIC FOOT	MINIMUM REQUIRED EQUIVALENT THICKNESS FOR FIRE-RESISTANCE RATING OF CLAY MASONRY PROTECTION ASSEMBLY, T_e (inches)			
		1 hour	2 hours	3 hours	4 hours			1 hour	2 hours	3 hours	4 hours
4 × 4 × $^1/_2$ wall thickness	120	1.44	2.72	3.76	4.68	4 double extra strong 0.674 wall thickness	120	1.26	2.55	3.60	4.52
	130	1.62	3.00	4.12	5.11		130	1.42	2.82	3.96	4.95
4 × 4 × $^3/_8$ wall thickness	120	1.56	2.84	3.88	4.78	4 extra strong 0.337 wall thickness	120	1.60	2.89	3.92	4.83
	130	1.74	3.12	4.23	5.21		130	1.77	3.16	4.28	5.25
4 × 4 × $^1/_4$ wall thickness	120	1.72	2.99	4.02	4.92	4 standard 0.237 wall thickness	120	1.74	3.02	4.05	4.95
	130	1.89	3.26	4.37	5.34		130	1.92	3.29	4.40	5.37
6 × 6 × $^1/_2$ wall thickness	120	1.33	2.58	3.62	4.52	5 double extra strong 0.750 wall thickness	120	1.17	2.44	3.48	4.40
	130	1.50	2.86	3.98	4.96		130	1.33	2.72	3.84	4.83
6 × 6 × $^3/_8$ wall thickness	120	1.48	2.74	3.76	4.67	5 extra strong 0.375 wall thickness	120	1.55	2.82	3.85	4.76
	130	1.65	3.01	4.13	5.10		130	1.72	3.09	4.21	5.18
6 × 6 × $^1/_4$ wall thickness	120	1.66	2.91	3.94	4.84	5 standard 0.258 wall thickness	120	1.71	2.97	4.00	4.90
	130	1.83	3.19	4.30	5.27		130	1.88	3.24	4.35	5.32
8 × 8 × $^1/_2$ wall thickness	120	1.27	2.50	3.52	4.42	6 double extra strong 0.864 wall thickness	120	1.04	2.28	3.32	4.23
	130	1.44	2.78	3.89	4.86		130	1.19	2.60	3.68	4.67
8 × 8 × $^3/_8$ wall thickness	120	1.43	2.67	3.69	4.59	6 extra strong 0.432 wall thickness	120	1.45	2.71	3.75	4.65
	130	1.60	2.95	4.05	5.02		130	1.62	2.99	4.10	5.08
8 × 8 × $^1/_4$ wall thickness	120	1.62	2.87	3.89	4.78	6 standard 0.280 wall thickness	120	1.65	2.91	3.94	4.84
	130	1.79	3.14	4.24	5.21		130	1.82	3.19	4.30	5.27

For SI: 1 inch = 25.4 mm, 1 pound per cubic foot = 16.02 kg/m^3.

TABLE 722.5.1(7)
MINIMUM COVER (inch) FOR STEEL COLUMNS ENCASED IN NORMAL-WEIGHT CONCRETE[a]
[FIGURE 722.5.1(6)(c)]

<table>
<tr><th rowspan="2">STRUCTURAL SHAPE</th><th colspan="5">FIRE-RESISTANCE RATING (hours)</th></tr>
<tr><th>1</th><th>1 1/2</th><th>2</th><th>3</th><th>4</th></tr>
<tr><td>W14 × 233</td><td rowspan="7">1</td><td rowspan="5">1</td><td rowspan="3">1</td><td rowspan="2">1 1/2</td><td>2</td></tr>
<tr><td>× 176</td><td rowspan="3">2 1/2</td></tr>
<tr><td>× 132</td><td rowspan="3">2</td></tr>
<tr><td>× 90</td><td rowspan="4">1 1/2</td></tr>
<tr><td>× 61</td><td rowspan="3">3</td></tr>
<tr><td>× 48</td><td rowspan="2">1 1/2</td><td rowspan="2">2 1/2</td></tr>
<tr><td>× 43</td></tr>
<tr><td>W12 × 152</td><td rowspan="5">1</td><td rowspan="2">1</td><td rowspan="2">1</td><td rowspan="3">2</td><td rowspan="2">2 1/2</td></tr>
<tr><td>× 96</td></tr>
<tr><td>× 65</td><td rowspan="3">1 1/2</td><td rowspan="3">1 1/2</td><td rowspan="3">3</td></tr>
<tr><td>× 50</td><td rowspan="2">2 1/2</td></tr>
<tr><td>× 40</td></tr>
<tr><td>W10 × 88</td><td>1</td><td rowspan="5">1 1/2</td><td rowspan="4">1 1/2</td><td>2</td><td rowspan="3">3</td></tr>
<tr><td>× 49</td><td rowspan="4">1</td><td rowspan="4">2 1/2</td></tr>
<tr><td>× 45</td></tr>
<tr><td>× 39</td><td rowspan="2">3 1/2</td></tr>
<tr><td>× 33</td><td>2</td></tr>
<tr><td>W8 × 67</td><td rowspan="6">1</td><td rowspan="2">1</td><td rowspan="3">1 1/2</td><td rowspan="3">2 1/2</td><td rowspan="2">3</td></tr>
<tr><td>× 58</td></tr>
<tr><td>× 48</td><td rowspan="4">1 1/2</td><td rowspan="3">3 1/2</td></tr>
<tr><td>× 31</td><td rowspan="3">2</td><td rowspan="3">3</td></tr>
<tr><td>× 21</td></tr>
<tr><td>× 18</td><td>4</td></tr>
<tr><td>W6 × 25</td><td rowspan="3">1</td><td>1 1/2</td><td>2</td><td rowspan="2">3</td><td>3 1/2</td></tr>
<tr><td>× 20</td><td rowspan="4">2</td><td rowspan="4">2 1/2</td><td rowspan="4">4</td></tr>
<tr><td>× 16</td><td rowspan="3">3 1/2</td></tr>
<tr><td>× 15</td><td rowspan="2">1 1/2</td></tr>
<tr><td>× 9</td></tr>
</table>

For SI: 1 inch = 25.4 mm.

a. The tabulated thicknesses are based on the assumed properties of normal-weight concrete given in Table 722.5.1(2).

TABLE 722.5.1(8)
MINIMUM COVER (inch) FOR STEEL COLUMNS ENCASED IN STRUCTURAL LIGHTWEIGHT CONCRETE[a]
[FIGURE 722.5.1(6)(c)]

<table>
<tr><th rowspan="2">STRUCTURAL SHAPE</th><th colspan="5">FIRE-RESISTANCE RATING (HOURS)</th></tr>
<tr><th>1</th><th>1 1/2</th><th>2</th><th>3</th><th>4</th></tr>
<tr><td>W14 × 233</td><td rowspan="5">1</td><td rowspan="5">1</td><td rowspan="4">1</td><td>1</td><td rowspan="2">1 1/2</td></tr>
<tr><td>× 193</td><td rowspan="3">1 1/2</td></tr>
<tr><td>× 74</td><td>2</td></tr>
<tr><td>× 61</td><td rowspan="2">2 1/2</td></tr>
<tr><td>× 43</td><td>1 1/2</td><td>2</td></tr>
<tr><td>W12 × 65</td><td rowspan="3">1</td><td rowspan="3">1</td><td rowspan="2">1</td><td>1 1/2</td><td>2</td></tr>
<tr><td>× 53</td><td rowspan="2">2</td><td rowspan="2">2 1/2</td></tr>
<tr><td>× 40</td><td>1 1/2</td></tr>
<tr><td>W10 × 112</td><td rowspan="4">1</td><td rowspan="4">1</td><td rowspan="3">1</td><td rowspan="2">1 1/2</td><td rowspan="2">2</td></tr>
<tr><td>× 88</td></tr>
<tr><td>× 60</td><td rowspan="2">2</td><td rowspan="2">2 1/2</td></tr>
<tr><td>× 33</td><td>1 1/2</td></tr>
<tr><td>W8 × 35</td><td rowspan="4">1</td><td rowspan="3">1</td><td rowspan="4">1 1/2</td><td rowspan="2">2</td><td>2 1/2</td></tr>
<tr><td>× 28</td><td rowspan="3">3</td></tr>
<tr><td>× 24</td><td rowspan="2">2 1/2</td></tr>
<tr><td>× 18</td><td>1 1/2</td></tr>
</table>

For SI: 1 inch = 25.4 mm.

a. The tabulated thicknesses are based on the assumed properties of structural lightweight concrete given in Table 722.5.1(2).

TABLE 722.5.1(9)
MINIMUM COVER (inch) FOR STEEL COLUMNS IN NORMAL-WEIGHT PRECAST COVERS[a]
[FIGURE 722.5.1 (6)(a)]

<table>
<tr><th rowspan="2">STRUCTURAL SHAPE</th><th colspan="5">FIRE-RESISTANCE RATING (hours)</th></tr>
<tr><th>1</th><th>1½</th><th>2</th><th>3</th><th>4</th></tr>
<tr><td>W14 × 233</td><td rowspan="8">1½</td><td rowspan="4">1½</td><td rowspan="2">1½</td><td rowspan="3">2½</td><td>3</td></tr>
<tr><td>× 211</td><td rowspan="4">3½</td></tr>
<tr><td>× 176</td><td rowspan="2">2</td></tr>
<tr><td>× 145</td><td rowspan="3">3</td></tr>
<tr><td>× 109</td><td rowspan="4">2</td><td rowspan="4">2½</td></tr>
<tr><td>× 99</td><td rowspan="2">4</td></tr>
<tr><td>× 61</td><td rowspan="2">3½</td></tr>
<tr><td>× 43</td><td>4½</td></tr>
<tr><td>W12 × 190</td><td rowspan="7">1½</td><td rowspan="4">1½</td><td>1½</td><td rowspan="2">2½</td><td rowspan="2">3½</td></tr>
<tr><td>× 152</td><td rowspan="3">2</td></tr>
<tr><td>× 120</td><td rowspan="2">3</td><td rowspan="3">4</td></tr>
<tr><td>× 96</td></tr>
<tr><td>× 87</td><td rowspan="3">2</td><td rowspan="3">2½</td><td rowspan="3">3½</td></tr>
<tr><td>× 58</td><td rowspan="2">4½</td></tr>
<tr><td>× 40</td></tr>
<tr><td>W10 × 112</td><td rowspan="5">1½</td><td rowspan="2">1½</td><td rowspan="2">2</td><td rowspan="3">3</td><td>3½</td></tr>
<tr><td>× 88</td><td rowspan="3">4</td></tr>
<tr><td>× 77</td><td rowspan="3">2</td><td rowspan="3">2½</td></tr>
<tr><td>× 54</td><td rowspan="2">3½</td></tr>
<tr><td>× 33</td><td>4½</td></tr>
<tr><td>W8 × 67</td><td rowspan="6">1½</td><td>1½</td><td>2</td><td>3</td><td rowspan="3">4</td></tr>
<tr><td>× 58</td><td rowspan="3">2</td><td rowspan="3">2½</td><td rowspan="4">3½</td></tr>
<tr><td>× 48</td></tr>
<tr><td>× 28</td><td rowspan="3">4½</td></tr>
<tr><td>× 21</td><td rowspan="2">2½</td><td rowspan="2">3</td></tr>
<tr><td>× 18</td><td>4</td></tr>
<tr><td>W6 × 25</td><td rowspan="3">1½</td><td>2</td><td>2½</td><td rowspan="2">3½</td><td rowspan="4">4½</td></tr>
<tr><td>× 20</td><td rowspan="4">2½</td><td rowspan="4">3</td></tr>
<tr><td>× 16</td><td rowspan="3">4</td></tr>
<tr><td>× 12</td><td rowspan="2">2</td></tr>
<tr><td>× 9</td><td>5</td></tr>
</table>

For SI: 1 inch = 25.4 mm.

a. The tabulated thicknesses are based on the assumed properties of normal-weight concrete given in Table 722.5.1(2).

TABLE 722.5.1(10)
MINIMUM COVER (inch) FOR STEEL COLUMNS IN STRUCTURAL LIGHTWEIGHT PRECAST COVERS[a]
[FIGURE 722.5.1 (6)(a)]

<table>
<tr><th rowspan="2">STRUCTURAL SHAPE</th><th colspan="5">FIRE-RESISTANCE RATING (hours)</th></tr>
<tr><th>1</th><th>1½</th><th>2</th><th>3</th><th>4</th></tr>
<tr><td>W14 × 233</td><td rowspan="8">1½</td><td rowspan="8">1½</td><td rowspan="5">1½</td><td rowspan="2">2</td><td>2½</td></tr>
<tr><td>× 176</td><td rowspan="5">3</td></tr>
<tr><td>× 145</td><td rowspan="5">2½</td></tr>
<tr><td>× 132</td></tr>
<tr><td>× 109</td></tr>
<tr><td>× 99</td><td rowspan="3">2</td></tr>
<tr><td>× 68</td><td rowspan="2">3½</td></tr>
<tr><td>× 43</td><td>3</td></tr>
<tr><td>W12 × 190</td><td rowspan="8">1½</td><td rowspan="8">1½</td><td rowspan="6">1½</td><td rowspan="3">2</td><td rowspan="2">2½</td></tr>
<tr><td>× 152</td></tr>
<tr><td>× 136</td><td rowspan="2">3</td></tr>
<tr><td>× 106</td><td rowspan="4">2½</td></tr>
<tr><td>× 96</td><td rowspan="4">3½</td></tr>
<tr><td>× 87</td></tr>
<tr><td>× 65</td><td rowspan="2">2</td></tr>
<tr><td>× 40</td><td>3</td></tr>
<tr><td>W10 × 112</td><td rowspan="7">1½</td><td rowspan="6">1½</td><td rowspan="3">1½</td><td>2</td><td rowspan="3">3</td></tr>
<tr><td>× 100</td><td rowspan="4">2½</td></tr>
<tr><td>× 88</td></tr>
<tr><td>× 77</td><td rowspan="4">2</td><td rowspan="4">3½</td></tr>
<tr><td>× 60</td></tr>
<tr><td>× 39</td><td rowspan="2">3</td></tr>
<tr><td>× 33</td><td>2</td></tr>
<tr><td>W8 × 67</td><td rowspan="5">1½</td><td rowspan="3">1½</td><td>1½</td><td>2½</td><td>3</td></tr>
<tr><td>× 48</td><td rowspan="3">2</td><td rowspan="4">3</td><td rowspan="3">3½</td></tr>
<tr><td>× 35</td></tr>
<tr><td>× 28</td><td rowspan="2">2</td></tr>
<tr><td>× 18</td><td>2½</td><td>4</td></tr>
<tr><td>W6 × 25</td><td rowspan="3">1½</td><td rowspan="3">2</td><td>2</td><td rowspan="2">3</td><td>3½</td></tr>
<tr><td>× 15</td><td rowspan="2">2½</td><td rowspan="2">4</td></tr>
<tr><td>× 9</td><td>3½</td></tr>
</table>

For SI: 1 inch = 25.4 mm.

a. The tabulated thicknesses are based on the assumed properties of structural lightweight concrete given in Table 722.5.1(2).

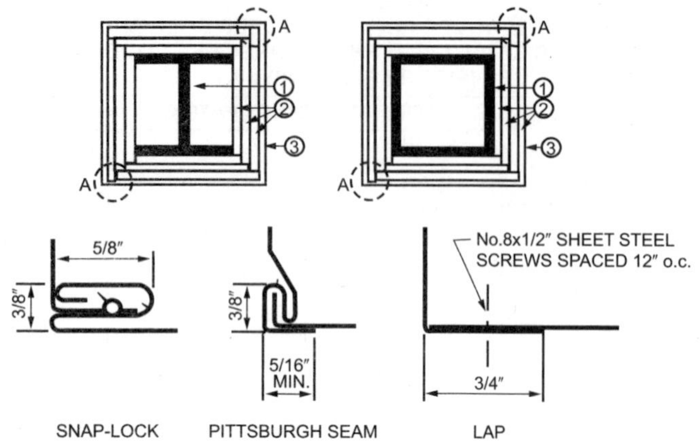

CORNER JOINT DETAILS (A)

For SI:1 inch = 25.4 mm, 1 foot = 305 mm.

1. Structural steel column, either wide flange or tubular shapes.
2. Type X gypsum board or gypsum panel products in accordance with ASTM C1177, C1178, C1278, C1396 or C1658. The total thickness of gypsum board or gypsum panel products calculated as *h* in Section 722.5.1.2 shall be applied vertically to an individual column using one of the following methods:
   1. As a single layer without horizontal joints.
   2. As multiple layers with horizontal joints not permitted in any layer.
   3. As multiple layers with horizontal joints staggered not less than 12 inches vertically between layers and not less than 8 feet vertically in any single layer. The total required thickness of gypsum board or gypsum panel products shall be determined on the basis of the specified fire-resistance rating and the weight-to-heated-perimeter ratio (W/D) of the column. For fire-resistance ratings of 2 hours or less, one of the required layers of gypsum board or gypsum panel product may be applied to the exterior of the sheet steel column covers with 1-inch long Type S screws spaced 1 inch from the wallboard edge and 8 inches on center. For such installations, 0.0149-inch minimum thickness galvanized steel corner beads with $1^1/_2$-inch legs shall be attached to the wallboard with Type S screws spaced 12 inches on center.
3. For fire-resistance ratings of 3 hours or less, the column covers shall be fabricated from 0.0239-inch minimum thickness galvanized or stainless steel. For 4-hour fire-resistance ratings, the column covers shall be fabricated from 0.0239-inch minimum thickness stainless steel. The column covers shall be erected with the Snap Lock or Pittsburgh joint details. For fire-resistance ratings of 2 hours or less, column covers fabricated from 0.0269-inch minimum thickness galvanized or stainless steel shall be permitted to be erected with lap joints. The lap joints shall be permitted to be located anywhere around the perimeter of the column cover. The lap joints shall be secured with $^1/_2$-inch-long No. 8 sheet metal screws spaced 12 inches on center. The column covers shall be provided with a minimum expansion clearance of $^1/_8$ inch per linear foot between the ends of the cover and any restraining construction.

FIGURE 722.5.1(2)
GYPSUM-PROTECTED STRUCTURAL STEEL COLUMNS WITH SHEET STEEL COLUMN COVERS

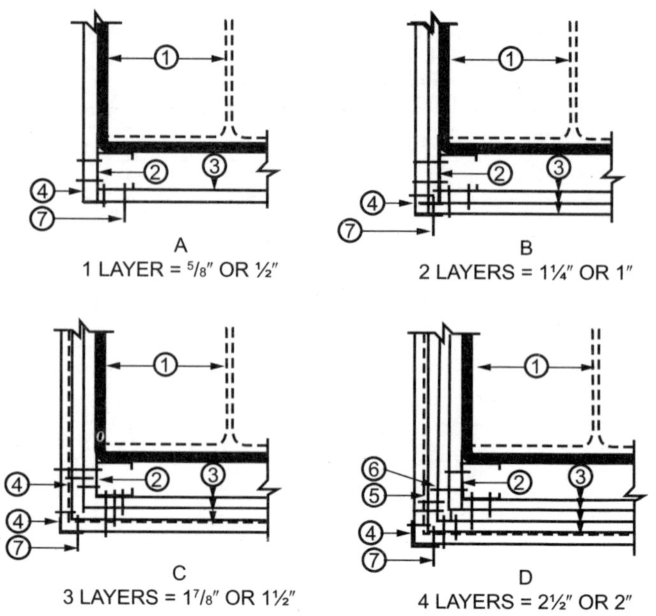

For SI:1 inch = 25.4 mm, 1 foot = -305 mm.

1. Structural steel column, either wide flange or tubular shapes.
2. $1^5/_8$-inch deep studs fabricated from 0.0179-inch minimum thickness galvanized steel with $1^5/_{16}$ or $1^7/_{16}$-inch legs. The length of the steel studs shall be $^1/_2$ inch less than the height of the assembly.
3. Type X gypsum board or gypsum panel products in accordance with ASTM C1177, C1178, C1278, C1396 or C1658. The total thickness of gypsum board or gypsum panel products calculated as *h* in Section 722.5.1.2 shall be applied vertically to an individual column using one of the following methods:
   1. As a single layer without horizontal joints.
   2. As multiple layers with horizontal joints not permitted in any layer.
   3. As multiple layers with horizontal joints staggered not less than 12 inches vertically between layers and not less than 8 feet vertically in any single layer. The total required thickness of gypsum board or gypsum panel products shall be determined on the basis of the specified fire-resistance rating and the weight-to-heated-perimeter ratio (W/D) of the column.
4. Galvanized 0.0149-inch minimum thickness steel corner beads with $1^1/_2$-inch legs attached to the gypsum board or gypsum panel products with 1-inch-long Type S screws spaced 12 inches on center.
5. No. 18 SWG steel tie wires spaced 24 inches on center.
6. Sheet metal angles with 2-inch legs fabricated from 0.0221-inch minimum thickness galvanized steel.
7. Type S screws, 1 inch long, shall be used for attaching the first layer of gypsum board or gypsum panel product to the steel studs and the third layer to the sheet metal angles at 24 inches on center. Type S screws $1^3/_4$ inches long shall be used for attaching the second layer of gypsum board or gypsum panel product to the steel studs and the fourth layer to the sheet metal angles at 12 inches on center. Type S screws $2^1/_4$ inches long shall be used for attaching the third layer of gypsum board or gypsum panel product to the steel studs at 12 inches on center.

FIGURE 722.5.1(3)
GYPSUM-PROTECTED STRUCTURAL STEEL COLUMNS WITH STEEL STUD/SCREW ATTACHMENT SYSTEM

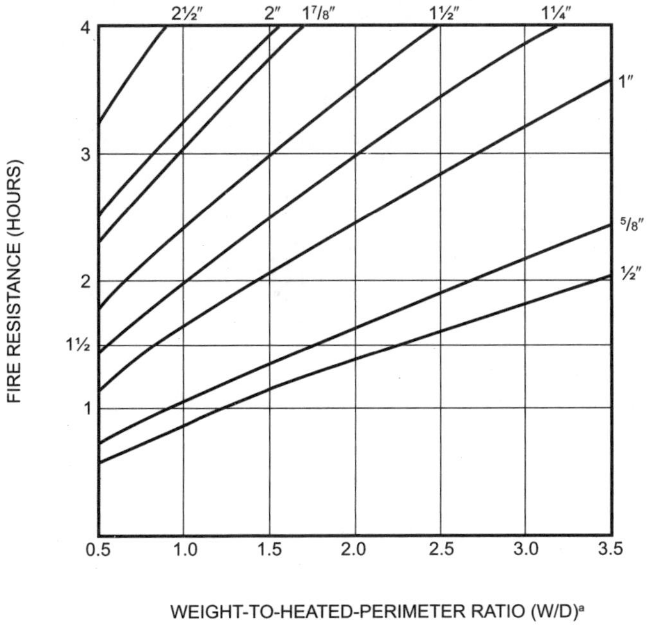

For SI:1 inch = 25.4 mm, 1 pound per linear foot/inch = 0.059 kg/m/mm.

a. The W/D ratios for typical wide flange columns are listed in Table 722.5.1(1). For other column shapes, the W/D ratios shall be determined in accordance with Section 722.5.1.1.

FIGURE 722.5.1(4)
FIRE RESISTANCE OF STRUCTURAL STEEL COLUMNS PROTECTED WITH VARIOUS THICKNESSES OF TYPE X GYPSUM WALLBOARD

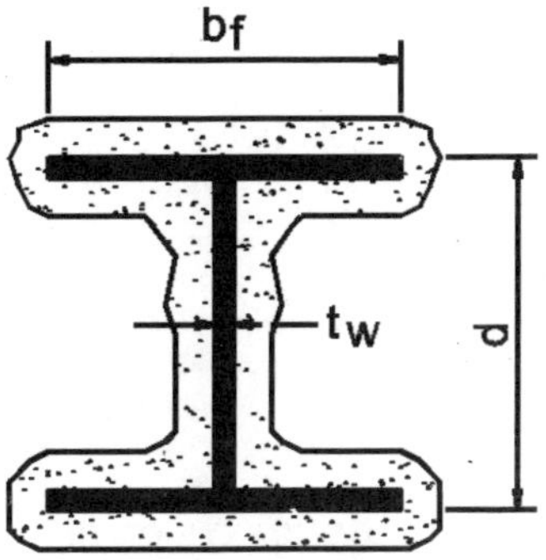

FIGURE 722.5.1(5)
WIDE FLANGE STRUCTURAL STEEL COLUMNS WITH SPRAYED FIRE-RESISTANT MATERIALS

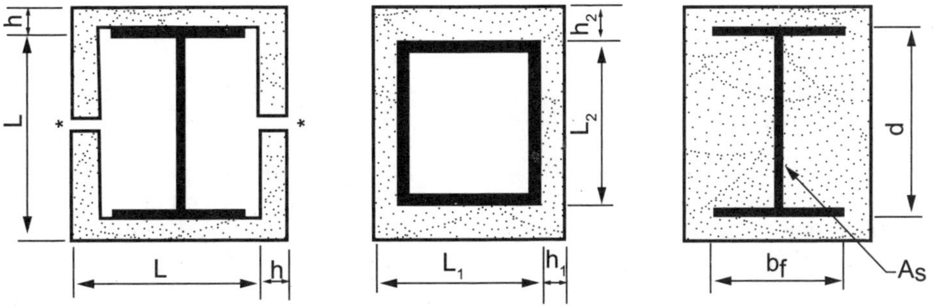

(a) PRECAST CONCRETE COLUMN COVERS

(b) CONCRETE ENCASED STRUCTURAL TUBE

(c) CONCRETE ENCASED WIDE-FLANGE SHAPE

a. Where the inside perimeter of the concrete protection is not square, L shall be taken as the average of L_1 and L_2. Where the thickness of concrete cover is not constant, h shall be taken of the average of h_1 and h_2.

b. Joints shall be protected with not less than a 1-inch thickness of ceramic fiber blanket but in no case less than one-half the thickness of the column cover (see Section 722.2.1.3).

FIGURE 722.5.1(6)
CONCRETE PROTECTED STRUCTURAL STEEL COLUMNS[a, b]

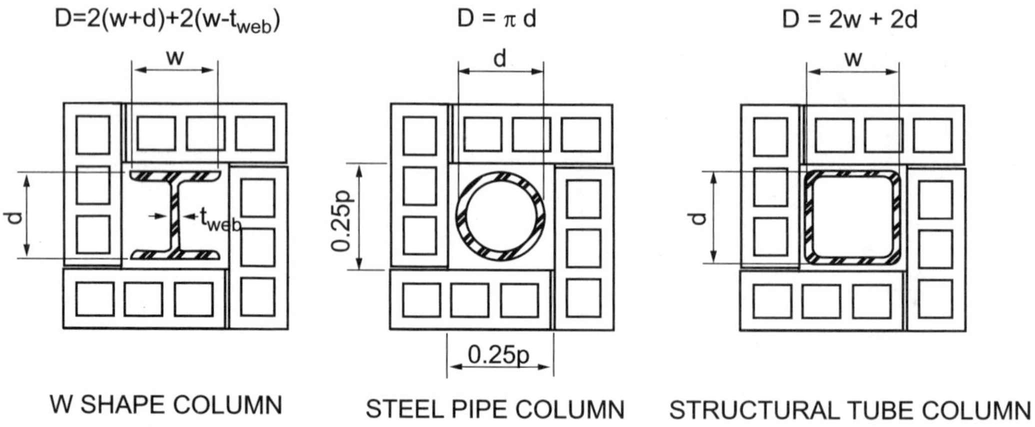

For SI: 1 inch = 25.4 mm.

d = Depth of a wide flange column, outside diameter of pipe column, or outside dimension of structural tubing column (inches).

t_{web} = Thickness of web of wide flange column (inches).

w = Width of flange of wide-flange column (inches).

FIGURE 722.5.1(7)
CONCRETE OR CLAY MASONRY PROTECTED STRUCTURAL STEEL COLUMNS

722.5.1.2 Gypsum wallboard protection. The *fire resistance* of structural steel columns with weight-to-heated-perimeter ratios (W/D) less than or equal to 3.65 and that are protected with Type X *gypsum wallboard* shall be permitted to be determined from the following expression:

$$R = 130\left[\frac{h(W'/D)}{2}\right]^{0.75} \quad \textbf{(Equation 7-12)}$$

where:

R = *Fire resistance* (minutes).

h = Total thickness of *gypsum wallboard* (inches).

D = Heated perimeter of the structural steel column (inches).

W' = Total weight of the structural steel column and *gypsum wallboard* protection (pounds per linear foot).

$W' = W + 50hD/144$.

722.5.1.2.1 Attachment. The *gypsum board* or *gypsum panel products* shall be supported as illustrated in either Figure 722.5.1(2) for *fire-resistance ratings* of 4 hours or less, or Figure 722.5.1(3) for *fire-resistance ratings* of 3 hours or less.

722.5.1.2.2 Gypsum wallboard equivalent to concrete. The determination of the *fire resistance* of structural steel columns from Figure 722.5.1(4) is permitted for various thicknesses of *gypsum wallboard* as a function of the weight-to-heated-perimeter ratio (W/D) of the column. For structural steel columns with weight-to-heated-perimeter ratios (W/D) greater than 3.65, the thickness of *gypsum wallboard* required for specified *fire-resistance ratings* shall be the same as the thickness determined for a W14 × 233 wide flange shape.

722.5.1.3 Sprayed fire-resistant materials. The *fire resistance* of wide-flange structural steel columns protected with sprayed fire-resistant materials, as illustrated in Figure 722.5.1(5), shall be permitted to be determined from the following expression:

$$R = [C_1(W/D) + C_2]h \quad \textbf{(Equation 7-13)}$$

where:

R = *Fire resistance* (minutes).

h = Thickness of sprayed fire-resistant material (inches).

D = Heated perimeter of the structural steel column (inches).

C_1 and C_2 = Material-dependent constants.

W = Weight of structural steel columns (pounds per linear foot).

The *fire resistance* of structural steel columns protected with intumescent or mastic fire-resistant coatings shall be determined on the basis of *fire-resistance* tests in accordance with Section 703.2.

722.5.1.3.1 Material-dependent constants. The material-dependent constants, C_1 and C_2, shall be determined for specific fire-resistant materials on the basis of standard fire endurance tests in accordance with Section 703.2. Unless evidence is submitted to the *building official* substantiating a broader application, this expression shall be limited to determining the *fire resistance* of structural steel columns with weight-to-heated-perimeter ratios (W/D) between the largest and smallest columns for which standard fire-resistance test results are available.

722.5.1.3.2 Identification. Sprayed fire-resistant materials shall be identified by density and thickness required for a given *fire-resistance rating*.

722.5.1.4 Concrete-protected columns. The *fire resistance* of structural steel columns protected with concrete, as illustrated in Figure 722.5.1(6) illustrations (a) and (b), shall be permitted to be determined from the following expression:

$$R = R_o(1 + 0.03m) \qquad \textbf{(Equation 7-14)}$$

where:

$$R_o = 10\,(W/D)^{0.7} + 17\,(h^{1.6}/k_c^{0.2}) \times [1 + 26\,\{H/p_c c_c h\,(L + h)\}^{0.8}]$$

As used in these expressions:

R = Fire endurance at equilibrium moisture conditions (minutes).

R_o = Fire endurance at zero moisture content (minutes).

m = Equilibrium moisture content of the concrete by volume (percent).

W = Average weight of the structural steel column (pounds per linear foot).

D = Heated perimeter of the structural steel column (inches).

h = Thickness of the concrete cover (inches).

k_c = Ambient temperature thermal conductivity of the concrete (Btu/hr ft °F).

H = Ambient temperature thermal capacity of the steel column = 0.11W (Btu/ ft °F).

p_c = Concrete density (pounds per cubic foot).

c_c = Ambient temperature specific heat of concrete (Btu/lb °F).

L = Interior dimension of one side of a square concrete box protection (inches).

722.5.1.4.1 Reentrant space filled. For wide-flange structural steel columns completely encased in concrete with all reentrant spaces filled Figure 722.5.1(6), illustration (c), the thermal capacity of the concrete within the reentrant spaces shall be permitted to be added to the thermal capacity of the steel column, as follows:

$$H = 0.11\,W + (p_c c_c/144)\,(b_f d - A_s) \qquad \textbf{(Equation 7-15)}$$

where:

b_f = Flange width of the structural steel column (inches).

d = Depth of the structural steel column (inches).

A_s = Cross-sectional area of the steel column (square inches).

722.5.1.4.2 Concrete properties unknown. If specific data on the properties of concrete are not available, the values given in Table 722.5.1(2) are permitted.

722.5.1.4.3 Minimum concrete cover. For structural steel column encased in concrete with all reentrant spaces filled, Figure 722.5.1(6), illustration (c) and Tables 722.5.1(7) and 722.5.1(8) indicate the thickness of concrete cover required for various *fire-resistance ratings* for typical wide-flange sections. The thicknesses of concrete indicated in these tables apply to structural steel columns larger than those specified.

722.5.1.4.4 Minimum precast concrete cover. For structural steel columns protected with precast concrete column covers as shown in Figure 722.5.1(6), illustration (a), Tables 722.5.1(9) and 722.5.1(10) indicate the thickness of the column covers required for various *fire-resistance ratings* for typical wide-flange shapes. The thicknesses of concrete given in these tables apply to structural steel columns larger than those specified.

722.5.1.4.5 Masonry protection. The *fire resistance* of structural steel columns protected with concrete *masonry units* or clay *masonry units* as illustrated in Figure 722.5.1(7) shall be permitted to be determined from the following expression:

$$R = 0.17\,(W/D)^{0.7} + [0.285\,(T_e^{1.6}/K^{0.2})]\,[1.0 + 42.7\,\{(A_s/d_m T_e)/(0.25p + T_e)\}^{0.8}] \qquad \textbf{(Equation 7-16)}$$

where:

R = *Fire-resistance rating* of column assembly (hours).

W = Average weight of structural steel column (pounds per foot).

D = Heated perimeter of structural steel column (inches) [see Figure 722.5.1(7)].

T_e = Equivalent thickness of concrete or clay *masonry unit* (inches) (see Table 722.3.2, Note a or Section 722.4.1).

K = Thermal conductivity of concrete or clay *masonry unit* (Btu/hr × ft × °F) [see Table 722.5.1(3)].

A_s = Cross-sectional area of structural steel column (square inches).

d_m = Density of the concrete or clay *masonry unit* (pounds per cubic foot).

p = Inner perimeter of concrete or clay masonry protection (inches) [see Figure 722.5.1(7)].

722.5.1.4.6 Equivalent concrete masonry thickness. For structural steel columns protected with concrete masonry, Table 722.5.1(5) gives the equivalent thickness of concrete masonry required for various *fire-resistance ratings* for typical column shapes. For structural steel columns protected with clay masonry, Table 722.5.1(6) gives the equivalent thickness of concrete masonry required for various *fire-resistance ratings* for typical column shapes.

722.5.2 Structural steel beams and girders. The *fire-resistance ratings* of structural steel beams and girders shall be based on the size of the element and the type of protection provided in accordance with this section.

722.5.2.1 Determination of fire resistance. These procedures establish a basis for determining resistance of structural steel beams and girders that differ in size from that specified in *approved* fire-resistance-rated assemblies as a function of the thickness of fire-resistant material and the weight (W) and heated perimeter (D) of the beam or girder. As used in these sections, W is the average weight of a *structural steel element* in pounds per linear foot (plf). The heated perimeter, D, is the inside perimeter of the fire-resistant material in inches as illustrated in Figure 722.5.2.

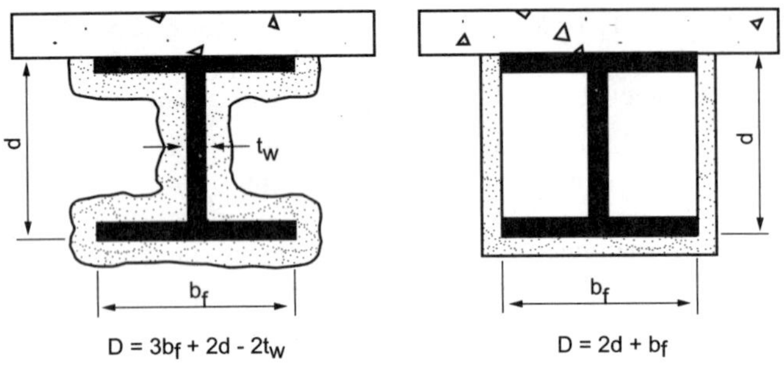

FIGURE 722.5.2
DETERMINATION OF THE HEATED PERIMETER OF STRUCTURAL STEEL BEAMS AND GIRDERS

722.5.2.1.1 Weight-to-heated perimeter. The weight-to-heated-perimeter ratios (W/D), for both contour and box fire-resistant protection profiles, for the wide flange shapes most often used as beams or girders are given in Table 722.5.1(4). For different shapes, the weight-to-heated-perimeter ratios (W/D) shall be determined in accordance with the definitions given in this section.

722.5.2.1.2 Beam and girder substitutions. Except as provided for in Section 722.5.2.2, structural steel beams in *approved* fire-resistance-rated assemblies shall be considered to be the minimum permissible size. Other beam or girder shapes shall be permitted to be substituted provided that the weight-to-heated-perimeter ratio (W/D) of the substitute beam is equal to or greater than that of the beam specified in the *approved* assembly.

722.5.2.2 Sprayed fire-resistant materials. The provisions in this section apply to structural steel beams and girders protected with sprayed fire-resistant materials. Larger or smaller beam and girder shapes shall be permitted to be substituted for beams specified in *approved* unrestrained or restrained fire-resistance-rated assemblies, provided that the thickness of the fire-resistant material is adjusted in accordance with the following expression:

$$h_2 = h_1 \, [(W_1 / D_1) + 0.60] / [(W_2 / D_2) + 0.60]$$

(Equation 7-17)

where:

h = Thickness of sprayed fire-resistant material in inches.

W = Weight of the structural steel beam or girder in pounds per linear foot.

D = Heated perimeter of the structural steel beam in inches.

Subscript 1 refers to the beam and fire-resistant material thickness in the *approved* assembly.

Subscript 2 refers to the substitute beam or girder and the required thickness of fire-resistant material.

The *fire resistance* of structural steel beams and girders protected with intumescent or mastic fire-resistant coatings shall be determined on the basis of fire-resistance tests in accordance with Section 703.2.

722.5.2.2.1 Minimum thickness. The use of Equation 7-17 is subject to the following conditions:

1. The weight-to-heated-perimeter ratio for the substitute beam or girder (W_2/D_2) shall be not less than 0.37.
2. The thickness of fire protection materials calculated for the substitute beam or girder (T_1) shall be not less than $^3/_8$ inch (9.5 mm).
3. The unrestrained or restrained beam rating shall be not less than 1 hour.
4. Where used to adjust the material thickness for a restrained beam, the use of this procedure is limited to structural steel sections classified as compact in accordance with AISC 360.

722.5.2.3 Structural steel trusses. The *fire resistance* of structural steel trusses protected with fire-resistant materials sprayed to each of the individual truss elements shall be permitted to be determined in accordance with this section. The thickness of the fire-resistant material shall be determined in accordance with Section 722.5.1.3. The weight-to-heated-perimeter ratio (W/D) of truss elements that can be simultaneously exposed to fire on all sides shall be determined on the same basis as columns, as specified in Section 722.5.1.1. The weight-to-heated-perimeter ratio (W/D) of truss elements that directly support floor or roof assembly shall be determined on the same basis as beams and girders, as specified in Section 722.5.2.1.

The *fire resistance* of structural steel trusses protected with intumescent or mastic fire-resistant coatings shall be determined on the basis of *fire resistance* tests in accordance with Section 703.2.

722.6 Wood assemblies. The provisions of this section contain procedures by which the *fire-resistance ratings* of wood assemblies are established by calculations.

722.6.1 General. This section contains procedures for calculating the *fire-resistance ratings* of walls, floor/ceiling and roof/ceiling assemblies based in part on the standard method of testing referenced in Section 703.2.

722.6.1.1 Maximum fire-resistance rating. *Fire-resistance ratings* calculated for assemblies using the methods in Section 722.6 shall be limited to not more than 1 hour.

722.6.1.2 Dissimilar membranes. Where dissimilar membranes are used on a wall assembly that requires consideration of fire exposure from both sides, the calculation shall be made from the least fire-resistant (weaker) side.

722.6.2 Walls, floors and roofs. These procedures apply to both load-bearing and nonload-bearing assemblies.

722.6.2.1 Fire-resistance rating of wood frame assemblies. The *fire-resistance rating* of a wood frame assembly is equal to the sum of the time assigned to the membrane on the fire-exposed side, the time assigned to the framing members and the time assigned for additional contribution by other protective measures such as insulation. The membrane on the unexposed side shall not be included in determining the *fire resistance* of the assembly.

722.6.2.2 Time assigned to membranes. Table 722.6.2(1) indicates the time assigned to membranes on the fire-exposed side.

722.6.2.3 Exterior walls. For an *exterior wall* with a *fire separation distance* greater than 10 feet (3048 mm), the wall is assigned a rating dependent on the interior membrane and the framing as described in Table 722.6.2(1) and Table 722.6.2(2). The membrane on the outside of the nonfire-exposed side of *exterior walls* with a *fire separation distance* greater than 10 feet (3048 mm) shall consist of sheathing, sheathing paper and siding as described in Table 722.6.2(3).

722.6.2.4 Floors and roofs. In the case of a floor or roof, the standard test provides only for testing for fire exposure from below. Except as noted in Section 703.2.3, floor or roof assemblies of wood framing shall have an upper membrane consisting of a subfloor and finished floor conforming to Table 722.6.2(4) or any other membrane that has a contribution to *fire resistance* of not less than 15 minutes in Table 722.6.2(1).

722.6.2.5 Additional protection. Table 722.6.2(5) indicates the time increments to be added to the *fire resistance* where glass fiber, rockwool, slag *mineral wool* or cellulose insulation is incorporated in the assembly.

722.6.2.6 Fastening. Fastening of wood frame assemblies and the fastening of membranes to the wood framing members shall be done in accordance with Chapter 23.

TABLE 722.6.2(1)
TIME ASSIGNED TO WALLBOARD MEMBRANES[a, b, c, d]

DESCRIPTION OF FINISH	TIME[e] (minutes)
$^{3}/_{8}$-inch wood structural panel bonded with exterior glue	5
$1^{5}/_{32}$-inch wood structural panel bonded with exterior glue	10
$1^{9}/_{32}$-inch wood structural panel bonded with exterior glue	15
$^{3}/_{8}$-inch gypsum wallboard	10
$^{1}/_{2}$-inch gypsum wallboard	15
$^{5}/_{8}$-inch gypsum wallboard	30
$^{1}/_{2}$-inch Type X gypsum wallboard	25
$^{5}/_{8}$-inch Type X gypsum wallboard	40
Double $^{3}/_{8}$-inch gypsum wallboard	25
$^{1}/_{2}$-inch + $^{3}/_{8}$-inch gypsum wallboard	35
Double $^{1}/_{2}$-inch gypsum wallboard	40

For SI: 1 inch = 25.4 mm.

a. These values apply only where membranes are installed on framing members that are spaced 16 inches o.c. or less.

b. Gypsum wallboard installed over framing or furring shall be installed so that all edges are supported, except $^{5}/_{8}$-inch Type X gypsum wallboard shall be permitted to be installed horizontally with the horizontal joints staggered 24 inches each side and unsupported but finished.

c. On wood frame floor/ceiling or roof/ceiling assemblies, gypsum board shall be installed with the long dimension perpendicular to framing members and shall have all joints finished.

d. The membrane on the unexposed side shall not be included in determining the fire resistance of the assembly. Where dissimilar membranes are used on a wall assembly, the calculation shall be made from the least fire-resistant (weaker) side.

e. The time assigned is not a finished rating.

TABLE 722.6.2(2)
TIME ASSIGNED FOR CONTRIBUTION OF WOOD FRAME [a, b, c]

DESCRIPTION	TIME ASSIGNED TO FRAME (minutes)
Wood studs 16 inches o.c.	20
Wood floor and roof joists 16 inches o.c.	10

For SI: 1 inch = 25.4 mm.

a. This table does not apply to studs or joists spaced more than 16 inches o.c.

b. All studs shall be nominal 2 × 4 and all joists shall have a nominal thickness of not less than 2 inches.

c. Allowable spans for joists shall be determined in accordance with Sections 2308.4.2.1, 2308.7.1 and 2308.7.2.

TABLE 722.6.2(3)
MEMBRANE[a] ON EXTERIOR FACE OF WOOD STUD WALLS

SHEATHING	PAPER	EXTERIOR FINISH
$^5/_8$-inch T & G lumber $^5/_{16}$-inch exterior glue *wood structural panel* $^1/_2$-inch gypsum wallboard $^5/_8$-inch gypsum wallboard $^1/_2$-inch fiberboard	Sheathing paper	Lumber siding Wood shingles and shakes $^1/_4$-inch *fiber-cement* lap, panel or shingle siding $^1/_4$-inch *wood structural panels*-exterior type $^1/_4$-inch hardboard Metal siding Stucco on metal lath Masonry veneer Vinyl siding
None	—	$^3/_8$-inch exterior-grade *wood structural panels*

For SI: 1 inch = 25.4 mm.

a. Any combination of sheathing, paper and exterior finish is permitted.

TABLE 722.6.2(4)
FLOORING OR ROOFING OVER WOOD FRAMING[a]

ASSEMBLY	STRUCTURAL MEMBERS	SUBFLOOR OR ROOF DECK	FINISHED FLOORING OR ROOFING
Floor	Wood	$^{15}/_{32}$-inch wood structural panels or $^{11}/_{16}$-inch T & G softwood	Hardwood or softwood flooring on building paper; resilient flooring; parquet floor; felted-synthetic fiber floor coverings, carpeting, or ceramic tile on $^1/_4$-inch-thick fiber-cement underlayment or $^3/_8$-inch-thick panel-type underlayment; ceramic tile on $1^1/_4$-inch mortar bed.
Roof	Wood	$^{15}/_{32}$-inch wood structural panels or $^{11}/_{16}$-inch T & G softwood	Finished roofing material with or without insulation

For SI: 1 inch = 25.4 mm.

a. This table applies only to wood joist construction. It is not applicable to wood truss construction.

TABLE 722.6.2(5)
TIME ASSIGNED FOR ADDITIONAL PROTECTION

DESCRIPTION OF ADDITIONAL PROTECTION	FIRE RESISTANCE (minutes)
Add to the fire-resistance rating of wood stud walls if the spaces between the studs are completely filled with glass fiber mineral wool batts weighing not less than 2 pounds per cubic foot (0.6 pound per square foot of wall surface) or rockwool or slag material wool batts weighing not less than 3.3 pounds per cubic foot (1 pound per square foot of wall surface), or cellulose insulation having a nominal density not less than 2.6 pounds per cubic foot.	15

For SI: 1 pound/cubic foot = 16.0185 kg/m^3.

722.7 Fire-resistance rating for mass timber. The required *fire resistance* of *mass timber* elements in Section 602.4 shall be determined in accordance with Section 703.2. The *fire-resistance rating* of *building elements* shall be as required in Tables 601 and 705.5 and as specified elsewhere in this code. The *fire-resistance rating* of the *mass timber* elements shall consist of the *fire resistance* of the unprotected element added to the protection time of the *noncombustible protection*.

722.7.1 Minimum required protection. Where required by Sections 602.4.1 through 602.4.3, *noncombustible protection* shall be provided for *mass timber building elements* in accordance with Table 722.7.1(1). The rating, in minutes, contributed by the *noncombustible protection* of *mass timber building elements,* components or assemblies, shall be established in accordance with Section 703.6. The protection contributions indicated in Table 722.7.1(2) shall be deemed to comply with this requirement where installed and fastened in accordance with Section 722.7.2.

722.7.2 Installation of gypsum board noncombustible protection. *Gypsum board* complying with Table 722.7.1(2) shall be installed in accordance with this section.

722.7.2.1 Interior surfaces. Layers of Type X *gypsum board* serving as *noncombustible protection* for *interior surfaces* of wall and ceiling assemblies determined in accordance with Table 722.7.1(1) shall be installed in accordance with the following:

1. Each layer shall be attached with Type S drywall screws of sufficient length to penetrate

the mass timber at least 1 inch (25 mm) when driven flush with the paper surface of the gypsum board.

Exception: The third layer, where determined necessary by Section 722.7, shall be permitted to be attached with 1-inch (25 mm) No. 6 Type S drywall screws to furring channels in accordance with AISI S220.

2. Screws for attaching the base layer shall be 12 inches (305 mm) on center in both directions.
3. Screws for each layer after the base layer shall be 12 inches (305 mm) on center in both directions and offset from the screws of the previous layers by 4 inches (102 mm) in both directions.
4. All panel edges of any layer shall be offset 18 inches (457 mm) from those of the previous layer.
5. All panel edges shall be attached with screws sized and offset as in Items 1 through 4 and placed at least 1 inch (25 mm) but not more than 2 inches (51 mm) from the panel edge.
6. All panels installed at wall-to-ceiling intersections shall be installed such that ceiling panels are installed first and the wall panels are installed after the ceiling panel has been installed and is fitted tight to the ceiling panel. Where multiple layers are required, each layer shall repeat this process.
7. All panels installed at a wall-to-wall intersection shall be installed such that the panels covering an exterior wall or a wall with a greater fire-resistance rating shall be installed first and the panels covering the other wall shall be fitted tight to the panel covering the first wall. Where multiple layers are required, each layer shall repeat this process.
8. Panel edges of the face layer shall be taped and finished with joint compound. Fastener heads shall be covered with joint compound.
9. Panel edges protecting mass timber elements adjacent to unprotected mass timber elements in accordance with Section 602.4.2.2 shall be covered with $1^1/_4$-inch (32 mm) metal corner bead and finished with joint compound.

722.7.2.2 Exterior surfaces. Layers of Type X *gypsum board* serving as *noncombustible protection* for the outside of the exterior *mass timber* walls determined in accordance with Table 722.7.1(1) shall be fastened 12 inches (305 mm) on center each way and 6 inches (152 mm) on center at all joints or ends. All panel edges shall be attached with fasteners located at least 1 inch (25 mm) but not more than 2 inches (51 mm) from the panel edge. Fasteners shall comply with one of the following:

1. Galvanized nails of minimum 12 gage with a $^7/_{16}$-inch (11 mm) head of sufficient length to penetrate the mass timber a minimum of 1 inch (25 mm).
2. Screws that comply with ASTM C1002 (Type S, W or G) of sufficient length to penetrate the mass timber a minimum of 1 inch (25 mm).

TABLE 722.7.1(1)
PROTECTION REQUIRED FROM NONCOMBUSTIBLE COVERING MATERIAL

REQUIRED FIRE-RESISTANCE RATING OF BUILDING ELEMENT PER TABLE 601 AND TABLE 705.5 (hours)	MINIMUM PROTECTION REQUIRED FROM NONCOMBUSTIBLE PROTECTION (minutes)
1	40
2	80
3 or more	120

TABLE 722.7.1(2)
PROTECTION PROVIDED BY NONCOMBUSTIBLE COVERING MATERIAL

NONCOMBUSTIBLE PROTECTION	PROTECTION CONTRIBUTION (minutes)
$^1/_2$-inch Type X gypsum board	25
$^5/_8$-inch Type X gypsum board	40

CHAPTER 8
INTERIOR FINISHES

User notes:

About this chapter: *Chapter 8 contains the performance requirements for controlling fire growth and smoke propagation within buildings by restricting interior finish and decorative materials. The provisions of this chapter require materials used as interior finishes and decorations to meet certain flame spread index or flame propagation criteria and smoke development criteria based on the relative fire hazard associated with the occupancy. The performance of the material is evaluated based on test standards.*

SECTION 801
SCOPE

801.1 Scope. The provisions of this chapter shall govern the use of materials used as *interior finishes*, *trim* and *decorative materials*.

SECTION 802
GENERAL

802.1 Interior wall and ceiling finish. The provisions of Section 803 shall limit the allowable fire performance and smoke development of *interior wall and ceiling finish* materials based on occupancy classification.

802.2 Interior floor finish. The provisions of Section 804 shall limit the allowable fire performance of *interior floor finish* materials based on occupancy classification.

[F] 802.3 Decorative materials and trim. *Decorative materials* and *trim* shall be restricted by combustibility, fire performance or flame propagation performance criteria in accordance with Section 806.

802.4 Applicability. For buildings in *flood hazard areas* as established in Section 1612.3, *interior finishes*, *trim* and *decorative materials* below the elevation required by Section 1612 shall be flood-damage-resistant materials.

802.5 Application. Combustible materials shall be permitted to be used as finish for walls, ceilings, floors and other *interior surfaces* of buildings.

802.6 Windows. Show windows in the *exterior walls* of the first *story* above *grade plane* shall be permitted to be of wood or of unprotected metal framing.

802.7 Foam plastics. Foam plastics shall not be used as *interior finish* except as provided in Section 803.4. Foam plastics shall not be used as interior *trim* except as provided in Section 806.5 or 2604.2. This section shall apply both to exposed foam plastics and to foam plastics used in conjunction with a textile or vinyl facing or cover.

SECTION 803
WALL AND CEILING FINISHES

803.1 General. *Interior wall and ceiling finish* materials shall be classified for fire performance and smoke development in accordance with Section 803.1.1 or 803.1.2, except as shown in Sections 803.1.3 through 803.15. Materials tested in accordance with Section 803.1.1 shall not be required to be tested in accordance with Section 803.1.2.

803.1.1 Interior wall and ceiling finish materials tested in accordance with NFPA 286. *Interior wall and ceiling finish* materials shall be classified in accordance with NFPA 286 and comply with Section 803.1.1.1. Materials complying with Section 803.1.1.1 shall be considered to also comply with the requirements of Class A.

803.1.1.1 Acceptance criteria for NFPA 286. The *interior finish* shall comply with the following:

1. During the 40 kW exposure, flames shall not spread to the ceiling.
2. The flame shall not spread to the outer extremity of the sample on any wall or ceiling.
3. Flashover, as defined in NFPA 286, shall not occur.
4. The peak heat release rate throughout the test shall not exceed 800 kW.
5. The total smoke released throughout the test shall not exceed 1,000 m^2.

803.1.2 Interior wall and ceiling finish materials tested in accordance with ASTM E84 or UL 723. *Interior wall and ceiling finish* materials shall be classified in accordance with ASTM E84 or UL 723. Such *interior finish* materials shall be grouped in the following classes in accordance with their *flame spread* and *smoke-developed indices*.

Class A = *Flame spread index* 0–25; *smoke-developed index* 0–450.

Class B = *Flame spread index* 26–75; *smoke developed index* 0–450.

Class C = *Flame spread index* 76–200; *smoke-developed index* 0–450.

Exception: Materials tested in accordance with Section 803.1.1 and as indicated in Sections 803.1.3 through 803.13.

803.1.3 Interior wall and ceiling finish materials with different requirements. The materials indicated in Sections 803.2 through 803.13 shall be tested as indicated in the corresponding sections.

803.2 Thickness exemption. Materials having a thickness less than 0.036 inch (0.9 mm) applied directly to the surface of walls or ceilings shall not be required to be tested.

803.3 Heavy timber exemption. Exposed portions of *building elements* complying with the requirements for buildings of heavy timber construction in Section 602.4 or Section 2304.11 shall not be subject to *interior finish* requirements except in *interior exit stairways*, interior exit *ramps*, and exit passageways.

803.4 Foam plastics. Foam plastics shall not be used as *interior finish* except as provided in Section 2603.9. This section shall apply both to exposed foam plastics and to foam plastics used in conjunction with a textile or vinyl facing or cover.

803.5 Textile wall coverings. Where used as interior wall finish materials, textile wall coverings, including materials having woven or nonwoven, napped, tufted, looped or similar surface and carpet and similar textile materials, shall be tested in the manner intended for use, using the product-mounting system, including adhesive, and shall comply with the requirements of one of the following: Section 803.1.1, 803.5.1 or 803.5.2.

803.5.1 Room corner test for textile wall coverings and expanded vinyl wall coverings. Textile wall coverings and *expanded vinyl wall coverings* shall meet the criteria of Section 803.5.1.1 when tested in the manner intended for use in accordance with the Method B protocol of NFPA 265 using the product-mounting system, including adhesive.

803.5.1.1 Acceptance criteria for NFPA 265. The *interior finish* shall comply with the following:

1. During the 40 kW exposure, flames shall not spread to the ceiling.
2. The flame shall not spread to the outer extremities of the samples on the 8-foot by 12-foot (203 by 305 mm) walls.
3. Flashover, as defined in NFPA 265, shall not occur.
4. The total smoke release throughout the test shall not exceed 1,000 m^2.

803.5.2 Acceptance criteria for textile and expanded vinyl wall or ceiling coverings tested to ASTM E84 or UL 723. Textile wall and ceiling coverings and expanded vinyl wall and ceiling coverings shall have a Class A *flame spread index* in accordance with ASTM E84 or UL 723 and be protected by an automatic sprinkler system installed in accordance with Section 903.3.1.1 or 903.3.1.2. Test specimen preparation and mounting shall be in accordance with ASTM E2404.

803.6 Textile ceiling coverings. Where used as interior ceiling finish materials, textile ceiling coverings, including materials having woven or nonwoven, napped, tufted, looped or similar surface and carpet and similar textile materials, shall be tested in the manner intended for use, using the product-mounting system, including adhesive, and shall comply with the requirements of Section 803.1.1 or 803.5.2.

803.7 Expanded vinyl wall coverings. Where used as interior wall finish materials, *expanded vinyl wall coverings* shall be tested in the manner intended for use, using the product-mounting system, including adhesive, and shall comply with the requirements of one of the following: Section 803.1.1, 803.5.1 or 803.5.2.

803.8 Expanded vinyl ceiling coverings. Where used as interior ceiling finish materials, expanded vinyl ceiling coverings shall be tested in the manner intended for use, using the product mounting system, including adhesive, and shall comply with the requirements of Section 803.1.1 or 803.5.2.

803.9 High-density polyethylene (HDPE) and polypropylene (PP). Where high-density polyethylene or polypropylene is used as an *interior finish*, it shall comply with Section 803.1.1.

803.10 Site-fabricated stretch systems. Where used as interior wall or interior ceiling finish materials, *site-fabricated stretch systems* containing all three components described in the definition in Chapter 2 shall be tested in the manner intended for use, and shall comply with the requirements of Section 803.1.1 or with the requirements of Class A in accordance with Section 803.1.2. If the materials are tested in accordance with ASTM E84 or UL 723, specimen preparation and mounting shall be in accordance with ASTM E2573.

803.11 Laminated products factory produced with a wood substrate. Laminated products factory produced with a wood substrate shall comply with one of the following:

1. The laminated product shall meet the criteria of Section 803.1.1.1 when tested in accordance with NFPA 286 using the product-mounting system, including adhesive, as described in Section 5.8 of NFPA 286.
2. The laminated product shall have a Class A, B, or C *flame spread index* and *smoke-developed index*, based on the requirements of Table 803.13, in accordance with ASTM E84 or UL 723. Test specimen preparation and mounting shall be in accordance with ASTM E2579.

803.12 Facings or wood veneers intended to be applied on site over a wood substrate. Facings or *veneers* intended to be applied on site over a wood substrate shall comply with one of the following:

1. The facing or *veneer* shall meet the criteria of Section 803.1.1.1 when tested in accordance with NFPA 286 using the product mounting system, including adhesive, as described in Section 5.9 of NFPA 286.
2. The facing or *veneer* shall have a Class A, B or C *flame spread index* and *smoke-developed index*, based on the requirements of Table 803.13, in accordance with ASTM E84 or UL 723. Test specimen preparation and mounting shall be in accordance with ASTM E2404.

803.13 Interior finish requirements based on occupancy. *Interior wall and ceiling finish* shall have a *flame spread* *index* not greater than that specified in Table 803.13 for the group and location designated. *Interior wall and ceiling finish* materials tested in accordance with NFPA 286 and meeting the acceptance criteria of Section 803.1.1.1, shall be permitted to be used where a Class A classification in accordance with ASTM E84 or UL 723 is required.

803.14 Stability. *Interior finish* materials regulated by this chapter shall be applied or otherwise fastened in such a manner that such materials will not readily become detached where subjected to room temperatures of 200°F (93°C) for not less than 30 minutes.

803.15 Application of interior finish materials to fire-resistance-rated or noncombustible *building elements*. Where *interior finish* materials are applied on walls, ceilings or structural elements required to have a *fire-resistance rating* or to be of noncombustible construction, these finish materials shall comply with the provisions of this section.

803.15.1 Direct attachment and furred construction. Where walls, ceilings or structural elements are required by any provision in this code to be of fire-resistance-rated or noncombustible construction, the *interior finish* material shall be applied directly against such construction or to furring strips not exceeding $1^3/_4$ inches (44 mm), applied directly against such surfaces.

803.15.1.1 Furred construction. If the *interior finish* material is applied to furring strips, the intervening spaces between such furring strips shall comply with one of the following:

1. Be filled with material that is inorganic or noncombustible.
2. Be filled with material that meets the requirements of a Class A material in accordance with Section 803.1.1 or 803.1.2.
3. Be fireblocked at not greater than 8 feet (2438 mm) in every direction in accordance with Section 718.

Exception: Compliance with Item 1, 2 or 3 is not required where the materials used to create the concealed space are noncombustible.

TABLE 803.13
INTERIOR WALL AND CEILING FINISH REQUIREMENTS BY OCCUPANCY[k]

GROUP	SPRINKLERED[l]			NONSPRINKLERED		
	Interior exit stairways and ramps and exit passageways[a, b]	Corridors and enclosure for exit access stairways and ramps	Rooms and enclosed spaces[c]	Interior exit stairways and ramps and exit passageways[a, b]	Corridors and enclosure for exit access stairways and ramps	Rooms and enclosed spaces[c]
A-1 & A-2	B	B	C	A	A[d]	B[e]
A-3[f], A-4, A-5	B	B	C	A	A[d]	C
B, E, M, R-1	B	C[m]	C	A	B	C
R-4	B	C	C	A	B	B
F	C	C	C	B	C	C
H	B	B	C[g]	A	A	B
I-1	B	C	C	A	B	B
I-2	B	B	B[h, i]	A	A	B
I-3	A	A[j]	C	A	A	B
I-4	B	B	B[h, i]	A	A	B
R-2	C	C	C	B	B	C
R-3	C	C	C	C	C	C
S	C	C	C	B	B	C
U	No restrictions			No restrictions		

For SI: 1 inch = 25.4 mm, 1 square foot = 0.0929 m².

a. Class C interior finish materials shall be permitted for wainscotting or paneling of not more than 1,000 square feet of applied surface area in the grade lobby where applied directly to a noncombustible base or over furring strips applied to a noncombustible base and fireblocked as required by Section 803.15.1.
b. In other than Group I-3 occupancies in buildings less than three stories above grade plane, Class B interior finish for nonsprinklered buildings and Class C interior finish for sprinklered buildings shall be permitted in interior exit stairways and ramps.
c. Requirements for rooms and enclosed spaces shall be based on spaces enclosed by partitions. Where a fire-resistance rating is required for structural elements, the enclosing partitions shall extend from the floor to the ceiling. Partitions that do not comply with this shall be considered to be enclosing spaces and the rooms or spaces on both sides shall be considered to be one room or space. In determining the applicable requirements for rooms and enclosed spaces, the specific occupancy thereof shall be the governing factor regardless of the group classification of the building or structure.
d. Lobby areas in Group A-1, A-2 and A-3 occupancies shall be not less than Class B materials.
e. Class C interior finish materials shall be permitted in places of assembly with an occupant load of 300 persons or less.
f. For places of religious worship, wood used for ornamental purposes, trusses, paneling or chancel furnishing shall be permitted.
g. Class B material is required where the building exceeds two stories.
h. Class C interior finish materials shall be permitted in administrative spaces.
i. Class C interior finish materials shall be permitted in rooms with a capacity of four persons or less.
j. Class B materials shall be permitted as wainscotting extending not more than 48 inches above the finished floor in corridors and exit access stairways and ramps.
k. Finish materials as provided for in other sections of this code.
l. Applies when protected by an automatic sprinkler system installed in accordance with Section 903.3.1.1 or 903.3.1.2.
m. Corridors in ambulatory care facilities shall be provided with Class A or B materials.

803.15.2 Set-out construction. Where walls and ceilings are required to be of fire-resistance-rated or noncombustible construction and walls are set out or ceilings are dropped distances greater than specified in Section 803.15.1, Class A finish materials, in accordance with Section 803.1.1 or 803.1.2, shall be used.

Exceptions:

1. Where *interior finish* materials are protected on both sides by an *automatic sprinkler system* in accordance with Section 903.3.1.1 or 903.3.1.2.
2. Where *interior finish* materials are attached to noncombustible backing or furring strips installed as specified in Section 803.15.1.1.
3. Where the combustible void is filled with a noncombustible material.

803.15.2.1 Hangers and assembly members. The hangers and assembly members of such dropped ceilings that are below the horizontal fire-resistance-rated floor or roof assemblies shall be of noncombustible materials. The construction of each set-out wall and horizontal fire-resistance-rated floor or roof assembly shall be of fire-resistance-rated construction as required elsewhere in this code.

Exception: In Types III and V construction, *fire-retardant-treated wood* shall be permitted for use as hangers and assembly members of dropped ceilings.

803.15.3 Heavy timber construction. Wall and ceiling finishes of all classes as permitted in this chapter that are installed directly against the wood decking or planking of heavy timber construction in Section 602.4.4.2 or 2304.11 or to wood furring strips applied directly to the wood decking or planking shall be fireblocked as specified in Section 803.15.1.1.

803.15.4 Materials. An interior wall or ceiling finish material that is not more than $^1/_4$ inch (6.4 mm) thick shall be applied directly onto the wall, ceiling or structural element without the use of furring strips and shall not be suspended away from the building element to which that finish material it is applied.

Exceptions:

1. Noncombustible *interior finish* materials.
2. Materials that meet the requirements of Class A materials in accordance with Section 803.1.1 or 803.1.2 where the qualifying tests were made with the material furred out from the noncombustible backing shall be permitted to be used with furring strips.
3. Materials that meet the requirements of Class A materials in accordance with Section 803.1.1 or 803.1.2 where the qualifying tests were made with the material suspended away from the noncombustible backing shall be permitted to be used suspended away from the building element.

SECTION 804
INTERIOR FLOOR FINISH

804.1 General. *Interior floor finish* and floor covering materials shall comply with Sections 804.2 through 804.4.2.

Exception: Floor finishes and coverings of a traditional type, such as wood, vinyl, linoleum or terrazzo, and resilient floor covering materials that are not composed of fibers.

804.2 Classification. *Interior floor finish* and floor covering materials required by Section 804.4.2 to be of Class I or II materials shall be classified in accordance with ASTM E648 or NFPA 253. The classification referred to herein corresponds to the classifications determined by ASTM E648 or NFPA 253 as follows: Class I, 0.45 watts/cm^2 or greater; Class II, 0.22 watts/cm^2 or greater.

804.3 Testing and identification. *Interior floor finish* and floor covering materials shall be tested by an agency in accordance with ASTM E648 or NFPA 253 and identified by a hang tag or other suitable method so as to identify the manufacturer or supplier and style, and shall indicate the *interior floor finish* or floor covering classification in accordance with Section 804.2. Carpet-type floor coverings shall be tested as proposed for use, including underlayment. Test reports confirming the information provided in the manufacturer's product identification shall be furnished to the building official on request.

804.4 Interior floor finish requirements. Interior floor covering materials shall comply with Sections 804.4.1 and 804.4.2 and *interior floor finish* materials shall comply with Section 804.4.2.

804.4.1 Test requirement. In all occupancies, interior floor covering materials shall comply with the requirements of the DOC FF-1 "pill test" (CPSC 16 CFR Part 1630) or with ASTM D2859.

804.4.2 Minimum critical radiant flux. In all occupancies, *interior floor finish* and floor covering materials in enclosures for *stairways* and *ramps*, exit passageways, *corridors* and rooms or spaces not separated from corridors by partitions extending from the floor to the underside of the ceiling shall withstand a minimum critical radiant flux. The minimum critical radiant flux shall be not less than Class I in Groups I-1, I-2 and I-3 and not less than Class II in Groups A, B, E, H, I-4, M, R-1, R-2 and S.

Exception: Where a building is equipped throughout with an automatic sprinkler system in accordance with Section 903.3.1.1 or 903.3.1.2, Class II materials are permitted in any area where Class I materials are required, and materials complying with DOC FF-1 "pill test" (CPSC 16 CFR Part 1630) or with ASTM D2859 are permitted in any area where Class II materials are required.

SECTION 805 COMBUSTIBLE MATERIALS IN TYPES I AND II CONSTRUCTION

805.1 Application. Combustible materials installed on or embedded in floors of buildings of Type I or II construction shall comply with Sections 805.1.1 through 805.1.3.

Exception: *Stages* and *platforms* constructed in accordance with Sections 410.2 and 410.3, respectively.

805.1.1 Subfloor construction. Floor sleepers, bucks and nailing blocks shall not be constructed of combustible materials, unless the space between the fire-resistance-rated floor assembly and the flooring is either solidly filled with noncombustible materials or fireblocked in accordance with Section 718, and provided that such open spaces shall not extend under or through permanent partitions or walls.

805.1.2 Wood finish flooring. Wood finish flooring is permitted to be attached directly to the embedded or fireblocked wood sleepers and shall be permitted where cemented directly to the top surface of fire-resistance-rated floor assemblies or directly to a wood subfloor attached to sleepers as provided for in Section 805.1.1.

805.1.3 Insulating boards. Combustible insulating boards not more than $^1/_2$ inch (12.7 mm) thick and covered with finish flooring are permitted where attached directly to a noncombustible floor assembly or to wood subflooring attached to sleepers as provided for in Section 805.1.1.

SECTION 806 DECORATIVE MATERIALS AND TRIM

[F] 806.1 General. The following requirements shall apply to all occupancies:

1. Furnishings or *decorative materials* of an *explosive* or highly flammable character shall not be used.
2. Fire-retardant coatings in existing buildings shall be maintained so as to retain the effectiveness of the treatment under service conditions encountered in actual use.
3. Furnishings or other objects shall not be placed to obstruct exits, access thereto, egress therefrom or visibility thereof.
4. The permissible amount of decorative vegetation and noncombustible *decorative materials* shall not be limited.

[F] 806.2 Combustible decorative materials. In Groups A, B, E, I, M and R-1 and in *dormitories* in Group R-2, curtains, draperies, fabric hangings and similar combustible *decorative materials* suspended from walls or ceilings shall comply with Section 806.4 and shall not exceed 10 percent of the specific wall or ceiling area to which such materials are attached.

Fixed or movable walls and partitions, paneling, wall pads and crash pads applied structurally or for decoration, acoustical correction, surface insulation or other purposes shall be considered to be *interior finish*, shall comply with Section 803 and shall not be considered to be *decorative materials* or furnishings.

Exceptions:

1. In auditoriums in Group A, the permissible amount of curtains, draperies, fabric hangings and similar combustible *decorative materials* suspended from walls or ceilings shall not exceed 75 percent of the aggregate wall area where the building is equipped throughout with an *approved automatic sprinkler system* in accordance with Section 903.3.1.1, and where the material is installed in accordance with Section 803.15 of this code.
2. In Group R-2 *dormitories*, within sleeping units and dwelling units, the permissible amount of curtains, draperies, fabric hangings and similar *decorative materials* suspended from walls or ceiling shall not exceed 50 percent of the aggregate wall areas where the building is equipped throughout with an *approved automatic sprinkler system* installed in accordance with Section 903.3.1.
3. In Group B and M occupancies, the amount of combustible *fabric partitions* suspended from the ceiling and not supported by the floor shall comply with Section 806.4 and shall not be limited.
4. The 10-percent limit shall not apply to curtains, draperies, fabric hangings and similar combustible *decorative materials* used as window coverings.

806.3 Occupancy-based requirements. Occupancy-based requirements for combustible *decorative materials*, other than decorative vegetation, not complying with Section 806.4 shall comply with Sections 807.5.1 through 807.5.6 of the *International Fire Code*.

[F] 806.4 Acceptance criteria and reports. Where required to exhibit improved fire performance, curtains, draperies, fabric hangings and similar combustible *decorative materials* suspended from walls or ceilings shall be tested by an *approved agency* and meet the flame propagation performance criteria of Test 1 or 2, as appropriate, of NFPA 701, or exhibit a maximum heat release rate of 100 kW when tested in accordance with NFPA 289, using the 20 kW ignition source. Reports of test results shall be prepared in accordance with the test method used and furnished to the *building official* on request.

[F] 806.5 Foam plastic. Foam plastic used as *trim* in any occupancy shall comply with Section 2604.2.

[F] 806.6 Pyroxylin plastic. Imitation leather or other material consisting of or coated with a pyroxylin or similarly hazardous base shall not be used in Group A occupancies.

[F] 806.7 Interior trim. Material, other than foam plastic used as interior *trim*, shall have a minimum Class C *flame spread* and *smoke-developed index* when tested in accordance with ASTM E84 or UL 723, as described in Section 803.1.2. Combustible *trim*, excluding handrails and guardrails, shall not exceed 10 percent of the specific wall or ceiling area to which it is attached.

[F] 806.8 Interior floor-wall base. *Interior floor-wall base* that is 6 inches (152 mm) or less in height shall be tested in accordance with Section 804.2 and shall be not less than Class II. Where a Class I floor finish is required, the floor-wall base shall be Class I.

Exception: Interior *trim* materials that comply with Section 806.7.

[F] 806.9 Combustible lockers. Where lockers constructed of combustible materials are used, the lockers shall be considered to be *interior finish* and shall comply with Section 803.

Exception: Lockers constructed entirely of wood and noncombustible materials shall be permitted to be used wherever interior finish materials are required to meet a Class C classification in accordance with Section 803.1.2.

SECTION 807
INSULATION

807.1 Insulation. Thermal and acoustical insulation shall comply with Section 720.

SECTION 808
ACOUSTICAL CEILING SYSTEMS

808.1 Acoustical ceiling systems. The quality, design, fabrication and erection of metal suspension systems for acoustical tile and lay-in panel ceilings in buildings or structures shall conform to generally accepted engineering practice, the provisions of this chapter and other applicable requirements of this code.

808.1.1 Materials and installation. Acoustical materials complying with the *interior finish* requirements of Section 803 shall be installed in accordance with the manufacturer's recommendations and applicable provisions for applying *interior finish*.

808.1.1.1 Suspended acoustical ceilings. Suspended acoustical ceiling systems shall be installed in accordance with the provisions of ASTM C635 and ASTM C636.

808.1.1.2 Fire-resistance-rated construction. Acoustical ceiling systems that are part of fire-resistance-rated construction shall be installed in the same manner used in the assembly tested and shall comply with the provisions of Chapter 7.

CHAPTER 9

FIRE PROTECTION AND LIFE SAFETY SYSTEMS

User note:

About this chapter: *Chapter 9 prescribes the minimum requirements for active fire protection equipment systems to perform the functions of detecting a fire, alerting the occupants or fire department of a fire emergency, mass notification, gas detection, controlling smoke and controlling or extinguishing the fire. Generally, the requirements are based on the occupancy, the height and the area of the building, because these are the factors that most affect fire-fighting capabilities and the relative hazard of a specific building or portion thereof. This chapter parallels and is substantially duplicated in Chapter 9 of the* International Fire Code®.

SECTION 901 GENERAL

901.1 Scope. The provisions of this chapter shall specify where fire protection and *life safety systems* are required and shall apply to the design, installation and operation of *fire protection* and *life safety systems*.

901.2 Fire protection systems. *Fire protection* and *life safety systems* shall be installed, repaired, operated and maintained in accordance with this code and the *International Fire Code*.

Any *fire protection or life safety system* for which an exception or reduction to the provisions of this code has been granted shall be considered to be a required system.

Exception: Any *fire protection or life safety system* or portion thereof not required by this code shall be permitted to be installed for partial or complete protection provided that such system meets the requirements of this code.

901.3 Modifications. Persons shall not remove or modify any *fire protection system* installed or maintained under the provisions of this code or the *International Fire Code* without approval by the *building official*.

901.4 Threads. Threads provided for fire department connections to sprinkler systems, standpipes, yard hydrants or any other fire hose connection shall be compatible with the connections used by the local fire department.

901.5 Acceptance tests. *Fire protection systems* shall be tested in accordance with the requirements of this code and the *International Fire Code*. Where required, the tests shall be conducted in the presence of the *building official*. Tests required by this code, the *International Fire Code* and the standards listed in this code shall be conducted at the expense of the owner or the owner's authorized agent. It shall be unlawful to occupy portions of a structure until the required *fire protection systems* within that portion of the structure have been tested and *approved*.

901.6 Supervisory service. Where required, *fire protection systems* shall be monitored by an approved supervising station in accordance with NFPA 72.

901.6.1 Automatic sprinkler systems. *Automatic sprinkler systems* shall be monitored by an *approved* supervising station.

Exceptions:

1. A supervising station is not required for *automatic sprinkler systems* protecting one- and two-family dwellings.
2. Limited area systems in accordance with Section 903.3.8.

[F] 901.6.2 Integrated testing. Where two or more fire protection or *life safety systems* are interconnected, the intended response of subordinate fire protection and *life safety systems* shall be verified when required testing of the initiating system is conducted. In addition, integrated testing shall be performed in accordance with Sections 901.6.2.1 and 901.6.2.2.

[F] 901.6.2.1 High-rise buildings. For high-rise buildings, integrated testing shall comply with NFPA 4, with an integrated test performed prior to issuance of the certificate of occupancy and at intervals not exceeding 10 years, unless otherwise specified by an integrated system test plan prepared in accordance with NFPA 4. If an equipment failure is detected during integrated testing, a repeat of the integrated test shall not be required, except as necessary to verify operation of fire protection or life safety functions that are initiated by equipment that was repaired or replaced.

[F] 901.6.2.2 Smoke control systems. Where a fire alarm system is integrated with a smoke control system as outlined in Section 909, integrated testing shall comply with NFPA 4, with an integrated test performed prior to issuance of the certificate of occupancy and at intervals not exceeding 10 years, unless otherwise specified by an integrated system test plan prepared in accordance with NFPA 4. If an equipment failure is detected during integrated testing, a repeat of the integrated test shall not be required, except as necessary to verify operation of fire protection or life safety functions that are initiated by equipment that was repaired or replaced.

901.6.3 Fire alarm systems. Fire alarm systems required by the provisions of Section 907.2 of this code and Sections 907.2 and 907.9 of the *International Fire Code* shall be monitored by an *approved* supervising station in accordance with Section 907.6.6 of this code.

Exceptions:

1. Single- and multiple-station smoke alarms required by Section 907.2.11.
2. Smoke detectors in Group I-3 occupancies.
3. Supervisory service is not required for *automatic sprinkler systems* in one- and two-family dwellings.

901.6.4 Group H. Supervision and monitoring of emergency alarm, detection and automatic fire-extinguishing systems in Group H occupancies shall be in accordance with the *International Fire Code*.

901.7 Fire areas. Where buildings, or portions thereof, are divided into *fire areas* so as not to exceed the limits established for requiring a *fire protection system* in accordance with this chapter, such *fire areas* shall be separated by *fire walls* constructed in accordance with Section 706, *fire barriers* constructed in accordance with Section 707, or *horizontal assemblies* constructed in accordance with Section 711, or a combination therof having a *fire-resistance rating* of not less than that determined in accordance with Section 707.3.10.

SECTION 902
FIRE PUMP AND RISER ROOM SIZE

[F] 902.1 Pump and riser room size. Where provided, fire pump rooms and *automatic sprinkler system* riser rooms shall be designed with adequate space for all equipment necessary for the installation, as defined by the manufacturer, with sufficient working room around the stationary equipment. Clearances around equipment to elements of permanent construction, including other installed equipment and appliances, shall be sufficient to allow inspection, service, repair or replacement without removing such elements of permanent construction or disabling the function of a required fire-resistance-rated assembly. Fire pump and *automatic sprinkler system* riser rooms shall be provided with doors and unobstructed passageways large enough to allow removal of the largest piece of equipment.

[F] 902.1.1 Access. Automatic sprinkler system risers, fire pumps and controllers shall be provided with ready access. Where located in a fire pump room or *automatic sprinkler system* riser room, the door shall be permitted to be locked provided that the key is available at all times.

[F] 902.1.2 Marking on access doors. Access doors for *automatic sprinkler system* riser rooms and fire pump rooms shall be labeled with an approved sign. The lettering shall be in contrasting color to the background. Letters shall have a minimum height of 2 inches (51 mm) with a minimum stroke of $^3/_8$ inch (10 mm).

[F] 902.1.3 Environment. Automatic sprinkler system riser rooms and fire pump rooms shall be maintained at a temperature of not less than 40°F (4°C). Heating units shall be permanently installed.

[F] 902.1.4 Lighting. Permanently installed artificial illumination shall be provided in the *automatic sprinkler system* riser rooms and fire pump rooms.

SECTION 903
AUTOMATIC SPRINKLER SYSTEMS

[F] 903.1 General. *Automatic sprinkler systems* shall comply with this section.

[F] 903.1.1 Alternative protection. Alternative *automatic fire-extinguishing systems* complying with Section 904 shall be permitted instead of automatic sprinkler protection where recognized by the applicable standard and *approved* by the fire code official.

[F] 903.2 Where required. Approved *automatic sprinkler systems* in new buildings and structures shall be provided in the locations described in Sections 903.2.1 through 903.2.12.

Exception: Spaces or areas in telecommunications buildings used exclusively for telecommunications equipment, associated electrical power distribution equipment, batteries and standby engines, provided that those spaces or areas are equipped throughout with an *automatic smoke detection system* in accordance with Section 907.2 and are separated from the remainder of the building by not less than 1-hour *fire barriers* constructed in accordance with Section 707 or not less than 2-hour *horizontal assemblies* constructed in accordance with Section 711, or both.

[F] 903.2.1 Group A. An *automatic sprinkler system* shall be provided throughout buildings and portions thereof used as Group A occupancies as provided in this section.

[F] 903.2.1.1 Group A-1. An *automatic sprinkler system* shall be provided throughout stories containing Group A-1 occupancies and throughout all stories from the Group A-1 occupancy to and including the *levels of exit discharge* serving that occupancy where one of the following conditions exists:

1. The *fire area* exceeds 12,000 square feet (1115 m^2).
2. The *fire area* has an *occupant load* of 300 or more.
3. The *fire area* is located on a floor other than a *level of exit discharge* serving such occupancies.
4. The *fire area* contains a multitheater complex.

[F] 903.2.1.2 Group A-2. An *automatic sprinkler system* shall be provided throughout stories containing Group A-2 occupancies and throughout all stories from the Group A-2 occupancy to and including the *levels of exit discharge* serving that occupancy where one of the following conditions exists:

1. The *fire area* exceeds 5,000 square feet (464 m^2).

2. The *fire area* has an *occupant load* of 100 or more.
3. The *fire area* is located on a floor other than a *level of exit discharge* serving such occupancies.

[F] 903.2.1.3 Group A-3. An *automatic sprinkler system* shall be provided throughout stories containing Group A-3 occupancies and throughout all stories from the Group A-3 occupancy to and including the *levels of exit discharge* serving that occupancy where one of the following conditions exists:

1. The *fire area* exceeds 12,000 square feet (1115 m^2).
2. The *fire area* has an *occupant load* of 300 or more.
3. The *fire area* is located on a floor other than a *level of exit discharge* serving such occupancies.

[F] 903.2.1.4 Group A-4. An *automatic sprinkler system* shall be provided throughout stories containing Group A-4 occupancies and throughout all stories from the Group A-4 occupancy to and including the *levels of exit discharge* serving that occupancy where one of the following conditions exists:

1. The *fire area* exceeds 12,000 square feet (1115 m^2).
2. The *fire area* has an *occupant load* of 300 or more.
3. The *fire area* is located on a floor other than a *level of exit discharge* serving such occupancies.

[F] 903.2.1.5 Group A-5. An *automatic sprinkler system* shall be provided for all enclosed Group A-5 accessory use areas in excess of 1,000 square feet (93 m^2).

[F] 903.2.1.5.1 Spaces under grandstands or bleachers. Enclosed spaces under *grandstands* or *bleachers* shall be equipped with an *automatic sprinkler system* in accordance with Section 903.3.1.1 where either of the following exist:

1. The enclosed area is 1,000 square feet (93 m^2) or less and is not constructed in accordance with Section 1030.1.1.1.
2. The enclosed area exceeds 1,000 square feet (93 m^2).

[F] 903.2.1.6 Assembly occupancies on roofs. Where an occupied roof has an assembly occupancy with an *occupant load* exceeding 100 for Group A-2 and 300 for other Group A occupancies, all floors between the occupied roof and the *level of exit discharge* shall be equipped with an *automatic sprinkler system* in accordance with Section 903.3.1.1 or 903.3.1.2.

Exception: *Open parking garages* of Type I or Type II construction.

903.2.1.7 Multiple fire areas. An *automatic sprinkler system* shall be provided where multiple *fire areas* of Group A-1, A-2, A-3 or A-4 occupancies share *exit* or exit access components and the combined *occupant load* of theses *fire areas* is 300 or more.

[F] 903.2.2 Ambulatory care facilities. An *automatic sprinkler system* shall be installed throughout the entire floor containing an *ambulatory care facility* where either of the following conditions exist at any time:

1. Four or more care recipients are *incapable of self-preservation*.
2. One or more care recipients that are *incapable of self-preservation* are located at other than the *level of exit discharge* serving such a facility.

In buildings where ambulatory care is provided on levels other than the *level of exit discharge*, an *automatic sprinkler system* shall be installed throughout the entire floor as well as all floors below where such care is provided, and all floors between the level of ambulatory care and the nearest *level of exit discharge*, the *level of exit discharge*, and all floors below the level of *exit discharge*.

Exception: Floors classified as an *open parking garage* are not required to be sprinklered.

[F] 903.2.3 Group E. An *automatic sprinkler system* shall be provided for Group E occupancies as follows:

1. Throughout all Group E *fire areas* greater than 12,000 square feet (1115 m^2) in area.
2. The Group E *fire area* is located on a floor other than a *level of exit discharge* serving such occupancies.

   Exception: In buildings where every classroom has not fewer than one exterior *exit* door at ground level, an *automatic sprinkler system* is not required in any area below the lowest *level of exit discharge* serving that area.

3. The Group E *fire area* has an *occupant load* of 300 or more.

[F] 903.2.4 Group F-1. An *automatic sprinkler system* shall be provided throughout all buildings containing a Group F-1 occupancy where one of the following conditions exists:

1. A Group F-1 *fire area* exceeds 12,000 square feet (1115 m^2).
2. A Group F-1 *fire area* is located more than three stories above *grade plane*.
3. The combined area of all Group F-1 *fire areas* on all floors, including any *mezzanines*, exceeds 24,000 square feet (2230 m^2).

[F] 903.2.4.1 Woodworking operations. An *automatic sprinkler system* shall be provided throughout all Group F-1 occupancy *fire areas* that contain woodworking operations in excess of 2,500 square feet (232 m^2) in area that generate finely divided combustible waste or use finely divided combustible materials.

[F] 903.2.4.2 Group F-1 distilled spirits. An *automatic sprinkler system* shall be provided throughout a Group F-1 *fire area* used for the manufacture of distilled spirits.

[F] 903.2.4.3 Group F-1 upholstered furniture or mattresses. An *automatic sprinkler system* shall be provided throughout a Group F-1 *fire area* that exceeds 2,500 square feet (232 m^2) used for the manufacture of upholstered furniture or mattresses.

[F] 903.2.5 Group H. *Automatic sprinkler systems* shall be provided in high-hazard occupancies as required in Sections 903.2.5.1 through 903.2.5.3.

[F] 903.2.5.1 General. An *automatic sprinkler system* shall be installed in Group H occupancies.

[F] 903.2.5.2 Group H-5 occupancies. An *automatic sprinkler system* shall be installed throughout buildings containing Group H-5 occupancies. The design of the sprinkler system shall be not less than that required by this code for the occupancy hazard classifications in accordance with Table 903.2.5.2.

Where the design area of the sprinkler system consists of a *corridor* protected by one row of sprinklers, the maximum number of sprinklers required to be calculated is 13.

[F] TABLE 903.2.5.2
GROUP H-5 SPRINKLER DESIGN CRITERIA

LOCATION	OCCUPANCY HAZARD CLASSIFICATION
Fabrication areas	Ordinary Hazard Group 2
Service corridors	Ordinary Hazard Group 2
Storage rooms without dispensing	Ordinary Hazard Group 2
Storage rooms with dispensing	Extra Hazard Group 2
Corridors	Ordinary Hazard Group 2

[F] 903.2.5.3 Pyroxylin plastics. An *automatic sprinkler system* shall be provided in buildings, or portions thereof, where cellulose nitrate film or pyroxylin plastics are manufactured, stored or handled in quantities exceeding 100 pounds (45 kg).

[F] 903.2.6 Group I. An *automatic sprinkler system* shall be provided throughout buildings with a Group I *fire area*.

Exceptions:

1. An *automatic sprinkler system* installed in accordance with Section 903.3.1.2 shall be permitted in Group I-1, Condition 1 facilities.
2. An *automatic sprinkler system* is not required where Group I-4 day care facilities are at the *level of exit discharge* and where every room where care is provided has not fewer than one exterior exit door.
3. In buildings where Group I-4 day care is provided on levels other than the *level of exit discharge*, an *automatic sprinkler system* in accordance with Section 903.3.1.1 shall be installed on the entire floor where care is provided, all floors between the level of care and the *level of exit discharge*, and all floors below the *level of exit discharge* other than areas classified as an open parking garage.

[F] 903.2.7 Group M. An *automatic sprinkler system* shall be provided throughout buildings containing a Group M occupancy where one of the following conditions exists:

1. A Group M *fire area* exceeds 12,000 square feet (1115 m^2).
2. A Group M *fire area* is located more than three stories above *grade plane*.
3. The combined area of all Group M *fire areas* on all floors, including any *mezzanines*, exceeds 24,000 square feet (2230 m^2).

[F] 903.2.7.1 High-piled storage. An *automatic sprinkler system* shall be provided in accordance with the *International Fire Code* in all buildings of Group M where storage of merchandise is in high-piled or rack storage arrays.

[F] 903.2.7.2 Group M upholstered furniture or mattresses. An *automatic sprinkler system* shall be provided throughout a Group M *fire area* where the area used for the display and sale of upholstered furniture or mattresses exceeds 5,000 square feet (464 m^2).

[F] 903.2.8 Group R. An *automatic sprinkler system* installed in accordance with Section 903.3 shall be provided throughout all buildings with a Group R *fire area*.

[F] 903.2.8.1 Group R-3. An *automatic sprinkler system* installed in accordance with Section 903.3.1.3 shall be permitted in Group R-3 occupancies.

[F] 903.2.8.2 Group R-4, Condition 1. An *automatic sprinkler system* installed in accordance with Section 903.3.1.3 shall be permitted in Group R-4, Condition 1 occupancies.

[F] 903.2.8.3 Group R-4, Condition 2. An *automatic sprinkler system* installed in accordance with Section 903.3.1.2 shall be permitted in Group R-4, Condition 2 occupancies.

[F] 903.2.8.4 Care facilities. An *automatic sprinkler system* installed in accordance with Section 903.3.1.3 shall be permitted in care facilities with five or fewer individuals in a single-family dwelling.

[F] 903.2.9 Group S-1. An *automatic sprinkler system* shall be provided throughout all buildings containing a Group S-1 occupancy where one of the following conditions exists:

1. A Group S-1 *fire area* exceeds 12,000 square feet (1115 m^2).
2. A Group S-1 *fire area* is located more than three stories above *grade plane*.
3. The combined area of all Group S-1 *fire areas* on all floors, including any *mezzanines*, exceeds 24,000 square feet (2230 m^2).

4. A Group S-1 *fire area* used for the storage of commercial motor vehicles where the *fire area* exceeds 5,000 square feet (464 m^2).

[F] 903.2.9.1 Repair garages. An *automatic sprinkler system* shall be provided throughout all buildings used as *repair garages* in accordance with Section 406, as shown:

1. Buildings having two or more *stories above grade plane*, including basements, with a *fire area* containing a *repair garage* exceeding 10,000 square feet (929 m^2).
2. Buildings not more than one *story above grade plane*, with a *fire area* containing a *repair garage* exceeding 12,000 square feet (1115 m^2).
3. Buildings with *repair garages* servicing vehicles parked in basements.
4. A Group S-1 *fire area* used for the repair of commercial motor vehicles where the *fire area* exceeds 5,000 square feet (464 m^2).

[F] 903.2.9.2 Bulk storage of tires. Buildings and structures where the area for the storage of tires exceeds 20,000 cubic feet (566 m^3) shall be equipped throughout with an *automatic sprinkler system* in accordance with Section 903.3.1.1.

[F] 903.2.9.3 Group S-1 Distilled spirits or wine. An *automatic sprinkler system* shall be provided throughout a Group S-1 *fire area* used for the bulk storage of distilled spirits or wine.

[F] 903.2.9.4 Group S-1 upholstered furniture and mattresses. An *automatic sprinkler system* shall be provided throughout a Group S-1 *fire area* where the area used for the storage of upholstered furniture or mattresses exceeds 2,500 square feet (232 m^2).

Exception: Self-service storage facilities not greater than one story above grade plane where all storage spaces can be accessed directly from the exterior.

[F] 903.2.10 Group S-2 parking garages. An *automatic sprinkler system* shall be provided throughout buildings classified as parking garages where any of the following conditions exists:

1. Where the fire area of the enclosed parking garage in accordance with Section 406.6 exceeds 12,000 square feet (1115 m^2).
2. Where the enclosed parking garage in accordance with Section 406.6 is located beneath other groups.

   Exception: Enclosed parking garages located beneath Group R-3 occupancies.
3. Where the *fire area* of the *open parking garage* in accordance with Section 406.5 exceeds 48,000 square feet (4460 m^2).

[F] 903.2.10.1 Commercial parking garages. An *automatic sprinkler system* shall be provided throughout buildings used for storage of commercial motor vehicles where the *fire area* exceeds 5,000 square feet (464 m^2).

[F] 903.2.10.2 Mechanical-access enclosed parking garages. An *approved automatic sprinkler system* shall be provided throughout buildings used for the storage of motor vehicles in a *mechanical-access enclosed parking garage*. The portion of the building that contains the *mechanical-access enclosed parking garage* shall be protected with a specially engineered *automatic sprinkler system*.

[F] 903.2.11 Specific building areas and hazards. In all occupancies other than Group U, an *automatic sprinkler system* shall be installed for building design or hazards in the locations set forth in Sections 903.2.11.1 through 903.2.11.6.

[F] 903.2.11.1 Stories without openings. An *automatic sprinkler system* shall be installed throughout all stories, including basements, of all buildings where the floor area exceeds 1,500 square feet (139.4 m^2) and where the *story* does not comply with the following criteria for *exterior wall* openings:

1. Openings below grade that lead directly to ground level by an exterior *stairway* complying with Section 1011 or an outside *ramp* complying with Section 1012. Openings shall be located in each 50 linear feet (15 240 mm), or fraction thereof, of *exterior wall* in the *story* on not fewer than one side. The required openings shall be distributed such that the lineal distance between adjacent openings does not exceed 50 feet (15 240 mm).
2. Openings entirely above the adjoining ground level totaling not less than 20 square feet (1.86 m^2) in each 50 linear feet (15 240 mm), or fraction thereof, of *exterior wall* in the *story* on not fewer than one side. The required openings shall be distributed such that the lineal distance between adjacent openings does not exceed 50 feet (15 240 mm). The height of the bottom of the clear opening shall not exceed 44 inches (1118 mm) measured from the floor.

[F] 903.2.11.1.1 Opening dimensions and access. Openings shall have a minimum dimension of not less than 30 inches (762 mm). Access to such openings shall be provided for the fire department from the exterior and shall not be obstructed in a manner such that fire fighting or rescue cannot be accomplished from the exterior.

[F] 903.2.11.1.2 Openings on one side only. Where openings in a *story* are provided on only one side and the opposite wall of such *story* is more than 75 feet (22 860 mm) from such openings, the *story* shall be equipped throughout with an *approved automatic sprinkler system*, or openings shall be provided on not fewer than two sides of the *story*.

[F] 903.2.11.1.3 Basements. Where any portion of a *basement* is located more than 75 feet (22 860 mm) from openings required by Section 903.2.11.1, or where walls, partitions or other obstructions are installed that restrict the application of water from hose streams, the *basement* shall be equipped throughout with an *approved automatic sprinkler system*.

[F] 903.2.11.2 Rubbish and linen chutes. An *automatic sprinkler system* shall be installed at the top of rubbish and linen chutes and in their terminal rooms. Chutes shall have additional sprinkler heads installed at alternate floors and at the lowest intake. Where a rubbish chute extends through a building more than one floor below the lowest intake, the extension shall have sprinklers installed that are recessed from the drop area of the chute and protected from freezing in accordance with Section 903.3.1.1. Such sprinklers shall be installed at alternate floors, beginning with the second level below the last intake and ending with the floor above the discharge. *Access* to sprinklers in chutes shall be provided for servicing.

[F] 903.2.11.3 Buildings 55 feet or more in height. An *automatic sprinkler system* shall be installed throughout buildings that have one or more stories with an *occupant load* of 30 or more located 55 feet (16 764 mm) or more above the lowest level of fire department vehicle access, measured to the finished floor.

Exception: Occupancies in Group F-2.

[F] 903.2.11.4 Ducts conveying hazardous exhausts. Where required by the *International Mechanical Code*, automatic sprinklers shall be provided in ducts conveying hazardous exhaust or flammable or combustible materials.

Exception: Ducts where the largest cross-sectional diameter of the duct is less than 10 inches (254 mm).

[F] 903.2.11.5 Commercial cooking operations. An *automatic sprinkler system* shall be installed in commercial kitchen exhaust hood and duct systems where an *automatic sprinkler system* is used to comply with Section 904.

[F] 903.2.11.6 Other required fire protection systems. In addition to the requirements of Section 903.2, the provisions indicated in Table 903.2.11.6 require the installation of a fire protection system for certain buildings and areas.

[[F] 903.2.12 During construction. *Automatic sprinkler systems* required during construction, alteration and demolition operations shall be provided in accordance with Chapter 33 of the *International Fire Code*.

[F] 903.3 Installation requirements. *Automatic sprinkler systems* shall be designed and installed in accordance with Sections 903.3.1 through 903.3.8.

[F] 903.3.1 Standards. Sprinkler systems shall be designed and installed in accordance with Section 903.3.1.1 unless otherwise permitted by Sections 903.3.1.2 and 903.3.1.3 and other chapters of this code, as applicable.

[F] 903.3.1.1 NFPA 13 sprinkler systems. Where the provisions of this code require that a building or portion thereof be equipped throughout with an *automatic sprinkler system* in accordance with this section, sprinklers shall be installed throughout in accordance with NFPA 13 except as provided in Sections 903.3.1.1.1 and 903.3.1.1.2.

[F] 903.3.1.1.1 Exempt locations. Automatic sprinklers shall not be required in the following rooms or areas where such rooms or areas are protected with an *approved* automatic fire detection system in accordance with Section 907.2 that will respond to visible or invisible particles of combustion. Sprinklers shall not be omitted from a room merely because it is damp, of fire-resistance-rated construction or contains electrical equipment.

1. A room where the application of water, or flame and water, constitutes a serious life or fire hazard.
2. A room or space where sprinklers are considered undesirable because of the nature of the contents, where *approved* by the fire code official.
3. Generator and transformer rooms separated from the remainder of the building by walls and floor/ceiling or roof/ceiling assemblies having a *fire-resistance rating* of not less than 2 hours.

[F]TABLE 903.2.11.6
ADDITIONAL REQUIRED PROTECTION SYSTEMS

SECTION	SUBJECT
402.5, 402.6.2	Covered and open mall buildings
403.3	High-rise buildings
404.3	Atriums
405.3	Underground structures
407.7	Group I-2
410.6	Stages
411.3	Special amusement buildings
412.2.4	Airport traffic control towers
412.3.6, 412.3.6.1, 412.5.6	Aircraft hangars
415.11.11	Group H-5 HPM exhaust ducts
416.5	Flammable finishes
417.4	Drying rooms
424.3	Play structures
428	Buildings containing laboratory suites
507	Unlimited area buildings
508.5.7	Live/work units
509.4	Incidental uses
1030.6.2.3	Smoke-protected assembly seating
IFC	Sprinkler system requirements as set forth in Section 903.2.11.6 of the *International Fire Code*

4. Rooms or areas that are of noncombustible construction with wholly noncombustible contents.
5. Fire service access elevator machine rooms and machinery spaces.
6. Machine rooms, machinery spaces, control rooms and control spaces associated with occupant evacuation elevators designed in accordance with Section 3008.

[F] 903.3.1.1.2 Bathrooms. In Group R occupancies sprinklers shall not be required in bathrooms that do not exceed 55 square feet (5 m^2) in area and are located within individual *dwelling units* or *sleeping units*, provided that walls and ceilings, including the walls and ceilings behind a shower enclosure or tub, are of noncombustible or limited-combustible materials with a 15-minute thermal barrier rating.

[F] 903.3.1.2 NFPA 13R sprinkler systems. *Automatic sprinkler systems* in Group R occupancies shall be permitted to be installed throughout in accordance with NFPA 13R where the Group R occupancy meets all of the following conditions:

1. Four stories or fewer above *grade plane*.
2. The floor level of the highest *story* is 30 feet (9144 mm) or less above the lowest level of fire department vehicle access.
3. The floor level of the lowest *story* is 30 feet (9144 mm) or less below the lowest level of fire department vehicle access.

The number of stories of Group R occupancies constructed in accordance with Sections 510.2 and 510.4 shall be measured from grade plane.

[F] 903.3.1.2.1 Balconies and decks. Sprinkler protection shall be provided for exterior balconies, decks and ground floor patios of *dwelling units* and *sleeping units* where either of the following conditions exists:

1. The building is of Type V construction, provided that there is a roof or deck above.
2. Exterior balconies, decks and ground floor patios of dwelling units and sleeping units are constructed in accordance with Section 705.2.3.1, Exception 3.

Sidewall sprinklers that are used to protect such areas shall be permitted to be located such that their deflectors are within 1 inch (25 mm) to 6 inches (152 mm) below the structural members and a maximum distance of 14 inches (356 mm) below the deck of the exterior balconies and decks that are constructed of open wood joist construction.

[F] 903.3.1.2.2 *Corridors* and balconies in the means of egress. Sprinkler protection shall be provided in *corridors* and for balconies in the *means of egress* where any of the following conditions apply:

1. Corridors with combustible floor or walls.
2. Corridors with an interior change of direction exceeding 45 degrees (0.79 rad).
3. Corridors that are less than 50 percent open to the outside atmosphere at the ends.
4. Open-ended corridors and associated exterior stairways and ramps as specified in Section 1027.6, Exception 3.
5. Egress balconies not complying with Sections 1021.2 and 1021.3.

[F] 903.3.1.2.3 Attics. *Attic* protection shall be provided as follows:

1. *Attics* that are used or intended for living purposes or storage shall be protected by an *automatic sprinkler system.*
2. Where fuel-fired equipment is installed in an unsprinklered *attic*, not fewer than one quick-response intermediate temperature sprinkler shall be installed above the equipment.
3. Where located in a building of Type III, Type IV or Type V construction designed in accordance with Section 510.2 or 510.4, attics not required by Item 1 to have sprinklers shall comply with one of the following if the roof assembly is located more than 55 feet (16 764 mm) above the lowest level of fire department vehicle access needed to meet the provisions in Section 503.
    - 3.1. Provide *automatic sprinkler system* protection.
    - 3.2. Construct the *attic* using noncombustible materials.
    - 3.3. Construct the *attic* using *fire-retardant-treated wood* complying with Section 2303.2.
    - 3.4. Fill the *attic* with noncombustible insulation.

The height of the roof assembly shall be determined by measuring the distance from the lowest required fire vehicle access road surface adjacent to the building to the eave of the highest pitched roof, the intersection of the highest roof to the exterior wall, or the top of the highest parapet, whichever yields the greatest distance. For the purpose of this measurement, required fire vehicle access roads shall include only those roads that are necessary for compliance with Section 503 of the *International Fire Code*:

4. Group R-4, Condition 2 occupancy *attics* not required by Item 1 to have sprinklers shall comply with one of the following:
    - 4.1. Provide *automatic sprinkler system* protection.
    - 4.2. Provide a heat detection system throughout the *attic* that is

arranged to activate the building fire alarm system.

4.3. Construct the *attic* using noncombustible materials.

4.4. Construct the *attic* using *fire-retardant-treated wood* complying with Section 2303.2.

4.5. Fill the *attic* with noncombustible insulation.

[F] 903.3.1.3 NFPA 13D sprinkler systems. *Automatic sprinkler systems* installed in one- and two-family *dwellings*; Group R-3; Group R-4, Condition 1; and *townhouses* shall be permitted to be installed throughout in accordance with NFPA 13D.

[F] 903.3.2 Quick-response and residential sprinklers. Where *automatic sprinkler systems* are required by this code, quick-response or residential automatic sprinklers shall be installed in all of the following areas in accordance with Section 903.3.1 and their listings:

1. Throughout all spaces within a *smoke compartment* containing care recipient *sleeping units* in Group I-2 in accordance with this code.
2. Throughout all spaces within a *smoke compartment* containing gas fireplace appliances and decorative gas appliances in Group I-2.
3. Throughout all spaces within a *smoke compartment* containing treatment rooms in *ambulatory care facilities*.
4. *Dwelling units* and *sleeping units* in Group I-1 and R occupancies.
5. Light-hazard occupancies as defined in NFPA 13.

[F] 903.3.3 Obstructed locations. Automatic sprinklers shall be installed with regard to obstructions that will delay activation or obstruct the water distribution pattern and shall be in accordance with the applicable *automatic sprinkler system* standard that is being used. Automatic sprinklers shall be installed in or under covered kiosks, displays, booths, concession stands, or equipment that exceeds 4 feet (1219 mm) in width. Not less than a 3-foot (914 mm) clearance shall be maintained between automatic sprinklers and the top of piles of *combustible fibers*.

Exception: Kitchen equipment under exhaust hoods protected with a fire-extinguishing system in accordance with Section 904.

[F] 903.3.4 Actuation. *Automatic sprinkler systems* shall be automatically actuated unless specifically provided for in this code.

[F] 903.3.5 Water supplies. Water supplies for *automatic sprinkler systems* shall comply with this section and the standards referenced in Section 903.3.1. The potable water supply shall be protected against backflow in accordance with the requirements of this section and the *International Plumbing Code*. For connections to public waterworks systems, the water supply test used for design of fire protection systems shall be adjusted to account for seasonal and daily pressure fluctuations based on information from the water supply authority and as approved by the fire code official.

[F] 903.3.5.1 Domestic services. Where the domestic service provides the water supply for the *automatic sprinkler system*, the supply shall be in accordance with this section.

[F] 903.3.5.2 Residential combination services. A single combination water supply shall be allowed provided that the domestic demand is added to the sprinkler demand as required by NFPA 13R.

[F] 903.3.6 Hose threads. Fire hose threads and fittings used in connection with *automatic sprinkler systems* shall be as prescribed by the fire code official.

[F] 903.3.7 Fire department connections. Fire department connections for *automatic sprinkler systems* shall be installed in accordance with Section 912.

[F] 903.3.8 Limited area sprinkler systems. Limited area sprinkler systems shall be in accordance with the standards listed in Section 903.3.1 except as provided in Sections 903.3.8.1 through 903.3.8.5.

903.3.8.1 Number of sprinklers. Limited area sprinkler systems shall not exceed six sprinklers in any single *fire area*.

903.3.8.2 Occupancy hazard classification. Only areas classified by NFPA 13 as Light Hazard or Ordinary Hazard Group 1 shall be permitted to be protected by limited area sprinkler systems.

903.3.8.3 Piping arrangement. Where a limited area sprinkler system is installed in a building with an automatic wet standpipe system, sprinklers shall be supplied by the standpipe system. Where a limited area sprinkler system is installed in a building without an automatic wet standpipe system, water shall be permitted to be supplied by the plumbing system provided that the plumbing system is capable of simultaneously supplying domestic and sprinkler demands.

903.3.8.4 Supervision. Control valves shall not be installed between the water supply and sprinklers unless the valves are of an *approved* indicating type that are supervised or secured in the open position.

903.3.8.5 Calculations. Hydraulic calculations in accordance with NFPA 13 shall be provided to demonstrate that the available water flow and pressure are adequate to supply all sprinklers installed in any single *fire area* with discharge densities corresponding to the hazard classification.

[F] 903.4 Sprinkler system supervision and alarms. Valves controlling the water supply for *automatic sprinkler systems*, pumps, tanks, water levels and temperatures, critical air pressures and waterflow switches on all sprinkler systems shall be electrically supervised by a *listed* fire alarm control unit.

Exceptions:

1. *Automatic sprinkler systems* protecting one- and two-family *dwellings*.

2. Limited area sprinkler systems in accordance with Section 903.3.8.
3. *Automatic sprinkler systems* installed in accordance with NFPA 13R where a common supply main is used to supply both domestic water and the *automatic sprinkler system*, and a separate shutoff valve for the *automatic sprinkler system* is not provided.
4. Jockey pump control valves that are sealed or locked in the open position.
5. Control valves to commercial kitchen hoods, paint spray booths or dip tanks that are sealed or locked in the open position.
6. Valves controlling the fuel supply to fire pump engines that are sealed or locked in the open position.
7. Trim valves to pressure switches in dry, preaction and deluge sprinkler systems that are sealed or locked in the open position.
8. Underground key or hub gate valves in roadway boxes.

[F] 903.4.1 Monitoring. Alarm, supervisory and trouble signals shall be distinctly different and shall be automatically transmitted to an *approved* supervising station or, where *approved* by the fire code official, shall sound an audible signal at a *constantly attended location*.

Exception: Backflow prevention device test valves located in limited area sprinkler system supply piping shall be locked in the open position. In occupancies required to be equipped with a fire alarm system, the backflow preventer valves shall be electrically supervised by a tamper switch installed in accordance with NFPA 72 and separately annunciated.

[F] 903.4.2 Alarms. An approved audible device, located on the exterior of the building in an approved location, shall be connected to each *automatic sprinkler system*. Such sprinkler waterflow alarm devices shall be activated by water flow equivalent to the flow of a single sprinkler of the smallest orifice size installed in the system. Where a fire alarm system is installed, actuation of the *automatic sprinkler system* shall actuate the building fire alarm system.

[F] 903.4.3 Floor control valves. *Approved* supervised indicating control valves shall be provided at the point of connection to the riser on each floor in high-rise buildings.

[F] 903.5 Testing and maintenance. Sprinkler systems shall be tested and maintained in accordance with the *International Fire Code*.

SECTION 904 ALTERNATIVE AUTOMATIC FIRE-EXTINGUISHING SYSTEMS

[F] 904.1 General. Automatic fire-extinguishing systems, other than *automatic sprinkler systems*, shall be designed, installed, inspected, tested and maintained in accordance with the provisions of this section and the applicable referenced standards.

[F] 904.2 Where permitted. Automatic fire-extinguishing systems installed as an alternative to the required *automatic sprinkler systems* of Section 903 shall be *approved* by the fire code official.

[F] 904.2.1 Restriction on using automatic sprinkler system exceptions or reductions. Automatic fire-extinguishing systems shall not be considered alternatives for the purposes of exceptions or reductions allowed for *automatic sprinkler systems* or by other requirements of this code.

[F] 904.2.2 Commercial hood and duct systems. Each required commercial kitchen exhaust hood and duct system required by Section 606 of the *International Fire Code* or Chapter 5 of the *International Mechanical Code* to have a Type I hood shall be protected with an approved automatic fire-extinguishing system installed in accordance with this code.

[F] 904.3 Installation. Automatic fire-extinguishing systems shall be installed in accordance with this section.

[F] 904.3.1 Electrical wiring. Electrical wiring shall be in accordance with NFPA 70.

[F] 904.3.2 Actuation. Automatic fire-extinguishing systems shall be automatically actuated and provided with a manual means of actuation in accordance with Section 904.11.1. Where more than one hazard could be simultaneously involved in fire due to their proximity, all hazards shall be protected by a single system designed to protect all hazards that could become involved.

Exception: Multiple systems shall be permitted to be installed if they are designed to operate simultaneously.

[F] 904.3.3 System interlocking. Automatic equipment interlocks with fuel shutoffs, ventilation controls, door closers, window shutters, conveyor openings, smoke and heat vents and other features necessary for proper operation of the fire-extinguishing system shall be provided as required by the design and installation standard utilized for the hazard.

[F] 904.3.4 Alarms and warning signs. Where alarms are required to indicate the operation of automatic fire-extinguishing systems, distinctive audible and visible alarms and warning signs shall be provided to warn of pending agent discharge. Where exposure to automatic-extinguishing agents poses a hazard to persons and a delay is required to ensure the evacuation of occupants before agent discharge, a separate warning signal shall be provided to alert occupants once agent discharge has begun. Audible signals shall be in accordance with Section 907.5.2.

[F] 904.3.5 Monitoring. Where a building fire alarm system is installed, automatic fire-extinguishing systems shall be monitored by the building fire alarm system in accordance with NFPA 72.

[F] 904.4 Inspection and testing. Automatic fire-extinguishing systems shall be inspected and tested in accordance with the provisions of this section prior to acceptance.

[F] 904.4.1 Inspection. Prior to conducting final acceptance tests, all of the following items shall be inspected:

1. Hazard specification for consistency with design hazard.
2. Type, location and spacing of automatic- and manual-initiating devices.
3. Size, placement and position of nozzles or discharge orifices.
4. Location and identification of audible and visible alarm devices.
5. Identification of devices with proper designations.
6. Operating instructions.

[F] 904.4.2 Alarm testing. Notification appliances, connections to fire alarm systems and connections to *approved* supervising stations shall be tested in accordance with this section and Section 907 to verify proper operation.

[F] 904.4.2.1 Audible and visible signals. The audibility and visibility of notification appliances signaling agent discharge or system operation, where required, shall be verified.

[F] 904.4.3 Monitor testing. Connections to protected premises and supervising station fire alarm systems shall be tested to verify proper identification and retransmission of alarms from automatic fire-extinguishing systems.

[F] 904.5 Wet-chemical systems. Wet-chemical extinguishing systems shall be installed, maintained, periodically inspected and tested in accordance with NFPA 17A and their listing. Records of inspections and testing shall be maintained.

[F] 904.6 Dry-chemical systems. Dry-chemical extinguishing systems shall be installed, maintained, periodically inspected and tested in accordance with NFPA 17 and their listing. Records of inspections and testing shall be maintained.

[F] 904.7 Foam systems. Foam-extinguishing systems shall be installed, maintained, periodically inspected and tested in accordance with NFPA 11 and NFPA 16 and their listing. Records of inspections and testing shall be maintained.

[F] 904.8 Carbon dioxide systems. Carbon dioxide extinguishing systems shall be installed, maintained, periodically inspected and tested in accordance with NFPA 12 and their listing. Records of inspections and testing shall be maintained.

[F] 904.9 Halon systems. Halogenated extinguishing systems shall be installed, maintained, periodically inspected and tested in accordance with NFPA 12A and their listing. Records of inspections and testing shall be maintained.

[F] 904.10 Clean-agent systems. Clean-agent fire-extinguishing systems shall be installed, maintained, periodically inspected and tested in accordance with NFPA 2001 and their listing. Records of inspections and testing shall be maintained.

[F] 904.11 Automatic water mist systems. *Automatic water mist systems* shall be permitted in applications that are consistent with the applicable listing or approvals and shall comply with Sections 904.11.1 through 904.11.3.

[F] 904.11.1 Design and installation requirements. *Automatic water mist systems* shall be designed and installed in accordance with Sections 904.11.1.1 through 904.11.1.4.

[F] 904.11.1.1 General. *Automatic water mist systems* shall be designed and installed in accordance with NFPA 750 and the manufacturer's instructions.

[F] 904.11.1.2 Actuation. *Automatic water mist systems* shall be automatically actuated.

[F] 904.11.1.3 Water supply protection. Connections to a potable water supply shall be protected against backflow in accordance with the *International Plumbing Code.*

[F] 904.11.1.4 Secondary water supply. Where a secondary water supply is required for an *automatic sprinkler system*, an *automatic water mist system* shall be provided with an *approved* secondary water supply.

[F] 904.11.2 Water mist system supervision and alarms. Supervision and alarms shall be provided as required for *automatic sprinkler systems* in accordance with Section 903.4.

[F] 904.11.2.1 Monitoring. Monitoring shall be provided as required for *automatic sprinkler systems* in accordance with Section 903.4.1.

[F] 904.11.2.2 Alarms. Alarms shall be provided as required for *automatic sprinkler systems* in accordance with Section 903.4.2.

[F] 904.11.2.3 Floor control valves. Floor control valves shall be provided as required for *automatic sprinkler systems* in accordance with Section 903.4.3.

[F] 904.11.3 Testing and maintenance. *Automatic water mist systems* shall be tested and maintained in accordance with the *International Fire Code.* *

[F] 904.12 Aerosol fire-extinguishing systems. Aerosol fire-extinguishing systems shall be installed, maintained, periodically inspected and tested in accordance with NFPA 2010, and their listing. **

Such devices and appurtenances shall be listed and installed in compliance with manufacturers' instructions.

[F] 904.13 Commercial cooking systems. The automatic fire-extinguishing system for commercial cooking systems shall be of a type recognized for protection of commercial cooking equipment and exhaust systems of the type and arrangement protected. Preengineered automatic dry- and wet-chemical extinguishing systems shall be tested in accordance with UL 300 and *listed* and *labeled* for the intended application. Other types of automatic fire-extinguishing systems shall be *listed* and *labeled* for specific use as protection for commercial cooking operations. The system shall be **

installed in accordance with this code, NFPA 96, its listing and the manufacturer's installation instructions. Automatic fire-extinguishing systems of the following types shall be installed in accordance with the referenced standard indicated, as follows:

1. Carbon dioxide extinguishing systems, NFPA 12.
2. *Automatic sprinkler systems*, NFPA 13.
3. Automatic water mist systems, NFPA 750.
4. Foam-water sprinkler system or foam-water spray systems, NFPA 16.
5. Dry-chemical extinguishing systems, NFPA 17.
6. Wet-chemical extinguishing systems, NFPA 17A.

Exception: Factory-built commercial cooking recirculating systems that are tested in accordance with UL 710B and *listed*, *labeled* and installed in accordance with Section 304.1 of the *International Mechanical Code*.

[F] 904.13.1 Manual system operation. A manual actuation device shall be located at or near a *means of egress* from the cooking area not less than 10 feet (3048 mm) and not more than 20 feet (6096 mm) from the kitchen exhaust system. The manual actuation device shall be installed not more than 48 inches (1200 mm) or less than 42 inches (1067 mm) above the floor and shall clearly identify the hazard protected. The manual actuation shall require a maximum force of 40 pounds (178 N) and a maximum movement of 14 inches (356 mm) to actuate the fire suppression system.

Exception: *Automatic sprinkler systems* shall not be required to be equipped with manual actuation means.

[F] 904.13.2 System interconnection. The actuation of the fire suppression system shall automatically shut down the fuel or electrical power supply to the cooking equipment. The fuel and electrical supply reset shall be manual.

[F] 904.13.3 Carbon dioxide systems. Where carbon dioxide systems are used, there shall be a nozzle at the top of the ventilating duct. Additional nozzles that are symmetrically arranged to give uniform distribution shall be installed within vertical ducts exceeding 20 feet (6096 mm) and horizontal ducts exceeding 50 feet (15 240 mm). *Dampers* shall be installed at either the top or the bottom of the duct and shall be arranged to operate automatically upon activation of the fire-extinguishing system. Where the *damper* is installed at the top of the duct, the top nozzle shall be immediately below the *damper*. Automatic carbon dioxide fire-extinguishing systems shall be sufficiently sized to protect against all hazards venting through a common duct simultaneously.

[F] 904.13.3.1 Ventilation system. Commercial-type cooking equipment protected by an automatic carbon dioxide-extinguishing system shall be arranged to shut off the ventilation system upon activation.

[F] 904.13.4 Special provisions for automatic sprinkler systems. *Automatic sprinkler systems* protecting commercial-type cooking equipment shall be supplied from a separate, indicating-type control valve that is identified. Access to the control valve shall be provided.

[F] 904.13.4.1 Listed sprinklers. Sprinklers used for the protection of fryers shall be tested in accordance with UL 199E, *listed* for that application and installed in accordance with their listing.

[F] 904.14 Domestic cooking facilities. Cooktops and ranges installed in the following occupancies shall be protected in accordance with Section 904.14.1: *

1. In Group I-1 occupancies where domestic cooking facilities are installed in accordance with Section 420.9.
2. In Group I-2 occupancies where domestic cooking facilities are installed in accordance with Section 407.2.7.
3. In Group R-2 college *dormitories* where domestic cooking facilities are installed in accordance with Section 420.11.

[F] 904.14.1 Protection from fire. Cooktops and ranges shall be protected in accordance with Section 904.14.1.1 or 904.14.1.2.

[F] 904.14.1.1 Automatic fire-extinguishing system. The domestic recirculating or exterior vented cooking hood provided over the cooktop or range shall be equipped with an approved automatic fire-extinguishing system complying with the following:

1. The automatic fire-extinguishing system shall be of a type recognized for protection of domestic cooking equipment. Preengineered automatic fire-extinguishing systems shall be listed and labeled in accordance with UL 300A and installed in accordance with the manufacturer's instructions.
2. Manual actuation of the fire-extinguishing system shall be provided in accordance with Section 904.13.1.
3. Interconnection of the fuel and electric power supply shall be in accordance with Section 904.13.2.

[F] 904.14.1.2 Ignition prevention. Cooktops and ranges shall include burners that have been tested and listed to prevent ignition of cooking oil with burners turned on to their maximum heat settings and allowed to operate for 30 minutes.

SECTION 905
STANDPIPE SYSTEMS

[F] 905.1 General. Standpipe systems shall be provided in new buildings and structures in accordance with Sections 905.2 through 905.11. In buildings used for high-piled combustible storage, fire protection shall be in accordance with the *International Fire Code*.

[F] 905.2 Installation standard. Standpipe systems shall be installed in accordance with this section and NFPA 14. Fire department connections for standpipe systems shall be in accordance with Section 912.

[F] 905.3 Required installations. Standpipe systems shall be installed where required by Sections 905.3.1 through 905.3.8. Standpipe systems are allowed to be combined with *automatic sprinkler systems*.

Exception: Standpipe systems are not required in Group R-3 occupancies.

[F] 905.3.1 Height. Class III standpipe systems shall be installed throughout buildings where any of the following conditions exist:

1. Four or more stories are above or below *grade plane*.
2. The floor level of the highest *story* is located more than 30 feet (9144 mm) above the lowest level of fire department vehicle access.
3. The floor level of the lowest *story* is located more than 30 feet (9144 mm) below the highest level of fire department vehicle access.

Exceptions:

1. Class I standpipes are allowed in buildings equipped throughout with an *automatic sprinkler system* in accordance with Section 903.3.1.1 or 903.3.1.2.
2. Class I standpipes are allowed in Group B and E occupancies.
3. Class I standpipes are allowed in parking garages.
4. Class I standpipes are allowed in basements equipped throughout with an *automatic sprinkler system*.
5. Class I standpipes are allowed in buildings where occupant-use hose lines will not be utilized by trained personnel or the fire department.
6. In determining the lowest level of fire department vehicle access, it shall not be required to consider either of the following:
   6.1. Recessed loading docks for four vehicles or less.
   6.2. Conditions where topography makes access from the fire department vehicle to the building impractical or impossible.

[F] 905.3.2 Group A. Class I automatic wet standpipes shall be provided in nonsprinklered Group A buildings having an *occupant load* exceeding 1,000 persons.

Exceptions:

1. Open-air-seating spaces without enclosed spaces.
2. Class I automatic dry and semiautomatic dry standpipes or manual wet standpipes are allowed in buildings that are not high-rise buildings.

[F] 905.3.3 Covered and open mall buildings. Covered mall and *open mall buildings* shall be equipped throughout with a standpipe system where required by Section 905.3.1. Mall buildings not required to be equipped with a standpipe system by Section 905.3.1 shall be equipped with Class I hose connections connected to the *automatic sprinkler system* sized to deliver water at 250 gallons per minute (946.4 L/min) at the hydraulically most remote hose connection while concurrently supplying the automatic sprinkler system demand. The standpipe system shall be designed to not exceed a 50 pounds per square inch (psi) (345 kPa) residual pressure loss with a flow of 250 gallons per minute (946.4 L/min) from the fire department connection to the hydraulically most remote hose connection. Hose connections shall be provided at each of the following locations:

1. Within the mall at the entrance to each *exit passageway* or *corridor*.
2. At each floor-level landing within *interior exit stairways* opening directly on the mall.
3. At exterior public entrances to the mall of a *covered mall building*.
4. At public entrances at the perimeter line of an *open mall building*.
5. At other locations as necessary so that the distance to reach all portions of a tenant space does not exceed 200 feet (60 960 mm) from a hose connection.

[F] 905.3.4 Stages. *Stages* greater than 1,000 square feet in area (93 m^2) shall be equipped with a Class III wet standpipe system with $1^1/_2$-inch and $2^1/_2$-inch (38 mm and 64 mm) hose connections on each side of the stage.

Exception: Where the building or area is equipped throughout with an *automatic sprinkler system*, a $1^1/_2$-inch (38 mm) hose connection shall be installed in accordance with NFPA 13 or in accordance with NFPA 14 for Class II or III standpipes.

[F] 905.3.4.1 Hose and cabinet. The $1^1/_2$-inch (38 mm) hose connections shall be equipped with sufficient lengths of $1^1/_2$-inch (38 mm) hose to provide fire protection for the stage area. Hose connections shall be equipped with an *approved* adjustable fog nozzle and be mounted in a cabinet or on a rack.

[F] 905.3.5 Underground buildings. Underground buildings shall be equipped throughout with a Class I automatic wet or manual wet standpipe system.

[F] 905.3.6 Helistops and heliports. Buildings with a rooftop *helistop* or *heliport* shall be equipped with a Class I or III standpipe system extended to the roof level on which the *helistop* or *heliport* is located in accordance with Section 2007.5 of the *International Fire Code.*

[F] 905.3.7 Marinas and boatyards. Standpipes in marinas and boatyards shall comply with Chapter 36 of the *International Fire Code*.

[F] 905.3.8 Landscaped roofs. Buildings or structures that have landscaped roofs and that are equipped with a standpipe system shall have the standpipe system extended to the roof level on which the landscaped roof is located.

[F] 905.4 Location of Class I standpipe hose connections. Class I standpipe hose connections shall be provided in all of the following locations:

1. In every required *interior exit stairway*, a hose connection shall be provided for each story above and below *grade plane*. Hose connections shall be located at the main floor landing unless otherwise *approved* by the fire code official.

   Exception: A single hose connection shall be permitted to be installed in the open corridor or open breezeway between open *stairs* that are not greater than 75 feet (22 860 mm) apart.

2. On each side of the wall adjacent to the exit opening of a *horizontal exit*.

   Exception: Where floor areas adjacent to a *horizontal exit* are reachable from an *interior exit stairway* hose connection by a 30-foot (9144 mm) hose stream from a nozzle attached to 100 feet (30 480 mm) of hose, a hose connection shall not be required at the *horizontal exit*.

3. In every *exit passageway*, at the entrance from the *exit passageway* to other areas of a building.

   Exception: Where floor areas adjacent to an *exit passageway* are reachable from an *interior exit stairway* hose connection by a 30-foot (9144 mm) hose stream from a nozzle attached to 100 feet (30 480 mm) of hose, a hose connection shall not be required at the entrance from the *exit passageway* to other areas of the building.

4. In covered mall buildings, adjacent to each exterior public entrance to the mall and adjacent to each entrance from an exit *passageway* or exit *corridor* to the mall. In *open mall buildings*, adjacent to each public entrance to the mall at the perimeter line and adjacent to each entrance from an exit *passageway* or *exit* corridor to the mall.

5. Where the roof has a slope less than 4 units vertical in 12 units horizontal (33.3-percent slope), a hose connection shall be located to serve the roof or at the highest landing of an *interior exit stairway* with access to the roof provided in accordance with Section 1011.12.

6. Where the most remote portion of a nonsprinklered floor or *story* is more than 150 feet (45 720 mm) from a hose connection or the most remote portion of a sprinklered floor or *story* is more than 200 feet (60 960 mm) from a hose connection, the fire code official is authorized to require that additional hose connections be provided in *approved* locations.

[F] 905.4.1 Protection. Risers and laterals of Class I standpipe systems not located within an *interior exit stairway* shall be protected by a degree of *fire resistance* equal to that required for vertical enclosures in the building in which they are located.

Exception: In buildings equipped throughout with an *approved automatic sprinkler system*, laterals that are not located within an *interior exit stairway* are not required to be enclosed within fire-resistance-rated construction.

[F] 905.4.2 Interconnection. In buildings where more than one standpipe is provided, the standpipes shall be interconnected in accordance with NFPA 14.

[F] 905.5 Location of Class II standpipe hose connections. Class II standpipe hose connections located so that all portions of the building are within 30 feet (9144 mm) of a nozzle attached to 100 feet (30 480 mm) of hose. Class II standpipe hose connections shall be located where they will have *ready access*.

[F] 905.5.1 Groups A-1 and A-2. In Group A-1 and A-2 occupancies having *occupant loads* exceeding 1,000 persons, hose connections shall be located on each side of any stage, on each side of the rear of the auditorium, on each side of the balcony and on each tier of dressing rooms.

[F] 905.5.2 Protection. Fire-resistance-rated protection of risers and laterals of Class II standpipe systems is not required.

[F] 905.5.3 Class II system 1-inch hose. A minimum 1-inch (25 mm) hose shall be allowed to be used for hose stations in light-hazard occupancies where investigated and *listed* for this service and where *approved* by the fire code official.

[F] 905.6 Location of Class III standpipe hose connections. Class III standpipe systems shall have hose connections located as required for Class I standpipes in Section 905.4 and shall have Class II hose connections as required in Section 905.5.

[F] 905.6.1 Protection. Risers and laterals of Class III standpipe systems shall be protected as required for Class I systems in accordance with Section 905.4.1.

[F] 905.6.2 Interconnection. In buildings where more than one Class III standpipe is provided, the standpipes shall be interconnected in accordance with NFPA 14.

[F] 905.7 Cabinets. Cabinets containing fire-fighting equipment such as standpipes, fire hoses, fire extinguishers or fire department valves shall not be blocked from use or obscured from view.

[F] 905.7.1 Cabinet equipment identification. Cabinets shall be identified in an *approved* manner by a permanently attached sign with letters not less than 2 inches (51 mm) high in a color that contrasts with the background color, indicating the equipment contained therein.

Exceptions:

1. Doors not large enough to accommodate a written sign shall be marked with a permanently attached pictogram of the equipment contained therein.
2. Doors that have either an *approved* visual identification clear glass panel or a complete glass door panel are not required to be marked.

[F] 905.7.2 Locking cabinet doors. Cabinets shall be unlocked.

Exceptions:

1. Visual identification panels of glass or other *approved* transparent frangible material that is easily broken and allows access.
2. *Approved* locking arrangements.
3. Group I-3 occupancies.

[F] 905.8 Dry standpipes. Dry standpipes shall not be installed.

Exception: Where subject to freezing and in accordance with NFPA 14.

[F] 905.9 Valve supervision. Valves controlling water supplies shall be supervised in the open position so that a change in the normal position of the valve will generate a supervisory signal at the supervising station required by Section 903.4. Where a fire alarm system is provided, a signal shall be transmitted to the control unit.

Exceptions:

1. Valves to underground key or hub valves in roadway boxes do not require supervision.
2. Valves locked in the normal position and inspected as provided in this code in buildings not equipped with a fire alarm system.

[F] 905.10 During construction. Standpipe systems required during construction and demolition operations shall be provided in accordance with Section 3311.

[F] 905.11 Locking standpipe outlet caps. The *fire code official* is authorized to require locking caps on the outlets on standpipes where the responding fire department carries key wrenches for the removal that are compatible with locking FDC connection caps.

SECTION 906
PORTABLE FIRE EXTINGUISHERS

[F] 906.1 Where required. Portable fire extinguishers shall be installed in all of the following locations:

1. In Group A, B, E, F, H, I, M, R-1, R-2, R-4 and S occupancies.

   Exceptions:

   1. In Group R-2 occupancies, portable fire extinguishers shall be required only in locations specified in Items 2 through 6 where each dwelling unit is provided with a portable fire extinguisher having a minimum rating of 1-A:10-B:C.
   2. In Group E occupancies. portable fire extinguishers shall be required only in locations specified in Items 2 through 6 where each classroom is provided with a portable fire extinguisher having a minimum rating of 2-A:20-B:C.
   3. In storage areas of Group S Occupancies where forklift, powered industrial truck or powered cart operators are the primary occupants, fixed extinguishers, as specified in NFPA 10, shall not be required where in accordance with all of the following:
      - 3.1. Use of vehicle-mounted extinguishers shall be approved by the fire code official.
      - 3.2. Each vehicle shall be equipped with a 10-pound, 40A:80B:C extinguisher affixed to the vehicle using a mounting bracket approved by the extinguisher manufacturer or the fire code official for vehicular use.
      - 3.3. Not less than two spare extinguishers of equal or greater rating shall be available on site to replace a discharged extinguisher.
      - 3.4. Vehicle operators shall be trained in the proper operation, use and inspection of extinguishers.
      - 3.5. Inspections of vehicle-mounted extinguishers shall be performed daily.
2. Within 30 feet (9144 mm) distance of travel from commercial cooking equipment and from domestic cooking equipment in Group I-1; I-2, Condition 1; and R-2 college *dormitory* occupancies.
3. In areas where flammable or *combustible liquids* are stored, used or dispensed.
4. On each floor of structures under construction, except Group R-3 occupancies, in accordance with Section 3315.1 of the *International Fire Code.*
5. Where required by the *International Fire Code* sections indicated in Table 906.1.
6. Special-hazard areas, including but not limited to laboratories, computer rooms and generator rooms, where required by the fire code official.

Exception: Portable fire extinguishers are not required at normally unmanned Group U occupancy buildings or structures where a portable fire extinguisher suitable to the hazard of the location is provided on the vehicle of visiting personnel.

[F] 906.2 General requirements. Portable fire extinguishers shall be selected and installed in accordance with this section and NFPA 10.

Exceptions:

1. The distance of travel to reach an extinguisher shall not apply to the spectator seating portions of Group A-5 occupancies.
2. In Group I-3, portable fire extinguishers shall be permitted to be located at staff locations.

[F] TABLE 906.1
ADDITIONAL REQUIRED PORTABLE FIRE EXTINGUISHERS IN THE INTERNATIONAL FIRE CODE

IFC SECTION	SUBJECT
303.5	Asphalt kettles
307.5	Open burning
308.1.3	Open flames—torches
309.4	Powered industrial trucks
1204.10	Portable Generators
2005.2	Aircraft towing vehicles
2005.3	Aircraft welding apparatus
2005.4	Aircraft fuel-servicing tank vehicles
2005.5	Aircraft hydrant fuel-servicing vehicles
2005.6	Aircraft fuel-dispensing stations
2007.7	Heliports and helistops
2108.4	Dry cleaning plants
2305.5	Motor fuel-dispensing facilities
2310.6.4	Marine motor fuel-dispensing facilities
2311.6	Repair garages
2404.4.1	Spray-finishing operations
2405.4.2	Dip-tank operations
2406.4.2	Powder-coating areas
2804.3	Lumberyards/woodworking facilities
2808.8	Recycling facilities
2809.5	Exterior lumber storage
2903.5	Organic-coating areas
3006.3	Industrial ovens
3107.9	Tents and membrane structures
3206.10	High-piled storage
3315.1	Buildings under construction or demolition
3318.3	Roofing operations
3408.2	Tire rebuilding/storage
3504.2.6	Welding and other hot work
3604.4	Marinas
3703.6	Combustible fibers
5703.2.1	Flammable and combustible liquids, general
5704.3.3.1	Indoor storage of flammable and combustible liquids
5704.3.7.5.2	Liquid storage rooms for flammable and combustible liquids
5705.4.9	Solvent distillation units
5706.2.7	Farms and construction sites—flammable and combustible liquids storage
5706.4.10.1	Bulk plants and terminals for flammable and combustible liquids
5706.5.4.5	Commercial, industrial, governmental or manufacturing establishments—fuel dispensing
5706.6.4	Tank vehicles for flammable and combustible liquids
5906.5.7	Flammable solids
6108.2	LP-gas

[F] 906.3 Size and distribution. The size and distribution of portable fire extinguishers shall be in accordance with Sections 906.3.1 through 906.3.4.

[F] TABLE 906.3(1)
FIRE EXTINGUISHERS FOR CLASS A FIRE HAZARDS

	LIGHT (Low) HAZARD OCCUPANCY	ORDINARY (Moderate) HAZARD OCCUPANCY	EXTRA (High) HAZARD OCCUPANCY
Minimum-rated single extinguisher	2-A[c]	2-A	4-A[a]
Maximum floor area per unit of A	3,000 square feet	1,500 square feet	1,000 square feet
Maximum floor area for extinguisher[b]	11,250 square feet	11,250 square feet	11,250 square feet
Maximum distance of travel to extinguisher	75 feet	75 feet	75 feet

For SI: 1 foot = 304.8 mm, 1 square foot = $0.0929 m^2$, 1 gallon = 3.785 L.

a. Two $2^1/_2$-gallon water-type extinguishers shall be deemed the equivalent of one 4-A rated extinguisher.

b. Annex E.3.3 of NFPA 10 provides more details concerning application of the maximum floor area criteria.

c. Two water-type extinguishers each with a 1-A rating shall be deemed the equivalent of one 2-A rated extinguisher for Light (Low) Hazard Occupancies.

[F] TABLE 906.3(2)
FIRE EXTINGUISHERS FOR FLAMMABLE OR COMBUSTIBLE LIQUIDS WITH DEPTHS LESS THAN OR EQUAL TO 0.25 INCH[a]

TYPE OF HAZARD	BASIC MINIMUM EXTINGUISHER RATING	MAXIMUM DISTANCE OF TRAVEL TO EXTINGUISHERS (feet)
Light (Low)	5-B	30
	10-B	50
Ordinary (Moderate)	10-B	30
	20-B	50
Extra (High)	40-B	30
	80-B	50

For SI: 1 inch = 25.4 mm, 1 foot = 304.8 mm.

a. For requirements on water-soluble flammable liquids and alternative sizing criteria, see Section 5.5 of NFPA 10.

[F] 906.3.1 Class A fire hazards. The minimum sizes and distribution of portable fire extinguishers for occupancies that involve primarily Class A fire hazards shall comply with Table 906.3(1).

[F] 906.3.2 Class B fire hazards. Portable fire extinguishers for occupancies involving flammable or combustible liquids with depths less than or equal to 0.25-inch (6.4 mm) shall be selected and placed in accordance with Table 906.3(2).

Portable fire extinguishers for occupancies involving flammable or combustible liquids with a depth of greater than 0.25-inch (6.4 mm) shall be selected and placed in accordance with NFPA 10.

[F] 906.3.3 Class C fire hazards. Portable fire extinguishers for Class C fire hazards shall be selected and placed on the basis of the anticipated Class A or B hazard.

[F] 906.3.4 Class D fire hazards. Portable fire extinguishers for occupancies involving combustible metals shall be selected and placed in accordance with NFPA 10.

[F] 906.4 Cooking equipment fires. Fire extinguishers provided for the protection of cooking equipment shall be of an *approved* type compatible with the automatic fire-extinguishing system agent. Cooking equipment involving solid fuels or vegetable or animal oils and fats shall be protected by a Class K-rated portable extinguisher in accordance with Sections 906.1, Item 2, 906.4.1 and 906.4.2 of the *International Fire Code*, as applicable.

[F] 906.5 Conspicuous location. Portable fire extinguishers shall be located in conspicuous locations where they will have *ready access* and be immediately available for use. These locations shall be along normal paths of travel, unless the fire code official determines that the hazard posed indicates the need for placement away from normal paths of travel.

[F] 906.6 Unobstructed and unobscured. Portable fire extinguishers shall not be obstructed or obscured from view. In rooms or areas in which visual obstruction cannot be completely avoided, means shall be provided to indicate the locations of extinguishers.

[F] 906.7 Hangers and brackets. Hand-held portable fire extinguishers, not housed in cabinets, shall be installed on the hangers or brackets supplied. Hangers or brackets shall be securely anchored to the mounting surface in accordance with the manufacturer's installation instructions.

[F] 906.8 Cabinets. Cabinets used to house portable fire extinguishers shall not be locked.

Exceptions:

1. Where portable fire extinguishers subject to malicious use or damage are provided with a means of ready access.
2. In Group I-3 occupancies and in mental health areas in Group I-2 occupancies, access to portable fire extinguishers shall be permitted to be locked or to be located in staff locations provided that the staff has keys.

[F] 906.9 Extinguisher installation. The installation of portable fire extinguishers shall be in accordance with Sections 906.9.1 through 906.9.3.

[F] 906.9.1 Extinguishers weighing 40 pounds or less. Portable fire extinguishers having a gross weight not exceeding 40 pounds (18 kg) shall be installed so that their tops are not more than 5 feet (1524 mm) above the floor.

[F] 906.9.2 Extinguishers weighing more than 40 pounds. Hand-held portable fire extinguishers having a gross weight exceeding 40 pounds (18 kg) shall be installed so that their tops are not more than 3.5 feet (1067 mm) above the floor.

[F] 906.9.3 Floor clearance. The clearance between the floor and the bottom of installed hand-held portable fire extinguishers shall be not less than 4 inches (102 mm).

[F] 906.10 Wheeled units. Wheeled fire extinguishers shall be conspicuously located in a designated location.

SECTION 907
FIRE ALARM AND DETECTION SYSTEMS

[F] 907.1 General. This section covers the application, installation, performance and maintenance of fire alarm systems and their components.

[F] 907.1.1 Construction documents. *Construction documents* for fire alarm systems shall be of sufficient clarity to indicate the location, nature and extent of the work proposed and show in detail that it will conform to the provisions of this code, the *International Fire Code*; and relevant laws, ordinances, rules and regulations, as determined by the fire code official.

[F] 907.1.2 Fire alarm shop drawings. Shop drawings for fire alarm systems shall be prepared in accordance with NFPA 72 and submitted for review and approval prior to system installation.

[F] 907.1.3 Equipment. Systems and components shall be *listed* and *approved* for the purpose for which they are installed.

[F] 907.2 Where required—new buildings and structures. An *approved* fire alarm system installed in accordance with the provisions of this code and NFPA 72 shall be provided in new buildings and structures in accordance with Sections 907.2.1 through 907.2.23 and provide occupant notification in accordance with Section 907.5, unless other requirements are provided by another section of this code.

Not fewer than one manual fire alarm box shall be provided in an *approved* location to initiate a fire *alarm signal* for fire alarm systems employing automatic fire detectors or waterflow detection devices. Where other sections of this code allow elimination of fire alarm boxes due to sprinklers, a single fire alarm box shall be installed.

Exceptions:

1. The manual fire alarm box is not required for fire alarm systems dedicated to elevator recall control and supervisory service.
2. The manual fire alarm box is not required for Group R-2 occupancies unless required by the *fire code official* to provide a means for fire watch personnel to initiate an alarm during a sprinkler system impairment event. Where provided, the manual fire alarm box shall not be located in an area that is open to the public.

[F] 907.2.1 Group A. A manual fire alarm system that activates the occupant notification system in accordance with Section 907.5 shall be installed in Group A occupancies where the *occupant load* due to the assembly occupancy is 300 or more, or where the Group A *occupant load* is more than 100 persons above or below the *lowest level of exit discharge*. Group A occupancies not separated from one another in accordance with Section

707.3.10 shall be considered as a single occupancy for the purposes of applying this section. Portions of Group E occupancies occupied for assembly purposes shall be provided with a fire alarm system as required for the Group E occupancy.

Exception: Manual fire alarm boxes are not required where the building is equipped throughout with an *automatic sprinkler system* installed in accordance with Section 903.3.1.1 and the occupant notification appliances will activate throughout the notification zones upon sprinkler water flow.

[F] 907.2.1.1 System initiation in Group A occupancies with an occupant load of 1,000 or more. Activation of the fire alarm in Group A occupancies with an *occupant load* of 1,000 or more shall initiate a signal using an emergency voice/alarm communications system in accordance with Section 907.5.2.2.

Exception: Where *approved*, the prerecorded announcement is allowed to be manually deactivated for a period of time, not to exceed 3 minutes, for the sole purpose of allowing a live voice announcement from an *approved, constantly attended location*.

[F] 907.2.1.2 Emergency voice/alarm communication captions. Stadiums, arenas and *grandstands* required to caption audible public announcements shall be in accordance with Section 907.5.2.2.4.

[F] 907.2.2 Group B. A manual fire alarm system, which activates the occupant notification system in accordance with Section 907.5, shall be installed in Group B occupancies where one of the following conditions exists:

1. The combined Group B *occupant load* of all floors is 500 or more.
2. The Group B *occupant load* is more than 100 persons above or below the lowest *level of exit discharge*.
3. The *fire area* contains an *ambulatory care facility*.

Exception: Manual fire alarm boxes are not required where the building is equipped throughout with an *automatic sprinkler system* installed in accordance with Section 903.3.1.1 and the occupant notification appliances will activate throughout the notification zones upon sprinkler water flow.

[F] 907.2.2.1 Ambulatory care facilities. *Fire areas* containing *ambulatory care facilities* shall be provided with an electronically supervised automatic smoke detection system installed within the ambulatory care facility and in *public use areas* outside of tenant spaces, including public *corridors* and elevator lobbies.

Exception: Buildings equipped throughout with an *automatic sprinkler system* in accordance with Section 903.3.1.1, provided that the occupant notification appliances will activate throughout the notification zones upon sprinkler waterflow.

[F] 907.2.3 Group E. A manual fire alarm system that initiates the occupant notification signal utilizing an emergency voice/alarm communication system meeting the requirements of Section 907.5.2.2 and installed in accordance with Section 907.6 shall be installed in Group E occupancies. Where *automatic sprinkler systems* or smoke detectors are installed, such systems or detectors shall be connected to the building fire alarm system.

Exceptions:

1. A manual fire alarm system shall not be required in Group E occupancies with an *occupant load* of 50 or less.
2. Emergency voice/alarm communication systems meeting the requirements of Section 907.5.2.2 and installed in accordance with Section 907.6 shall not be required in Group E occupancies with *occupant loads* of 100 or less, provided that activation of the manual fire alarm system initiates an *approved* occupant notification signal in accordance with Section 907.5.
3. Manual fire alarm boxes shall not be required in Group E occupancies where all of the following apply:
   - 3.1. Interior *corridors* are protected by smoke detectors.
   - 3.2. Auditoriums, cafeterias, gymnasiums and similar areas are protected by *heat detectors* or other *approved* detection devices.
   - 3.3. Shops and laboratories involving dusts or vapors are protected by *heat detectors* or other *approved* detection devices.
   - 3.4 Manual activation is provided from a normally occupied location.
4. Manual fire alarm boxes shall not be required in Group E occupancies where all of the following apply:
   - 4.1. The building is equipped throughout with an *approved automatic sprinkler system* installed in accordance with Section 903.3.1.1.
   - 4.2. The emergency voice/alarm communication system will activate on sprinkler waterflow.
   - 4.3. Manual activation is provided from a normally occupied location.

[F] 907.2.4 Group F. A manual fire alarm system that activates the occupant notification system in accordance with Section 907.5 shall be installed in Group F occupancies where both of the following conditions exist:

1. The Group F occupancy is two or more *stories* in height.

2. The Group F occupancy has a combined *occupant load* of 500 or more above or below the lowest *level of exit discharge*.

Exception: Manual fire alarm boxes are not required where the building is equipped throughout with an *automatic sprinkler system* installed in accordance with Section 903.3.1.1 and the occupant notification appliances will activate throughout the notification zones upon sprinkler water flow.

[F] 907.2.5 Group H. A manual fire alarm system that activates the occupant notification system in accordance with Section 907.5 shall be installed in Group H-5 occupancies and in occupancies used for the manufacture of organic coatings. An automatic smoke detection system shall be installed for *highly toxic* gases, organic peroxides and oxidizers in accordance with Chapters 60, 62 and 63, respectively, of the *International Fire Code*.

[F] 907.2.6 Group I. A manual fire alarm system that activates the occupant notification system in accordance with Section 907.5 shall be installed in Group I occupancies. An automatic smoke detection system that activates the occupant notification system in accordance with Section 907.5 shall be provided in accordance with Sections 907.2.6.1, 907.2.6.2 and 907.2.6.3.3.

Exceptions:

1. Manual fire alarm boxes in sleeping units of Group I-1 and I-2 occupancies shall not be required at *exits* if located at all care providers' control stations or other constantly attended staff locations, provided that such manual fire alarm boxes are visible and provided with *ready access*, and the distances of travel required in Section 907.4.2.1 are not exceeded.
2. Occupant notification systems are not required to be activated where private mode signaling installed in accordance with NFPA 72 is *approved* by the fire code official and staff evacuation responsibilities are included in the fire safety and evacuation plan required by Section 404 of the *International Fire Code*.

[F] 907.2.6.1 Group I-1. In Group I-1 occupancies, an automatic smoke detection system shall be installed in *corridors*, waiting areas open to corridors and *habitable spaces* other than *sleeping units* and kitchens. The system shall be activated in accordance with Section 907.5.

Exceptions:

1. For Group I-1, Condition 1 occupancies, smoke detection in *habitable spaces* is not required where the facility is equipped throughout with an *automatic sprinkler system* installed in accordance with Section 903.3.1.1.
2. Smoke detection is not required for exterior balconies.

[F] 907.2.6.1.1 Smoke alarms. Single- and multiple-station smoke alarms shall be installed in accordance with Section 907.2.11.

[F] 907.2.6.2 Group I-2. An automatic smoke detection system shall be installed in *corridors* in Group I-2, Condition 1 facilities and spaces permitted to be open to the corridors by Section 407.2. The system shall be activated in accordance with Section 907.4. Group I-2, Condition 2 occupancies shall be equipped with an automatic smoke detection system as required in Section 407.

Exceptions:

1. Corridor smoke detection is not required in *smoke compartments* that contain sleeping units where such units are provided with smoke detectors that comply with UL 268. Such detectors shall provide a visual display on the corridor side of each sleeping unit and shall provide an audible and visual alarm at the care providers' station attending each unit.
2. Corridor smoke detection is not required in *smoke compartments* that contain sleeping units where sleeping unit doors are equipped with automatic door-closing devices with integral smoke detectors on the unit sides installed in accordance with their listing, provided that the integral detectors perform the required alerting function.

[F] 907.2.6.3 Group I-3 occupancies. Group I-3 occupancies shall be equipped with a manual fire alarm system and automatic smoke detection system installed for alerting staff.

[F] 907.2.6.3.1 System initiation. Actuation of an automatic fire-extinguishing system, *automatic sprinkler system*, a manual fire alarm box or a fire detector shall initiate an approved fire *alarm signal* that automatically notifies staff.

[F] 907.2.6.3.2 Manual fire alarm boxes. Manual fire alarm boxes are not required to be located in accordance with Section 907.4.2 where the fire alarm boxes are provided at staff-attended locations having direct supervision over areas where manual fire alarm boxes have been omitted.

[F] 907.2.6.3.2.1 Manual fire alarm boxes in detainee areas. Manual fire alarm boxes are allowed to be locked in areas occupied by detainees, provided that staff members are present within the subject area and have keys readily available to operate the manual fire alarm boxes.

[F] 907.2.6.3.3 Automatic smoke detection system. An automatic smoke detection system shall be installed throughout resident housing areas, including *sleeping units* and contiguous day rooms, group activity spaces and other common spaces normally open to residents.

Exceptions:

1. Other *approved* smoke detection arrangements providing equivalent protection, including, but not limited to, placing detectors in exhaust ducts from cells or behind protective guards *listed* for the purpose, are allowed where necessary to prevent damage or tampering.
2. *Sleeping units* in Use Conditions 2 and 3 as described in Section 308.
3. Smoke detectors are not required in *sleeping units* with four or fewer occupants in *smoke compartments* that are equipped throughout with an *automatic sprinkler system* installed in accordance with Section 903.3.1.1.

[F] 907.2.7 Group M. A manual fire alarm system that activates the occupant notification system in accordance with Section 907.5 shall be installed in Group M occupancies where one of the following conditions exists:

1. The combined Group M *occupant load* of all floors is 500 or more persons.
2. The Group M *occupant load* is more than 100 persons above or below the lowest *level of exit discharge*.

Exceptions:

1. A manual fire alarm system is not required in *covered or open mall buildings* complying with Section 402.
2. Manual fire alarm boxes are not required where the building is equipped throughout with an *automatic sprinkler system* installed in accordance with Section 903.3.1.1 and the occupant notification appliances will automatically activate throughout the notification zones upon sprinkler water flow.

[F] 907.2.7.1 Occupant notification. During times that the building is occupied, the initiation of a signal from a manual fire alarm box or from a waterflow switch shall not be required to activate the alarm notification appliances when an *alarm signal* is activated at a *constantly attended location* from which evacuation instructions shall be initiated over an emergency voice/alarm communication system installed in accordance with Section 907.5.2.2.

[F] 907.2.8 Group R-1. Fire alarm systems and smoke alarms shall be installed in Group R-1 occupancies as required in Sections 907.2.8.1 through 907.2.8.3.

[F] 907.2.8.1 Manual fire alarm system. A manual fire alarm system that activates the occupant notification system in accordance with Section 907.5 shall be installed in Group R- 1 occupancies.

Exceptions:

1. A manual fire alarm system is not required in buildings not more than two *stories* in height where all individual *sleeping units* and contiguous *attic* and crawl spaces to those units are separated from each other and public or common areas by not less than 1-hour *fire partitions* and each individual *sleeping unit* has an *exit* directly to a *public way, egress court* or *yard*.
2. Manual fire alarm boxes are not required throughout the building where all of the following conditions are met:
   - 2.1. The building is equipped throughout with an *automatic sprinkler system* installed in accordance with Section 903.3.1.1 or 903.3.1.2.
   - 2.2. The notification appliances will activate upon sprinkler water flow.
   - 2.3. Not fewer than one manual fire alarm box is installed at an *approved* location.

[F] 907.2.8.2 Automatic smoke detection system. An automatic smoke detection system that activates the occupant notification system in accordance with Section 907.5 shall be installed throughout all interior *corridors* serving *sleeping units*.

Exception: An automatic smoke detection system is not required in buildings that do not have interior *corridors* serving *sleeping units* and where each *sleeping unit* has a *means of egress* door opening directly to an *exit* or to an exterior *exit access* that leads directly to an *exit*.

[F] 907.2.8.3 Smoke alarms. Single- and multiple-station smoke alarms shall be installed in accordance with Section 907.2.11.

[F] 907.2.9 Group R-2. Fire alarm systems and smoke alarms shall be installed in Group R-2 occupancies as required in Sections 907.2.9.1 through 907.2.9.3.

[F] 907.2.9.1 Manual fire alarm system. A manual fire alarm system that activates the occupant notification system in accordance with Section 907.5 shall be installed in Group R-2 occupancies where any of the following conditions apply:

1. Any *dwelling unit* or *sleeping unit* is located three or more *stories* above the lowest *level of exit discharge*.
2. Any *dwelling unit* or *sleeping unit* is located more than one *story* below the highest *level of exit discharge* of *exits* serving the *dwelling unit* or *sleeping unit*.

3. The building contains more than 16 *dwelling units* or *sleeping units*.

Exceptions:

1. A fire alarm system is not required in buildings not more than two stories in height where all *dwelling units* or *sleeping units* and contiguous *attic* and crawl spaces are separated from each other and public or common areas by not less than 1-hour *fire partitions* and each *dwelling unit* or *sleeping unit* has an *exit* directly to a *public way, egress court* or *yard*.
2. Manual fire alarm boxes are not required where the building is equipped throughout with an *automatic sprinkler system* installed in accordance with Section 903.3.1.1 or 903.3.1.2 and the occupant notification appliances will automatically activate throughout the notification zones upon a sprinkler water flow.
3. A fire alarm system is not required in buildings that do not have interior *corridors* serving dwelling units and are protected by an approved *automatic sprinkler system* installed in accordance with Section 903.3.1.1 or 903.3.1.2, provided that *dwelling units* either have a *means of egress* door opening directly to an exterior *exit access* that leads directly to the exits or are served by open-ended *corridors* designed in accordance with Section 1027.6, Exception 3.

[F] 907.2.9.2 Smoke alarms. Single- and multiple-station smoke alarms shall be installed in accordance with Section 907.2.11.

[F] 907.2.9.3 Group R-2 college and university buildings. An automatic smoke detection system that activates the occupant notification system in accordance with Section 907.5 shall be installed in Group R-2 occupancies operated by a college or university for student or staff housing in all of the following locations:

1. Common spaces outside of *dwelling units* and *sleeping units*.
2. Laundry rooms, mechanical equipment rooms and storage rooms.
3. All interior *corridors* serving *sleeping units* or *dwelling units*.

   Exception: An automatic smoke detection system is not required in buildings that do not have interior *corridors* serving *sleeping units* or *dwelling units* and where each *sleeping unit* or *dwelling unit* either has a *means of egress* door opening directly to an exterior *exit access* that leads directly to an *exit* or a *means of egress* door opening directly to an *exit*.

Required smoke alarms in *dwelling units* and *sleeping units* in Group R-2 occupancies operated by a college or university for student or staff housing shall be interconnected with the fire alarm system in accordance with NFPA 72.

[F] 907.2.10 Group S. A manual fire alarm system that activates the occupant notification system in accordance with Section 907.5 shall be installed in Group S public- and self-storage occupancies three stories or greater in height for interior corridors and interior common areas. Visible notification appliances are not required within storage units.

Exception: Manual fire alarm boxes are not required where the building is equipped throughout with an automatic sprinkler system installed in accordance with Section 903.3.1.1, and the occupant notification appliances will activate throughout the notification zones upon sprinkler water flow.

[F] 907.2.11 Single- and multiple-station smoke alarms. *Listed* single- and multiple-station smoke alarms complying with UL 217 shall be installed in accordance with Sections 907.2.11.1 through 907.2.11.7 and NFPA 72.

[F] 907.2.11.1 Group R-1. Single- or multiple-station smoke alarms shall be installed in all of the following locations in Group R-1:

1. In sleeping areas.
2. In every room in the path of the *means of egress* from the sleeping area to the door leading from the *sleeping unit*.
3. In each *story* within the *sleeping unit*, including basements. For *sleeping units* with split levels and without an intervening door between the adjacent levels, a smoke alarm installed on the upper level shall suffice for the adjacent lower level provided that the lower level is less than one full *story* below the upper level.

[F] 907.2.11.2 Groups R-2, R-3, R-4 and I-1. Single- or multiple-station smoke alarms shall be installed and maintained in Groups R-2, R-3, R-4 and I-1 regardless of *occupant load* at all of the following locations:

1. On the ceiling or wall outside of each separate sleeping area in the immediate vicinity of bedrooms.
2. In each room used for sleeping purposes.
3. In each *story* within a *dwelling unit*, including basements but not including crawl spaces and uninhabitable *attics*. In *dwellings* or *dwelling units* with split levels and without an intervening door between the adjacent levels, a smoke alarm installed on the upper level shall suffice for the adjacent lower level provided that the lower level is less than one full *story* below the upper level.

[F] 907.2.11.3 Installation near cooking appliances. Smoke alarms shall not be installed in the following locations unless this would prevent placement of a smoke alarm in a location required by Section 907.2.11.1 or 907.2.11.2:

1. Ionization smoke alarms shall not be installed less than 20 feet (6096 mm) horizontally from a permanently installed cooking appliance.
2. Ionization smoke alarms with an alarm-silencing switch shall not be installed less than 10 feet (3048 mm) horizontally from a permanently installed cooking appliance.
3. Photoelectric smoke alarms shall not be installed less than 6 feet (1829 mm) horizontally from a permanently installed cooking appliance.

[F] 907.2.11.4 Installation near bathrooms. Smoke alarms shall be installed not less than 3 feet (914 mm) horizontally from the door or opening of a bathroom that contains a bathtub or shower unless this would prevent placement of a smoke alarm required by Section 907.2.11.1 or 907.2.11.2.

[F] 907.2.11.5 Interconnection. Where more than one smoke alarm is required to be installed within an individual *dwelling unit* or *sleeping unit* in Group R or I-1 occupancies, the smoke alarms shall be interconnected in such a manner that the activation of one alarm will activate all of the alarms in the individual unit. Physical interconnection of smoke alarms shall not be required where *listed* wireless alarms are installed and all alarms sound upon activation of one alarm. The alarm shall be clearly audible in all bedrooms over background noise levels with all intervening doors closed.

[F] 907.2.11.6 Power source. In new construction, required smoke alarms shall receive their primary power from the building wiring where such wiring is served from a commercial source and shall be equipped with a battery backup. Smoke alarms with integral strobes that are not equipped with battery backup shall be connected to an emergency electrical system in accordance with Section 2702. Smoke alarms shall emit a signal when the batteries are low. Wiring shall be permanent and without a disconnecting switch other than as required for overcurrent protection.

Exception: Smoke alarms are not required to be equipped with battery backup where they are connected to an emergency electrical system that complies with Section 2702.

[F] 907.2.11.7 Smoke detection system. Smoke detectors listed in accordance with UL 268 and provided as part of the building *fire alarm system* shall be an acceptable alternative to single- and multiple-station *smoke alarms* and shall comply with the following:

1. The *fire alarm* system shall comply with all applicable requirements in Section 907.
2. Activation of a smoke detector in a *dwelling unit* or *sleeping unit* shall initiate alarm notification in the *dwelling unit* or *sleeping unit* in accordance with Section 907.5.2.
3. Activation of a smoke detector in a *dwelling unit* or *sleeping unit* shall not activate alarm notification appliances outside of the *dwelling unit* or *sleeping unit*, provided that a supervisory signal is generated and monitored in accordance with Section 907.6.6.

[F] 907.2.12 Special amusement buildings. An automatic smoke detection system shall be provided in *special amusement buildings* in accordance with Sections 907.2.12.1 through 907.2.12.3.

[F] 907.2.12.1 Alarm. Activation of any single smoke detector, the *automatic sprinkler system* or any other automatic fire detection device shall immediately activate an audible and visible alarm at the building at a *constantly attended location* from which emergency action can be initiated, including the capability of manual initiation of requirements in Section 907.2.12.2.

[F] 907.2.12.2 System response. The activation of two or more smoke detectors, a single smoke detector equipped with an *alarm verification feature*, the *automatic sprinkler system* or other *approved* fire detection device shall automatically do all of the following:

1. Cause illumination of the *means of egress* with light of not less than 1 footcandle (11 lux) at the walking surface level.
2. Stop any conflicting or confusing sounds and visual distractions.
3. Activate an *approved* directional exit marking that will become apparent in an emergency.
4. Activate a prerecorded message, audible throughout the *special amusement building*, instructing patrons to proceed to the nearest exit. Alarm signals used in conjunction with the prerecorded message shall produce a sound that is distinctive from other sounds used during normal operation.

[F] 907.2.12.3 Emergency voice/alarm communication system. An emergency voice/alarm communication system, which is allowed to serve as a public address system, shall be installed in accordance with Section 907.5.2.2 and be audible throughout the entire *special amusement building*.

[F] 907.2.13 High-rise buildings. High-rise buildings shall be provided with an automatic smoke detection system in accordance with Section 907.2.13.1, a fire department communication system in accordance with Section 907.2.13.2 and an emergency voice/alarm communication system in accordance with Section 907.5.2.2.

Exceptions:

1. Airport traffic control towers in accordance with Sections 412 and 907.2.22.

2. *Open parking garages* in accordance with Section 406.5.
3. Buildings with an occupancy in Group A-5 in accordance with Section 303.1.
4. Low-hazard special occupancies in accordance with Section 503.1.1.
5. Buildings with an occupancy in Group H-1, H-2 or H-3 in accordance with Section 415.
6. In Group I-1 and I-2 occupancies, the alarm shall sound at a *constantly attended location* and occupant notification shall be broadcast by the emergency voice/alarm communication system.

[F] 907.2.13.1 Automatic smoke detection. Automatic smoke detection in high-rise buildings shall be in accordance with Sections 907.2.13.1.1 and 907.2.13.1.2.

[F] 907.2.13.1.1 Area smoke detection. Area smoke detectors shall be provided in accordance with this section. Smoke detectors shall be connected to an automatic fire alarm system. The activation of any detector required by this section shall activate the emergency voice/alarm communication system in accordance with Section 907.5.2.2. In addition to smoke detectors required by Sections 907.2.1 through 907.2.9, smoke detectors shall be located as follows:

1. In each mechanical equipment, electrical, transformer, telephone equipment or similar room that is not provided with sprinkler protection.
2. In each elevator machine room, machinery space, control room and control space and in elevator lobbies.

[F] 907.2.13.1.2 Duct smoke detection. Duct smoke detectors complying with Section 907.3.1 shall be located as follows:

1. In the main return air and exhaust air plenum of each air-conditioning system having a capacity greater than 2,000 cubic feet per minute (cfm) (0.94 m^3/s). Such detectors shall be located in a serviceable area downstream of the last duct inlet.
2. At each connection to a vertical duct or riser serving two or more stories from a return air duct or plenum of an air-conditioning system. In Group R-1 and R-2 occupancies, a smoke detector is allowed to be used in each return air riser carrying not more than 5,000 cfm (2.4 m^3/s) and serving not more than 10 air-inlet openings.

[F] 907.2.13.2 Fire department communication system. Where a wired communication system is *approved* in lieu of an in-building two-way emergency responder communication coverage system in accordance with Section 510 of the *International Fire Code*, the wired fire department communication system shall be designed and installed in accordance with NFPA 72 and shall operate between a fire command center complying with Section 911, elevators, elevator lobbies, emergency and standby power rooms, fire pump rooms, *areas of refuge* and inside *interior exit stairways*. The fire department communication device shall be provided at each floor level within the *interior exit stairway*.

[F] 907.2.13.3 Multiple-channel voice evacuation. In buildings with an occupied floor more than 120 feet (36 576 mm) above the lowest level of fire department vehicle access, voice evacuation systems for high-rise buildings shall be multiple-channel systems.

[F] 907.2.14 Atriums connecting more than two stories. A fire alarm system shall be installed in occupancies with an *atrium* that connects more than two *stories*, with smoke detection installed in locations required by a rational analysis in Section 909.4 and in accordance with the system operation requirements in Section 909.17. The system shall be activated in accordance with Section 907.5. Such occupancies in Group A, E or M shall be provided with an emergency voice/alarm communication system complying with the requirements of Section 907.5.2.2.

[F] 907.2.15 High-piled combustible storage areas. An automatic smoke detection system shall be installed throughout high-piled combustible storage areas where required by Section 3206.5 of the *International Fire Code*.

[F] 907.2.16 Aerosol storage uses. *Aerosol product* rooms and general-purpose warehouses containing aerosol products shall be provided with an *approved* manual fire alarm system where required by the *International Fire Code*.

[F] 907.2.17 Lumber, wood structural panel and veneer mills. Lumber, *wood structural panel* and *veneer* mills shall be provided with a manual fire alarm system.

[F] 907.2.18 Underground buildings with smoke control systems. Where a smoke control system is installed in an underground building in accordance with this code, automatic smoke detectors shall be provided in accordance with Section 907.2.18.1.

[F] 907.2.18.1 Smoke detectors. Not fewer than one smoke detector *listed* for the intended purpose shall be installed in all of the following areas:

1. Mechanical equipment, electrical, transformer, telephone equipment, elevator machine or similar rooms.
2. Elevator lobbies.
3. The main return and exhaust air plenum of each air-conditioning system serving more than one *story* and located in a serviceable area downstream of the last duct inlet.

4. Each connection to a vertical duct or riser serving two or more floors from return air ducts or plenums of heating, ventilating and air-conditioning systems, except that in Group R occupancies, a *listed* smoke detector is allowed to be used in each return air riser carrying not more than 5,000 cubic feet per minute (2.4 m^3/s) and serving not more than 10 air-inlet openings.

[F] 907.2.18.2 Alarm required. Activation of the smoke control system shall activate an audible alarm at a *constantly attended location.*

[F] 907.2.19 Deep underground buildings. Where the lowest level of a structure is more than 60 feet (18 288 mm) below the finished floor of the lowest *level of exit discharge*, the structure shall be equipped throughout with a manual fire alarm system, including an emergency voice/alarm communication system installed in accordance with Section 907.5.2.2.

[F] 907.2.20 Covered and open mall buildings. Where the total floor area exceeds 50,000 square feet (4645 m^2) within either a *covered mall building* or within the perimeter line of an *open mall building*, an emergency voice/alarm communication system shall be provided. Access to emergency voice/alarm communication systems serving a mall, required or otherwise, shall be provided for the fire department. The system shall be provided in accordance with Section 907.5.2.2.

[F] 907.2.21 Residential aircraft hangars. Not fewer than one single-station smoke alarm shall be installed within a *residential aircraft hangar* as defined in Chapter 2 and shall be interconnected into the residential smoke alarm or other sounding device to provide an alarm that will be audible in all sleeping areas of the *dwelling*.

[F] 907.2.22 Airport traffic control towers. An automatic smoke detection system that activates the occupant notification system in accordance with Section 907.5 shall be provided in airport control towers in accordance with Sections 907.2.22.1 and 907.2.22.2.

Exception: Audible appliances shall not be installed within the control tower cab.

[F] 907.2.22.1 Airport traffic control towers with multiple exits and automatic sprinklers. Airport traffic control towers with multiple *exits* and equipped throughout with an *automatic sprinkler system* in accordance with Section 903.3.1.1 shall be provided with smoke detectors in all of the following locations:

1. Airport traffic control cab.
2. Electrical and mechanical equipment rooms.
3. Airport terminal radar and electronics rooms.
4. Outside each opening into *interior exit stairways*.
5. Along the single *means of egress* permitted from observation levels.
6. Outside each opening into the single *means of egress* permitted from observation levels.

[F] 907.2.22.2 Other airport traffic control towers. Airport traffic control towers with a single exit or where sprinklers are not installed throughout shall be provided with smoke detectors in all of the following locations:

1. Airport traffic control cab.
2. Electrical and mechanical equipment rooms.
3. Airport terminal radar and electronics rooms.
4. Office spaces incidental to the tower operation.
5. Lounges for employees, including sanitary facilities.
6. *Means of egress.*
7. Utility *shafts* where access to smoke detectors can be provided.

[F] 907.2.23 Energy storage systems. An automatic smoke detection system or radiant-energy detection system shall be installed in rooms, areas and walk-in units containing energy storage systems as required in Section 1207.5.4 of the *International Fire Code*.

[F] 907.3 Fire safety functions. Automatic fire detectors utilized for the purpose of performing fire safety functions shall be connected to the building's fire alarm control unit where a fire alarm system is required by Section 907.2. Detectors shall, upon actuation, perform the intended function and activate the alarm notification appliances or activate a visible and audible supervisory signal at a *constantly attended location*. In buildings not equipped with a fire alarm system, the automatic fire detector shall be powered by normal electrical service and, upon actuation, perform the intended function. The detectors shall be located in accordance with NFPA 72.

[F] 907.3.1 Duct smoke detectors. Smoke detectors installed in ducts shall be *listed* for the air velocity, temperature and humidity present in the duct. Duct smoke detectors shall be connected to the building's fire alarm control unit where a fire alarm system is required by Section 907.2. Activation of a duct smoke detector shall initiate a visible and audible supervisory signal at a *constantly attended location* and shall perform the intended fire safety function in accordance with this code and the *International Mechanical Code*. In facilities that are required to be monitored by a supervising station, duct smoke detectors shall report only as a supervisory signal and not as a fire alarm. They shall not be used as a substitute for required open area detection.

Exceptions:

1. The supervisory signal at a *constantly attended location* is not required where duct smoke detectors activate the building's alarm notification appliances.
2. In occupancies not required to be equipped with a fire alarm system, actuation of a smoke detector shall activate a visible and an audible signal in an *approved* location. Smoke detector trouble conditions shall activate a visible or

audible signal in an *approved* location and shall be identified as air duct detector trouble.

[F] 907.3.2 Special locking systems. Where special locking systems are installed on *means of egress* doors in accordance with Sections 1010.2.13 or 1010.2.14, an automatic detection system shall be installed as required by that section.

[F] 907.3.3 Elevator emergency operation. Automatic fire detectors installed for elevator emergency operation shall be installed in accordance with the provisions of ASME A17.1/CSA B44 and NFPA 72.

[F] 907.3.4 Wiring. The wiring to the auxiliary devices and equipment used to accomplish the fire safety functions shall be monitored for integrity in accordance with NFPA 72.

[F] 907.4 Initiating devices. Where a fire alarm system is required by another section of this code, occupant notification in accordance with Section 907.5 shall be initiated by one or more of the following. Initiating devices shall be installed in accordance with Sections 907.4.1 through 907.4.3.1.

1. Manual fire alarm boxes.
2. Automatic fire detectors.
3. *Automatic sprinkler system* waterflow devices.
4. Automatic fire-extinguishing systems.

[F] 907.4.1 Protection of fire alarm control unit. In areas that are not continuously occupied, a single smoke detector shall be provided at the location of each fire alarm control unit, notification appliance circuit power extenders, and supervising station transmitting equipment.

Exception: Where ambient conditions prohibit installation of a smoke detector, a *heat detector* shall be permitted.

[F] 907.4.2 Manual fire alarm boxes. Where a manual fire alarm system is required by another section of this code, it shall be activated by fire alarm boxes installed in accordance with Sections 907.4.2.1 through 907.4.2.6.

[F] 907.4.2.1 Location. Manual fire alarm boxes shall be located not more than 5 feet (1524 mm) from the entrance to each *exit*. In buildings not protected by an *automatic sprinkler system* in accordance with Section 903.3.1.1 or 903.3.1.2, additional manual fire alarm boxes shall be located so that the distance of travel to the nearest box does not exceed 200 feet (60 960 mm).

[F] 907.4.2.2 Height. The height of the manual fire alarm boxes shall be not less than 42 inches (1067 mm) and not more than 48 inches (1372 mm) measured vertically, from the floor level to the activating handle or lever of the box.

[F] 907.4.2.3 Color. Manual fire alarm boxes shall be red in color.

[F] 907.4.2.4 Signs. Where fire alarm systems are not monitored by an *approved* supervising station in accordance with Section 907.6.6, an *approved* permanent sign shall be installed adjacent to each manual fire alarm box that reads: WHEN ALARM SOUNDS CALL FIRE DEPARTMENT.

Exception: Where the manufacturer has permanently provided this information on the manual fire alarm box.

[F] 907.4.2.5 Protective covers. The fire code official is authorized to require the installation of *listed* manual fire alarm box protective covers to prevent malicious false alarms or to provide the manual fire alarm box with protection from physical damage. The protective cover shall be transparent or red in color with a transparent face to permit visibility of the manual fire alarm box. Each cover shall include proper operating instructions. A protective cover that emits a local *alarm signal* shall not be installed unless *approved*. Protective covers shall not project more than that permitted by Section 1003.3.3.

[F] 907.4.2.6 Unobstructed and unobscured. Manual fire alarm boxes shall be provided with ready access, unobstructed, unobscured and visible at all times.

[F] 907.4.3 Automatic smoke detection. Where an automatic smoke detection system is required, it shall utilize smoke detectors unless ambient conditions prohibit such an installation. In spaces where smoke detectors cannot be utilized due to ambient conditions, *approved* automatic *heat detectors* shall be permitted.

[F] 907.4.3.1 Automatic sprinkler system. For conditions other than specific fire safety functions noted in Section 907.3, in areas where ambient conditions prohibit the installation of smoke detectors, an *automatic sprinkler system* installed in such areas in accordance with Section 903.3.1.1 or 903.3.1.2 and that is connected to the fire alarm system shall be *approved* as automatic heat detection.

[F] 907.5 Occupant notification. Occupant notification by fire alarms shall be in accordance with Sections 907.5.1 through 907.5.2.3.3. Occupant notification by smoke alarms in Group R-1 and R-2 occupancies shall comply with Section 907.5.2.1.3.2.

907.5.1 Alarm activation and annunciation. Upon activation, fire alarm systems shall initiate occupant notification and shall annunciate at the fire alarm control unit, or where allowed elsewhere by Section 907, at a *constantly attended location*.

907.5.1.1 Presignal feature. A presignal feature shall be provided only where *approved*. The presignal shall be annunciated at an approved, *constantly attended location,* having the capability to activate the occupant notification system in the event of fire or other emergency.

[F] 907.5.2 Alarm notification appliances. Alarm notification appliances shall be provided and shall be *listed* for their purpose.

[F] 907.5.2.1 Audible alarms. Audible alarm notification appliances shall be provided and emit a distinctive sound that is not to be used for any purpose other than that of a fire alarm.

Exceptions:

1. Audible alarm notification appliances are not required in critical care areas of Group I-2, Condition 2 occupancies that are in compliance with Section 907.2.6, Exception 2.
2. A visible *alarm notification appliance* installed in a nurses' control station or other continuously attended staff location in a Group I-2, Condition 2 suite shall be an acceptable alternative to the installation of audible alarm notification appliances throughout a suite or unit in Group I-2, Condition 2 occupancies that are in compliance with Section 907.2.6, Exception 2.
3. Where provided, audible notification appliances located in each enclosed occupant evacuation elevator lobby in accordance with Section 3008.9.1 shall be connected to a separate notification zone for manual paging only.

[F] 907.5.2.1.1 Average sound pressure. The audible alarm notification appliances shall provide a sound pressure level of 15 decibels (dBA) above the *average ambient sound level* or 5 dBA above the maximum sound level having a duration of not less than 60 seconds, whichever is greater, in every *occupiable space* within the building.

[F] 907.5.2.1.2 Maximum sound pressure. The total sound pressure level produced by combining the ambient sound pressure level with all audible notification appliances operating shall not exceed 110 dBA at the minimum hearing distance from the audible appliance. Where the average ambient noise is greater than 105 dBA, visible alarm notification appliances shall be provided in accordance with NFPA 72 and audible alarm notification appliances shall not be required.

[F] 907.5.2.1.3 Audible signal frequency in Group R-1 and R-2 sleeping rooms. Audible signal frequency in Group R-1 and R-2 occupancies shall be in accordance with Sections 907.5.2.1.3.1 and 907.5.2.1.3.2.

[F] 907.5.2.1.3.1 Fire alarm system signal. In sleeping rooms of Group R-1 and R-2 occupancies, the audible alarm activated by a fire alarm system shall be a 520-Hz low-frequency signal complying with NFPA 72.

[F] 907.5.2.1.3.2 Smoke alarm signal in sleeping rooms. In sleeping rooms of Group R-1 and R-2 occupancies that are required by Section 907.2.8 or 907.2.9 to have a fire alarm system, the audible *alarm signal* activated by single- or multiple-station smoke alarms in the *dwelling unit* or *sleeping unit* shall be a 520-Hz signal complying with NFPA 72. Where a sleeping room smoke alarm is unable to produce a 520-Hz signal, the 520-Hz *alarm signal* shall be provided by a *listed* notification appliance or a smoke detector with an integral 520-Hz sounder.

[F] 907.5.2.2 Emergency voice/alarm communication systems. Emergency voice/alarm communication systems required by this code shall be designed and installed in accordance with NFPA 72. The operation of any automatic fire detector, sprinkler waterflow device or manual fire alarm box shall automatically sound an alert tone followed by voice instructions giving *approved* information and directions for a general or staged evacuation in accordance with the building's fire safety and evacuation plans required by Section 404 of the *International Fire Code*. In high-rise buildings, the system shall operate on at least the alarming floor, the floor above and the floor below. Speakers shall be provided throughout the building by paging zones. At a minimum, paging zones shall be provided as follows:

1. Elevator groups.
2. *Interior exit stairways.*
3. Each floor.
4. *Areas of refuge* as defined in Chapter 2.

Exception: In Group I-1 and I-2 occupancies, the alarm shall sound in a constantly attended area and a general occupant notification shall be broadcast over the overhead page.

[F] 907.5.2.2.1 Manual override. A manual override for emergency voice communication shall be provided on a selective and all-call basis for all paging zones.

[F] 907.5.2.2.2 Live voice messages. The emergency voice/alarm communication system shall have the capability to broadcast live voice messages by paging zones on a selective and all-call basis.

[F] 907.5.2.2.3 Alternative uses. The emergency voice/alarm communication system shall be allowed to be used for other announcements, provided that the manual fire alarm use takes precedence over any other use.

[F] 907.5.2.2.4 Emergency voice/alarm communication captions. Where stadiums, arenas and *grandstands* have 15,000 fixed seats or more and provide audible public announcements, the emergency/voice alarm communication system shall provide prerecorded or real-time captions. Prerecorded or live emergency captions shall be from an *approved* location constantly attended by personnel trained to respond to an emergency.

[F] 907.5.2.2.5 Standby power. Emergency voice/alarm communications systems shall be provided with standby power in accordance with Section 2702.

[F] 907.5.2.3 Visible alarms. Visible alarm notification appliances shall be provided in accordance with Sections 907.5.2.3.1 through 907.5.2.3.3.

Exceptions:

1. Visible alarm notification appliances are not required in *alterations*, except where an existing fire alarm system is upgraded or replaced, or a new fire alarm system is installed.
2. Visible alarm notification appliances shall not be required in *exits* as defined in Chapter 2.
3. Visible alarm notification appliances shall not be required in elevator cars.
4. Visual alarm notification appliances are not required in critical care areas of Group I-2, Condition 2 occupancies that are in compliance with Section 907.2.6, Exception 2.
5. A visible *alarm notification appliance* installed in a nurses' control station or other continuously attended staff location in a Group I-2, Condition 2 suite shall be an acceptable alternative to the installation of visible alarm notification appliances throughout the suite or unit in Group I-2, Condition 2 occupancies that are in compliance with Section 907.2.6, Exception 2.

[F] 907.5.2.3.1 Public use areas and common use areas. Visible alarm notification appliances shall be provided in *public use areas* and *common use areas.*

Exception: Where *employee work areas* have audible alarm coverage, the notification appliance circuits serving the *employee work areas* shall be initially designed with not less than 20-percent spare capacity to account for the potential of adding visible notification appliances in the future to accommodate hearing-impaired employee(s).

[F] 907.5.2.3.2 Groups I-1 and R-1. *Habitable spaces* in *dwelling units* and *sleeping units* in Group I-1 and R-1 occupancies in accordance with Table 907.5.2.3.2 shall be provided with visible alarm notification. Visible alarms shall be activated by the in-room smoke alarm and the building fire alarm system.

[F] 907.5.2.3.3 Group R-2. In Group R-2 occupancies required by Section 907 to have a fire alarm system, each *story* that contains *dwelling units* and *sleeping units* shall be provided with the capability to support future visible alarm notification appliances in accordance with Chapter 11 of ICC A117.1. Such capability shall accommodate wired or wireless equipment.

[F] 907.5.2.3.3.1 Wired equipment. Where wired equipment is used to comply with the future capability required by Section 907.5.2.3.3, the system shall include one of the following capabilities:

1. The replacement of audible appliances with combination audible/visible appliances or additional visible notification appliances.
2. The future extension of the existing wiring from the unit smoke alarm locations to required locations for visible appliances.
3. For wired equipment, the fire alarm power supply and circuits shall have not less than 5-percent excess capacity to accommodate the future addition of visible alarm notification appliances, and a single access point to such circuits shall be available on every story. Such circuits shall not be required to be extended beyond a single access point on a story. The fire alarm system shop drawings required by Section 907.1.2 shall include the power supply and circuit documentation to accommodate the future addition of visible notification appliances.

[F] 907.6 Installation and monitoring. A fire alarm system shall be installed and monitored in accordance with Sections 907.6.1 through 907.6.6.3 and NFPA 72.

[F] 907.6.1 Wiring. Wiring shall comply with the requirements of NFPA 70 and NFPA 72. Wireless protection systems utilizing radio-frequency transmitting devices shall comply with the special requirements for supervision of low-power wireless systems in NFPA 72.

[F] 907.6.2 Power supply. The primary and secondary power supply for the fire alarm system shall be provided in accordance with NFPA 72.

Exception: Back-up power for single-station and multiple-station smoke alarms as required in Section 907.2.11.6.

[F] TABLE 907.5.2.3.2
VISIBLE ALARMS

NUMBER OF SLEEPING UNITS	SLEEPING ACCOMMODATIONS WITH VISIBLE ALARMS
6 to 25	2
26 to 50	4
51 to 75	7
76 to 100	9
101 to 150	12
151 to 200	14
201 to 300	17
301 to 400	20
401 to 500	22
501 to 1,000	5% of total
1,001 and over	50 plus 3 for each 100 over 1,000

[F] 907.6.3 Initiating device identification. The fire alarm system shall identify the specific *initiating device* address, location, device type, floor level where applicable and status including indication of normal, alarm, trouble and supervisory status, as appropriate.

Exceptions:

1. Fire alarm systems in single-*story* buildings less than 22,500 square feet (2090 m^2) in area.
2. Fire alarm systems that only include manual fire alarm boxes, waterflow initiating devices and not more than 10 additional alarm-initiating devices.
3. Special initiating devices that do not support individual device identification.
4. Fire alarm systems or devices that are replacing existing equipment.

[F] 907.6.3.1 Annunciation. The *initiating device* status shall be annunciated at an *approved* on-site location.

[F] 907.6.4 Zones. Each floor shall be zoned separately and a zone shall not exceed 22,500 square feet (2090 m^2). The length of any zone shall not exceed 300 feet (91 440 mm) in any direction.

Exception: *Automatic sprinkler system* zones shall not exceed the area permitted by NFPA 13.

[F] 907.6.4.1 Zoning indicator panel. A zoning indicator panel and the associated controls shall be provided in an *approved* location. The visual zone indication shall lock in until the system is reset and shall not be canceled by the operation of an audible-alarm silencing switch.

[F] 907.6.4.2 High-rise buildings. In high-rise buildings, a separate zone by floor shall be provided for each of the following types of alarm-initiating devices where provided:

1. Smoke detectors.
2. Sprinkler waterflow devices.
3. Manual fire alarm boxes.
4. Other *approved* types of automatic *fire protection systems*.

[F] 907.6.5 Access. Access shall be provided to each fire alarm device and notification appliance for periodic inspection, maintenance and testing.

[F] 907.6.6 Monitoring. Fire alarm systems required by this chapter or by the *International Fire Code* shall be monitored by an *approved* supervising station in accordance with NFPA 72.

Exception: Monitoring by a supervising station is not required for:

1. Single- and multiple-station smoke alarms required by Section 907.2.11.
2. Smoke detectors in Group I-3 occupancies.
3. *Automatic sprinkler systems* in one- and two-family dwellings.

[F] 907.6.6.1 Transmission of alarm signals. Transmission of alarm signals to a supervising station shall be in accordance with NFPA 72.

[F] 907.6.6.2 MIY Monitoring. Direct transmission of alarms associated with monitor it yourself (MIY) transmitters to a public safety answering point (PSAP) shall not be permitted unless *approved* by the *fire code official*.

[F] 907.6.6.3 Termination of monitoring service. Termination of fire alarm monitoring services shall be in accordance with Section 901.9 of the *International Fire Code*.

[F] 907.7 Acceptance tests and completion. Upon completion of the installation, the fire alarm system and all fire alarm components shall be tested in accordance with NFPA 72.

[F] 907.7.1 Single- and multiple-station alarm devices. When the installation of the alarm devices is complete, each device and interconnecting wiring for multiple-station alarm devices shall be tested in accordance with the smoke alarm provisions of NFPA 72.

[F] 907.7.2 Record of completion. A record of completion in accordance with NFPA 72 verifying that the system has been installed and tested in accordance with the *approved* plans and specifications shall be provided.

[F] 907.7.3 Instructions. Operating, testing and maintenance instructions and *record drawings* ("as-builts") and equipment specifications shall be provided at an *approved* location.

[F] 907.8 Inspection, testing and maintenance. The maintenance and testing schedules and procedures for fire alarm and fire detection systems shall be in accordance with Section 907.8 of the *International Fire Code*.

SECTION 908
EMERGENCY ALARM SYSTEMS

[F] 908.1 Group H occupancies. Emergency alarms for the detection and notification of an emergency condition in Group H occupancies shall be provided in accordance with Section 415.5.

[F] 908.2 Group H-5 occupancy. Emergency alarms for notification of an emergency condition in an HPM facility shall be provided as required in Section 415.11.4.

[F] 908.3 Fire alarm system interface. Where an emergency alarm system is interfaced with a building's fire alarm system, the signal produced at the fire alarm control unit shall be a supervisory signal.

SECTION 909
SMOKE CONTROL SYSTEMS

[F] 909.1 Scope and purpose. This section applies to mechanical or passive smoke control systems where they are required by other provisions of this code. The purpose of this section is to establish minimum requirements for the design, installation and acceptance testing of smoke control systems that are intended to provide a tenable environment for the evacuation or relocation of occupants. These provisions are not intended for the preservation of contents, the timely restoration of operations or for assistance in fire suppression or overhaul activities. Smoke control systems regulated by this section serve a different purpose than the smoke- and heat-removal provisions found in Section 910. Mechanical smoke control systems shall not be considered exhaust systems under Chapter 5 of the *International Mechanical Code*.

[F] 909.2 General design requirements. Buildings, structures or parts thereof required by this code to have a smoke control system or systems shall have such systems designed in accordance with the applicable requirements of Section 909 and the generally accepted and well-established principles of engineering relevant to the design. The *construction documents* shall include sufficient information and detail to adequately describe the elements of the design necessary for the proper implementation of the smoke control systems. These documents shall be accompanied by sufficient information and analysis to demonstrate compliance with these provisions.

[F] 909.3 Special inspection and test requirements. In addition to the ordinary inspection and test requirements that buildings, structures and parts thereof are required to undergo, smoke control systems subject to the provisions of Section 909 shall undergo *special inspections* and tests sufficient to verify the proper commissioning of the smoke control design in its final installed condition. The design submission accompanying the *construction documents* shall clearly detail procedures and methods to be used and the items subject to such inspections and tests. Such commissioning shall be in accordance with generally accepted engineering practice and, where possible, based on published standards for the particular testing involved. The *special inspections* and tests required by this section shall be conducted under the same terms in Section 1704.

[F] 909.4 Analysis. A rational analysis supporting the types of smoke control systems to be employed, their methods of operation, the systems supporting them and the methods of construction to be utilized shall accompany the submitted *construction documents* and shall include, but not be limited to, the items indicated in Sections 909.4.1 through 909.4.7.

[F] 909.4.1 Stack effect. The system shall be designed such that the maximum probable normal or reverse stack effect will not adversely interfere with the system's capabilities. In determining the maximum probable stack effect, altitude, elevation, weather history and interior temperatures shall be used.

[F] 909.4.2 Temperature effect of fire. Buoyancy and expansion caused by the design fire in accordance with Section 909.9 shall be analyzed. The system shall be designed such that these effects do not adversely interfere with the system's capabilities.

[F] 909.4.3 Wind effect. The design shall consider the adverse effects of wind. Such consideration shall be consistent with the wind-loading provisions of Chapter 16.

[F] 909.4.4 HVAC systems. The design shall consider the effects of the heating, ventilating and air-conditioning (HVAC) systems on both smoke and fire transport. The analysis shall include all permutations of systems status. The design shall consider the effects of the fire on the HVAC systems.

[F] 909.4.5 Climate. The design shall consider the effects of low temperatures on systems, property and occupants. Air inlets and exhausts shall be located so as to prevent snow or ice blockage.

[F] 909.4.6 Duration of operation. All portions of active or engineered smoke control systems shall be capable of continued operation after detection of the fire event for a period of not less than either 20 minutes or 1.5 times the calculated egress time, whichever is greater.

909.4.7 Smoke control system interaction. The design shall consider the interaction effects of the operation of multiple smoke control systems for all design scenarios.

[F] 909.5 Smoke barrier construction. *Smoke barriers* required for passive smoke control and a smoke control system using the pressurization method shall comply with Section 709. The maximum allowable leakage area shall be the aggregate area calculated using the following leakage area ratios:

1. Walls $A/A_w = 0.00100$
2. Interior *exit stairways* and *ramps* and *exit passageways*: $A/A_w = 0.00035$
3. Enclosed *exit access stairways* and *ramps* and all other *shafts*: $A/A_w = 0.00150$
4. Floors and roofs: $A/A_F = 0.00050$

where:

A = Total leakage area, square feet (m^2).

A_F = Unit floor or roof area of barrier, square feet (m^2).

A_w = Unit wall area of barrier, square feet (m^2).

The leakage area ratios shown do not include openings due to gaps around doors and operable windows. The total leakage area of the *smoke barrier* shall be determined in accordance with Section 909.5.1 and tested in accordance with Section 909.5.2.

[F] 909.5.1 Total leakage area. Total leakage area of the barrier is the product of the *smoke barrier* gross area multiplied by the allowable leakage area ratio, plus the area of other openings such as gaps around doors and operable windows.

[F] 909.5.2 Testing of leakage area. Compliance with the maximum total leakage area shall be determined by achieving the minimum air pressure difference across the barrier with the system in the smoke control mode for mechanical smoke control systems utilizing the pressurization method. Compliance with the maximum total leakage area of passive smoke control systems shall be verified through methods such as door fan testing or other methods, as *approved* by the fire code official.

[F] 909.5.3 Opening protection. Openings in *smoke barriers* shall be protected by automatic-closing devices actuated by the required controls for the mechanical smoke control system. Door openings shall be protected by *fire door assemblies* complying with Section 716.

Exceptions:

1. Passive smoke control systems with automatic-closing devices actuated by spot-type smoke detectors *listed* for releasing service installed in accordance with Section 907.3.
2. Fixed openings between smoke zones that are protected utilizing the airflow method.
3. In Group I-1, Condition 2; Group I-2; and *ambulatory care facilities*, where a pair of opposite-swinging doors are installed across a *corridor* in accordance with Section 909.5.3.1, the doors shall not be required to be protected in accordance with Section 716. The doors shall be close-fitting within operational tolerances and shall not have a center mullion or undercuts in excess of $^{3}/_{4}$ inch (19.1 mm), louvers or grilles. The doors shall have head and jamb stops and astragals or rabbets at meeting edges and, where permitted by the door manufacturer's listing, positive-latching devices are not required.
4. In Group I-2 and *ambulatory care facilities*, where such doors are special-purpose horizontal sliding, accordion or folding door assemblies installed in accordance with Section 1010.3.3 and are automatic closing by smoke detection in accordance with Section 716.2.6.5.
5. Group I-3.
6. Openings between smoke zones with clear ceiling heights of 14 feet (4267 mm) or greater and bank-down capacity of greater than 20 minutes as determined by the design fire size.

909.5.3.1 Group I-1, Condition 2; Group I-2; and ambulatory care facilities. In Group I-1, Condition 2; Group I-2; and *ambulatory care facilities*, where doors are installed across a *corridor*, the doors shall be automatic closing by smoke detection in accordance with Section 716.2.6.5 and shall have a vision panel with fire-protection-rated glazing materials in fire protection-rated frames, the area of which shall not exceed that tested.

[F] 909.5.3.2 Ducts and air transfer openings. Ducts and air transfer openings are required to be protected with a minimum Class II, 250°F (121°C) *smoke damper* complying with Section 717.

[F] 909.6 Pressurization method. The primary mechanical means of controlling smoke shall be by pressure differences across *smoke barriers*. Maintenance of a tenable environment is not required in the smoke control zone of fire origin.

[F] 909.6.1 Minimum pressure difference. The pressure difference across a *smoke barrier* used to separate smoke zones shall be not less than 0.05-inch water gage (0.0124 kPa) in fully sprinklered buildings.

In buildings permitted to be other than fully sprinklered, the smoke control system shall be designed to achieve pressure differences not less than two times the maximum calculated pressure difference produced by the design fire.

[F] 909.6.2 Maximum pressure difference. The maximum air pressure difference across a *smoke barrier* shall be determined by required door-opening or closing forces. The actual force required to open exit doors when the system is in the smoke control mode shall be in accordance with Section 1010.1.3. Opening and closing forces for other doors shall be determined by standard engineering methods for the resolution of forces and reactions. The calculated force to set a side-hinged, swinging door in motion shall be determined by:

$$F = F_{dc} + K(WA\Delta P)/2(W-d) \qquad \textbf{(Equation 9-1)}$$

where:

A = Door area, square feet (m^2).

d = Distance from door handle to latch edge of door, feet (m).

F = Total door opening force, pounds (N).

F_{dc} = Force required to overcome closing device, pounds (N).

K = Coefficient 5.2 (1.0).

W = Door width, feet (m).

ΔP = Design pressure difference, inches of water (Pa).

[F] 909.6.3 Pressurized stairways and elevator hoistways. Where *stairways* or elevator hoistways are pressurized, such pressurization systems shall comply with Section 909 as smoke control systems, in addition to the requirements of Sections 909.20 of this code and 909.21 of the *International Fire Code*.

[F] 909.7 Airflow design method. Where *approved* by the fire code official, smoke migration through openings fixed in a permanently open position, which are located between smoke control zones by the use of the airflow method, shall be permitted. The design airflow shall be in accordance with this section. Airflow shall be directed to limit smoke migration from the fire zone. The geometry of openings shall be considered to prevent flow reversal from turbulent effects. Smoke control systems using the airflow method shall be designed in accordance with NFPA 92.

[F] 909.7.1 Prohibited conditions. This method shall not be employed where either the quantity of air or the velocity of the airflow will adversely affect other portions of the smoke control system, unduly intensify the fire, disrupt plume dynamics or interfere with exiting. Airflow toward the fire shall not exceed 200 feet per minute (1.02 m/s). Where the calculated airflow exceeds this limit, the airflow method shall not be used.

[F] 909.8 Exhaust method. Where *approved* by the fire code official, mechanical smoke control for large enclosed volumes, such as in *atriums* or malls, shall be permitted to utilize the exhaust method. Smoke control systems using the exhaust method shall be designed in accordance with NFPA 92.

[F] 909.8.1 Smoke layer. The height of the lowest horizontal surface of the smoke layer interface shall be maintained not less than 6 feet (1829 mm) above a walking surface that forms a portion of a required egress system within the smoke zone.

[F] 909.9 Design fire. The design fire shall be based on a rational analysis performed by the *registered design professional* and *approved* by the fire code official. The design fire shall be based on the analysis in accordance with Section 909.4 and this section.

[F] 909.9.1 Factors considered. The engineering analysis shall include the characteristics of the fuel, fuel load, effects included by the fire and whether the fire is likely to be steady or unsteady.

[F] 909.9.2 Design fire fuel. Determination of the design fire shall include consideration of the type of fuel, fuel spacing and configuration.

[F] 909.9.3 Heat-release assumptions. The analysis shall make use of best available data from *approved* sources and shall not be based on excessively stringent limitations of combustible material.

[F] 909.9.4 Sprinkler effectiveness assumptions. A documented engineering analysis shall be provided for conditions that assume fire growth is halted at the time of sprinkler activation.

[F] 909.10 Equipment. Equipment including, but not limited to, fans, ducts, automatic *dampers* and balance *dampers*, shall be suitable for its intended use, suitable for the probable exposure temperatures that the rational analysis indicates and as *approved* by the fire code official.

[F] 909.10.1 Exhaust fans. Components of exhaust fans shall be rated and certified by the manufacturer for the probable temperature rise to which the components will be exposed. This temperature rise shall be computed by:

$$T_s = (Q_c/mc) + (T_a) \qquad \textbf{(Equation 9-2)}$$

where:

c = Specific heat of smoke at smoke layer temperature, Btu/lb°F (kJ/kg × K).

m = Exhaust rate, pounds per second (kg/s).

Q_c = Convective heat output of fire, Btu/s (kW).

T_a = Ambient temperature, °F (K).

T_s = Smoke temperature, °F (K).

Exception: Reduced T_s as calculated based on the assurance of adequate dilution air.

[F] 909.10.2 Ducts. Duct materials and joints shall be capable of withstanding the probable temperatures and pressures to which they are exposed as determined in accordance with Section 909.10.1. Ducts shall be constructed and supported in accordance with the *International Mechanical Code*. Ducts shall be leak tested to 1.5 times the maximum design pressure in accordance with nationally accepted practices. Measured leakage shall not exceed 5 percent of design flow. Results of such testing shall be a part of the documentation procedure. Ducts shall be supported directly from fire-resistance-rated structural elements of the building by substantial, noncombustible supports.

Exception: Flexible connections, for the purpose of vibration isolation, complying with the *International Mechanical Code* and that are constructed of *approved* fire-resistance-rated materials.

[F] 909.10.3 Equipment, inlets and outlets. Equipment shall be located so as to not expose uninvolved portions of the building to an additional fire hazard. Outside air inlets shall be located so as to minimize the potential for introducing smoke or flame into the building. Exhaust outlets shall be so located as to minimize reintroduction of smoke into the building and to limit exposure of the building or adjacent buildings to an additional fire hazard.

[F] 909.10.4 Automatic dampers. Automatic *dampers*, regardless of the purpose for which they are installed within the smoke control system, shall be listed and conform to the requirements of *approved*, recognized standards.

[F] 909.10.5 Fans. In addition to other requirements, belt-driven fans shall have 1.5 times the number of belts required for the design duty, with the minimum number of belts being two. Fans shall be selected for stable performance based on normal temperature and, where applicable, elevated temperature. Calculations and manufacturer's fan curves shall be part of the documentation procedures. Fans shall be supported and restrained by noncombustible devices in accordance with the requirements of Chapter 16.

Motors driving fans shall not be operated beyond their nameplate horsepower (kilowatts), as determined from measurement of actual current draw, and shall have a minimum service factor of 1.15.

[F] 909.11 Standby power. Smoke control systems shall be provided with standby power in accordance with Section 2702.

909.11.1 Equipment room. The standby power source and its transfer switches shall be in a room separate from the normal power transformers and switch gears and ventilated directly to and from the exterior. The room shall be enclosed with not less than 1-hour *fire barriers* constructed in accordance with Section 707 or *horizontal assemblies* constructed in accordance with Section 711, or both.

[F] 909.11.2 Power sources and power surges. Elements of the smoke control system relying on volatile memories or the like shall be supplied with uninterruptable power sources of sufficient duration to span 15-minute primary power interruption. Elements of the smoke control system susceptible to power surges shall be suitably protected by conditioners, suppressors or other *approved* means.

[F] 909.12 Detection and control systems. Fire detection systems providing control input or output signals to mechanical smoke control systems or elements thereof shall comply with the requirements of Section 907. Such systems shall be equipped with a control unit complying with UL 864 and *listed* as smoke control equipment.

909.12.1 Verification. Control systems for mechanical smoke control systems shall include provisions for verification. Verification shall include positive confirmation of actuation, testing, manual override and the presence of power downstream of all disconnects. A preprogrammed weekly test sequence shall report abnormal conditions audibly, visually and by printed report. The preprogrammed weekly test shall operate all devices, equipment and components used for smoke control.

Exception: Where verification of individual components tested through the preprogrammed weekly testing sequence will interfere with, and produce unwanted effects to, normal building operation, such individual components are permitted to be bypassed from the preprogrammed weekly testing, where *approved* by the building official and in accordance with both of the following:

1. Where the operation of components is bypassed from the preprogrammed weekly test, presence of power downstream of all disconnects shall be verified weekly by a *listed* control unit.
2. Testing of all components bypassed from the preprogrammed weekly test shall be in accordance with Section 909.20.6 of the *International Fire Code*.

[F] 909.12.2 Wiring. In addition to meeting requirements of NFPA 70, all wiring, regardless of voltage, shall be fully enclosed within continuous raceways.

[F] 909.12.3 Activation. Smoke control systems shall be activated in accordance with this section.

[F] 909.12.3.1 Pressurization, airflow or exhaust method. Mechanical smoke control systems using the pressurization, airflow or exhaust method shall have completely automatic control.

[F] 909.12.3.2 Passive method. Passive smoke control systems actuated by *approved* spot-type detectors *listed* for releasing service shall be permitted.

[F] 909.12.4 Automatic control. Where completely automatic control is required or used, the automatic-control sequences shall be initiated from an appropriately zoned *automatic sprinkler system* complying with Section 903.3.1.1, manual controls provided with *ready access* for the fire department and any smoke detectors required by engineering analysis.

[F] 909.13 Control air tubing. Control air tubing shall be of sufficient size to meet the required response times. Tubing shall be flushed clean and dry prior to final connections and shall be adequately supported and protected from damage. Tubing passing through concrete or masonry shall be sleeved and protected from abrasion and electrolytic action.

[F] 909.13.1 Materials. Control-air tubing shall be hard-drawn copper, Type L, ACR in accordance with ASTM B42, ASTM B43, ASTM B68/B68M, ASTM B88, ASTM B251 and ASTM B280. Fittings shall be wrought copper or brass, solder type in accordance with ASME B16.18 or ASME B16.22. Changes in direction shall be made with appropriate tool bends. Brass compression-type fittings shall be used at final connection to devices; other joints shall be brazed using a BCuP-5 brazing alloy with solidus above 1,100°F (593°C) and liquids below 1,500°F (816°C). Brazing flux shall be used on copper-to-brass joints only.

Exception: Nonmetallic tubing used within control panels and at the final connection to devices provided that all of the following conditions are met:

1. Tubing shall comply with the requirements of Section 602.2.1.3 of the *International Mechanical Code*.
2. Tubing and connected devices shall be completely enclosed within a galvanized or paint-grade steel enclosure having a minimum thickness of 0.0296 inch (0.7534 mm) (No. 22 gage). Entry to the enclosure shall be by copper tubing with a protective grommet of neoprene or Teflon or by suitable brass compression to male barbed adapter.
3. Tubing shall be identified by appropriately documented coding.
4. Tubing shall be neatly tied and supported within the enclosure. Tubing bridging cabinets and doors or moveable devices shall be of sufficient length to avoid tension and excessive stress. Tubing shall be protected against abrasion. Tubing connected to devices on doors shall be fastened along hinges.

[F] 909.13.2 Isolation from other functions. Control tubing serving other than smoke control functions shall be isolated by automatic isolation valves or shall be an independent system.

[F] 909.13.3 Testing. Control air tubing shall be tested at three times the operating pressure for not less than 30 minutes without any noticeable loss in gauge pressure prior to final connection to devices.

[F] 909.14 Marking and identification. The detection and control systems shall be clearly marked at all junctions, accesses and terminations.

[F] 909.15 Control diagrams. Identical control diagrams showing all devices in the system and identifying their location and function shall be maintained current and kept on file with the fire code official, the fire department and in the fire command center in a format and manner *approved* by the *fire code official*.

[F] 909.16 Fire fighter's smoke control panel. A fire fighter's smoke control panel for fire department emergency response purposes only shall be provided and shall include manual control or override of automatic control for mechanical smoke control systems. The panel shall be located in a fire command center complying with Section 911 in high-rise buildings or buildings with smoke-protected assembly seating. In all other buildings, the fire fighter's smoke control panel shall be installed in an *approved* location adjacent to the fire alarm control panel. The fire fighter's smoke control panel shall comply with Sections 909.16.1 through 909.16.3.

[F] 909.16.1 Smoke control systems. Fans within the building shall be shown on the fire fighter's control panel. A clear indication of the direction of airflow and the relationship of components shall be displayed. Status indicators shall be provided for all smoke control equipment, annunciated by fan and zone, and by pilot-lamp-type indicators as follows:

1. Fans, *dampers* and other operating equipment in their normal status—WHITE.
2. Fans, *dampers* and other operating equipment in their off or closed status—RED.
3. Fans, *dampers* and other operating equipment in their on or open status—GREEN.
4. Fans, *dampers* and other operating equipment in a fault status—YELLOW/AMBER.

[F] 909.16.2 Smoke control panel. The fire fighter's control panel shall provide control capability over the complete smoke control system equipment within the building as follows:

1. ON-AUTO-OFF control over each individual piece of operating smoke control equipment that can be controlled from other sources within the building. This includes *stairway* pressurization fans; smoke exhaust fans; supply, return and exhaust fans; elevator *shaft* fans and other operating equipment used or intended for smoke control purposes.
2. OPEN-AUTO-CLOSE control over individual *dampers* relating to smoke control and that are controlled from other sources within the building.
3. ON-OFF or OPEN-CLOSE control over smoke control and other critical equipment associated with a fire or smoke emergency and that can only be controlled from the fire fighter's control panel.

Exceptions:

1. Complex systems, where *approved*, where the controls and indicators are combined to control and indicate all elements of a single smoke zone as a unit.
2. Complex systems, where *approved*, where the control is accomplished by computer interface using *approved*, plain English commands.

[F] 909.16.3 Control action and priorities. The fire-fighter's control panel actions shall be as follows:

1. ON-OFF and OPEN-CLOSE control actions shall have the highest priority of any control point within the building. Once issued from the fire fighter's control panel, automatic or manual control from any other control point within the building shall not contradict the control action. Where automatic means are provided to interrupt normal, nonemergency equipment operation or produce a specific result to safeguard the building or equipment including, but not limited to, duct freezestats, duct smoke detectors, high-temperature cutouts, temperature-actuated linkage and similar devices, such means shall be capable of being overridden by the fire fighter's control panel. The last control action as indicated by each fire fighter's control panel switch position shall prevail. Control actions shall not require the smoke control system to assume more than one configuration at any one time.

   Exception: Power disconnects required by NFPA 70.

2. Only the AUTO position of each three-position fire-fighter's control panel switch shall allow automatic or manual control action from other control points within the building. The AUTO position shall be the NORMAL, nonemergency, building control position. Where a fire fighter's control panel is in the AUTO position, the actual status of the device (on, off, open, closed) shall continue to be indicated by the status indicator described in Section 909.16.1. Where directed by an automatic signal to assume an emergency condition, the NORMAL position shall become the emergency condition for that device or group of devices within the zone. Control actions shall not require the smoke control system to assume more than one configuration at any one time.

[F] 909.17 System response time. Smoke-control system activation shall be initiated immediately after receipt of an appropriate automatic or manual activation command. Smoke control systems shall activate individual components (such as *dampers* and fans) in the sequence necessary to prevent physical damage to the fans, *dampers*, ducts and other equipment. For purposes of smoke control, the fire fighter's control panel response time shall be the same for automatic or manual smoke control action initiated from any other building control point. The total response time, including that necessary for detection, shutdown of operating equipment and smoke control system startup, shall allow for full operational mode to be achieved before the conditions in the space exceed the design smoke condition. Upon receipt of an alarm condition at the fire alarm control panel, fans, dampers and automatic doors shall have achieved their proper operating state and the final status shall be indicated at the smoke control panel within 90 seconds. The system response time for each component and their sequential relationships shall be detailed in the required rational analysis and verification of their installed condition reported in the required final report.

[F] 909.18 Acceptance testing. Devices, equipment, components and sequences shall be individually tested. These tests, in addition to those required by other provisions of this code, shall consist of determination of function, sequence and, where applicable, capacity of their installed condition.

[F] 909.18.1 Detection devices. Smoke or fire detectors that are a part of a smoke control system shall be tested in accordance with Chapter 9 in their installed condition. Where applicable, this testing shall include verification of airflow in both minimum and maximum conditions.

[F] 909.18.2 Ducts. Ducts that are part of a smoke control system shall be traversed using generally accepted practices to determine actual air quantities.

[F] 909.18.3 Dampers. *Dampers* shall be tested for function in their installed condition.

[F] 909.18.4 Inlets and outlets. Inlets and outlets shall be read using generally accepted practices to determine air quantities.

[F] 909.18.5 Fans. Fans shall be examined for correct rotation. Measurements of voltage, amperage, revolutions per minute (rpm) and belt tension shall be made.

[F] 909.18.6 Smoke barriers. Measurements using inclined manometers or other *approved* calibrated measuring devices shall be made of the pressure differences across *smoke barriers*. Such measurements shall be conducted for each possible smoke control condition.

[F] 909.18.7 Controls. Each smoke zone equipped with an automatic-initiation device shall be put into operation by the actuation of one such device. Each additional device within the zone shall be verified to cause the same sequence without requiring the operation of fan motors in order to prevent damage. Control sequences shall be verified throughout the system, including verification of override from the fire fighter's control panel and simulation of standby power conditions.

[F] 909.18.8 Testing for smoke control. Smoke control systems shall be tested by a *special inspector* in accordance with Section 1705.19 .

[F] 909.18.8.1 Scope of testing. Testing shall be conducted in accordance with the following:

1. During erection of ductwork and prior to concealment for the purposes of leakage testing and recording of device location.
2. Prior to occupancy and after sufficient completion for the purposes of pressure-difference testing, flow measurements, and detection and control verification.

[F] 909.18.8.2 Qualifications. *Approved* agencies for smoke control testing shall have expertise in fire protection engineering, mechanical engineering and certification as air balancers.

[F] 909.18.8.3 Reports. A complete report of testing shall be prepared by the *approved* agency. The report shall include identification of all devices by manufacturer, nameplate data, design values, measured values and identification tag or *mark*. The report shall be reviewed by the responsible *registered design professional* and, when satisfied that the design intent has been achieved, the responsible *registered design professional* shall sign, seal and date the report.

[F] 909.18.8.3.1 Report filing. A copy of the final report shall be filed with the fire code official and an identical copy shall be maintained in an *approved* location at the building.

[F] 909.18.9 Identification and documentation. Charts, drawings and other documents identifying and locating each component of the smoke control system, and describing its proper function and maintenance requirements, shall be maintained on file at the building as an attachment to the report required by Section 909.18.8.3. Devices shall have an *approved* identifying tag or *mark* on them consistent with the other required documentation and shall be dated indicating the last time they were successfully tested and by whom.

[F] 909.19 System acceptance. Buildings, or portions thereof, required by this code to comply with this section shall not be issued a certificate of occupancy until such time that the fire code official determines that the provisions of this section have been fully complied with and that the fire department has received satisfactory instruction on the operation, both automatic and manual, of the system and a written maintenance program complying with the requirements of Section 909.20.1 of the *International Fire Code* has been submitted and approved by the fire code official.

Exception: In buildings of phased construction, a temporary certificate of occupancy, as *approved* by the fire code official, shall be allowed provided that those portions of the building to be occupied meet the requirements of this section and that the remainder does not pose a significant hazard to the safety of the proposed occupants or adjacent buildings.

909.20 Smokeproof enclosures. Where required by Section 1023.12, a *smokeproof enclosure* shall be constructed in accordance with this section. A *smokeproof enclosure* shall consist of an *interior exit stairway* or *ramp* that is enclosed in accordance with the applicable provisions of Section 1023 and an open exterior balcony, ventilated vestibule or pressurized *stair* and pressurized entrance vestibule meeting the requirements of this section. Where access to the roof is required by the *International Fire Code*, such access shall be from the *smokeproof enclosure* where a *smokeproof enclosure* is required.

909.20.1 Access. Access to the *stairway* or *ramp* shall be by way of a vestibule or an open exterior balcony. The minimum dimension of the vestibule shall be not less than the required width of the *corridor* leading to the vestibule but shall not have a width of less than 44 inches (1118 mm) and shall not have a length of less than 72 inches (1829 mm) in the direction of egress travel.

909.20.2 Construction. The *smokeproof enclosure* shall be separated from the remainder of the building by not less than 2-hour *fire barriers* constructed in accordance with Section 707 or *horizontal assemblies* constructed in accordance with Section 711, or both. Openings are not permitted other than the required *means of egress* doors. The vestibule shall be separated from the *stairway* or *ramp* by not less than 2-hour *fire barriers* constructed in accordance with Section 707 or *horizontal assemblies* constructed in accordance with Section 711, or both. The open exterior balcony shall be constructed in accordance with the *fire-resistance rating* requirements for floor assemblies.

909.20.2.1 Door closers. Doors in a *smokeproof enclosure* shall be self- or automatic closing by actuation of a smoke detector in accordance with Section 716.2.6.6 and shall be installed at the floor-side entrance to the *smokeproof enclosure*. The actuation of the smoke detector on any door shall activate the closing devices on all doors in the *smokeproof enclosure* at all levels. Smoke detectors shall be installed in accordance with Section 907.3.

909.20.3 Natural ventilation alternative. The provisions of Sections 909.20.3.1 through 909.20.3.3 shall apply to ventilation of *smokeproof enclosures* by natural means.

909.20.3.1 Balcony doors. Where access to the *stairway* or *ramp* is by way of an open exterior balcony, the door assembly into the enclosure shall be a *fire door assembly* in accordance with Section 716.

909.20.3.2 Vestibule doors. Where access to the *stairway* or *ramp* is by way of a vestibule, the door assembly into the vestibule shall be a *fire door assembly* complying with Section 716. The door assembly from the vestibule to the *stairway* shall have not less than a 20-minute *fire protection rating* complying with Section 716.

909.20.3.3 Vestibule ventilation. Each vestibule shall have a minimum net area of 16 square feet (1.5 m^2) of opening in a wall facing an outer *court, yard* or *public way* that is not less than 20 feet (6096 mm) in width.

909.20.4 Mechanical ventilation alternative. The provisions of Sections 909.20.4.1 through 909.20.4.4 shall apply to ventilation of *smokeproof enclosures* by mechanical means.

909.20.4.1 Vestibule doors. The door assembly from the building into the vestibule shall be a *fire door assembly* complying with Section 716.2.2.1. The door assembly from the vestibule to the *stairway* or *ramp* shall not have less than a 20-minute *fire protection rating* and shall meet the requirements for a smoke door assembly in accordance with Section 716.2.2.1. The door shall be installed in accordance with NFPA 105.

909.20.4.2 Vestibule ventilation. The vestibule shall be supplied with not less than one air change per minute and the exhaust shall be not less than 150 percent of supply. Supply air shall enter and exhaust air shall discharge from the vestibule through separate, tightly constructed ducts used only for that purpose. Supply air shall enter the vestibule within 6 inches (152 mm) of the floor level. The top of the exhaust register shall be located at the top of the smoke trap but not more than 6 inches (152 mm) down from the top of the trap, and shall be entirely within the smoke trap area. Doors in the open position shall not obstruct duct openings. Duct openings with controlling *dampers* are permitted where necessary to meet the design requirements, but *dampers* are not otherwise required.

909.20.4.2.1 Engineered ventilation system. Where a specially engineered system is used, the system shall exhaust a quantity of air equal to not less than 90 air changes per hour from any vestibule in emergency operation mode and shall be sized to handle three vestibules simultaneously. Smoke detectors shall be located at the floor-side entrance to each vestibule and shall activate the system for the affected vestibule. Smoke detectors shall be installed in accordance with Section 907.3.

909.20.4.3 Smoke trap. The vestibule ceiling shall be not less than 20 inches (508 mm) higher than the door opening into the vestibule to serve as a smoke and heat trap and to provide an upward-moving air column. The height shall not be decreased unless *approved* and justified by design and test.

909.20.4.4 Stairway or ramp shaft air movement system. The *stairway* or *ramp shaft* shall be provided with a dampered relief opening and supplied with sufficient air to maintain a minimum positive pressure of 0.10 inch of water (25 Pa) in the *shaft* relative to the vestibule with all doors closed.

909.20.5 Stairway and ramp pressurization alternative. Where the building is equipped throughout with an *automatic sprinkler system* in accordance with Section 903.3.1.1, the vestibule is not required, provided that each *interior exit stairway* or *ramp* is pressurized to not less than 0.10 inch of water (25 Pa) and not more than 0.35 inches of water (87 Pa) in the *shaft* relative to the building measured with all *interior exit stairway* and *ramp* doors closed under maximum anticipated conditions of stack effect and wind effect.

909.20.6 Pressurized stair and vestibule alternative. The provisions of Sections 909.20.6.1 through 909.20.6.3 shall apply to *smokeproof enclosures* using a pressurized *stair* and pressurized entrance vestibule.

909.20.6.1 Vestibule doors. The door assembly from the building into the vestibule shall be a *fire door assembly* complying with Section 716.2.2.1. The door assembly from the vestibule to the *stairway* shall have not less than a 20-minute *fire protection rating* and meet the requirements for a smoke door assembly in accordance with Section 716.2.2.1. The door shall be installed in accordance with NFPA 105.

909.20.6.2 Pressure difference. The stair enclosure shall be pressurized to not less than 0.05 inch of water gage (12.44 Pa) positive pressure relative to the vestibule with all *stairway* doors closed under the maximum anticipated stack pressures. The vestibule, with doors closed, shall have not less than 0.05 inch of water gage (12.44 Pa) positive pressure relative to the fire floor. The pressure difference across doors shall not exceed 30 pounds (133-N) maximum force to begin opening the door.

909.20.6.3 Dampered relief opening. A controlled relief vent having the capacity to discharge not less than 2,500 cubic feet per minute (1180 L/s) of air at the design pressure difference shall be located in the upper portion of the pressurized exit enclosure.

909.20.7 Ventilating equipment. The activation of ventilating equipment required by the alternatives in Sections 909.20.4, 909.20.5 and 909.20.6 shall be by smoke detectors installed at each floor level at an *approved* location at the entrance to the *smokeproof enclosure*. When the closing device for the *stairway* and *ramp shaft* and vestibule doors is activated by smoke detection or power failure, the mechanical equipment shall activate and operate at the required performance levels. Smoke detectors shall be installed in accordance with Section 907.3.

909.20.7.1 Ventilation systems. *Smokeproof enclosure* ventilation systems shall be independent of other building ventilation systems. The equipment, control wiring, power wiring and ductwork shall comply with one of the following:

1. Equipment, control wiring, power wiring and ductwork shall be located exterior to the building and directly connected to the *smokeproof enclosure* or connected to the *smokeproof enclosure* by ductwork enclosed by not less than 2-hour *fire barriers* constructed in accordance with Section 707 or *horizontal assemblies* constructed in accordance with Section 711, or both.
2. Equipment, control wiring, power wiring and ductwork shall be located within the *smokeproof enclosure* with intake or exhaust directly from and to the outside or through ductwork enclosed by not less than 2-hour *fire barriers* constructed in accordance with Section 707 or *horizontal assemblies* constructed in accordance with Section 711, or both.
3. Equipment, control wiring, power wiring and ductwork shall be located within the building if separated from the remainder of the building, including other mechanical equipment, by not less than 2-hour *fire barriers* constructed in accordance with Section 707 or *horizontal assemblies* constructed in accordance with Section 711, or both.

Exception:

1. Control wiring and power wiring located outside of a 2-hour fire barrier construction shall be protected using any one of the following methods:
   - 1.1. Cables used for survivability of required critical circuits shall be listed in accordance with UL 2196 and shall have a *fire-resistance rating* of not less than 2 hours.
   - 1.2. Where encased with not less than 2 inches (51 mm) of concrete.
   - 1.3. *Electrical circuit protective systems* shall have a *fire-resistance rating* of not less than 2 hours. *Electrical circuit protective systems* shall be installed in accordance with their listing requirements.

909.20.7.2 Standby power. Mechanical vestibule and *stairway* and *ramp shaft* ventilation systems and automatic fire detection systems shall be provided with standby power in accordance with Section 2702.

909.20.7.3 Acceptance and testing. Before the mechanical equipment is *approved*, the system shall be tested in the presence of the *building official* to confirm that the system is operating in compliance with these requirements.

909.21 Elevator hoistway pressurization alternative. Where elevator hoistway pressurization is provided in lieu of required enclosed elevator lobbies, the pressurization system shall comply with Sections 909.21.1 through 909.21.11.

909.21.1 Pressurization requirements. Elevator hoistways shall be pressurized to maintain a minimum positive pressure of 0.10 inch of water (25 Pa) and a maximum positive pressure of 0.25 inch of water (67 Pa) with respect to adjacent occupied space on all floors. This pressure shall be measured at the midpoint of each hoistway door, with all elevator cars at the floor of recall and all hoistway doors on the floor of recall open and all other hoistway doors closed. The pressure differentials shall be measured between the hoistway and the adjacent elevator landing. The opening and closing of hoistway doors at each level must be demonstrated during this test. The supply air intake shall be from an outside, uncontaminated source located a minimum distance of 20 feet (6096 mm) from any air exhaust system or outlet.

Exceptions:

1. On floors containing only Group R occupancies, the pressure differential is permitted to be measured between the hoistway and a *dwelling unit* or *sleeping unit*.
2. Where an elevator opens into a lobby enclosed in accordance with Section 3007.6 or 3008.6, the pressure differential is permitted to be measured between the hoistway and the space immediately outside the door(s) from the floor to the enclosed lobby.
3. The pressure differential is permitted to be measured relative to the outdoor atmosphere on floors other than the following:
   - 3.1. The fire floor.
   - 3.2. The two floors immediately below the fire floor.
   - 3.3. The floor immediately above the fire floor.
4. The minimum positive pressure of 0.10 inch of water (25 Pa) and a maximum positive pressure

of 0.25 inch of water (67 Pa) with respect to occupied floors are not required at the floor of recall with the doors open.

909.21.1.1 Use of ventilation systems. Ventilation systems, other than hoistway supply air systems, are permitted to be used to exhaust air from adjacent spaces on the fire floor, two floors immediately below and one floor immediately above the fire floor to the building's exterior where necessary to maintain positive pressure relationships as required in Section 909.21.1 during operation of the elevator *shaft* pressurization system.

909.21.2 Rational analysis. A rational analysis complying with Section 909.4 shall be submitted with the *construction documents*.

909.21.3 Ducts for system. Any duct system that is part of the pressurization system shall be protected with the same *fire-resistance rating* as required for the elevator *shaft* enclosure.

909.21.4 Fan system. The fan system provided for the pressurization system shall be as required by Sections 909.21.4.1 through 909.21.4.4.

909.21.4.1 Fire resistance. Where located within the building, the fan system that provides the pressurization shall be protected with the same *fire-resistance rating* required for the elevator *shaft* enclosure.

909.21.4.2 Smoke detection. The fan system shall be equipped with a smoke detector that will automatically shut down the fan system when smoke is detected within the system.

909.21.4.3 Separate systems. A separate fan system shall be used for each elevator hoistway.

909.21.4.4 Fan capacity. The supply fan shall be either adjustable with a capacity of not less than 1,000 cubic feet per minute (0.4719 m^3/s) per door, or that specified by a *registered design professional* to meet the requirements of a designed pressurization system.

909.21.5 Standby power. The pressurization system shall be provided with standby power in accordance with Section 2702.

909.21.6 Activation of pressurization system. The elevator pressurization system shall be activated upon activation of either the building fire alarm system or the elevator lobby smoke detectors. Where both a building fire alarm system and elevator lobby smoke detectors are present, each shall be independently capable of activating the pressurization system.

909.21.7 Testing. Testing for performance shall be required in accordance with Section 909.18.8. System acceptance shall be in accordance with Section 909.19.

909.21.8 Marking and identification. Detection and control systems shall be marked in accordance with Section 909.14.

909.21.9 Control diagrams. Control diagrams shall be provided in accordance with Section 909.15.

909.21.10 Control panel. A control panel complying with Section 909.16 shall be provided.

909.21.11 System response time. Hoistway pressurization systems shall comply with the requirements for smoke control system response time in Section 909.17.

SECTION 910
SMOKE AND HEAT REMOVAL

[F] 910.1 General. Where required by this code, smoke and heat vents or mechanical smoke removal systems shall conform to the requirements of this section.

[F] 910.2 Where required. Smoke and heat vents or a mechanical smoke removal system shall be installed as required by Sections 910.2.1 and 910.2.2.

Exceptions:

1. Frozen food warehouses used solely for storage of Class I and II commodities where protected by an *approved automatic sprinkler system*.
2. Smoke and heat removal shall not be required in areas of buildings equipped with early suppression fast-response (ESFR) sprinklers.
3. Smoke and heat removal shall not be required in areas of buildings equipped with control mode special application sprinklers with a response time index of 50 $(m \times s)^{1/2}$ or less that are listed to control a fire in stored commodities with 12 or fewer sprinklers.

910.2.1 Group F-1 or S-1. Smoke and heat vents installed in accordance with Section 910.3 or a mechanical smoke removal system installed in accordance with Section 910.4 shall be installed in buildings and portions thereof used as a Group F-1 or S-1 occupancy having more than 50,000 square feet (4645 m^2) of undivided area. In occupied portions of a building equipped throughout with an *automatic sprinkler system* in accordance with Section 903.3.1.1 where the upper surface of the *story* is not a roof assembly, a mechanical smoke removal system in accordance with Section 910.4 shall be installed.

Exception: Group S-1 aircraft repair hangars.

[F] 910.2.2 High-piled combustible storage. Smoke and heat removal required by Table 3206.2 of the *International Fire Code* for buildings and portions thereof containing high-piled combustible storage shall be installed in accordance with Section 910.3 in unsprinklered buildings. In buildings and portions thereof containing high-piled combustible storage equipped throughout with an *automatic sprinkler system* in accordance with Section 903.3.1.1, a smoke and heat removal system shall be installed in accordance with Section 910.3 or 910.4. In occupied portions of a building equipped throughout with an *automatic sprinkler system* in accordance with Section 903.3.1.1, where the upper surface of the *story* is not a roof assembly, a mechanical smoke removal system in accordance with Section 910.4 shall be installed.

[F] 910.3 Smoke and heat vents. The design and installation of smoke and heat vents shall be in accordance with Sections 910.3.1 through 910.3.3.

[F] 910.3.1 Listing and labeling. Smoke and heat vents shall be *listed* and labeled to indicate compliance with UL 793 or FM 4430.

[F] 910.3.2 Smoke and heat vent locations. Smoke and heat vents shall be located 20 feet (6096 mm) or more from adjacent *lot lines* and *fire walls* and 10 feet (3048 mm) or more from *fire barriers*. Vents shall be uniformly located within the roof in the areas of the building where the vents are required to be installed by Section 910.2 with consideration given to roof pitch, sprinkler location and structural members.

910.3.3 Smoke and heat vents area. The required aggregate area of smoke and heat vents shall be calculated as follows:

For buildings equipped throughout with an *automatic sprinkler system* in accordance with Section 903.3.1.1:

$A_{VR} = V/9000$ **(Equation 9-3)**

where:

A_{VR} = The required aggregate vent area (ft^2).

V = Volume (ft^3) of the area that requires smoke removal.

For unsprinklered buildings:

$A_{VR} = A_{FA}/50$ **(Equation 9-4)**

where:

A_{VR} = The required aggregate vent area (ft^2).

A_{FA} = The area of the floor in the area that requires smoke removal.

[F] 910.3.4 Vent operation. Smoke and heat vents shall be capable of being operated by *approved* automatic and manual means.

[F] 910.3.5 Fusible link temperature rating. Where vents are installed in areas provided with automatic fire sprinklers and the vents operate by fusible link, the fusible link shall have a temperature rating of 360°F (182°C).

[F] 910.4 Mechanical smoke removal systems. Mechanical smoke removal systems shall be designed and installed in accordance with Sections 910.4.1 through 910.4.7.

910.4.1 Automatic sprinklers required. The building shall be equipped throughout with an *approved automatic sprinkler system* in accordance with Section 903.3.1.1.

910.4.2 Exhaust fan construction. Exhaust fans that are part of a mechanical smoke removal system shall be rated for operation at 221°F (105°C). Exhaust fan motors shall be located outside of the exhaust fan air stream.

910.4.3 System design criteria. The mechanical smoke removal system shall be sized to exhaust the building at a minimum rate of two air changes per hour based on the volume of the building or portion thereof without contents. The capacity of each exhaust fan shall not exceed 30,000 cubic feet per minute (14.2 m^3/s).

910.4.3.1 Makeup air. Makeup air openings shall be provided within 6 feet (1829 mm) of the floor level. Operation of makeup air openings shall be manual or automatic. The minimum gross area of makeup air inlets shall be 8 square feet per 1,000 cubic feet per minute (0.74 m^2 per 0.4719 m^3/s) of smoke exhaust.

910.4.4 Activation. The mechanical smoke removal system shall be activated by manual controls only.

910.4.5 Manual control location. Manual controls shall be located where they are able to be accessed by the fire service from an exterior door of the building and separated from the remainder of the building by not less than 1-hour *fire barriers* constructed in accordance with Section 707 or *horizontal assemblies* constructed in accordance with Section 711, or both.

[F] 910.4.6 Control wiring. Wiring for operation and control of mechanical smoke removal systems shall be connected ahead of the main disconnect in accordance with Section 701.12E of NFPA 70 and be protected against interior fire exposure to temperatures in excess of 1,000°F (538°C) for a period of not less than 15 minutes.

[F] 910.4.7 Controls. Where building air-handling and mechanical smoke removal systems are combined or where independent building air-handling systems are provided, fans shall automatically shut down in accordance with the *International Mechanical Code*. The manual controls provided for the smoke removal system shall have the capability to override the automatic shutdown of fans that are part of the smoke removal system.

910.5 Maintenance. Smoke and heat vents and mechanical smoke removal systems shall be maintained in accordance with the *International Fire Code*.

SECTION 911
FIRE COMMAND CENTER

[F] 911.1 General. Where required by other sections of this code, in buildings classified as high-rise buildings by this code and in all F-1 and S-1 occupancies with a building footprint of over 500,000 square feet (46 452 m^2), a fire command center for fire department operations shall be provided and shall comply with Sections 911.1.1 through 911.1.7.

[F] 911.1.1 Location and access. The location and access to the fire command center shall be *approved* by the *fire code official*.

[F] 911.1.2 Separation. The fire command center shall be separated from the remainder of the building by not less than a 1-hour *fire barrier* constructed in accordance with Section 707 or *horizontal assembly* constructed in accordance with Section 711, or both.

[F] 911.1.3 Size. The fire command center shall be not less than 0.015 percent of the total building area of the facility served or 200 square feet (19 m^2) in area, whichever is greater, with a minimum dimension of 0.7 times the square root of the room area or 10 feet (3048 mm), whichever is greater. Where a fire command is required for Group F-1 and S-1 occupancies with a building footprint greater than 500,000 square feet (46 452 m^2) in area, the fire command center shall have a minimum size of 96 square feet (9 m^2) with a minimum dimension of 8 feet (2348 mm) where *approved* by the *fire code official*.

[F] 911.1.4 Layout approval. A layout of the fire command center and all features required by this section to be contained therein shall be submitted for approval prior to installation.

[F] 911.1.5 Storage. Storage unrelated to operation of the fire command center shall be prohibited.

[F] 911.1.6 Required features. The fire command center shall comply with NFPA 72 and shall contain all of the following features:

1. The emergency voice/alarm communication system control unit.
2. The fire department communications system.
3. Fire detection and alarm system *annunciator*.
4. *Annunciator* unit visually indicating the location of the elevators and whether they are operational.
5. Status indicators and controls for air distribution systems.
6. The fire fighter's control panel required by Section 909.16 for smoke control systems installed in the building.
7. Controls for unlocking *interior exit stairway* doors simultaneously.
8. Sprinkler valve and waterflow detector display panels.
9. Emergency and standby power status indicators.
10. A telephone for fire department use with controlled access to the public telephone system.
11. Fire pump status indicators.
12. Schematic building plans indicating the typical floor plan and detailing the building core, *means of egress*, fire protection systems, fire fighter air replenishment system, fire-fighting equipment and fire department access and the location of *fire walls, fire barriers, fire partitions, smoke barriers* and *smoke partitions*.
13. An *approved* Building Information Card that contains, but is not limited to, the following information:
    - 13.1. General building information that includes: property name, address, the number of floors in the building above and below grade, use and occupancy classification (for mixed uses, identify the different types of occupancies on each floor), and the estimated building population during the day, night and weekend.
    - 13.2. Building emergency contact information that includes: a list of the building's emergency contacts including but not limited to building manager and building engineer and their respective work phone number, cell phone number, e-mail address.
    - 13.3. Building construction information that includes: the type of building construction including but not limited to floors, walls, columns, and roof assembly.
    - 13.4. *Exit access* and *exit stairway* information that includes: number of *exit access* and *exit stairways* in the building, each *exit access* and *exit stairway* designation and floors served, location where each *exit access* and *exit stairway* discharges, *interior exit stairways* that are pressurized, *exit stairways* provided with emergency lighting, each *exit stairway* that allows reentry, *exit stairways* providing roof access; elevator information that includes: number of elevator banks, elevator bank designation, elevator car numbers and respective floors that they serve; location of elevator machine rooms, control rooms and control spaces; location of sky lobby, location of freight elevator banks.
    - 13.5. Building services and system information that includes: location of mechanical rooms, location of building management system, location and capacity of all fuel oil tanks, location of emergency generator, location of natural gas service.
    - 13.6. Fire protection system information that includes: location of standpipes, location of fire pump room, location of fire department connections, floors protected by automatic sprinklers, location of different types of *automatic sprinkler systems* installed including, but not limited to, dry, wet and pre-action.
    - 13.7 Hazardous material information that includes: location of hazardous material, quantity of hazardous material.
14. Work table.
15. Generator supervision devices, manual start and transfer features.
16. Public address system, where specifically required by other sections of this code.
17. Elevator fire recall switch in accordance with ASME A17.1/CSA B44.
18. Elevator emergency or standby power selector switch(es), where emergency or standby power is provided.

[F] 911.1.7 Fire command center identification. The *fire command center* shall be identified by a permanent easily visible sign reading "FIRE COMMAND CENTER" located on the door to the *fire command center*.

SECTION 912
FIRE DEPARTMENT CONNECTIONS

[F] 912.1 Installation. Fire department connections shall be installed in accordance with the NFPA standard applicable to the system design and shall comply with Sections 912.2 through 912.6.

[F] 912.2 Location. With respect to hydrants, driveways, buildings and landscaping, fire department connections shall be so located that fire apparatus and hose connected to supply the system will not obstruct access to the buildings for other fire apparatus. The location of fire department connections shall be *approved* by the *fire code official*.

[F] 912.2.1 Visible location. Fire department connections shall be located on the street side of buildings or facing approved fire apparatus access roads, fully visible and recognizable from the street, fire apparatus access road or nearest point of fire department vehicle access or as otherwise *approved* by the *fire code official*.

[F] 912.2.2 Existing buildings. On existing buildings, wherever the fire department connection is not visible to approaching fire apparatus, the fire department connection shall be indicated by an *approved* sign mounted on the street front or on the side of the building. Such sign shall have the letters "FDC" not less than 6 inches (152 mm) high and words in letters not less than 2 inches (51 mm) high or an arrow to indicate the location. Such signs shall be subject to the approval of the *fire code official*.

[F] 912.3 Fire hose threads. Fire hose threads used in connection with standpipe systems shall be *approved* and shall be compatible with fire department hose threads.

[F] 912.4 Access. Immediate access to fire department connections shall be maintained at all times and without obstruction by fences, bushes, trees, walls or any other fixed or moveable object. Access to fire department connections shall be *approved* by the *fire code official*.

Exception: Fences, where provided with an access gate equipped with a sign complying with the legend requirements of this section and a means of emergency operation. The gate and the means of emergency operation shall be *approved* by the *fire code official* and maintained operational at all times.

[F] 912.4.1 Locking fire department connection caps. The fire code official is authorized to require locking caps on fire department connections for water-based *fire protection systems* where the responding fire department carries appropriate key wrenches for removal.

[F] 912.4.2 Clear space around connections. A working space of not less than 36 inches (762 mm) in width, 36 inches (914 mm) in depth and 78 inches (1981 mm) in height shall be provided and maintained in front of and to the sides of wall-mounted fire department connections and around the circumference of free-standing fire department connections, except as otherwise required or *approved* by the *fire code official*.

[F] 912.4.3 Physical protection. Where fire department connections are subject to impact by a motor vehicle, vehicle impact protection shall be provided in accordance with Section 312 of the *International Fire Code*.

[F] 912.5 Signs. A metal sign with raised letters not less than 1 inch (25 mm) in size shall be mounted on all fire department connections serving automatic sprinklers, standpipes or fire pump connections. Such signs shall read: "AUTOMATIC SPRINKLERS," "STANDPIPES," or "TEST CONNECTION," or a combination thereof as applicable. Where the fire department connection does not serve the entire building, a sign shall be provided indicating the portions of the building served.

[P] 912.6 Backflow protection. The potable water supply to automatic sprinkler and standpipe systems shall be protected against backflow as required by the *International Plumbing Code*.

SECTION 913
FIRE PUMPS

[F] 913.1 General. Where provided, fire pumps for fire protection systems shall be installed in accordance with this section and NFPA 20.

Exception: Pumps for automatic sprinkler systems installed in accordance with Section 903.3.1.3, or Section P2904 of the *International Residential Code*.

[F] 913.2 Protection against interruption of service. The fire pump, driver and controller shall be protected in accordance with NFPA 20 against possible interruption of service through damage caused by explosion, fire, *flood*, earthquake, rodents, insects, windstorm, freezing, vandalism and other adverse conditions.

913.2.1 Protection of fire pump rooms. Fire pumps shall be located in rooms that are separated from all other areas of the building by 2-hour *fire barriers* constructed in accordance with Section 707 or 2-hour *horizontal assemblies* constructed in accordance with Section 711, or both.

Exceptions:

1. In other than high-rise buildings, separation by 1-hour *fire barriers* constructed in accordance with Section 707 or 1-hour *horizontal assemblies* constructed in accordance with Section 711, or both, shall be permitted in buildings equipped throughout with an *automatic sprinkler system* in accordance with Section 903.3.1.1 or 903.3.1.2.
2. Separation is not required for fire pumps physically separated in accordance with NFPA 20.

[F] 913.2.2 Circuits supplying fire pumps. Cables used for survivability of circuits supplying fire pumps shall be protected using one of the following methods:

1. Cables used for survivability of required critical circuits shall be *listed* in accordance with UL 2196 and shall have a *fire-resistance rating* of not less than 1 hour.
2. *Electrical circuit protective systems* shall have a *fire-resistance rating* of not less than 1 hour. *Electrical circuit protective systems* shall be installed in accordance with their listing requirements.
3. Construction having a *fire-resistance rating* of not less than 1 hour.
4. The cable or raceway is encased in a minimum of 2 inches (51 mm) of concrete.

Exception: This section shall not apply to cables, or portions of cables, located within a fire pump room or generator room which is separated from the remainder of the occupancy with *fire-resistance-rated* construction.

[F] 913.3 Temperature of pump room. Suitable means shall be provided for maintaining the temperature of a pump room or pump house, where required, above 40°F (5°C).

[F] 913.3.1 Engine manufacturer's recommendation. Temperature of the pump room, pump house or area where engines are installed shall never be less than the minimum recommended by the engine manufacturer. The engine manufacturer's recommendations for oil heaters shall be followed.

[F] 913.4 Valve supervision. Where provided, the fire pump suction, discharge and bypass valves, and isolation valves on the backflow prevention device or assembly shall be supervised open by one of the following methods:

1. Central-station, proprietary or remote-station signaling service.
2. Local signaling service that will cause the sounding of an audible signal at a *constantly attended location*.
3. Locking valves open.
4. Sealing of valves and *approved* weekly recorded inspection where valves are located within fenced enclosures under the control of the owner.

[F] 913.4.1 Test outlet valve supervision. Fire pump test outlet valves shall be supervised in the closed position.

[F] 913.5 Acceptance test. Acceptance testing shall be done in accordance with the requirements of NFPA 20.

SECTION 914
EMERGENCY RESPONDER SAFETY FEATURES

[F] 914.1 Shaftway markings. Vertical *shafts* shall be identified as required by Sections 914.1.1 and 914.1.2.

[F] 914.1.1 Exterior access to shaftways. Outside openings accessible to the fire department and that open directly on a hoistway or shaftway communicating between two or more floors in a building shall be plainly marked with the word "SHAFTWAY" in red letters not less than 6 inches (152 mm) high on a white background. Such warning signs shall be placed so as to be readily discernible from the outside of the building.

[F] 914.1.2 Interior access to shaftways. Door or window openings to a hoistway or shaftway from the interior of the building shall be plainly marked with the word "SHAFTWAY" in red letters not less than 6 inches (152 mm) high on a white background. Such warning signs shall be placed so as to be readily discernible.

Exception: Markings shall not be required on shaftway openings that are readily discernible as openings onto a shaftway by the construction or arrangement.

[F] 914.2 Equipment room identification. Fire protection equipment shall be identified in an *approved* manner. Rooms containing controls for air-conditioning systems, sprinkler risers and valves or other fire detection, suppression or control elements shall be identified for the use of the fire department. *Approved* signs required to identify fire protection equipment and equipment location shall be constructed of durable materials, permanently installed and readily visible.

SECTION 915
CARBON MONOXIDE DETECTION

[F] 915.1 General. Carbon monoxide detection shall be installed in new buildings in accordance with Sections 915.1.1 through 915.6. Carbon monoxide detection shall be installed in existing buildings in accordance with Chapter 11 of the *International Fire Code.*

[F] 915.1.1 Where required. Carbon monoxide detection shall be provided in Group I-1, I-2, I-4 and R occupancies and in classrooms in Group E occupancies in the locations specified in Section 915.2 where any of the conditions in Sections 915.1.2 through 915.1.6 exist.

[F] 915.1.2 Fuel-burning appliances and fuel-burning fireplaces. Carbon monoxide detection shall be provided in *dwelling units, sleeping units* and classrooms that contain a fuel-burning appliance or a fuel-burning fireplace.

[F] 915.1.3 Fuel burning, forced-air furnaces. Carbon monoxide detection shall be provided in dwelling units, sleeping units and classrooms served by a fuel-burning, forced-air furnace.

Exception: Carbon monoxide detection shall not be required in *dwelling units, sleeping units* and classrooms if a carbon monoxide detector is provided in the first room or area served by each main duct leaving the furnace, and the carbon monoxide alarm signals are automatically transmitted to an approved location.

[F] 915.1.4 Fuel-burning appliances outside of dwelling units, sleeping units and classrooms. Carbon monoxide detection shall be provided in *dwelling units,* *sleeping units* and classrooms located in buildings that contain fuel-burning appliances or fuel-burning fireplaces.

Exceptions:

1. Carbon monoxide detection shall not be required in *dwelling units, sleeping units* and classrooms without communicating openings between the fuel-burning appliance or fuel-burning fireplace and the *dwelling unit, sleeping unit* or classroom.
2. Carbon monoxide detection shall not be required in *dwelling units, sleeping units* and classrooms where a carbon monoxide detector is provided in one of the following locations:
   - 2.1. In an *approved* location between the fuel-burning appliance or fuel-burning fireplace and the *dwelling unit, sleeping unit* or classroom.
   - 2.2. On the ceiling of the room containing the fuel-burning appliance or fuel-burning fireplace.

[F] 915.1.5 Private garages. Carbon monoxide detection shall be provided in *dwelling units, sleeping units* and classrooms in buildings with attached *private garages*.

Exceptions:

1. Carbon monoxide detection shall not be required in dwelling units, sleeping units and classrooms without communicating openings between the *private garage* and the *dwelling unit, sleeping unit* or classroom.
2. Carbon monoxide detection shall not be required in *dwelling units, sleeping units* and classrooms located more than one *story* above or below a *private garage*.
3. Carbon monoxide detection shall not be required where the *private garage* connects to the building through an *open-ended corridor*.
4. Where a carbon monoxide detector is provided in an *approved* location between openings to a *private garage* and *dwelling units, sleeping units* or classrooms.

[F] 915.1.6 Exempt garages. For determining compliance with Section 915.1.5, an *open parking garage* complying with Section 406.5 or an enclosed parking garage complying with Section 406.6 shall not be considered a *private garage*.

[F] 915.2 Locations. Where required by Section 915.1.1, carbon monoxide detection shall be installed in the locations specified in Sections 915.2.1 through 915.2.3.

[F] 915.2.1 Dwelling units. Carbon monoxide detection shall be installed in *dwelling units* outside of each separate sleeping area in the immediate vicinity of the bedrooms. Where a fuel-burning appliance is located within a bedroom or its attached bathroom, carbon monoxide detection shall be installed within the bedroom.

[F] 915.2.2 Sleeping units. Carbon monoxide detection shall be installed in *sleeping units*.

Exception: Carbon monoxide detection shall be allowed to be installed outside of each separate sleeping area in the immediate vicinity of the *sleeping unit* where the *sleeping unit* or its attached bathroom does not contain a fuel-burning appliance and is not served by a forced air furnace.

[F] 915.2.3 Group E occupancies. Carbon monoxide detectors shall be installed in classrooms in Group E occupancies. Carbon monoxide alarm signals shall be automatically transmitted to an on-site location that is staffed by school personnel.

Exception: Carbon monoxide alarm signals shall not be required to be automatically transmitted to an on-site location that is staffed by school personnel in Group E occupancies with an *occupant load* of 30 or less.

[F] 915.3 Carbon monoxide detection. Carbon monoxide detection required by Sections 915.1 through 915.2.3 shall be provided by carbon monoxide alarms complying with Section 915.4 or carbon monoxide detection systems complying with Section 915.5.

[F] 915.4 Carbon monoxide alarms. Carbon monoxide alarms shall comply with Sections 915.4.1 through 915.4.4.

[F] 915.4.1 Power source. Carbon monoxide alarms shall receive their primary power from the building wiring where such wiring is served from a commercial source, and when primary power is interrupted, shall receive power from a battery. Wiring shall be permanent and without a disconnecting switch other than that required for overcurrent protection.

Exception: Where installed in buildings without commercial power, battery-powered carbon monoxide alarms shall be an acceptable alternative.

[F] 915.4.2 Listings. Carbon monoxide alarms shall be listed in accordance with UL 2034.

[F] 915.4.3 Locations. Carbon monoxide alarms shall only be installed in *dwelling units* and in *sleeping units*. They shall not be installed in locations where the code requires carbon monoxide detectors to be used.

[F] 915.4.4 Combination alarms. Combination carbon monoxide/smoke alarms shall be an acceptable alternative to carbon monoxide alarms. Combination carbon monoxide/smoke alarms shall be listed in accordance with UL 217 and UL 2034.

[F] 915.5 Carbon monoxide detection systems. Carbon monoxide detection systems shall be an acceptable alternative to carbon monoxide alarms and shall comply with Sections 915.5.1 through 915.5.3.

[F] 915.5.1 General. Carbon monoxide detection systems shall comply with NFPA 720. Carbon monoxide detectors shall be *listed* in accordance with UL 2075.

[F] 915.5.2 Locations. Carbon monoxide detectors shall be installed in the locations specified in Section 915.2.

These locations supersede the locations specified in NFPA 720.

[F] 915.5.3 Combination detectors. Combination carbon monoxide/smoke detectors installed in carbon monoxide detection systems shall be an acceptable alternative to carbon monoxide detectors, provided that they are listed in accordance with UL 268 and UL 2075.

[F] 915.6 Maintenance. Carbon monoxide alarms and carbon monoxide detection systems shall be maintained in accordance with the *International Fire Code*.

SECTION 916
GAS DETECTION SYSTEMS

[F] 916.1 Gas detection systems. Gas detection systems required by this code shall comply with Sections 916.2 through 916.11.

[F] 916.2 Permits. Permits shall be required as set forth in Section 105.6.10 of the *International Fire Code*.

[F] 916.2.1 Construction documents. Documentation of the gas detection system design and equipment to be used that demonstrates compliance with the requirements of this code and the *International Fire Code* shall be provided with the application for permit.

[F] 916.3 Equipment. Gas detection system equipment shall be designed for use with the gases being detected and shall be installed in accordance with manufacturer's instructions.

[F] 916.4 Power connections. Gas detection systems shall be permanently connected to the building electrical power supply or shall be permitted to be cord connected to an unswitched receptacle using an *approved* restraining means that secures the plug to the receptacle.

[F] 916.5 Emergency and standby power. Standby or emergency power shall be provided or the gas detection system shall initiate a trouble signal at an *approved* location if the power supply is interrupted.

[F] 916.6 Sensor locations. Sensors shall be installed in *approved* locations where leaking gases are expected to accumulate.

[F] 916.7 Gas sampling. Gas sampling shall be performed continuously. Sample analysis shall be processed immediately after sampling, except as follows:

1. For HPM gases, sample analysis shall be performed at intervals not exceeding 30 minutes.
2. For toxic gases that are not HPM, sample analysis shall be performed at intervals not exceeding 5 minutes in accordance with Section 6004.2.2.7 of the *International Fire Code*.
3. Where a less frequent or delayed sampling interval is *approved*.

[F] 916.8 System activation. A gas detection alarm shall be initiated where any sensor detects a concentration of gas exceeding the following thresholds:

1. For flammable gases, a gas concentration exceeding 25 percent of the lower flammability limit (LFL).
2. For nonflammable gases, a gas concentration exceeding one-half of the IDLH, unless a different threshold is specified by the section of this code requiring a gas detection system.

Upon activation of a gas detection alarm, alarm signals or other required responses shall be as specified by the section of this code requiring a gas detection system. Audible and visible alarm signals associated with a gas detection alarm shall be distinct from fire alarm and carbon monoxide alarm signals.

[F] 916.9 Signage. Signs shall be provided adjacent to gas detection system alarm signaling devices that advise occupants of the nature of the signals and actions to take in response to the signal.

[F] 916.10 Fire alarm system connections. Gas sensors and gas detection systems shall not be connected to fire alarm systems unless approved and connected in accordance with the fire alarm equipment manufacturer's instructions.

[F] 916.11 Inspection, testing and sensor calibration. Gas detection systems and sensors shall be inspected, tested and calibrated in accordance with the *International Fire Code*.

SECTION 917
MASS NOTIFICATION SYSTEMS

[F] 917.1 College and university campuses. Prior to construction of a new building requiring a fire alarm system on a multiple-building college or university campus having a cumulative building *occupant load* of 1,000 or more, a mass notification risk analysis shall be conducted in accordance with NFPA 72. Where the risk analysis determines a need for mass notification, an *approved* mass notification system shall be provided in accordance with the findings of the risk analysis.

SECTION 918
EMERGENCY RESPONDER COMMUNICATION COVERAGE

[F] 918.1 General. In-building two-way emergency responder communication coverage shall be provided in all new buildings in accordance with Section 510 of the *International Fire Code*.

CHAPTER 10

MEANS OF EGRESS

User notes:

About this chapter: *Chapter 10 provides the general criteria for designing the means of egress established as the primary method for protection of people in buildings by allowing timely relocation or evacuation of building occupants. Both prescriptive and performance language is utilized in this chapter to provide for a basic approach in the determination of a safe exiting system for all occupancies. It addresses all portions of the egress system (exit access, exits and exit discharge) and includes design requirements as well as provisions regulating individual components. The requirements detail the size, arrangement, number and protection of means of egress components. Functional and operational characteristics that will permit the safe use of components without special knowledge or effort are specified.*

The means of egress protection requirements work in coordination with other sections of the code, such as protection of vertical openings (see Chapter 7), interior finish (see Chapter 8), fire suppression and detection systems (see Chapter 9) and numerous others, all having an impact on life safety. Chapter 10 is subdivided into four main sections: general (Sections 1003–1015), exit access (Sections 1016–1021), exit (Sections 1022–1027) and exit discharge (Sections 1028–1029). Special allowances for the unique requirements for assembly spaces (Section 1030) and emergency escape and rescue openings (Section 1031) complete the chapter. Chapter 10 of this code is duplicated in Chapter 10 of the International Fire Code®*; however, the* International Fire Code *contains one additional section on maintenance of the means of egress system in existing buildings.*

Section 1010 was extensively reorganized for the 2021 edition. For complete information, see the moved sections table in the preface information for the International Building Code.

SECTION 1001 ADMINISTRATION

1001.1 General. Buildings or portions thereof shall be provided with a *means of egress* system as required by this chapter. The provisions of this chapter shall control the design, construction and arrangement of *means of egress* components required to provide an *approved means of egress* from structures and portions thereof.

1001.2 Minimum requirements. It shall be unlawful to alter a building or structure in a manner that will reduce the number of *exits* or the minimum width or required capacity of the *means of egress* to less than required by this code.

[F] SECTION 1002 MAINTENANCE AND PLANS

[F] 1002.1 Maintenance. *Means of egress* shall be maintained in accordance with the *International Fire Code*.

[F] 1002.2 Fire safety and evacuation plans. Fire safety and evacuation plans shall be provided for all occupancies and buildings where required by the *International Fire Code*. Such fire safety and evacuation plans shall comply with the applicable provisions of Sections 401.2 and 404 of the *International Fire Code*.

SECTION 1003 GENERAL MEANS OF EGRESS

1003.1 Applicability. The general requirements specified in Sections 1003 through 1015 shall apply to all three elements of the *means of egress* system, in addition to those specific requirements for the *exit access*, the *exit* and the *exit discharge* detailed elsewhere in this chapter.

1003.2 Ceiling height. The *means of egress* shall have a ceiling height of not less than 7 feet 6 inches (2286 mm) above the finished floor.

Exceptions:

1. Sloped ceilings in accordance with Section 1208.2.
2. Ceilings of *dwelling units* and *sleeping units* within residential occupancies in accordance with Section 1208.2.
3. Allowable projections in accordance with Section 1003.3.
4. *Stair* headroom in accordance with Section 1011.3.
5. Door height in accordance with Section 1010.1.1.
6. *Ramp* headroom in accordance with Section 1012.5.2.
7. The clear height of floor levels in vehicular and pedestrian traffic areas of public and private parking garages in accordance with Section 406.2.2.
8. Areas above and below *mezzanine* floors in accordance with Section 505.2.

1003.3 Protruding objects. Protruding objects on *circulation paths* shall comply with the requirements of Sections 1003.3.1 through 1003.3.4.

1003.3.1 Headroom. Protruding objects are permitted to extend below the minimum ceiling height required by Section 1003.2 where a minimum headroom of 80 inches

(2032 mm) is provided over any circulation paths, including walks, *corridors*, *aisles* and passageways. Not more than 50 percent of the ceiling area of a *means of egress* shall be reduced in height by protruding objects.

Exception: Door closers and stops shall not reduce headroom to less than 78 inches (1981 mm).

A barrier shall be provided where the vertical clearance above a *circulation path* is less than 80 inches (2032 mm) high above the finished floor. The leading edge of such a barrier shall be located 27 inches (686 mm) maximum above the finished floor.

1003.3.2 Post-mounted objects. A free-standing object mounted on a post or pylon shall not overhang that post or pylon more than 4 inches (102 mm) where the lowest point of the leading edge is more than 27 inches (686 mm) and less than 80 inches (2032 mm) above the finished floor. Where a sign or other obstruction is mounted between posts or pylons and the clear distance between the posts or pylons is greater than 12 inches (305 mm), the lowest edge of such sign or obstruction shall be 27 inches (686 mm) maximum or 80 inches (2032 mm) minimum above the finished floor or ground.

Exception: These requirements shall not apply to sloping portions of *handrails* between the top and bottom riser of *stairs* and above the *ramp* run.

1003.3.3 Horizontal projections. Objects with leading edges more than 27 inches (685 mm) and not more than 80 inches (2030 mm) above the finished floor shall not project horizontally more than 4 inches (102 mm) into the *circulation path*.

Exception: *Handrails* are permitted to protrude $4^1/_2$ inches (114 mm) from the wall or *guard*.

1003.3.4 Clear width. Protruding objects shall not reduce the minimum clear width of *accessible routes*.

1003.4 Slip-resistant surface. Circulation paths of the *means of egress* shall have a slip-resistant surface and be securely attached.

1003.5 Elevation change. Where changes in elevation of less than 12 inches (305 mm) exist in the *means of egress*, sloped surfaces shall be used. Where the slope is greater than one unit vertical in 20 units horizontal (5-percent slope), *ramps* complying with Section 1012 shall be used. Where the difference in elevation is 6 inches (152 mm) or less, the *ramp* shall be equipped with either *handrails* or floor finish materials that contrast with adjacent floor finish materials.

Exceptions:

1. Steps at exterior doors complying with Section 1010.1.4.
2. A *stair* with a single riser or with two risers and a tread is permitted at locations not required to be *accessible* by Chapter 11 where the risers and treads comply with Section 1011.5, the minimum depth of the tread is 13 inches (330 mm) and not less than one *handrail* complying with Section 1014 is provided within 30 inches (762 mm) of the centerline of the normal path of egress travel on the *stair*.
3. A step is permitted in *aisles* serving seating that has a difference in elevation less than 12 inches (305 mm) at locations not required to be *accessible* by Chapter 11, provided that the risers and treads comply with Section 1030.14 and the *aisle* is provided with a *handrail* complying with Section 1030.16.

Throughout a *story* in a Group I-2 occupancy, any change in elevation in portions of the *means of egress* that serve nonambulatory persons shall be by means of a *ramp* or sloped walkway.

1003.6 Means of egress continuity. The path of egress travel along a *means of egress* shall not be interrupted by a building element other than a *means of egress* component as specified in this chapter. Obstructions shall not be placed in the minimum width or required capacity of a *means of egress* component except projections permitted by this chapter. The minimum width or required capacity of a *means of egress* system shall not be diminished along the path of egress travel.

1003.7 Elevators, escalators and moving walks. Elevators, escalators and moving walks shall not be used as a component of a required *means of egress* from any other part of the building.

Exception: Elevators used as an *accessible* means of egress in accordance with Section 1009.4.

SECTION 1004
OCCUPANT LOAD

1004.1 Design occupant load. In determining *means of egress* requirements, the number of occupants for whom *means of egress* facilities are provided shall be determined in accordance with this section.

1004.2 Cumulative occupant loads. Where the path of egress travel includes intervening rooms, areas or spaces, cumulative *occupant loads* shall be determined in accordance with this section.

1004.2.1 Intervening spaces or accessory areas. Where occupants egress from one or more rooms, areas or spaces through others, the design *occupant load* shall be the combined *occupant load* of interconnected accessory or intervening spaces. Design of egress path capacity shall be based on the cumulative portion of *occupant loads* of all rooms, areas or spaces to that point along the path of egress travel.

1004.2.2 Adjacent levels for mezzanines. That portion of the *occupant load* of a *mezzanine* with required egress through a room, area or space on an adjacent level shall be added to the *occupant load* of that room, area or space.

1004.2.3 Adjacent stories. Other than for the egress components designed for convergence in accordance with Section 1005.6, the *occupant load* from separate stories shall not be added.

1004.3 Multiple function occupant load. Where an area under consideration contains multiple functions having different *occupant load* factors, the design *occupant load* for such area shall be based on the floor area of each function calculated independently.

1004.4 Multiple occupancies. Where a building contains two or more occupancies, the *means of egress* requirements shall apply to each portion of the building based on the occupancy of that space. Where two or more occupancies utilize portions of the same *means of egress* system, those egress components shall meet the more stringent requirements of all occupancies that are served.

1004.5 Areas without fixed seating. The number of occupants shall be computed at the rate of one occupant per unit of area as prescribed in Table 1004.5. For areas without *fixed seating*, the *occupant load* shall be not less than that number determined by dividing the floor area under consideration by the *occupant load* factor assigned to the function of the space as set forth in Table 1004.5. Where an intended function is not *listed* in Table 1004.5, the *building official* shall establish a function based on a *listed* function that most nearly resembles the intended function.

Exception: Where *approved* by the *building official*, the actual number of occupants for whom each occupied space, floor or building is designed, although less than those determined by calculation, shall be permitted to be used in the determination of the design *occupant load*.

1004.5.1 Increased occupant load. The *occupant load* permitted in any building, or portion thereof, is permitted to be increased from that number established for the occupancies in Table 1004.5, provided that all other requirements of the code are met based on such modified number and the *occupant load* does not exceed one occupant per 7 square feet (0.65 m2) of occupiable floor space. Where required by the *building official*, an *approved aisle*, seating or fixed equipment diagram substantiating any increase in *occupant load* shall be submitted. Where required by the *building official*, such diagram shall be posted.

1004.6 Fixed seating. For areas having *fixed seats* and *aisles*, the *occupant load* shall be determined by the number of *fixed seats* installed therein. The *occupant load* for areas in which *fixed seating* is not installed, such as waiting spaces, shall be determined in accordance with Section 1004.5 and added to the number of *fixed seats*.

The *occupant load* of *wheelchair spaces* and the associated companion seat shall be based on one occupant for each *wheelchair space* and one occupant for the associated companion seat provided in accordance with Section 1109.2.3.

For areas having *fixed seating* without dividing arms, the *occupant load* shall be not less than the number of seats based on one person for each 18 inches (457 mm) of seating length.

The *occupant load* of seating booths shall be based on one person for each 24 inches (610 mm) of booth seat length measured at the backrest of the seating booth.

TABLE 1004.5
MAXIMUM FLOOR AREA ALLOWANCES PER OCCUPANT

FUNCTION OF SPACE	OCCUPANT LOAD FACTOR[a]
Accessory storage areas, mechanical equipment room	300 gross
Agricultural building	300 gross
Aircraft hangars	500 gross
Airport terminal	
Baggage claim	20 gross
Baggage handling	300 gross
Concourse	100 gross
Waiting areas	15 gross
Assembly	
Gaming floors (keno, slots, etc.)	11 gross
Exhibit gallery and museum	30 net
Assembly with fixed seats	See Section 1004.6
Assembly without fixed seats	
Concentrated (chairs only—not fixed)	7 net
Standing space	5 net
Unconcentrated (tables and chairs)	15 net
Bowling centers, allow 5 persons for each lane including 15 feet of runway, and for additional areas	7 net
Business areas	150 gross
Concentrated business use areas	See Section 1004.8
Courtrooms—other than fixed seating areas	40 net
Day care	35 net
Dormitories	50 gross
Educational	
Classroom area	20 net
Shops and other vocational room areas	50 net
Exercise rooms	50 gross
Group H-5 fabrication and manufacturing areas	200 gross
Industrial areas	100 gross
Institutional areas	
Inpatient treatment areas	240 gross
Outpatient areas	100 gross
Sleeping areas	120 gross
Kitchens, commercial	200 gross
Library	
Reading rooms	50 net
Stack area	100 gross
Locker rooms	50 gross
Mall buildings—covered and open	See Section 402.8.2
Mercantile	60 gross
Storage, stock, shipping areas	300 gross
Parking garages	200 gross
Residential	200 gross
Skating rinks, swimming pools	
Rink and pool	50 gross
Decks	15 gross
Stages and platforms	15 net
Warehouses	500 gross

For SI: 1 foot = 304.8 mm, 1 square foot = 0.0929 m^2.

a. Floor area in square feet per occupant.

1004.7 Outdoor areas. *Yards*, patios, occupied roofs, *courts* and similar outdoor areas accessible to and usable by the building occupants shall be provided with *means of egress* as required by this chapter. The *occupant load* of such outdoor areas shall be assigned by the *building official* in accordance with the anticipated use. Where outdoor areas are to be used by persons in addition to the occupants of the building, and the path of egress travel from the outdoor areas passes through the building, *means of egress* requirements for the building shall be based on the sum of the *occupant loads* of the building plus the outdoor areas.

Exceptions:

1. Outdoor areas used exclusively for service of the building need only have one *means of egress*.
2. Both outdoor areas associated with Group R-3 and individual dwelling units of Group R-2.

1004.8 Concentrated business use areas. The *occupant load* factor for concentrated business use shall be applied to telephone call centers, trading floors, electronic data processing centers and similar business use areas with a higher density of occupants than would normally be expected in a typical business occupancy environment. Where approved by the *building official*, the *occupant load* for concentrated business use areas shall be the actual *occupant load*, but not less than one occupant per 50 square feet (4.65 m^2) of gross occupiable floor space.

1004.9 Posting of occupant load. Every room or space that is an assembly occupancy shall have the *occupant load* of the room or space posted in a conspicuous place, near the main *exit* or *exit access doorway* from the room or space, for the intended configurations. Posted signs shall be of an approved legible permanent design and shall be maintained by the owner or the owner's authorized agent.

SECTION 1005
MEANS OF EGRESS SIZING

1005.1 General. All portions of the *means of egress* system shall be sized in accordance with this section.

Exception: *Aisles* and *aisle accessways* in rooms or spaces used for assembly purposes complying with Section 1030.

1005.2 Minimum width based on component. The minimum width, in inches (mm), of any *means of egress* components shall be not less than that specified for such component, elsewhere in this code.

1005.3 Required capacity based on occupant load. The required capacity, in inches (mm), of the *means of egress* for any room, area, space or *story* shall be not less than that determined in accordance with Sections 1005.3.1 and 1005.3.2.

1005.3.1 Stairways. The capacity, in inches, of means of egress *stairways* shall be calculated by multiplying the *occupant load* served by such *stairways* by a *means of egress* capacity factor of 0.3 inch (7.6 mm) per occupant. Where *stairways* serve more than one *story*, only the *occupant load* of each *story* considered individually shall be used in calculating the required capacity of the *stairways* serving that *story*.

Exceptions:

1. For other than Group H and I-2 occupancies, the capacity, in inches, of means of egress *stairways* shall be calculated by multiplying the *occupant load* served by such *stairways* by a means of egress capacity factor of 0.2 inch (5.1 mm) per occupant in buildings equipped throughout with an *automatic sprinkler system* installed in accordance with Section 903.3.1.1 or 903.3.1.2 and an *emergency voice/alarm communication system* in accordance with Section 907.5.2.2.
2. Facilities with *smoke-protected assembly seating* shall be permitted to use the capacity factors in Table 1030.6.2 indicated for stepped *aisles* for *exit access* or exit *stairways* where the entire path for *means of egress* from the seating to the *exit discharge* is provided with a smoke control system complying with Section 909.
3. Facilities with *open-air assembly seating* shall be permitted to the capacity factors in Section 1030.6.3 indicated for stepped *aisles* for *exit access* or *exit stairways* where the entire path for *means of egress* from the seating to the *exit discharge* is open to the outdoors.

1005.3.2 Other egress components. The capacity, in inches, of *means of egress* components other than *stairways* shall be calculated by multiplying the *occupant load* served by such component by a means of egress capacity factor of 0.2 inch (5.1 mm) per occupant.

Exceptions:

1. For other than Group H and I-2 occupancies, the capacity, in inches, of *means of egress* components other than *stairways* shall be calculated by multiplying the *occupant load* served by such component by a means of egress capacity factor of 0.15 inch (3.8 mm) per occupant in buildings equipped throughout with an *automatic sprinkler system* installed in accordance with Section 903.3.1.1 or 903.3.1.2 and an emergency voice/alarm communication system in accordance with Section 907.5.2.2.
2. Facilities with *smoke-protected assembly seating* shall be permitted to use the capacity factors in Table 1030.6.2 indicated for level or ramped *aisles* for *means of egress* components other than *stairways* where the entire path for *means of egress* from the seating to the *exit discharge* is provided with a smoke control system complying with Section 909.
3. Facilities with *open-air assembly seating* shall be permitted to the capacity factors in Section 1030.6.3 indicated for level or ramped *aisles* for *means of egress* components other than

stairways where the entire path for *means of egress* from the seating to the *exit discharge* is open to the outdoors.

1005.4 Continuity. The minimum width or required capacity of the *means of egress* required from any *story* of a building shall not be reduced along the path of egress travel until arrival at the *public way*.

1005.5 Distribution of minimum width and required capacity. Where more than one *exit*, or access to more than one *exit*, is required, the *means of egress* shall be configured such that the loss of any one *exit*, or access to one *exit*, shall not reduce the available capacity or width to less than 50 percent of the required capacity or width.

1005.6 Egress convergence. Where the *means of egress* from stories above and below converge at an intermediate level, the capacity of the *means of egress* from the point of convergence shall be not less than the largest minimum width or the sum of the required capacities for the *stairways* or *ramps* serving the two adjacent stories, whichever is larger.

1005.7 Encroachment. Encroachments into the required *means of egress* width shall be in accordance with the provisions of this section.

1005.7.1 Doors. Doors, when fully opened, shall not reduce the required width by more than 7 inches (178 mm). Doors in any position shall not reduce the required width by more than one-half.

Exceptions:

1. Surface-mounted latch release hardware shall be exempt from inclusion in the 7-inch maximum (178 mm) encroachment where both of the following conditions exist:
   - 1.1. The hardware is mounted to the side of the door facing away from the adjacent wall where the door is in the open position.
   - 1.2. The hardware is mounted not less than 34 inches (865 mm) nor more than 48 inches (1219 mm) above the finished floor.
2. The restrictions on door swing shall not apply to doors within individual *dwelling units* and *sleeping units* of Group R-2 occupancies and *dwelling units* of Group R-3 occupancies.

1005.7.2 Other projections. *Handrail* projections shall be in accordance with the provisions of Section 1014.8. Other nonstructural projections such as *trim* and similar decorative features shall be permitted to project into the required width not more than $1^1/_2$ inches (38 mm) on each side.

Exception: Projections are permitted in *corridors* within Group I-2, Condition 1 in accordance with Section 407.4.3.

1005.7.3 Protruding objects. Protruding objects shall comply with the applicable requirements of Section 1003.3.

SECTION 1006
NUMBER OF EXITS AND EXIT ACCESS DOORWAYS

1006.1 General. The number of *exits* or *exit access doorways* required within the *means of egress* system shall comply with the provisions of Section 1006.2 for spaces, including *mezzanines*, and Section 1006.3 for *stories* or occupied roofs.

1006.2 Egress from spaces. Rooms, areas or spaces, including *mezzanines*, within a *story* or *basement* shall be provided with the number of *exits* or access to *exits* in accordance with this section.

1006.2.1 Egress based on occupant load and common path of egress travel distance. Two *exits* or *exit access doorways* from any space shall be provided where the design *occupant load* or the *common path of egress* travel distance exceeds the values *listed* in Table 1006.2.1. The cumulative *occupant load* from adjacent rooms, areas or spaces shall be determined in accordance with Section 1004.2.

Exceptions:

1. The number of *exits* from foyers, lobbies, vestibules or similar spaces need not be based on cumulative *occupant loads* for areas discharging through such spaces, but the capacity of the *exits* from such spaces shall be based on applicable cumulative *occupant loads*.
2. *Care suites* in Group I-2 occupancies complying with Section 407.4.
3. Unoccupied mechanical rooms and *penthouses* are not required to comply with the common path of egress travel distance measurement.

1006.2.1.1 Three or more exits or exit access doorways. Three *exits* or *exit access doorways* shall be provided from any space with an *occupant load* of 501 to 1,000. Four *exits* or *exit access doorways* shall be provided from any space with an *occupant load* greater than 1,000.

1006.2.2 Egress based on use. The numbers, configuration and types of components of *exits* or access to *exits* shall be provided in the uses described in Sections 1006.2.2.1 through 1006.2.2.6.

1006.2.2.1 Boiler, incinerator and furnace rooms. Two *exit access doorways* are required in boiler, incinerator and furnace rooms where the area is over 500 square feet (46 m^2) and any fuel-fired equipment exceeds 400,000 British thermal units (Btu) (422 000 KJ) input capacity. Where two *exit access doorways* are required, one is permitted to be a fixed ladder or an *alternating tread device*. *Exit access doorways* shall be separated by a horizontal distance equal to one-half the length of the maximum overall diagonal dimension of the room.

1006.2.2.2 Refrigeration machinery rooms. Machinery rooms larger than 1,000 square feet (93 m^2) shall have not less than two *exits* or exit access doorways. Where two *exit access doorways* are required, one such doorway is permitted to be served by a fixed ladder or an *alternating tread device*. *Exit access doorways* shall be separated by a horizontal distance equal to one-half the maximum horizontal dimension of the room.

All portions of machinery rooms shall be within 150 feet (45 720 mm) of an *exit* or *exit access doorway*. An increase in *exit access* travel distance is permitted in accordance with Section 1017.1.

Exit and *exit access doorways* shall swing in the direction of egress travel and shall be equipped with *panic hardware*, regardless of the *occupant load* served. Exit and *exit access doorways* shall be tight fitting and *self-closing*.

1006.2.2.3 Refrigerated rooms or spaces. Rooms or spaces having a floor area larger than 1,000 square feet (93 m^2), containing a refrigerant evaporator and maintained at a temperature below 68°F (20°C), shall have access to not less than two *exits* or exit access doorways.

Exit access travel distance shall be determined as specified in Section 1017.1, but all portions of a refrigerated room or space shall be within 150 feet (45 720 mm) of an *exit* or *exit access doorway* where such rooms are not protected by an *approved automatic sprinkler system*. Egress is allowed through adjoining refrigerated rooms or spaces.

Exception: Where using refrigerants in quantities limited to the amounts based on the volume set forth in the *International Mechanical Code*.

1006.2.2.4 Electrical rooms. T he location and number of *exit* or *exit access doorways* shall be provided for electrical rooms in accordance with Section 110.26 of NFPA 70 for electrical equipment rated 1,000 volts or less, and Section 110.33 of NFPA 70 for electrical equipment rated over 1,000 volts. Panic hardware shall be provided where required in accordance with Section 1010.2.9.2.

1006.2.2.5 Vehicular ramps. Vehicular *ramps* shall not be considered as an *exit access ramp* unless pedestrian facilities are provided.

1006.2.2.6 Groups R-3 and R-4. Where Group R-3 occupancies are permitted by Section 903.2.8 to be protected by an *automatic sprinkler system* installed in accordance with Section 903.3.1.3, the *exit access* travel distance for Group R-3 shall be not more than 125 feet (38 100 mm). Where Group R-4 occupancies are permitted by Section 903.2.8 to be protected by an *automatic sprinkler system* installed in accordance with

TABLE 1006.2.1
SPACES WITH ONE EXIT OR EXIT ACCESS DOORWAY

OCCUPANCY	MAXIMUM OCCUPANT LOAD OF SPACE	MAXIMUM COMMON PATH OF EGRESS TRAVEL DISTANCE (feet)		
		Without Sprinkler System (feet)		With Sprinkler System (feet)
		Occupant Load		
		OL ≤ 30	OL > 30	
A[c], E, M	49	75	75	75[a]
B	49	100	75	100[a]
F	49	75	75	100[a]
H-1, H-2, H-3	3	NP	NP	25[b]
H-4, H-5	10	NP	NP	75[b]
I-1, I-2[d], I-4	10	NP	NP	75[a]
I-3	10	NP	NP	100[a]
R-1	10	NP	NP	75[a]
R-2	20	NP	NP	125[a]
R-3[e]	20	NP	NP	125[a, g]
R-4[e]	20	NP	NP	125[a, g]
S[f]	29	100	75	100[a]
U	49	100	75	75[a]

For SI: 1 foot = 304.8 mm.

NP = Not Permitted.

a. Buildings equipped throughout with an automatic sprinkler system in accordance with Section 903.3.1.1 or 903.3.1.2. See Section 903 for occupancies where automatic sprinkler systems are permitted in accordance with Section 903.3.1.2.

b. Group H occupancies equipped throughout with an automatic sprinkler system in accordance with Section 903.2.5.

c. For a room or space used for assembly purposes having fixed seating, see Section 1030.8.

d. For the travel distance limitations in Group I-2, see Section 407.4.

e. The common path of egress travel distance shall only apply in a Group R-3 occupancy located in a mixed occupancy building.

f. The length of common path of egress travel distance in a Group S-2 open parking garage shall be not more than 100 feet.

g. For the travel distance limitations in Groups R-3 and R-4 equipped throughout with an automatic sprinkler system in accordance with Section 903.3.1.3, see Section 1006.2.2.6.

Section 903.3.1.3, the *exit access* travel distance for Group R-4 shall be not more than 75 feet (22 860 mm).

1006.3 Egress from stories or occupied roofs. The *means of egress* system serving any *story* or occupied roof shall be provided with the number of separate and distinct *exits* or access to *exits* based on the aggregate *occupant load* served in accordance with this section.

1006.3.1 Occupant load. Where *stairways* serve more than one *story*, or more than one *story* and an occupied roof, only the *occupant load* of each *story* or occupied roof, considered individually, shall be used when calculating the required number of *exits* or access to *exits* serving that *story*.

1006.3.2 Path of egress travel. The path of egress travel to an *exit* shall not pass through more than one adjacent *story*.

Exception: The path of egress travel to an *exit* shall be permitted to pass through more than one adjacent *story* in any of the following:

1. In Group R-1, R-2 or R-3 occupancies, *exit access stairways* and *ramps* connecting four stories or less serving and contained within an individual dwelling unit, sleeping unit or live/work unit.
2. *Exit access stairways* serving and contained within a Group R-3 congregate residence or a Group R-4 facility.
3. *Exit access stairways* and *ramps* within an *atrium* complying with Section 404.
4. *Exit access stairways* and *ramps* in *open parking garages* that serve only the parking garage.
5. *Exit access stairways* and *ramps* serving *open-air assembly seating* complying with the exit access travel distance requirements of Section 1030.7.
6. *Exit access stairways* and *ramps* between the balcony, gallery or press box and the main assembly floor in occupancies such as theaters, *places of religious worship*, auditoriums and sports facilities.
7. Exterior *exit access stairways* and *ramps* between occupied roofs.

1006.3.3 Egress based on occupant load. Each *story* and occupied roof shall have the minimum number of separate and distinct *exits*, or access to *exits*, as specified in Table 1006.3.3. A single *exit* or access to a single *exit* shall be permitted in accordance with Section 1006.3.4. The required number of *exits*, or *exit access stairways* or *ramps* providing access to *exits*, from any *story* or occupied roof shall be maintained until arrival at the *exit discharge* or a *public way*.

TABLE 1006.3.3
MINIMUM NUMBER OF EXITS OR ACCESS TO EXITS PER STORY

OCCUPANT LOAD PER STORY	MINIMUM NUMBER OF EXITS OR ACCESS TO EXITS FROM STORY
1-500	2
501-1,000	3
More than 1,000	4

1006.3.4 Single exits. A single *exit* or access to a single *exit* shall be permitted from any *story* or occupied roof where one of the following conditions exists:

1. The *occupant load*, number of *dwelling units* and exit access travel distance do not exceed the values in Table 1006.3.4(1) or 1006.3.4(2).
2. Rooms, areas and spaces complying with Section 1006.2.1 with *exits* that discharge directly to the exterior at the *level of exit discharge*, are permitted to have one *exit* or access to a single *exit*.
3. Parking garages where vehicles are mechanically parked shall be permitted to have one *exit* or access to a single *exit*.
4. Group R-3 and R-4 occupancies shall be permitted to have one *exit* or access to a single *exit*.
5. Individual single-story or multistory *dwelling units* shall be permitted to have a single *exit* or access to a single *exit* from the *dwelling unit* provided that both of the following criteria are met:
   - 5.1. The *dwelling unit* complies with Section 1006.2.1 as a space with one *means of egress*.
   - 5.2. Either the exit from the *dwelling unit* discharges directly to the exterior at the *level of exit discharge*, or the *exit access* outside the *dwelling unit's* entrance door provides access to not less than two *approved* independent *exits*.

1006.3.4.1 Mixed occupancies. Where one *exit*, or *exit access stairway* or *ramp* providing access to exits at other stories, is permitted to serve individual *stories*, mixed occupancies shall be permitted to be served by single *exits* provided that each individual occupancy complies with the applicable requirements of Table 1006.3.4(1) or 1006.3.4(2) for that occupancy. Where applicable, cumulative *occupant loads* from adjacent occupancies shall be considered to be in accordance with the provisions of Section 1004.1. In each *story* of a mixed occupancy building, the maximum number of occupants served by a single exit shall be such that the sum of the ratios of the calculated number of occupants of the space divided by the allowable number of occupants indicated in Table 1006.3.4(2) for each occupancy does not exceed one. Where *dwelling units* are located on a story with other occupancies, the actual number of *dwelling units* divided by four plus the ratio from the other occupancy does not exceed one.

**TABLE 1006.3.4(1)
STORIES WITH ONE EXIT OR ACCESS TO ONE EXIT FOR R-2 OCCUPANCIES**

STORY	OCCUPANCY	MAXIMUM NUMBER OF DWELLING UNITS	MAXIMUM EXIT ACCESS TRAVEL DISTANCE
Basement, first, second or third story above grade plane	R-2[a, b]	4 dwelling units	125 feet
Fourth story above grade plane and higher	NP	NA	NA

For SI: 1 foot = 304.8 mm.
NP = Not Permitted.
NA = Not Applicable.
a. Buildings classified as Group R-2 equipped throughout with an automatic sprinkler system in accordance with Section 903.3.1.1 or 903.3.1.2 and provided with emergency escape and rescue openings in accordance with Section 1031.
b. This table is used for R-2 occupancies consisting of dwelling units. For R-2 occupancies consisting of sleeping units, use Table 1006.3.4(2).

**TABLE 1006.3.4(2)
STORIES WITH ONE EXIT OR ACCESS TO ONE EXIT FOR OTHER OCCUPANCIES**

STORY	OCCUPANCY	MAXIMUM OCCUPANT LOAD PER STORY	MAXIMUM EXIT ACCESS TRAVEL DISTANCE (feet)
First story above or below grade plane	A, B[b], E, F[b], M, U	49	75
	H-2, H-3	3	25
	H-4, H-5, I, R-1, R-2[a, c]	10	75
	S[b, d]	29	75
Second story above grade plane	B, F, M, S[d]	29	75
Third story above grade plane and higher	NP	NA	NA

For SI: 1 foot = 304.8 mm.
NP = Not Permitted.
NA = Not Applicable.
a. Buildings classified as Group R-2 equipped throughout with an automatic sprinkler system in accordance with Section 903.3.1.1 or 903.3.1.2 and provided with emergency escape and rescue openings in accordance with Section 1031.
b. Group B, F and S occupancies in buildings equipped throughout with an automatic sprinkler system in accordance with Section 903.3.1.1 shall have a maximum exit access travel distance of 100 feet.
c. This table is used for R-2 occupancies consisting of sleeping units. For R-2 occupancies consisting of dwelling units, use Table 1006.3.4(1).
d. The length of exit access travel distance in a Group S-2 open parking garage shall be not more than 100 feet.

SECTION 1007 EXIT AND EXIT ACCESS DOORWAY CONFIGURATION

1007.1 General. *Exits*, *exit access doorways*, and *exit access stairways* and *ramps* serving spaces, including individual building *stories*, shall be separated in accordance with the provisions of this section.

1007.1.1 Two exits or exit access doorways. Where two *exits*, *exit access doorways*, *exit access stairways* or *ramps*, or any combination thereof, are required from any portion of the *exit access*, they shall be placed a distance apart equal to not less than one-half of the length of the maximum overall diagonal dimension of the building or area to be served measured in a straight line between them. Interlocking or *scissor stairways* shall be counted as one *exit stairway*.

Exceptions:

1. Where interior *exit stairways* or *ramps* are interconnected by a 1-hour fire-resistance-rated *corridor* conforming to the requirements of Section 1020, the required exit separation shall be measured along the shortest direct line of travel within the corridor.
2. Where a building is equipped throughout with an *automatic sprinkler system* in accordance with Section 903.3.1.1 or 903.3.1.2, the separation distance shall be not less than one-third of the length of the maximum overall diagonal dimension of the area served.

1007.1.1.1 Measurement point. The separation distance required in Section 1007.1.1 shall be measured in accordance with the following:

1. The separation distance to *exit* or *exit access doorways* shall be measured to any point along the width of the doorway.
2. The separation distance to *exit access stairways* shall be measured to the closest riser.
3. The separation distance to *exit access ramps* shall be measured to the start of the *ramp* run.

1007.1.2 Three or more exits or exit access doorways. Where access to three or more *exits* is required, not less than two *exit* or *exit access doorways* shall be arranged in accordance with the provisions of Section 1007.1.1. Additional required *exit* or *exit access doorways* shall be arranged a reasonable distance apart so that if one becomes blocked, the others will be available.

1007.1.3 Remoteness of exit access stairways or ramps. Where two *exit access stairways* or *ramps* provide the required *means of egress* to *exits* at another *story*, the required separation distance shall be maintained for all portions of such *exit access stairways* or *ramps*.

1007.1.3.1 Three or more exit access stairways or ramps. Where more than two *exit access stairways* or *ramps* provide the required *means of egress*, not less than two shall be arranged in accordance with Section 1007.1.3.

SECTION 1008 MEANS OF EGRESS ILLUMINATION

1008.1 Means of egress illumination. Illumination shall be provided in the *means of egress* in accordance with Section 1008.2. Under emergency power, *means of egress* illumination shall comply with Section 1008.3.

1008.2 Illumination required. The *means of egress* serving a room or space shall be illuminated at all times that the room or space is occupied.

Exceptions:

1. Occupancies in Group U.
2. *Aisle accessways* in Group A.
3. *Dwelling units* and *sleeping units* in Groups R-1, R-2 and R-3.
4. *Sleeping units* of Group I occupancies.

1008.2.1 Illumination level under normal power. The *means of egress* illumination level shall be not less than 1 footcandle (11 lux) at the walking surface. Along *exit access stairways*, exit stairways and at their required landings, the illumination level shall not be less than 10 footcandles (108 lux) at the walking surface when the *stairway* is in use.

Exception: For auditoriums, theaters, concert or opera halls and similar assembly occupancies, the illumination at the walking surface is permitted to be reduced during performances by one of the following methods provided that the required illumination is automatically restored upon activation of a premises' *fire alarm system*:

1. Externally illuminated walking surfaces shall be permitted to be illuminated to not less than 0.2 footcandle (2.15 lux).
2. Steps, landings and the sides of *ramps* shall be permitted to be marked with *self-luminous* materials in accordance with Sections 1025.2.1, 1025.2.2 and 1025.2.4 by systems *listed* in accordance with UL 1994.

1008.2.2 Group I-2. In Group I-2 occupancies where two or more *exits* are required, on the exterior landings required by Section 1010.1.5, means of egress illumination levels for the exit discharge shall be provided such that failure of a single lamp in a luminaire shall not reduce the illumination level on that landing to less than 1 footcandle (11 lux).

1008.2.3 Exit discharge. Illumination shall be provided along the path of travel for the exit discharge from each exit to the *public way*.

Exception: Illumination shall not be required where the path of the exit discharge meets both of the following requirements:

1. The path of exit discharge is illuminated from the exit to a safe dispersal area complying with Section 1028.5.
2. A dispersal area shall be illuminated to a level not less than 1 footcandle (11 lux) at the walking surface.

1008.3 Emergency power for illumination. The power supply for *means of egress* illumination shall normally be provided by the premises' electrical supply.

1008.3.1 General. In the event of power supply failure in rooms and spaces that require two or more *exits* or access to exits, an emergency electrical system shall automatically illuminate all of the following areas:

1. *Aisles.*
2. *Corridors.*
3. *Exit access stairways* and *ramps.*

1008.3.2 Buildings. In the event of power supply failure in buildings that require two or more *exits* or access to exits, an emergency electrical system shall automatically illuminate all of the following areas:

1. *Interior exit access stairways* and *ramps.*
2. *Interior* and *exterior exit stairways* and *ramps.*
3. *Exit passageways.*
4. Vestibules and areas on the level of discharge used for *exit discharge* in accordance with Section 1028.2.
5. Exterior landings as required by Section 1010.1.5 for *exit doorways* that lead directly to the *exit discharge.*

1008.3.3 Rooms and spaces. In the event of power supply failure, an emergency electrical system shall automatically illuminate all of the following areas:

1. Electrical equipment rooms.
2. Fire command centers.
3. Fire pump rooms.
4. Generator rooms.
5. Public restrooms with an area greater than 300 square feet (27.87 m^2).

1008.3.4 Duration. The emergency power system shall provide power for a duration of not less than 90 minutes and shall consist of storage batteries, unit equipment or an on-site generator. The installation of the emergency power system shall be in accordance with Section 2702.

1008.3.5 Illumination level under emergency power. Emergency lighting facilities shall be arranged to provide initial illumination that is not less than an average of 1 footcandle (11 lux) and a minimum at any point of 0.1 footcandle (1 lux) measured along the path of egress at floor level. Illumination levels shall be permitted to decline to 0.6 footcandle (6 lux) average and a minimum at any point of 0.06 footcandle (0.6 lux) at the end of the emergency lighting time duration. A maximum-to-minimum illumination uniformity ratio of 40 to 1 shall not be exceeded. In Group I-2 occupancies, failure of a single lamp in a luminaire shall not reduce the illumination level to less than 0.2 footcandle (2.2 lux).

SECTION 1009
ACCESSIBLE MEANS OF EGRESS

1009.1 Accessible means of egress required. *Accessible* means of egress shall comply with this section. *Accessible* spaces shall be provided with not less than one accessible means of egress. Where more than one *means of egress* is required by Section 1006.2 or 1006.3 from any accessible space, each *accessible* portion of the space shall be served by not less than two accessible means of egress.

Exceptions:

1. One *accessible means of egress* is required from an *accessible mezzanine* level in accordance with Section 1009.3, 1009.4 or 1009.5.
2. In assembly areas with ramped *aisles* or stepped *aisles*, one *accessible means of egress* is permitted where the *common path of egress travel* is *accessible* and meets the requirements in Section 1030.8.

1009.2 Continuity and components. Each required *accessible means of egress* shall be continuous to a *public way* and shall consist of one or more of the following components:

1. *Accessible routes* complying with Section 1104.
2. *Interior exit stairways* complying with Sections 1009.3 and 1023.
3. *Exit access stairways* complying with Sections 1009.3 and 1019.3 or 1019.4.
4. *Exterior exit stairways* complying with Sections 1009.3 and 1027 and serving levels other than the *level of exit discharge*.
5. Elevators complying with Section 1009.4.
6. Platform lifts complying with Section 1009.5.
7. *Horizontal exits* complying with Section 1026.
8. *Ramps* complying with Section 1012.
9. *Areas of refuge* complying with Section 1009.6.
10. Exterior areas for assisted rescue complying with Section 1009.7 serving *exits* at the *level of exit discharge*.

1009.2.1 Elevators required. In buildings where a required accessible floor or occupied roof is four or more stories above or below a *level of exit discharge*, not less than one required *accessible means of egress* shall be an elevator complying with Section 1009.4.

Exceptions:

1. In buildings equipped throughout with an *automatic sprinkler system* installed in accordance with Section 903.3.1.1 or 903.3.1.2, the elevator shall not be required on floors provided with a *horizontal exit* and located at or above the *levels of exit discharge*.
2. In buildings equipped throughout with an *automatic sprinkler system* installed in accordance with Section 903.3.1.1 or 903.3.1.2, the elevator shall not be required on floors provided with a *ramp* conforming to the provisions of Section 1012.

1009.3 Stairways. In order to be considered part of an *accessible means of egress*, a *stairway* between *stories* shall comply with Sections 1009.3.1 through 1009.3.3.

1009.3.1 Exit access stairways. *Exit access stairways* that connect levels in the same *story* are not permitted as part of an *accessible means of egress*.

Exception: *Exit access stairways* providing *means of egress* from *mezzanines* are permitted as part of an *accessible means of egress*.

1009.3.2 Stairway width. *Stairways* shall have a clear width of 48 inches (1219 mm) minimum between *handrails*.

Exceptions:

1. The clear width of 48 inches (1219 mm) between *handrails* is not required in buildings equipped throughout with an *automatic sprinkler system* installed in accordance with Section 903.3.1.1 or 903.3.1.2.
2. The clear width of 48 inches (1219 mm) between *handrails* is not required for *stairways* accessed from a refuge area in conjunction with a *horizontal exit*.

1009.3.3 Area of refuge. *Stairways* shall either incorporate an *area of refuge* within an enlarged floor-level landing or shall be accessed from an *area of refuge* complying with Section 1009.6.

Exceptions:

1. *Areas of refuge* are not required at *exit access stairways* where two-way communication is provided at the elevator landing in accordance with Section 1009.8.
2. *Areas of refuge* are not required at *stairways* in buildings equipped throughout with an *automatic sprinkler system* installed in accordance with Section 903.3.1.1 or 903.3.1.2.
3. *Areas of refuge* are not required at *stairways* serving *open parking garages*.
4. *Areas of refuge* are not required for *smoke-protected or open-air assembly seating* areas

complying with Sections 1030.6.2 and 1030.6.3.

5. *Areas of refuge* are not required at *stairways* in Group R-2 occupancies.
6. *Areas of refuge* are not required for *stairways* accessed from a refuge area in conjunction with a *horizontal exit.*

1009.4 Elevators. In order to be considered part of an *accessible means of egress*, an elevator shall comply with Sections 1009.4.1 and 1009.4.2.

1009.4.1 Standby power. The elevator shall meet the emergency operation and signaling device requirements of Section 2.27 of ASME A17.1/CSA B44. Standby power shall be provided in accordance with Chapter 27 and Section 3003.

1009.4.2 Area of refuge. The elevator shall be accessed from an *area of refuge* complying with Section 1009.6.

Exceptions:

1. *Areas of refuge* are not required at the elevator in *open parking garages*.
2. *Areas of refuge* are not required in buildings and facilities equipped throughout with an *automatic sprinkler system* installed in accordance with Section 903.3.1.1 or 903.3.1.2.
3. *Areas of refuge* are not required at elevators not required to be located in a *shaft* in accordance with Section 712.
4. *Areas of refuge* are not required at elevators serving *smoke-protected or open-air assembly seating* areas complying with Sections 1030.6.2 and 1030.6.3.
5. *Areas of refuge* are not required for elevators accessed from a refuge area in conjunction with a *horizontal exit*.

1009.5 Platform lifts. Platform lifts shall be permitted to serve as part of an *accessible means of egress* where allowed as part of a required *accessible route* in Section 1110.9 except for Item 10. Standby power for the platform lift shall be provided in accordance with Chapter 27.

1009.6 Areas of refuge. Every required *area of refuge* shall be accessible from the space it serves by an *accessible means of egress*.

1009.6.1 Travel distance. The maximum travel distance from any accessible space to an *area of refuge* shall not exceed the *exit access* travel distance permitted for the occupancy in accordance with Section 1017.1.

1009.6.2 Stairway or elevator access. Every required *area of refuge* shall have *direct access* to a *stairway* complying with Sections 1009.3 and 1023 or an elevator complying with Section 1009.4.

Exception: An interior area of refuge at the level of exit discharge that provides direct access to an exterior exit door.

1009.6.3 Size. Each *area of refuge* shall be sized to accommodate one *wheelchair space* of 30 inches by 52 inches (762 mm by 1320 mm) for each 200 occupants or portion thereof, based on the *occupant load* of the *area of refuge* and areas served by the *area of refuge*. Such *wheelchair spaces* shall not reduce the *means of egress* minimum width or required capacity. Access to any of the required *wheelchair spaces* in an *area of refuge* shall not be obstructed by more than one adjoining *wheelchair space*.

1009.6.4 Separation. Each *area of refuge* shall be separated from the remainder of the *story* by a *smoke barrier* complying with Section 709 or a *horizontal exit* complying with Section 1026. Each *area of refuge* shall be designed to minimize the intrusion of smoke.

Exceptions:

1. *Areas of refuge* located within an enclosure for *interior exit stairways* complying with Section 1023.
2. *Areas of refuge* in outdoor facilities where *exit access* is essentially open to the outside.

1009.6.5 Two-way communication. *Areas of refuge* shall be provided with a two-way communication system complying with Sections 1009.8.1 and 1009.8.2.

1009.7 Exterior areas for assisted rescue. Exterior areas for assisted rescue shall be accessed by an *accessible route* from the area served.

Where the *exit discharge* does not include an *accessible route* from an exit located on the *level of exit discharge* to a *public way*, an exterior area of assisted rescue shall be provided on the exterior landing in accordance with Sections 1009.7.1 through 1009.7.4.

1009.7.1 Size. Each exterior area for assisted rescue shall be sized to accommodate *wheelchair spaces* in accordance with Section 1009.6.3.

1009.7.2 Separation. Exterior walls separating the exterior area of assisted rescue from the interior of the building shall have a minimum *fire-resistance rating* of 1 hour, rated for exposure to fire from the inside. The fire-resistance-rated *exterior wall* construction shall extend horizontally not less than 10 feet (3048 mm) beyond the landing on either side of the landing or equivalent fire-resistance-rated construction is permitted to extend out perpendicular to the *exterior wall* not less than 4 feet (1219 mm) on the side of the landing. The *fire-resistance-rated* construction shall extend vertically from the ground to a point not less than 10 feet (3048 mm) above the floor level of the area for assisted rescue or to the roof line, whichever is lower. Openings within such *fire-resistance-rated exterior walls* shall be protected in accordance with Section 716.

Exception: The *fire-resistance rating* and opening protectives are not required in the *exterior wall* where the building is equipped throughout with an *automatic sprinkler system* installed in accordance with Section 903.3.1.1 or 903.3.1.2.

1009.7.3 Openness. The exterior area for assisted rescue shall be open to the outside air. The sides other than the separation walls shall be not less than 50 percent open, and the open area shall be distributed so as to minimize the accumulation of smoke or toxic gases.

1009.7.4 Stairways. *Stairways* that are part of the *means of egress* for the exterior area for assisted rescue shall provide a minimum clear width of 48 inches (1219 mm) between *handrails*.

Exception: The minimum clear width of 48 inches (1219 mm) between *handrails* is not required at *stairways* serving buildings equipped throughout with an *automatic sprinkler system* installed in accordance with Section 903.3.1.1 or 903.3.1.2.

1009.8 Two-way communication. A two-way communication system complying with Sections 1009.8.1 and 1009.8.2 shall be provided at the landing serving each elevator or bank of elevators on each accessible floor that is one or more stories above or below the *level of exit discharge*.

Exceptions:

1. Two-way communication systems are not required at the landing serving each elevator or bank of elevators where the two-way communication system is provided within *areas of refuge* in accordance with Section 1009.6.5.
2. Two-way communication systems are not required on floors provided with *ramps* conforming to the provisions of Section 1012.
3. Two-way communication systems are not required at the landings serving only service elevators that are not designated as part of the *accessible means of egress* or serve as part of the required *accessible route* into a facility.
4. Two-way communication systems are not required at the landings serving only freight elevators.
5. Two-way communication systems are not required at the landing serving a private residence elevator.
6. Two-way communication systems are not required in Group I-2 or I-3 facilities.

1009.8.1 System requirements. Two-way communication systems shall provide communication between each required location and the *fire command center* or a central control point location *approved* by the fire department. Where the central control point is not a *constantly attended location*, the two-way communication system shall have timed, automatic telephone dial-out capability that provides two-way communication with an approved supervising station or 9-1-1. The two-way communication system shall include both audible and visible signals.

1009.8.2 Directions. Directions for the use of the two-way communication system, instructions for summoning assistance via the two-way communication system and written identification of the location shall be posted adjacent to the two-way communication system. Signage shall comply with the ICC A117.1 requirements for visual characters.

1009.9 Signage. Signage indicating special accessibility provisions shall be provided as shown:

1. Each door providing access to an *area of refuge* from an adjacent floor area shall be identified by a sign stating, "AREA OF REFUGE."
2. Each door providing access to an exterior area for assisted rescue shall be identified by a sign stating, "EXTERIOR AREA FOR ASSISTED RESCUE."

Signage shall comply with the ICC A117.1 requirements for visual characters and include the International Symbol of Accessibility. Where exit sign illumination is required by Section 1013.3, the signs shall be illuminated. Additionally, visual characters, raised character and braille signage complying with ICC A117.1 shall be located at each door to an *area of refuge* and exterior area for assisted rescue in accordance with Section 1013.4.

1009.10 Directional signage. Directional signage indicating the location of all other *means of egress* and which of those are *accessible* means of egress shall be provided at the following:

1. At *exits* serving a required accessible space but not providing an *approved accessible means of egress*.
2. At elevator landings.
3. Within *areas of refuge*.

1009.11 Instructions. In *areas of refuge* and exterior areas for assisted rescue, instructions on the use of the area under emergency conditions shall be posted. Signage shall comply with the ICC A117.1 requirements for visual characters. The instructions shall include all of the following:

1. Persons able to use the *exit stairway* do so as soon as possible, unless they are assisting others.
2. Information on planned availability of assistance in the use of *stairs* or supervised operation of elevators and how to summon such assistance.
3. Directions for use of the two-way communication system where provided.

SECTION 1010
DOORS, GATES AND TURNSTILES

1010.1 General. Doors in the *means of egress* shall comply with the requirements of Sections 1010.1.1 through 1010.3.4. Exterior *exit* doors shall also comply with the requirements of Section 1022.2. Gates in the *means of egress* shall comply with the requirements of Sections 1010.4 and 1010.4.1. Turnstiles in the *means of egress* shall comply with the requirements of Sections 1010.5 through 1010.5.4.

Doors, gates and turnstiles provided for egress purposes in numbers greater than required by this code shall comply with the requirements of this section.

Doors in the *means of egress* shall be readily distinguishable from the adjacent construction and finishes such that the doors are easily recognizable as doors. Mirrors or similar reflecting materials shall not be used on *means of egress* doors. *Means of egress* doors shall not be concealed by curtains, drapes, decorations or similar materials.

1010.1.1 Size of doors. The required capacity of each door opening shall be sufficient for the *occupant load* thereof and shall provide a minimum clear opening width of 32 inches (813 mm). The clear opening width of doorways with swinging doors shall be measured between the face of the door and the stop, with the door open 90 degrees (1.57 rad). Where this section requires a minimum clear opening width of 32 inches (813 mm) and a door opening includes two door leaves without a mullion, one leaf shall provide a minimum clear opening width of 32 inches (813 mm). In Group I-2, doors serving as means of egress doors where used for the movement of beds shall provide a minimum clear opening width of $41^1/_2$ inches (1054 mm). The minimum clear opening height of doors shall be not less than 80 inches (2032 mm).

Exceptions:

1. In Group R-2 and R-3 *dwelling and sleeping units* that are not required to be an *Accessible unit*, *Type A unit* or *Type B unit*, the minimum width shall not apply to door openings that are not part of the required *means of egress*.
2. In Group I-3, door openings to resident *sleeping units* that are not required to be an *Accessible unit* shall have a minimum clear opening width of 28 inches (711 mm).
3. Door openings to storage closets less than 10 square feet (0.93 m^2) in area shall not be limited by the minimum clear opening width.
4. The maximum width of door leaves in revolving doors that comply with Section 1010.3.1 shall not be limited.
5. The maximum width of door leaves in *power-operated doors* that comply with Section 1010.3.2 shall not be limited.
6. Door openings within a *dwelling unit* or *sleeping unit* shall have a minimum clear opening height of 78 inches (1981 mm).
7. In *dwelling and sleeping units* that are not required to be *Accessible*, Type A or *Type B units*, exterior door openings other than the required *exit* door shall have a minimum clear opening height of 76 inches (1930 mm).
8. In Groups I-1, R-2, R-3 and R-4, in *dwelling and sleeping units* that are not required to be Accessible, Type A or *Type B units*, the minimum clear opening widths shall not apply to interior egress doors.
9. Door openings required to be *accessible* within *Type B units* intended for user passage shall have a minimum clear opening width of 31.75 inches (806 mm).
10. Doors to walk-in freezers and coolers less than 1,000 square feet (93 m^2) in area shall have a maximum width of 60 inches (1524 mm) nominal.
11. Doors serving nonaccessible single-user shower or sauna compartments, toilet stalls or dressing, fitting or changing rooms shall have a minimum clear opening width of 20 inches (508 mm).

1010.1.1.1 Projections into clear opening. There shall not be projections into the required clear opening width lower than 34 inches (864 mm) above the floor or ground. Projections into the clear opening width between 34 inches (864 mm) and 80 inches (2032 mm) above the floor or ground shall not exceed 4 inches (102 mm).

Exception: Door closers, overhead door stops, power door operators, and electromagnetic door locks shall be permitted to be 78 inches (1980 mm) minimum above the floor.

1010.1.2 Egress door types. Egress doors shall be of the side-hinged swinging door, pivoted door, or *balanced door* types.

Exceptions:

1. *Private garages*, office areas, factory and storage areas with an *occupant load* of 10 or less.
2. Group I-3 occupancies used as a place of detention.
3. Critical or intensive care patient rooms within suites of health care facilities.
4. Doors within or serving a single *dwelling unit* in Groups R-2 and R-3.
5. In other than Group H occupancies, revolving doors complying with Section 1010.3.1.
6. In other than Group H occupancies, special purpose horizontal sliding, accordion or folding door assemblies complying with Section 1010.3.3.
7. *Power-operated* doors in accordance with Section 1010.3.2.
8. Doors serving a bathroom within an individual *sleeping unit* in Group R-1.
9. In other than Group H occupancies, manually operated horizontal sliding doors are permitted in a *means of egress* from spaces with an *occupant load* of 10 or less.

1010.1.2.1 Direction of swing. Side-hinged swinging doors, pivoted doors and balanced doors shall swing in the direction of egress travel where serving a room or area containing an *occupant* load of 50 or more persons or a Group H occupancy.

1010.1.3 Forces to unlatch and open doors. The forces to unlatch doors shall comply with the following:

1. Where door hardware operates by push or pull, the operational force to unlatch the door shall not exceed 15 pounds (67 N).
2. Where door hardware operates by rotation, the operational force to unlatch the door shall not exceed 28 inch-pounds (315 N-cm).

The force to open doors shall comply with the following:

1. For interior swinging egress doors that are manually operated, other than doors required to be fire rated, the force for pushing or pulling open the door shall not exceed 5 pounds (22 N).
2. For other swinging doors, sliding doors or folding doors, and doors required to be fire rated, the door shall require not more than a 30-pound (133 N) force to be set in motion and shall move to a full-open position when subjected to not more than a 15-pound (67 N) force.

1010.1.3.1 Location of applied forces. Forces shall be applied to the latch side of the door.

1010.1.3.2 Manual horizontal sliding doors. Where a manual horizontal sliding door is required to latch, the latch or other mechanism shall prevent the door from rebounding into a partially open position when the door is closed.

1010.1.4 Floor elevation. There shall be a floor or landing on each side of a door. Such floor or landing shall be at the same elevation on each side of the door. Landings shall be level except for exterior landings, which are permitted to have a slope not to exceed 0.25 unit vertical in 12 units horizontal (2-percent slope).

Exceptions:

1. At doors serving individual dwelling units or sleeping units in Groups R-2 and R-3 : a door is permitted to open at the top step of an interior *flight* of stairs, provided that the door does not swing over the top step.
2. At exterior doors serving Groups F, H, R-2 and S and where such doors are not part of an accessible route, the landing at an exterior door shall not be more than 7 inches (178 mm) below the landing on the egress side of the door, provided that the door, other than an exterior storm or screen door, does not swing over the landing.
3. At exterior doors serving Group U and individual dwelling units and sleeping units in Groups R-2 and R-3, and where such units are not required to be Accessible units, Type A units or Type B units, the landing at an exterior doorway shall be not more than $7^3/_4$ inches (197 mm) below the landing on the egress side of the door. Such doors, including storm or screen doors, shall be permitted to swing over either landing.
4. Variations in elevation due to differences in finish materials, but not more than $^1/_2$ inch (12.7 mm).
5. Exterior decks, patios or balconies that are part of Type B *dwelling units* or sleeping units, that have impervious surfaces and that are not more than 4 inches (102 mm) below the finished floor level of the adjacent interior space of the *dwelling unit* or *sleeping unit.*
6. Doors serving equipment spaces not required to be *accessible* in accordance with Section 1103.2.9 and serving an *occupant load* of five or less shall be permitted to have a landing on one side to be not more than 7 inches (178 mm) above or below the landing on the egress side of the door.

1010.1.5 Landings at doors. Landings shall have a width not less than the width of the *stairway* or the door, whichever is greater. Doors in the fully open position shall not reduce a required dimension by more than 7 inches (178 mm). Where a landing serves an *occupant load* of 50 or more, doors in any position shall not reduce the landing to less than one-half its required width. Landings shall have a length measured in the direction of travel of not less than 44 inches (1118 mm).

Exception: Landing length in the direction of travel in Groups R-3 and U and within individual units of Group R-2 need not exceed 36 inches (914 mm).

1010.1.6 Thresholds. Thresholds at doorways shall not exceed $^3/_4$ inch (19.1 mm) in height above the finished floor or landing for sliding doors serving *dwelling units* or $^1/_2$ inch (12.7 mm) above the finished floor or landing for other doors. Raised thresholds and floor level changes greater than $^1/_4$ inch (6.4 mm) at doorways shall be beveled with a slope not greater than one unit vertical in two units horizontal (50-percent slope).

Exceptions:

1. In occupancy Group R-2 or R-3, threshold heights for sliding and side-hinged exterior doors shall be permitted to be up to $7^3/_4$ inches (197 mm) in height if all of the following apply:
   1.1. The door is not part of the required *means of egress.*
   1.2. The door is not part of an *accessible route* as required by Chapter 11.
   1.3. The door is not part of an *Accessible unit, Type A unit* or *Type B unit.*
2. In *Type B units*, where Exception 5 to Section 1010.1.4 permits a 4-inch (102 mm) elevation change at the door, the threshold height on the exterior side of the door shall not exceed $4^3/_4$ inches (120 mm) in height above the exterior deck, patio or balcony for sliding doors or $4^1/_2$ inches (114 mm) above the exterior deck, patio or balcony for other doors.

1010.1.7 Door arrangement. Space between two doors in a series shall be 48 inches (1219 mm) minimum plus the width of a door swinging into the space. Doors in a series shall swing either in the same direction or away from the space between the doors.

Exceptions:

1. The minimum distance between horizontal sliding *power-operated* doors in a series shall be 48 inches (1219 mm).

2. Storm and screen doors serving individual *dwelling units* in Groups R-2 and R-3 need not be spaced 48 inches (1219 mm) from the other door.
3. Doors within individual *dwelling units* in Groups R-2 and R-3 other than within Type A *dwelling units*.

1010.2 Door operations. Except as specifically permitted by this section, egress doors shall be readily openable from the egress side without the use of a key or special knowledge or effort.

1010.2.1 Unlatching. The unlatching of any door or leaf for egress shall require not more than one motion in a single linear or rotational direction to release all latching and all locking devices.

Exceptions:

1. Places of detention or restraint.
2. Where manually operated bolt locks are permitted by Section 1010.2.5.
3. Doors with automatic flush bolts as permitted by Section 1010.2.4, Item 4.
4. Doors from individual *dwelling units* and *sleeping units* of Group R occupancies as permitted by Section 1010.2.4, Item 5.

1010.2.2 Hardware. Door handles, pulls, latches, locks and other operating devices on doors required to be *accessible* by Chapter 11 shall not require tight grasping, tight pinching or twisting of the wrist to operate.

1010.2.3 Hardware height. Door handles, pulls, latches, locks and other operating devices shall be installed 34 inches (864 mm) minimum and 48 inches (1219 mm) maximum above the finished floor. Locks used only for security purposes and not used for normal operation are permitted at any height.

Exception: Access doors or gates in barrier walls and fences protecting pools, spas and hot tubs shall be permitted to have operable parts of the latch release on self-latching devices at 54 inches (1370 mm) maximum above the finished floor or ground, provided that the self-latching devices are not also self-locking devices operated by means of a key, electronic opener or integral combination lock.

1010.2.4 Locks and latches. Locks and latches shall be permitted to prevent operation of doors where any of the following exist:

1. Places of detention or restraint.
2. In Group I-1, Condition 2 and Group I-2 occupancies where the clinical needs of persons receiving care require containment or where persons receiving care pose a security threat, provided that all clinical staff can readily unlock doors at all times, and all such locks are keyed to keys carried by all clinical staff at all times or all clinical staff have the codes or other means necessary to operate the locks at all times.
3. In buildings in occupancy Group A having an *occupant load* of 300 or less, Groups B, F, M and S, and in *places of religious worship*, the main door or doors are permitted to be equipped with key-operated locking devices from the egress side provided:
   3.1. The locking device is readily distinguishable as locked.
   3.2. A readily visible durable sign is posted on the egress side on or adjacent to the door stating: THIS DOOR TO REMAIN UNLOCKED WHEN THIS SPACE IS OCCUPIED. The sign shall be in letters 1 inch (25 mm) high on a contrasting background.
   3.3. The use of the key-operated locking device is revocable by the *building official* for due cause.
4. Where egress doors are used in pairs, *approved* automatic flush bolts shall be permitted to be used, provided that the door leaf having the automatic flush bolts does not have a doorknob or surface-mounted hardware.
5. Doors from individual *dwelling* or *sleeping units* of Group R occupancies having an *occupant load* of 10 or less are permitted to be equipped with a night latch, dead bolt or security chain, provided such devices are openable from the inside without the use of a key or tool.
6. *Fire doors* after the minimum elevated temperature has disabled the unlatching mechanism in accordance with *listed fire door* test procedures.
7. Doors serving roofs not intended to be occupied shall be permitted to be locked preventing entry to the building from the roof.
8. Other than egress *courts*, where occupants must egress from an exterior space through the building for *means of egress*, exit access doors shall be permitted to be equipped with an approved locking device where installed and operated in accordance with all of the following:
   8.1. The maximum *occupant load* shall be posted where required by Section 1004.9. Such signage shall be permanently affixed inside the building and shall be posted in a conspicuous space near all the exit access doorways.
   8.2. A weatherproof telephone or two-way communication system installed in accordance with Sections 1009.8.1 and 1009.8.2 shall be located adjacent to not less than one required exit access door on the exterior side.
   8.3. The egress door locking device is readily distinguishable as locked and shall be a key-operated locking device.

8.4. A clear window or glazed door opening, not less than 5 square feet (0.46 m^2) in area, shall be provided at each exit access door to determine if there are occupants using the outdoor area.

8.5. A readily visible, durable sign shall be posted on the interior side on or adjacent to each locked required exit access door serving the exterior area stating, "THIS DOOR TO REMAIN UNLOCKED WHEN THE OUTDOOR AREA IS OCCUPIED." The letters on the sign shall be not less than 1 inch (25.4 mm) high on a contrasting background.

8.6. The *occupant load* of the occupied exterior area shall not exceed 300 occupants in accordance with Section 1004.

9. Locking devices are permitted on doors to balconies, decks or other exterior spaces serving individual dwelling or sleeping units.

10. Locking devices are permitted on doors to balconies, decks or other exterior spaces of 250 square feet (23.23 m^2) or less serving a private office space.

1010.2.5 Bolt locks. Manually operated flush bolts or surface bolts are not permitted.

Exceptions:

1. On doors not required for egress in individual *dwelling units* or *sleeping units.*
2. Where a pair of doors serves a storage or equipment room, manually operated edge- or surface-mounted bolts are permitted on the inactive leaf.
3. Where a pair of doors serves an *occupant load* of less than 50 persons in a Group B, F or S occupancy, manually operated edge- or surface-mounted bolts are permitted on the inactive leaf. The inactive leaf shall not contain doorknobs, *panic* bars or similar operating hardware.
4. Where a pair of doors serves a Group B, F or S occupancy, manually operated edge- or surface-mounted bolts are permitted on the inactive leaf provided that such inactive leaf is not needed to meet egress capacity requirements and the building is equipped throughout with an *automatic sprinkler system* in accordance with Section 903.3.1.1. The inactive leaf shall not contain doorknobs, *panic* bars or similar operating hardware.
5. Where a pair of doors serves patient care rooms in Group I-2 occupancies, self-latching edge- or surface-mounted bolts are permitted on the inactive leaf provided that the inactive leaf is not needed to meet egress capacity requirements and the inactive leaf shall not contain doorknobs, *panic* bars or similar operating hardware.

1010.2.6 Closet doors. Closet doors that latch in the closed position shall be openable from inside the closet.

1010.2.7 Stairway doors. Interior *stairway* means of egress doors shall be openable from both sides without the use of a key or special knowledge or effort.

Exceptions:

1. *Stairway* discharge doors shall be openable from the egress side and shall only be locked from the opposite side.
2. This section shall not apply to doors arranged in accordance with Section 403.5.3.
3. *Stairway* exit doors are permitted to be locked from the side opposite the egress side, provided that they are openable from the egress side and capable of being unlocked simultaneously without unlatching upon a signal from the *fire command center*, if present, or a signal by emergency personnel from a single location inside the main entrance to the building.
4. *Stairway exit* doors shall be openable from the egress side and shall only be locked from the opposite side in Group B, F, M and S occupancies where the only interior access to the tenant space is from a single *exit stairway* where permitted in Section 1006.3.4.
5. *Stairway* exit doors shall be openable from the egress side and shall only be locked from the opposite side in Group R-2 occupancies where the only interior access to the *dwelling unit* is from a single exit *stairway* where permitted in Section 1006.3.4.

1010.2.8 Locking arrangements in educational occupancies. In Group E occupancies, Group B educational occupancies and Group I-4 occupancies, egress doors from classrooms, offices and other occupied rooms with locking arrangements designed to keep intruders from entering the room shall comply with all of the following conditions:

1. The door shall be capable of being unlocked from outside the room with a key or other *approved* means.
2. The door shall be openable from within the room in accordance with Section 1010.2.
3. Modifications shall not be made to listed *panic hardware*, *fire door* hardware or door closers.
4. Modifications to *fire door assemblies* shall be in accordance with NFPA 80.

Remote locking or unlocking of doors from an approved location shall be permitted in addition to the unlocking operation in Item 1.

1010.2.9 Panic and fire exit hardware. Swinging doors serving a Group H occupancy and swinging doors serving rooms or spaces with an *occupant load* of 50 or more in a

Group A or E occupancy shall not be provided with a latch or lock other than *panic hardware* or *fire exit hardware*.

Exceptions:

1. A main exit of a Group A occupancy shall be permitted to have locking devices in accordance with Section 1010.2.4, Item 3.
2. Doors provided with *panic hardware* or *fire exit hardware* and serving a Group A or E occupancy shall be permitted to be electrically locked in accordance with Section 1010.2.11 or 1010.2.12.
3. Exit access doors serving occupied exterior areas shall be permitted to be locked in accordance with Section 1010.2.4, Item 8.
4. Courtrooms shall be permitted to be locked in accordance with Section 1010.2.13, Item 3.

1010.2.9.1 Refrigeration machinery room. Refrigeration machinery rooms larger than 1,000 square feet (93 m^2) shall have not less than two exit or exit access doorways that swing in the direction of egress travel and shall be equipped with *panic hardware* or *fire exit hardware*.

1010.2.9.2 Rooms with electrical equipment. Exit or exit access doors serving transformer vaults, rooms designated for batteries or energy storage systems, or modular data centers shall be equipped with panic hardware or fire exit hardware. Rooms containing electrical equipment rated 800 amperes or more that contain overcurrent devices, switching devices or control devices and where the exit or exit access door is less than 25 feet (7620 mm) from the equipment working space as required by NFPA 70, such doors shall not be provided with a latch or lock other than *panic hardware* or *fire exit hardware*. The doors shall swing in the direction of egress travel.

1010.2.9.3 Installation. Where panic or *fire exit hardware* is installed, it shall comply with the following:

1. *Panic hardware* shall be *listed* in accordance with UL 305.
2. *Fire exit hardware* shall be *listed* in accordance with UL 10C and UL 305.
3. The actuating portion of the releasing device shall extend not less than one-half of the door leaf width.
4. The maximum unlatching force shall not exceed 15 pounds (67 N).

1010.2.9.4 Balanced doors. If *balanced doors* are used and *panic hardware* is required, the *panic hardware* shall be the push-pad type and the pad shall not extend more than one-half the width of the door measured from the latch side.

1010.2.10 Monitored or recorded egress. Where electrical systems that monitor or record egress activity are incorporated, the locking system shall comply with Section 1010.2.11, 1010.2.12, 1010.2.13, 1010.2.14 or 1010.2.15 or shall be readily openable from the egress side without the use of a key or special knowledge or effort.

1010.2.11 Door hardware release of electrically locked egress doors. Door hardware release of electric locking systems shall be permitted on doors in the *means of egress* in any occupancy except Group H where installed and operated in accordance with all of the following:

1. The door hardware that is affixed to the door leaf has an obvious method of operation that is readily operated under all lighting conditions.
2. The door hardware is capable of being operated with one hand and shall comply with Section 1010.2.1.
3. Operation of the door hardware directly interrupts the power to the electric lock and unlocks the door immediately.
4. Loss of power to the electric locking system automatically unlocks the door.
5. Where *panic* or *fire exit hardware* is required by Section 1010.2.9, operation of the *panic* or *fire exit hardware* also releases the electric lock.
6. The locking system units shall be *listed* in accordance with UL 294.

1010.2.12 Sensor release of electrically locked egress doors. Sensor release of electric locking systems shall be permitted on doors located in the *means of egress* in any occupancy except Group H where installed and operated in accordance with all of the following criteria:

1. The sensor shall be installed on the egress side, arranged to detect an occupant approaching the doors, and shall cause the electric locking system to unlock.
2. The electric locks shall be arranged to unlock by a signal from or loss of power to the sensor.
3. Loss of power to the lock or locking system shall automatically unlock the electric locks.
4. The doors shall be arranged to unlock from a manual unlocking device located 40 inches to 48 inches (1016 mm to 1219 mm) vertically above the floor and within 5 feet (1524 mm) of the secured doors. Ready access shall be provided to the manual unlocking device and the device shall be clearly identified by a sign that reads "PUSH TO EXIT." When operated, the manual unlocking device shall result in direct interruption of power to the electric lock—independent of other electronics—and the electric lock shall remain unlocked for not less than 30 seconds.
5. Activation of the building *fire alarm system*, where provided, shall automatically unlock the electric lock, and the electric lock shall remain unlocked until the *fire alarm system* has been reset.
6. Activation of the building *automatic sprinkler system* or fire detection system, where provided,

shall automatically unlock the electric lock. The electric lock shall remain unlocked until the *fire alarm system* has been reset.

7. Emergency lighting shall be provided on the egress side of the door.
8. The door locking system units shall be *listed* in accordance with UL 294.

1010.2.13 Delayed egress. Delayed egress locking systems shall be permitted to be installed on doors serving the following occupancies in buildings that are equipped throughout with an *automatic sprinkler system* in accordance with Section 903.3.1.1 or an *approved automatic smoke* or *heat detection system* installed in accordance with Section 907.

1. Group B, F, I, M, R, S and U occupancies.
2. Group E classrooms with an *occupant load* of less than 50.
3. In courtrooms in Group A-3 and B occupancies, delayed egress locking systems shall be permitted to be installed on exit or *exit access* doors, other than the main exit or *exit access* door, in buildings that are equipped throughout with an *automatic sprinkler system* in accordance with Section 903.3.1.1.

1010.2.13.1 Delayed egress locking system. The delayed egress locking system shall be installed and operated in accordance with all of the following:

1. The delay electronics of the delayed egress locking system shall deactivate upon actuation of the *automatic sprinkler system* or *automatic fire detection system*, allowing immediate free egress.
2. The delay electronics of the delayed egress locking system shall deactivate upon loss of power controlling the lock or lock mechanism, allowing immediate free egress.
3. The delayed egress locking system shall have the capability of being deactivated at the *fire command center* and other *approved* locations.
4. An attempt to egress shall initiate an irreversible process that shall allow such egress in not more than 15 seconds when a physical effort to exit is applied to the egress side door hardware for not more than 3 seconds. Initiation of the irreversible process shall activate an audible signal in the vicinity of the door. Once the delay electronics have been deactivated, rearming the delay electronics shall be by manual means only.

   Exception: Where *approved*, a delay of not more than 30 seconds is permitted on a delayed egress door.

5. The egress path from any point shall not pass through more than one delayed egress locking system.

   Exceptions:

   1. In Group I-1, Condition 2, Group I-2 or I-3 occupancies, the egress path from any point in the building shall pass through not more than two delayed egress locking systems provided that the combined delay does not exceed 30 seconds.
   2. In Group I-1, Condition 1 or Group I-4 occupancies, the egress path from any point in the building shall pass through not more than two delayed egress locking systems provided the combined delay does not exceed 30 seconds and the building is equipped throughout with an *automatic sprinkler system* in accordance with Section 903.3.1.1.

6. A sign shall be provided on the door and shall be located above and within 12 inches (305 mm) of the door exit hardware:

   Exception: Where *approved*, in Group I occupancies, the installation of a sign is not required where care recipients who because of clinical needs require restraint or containment as part of the function of the treatment area.

   6.1. For doors that swing in the direction of egress, the sign shall read, "PUSH UNTIL ALARM SOUNDS. DOOR CAN BE OPENED IN 15 [30] SECONDS."

   6.2. For doors that swing in the opposite direction of egress, the sign shall read, "PULL UNTIL ALARM SOUNDS. DOOR CAN BE OPENED IN 15 [30] SECONDS."

   6.3. The sign shall comply with the visual character requirements in ICC A117.1.

7. Emergency lighting shall be provided on the egress side of the door.
8. The delayed egress locking system units shall be *listed* in accordance with UL 294.

1010.2.14 Controlled egress doors in Groups I-1 and I-2. Electric locking systems, including electro-mechanical locking systems and electromagnetic locking systems, shall be permitted to be locked in the *means of egress* in Group I-1 or I-2 occupancies where the clinical needs of persons receiving care require their containment. Controlled egress doors shall be permitted in such occupancies where the building is equipped throughout with an *automatic sprinkler system* in accordance with Section 903.3.1.1 or an *approved automatic smoke detection system* installed in accordance with Section 907, provided that the doors are installed and operate in accordance with all of the following:

1. The door locks shall unlock on actuation of the *automatic sprinkler system* or *automatic smoke detection system*.
2. The door locks shall unlock on loss of power controlling the lock or lock mechanism.
3. The door locking system shall be installed to have the capability of being unlocked by a switch located at the *fire command center*, a nursing station or other *approved* location. The switch shall directly break power to the lock.
4. A building occupant shall not be required to pass through more than one door equipped with a controlled egress locking system before entering an *exit*.
5. The procedures for unlocking the doors shall be described and *approved* as part of the emergency planning and preparedness required by Chapter 4 of the *International Fire Code*.
6. All clinical staff shall have the keys, codes or other means necessary to operate the locking systems.
7. Emergency lighting shall be provided at the door.
8. The door locking system units shall be *listed* in accordance with UL 294.

Exceptions:

1. Items 1 through 4 shall not apply to doors to areas occupied by persons who, because of clinical needs, require restraint or containment as part of the function of a psychiatric or cognitive treatment area.
2. Items 1 through 4 shall not apply to doors to areas where a *listed* egress control system is utilized to reduce the risk of child abduction from nursery and obstetric areas of a Group I-2 *hospital*.

1010.2.15 Locking arrangements in buildings within correctional facilities. In *buildings* within correctional and detention facilities, doors in *means of egress* serving rooms or spaces occupied by persons whose movements are controlled for security reasons shall be permitted to be locked where equipped with egress control devices that shall unlock manually and by not less than one of the following means:

1. Activation of an *automatic sprinkler system* installed in accordance with Section 903.3.1.1.
2. Activation of an *approved manual fire alarm box*.
3. A signal from a *constantly attended location*.

1010.3 Special doors. Special doors and security grilles shall comply with the requirements of Sections 1010.3.1 through 1010.3.4.

1010.3.1 Revolving doors. Revolving doors shall comply with the following:

1. Revolving doors shall comply with BHMA A156.27 and shall be installed in accordance with the manufacturer's instructions.
2. Each revolving door shall be capable of *breakout* in accordance with BHMA A156.27 and shall provide an aggregate width of not less than 36 inches (914 mm).
3. A revolving door shall not be located within 10 feet (3048 mm) of the foot or top of *stairways* or escalators. A dispersal area shall be provided between the *stairways* or escalators and the revolving doors.
4. The revolutions per minute (rpm) for a revolving door shall not exceed the maximum rpm as specified in BHMA A156.27. Manual revolving doors shall comply with Table 1010.3.1(1). Automatic or *power-operated* revolving doors shall comply with Table 1010.3.1(2).
5. An emergency stop switch shall be provided near each entry point of power or automatic operated revolving doors within 48 inches (1219 mm) of the door and between 34 inches (864 mm) and 48 inches (1219 mm) above the floor. The activation area of the emergency stop switch button shall be not less than 1 inch (25 mm) in diameter and shall be red.
6. Each revolving door shall have a side-hinged swinging door that complies with Section 1010.1 in the same wall and within 10 feet (3048 mm) of the revolving door.
7. Revolving doors shall not be part of an *accessible route* required by Section 1009 and Chapter 11.

TABLE 1010.3.1(1)
MAXIMUM DOOR SPEED MANUAL REVOLVING DOORS

REVOLVING DOOR MAXIMUM NOMINAL DIAMETER (FT-IN)	MAXIMUM ALLOWABLE REVOLVING DOOR SPEED (RPM)
6-0	12
7-0	11
8-0	10
9-0	9
10-0	8

For SI: 1 inch = 25.4 mm, 1 foot = 304.8 mm.

TABLE 1010.3.1(2)
MAXIMUM DOOR SPEED AUTOMATIC OR POWER-OPERATED REVOLVING DOORS

REVOLVING DOOR MAXIMUM NOMINAL DIAMETER (FT-IN)	MAXIMUM ALLOWABLE REVOLVING DOOR SPEED (RPM)
8-0	7.2
9-0	6.4
10-0	5.7
11-0	5.2
12-0	4.8
12-6	4.6
14-0	4.1
16-0	3.6
17-0	3.4
18-0	3.2
20-0	2.9
24-0	2.4

For SI: 1 inch = 25.4 mm, 1 foot = 304.8 mm.

1010.3.1.1 Egress component. A revolving door used as a component of a *means of egress* shall comply with Section 1010.3.1 and the following three conditions:

1. Revolving doors shall not be given credit for more than 50 percent of the minimum width or required capacity.
2. Each revolving door shall be credited with a capacity based on not more than a 50-person *occupant load*.
3. Each revolving door shall provide for egress in accordance with BHMA A156.27 with a *breakout* force of not more than 130 pounds (578 N).

1010.3.1.2 Other than egress component. A revolving door used as other than a component of a *means of egress* shall comply with Section 1010.3.1. The *breakout* force of a revolving door not used as a component of a *means of egress* shall not be more than 180 pounds (801 N).

Exception: A *breakout* force in excess of 180 pounds (801 N) is permitted if the *breakout* force is reduced to not more than 130 pounds (578 N) when not less than one of the following conditions is satisfied:

1. There is a power failure or power is removed to the device holding the door wings in position.
2. There is an actuation of the *automatic sprinkler system* where such system is provided.
3. There is an actuation of a smoke detection system that is installed in accordance with Section 907 to provide coverage in areas within the building that are within 75 feet (22 860 mm) of the revolving doors.
4. There is an actuation of a manual control switch, in an *approved* location and clearly identified, that reduces the *breakout* force to not more than 130 pounds (578 N).

1010.3.2 Power-operated doors. Where means of egress doors are operated or assisted by power, the design shall be such that in the event of power failure, the door is capable of being opened manually to permit means of egress travel or closed where necessary to safeguard *means of egress*. The forces required to open these doors manually shall not exceed those specified in Section 1010.1.3, except that the force to set the door in motion shall not exceed 50 pounds (220 N). The door shall be capable of opening from any position to the full width of the opening in which such door is installed when a force is applied to the door on the side from which egress is made. *Power-operated* swinging doors, *power-operated* sliding doors and *power-operated* folding doors shall comply with BHMA A156.10. *Power-assisted* swinging doors and low-energy *power-operated* swinging doors shall comply with BHMA A156.19. Low-energy *power-operated* sliding *doors* and low-energy *power-operated* folding *doors* shall comply with BHMA A156.38.

Exceptions:

1. Occupancies in Group I-3.
2. Special purpose horizontal sliding, accordion or folding doors complying with Section 1010.3.3.
3. For a biparting door in the emergency *breakout* mode, a door leaf located within a multiple-leaf opening shall be exempt from the minimum 32-inch (813 mm) single-leaf requirement of Section 1010.1.1, provided that a minimum 32-inch (813 mm) clear opening is provided when the two biparting leaves meeting in the center are broken out.

1010.3.3 Special purpose horizontal sliding, accordion or folding doors. In other than Group H occupancies, special purpose horizontal sliding, accordion or folding door assemblies permitted to be a component of a *means of egress* in accordance with Exception 6 to Section 1010.1.2 shall comply with all of the following criteria:

1. The doors shall be power operated and shall be capable of being operated manually in the event of power failure.
2. The doors shall be openable by a simple method without special knowledge or effort from the egress side or sides.
3. The force required to operate the door shall not exceed 30 pounds (133 N) to set the door in motion and 15 pounds (67 N) to close the door or open it to the minimum required width.
4. The door shall be openable with a force not to exceed 15 pounds (67 N) when a force of 250 pounds (1100 N) is applied perpendicular to the door adjacent to the operating device.

5. The door assembly shall comply with the applicable *fire protection rating* and, where rated, shall be *self-closing* or automatic closing by smoke detection in accordance with Section 716.2.6.6, shall be installed in accordance with NFPA 80 and shall comply with Section 716.
6. The door assembly shall have an integrated standby power supply.
7. The door assembly power supply shall be electrically supervised.
8. The door shall open to the minimum required width within 10 seconds after activation of the operating device.

1010.3.4 Security grilles. In Groups B, F, M and S, horizontal sliding or vertical security grilles are permitted at the main *exit* and shall be openable from the inside without the use of a key or special knowledge or effort during periods that the space is occupied. The grilles shall remain secured in the full-open position during the period of occupancy by the general public. Where two or more exits or access to exits are required, not more than one-half of the *exits* or *exit access doorways* shall be equipped with horizontal sliding or vertical security grilles.

1010.4 Gates. Gates serving the *means of egress* system shall comply with the requirements of this section. Gates used as a component in a *means of egress* shall conform to the applicable requirements for doors.

Exception: Horizontal sliding or swinging gates exceeding the 4-foot (1219 mm) maximum leaf width limitation are permitted in fences and walls surrounding a stadium.

1010.4.1 Stadiums. *Panic hardware* is not required on gates surrounding stadiums where such gates are under constant immediate supervision while the public is present, and where safe dispersal areas based on 3 square feet (0.28 m2) per occupant are located between the fence and enclosed space. Such required safe dispersal areas shall not be located less than 50 feet (15 240 mm) from the enclosed space. See Section 1028.5 for *means of egress* from safe dispersal areas.

1010.5 Turnstiles and similar devices. Turnstiles or similar devices that restrict travel to one direction shall not be placed so as to obstruct any required *means of egress*, except where permitted in accordance with Sections 1010.5.1, 1010.5.2 and 1010.5.3.

1010.5.1 Capacity. Each turnstile or similar device shall be credited with a capacity based on not more than a 50-person *occupant load* where all of the following provisions are met:

1. Each device shall turn free in the direction of egress travel when primary power is lost and on the manual release by an employee in the area.
2. Such devices are not given credit for more than 50 percent of the required egress capacity or width.
3. Each device is not more than 39 inches (991 mm) high.
4. Each device has not less than $16^1/_2$ inches (419 mm) clear width at and below a height of 39 inches (991 mm) and not less than 22 inches (559 mm) clear width at heights above 39 inches (991 mm).

1010.5.1.1 Clear width. Where located as part of an *accessible* route, turnstiles shall have not less than 36 inches (914 mm) clear width at and below a height of 34 inches (864 mm), not less than 32 inches (813 mm) clear width between 34 inches (864 mm) and 80 inches (2032 mm) and shall consist of a mechanism other than a revolving device.

1010.5.2 Security access turnstiles. Security access turnstiles that inhibit travel in the direction of egress utilizing a physical barrier shall be permitted to be considered as a component of the *means of egress*, provided that all of the following criteria are met:

1. The building is protected throughout by an approved, supervised *automatic sprinkler system* in accordance with Section 903.3.1.1.
2. Each security access turnstile lane configuration has a minimum clear passage width of 22 inches (559 mm).
3. Any security access turnstile lane configuration providing a clear passage width of less than 32 inches (810 mm) shall be credited with a maximum egress capacity of 50 persons.
4. Any security access turnstile lane configuration providing a clear passage width of 32 inches (810 mm) or more shall be credited with a maximum egress capacity as calculated in accordance with Section 1005.
5. Each secured physical barrier shall automatically retract or swing to an unobstructed open position in the direction of egress, under each of the following conditions:
   - 5.1. Upon loss of power to the turnstile or any part of the access control system that secures the physical barrier.
   - 5.2. Upon actuation of a clearly identified manual release device with ready access that results in direct interruption of power to each secured physical barrier, after which such barriers remain in the open position for not less than 30 seconds. The manual release device shall be positioned at one of the following locations:
     - 5.2.1. On the egress side of each security access turnstile lane.
     - 5.2.2. At an *approved* location where it can be actuated by an employee assigned to the area at all times that the building is occupied.
   - 5.3. Upon actuation of the building *fire alarm system*, if provided, after which the physi-

cal barrier remains in the open position until the *fire alarm system* is manually reset.

Exception: Actuation of a manual fire alarm box.

5.4. Upon actuation of the building automatic sprinkler system or fire detection system, after which the physical barrier remains in the open position until the *fire alarm system* is manually reset.

1010.5.3 High turnstile. Turnstiles more than 39 inches (991 mm) high shall meet the requirements for revolving doors or the requirements of Section 1010.5.2 for security access turnstiles.

1010.5.4 Additional door. Where serving an *occupant load* greater than 300, each turnstile that is not portable shall have a side-hinged swinging door that conforms to Section 1010.1 within 50 feet (15 240 mm).

Exception: A side-hinged swinging door is not required at security access turnstiles that comply with Section 1010.5.2.

SECTION 1011 STAIRWAYS

1011.1 General. *Stairways* serving occupied portions of a building shall comply with the requirements of Sections 1011.2 through 1011.13. *Alternating tread devices* shall comply with Section 1011.14. Ship's ladders shall comply with Section 1011.15. Ladders shall comply with Section 1011.16.

Exception: Within rooms or spaces used for assembly purposes, stepped *aisles* shall comply with Section 1030.

1011.2 Width and capacity. The required capacity of *stairways* shall be determined as specified in Section 1005.1, but the minimum width shall be not less than 44 inches (1118 mm). See Section 1009.3 for *accessible means of egress stairways*.

Exceptions:

1. *Stairways* serving an *occupant load* of less than 50 shall have a width of not less than 36 inches (914 mm).
2. *Spiral stairways* as provided for in Section 1011.10.
3. Where an incline platform lift or *stairway* chairlift is installed on *stairways* serving occupancies in Group R-3, or within *dwelling units* in occupancies in Group R-2, a clear passage width not less than 20 inches (508 mm) shall be provided. Where the seat and platform can be folded when not in use, the distance shall be measured from the folded position.

1011.3 Headroom. *Stairways* shall have a headroom clearance of not less than 80 inches (2032 mm) measured vertically from a line connecting the edge of the *nosings*. Such headroom shall be continuous above the *stairway* to the point where the line intersects the landing below, one tread depth beyond the bottom riser. The minimum clearance shall be maintained the full width of the *stairway* and landing.

Exceptions:

1. *Spiral stairways* complying with Section 1011.10 are permitted a 78-inch (1981 mm) headroom clearance.
2. In Group R-3 occupancies; within *dwelling units* in Group R-2 occupancies; and in Group U occupancies that are accessory to a Group R-3 occupancy or accessory to individual *dwelling units* in Group R-2 occupancies; where the *nosings* of treads at the side of a *flight* extend under the edge of a floor opening through which the *stair* passes, the floor opening shall be allowed to project horizontally into the required headroom not more than $4^3/_4$ inches (121 mm).

1011.4 Walkline. The walkline across *winder* treads shall be concentric to the direction of travel through the turn and located 12 inches (305 mm) from the side where the *winders* are narrower. The 12-inch (305 mm) dimension shall be measured from the widest point of the clear *stair* width at the walking surface of the *winder*. Where *winders* are adjacent within the *flight*, the point of the widest clear *stair* width of the adjacent *winders* shall be used.

1011.5 Stair treads and risers. *Stair* treads and risers shall comply with Sections 1011.5.1 through 1011.5.5.3.

1011.5.1 Dimension reference surfaces. For the purpose of this section, all dimensions are exclusive of carpets, rugs or runners.

1011.5.2 Riser height and tread depth. *Stair* riser heights shall be 7 inches (178 mm) maximum and 4 inches (102 mm) minimum. The riser height shall be measured vertically between the *nosings* of adjacent treads or between the *stairway* landing and the adjacent tread. Rectangular tread depths shall be 11 inches (279 mm) minimum measured horizontally between the vertical planes of the foremost projection of adjacent treads and at a right angle to the tread's *nosing*. *Winder* treads shall have a minimum tread depth of 11 inches (279 mm) between the vertical planes of the foremost projection of adjacent treads at the intersections with the walkline and a minimum tread depth of 10 inches (254 mm) within the clear width of the stair.

Exceptions:

1. *Spiral stairways* in accordance with Section 1011.10.
2. *Stairways* connecting stepped *aisles* to cross *aisles* or concourses shall be permitted to use the riser/tread dimension in Section 1030.14.2.
3. In Group R-3 occupancies; within *dwelling units* in Group R-2 occupancies; and in Group U occupancies that are accessory to a Group R-3 occupancy or accessory to individual *dwelling units* in Group R-2 occupancies; the maximum riser height shall be $7^3/_4$ inches (197

mm); the minimum tread depth shall be 10 inches (254 mm); the minimum *winder* tread depth at the walkline shall be 10 inches (254 mm); and the minimum *winder* tread depth shall be 6 inches (152 mm). A *nosing* projection not less than $^3/_4$ inch (19.1 mm) but not more than $1^1/_4$ inches (32 mm) shall be provided on *stairways* with solid risers where the tread depth is less than 11 inches (279 mm).

4. See Section 503.1 of the *International Existing Building Code* for the replacement of existing *stairways*.
5. In Group I-3 facilities, *stairways* providing access to guard towers, observation stations and control rooms, not more than 250 square feet (23 m^2) in area, shall be permitted to have a maximum riser height of 8 inches (203 mm) and a minimum tread depth of 9 inches (229 mm).

1011.5.3 Winder treads. *Winder* treads are not permitted in *means of egress stairways* except within a *dwelling unit*.

Exceptions:

1. Curved *stairways* in accordance with Section 1011.9.
2. *Spiral stairways* in accordance with Section 1011.10.

1011.5.4 Dimensional uniformity. *Stair* treads and risers shall be of uniform size and shape. The tolerance between the largest and smallest riser height or between the largest and smallest tread depth shall not exceed $^3/_8$ inch (9.5 mm) in any *flight* of *stairs*. The greatest *winder* tread depth at the walkline within any *flight of stairs* shall not exceed the smallest by more than $^3/_8$ inch (9.5 mm).

Exceptions:

1. *Stairways* connecting stepped *aisles* to cross *aisles* or concourses shall be permitted to comply with the dimensional nonuniformity in Section 1030.14.2.
2. Consistently shaped *winders*, complying with Section 1011.5, differing from rectangular treads in the same *flight* of *stairs*.
3. Nonuniform riser dimension complying with Section 1011.5.4.1.

1011.5.4.1 Nonuniform height risers. Where the bottom or top riser adjoins a sloping *public way*, walkway or driveway having an established grade and serving as a landing, the bottom or top riser is permitted to be reduced along the slope to less than 4 inches (102 mm) in height, with the variation in height of the bottom or top riser not to exceed one unit vertical in 12 units horizontal (8-percent slope) of *stair* width. The *nosings* or leading edges of treads at such nonuniform height risers shall have a distinctive marking stripe, different from any other *nosing* marking provided on the *stair flight*. The distinctive marking stripe shall be visible in descent of the *stair* and shall have a slip-resistant surface. Marking stripes shall have a width of not less than 1 inch (25 mm) but not more than 2 inches (51 mm).

1011.5.5 Nosing and riser profile. *Nosings* shall have a curvature or bevel of not less than $^1/_{16}$ inch (1.6 mm) but not more than $^9/_{16}$ inch (14.3 mm) from the foremost projection of the tread. Risers shall be solid and vertical or sloped under the tread above from the underside of the *nosing* above at an angle not more than 30 degrees (0.52 rad) from the vertical.

1011.5.5.1 Nosing projection size. The leading edge (*nosings*) of treads shall project not more than $1^1/_4$ inches (32 mm) beyond the tread below.

1011.5.5.2 Nosing projection uniformity. *Nosing* projections of the leading edges shall be of uniform size, including the projections of the *nosing's* leading edge of the floor at the top of a *flight*.

1011.5.5.3 Solid risers. Risers shall be solid.

Exceptions:

1. Solid risers are not required for *stairways* that are not required to comply with Section 1009.3, provided that the opening between treads does not permit the passage of a sphere with a diameter of 4 inches (102 mm).
2. Solid risers are not required for occupancies in Group I-3 or in Group F, H and S occupancies other than areas accessible to the public. The size of the opening in the riser is not restricted.
3. Solid risers are not required for *spiral stairways* constructed in accordance with Section 1011.10.

1011.6 Stairway landings. There shall be a floor or landing at the top and bottom of each *stairway*. The width of landings, measured perpendicularly to the direction of travel, shall be not less than the width of *stairways* served. Every landing shall have a minimum depth, measured parallel to the direction of travel, equal to the width of the *stairway* or 48 inches (1219 mm), whichever is less. Doors opening onto a landing shall not reduce the landing to less than one-half the required width. When fully open, the door shall not project more than 7 inches (178 mm) into the required width of a landing. Where wheelchair spaces are required on the *stairway* landing in accordance with Section 1009.6.3, the *wheelchair space* shall not be located in the required width of the landing and doors shall not swing over the *wheelchair spaces*.

Exceptions:

1. Where *stairways* connect stepped *aisles* to cross *aisles* or concourses, *stairway* landings are not required at the transition between *stairways* and stepped *aisles* constructed in accordance with Section 1030.

2. Where curved *stairways* of constant radius have intermediate landings, the landing depth shall be measured horizontally between the intersection of the walkline of the lower *flight* at the landing *nosing* and the intersection of the walkline of the upper *flight* at the *nosing* of the lowest tread of the upper *flight*.
3. Where a landing turns 90 degrees (1.57 rad) or more, the minimum landing depth in accordance with this section shall not be required where the landing provided is not less than that described by an arc with a radius equal to the width of the *flight* served.

1011.7 Stairway construction. *Stairways* shall be built of materials consistent with the types permitted for the type of construction of the building.

Exceptions:

1. Wood handrails shall be permitted in all types of construction.
2. Interior exit stairways in accordance with Section 510.2.

1011.7.1 Stairway walking surface. The walking surface of treads and landings of a *stairway* shall not be sloped steeper than one unit vertical in 48 units horizontal (2-percent slope) in any direction. *Stairway* treads and landings shall have a solid surface. Finish floor surfaces shall be securely attached.

Exceptions:

1. Openings in *stair* walking surfaces shall be a size that does not permit the passage of $^1/_2$-inch-diameter (12.7 mm) sphere. Elongated openings shall be placed so that the long dimension is perpendicular to the direction of travel.
2. In Group F, H and S occupancies, other than areas of parking structures accessible to the public, openings in treads and landings shall not be prohibited provided that a sphere with a diameter of $1^1/_8$ inches (29 mm) cannot pass through the opening.

1011.7.2 Outdoor conditions. Outdoor *stairways* and outdoor approaches to *stairways* shall be designed so that water will not accumulate on walking surfaces.

1011.7.3 Enclosures under interior stairways. The walls and soffits within enclosed usable spaces under enclosed and unenclosed *stairways* shall be protected by 1-hour fire-resistance-rated construction or the *fire-resistance rating* of the *stairway* enclosure, whichever is greater. Access to the enclosed space shall not be directly from within the *stairway* enclosure.

Exception: Spaces under *stairways* serving and contained within a single residential *dwelling unit* in Group R-2 or R-3 shall be permitted to be protected on the enclosed side with $^1/_2$-inch (12.7 mm) *gypsum board*.

1011.7.4 Enclosures under exterior stairways. There shall not be enclosed usable space under exterior exit *stairways* unless the space is completely enclosed in 1-hour fire-resistance-rated construction. The open space under exterior *stairways* shall not be used for any purpose.

1011.8 Vertical rise. A *flight* of *stairs* shall not have a vertical rise greater than 12 feet (3658 mm) between floor levels or landings.

Exception: Spiral *stairways* used as a *means of egress* from *technical production areas*.

1011.9 Curved stairways. Curved *stairways* with *winder* treads shall have treads and risers in accordance with Section 1011.5 and the smallest radius shall be not less than twice the minimum width or required capacity of the stairway.

Exception: The radius restriction shall not apply to curved *stairways* in Group R-3 and within individual dwelling units in Group R-2.

1011.10 Spiral stairways. *Spiral stairways* are permitted to be used as a component in the *means of egress* only within *dwelling units* or from a space not more than 250 square feet (23 m^2) in area and serving not more than five occupants, or from *technical production areas* in accordance with Section 410.5.

A *spiral stairway* shall have a $6^3/_4$-inch (171 mm) minimum clear tread depth at a point 12 inches (305 mm) from the narrow edge. The risers shall be sufficient to provide a headroom of 78 inches (1981 mm) minimum, but riser height shall not be more than $9^1/_2$ inches (241 mm). The minimum *stairway* clear width at and below the *handrail* shall be 26 inches (660 mm).

1011.11 Handrails. *Flights of stairways* shall have *handrails* on each side and shall comply with Section 1014. Where glass is used to provide the *handrail*, the *handrail* shall comply with Section 2407.

Exceptions:

1. *Flights of stairways* within *dwelling units* and *flights of spiral stairways* are permitted to have a *handrail* on one side only.
2. Decks, patios and walkways that have a single change in elevation where the landing depth on each side of the change of elevation is greater than what is required for a landing do not require *handrails*.
3. In Group R-3 occupancies, a change in elevation consisting of a single riser at an entrance or egress door does not require *handrails*.
4. Changes in room elevations of three or fewer risers within *dwelling units* and *sleeping units* in Groups R-2 and R-3 do not require *handrails*.
5. Where a platform lift is in a stationary position and the floor of the platform lift serves as the upper landing of a *stairway*, *handrails* shall not be

required on the *stairway*, provided that all of the following criteria are met:

5.1. The *stairway* contains not more than two risers.

5.2. A handhold, positioned horizontally or vertically, is located on one side of the *stairway* adjacent to the top landing.

5.3. The handhold is located not less than 34 inches (864 mm) and not more than 42 inches (1067 mm) above the bottom landing of the *stairway*.

5.4. The handhold gripping surface complies with Section 1014.3, and is not less than 4.5 inches (114 mm) in length.

1011.12 Stairway to roof. In buildings four or more stories above grade plane, one *stairway* shall extend to the roof surface unless the roof has a slope steeper than four units vertical in 12 units horizontal (33-percent slope).

Exception: Other than where required by Section 1011.12.1, in buildings without an occupied roof access to the roof from the top *story* shall be permitted to be by an *alternating tread device*, a ships ladder or a permanent ladder.

1011.12.1 Stairway to elevator equipment. Roofs and *penthouses* containing elevator equipment that must be accessed for maintenance are required to be accessed by a *stairway*.

1011.12.2 Roof access. Where a *stairway* is provided to a roof, access to the roof shall be provided through a *penthouse* complying with Section 1511.2.

Exception: In buildings without an occupied roof, access to the roof shall be permitted to be a roof hatch or trap door not less than 16 square feet (1.5 m^2) in area and having a minimum dimension of 2 feet (610 mm).

1011.13 Guards. *Guards* shall be provided along *stairways* and landings where required by Section 1015 and shall be constructed in accordance with Section 1015. Where the roof hatch opening providing the required access is located within 10 feet (3049 mm) of the roof edge, such roof access or roof edge shall be protected by *guards* installed in accordance with Section 1015.

1011.14 Alternating tread devices. *Alternating tread devices* are limited to an element of a *means of egress* in buildings of Groups F, H and S from a *mezzanine* not more than 250 square feet (23 m^2) in area and that serves not more than five occupants; in buildings of Group I-3 from a guard tower, observation station or control room not more than 250 square feet (23 m^2) in area and for access to unoccupied roofs. *Alternating tread devices* used as a *means of egress* shall not have a rise greater than 20 feet (6096 mm) between floor levels or landings.

1011.14.1 Handrails of alternating tread devices. *Handrails* shall be provided on both sides of *alternating tread devices* and shall comply with Section 1014.

1011.14.2 Treads of alternating tread devices. *Alternating tread devices* shall have a minimum tread depth of 5 inches (127 mm), a minimum projected tread depth of $8^1/_2$ inches (216 mm), a minimum tread width of 7 inches (178 mm) and a maximum riser height of $9^1/_2$ inches (241 mm). The tread depth shall be measured horizontally between the vertical planes of the foremost projections of adjacent treads. The riser height shall be measured vertically between the leading edges of adjacent treads. The riser height and tread depth provided shall result in an angle of ascent from the horizontal of between 50 and 70 degrees (0.87 and 1.22 rad). The initial tread of the device shall begin at the same elevation as the platform, landing or floor surface.

Exception: *Alternating tread devices* used as an element of a *means of egress* in buildings from a *mezzanine* area not more than 250 square feet (23 m^2) in area that serves not more than five occupants shall have a minimum tread depth of 3 inches (76 mm) with a minimum projected tread depth of $10^1/_2$ inches (267 mm). The rise to the next alternating tread surface shall not exceed 8 inches (203 mm).

1011.15 Ship's ladders. Ship's ladders are permitted to be used in Group I-3 as a component of a *means of egress* to and from control rooms or elevated facility observation stations not more than 250 square feet (23 m^2) with not more than three occupants and for access to unoccupied roofs. The minimum clear width at and below the *handrails* shall be 20 inches (508 mm). Ship's ladders shall be designed for the live loads indicated in Section 1607.17.

1011.15.1 Handrails of ship's ladders. *Handrails* shall be provided on both sides of ship's ladders.

1011.15.2 Treads of ship's ladders. Ship's ladders shall have a minimum tread depth of 5 inches (127 mm). The tread shall be projected such that the total of the tread depth plus the *nosing* projection is not less than $8^1/_2$ inches (216 mm). The maximum riser height shall be $9^1/_2$ inches (241 mm).

1011.16 Ladders. Permanent ladders shall not serve as a part of the *means of egress* from occupied spaces within a building. Permanent ladders shall be constructed in accordance with Section 306.5 of the *International Mechanical Code* and designed for the live loads indicated in Section 1607.17. Permanent ladders shall be permitted to provide access to the following areas:

1. Spaces frequented only by personnel for maintenance, repair or monitoring of equipment.
2. Nonoccupiable spaces accessed only by catwalks, crawl spaces, freight elevators or very narrow passageways.
3. Raised areas used primarily for purposes of security, life safety or fire safety including, but not limited to, observation galleries, prison guard towers, fire towers or lifeguard stands.
4. Elevated levels in Group U not open to the general public.

5. Nonoccupied roofs that are not required to have *stairway* access in accordance with Section 1011.12.1.
6. Where permitted to access equipment and appliances in accordance with Section 306.5 of the *International Mechanical Code*.

SECTION 1012
RAMPS

1012.1 Scope. The provisions of this section shall apply to *ramps* used as a component of a *means of egress*.

Exceptions:

1. Ramped *aisles* within assembly rooms or spaces shall comply with the provisions in Section 1030.
2. Curb ramps shall comply with ICC A117.1.
3. Vehicle ramps in parking garages for pedestrian *exit access* shall not be required to comply with Sections 1012.3 through 1012.10 where they are not an *accessible* route serving accessible parking spaces, other required accessible elements or part of an *accessible means of egress*.

1012.2 Slope. *Ramps* used as part of a *means of egress* shall have a running slope not steeper than 1 unit vertical in 12 units horizontal (8.3-percent slope). The slope of other pedestrian *ramps* shall not be steeper than 1 unit vertical in 8 units horizontal (12.5-percent slope).

1012.3 Cross slope. The slope measured perpendicular to the direction of travel of a *ramp* shall not be steeper than one unit vertical in 48 units horizontal (2-percent slope).

1012.4 Vertical rise. The rise for any *ramp* run shall be 30 inches (762 mm) maximum.

1012.5 Minimum dimensions. The minimum dimensions of *means of egress ramps* shall comply with Sections 1012.5.1 through 1012.5.3.

1012.5.1 Width and capacity. The minimum width and required capacity of a *means of egress ramp* shall be not less than that required for *corridors* by Section 1020.3. The clear width of a *ramp* between *handrails*, if provided, or other permissible projections shall be 36 inches (914 mm) minimum.

1012.5.2 Headroom. The minimum headroom in all parts of the *means of egress ramp* shall be not less than 80 inches (2032 mm) above the finished floor of the *ramp* run and any intermediate landings. The minimum clearance shall be maintained for the full width of the *ramp* and landing.

1012.5.3 Restrictions. *Means of egress ramps* shall not reduce in width in the direction of egress travel. Projections into the required *ramp* and landing width are prohibited. Doors opening onto a landing shall not reduce the clear width to less than 42 inches (1067 mm).

1012.6 Landings. *Ramps* shall have landings at the bottom and top of each *ramp*, points of turning, entrance, *exits* and at doors. Landings shall comply with Sections 1012.6.1 through 1012.6.5.

1012.6.1 Slope. Landings shall have a slope not steeper than one unit vertical in 48 units horizontal (2-percent slope) in any direction. Changes in level are not permitted.

1012.6.2 Width. The landing width shall be not less than the width of the widest *ramp* run adjoining the landing.

1012.6.3 Length. The landing length shall be 60 inches (1525 mm) minimum.

Exceptions:

1. In Group R-2 and R-3 individual dwelling and *sleeping units* that are not required to be *Accessible units*, *Type A units* or *Type B units* in accordance with Section 1108, landings are permitted to be 36 inches (914 mm) minimum.
2. Where the *ramp* is not a part of an *accessible* route, the length of the landing shall not be required to be more than 48 inches (1219 mm) in the direction of travel.

1012.6.4 Change in direction. Where changes in direction of travel occur at landings provided between *ramp* runs, the landing shall be 60 inches by 60 inches (1524 mm by 1524 mm) minimum.

Exception: In Group R-2 and R-3 individual *dwelling* or *sleeping units* that are not required to be *Accessible units*, *Type A units* or *Type B units* in accordance with Section 1108, landings are permitted to be 36 inches by 36 inches (914 mm by 914 mm) minimum.

1012.6.5 Doorways. Where doorways are located adjacent to a *ramp* landing, maneuvering clearances required by *ICC A117.1* are permitted to overlap the required landing area.

1012.7 Ramp construction. *Ramps* shall be built of materials consistent with the types permitted for the type of construction of the building, except that wood *handrails* shall be permitted for all types of construction.

1012.7.1 Ramp surface. The surface of *ramps* shall be of slip-resistant materials that are securely attached.

1012.7.2 Outdoor conditions. Outdoor *ramps* and outdoor approaches to *ramps* shall be designed so that water will not accumulate on walking surfaces.

1012.8 Handrails. *Ramps* with a rise greater than 6 inches (152 mm) shall have *handrails* on both sides. *Handrails* shall comply with Section 1014.

1012.9 Guards. *Guards* shall be provided where required by Section 1015 and shall be constructed in accordance with Section 1015.

1012.10 Edge protection. Edge protection complying with Section 1012.10.1 or 1012.10.2 shall be provided on each side of *ramp* runs and at each side of *ramp* landings.

Exceptions:

1. Edge protection is not required on *ramps* that are not required to have *handrails*, provided they have flared sides that comply with the *ICC A117.1* curb *ramp* provisions.
2. Edge protection is not required on the sides of *ramp* landings serving an adjoining *ramp* run or *stairway*.
3. Edge protection is not required on the sides of *ramp* landings having a vertical dropoff of not more than $^1/_2$ inch (12.7 mm) within 10 inches (254 mm) horizontally of the required landing area.

1012.10.1 Curb, rail, wall or barrier. A curb, rail, wall or barrier shall be provided to serve as edge protection. A curb shall be not less than 4 inches (102 mm) in height. Barriers shall be constructed so that the barrier prevents the passage of a 4-inch-diameter (102 mm) sphere, where any portion of the sphere is within 4 inches (102 mm) of the floor or ground surface.

1012.10.2 Extended floor or ground surface. The floor or ground surface of the *ramp* run or landing shall extend 12 inches (305 mm) minimum beyond the inside face of a *handrail* complying with Section 1014.

SECTION 1013
EXIT SIGNS

1013.1 Where required. Exits and *exit access* doors shall be marked by an *approved* exit sign readily visible from any direction of egress travel. The path of egress travel to *exits* and within *exits* shall be marked by readily visible exit signs to clearly indicate the direction of egress travel in cases where the exit or the path of egress travel is not immediately visible to the occupants. Intervening means of egress doors within *exits* shall be marked by exit signs. Exit sign placement shall be such that any point in an *exit access corridor* or *exit passageway* is within 100 feet (30 480 mm) or the *listed* viewing distance of the sign, whichever is less, from the nearest visible *exit* sign.

Exceptions:

1. Exit signs are not required in rooms or areas that require only one exit or *exit access*.
2. Main exterior exit doors or gates that are obviously and clearly identifiable as *exits* need not have exit signs where *approved* by the building official.
3. Exit signs are not required in occupancies in Group U and individual *sleeping units* or *dwelling units* in Group R-1, R-2 or R-3.
4. Exit signs are not required in dayrooms, sleeping rooms or *dormitories* in occupancies in Group I-3.
5. In occupancies in Groups A-4 and A-5, exit signs are not required on the seating side of vomitories or openings into seating areas where exit signs are provided in the concourse that are readily apparent from the vomitories. Egress lighting is provided to identify each vomitory or opening within the seating area in an emergency.

1013.2 Low-level exit signs in Group R-1. Where exit signs are required in Group R-1 occupancies by Section 1013.1, additional low-level exit signs shall be provided in all areas serving guest rooms in Group R-1 occupancies and shall comply with Section 1013.5.

The bottom of the sign shall be not less than 10 inches (254 mm) nor more than 18 inches (455 mm) above the floor level. The sign shall be flush mounted to the door or wall. Where mounted on the wall, the edge of the sign shall be within 4 inches (102 mm) of the door frame on the latch side.

1013.3 Illumination. Exit signs shall be internally or externally illuminated.

Exception: Tactile signs required by Section 1013.4 need not be provided with illumination.

1013.4 Raised character and braille exit signs. Where exit signs are provided at an *area of refuge* with *direct access* to a *stairway*, an exterior area for assisted rescue, an *exit stairway* or *ramp*, an *exit passageway, a horizontal exit* and the *exit discharge*, a sign stating "EXIT" in visual characters, raised characters and braille and complying with ICC A117.1 shall be provided.

1013.5 Internally illuminated exit signs. Electrically powered, *self-luminous* and *photoluminescent* exit signs shall be *listed* and labeled in accordance with UL 924 and shall be installed in accordance with the manufacturer's instructions and Chapter 27. Exit signs shall be illuminated at all times.

1013.6 Externally illuminated exit signs. Externally illuminated exit signs shall comply with Sections 1013.6.1 through 1013.6.3.

1013.6.1 Graphics. Every exit sign and directional exit sign shall have plainly legible letters not less than 6 inches (152 mm) high with the principal strokes of the letters not less than $^3/_4$ inch (19.1 mm) wide. The word "EXIT" shall have letters having a width not less than 2 inches (51 mm) wide, except the letter "I," and the minimum spacing between letters shall be not less than $^3/_8$ inch (9.5 mm). Signs larger than the minimum established in this section shall have letter widths, strokes and spacing in proportion to their height.

The word "EXIT" shall be in high contrast with the background and shall be clearly discernible when the means of exit sign illumination is or is not energized. If a chevron directional indicator is provided as part of the exit sign, the construction shall be such that the direction of the chevron directional indicator cannot be readily changed.

1013.6.2 Exit sign illumination. The face of an exit sign illuminated from an external source shall have an intensity of not less than 5 footcandles (54 lux).

1013.6.3 Power source. Exit signs shall be illuminated at all times. To ensure continued illumination for a duration of not less than 90 minutes in case of primary power loss, the sign illumination means shall be connected to an emergency power system provided from storage batteries, unit equipment or an on-site generator. The installation of the emergency power system shall be in accordance with Chapter 27. Group I-2, Condition 2 exit sign illumination shall not be provided by unit equipment batteries only.

Exception: *Approved* exit sign illumination types that provide continuous illumination independent of external power sources for a duration of not less than 90 minutes, in case of primary power loss, are not required to be connected to an emergency electrical system.

SECTION 1014
HANDRAILS

1014.1 Where required. *Handrails* serving *flights of stairways*, *ramps*, stepped *aisles* and ramped *aisles* shall be adequate in strength and attachment in accordance with Section 1607.9. *Handrails* required for flights of *stairways* by Section 1011.11 shall comply with Sections 1014.2 through 1014.9. *Handrails* required for *ramps* by Section 1012.8 shall comply with Sections 1014.2 through 1014.8. *Handrails* for stepped *aisles* and ramped *aisles* required by Section 1030.16 shall comply with Sections 1014.2 through 1014.8.

1014.2 Height. *Handrail* height, measured above *stair* tread *nosings*, or finish surface of *ramp* slope, shall be uniform, not less than 34 inches (864 mm) and not more than 38 inches (965 mm). *Handrail* height of *alternating tread devices* and ships ladders, measured above tread *nosings*, shall be uniform, not less than 30 inches (762 mm) and not more than 34 inches (864 mm).

Exceptions:

1. Where *handrail* fittings or bendings are used to provide continuous transition between flights, the fittings or bendings shall be permitted to exceed the maximum height.
2. In Group R-3 occupancies; within *dwelling units* in Group R-2 occupancies; and in Group U occupancies that are associated with a Group R-3 occupancy or associated with individual *dwelling units* in Group R-2 occupancies; where *handrail* fittings or bendings are used to provide continuous transition between flights, transition at *winder* treads, transition from *handrail* to guard, or where used at the start of a *flight*, the *handrail* height at the fittings or bendings shall be permitted to exceed the maximum height.
3. *Handrails* on top of a *guard* where permitted along stepped *aisles* and ramped *aisles* in accordance with Section 1030.16.

1014.3 Handrail graspability. Required *handrails* shall comply with Section 1014.3.1 or shall provide equivalent graspability.

Exception: In Group R-3 occupancies; within *dwelling units* in Group R-2 occupancies; and in Group U occupancies that are accessory to a Group R-3 occupancy or accessory to individual *dwelling units* in Group R-2 occupancies; *handrails* shall be Type I in accordance with Section 1014.3.1, Type II in accordance with Section 1014.3.2 or shall provide equivalent graspability.

1014.3.1 Type I. *Handrails* with a circular cross section shall have an outside diameter of not less than $1^1/_4$ inches (32 mm) and not greater than 2 inches (51 mm). Where the *handrail* is not circular, it shall have a perimeter dimension of not less than 4 inches (102 mm) and not greater than $6^1/_4$ inches (160 mm) with a maximum cross-sectional dimension of $2^1/_4$ inches (57 mm) and minimum cross-sectional dimension of 1 inch (25 mm). Edges shall have a minimum radius of 0.01 inch (0.25 mm).

1014.3.2 Type II. *Handrails* with a perimeter greater than $6^1/_4$ inches (160 mm) shall provide a graspable finger recess area on both sides of the profile. The finger recess shall begin within a distance of $^3/_4$ inch (19 mm) measured vertically from the tallest portion of the profile and achieve a depth of not less than $^5/_{16}$ inch (8 mm) within $^7/_8$ inch (22 mm) below the widest portion of the profile. This required depth shall continue for not less than $^3/_8$ inch (10 mm) to a level that is not less than $1^3/_4$ inches (45 mm) below the tallest portion of the profile. The width of the *handrail* above the recess shall be not less than $1^1/_4$ inches (32 mm) to not greater than $2^3/_4$ inches (70 mm). Edges shall have a minimum radius of 0.01 inch (0.25 mm).

1014.4 Continuity. *Handrail* gripping surfaces shall be continuous, without interruption by newel posts or other obstructions.

Exceptions:

1. *Handrails* within *dwelling units* are permitted to be interrupted by a newel post at a turn or landing.
2. Within a *dwelling unit*, the use of a volute, turnout, starting easing or starting newel is allowed over the lowest tread.
3. Handrail brackets or balusters attached to the bottom surface of the *handrail* that do not project horizontally beyond the sides of the *handrail* within $1^1/_2$ inches (38 mm) of the bottom of the *handrail* shall not be considered obstructions. For each $^1/_2$ inch (12.7 mm) of additional *handrail* perimeter dimension above 4 inches (102 mm), the vertical clearance dimension of $1^1/_2$ inches (38 mm) shall be permitted to be reduced by $^1/_8$ inch (3.2 mm).
4. Where *handrails* are provided along walking surfaces with slopes not steeper than 1:20, the bottoms of the *handrail* gripping surfaces shall be permitted to be obstructed along their entire length

where they are integral to crash rails or bumper *guards*.

5. *Handrails* serving stepped *aisles* or ramped *aisles* are permitted to be discontinuous in accordance with Section 1030.16.1.

1014.5 Fittings. *Handrails* shall not rotate within their fittings.

1014.6 Handrail extensions. *Handrails* shall return to a wall, *guard* or the walking surface or shall be continuous to the *handrail* of an adjacent *flight of stairs* or *ramp* run. Where *handrails* are not continuous between flights, the *handrails* shall extend horizontally not less than 12 inches (305 mm) beyond the top riser and continue to slope for the depth of one tread beyond the bottom riser. At *ramps* where *handrails* are not continuous between runs, the *handrails* shall extend horizontally above the landing 12 inches (305 mm) minimum beyond the top and bottom of *ramp* runs. The extensions of *handrails* shall be in the same direction of the flights of *stairs* at *stairways* and the *ramp* runs at *ramps*.

Exceptions:

1. *Handrails* within a *dwelling unit* that is not required to be *accessible* need extend only from the top riser to the bottom riser.
2. *Handrails* serving *aisles* in rooms or spaces used for assembly purposes are permitted to comply with the *handrail* extensions in accordance with Section 1030.16.
3. *Handrails* for *alternating tread devices* and ships ladders are permitted to terminate at a location vertically above the top and bottom risers. *Handrails* for *alternating tread devices* are not required to be continuous between flights or to extend beyond the top or bottom risers.

1014.7 Clearance. Clear space between a *handrail* and a wall or other surface shall be not less than $1^1/_2$ inches (38 mm). A *handrail* and a wall or other surface adjacent to the *handrail* shall be free of any sharp or abrasive elements.

1014.8 Projections. On *ramps* and on ramped *aisles* that are part of an *accessible* route, the clear width between *handrails* shall be 36 inches (914 mm) minimum. Projections into the required width of *aisles*, *stairways* and *ramps* at each side shall not exceed $4^1/_2$ inches (114 mm) at or below the *handrail* height. Projections into the required width shall not be limited above the minimum headroom height required in Section 1011.3. Projections due to intermediate *handrails* shall not constitute a reduction in the egress width. Where a pair of intermediate *handrails* are provided within the *stairway* width without a walking surface between the pair of intermediate *handrails* and the distance between the pair of intermediate *handrails* is greater than 6 inches (152 mm), the available egress width shall be reduced by the distance between the closest edges of each such intermediate pair of *handrails* that is greater than 6 inches (152 mm).

1014.9 Intermediate handrails. *Stairways* shall have intermediate *handrails* located in such a manner that all portions of the *stairway* minimum width or required capacity are within 30 inches (762 mm) of a *handrail*. On monumental *stairs*, *handrails* shall be located along the most direct path of egress travel.

SECTION 1015 GUARDS

1015.1 General. *Guards* shall comply with the provisions of Sections 1015.2 through 1015.7. Operable windows with sills located more than 72 inches (1829 mm) above finished grade or other surface below shall comply with Section 1015.8.

1015.2 Where required. *Guards* shall be located along open-sided walking surfaces, including *mezzanines*, equipment platforms, *aisles*, *stairs*, *ramps* and landings that are located more than 30 inches (762 mm) measured vertically to the floor or grade below at any point within 36 inches (914 mm) horizontally to the edge of the open side. *Guards* shall be adequate in strength and attachment in accordance with Section 1607.9.

Exceptions: *Guards* are not required for the following locations:

1. On the loading side of loading docks or piers.
2. On the audience side of *stages* and raised *platforms*, including *stairs* leading up to the *stage* and raised *platforms*.
3. On raised *stage* and *platform* floor areas, such as runways, *ramps* and side *stages* used for entertainment or presentations.
4. At vertical openings in the performance area of *stages* and *platforms*.
5. At elevated walking surfaces appurtenant to *stages* and *platforms* for access to and utilization of special lighting or equipment.
6. Along vehicle service pits not accessible to the public.
7. In assembly seating areas at cross *aisles* in accordance with Section 1030.17.2.
8. On the loading side of station platforms on fixed guideway transit or passenger rail systems.

1015.2.1 Glazing. Where glass is used to provide a *guard* or as a portion of the *guard* system, the *guard* shall comply with Section 2407. Where the glazing provided does not meet the strength and attachment requirements of Section 1607.9, complying *guards* shall be located along glazed sides of open-sided walking surfaces.

1015.3 Height. Required *guards* shall be not less than 42 inches (1067 mm) high, measured vertically as follows:

1. From the adjacent walking surfaces.
2. On *stairways* and stepped *aisles*, from the line connecting the leading edges of the tread *nosings*.
3. On *ramps* and ramped *aisles*, from the *ramp* surface at the guard.

Exceptions:

1. For occupancies in Group R-3 not more than three stories above grade in height and within individual

dwelling units in occupancies in Group R-2 not more than three stories above grade in height with separate *means of egress*, required *guards* shall be not less than 36 inches (914 mm) in height measured vertically above the adjacent walking surfaces.

2. For occupancies in Group R-3, and within individual *dwelling units* in occupancies in Group R-2, *guards* on the open sides of *stairs* shall have a height not less than 34 inches (864 mm) measured vertically from a line connecting the leading edges of the treads.
3. For occupancies in Group R-3, and within individual *dwelling units* in occupancies in Group R-2, where the top of the *guard* serves as a *handrail* on the open sides of *stairs*, the top of the *guard* shall be not less than 34 inches (864 mm) and not more than 38 inches (965 mm) measured vertically from a line connecting the leading edges of the treads.
4. The *guard* height in assembly seating areas shall comply with Section 1030.17 as applicable.
5. Along *alternating tread devices* and ships ladders, *guards* where the top rail serves as a *handrail* shall have height not less than 30 inches (762 mm) and not more than 34 inches (864 mm), measured vertically from the leading edge of the device tread *nosing*.
6. In Group F occupancies where *exit access stairways* serve fewer than three stories and such *stairways* are not open to the public, and where the top of the *guard* also serves as a *handrail*, the top of the *guard* shall be not less than 34 inches (864 mm) and not more than 38 inches (965 mm) measured vertically from a line connecting the leading edges of the treads.

1015.4 Opening limitations. Required *guards* shall not have openings that allow passage of a sphere 4 inches (102 mm) in diameter from the walking surface to the required *guard* height.

Exceptions:

1. From a height of 36 inches (914 mm) to 42 inches (1067 mm), *guards* shall not have openings that allow passage of a sphere $4^3/_8$ inches (111 mm) in diameter.
2. The triangular openings at the open sides of a *stair*, formed by the riser, tread and bottom rail shall not allow passage of a sphere 6 inches (152 mm) in diameter.
3. At elevated walking surfaces for access to and use of electrical, mechanical or plumbing systems or equipment, *guards* shall not have openings that allow passage of a sphere 21 inches (533 mm) in diameter.
4. In areas that are not open to the public within occupancies in Group I-3, F, H or S, and for *alternating tread devices* and ships ladders, *guards* shall not have openings that allow passage of a sphere 21 inches (533 mm) in diameter.
5. In assembly seating areas, *guards* required at the end of *aisles* in accordance with Section 1030.17.4 shall not have openings that allow passage of a sphere 4 inches (102 mm) in diameter up to a height of 26 inches (660 mm). From a height of 26 inches (660 mm) to 42 inches (1067 mm) above the adjacent walking surfaces, *guards* shall not have openings that allow passage of a sphere 8 inches (203 mm) in diameter.
6. Within individual *dwelling units* and *sleeping units* in Group R-2 and R-3 occupancies, *guards* on the open sides of *stairs* shall not have openings that allow passage of a sphere $4^3/_8$ (111 mm) inches in diameter.

1015.5 Screen porches. Porches and decks that are enclosed with insect screening shall be provided with *guards* where the walking surface is located more than 30 inches (762 mm) above the floor or grade below.

1015.6 Mechanical equipment, systems and devices. *Guards* shall be provided where various components that require service are located within 10 feet (3048 mm) of a roof edge or open side of a walking surface and such edge or open side is located more than 30 inches (762 mm) above the floor, roof or grade below. The *guard* shall extend not less than 30 inches (762 mm) beyond each end of such components. The *guard* shall be constructed so as to prevent the passage of a sphere 21 inches (533 mm) in diameter.

Exception: *Guards* are not required where personal fall arrest anchorage connector devices that comply with ANSI/ASSE Z 359.1 are installed.

1015.7 Roof access. *Guards* shall be provided where the roof hatch opening is located within 10 feet (3048 mm) of a roof edge or open side of a walking surface and such edge or open side is located more than 30 inches (762 mm) above the floor, roof or grade below. The *guard* shall extend not less than 30 inches (762 mm) beyond each end of the hatch parallel to the roof edge. The *guard* shall be constructed so as to prevent the passage of a sphere 21 inches (533 mm) in diameter.

Exception: *Guards* are not required where personal fall arrest anchorage connector devices that comply with ANSI/ASSE Z 359.1 are installed.

1015.8 Window openings. Windows in Group R-2 and R-3 buildings including *dwelling units*, where the bottom of the clear opening of an operable window is located less than 36 inches (914 mm) above the finished floor and more than 72 inches (1829 mm) above the finished grade or other surface below on the exterior of the building, shall comply with one of the following:

1. Operable windows where the top of the sill of the opening is located more than 75 feet (22 860 mm) above the finished grade or other surface below and that are provided with window fall prevention devices that comply with ASTM F2006.

2. Operable windows where the openings will not allow a 4-inch-diameter (102 mm) sphere to pass through the opening when the window is in its largest opened position.
3. Operable windows where the openings are provided with window fall prevention devices that comply with ASTM F2090.
4. Operable windows that are provided with window opening control devices that comply with Section 1015.8.1.

1015.8.1 Window opening control devices. Window opening control devices shall comply with ASTM F2090. The window opening control device, after operation to release the control device allowing the window to fully open, shall not reduce the minimum net clear opening area of the window unit to less than the area required by Section 1031.3.1.

SECTION 1016
EXIT ACCESS

1016.1 General. The *exit access* shall comply with the applicable provisions of Sections 1003 through 1015. *Exit access* arrangement shall comply with Sections 1016 through 1021.

1016.2 Egress through intervening spaces. Egress through intervening spaces shall comply with this section.

1. *Exit access* through an enclosed elevator lobby is permitted. Where access to two or more exits or exit access doorways is required in Section 1006.2.1, access to not less than one of the required *exits* shall be provided without travel through the enclosed elevator lobbies required by Section 3006. Where the path of *exit access* travel passes through an enclosed elevator lobby, the level of protection required for the enclosed elevator lobby is not required to be extended to the *exit* unless direct access to an *exit* is required by other sections of this code.
2. Egress from a room or space shall not pass through adjoining or intervening rooms or areas, except where such adjoining rooms or areas and the area served are accessory to one or the other, are not a Group H occupancy and provide a discernible path of egress travel to an exit.

   Exception: *Means of egress* are not prohibited through adjoining or intervening rooms or spaces in a Group H, S or F occupancy where the adjoining or intervening rooms or spaces are the same or a lesser hazard occupancy group.
3. An *exit access* shall not pass through a room that can be locked to prevent egress.
4. *Means of egress* from *dwelling units* or sleeping areas shall not lead through other sleeping areas, toilet rooms or bathrooms.
5. Egress shall not pass through kitchens, storage rooms, closets or spaces used for similar purposes.

   Exceptions:

   1. *Means of egress* are not prohibited through a kitchen area serving adjoining rooms constituting part of the same *dwelling unit* or *sleeping unit.*
   2. *Means of egress* are not prohibited through stockrooms in Group M occupancies where all of the following are met:

      2.1. The stock is of the same hazard classification as that found in the main retail area.

      2.2. Not more than 50 percent of the *exit access* is through the stockroom.

      2.3. The stockroom is not subject to locking from the egress side.

      2.4. There is a demarcated, minimum 44-inch-wide (1118 mm) *aisle* defined by full- or partial-height fixed walls or similar construction that will maintain the required width and lead directly from the retail area to the exit without obstructions.

1016.2.1 Multiple tenants. Where more than one tenant occupies any one floor of a building or structure, each tenant space, *dwelling unit* and *sleeping unit* shall be provided with access to the required *exits* without passing through adjacent tenant spaces, *dwelling units* and *sleeping units*.

Exception: The *means of egress* from a smaller tenant space shall not be prohibited from passing through a larger adjoining tenant space where such rooms or spaces of the smaller tenant occupy less than 10 percent of the area of the larger tenant space through which they pass; are the same or similar occupancy group; a discernible path of egress travel to an *exit* is provided; and the *means of egress* into the adjoining space is not subject to locking from the egress side. A required *means of egress* serving the larger tenant space shall not pass through the smaller tenant space or spaces.

SECTION 1017
EXIT ACCESS TRAVEL DISTANCE

1017.1 General. Travel distance within the *exit access* portion of the *means of egress* system shall be in accordance with this section.

1017.2 Limitations. *Exit access* travel distance shall not exceed the values given in Table 1017.2.

TABLE 1017.2
EXIT ACCESS TRAVEL DISTANCE[a]

OCCUPANCY	WITHOUT SPRINKLER SYSTEM (feet)	WITH SPRINKLER SYSTEM (feet)
A, E, F-1, M, R, S-1	200[e]	250[b]
I-1	Not Permitted	250[b]
B	200	300[c]
F-2, S-2, U	300	400[c]
H-1	Not Permitted	75[d]
H-2	Not Permitted	100[d]
H-3	Not Permitted	150[d]
H-4	Not Permitted	175[d]
H-5	Not Permitted	200[c]
I-2, I-3	Not Permitted	200[c]
I-4	150	200[c]

For SI: 1 foot = 304.8 mm.

a. See the following sections for modifications to *exit access* travel distance requirements:
   - Section 402.8: For the distance limitation in malls.
   - Section 407.4: For the distance limitation in Group I-2.
   - Sections 408.6.1 and 408.8.1: For the distance limitations in Group I-3.
   - Section 411.2: For the distance limitation in special amusement areas.
   - Section 412.6: For the distance limitations in aircraft manufacturing facilities.
   - Section 1006.2.2.2: For the distance limitation in refrigeration machinery rooms.
   - Section 1006.2.2.3: For the distance limitation in refrigerated rooms and spaces.
   - Section 1006.3.4: For buildings with one exit.
   - Section 1017.2.2: For increased distance limitation in Groups F-1 and S-1.
   - Section 1030.7: For increased limitation in assembly seating.
   - Section 3103.4: For temporary structures.
   - Section 3104.9: For pedestrian walkways.

b. Buildings equipped throughout with an automatic sprinkler system in accordance with Section 903.3.1.1 or 903.3.1.2. See Section 903 for occupancies where automatic sprinkler systems are permitted in accordance with Section 903.3.1.2.

c. Buildings equipped throughout with an automatic sprinkler system in accordance with Section 903.3.1.1.

d. Group H occupancies equipped throughout with an automatic sprinkler system in accordance with Section 903.2.5.1.

e. Group R-3 and R-4 buildings equipped throughout with an automatic sprinkler system in accordance with Section 903.3.1.3. See Section 903.2.8 for occupancies where automatic sprinkler systems are permitted in accordance with Section 903.3.1.3.

1017.2.1 Exterior egress balcony increase. *Exit access* travel distances specified in Table 1017.2 shall be increased up to an additional 100 feet (30 480 mm) provided that the last portion of the *exit access* leading to the exit occurs on an exterior egress balcony constructed in accordance with Section 1021. The length of such balcony shall be not less than the amount of the increase taken.

1017.2.2 Groups F-1 and S-1 increase. The maximum *exit access* travel distance shall be 400 feet (122 m) in Group F-1 or S-1 occupancies where all of the following conditions are met:

1. The portion of the building classified as Group F-1 or S-1 is limited to one *story* in height.
2. The minimum height from the finished floor to the bottom of the ceiling or roof slab or deck is 24 feet (7315 mm).
3. The building is equipped throughout with an *automatic sprinkler system* in accordance with Section 903.3.1.1.

1017.3 Measurement. *Exit access* travel distance shall be measured from the most remote point of each room, area or space along the natural and unobstructed path of horizontal and vertical egress travel to the entrance to an *exit*. Where more than one exit is required, *exit* access travel distance shall be measured to the nearest exit.

Exceptions:

1. In *open parking garages*, *exit access* travel distance is permitted to be measured to the closest riser of an *exit access stairway* or the closest slope of an *exit access ramp*.
2. In smoke protected seating and open air assembly seating, exit access travel distance shall be measured in accordance with Section 1030.7.

1017.3.1 Exit access stairways and ramps. Travel distance on *exit access stairways* or *ramps* shall be included in the *exit access* travel distance measurement. The measurement along *stairways* shall be made on a plane parallel and tangent to the *stair* tread *nosings* in the center of the *stair* and landings. The measurement along *ramps* shall be made on the walking surface in the center of the *ramp* and landings.

1017.3.2 Atriums. Exit acess travel distance for areas open to an *atrium* shall comply with the requirements of Sections 1017.3.2.1 through 1017.3.2.3.

1017.3.2.1 Egress not through the atrium. Where required access to the exits is not through the *atrium*, exit access travel distance shall comply with Section 1017.2.

1017.3.2.2 Exit access travel distance at the level of exit discharge. Where the path of egress travel is through an *atrium* space, exit access travel distance at the *level of exit discharge* shall be determined in accordance with Section 1017.2.

1017.3.2.3 Exit access travel distance at other than the level of exit discharge. Where the path of egress travel is not at the *level of exit discharge* from the *atrium*, that portion of the total permitted exit access travel distance that occurs within the *atrium* shall be not greater than 200 feet (60 960 mm).

SECTION 1018 AISLES

1018.1 General. *Aisles* and *aisle accessways* serving as a portion of the *exit access* in the *means of egress* system shall comply with the requirements of this section. *Aisles* or *aisle accessways* shall be provided from all occupied portions of the *exit access* that contain seats, tables, furnishings, displays and similar fixtures or equipment. The minimum width or required capacity of *aisles* shall be unobstructed.

Exception: Encroachments complying with Section 1005.7.

1018.2 Aisles in assembly spaces. *Aisles* and *aisle accessways* serving a room or space used for assembly purposes shall comply with Section 1030.

1018.3 Aisles in Groups B and M. In Group B and M occupancies, the minimum clear aisle width shall be determined by Section 1005.1 for the *occupant load* served, but shall be not less than that required for *corridors* by Section 1020.3.

Exception: Nonpublic *aisles* serving less than 50 people and not required to be *accessible* by Chapter 11 need not exceed 28 inches (711 mm) in width.

1018.4 Aisle accessways in Group M. An *aisle accessway* shall be provided on not less than one side of each element within the *merchandise pad*. The minimum clear width for an *aisle accessway* not required to be *accessible* shall be 30 inches (762 mm). The required clear width of the *aisle accessway* shall be measured perpendicular to the elements and merchandise within the *merchandise pad*. The 30-inch (762 mm) minimum clear width shall be maintained to provide a path to an adjacent *aisle* or *aisle accessway*. The *common path of egress travel* shall not exceed 30 feet (9144 mm) from any point in the *merchandise pad*.

Exception: For areas serving not more than 50 occupants, the *common path of egress travel* shall not exceed 75 feet (22 860 mm).

1018.5 Aisles in other than assembly spaces and Groups B and M. In other than rooms or spaces used for assembly purposes and Group B and M occupancies, the minimum clear aisle capacity shall be determined by Section 1005.1 for the *occupant load* served, but the width shall be not less than that required for *corridors* by Section 1020.3.

Exception: Nonpublic *aisles* serving less than 50 people and not required to be *accessible* by Chapter 11 need not exceed 28 inches (711 mm) in width.

SECTION 1019 EXIT ACCESS STAIRWAYS AND RAMPS

1019.1 General. *Exit access stairways* and *ramps* serving as an *exit access* component in a *means of egress* system shall comply with the requirements of this section. The number of stories connected by *exit access stairways* and *ramps* shall include *basements*, but not *mezzanines*.

1019.2 All occupancies. *Exit access stairways* and *ramps* that serve floor levels within a single *story* are not required to be enclosed.

1019.3 Occupancies other than Groups I-2 and I-3. In other than Group I-2 and I-3 occupancies, floor openings containing *exit access stairways* or *ramps* shall be enclosed with a shaft enclosure constructed in accordance with Section 713.

Exceptions:

1. *Exit access stairways* and *ramps* that serve or atmospherically communicate between only two adjacent stories. Such interconnected stories shall not be open to other stories.
2. In Group R-1, R-2 or R-3 occupancies, *exit access stairways* and *ramps* connecting four stories or less serving and contained within an individual dwelling unit or sleeping unit or live/work unit.
3. *Exit access stairways* serving and contained within a Group R-3 congregate residence or a Group R-4 facility are not required to be enclosed.
4. *Exit access stairways* and *ramps* in buildings equipped throughout with an automatic sprinkler system in accordance with Section 903.3.1.1, where the area of the vertical opening between stories does not exceed twice the horizontal projected area of the stairway or *ramp* and the opening is protected by a draft curtain and closely spaced sprinklers in accordance with NFPA 13. In other than Group B and M occupancies, this provision is limited to openings that do not connect more than four stories.
5. *Exit access stairways* and *ramps* within an *atrium* complying with the provisions of Section 404.
6. *Exit access stairways* and *ramps* in *open parking garages* that serve only the parking garage.
7. *Exit access stairways* and *ramps* serving smoke-protected or *open-air assembly seating* complying with the exit access travel distance requirements of Section 1030.7.
8. *Exit access stairways* and *ramps* between the balcony, gallery or press box and the main assembly floor in occupancies such as theaters, *places of religious worship*, auditoriums and sports facilities.
9. Exterior *exit access stairways* or *ramps* between occupied roofs.

1019.4 Group I-2 and I-3 occupancies. In Group I-2 and I-3 occupancies, floor openings between stories containing *exit access stairways* or *ramps* are required to be enclosed with a shaft enclosure constructed in accordance with Section 713.

Exception: In Group I-3 occupancies, *exit access stairways* or *ramps* constructed in accordance with Section 408 are not required to be enclosed.

SECTION 1020 CORRIDORS

1020.1 General. Corridors serving as an exit access component in a *means of egress* system shall comply with the requirements of Sections 1020.2 through 1020.7.

1020.2 Construction. Corridors shall be fire-resistance rated in accordance with Table 1020.2. The *corridor* walls required to be fire-resistance rated shall comply with Section 708 for *fire partitions*.

Exceptions:

1. A *fire-resistance rating* is not required for *corridors* in an occupancy in Group E where each room that is used for instruction has not less than one door opening directly to the exterior and rooms for assembly purposes have not less than one-half of the required means of egress doors opening directly to the exterior. Exterior doors specified in this exception are required to be at ground level.
2. A *fire-resistance rating* is not required for *corridors* contained within a *dwelling unit* or *sleeping unit* in an occupancy in Groups I-1 and R.
3. A *fire-resistance rating* is not required for *corridors* in *open parking garages*.
4. A *fire-resistance rating* is not required for *corridors* in an occupancy in Group B that is a space requiring only a single *means of egress* complying with Section 1006.2.
5. *Corridors* adjacent to the *exterior walls* of buildings shall be permitted to have unprotected openings on unrated *exterior walls* where unrated walls are permitted by Table 705.5 and unprotected openings are permitted by Table 705.8.

TABLE 1020.2 CORRIDOR FIRE-RESISTANCE RATING

OCCUPANCY	OCCUPANT LOAD SERVED BY CORRIDOR	REQUIRED FIRE-RESISTANCE RATING (hours)	
		Without sprinkler system	With sprinkler system
H-1, H-2, H-3	All	Not Permitted	1[c]
H-4, H-5	Greater than 30	Not Permitted	1[c]
A, B, E, F, M, S, U	Greater than 30	1	0
R	Greater than 10	Not Permitted	0.5[c]/1[d]
I-2[a]	All	Not Permitted	0
I-1, I-3	All	Not Permitted	1[b, c]
I-4	All	1	0

a. For requirements for occupancies in Group I-2, see Sections 407.2 and 407.3.

b. For a reduction in the fire-resistance rating for occupancies in Group I-3, see Section 408.8.

c. Buildings equipped throughout with an automatic sprinkler system in accordance with Section 903.3.1.1 or 903.3.1.2 where allowed.

d. Group R-3 and R-4 buildings equipped throughout with an automatic sprinkler system in accordance with Section 903.3.1.3. See Section 903.2.8 for occupancies where automatic sprinkler systems are permitted in accordance with Section 903.3.1.3.

1020.2.1 Hoistway opening protection. Elevator hoistway openings shall be protected in accordance with Section 3006.2.1.

1020.3 Width and capacity. The required capacity of *corridors* shall be determined as specified in Section 1005.1, but the minimum width shall be not less than that specified in Table 1020.3.

Exception: In Group I-2 occupancies, *corridors* are not required to have a clear width of 96 inches (2438 mm) in areas where there will not be stretcher or bed movement for access to care or as part of the *defend-in-place* strategy.

TABLE 1020.3 MINIMUM CORRIDOR WIDTH

OCCUPANCY	MINIMUM WIDTH (inches)
Any facility not listed in this table	44
Access to and utilization of mechanical, plumbing or electrical systems or equipment	24
With an occupant load of less than 50	36
Within a dwelling unit	36
In Group E with a corridor having an occupant load of 100 or more	72
In corridors and areas serving stretcher traffic in ambulatory care facilities	72
Group I-2 in areas where required for bed movement	96

For SI: 1 inch = 25.4 mm.

1020.4 Obstruction. The minimum width or required capacity of *corridors* shall be unobstructed.

Exception: Encroachments complying with Section 1005.7.

1020.5 Dead ends. Where more than one *exit* or *exit access doorway* is required, the *exit access* shall be arranged such that dead-end *corridors* do not exceed 20 feet (6096 mm) in length.

Exceptions:

1. In Group I-3, Condition 2, 3 or 4, occupancies, the dead end in a *corridor* shall not exceed 50 feet (15 240 mm).
2. In occupancies in Groups B, E, F, I-1, M, R-1, R-2, S and U, where the building is equipped throughout with an *automatic sprinkler system* in accordance with Section 903.3.1.1, the length of the dead-end *corridors* shall not exceed 50 feet (15 240 mm).
3. A dead-end *corridor* shall not be limited in length where the length of the dead-end *corridor* is less than 2.5 times the least width of the dead-end *corridor*.
4. In Group I-2, Condition 2 occupancies, the length of dead-end *corridors* that do not serve patient rooms or patient treatment spaces shall not exceed 30 feet (9144 mm).

1020.6 Air movement in corridors. *Corridors* shall not serve as supply, return, exhaust, relief or ventilation air ducts.

Exceptions:

1. Use of a *corridor* as a source of makeup air for exhaust systems in rooms that open directly onto such corridors, including toilet rooms, bathrooms, dressing rooms, smoking lounges and janitor closets, shall be permitted, provided that each such *corridor* is directly supplied with outdoor air at a rate greater than the rate of makeup air taken from the corridor.
2. Where located within a *dwelling unit*, the use of *corridors* for conveying return air shall not be prohibited.
3. Where located within tenant spaces of 1,000 square feet (93 m^2) or less in area, utilization of *corridors* for conveying return air is permitted.
4. Transfer air movement required to maintain the pressurization difference within health care facilities in accordance with ASHRAE 170.

1020.6.1 Corridor ceiling. Use of the space between the corridor ceiling and the floor or roof structure above as a return air plenum is permitted for one or more of the following conditions:

1. The *corridor* is not required to be of *fire-resistance-rated* construction.
2. The *corridor* is separated from the plenum by *fire-resistance-rated* construction.
3. The air-handling system serving the *corridor* is shut down upon activation of the air-handling unit smoke detectors required by the *International Mechanical Code*.
4. The air-handling system serving the *corridor* is shut down upon detection of sprinkler water flow where the building is equipped throughout with an automatic sprinkler system.
5. The space between the corridor ceiling and the floor or roof structure above the *corridor* is used as a component of an *approved* engineered smoke control system.

1020.7 Corridor continuity. *Fire-resistance-rated corridors* shall be continuous from the point of entry to an *exit*, and shall not be interrupted by intervening rooms. Where the path of egress travel within a *fire-resistance-rated corridor* to the *exit* includes travel along unenclosed *exit access stairways* or *ramps*, the *fire-resistance rating* shall be continuous for the length of the *stairway* or *ramp* and for the length of the connecting *corridor* on the adjacent floor leading to the exit.

Exceptions:

1. Foyers, lobbies or reception rooms constructed as required for *corridors* shall not be construed as intervening rooms.
2. Enclosed elevator lobbies as permitted by Item 1 of Section 1016.2 shall not be construed as intervening rooms.

SECTION 1021 EGRESS BALCONIES

1021.1 General. Balconies used for egress purposes shall conform to the same requirements as *corridors* for minimum width, required capacity, headroom, dead ends and projections.

1021.2 Wall separation. Exterior egress balconies shall be separated from the interior of the building by walls and opening protectives as required for *corridors*.

Exception: Separation is not required where the exterior egress balcony is served by not less than two *stairways* and a dead-end travel condition does not require travel past an unprotected opening to reach a *stairway*.

1021.3 Openness. The long side of an egress balcony shall be not less than 50 percent open, and the open area above the *guards* shall be so distributed as to minimize the accumulation of smoke or toxic gases.

1021.4 Location. Exterior egress balconies shall have a minimum *fire separation distance* of 10 feet (3048 mm) measured at right angles from the exterior edge of the egress balcony to the following:

1. Adjacent *lot lines*.
2. Other portions of the building.
3. Other buildings on the same lot unless the adjacent building *exterior walls* and openings are protected in accordance with Section 705 based on *fire separation distance*.

For the purposes of this section, other portions of the building shall be treated as separate buildings.

SECTION 1022 EXITS

1022.1 General. *Exits* shall comply with Sections 1022 through 1027 and the applicable requirements of Sections 1003 through 1015. An *exit* shall not be used for any purpose that interferes with its function as a *means of egress*. Once a given level of *exit* protection is achieved, such level of protection shall not be reduced until arrival at the *exit discharge*. Exits shall be continuous from the point of entry into the *exit* to the *exit discharge*.

1022.2 Exterior exit doors. Buildings or structures used for human occupancy shall have not less than one exterior door that meets the requirements of Section 1010.1.1.

1022.2.1 Detailed requirements. Exterior *exit* doors shall comply with the applicable requirements of Section 1010.1.

1022.2.2 Arrangement. Exterior *exit* doors shall lead directly to the *exit discharge* or the *public way*.

SECTION 1023 INTERIOR EXIT STAIRWAYS AND RAMPS

1023.1 General. *Interior exit stairways* and *ramps* serving as an exit component in a *means of egress* system shall comply with the requirements of this section. *Interior exit stairways* and *ramps* shall be enclosed and lead directly to the exterior of the building or shall be extended to the exterior of the building with an *exit passageway* conforming to the requirements of Section 1024, except as permitted in Section 1028.2. An *interior exit stairway* or *ramp* shall not be used for any purpose other than as a *means of egress* and a *circulation path.*

1023.2 Construction. Enclosures for interior exit *stairways* and *ramps* shall be constructed as *fire barriers* in accordance with Section 707 or *horizontal assemblies* constructed in accordance with Section 711, or both. *Interior exit stairway* and *ramp* enclosures shall have a *fire-resistance rating* of not less than 2 hours where connecting four stories or more and not less than 1 hour where connecting less than four stories. The number of stories connected by the interior exit *stairways* or *ramps* shall include any *basements*, but not any *mezzanines*. Enclosures for interior *exit stairways* and *ramps* shall have a *fire-resistance rating* not less than the floor assembly penetrated, but need not exceed 2 hours.

Exceptions:

1. *Interior exit stairways* and *ramps* in Group I-3 occupancies in accordance with the provisions of Section 408.3.8.
2. *Interior exit stairways* within an *atrium* enclosed in accordance with Section 404.6.
3. Interior exit stairways in accordance with Section 510.2.

1023.3 Termination. *Interior exit stairways* and *ramps* shall terminate at an *exit discharge* or a *public way.*

Exception: A combination of *interior exit stairways, interior exit ramps* and *exit passageways*, constructed in accordance with Sections 1023.2, 1023.3.1 and 1024, respectively, and forming a continuous protected enclosure, shall be permitted to extend an *interior exit stairway* or *ramp* to the *exit discharge* or a *public way.*

1023.3.1 Extension. Where *interior exit stairways* and *ramps* are extended to an *exit discharge* or a *public way* by an *exit passageway*, the *interior exit stairway* and *ramp* shall be separated from the *exit passageway* by a *fire barrier* constructed in accordance with Section 707 or a *horizontal assembly* constructed in accordance with Section 711, or both. The *fire-resistance rating* shall be not less than that required for the *interior exit stairway* and *ramp*. A *fire door assembly* complying with Section 716 shall be installed in the *fire barrier* to provide a *means of egress* from the *interior exit stairway* and *ramp* to the *exit passageway*. Openings in the *fire barrier* other than the *fire door assembly* are prohibited. Penetrations of the *fire barrier* are prohibited.

Exceptions:

1. Penetrations of the *fire barrier* in accordance with Section 1023.5 shall be permitted.
2. Separation between an *interior exit stairway* or *ramp* and the *exit passageway* extension shall not be required where there are no openings into the *exit passageway* extension.
3. Separation between an *interior exit stairway* or *ramp* and the *exit passageway* extension shall not be required where the *interior exit stairway* and the *exit passageway* extension are pressurized in accordance with Section 909.20.5.

1023.4 Openings. *Interior exit stairway* and *ramp* opening protectives shall be in accordance with the requirements of Section 716.

Openings in *interior exit stairways* and *ramps* other than unprotected exterior openings shall be limited to those required for *exit access* to the enclosure from normally occupied spaces and for egress from the enclosure.

Elevators shall not open into *interior exit stairways* and *ramps.*

1023.5 Penetrations. Penetrations into or through interior exit *stairways* and *ramps* are prohibited except for the following:

1. Equipment and ductwork necessary for independent ventilation or pressurization.
2. *Fire protection systems.*
3. Security systems.
4. Two-way communication systems.
5. Electrical raceway for fire department communication systems.
6. Electrical raceway serving the *interior exit stairway* and *ramp* and terminating at a steel box not exceeding 16 square inches (0.010 m^2).
7. Structural elements supporting the *interior exit stairway* or *ramp* or enclosure, such as beams or joists.

Such penetrations shall be protected in accordance with Section 714. There shall not be penetrations or communication openings, whether protected or not, between adjacent interior exit *stairways* and *ramps.*

Exception: *Membrane penetrations* shall be permitted on the outside of the *interior exit stairway* and *ramp*. Such penetrations shall be protected in accordance with Section 714.4.2.

1023.6 Ventilation. Equipment and ductwork for *interior exit stairway* and *ramp* ventilation as permitted by Section 1023.5 shall comply with one of the following items:

1. Such equipment and ductwork shall be located exterior to the building and shall be directly connected to the *interior exit stairway* and *ramp* by ductwork enclosed in construction as required for *shafts*.
2. Where such equipment and ductwork is located within the *interior exit stairway* and *ramp*, the intake air shall be taken directly from the outdoors and the exhaust air shall be discharged directly to the outdoors, or such air shall be conveyed through ducts enclosed in construction as required for *shafts*.
3. Where located within the building, such equipment and ductwork shall be separated from the remainder of the building, including other mechanical equipment, with construction as required for *shafts*.

In each case, openings into the fire-resistance-rated construction shall be limited to those needed for maintenance and operation and shall be protected by opening protectives in accordance with Section 716 for shaft enclosures.

The *interior exit stairway* and *ramp* ventilation systems shall be independent of other building ventilation systems.

1023.7 Interior exit stairway and ramp exterior walls. *Exterior walls* of the *interior exit stairway* or *ramp* shall comply with the requirements of Section 705 for *exterior walls*. Where nonrated walls or unprotected openings enclose the exterior of the *stairway* or *ramps* and the walls or openings are exposed by other parts of the building at an angle of less than 180 degrees (3.14 rad), the building *exterior walls* within 10 feet (3048 mm) horizontally of a nonrated wall or unprotected opening shall have a *fire-resistance rating* of not less than 1 hour. Openings within such *exterior* walls shall be protected by opening protectives having a *fire protection rating* of not less than $^{3}/_{4}$ hour. This construction shall extend vertically from the ground to a point 10 feet (3048 mm) above the topmost landing of the *stairway* or *ramp*, or to the roof line, whichever is lower.

1023.8 Barrier at level of exit discharge. An *interior exit stairway* and *ramp* shall not continue below its *level of exit discharge* unless an *approved* barrier is provided at the *level of exit discharge* to prevent persons from unintentionally continuing into levels below. Directional exit signs shall be provided as specified in Section 1013.

1023.9 Stairway identification signs. A sign shall be provided at each floor landing in an *interior exit stairway* and *ramp* connecting more than three stories designating the floor level, the terminus of the top and bottom of the *interior exit stairway* and *ramp* and the identification of the *stairway* or *ramp*. The signage shall state the story of and direction to the *exit discharge*, and the availability of roof access from the *interior exit stairway* and *ramp* for the fire department. The bottom of the sign shall be located not less than 5 feet (1524 mm) above the floor landing in a position that is readily visible when the doors are in the open and closed positions.

1023.9.1 Signage requirements. *Stairway* identification signs shall comply with all of the following requirements:

1. The signs shall be a minimum size of 18 inches (457 mm) by 12 inches (305 mm).
2. The letters designating the identification of the *interior exit stairway* and *ramp* shall be not less than $1^{1}/_{2}$ inches (38 mm) in height.
3. The number designating the floor level shall be not less than 5 inches (127 mm) in height and located in the center of the sign.
4. Other lettering and numbers shall be not less than 1 inch (25 mm) in height.
5. Characters and their background shall have a nonglare finish. Characters shall contrast with their background, with either light characters on a dark background or dark characters on a light background.
6. Where signs required by Section 1023.9 are installed in the interior exit *stairways* and *ramps* of buildings subject to Section 1025, the signs shall be made of the same materials as required by Section 1025.4.

1023.10 Elevator lobby identification signs. At landings in interior exit *stairways* where two or more doors lead to the floor level, any door with direct access to an enclosed elevator lobby shall be identified by signage located on the door or directly adjacent to the door stating "Elevator Lobby." Signage shall be in accordance with Section 1023.9.1, Items 4, 5 and 6.

1023.11 Tactile floor-level signs. Where floor level signs are provided in *interior exit stairways* and *ramps*, a floor-level sign identifying the floor level in visual characters, raised characters and braille complying with ICC A117.1 shall be located at each floor-level landing adjacent to the door leading from the *interior exit stairway* and *ramp* into the corridor.

1023.12 Smokeproof enclosures. Where required by Section 403.5.4, 405.7.2 or 412.2.2.1, interior exit *stairways* and *ramps* shall be *smokeproof enclosures* in accordance with Section 909.20.

1023.12.1 Termination and extension. A *smokeproof enclosure* shall terminate at an *exit discharge* or a *public way*. The *smokeproof enclosure* shall be permitted to be extended by an *exit passageway* in accordance with Section 1023.3. The *exit passageway* shall be without openings other than the *fire door assembly* required by Section 1023.3.1 and those necessary for egress from the *exit passageway*. The *exit passageway* shall be separated from the remainder of the building by 2-hour *fire barriers* constructed in accordance with Section 707 or *horizontal assemblies* constructed in accordance with Section 711, or both.

Exceptions:

1. Openings in the *exit passageway* serving a *smokeproof enclosure* are permitted where the *exit passageway* is protected and pressurized in

the same manner as the *smokeproof enclosure*, and openings are protected as required for access from other floors.

2. The *fire barrier* separating the *smokeproof enclosure* from the *exit passageway* is not required, provided that the *exit passageway* is protected and pressurized in the same manner as the *smokeproof enclosure*.
3. A *smokeproof enclosure* shall be permitted to egress through areas on the *level of exit discharge* or vestibules as permitted by Section 1028.

1023.12.2 Enclosure access. Access to the *stairway* or *ramp* within a *smokeproof enclosure* shall be by way of a vestibule or an open exterior balcony.

Exception: Access is not required by way of a vestibule or exterior balcony for *stairways* and *ramps* using the pressurization alternative complying with Section 909.20.5.

1023.13 Standpipes. Standpipes and standpipe hose connections shall be provided where required by Sections 905.3 and 905.4.

SECTION 1024
EXIT PASSAGEWAYS

1024.1 General. Exit passageways serving as an exit component in a *means of egress* system shall comply with the requirements of this section. An *exit passageway* shall not be used for any purpose other than as a *means of egress* and a *circulation path*.

1024.2 Width and capacity. The required capacity of exit passageways shall be determined as specified in Section 1005.1 but the minimum width shall be not less than 44 inches (1118 mm), except that exit passageways serving an *occupant load* of less than 50 shall be not less than 36 inches (914 mm) in width. The minimum width or required capacity of exit passageways shall be unobstructed.

Exception: Encroachments complying with Section 1005.7.

1024.3 Construction. *Exit passageway* enclosures shall have walls, floors and ceilings of not less than a 1-hour *fire-resistance rating*, and not less than that required for any connecting *interior exit stairway* or *ramp*. Exit passageways shall be constructed as *fire barriers* in accordance with Section 707 or *horizontal assemblies* constructed in accordance with Section 711, or both.

1024.4 Termination. Exit passageways on the *level of exit discharge* shall terminate at an *exit discharge*. Exit passageways on other levels shall terminate at an exit.

1024.5 Openings. *Exit passageway* opening protectives shall be in accordance with the requirements of Section 716.

Except as permitted in Section 402.8.7, openings in exit passageways other than unprotected exterior openings shall be limited to those necessary for *exit access* to the *exit passageway* from normally occupied spaces and for egress from the *exit passageway*.

Where an *interior exit stairway* or *ramp* is extended to an *exit discharge* or a *public way* by an *exit passageway*, the *exit passageway* shall comply with Section 1023.3.1.

Elevators shall not open into an *exit passageway*.

1024.6 Penetrations. Penetrations into or through an *exit passageway* are prohibited except for the following:

1. Equipment and ductwork necessary for independent ventilation or pressurization.
2. Fire protection systems.
3. Security systems.
4. Two-way communication systems.
5. Electrical raceway for fire department communication.
6. Electrical raceway serving the *exit passageway* and terminating at a steel box not exceeding 16 square inches (0.010 m^2).

Such penetrations shall be protected in accordance with Section 714. There shall not be penetrations or communicating openings, whether protected or not, between adjacent exit passageways.

Exception: *Membrane penetrations* shall be permitted on the outside of the *exit passageway*. Such penetrations shall be protected in accordance with Section 714.4.2.

1024.7 Ventilation. Equipment and ductwork for *exit passageway* ventilation as permitted by Section 1024.6 shall comply with one of the following:

1. The equipment and ductwork shall be located exterior to the building and shall be directly connected to the *exit passageway* by ductwork enclosed in construction as required for *shafts*.
2. Where the equipment and ductwork is located within the *exit passageway*, the intake air shall be taken directly from the outdoors and the exhaust air shall be discharged directly to the outdoors, or the air shall be conveyed through ducts enclosed in construction as required for *shafts*.
3. Where located within the building, the equipment and ductwork shall be separated from the remainder of the building, including other mechanical equipment, with construction as required for *shafts*.

In each case, openings into the fire-resistance-rated construction shall be limited to those needed for maintenance and operation and shall be protected by opening protectives in accordance with Section 716 for shaft enclosures.

Exit passageway ventilation systems shall be independent of other building ventilation systems.

1024.8 Exit passageway exterior walls. *Exterior walls* of the *exit passageway* shall comply with Section 705. Where nonrated walls or unprotected openings enclose the exterior of the *exit passageway* and the walls or openings are exposed by other parts of the building at an angle of less than 180 degrees (3.14 rad), the building *exterior walls* within 10 feet (3048 mm) horizontally of a nonrated wall or unprotected opening shall have a *fire-resistance rating* of not less than 1 hour. Openings within such *exterior walls* shall be protected by opening protectives having a *fire protection rating* of not less than $^3/_4$ hour. This construction shall extend vertically from the ground to a point 10 feet (3048 mm) above the floor of the *exit passageway*, or to the roof line, whichever is lower.

1024.9 Standpipes. Standpipes and standpipe hose connections shall be provided where required by Sections 905.3 and 905.4.

SECTION 1025
LUMINOUS EGRESS PATH MARKINGS

1025.1 General. *Approved* luminous egress path markings delineating the exit path shall be provided in *high-rise buildings* of Group A, B, E, I-1, M or R-1 occupancies in accordance with this section.

Exception: Luminous egress path markings shall not be required on the *level of exit discharge* in lobbies that serve as part of the exit path in accordance with Section 1028.2, Exception 1.

1025.2 Markings within exit components. Egress path markings shall be provided in *interior exit stairways*, interior exit *ramps* and exit passageways, in accordance with Sections 1025.2.1 through 1025.2.6.3.

1025.2.1 Steps. A solid and continuous stripe shall be applied to the horizontal leading edge of each step and shall extend for the full length of the step. Outlining stripes shall have a minimum horizontal width of 1 inch (25 mm) and a maximum width of 2 inches (51 mm). The leading edge of the stripe shall be placed not more than $^1/_2$ inch (12.7 mm) from the leading edge of the step and the stripe shall not overlap the leading edge of the step by not more than $^1/_2$ inch (12.7 mm) down the vertical face of the step.

Exception: The minimum width of 1 inch (25 mm) shall not apply to outlining stripes *listed* in accordance with UL 1994.

1025.2.2 Landings. The leading edge of landings shall be marked with a stripe consistent with the dimensional requirements for steps.

1025.2.3 Handrails. *Handrails* and *handrail* extensions shall be marked with a solid and continuous stripe having a minimum width of 1 inch (25 mm). The stripe shall be placed on the top surface of the *handrail* for the entire length of the *handrail*, including extensions and newel post caps. Where *handrails* or *handrail* extensions bend or turn corners, the stripe shall not have a gap of more than 4 inches (102 mm).

Exception: The minimum width of 1 inch (25 mm) shall not apply to outlining stripes *listed* in accordance with UL 1994.

1025.2.4 Perimeter demarcation lines. *Stair* landings and other floor areas within *interior exit stairways*, interior exit *ramps* and exit passageways, with the exception of the sides of steps, shall be provided with solid and continuous demarcation lines on the floor or on the walls or a combination of both. The stripes shall be 1 to 2 inches (25 mm to 51 mm) wide with interruptions not exceeding 4 inches (102 mm).

Exception: The minimum width of 1 inch (25 mm) shall not apply to outlining stripes *listed* in accordance with UL 1994.

1025.2.4.1 Floor-mounted demarcation lines. Perimeter demarcation lines shall be placed within 4 inches (102 mm) of the wall and shall extend to within 2 inches (51 mm) of the markings on the leading edge of landings. The demarcation lines shall continue across the floor in front of all doors.

Exception: Demarcation lines shall not extend in front of exit discharge doors that lead out of an exit and through which occupants must travel to complete the exit path.

1025.2.4.2 Wall-mounted demarcation lines. Perimeter demarcation lines shall be placed on the wall with the bottom edge of the stripe not more than 4 inches (102 mm) above the finished floor. At the top or bottom of the *stairs*, demarcation lines shall drop vertically to the floor within 2 inches (51 mm) of the step or landing edge. Demarcation lines on walls shall transition vertically to the floor and then extend across the floor where a line on the floor is the only practical method of outlining the path. Where the wall line is broken by a door, demarcation lines on walls shall continue across the face of the door or transition to the floor and extend across the floor in front of such door.

Exception: Demarcation lines shall not extend in front of exit discharge doors that lead out of an exit and through which occupants must travel to complete the exit path.

1025.2.4.3 Transition. Where a wall-mounted demarcation line transitions to a floor-mounted demarcation line, or vice versa, the wall-mounted demarcation line shall drop vertically to the floor to meet a complimentary extension of the floor-mounted demarcation line, thus forming a continuous marking.

1025.2.5 Obstacles. Obstacles at or below 6 feet 6 inches (1981 mm) in height and projecting more than 4 inches (102 mm) into the egress path shall be outlined with markings not less than 1 inch (25 mm) in width comprised of a pattern of alternating equal bands, of luminous material and black, with the alternating bands not more than 2 inches (51 mm) thick and angled at 45 degrees (0.79 rad). Obstacles shall include, but are not limited to, standpipes, hose cabinets, wall projections and restricted height areas. However, such markings shall not conceal any required information or indicators including but not limited to instructions to occupants for the use of standpipes.

Exception: The minimum width of 1 inch (25 mm) shall not apply to markings *listed* in accordance with UL 1994.

1025.2.6 Doors within the exit path. Doors through which occupants must pass in order to complete the exit path shall be provided with markings complying with Sections 1025.2.6.1 through 1025.2.6.3.

1025.2.6.1 Emergency exit symbol. The doors shall be identified by a low-location luminous emergency exit symbol complying with NFPA 170. The exit symbol shall be not less than 4 inches (102 mm) in height and shall be mounted on the door, centered horizontally, with the top of the symbol not higher than 18 inches (457 mm) above the finished floor.

1025.2.6.2 Door hardware markings. Door hardware shall be marked with not less than 16 square inches (10323 $mm^{2)}$ of luminous material. This marking shall be located behind, immediately adjacent to, or on the door handle or escutcheon. Where a *panic* bar is installed, such material shall be not less than 1 inch (25 mm) wide for the entire length of the actuating bar or touchpad.

1025.2.6.3 Door frame markings. The top and sides of the door frame shall be marked with a solid and continuous 1-inch- to 2-inch-wide (25 mm to 51 mm) stripe. Where the door molding does not provide sufficient flat surface on which to locate the stripe, the stripe shall be permitted to be located on the wall surrounding the frame.

1025.3 Uniformity. Placement and dimensions of markings shall be consistent and uniform throughout the same enclosure.

1025.4 Self-luminous and photoluminescent. Luminous egress path markings shall be permitted to be made of any material, including paint, provided that an electrical charge is not required to maintain the required luminance. Such materials shall include, but not be limited to, *self-luminous* materials and *photoluminescent* materials. Materials shall comply with either of the following standards:

1. UL 1994.
2. ASTM E2072, except that the charging source shall be 1 footcandle (11 lux) of fluorescent illumination for 60 minutes, and the minimum luminance shall be 30 milicandelas per square meter at 10 minutes and 5 milicandelas per square meter after 90 minutes.

1025.5 Illumination. Where *photoluminescent* exit path markings are installed, they shall be provided with not less than 1 footcandle (11 lux) of illumination for not less than 60 minutes prior to periods when the building is occupied and continuously during occupancy.

SECTION 1026
HORIZONTAL EXITS

1026.1 General. Horizontal *exits* serving as an *exit* in a *means of egress* system shall comply with the requirements of this section. A *horizontal exit* shall not serve as the only exit from a portion of a building, and where two or more *exits* are required, not more than one-half of the total number of *exits* or total exit minimum width or required capacity shall be horizontal *exits*.

Exceptions:

1. Horizontal *exits* are permitted to comprise two-thirds of the required *exits* from any building or floor area for occupancies in Group I-2.
2. Horizontal *exits* are permitted to comprise 100 percent of the *exits* required for occupancies in Group I-3. Not less than 6 square feet (0.6 m2) of accessible space per occupant shall be provided on each side of the *horizontal exit* for the total number of people in adjoining compartments.

1026.2 Separation. The separation between buildings or refuge areas connected by a *horizontal exit* shall be provided by a *fire wall* complying with Section 706; or by a *fire barrier* complying with Section 707 or a *horizontal assembly* complying with Section 711, or both. The minimum *fire-resistance rating* of the separation shall be 2 hours. Opening protectives in horizontal *exits* shall also comply with Section 716. Duct and air transfer openings in a *fire wall* or *fire barrier* that serves as a *horizontal exit* shall also comply with Section 717. The *horizontal exit* separation shall extend vertically through all levels of the building unless floor assemblies have a *fire-resistance rating* of not less than 2 hours and do not have unprotected openings.

Exception: A *fire-resistance rating* is not required at horizontal *exits* between a building area and an above-grade *pedestrian walkway* constructed in accordance with Section 3104, provided that the distance between connected buildings is more than 20 feet (6096 mm).

Horizontal exits constructed as *fire barriers* shall be continuous from *exterior wall* to *exterior wall* so as to divide completely the floor served by the *horizontal exit*.

1026.3 Opening protectives. *Fire doors* in horizontal *exits* shall be *self-closing* or automatic-closing when activated by a smoke detector in accordance with Section 716.2.6.6. Doors, where located in a cross-*corridor* condition, shall be automatic-closing by activation of a smoke detector installed in accordance with Section 716.2.6.6.

1026.4 Refuge area. The refuge area of a *horizontal exit* shall be a space occupied by the same tenant or a public area and each such refuge area shall be adequate to accommodate the original *occupant load* of the refuge area plus the *occupant load* anticipated from the adjoining compartment. The anticipated *occupant load* from the adjoining compartment shall be based on the capacity of the *horizontal exit doors* entering the refuge area or the total *occupant load* of the adjoining compartment, whichever is less.

1026.4.1 Capacity. The capacity of the refuge area shall be computed based on a *net floor area* allowance of 3 square feet (0.2787 m^2) for each occupant to be accommodated therein. Where the *horizontal exit* also forms a *smoke compartment*, the capacity of the refuge area for Group I-1, I-2 and I-3 occupancies and Group B ambulatory care facilities shall comply with Sections 407.5.3, 408.6.2, 420.6.1 and 422.3.2 as applicable.

1026.4.2 Number of exits. The refuge area into which a *horizontal exit* leads shall be provided with *exits* adequate to meet the occupant requirements of this chapter, but not including the added *occupant load* imposed by persons entering the refuge area through horizontal *exits* from other areas. Not less than one refuge area exit shall lead directly to the exterior or to an *interior exit stairway* or ramp.

Exception: The adjoining compartment shall not be required to have a *stairway* or door leading directly outside, provided that the refuge area into which a *horizontal exit* leads has *stairways* or doors leading directly outside and are so arranged that egress shall not require the occupants to return through the compartment from which egress originates.

1026.5 Standpipes. Standpipes and standpipe hose connections shall be provided where required by Sections 905.3 and 905.4.

SECTION 1027
EXTERIOR EXIT STAIRWAYS AND RAMPS

1027.1 General. Exterior exit stairways and ramps serving as an exit component in a *means of egress* system shall comply with the requirements of this section.

1027.2 Use in a means of egress. *Exterior exit stairways* shall not be used as an element of a required *means of egress* for Group I-2 occupancies. For occupancies in other than Group I-2, *exterior exit stairways* and *ramps* shall be permitted as an element of a required *means of egress* for buildings not exceeding six stories above grade plane or that are not *high-rise buildings*.

1027.3 Open side. *Exterior exit stairways* and *ramps* serving as an element of a required *means of egress* shall be open on not less than one side, except for required structural columns, beams, *handrails* and *guards*. An open side shall have not less than 35 square feet (3.3 m^2) of aggregate open area adjacent to each floor level and the level of each intermediate landing. The required open area shall be located not less than 42 inches (1067 mm) above the adjacent floor or landing level.

1027.4 Side yards. The open areas adjoining *exterior exit stairways* or *ramps* shall be either *yards*, *courts* or *public ways*; the remaining sides are permitted to be enclosed by the *exterior walls* of the building.

1027.5 Location. *Exterior exit stairways* and *ramps* shall have a minimum *fire separation distance* of 10 feet (3048 mm) measured at right angles from the exterior edge of the *stairway* or *ramps*, including landings, to:

1. Adjacent *lot lines*.
2. Other portions of the building.
3. Other buildings on the same lot unless the adjacent building *exterior walls* and openings are protected in accordance with Section 705 based on *fire separation distance*.

For the purposes of this section, other portions of the building shall be treated as separate buildings.

Exception: *Exterior exit stairways* and *ramps* serving individual *dwelling units* of Group R-3 shall have a minimum *fire separation distance* of 5 feet (1525 mm).

1027.6 Exterior exit stairway and ramp protection. *Exterior exit stairways* and *ramps* shall be separated from the interior of the building as required in Section 1023.2. Openings shall be limited to those necessary for egress from normally occupied spaces. Where a vertical plane projecting from the edge of an *exterior exit stairway* or *ramp* and landings is exposed by other parts of the building at an angle of less than 180 degrees (3.14 rad), the *exterior wall* shall be rated in accordance with Section 1023.7.

Exceptions:

1. Separation from the interior of the building is not required for occupancies, other than those in Group R-1 or R-2, in buildings that are not more than two stories above grade plane where a *level of exit discharge* serving such occupancies is the first *story above grade plane*.
2. Separation from the interior of the building is not required where the *exterior exit stairway* or *ramp* is served by an *exterior exit ramp* or balcony that connects two remote exterior exit *stairways* or other *approved exits* with a perimeter that is not less than 50 percent open. To be considered open, the opening shall be not less than 50 percent of the height of the enclosing wall, with the top of the openings not less than 7 feet (2134 mm) above the top of the balcony.
3. Separation from the open-ended *corridor* of the building is not required for *exterior exit stairways* or *ramps*, provided that Items 3.1 through 3.5 are met:
    - 3.1. The building, including open-ended *corridors*, and *stairways* and *ramps*, shall be equipped throughout with an *automatic sprinkler system* in accordance with Section 903.3.1.1 or 903.3.1.2.
    - 3.2. The open-ended *corridors* comply with Section 1020.
    - 3.3. The open-ended *corridors* are connected on each end to an *exterior exit stairway* or *ramp* complying with Section 1027.

3.4. The *exterior walls* and openings adjacent to the *exterior exit stairway* or *ramp* comply with Section 1023.7.

3.5. At any location in an open-ended *corridor* where a change of direction exceeding 45 degrees (0.79 rad) occurs, a clear opening of not less than 35 square feet (3.3 m^2) or an *exterior stairway* or *ramp* shall be provided. Where clear openings are provided, they shall be located so as to minimize the accumulation of smoke or toxic gases.

4. In Group R-3 occupancies not more than four stories in height, *exterior exit stairways* and *ramps* serving individual *dwelling units* are not required to be separated from the interior of the building where the *exterior exit stairway* or *ramp* discharges directly to grade.

SECTION 1028
EXIT DISCHARGE

1028.1 General. The exit discharge shall comply with Sections 1028 and 1029 and the applicable requirements of Sections 1003 through 1015.

1028.2 Exit discharge. *Exits* shall discharge directly to the exterior of the building. The *exit discharge* shall be at grade or shall provide a direct path of egress travel to grade. The *exit discharge* shall not reenter a building. The combined use of Exceptions 1 and 2 shall not exceed 50 percent of the number and minimum width or required capacity of the required *exits*.

Exceptions:

1. Not more than 50 percent of the number and minimum width or required capacity of *interior exit stairways* and *ramps* is permitted to egress through areas, including *atriums*, on the level of discharge provided that all of the following conditions are met:

   1.1. Discharge of *interior exit stairways* and *ramps* shall be provided with a free and unobstructed path of travel to an exterior *exit* door and such *exit* is readily visible and identifiable from the point of termination of the enclosure.

   1.2. The entire area of the *level of exit discharge* is separated from areas below by construction conforming to the *fire-resistance rating* for the enclosure.

   1.3. The egress path from the *interior exit stairway* and *ramp* on the *level of exit discharge* is protected throughout by an *approved automatic sprinkler system*. Portions of the *level of exit discharge* with access to the egress path shall be either equipped throughout with an *automatic sprinkler system* installed in accordance with Section 903.3.1.1 or 903.3.1.2, or separated from the egress path in accordance with the requirements for the enclosure of *interior exit stairways* or *ramps*.

   1.4. Where a required *interior exit stairway* or *ramp* and an *exit access stairway* or *ramp* serve the same floor level and terminate at the same *level of exit discharge*, the termination of the *exit access stairway* or *ramp* and the *exit discharge* door of the *interior exit stairway* or *ramp* shall be separated by a distance of not less than 30 feet (9144 mm) or not less than one-fourth the length of the maximum overall diagonal dimension of the building, whichever is less. The distance shall be measured in a straight line between the *exit discharge* door from the *interior exit stairway* or *ramp* and the last tread of the *exit access stairway* or termination of slope of the *exit access ramp*.

2. Not more than 50 percent of the number and minimum width or required capacity of the *interior exit stairways* and *ramps* is permitted to egress through a vestibule provided that all of the following conditions are met:

   2.1. The entire area of the vestibule is separated from areas below by construction conforming to the *fire-resistance rating* of the *interior exit stairway* or *ramp enclosure*.

   2.2. The depth from the exterior of the building is not greater than 10 feet (3048 mm) and the length is not greater than 30 feet (9144 mm).

   2.3. The area is separated from the remainder of the *level of exit discharge* by a *fire partition* constructed in accordance with Section 708.

   Exception: The maximum transmitted temperature rise is not required.

   2.4. The area is used only for *means of egress* and *exits* directly to the outside.

3. *Horizontal exits* complying with Section 1026 shall not be required to discharge directly to the exterior of the building.

1028.3 Exit discharge width or capacity. The minimum width or required capacity of the *exit discharge* shall be not less than the minimum width or required capacity of the *exits* being served.

1028.4 Exit discharge components. *Exit discharge* components shall be sufficiently open to the exterior so as to minimize the accumulation of smoke and toxic gases. *

1028.5 Access to a public way. The *exit discharge* shall provide a direct and unobstructed access to a *public way*.

Exception: Where access to a *public way* cannot be provided, a safe dispersal area shall be provided where all of the following are met:

1. The area shall be of a size to accommodate not less than 5 square feet (0.46 m^2) for each person.
2. The area shall be located on the same lot not less than 50 feet (15 240 mm) away from the building requiring egress.
3. The area shall be permanently maintained and identified as a safe dispersal area.
4. The area shall be provided with a safe and unobstructed path of travel from the building.

**

SECTION 1029
EGRESS COURTS

1029.1 General. *Egress courts* serving as an *exit discharge* component in the *means of egress* system shall comply with the requirements in this section.

1029.2 Width or capacity. The required capacity of egress courts shall be determined as specified in Section 1005.1, but the minimum width shall be not less than 44 inches (1118 mm), except as specified herein. Egress courts serving Group R-3 and U occupancies shall be not less than 36 inches (914 mm) in width. The required capacity and width of egress courts shall be unobstructed to a height of 7 feet (2134 mm). The width of the egress court shall be not less than the required capacity.

Exception: Encroachments complying with Section 1005.7.

1029.3 Construction and openings. Where an *egress court* serving a building or portion thereof is less than 10 feet (3048 mm) in width, the *egress court* walls shall have not less than 1-hour *fire-resistance-rated* construction for a distance of 10 feet (3048 mm) above the floor of the *egress court*. Openings within such walls shall be protected by opening protectives having a *fire protection rating* of not less than $^3/_4$ hour.

Exceptions:

1. *Egress courts* serving an *occupant load* of less than 10.
2. *Egress courts* serving Group R-3.

SECTION 1030
ASSEMBLY

1030.1 General. A room or space used for assembly purposes that contains seats, tables, displays, equipment or other material shall comply with this section.

1030.1.1 Bleachers. *Bleachers*, *grandstands* and *folding and telescopic seating*, that are not building elements, shall comply with ICC 300.

1030.1.1.1 Spaces under grandstands and bleachers. Spaces under *grandstands* or *bleachers* shall be separated by *fire barriers* complying with Section 707 and *horizontal assemblies* complying with Section 711 with not less than 1-hour *fire-resistance-rated* construction.

Exceptions:

1. Ticket booths less than 100 square feet (9.29 m^2) in area.
2. Toilet rooms.
3. Other accessory use areas 1,000 square feet (92.9 m^2) or less in area and equipped with an *automatic sprinkler system* in accordance with Section 903.3.1.1.

1030.2 Assembly main exit. A building, room or space used for assembly purposes that has an *occupant load* of greater than 300 and is provided with a main *exit*, that main *exit* shall be of sufficient capacity to accommodate not less than one-half of the *occupant load*, but such capacity shall be not less than the total required capacity of all *means of egress* leading to the *exit*. Where the building is classified as a Group A occupancy, the main *exit* shall front on not less than one street or an unoccupied space of not less than 10 feet (3048 mm) in width that adjoins a street or *public way*. In a building, room or space used for assembly purposes where there is not a well-defined main *exit* or where multiple main *exits* are provided, *exits* shall be permitted to be distributed around the perimeter of the building provided that the total capacity of egress is not less than 100 percent of the required capacity.

1030.3 Assembly other exits. In addition to having access to a main *exit*, each level in a building used for assembly purposes having an *occupant load* greater than 300 and provided with a main *exit*, shall be provided with additional *means of egress* that shall provide an egress capacity for not less than one-half of the total *occupant load* served by that level and shall comply with Section 1007.1. In a building used for assembly purposes where there is not a well-defined main *exit* or where multiple main *exits* are provided, *exits* for each level shall be permitted to be distributed around the perimeter of the building, provided that the total width of egress is not less than 100 percent of the required width.

1030.4 Foyers and lobbies. In Group A-1 occupancies, where persons are admitted to the building at times when seats are not available, such persons shall be allowed to wait in a lobby or similar space, provided that such lobby or similar space shall not encroach on the minimum width or required capacity of the *means of egress*. Such foyer, if not directly connected to a public street by all the main entrances or *exits*, shall have a straight and unobstructed corridor or path of travel to every such main entrance or *exit*.

1030.5 Interior balcony and gallery means of egress. For balconies, galleries or press boxes having a seating capacity of 50 or more located in a building, room or space used for assembly purposes, not less than two *means of egress* shall be provided, with one from each side of every balcony, gallery or press box.

1030.6 Capacity of aisle for assembly. The required capacity of *aisles* shall be not less than that determined in accordance with Section 1030.6.1 where *smoke-protected assembly seating* is not provided, Section 1030.6.2 where *smoke-protected assembly seating* is provided and Section 1030.6.3 where *open-air assembly seating* is provided.

1030.6.1 Without smoke protection. The required capacity in inches (mm) of the *aisles* for assembly seating without smoke protection shall be not less than the *occupant load* served by the egress element in accordance with all of the following, as applicable:

1. Not less than 0.3 inch (7.6 mm) of aisle capacity for each occupant served shall be provided on stepped *aisles* having riser heights 7 inches (178 mm) or less and tread depths 11 inches (279 mm) or greater, measured horizontally between tread *nosings*.
2. Not less than 0.005 inch (0.127 mm) of additional aisle capacity for each occupant shall be provided for each 0.10 inch (2.5 mm) of riser height above 7 inches (178 mm).
3. Where egress requires stepped aisle descent, not less than 0.075 inch (1.9 mm) of additional aisle capacity for each occupant shall be provided on those portions of aisle capacity that do not have a *handrail* within a horizontal distance of 30 inches (762 mm).
4. Ramped *aisles*, where slopes are steeper than one unit vertical in 12 units horizontal (8-percent slope), shall have not less than 0.22 inch (5.6 mm) of clear aisle capacity for each occupant served. Level or ramped *aisles*, where slopes are not steeper than one unit vertical in 12 units horizontal (8-percent slope), shall have not less than 0.20 inch (5.1 mm) of clear aisle capacity for each occupant served.

1030.6.2 Smoke-protected assembly seating. The required capacity in inches (mm) of the *aisle* for *smoke-protected assembly seating* shall be not less than the *occupant load* served by the egress element multiplied by the appropriate factor in Table 1030.6.2. The total number of seats specified shall be those within the space exposed to the same smoke-protected environment. Interpolation is permitted between the specific values shown. A life safety evaluation, complying with NFPA 101, shall be done for a facility utilizing the reduced width requirements of Table 1030.6.2 for *smoke-protected assembly seating*.

1030.6.2.1 Smoke control. *Aisles* and *aisle accessways* serving a *smoke-protected assembly seating* area shall be provided with a smoke control system complying with Section 909 or natural ventilation designed to maintain the smoke level not less than 6 feet (1829 mm) above the floor of the *means of egress*.

1030.6.2.2 Roof height. A *smoke-protected assembly seating* area with a roof shall have the lowest portion of the *roof deck* not less than 15 feet (4572 mm) above the highest *aisle* or *aisle accessway*.

Exception: A roof canopy in an outdoor stadium shall be permitted to be less than 15 feet (4572 mm) above the highest *aisle* or *aisle accessway* provided that there are no objects less than 80 inches (2032 mm) above the highest *aisle* or *aisle accessway*.

1030.6.2.3 Automatic sprinklers. Enclosed areas with walls and ceilings in buildings or structures containing *smoke-protected assembly seating* shall be protected with an *approved automatic sprinkler system* in accordance with Section 903.3.1.1.

Exceptions:

1. The floor area used for contests, performances or entertainment provided that the roof construction is more than 50 feet (15 240 mm) above the floor level and the use is restricted to low fire hazard uses.
2. Press boxes and storage facilities less than 1,000 square feet (93 m^2) in area.

1030.6.3 Open-air assembly seating. In *open-air assembly seating*, the required capacity in inches (mm) of *aisles* shall be not less than the total *occupant load* served by the egress element multiplied by 0.08 (2.0 mm) where egress is by stepped *aisle* and multiplied by 0.06 (1.52 mm) where egress is by level *aisles* and ramped *aisles*.

Exception: The required capacity in inches (mm) of *aisles* shall be permitted to comply with Section 1030.6.2 for the number of seats in the *open-air assembly seating* where Section 1030.6.2 permits less capacity.

TABLE 1030.6.2
CAPACITY FOR AISLES FOR SMOKE-PROTECTED ASSEMBLY

TOTAL NUMBER OF SEATS IN THE SMOKE-PROTECTED ASSEMBLY SEATING	INCHES OF CAPACITY PER SEAT SERVED			
	Stepped aisles with handrails within 30 inches	Stepped aisles without handrails within 30 inches	Level aisles or ramped aisles not steeper than 1 in 10 in slope	Ramped aisles steeper than 1 in 10 in slope
Equal to or less than 5,000	0.200	0.250	0.150	0.165
10,000	0.130	0.163	0.100	0.110
15,000	0.096	0.120	0.070	0.077
20,000	0.076	0.095	0.056	0.062
Equal to or greater than 25,000	0.060	0.075	0.044	0.048

For SI: 1 inch = 25.4 mm.

1030.6.3.1 Automatic sprinklers. Enclosed areas with walls and ceilings in buildings or structures containing *open-air assembly seating* shall be protected with an approved automatic sprinkler system in accordance with Section 903.3.1.1.

Exceptions:

1. The floor area used for contests, performances or entertainment, provided that the roof construction is more than 50 feet (15 240 mm) above the floor level and the use is restricted to low fire hazard uses.
2. Press boxes and storage facilities less than 1,000 square feet (93 m^2) in area.
3. Open-air assembly seating facilities where seating and the means of egress in the seating area are essentially open to the outside.

1030.7 Travel distance. The *exit access* travel distance shall comply with Section 1017. Where *aisles* are provided for seating, the distance shall be measured along the *aisles* and *aisle accessways* without travel over or on the seats.

Exceptions:

1. In facilities with *smoke-protected assembly seating*, the total *exit access* travel distance shall be not greater than 400 feet (122 m). That portion of the total permitted *exit access* travel distance from each seat to the nearest entrance to a vomitory or concourse shall not exceed 200 feet (60 960 mm). The portion of the total permitted *exit access* travel distance from the entrance to the vomitory or concourse to one of the following shall not exceed 200 feet (60 960 mm):
   1.1. The closest riser of an *exit access stairway*.
   1.2. The closest slope of an *exit access ramp*.
   1.3. An exit.
2. In facilities with *open-air assembly seating* of Type III, IV or V construction, the total *exit access* travel distance to one of the following shall not exceed 400 feet (122 m):
   2.1. The closest riser of an *exit access stairway*.
   2.2. The closest slope of an *exit access* ramp.
   2.3. An exit.
3. In facilities with *open-air assembly seating* of Type I or II construction, the total *exit access* travel distance shall not be limited.

1030.8 Common path of egress travel. The *common path of egress travel* shall not exceed 30 feet (9144 mm) from any seat to a point where an occupant has a choice of two paths of egress travel to two *exits*.

Exceptions:

1. For areas serving less than 50 occupants, the *common path of egress travel* shall not exceed 75 feet (22 860 mm).
2. For *smoke-protected* or *open-air assembly seating*, the *common path of egress travel* shall not exceed 50 feet (15 240 mm).

1030.8.1 Path through adjacent row. Where one of the two paths of travel is across the *aisle* through a row of seats to another *aisle*, there shall be not more than 24 seats between the two aisles, and the minimum clear width between rows for the row between the two *aisles* shall be 12 inches (305 mm) plus 0.6 inch (15.2 mm) for each additional seat above seven in the row between aisles.

Exception: For *smoke-protected* or *open-air assembly seating* there shall be not more than 40 seats between the two *aisles* and the minimum clear width shall be 12 inches (305 mm) plus 0.3 inch (7.6 mm) for each additional seat.

1030.9 Assembly aisles are required. Every occupied portion of any building, room or space used for assembly purposes that contains seats, tables, displays, similar fixtures or equipment shall be provided with *aisles* leading to *exits* or *exit access doorways* in accordance with this section.

1030.9.1 Minimum aisle width. The minimum clear width for *aisles* shall comply with one of the following:

1. Forty-eight inches (1219 mm) for stepped *aisles* having seating on both sides.

   Exception: Thirty-six inches (914 mm) where the stepped *aisles* serve less than 50 seats.
2. Thirty-six inches (914 mm) for stepped *aisles* having seating on only one side.

   Exception: Twenty-three inches (584 mm) between a stepped aisle *handrail* and seating where a stepped *aisle* does not serve more than five rows on one side.
3. Twenty-three inches (584 mm) between a stepped aisle *handrail* or *guard* and seating where the stepped *aisle* is subdivided by a mid-aisle *handrail.*
4. Forty-two inches (1067 mm) for level or ramped *aisles* having seating on both sides.

   Exceptions:

   1. Thirty-six inches (914 mm) where the *aisle* serves less than 50 seats.
   2. Thirty inches (762 mm) where the *aisle* serves less than 15 seats and does not serve as part of an *accessible* route.
5. Thirty-six inches (914 mm) for level or ramped *aisles* having seating on only one side.

   Exception: Thirty inches (762 mm) where the *aisle* serves fewer than 15 seats and does not serve as part of an *accessible* route.

1030.9.2 Aisle catchment area. The *aisle* shall provide sufficient capacity for the number of persons accommodated by the catchment area served by the *aisle.* The catchment area served by an *aisle* is that portion of the total space served by that section of the *aisle*. In establishing catchment areas, the assumption shall be made that there is a balanced use of all *means of egress*, with the number of persons in proportion to egress capacity.

1030.9.3 Converging aisles. Where *aisles* converge to form a single path of egress travel, the required capacity of that path shall be not less than the combined required capacity of the converging aisles.

1030.9.4 Uniform width and capacity. Those portions of *aisles*, where egress is possible in either of two directions, shall be uniform in minimum width or required capacity.

1030.9.5 Dead-end aisles. Each end of an *aisle* shall be continuous to a cross *aisle*, foyer, doorway, vomitory, concourse or *stairway* in accordance with Section 1030.9.7 having access to an *exit*.

Exceptions:

1. Dead-end *aisles* shall be not greater than 20 feet (6096 mm) in length.
2. Dead-end *aisles* longer than 16 rows are permitted where seats beyond the 16th row dead-end *aisle* are not more than 24 seats from another *aisle*, measured along a row of seats having a minimum clear width of 12 inches (305 mm) plus 0.6 inch (15.2 mm) for each additional seat above seven in the row where seats have backrests or beyond 10 where seats are without backrests in the row.
3. For *smoke-protected* or *open-air assembly seating*, the dead-end *aisle* length of vertical *aisles* shall not exceed a distance of 21 rows.
4. For *smoke-protected* or *open-air assembly seating*, a longer dead-end *aisle* is permitted where seats beyond the 21-row dead-end *aisle* are not more than 40 seats from another *aisle*, measured along a row of seats having an *aisle* accessway with a minimum clear width of 12 inches (305 mm) plus 0.3 inch (7.6 mm) for each additional seat above seven in the row where seats have backrests or beyond 10 where seats are without backrests in the row.

1030.9.6 Aisle measurement. The clear width for *aisles* shall be measured to walls, edges of seating and tread edges except for permitted projections.

Exception: The clear width of *aisles* adjacent to seating at tables shall be permitted to be measured in accordance with Section 1030.13.1.

1030.9.6.1 Assembly aisle obstructions. There shall not be obstructions in the minimum width or required capacity of *aisles*.

Exception: *Handrails* are permitted to project into the required width of stepped *aisles* and ramped *aisles* in accordance with Section 1014.8.

1030.9.7 Stairways connecting to stepped aisles. A *stairway* that connects a stepped *aisle* to a cross *aisle* or concourse shall be permitted to comply with the assembly aisle walking surface requirements of Section 1030.14. Transitions between *stairways* and stepped *aisles* shall comply with Section 1030.10.

1030.9.8 Stairways connecting to vomitories. A *stairway* that connects a vomitory to a cross *aisle* or concourse shall be permitted to comply with the assembly aisle walking surface requirements of Section 1030.14. Transitions between *stairways* and stepped *aisles* shall comply with Section 1030.10.

1030.10 Transitions. Transitions between *stairways* and stepped *aisles* shall comply with either Section 1030.10.1 or 1030.10.2.

1030.10.1 Transitions to stairways that maintain stepped aisle riser and tread dimensions. Stepped *aisles*, transitions and *stairways* that maintain the stepped aisle riser and tread dimensions shall comply with Section 1030.14 as one *exit access* component.

1030.10.2 Transitions to stairways that do not maintain stepped aisle riser and tread dimensions. Transitions between *stairways* and stepped *aisles* having different riser and tread dimensions shall comply with Sections 1030.10.2.1 through 1030.10.3.

1030.10.2.1 Stairways and stepped aisles in a straight run. Where *stairways* and stepped *aisles* are in a straight run, transitions shall have one of the following:

1. A depth of not less than 22 inches (559 mm) where the treads on the descending side of the transition have greater depth.
2. A depth of not less than 30 inches (762 mm) where the treads on the descending side of the transition have lesser depth.

1030.10.2.2 Stairways that change direction from stepped aisles. Transitions where the *stairway* changes direction from the stepped *aisle* shall have a minimum depth of 11 inches (280 mm) or the stepped aisle tread depth, whichever is greater, between the stepped *aisle* and *stairway*.

1030.10.3 Transition marking. A distinctive marking stripe shall be provided at each *nosing* or leading edge adjacent to the transition. Such stripe shall be not less than 1 inch (25 mm), and not more than 2 inches (51 mm), wide. The edge marking stripe shall be distinctively different from the stepped *aisle* contrasting marking stripe.

1030.11 Stepped aisles at vomitories. Stepped *aisles* that change direction at vomitories shall comply with Section 1030.11.1 Transitions between a stepped *aisle* above a vomitory and a stepped *aisle* to the side of a vomitory shall comply with Section 1030.11.2.

1030.11.1 Stepped aisles that change direction at vomitories. Stepped aisle treads where the stepped *aisle* changes direction at a vomitory shall have a depth of not less than 11 inches (280 mm) or the stepped aisle tread depth, whichever is greater. The height of a stepped aisle tread above a transition at a vomitory shall comply with Section 1030.14.2.2.

1030.11.2 Stepped aisle transitions at the top of vomitories. Transitions between the stepped *aisle* above a vomitory and stepped *aisles* to the side of a vomitory shall have a depth of not less than 11 inches (280 mm) or the stepped aisle tread depth, whichever is greater.

1030.12 Construction. *Aisles*, stepped *aisles* and ramped *aisles* shall be built of materials consistent with the types permitted for the type of construction of the building.

Exception: Wood *handrails* shall be permitted for all types of construction.

1030.12.1 Walking surface. The surface of *aisles*, stepped *aisles* and ramped *aisles* shall be of slip-resistant materials that are securely attached. The surface for stepped *aisles* shall comply with Section 1011.7.1.

1030.12.2 Outdoor conditions. Outdoor *aisles*, stepped *aisles* and ramped *aisles* and outdoor approaches to *aisles*, stepped *aisles* and ramped *aisles* shall be designed so that water will not accumulate on the walking surface.

1030.13 Aisle accessways. *Aisle accessways* for seating at tables shall comply with Section 1030.13.1. *Aisle accessways* for seating in rows shall comply with Section 1030.13.2.

1030.13.1 Seating at tables. Where seating is located at a table or counter and is adjacent to an *aisle* or *aisle accessway*, the measurement of required clear width of the *aisle* or *aisle accessway* shall be made to a line 19 inches (483 mm) away from and parallel to the edge of the table or counter. The 19-inch (483 mm) distance shall be measured perpendicular to the side of the table or counter. In the case of other side boundaries for *aisles* or *aisle accessways*, the clear width shall be measured to walls, edges of seating and tread edges.

Exception: Where tables or counters are served by fixed seats, the width of the *aisle* or *aisle accessway* shall be measured from the back of the seat.

1030.13.1.1 Aisle accessway capacity and width for seating at tables. *Aisle accessways* serving arrangements of seating at tables or counters shall comply with the capacity requirements of Section 1005.1 but shall not have less than 12 inches (305 mm) of width plus $^1/_2$ inch (12.7 mm) of width for each additional 1 foot (305 mm), or fraction thereof, beyond 12 feet (3658 mm) of aisle accessway length measured from the center of the seat farthest from an *aisle*.

Exception: Portions of an *aisle accessway* having a length not exceeding 6 feet (1829 mm) and used by a total of not more than four persons.

1030.13.1.2 Seating at table aisle accessway length. The length of travel along the *aisle accessway* shall not exceed 30 feet (9144 mm) from any seat to the point where a person has a choice of two or more paths of egress travel to separate *exits*.

1030.13.2 Clear width of aisle accessways serving seating in rows. Where seating rows have 14 or fewer seats, the minimum clear aisle accessway width shall be not less than 12 inches (305 mm) measured as the clear horizontal distance from the back of the row ahead and the nearest projection of the row behind. Where chairs have automatic or self-rising seats, the measurement shall be made with seats in the raised position. Where any chair in the row does not have an automatic or self-rising seat, the measurements shall be made with the seat in the down position. For seats with folding tablet arms, row spacing shall be determined with the tablet arm in the used position.

Exception: For seats with folding tablet arms, row spacing is permitted to be determined with the tablet arm in the stored position where the tablet arm when raised manually to vertical position in one motion automatically returns to the stored position by force of gravity.

1030.13.2.1 Dual access. For rows of seating served by *aisles* or doorways at both ends, there shall be not more than 100 seats per row. The minimum clear width of 12 inches (305 mm) between rows shall be increased by 0.3 inch (7.6 mm) for every additional seat beyond 14 seats where seats have backrests or beyond 21 where seats are without backrests. The minimum clear width is not required to exceed 22 inches (559 mm).

Exception: For *smoke-protected* or *open-air assembly seating*, the row length limits for a 12-inch-wide (305 mm) *aisle accessway*, beyond which the aisle accessway minimum clear width shall be increased, are in Table 1030.13.2.1.

1030.13.2.2 Single access. For rows of seating served by an *aisle* or doorway at only one end of the row, the minimum clear width of 12 inches (305 mm) between rows shall be increased by 0.6 inch (15.2 mm) for every additional seat beyond seven seats where seats have backrests or beyond 10 where seats are without backrests. The minimum clear width is not required to exceed 22 inches (559 mm).

Exception: For *smoke-protected* or *open-air assembly seating*, the row length limits for a 12-inch-wide (305 mm) *aisle accessway*, beyond which the aisle accessway minimum clear width shall be increased, are in Table 1030.13.2.1.

1030.14 Assembly aisle walking surfaces. Ramped *aisles* shall comply with Sections 1030.14.1 through 1030.14.1.3. Stepped *aisles* shall comply with Sections 1030.14.2 through 1030.14.2.4.

1030.14.1 Ramped aisles. *Aisles* that are sloped more than 1 unit vertical in 20 units horizontal (5-percent slope) shall be considered to be a ramped *aisle*. Ramped *aisles* that serve as part of an *accessible route* in accordance with Sections 1009 and 1109.2 shall have a maximum slope of 1 unit vertical in 12 units horizontal (8-percent slope). The slope of other ramped *aisles* shall not exceed 1 unit vertical in 8 units horizontal (12.5-percent slope).

**TABLE 1030.13.2.1
SMOKE-PROTECTED OR OPEN-AIR ASSEMBLY AISLE ACCESSWAYS**

NUMBER OF SEATS IN THE SMOKE-PROTECTED OR OPEN-AIR ASSEMBLY SEATING	MAXIMUM NUMBER OF SEATS PER ROW PERMITTED TO HAVE A MINIMUM 12-INCH CLEAR WIDTH AISLE ACCESSWAY			
	Aisle or doorway at both ends of row		Aisle or doorway at one end of row only	
	Seats with backrests	Seats without backrests	Seats with backrests	Seats without backrests
Less than 4,000	14	21	7	10
4,000 to 6,999	15	22	7	10
7,000 to 9,999	16	23	8	11
10,000 to 12,999	17	24	8	11
13,000 to 15,999	18	25	9	12
16,000 to 18,999	19	26	9	12
19,000 to 21,999	20	27	10	13
22,000 and greater	21	28	11	14

For SI: 1 inch = 25.4 mm.

1030.14.1.1 Cross slope. The slope measured perpendicular to the direction of travel of a ramped *aisle* shall not be steeper than 1 unit vertical in 48 units horizontal (2-percent slope).

1030.14.1.2 Landings. Ramped *aisles* shall have landings in accordance with Sections 1012.6 through 1012.6.5. Landings for ramped *aisles* shall be permitted to overlap required *aisles* or cross *aisles*.

1030.14.1.3 Edge protection. Ramped *aisles* shall have edge protection in accordance with Sections 1012.10 and 1012.10.1.

Exception: In assembly spaces with *fixed seating*, edge protection is not required on the sides of ramped *aisles* where the ramped *aisles* provide access to the adjacent seating and *aisle accessways*.

1030.14.2 Stepped aisles. *Aisles* with a slope exceeding one unit vertical in eight units horizontal (12.5-percent slope) shall consist of a series of risers and treads that extends across the full width of *aisles* and complies with Sections 1030.14.2.1 through 1030.14.2.4.

1030.14.2.1 Treads. Tread depths shall be not less than 11 inches (279 mm) and shall have dimensional uniformity.

Exception: The tolerance between adjacent treads shall not exceed $^{3}/_{16}$ inch (4.8 mm).

1030.14.2.2 **Risers.** Where the gradient of stepped *aisles* is to be the same as the gradient of adjoining seating areas, the riser height shall be not less than 4 inches (102 mm) nor more than 8 inches (203 mm) and shall be uniform within each *flight*.

Exceptions:

1. Riser height nonuniformity shall be limited to the extent necessitated by changes in the gradient of the adjoining seating area to maintain adequate sightlines. Where nonuniformities exceed $^{3}/_{16}$ inch (4.8 mm) between adjacent risers, the exact location of such nonuniformities shall be indicated with a distinctive marking stripe on each tread at the *nosing* or leading edge adjacent to the nonuniform risers. Such stripe shall be not less than 1 inch (25 mm), and not more than 2 inches (51 mm), wide. The edge marking stripe shall be distinctively different from the contrasting marking stripe.
2. Riser heights not exceeding 9 inches (229 mm) shall be permitted where they are necessitated by the slope of the adjacent seating areas to maintain sightlines.

1030.14.2.2.1 Construction tolerances. The tolerance between adjacent risers on a stepped *aisle* that were designed to be equal height shall not exceed $^{3}/_{16}$ inch (4.8 mm). Where the stepped *aisle* is designed in accordance with Exception 1 of Section 1030.14.2.2, the stepped *aisle* shall be constructed so that each riser of unequal height, determined in the direction of descent, is not more than $^{3}/_{8}$ inch (9.5 mm) in height different from adjacent risers where stepped *aisle* treads are less than 22 inches (560 mm) in depth and $^{3}/_{4}$ inch (19.1 mm) in height different from adjacent risers where stepped *aisle* treads are 22 inches (560 mm) or greater in depth.

1030.14.2.3 Tread contrasting marking stripe. A contrasting marking stripe shall be provided on each tread at the *nosing* or leading edge such that the location of each tread is readily apparent when viewed in descent. Such stripe shall be not less than 1 inch (25 mm) and not more than 2 inches (51 mm) wide.

Exception: The contrasting marking stripe is permitted to be omitted where tread surfaces are such that the location of each tread is readily apparent when viewed in descent.

1030.14.2.4 Nosing and profile. *Nosing* and riser profile shall comply with Sections 1011.5.5 through 1011.5.5.3.

1030.15 Seat stability. In a building, room or space used for assembly purposes, the seats shall be securely fastened to the floor.

Exceptions:

1. In a building, room or space used for assembly purposes or portions thereof without ramped or tiered floors for seating and with 200 or fewer seats, the seats shall not be required to be fastened to the floor.
2. In a building, room or space used for assembly purposes or portions thereof with seating at tables and without ramped or tiered floors for seating, the seats shall not be required to be fastened to the floor.
3. In a building, room or space used for assembly purposes or portions thereof without ramped or tiered floors for seating and with greater than 200 seats, the seats shall be fastened together in groups of not less than three or the seats shall be securely fastened to the floor.
4. In a building, room or space used for assembly purposes where flexibility of the seating arrangement is an integral part of the design and function of the space and seating is on tiered levels, not more than 200 seats shall not be required to be fastened to the floor. Plans showing seating, tiers and *aisles* shall be submitted for approval.
5. Groups of seats within a building, room or space used for assembly purposes separated from other seating by railings, *guards*, partial height walls or similar barriers with level floors and having not more than 14 seats per group shall not be required to be fastened to the floor.
6. Seats intended for musicians or other performers and separated by railings, *guards*, partial height walls or similar barriers shall not be required to be fastened to the floor.

1030.16 Handrails. Ramped *aisles* having a slope exceeding one unit vertical in 15 units horizontal (6.7-percent slope) and stepped *aisles* shall be provided with *handrails* in compliance with Section 1014 located either at one or both sides of the *aisle* or within the *aisle* width. Where stepped *aisles* have seating on one side and the *aisle* width is 74 inches (1880 mm) or greater, two *handrails* are required. Where two *handrails* are required, one of the *handrails* shall be within 30 inches (762 mm) horizontally of the stepped *aisle*.

Exceptions:

1. *Handrails* are not required for ramped *aisles* with seating on both sides.
2. *Handrails* are not required where, at the side of the *aisle*, there is a *guard* with a top surface that complies with the graspability requirements of *handrails* in accordance with Section 1014.3.
3. *Handrail* extensions are not required at the top and bottom of stepped *aisles* and ramped *aisles* to permit crossovers within the *aisles*.

1030.16.1 Discontinuous mid-aisle handrails. Where there is seating on both sides of the *aisle*, the mid-aisle *handrails* shall be discontinuous. Where a stepped *aisle* is required to have two *handrails*, the mid-aisle *handrails* shall be discontinuous. Gaps or breaks shall be provided at intervals not exceeding five rows to facilitate access to seating and to permit crossing from one side of the *aisle* to the other. These gaps or breaks shall have a clear width of not less than 22 inches (559 mm) and not greater than 36 inches (914 mm), measured horizontally, and the mid-aisle *handrail* shall have rounded terminations or bends.

1030.16.2 Handrail termination. *Handrails* located on the side of stepped *aisles* shall return to a wall, *guard* or the walking surface or shall be continuous to the *handrail* of an adjacent stepped *aisle flight*.

1030.16.3 Mid-aisle termination. Mid-aisle *handrails* shall not extend beyond the lowest riser and shall terminate within 18 inches (381 mm), measured horizontally, from the lowest riser. *Handrail* extensions are not required.

Exception: Mid-aisle *handrails* shall be permitted to extend beyond the lowest riser where the *handrail* extensions do not obstruct the width of the cross *aisle*.

1030.16.4 Rails. Where mid-aisle *handrails* are provided in stepped *aisles*, there shall be an additional rail located approximately 12 inches (305 mm) below the *handrail*. The rail shall be adequate in strength and attachment in accordance with Section 1607.9.1.2.

1030.17 Assembly guards. *Guards* adjacent to seating in a building, room or space used for assembly purposes shall be provided where required by Section 1015 and shall be constructed in accordance with Section 1015 except where provided in accordance with Sections 1030.17.1 through 1030.17.4. At *bleachers*, *grandstands* and *folding and telescopic seating*, *guards* must be provided where required by ICC 300 and Section 1030.17.1.

1030.17.1 Perimeter guards. Perimeter *guards* shall be provided where the footboards or walking surface of seating facilities are more than 30 inches (762 mm) above the floor or grade below. Where the seatboards are adjacent to the perimeter, *guard* height shall be 42 inches (1067 mm) high minimum, measured from the seatboard. Where the seats are self-rising, *guard* height shall be 42 inches (1067 mm) high minimum, measured from the floor surface. Where there is an *aisle* between the seating and the perimeter, the *guard* height shall be measured in accordance with Section 1015.3.

Exceptions:

1. *Guards* that impact sightlines shall be permitted to comply with Section 1030.17.3.
2. *Bleachers*, *grandstands* and *folding and telescopic seating* shall not be required to have

perimeter *guards* where the seating is located adjacent to a wall and the space between the wall and the seating is less than 4 inches (102 mm).

1030.17.2 Cross aisles. Cross *aisles* located more than 30 inches (762 mm) above the floor or grade below shall have *guards* in accordance with Section 1015.

Where an elevation change of 30 inches (762 mm) or less occurs between a cross *aisle* and the adjacent floor or grade below, *guards* not less than 26 inches (660 mm) above the *aisle* floor shall be provided.

Exception: Where the backs of seats on the front of the cross *aisle* project 24 inches (610 mm) or more above the adjacent floor of the *aisle*, a *guard* need not be provided.

1030.17.3 Sightline-constrained guard heights. Unless subject to the requirements of Section 1030.17.4, a fascia or railing system in accordance with the *guard* requirements of Section 1015 and having a minimum height of 26 inches (660 mm) shall be provided where the floor or footboard elevation is more than 30 inches (762 mm) above the floor or grade below and the fascia or railing would otherwise interfere with the sightlines of immediately adjacent seating.

1030.17.4 Guards at the end of aisles. A fascia or railing system complying with the *guard* requirements of Section 1015 shall be provided for the full width of the *aisle* where the foot of the *aisle* is more than 30 inches (762 mm) above the floor or grade below. The fascia or railing shall be not less than 36 inches (914 mm) high and shall provide not less than 42 inches (1067 mm) measured diagonally between the top of the rail and the *nosing* of the nearest tread.

SECTION 1031
EMERGENCY ESCAPE AND RESCUE

1031.1 General. *Emergency escape and rescue openings* shall comply with the requirements of this section.

1031.2 Where required. In addition to the *means of egress* required by this chapter, *emergency escape and rescue openings* shall be provided in the following occupancies:

1. Group R-2 occupancies located in stories with only one *exit* or *access* to only one *exit* as permitted by Tables 1006.3.4(1) and 1006.3.4(2).
2. Group R-3 and R-4 occupancies.

Basements and sleeping rooms below the fourth *story above grade plane* shall have not fewer than one *emergency escape and rescue opening* in accordance with this section. Where *basements* contain one or more sleeping rooms, an *emergency escape and rescue opening* shall be required in each sleeping room, but shall not be required in adjoining areas of the *basement*. Such openings shall open directly into a *public way* or to a *yard* or *court* that opens to a *public way*.

Exceptions:

1. *Basements* with a ceiling height of less than 80 inches (2032 mm) shall not be required to have *emergency escape and rescue openings*.
2. *Emergency escape and rescue openings* are not required from *basements* or sleeping rooms that have an *exit* door or *exit access* door that opens directly into a *public way* or to a *yard*, court or exterior egress balcony that opens to a *public way*.
3. *Basements* without *habitable spaces* and having not more than 200 square feet (18.6 m^2) in floor area shall not be required to have *emergency escape and rescue openings*.
4. *Storm shelters* are not required to comply with this section where the shelter is constructed in accordance with ICC 500.
5. Within individual *dwelling* and *sleeping units* in Groups R-2 and R-3, where the building is equipped throughout with an *automatic sprinkler system* installed in accordance with Section 903.3.1.1, 903.3.1.2 or 903.3.1.3, *sleeping rooms* in *basements* shall not be required to have *emergency escape and rescue openings* provided that the basement has one of the following:
   5.1. One *means of egress* and one *emergency escape and rescue opening*.
   5.2. Two *means of egress*.

1031.2.1 Operational constraints and opening control devices. *Emergency escape and rescue openings* shall be operational from inside the room without the use of keys or tools. Window-opening control devices complying with ASTM F2090 shall be permitted for use on windows serving as a required *emergency escape and rescue opening*.

1031.3 Emergency escape and rescue openings. *Emergency escape and rescue openings* shall comply with Sections 1031.3.1 through 1031.3.3.

1031.3.1 Minimum size. *Emergency escape and rescue openings* shall have a minimum net clear opening of 5.7 square feet (0.53 m^2).

Exception: The minimum net clear opening for grade-floor *emergency escape and rescue openings* shall be 5 square feet (0.46 m^2).

1031.3.2 Minimum dimensions. The minimum net clear opening height dimension shall be 24 inches (610 mm). The minimum net clear opening width dimension shall be 20 inches (508 mm). The net clear opening dimensions shall be the result of normal operation of the opening.

1031.3.3 Maximum height from floor. *Emergency escape and rescue openings* shall have the bottom of the clear opening not greater than 44 inches (1118 mm) measured from the floor.

1031.4 Emergency escape and rescue doors. Where a door is provided as the required *emergency escape and rescue opening*, it shall be a swinging door or a sliding door.

1031.5 Area wells. An *emergency escape and rescue opening* with the bottom of the clear opening below the adjacent grade shall be provided with an area well in accordance with Sections 1031.5.1 through 1031.5.3.

1031.5.1 Minimum size. The minimum horizontal area of the area well shall be 9 square feet (0.84 m^2), with a horizontal projection and width of not less than 36 inches (914 mm). The area well shall allow the *emergency escape and rescue opening* to be fully opened.

Exception: The ladder or steps required by Section 1031.5.2 shall be permitted to encroach not more than 6 inches (152 mm) into the required dimensions of the area well.

1031.5.2 Ladders or steps. Area wells with a vertical depth of more than 44 inches (1118 mm) shall be equipped with an *approved* permanently affixed ladder or steps. The ladder or steps shall not be obstructed by the *emergency escape and rescue opening* when the window or door is in the open position. Ladders or steps required by this section shall not be required to comply with Section 1011.

1031.5.2.1 Ladders. Ladders or rungs shall have an inside width of at least 12 inches (305 mm), shall project at least 3 inches (76 mm) from the wall and shall be spaced not more than 18 inches (457 mm) on center (o.c.) vertically for the full height of the area well.

1031.5.2.2 Steps. Steps shall have an inside width of not less than 12 inches (305 mm), shall have treads greater than 5 inches (127 mm) in depth and a riser height not greater than 18 inches (457 mm) for the full height of the area well.

1031.5.3 Drainage. Area wells shall be designed for proper drainage by connecting to the building's foundation drainage system required by Section 1805.

Exception: A drainage system for area wells is not required where the foundation is on well-drained soil or sand-gravel mixture soils in accordance with the United Soil Classification System, Group I Soils, in accordance with Section 1803.5.1.

1031.6 Bars, grilles, covers and screens. Where bars, grilles, covers, screens or similar devices are placed over *emergency escape and rescue openings* or area wells that serve such openings, the minimum net clear opening size shall comply with Sections 1031.3 and 1031.5. Such devices shall be releasable or removable from the inside without the use of a key, tool or force greater than that which is required for normal operation of the *emergency escape and rescue opening*.

CHAPTER 11
ACCESSIBILITY

User note:

About this chapter: *Chapter 11 contains provisions that set forth requirements for accessibility of buildings and their associated sites and facilities for people with physical disabilities. The fundamental philosophy of the code on the subject of accessibility is that everything is required to be accessible. This is reflected in the basic applicability requirement (see Section 1103.1). The code's scoping requirements then address the conditions under which accessibility is not required in terms of exceptions to this general mandate. While the IBC contains scoping provisions for accessibility (for example, what, where and how many), ICC A117.1, Accessible and Usable Buildings and Facilities, is the referenced standard for the technical provisions (in other words, how). Accessibility criteria for existing buildings are addressed in the* International Existing Building Code®. *The* International Residential Code® *references Chapter 11 for accessibility provisions; therefore, this chapter may be applicable to housing covered under the* International Residential Code. *The provisions in the I-Codes are intended to meet or exceed the requirements in the federal accessibility requirement found in the Americans with Disabilities Act and the Fair Housing Act.*

There are many accessibility issues that not only benefit people with disabilities, but also provide a tangible benefit to people without disabilities. This type of requirement can be set forth in the code as generally applicable without necessarily identifying it specifically as an accessibility-related issue. Such a requirement would then be considered as having been "mainstreamed." For example, visible alarms are located in Chapter 9 and accessible means of egress and ramp requirements are addressed in Chapter 10.

SECTION 1101 GENERAL

1101.1 Scope. The provisions of this chapter shall control the design and construction of facilities for accessibility for individuals with disabilities.

SECTION 1102 COMPLIANCE

1102.1 Design. Buildings and facilities shall be designed and constructed to be *accessible* in accordance with this code and ICC A117.1.

SECTION 1103 SCOPING REQUIREMENTS

1103.1 Where required. *Sites*, buildings, *structures, facilities*, elements and spaces, temporary or permanent, shall be accessible to individuals with disabilities.

1103.2 General exceptions. *Sites*, buildings, *structures, facilities*, elements and spaces shall be exempt from this chapter to the extent specified in this section.

1103.2.1 Specific requirements. *Accessibility* is not required in buildings and *facilities*, or portions thereof, to the extent permitted by Sections 1104 through 1112.

1103.2.2 Employee work areas. Spaces and elements within *employee work areas* shall only be required to comply with Sections 907.5.2.3.1, 1009 and 1104.3.1 and shall be designed and constructed so that individuals with disabilities can approach, enter and exit the work area. Work areas, or portions of work areas, other than raised courtroom stations in accordance with Section 1109.4.1.4, that are less than 300 square feet (30 m^2) in area and located 7 inches (178 mm) or more above or below the ground or finished floor where the change in elevation is essential to the function of the space shall be exempt from all requirements.

1103.2.3 Detached dwellings. Detached one- and two-family *dwellings*, their accessory structures and their associated *sites* and *facilities* are not required to comply with this chapter.

1103.2.4 Utility buildings. Group U occupancies are not required to comply with this chapter other than the following:

1. In agricultural buildings, access is required to paved work areas and areas open to the general public.
2. *Private garages* or carports that contain required accessible parking.

1103.2.5 Construction sites. Structures, *sites* and equipment directly associated with the actual processes of construction including, but not limited to, scaffolding, bridging, materials hoists, materials storage or construction trailers are not required to comply with this chapter.

1103.2.6 Raised areas. Raised areas used primarily for purposes of security, life safety or fire safety including, but not limited to, observation galleries, prison guard towers, fire towers or lifeguard stands are not required to comply with this chapter.

1103.2.7 Limited access spaces. Spaces accessed only by ladders, catwalks, crawl spaces, freight elevators or very narrow passageways are not required to comply with this chapter.

1103.2.8 Areas in places of religious worship. Raised or lowered areas, or portions of areas, in *places of religious worship* that are less than 300 square feet (30 m^2) in area and located 7 inches (178 mm) or more above or below the finished floor and used primarily for the performance of religious ceremonies are not required to comply with this chapter.

1103.2.9 Equipment spaces. Spaces frequented only by service personnel for maintenance, repair or occasional monitoring of equipment are not required to comply with this chapter.

1103.2.10 Highway tollbooths. Highway tollbooths where the access is provided only by bridges above the vehicular traffic or underground tunnels are not required to comply with this chapter.

1103.2.11 Residential Group R-1. Buildings of Group R-1 containing not more than five *sleeping units* for rent or hire that are also occupied as the residence of the proprietor are not required to comply with this chapter.

1103.2.12 Day care facilities. Where a day care facility is part of a *dwelling unit*, only the portion of the structure utilized for the day care facility is required to comply with this chapter.

1103.2.13 Detention and correctional facilities. In detention and correctional facilities, common use areas that are used only by inmates or detainees and security personnel, and that do not serve holding *cells* or housing *cells* required to be *Accessible units*, are not required to comply with this chapter.

1103.2.14 Walk-in coolers and freezers. Walk-in cooler and freezer equipment accessed only from *employee work areas* is not required to comply with this chapter.

SECTION 1104
ACCESSIBLE ROUTE

1104.1 Site arrival points. At least one *accessible* route within the *site* shall be provided from public transportation stops, accessible parking, accessible passenger loading zones, and public streets or sidewalks to the accessible building entrance served.

Exception: Other than in buildings or *facilities* containing or serving *Type B units*, an *accessible route* shall not be required between *site* arrival points and the building or *facility* entrance if the only means of access between them is a vehicular way not providing for pedestrian access.

1104.2 Within a site. At least one *accessible* route shall connect accessible buildings, accessible facilities, accessible elements and accessible spaces that are on the same *site*.

Exceptions:

1. An *accessible* route is not required between accessible buildings, accessible facilities, accessible elements and accessible spaces that have, as the only means of access between them, a vehicular way not providing for pedestrian access.
2. An *accessible route* to recreational facilities shall only be required to the extent specified in Section 1111.

1104.3 Connected spaces. Where a building or portion of a building is required to be *accessible*, at least one *accessible* route shall be provided to each portion of the building, to accessible building entrances connecting accessible *pedestrian walkways* and to the *public way*.

Exceptions:

1. *Stories* and *mezzanines* exempted by Section 1104.4.
2. In a building, room or space used for assembly purposes with *fixed seating*, an *accessible route* shall not be required to serve levels where *wheelchair spaces* are not provided.
3. Vertical access to elevated employee work stations within a courtroom complying with Section 1109.4.1.4.
4. An *accessible route* to recreational facilities shall only be required to the extent specified in Section 1111.

1104.3.1 Employee work areas. *Common use circulation paths* within *employee work areas* shall be *accessible routes*.

Exceptions:

1. *Common use circulation paths*, located within *employee work areas* that are less than 1,000 square feet (93 m^2) in size and defined by permanently installed partitions, counters, casework or furnishings, shall not be required to be *accessible routes*.
2. *Common use circulation paths*, located within *employee work areas*, that are an integral component of equipment, shall not be required to be *accessible routes*.
3. *Common use circulation paths*, located within exterior *employee work areas* that are fully exposed to the weather, shall not be required to be *accessible routes*.

1104.3.2 Press boxes. Press boxes in a building, room or space used for assembly purposes shall be on an *accessible route*.

Exceptions:

1. An *accessible route* shall not be required to press boxes in *bleachers* that have a single point of entry from the *bleachers*, provided that the aggregate area of all press boxes for each playing field is not more than 500 square feet (46 m^2).
2. An *accessible route* shall not be required to free-standing press boxes that are more than 12 feet (3660 mm) above grade provided that the aggregate area of all press boxes for each playing field is not more than 500 square feet (46 m^2).

1104.4 Multistory buildings and facilities. At least one *accessible* route shall connect each accessible *story, mezza-*

nine and occupied roofs in multilevel buildings and *facilities*.

Exceptions:

1. An *accessible* route is not required to *stories, mezzanines* and occupied roofs that have an aggregate area of not more than 3,000 square feet (278.7 m^2) and are located above and below accessible levels. This exception shall not apply to:
   - 1.1. Multiple tenant facilities of Group M occupancies containing five or more tenant spaces used for the sales or rental of goods and where at least one such tenant space is located on a floor level above or below the accessible levels.
   - 1.2. *Stories* or *mezzanines* containing offices of health care providers (Group B or I).
   - 1.3. Passenger transportation facilities and airports (Group A-3 or B).
   - 1.4. Government buildings.
   - 1.5. Structures with four or more dwelling units.
2. *Stories*, *mezzanines* or occupied roofs that do not contain accessible elements or other spaces as determined by Section 1108 or 1109 are not required to be served by an accessible route from an *accessible* level.
3. In air traffic control towers, an *accessible route* is not required to serve the cab and the floor immediately below the cab.
4. Where a two-story building or facility has one *story* or *mezzanine* with an *occupant load* of five or fewer persons that does not contain *public use* space, that *story* or *mezzanine* shall not be required to be connected by an *accessible route* to the *story* above or below.

1104.5 Location. *Accessible routes* shall coincide with or be located in the same area as a general *circulation path*. Where the *circulation path* is interior, the *accessible route* shall be interior. Where only one *accessible route* is provided, the *accessible route* shall not pass through kitchens, storage rooms, restrooms, closets or similar spaces.

Exceptions:

1. *Accessible routes* from parking garages contained within and serving *Type B units* are not required to be interior.
2. A single *accessible route* is permitted to pass through a kitchen or storage room in an *Accessible unit, Type A unit* or *Type B unit*.

1104.6 Security barriers. Security barriers including, but not limited to, security bollards and security check points shall not obstruct a required *accessible route* or accessible means of egress.

Exception: Where security barriers incorporate elements that cannot comply with these requirements, such as certain metal detectors, fluoroscopes or other similar devices, the *accessible route* shall be permitted to be provided adjacent to security screening devices. The *accessible route* shall permit persons with disabilities passing around security barriers to maintain visual contact with their personal items to the same extent provided others passing through the security barrier.

SECTION 1105
ACCESSIBLE ENTRANCES

1105.1 Public entrances. In addition to accessible entrances required by Sections 1105.1.2 through 1105.1.8, at least 60 percent of all *public entrances* shall be *accessible.*

Exceptions:

1. An accessible entrance is not required to areas not required to be *accessible*.
2. Loading and *service entrances* that are not the only entrance to a tenant space.

1105.1.1 Automatic doors. In facilities with the occupancies and building *occupant loads* indicated in Table 1105.1.1, *public entrances* that are required to be *accessible* shall have one door be either a full *power-operated* door or a *low-energy power-operated door*. Where the *public entrance* includes a vestibule, at least one door into and one door out of the vestibule shall meet the requirements of this section.

TABLE 1105.1.1
PUBLIC ENTRANCE WITH POWER-OPERATED DOOR[a]

OCCUPANCY	BUILDING OCCUPANT LOAD GREATER THAN
A-1, A-2, A-3, A-4	300
B, M, R-1	500

a. In mixed-use facilities where the total sum of the building occupant load is greater than those listed, the most restrictive building occupant load shall apply.

1105.1.2 Parking garage entrances. Where provided, direct access for pedestrians from parking structures to buildings or facility entrances shall be *accessible.*

1105.1.3 Entrances from tunnels or elevated walkways. Where direct access is provided for pedestrians from a pedestrian tunnel or elevated walkway to a building or facility, at least one entrance to the building or facility from each tunnel or walkway shall be *accessible.*

1105.1.4 Restricted entrances. Where *restricted entrances* are provided to a building or facility, at least one *restricted entrance* to the building or facility shall be *accessible.*

1105.1.5 Entrances for inmates or detainees. Where entrances used only by inmates or detainees and security personnel are provided at judicial facilities, detention facilities or correctional facilities, at least one such entrance shall be *accessible.*

1105.1.6 Service entrances. If a *service entrance* is the only entrance to a building or a tenant space in a facility, that entrance shall be *accessible.*

1105.1.7 Tenant spaces. At least one accessible entrance shall be provided to each tenant in a facility.

Exception: An accessible entrance is not required to *self-service storage facilities* that are not required to be *accessible*.

1105.1.8 Dwelling units and sleeping units. At least one accessible entrance shall be provided to each *dwelling unit* and *sleeping unit* in a facility.

Exception: An accessible entrance is not required to *dwelling units* and *sleeping units* that are not required to be *Accessible units, Type A units* or *Type B units*.

SECTION 1106 PARKING AND PASSENGER LOADING FACILITIES

1106.1 General. Parking shall comply with Sections 1106.2 through 1106.8. Passenger loading zones shall comply with Section 1106.9.

1106.2 Required. Where parking is provided, accessible parking spaces shall be provided in compliance with Table 1106.2, except as required by Sections 1106.3 through 1106.5. Where more than one parking facility is provided on a site, the number of parking spaces required to be *accessible* shall be calculated separately for each parking facility.

Exception: This section does not apply to parking spaces used exclusively for buses, trucks, other delivery vehicles, law enforcement vehicles or vehicular impound and motor pools where lots accessed by the public are provided with an accessible passenger loading zone.

TABLE 1106.2 ACCESSIBLE PARKING SPACES

TOTAL PARKING SPACES PROVIDED IN PARKING FACILITIES	REQUIRED MINIMUM NUMBER OF ACCESSIBLE SPACES
1 to 25	1
26 to 50	2
51 to 75	3
76 to 100	4
101 to 150	5
151 to 200	6
201 to 300	7
301 to 400	8
401 to 500	9
501 to 1,000	2% of total
1,001 and over	20, plus one for each 100, or fraction thereof, over 1,000

1106.3 Groups I-1, R-1, R-2, R-3 and R-4. *Accessible* parking spaces shall be provided in Group I-1, R-1, R-2, R-3 and R-4 occupancies in accordance with Items 1 through 4 as applicable.

1. In Group R-2, R-3 and R-4 occupancies that are required to have Accessible, Type A or *Type B dwelling units* or *sleeping units*, at least 2 percent, but not less than one, of each type of parking space provided shall be accessible.
2. In Group I-1 and R-1 occupancies, accessible parking shall be provided in accordance with Table 1106.2.
3. Where at least one parking space is provided for each *dwelling unit* or *sleeping unit*, at least one *accessible* parking space shall be provided for each *Accessible* and *Type A unit*.
4. Where parking is provided within or beneath a building, *accessible* parking spaces shall be provided within or beneath the building.

1106.4 Hospital outpatient facilities. At least 10 percent, but not less than one, of care recipient and visitor parking spaces provided to serve *hospital* outpatient facilities shall be *accessible*.

1106.5 Rehabilitation facilities and outpatient physical therapy facilities. At least 20 percent, but not less than one, of the portion of care recipient and visitor parking spaces serving rehabilitation facilities specializing in treating conditions that affect mobility and outpatient physical therapy facilities shall be *accessible*.

1106.6 Van spaces. For every six or fraction of six accessible parking spaces, at least one shall be a van-accessible parking space.

Exception: In Group U *private garages* that serve Group R-2 and R-3 occupancies, van-accessible spaces shall be permitted to have vehicular routes, entrances, parking spaces and access aisles with a minimum vertical clearance of 7 feet (2134 mm).

1106.7 Location. Accessible parking spaces shall be located on the shortest accessible route of travel from adjacent parking to an accessible building entrance. In parking facilities that do not serve a particular building, accessible parking spaces shall be located on the shortest route to an accessible pedestrian entrance to the parking facility. Where buildings have multiple accessible entrances with adjacent parking, *accessible* parking spaces shall be dispersed and located near the accessible entrances.

Exceptions:

1. In multilevel parking structures, van-accessible parking spaces are permitted on one level.
2. *Accessible* parking spaces shall be permitted to be located in different parking facilities if substantially equivalent or greater accessibility is provided in terms of distance from an *accessible* entrance or entrances, parking fee and user convenience.

1106.8 Parking meters and pay stations. Where parking meters and pay stations serve accessible parking spaces, such parking meters and pay stations shall be *accessible*.

1106.9 Passenger loading zones. Passenger loading zones shall be *accessible*.

1106.9.1 Continuous loading zones. Where passenger loading zones are provided, one passenger loading zone in every continuous 100 linear feet (30.4 m) maximum of loading zone space shall be *accessible*.

1106.9.2 Medical facilities. A passenger loading zone shall be provided at an accessible entrance to licensed medical and long-term care facilities where people receive physical or medical treatment or care and where the period of stay exceeds 24 hours.

1106.9.3 Valet parking. A passenger loading zone shall be provided at valet parking services.

1106.9.4 Mechanical access parking garages. Mechanical access parking garages shall provide at least one passenger loading zone at vehicle drop-off and vehicle pick-up areas.

SECTION 1107 MOTOR-VEHICLE-RELATED FACILITIES

1107.1 General. Electrical vehicle charging stations shall comply with Section 1107.2. Fuel-dispensing systems shall comply with Section 1107.3.

1107.2 Electrical vehicle charging stations. Electrical vehicle charging stations shall comply with Sections 1107.2.1 and 1107.2.2.

Exception: Electrical vehicle charging stations provided to serve Group R-2, R-3 and R-4 occupancies are not required to comply with this section.

1107.2.1 Number of accessible vehicle spaces. Not less than 5 percent of vehicle spaces on the site served by electrical vehicle charging systems, but not fewer than one for each type of electric vehicle charging system, shall be accessible.

1107.2.2 Vehicle space size. Accessible vehicle spaces shall comply with the requirements for a van accessible parking space that is 132 inches (3350 mm) minimum in width with an adjoining access aisle that is 60 inches (1525 mm) minimum in width.

** **1107.3 Fuel-dispensing systems.** Fuel-dispensing systems shall be *accessible*.

SECTION 1108 DWELLING UNITS AND SLEEPING UNITS

1108.1 General. In addition to the other requirements of this chapter, occupancies having *dwelling units* or *sleeping units* shall be provided with accessible features in accordance with this section.

1108.2 Design. *Dwelling units* and *sleeping units* that are required to be *Accessible units, Type A units* and *Type B units* shall comply with the applicable portions of Chapter 11 of ICC A117.1. Units required to be *Type A units* are permitted to be designed and constructed as *Accessible units*. Units required to be *Type B units* are permitted to be designed and constructed as *Accessible units* or as *Type A units*.

1108.3 Accessible spaces. Rooms and spaces available to the general public or available for use by residents and serving Accessible units, *Type A units* or *Type B units* shall be *accessible*. *Accessible* spaces shall include toilet and bathing rooms, kitchen, living and dining areas and any exterior spaces, including patios, terraces and balconies.

Exceptions:

1. *Stories* and *mezzanines* exempted by Section 1108.4.
2. Recreational facilities in accordance with Section 1111.2.
3. Exterior decks, patios or balconies that are part of *Type B units* and have impervious surfaces, and that are not more than 4 inches (102 mm) below the finished floor level of the adjacent interior space of the unit.

1108.4 Accessible route. Not fewer than one accessible route shall connect accessible building or facility entrances with the primary entrance of each *Accessible unit, Type A unit* and *Type B unit* within the building or facility and with those exterior and interior spaces and facilities that serve the units.

Exceptions:

1. If due to circumstances outside the control of the owner, either the slope of the finished ground level between accessible facilities and buildings exceeds one unit vertical in 12 units horizontal (1:12), or where physical barriers or legal restrictions prevent the installation of an *accessible route*, a vehicular route with parking that complies with Section 1106 at each *public* or common use facility or building is permitted in place of the *accessible route*.
2. In Group I-3 facilities, an *accessible route* is not required to connect *stories* or *mezzanines* where *Accessible units*, all common use areas serving *Accessible units* and all *public use* areas are on an *accessible route*.
3. In Group R-2 facilities with *Type A units* complying with Section 1108.6.2.2.1, an *accessible route* is not required to connect *stories* or *mezzanines* where *Type A units*, all common use areas serving *Type A units* and all public use areas are on an *accessible route*.
4. In other than Group R-2 *dormitory* housing provided by places of education, in Group R-2 facilities with *Accessible units* complying with Section 1108.6.2.3.1, an *accessible route* is not required to connect stories or *mezzanines* where *Accessible units*, all common use areas serving *Accessible units* and all *public use* areas are on an *accessible* route.
5. In Group R-1, an *accessible route* is not required to connect *stories* or *mezzanines* within individual units, provided the *accessible* level meets the provisions for *Accessible units* and sleeping accommodations for two persons minimum and a toilet facility are provided on that level.

6. In congregate residences in Groups R-3 and R-4, an *accessible route* is not required to connect *stories* or *mezzanines* where *Accessible units* or *Type B units*, all common use areas serving *Accessible units* and *Type B units* and all *public use* areas serving *Accessible units* and *Type B units* are on an *accessible route*.
7. An *accessible route* between *stories* is not required where *Type B units* are exempted by Section 1108.7.

1108.5 Group I. *Accessible units* and *Type B units* shall be provided in Group I occupancies in accordance with Sections 1108.5.1 through 1108.5.5.

1108.5.1 Group I-1. *Accessible units* and *Type B units* shall be provided in Group I-1 occupancies in accordance with Sections 1108.5.1.1 and 1108.5.1.3.

1108.5.1.1 Accessible units. In Group I-1, Condition 1, at least 4 percent, but not less than one, of the *dwelling units* and *sleeping units* shall be *Accessible units*. Accessible dwelling units and sleeping units shall be dispersed among the various classes of units.

Exceptions:

1. Water closets shall not be required to comply with ICC A117.1 where such water closets comply with Section 1110.2.2, in not more than 50 percent of the *Accessible units*.
2. Roll-in-type showers shall not be required to comply with ICC A117.1 where roll-in-type showers comply with Section 1110.2.3, in not more than 50 percent of the *Accessible units*.

1108.5.1.2 Accessible units in Group I-1, Condition 2. In Group I-1, Condition 2, at least 10 percent, but not less than one, of the dwelling units and sleeping units shall be *Accessible units*. Accessible dwelling units and sleeping units shall be dispersed among the various classes of units.

Exceptions:

1. Water closets shall not be required to comply with ICC A117.1 where such water closets comply with Section 1110.2.2, in not more than 50 percent of the *Accessible units*.
2. Roll-in-type showers shall not be required to comply with ICC A117.1 where roll-in-type showers comply with Section 1110.2.3, in not more than 50 percent of the *Accessible units*.

1108.5.1.3 Type B units. In structures with four or more *dwelling units* or *sleeping units intended to be occupied as a residence*, every *dwelling unit* and *sleeping unit intended to be occupied as a residence* shall be a *Type B unit*.

Exception: The number of *Type B units* is permitted to be reduced in accordance with Section 1108.7.

1108.5.2 Group I-2 nursing homes. *Accessible units* and *Type B units* shall be provided in *nursing homes* of Group I-2, Condition 1 occupancies in accordance with Sections 1108.5.2.1 and 1108.5.2.2.

1108.5.2.1 Accessible units. At least 50 percent but not less than one of each type of the *dwelling units* and *sleeping units* shall be *Accessible units*.

Exceptions:

1. Water closets shall not be required to comply with ICC A117.1 where such water closets comply with Section 1110.2.2, in not more than 90 percent of the *Accessible units*.
2. Roll-in-type showers shall not be required to comply with ICC A117.1 where roll-in-type showers comply with Section 1110.2.3, in not more than 90 percent of the *Accessible units*.

1108.5.2.2 Type B units. In structures with four or more *dwelling units* or *sleeping units intended to be occupied as a residence*, every *dwelling unit* and *sleeping unit intended to be occupied as a residence* shall be a *Type B unit*.

Exception: The number of *Type B units* is permitted to be reduced in accordance with Section 1108.7.

1108.5.3 Group I-2 hospitals. *Accessible units* and *Type B units* shall be provided in general-purpose *hospitals*, psychiatric facilities and *detoxification facilities* of Group I-2 occupancies in accordance with Sections 1108.5.3.1 and 1108.5.3.2.

1108.5.3.1 Accessible units. At least 10 percent, but not less than one, of the *dwelling units* and *sleeping units* shall be *Accessible units*.

Exception: Entry doors to *Accessible dwelling units* or *sleeping units* shall not be required to provide the maneuvering clearance beyond the latch side of the door.

1108.5.3.2 Type B units. In structures with four or more *dwelling units* or *sleeping units intended to be occupied as a residence*, every *dwelling unit* and *sleeping unit intended to be occupied as a residence* shall be a *Type B unit*.

Exception: The number of *Type B units* is permitted to be reduced in accordance with Section 1108.7.

1108.5.4 Group I-2 rehabilitation facilities. In *hospitals* and rehabilitation facilities of Group I-2 occupancies that specialize in treating conditions that affect mobility, or units within either that specialize in treating conditions that affect mobility, 100 percent of the *dwelling units* and *sleeping units* shall be *Accessible units*.

Exceptions:

1. Water closets shall not be required to comply with ICC A117.1 where such water closets comply with Section 1110.2.2, in not more than 50 percent of *Accessible units*.

2. Roll-in-type showers shall not be required to comply with ICC A117.1 where roll-in-type showers comply with Section 1110.2.3, in not more than 50 percent of *Accessible units*.

1108.5.5 Group I-3. *Accessible units* shall be provided in Group I-3 occupancies in accordance with Sections 1108.5.5.1 through 1108.5.5.3.

1108.5.5.1 Group I-3 sleeping units. In Group I-3 occupancies, at least 3 percent of the total number of *sleeping units* in the facility, but not less than one unit in each classification level, shall be *Accessible units*.

1108.5.5.2 Special holding cells and special housing cells or rooms. In addition to the *Accessible units* required by Section 1108.5.5.1, where special holding *cells* or special housing *cells* or rooms are provided, at least one serving each purpose shall be an *Accessible unit*. *Cells* or rooms subject to this requirement include, but are not limited to, those used for purposes of orientation, protective custody, administrative or disciplinary detention or segregation, detoxification and medical isolation.

Exception: *Cells* or rooms specially designed without protrusions and that are used solely for purposes of suicide prevention shall not be required to include grab bars.

1108.5.5.3 Medical care facilities. Patient *sleeping units* or *cells* required to be *Accessible units* in *medical care* facilities shall be provided in addition to any medical isolation *cells* required to comply with Section 1108.5.5.2.

1108.6 Group R. *Accessible units*, *Type A units* and *Type B units* shall be provided in Group R occupancies in accordance with Sections 1108.6.1 through 1108.6.4.

1108.6.1 Group R-1. *Accessible units* and *Type B units* shall be provided in Group R-1 occupancies in accordance with Sections 1108.6.1.1 and 1108.6.1.2.

1108.6.1.1 Accessible units. *Accessible dwelling units* and *sleeping units* shall be provided in accordance with Table 1108.6.1.1. On a multiple-building site, where structures contain more than 50 *dwelling units* or *sleeping units*, the number of *Accessible units* shall be determined per structure. On a multiple-building site, where structures contain 50 or fewer *dwelling units* or *sleeping units*, all *dwelling units* and *sleeping units* on a site shall be considered to determine the total number of *Accessible units*. *Accessible units* shall be dispersed among the various classes of units.

1108.6.1.2 Type B units. In structures with four or more *dwelling units* or *sleeping units intended to be occupied as a residence*, every *dwelling unit* and *sleeping unit intended to be occupied as a residence* shall be a *Type B unit*.

Exception: The number of *Type B units* is permitted to be reduced in accordance with Section 1108.7.

1108.6.2 Group R-2. *Accessible units, Type A units* and *Type B units* shall be provided in Group R-2 occupancies in accordance with Sections 1108.6.2.1 through 1108.6.2.3.

1108.6.2.1 Live/work units. In *live/work units* constructed in accordance with Section 419, the nonresidential portion is required to be *accessible*. In a structure where there are four or more *live/work units intended to be occupied as a residence*, the residential portion of the *live/work unit* shall be a *Type B unit*.

Exception: The number of *Type B units* is permitted to be reduced in accordance with Section 1108.7.

1108.6.2.2 Apartment houses, monasteries and convents. *Type A units* and *Type B units* shall be provided in apartment houses, monasteries and convents in accordance with Sections 1108.6.2.2.1 and 1108.6.2.2.2. Bedrooms in monasteries and convents shall be counted as units for the purpose of determining the number of units. Where the bedrooms are grouped in *sleeping units*, only one bedroom in each *sleeping unit* shall count toward the number of required *Type A units*.

TABLE 1108.6.1.1
ACCESSIBLE DWELLING UNITS AND SLEEPING UNITS

TOTAL NUMBER OF UNITS PROVIDED	MINIMUM REQUIRED NUMBER OF ACCESSIBLE UNITS WITHOUT ROLL-IN SHOWERS	MINIMUM REQUIRED NUMBER OF ACCESSIBLE UNITS WITH ROLL-IN SHOWERS	TOTAL NUMBER OF REQUIRED ACCESSIBLE UNITS
1 to 25	1	0	1
26 to 50	2	0	2
51 to 75	3	1	4
76 to 100	4	1	5
101 to 150	5	2	7
151 to 200	6	2	8
201 to 300	7	3	10
301 to 400	8	4	12
401 to 500	9	4	13
501 to 1,000	2% of total	1% of total	3% of total
Over 1,000	20, plus 1 for each 100, or fraction thereof, over 1,000	10 plus 1 for each 100, or fraction thereof, over 1,000	30 plus 2 for each 100, or fraction thereof, over 1,000

1108.6.2.2.1 Type A units. In Group R-2 occupancies containing more than 20 *dwelling units* or *sleeping units*, at least 2 percent but not less than one of the units shall be a *Type A unit*. All Group R-2 units on a site shall be considered to determine the total number of units and the required number of *Type A units*. *Type A units* shall be dispersed among the various classes of units.

Exceptions:

1. The number of *Type A units* is permitted to be reduced in accordance with Section 1108.7.
2. *Existing structures* on a site shall not contribute to the total number of units on a site.

1108.6.2.2.2 Type B units. Where there are four or more *dwelling units* or *sleeping units intended to be occupied as a residence* in a single structure, every *dwelling unit* and *sleeping unit intended to be occupied as a residence* shall be a *Type B unit*.

Exception: The number of *Type B units* is permitted to be reduced in accordance with Section 1108.7.

1108.6.2.3 Group R-2 other than live/work units, apartment houses, monasteries and convents. In Group R-2 occupancies, other than *live/work units*, apartment houses, monasteries and convents falling within the scope of Sections 1108.6.2.1 and 1108.6.2.2, *Accessible units* and *Type B units* shall be provided in accordance with Sections 1108.6.2.3.1 and 1108.6.2.3.2. Bedrooms within *congregate living facilities*, *dormitories*, sororities, fraternities and *boarding houses* shall be counted as *sleeping units* for the purpose of determining the number of units. Where the *bedrooms* are grouped into *dwelling* or *sleeping units*, only one *bedroom* in each *dwelling* or *sleeping unit* shall be permitted to count toward the number of required *Accessible units*.

1108.6.2.3.1 Accessible units. *Accessible dwelling units* and *sleeping units* shall be provided in accordance with Table 1108.6.1.1.

1108.6.2.3.2 Type B units. Where there are four or more *dwelling units* or *sleeping units intended to be occupied as a residence* in a single structure, every *dwelling unit* and every *sleeping unit intended to be occupied as a residence* shall be a *Type B unit*.

Exception: The number of *Type B units* is permitted to be reduced in accordance with Section 1108.7.

1108.6.3 Group R-3. In Group R-3 occupancies where there are four or more *dwelling units* or *sleeping units intended to be occupied as a residence* in a single structure, every *dwelling unit* and *sleeping unit intended to be occupied as a residence* shall be a *Type B unit*. Bedrooms within *congregate living facilities*, *dormitories*, sororities, fraternities, and *boarding houses* shall be counted as *sleeping units* for the purpose of determining the number of units.

Exception: The number of *Type B units* is permitted to be reduced in accordance with Section 1108.7.

1108.6.4 Group R-4. *Accessible units* and *Type B units* shall be provided in Group R-4 occupancies in accordance with Sections 1108.6.4.1 and 1108.6.4.2. Bedrooms in Group R-4 facilities shall be counted as *sleeping units* for the purpose of determining the number of units.

1108.6.4.1 Accessible units. In Group R-4, Condition 1, at least one of the *sleeping units* shall be an *Accessible unit*. In Group R-4, Condition 2, at least two of the *sleeping units* shall be an *Accessible unit*.

1108.6.4.2 Type B units. In structures with four or more *sleeping units intended to be occupied as a residence*, every *sleeping unit intended to be occupied as a residence* shall be a *Type B unit*.

Exception: The number of *Type B units* is permitted to be reduced in accordance with Section 1108.7.

1108.7 General exceptions. Where specifically permitted by Section 1108.5 or 1108.6, the required number of *Type A units* and *Type B units* is permitted to be reduced in accordance with Sections 1108.7.1 through 1108.7.5.

1108.7.1 Structures without elevator service. Where elevator service is not provided in a structure, only the *dwelling units* and *sleeping units* that are located on stories indicated in Sections 1108.7.1.1 and 1108.7.1.2 are required to be *Type A units* and *Type B units*, respectively. The number of *Type A units* shall be determined in accordance with Section 1108.6.2.2.1.

1108.7.1.1 One story with Type B units required. At least one *story* containing *dwelling units* or *sleeping units intended to be occupied as a residence* shall be provided with an accessible entrance from the exterior of the structure and all units *intended to be occupied as a residence* on that *story* shall be *Type B units*.

1108.7.1.2 Additional stories with Type B units. Where stories have entrances not included in determining compliance with Section 1108.7.1.1, and such entrances are proximate to arrival points intended to serve units on that *story*, as indicated in Items 1 and 2, all *dwelling units* and *sleeping units intended to be occupied as a residence* served by that entrance on that *story* shall be *Type B units*.

1. Where the slopes of the undisturbed site measured between the planned entrance and all vehicular or pedestrian arrival points within 50 feet (15 240 mm) of the planned entrance are 10 percent or less.
2. Where the slopes of the planned finished grade measured between the entrance and all vehicular or pedestrian arrival points within 50 feet (15 240 mm) of the planned entrance are 10 percent or less.

Where arrival points are not within 50 feet (15 240 mm) of the entrance, the closest arrival point shall be used to determine access unless that arrival point serves the *story* required by Section 1108.7.1.1.

1108.7.2 Multistory units. A *multistory dwelling unit* or *sleeping unit* that is not provided with elevator service is not required to be a *Type B unit*. Where a *multistory unit* is provided with external elevator service to only one floor, the floor provided with elevator service shall be the primary entry to the unit, shall comply with the requirements for a *Type B unit* and, where provided within the unit, a living area, a kitchen and a toilet facility shall be provided on that floor.

1108.7.3 Elevator service to the lowest story with units. Where elevator service in the building provides an *accessible route* only to the lowest *story* containing *dwelling units* or *sleeping units intended to be occupied as a residence*, only the units on that *story* that are *intended to be occupied as a residence* are required to be *Type B units*.

1108.7.4 Site impracticality. On a site with multiple nonelevator buildings, the number of units required by Section 1108.7.1 to be *Type B units* is permitted to be reduced to a percentage that is equal to the percentage of the entire site having grades, prior to development, that are less than 10 percent, provided that all of the following conditions are met:

1. Not less than 20 percent of the units required by Section 1108.7.1 on the site are *Type B units*.
2. Units required by Section 1108.7.1, where the slope between the building entrance serving the units on that *story* and a pedestrian or vehicular arrival point is not greater than 8.33 percent, are *Type B units*.
3. Units required by Section 1108.7.1, where an elevated walkway is planned between a building entrance serving the units on that *story* and a pedestrian or vehicular arrival point and the slope between them is 10 percent or less, are *Type B units*.
4. Units served by an elevator in accordance with Section 1108.7.3 are *Type B units*.

1108.7.5 Flood hazard areas. *Type A units* and *Type B units* shall not be required for buildings without elevator service that are located in *flood hazard areas* as established in Section 1612.3, where the minimum required elevation of the *lowest floor* or lowest supporting horizontal structural member, as applicable, results in all of the following:

1. A difference in elevation between the minimum required floor elevation at the primary entrances and vehicular and pedestrian arrival points within 50 feet (15 240 mm) exceeding 30 inches (762 mm).
2. A slope exceeding 10 percent between the minimum required floor elevation at the primary entrances and vehicular and pedestrian arrival points within 50 feet (15 240 mm).

Where such arrival points are not within 50 feet (15 240 mm) of the primary entrances, the closest arrival points shall be used.

SECTION 1109
SPECIAL OCCUPANCIES

1109.1 General. In addition to the other requirements of this chapter, the requirements of Sections 1109.2 through 1109.4 shall apply to specific occupancies.

1109.2 Assembly area seating. A building, room or space used for assembly purposes with *fixed seating* shall comply with Sections 1109.2.1 through 1109.2.5. Lawn seating shall comply with Section 1109.2.6. Assistive listening systems shall comply with Section 1109.2.7. Performance areas viewed from assembly seating areas shall comply with Section 1109.2.8. Dining areas shall comply with Section 1109.2.9.

1109.2.1 Services. If a service or facility is provided in an area that is not *accessible*, the same service or facility shall be provided on an accessible level and shall be *accessible*.

1109.2.2 Wheelchair spaces. In rooms and spaces used for assembly purposes with *fixed seating*, accessible *wheelchair spaces* shall be provided in accordance with Sections 1109.2.2.1 through 1109.2.2.3.

1109.2.2.1 General seating. *Wheelchair spaces* shall be provided in accordance with Table 1109.2.2.1.

TABLE 1109.2.2.1
ACCESSIBLE WHEELCHAIR SPACES

CAPACITY OF SEATING IN ASSEMBLY AREAS	MINIMUM REQUIRED NUMBER OF WHEELCHAIR SPACES
4 to 25	1
26 to 50	2
51 to 100	4
101 to 300	5
301 to 500	6
501 to 5,000	6, plus 1 for each 150, or fraction thereof, between 501 through 5,000
5,001 and over	36 plus 1 for each 200, or fraction thereof, over 5,000

1109.2.2.2 Luxury boxes, club boxes and suites. In each luxury box, club box and suite within arenas, stadiums and *grandstands, wheelchair spaces* shall be provided in accordance with Table 1109.2.2.1.

1109.2.2.3 Other boxes. In boxes other than those required to comply with Section 1109.2.2.2, the total number of *wheelchair spaces* provided shall be determined in accordance with Table 1109.2.2.1. *Wheelchair spaces* shall be located in not less than 20 percent of all boxes provided.

1109.2.3 Companion seats. At least one companion seat shall be provided for each *wheelchair space* required by Sections 1109.2.2.1 through 1109.2.2.3.

1109.2.4 Dispersion of wheelchair spaces in multilevel assembly seating areas. In *multilevel assembly seating* areas, *wheelchair spaces* shall be provided on the main floor level and on one of each two additional floor or *mezzanine* levels. *Wheelchair spaces* shall be provided in each luxury box, club box and suite within assembly facilities.

Exceptions:

1. In *multilevel assembly seating* areas utilized for worship services where the second floor or *mezzanine* level contains 25 percent or less of the total seating capacity, *wheelchair spaces* shall be permitted to all be located on the main level.
2. In *multilevel assembly seating* areas where the second floor or *mezzanine* level provides 25 percent or less of the total seating capacity and 300 or fewer seats, all *wheelchair spaces* shall be permitted to be located on the main level.
3. *Wheelchair spaces* in team or player seating serving *areas of sport activity* are not required to be dispersed.

1109.2.5 Designated aisle seats. At least 5 percent, but not less than one, of the total number of aisle seats provided shall be designated aisle seats and shall be the aisle seats located closest to *accessible routes*.

Exception: Designated aisle seats are not required in team or player seating serving *areas of sport activity*.

1109.2.6 Lawn seating. Lawn seating areas and exterior overflow seating areas, where fixed seats are not provided, shall connect to an *accessible route*.

1109.2.7 Assistive listening systems. Each building, room or space used for assembly purposes where audible communications are integral to the use of the space shall have an assistive listening system.

Exception: Other than in courtrooms, an assistive listening system is not required where there is no audio amplification system.

1109.2.7.1 Receivers. The number and type of receivers shall be provided for assistive listening systems in accordance with Table 1109.2.7.1.

Exceptions:

1. Where a building contains more than one room or space used for assembly purposes, the total number of required receivers shall be permitted to be calculated based on the total number of seats in the building, provided that all receivers are usable with all systems and if the rooms or spaces used for assembly purposes required to provide assistive listening are under one management.
2. Where all seats in a building, room or space used for assembly purposes are served by an induction loop assistive listening system, the minimum number of receivers required by Table 1109.2.7.1 to be hearing-aid compatible shall not be required.

1109.2.7.2 Ticket windows. Where ticket windows are provided in stadiums and arenas, at least one window at each location shall have an assistive listening system.

1109.2.7.3 Public address systems. Where stadiums, arenas and *grandstands* have 15,000 fixed seats or more and provide audible public announcements, they shall also provide prerecorded or real-time captions of those audible public announcements.

1109.2.8 Performance areas. An *accessible route* shall directly connect the performance area to the assembly seating area where a *circulation path* directly connects a performance area to an assembly seating area. An *accessible route* shall be provided from performance areas to ancillary areas or facilities used by performers.

1109.2.9 Dining and drinking areas. In dining and drinking areas, all interior and exterior floor areas shall be *accessible* and be on an accessible route.

Exceptions:

1. An accessible route between *accessible* levels and stories above or below is not required where permitted by Section 1104.4, Exception 1.

TABLE 1109.2.7.1
RECEIVERS FOR ASSISTIVE LISTENING SYSTEMS

CAPACITY OF SEATING IN ASSEMBLY AREAS	MINIMUM REQUIRED NUMBER OF RECEIVERS	MINIMUM NUMBER OF RECEIVERS TO BE HEARING-AID COMPATIBLE
50 or less	2	2
51 to 200	2, plus 1 per 25 seats over 50 seats*	2
201 to 500	2, plus 1 per 25 seats over 50 seats*	1 per 4 receivers*
501 to 1,000	20, plus 1 per 33 seats over 500 seats*	1 per 4 receivers*
1,001 to 2,000	35, plus 1 per 50 seats over 1,000 seats*	1 per 4 receivers*
Over 2,000	55, plus 1 per 100 seats over 2,000 seats*	1 per 4 receivers*

Note: * = or fraction thereof

2. An accessible route to dining and drinking areas in a *mezzanine* is not required, provided that the *mezzanine* contains less than 25 percent of the total combined area for dining and drinking and the same services, and decor are provided in the *accessible* area.
3. In sports facilities, tiered dining areas providing seating required to be *accessible* shall be required to have *accessible routes* serving at least 25 percent of the dining area, provided that *accessible routes* serve *accessible* seating and where each tier is provided with the same services.
4. Employee-only work areas shall comply with Sections 1103.2.2 and 1104.3.1.

1109.2.9.1 Dining surfaces. Where dining surfaces for the consumption of food or drink are provided, at least 5 percent, but not less than one, of the dining surfaces for the seating and standing spaces shall be *accessible* and be distributed throughout the facility and located on a level accessed by an *accessible route*.

1109.3 Self-service storage facilities. *Self-service storage facilities* shall provide accessible individual self-storage spaces in accordance with Table 1109.3.

TABLE 1109.3
ACCESSIBLE SELF-SERVICE STORAGE FACILITIES

TOTAL SPACES IN FACILITY	MINIMUM NUMBER OF REQUIRED ACCESSIBLE SPACES
1 to 200	5%, but not less than 1
Over 200	10, plus 2% of total number of units over 200

1109.3.1 Dispersion. Accessible individual self-service storage spaces shall be dispersed throughout the various classes of spaces provided. Where more classes of spaces are provided than the number of required accessible spaces, the number of *accessible* spaces shall not be required to exceed that required by Table 1109.3. *Accessible* spaces are permitted to be dispersed in a single building of a multiple-building facility.

1109.4 Judicial facilities. Judicial facilities shall comply with Sections 1109.4.1 and 1109.4.2.

1109.4.1 Courtrooms. Each courtroom shall be accessible and comply with Sections 1109.4.1.1 through 1109.4.1.5.

1109.4.1.1 Jury box. A *wheelchair space* shall be provided within the jury box.

Exception: Adjacent companion seating is not required.

1109.4.1.2 Gallery seating. *Wheelchair spaces* shall be provided in accordance with Table 1109.2.2.1. Designated aisle seats shall be provided in accordance with Section 1109.2.5.

1109.4.1.3 Assistive listening systems. An assistive listening system must be provided. Receivers shall be provided for the assistive listening system in accordance with Section 1109.2.7.1.

1109.4.1.4 Employee work stations. The judge's bench, clerk's station, bailiff's station, deputy clerk's station and court reporter's station shall be located on an *accessible route*. The vertical access to elevated employee work stations within a courtroom is not required at the time of initial construction, provided a *ramp*, lift or elevator can be installed without requiring reconfiguration or extension of the courtroom or extension of the electrical system.

1109.4.1.5 Other work stations. The litigant's and counsel stations, including the lectern, shall be *accessible*.

1109.4.2 Holding cells. Central holding *cells* and court-floor holding *cells* shall comply with Sections 1109.4.2.1 and 1109.4.2.2.

1109.4.2.1 Central holding cells. Where separate central holding *cells* are provided for adult males, juvenile males, adult females or juvenile females, one of each type shall be *accessible*. Where central holding *cells* are provided and are not separated by age or sex, at least one accessible cell shall be provided.

1109.4.2.2 Court-floor holding cells. Where separate court-floor holding *cells* are provided for adult males, juvenile males, adult females or juvenile females, each courtroom shall be served by one accessible cell of each type. Where court-floor holding *cells* are provided and are not separated by age or sex, courtrooms shall be served by at least one accessible cell. *Accessible cells* shall be permitted to serve more than one courtroom.

SECTION 1110
OTHER FEATURES AND FACILITIES

1110.1 General. *Accessible* building features and facilities shall be provided in accordance with Sections 1110.2 through 1110.16.

Exception: *Accessible units, Type A units* and *Type B units* shall comply with Chapter 10 of ICC A117.1.

1110.2 Toilet and bathing facilities. Each toilet room and bathing room shall be *accessible*. Where a floor level is not required to be connected by an *accessible route*, the only toilet rooms or bathing rooms provided within the facility shall not be located on the inaccessible floor. Except as provided for in Sections 1110.2.4 and 1110.2.5, at least one of each type of fixture, element, control or dispenser in each accessible toilet room and bathing room shall be *accessible*.

Exceptions:

1. Toilet rooms or bathing rooms accessed only through a private office, not for *common* or *public use* and intended for use by a single occupant, shall be permitted to comply with the specific exceptions in ICC A117.1.

2. This section is not applicable to toilet and bathing rooms that serve *dwelling units* or *sleeping units* that are not required to be *accessible* by Section 1108.
3. Where multiple single-user toilet rooms or bathing rooms are clustered at a single location, at least 50 percent but not less than one room for each use at each cluster shall be *accessible*.
4. Where no more than one urinal is provided in a toilet room or bathing room, the urinal is not required to be *accessible*.
5. Toilet rooms or bathing rooms that are part of critical care or intensive care patient sleeping rooms serving *Accessible units* are not required to be *accessible*.
6. Toilet rooms or bathing rooms designed for bariatrics patients are not required to comply with the toilet room and bathing room requirement in ICC A117.1. The *sleeping units* served by bariatrics toilet or bathing rooms shall not count toward the required number of Accessible *sleeping units*.
7. Where permitted in Section 1108, in toilet rooms or bathrooms serving Accessible units, water closets designed for assisted toileting shall comply with Section 1110.2.2.
8. Where permitted in Section 1108, in bathrooms serving Accessible units, showers designed for assisted bathing shall comply with Section 1110.2.3.
9. Where toilet facilities are primarily for children's use, required *accessible* water closets, toilet compartments and lavatories shall be permitted to comply with children's provision of ICC A117.1.

1110.2.1 Family or assisted-use toilet and bathing rooms. In assembly and mercantile occupancies, an accessible family or assisted-use toilet room shall be provided where an aggregate of six or more male and female water closets is required. In buildings of mixed occupancy, only those water closets required for the assembly or mercantile occupancy shall be used to determine the family or assisted-use toilet room requirement. In recreational facilities where separate-sex bathing rooms are provided, an accessible family or assisted-use bathing room shall be provided. Fixtures located within family or assisted-use toilet and bathing rooms shall be included in determining the number of fixtures provided in an occupancy.

Exception: Where each separate-sex bathing room has only one shower or bathtub fixture, a family or assisted-use bathing room is not required.

1110.2.1.1 Standard. Family or assisted-use toilet and bathing rooms shall comply with Sections 1110.2.1.2 through 1110.2.1.6.

1110.2.1.2 Family or assisted-use toilet rooms. Family or assisted-use toilet rooms shall include only one water closet and only one lavatory. A family or assisted-use bathing room in accordance with Section 1110.2.1.3 shall be considered to be a family or assisted-use toilet room.

Exception: The following additional fixtures shall be permitted in a family or assisted-use toilet room:

1. A urinal.
2. A child-height water closet.
3. A child-height lavatory.

1110.2.1.3 Family or assisted-use bathing rooms. Family or assisted-use bathing rooms shall include only one shower or bathtub fixture. Family or assisted-use bathing rooms shall also include one water closet and one lavatory. Where storage facilities are provided for separate-sex bathing rooms, accessible storage facilities shall be provided for family or assisted-use bathing rooms.

1110.2.1.4 Location. Family or assisted-use toilet and bathing rooms shall be located on an *accessible route*. Family or assisted-use toilet rooms shall be located not more than one *story* above or below separate-sex toilet rooms. The *accessible route* from any separate-sex toilet room to a family or assisted-use toilet room shall not exceed 500 feet (152 m).

1110.2.1.5 Prohibited location. In passenger transportation facilities and airports, the *accessible route* from separate-sex toilet rooms to a family or assisted-use toilet room shall not pass through security checkpoints.

1110.2.1.6 Privacy. Doors to family or assisted-use toilet and bathing rooms shall be securable from within the room and be provided with an "occupied" indicator.

1110.2.2 Water closets designed for assisted toileting. Water closets designed for assisted toileting shall comply with Sections 1110.2.2.1 through 1110.2.2.6.

1110.2.2.1 Location. The centerline of the water closet shall be not less than 24 inches (610 mm) and not greater than 26 inches (660 mm) from one side of the required clearance.

1110.2.2.2 Clearance. Clearance around the water closet shall comply with Sections 1110.2.2.2.1 through 1110.2.2.2.3.

1110.2.2.2.1 Clearance width. Clearance around a water closet shall be not less than 66 inches (1675 mm) in width, measured perpendicularly from the side of the clearance that is not less than 24 inches (610 mm) and not greater than 26 inches (660 mm) from the water closet centerline.

1110.2.2.2.2 Clearance depth. Clearance around the water closet shall be not less than 78 inches (1980 mm) in depth, measured perpendicularly from the rear wall

1110.2.2.2.3 Clearance overlap. The required clearance around the water closet shall permit overlaps per ICC A117.1, Section 604.3.3

1110.2.2.3 Height. The height of the water closet seats shall comply with ICC A117.1, Section 604.4.

1110.2.2.4 Swing-up grab bars. Swing-up grab bars shall comply with ICC A117.1, Sections 609.2 and 609.8. Swing-up grab bars shall be provided on both sides of the water closet and shall comply with all of the following:

1. The centerline of the grab bar shall be not less than 14 inches (356 mm) and not greater than 16 inches (405 mm) from the centerline of the water closet.
2. The length of the grab bar is not less than 36 inches (915 mm) in length, measured from the rear wall to the end of the grab bar.
3. The top of the grab bar in the down position is not less than 30 inches (760 mm) and not greater than34 inches (865 mm) above the floor.

1110.2.2.5 Flush controls. Flush controls shall comply with ICC A117.1, Section 604.6.

1110.2.2.6 Dispensers Toilet paper dispensers shall be mounted on at least one of the swing-up grab bars and the outlet of the dispenser shall be located not less than 24 inches (610 mm) and not greater than 36 inches (915 mm) from the rear wall.

1110.2.3 Standard roll-in-type shower compartment designed for assisted bathing. Standard roll-in-type shower compartments designed for assisted bathing shall comply with Sections 1110.2.3.1 through 1110.2.3.9.

1110.2.3.1 Size. Standard roll-in-type shower compartments shall have a clear inside dimension of not less than 60 inches (1525 mm) in width and 30 inches (760 mm) in depth, measured at the center point of opposing sides. An entry not less than 60 inches (1525 mm) in width shall be provided.

1110.2.3.2 Clearance. A clearance of not less than 60 inches (1525 mm) in length adjacent to the 60-inch (1525 mm) width of the open face of the shower compartment, and not less than 30 inches (760 mm) in depth, shall be provided.

Exceptions:

1. A lavatory complying with ICC A117.1, Section 606 shall be permitted at one end of the clearance.
2. Where the shower compartment exceeds minimum sizes, the clear floor space shall be placed adjacent to the grab bars and not less than 30 inches (762 mm) from the back wall.

1110.2.3.3 Grab bars. Grab bars shall comply with ICC A117.1, Section 609 and shall be provided in accordance with Sections 1110.2.3.3.1 and 1110.2.3.3.2. In standard roll-in-type shower compartments, grab bars shall be provided on three walls. Where multiple grab bars are used, required horizontal grab bars shall be installed at the same height above the floor. Grab bars can be separate bars or one continuous bar.

1110.2.3.3.1 Back-wall grab bar. The back-wall grab bar shall extend the length of the back wall and extend within 6 inches (150 mm) maximum from the two adjacent sidewalls.

Exception: The back-wall grab bar shall not be required to exceed 48 inches (1220 mm) in length. The rear grab bar shall be located with one end within 6 inches maximum of a sidewall with a grab bar complying with Section 1110.2.3.3.2.

1110.2.3.3.2 Sidewall grab bars. The sidewall grab bars shall extend the length of the wall and extend within 6 inches (150 mm) of the adjacent back wall.

Exceptions:

1. The sidewall grab bar shall not be required to exceed 30 inches (760 mm) in length. The side grab bar shall be located with one end within 6 inches (152 mm) of the back wall with a grab bar complying with Section 1110.2.3.3.1.
2. Where the sidewalls are located 72 inches (1830 mm) or greater apart, a grab bar is not required on one of the sidewalls.

1110.2.3.4 Seats. Wall-mounted folding seats shall not be installed.

1110.2.3.5 Controls and hand showers. In standard roll-in-type showers, the controls and hand shower shall be located not less than 38 inches (965 mm) and not greater than 48 inches (1220 mm) above the shower floor. Controls shall be located to facilitate caregiver access.

1110.2.3.6 Hand showers. Hand showers shall comply with ICC A117.1, Section 608.5.

1110.2.3.7 Thresholds. Thresholds shall comply with ICC A117.1, Section 608.6.

1110.2.3.8 Shower enclosures. Shower compartment enclosures for shower compartments shall comply with ICC A117.1, Section 608.7.

1110.2.3.9 Water temperature. Water temperature shall comply with ICC A117.1, Section 608.8.

1110.2.4 Water closet compartment. Where water closet compartments are provided in a toilet room or bathing room, at least 5 percent of the total number of compartments shall be wheelchair accessible. Where the combined total water closet compartments and urinals provided in a toilet room or bathing room is six or more, at least 5 percent of the total number of compartments shall be ambulatory accessible, provided in addition to the wheelchair-accessible compartment.

1110.2.5 Lavatories. Where lavatories are provided, at least 5 percent, but not less than one, shall be *accessible*. Where an accessible lavatory is located within the accessible water closet compartment at least one additional accessible lavatory shall be provided in the multicompartment toilet room outside the water closet compartment. Where the total lavatories provided in a toilet room or bathing facility is six or more, at least one lavatory with enhanced reach ranges shall be provided.

1110.3 Sinks. Where sinks are provided, at least 5 percent but not less than one provided in accessible spaces shall be *accessible*.

Exception: Mop or service sinks are not required to be *accessible*.

1110.4 Kitchens and kitchenettes. Where kitchens and kitchenettes are provided in accessible spaces or rooms, they shall be *accessible*.

1110.5 Drinking fountains. Where drinking fountains are provided on an exterior site, on a floor or within a secured area, the drinking fountains shall be provided in accordance with Sections 1110.5.1 and 1110.5.2.

1110.5.1 Minimum number. Not fewer than two drinking fountains shall be provided. One drinking fountain shall comply with the requirements for people who use a wheelchair and one drinking fountain shall comply with the requirements for standing persons.

Exceptions:

1. A single drinking fountain with two separate spouts that complies with the requirements for people who use a wheelchair and standing persons shall be permitted to be substituted for two separate drinking fountains.
2. Where drinking fountains are primarily for children's use, drinking fountains for people using wheelchairs shall be permitted to comply with the children's provisions in ICC A117.1 and drinking fountains for standing children shall be permitted to provide the spout at 30 inches (762 mm) minimum above the floor.

1110.5.2 More than the minimum number. Where more than the minimum number of drinking fountains specified in Section 1110.5.1 is provided, 50 percent of the total number of drinking fountains provided shall comply with the requirements for persons who use a wheelchair and 50 percent of the total number of drinking fountains provided shall comply with the requirements for standing persons.

Exceptions:

1. Where 50 percent of the drinking fountains yields a fraction, 50 percent shall be permitted to be rounded up or down, provided that the total number of drinking fountains complying with this section equals 100 percent of the drinking fountains.
2. Where drinking fountains are primarily for children's use, drinking fountains for people using wheelchairs shall be permitted to comply with the children's provisions in ICC A117.1 and drinking fountains for standing children shall be permitted to provide the spout at 30 inches (762 mm) minimum above the floor.

1110.6 Bottle-filling stations. Where bottle-filling stations are provided, they shall be *accessible*.

Exception: Bottle-filling stations over drinking fountains for standing persons are not required to be accessible, provided that bottle-filling stations are also located over the drinking fountains for persons using wheelchairs.

1110.7 Saunas and steam rooms. Where provided, saunas and steam rooms shall be *accessible*.

Exception: Where saunas or steam rooms are clustered at a single location, at least 5 percent of the saunas and steam rooms, but not less than one, of each type in each cluster shall be *accessible*.

1110.8 Elevators. Passenger elevators on an *accessible route* shall be *accessible* and comply with Chapter 30.

1110.9 Lifts. Platform (wheelchair) lifts are permitted to be a part of a required *accessible route* in new construction where indicated in Items 1 through 10. Platform (wheelchair) lifts shall be installed in accordance with ASME A18.1.

1. An *accessible route* to a performing area and speaker platforms.
2. An *accessible route* to *wheelchair spaces* required to comply with the wheelchair space dispersion requirements of Sections 1109.2.2 through 1109.2.6.
3. An *accessible route* to spaces that are not open to the general public with an *occupant load* of not more than five.
4. An *accessible route* within an individual *dwelling unit* or *sleeping unit* required to be an *Accessible unit, Type A unit* or *Type B unit*.
5. An *accessible route* to jury boxes and witness stands; raised courtroom stations including judges' benches, clerks' stations, bailiffs' stations, deputy clerks' stations and court reporters' stations; and to depressed areas such as the well of the court.
6. An *accessible route* to load and unload areas serving amusement rides.
7. An *accessible route* to play components or *soft contained play structures*.
8. An *accessible route* to team or player seating areas serving *areas of sport activity*.
9. An *accessible route* instead of gangways serving recreational boating facilities and fishing piers and platforms.
10. An *accessible route* where existing exterior site constraints make use of a *ramp* or elevator infeasible.

1110.10 Storage. Where fixed or built-in storage elements such as cabinets, coat hooks, shelves, medicine cabinets, lockers, closets and drawers are provided in required accessible spaces, at least 5 percent, but not less than one of each type shall be *accessible*.

1110.10.1 Equity. *Accessible* facilities and spaces shall be provided with the same storage elements as provided in the similar nonaccessible facilities and spaces.

1110.10.2 Shelving and display units. Self-service shelves and display units shall be located on an *accessible route*. Such shelving and display units shall not be required to comply with reach-range provisions.

1110.11 Detectable warnings. Passenger transit platform edges bordering a drop-off and not protected by platform screens or *guards* shall have a *detectable warning*.

Exception: *Detectable warnings* are not required at bus stops.

1110.12 Seating at tables, counters and work surfaces. Where seating or standing space at fixed or built-in tables, counters or work surfaces is provided in accessible spaces, at least 5 percent of the seating and standing spaces, but not less than one, shall be *accessible*.

Exception: Check-writing surfaces at check-out aisles not required to comply with Section 1110.13.1 are not required to be *accessible*.

1110.12.1 Dispersion. *Accessible* fixed or built-in seating at tables, counters or work surfaces shall be distributed throughout the space or facility containing such elements and located on a level accessed by an *accessible route*.

1110.12.2 Visiting areas. Visiting areas in judicial facilities and Group I-3 shall comply with Sections 1110.12.2.1 and 1110.12.2.2.

1110.12.2.1 Cubicles and counters. At least 5 percent, but not less than one of the cubicles, shall be *accessible* on both the visitor and detainee sides. Where counters are provided, at least one shall be *accessible* on both the visitor and detainee sides.

Exception: This requirement shall not apply to the detainee side of cubicles or counters at noncontact visiting areas not serving *Accessible unit* holding *cells*.

1110.12.2.2 Partitions. Where solid partitions or security glazing separate visitors from detainees, at least one of each type of cubicle or counter partition shall be *accessible*.

1110.13 Service facilities. Service facilities shall provide for accessible features in accordance with Sections 1110.14 through 1110.13.4.

* **1110.13.1 Check-out aisles.** Where check-out aisles are provided, accessible check-out aisles shall be provided in accordance with Table 1110.13.1. Where check-out aisles serve different functions, accessible check-out aisles shall be provided in accordance with Table 1110.13.1 for each function. Where check-out aisles are dispersed throughout the building or facility, accessible check-out aisles shall also be dispersed. Traffic control devices, security devices and turnstiles located in accessible check-out aisles or lanes shall be *accessible*.

Exception: Where the public use area is under 5,000 square feet (465 m^2) not more than one accessible check-out aisle shall be required.

TABLE 1110.13.1
ACCESSIBLE CHECK-OUT AISLES

TOTAL CHECK-OUT AISLES OF EACH FUNCTION	MINIMUM NUMBER OF ACCESSIBLE CHECK-OUT AISLES OF EACH FUNCTION
1 to 4	1
5 to 8	2
9 to 15	3
Over 15	3, plus 20% of additional aisles

1110.13.2 Sales and service counters and windows. Where counters or windows are provided for sale or distribution of goods or services, at least one of each type of counter and window provided shall be *accessible*. Where such counters or windows are dispersed throughout the building or facility, accessible counters or windows shall also be dispersed.

1110.13.3 Food service lines. Food service lines shall be *accessible*. Where self-service shelves are provided, at least 50 percent, but not less than one, of each type provided shall be *accessible*.

1110.13.4 Queue and waiting lines. Queue and waiting lines servicing accessible counters or check-out aisles shall be *accessible*. **

1110.14 Dressing, fitting and locker rooms. Where dressing rooms, fitting rooms or locker rooms are provided, at least 5 percent, but not less than one, of each type of use in each cluster provided shall be *accessible*.

1110.15 Controls, operating mechanisms and hardware. Controls, operating mechanisms and hardware intended for operation by the occupant, including switches that control lighting and ventilation and electrical convenience outlets, in accessible spaces, along accessible routes or as parts of *accessible* elements shall be *accessible*.

Exceptions:

1. Operable parts that are intended for use only by service or maintenance personnel shall not be required to be *accessible*.
2. Access doors or gates in barrier walls and fences protecting pools, spas and hot tubs shall be permitted to comply with Section 1010.2.3.
3. Operable parts exempted in accordance with ICC A117.1 are not required to be accessible. *

1110.16 Gaming machines and gaming tables. At least two percent of the total, but not fewer than one, of each *gaming machine type* and *gaming table type* shall be *accessible*. Where multiple *gaming areas* occur, accessible gaming machines and gaming tables shall be distributed throughout.

SECTION 1111
RECREATIONAL FACILITIES

1111.1 General. Recreational facilities shall be provided with accessible features in accordance with Sections 1111.2 through 1111.4.

1111.2 Facilities serving Group R-2, R-3 and R-4 occupancies. Recreational facilities that serve Group R-2, R-3 and Group R-4 occupancies shall comply with Sections 1111.2.1 through 1111.2.3, as applicable.

1111.2.1 Facilities serving Accessible units. In Group R-2 and R-4 occupancies where recreational facilities serve *Accessible units*, every recreational facility of each type serving *Accessible units* shall be *accessible*.

1111.2.2 Facilities serving Type A and Type B units in a single building. In Group R-2, R-3 and R-4 occupancies where recreational facilities serve a single building containing *Type A units* or *Type B units*, 25 percent, but not less than one, of each type of recreational facility shall be *accessible*. Every recreational facility of each type on a site shall be considered to determine the total number of each type that is required to be *accessible*.

1111.2.3 Facilities serving Type A and Type B units in multiple buildings. In Group R-2, R-3 and R-4 occupancies on a single site where multiple buildings containing *Type A units* or *Type B units* are served by recreational facilities, 25 percent, but not less than one, of each type of recreational facility serving each building shall be *accessible*. The total number of each type of recreational facility that is required to be *accessible* shall be determined by considering every recreational facility of each type serving each building on the site.

1111.3 Other occupancies. Recreational facilities not falling within the purview of Section 1111.2 shall be *accessible*.

1111.4 Recreational facilities. Recreational facilities shall be *accessible* and shall be on an *accessible route* to the extent specified in this section.

1111.4.1 Area of sport activity. Each *area of sport activity* shall be on an *accessible route* and shall not be required to be *accessible* except as provided for in Sections 1111.4.2 through 1111.4.15.

1111.4.2 Team or player seating. At least one wheelchair space shall be provided in team or player seating areas serving *areas of sport activity*.

Exception: Wheelchair spaces shall not be required in team or player seating areas serving bowling lanes that are not required to be *accessible* in accordance with Section 1111.4.3.

1111.4.3 Bowling lanes. An *accessible route* shall be provided to at least 5 percent, but not less than one, of each type of bowling lane.

1111.4.4 Court sports. In court sports, at least one *accessible route* shall directly connect both sides of the court.

1111.4.5 Raised boxing or wrestling rings. Raised boxing or wrestling rings are not required to be *accessible* or to be on an *accessible route*.

1111.4.6 Raised refereeing, judging and scoring areas. Raised structures used solely for refereeing, judging or scoring a sport are not required to be *accessible* or to be on an *accessible route*.

1111.4.7 Animal containment areas. Animal containment areas that are not within public use areas are not required to be *accessible* or to be on an *accessible route*.

1111.4.8 Amusement rides. Amusement rides that move persons through a fixed course within a defined area shall comply with Sections 1111.4.8.1 through 1111.4.8.3.

Exception: Mobile or portable amusement rides shall not be required to be *accessible*.

1111.4.8.1 Load and unload areas. Load and unload areas serving amusement rides shall be *accessible* and be on an *accessible route*. Where load and unload areas have more than one loading or unloading position, at least one loading and unloading position shall be on an *accessible route*.

1111.4.8.2 Wheelchair spaces, ride seats designed for transfer and transfer devices. Where amusement rides are in the load and unload position, the following shall be on an *accessible route*.

1. The position serving a wheelchair space.
2. Amusement ride seats designed for transfer.
3. Transfer devices.

1111.4.8.3 Minimum number. Amusement rides shall provide at least one wheelchair space, amusement ride seat designed for transfer or transfer device.

Exceptions:

1. Amusement rides that are controlled or operated by the rider are not required to comply with this section.
2. Amusement rides designed primarily for children, where children are assisted on and off the ride by an adult, are not required to comply with this section.
3. Amusement rides that do not provide seats that are built-in or mechanically fastened shall not be required to comply with this section.

1111.4.9 Recreational boating facilities. Boat slips required to be *accessible* by Sections 1111.4.9.1 and 1111.4.9.2 and boarding piers at boat launch ramps required to be *accessible* by Section 1111.4.9.3 shall be on an *accessible route*.

1111.4.9.1 Boat slips. Accessible boat slips shall be provided in accordance with Table 1111.4.9.1. All units on the site shall be combined to determine the number of *accessible* boat slips required. Where the number of boat slips is not identified, each 40 feet (12 m) of boat slip edge provided along the perimeter of the pier shall be counted as one boat slip for the purpose of this section.

Exception: Boat slips not designed for embarking or disembarking are not required to be *accessible* or be on an *accessible route*.

TABLE 1111.4.9.1
BOAT SLIPS

TOTAL NUMBER OF BOAT SLIPS PROVIDED	MINIMUM NUMBER OF REQUIRED ACCESSIBLE BOAT SLIPS
1 to 25	1
26 to 50	2
51 to 100	3
101 to 150	4
151 to 300	5
301 to 400	6
401 to 500	7
501 to 600	8
601 to 700	9
701 to 800	10
801 to 900	11
901 to 1,000	12
1,001 and over	12, plus 1 for every 100, or fraction thereof, over 1,000

1111.4.9.2 Dispersion. Accessible boat slips shall be dispersed throughout the various types of boat slips provided. Where the minimum number of *accessible* boat slips has been met, further dispersion shall not be required.

1111.4.9.3 Boarding piers at boat launch ramps. Where boarding piers are provided at boat launch ramps, at least 5 percent, but not less than one, of the boarding piers shall be *accessible*.

1111.4.10 Exercise machines and equipment. At least one of each type of exercise machine and equipment shall be on an *accessible route*.

1111.4.11 Fishing piers and platforms. Fishing piers and platforms shall be *accessible* and be on an *accessible route*.

1111.4.12 Miniature golf facilities. Miniature golf facilities shall comply with Sections 1111.4.12.1 through 1111.4.12.3.

1111.4.12.1 Minimum number. At least 50 percent of holes on miniature golf courses shall be *accessible*.

1111.4.12.2 Miniature golf course configuration. Miniature golf courses shall be configured so that the accessible holes are consecutive. Miniature golf courses shall provide an accessible route from the last *accessible* hole to the course entrance or exit without requiring travel through any other holes on the course.

Exception: One break in the sequence of consecutive holes shall be permitted provided that the last hole on the miniature golf course is the last hole in the sequence.

1111.4.12.3 Accessible route. Holes required to comply with Section 1111.4.12.1, including the start of play, shall be on an *accessible route*.

1111.4.13 Play areas. Play areas containing play components designed and constructed for children shall be located on an *accessible route*.

1111.4.14 Swimming pools, wading pools, cold baths, hot tubs and spas. *Swimming pools*, wading pools, cold baths, hot tubs and spas shall be *accessible* and be on an accessible route.

Exceptions:

1. Catch pools or a designated section of a pool used as a terminus for a water slide flume shall not be required to provide an *accessible* means of entry, provided that a portion of the catch pool edge is on an *accessible route*.
2. Where spas, cold baths or hot tubs are provided in a cluster, at least 5 percent, but not less than one of each type of spa, cold bath or hot tub in each cluster, shall be accessible and be on an *accessible route*.
3. *Swimming pools*, wading pools, spas, cold baths and hot tubs that are required to be *accessible* by Sections 1111.2.2 and 1111.2.3 are not required to provide *accessible* means of entry into the water.

1111.4.14.1 Raised diving boards and diving platforms. Raised diving boards and diving platforms are not required to be *accessible* or to be on an *accessible route*.

1111.4.14.2 Water slides. Water slides are not required to be *accessible* or to be on an *accessible route*.

1111.4.15 Shooting facilities with firing positions. Where shooting facilities with firing positions are designed and constructed at a site, at least 5 percent, but not less than one, of each type of firing position shall be *accessible* and be on an *accessible route*.

Exception: Shooting facilities with firing positions on free-standing platforms that are elevated more than 12 feet (3660 mm) above grade, provided that the aggregate area of the elevated firing positions is not more than 500 square feet (46 m^2), are not required to be accessible.

SECTION 1112
SIGNAGE

1112.1 Signs. Required accessible elements shall be identified by the International Symbol of Accessibility at the following locations.

1. Accessible parking spaces required by Section 1106.2.

   Exception: Where the total number of parking spaces provided is four or less, identification of accessible parking spaces is not required.

2. *Accessible* parking spaces required by Section 1106.3.

   Exception: In Group I-1, R-2, R-3 and R-4 facilities, where parking spaces are assigned to specific *dwelling units* or *sleeping units*, identification of *accessible* parking spaces is not required.

3. *Accessible* passenger loading zones.
4. *Accessible* toilet or bathing rooms where not all toilet or bathing rooms are *accessible.*
5. *Accessible* entrances where not all entrances are *accessible.*
6. *Accessible* check-out aisles where not all aisles are *accessible*. The sign, where provided, shall be above the check-out aisle in the same location as the check-out aisle number or type of check-out identification.
7. *Accessible* dressing, fitting and locker rooms where not all such rooms are *accessible.*
8. *Accessible areas of refuge* in accordance with Section 1009.9.
9. Exterior areas for assisted rescue in accordance with Section 1009.9.
10. In recreational facilities, lockers that are required to be *accessible* in accordance with Section 1110.10.

1112.2 Signs identifying toilet or bathing rooms. Signs required in Section 403.4 of the *International Plumbing Code* identifying toilet rooms and bathing rooms shall be visual characters, raised characters and braille complying with ICC A117.1. Where pictograms are provided as designations for toilet rooms and bathing rooms, the pictograms shall have visual characters, raised characters and braille complying with ICC A117.1.

1112.3 Directional signage. Directional signage indicating the route to the nearest like accessible element shall be provided at the following locations. These directional signs shall include the International Symbol of Accessibility and sign characters shall meet the visual character requirements in accordance with ICC A117.1.

1. Inaccessible building entrances.
2. Inaccessible public toilets and bathing facilities.
3. Elevators not serving an accessible route.
4. At each separate-sex toilet and bathing room indicating the location of the nearest family/assisted use toilet or bathing room where provided in accordance with Section 1110.2.1.
5. At *exits* and *exit stairways* serving a required *accessible* space, but not providing an *approved* accessible means of egress, signage shall be provided in accordance with Section 1009.10.
6. Where drinking fountains for persons using wheelchairs and drinking fountains for standing persons are not located adjacent to each other, directional signage shall be provided indicating the location of the other drinking fountains.

1112.4 Other signs. Signage indicating special accessibility provisions shall be provided as shown.

1. Each assembly area required to comply with Section 1109.2.7 shall provide a sign notifying patrons of the availability of assistive listening systems. The sign shall comply with ICC A117.1 requirements for visual characters and include the International Symbol of Access for Hearing Loss.

   Exception: Where ticket offices or windows are provided, signs are not required at each assembly area provided that signs are displayed at each ticket office or window informing patrons of the availability of assistive listening systems.

2. At each door to an *area of refuge* providing *direct access* to a *stairway*, exterior area for assisted rescue, exit *stairway, exit passageway* or *exit discharge*, signage shall be provided in accordance with Section 1013.4.
3. At *areas of refuge*, signage shall be provided in accordance with Section 1009.11.
4. At exterior areas for assisted rescue, signage shall be provided in accordance with Section 1009.11.
5. At two-way communication systems, signage shall be provided in accordance with Section 1009.8.2.
6. In *interior exit stairways* and *ramps*, floor level signage shall be provided in accordance with Section 1023.9.
7. Signs identifying the type of access provided on amusement rides required to be *accessible* by Section 1111.4.8 shall be provided at entries to queues and waiting lines. In addition, where accessible unload areas also serve as accessible load areas, signs indicating the location of the accessible load and unload areas shall be provided at entries to queues and waiting lines. These directional sign characters shall meet the visual character requirements in accordance with ICC A117.1.

1112.5 Variable message signs. Where provided in the locations in Sections 1112.5.1 and 1112.5.2, variable message signs shall comply with the variable message sign requirements of ICC A117.1.

1112.5.1 Transportation facilities. Where provided in transportation facilities, variable message signs conveying transportation-related information shall comply with Section 1112.5.

1112.5.2 Emergency shelters. Where provided in buildings that are designated as emergency shelters, variable message signs conveying emergency-related information shall comply with Section 1112.5.

Exception: Where equivalent information is provided in an audible manner, variable message signs are not required to comply with ICC A117.1.

CHAPTER 12

INTERIOR ENVIRONMENT

User note:

About this chapter: *Chapter 12 provides minimum provisions for the interior of buildings—the occupied environment. Ventilation, lighting, and space heating are directly regulated in this chapter and in conjunction with the* International Mechanical Code® *and the* International Energy Conservation Code®. *Minimum room size, maximum room-to-room sound transmission and classroom acoustics are set for educational occupancies.*

SECTION 1201
GENERAL

1201.1 Scope. The provisions of this chapter shall govern ventilation, temperature control, lighting, *yards* and *courts*, sound transmission, room dimensions, surrounding materials and rodentproofing associated with the interior spaces of buildings.

SECTION 1202
VENTILATION

1202.1 General. Buildings shall be provided with natural ventilation in accordance with Section 1202.5, or mechanical ventilation in accordance with the *International Mechanical Code.*

Where the air infiltration rate in a *dwelling unit* is less than 5 air changes per hour where tested with a blower door at a pressure 0.2 inch w.c. (50 Pa) in accordance with Section R402.4.1.2 of the *International Energy Conservation Code—Residential Provisions*, the *dwelling unit* shall be ventilated by mechanical means in accordance with Section 403 of the *International Mechanical Code. Ambulatory care facilities* and Group I-2 occupancies shall be ventilated by mechanical means in accordance with Section 407 of the *International Mechanical Code.*

1202.2 Roof ventilation. Roof assemblies shall be ventilated in accordance with this section or shall comply with Section 1202.3.

1202.2.1 Ventilated attics and rafter spaces. Enclosed *attics* and enclosed rafter spaces formed where ceilings are applied directly to the underside of roof framing members shall have cross ventilation for each separate space by ventilation openings protected against the entrance of rain and snow. Blocking and bridging shall be arranged so as not to interfere with the movement of air. An airspace of not less than 1 inch (25 mm) shall be provided between the insulation and the roof sheathing. The net free ventilating area shall be not less than $^1/_{150}$ of the area of the space ventilated. Ventilators shall be installed in accordance with manufacturer's installation instructions.

Exception: The net free cross-ventilation area shall be permitted to be reduced to $^1/_{300}$ provided both of the following conditions are met:

1. In Climate Zones 6, 7 and 8, a Class I or II vapor retarder is installed on the warm-in-winter side of the ceiling.
2. At least 40 percent and not more than 50 percent of the required venting area is provided by ventilators located in the upper portion of the *attic* or rafter space. Upper ventilators shall be located not more than 3 feet (914 mm) below the ridge or highest point of the space, measured vertically, with the balance of the *ventilation* provided by eave or cornice vents. Where the location of wall or roof framing members conflicts with the installation of upper ventilators, installation more than 3 feet (914 mm) below the ridge or highest point of the space shall be permitted.

1202.2.2 Openings into attic. Exterior openings into the *attic* space of any building intended for human occupancy shall be protected to prevent the entry of birds, squirrels, rodents, snakes and other similar creatures. Openings for ventilation having a least dimension of not less than $^1/_{16}$ inch (1.6 mm) and not more than $^1/_4$ inch (6.4 mm) shall be permitted. Openings for ventilation having a least dimension larger than $^1/_4$ inch (6.4 mm) shall be provided with corrosion-resistant wire cloth screening, hardware cloth, perforated vinyl or similar material with openings having a least dimension of not less than $^1/_{16}$ inch (1.6 mm) and not more than $^1/_4$ inch (6.4 mm). Where combustion air is obtained from an *attic* area, it shall be in accordance with Chapter 7 of the *International Mechanical Code.*

1202.3 Unvented attic and unvented enclosed rafter assemblies. Unvented *attics* and unvented enclosed roof framing assemblies created by ceilings applied directly to the underside of the roof framing members/rafters and the structural roof sheathing at the top of the roof framing members shall be permitted where all of the following conditions are met:

1. The unvented *attic* space is completely within the *building thermal envelope.*
2. No interior Class I vapor retarders are installed on the ceiling side (*attic* floor) of the unvented *attic* assembly or on the ceiling side of the unvented enclosed roof framing assembly.
3. Where wood shingles or shakes are used, not less than a $^1/_4$-inch (6.4 mm) vented airspace separates the shingles or shakes and the roofing *underlayment* above the structural sheathing.

4. In Climate Zones 5, 6, 7 and 8, any *air-impermeable insulation* shall be a Class II vapor retarder or shall have a Class II vapor retarder coating or covering in direct contact with the underside of the insulation.

5. Insulation shall comply with either Item 5.1 or 5.2, and additionally Item 5.3.

   5.1. Item 5.1.1, 5.1.2, 5.1.3 or 5.1.4 shall be met, depending on the air permeability of the insulation directly under the structural roof sheathing.

      5.1.1. Where only *air-impermeable insulation* is provided, it shall be applied in direct contact with the underside of the structural roof sheathing.

      5.1.2. Where air-permeable insulation is provided inside the building thermal envelope, it shall be installed in accordance with Item 5.1.1. In addition to the air-permeable insulation installed directly below the structural sheathing, rigid board or sheet insulation shall be installed directly above the structural roof sheathing in accordance with the *R*-values in Table 1202.3 for condensation control.

      5.1.3. Where both air-impermeable and air-permeable insulation are provided, the *air-impermeable insulation* shall be applied in direct contact with the underside of the structural roof sheathing in accordance with Item 5.1.1 and shall be in accordance with the R-values in Table 1202.3 for condensation control. The *air-permeable insulation* shall be installed directly under the *air-impermeable insulation*.

      5.1.4. Alternatively, sufficient rigid board or sheet insulation shall be installed directly above the structural roof sheathing to maintain the monthly average temperature of the underside of the structural roof sheathing above 45°F (7°C). For calculation purposes, an interior air temperature of 68°F (20°C) is assumed and the exterior air temperature is assumed to be the monthly average outside air temperature of the three coldest months.

   5.2. In Climate Zones 1, 2 and 3, air-permeable insulation installed in unvented attics shall meet the following requirements:

      5.2.1. A vapor diffusion port shall be installed not more than 12 inches (305 mm) from the highest point of the roof, measured vertically from the highest point of the roof to the lower edge of the port.

      5.2.2. The port area shall be greater than or equal to $^{1}/_{600}$ of the ceiling area. Where there are multiple ports in the attic, the sum of the port areas shall be greater than or equal to the area requirement.

      5.2.3. The vapor permeable membrane in the vapor diffusion port shall have a vapor permeance rating of greater than or equal to 20 perms when tested in accordance with Procedure A of ASTM E96.

      5.2.4. The vapor diffusion port shall serve as an air barrier between the attic and the exterior of the building.

      5.2.5. The vapor diffusion port shall protect the attic against the entrance of rain and snow.

      5.2.6. Framing members and blocking shall not block the free flow of water vapor to the port. Not less than a 2-inch (50 mm) space shall be provided between any blocking and the roof sheathing. Air-permeable insulation shall be permitted within that space.

TABLE 1202.3
INSULATION FOR CONDENSATION CONTROL

CLIMATE ZONE	MINIMUM *R*-VALUE OF AIR-IMPERMEABLE INSULATION[a]
2B and 3B tile roof only	0 (none required)
1, 2A, 2B, 3A, 3B, 3C	R-5
4C	R-10
4A, 4B	R-15
5	R-20
6	R-25
7	R-30
8	R-35

a. Contributes to, but does not supersede, thermal resistance requirements for attic and roof assemblies in Section C402.2.1 of the *International Energy Conservation Code*.

5.2.7. The roof slope shall be greater than or equal to 3 units vertical in 12 units horizontal (3:12).

5.2.8. Where only air-permeable insulation is used, it shall be installed directly below the structural roof sheathing, on top the attic floor, or on top of the ceiling.

5.2.9. Where only air-permeable insulation is used and is installed directly below the structural roof sheathing, air shall be supplied at a flow rate greater than or equal to 50 cubic feet per minute (23.6 L/s) per 1,000 square feet (93 m^2) of ceiling.

5.3. The air shall be supplied from ductwork providing supply air to the occupiable space when the conditioning system is operating. Alternatively, the air shall be supplied by a supply fan when the conditioning system is operating. Where preformed insulation board is used as the air-impermeable insulation layer, it shall be sealed at the perimeter of each individual sheet interior surface to form a continuous layer.

Exceptions:

1. Section 1202.3 does not apply to special use structures or enclosures such as swimming pool enclosures, data processing centers, hospitals or art galleries.
2. Section 1202.3 does not apply to enclosures in Climate Zones 5 through 8 that are humidified beyond 35 percent during the three coldest months.

1202.4 Under-floor ventilation. The space between the bottom of the floor joists and the earth under any building except spaces occupied by basements or cellars shall be provided with ventilation in accordance with Section 1202.4.1, 1202.4.2 or 1202.4.3.

1202.4.1 Ventilation openings. Ventilation openings through foundation walls shall be provided. The openings shall be placed so as to provide cross ventilation of the under-floor space. The net area of ventilation openings shall be in accordance with Section 1202.4.1.1 or 1202.4.1.2. Ventilation openings shall be covered for their height and width with any of the following materials, provided that the least dimension of the covering shall be not greater than $^1/_4$ inch (6.4 mm):

1. Perforated sheet metal plates not less than 0.070 inch (1.8 mm) thick.
2. Expanded sheet metal plates not less than 0.047 inch (1.2 mm) thick.
3. Cast-iron grilles or gratings.
4. Extruded load-bearing vents.
5. Hardware cloth of 0.035-inch (0.89 mm) wire or heavier.
6. Corrosion-resistant wire mesh, with the least dimension not greater than $^1/_8$ inch (3.2 mm).
7. Operable louvres, where ventilation is provided in accordance with Section 1202.4.1.2.

1202.4.1.1 Ventilation area for crawl spaces with open earth floors. The net area of ventilation openings for crawl spaces with uncovered earth floors shall be not less than 1 square foot for each 150 square feet (0.67 m^2 for each 100 m^2) of crawl space area.

1202.4.1.2 Ventilation area for crawl spaces with covered floors. The net area of ventilation openings for crawl spaces with the ground surface covered with a Class I vapor retarder shall be not less than 1 square foot for each 1,500 square feet (0.67 m^2 for each 1000 m^2) of crawl space area.

1202.4.2 Ventilation in cold climates. In extremely cold climates, where a ventilation opening will cause a detrimental loss of energy, ventilation openings to the interior of the structure shall be provided.

1202.4.3 Mechanical ventilation. Mechanical ventilation shall be provided to crawl spaces where the ground surface is covered with a Class I vapor retarder. Ventilation shall be in accordance with Section 1202.4.3.1 or 1202.4.3.2.

1202.4.3.1 Continuous mechanical ventilation. Continuously operated mechanical ventilation shall be provided at a rate of 1.0 cubic foot per minute (cfm) for each 50 square feet (1.02 L/s for each 10 m^2) of crawl space ground surface area and the ground surface shall be covered with a Class I vapor retarder.

1202.4.3.2 Conditioned space. The crawl space shall be conditioned in accordance with the *International Mechanical Code* and the walls of the crawl space shall be insulated in accordance with the *International Energy Conservation Code.*

1202.4.4 Flood hazard areas. For buildings in *flood hazard areas* as established in Section 1612.3, the openings for under-floor ventilation shall be deemed as meeting the flood opening requirements of ASCE 24 provided that the ventilation openings are designed and installed in accordance with ASCE 24.

1202.5 Natural ventilation. Natural *ventilation* of an occupied space shall be through windows, doors, louvers or other openings to the outdoors. The operating mechanism for such openings shall be provided with ready access so that the openings are readily controllable by the building occupants.

1202.5.1 Ventilation area required. The openable area of the openings to the outdoors shall be not less than 4 percent of the floor area being ventilated.

1202.5.1.1 Adjoining spaces. Where rooms and spaces without openings to the outdoors are ventilated through an adjoining room, the opening to the adjoining room shall be unobstructed and shall have an area of not less than 8 percent of the floor area of the interior room or space, but not less than 25 square feet (2.3

m^2). The openable area of the openings to the outdoors shall be based on the total floor area being ventilated.

Exception: Exterior openings required for *ventilation* shall be allowed to open into a *sunroom* with *thermal isolation* or a patio cover provided that the openable area between the *sunroom* addition or patio cover and the interior room shall have an area of not less than 8 percent of the floor area of the interior room or space, but not less than 20 square feet (1.86 m^2). The openable area of the openings to the outdoors shall be based on the total floor area being ventilated.

1202.5.1.2 Openings below grade. Where openings below grade provide required natural *ventilation*, the outside horizontal clear space measured perpendicular to the opening shall be one and one-half times the depth of the opening. The depth of the opening shall be measured from the average adjoining ground level to the bottom of the opening.

1202.5.2 Contaminants exhausted. Contaminant sources in naturally ventilated spaces shall be removed in accordance with the *International Mechanical Code* and the *International Fire Code*.

1202.5.2.1 Bathrooms. Rooms containing bathtubs, showers, spas and similar bathing fixtures shall be mechanically ventilated in accordance with the *International Mechanical Code*.

1202.5.3 Openings on yards or courts. Where natural *ventilation* is to be provided by openings onto *yards* or *courts*, such *yards* or *courts* shall comply with Section 1205.

1202.6 Other ventilation and exhaust systems. *Ventilation* and exhaust systems for occupancies and operations involving flammable or combustible hazards or other contaminant sources as covered in the *International Mechanical Code* or the *International Fire Code* shall be provided as required by both codes.

SECTION 1203 TEMPERATURE CONTROL

1203.1 Equipment and systems. Interior spaces intended for human occupancy shall be provided with active or passive space heating systems capable of maintaining an indoor temperature of not less than 68°F (20°C) at a point 3 feet (914 mm) above the floor on the design heating day.

Exceptions: Space heating systems are not required for:

1. Interior spaces where the primary purpose of the space is not associated with human comfort.
2. Group F, H, S or U occupancies.

SECTION 1204 LIGHTING

1204.1 General. Every space intended for human occupancy shall be provided with natural light by means of exterior glazed openings in accordance with Section 1204.2 or shall be provided with artificial light in accordance with Section 1204.3. Exterior glazed openings shall open directly onto a *public way* or onto a *yard* or *court* in accordance with Section 1205.

1204.2 Natural light. The minimum net glazed area shall be not less than 8 percent of the floor area of the room served.

1204.2.1 Adjoining spaces. For the purpose of natural lighting, any room is permitted to be considered as a portion of an adjoining room where one-half of the area of the common wall is open and unobstructed and provides an opening of not less than one-tenth of the floor area of the interior room or 25 square feet (2.32 m^2), whichever is greater.

Exception: Openings required for natural light shall be permitted to open into a *sunroom* with *thermal isolation* or a patio cover where the common wall provides a glazed area of not less than one-tenth of the floor area of the interior room or 20 square feet (1.86 m^2), whichever is greater.

1204.2.2 Exterior openings. Exterior openings required by Section 1204.2 for natural light shall open directly onto a *public way*, *yard* or *court*, as set forth in Section 1205.

Exceptions:

1. Required exterior openings are permitted to open into a roofed porch where the porch meets all of the following criteria:
   - 1.1. Abuts a *public way*, *yard* or *court*.
   - 1.2. Has a ceiling height of not less than 7 feet (2134 mm).
   - 1.3. Has a longer side at least 65 percent open and unobstructed.
2. Skylights are not required to open directly onto a *public way*, *yard* or *court*.

1204.3 Artificial light. Artificial light shall be provided that is adequate to provide an average illumination of 10 footcandles (107 lux) over the area of the room at a height of 30 inches (762 mm) above the floor level.

1204.4 Stairway illumination. *Stairways* within *dwelling units* and *exterior stairways* serving a *dwelling unit* shall have an illumination level on tread runs of not less than 1 footcandle (11 lux). *Stairways* in other occupancies shall be governed by Chapter 10.

1204.4.1 Controls. The control for activation of the required *stairway* lighting shall be in accordance with NFPA 70.

1204.5 Emergency egress lighting. The *means of egress* shall be illuminated in accordance with Section 1008.1.

SECTION 1205 YARDS OR COURTS

1205.1 General. This section shall apply to *yards* and *courts* adjacent to exterior openings that provide natural light or ventilation. Such *yards* and *courts* shall be on the same *lot* as the building.

1205.2 Yards. *Yards* shall be not less than 3 feet (914 mm) in width for buildings two *stories* or less above *grade plane*. For buildings more than two *stories above grade plane*, the minimum width of the *yard* shall be increased at the rate of 1 foot (305 mm) for each additional *story*. For buildings exceeding 14 *stories above grade plane*, the required width of the *yard* shall be computed on the basis of 14 *stories above grade plane*.

1205.3 Courts. *Courts* shall be not less than 3 feet (914 mm) in width. *Courts* having windows opening on opposite sides shall be not less than 6 feet (1829 mm) in width. *Courts* shall be not less than 10 feet (3048 mm) in length unless bounded on one end by a *public way* or *yard*. For buildings more than two *stories above grade plane*, the *court* shall be increased 1 foot (305 mm) in width and 2 feet (610 mm) in length for each additional *story*. For buildings exceeding 14 *stories above grade plane*, the required dimensions shall be computed on the basis of 14 *stories above grade plane*.

1205.3.1 Court access. Access shall be provided to the bottom of *courts* for cleaning purposes.

1205.3.2 Air intake. *Courts* more than two *stories* in height shall be provided with a horizontal air intake at the bottom not less than 10 square feet (0.93 m^2) in area and leading to the exterior of the building unless abutting a *yard* or *public way*.

1205.3.3 Court drainage. The bottom of every *court* shall be properly graded and drained to a public sewer or other *approved* disposal system complying with the *International Plumbing Code*.

SECTION 1206
SOUND TRANSMISSION

1206.1 Scope. This section shall apply to common interior walls, partitions and floor/ceiling assemblies between adjacent *dwelling units* and *sleeping units* or between *dwelling units* and *sleeping units* and adjacent public areas.

1206.2 Airborne sound. Walls, partitions and floor-ceiling assemblies separating *dwelling units* and *sleeping units* from each other or from public or service areas shall have a sound transmission class of not less than 50 where tested in accordance with ASTM E90, or have a Normalized Noise Isolation Class (NNIC) rating of not less than 45 if field tested, in accordance with ASTM E336 for airborne noise. Alternatively, the sound transmission class of walls, partitions and floor-ceiling assemblies shall be established by engineering analysis based on a comparison of walls, partitions and floor-ceiling assemblies having sound transmission class ratings as determined by the test procedures set forth in ASTM E90. Penetrations or openings in construction assemblies for piping; electrical devices; recessed cabinets; bathtubs; soffits; or heating, ventilating or exhaust ducts shall be sealed, lined, insulated or otherwise treated to maintain the required ratings. This requirement shall not apply to entrance doors; however, such doors shall be tight fitting to the frame and sill.

1206.2.1 Masonry. The sound transmission class of concrete masonry and clay masonry assemblies shall be calculated in accordance with TMS 302 or determined through testing in accordance with ASTM E90.

1206.3 Structure-borne sound. Floor-ceiling assemblies between *dwelling units* and *sleeping units* or between a *dwelling unit* or *sleeping unit* and a public or service area within the structure shall have an impact insulation class rating of not less than 50 where tested in accordance with ASTM E492, or have a Normalized Impact Sound Rating (NISR) of not less than 45 if field tested in accordance with ASTM E1007. Alternatively, the impact insulation class of floor-ceiling assemblies shall be established by engineering analysis based on a comparison of floor-ceiling assemblies having impact insulation class ratings as determined by the test procedures in ASTM E492.

SECTION 1207
ENHANCED CLASSROOM ACOUSTICS

1207.1 General. Enhanced classroom acoustics, where required by this section, shall comply with Section 808 of ICC A117.1.

1207.2 Where required. In Group E occupancies, enhanced classroom acoustics shall be provided in all classrooms with a volume of 20,000 cubic feet (566 m^3) or less.

SECTION 1208
INTERIOR SPACE DIMENSIONS

1208.1 Minimum room widths. *Habitable* spaces, other than a kitchen, shall be not less than 7 feet (2134 mm) in any plan dimension. Kitchens shall have a clear passageway of not less than 3 feet (914 mm) between counter fronts and appliances or counter fronts and walls.

1208.2 Minimum ceiling heights. *Occupiable spaces*, *habitable spaces* and corridors shall have a ceiling height of not less than 7 feet 6 inches (2286 mm) above the finished floor. Bathrooms, toilet rooms, kitchens, storage rooms and laundry rooms shall have a ceiling height of not less than 7 feet (2134 mm) above the finished floor.

Exceptions:

1. In one- and two-family *dwellings*, beams or girders spaced not less than 4 feet (1219 mm) on center shall be permitted to project not more than 6 inches (152 mm) below the required ceiling height.
2. If any room in a building has a sloped ceiling, the prescribed ceiling height for the room is required in one-half the area thereof. Any portion of the room measuring less than 5 feet (1524 mm) from the finished floor to the ceiling shall not be included in any computation of the minimum area thereof.
3. The height of *mezzanines* and spaces below *mezzanines* shall be in accordance with Section 505.2.

4. Corridors contained within a *dwelling unit* or *sleeping unit* in a Group R occupancy shall have a ceiling height of not less than 7 feet (2134 mm) above the finished floor.

1208.2.1 Furred ceiling. Any room with a furred ceiling shall be required to have the minimum ceiling height in two-thirds of the area thereof, but in no case shall the height of the furred ceiling be less than 7 feet (2134 mm).

1208.3 Room area. Every *dwelling unit* shall have not less than one room that shall have not less than 120 square feet (11.2 m^2) of *net floor area*. Other habitable rooms shall have a *net floor area* of not less than 70 square feet (6.5 m^2).

Exception: Kitchens are not required to be of a minimum floor area.

1208.4 Efficiency dwelling units. *Efficiency dwelling units* shall conform to the requirements of the code except as modified herein:

1. The unit shall have a living room of not less than 190 square feet (17.7 m^2) of floor area.
2. The unit shall be provided with a separate closet.
3. For other than *Accessible*, Type A and Type B dwelling units, the unit shall be provided with a kitchen sink, cooking appliance and refrigerator, each having a clear working space of not less than 30 inches (762 mm) in front. Light and *ventilation* conforming to this code shall be provided.
4. The unit shall be provided with a separate bathroom containing a water closet, lavatory and bathtub or shower.

SECTION 1209
ACCESS TO UNOCCUPIED SPACES

1209.1 Crawl spaces. Crawl spaces shall be provided with not less than one access opening that shall be not less than 18 inches by 24 inches (457 mm by 610 mm).

1209.2 Attic spaces. An opening not less than 20 inches by 30 inches (559 mm by 762 mm) shall be provided to any *attic* area having a clear height of over 30 inches (762 mm). Clear headroom of not less than 30 inches (762 mm) shall be provided in the *attic* space at or above the access opening.

1209.3 Mechanical appliances. Access to mechanical appliances installed in under-floor areas, in *attic* spaces and on roofs or elevated structures shall be in accordance with the *International Mechanical Code.*

SECTION 1210
TOILET AND BATHROOM REQUIREMENTS

[P] 1210.1 Required fixtures. The number and type of plumbing fixtures provided in any occupancy shall comply with Chapter 29.

[P] 1210.2 Finish materials. Walls, floors and partitions in toilet and bathrooms shall comply with Sections 1210.2.1 through 1210.2.4.

[P] 1210.2.1 Floors and wall bases. In other than *dwelling units*, toilet, bathing and shower room floor finish materials shall have a smooth, hard, nonabsorbent surface. The intersections of such floors with walls shall have a smooth, hard, nonabsorbent vertical base that extends upward onto the walls not less than 4 inches (102 mm).

[P] 1210.2.2 Walls and partitions. Walls and partitions within 2 feet (610 mm) of service sinks, urinals and water closets shall have a smooth, hard, nonabsorbent surface, to a height of not less than 4 feet (1219 mm) above the floor, and except for structural elements, the materials used in such walls shall be of a type that is not adversely affected by moisture.

Exception: This section does not apply to the following buildings and spaces:

1. Dwelling units and *sleeping units.*
2. Toilet rooms that are not accessible to the public and that have not more than one water closet.

Accessories such as grab bars, towel bars, paper dispensers and soap dishes, provided on or within walls, shall be installed and sealed to protect structural elements from moisture.

[P] 1210.2.3 Showers. Shower compartments and walls above bathtubs with installed shower heads shall be finished with a smooth, nonabsorbent surface to a height not less than 72 inches (1829 mm) above the drain inlet.

[P] 1210.2.4 Waterproof joints. Built-in tubs with showers shall have waterproof joints between the tub and adjacent wall.

[P] 1210.3 Privacy. Public restrooms shall be visually screened from outside entry or exit doorways to ensure user privacy within the restroom. This provision shall also apply where mirrors would compromise personal privacy. Privacy at water closets and urinals shall be provided in accordance with Sections 1210.3.1 and 1210.3.2.

Exception: Visual screening shall not be required for single-occupant toilet rooms with a lockable door.

[P] 1210.3.1 Water closet compartment. Each water closet utilized by the public or employees shall occupy a separate compartment with walls or partitions and a door enclosing the fixtures to ensure privacy.

Exceptions:

1. Water closet compartments shall not be required in a single-occupant toilet room with a lockable door.
2. Toilet rooms located in child day care facilities and containing two or more water closets shall be permitted to have one water closet without an enclosing compartment.
3. This provision is not applicable to toilet areas located within Group I-3 occupancy housing areas.

[P] 1210.3.2 Urinal partitions. Each urinal utilized by the public or employees shall occupy a separate area with walls or partitions to provide privacy. The walls or partitions shall begin at a height not more than 12 inches (305 mm) from and extend not less than 60 inches (1524 mm) above the finished floor surface. The walls or partitions shall extend from the wall surface at each side of the urinal not less than 18 inches (457 mm) or to a point not less than 6 inches (152 mm) beyond the outermost front lip of the urinal measured from the finished backwall surface, whichever is greater.

Exceptions:

1. Urinal partitions shall not be required in a single-occupant or family or assisted-use toilet room with a lockable door.
2. Toilet rooms located in child day care facilities and containing two or more urinals shall be permitted to have one urinal without partitions.

CHAPTER 13

ENERGY EFFICIENCY

User notes:

About this chapter: *The purpose of Chapter 13 is to provide minimum design requirements that will promote efficient energy utilization in buildings. The requirements address the building thermal envelope, the selection of HVAC and lighting equipment, and the installation of controls that dictate efficient operation. Also promoted is efficient use of power where providing water and lighting to the users of the space.*

Code development reminder: *Code change proposals to this chapter will be considered by the International Energy Conservation Code Development Committee during the 2022 (Group B) Code Development Cycle.*

SECTION 1301
GENERAL

[E] 1301.1 Scope. This chapter governs the design and construction of buildings for energy efficiency.

[E] 1301.1.1 Criteria. Buildings shall be designed and constructed in accordance with the *International Energy Conservation Code.*

CHAPTER 14

EXTERIOR WALLS

User notes:

About this chapter: *Chapter 14 addresses requirements for exterior walls of buildings. Minimum standards for wall covering materials, such as material performance and fire resistance, installation of wall coverings and the ability of the wall to provide weather protection are provided. This chapter also contains limitations on the areas and heights of combustible wall coverings based on fire separation distances, radiant heat exposure and surface burning characteristics.*

Code development reminder: *Code change proposals to sections preceded by the designation [BS] will be considered by the IBC—Structural Code Development Committee during the 2022 (Group B) Code Development Cycle.*

SECTION 1401
GENERAL

1401.1 Scope. The provisions of this chapter shall establish the minimum requirements for *exterior walls*; *exterior wall coverings*; *exterior wall* openings; exterior windows and doors; and architectural *trim*.

SECTION 1402
PERFORMANCE REQUIREMENTS

1402.1 General. The provisions of this section shall apply to *exterior walls*, wall coverings and components thereof.

1402.2 Weather protection. *Exterior walls* shall provide the building with a weather-resistant *exterior wall envelope*. The *exterior wall envelope* shall include flashing, as described in Section 1404.4. The *exterior wall envelope* shall be designed and constructed in such a manner as to prevent the accumulation of water within the wall assembly by providing a *water-resistive barrier* behind the exterior *veneer*, as described in Section 1403.2, and a means for draining water that enters the assembly to the exterior. Protection against condensation in the *exterior wall* assembly shall be provided in accordance with Section 1404.3.

Exceptions:

1. A weather-resistant *exterior wall envelope* shall not be required over concrete or masonry walls designed in accordance with Chapters 19 and 21, respectively.
2. Compliance with the requirements for a means of drainage, and the requirements of Sections 1403.2 and 1404.4, shall not be required for an *exterior wall envelope* that has been demonstrated through testing to resist wind-driven rain, including joints, penetrations and intersections with dissimilar materials, in accordance with ASTM E331 under the following conditions:

The *exterior wall envelope* design shall be considered to resist wind-driven rain where the results of testing indicate that water did not penetrate control joints in the *exterior wall envelope*, joints at the perimeter of openings or intersections of terminations with dissimilar materials.

2.1. *Exterior wall envelope* test assemblies shall include not fewer than one opening, one control joint, one wall/eave interface and one wall sill. Tested openings and penetrations shall be representative of the intended end-use configuration.

2.2. *Exterior wall envelope* test assemblies shall be not less than 4 feet by 8 feet (1219 mm by 2438 mm) in size.

2.3. *Exterior wall envelope* assemblies shall be tested at a minimum differential pressure of 6.24 pounds per square foot (0.297 kN/m^2).

2.4. *Exterior wall envelope* assemblies shall be subjected to a minimum test exposure duration of 2 hours.

3. *Exterior insulation and finish systems* (EIFS) complying with Section 1407.4.1.

[BS] 1402.3 Structural. *Exterior walls*, and the associated openings, shall be designed and constructed to resist safely the superimposed *loads* required by Chapter 16.

1402.4 Fire resistance. *Exterior walls* shall be fire-resistance rated as required by other sections of this code with opening protection as required by Chapter 7.

1402.5 Water-resistive barriers. *Exterior walls* on buildings of Type I, II, III or IV construction that are greater than 40 feet (12 192 mm) in height above grade plane and contain a combustible *water-resistive barrier* shall be tested in accordance with and comply with the acceptance criteria of NFPA 285.Combustibility shall be determined in accordance with Section 703.3. For the purposes of this section, *fenestration* products, flashing of *fenestration* products and *water-resistive-barrier* flashing and accessories at other locations, including through wall flashings, shall not be considered part of the *water-resistive barrier*.

Exceptions:

1. Walls in which the *water-resistive barrier* is the only combustible component and the *exterior wall*

has a wall covering of brick, concrete, stone, terra cotta, stucco or steel with minimum thicknesses in accordance with Table 1404.2.

2. Walls in which the *water-resistive barrier* is the only combustible component and the *water-resistive barrier* complies with the following:

    2.1. A peak heat release rate of less than 150 kW/m^2, a total heat release of less than 20 MJ/m^2 and an effective heat of combustion of less than 18 MJ/kg when tested on specimens at the thickness intended for use, in accordance with ASTM E1354, in the horizontal orientation and at an incident radiant heat flux of 50 kW/m^2.

    2.2. A flame spread index of 25 or less and a smoke-developed index of 450 or less as determined in accordance with ASTM E84 or UL 723, with test specimen preparation and mounting in accordance with ASTM E2404.

[BS] 1402.6 Flood resistance. For buildings in *flood hazard areas* as established in Section 1612.3, *exterior walls* extending below the elevation required by Section 1612 shall be constructed with flood-damage-resistant materials.

[BS] 1402.7 Flood resistance for coastal high-hazard areas and coastal A zones. For buildings in *coastal high-hazard areas* and coastal A zones as established in Section 1612.3, electrical, mechanical and plumbing system components shall not be mounted on or penetrate through *exterior walls* that are designed to break away under *flood loads*.

SECTION 1403
MATERIALS

1403.1 General. Materials used for the construction of *exterior walls* shall comply with the provisions of this section. Materials not prescribed herein shall be permitted, provided that any such alternative has been *approved*.

1403.2 Water-resistive barrier. Not fewer than one layer of *water-resistive barrier* material shall be attached to the studs or sheathing, with flashing as described in Section 1404.4, in such a manner as to provide a continuous *water-resistive barrier* behind the exterior wall *veneer*.

Water-resistive barriers shall comply with one of the following:

1. No. 15 felt complying with ASTM D226, Type 1.
2. ASTM E2556, Type I or II.
3. ASTM E331 in accordance with Section 1402.2.
4. Other approved materials installed in accordance with the manufacturer's installation instructions.

[BS] 1403.3 Wood. *Exterior walls* of wood construction shall be designed and constructed in accordance with Chapter 23.

[BS] 1403.3.1 Basic hardboard. Basic *hardboard* shall conform to the requirements of ANSI A135.4.

[BS] 1403.3.2 Hardboard siding. *Hardboard* siding shall conform to the requirements of ANSI A135.6 and, where used structurally, shall be so identified by the *label* of an *approved* agency.

[BS] 1403.4 Masonry. *Exterior walls* of masonry construction shall be designed and constructed in accordance with this section and Chapter 21. *Masonry units*, *mortar* and metal accessories used in anchored and adhered *veneer* shall meet the physical requirements of Chapter 21. The backing of anchored and adhered *veneer* shall be of concrete, masonry, steel framing or wood framing. Continuous insulation meeting the applicable requirements of this code shall be permitted between the backing and the masonry *veneer*.

[BS] 1403.5 Metal. *Exterior walls* constructed of cold-formed or structural steel shall be designed in accordance with Chapter 22. *Exterior walls* constructed of aluminum shall be designed in accordance with Chapter 20.

[BS] 1403.5.1 Aluminum siding. Aluminum siding shall conform to the requirements of AAMA 1402.

[BS] 1403.5.2 Cold-rolled copper. Copper shall conform to the requirements of ASTM B370.

[BS] 1403.5.3 Lead-coated copper. Lead-coated copper shall conform to the requirements of ASTM B101.

[BS] 1403.6 Concrete. *Exterior walls* of concrete construction shall be designed and constructed in accordance with Chapter 19.

[BS] 1403.7 Glass-unit masonry *Exterior walls* of glass-unit masonry shall be designed and constructed in accordance with Chapter 21.

1403.8 Plastics. Plastic panel, apron or spandrel walls as defined in this code shall not be limited in thickness, provided that such plastics and their assemblies conform to the requirements of Chapter 26 and are constructed of *approved* weather-resistant materials of adequate strength to resist the wind *loads* for cladding specified in Chapter 16.

1403.9 Vinyl siding. Vinyl siding shall be certified and labeled as conforming to the requirements of ASTM D3679 by an *approved* quality control agency.

1403.10 Fiber-cement siding. *Fiber-cement siding* shall conform to the requirements of ASTM C1186, Type A (or ISO 8336, Category A), and shall be so identified on labeling listing an *approved* quality control agency.

1403.11 Exterior insulation and finish systems. *Exterior insulation and finish systems* (EIFS) and *exterior insulation and finish systems (EIFS) with drainage* shall comply with Section 1407.

1403.12 Polypropylene siding. *Polypropylene siding* shall be certified and labeled as conforming to the requirements of ASTM D7254 and those of Section 1403.12.1 or 1403.12.2 by an approved quality control agency. *Polypropylene siding* shall be installed in accordance with the requirements of Section 1404.18 and in accordance with the manufacturer's instructions. *Polypropylene siding* shall be secured to the building so as to provide weather protection for the *exterior walls* of the building.

1403.12.1 Flame spread index. The certification of the *flame spread index* shall be accompanied by a test report stating that all portions of the test specimen ahead of the flame front remained in position during the test in accordance with ASTM E84 or UL 723.

1403.12.2 Fire separation distance. The *fire separation distance* between a building with *polypropylene siding* and the adjacent building shall be not less than 10 feet (3048 mm).

1403.13 Foam plastic insulation. Foam plastic insulation used in *exterior wall covering* assemblies shall comply with Chapter 26.

1403.14 Attachments through insulation. *Exterior wall coverings* attached to the building structure through foam plastic insulating sheathing shall comply with the attachment requirements of Section 2603.11, 2603.12, or 2603.13.

SECTION 1404
INSTALLATION OF WALL COVERINGS

1404.1 General. *Exterior wall coverings* shall be designed and constructed in accordance with the applicable provisions of this section.

1404.2 Weather protection. *Exterior walls* shall provide weather protection for the building. The materials of the minimum nominal thickness specified in Table 1404.2 shall be acceptable as *approved* weather coverings.

1404.3 Vapor retarders. Vapor retarder materials shall be classified in accordance with Table 1404.3(1). A vapor retarder shall be provided on the interior side of frame walls in accordance with Tables 1404.3(2) and 1404.3(3), or an approved design using accepted engineering practice for hygrothermal analysis. The appropriate climate zone shall be selected in accordance with Chapter 3 of the International Energy Conservation Code.

TABLE 1404.3(1)
VAPOR RETARDER MATERIALS AND CLASSES

VAPOR RETARDER CLASS	ACCEPTABLE MATERIALS
I	Sheet polyethylene, nonperforated aluminum foil, or other approved materials with a perm rating of less than or equal to 0.1
II	Kraft-faced fiberglass batts or vapor retarder paint or other approved materials, applied in accordance with the manufacturer's instructions for a perm rating greater than 0.1 and less than or equal to 1.0
III	Latex paint, enamel paint, or other approved materials, applied in accordance with the manufacturer's instructions for a perm rating of greater than 1.0 and less than or equal to 10

TABLE 1404.2
MINIMUM THICKNESS OF WEATHER COVERINGS

COVERING TYPE	MINIMUM THICKNESS (inches)
Adhered masonry veneer	0.25
Aluminum siding	0.019
Anchored masonry veneer	
Stone (natural)	2.0
Architectural cast stone	2.5
Other	2.0
Asbestos-cement boards	0.125
Asbestos shingles	0.156
Cold-rolled copper[d]	0.0216 nominal
Copper shingles[d]	0.0162 nominal
Exterior plywood (with sheathing)	0.313
Exterior plywood (without sheathing)	See Section 2304.6
Fiber cement lap siding	0.25[c]
Fiber cement panel siding	0.25[c]
Fiberboard siding	0.5
Glass-fiber reinforced concrete panels	0.375
Hardboard siding[c]	0.25
High-yield copper[d]	0.0162 nominal
Lead-coated copper[d]	0.0216 nominal
Lead-coated high-yield copper	0.0162 nominal
Marble slabs	1
Particleboard (with sheathing)	See Section 2304.6
Particleboard (without sheathing)	See Section 2304.6
Porcelain tile	0.125 nominal
Steel (approved corrosion resistant)	0.0149
Structural glass	0.344
Stucco or exterior cement plaster	
Three-coat work over:	
Metal plaster base	0.875[b]
Unit masonry	0.625[b]
Cast-in-place or precast concrete	0.625[b]
Two-coat work over:	
Unit masonry	0.5[b]
Cast-in-place or precast concrete	0.375[b]
Terra cotta (anchored)	1
Terra cotta (adhered)	0.25
Vinyl siding	0.035
Wood shingles	0.375
Wood siding (without sheathing)[a]	0.5

For SI: 1 inch = 25.4 mm, 1 ounce = 28.35 g, 1 square foot = 0.093 m^2.

a. Wood siding of thicknesses less than 0.5 inch shall be placed over sheathing that conforms to Section 2304.6.

b. Exclusive of texture.

c. As measured at the bottom of decorative grooves.

d. 16 ounces per square foot for cold-rolled copper and lead-coated copper, 12 ounces per square foot for copper shingles, high-yield copper and lead-coated high-yield copper.

TABLE 1404.3(2)
VAPOR RETARDER OPTIONS

CLIMATE ZONE	VAPOR RETARDER CLASS		
	I	II	III[a]
1, 2	Not permitted	Not Permitted	Permitted
3, 4 (except Marine 4)	Not permitted	Permitted	Permitted
Marine 4, 5, 6, 7, 8	Permitted	Permitted	See Table 1404.3(3)

a. See also Section 1404.3.2.

TABLE 1404.3(3)
CLASS III VAPOR RETARDERS

ZONE	CLASS III VAPOR RETARDERS PERMITTED FOR:[a, b]
4	Vented cladding over wood structural panels Vented cladding over fiberboard Vented cladding over gypsum Continuous insulation with *R*-value ≥ R-2.5 over 2 × 4 wall Continuous insulation with *R*-value ≥ R-3.75 over 2 × 6 wall
5	Vented cladding over wood structural panels Vented cladding over fiberboard Vented cladding over gypsum Continuous insulation with *R*-value ≥ R-5 over 2 × 4 wall Continuous insulation with *R*-value ≥ R-7.5 over 2 × 6 wall
6	Vented cladding over fiberboard Vented cladding over gypsum Continuous insulation with *R*-value ≥ R-7.5 over 2 × 4 wall Continuous insulation with *R*-value ≥ R-11.25 over 2 × 6 wall
7	Continuous insulation with *R*-value ≥ R-10 over 2 × 4 wall Continuous insulation with *R*-value ≥ R-15 over 2 × 6 wall
8	Continuous insulation with *R*-value ≥ R-12.5 over 2 × 4 wall Continuous insulation with *R*-value ≥ R-20 over 2 × 6 wall

a. Vented cladding shall include vinyl lap siding, polypropylene, or horizontal aluminum siding, brick veneer with airspace as specified in this code, and other approved vented claddings.

b. The requirements in this table apply only to insulation used to control moisture in order to permit the use of Class III vapor retarders. The insulation materials used to satisfy this option also contribute to but do not supersede the thermal envelope requirements of the *International Energy Conservation Code*.

1404.3.1 Class I and II vapor retarders. Where a Class II vapor retarder is used in combination with foam plastic insulating sheathing installed as continuous insulation on the exterior side of frame walls, the continuous insulation shall comply with Table 1404.3.1 and the Class II vapor retarder shall have a vapor permeance greater than 1 perm when measured by ASTM E96 water method (Procedure B). Use of a Class I interior vapor retarder in frame walls with a Class I vapor retarder on the exterior side shall require an approved design.

Exceptions:

1. Basement walls.
2. Below-grade portion of any wall.
3. Construction where accumulation, condensation or freezing of moisture will not damage the materials.
4. Class I and II vapor retarders with vapor permeance greater than 1 perm when measured by ASTM E96 water method (Procedure B) shall be allowed on the interior side of any frame wall in all climate zones.

TABLE 1404.3.1
CONTINUOUS INSULATION WITH CLASS II VAPOR RETARDER

CLIMATE ZONE	PERMITTED CONDITIONS[a]
3	Continuous insulation with *R*-value ≥ R-2
4, 5, 6	Continuous insulation with *R*-value ≥ R-3 over 2 × 4 wall Continuous insulation with *R*-value ≥ R-5 over 2 × 6 wall
7	Continuous insulation with *R*-value ≥ R-5 over 2 × 4 wall Continuous insulation with *R*-value ≥ R-7.5 over 2 × 6 wall
8	Continuous insulation with *R*-value ≥ R-7.5 over 2 × 4 wall Continuous insulation with *R*-value ≥ R-10 over 2 × 6 wall

a. In addition to the vapor retarder, spray foam with a maximum permeance of 1.5 perms at the installed thickness, applied to the interior cavity side of wood structural panels, fiberboard, insulating sheathing or gypsum is deemed to comply with the continuous insulation requirement only for the moisture control purposes of this table where the spray foam *R*-value plus any continuous insulation *R*-value provided equals or exceeds the specified continuous insulation *R*-value.

1404.3.2 Class III vapor retarders. Only Class III vapor retarders shall be used on the interior side of frame walls where foam plastic insulating sheathing with a perm rating of less than 1 is applied in accordance with Table 1404.3(3) on the exterior side of the frame wall.

1404.3.2.1 Spray foam plastic insulation for moisture control with Class III vapor retarders. For purposes of compliance with Table 1404.3(3), spray foam with a maximum permeance of 1.5 perms at the installed thickness applied to the interior cavity side of wood structural panels, fiberboard, insulating sheathing or gypsum shall be deemed to meet the continuous insulation *R*-value requirement where the spray foam *R*-value meets or exceeds the specified continuous insulation *R*-value.

1404.3.2.1.1 Hybrid insulation for moisture control with Class III vapor retarders. For the purposes of compliance with Table 1404.3(3), the combined *R*-values of spray foam plastic insulation and continuous insulation shall be permitted to be counted toward the continuous *R*-value requirement.

1404.4 Flashing. Flashing shall be installed in such a manner so as to prevent moisture from entering the wall or to redirect that moisture to the surface of the exterior wall finish or to a *water-resistive barrier* complying with Section 1403.2 and that is part of a means of drainage complying with Section 1402.2. Flashing shall be installed at the perimeters of exterior door and window assemblies, penetrations and terminations of *exterior wall* assemblies, *exterior wall* intersections with roofs, chimneys, porches, decks, balconies and similar projections and at built-in gutters and similar locations where moisture could enter the wall. Flashing with projecting flanges shall be installed on both sides and the ends of copings, under sills and continuously above projecting trim. Where self-adhered membranes are used as flashings of *fenestration* in wall assemblies, those self-adhered flashings shall comply with AAMA 711. Where fluid applied membranes are used as flashing for *exterior wall* openings, those fluid applied membrane flashings shall comply with AAMA 714.

1404.4.1 Exterior wall pockets. In *exterior walls* of buildings or structures, wall pockets or crevices in which moisture can accumulate shall be avoided or protected with caps or drips, or other *approved* means shall be provided to prevent water damage.

1404.4.2 Masonry. Flashing and weep holes in anchored *veneer* designed in accordance with Section 1404.6 shall be located not more than 10 inches (245 mm) above finished ground level above the foundation wall or slab. At other points of support including structural floors, shelf angles and lintels, flashing and weep holes shall be located in the first course of masonry above the support.

1404.5 Wood veneers. Wood *veneers* on *exterior walls* of buildings of Types I, II, III and IV construction shall be not less than 1 inch (25 mm) nominal thickness, 0.438-inch (11.1 mm) exterior *hardboard* siding or 0.375-inch (9.5 mm) exterior-type *wood structural panels* or particleboard and shall conform to the following:

1. The *veneer* shall not exceed 40 feet (12 190 mm) in height above grade. Where *fire-retardant-treated wood* is used, the height shall not exceed 60 feet (18 290 mm) in height above grade.
2. The *veneer* is attached to or furred from a noncombustible backing that is fire-resistance rated as required by other provisions of this code.
3. Where open or spaced wood *veneers* (without concealed spaces) are used, they shall not project more than 24 inches (610 mm) from the building wall.

[BS] 1404.6 Anchored masonry veneer. *Anchored masonry veneer* shall comply with the provisions of Sections 1404.6 through 1404.9 and Sections 12.1 and 12.2 of TMS 402.

[BS] 1404.6.1 Tolerances. *Anchored masonry veneers* in accordance with Chapter 14 are not required to meet the tolerances in Article 3.3 F1 of TMS 602.

[BS] 1404.6.2 Seismic requirements. *Anchored masonry veneer* located in *Seismic Design Category* C, D, E or F shall conform to the requirements of Section 12.2.2.11 of TMS 402.

[BS] 1404.7 Stone veneer. Anchored stone *veneer* units not exceeding 10 inches (254 mm) in thickness shall be anchored directly to masonry, concrete or to stud construction by one of the following methods:

1. With concrete or masonry backing, anchor ties shall be not less than 0.1055-inch (2.68 mm) corrosion-resistant wire, or *approved* equal, formed beyond the base of the backing. The legs of the loops shall be not less than 6 inches (152 mm) in length bent at right angles and laid in the *mortar* joint, and spaced so that the eyes or loops are 12 inches (305 mm) maximum on center in both directions. There shall be provided not less than a 0.1055-inch (2.68 mm) corrosion-resistant wire tie, or *approved* equal, threaded through the exposed loops for every 2 square feet (0.2 m^2) of stone *veneer*. This tie shall be a loop having legs not less than 15 inches (381 mm) in length bent so that the tie will lie in the stone *veneer mortar* joint. The last 2 inches (51 mm) of each wire leg shall have a right-angle bend. One-inch (25 mm) minimum thickness of cement grout shall be placed between the backing and the stone *veneer*.
2. With wood stud backing, a 2-inch by 2-inch (51 by 51 mm) 0.0625-inch (1.59 mm) zinc-coated or nonmetallic coated wire mesh with two layers of *water-resistive barrier* in accordance with Section 1403.2 shall be applied directly to wood studs spaced not more than 16 inches (406 mm) on center. On studs, the mesh shall be attached with 2-inch-long (51 mm) corrosion-resistant steel wire furring nails at 4 inches (102 mm) on center providing a minimum 1.125-inch (29 mm) penetration into each stud and with 8d annular threaded nails at 8 inches (203 mm) on center. into top and bottom plates or with equivalent wire ties. There shall be not less than a 0.1055-inch (2.68 mm) zinc-coated or nonmetallic coated wire, or approved equal, attached to the stud with not smaller than an 8d (0.120 in. diameter) annular threaded nail for every 2 square feet (0.2 m^2) of stone *veneer*. This tie shall be a loop having legs not less than 15 inches (381 mm) in length, so bent that the tie will lie in the stone *veneer mortar* joint. The last 2 inches (51 mm) of each wire leg shall have a right-angle bend. One-inch (25 mm) minimum thickness of cement grout shall be placed between the backing and the stone *veneer*.
3. With cold-formed steel stud backing, a 2-inch by 2-inch (51 by 51 mm) 0.0625-inch (1.59 mm) zinc-coated or nonmetallic coated wire mesh with two layers of *water-resistive barrier* in accordance with Section 1403.2 shall be applied directly to steel studs spaced a not more than 16 inches (406 mm) on center. The mesh shall be attached with corrosion-resistant #8 self-drilling, tapping screws at 4 inches (102 mm) on center, and at 8 inches (203 mm) on center into top and bottom tracks or with equivalent wire ties. Screws shall extend through the steel connection not fewer than three exposed threads. There shall be not less than a 0.1055-inch (2.68 mm) corrosion-resistant wire, or approved equal, attached to the stud with not smaller than a #8 self-drilling, tapping screw extending through the steel framing not fewer than three exposed threads for every 2 square feet (0.2 m^2) of stone *veneer*. This tie shall be a loop having legs not less than 15 inches (381 mm) in length, so bent that the tie will lie in the stone *veneer mortar* joint. The last 2 inches (51 mm) of each wire leg shall have a right-angle bend. Cement grout not less than 1 inch (25 mm) in thickness shall

be placed between the backing and the stone *veneer*. The cold-formed steel framing members shall have a minimum bare steel thickness of 0.0428 inches (1.087 mm).

[BS] 1404.8 Slab-type veneer. Anchored slab-type *veneer* units not exceeding 2 inches (51 mm) in thickness shall be anchored directly to masonry, concrete or *light-frame construction*. For *veneer* units of marble, travertine, granite or other stone units of slab form, ties of corrosion-resistant dowels in drilled holes shall be located in the middle third of the edge of the units, spaced not more than 24 inches (610 mm) apart around the periphery of each unit with not less than four ties per *veneer* unit. Units shall not exceed 20 square feet (1.9 m^2) in area. If the dowels are not tight fitting, the holes shall be drilled not more than 0.063 inch (1.6 mm) larger in diameter than the dowel, with the hole countersunk to a diameter and depth equal to twice the diameter of the dowel in order to provide a tight-fitting key of cement *mortar* at the dowel locations where the *mortar* in the joint has set. *Veneer* ties shall be corrosion-resistant metal capable of resisting, in tension or compression, a force equal to two times the weight of the attached *veneer*. If made of sheet metal, *veneer* ties shall be not smaller in area than 0.0336 by 1 inch (0.853 by 25 mm) or, if made of wire, not smaller in diameter than 0.1483-inch (3.76 mm) wire.

[BS] 1404.9 Terra cotta. Anchored terra cotta or ceramic units not less than $1^5/_8$ inches (41 mm) thick shall be anchored directly to masonry, concrete or stud construction. Tied terra cotta or ceramic *veneer* units shall be not less than $1^5/_8$ inches (41 mm) thick with projecting dovetail webs on the back surface spaced approximately 8 inches (203 mm) on center. The facing shall be tied to the backing wall with corrosion-resistant metal anchors of not less than No. 8 gage wire installed at the top of each piece in horizontal *bed joints* not less than 12 inches (305 mm) nor more than 18 inches (457 mm) on center; these anchors shall be secured to $^1/_4$-inch (6.4 mm) corrosion-resistant pencil rods that pass through the vertical aligned loop anchors in the backing wall. The *veneer* ties shall have sufficient strength to support the full weight of the *veneer* in tension. The facing shall be set with not less than a 2-inch (51 mm) space from the backing wall and the space shall be filled solidly with Portland cement grout and pea gravel. Immediately prior to setting, the backing wall and the facing shall be drenched with clean water and shall be distinctly damp when the grout is poured.

[BS] 1404.10 Adhered masonry veneer. *Adhered masonry veneer* shall comply with the applicable requirements in this section and Sections 12.1 and 12.3 of TMS 402.

[BS] 1404.10.1 Exterior adhered masonry veneer. Exterior *adhered masonry veneer* shall be installed in accordance with Section 1404.10 and the manufacturer's instructions.

[BS] 1404.10.1.1 Water-resistive barriers. *Water-resistive barriers* shall be installed as required in Section 2510.6.

[BS] 1404.10.1.2 Flashing. Flashing shall comply with the applicable requirements of Sections 1404.4 and 1404.10.1.2.1.

[BS] 1404.10.1.2.1 Flashing at foundation. A corrosion-resistant screed or flashing of a minimum 0.019-inch (0.48 mm) or 26 gage galvanized or plastic with a minimum vertical attachment flange of $3^1/_2$ inches (89 mm) shall be installed to extend not less than 1 inch (25 mm) below the foundation plate line on exterior stud walls in accordance with Section 1404.4. The *water-resistive barrier* shall lap over the exterior of the attachment flange of the screed or flashing.

[BS] 1404.10.1.3 Clearances. On exterior stud walls, *adhered masonry veneer* shall be installed not less than 4 inches (102 mm) above the earth, or not less than 2 inches (51 mm) above paved areas, or not less than $^1/_2$ inch (12.7 mm) above exterior walking surfaces that are supported by the same foundation that supports the *exterior wall*.

[BS] 1404.10.1.4 Adhered masonry veneer installed with lath and mortar. Exterior *adhered masonry veneer* installed with lath and *mortar* shall comply with the following.

[BS] 1404.10.1.4.1 Lathing. Lathing shall comply with the requirements of Section 2510.

[BS] 1404.10.1.4.2 Scratch coat. A nominal $^1/_2$-inch-thick (12.7 mm) layer of *mortar* complying with the material requirements of Sections 2103 and 2512.2 shall be applied, encapsulating the lathing. The surface of this *mortar* shall be scored horizontally, resulting in a scratch coat.

[BS] 1404.10.1.4.3 Adhering veneer. The masonry *veneer* units shall be adhered to the *mortar* scratch coat with a nominal $^1/_2$-inch-thick (12.7 mm) setting bed of *mortar* complying with Sections 2103 and 2512.2 applied to create a full setting bed for the back of the masonry *veneer* units. The masonry *veneer* units shall be worked into the setting bed resulting in a nominal $^3/_8$-inch (9.5 mm) setting bed after the masonry *veneer* units are applied.

[BS] 1404.10.1.5 Adhered masonry veneer applied directly to masonry and concrete. *Adhered masonry veneer* applied directly to masonry or concrete shall comply with the applicable requirements of Section 1404.10 and with the requirements of Section 1404.10.1.4 or 2510.7.

[BS] 1404.10.1.6 Cold weather construction. Cold weather construction of *adhered masonry veneer* shall comply with the requirements of Sections 2104 and 2512.4.

[BS] 1404.10.1.7 Hot weather construction. Hot weather construction of *adhered masonry veneer* shall comply with the requirements of Section 2104.

[BS] 1404.10.2 Exterior adhered masonry veneers—porcelain tile. Adhered units weighing more than 3.5 pounds per square foot (0.17 kN/m^2) shall not exceed 48 inches (1219 mm) in any face dimension nor more than 9 square feet (0.8 m^2) in total face area and shall not weigh more than 6 pounds per square foot (0.29 kN/m^2).

Adhered units weighing less than or equal to 3.5 pounds per square foot (0.17 kN/m^2) shall not exceed 72 inches (1829 mm) in any face dimension nor more than 17.5 square feet (1.6 m^2) in total face area. Porcelain tile shall be adhered to an approved backing system.

[BS] 1404.10.3 Interior adhered masonry veneers. Interior *adhered masonry veneers* shall have a maximum weight of 20 psf (0.958 kg/m^2) and shall be installed in accordance with Section 1404.10. Where the interior *adhered masonry veneer* is supported by wood construction, the supporting members shall be designed to limit deflection to $^1/_{600}$ of the span of the supporting members.

[BS] 1404.11 Metal veneers. *Veneers* of metal shall be fabricated from *approved* corrosion-resistant materials or shall be protected front and back with porcelain enamel, or otherwise be treated to render the metal resistant to corrosion. Such *veneers* shall be not less than 0.0149-inch (0.378 mm) nominal thickness sheet steel mounted on wood or metal furring strips or approved sheathing on *light-frame construction*.

[BS] 1404.11.1 Attachment. Exterior metal *veneer* shall be securely attached to the supporting masonry or framing members with corrosion-resistant fastenings, metal ties or by other *approved* devices or methods. The spacing of the fastenings or ties shall not exceed 24 inches (610 mm) either vertically or horizontally, but where units exceed 4 square feet (0.4 m^2) in area there shall be not less than four attachments per unit. The metal attachments shall have a cross-sectional area not less than provided by W 1.7 wire. Such attachments and their supports shall be designed and constructed to resist the wind *loads* as specified in Section 1609 for components and cladding.

1404.11.2 Weather protection. Metal supports for exterior metal *veneer* shall be protected by painting, galvanizing or by other equivalent coating or treatment. Wood studs, furring strips or other wood supports for exterior metal *veneer* shall be *approved* pressure-treated wood or protected as required in Section 1402.2. Joints and edges exposed to the weather shall be caulked with *approved* durable waterproofing material or by other *approved* means to prevent penetration of moisture.

1404.11.3 Backup. Masonry backup shall not be required for metal *veneer* unless required by the fire-resistance requirements of this code.

1404.11.4 Grounding. Grounding of metal *veneers* on buildings shall comply with the requirements of Chapter 27 of this code.

[BS] 1404.12 Glass veneer. The area of a single section of thin exterior structural glass *veneer* shall not exceed 10 square feet (0.93 m^2) where that section is not more than 15 feet (4572 mm) above the level of the sidewalk or grade level directly below, and shall not exceed 6 square feet (0.56 m^2) where it is more than 15 feet (4572 mm) above that level.

[BS] 1404.12.1 Length and height. The length or height of any section of thin exterior structural glass *veneer* shall not exceed 48 inches (1219 mm).

[BS] 1404.12.2 Thickness. The thickness of thin exterior structural glass *veneer* shall be not less than 0.344 inch (8.7 mm).

[BS] 1404.12.3 Application. Thin exterior structural glass *veneer* shall be set only after backing is thoroughly dry and after application of an *approved* bond coat uniformly over the entire surface of the backing so as to effectively seal the surface. Glass shall be set in place with an *approved* mastic cement in sufficient quantity so that not less than 50 percent of the area of each glass unit is directly bonded to the backing by mastic not less than $^1/_4$ inch (6.4 mm) thick and not more than $^5/_8$ inch (15.9 mm) thick. The bond coat and mastic shall be evaluated for compatibility and shall bond firmly together.

[BS] 1404.12.4 Installation at sidewalk level. Where glass extends to a sidewalk surface, each section shall rest in an *approved* metal molding, and be set not less than $^1/_4$ inch (6.4 mm) above the highest point of the sidewalk. The space between the molding and the sidewalk shall be thoroughly caulked and made watertight.

[BS] 1404.12.4.1 Installation above sidewalk level. Where thin exterior structural glass *veneer* is installed above the level of the top of a bulkhead facing, or at a level more than 36 inches (914 mm) above the sidewalk level, the mastic cement binding shall be supplemented with *approved* nonferrous metal shelf angles located in the horizontal joints in every course. Such shelf angles shall be not less than 0.0478-inch (1.2 mm) thick and not less than 2 inches (51 mm) long and shall be spaced at *approved* intervals, with not less than two angles for each glass unit. Shelf angles shall be secured to the wall or backing with expansion bolts, toggle bolts or by other *approved* methods.

[BS] 1404.12.5 Joints. Unless otherwise specifically *approved* by the *building official*, abutting edges of thin exterior structural glass *veneer* shall be ground square. Mitered joints shall not be used except where specifically *approved* for wide angles. Joints shall be uniformly buttered with an *approved* jointing compound and horizontal joints shall be held to not less than 0.063 inch (1.6 mm) by an *approved* nonrigid substance or device. Where thin exterior structural glass *veneer* abuts nonresilient material at sides or top, expansion joints not less than $^1/_4$ inch (6.4 mm) wide shall be provided.

[BS] 1404.12.6 Mechanical fastenings. Thin exterior structural glass *veneer* installed above the level of the heads of show windows and *veneer* installed more than 12 feet (3658 mm) above sidewalk level shall, in addition to the mastic cement and shelf angles, be held in place by the use of fastenings at each vertical or horizontal edge, or at the four corners of each glass unit. Fastenings shall be secured to the wall or backing with expansion bolts, toggle bolts or by other methods. Fastenings shall be so designed as to hold the glass *veneer* in a vertical plane independent of the mastic cement. Shelf angles providing both support and fastenings shall be permitted.

[BS] 1404.12.7 Flashing. Exposed edges of thin exterior structural glass *veneer* shall be flashed with overlapping corrosion-resistant metal flashing and caulked with a waterproof compound in a manner to effectively prevent the entrance of moisture between the glass *veneer* and the backing.

1404.13 Exterior windows and doors. Windows and doors installed in *exterior walls* shall conform to the testing and performance requirements of Section 1709.5.

1404.13.1 Installation. Windows and doors shall be installed in accordance with *approved* manufacturer's instructions. Fastener size and spacing shall be provided in such instructions and shall be calculated based on maximum *loads* and spacing used in the tests.

[BS] 1404.14 Vinyl siding. Vinyl siding conforming to the requirements of this section and complying with ASTM D3679 shall be permitted on *exterior walls* where the design wind pressure determined in accordance with Section 1609 does not exceed 30 pounds per square foot (1.44 kN/m^2). Where the design wind pressure exceeds 30 pounds per square foot (1.44 kN/m^2), tests or calculations indicating compliance with Chapter 16 shall be submitted. Vinyl siding shall be secured to the building so as to provide weather protection for the *exterior walls* of the building.

[BS] 1404.14.1 Application. The siding shall be applied over sheathing or materials listed in Section 2304.6. Siding shall be applied to conform to the *water-resistive barrier* requirements in Section 1402. Siding and accessories shall be installed in accordance with the *approved* manufacturer's instructions.

1404.14.1.1 Fasteners and fastener penetration for wood construction. Unless otherwise specified in the approved manufacturer's instructions, nails used to fasten the siding and accessories shall be corrosion resistant and have not less than a 0.313-inch (7.9 mm) head diameter and $^1/_8$-inch (3.18 mm) shank diameter. The penetration into nailable substrate shall be not less than $1^1/_4$ inches (32 mm).

1404.14.1.2 Fasteners and fastener penetration for cold-formed steel light-fame construction. For cold-formed steel light-frame construction, corrosion-resistant fasteners shall be used. Screw fasteners shall penetrate through the steel with not fewer than three exposed threads. Other fasteners shall be installed in accordance with the approved construction documents and manufacturer's instructions.

1404.14.1.3 Fastener spacing. Unless specified otherwise by the approved manufacturer's instructions, fasteners shall be installed in the middle third of the slots of the nail hem and spacing between fasteners shall be not greater than 16 inches (406 mm) for horizontal siding and 12 inches (305 mm) for vertical siding.

[BS] 1404.15 Cement plaster. *Cement plaster* applied to *exterior walls* shall conform to the requirements specified in Chapter 25.

[BS] 1404.16 Fiber-cement siding. *Fiber-cement siding* complying with Section 1403.10 shall be permitted on *exterior walls* of Types I, II, III, IV and V construction for wind pressure resistance or wind speed exposures as indicated by the manufacturer's listing and *label* and *approved* installation instructions. Where specified, the siding shall be installed over sheathing or materials listed in Section 2304.6 and shall be installed to conform to the *water-resistive barrier* requirements in Section 1402. Siding and accessories shall be installed in accordance with *approved* manufacturer's instructions. Unless otherwise specified in the *approved* manufacturer's instructions, nails used to fasten the siding to wood studs shall be corrosion-resistant round head smooth shank and shall be long enough to penetrate the studs not less than 1 inch (25 mm). For cold-formed steel *light-frame construction*, corrosion-resistant fasteners shall be used. Screw fasteners shall penetrate the cold-formed steel framing not fewer than three exposed full threads. Other fasteners shall be installed in accordance with the approved construction documents and manufacturer's instructions.

[BS] 1404.16.1 Panel siding. *Fiber-cement* panels shall comply with the requirements of ASTM C1186, Type A, minimum Grade II (or ISO 8336, Category A, minimum Class 2). Panels shall be installed with the long dimension either parallel or perpendicular to framing. Vertical and horizontal joints shall occur over framing members and shall be protected with caulking, with battens or flashing, or be vertical or horizontal shiplap or otherwise designed to comply with Section 1402.2. Panel siding shall be installed with fasteners in accordance with the *approved* manufacturer's instructions.

[BS] 1404.16.2 Lap siding. *Fiber-cement* lap siding having a maximum width of 12 inches (305 mm) shall comply with the requirements of ASTM C1186, Type A, minimum Grade II (or ISO 8336, Category A, minimum Class 2). Lap siding shall be lapped not less than $1^1/_4$ inches (32 mm) and lap siding not having tongue-and-groove end joints shall have the ends protected with caulking, covered with an H-section joint cover, located over a strip of flashing or shall be otherwise designed to comply with Section 1402.2. Lap siding courses shall be installed with the fastener heads exposed or concealed in accordance with the *approved* manufacturer's instructions.

[BS] 1404.17 Fastening. Weather boarding and wall coverings shall be securely fastened with aluminum, copper, zinc, zinc-coated or other *approved* corrosion-resistant fasteners in accordance with the nailing schedule in Table 2304.10.2 or the *approved* manufacturer's instructions. Shingles and other weather coverings shall be attached with appropriate standard-shingle nails to furring strips securely nailed to studs, or with *approved* mechanically bonding nails, except where sheathing is of wood not less than 1-inch (25 mm) nominal thickness or of *wood structural panels* as specified in Table 2308.6.3(3).

[BS] 1404.18 Polypropylene siding. *Polypropylene siding* conforming to the requirements of this section and complying with Section 1403.12 shall be limited to *exterior walls* located in areas where the wind speed specified in Chapter 16 does not exceed 100 miles per hour (45 m/s) and the building height is less than or equal to 40 feet (12 192 mm) in Exposure C. Where construction is located in areas where the basic wind speed exceeds 100 miles per hour (45 m/s), or building heights are in excess of 40 feet (12 192 mm), tests or calculations indicating compliance with Chapter 16 shall be submitted. *Polypropylene siding* shall be installed in accordance with the manufacturer's instructions. *Polypropylene siding* shall be secured to the building so as to provide weather protection for the *exterior walls* of the building.

SECTION 1405
COMBUSTIBLE MATERIALS ON THE EXTERIOR SIDE OF EXTERIOR WALLS

1405.1 Combustible exterior wall coverings. Combustible *exterior wall coverings* shall comply with this section.

Exception: Plastics complying with Chapter 26.

1405.1.1 Types I, II, III and IV construction. On buildings of Types I, II, III and IV construction, *exterior wall coverings* shall be permitted to be constructed of combustible materials, complying with the following limitations:

1. Combustible *exterior wall coverings* shall not exceed 10 percent of an *exterior wall* surface area where the *fire separation distance* is 5 feet (1524 mm) or less.
2. Combustible *exterior wall coverings* shall be limited to 40 feet (12 192 mm) in height above *grade plane*.
3. Combustible *exterior wall coverings* constructed of *fire-retardant-treated wood* complying with Section 2303.2 for exterior installation shall not be limited in wall surface area where the *fire separation distance* is 5 feet (1524 mm) or less and shall be permitted up to 60 feet (18 288 mm) in height above *grade plane* regardless of the *fire separation distance*.
4. Wood *veneers* shall comply with Section 1404.5.

1405.1.1.1 Ignition resistance. Where permitted by Section 1405.1.1, combustible *exterior wall coverings* shall be tested in accordance with NFPA 268.

Exceptions:

1. Wood or wood-based products.
2. Other combustible materials covered with an exterior weather covering, other than vinyl sidings, included in and complying with the thickness requirements of Table 1404.2.
3. Aluminum having a minimum thickness of 0.019 inch (0.48 mm).

1405.1.1.1.1 Fire separation 5 feet or less. Where installed on *exterior walls* having a *fire separation distance* of 5 feet (1524 mm) or less, combustible *exterior wall coverings* shall not exhibit sustained flaming as defined in NFPA 268.

1405.1.1.1.2 Fire separation greater than 5 feet. For *fire separation distances* greater than 5 feet (1524 mm), any *exterior wall covering* shall be permitted that has been exposed to a reduced level of incident radiant heat flux in accordance with the NFPA 268 test method without exhibiting sustained flaming. The minimum *fire separation distance* required for the *exterior wall covering* shall be determined from Table 1405.1.1.1.2 based on the maximum tolerable level of incident radiant heat flux that does not cause sustained flaming of the *exterior wall covering*.

TABLE 1405.1.1.1.2
MINIMUM FIRE SEPARATION FOR COMBUSTIBLE EXTERIOR WALL COVERINGS

FIRE SEPARATION DISTANCE (feet)	TOLERABLE LEVEL INCIDENT RADIANT HEAT ENERGY (kW/m^2)
5	12.5
6	11.8
7	11.0
8	10.3
9	9.6
10	8.9
11	8.3
12	7.7
13	7.2
14	6.7
15	6.3
16	5.9
17	5.5
18	5.2
19	4.9
20	4.6
21	4.4
22	4.1
23	3.9
24	3.7
25	3.5

For SI: 1 foot = 304.8 mm, 1 Btu/H^2 × °F = 0.0057 kW/m^2 × K.

1405.1.2 Location. Combustible *exterior wall coverings* located along the top of *exterior walls* shall be completely backed up by the *exterior wall* and shall not extend over or above the top of the *exterior wall*.

1405.1.3 Fireblocking. Where the combustible *exterior wall covering* is furred out from the *exterior wall* and forms a solid surface, the distance between the back of the *exterior wall covering* and the *exterior wall* shall not exceed $1^5/_8$ inches (41 mm). The concealed space thereby created shall be fireblocked in accordance with Section 718.

Exception: The distance between the back of the *exterior wall covering* and the *exterior wall* shall be permitted to exceed $1^5/_8$ inches (41 mm) where the concealed space is not required to be fireblocked by Section 718.

SECTION 1406
METAL COMPOSITE MATERIALS (MCM)

1406.1 General. The provisions of this section shall govern the materials, construction and quality of *metal composite materials (MCM)* for use as *exterior wall coverings* in addition to other applicable requirements of Chapters 14 and 16.

1406.2 Exterior wall covering. MCM used as *exterior wall* covering or as elements of balconies and similar projections and bay and oriel windows to provide cladding or weather resistance shall comply with Sections 1406.4 through 1406.13 .

1406.3 Architectural trim and embellishments. MCM used as architectural trim or embellishments shall comply with Sections 1406.7 through 1406.13 .

1406.4 Structural design. MCM systems shall be designed and constructed to resist wind *loads* as required by Chapter 16 for components and cladding.

1406.5 Approval. Results of *approved* tests or an engineering analysis shall be submitted to the *building official* to verify compliance with the requirements of Chapter 16 for wind *loads*.

1406.6 Weather resistance. MCM systems shall comply with Section 1402 and shall be designed and constructed to resist wind and rain in accordance with this section and the manufacturer's installation instructions.

1406.7 Durability. MCM systems shall be constructed of *approved* materials that maintain the performance characteristics required in Section 1406 for the duration of use.

1406.8 Fire-resistance rating. Where MCM systems are used on *exterior walls* required to have a *fire-resistance rating* in accordance with Section 705, evidence shall be submitted to the *building official* that the required *fire-resistance rating* is maintained.

Exception: MCM systems that are part of an *exterior wall envelope* not containing foam plastic insulation and are installed on the outer surface of a fire-resistance-rated *exterior wall* in a manner such that the attachments do not penetrate through the entire *exterior wall* assembly, shall not be required to comply with this section.

1406.9 Surface-burning characteristics. Unless otherwise specified, MCM shall have a *flame spread index* of 75 or less and a *smoke-developed index* of 450 or less when tested in the maximum thickness intended for use in accordance with ASTM E84 or UL 723.

1406.10 Types I, II, III and IV construction. Where installed on buildings of Types I, II, III and IV construction, *metal composite material* (MCM) shall comply with Sections 1406.10.1 and 1406.10.2 for installations up to 40 feet (12 192 mm) above *grade plane*. Where installed on buildings of Types I, II, III and IV construction, MCMs and MCM systems shall comply with Sections 1406.10.1 through 1406.10.3, for installations greater than 40 feet (12 192 mm) above *grade plane*.

1406.10.1 Surface-burning characteristics. MCM shall have a *flame spread index* of not more than 25 and a *smoke-developed index* of not more than 450 when tested in the maximum thickness intended for use in accordance with ASTM E84 or UL 723.

1406.10.2 Thermal barriers. MCM shall be separated from the interior of a building by an approved thermal barrier consisting of $^1/_2$-inch (12.7 mm) *gypsum wallboard* or material that is tested in accordance with and meets the acceptance criteria of both the Temperature Transmission Fire Test and the Integrity Fire Test of NFPA 275.

Exceptions:

1. The MCM system is specifically approved based on tests conducted in accordance with NFPA 286 and with the acceptance criteria of Section 803.1.1.1, UL 1040 or UL 1715. Such testing shall be performed with the MCM in the maximum thickness intended for use. The MCM system shall include seams, joints and other typical details used in the installation and shall be tested in the manner intended for use.
2. The MCM is used as elements of balconies and similar projections, architectural trim or embellishments.

1406.10.3 Full-scale tests. The MCM system shall be tested in accordance with, and comply with, the acceptance criteria of NFPA 285. Such testing shall be performed on the MCM system with the MCM in the maximum thickness intended for use.

1406.11 Type V construction. MCM shall be permitted to be installed on buildings of Type V construction.

1406.12 Foam plastic insulation. Where MCM systems are included in an exterior wall envelope containing foam plastic insulation, the exterior wall envelope shall also comply with the requirements of Section 2603.

1406.13 Labeling. MCM shall be labeled in accordance with Section 1703.5.

SECTION 1407
EXTERIOR INSULATION AND FINISH SYSTEMS (EIFS)

1407.1 General. The provisions of this section shall govern the materials, construction and quality of *exterior insulation* *and finish systems* (EIFS) for use as *exterior wall coverings* in addition to other applicable requirements of Chapters 7, 14, 16, 17 and 26.

1407.2 Performance characteristics. EIFS shall be constructed such that it meets the performance characteristics required in ASTM E2568.

[BS] 1407.3 Structural design. The underlying structural framing and substrate shall be designed and constructed to resist *loads* as required by Chapter 16.

1407.4 Weather resistance. EIFS shall comply with Section 1402 and shall be designed and constructed to resist wind and rain in accordance with this section and the manufacturer's application instructions.

1407.4.1 EIFS with drainage. EIFS with drainage shall have an average minimum drainage efficiency of 90 percent when tested in accordance the requirements of ASTM E2273 and is required on framed walls of Type V construction, Group R1, R2, R3 and R4 occupancies.

1407.4.1.1 Water-resistive barrier. For EIFS with drainage, the *water-resistive barrier* shall comply with Section 1403.2 or ASTM E2570.

1407.5 Installation. Installation of the EIFS and EIFS with drainage shall be in accordance with the EIFS manufacturer's instructions.

1407.6 Special inspections. EIFS installations shall comply with the provisions of Sections 1704.2 and 1705.17 .

SECTION 1408
HIGH-PRESSURE DECORATIVE EXTERIOR-GRADE COMPACT LAMINATES (HPL)

1408.1 General. The provisions of this section shall govern the materials, construction and quality of High-Pressure Decorative *Exterior-Grade Compact Laminates (HPL)* for use as *exterior wall coverings* in addition to other applicable requirements of Chapters 14 and 16.

1408.2 Exterior wall covering. HPL used as *exterior wall covering* or as elements of balconies and similar projections and bay and oriel windows to provide cladding or weather resistance shall comply with Sections 1408.4 through 1408.14.

1408.3 Architectural trim and embellishments. HPL used as architectural trim or embellishments shall comply with Sections 1408.7 through 1408.14.

[BS] 1408.4 Structural design. HPL systems shall be designed and constructed to resist wind *loads* as required by Chapter 16 for components and cladding.

1408.5 Approval. Results of approved tests or an engineering analysis shall be submitted to the building official to verify compliance with the requirements of Chapter 16 for wind *loads*.

1408.6 Weather resistance. HPL systems shall comply with Section 1402 and shall be designed and constructed to resist wind and rain in accordance with this section and the manufacturer's instructions.

1408.7 Durability. HPL systems shall be constructed of approved materials that maintain the performance characteristics required in Section 1408 for the duration of use.

1408.8 Fire-resistance rating. Where HPL systems are used on *exterior walls* required to have a *fire-resistance rating* in accordance with Section 705, evidence shall be submitted to the building official that the required *fire-resistance rating* is maintained.

Exception: HPL systems not containing foam plastic insulation, which are installed on the outer surface of a fire-resistance-rated *exterior wall* in a manner such that the attachments do not penetrate through the entire *exterior wall* assembly, shall not be required to comply with this section.

1408.9 Surface-burning characteristics. Unless otherwise specified, HPL shall have a *flame spread index* of 75 or less and a *smoke-developed index* of 450 or less when tested in the minimum and maximum thicknesses intended for use in accordance with ASTM E84 or UL 723.

1408.10 Types I, II, III and IV construction. Where installed on buildings of Types I, II, III and IV construction, HPL systems shall comply with Sections 1408.10.1 through 1408.10.4, or Section 1408.11.

1408.10.1 Surface-burning characteristics. HPL shall have a *flame spread index* of not more than 25 and a *smoke-developed index* of not more than 450 when tested in the minimum and maximum thicknesses intended for use in accordance with ASTM E84 or UL 723.

1408.10.2 Thermal barriers. HPL shall be separated from the interior of a building by an approved thermal barrier consisting of $^1/_2$-inch (12.7 mm) *gypsum wallboard* or a material that is tested in accordance with and meets the acceptance criteria of both the Temperature Transmission Fire Test and the Integrity Fire Test of NFPA 275.

1408.10.3 Thermal barrier not required. The thermal barrier specified for HPL in Section 1408.10.2 is not required where:

1. The HPL system is specifically approved based on tests conducted in accordance with NFPA 286, and with the acceptance criteria of Section 803.1.1.1, or with UL 1040 or UL 1715. Such testing shall be performed with the HPL in the minimum and maximum thicknesses intended for use. The HPL system shall include seams, joints and other typical details used in the installation and shall be tested in the manner intended for use.
2. The HPL is used as elements of balconies and similar projections, architectural trim or embellishments.

1408.10.4 Full-scale tests. The HPL system shall be tested in accordance with, and comply with, the acceptance criteria of NFPA 285. Such testing shall be performed on the HPL system with the HPL in the minimum and maximum thicknesses intended for use.

1408.11 Alternate conditions. HPL and HPL systems shall not be required to comply with Sections 1408.10.1 through 1408.10.4 provided that such systems comply with Section 1408.11.1.

1408.11.1 Installations up to 40 feet in height. HPL shall be permitted to be installed up to 40 feet (12 190 mm) in height above *grade plane* where installed in accordance with Section 1408.11.1.1 or 1408.11.1.2.

1408.11.1.1 Fire separation distance of 5 feet or less. Where the *fire separation distance* is 5 feet (1524 mm) or less, the area of HPL shall not exceed 10 percent of the *exterior wall* surface.

1408.11.1.2 Fire separation distance greater than 5 feet. Where the *fire separation distance* is greater than 5 feet (1524 mm), the area of *exterior wall* surface coverage using HPL shall not be limited.

1408.12 Type V construction. HPL shall be permitted to be installed on buildings of Type V construction.

1408.13 Foam plastic insulation. HPL systems containing foam plastic insulation shall comply with the requirements of Section 2603.

1408.14 Labeling. HPL shall be labeled in accordance with Section 1703.5.

SECTION 1409
PLASTIC COMPOSITE DECKING

1409.1 Plastic composite decking. Exterior deck boards, *stair* treads, *handrails* and *guards* constructed of plastic composites, including plastic lumber, shall comply with Section 2612.

CHAPTER 15
ROOF ASSEMBLIES AND ROOFTOP STRUCTURES

User notes:

About this chapter: *Chapter 15 provides minimum requirements for the design and construction of roof assemblies and rooftop structures. The criteria address the weather-protective barrier at the roof and, in most circumstances, a fire-resistant barrier. The chapter is largely prescriptive in nature and is based on decades of experience with various traditional materials, but it also recognizes newer products. Section 1511 addresses rooftop structures, which include penthouses, tanks, towers and spires. Rooftop penthouses larger than prescribed in this chapter must be treated as a story under Chapter 5.*

Code development reminder: *Code change proposals to sections preceded by the designation [BF], [BG] or [P] will be considered by one of the code development committees meeting during the 2021 (Group A) Code Development Cycle. All other code change proposals will be considered by the IBC—Structural Code Development Committee during the 2022 (Group B) Code Development Cycle.*

SECTION 1501
GENERAL

1501.1 Scope. The provisions of this chapter shall govern the design, materials, construction and quality of *roof assemblies*, and *rooftop structures*.

SECTION 1502
ROOF DRAINAGE

[P] 1502.1 General. Design and installation of roof drainage systems shall comply with this section, Section 1611 of this code and Chapter 11 of the *International Plumbing Code*.

[P] 1502.2 Secondary (emergency overflow) drains or scuppers. Where roof drains are required, secondary (emergency overflow) roof drains or scuppers shall be provided where the roof perimeter construction extends above the roof in such a manner that water will be entrapped if the primary drains allow buildup for any reason. The installation and sizing of secondary emergency overflow drains, leaders and conductors shall comply with Section 1611 of this code and Chapter 11 of the *International Plumbing Code*.

1502.3 Scuppers. Where *scuppers* are used for secondary (emergency overflow) roof drainage, the quantity, size, location and inlet elevation of the *scuppers* shall be sized to prevent the depth of ponding water from exceeding that for which the roof was designed as determined by Section 1611.1. *Scuppers* shall not have an opening dimension of less than 4 inches (102 mm). The flow through the primary system shall not be considered when locating and sizing *scuppers*.

1502.4 Gutters. Gutters and leaders placed on the outside of buildings, other than Group R-3, *private garages* and buildings of Type V construction, shall be of noncombustible material or not less than Schedule 40 plastic pipe.

SECTION 1503
WEATHER PROTECTION

1503.1 General. *Roof decks* shall be covered with *approved roof coverings* secured to the building or structure in accordance with the provisions of this chapter. *Roof coverings* shall be designed in accordance with this code, and installed in accordance with this code and the manufacturer's *approved* instructions.

1503.2 Flashing. Flashing shall be installed in such a manner so as to prevent water from entering the wall and roof through joints in copings, through moisture-permeable materials and at intersections with *parapet walls* and other penetrations through the roof plane.

1503.2.1 Locations. Flashing shall be installed at wall and roof intersections, at gutters, wherever there is a change in roof slope or direction and around roof openings. Where flashing is of metal, the metal shall be corrosion resistant with a thickness of not less than 0.019 inch (0.483 mm) (No. 26 galvanized sheet).

1503.3 Parapet walls. Parapet walls shall be coped or covered in accordance with Sections 1503.3.1 and 1503.3.2. The top surface of the parapet wall shall provide positive drainage.

1503.3.1 Fire-resistance-rated parapet walls. *Parapet walls* required by Section 705.11 shall be coped or covered with weatherproof materials of a width not less than the thickness of the *parapet wall* such that the *fire-resistance rating* of the wall is not decreased.

1503.3.2 Other parapet walls. *Parapet walls* meeting one of the exceptions in Section 705.11 shall be coped or covered with weatherproof materials of a width not less than the thickness of the *parapet wall*.

1503.4 Attic and rafter ventilation. Intake and exhaust vents shall be provided in accordance with Section 1202.2 and the vent product manufacturer's installation instructions.

1503.5 Crickets and saddles. A cricket or saddle shall be installed on the ridge side of any chimney or penetration greater than 30 inches (762 mm) wide as measured perpendicular to the slope. Cricket or saddle coverings shall be sheet metal or of the same material as the *roof covering*.

Exception: *Unit skylights* installed in accordance with Section 2405.5 and flashed in accordance with the manufacturer's instructions shall be permitted to be installed without a cricket or saddle.

SECTION 1504
PERFORMANCE REQUIREMENTS

1504.1 Wind resistance of roofs. *Roof decks* and *roof coverings* shall be designed for wind *loads* in accordance with Chapter 16 and Sections 1504.2, 1504.3, 1504.4 and 1504.5

1504.2 Wind resistance of asphalt shingles. Asphalt shingles shall be tested in accordance with ASTM D7158. Asphalt shingles shall meet the classification requirements of Table 1504.2 for the appropriate maximum basic wind speed. Asphalt shingle packaging shall bear a label to indicate compliance with ASTM D7158 and the required classification in Table 1504.2.

Exception: Asphalt shingles not included in the scope of ASTM D7158 shall be tested and labeled in accordance with ASTM D3161. Asphalt shingle packaging shall bear a label to indicate compliance with ASTM D3161 and the required classification in Table 1504.2.

1504.3 Wind resistance of clay and concrete tile. Wind *loads* on clay and concrete tile roof coverings shall be in accordance with Section 1609.5.

1504.3.1 Testing. Testing of concrete and clay roof tiles shall be in accordance with Sections 1504.3.1.1, 1504.3.1.2 and 1504.3.1.3.

1504.3.1.1 Overturning resistance. Concrete and clay roof tiles shall be tested to determine their resistance to overturning due to wind in accordance with Chapter 15 and either SBCCI SSTD 11 or ASTM C1568.

1504.3.1.2 Wind tunnel testing. Where concrete and clay roof tiles do not satisfy the limitations in Chapter 16 for rigid tile, a wind tunnel test shall be used to determine the wind characteristics of the concrete or clay tile *roof covering* in accordance with Chapter 15 and either SBCCI SSTD 11 or ASTM C1569.

1504.3.1.3 Air permeability testing. The lift coefficient for concrete and clay tile shall be 0.2 or shall be determined in accordance with SBCCI SSTD 11 or ASTM C1570.

1504.4 Wind resistance of nonballasted roofs. *Roof coverings* installed on roofs in accordance with Section 1507 that are mechanically attached or adhered to the *roof deck* shall be designed to resist the design wind *load* pressures for components and cladding in accordance with Section 1609.5.2. The wind *load* on the *roof covering* shall be permitted to be determined using *allowable stress design*.

1504.4.1 Other roof systems. Built-up, modified bitumen, fully adhered or mechanically attached single-ply roof systems, metal panel roof systems applied to a solid or closely fitted deck and other types of membrane *roof coverings* shall be tested in accordance with FM 4474, UL 580 or UL 1897.

1504.4.2 Structural metal panel roof systems. Where the *metal roof panel* functions as the *roof deck* and *roof covering* and it provides both weather protection and support for *loads*, the structural metal panel roof system shall comply with this section. Structural standing-seam metal panel roof systems shall be tested in accordance with ASTM E1592 or FM 4474. Structural through-fastened metal panel roof systems shall be tested in accordance with ASTM E1592, FM 4474 or UL 580.

Exceptions:

1. Metal roofs constructed of cold-formed steel shall be permitted to be designed and tested in accordance with the applicable referenced structural design standard in Section 2210.1.
2. Metal roofs constructed of aluminum shall be permitted to be designed and tested in accordance with the applicable referenced structural design standard in Section 2002.1.

1504.4.3 Metal roof shingles. *Metal roof shingles* applied to a solid or closely fitted deck shall be tested in accordance with ASTM D3161, FM 4474, UL 580 or UL 1897. *Metal roof shingles* tested in accordance with ASTM D3161 shall meet the classification requirements of Table 1504.2 for the appropriate maximum basic wind speed and the metal shingle packaging shall bear a label to indicate compliance with ASTM D3161 and the required classification in Table 1504.2.

TABLE 1504.2
CLASSIFICATION OF STEEP SLOPE ROOF SHINGLES TESTED IN ACCORDANCE WITH ASTM D3161OR D7158

MAXIMUM BASIC WIND SPEED, *V*, FROM FIGURES 1609.3(1)–(8) OR ASCE 7(mph)	MAXIMUM ALLOWABLE STRESS DESIGN WIND SPEED, V_{asd}, FROM Table 1609.3.1 (mph)	ASTM D7158[a] CLASSIFICATION	ASTM D3161 or UL 7103 CLASSIFICATION
110	85	D, G or H	A, D or F
116	90	D, G or H	A, D or F
129	100	G or H	A, D or F
142	110	G or H	F
155	120	G or H	F
168	130	H	F
181	140	H	F
194	150	H	F

For SI: 1 foot = 304.8 mm; 1 mph = 0.447 m/s.

a. The standard calculations contained in ASTM D7158 assume Exposure Category B or C and building height of 60 feet or less. Additional calculations are required for conditions outside of these assumptions.

1504.5 Ballasted low-slope single-ply roof systems. Ballasted low-slope (roof slope < 2:12) single-ply roof system coverings installed in accordance with Section 1507.12 shall be designed in accordance with ANSI/SPRI RP-4.

1504.6 Edge systems for low-slope roofs. Metal edge systems, except gutters and counterflashing, installed on built-up, modified bitumen and single-ply roof systems having a slope less than 2 units vertical in 12 units horizontal (2:12) shall be designed and installed for wind *loads* in accordance with Chapter 16 and tested for resistance in accordance with Test Methods RE-1, RE-2 and RE-3 of ANSI/SPRI ES-1, except basic design *wind speed*, V, shall be determined from Figures 1609.3(1) through 1609.3(12) as applicable.

1504.6.1 Gutter securement for low-slope roofs. Gutters that are used to secure the perimeter edge of the roof membrane on low-slope (less than 2:12 slope) built-up, modified bitumen, and single-ply roofs, shall be designed, constructed and installed to resist wind loads in accordance with Section 1609 and shall be tested in accordance with Test Methods G-1 and G-2 of SPRI GT-1.

1504.7 Physical properties. *Roof coverings* installed on low-slope roofs (roof slope < 2:12) in accordance with Section 1507 shall demonstrate physical integrity over the working life of the roof based on 2,000 hours of exposure to accelerated weathering tests conducted in accordance with ASTM G152, ASTM G154 or ASTM G155. Those *roof coverings* that are subject to cyclical flexural response due to wind *loads* shall not demonstrate any significant loss of tensile strength for unreinforced membranes or breaking strength for reinforced membranes when tested as herein required.

1504.8 Impact resistance. *Roof coverings* installed on low-slope roofs (roof slope < 2:12) in accordance with Section 1507 shall resist impact damage based on the results of tests conducted in accordance with ASTM D3746, ASTM D4272 or the "Resistance to Foot Traffic Test" in FM 4470.

1504.9 Wind resistance of aggregate-surfaced roofs. Parapets shall be provided for aggregate surfaced roofs and shall comply with Table 1504.9.

SECTION 1505
FIRE CLASSIFICATION

[BF] 1505.1 General. *Roof assemblies* shall be divided into the classes defined in this section. Class A, B and C *roof assemblies* and *roof coverings* required to be listed by this section shall be tested in accordance with ASTM E108 or UL 790. In addition, *fire-retardant-treated wood roof coverings* shall be tested in accordance with ASTM D2898. The minimum *roof coverings* installed on buildings shall comply with Table 1505.1 based on the type of construction of the building.

Exception: *Skylights and sloped glazing* that comply with Chapter 24 or Section 2610.

TABLE 1504.9
MINIMUM REQUIRED PARAPET HEIGHT (INCHES) FOR AGGREGATE SURFACED ROOFS[a, b, c]

AGGREGATE SIZE	MEAN ROOF HEIGHT (ft)	WIND EXPOSURE AND BASIC DESIGN WIND SPEED (MPH)																	
		Exposure B									Exposure C[d]								
		≤ 95	100	105	110	115	120	130	140	150	≤ 95	100	105	110	115	120	130	140	150
ASTM D1863 (No. 7 or No. 67)	15	2	2	2	2	12	12	16	20	24	2	13	15	18	20	23	27	32	37
	20	2	2	2	2	12	14	18	22	26	12	15	17	19	22	24	29	34	39
	30	2	2	2	13	15	17	21	25	30	14	17	19	22	24	27	32	37	42
	50	12	12	14	16	18	21	25	30	35	17	19	22	25	28	30	36	41	47
	100	14	16	19	21	24	27	32	37	42	21	24	26	29	32	35	41	47	53
	150	17	19	22	25	27	30	36	41	46	23	26	29	32	35	38	44	50	56
ASTM D1863 (No. 6)	15	2	2	2	2	12	12	12	15	18	2	2	2	13	15	17	22	26	30
	20	2	2	2	2	12	12	13	17	21	2	2	12	15	17	19	23	28	32
	30	2	2	2	2	12	12	16	20	24	2	12	14	17	19	21	26	31	35
	50	12	12	12	12	14	16	20	24	28	12	15	17	19	22	24	29	34	39
	100	12	12	14	16	19	21	26	30	35	16	18	21	24	26	29	34	39	45
	150	12	14	17	19	22	24	29	34	39	18	21	23	26	29	32	37	43	48

For SI: 1 inch = 25.4 mm; 1 foot = 304.8 mm; 1 mile per hour = 0.447 m/s.

a. Interpolation shall be permitted for mean roof height and parapet height.

b. Basic design wind speed, *V*, and wind exposure shall be determined in accordance with Section 1609.

c. Where the minimum required parapet height is indicated to be 2 inches (51 mm), a gravel stop shall be permitted and shall extend not less than 2 inches (51 mm) from the roof surface and not less than the height of the aggregate.

d. For Exposure D, add 8 inches (203 mm) to the parapet height required for Exposure C and the parapet height shall not be less than 12 inches (305 mm).

TABLE 1505.1
MINIMUM ROOF COVERING CLASSIFICATION FOR TYPES OF CONSTRUCTION[a, b]

IA	IB	IIA	IIB	IIIA	IIIB	IV	VA	VB
B	B	B	C[c]	B	C[c]	B	B	C[c]

For SI: 1 foot = 304.8 mm, 1 square foot = 0.0929 m^2.

a. Unless otherwise required in accordance with the *International Wildland-Urban Interface Code* or due to the location of the building within a fire district in accordance with Appendix D.

b. Nonclassified roof coverings shall be permitted on buildings of Group R-3 and Group U occupancies, where there is a minimum fire-separation distance of 6 feet measured from the leading edge of the roof.

c. Buildings that are not more than two stories above grade plane and having not more than 6,000 square feet of projected roof area and where there is a minimum 10-foot fire-separation distance from the leading edge of the roof to a lot line on all sides of the building, except for street fronts or public ways, shall be permitted to have roofs of No. 1 cedar or redwood shakes and No. 1 shingles constructed in accordance with Section 1505.7.

[BF] 1505.2 Class A roof assemblies. Class A *roof assemblies* are those that are effective against severe fire test exposure. Class A *roof assemblies* and *roof coverings* shall be *listed* and identified as Class A by an *approved* testing agency. Class A *roof assemblies* shall be permitted for use in buildings or structures of all types of construction.

Exceptions:

1. Class A *roof assemblies* include those with coverings of brick, masonry or an exposed concrete roof deck.
2. Class A *roof assemblies* also include ferrous or copper shingles or sheets, metal sheets and shingles, clay or concrete roof tile or slate installed on noncombustible decks or ferrous, copper or metal sheets installed without a roof deck on noncombustible framing.
3. Class A *roof assemblies* include minimum 16 ounce per square foot (0.0416 kg/m^2) copper sheets installed over combustible decks.
4. Class A *roof assemblies* include slate installed over ASTM D226, Type II *underlayment* over combustible decks.

[BF] 1505.3 Class B roof assemblies. Class B *roof assemblies* are those that are effective against moderate fire-test exposure. Class B *roof assemblies* and *roof coverings* shall be *listed* and identified as Class B by an *approved* testing agency.

[BF] 1505.4 Class C roof assemblies. Class C *roof assemblies* are those that are effective against light fire-test exposure. Class C *roof assemblies* and *roof coverings* shall be *listed* and identified as Class C by an *approved* testing agency.

[BF] 1505.5 Nonclassified roofing. Nonclassified roofing is *approved* material that is not *listed* as a Class A, B or C *roof covering*.

[BF] 1505.6 Fire-retardant-treated wood shingles and shakes. *Fire-retardant-treated wood* shakes and shingles shall be treated by impregnation with chemicals by the full-cell vacuum-pressure process, in accordance with AWPA C1. Each bundle shall be marked to identify the manufactured unit and the manufacturer, and shall be *labeled* to identify the classification of the material in accordance with the testing required in Section 1505.1, the treating company and the quality control agency.

[BF] 1505.7 Special purpose roofs. Special purpose wood shingle or wood shake roofing shall conform to the grading and application requirements of Section 1507.8 or 1507.9. In addition, an *underlayment* of $^5/_8$-inch (15.9 mm) Type X water-resistant gypsum backing board or *gypsum sheathing* shall be placed under minimum nominal $^1/_2$-inch-thick (12.7 mm) *wood structural panel* solid sheathing or 1-inch (25 mm) nominal spaced sheathing.

[BF] 1505.8 Building-integrated photovoltaic (BIPV) products. *BIPV products* installed as the roof covering shall be tested, *listed* and *labeled* for fire classification in accordance with Section 1505.1.

[BF] 1505.9 Rooftop mounted photovoltaic (PV) panel systems. Rooftop mounted *photovoltaic (PV) panel systems* shall be tested, *listed* and identified with a fire classification in accordance with UL 2703. Listed systems shall be installed in accordance with the manufacturer's installation instructions and their listing. The fire classification shall comply with Table 1505.1 based on the type of construction of the building.

[BF] 1505.10 Landscaped roofs. Landscaped roofs shall comply with Sections 1505.1 and 1507.15 and shall be installed in accordance with ANSI/SPRI VF-1.

SECTION 1506 MATERIALS

1506.1 Scope. The requirements set forth in this section shall apply to the application of roof-covering materials specified herein. *Roof coverings* shall be applied in accordance with this chapter and the *roof covering* listing as required by Section 1505. Installation of *roof coverings* shall comply with the applicable provisions of Section 1507.

1506.2 Material specifications and physical characteristics. Roof-covering materials shall conform to the applicable standards listed in this chapter.

1506.3 Product identification. Roof-covering materials shall be delivered in packages bearing the manufacturer's identifying marks and *approved* testing agency labels required in accordance with Section 1505. Bulk shipments of materials shall be accompanied with the same information issued in the form of a certificate or on a bill of lading by the manufacturer.

SECTION 1507 REQUIREMENTS FOR ROOF COVERINGS

1507.1 Scope. *Roof coverings* shall be applied in accordance with the applicable provisions of this section and the manufacturer's installation instructions.

1507.1.1 Underlayment. Underlayment for asphalt shingles, clay and concrete tile, metal roof shingles, mineral-surfaced roll roofing, slate and slate-type shingles, wood shingles, wood shakes, metal roof panels and *photovoltaic* *shingles* shall conform to the applicable standards listed in this chapter. Underlayment materials required to comply with ASTM D226, D1970, D4869 and D6757 shall bear a label indicating compliance with the standard designation and, if applicable, type classification indicated in Table 1507.1.1(1). Underlayment shall be applied in accordance with Table 1507.1.1(2). Underlayment shall be attached in accordance with Table 1507.1.1(3).

Exceptions:

1. As an alternative, a minimum 4-inch-wide (102 mm) strip of self-adhering polymer modified bitumen membrane complying with ASTM D1970 and installed in accordance with the manufacturer's installation instructions for the deck material shall be applied over all joints in the roof decking. An approved underlayment for the applicable roof covering for design wind speeds less than 120 mph (54 m/s) shall be applied over the 4-inch-wide (102 mm) membrane strips.
2. As an alternative, two layers of underlayment complying with ASTM D226 Type II or ASTM D4869 Type IV shall be permitted to be installed as follows: Apply a 19-inch (483 mm) strip of underlayment parallel with the eave. Starting at the eave, apply 36-inch-wide (914 mm) strips of underlayment felt, overlapping successive sheets 19 inches (483 mm). The underlayment shall be attached with corrosion-resistant fasteners in a grid pattern of 12 inches (305 mm) between side laps with a 6-inch (152 mm) spacing at side and end laps. End laps shall be 4 inches (102 mm) and shall be offset by 6 feet (1829 mm). Underlayment shall be attached using metal or plastic cap nails with a nominal cap diameter of not less than 1 inch (25.4 mm). Metal caps shall have a thickness of not less than 32-gage sheet metal. Power-driven metal caps shall have a thickness of not less than 0.010 inch (0.254 mm). Thickness of the outside edge of plastic caps shall be not less than 0.035 inch (0.89 mm). The cap nail shank shall be not less than 0.083 inch (2.1 mm) for ring shank cap nails and 0.091 inch (2.3 mm) for smooth shank cap nails. The cap nail shank shall have a length sufficient to penetrate through the roof sheathing or not less than $^{3}/_{4}$ inch (19.1 mm) into the roof sheathing.
3. Structural metal panels that do not require a substrate or underlayment.

1507.1.2 Ice barriers. In areas where there has been a history of ice forming along the eaves causing a backup of water, an ice barrier shall be installed for asphalt shingles, *metal roof shingles*, mineral-surfaced roll roofing, slate and slate-type shingles, wood shingles, and wood shakes. The ice barrier shall consist of not less than two layers of *underlayment* cemented together, or a self-adhering polymer modified bitumen sheet shall be used in place of normal *underlayment* and extend from the lowest edges of all roof surfaces to a point not less than 24 inches (610 mm) inside the *exterior wall* line of the building.

Exception: Detached accessory structures that do not contain conditioned floor area.

TABLE 1507.1.1(1)
UNDERLAYMENT TYPES

ROOF COVERING	SECTION	MAXIMUM BASIC DESIGN WIND SPEED, V < 140 MPH	MAXIMUM BASIC DESIGN WIND SPEED, V ≥ 140 MPH
Asphalt shingles	1507.2	ASTM D226 Type I or II ASTM D4869 Type I, II, III or IV ASTM D6757	ASTM D226 Type II ASTM D4869 Type IV ASTM D6757
Clay and concrete tiles	1507.3	ASTM D226 Type II ASTM D2626 Type I ASTM D6380 Class M mineral surfaced roll roofing	ASTM D226 Type II ASTM D2626 Type I ASTM D6380 Class M mineral surfaced roll roofing
Metal roof panels	1507.4	Manufacturer's instructions	ASTM D226 Type II ASTM D4869 Type IV
Metal roof shingles	1507.5	ASTM D226 Type I or II ASTM D4869 Type I, II, III or IV	ASTM D226 Type II ASTM D4869 Type IV
Mineral-surfaced roll roofing	1507.6	ASTM D226 Type I or II ASTM D4869 Type I, II, III or IV	ASTM D226 Type II ASTM D4869 Type IV
Slate shingles	1507.7	ASTM D226 Type II ASTM D4869 Type III or IV	ASTM D226 Type II ASTM D4869 Type IV
Wood shingles	1507.8	ASTM D226 Type I or II ASTM D4869 Type I, II, III or IV	ASTM D226 Type II ASTM D4869 Type IV
Wood shakes	1507.9	ASTM D226 Type I or II ASTM D4869 Type I, II, III or IV	ASTM D226 Type II ASTM D4869 Type IV
Photovoltaic shingles	1507.16	ASTM D226 Type I or II ASTM D4869 Type I, II, III or IV ASTM D6757	ASTM D226 Type II ASTM D4869 Type IV ASTM D6757

TABLE 1507.1.1(2)
UNDERLAYMENT APPLICATION

ROOF COVERING	SECTION	MAXIMUM BASIC DESIGN WIND SPEED, $V < 140$ MPH	MAXIMUM BASIC DESIGN WIND SPEED, $V \geq 140$ MPH
Asphalt shingles	1507.2	For roof slopes from 2 units vertical in 12 units horizontal (2:12), up to 4 units vertical in 12 units horizontal (4:12), underlayment shall be two layers applied as follows: Apply a 19-inch strip of underlayment felt parallel to and starting at the eaves. Starting at the eave, apply 36-inch-wide sheets of underlayment, overlapping successive sheets 19 inches. End laps shall be 4 inches and shall be offset by 6 feet. Distortions in the underlayment shall not interfere with the ability of the shingles to seal. For roof slopes of 4 units vertical in 12 units horizontal (4:12) or greater, underlayment shall be one layer applied as follows: Underlayment shall be applied shingle fashion, parallel to and starting from the eave and lapped 2 inches, Distortions in the underlayment shall not interfere with the ability of the shingles to seal. End laps shall be 4 inches and shall be offset by 6 feet.	Same as Maximum Basic Design Wind Speed, $V < 140$ mph except all laps shall be not less than 4 inches
Clay and concrete tile	1507.3	For roof slopes from $2^1/_2$ units vertical in 12 units horizontal ($2^1/_2$:12), up to 4 units vertical in 12 units horizontal (4:12), underlayment shall be not fewer than two layers applied as follows: Starting at the eave, a 19-inch strip of underlayment shall be applied parallel with the eave. Starting at the eave, a 36-inch-wide strip of underlayment felt shall be applied, overlapping successive sheets 19 inches. End laps shall be 4 inches and shall be offset by 6 feet. For roof slopes of 4 units vertical in 12 units horizontal (4:12) or greater, underlayment shall be one layer applied as follows: Underlayment shall be applied shingle fashion, parallel to and starting from the eave and lapped 2 inches. End laps shall be 4 inches and shall be offset by 6 feet.	Same as Maximum Basic Design Wind Speed, $V < 140$ mph except all laps shall be not less than 4 inches
Metal roof panels	1507.4	Apply in accordance with the manufacturer's installation instructions	For roof slopes from 2 units vertical in 12 units horizontal (2:12), up to 4 units vertical in 12 units horizontal (4:12), underlayment shall be two layers applied as follows: Apply a 19-inch strip of underlayment felt parallel to and starting at the eaves. Starting at the eave, apply 36-inch-wide sheets of underlayment, overlapping successive sheets 19 inches. End laps shall be 4 inches and shall be offset by 6 feet. For roof slopes of 4 units vertical in 12 units horizontal (4:12) or greater, underlayment shall be one layer applied as follows: Underlayment shall be applied shingle fashion, parallel to and starting from the eave and lapped 4 inches. End laps shall be 4 inches and shall be offset by 6 feet.
Metal roof shingles	1507.5		
Mineral-surfaced roll roofing	1507.6		
Slate shingles	1507.7		
Wood shingles	1507.8		
Wood shakes	1507.9		
Photovoltaic shingles	1507.16	For roof slopes from 3 units vertical in 12 units horizontal (3:12), up to 4 units vertical in 12 units horizontal (4:12), underlayment shall be two layers applied as follows: Apply a 19-inch strip of underlayment felt parallel to and starting at the eaves. Starting at the eave, apply 36-inch-wide sheets of underlayment, overlapping successive sheets 19 inches. End laps shall be 4 inches and shall be offset by 6 feet. Distortions in the underlayment shall not interfere with the ability of the shingles to seal. For roof slopes of 4 units vertical in 12 units horizontal (4:12) or greater, underlayment shall be one layer applied as follows: Underlayment shall be applied shingle fashion, parallel to and starting from the eave and lapped 2 inches. Distortions in the underlayment shall not interfere with the ability of the shingles to seal. End laps shall be 4 inches and shall be offset by 6 feet.	Same as Maximum Basic Design Wind Speed, $V < 140$ mph except all laps shall be not less than 4 inches

For SI: 1 inch = 25.4 mm, 1 foot = 304.8 mm; 1 mile per hour = 0.447 m/s.

TABLE 1507.1.1(3)
UNDERLAYMENT ATTACHMENT

ROOF COVERING	SECTION	MAXIMUM BASIC DESIGN WIND SPEED, *V* < 140 MPH	MAXIMUM BASIC DESIGN WIND SPEED, *V* ≥ 140 MPH
Asphalt shingles	1507.2	Fastened sufficiently to hold in place	The underlayment shall be attached with corrosion-resistant fasteners in a grid pattern of 12 inches between side laps with a 6-inch spacing at side and end laps. Underlayment shall be attached using metal or plastic cap nails or cap staples with a nominal cap diameter of not less than 1 inch. Metal caps shall have a thickness of not less than 32-gage (0.0134 inch) sheet metal. Power-driven metal caps shall have a minimum thickness of 0.010 inch. Minimum thickness of the outside edge of plastic caps shall be 0.035 inch. The cap nail shank shall be not less than 0.083 inch for ring shank cap nails and 0.091 inch for smooth shank cap nails. Staples shall be not less than 21 gage (0.032 inch). The cap nail shank and cap staple legs shall have a length sufficient to penetrate through the roof sheathing or not less than $^{3}/_{4}$ inch into the roof sheathing.
Clay and concrete tile	1507.3		
Photovoltaic shingles	1507.16		
Metal roof panels	1507.4	Manufacturer's installation instructions	The underlayment shall be attached with corrosion-resistant fasteners in a grid pattern of 12 inches between side laps with a 6-inch spacing at side and end laps. Underlayment shall be attached using metal or plastic cap nails or cap staples with a nominal cap diameter of not less than 1 inch. Metal caps shall have a thickness of not less than 32-gage sheet metal. Power-driven metal caps shall have a minimum thickness of 0.010 inch. Minimum thickness of the outside edge of plastic caps shall be 0.035 inch. The cap nail shank shall be not less than 0.083 inch for ring shank cap nails and 0.091 inch for smooth shank cap nails. Staples shall be not less than 21 gage. The cap nail shank and cap staple legs shall have a length sufficient to penetrate through the roof sheathing or not less than $^{3}/_{4}$ inch into the roof sheathing.
Metal roof shingles	1507.5		
Mineral-surfaced roll roofing	1507.6		
Slate shingles	1507.7		
Wood shingles	1507.8		
Wood shakes	1507.9		

For SI: 1 inch = 25.4 mm; 1 mile per hour = 0.447 m/s.

1507.2 Asphalt shingles. The installation of asphalt shingles shall comply with the provisions of this section.

1507.2.1 Deck requirements. Asphalt shingles shall be fastened to solidly sheathed decks.

1507.2.2 Slope. Asphalt shingles shall only be used on roof slopes of 2 units vertical in 12 units horizontal (17-percent slope) or greater. For roof slopes from 2 units vertical in 12 units horizontal (17-percent slope) up to 4 units vertical in 12 units horizontal (33-percent slope), double *underlayment* application is required in accordance with Section 1507.2.8.

1507.2.3 Underlayment. *Underlayment* shall comply with Section 1507.1.1.

1507.2.4 Asphalt shingles. Asphalt shingles shall comply with ASTM D3462.

1507.2.5 Fasteners. Fasteners for asphalt shingles shall be galvanized, stainless steel, aluminum or copper roofing nails, minimum 12-gage [0.105 inch (2.67 mm)] shank with a minimum $^{3}/_{8}$-inch-diameter (9.5 mm) head, of a length to penetrate through the roofing materials and not less than $^{3}/_{4}$ inch (19.1 mm) into the roof sheathing. Where the roof sheathing is less than $^{3}/_{4}$ inch (19.1 mm) thick, the nails shall penetrate through the sheathing. Fasteners shall comply with ASTM F1667.

1507.2.6 Attachment. Asphalt shingles shall have the minimum number of fasteners required by the manufacturer, but not less than four fasteners per strip shingle or two fasteners per individual shingle. Where the roof slope exceeds 21 units vertical in 12 units horizontal (21:12), shingles shall be installed as required by the manufacturer.

1507.2.7 Ice barrier. Where required, ice barriers shall comply with Section 1507.1.2.

1507.2.8 Flashings. Flashing for asphalt shingles shall comply with this section. Flashing shall be applied in accordance with this section and the asphalt shingle manufacturer's printed instructions.

1507.2.8.1 Base and cap flashing. Base and cap flashing shall be installed in accordance with the manufacturer's instructions. Base flashing shall be of either corrosion-resistant metal of minimum nominal 0.019-inch (0.483 mm) thickness or mineral-surfaced roll roofing weighing not less than 77 pounds per 100 square feet (3.76 kg/m^2). Cap flashing shall be corrosion-resistant metal of minimum nominal 0.019-inch (0.483 mm) thickness.

1507.2.8.2 Valleys. Valley linings shall be installed in accordance with the manufacturer's instructions before applying shingles. Valley linings of the following types shall be permitted:

1. For open valleys (valley lining exposed) lined with metal, the valley lining shall be not less than 24 inches (610 mm) wide and of any of

the corrosion-resistant metals in Table 1507.2.8.2.

2. For open valleys, valley lining of two plies of mineral-surfaced roll roofing complying with ASTM D3909 or ASTM D6380 shall be permitted. The bottom layer shall be 18 inches (457 mm) and the top layer not less than 36 inches (914 mm) wide.
3. For closed valleys (valleys covered with shingles), valley lining of one ply of smooth roll roofing complying with ASTM D6380, and not less than 36 inches (914 mm) wide or types as described in Item 1 or 2 above shall be permitted. Self-adhering polymer modified bitumen *underlayment* bearing a label indicating compliance with ASTM D1970 shall be permitted in lieu of the lining material.

1507.2.8.3 Drip edge. A drip edge shall be provided at eaves and rake edges of shingle roofs. Adjacent segments of the drip edge shall be lapped not less than 2 inches (51 mm). The vertical leg of drip edges shall be not less than $1^1/_2$ inches (38 mm) in width and shall extend not less than $^1/_4$ inch (6.4 mm) below sheathing. The drip edge shall extend back on the roof not less than 2 inches (51 mm). *Underlayment* shall be installed over drip edges along eaves. Drip edges shall be installed over *underlayment* along rake edges. Drip edges shall be mechanically fastened at intervals not greater than 12 inches (305 mm) on center.

1507.3 Clay and concrete tile. The installation of clay and concrete tile shall comply with the provisions of this section.

1507.3.1 Deck requirements. Concrete and clay tile shall be installed only over solid sheathing.

Exception: Spaced lumber sheathing shall be permitted in *Seismic Design Categories* A, B and C.

1507.3.2 Deck slope. Clay and concrete roof tile shall be installed on roof slopes of $2^1/_2$ units vertical in 12 units horizontal (21-percent slope) or greater. For roof slopes from $2^1/_2$ units vertical in 12 units horizontal (21-percent slope) to 4 units vertical in 12 units horizontal (33-percent slope), double *underlayment* application is required in accordance with Section 1507.3.3.

1507.3.3 Underlayment. Unless otherwise noted, required *underlayment* shall conform to: ASTM D226, Type II; ASTM D2626 or ASTM D6380, Class M mineral-surfaced roll roofing.

1507.3.4 Clay tile. Clay roof tile shall comply with ASTM C1167.

1507.3.5 Concrete tile. Concrete roof tile shall comply with ASTM C1492.

1507.3.6 Fasteners. Tile fasteners shall be corrosion resistant and not less than 11-gage, [0.120 inch (3 mm)], $^5/_{16}$-inch (8.0 mm) head, and of sufficient length to penetrate the deck not less than $^3/_4$ inch (19.1 mm) or through the thickness of the deck, whichever is less. Attaching wire for clay or concrete tile shall not be smaller than 0.083 inch (2.1 mm). Perimeter fastening areas include three tile courses but not less than 36 inches (914 mm) from either side of hips or ridges and edges of eaves and *gable* rakes.

1507.3.7 Attachment. Clay and concrete roof tiles shall be fastened in accordance with Table 1507.3.7.

1507.3.8 Application. Tile shall be applied according to the manufacturer's installation instructions, based on the following:

1. Climatic conditions.
2. Roof slope.
3. *Underlayment* system.
4. Type of tile being installed.

TABLE 1507.2.8.2
VALLEY LINING MATERIAL

MATERIAL	MINIMUM THICKNESS	GAGE	WEIGHT
Aluminum	0.024 in.	—	—
Cold-rolled copper	0.0216 in.	—	ASTM B370, 16 oz. per square ft.
Copper	—	—	16 oz
Galvanized steel	0.0179 in.	26 (zinc-coated G90)	—
High-yield copper	0.0162 in.	—	ASTM B370, 12 oz. per square ft.
Lead	—	—	2.5 pounds
Lead-coated copper	0.0216 in.	—	ASTM B101, 16 oz. per square ft.
Lead-coated high-yield copper	0.0162 in.	—	ASTM B101, 12 oz. per square ft.
Painted terne	—	—	20 pounds
Stainless steel	—	28	—
Zinc alloy	0.027 in.	—	—

For SI: 1 inch = 25.4 mm, 1 pound = 0.454 kg, 1 ounce = 28.35 g, 1 square foot = 0.0929 m^2.

TABLE 1507.3.7 CLAY AND CONCRETE TILE ATTACHMENT[a, b, c]

GENERAL—CLAY OR CONCRETE ROOF TILE

Maximum Allowable Stress Design Wind Speed, V_{asd}[f] (mph)	Mean roof height (feet)	Roof slope < 3:12	Roof slope 3:12 and over
85	0-60	One fastener per tile. Flat tile without vertical laps, two fasteners per tile.	Two fasteners per tile. Only one fastener on slopes of 7:12 and less for tiles with installed weight exceeding 7.5 lbs./sq. ft. having a width not more than 16 inches.
100	0-40		
100	> 40-60	The head of all tiles shall be nailed. The nose of all eave tiles shall be fastened with approved clips. Rake tiles shall be nailed with two nails. The nose of all ridge, hip and rake tiles shall be set in a bead of roofer's mastic.	
110	0-60	The fastening system shall resist the wind forces in Section 1609.5.3.	
120	0-60	The fastening system shall resist the wind forces in Section 1609.5.3.	
130	0-60	The fastening system shall resist the wind forces in Section 1609.5.3.	
All	> 60	The fastening system shall resist the wind forces in Section 1609.5.3.	

INTERLOCKING CLAY OR CONCRETE ROOF TILE WITH PROJECTING ANCHOR LUGS[d, e] (Installations on spaced/solid sheathing with battens or spaced sheathing)

Maximum Allowable Stress Design Wind Speed, V_{asd}[f](mph)	Mean roof height (feet)	Roof slope < 5:12	Roof slope 5:12 < 12:12	Roof slope 12:12 and over
85	0-60	Fasteners are not required. Tiles with installed weight less than 9 lbs./sq. ft. require not fewer than one fastener per tile.	One fastener per tile every other row. Perimeter tiles require one fastener. Tiles with installed weight less than 9 lbs./sq. ft. require not fewer than one fastener per tile.	One fastener required for every tile. Tiles with installed weight less than 9 lbs./sq. ft. require not fewer than one fastener per tile.
100	0-40			
100	> 40-60	The head of all tiles shall be nailed. The nose of all eave tiles shall be fastened with approved clips. Rake tiles shall be nailed with two nails. The nose of all ridge, hip and rake tiles shall be set in a bead of roofer's mastic.		
110	0-60	The fastening system shall resist the wind forces in Section 1609.5.3.		
120	0-60	The fastening system shall resist the wind forces in Section 1609.5.3.		
130	0-60	The fastening system shall resist the wind forces in Section 1609.5.3.		
All	> 60	The fastening system shall resist the wind forces in Section 1609.5.3.		

INTERLOCKING CLAY OR CONCRETE ROOF TILE WITH PROJECTING ANCHOR LUGS (Installations on solid sheathing without battens)

Maximum Allowable Stress Wind Speed, V_{asd}[f](mph)	Mean roof height (feet)	All roof slopes
85	0-60	One fastener per tile.
100	0-40	One fastener per tile.
100	> 40-60	The head of all tiles shall be nailed. The nose of all eave tiles shall be fastened with approved clips. Rake tiles shall be nailed with two nails. The nose of all ridge, hip and rake tiles shall be set in a bead of roofer's mastic.
110	0-60	The fastening system shall resist the wind forces in Section 1609.5.3.
120	0-60	The fastening system shall resist the wind forces in Section 1609.5.3.
130	0-60	The fastening system shall resist the wind forces in Section 1609.5.3.
All	> 60	The fastening system shall resist the wind forces in Section 1609.5.3.

For SI: 1 inch = 25.4 mm, 1 foot = 304.8 mm, 1 mile per hour = 0.447 m/s, 1 pound per square foot = 4.882 kg/m^2.

a. Minimum fastener size. Corrosion-resistant nails not less than No. 11 gage with $^5/_{16}$-inch head. Fasteners shall be long enough to penetrate into the sheathing $^3/_4$ inch or through the thickness of the sheathing, whichever is less. Attaching wire for clay and concrete tile shall not be smaller than 0.083 inch.

b. Snow areas. Not fewer than two fasteners per tile are required or battens and one fastener.

c. Roof slopes greater than 24:12. The nose of all tiles shall be securely fastened.

d. Horizontal battens. Battens shall be not less than 1 inch by 2 inches nominal. Provisions shall be made for drainage by a riser of not less than $^1/_8$ inch at each nail or by 4-foot-long battens with not less than a $^1/_2$-inch separation between battens. Horizontal battens are required for slopes over 7:12.

e. Perimeter fastening areas include three tile courses but not less than 36 inches from either side of hips or ridges and edges of eaves and gable rakes.

f. V_{asd} shall be determined in accordance with Section 1609.3.1.

1507.3.9 Flashing. At the juncture of the roof vertical surfaces, flashing and counterflashing shall be provided in accordance with the manufacturer's installation instructions, and where of metal, shall be not less than 0.019-inch (0.48 mm) (No. 26 galvanized sheet gage) corrosion-resistant metal. The valley flashing shall extend not less than 11 inches (279 mm) from the centerline each way and have a splash diverter rib not less than 1 inch (25 mm) high at the flow line formed as part of the flashing. Sections of flashing shall have an end lap of not less than 4 inches (102 mm). For roof slopes of three units vertical in 12 units horizontal (25-percent slope) and over, the valley flashing shall have a 36-inch-wide (914 mm) *underlayment* of either one layer of Type I *underlayment* running the full length of the valley, or a self-adhering polymer-modified bitumen sheet bearing a label indicating compliance with ASTM D1970, in addition to other required *underlayment*. In areas where the average daily temperature in January is 25°F (-4°C) or less or where there is a possibility of ice forming along the eaves causing a backup of water, the metal valley flashing *underlayment* shall be solid cemented to the roofing *underlayment* for slopes under seven units vertical in 12 units horizontal (58-percent slope) or self-adhering polymer-modified bitumen sheet shall be installed.

1507.4 Metal roof panels. The installation of *metal roof panels* shall comply with the provisions of this section.

1507.4.1 Deck requirements. *Metal roof panel roof coverings* shall be applied to a solid or closely fitted deck, except where the *roof covering* is specifically designed to be applied to spaced supports.

1507.4.2 Deck slope. Minimum slopes for *metal roof panels* shall comply with the following:

1. The minimum slope for lapped, nonsoldered seam *metal roof panels* without applied lap sealant shall be three units vertical in 12 units horizontal (25-percent slope).
2. The minimum slope for lapped, nonsoldered seam *metal roof panels* with applied lap sealant shall be one-half unit vertical in 12 units horizontal (4-percent slope). Lap sealants shall be applied in accordance with the *approved* manufacturer's installation instructions.
3. The minimum slope for standing-seam *metal roof panel* systems shall be one-quarter unit vertical in 12 units horizontal (2-percent slope).

1507.4.3 Material standards. Metal-sheet *roof covering* systems that incorporate supporting structural members shall be designed in accordance with Chapter 22. Metal-sheet *roof coverings* installed over structural decking shall comply with Table 1507.4.3(1). The materials used for metal-sheet *roof coverings* shall be naturally corrosion resistant or provided with *corrosion resistance* in accordance with the standards and minimum thicknesses shown in Table 1507.4.3(2).

1507.4.4 Attachment. *Metal roof panels* shall be secured to the supports in accordance with the *approved* manufacturer's fasteners. In the absence of manufacturer recommendations, the following fasteners shall be used:

1. Galvanized fasteners shall be used for steel roofs.
2. Copper, brass, bronze, copper alloy or 300 series stainless-steel fasteners shall be used for copper roofs.
3. Stainless-steel fasteners are acceptable for all types of metal roofs.
4. Aluminum fasteners are acceptable for aluminum roofs attached to aluminum supports.

TABLE 1507.4.3(1)
METAL ROOF COVERINGS

ROOF COVERING TYPE	STANDARD APPLICATION RATE/THICKNESS
Aluminum	ASTM B209, 0.024 inch minimum thickness for roll-formed panels and 0.019 inch minimum thickness for press-formed shingles.
Aluminum-zinc alloy coated steel	ASTM A792 AZ 50
Cold-rolled copper	ASTM B370 minimum 16 oz./sq. ft. and 12 oz./sq. ft. high yield copper for metal-sheet roof covering systems: 12 oz./sq. ft. for preformed metal shingle systems.
Copper	16 oz./sq. ft. for metal-sheet roof-covering systems; 12 oz./sq. ft. for preformed metal shingle systems.
Galvanized steel	ASTM A653 G90 zinc-coated[a].
Hard lead	2 lbs./sq. ft.
Lead-coated copper	ASTM B101
Prepainted steel	ASTM A755
Soft lead	3 lbs./sq. ft.
Stainless steel	ASTM A240, 300 Series Alloys
Steel	ASTM A924
Terne and terne-coated stainless	Terne coating of 40 lbs. per double base box, field painted where applicable in accordance with manufacturer's installation instructions.
Zinc	0.027 inch minimum thickness; 99.995% electrolytic high-grade zinc with alloy additives of copper (0.08% - 0.20%), titanium (0.07% - 0.12%) and aluminum (0.015%).

For SI: 1 ounce per square foot = 0.305 kg/m^2, 1 pound per square foot = 4.882 kg/m^2, 1 inch = 25.4 mm, 1 pound = 0.454 kg.

a. For Group U buildings, the minimum coating thickness for ASTM A653 galvanized steel roofing shall be G60.

TABLE 1507.4.3(2)
MINIMUM CORROSION RESISTANCE

55% Aluminum-zinc alloy coated steel	ASTM A792 AZ 50
5% Aluminum alloy-coated steel	ASTM A875 GF60
Aluminum-coated steel	ASTM A463 T2 65
Galvanized steel	ASTM A653 G90
Prepainted steel	ASTM A755[a]

a. Paint systems in accordance with ASTM A755 shall be applied over steel products with corrosion-resistant coatings complying with ASTM A463, ASTM A653, ASTM A792 or ASTM A875.

1507.4.5 Underlayment and high wind. *Underlayment* shall comply with Section 1507.1.1.

1507.5 Metal roof shingles. The installation of *metal roof shingles* shall comply with the provisions of this section.

1507.5.1 Deck requirements. *Metal roof shingles* shall be applied to a solid or closely fitted deck, except where the *roof covering* is specifically designed to be applied to spaced sheathing.

1507.5.2 Deck slope. *Metal roof shingles* shall not be installed on roof slopes below three units vertical in 12 units horizontal (25-percent slope).

1507.5.3 Underlayment. Underlayment shall comply with Section 1507.1.1.

1507.5.4 Ice barrier. Where required, ice barriers shall comply with Section 1507.1.2.

1507.5.5 Material standards. *Metal roof shingle roof coverings* shall comply with Table 1507.4.3(1). The materials used for metal-roof shingle *roof coverings* shall be naturally corrosion resistant or provided with *corrosion resistance* in accordance with the standards and minimum thicknesses specified in the standards listed in Table 1507.4.3(2).

1507.5.6 Attachment. *Metal roof shingles* shall be secured to the roof in accordance with the *approved* manufacturer's installation instructions.

1507.5.7 Flashing. Roof valley flashing shall be of corrosion-resistant metal of the same material as the *roof covering* or shall comply with the standards in Table 1507.4.3(1). The valley flashing shall extend not less than 8 inches (203 mm) from the centerline each way and shall have a splash diverter rib not less than $^{3}/_{4}$ inch (19.1 mm) high at the flow line formed as part of the flashing. Sections of flashing shall have an end lap of not less than 4 inches (102 mm). In areas where the average daily temperature in January is 25°F (-4°C) or less or where there is a possibility of ice forming along the eaves causing a backup of water, the metal valley flashing shall have a 36-inch-wide (914 mm) *underlayment* directly under it consisting of either one layer of *underlayment* running the full length of the valley or a self-adhering polymer-modified bitumen sheet bearing a label indicating compliance with ASTM D1970, in addition to *underlayment* required for *metal roof shingles*. The metal valley flashing *underlayment* shall be solidly cemented to the roofing *underlayment* for roof slopes under seven units vertical in 12 units horizontal (58-percent slope) or self-adhering polymer-modified bitumen sheet shall be installed.

1507.6 Mineral-surfaced roll roofing. The installation of mineral-surfaced roll roofing shall comply with this section.

1507.6.1 Deck requirements. Mineral-surfaced roll roofing shall be fastened to solidly sheathed roofs.

1507.6.2 Deck slope. Mineral-surfaced roll roofing shall not be applied on roof slopes below one unit vertical in 12 units horizontal (8-percent slope).

1507.6.3 Underlayment. *Underlayment* shall comply with Section 1507.1.1.

1507.6.4 Ice barrier. Where required, ice barriers shall comply with Section 1507.1.2.

1507.6.5 Material standards. Mineral-surfaced roll roofing shall conform to ASTM D3909 or ASTM D6380.

1507.7 Slate shingles. The installation of slate shingles shall comply with the provisions of this section.

1507.7.1 Deck requirements. Slate shingles shall be fastened to solidly sheathed roofs.

1507.7.2 Deck slope. Slate shingles shall only be used on slopes of four units vertical in 12 units horizontal (4:12) or greater.

1507.7.3 Underlayment. *Underlayment* shall comply with Section 1507.1.1.

1507.7.4 Ice barrier. Where required, ice barriers shall comply with Section 1507.1.2.

1507.7.5 Material standards. Slate shingles shall comply with ASTM C406.

1507.7.6 Application. Minimum headlap for slate shingles shall be in accordance with Table 1507.7.6. Slate shingles shall be secured to the roof with two fasteners per slate.

TABLE 1507.7.6
SLATE SHINGLE HEADLAP

SLOPE	HEADLAP (inches)
4:12 < slope < 8:12	4
8:12 < slope < 20:12	3
slope ≥ 20:12	2

For SI: 1 inch = 25.4 mm.

1507.7.7 Flashing. Flashing and counterflashing shall be made with sheet metal. Valley flashing shall be not less than 15 inches (381 mm) wide. Valley and flashing metal shall be a minimum uncoated thickness of 0.0179-inch (0.455 mm) zinc-coated G90. Chimneys, stucco or brick walls shall have not fewer than two plies of felt for a cap flashing consisting of a 4-inch-wide (102 mm) strip of felt set in plastic cement and extending 1 inch (25 mm) above the first felt and a top coating of plastic cement. The felt shall extend over the base flashing 2 inches (51 mm).

1507.8 Wood shingles. The installation of wood shingles shall comply with the provisions of this section and Table 1507.8.

1507.8.1 Deck requirements. Wood shingles shall be installed on solid or spaced sheathing. Where spaced sheathing is used, sheathing boards shall be not less than 1-inch by 4-inch (25 mm by 102 mm) nominal dimensions and shall be spaced on centers equal to the weather exposure to coincide with the placement of fasteners.

1507.8.1.1 Solid sheathing required. Solid sheathing is required in areas where the average daily temperature in January is 25°F (-4°C) or less or where there is a possibility of ice forming along the eaves causing a backup of water.

TABLE 1507.8
WOOD SHINGLE AND SHAKE INSTALLATION

ROOF ITEM	WOOD SHINGLES	WOOD SHAKES
1. Roof slope	Wood shingles shall be installed on slopes of not less than 3 units vertical in 12 units horizontal (3:12).	Wood shakes shall be installed on slopes of not less than 4 units vertical in 12 units horizontal (4:12).
2. Deck requirement		
Temperate climate	Shingles shall be applied to roofs with solid or spaced sheathing. Where spaced sheathing is used, sheathing boards shall be not less than 1″ × 4″ nominal dimensions and shall be spaced on centers equal to the weather exposure to coincide with the placement of fasteners.	Shakes shall be applied to roofs with solid or spaced sheathing. Where spaced sheathing is used, sheathing boards shall be not less than 1″ × 4″ nominal dimensions and shall be spaced on centers equal to the weather exposure to coincide with the placement of fasteners. Where 1″ × 4″ spaced sheathing is installed at 10 inches, boards must be installed between the sheathing boards.
In areas where the average daily temperature in January is 25°F or less or where there is a possibility of ice forming along the eaves causing a backup of water.	Solid sheathing is required.	Solid sheathing is required.
3. Interlayment	No requirements.	Interlayment shall comply with ASTM D226, Type 1.
4. Underlayment		
Temperate climate	Underlayment shall comply with Section 1507.1.1.	Underlayment shall comply with Section 1507.1.1.
5. Application		
Attachment	Fasteners for wood shingles shall be hot-dipped galvanized or Type 304 (Type 316 for coastal areas) stainless steel with a minimum penetration of 0.75 inch into the sheathing. For sheathing less than 0.5 inch thick, the fasteners shall extend through the sheathing.	Fasteners for wood shakes shall be hot-dipped galvanized or Type 304 (Type 316 for coastal areas) with a minimum penetration of 0.75 inch into the sheathing. For sheathing less than 0.5 inch thick, the fasteners shall extend through the sheathing.
No. of fasteners	Two per shingle.	Two per shake.
Exposure	Weather exposures shall not exceed those set forth in Table 1507.8.7.	Weather exposures shall not exceed those set forth in Table 1507.9.8.
Method	Shingles shall be laid with a side lap of not less than 1.5 inches between joints in courses, and no two joints in any three adjacent courses shall be in direct alignment. Spacing between shingles shall be 0.25 to 0.375 inch.	Shakes shall be laid with a side lap of not less than 1.5 inches between joints in adjacent courses. Spacing between shakes shall not be less than 0.375 inch or more than 0.625 inch for shakes and taper sawn shakes of naturally durable wood and shall be 0.25 to 0.375 inch for preservative-treated taper sawn shakes.
Flashing	In accordance with Section 1507.8.8.	In accordance with Section 1507.9.9.

For SI: 1 inch = 25.4 mm, °C = [(°F) - 32]/1.8.

1507.8.2 Deck slope. Wood shingles shall be installed on slopes of not less than three units vertical in 12 units horizontal (25-percent slope).

1507.8.3 Underlayment. *Underlayment* shall comply with Section 1507.1.1.

1507.8.4 Ice barrier. Where required, ice barriers shall comply with Section 1507.1.2.

1507.8.5 Material standards. Wood shingles shall be of *naturally durable wood* and comply with the requirements of Table 1507.8.5.

1507.8.6 Attachment. Fasteners for wood shingles shall be corrosion resistant with a minimum penetration of $^{3}/_{4}$ inch (19.1 mm) into the sheathing. For sheathing less than $^{1}/_{2}$ inch (12.7 mm) in thickness, the fasteners shall extend through the sheathing. Each shingle shall be attached with not fewer than two fasteners.

TABLE 1507.8.5
WOOD SHINGLE MATERIAL REQUIREMENTS

MATERIAL	APPLICABLE MINIMUM GRADES	GRADING RULES
Wood shingles of naturally durable wood	1, 2 or 3	CSSB

CSSB = Cedar Shake and Shingle Bureau

1507.8.7 Application. Wood shingles shall be laid with a side lap not less than $1^{1}/_{2}$ inches (38 mm) between joints in adjacent courses, and not be in direct alignment in alternate courses. Spacing between shingles shall be $^{1}/_{4}$ to $^{3}/_{8}$ inch (6.4 to 9.5 mm). Weather exposure for wood shingles shall not exceed that set in Table 1507.8.7.

TABLE 1507.8.7
WOOD SHINGLE WEATHER EXPOSURE AND ROOF SLOPE

ROOFING MATERIAL	LENGTH (inches)	GRADE	EXPOSURE (inches)	
			3:12 pitch to < 4:12	4:12 pitch or steeper
Shingles of naturally durable wood	16	No. 1	3.75	5
		No. 2	3.5	4
		No. 3	3	3.5
	18	No. 1	4.25	5.5
		No. 2	4	4.5
		No. 3	3.5	4
	24	No. 1	5.75	7.5
		No. 2	5.5	6.5
		No. 3	5	5.5

For SI: 1 inch = 25.4 mm.

1507.8.8 Flashing. At the juncture of the roof and vertical surfaces, flashing and counterflashing shall be provided in accordance with the manufacturer's installation instructions, and where of metal, shall be not less than 0.019-inch (0.48 mm) (No. 26 galvanized sheet gage) corrosion-resistant metal. The valley flashing shall extend not less than 11 inches (279 mm) from the centerline each way and have a splash diverter rib not less than 1 inch (25 mm) high at the flow line formed as part of the flashing. Sections of flashing shall have an end lap of not less than 4 inches (102 mm). For roof slopes of three units vertical in 12 units horizontal (25-percent slope) and over, the valley flashing shall have a 36-inch-wide (914 mm) *underlayment* of either one layer of Type I *underlayment* running the full length of the valley or a self-adhering polymer-modified bitumen sheet bearing a label indicating compliance with ASTM D1970, in addition to other required *underlayment*. In areas where the average daily temperature in January is 25°F (-4°C) or less or where there is a possibility of ice forming along the eaves causing a backup of water, the metal valley flashing *underlayment* shall be solidly cemented to the roofing *underlayment* for slopes under seven units vertical in 12 units horizontal (58-percent slope) or self-adhering polymer-modified bitumen sheet shall be installed.

1507.8.9 Label required. Each bundle of shingles shall be identified by a label of an approved grading or inspection bureau or agency.

1507.9 Wood shakes. The installation of wood shakes shall comply with the provisions of this section and Table 1507.8.

1507.9.1 Deck requirements. Wood shakes shall only be used on solid or spaced sheathing. Where spaced sheathing is used, sheathing boards shall be not less than 1-inch by 4-inch (25 mm by 102 mm) nominal dimensions and shall be spaced on centers equal to the weather exposure to coincide with the placement of fasteners. Where 1-inch by 4-inch (25 mm by 102 mm) spaced sheathing is installed at 10 inches (254 mm) on center, additional 1-inch by 4-inch (25 mm by 102 mm) boards shall be installed between the sheathing boards.

1507.9.1.1 Solid sheathing required. Solid sheathing is required in areas where the average daily temperature in January is 25°F (-4°C) or less or where there is a possibility of ice forming along the eaves causing a backup of water.

1507.9.2 Deck slope. Wood shakes shall only be used on slopes of not less than 4 units vertical in 12 units horizontal (33-percent slope).

1507.9.3 Underlayment. *Underlayment* shall comply with Section 1507.1.1.

1507.9.4 Ice barrier. Where required, ice barriers shall comply with Section 1507.1.2.

1507.9.5 Interlayment. *Interlayment* shall comply with ASTM D226, Type I.

1507.9.6 Material standards. Wood shakes shall comply with the requirements of Table 1507.9.6.

TABLE 1507.9.6
WOOD SHAKE MATERIAL REQUIREMENTS

MATERIAL	MINIMUM GRADES	APPLICABLE GRADING RULES
Wood shakes of naturally durable wood	1	CSSB
Taper sawn shakes of naturally durable wood	1 or 2	CSSB
Preservative-treated shakes and shingles of naturally durable wood	1	CSSB
Fire-retardant-treated shakes and shingles of naturally durable wood	1	CSSB
Preservative-treated taper sawn shakes of Southern pine treated in accordance with AWPA U1 (Commodity Specification A, Special Requirement 4.6)	1 or 2	TFS

CSSB = Cedar Shake and Shingle Bureau.
TFS = Forest Products Laboratory of the Texas Forest Services.

1507.9.7 Attachment. Fasteners for wood shakes shall be corrosion resistant with a minimum penetration of $^3/_4$ inch (19.1 mm) into the sheathing. For sheathing less than $^1/_2$ inch (12.7 mm) in thickness, the fasteners shall extend through the sheathing. Each shake shall be attached with not fewer than two fasteners.

1507.9.8 Application. Wood shakes shall be laid with a side lap not less than $1^1/_2$ inches (38 mm) between joints in adjacent courses. Spacing between shakes in the same course shall be $^3/_8$ to $^5/_8$ inch (9.5 to 15.9 mm) for shakes and taper sawn shakes of *naturally durable wood* and shall be $^1/_4$ to $^3/_8$ inch (6.4 to 9.5 mm) for preservative taper sawn shakes. Weather exposure for wood shakes shall not exceed those set in Table 1507.9.8.

1507.9.9 Flashing. At the juncture of the roof and vertical surfaces, flashing and counterflashing shall be provided in accordance with the manufacturer's installation instructions, and where of metal, shall be not less than 0.019-inch (0.48 mm) (No. 26 galvanized sheet gage) corrosion-resistant metal. The valley flashing shall extend not less than 11 inches (279 mm) from the centerline each way and have a splash diverter rib not less than 1 inch (25 mm) high at the flow line formed as part of the flashing. Sections of flashing shall have an end lap of not less than 4 inches (102 mm). For roof slopes of three units vertical in 12 units horizontal (25-percent slope) and over, the valley flashing shall have a 36-inch-wide (914 mm) *underlayment* of either one layer of Type I *underlayment* running the full length of the valley or a self-adhering polymer-modified bitumen sheet bearing a label indicating compliance with ASTM D1970, in addition to other required *underlayment*. In areas where the average daily temperature in January is 25°F (-4°C) or less or where there is a possibility of ice forming along the eaves causing a backup of water, the metal valley flashing *underlayment* shall be solidly cemented to the roofing *underlayment* for slopes under seven units vertical in 12 units horizontal (58-percent slope) or self-adhering polymer-modified bitumen sheet shall be installed.

1507.9.10 Label required. Each bundle of shakes shall be identified by a label of an approved grading or inspection bureau or agency.

1507.10 Built-up roofs. The installation of built-up roofs shall comply with the provisions of this section.

1507.10.1 Slope. Built-up roofs shall have a design slope of not less than $^1/_4$ unit vertical in 12 units horizontal (2-percent slope) for drainage, except for coal-tar built-up roofs that shall have a design slope of not less than $^1/_8$ unit vertical in 12 units horizontal (1-percent slope).

1507.10.2 Material standards. *Built-up roof covering* materials shall comply with the standards in Table 1507.10.2 or UL 55A.

1507.11 Modified bitumen roofing. The installation of modified bitumen roofing shall comply with the provisions of this section.

1507.11.1 Slope. Modified bitumen roofing shall have a design slope of not less than $^1/_4$ unit vertical in 12 units horizontal (2-percent slope) for drainage.

1507.11.2 Material standards. Modified bitumen roofing materials shall comply with ASTM D6162, ASTM D6163, ASTM D6164, ASTM D6222, ASTM D6223, ASTM D6298 or ASTM D6509.

1507.11.2.1 Base sheet. A base sheet that complies with the requirements of Section 1507.11.2, ASTM D1970 or ASTM D4601 shall be permitted to be used with a modified bitumen cap sheet.

TABLE 1507.10.2
BUILT-UP ROOFING MATERIAL STANDARDS

MATERIAL STANDARD	STANDARD
Acrylic coatings used in roofing	ASTM D6083
Aggregate surfacing	ASTM D1863
Asphalt adhesive used in roofing	ASTM D3747
Asphalt cements used in roofing	ASTM D2822; D3019; D4586
Asphalt-coated glass fiber base sheet	ASTM D4601
Asphalt coatings used in roofing	ASTM D1227; D2823; D2824; D4479
Asphalt glass felt	ASTM D2178
Asphalt primer used in roofing	ASTM D41
Asphalt-saturated and asphalt-coated organic felt base sheet	ASTM D2626
Asphalt-saturated organic felt (perforated)	ASTM D226
Asphalt used in roofing	ASTM D312
Coal-tar cements used in roofing	ASTM D4022; D5643
Coal-tar saturated organic felt	ASTM D227
Coal-tar pitch used in roofing	ASTM D450; Type I or II
Coal-tar primer used in roofing, dampproofing and waterproofing	ASTM D43
Glass mat, coal tar	ASTM D4990
Glass mat, venting type	ASTM D4897
Mineral-surfaced inorganic cap sheet	ASTM D3909
Thermoplastic fabrics used in roofing	ASTM D5665, D5726

TABLE 1507.9.8
WOOD SHAKE WEATHER EXPOSURE AND ROOF SLOPE

ROOFING MATERIAL	LENGTH (inches)	GRADE	EXPOSURE (inches) 4:12 PITCH OR STEEPER
Shakes of naturally durable wood	18	No. 1	7.5
	24	No. 1	10[a]
Preservative-treated taper sawn shakes of Southern yellow pine	18	No. 1	7.5
	24	No. 1	10
	18	No. 2	5.5
	24	No. 2	7.5
Taper sawn shakes of naturally durable wood	18	No. 1	7.5
	24	No. 1	10
	18	No. 2	5.5
	24	No. 2	7.5

For SI: 1 inch = 25.4 mm.

a. For 24-inch by 0.375-inch handsplit shakes, the maximum exposure is 7.5 inches.

1507.12 Single-ply roofing. The installation of single-ply roofing shall comply with the provisions of this section.

1507.12.1 Slope. Single-ply membrane roofs shall have a design slope of not less than $^1/_4$ unit vertical in 12 units horizontal (2-percent slope) for drainage.

1507.12.2 Material standards. Single-ply roof coverings shall comply with the material standards in Table 1507.12.2.

TABLE 1507.12.2
SINGLE-PLY ROOFING MATERIAL STANDARDS

MATERIAL	MATERIAL STANDARD
Chlorosulfonated polyethylene (CSPE) or polyisobutylene (PIB)	ASTM D5019
Ethylene propylene diene monomer (EPDM)	ASTM D4637
Ketone Ethylene Ester (KEE)	ASTM D6754
Polyvinyl Chloride (PVC) or (PVC/KEE)	ASTM D4434
Thermoplastic polyolefin (TPO)	ASTM D6878

1507.12.3 Ballasted low-slope roofs. Ballasted low-slope roofs (roof slope < 2:12) shall be installed in accordance with this section and Section 1504.5. Stone used as *ballast* shall comply with ASTM D448 or ASTM D7655.

1507.13 Sprayed polyurethane foam roofing. The installation of sprayed polyurethane foam roofing shall comply with the provisions of this section.

1507.13.1 Slope. Sprayed polyurethane foam roofs shall have a design slope of not less than $^1/_4$ unit vertical in 12 units horizontal (2-percent slope) for drainage.

1507.13.2 Material standards. Spray-applied polyurethane foam insulation shall comply with ASTM C1029 Type III or IV or ASTM D7425.

1507.13.3 Application. Foamed-in-place roof insulation shall be installed in accordance with the manufacturer's instructions. A liquid-applied protective coating that complies with Table 1507.13.3 shall be applied not less than 2 hours nor more than 72 hours following the application of the foam.

TABLE 1507.13.3
PROTECTIVE COATING MATERIAL STANDARDS

MATERIAL	STANDARD
Acrylic coating	ASTM D6083
Silicone coating	ASTM D6694
Moisture-cured polyurethane coating	ASTM D6947

1507.13.4 Foam plastics. Foam plastic materials and installation shall comply with Chapter 26.

1507.14 Liquid-applied roofing. The installation of liquid-applied roofing shall comply with the provisions of this section.

1507.14.1 Slope. Liquid-applied roofing shall have a design slope of not less than $^1/_4$ unit vertical in 12 units horizontal (2-percent slope).

1507.14.2 Material standards. Liquid-applied roofing shall comply with ASTM C836, ASTM C957 or ASTM D3468.

1507.15 Vegetative roofs and landscaped roofs. *Vegetative roofs* and landscaped roofs shall comply with the requirements of this chapter, Section 1607.14.2.2 and the *International Fire Code*.

[BF] 1507.15.1 Structural fire resistance. The structural frame and roof construction supporting the load imposed on the roof by the *vegetative roof* or landscaped roofs shall comply with the requirements of Table 601.

1507.16 Photovoltaic shingles. The installation of *photovoltaic shingles* shall comply with the provisions of this section.

1507.16.1 Deck requirements. *Photovoltaic shingles* shall be applied to a solid or closely fitted deck, except where the shingles are specifically designed to be applied over spaced sheathing.

1507.16.2 Deck slope. *Photovoltaic shingles* shall be installed on roof slopes of not less than 2 units vertical in 12 units horizontal (2:12).

1507.16.3 Underlayment. *Underlayment* shall comply with Section 1507.1.1.

1507.16.4 Ice barrier. Where required, ice barriers shall comply with Section 1507.1.2.

1507.16.5 Fasteners. Fasteners for *photovoltaic shingles* shall be galvanized, stainless steel, aluminum or copper roofing nails, minimum 12-gage [0.105 inch (2.67 mm)] shank with a minimum $^3/_8$-inch-diameter (9.5 mm) head, of a length to penetrate through the roofing materials and not less than $^3/_4$ inch (19.1 mm) into the roof sheathing. Where the roof sheathing is less than $^3/_4$ inch (19.1 mm) thick, the nails shall penetrate through the sheathing. Fasteners shall comply with ASTM F1667.

1507.16.6 Material standards. *Photovoltaic shingles* shall be *listed* and labeled in accordance with UL 7103 or with both UL 61730-1 and UL 61730-2.

1507.16.7 Attachment. *Photovoltaic shingles* shall be attached in accordance with the manufacturer's installation instructions.

1507.16.8 Wind resistance. *Photovoltaic shingles* shall comply with the classification requirements of Table 1504.2 for the appropriate maximum nominal design wind speed.

1507.17 Building-integrated photovoltaic roof panels. The installation of building-integrated photovoltaic (*BIPV*) roof panels shall comply with the provisions of this section.

1507.17.1 Deck requirements. *BIPV* roof panels shall be applied to a solid or closely fitted deck, except where the *roof covering* is specifically designed to be applied over spaced sheathing.

1507.17.2 Deck slope. BIPV roof panels shall be used only on roof slopes of 2 units vertical in 12 units horizontal (2:12) or greater.

1507.17.3 Underlayment. *Underlayment* shall comply with ASTM D226, ASTM D4869 or ASTM D6757.

1507.17.4 Underlayment application. *Underlayment* shall be applied *shingle fashion*, parallel to and starting from the eave, lapped 2 inches (51 mm) and fastened sufficiently to hold in place.

1507.17.4.1 High-wind attachment. *Underlayment* applied in areas subject to high winds [V_{asd} greater than 110 mph (49 m/s) as determined in accordance with Section 1609.3.1] shall be applied in accordance with the manufacturer's instructions. Fasteners shall be applied along the overlap at not more than 36 inches (914 mm) on center. *Underlayment* installed where V_{asd} is not less than 120 mph (54 m/s) shall comply with ASTM D226, Type III, ASTM D4869, Type IV or ASTM D6757. The *underlayment* shall be attached in a grid pattern of 12 inches (305 mm) between side laps with a 6-inch (152 mm) spacing at the side laps. The *underlayment* shall be applied in accordance with Section 1507.1.1 except all laps shall be not less than 4 inches (102 mm). *Underlayment* shall be attached using cap nails or cap staples. Caps shall be metal or plastic with a nominal head diameter of not less than 1 inch (25.4 mm). Metal caps shall have a thickness of not less than 0.010 inch (0.25 mm). Power-driven metal caps shall have a thickness of not less than 0.010 inch (0.25 mm). Thickness of the outside edge of plastic caps shall be not less than 0.035 inch (0.89 mm). The cap nail shank shall be not less than 0.083 inch (2.11 mm) for ring shank cap nails and 0.091 inch (2.31 mm) for smooth shank cap nails. Staple gage shall be not less than 21 gage [0.0.2 inch (0.81 mm)]. Cap nail shank and cap staple legs shall have a length sufficient to penetrate through-the-roof sheathing or not less than $^{3}/_{4}$ inch (19.1 mm) into the roof sheathing.

Exception: As an alternative, adhered *underlayment* complying with ASTM D1970 shall be permitted.

1507.17.4.2 Ice barrier. In areas where there has been a history of ice forming along the eaves causing a back-up of water, an ice barrier consisting of not fewer than two layers of *underlayment* cemented together or of a self-adhering polymer modified bitumen sheet shall be used instead of normal *underlayment* and extend from the lowest edges of all roof surfaces to a point not less than 24 inches (610 mm) inside the *exterior wall* line of the building.

Exception: Detached accessory structures that do not contain conditioned floor area.

1507.17.5 Material standards. BIPV roof panels shall be listed and labeled in accordance with UL 7103 or with both UL 61730-1 and UL 61730-2.

1507.17.6 Attachment. *BIPV* roof panels shall be attached in accordance with the manufacturer's installation instructions.

SECTION 1508
ROOF INSULATION

[BF] 1508.1 General. The use of above-deck thermal insulation shall be permitted provided that such insulation is covered with an approved *roof covering* and passes the tests of NFPA 276 or UL 1256 when tested as an assembly.

Exceptions:

1. Foam plastic roof insulation shall conform to the material and installation requirements of Chapter 26.
2. Where a concrete or composite metal and concrete roof deck is used and the above-deck thermal insulation is covered with an approved *roof covering*.

[BF] 1508.2 Material standards. Above-deck thermal insulation board shall comply with the standards in Table 1508.2.

[BF] TABLE 1508.2
MATERIAL STANDARDS FOR ROOF INSULATION

Cellular glass board	ASTM C552
Composite boards	ASTM C1289, Type III, IV, V or VII
Expanded polystyrene	ASTM C578
Extruded polystyrene	ASTM C578
Fiber-reinforced gypsum board	ASTM C1278
Glass-faced gypsum board	ASTM C1177
High-density polyisocyanurate board	ASTM C1289, Type II, Class 4
Mineral fiber insulation board	ASTM C726
Perlite board	ASTM C728
Polyisocyanurate board	ASTM C1289, Type I or II
Wood fiberboard	ASTM C208, Type II

SECTION 1509
ROOF COATINGS

1509.1 General. The installation of a *roof coating* on a *roof covering* shall comply with the requirements of Section 1505 and this section.

1509.2 Material standards. Roof coating materials shall comply with the standards in Table 1509.2.

TABLE 1509.2
ROOF COATING MATERIAL STANDARDS

MATERIAL	STANDARD
Acrylic coating	ASTM D6083
Asphaltic emulsion coating	ASTM D1227
Asphalt coating	ASTM D2823
Asphalt roof coating	ASTM D4479
Aluminum-pigmented asphalt coating	ASTM D2824
Silicone coating	ASTM D6694
Moisture-cured polyurethane coating	ASTM D6947

SECTION 1510
RADIANT BARRIERS INSTALLED ABOVE DECK

[BF] 1510.1 General. A *radiant barrier* installed above a deck shall comply with Sections 1510.2 through 1510.4.

[BF] 1510.2 Fire testing. *Radiant barriers* shall be permitted for use above decks where the *radiant barrier* is covered with an *approved roof covering* and the system consisting of the *radiant barrier* and the *roof covering* complies with the requirements of either FM 4450 or UL 1256.

[BF] 1510.3 Installation. The low *emittance* surface of the *radiant barrier* shall face the continuous airspace between the *radiant barrier* and the roof covering.

[BF] 1510.4 Material standards. A *radiant barrier* installed above a deck shall comply with ASTM C1313/1313M.

SECTION 1511
ROOFTOP STRUCTURES

[BG] 1511.1 General. The provisions of this section shall govern the construction of *rooftop structures*.

1511.1.1 Area limitation. The aggregate area of *penthouses* and other enclosed *rooftop structures* shall not exceed one-third the area of the supporting roof deck. Such *penthouses* and other enclosed *rooftop structures* shall not be required to be included in determining the *building area* or number of stories as regulated by Section 503.1. The area of such *penthouses* shall not be included in determining the *fire area* specified in Section 901.7.

[BG] 1511.2 Penthouses. *Penthouses* in compliance with Sections 1511.2.1 through 1511.2.4shall be considered as a portion of the *story* directly below the *roof deck* on which such *penthouses* are located. Other *penthouses* shall be considered as an additional *story* of the building.

[BG] 1511.2.1 Height above roof deck. *Penthouses* constructed on buildings of other than Type I construction shall not exceed 18 feet (5486 mm) in height above the *roof deck* as measured to the average height of the roof of the *penthouse*. *Penthouses* located on the roof of buildings of Type I construction shall not be limited in height.

Exception: Where used to enclose tanks or elevators that travel to the roof level, penthouses shall be permitted to have a maximum height of 28 feet (8534 mm) above the roof deck.

[BG] 1511.2.2 Use limitations. *Penthouses* shall not be used for purposes other than the shelter of mechanical or electrical equipment, tanks, elevators and related machinery, stairways or vertical *shaft* openings in the roof assembly, including ancillary spaces used to access elevators and stairways.

[BG] 1511.2.3 Weather protection. Provisions such as louvers, louver blades or flashing shall be made to protect the mechanical and electrical equipment and the building interior from the elements.

[BG] 1511.2.4 Type of construction. Penthouses shall be constructed of building elements as required for the type of construction of the building on which such penthouses are built.

Exceptions:

1. On buildings of Type I construction, the *exterior walls* and roofs of *penthouses* with a *fire separation distance* greater than 5 feet (1524 mm) and less than 20 feet (6096 mm) shall be permitted to have not less than a 1-hour *fire-resistance rating*. The *exterior walls* and roofs of *penthouses* with a *fire separation distance* of 20 feet (6096 mm) or greater shall not be required to have a *fire-resistance rating*.
2. On buildings of Type I construction two stories or less in height above *grade plane* or of Type II construction, the *exterior walls* and roofs of *penthouses* with a *fire separation distance* greater than 5 feet (1524 mm) and less than 20 feet (6096 mm) shall be permitted to have not less than a 1-hour *fire-resistance rating* or a lesser *fire-resistance rating* as required by Table 705.5 and be constructed of *fire-retardant-treated wood*. The *exterior walls* and roofs of *penthouses* with a *fire separation distance* of 20 feet (6096 mm) or greater shall be permitted to be constructed of *fire-retardant-treated wood* and shall not be required to have a *fire-resistance rating*. Interior framing and walls shall be permitted to be constructed of *fire-retardant-treated wood*.
3. On buildings of Type III, IV or V construction, the *exterior walls* of *penthouses* with a *fire separation distance* greater than 5 feet (1524 mm) and less than 20 feet (6096 mm) shall be permitted to have not less than a 1-hour *fire-resistance rating* or a lesser *fire-resistance rating* as required by Table 705.5. On buildings of Type III, IV or VA construction, the *exterior walls* of *penthouses* with a *fire separation distance* of 20 feet (6096 mm) or greater shall be permitted to be of heavy timber construction complying with Sections 602.4 and 2304.11 or noncombustible construction or *fire-retardant-treated wood* and shall not be required to have a *fire-resistance rating*.

[BG] 1511.3 Tanks. Tanks having a capacity of more than 500 gallons (1893 L) located on the *roof deck* of a building shall be supported on masonry, reinforced concrete, steel or heavy timber construction complying with Section 2304.11 provided that, where such supports are located in the building above the lowest story, the support shall be fire-resistance rated as required for Type IA construction.

[BG] 1511.3.1 Valve and drain. In the bottom or on the side near the bottom of the tank, a pipe or outlet, fitted with a suitable quick-opening valve for discharging the contents into a drain in an emergency shall be provided.

[BG] 1511.3.2 Location. Tanks shall not be placed over or near a *stairway* or an elevator *shaft*, unless there is a solid roof or floor underneath the tank.

[BG] 1511.3.3 Tank cover. Unenclosed roof tanks shall have covers sloping toward the perimeter of the tanks.

[BG] 1511.4 Cooling towers. Cooling towers located on the *roof deck* of a building and greater than 250 square feet (23.2 m^2) in base area or greater than 15 feet (4572 mm) in height above the *roof deck*, as measured to the highest point on the cooling tower, where the roof is greater than 50 feet (15 240 mm) in height above *grade plane* shall be constructed of noncombustible materials. The base area of cooling towers shall not exceed one-third the area of the supporting *roof deck*.

Exception: Drip boards and the enclosing construction shall be permitted to be of wood not less than 1 inch (25 mm) nominal thickness, provided that the wood is covered on the exterior of the tower with noncombustible material.

[BG] 1511.5 Towers, spires, domes and cupolas. Towers, spires, domes and cupolas shall be of a type of construction having *fire-resistance ratings* not less than required for the building on top of which such tower, spire, dome or cupola is built. Towers, spires, domes and cupolas greater than 85 feet (25 908 mm) in height above *grade plane* as measured to the highest point on such structures, and either greater than 200 square feet (18.6 m^2) in horizontal area or used for any purpose other than a belfry or an architectural embellishment, shall be constructed of and supported on Type I or II construction.

[BG] 1511.5.1 Noncombustible construction required. Towers, spires, domes and cupolas greater than 60 feet (18 288 mm) in height above the highest point at which such structure contacts the roof as measured to the highest point on such structure, or that exceeds 200 square feet (18.6 m^2) in area at any horizontal section, or which is intended to be used for any purpose other than a belfry or architectural embellishment, or is located on the top of a building greater than 50 feet (1524 mm) in *building height* shall be constructed of and supported by noncombustible materials and shall be separated from the building below by construction having a *fire-resistance rating* of not less than 1.5 hours with openings protected in accordance with Section 711. Such structures located on the top of a building greater than 50 feet (15 240 mm) in *building height* shall be supported by noncombustible construction.

[BG] 1511.5.2 Towers and spires. Enclosed towers and spires shall have *exterior walls* constructed as required for the building on top of which such towers and spires are built. The *roof covering* of spires shall be not less than the same class of *roof covering* required for the building on top of which the spire is located.

[BG] 1511.6 Mechanical equipment screens. *Mechanical equipment screens* shall be constructed of the materials specified for the *exterior walls* in accordance with the type of construction of the building. Where the *fire separation distance* is greater than 5 feet (1524 mm), *mechanical equipment screens* shall not be required to comply with the *fire-resistance rating* requirements.

[BG] 1511.6.1 Height limitations. *Mechanical equipment screens* shall not exceed 18 feet (5486 mm) in height above the *roof deck*, as measured to the highest point on the *mechanical equipment screen*.

Exception: Where located on buildings of Type IA construction, the height of *mechanical equipment screens* shall not be limited.

[BG] 1511.6.2 Type I, II, III or IV construction. Regardless of the requirements in Section 1511.6, *mechanical equipment screens* that are located on the *roof decks* of buildings of Type I, II, III or IV construction shall be permitted to be constructed of combustible materials in accordance with any one of the following limitations:

1. The *fire separation distance* shall be not less than 20 feet (6096 mm) and the height of the *mechanical equipment screen* above the roof deck shall not exceed 4 feet (1219 mm) as measured to the highest point on the *mechanical equipment screen*.
2. The *fire separation distance* shall be not less than 20 feet (6096 mm) and the *mechanical equipment screen* shall be constructed of *fire-retardant-treated wood* complying with Section 2303.2 for exterior installation.
3. Where exterior wall covering panels are used, the panels shall have a *flame spread index* of 25 or less when tested in the minimum and maximum thicknesses intended for use, with each face tested independently in accordance with ASTM E84 or UL 723. The panels shall be tested in the minimum and maximum thicknesses intended for use in accordance with, and shall comply with the acceptance criteria of, NFPA 285 and shall be installed as tested. Where the panels are tested as part of an *exterior wall* assembly in accordance with NFPA 285, the panels shall be installed on the face of the *mechanical equipment screen* supporting structure in the same manner as they were installed on the tested *exterior wall* assembly.

[BG] 1511.6.3 Type V construction. The height of mechanical equipment screens located on the *roof decks* of buildings of Type V construction, as measured from *grade plane* to the highest point on the *mechanical equipment screen*, shall be permitted to exceed the maximum *building height* allowed for the building by other provisions of this code where complying with any one of the following limitations, provided that the *fire separation distance* is greater than 5 feet (1524 mm):

1. Where the *fire separation distance* is not less than 20 feet (6096 mm), the height above *grade plane* of the *mechanical equipment screen* shall not exceed 4 feet (1219 mm) more than the maximum *building height* allowed.
2. The *mechanical equipment screen* shall be constructed of noncombustible materials.

3. The *mechanical equipment screen* shall be constructed of *fire-retardant-treated wood* complying with Section 2303.2 for exterior installation.
4. Where the *fire separation distance* is not less than 20 feet (6096 mm), the *mechanical equipment screen* shall be constructed of materials having a *flame spread index* of 25 or less when tested in the minimum and maximum thicknesses intended for use with each face tested independently in accordance with ASTM E84 or UL 723.

[BG] 1511.7 Other rooftop structures. *Rooftop structures* not regulated by Sections 1511.2 through 1511.6 shall comply with Sections 1511.7.1 through 1511.7.5, as applicable.

[BG] 1511.7.1 Aerial supports. Aerial supports shall be constructed of noncombustible materials.

Exception: Aerial supports not greater than 12 feet (3658 mm) in height as measured from the *roof deck* to the highest point on the aerial supports shall be permitted to be constructed of combustible materials.

[BG] 1511.7.2 Bulkheads. Bulkheads used for the shelter of mechanical or electrical equipment or vertical *shaft* openings in the *roof assembly* shall comply with Section 1511.2 as *penthouses*. Bulkheads used for any other purpose shall be considered as an additional story of the building.

[BG] 1511.7.3 Dormers. Dormers shall be of the same type of construction as required for the roof in which such dormers are located or the *exterior walls* of the building.

[BG] 1511.7.4 Fences. Fences and similar structures shall comply with Section 1511.6 as *mechanical equipment screens*.

[BG] 1511.7.5 Flagpoles. Flagpoles and similar structures shall not be required to be constructed of noncombustible materials and shall not be limited in height or number.

[BG] 1511.8 Structural fire resistance. The structural frame and roof construction supporting *loads* imposed upon the roof by any *rooftop structure* shall comply with the requirements of Table 601. The fire-resistance reduction permitted by Table 601, Note a, shall not apply to roofs containing *rooftop structures*.

SECTION 1512
REROOFING

1512.1 General. Materials and methods of application used for recovering or replacing an existing *roof covering* shall comply with the requirements of Chapter 15.

Exceptions:

1. *Roof replacement* or *roof recover* of existing low-slope *roof coverings* shall not be required to meet the minimum design slope requirement of $^1/_4$ unit vertical in 12 units horizontal (2-percent slope) in Section 1507 for roofs that provide *positive roof drainage*.
2. Recovering or replacing an existing *roof covering* shall not be required to meet the requirement for secondary (emergency overflow) drains or *scuppers* in Section 1502.2 for roofs that provide for *positive roof drainage*. For the purposes of this exception, existing secondary drainage or *scupper* systems required in accordance with this code shall not be removed unless they are replaced by secondary drains or *scuppers* designed and installed in accordance with Section 1502.2.

* **1512.2 Roof replacement.** *Roof replacement* shall include the removal of all existing layers of *roof assembly* materials down to the *roof deck*.

Exception: Where the existing *roof assembly* includes an ice barrier membrane that is adhered to the *roof deck*, the existing ice barrier membrane shall be permitted to remain in place and covered with an additional layer of ice barrier membrane in accordance with Section 1507.

1512.2.1 Roof recover. The installation of a new roof covering over an existing roof covering shall be permitted where any of the following conditions occur:

1. Where the new roof covering is installed in accordance with the roof covering manufacturer's approved instructions.
2. Complete and separate roofing systems, such as standing-seam *metal roof panel* systems, that are designed to transmit the roof *loads* directly to the building's structural system and that do not rely on existing roofs and roof coverings for support, shall not require the removal of existing roof coverings.
3. Metal panel, metal shingle and concrete and clay tile roof coverings shall be permitted to be installed over existing wood shake roofs when applied in accordance with Section 1512.3.
4. The application of a new protective roof coating over an existing protective roof coating, *metal roof panel*, built-up roof, spray polyurethane foam roofing system, *metal roof shingles*, mineral-surfaced roll roofing, modified bitumen roofing or thermoset and thermoplastic single-ply roofing shall be permitted without tear off of existing roof coverings.

1512.2.1.1 Exceptions. A *roof recover* shall not be permitted where any of the following conditions occur:

1. Where the existing roof or *roof covering* is water soaked or has deteriorated to the point that the existing roof or *roof covering* is not adequate as a base for additional roofing.
2. Where the existing *roof covering* is slate, clay, cement or asbestos-cement tile.
3. Where the existing roof has two or more applications of any type of *roof covering*.

1512.3 Roof recovering. Where the application of a new *roof covering* over wood shingle or shake roofs creates a combustible concealed space, the entire existing surface shall be covered with *gypsum board*, mineral fiber, glass fiber or other *approved* materials securely fastened in place.

1512.4 Reinstallation of materials. Existing slate, clay or cement tile shall be permitted for reinstallation, except that damaged, cracked or broken slate or tile shall not be reinstalled. Existing vent flashing, metal edgings, drain outlets, collars and metal counterflashings shall not be reinstalled where rusted, damaged or deteriorated. Existing *ballast* that is damaged, cracked or broken shall not be reinstalled. Existing aggregate surfacing materials from built-up roofs shall not be reinstalled.

1512.5 Flashings. Flashings shall be reconstructed in accordance with *approved* manufacturer's installation instructions. Metal flashing to which bituminous materials are to be adhered shall be primed prior to installation.

CHAPTER 16
STRUCTURAL DESIGN

User notes:

About this chapter: *Chapter 16 establishes minimum design requirements so that the structural components of buildings are proportioned to resist the loads that are likely to be encountered. In addition, this chapter assigns buildings and structures to risk categories that are indicative of their intended use. The loads specified herein along with the required load combinations have been established through research and service performance of buildings and structures. The application of these loads and adherence to the serviceability criteria enhance the protection of life and property.*

Code development reminder: *Code change proposals to this chapter will be considered by the IBC—Structural Code Development Committee during the 2022 (Group B) Code Development Cycle.*

SECTION 1601
GENERAL

1601.1 Scope. The provisions of this chapter shall govern the structural design of buildings, structures and portions thereof regulated by this code.

SECTION 1602
NOTATIONS

1602.1 Notations. The following notations are used in this chapter:

D = Dead load.

D_i = Weight of ice in accordance with Chapter 10 of ASCE 7.

E = Combined effect of horizontal and vertical earthquake induced forces as defined in Section 12.4 of ASCE 7.

F = Load due to fluids with well-defined pressures and maximum heights.

F_a = Flood load in accordance with Chapter 5 of ASCE 7.

H = Load due to lateral earth pressures, ground water pressure or pressure of bulk materials.

L = Live load.

L_r = Roof live load.

R = Rain load.

S = Snow load.

T = Cumulative effects of self-straining load forces and effects.

V_{asd} = Allowable stress design wind speed, miles per hour (mph) (km/hr) where applicable.

V = Basic design wind speeds, miles per hour (mph) (km/hr) determined from Figures 1609.3(1) through 1609.3(12) or ASCE 7.

W = Load due to wind pressure.

W_i = Wind-on-ice in accordance with Chapter 10 of ASCE 7.

SECTION 1603
CONSTRUCTION DOCUMENTS

1603.1 General. *Construction documents* shall show the size, section and relative locations of structural members with floor levels, column centers and offsets dimensioned. The design loads and other information pertinent to the structural design required by Sections 1603.1.1 through 1603.1.9 shall be indicated on the *construction documents.*

Exception: *Construction documents* for buildings constructed in accordance with the *conventional light-frame construction* provisions of Section 2308 shall indicate the following structural design information:

1. Floor and roof dead and live loads.
2. Ground snow load, p_g.
3. Basic design wind speed, V, miles per hour (mph) (km/hr) and allowable stress design wind speed, V_{asd}, as determined in accordance with Section 1609.3.1 and wind exposure.
4. *Seismic design category* and *site class.*
5. Flood design data, if located in *flood hazard areas* established in Section 1612.3.
6. Design load-bearing values of soils.
7. Rain load data.

1603.1.1 Floor live load. The uniformly distributed, concentrated and impact floor *live load* used in the design shall be indicated for floor areas. Use of *live load* reduction in accordance with Section 1607.12 shall be indicated for each type of *live load* used in the design.

1603.1.2 Roof live load. The *roof live load* used in the design shall be indicated for roof areas (Section 1607.14).

1603.1.3 Roof snow load data. The ground snow *load*, p_g, shall be indicated. In areas where the ground snow *load*, p_g, exceeds 10 pounds per square foot (psf) (0.479 kN/m^2), the following additional information shall also be provided, regardless of whether snow *loads* govern the design of the roof:

1. Flat-roof snow *load*, p_f.
2. Snow exposure factor, C_e.
3. Snow *load* importance factor, I_s.
4. Thermal factor, C_t.
5. Slope factor(s), C_s.

6. Drift surcharge load(s), p_d, where the sum of p_d and p_f exceeds 20 psf (0.96 kN/m^2).
7. Width of snow drift(s), *w*.

1603.1.4 Wind design data. The following information related to wind *loads* shall be shown, regardless of whether wind *loads* govern the design of the lateral force-resisting system of the structure:

1. Basic design *wind speed*, *V*, miles per hour and *allowable stress design wind speed*, V_{asd}, as determined in accordance with Section 1609.3.1.
2. *Risk category*.
3. Wind exposure. Applicable wind direction if more than one wind exposure is utilized.
4. Applicable internal pressure coefficient.
5. Design wind pressures and their applicable zones with dimensions to be used for exterior component and cladding materials not specifically designed by the *registered design professional* responsible for the design of the structure, pounds per square foot (kN/m^2).

1603.1.5 Earthquake design data. The following information related to seismic *loads* shall be shown, regardless of whether seismic *loads* govern the design of the lateral force-resisting system of the structure:

1. *Risk category*.
2. Seismic importance factor, I_e.
3. Mapped spectral response acceleration parameters, S_S and S_1.
4. *Site class*.
5. Design spectral response acceleration parameters, S_{DS} and S_{D1}.
6. *Seismic design category*.
7. Basic seismic force-resisting system(s).
8. Design base shear(s).
9. Seismic response coefficient(s), *CS*.
10. Response modification coefficient(s), *R*.
11. Analysis procedure used.

1603.1.6 Geotechnical information. The design load-bearing values of soils shall be shown on the *construction documents*.

1603.1.7 Flood design data. For buildings located in whole or in part in *flood hazard areas* as established in Section 1612.3, the documentation pertaining to design, if required in Section 1612.4, shall be included and the following information, referenced to the datum on the community's *Flood Insurance Rate Map* (FIRM), shall be shown, regardless of whether *flood loads* govern the design of the building:

1. *Flood design* class assigned according to ASCE 24.
2. In *flood hazard areas* other than *coastal high hazard areas* or *coastal A zones*, the elevation of the proposed *lowest floor*, including the basement.
3. In *flood hazard areas* other than *coastal high hazard areas* or *coastal A zones*, the elevation to which any nonresidential building will be dry floodproofed.
4. In *coastal high hazard areas* and *coastal A zones*, the proposed elevation of the bottom of the lowest horizontal structural member of the *lowest floor*, including the basement.

1603.1.8 Special loads. Special *loads* that are applicable to the design of the building, structure or portions thereof, including but not limited to the *loads* of machinery or equipment, and that are greater than specified floor and roof *loads* shall be specified by their descriptions and locations.

1603.1.8.1 Photovoltaic panel systems. The *dead load* of rooftop-mounted *photovoltaic panel systems*, including rack support systems, shall be indicated on the *construction documents*.

1603.1.9 Roof rain load data. Rain intensity, *i* (in/hr) (cm/hr), shall be shown regardless of whether rain *loads* govern the design.

SECTION 1604
GENERAL DESIGN REQUIREMENTS

1604.1 General. Building, structures and parts thereof shall be designed and constructed in accordance with *strength* design, *load and resistance factor* design, *allowable stress design*, empirical design or conventional construction methods, as permitted by the applicable material chapters and referenced standards.

1604.2 Strength. Buildings and *other structures*, and parts thereof, shall be designed and constructed to support safely the *factored loads* in load combinations defined in this code without exceeding the appropriate strength *limit states* for the materials of construction. Alternatively, buildings and *other structures*, and parts thereof, shall be designed and constructed to support safely the *nominal loads* in load combinations defined in this code without exceeding the appropriate specified allowable stresses for the materials of construction.

Loads and forces for occupancies or uses not covered in this chapter shall be subject to the approval of the *building official*.

1604.3 Serviceability. Structural systems and members thereof shall be designed to have adequate stiffness to limit deflections as indicated in Table 1604.3.

1604.3.1 Deflections. The deflections of structural members shall not exceed the more restrictive of the limitations of Sections 1604.3.2 through 1604.3.5 or that permitted by Table 1604.3.

1604.3.2 Reinforced concrete. The deflection of reinforced concrete structural members shall not exceed that permitted by ACI 318.

1604.3.3 Steel. The deflection of steel structural members shall not exceed that permitted by AISC 360, AISI S100, ASCE 8, SJI 100 or SJI 200, as applicable.

TABLE 1604.3
DEFLECTION LIMITS[a, b, c, h, i]

CONSTRUCTION	*L* or L_r	*S* or *W*[f]	*D* + *L*[d, g]
Roof members:[e]			
Supporting plaster or stucco ceiling	l/360	l/360	l/240
Supporting nonplaster ceiling	l/240	l/240	l/180
Not supporting ceiling	l/180	l/180	l/120
Floor members	l/360	—	l/240
Exterior walls:			
With plaster or stucco finishes	—	l/360	—
With other brittle finishes	—	l/240	—
With flexible finishes	—	l/120	—
Interior partitions:[b]			
With plaster or stucco finishes	l/360	—	—
With other brittle finishes	l/240	—	—
With flexible finishes	l/120	—	—
Farm buildings	—	—	l/180
Greenhouses	—	—	l/120

For SI: 1 foot = 304.8 mm.

a. For structural roofing and siding made of formed metal sheets, the total load deflection shall not exceed l/60. For secondary roof structural members supporting formed metal roofing, the live load deflection shall not exceed l/150. For secondary wall members supporting formed metal siding, the design wind load deflection shall not exceed l/90. For roofs, this exception only applies when the metal sheets have no roof covering.

b. Flexible, folding and portable partitions are not governed by the provisions of this section. The deflection criterion for interior partitions is based on the horizontal load defined in Section 1607.16.

c. See Section 2403 for glass supports.

d. The deflection limit for the $D + (L + L_r)$ load combination only applies to the deflection due to the creep component of long-term dead load deflection plus the short-term live load deflection. For lumber, structural glued laminated timber, prefabricated wood I-joists and structural composite lumber members that are dry at time of installation and used under dry conditions in accordance with the ANSI/AWC NDS, the creep component of the long-term deflection shall be permitted to be estimated as the immediate dead load deflection resulting from 0.5*D*. For lumber and glued laminated timber members installed or used at all other moisture conditions or cross laminated timber and wood structural panels that are dry at time of installation and used under dry conditions in accordance with the ANSI/AWC NDS, the creep component of the long-term deflection is permitted to be estimated as the immediate dead load deflection resulting from *D*. The value of 0.5*D* shall not be used in combination with ANSI/AWC NDS provisions for long-term loading.

e. The preceding deflections do not ensure against ponding. Roofs that do not have sufficient slope or camber to ensure adequate drainage shall be investigated for ponding. See Chapter 8 of ASCE 7.

f. The wind load shall be permitted to be taken as 0.42 times the "component and cladding" loads or directly calculated using the 10-year mean return interval wind speed for the purpose of determining deflection limits in Table 1604.3. Where framing members support glass, the deflection limit therein shall not exceed that specified in Section 1604.3.7

g. For steel structural members, the deflection due to creep component of long-term dead load shall be permitted to be taken as zero.

h. For aluminum structural members or aluminum panels used in skylights and sloped glazing framing, roofs or walls of sunroom additions or patio covers not supporting edge of glass or aluminum sandwich panels, the total load deflection shall not exceed l/60. For continuous aluminum structural members supporting edge of glass, the total load deflection shall not exceed l/175 for each glass lite or l/60 for the entire length of the member, whichever is more stringent. For aluminum sandwich panels used in roofs or walls of sunroom additions or patio covers, the total load deflection shall not exceed l/120.

i. l = Length of the member between supports. For cantilever members, l shall be taken as twice the length of the cantilever.

1604.3.4 Masonry. The deflection of masonry structural members shall not exceed that permitted by TMS 402.

1604.3.5 Aluminum. The deflection of aluminum structural members shall not exceed that permitted by AA ADM.

1604.3.6 Limits. The deflection limits of Section 1604.3.1 shall be used unless more restrictive deflection limits are required by a referenced standard for the element or finish material.

1604.3.7 Framing supporting glass. The deflection of framing members supporting glass subjected to 0.6 times the "component and cladding" wind loads shall not exceed either of the following:

1. $^1/_{175}$ of the length of span of the framing member, for framing members having a length not more than 13 feet 6 inches (4115 mm).
2. $^1/_{240}$ of the length of span of the framing member + $^1/_4$ inch (6.4 mm), for framing members having a length greater than 13 feet 6 inches (4115 mm).

1604.4 Analysis. *Load effects* on structural members and their connections shall be determined by methods of structural analysis that take into account equilibrium, general stability, geometric compatibility and both short- and long-term material properties.

Members that tend to accumulate residual deformations under repeated service *loads* shall have included in their analysis the effects of added deformations expected to occur during their service life.

Any system or method of construction to be used shall be based on a rational analysis in accordance with well-established principles of mechanics. Such analysis shall result in a system that provides a complete *load* path capable of transferring *loads* from their point of origin to the load-resisting elements.

The total lateral force shall be distributed to the various vertical elements of the lateral force-resisting system in proportion to their rigidities, considering the rigidity of the horizontal bracing system or *diaphragm*. Rigid elements assumed not to be a part of the lateral force-resisting system are permitted to be incorporated into buildings provided that their effect on the action of the system is considered and provided for in the design. A *diaphragm* is rigid for the purpose of distribution of *story* shear and torsional moment when the lateral deformation of the *diaphragm* is less than or equal to two times the average *story* drift. Where required by ASCE 7, provisions shall be made for the increased forces induced on resisting elements of the structural system resulting from torsion due to eccentricity between the center of application of the lateral forces and the center of rigidity of the lateral force-resisting system.

Every structure shall be designed to resist the effects caused by the forces specified in this chapter, including overturning, uplift and sliding. Where sliding is used to isolate the elements, the effects of friction between sliding elements shall be included as a force.

1604.5 Risk category. Each building and structure shall be assigned a *risk category* in accordance with Table 1604.5. Where a referenced standard specifies an occupancy category, the *risk category* shall not be taken as lower than the occupancy category specified therein. Where a referenced standard specifies that the assignment of a *risk category* be in accordance with ASCE 7, Table 1.5-1, Table 1604.5 shall be used in lieu of ASCE 7, Table 1.5-1.

Exception: The assignment of buildings and structures to Tsunami *Risk Categories* III and IV is permitted to be in accordance with Section 6.4 of ASCE 7.

1604.5.1 Multiple occupancies. Where a building or structure is occupied by two or more occupancies not included in the same *risk category*, it shall be assigned the classification of the highest *risk category* corresponding to the various occupancies. Where buildings or structures have two or more portions that are structurally separated, each portion shall be separately classified. Where a separated portion of a building or structure provides required access to, required egress from or shares life safety components with another portion having a higher *risk category*, both portions shall be assigned to the higher *risk category*.

Exception: Where a *storm shelter* designed and constructed in accordance with ICC 500 is provided in a building, structure or portion thereof normally occupied for other purposes, the *risk category* for the normal occupancy of the building shall apply unless the *storm shelter* is a designated emergency shelter in accordance with Table 1604.5.

1604.6 In-situ load tests. The *building official* is authorized to require an engineering analysis or a load test, or both, of any construction whenever there is reason to question the safety of the construction for the intended occupancy. Engineering analysis and load tests shall be conducted in accordance with Section 1708.

1604.7 Preconstruction load tests. Materials and methods of construction that are not capable of being designed by *approved* engineering analysis or that do not comply with the applicable referenced standards, or alternative test procedures in accordance with Section 1707, shall be load tested in accordance with Section 1709.

1604.8 Anchorage. Buildings and *other structures*, and portions thereof, shall be provided with anchorage in accordance with Sections 1604.8.1 through 1604.8.3, as applicable.

1604.8.1 General. Anchorage of the roof to walls and columns, and of walls and columns to foundations, shall be provided to resist the uplift and sliding forces that result from the application of the prescribed *loads*.

1604.8.2 Structural walls. Walls that provide vertical load-bearing resistance or lateral shear resistance for a portion of the structure shall be anchored to the roof and to all floors and members that provide lateral support for the wall or that are supported by the wall. The connections shall be capable of resisting the horizontal forces specified in Section 1.4.4 of ASCE 7 for walls of structures assigned to *Seismic Design Category* A and to Section 12.11 of ASCE 7 for walls of structures assigned to all other *seismic design categories*. Required anchors in masonry walls of hollow units or *cavity walls* shall be embedded in a reinforced grouted structural element of the wall. See Sections 1609 for wind design requirements and 1613 for earthquake design requirements.

1604.8.3 Decks. Where supported by attachment to an *exterior wall*, decks shall be positively anchored to the primary structure and designed for both vertical and lateral loads as applicable. Such attachment shall not be accomplished by the use of toenails or nails subject to withdrawal. Where positive connection to the primary building structure cannot be verified during inspection, decks shall be self-supporting. Connections of decks with cantilevered framing members to *exterior walls* or other framing members shall be designed for both of the following:

1. The reactions resulting from the *dead load* and *live load* specified in Table 1607.1, or the snow *load* specified in Section 1608, in accordance with Section 1605, acting on all portions of the deck.
2. The reactions resulting from the *dead load* and *live load* specified in Table 1607.1, or the snow *load* specified in Section 1608, in accordance with Section 1605, acting on the cantilevered portion of the deck, and no *live load* or snow *load* on the remaining portion of the deck.

1604.9 Wind and seismic detailing. Lateral force-resisting systems shall meet seismic detailing requirements and limitations prescribed in this code and ASCE 7 Chapters 11, 12, 13, 15, 17 and 18 as applicable, even where wind *load effects* are greater than seismic *load effects*.

Exception: References within ASCE 7 to Chapter 14 shall not apply, except as specifically required herein.

1604.10 Loads on storm shelters. *Loads* and load combinations on storm shelters shall be determined in accordance with ICC 500.

TABLE 1604.5
RISK CATEGORY OF BUILDINGS AND OTHER STRUCTURES

RISK CATEGORY	NATURE OF OCCUPANCY
I	Buildings and other structures that represent a low hazard to human life in the event of failure, including but not limited to: • Agricultural facilities. • Certain temporary facilities. • Minor storage facilities.
II	Buildings and other structures except those listed in Risk Categories I, III and IV.

III	Buildings and other structures that represent a substantial hazard to human life in the event of failure, including but not limited to:
	• Buildings and other structures whose primary occupancy is public assembly with an occupant load greater than 300.
	• Buildings and other structures containing one or more public assembly spaces, each having an occupant load greater than 300 and a cumulative occupant load of the public assembly spaces of greater than 2,500.
	• Buildings and other structures containing Group E or Group I-4 occupancies or combination therof, with an occupant load greater than 250.
	• Buildings and other structures containing educational occupancies for students above the 12th grade with an occupant load greater than 500.
	• Group I-2, Condition 1 occupancies with 50 or more care recipients.
	• Group I-2, Condition 2 occupancies not having emergency surgery or emergency treatment facilities.
	• Group I-3 occupancies.
	• Any other occupancy with an occupant load greater than 5,000.[a]
	• Power-generating stations, water treatment facilities for potable water, wastewater treatment facilities and other public utility facilities not included in Risk Category IV.
	• Buildings and other structures not included in Risk Category IV containing quantities of toxic or explosive materials that:
	    • Exceed maximum allowable quantities per control area as given in Table 307.1(1) or 307.1(2) or per outdoor control area in accordance with the *International Fire Code*; and
	    • Are sufficient to pose a threat to the public if released.[b]

IV	Buildings and other structures designated as essential facilities, including but not limited to:
	• Group I-2, Condition 2 occupancies having emergency surgery or emergency treatment facilities.
	• Ambulatory care facilities having emergency surgery or emergency treatment facilities.
	• Fire, rescue, ambulance and police stations and emergency vehicle garages
	• Designated earthquake, hurricane or other emergency shelters.
	• Designated emergency preparedness, communications and operations centers and other facilities required for emergency response.
	• Power-generating stations and other public utility facilities required as emergency backup facilities for Risk Category IV structures.
	• Buildings and other structures containing quantities of highly toxic materials that:
	    • Exceed maximum allowable quantities per control area as given in Table 307.1(2) or per outdoor control area in accordance with the *International Fire Code*; and
	    • Are sufficient to pose a threat to the public if released.[b]
	• Aviation control towers, air traffic control centers and emergency aircraft hangars.
	• Buildings and other structures having critical national defense functions.
	• Water storage facilities and pump structures required to maintain water pressure for fire suppression.

a. For purposes of occupant load calculation, occupancies required by Table 1004.5 to use gross floor area calculations shall be permitted to use net floor areas to determine the total occupant load.

b. Where approved by the building official, the classification of buildings and other structures as Risk Category III or IV based on their quantities of toxic, highly toxic or explosive materials is permitted to be reduced to Risk Category II, provided that it can be demonstrated by a hazard assessment in accordance with Section 1.5.3 of ASCE 7 that a release of the toxic, highly toxic or explosive materials is not sufficient to pose a threat to the public.

SECTION 1605
LOAD COMBINATIONS

1605.1 General. Buildings and *other structures* and portions thereof shall be designed to resist the strength load combinations specified in ASCE 7, Section 2.3, the *allowable stress design* load combinations specified in ASCE 7, Section 2.4, or the alternative *allowable stress design* load combinations of Section 1605.2.

Exceptions:

1. The modifications to load combinations of ASCE 7 Section 2.3, ASCE 7 Section 2.4, and Section 1605.2 specified in ASCE 7 Chapters 18 and 19 shall apply.
2. Where the allowable stress design load combinations of ASCE 7 Section 2.4 are used, flat roof snow loads of 30 pounds per square foot (1.44 kN/m^2) and roof live loads of 30 pounds per square foot (1.44 kN/m^2) or less need not be combined with seismic load. Where flat roof snow loads exceed 30 pounds per square foot (1.44 kN/m^2), 20 percent shall be combined with seismic loads.
3. Where the allowable stress design load combinations of ASCE 7 Section 2.4 are used, crane hook loads need not be combined with *roof live loads* or with more than three-fourths of the snow load or one-half of the wind loads.

1605.1.1 Stability. Regardless of which load combinations are used to design for strength, where overall structure stability (such as stability against overturning, sliding, or buoyancy) is being verified, use of the load combinations specified in Section 2.3 or 2.4 of ASCE 7, and in Section 1605.2 shall be permitted. Where the load combinations specified in ASCE 7, Section 2.3 are used, strength reduction factors applicable to soil resistance shall be provided by a registered design professional. The stability of retaining walls shall be verified in accordance with Section 1807.2.3.

1605.2 Alternative allowable stress design load combinations. In lieu of the load combinations in ASCE 7, Section 2.4, structures and portions thereof shall be permitted to be designed for the most critical effects resulting from the following combinations. Where using these alternative allowable stress load combinations that include wind or seismic *loads*, allowable stresses are permitted to be increased or load combinations reduced where permitted by the material chapter of this code or the referenced standards. For load combinations that include the counteracting effects of dead and wind *loads*, only two-thirds of the minimum *dead load* likely to be in place during a design wind event shall be used. Where using these alternative load combinations to evaluate sliding, overturning and soil bearing at the soil-structure interface, the reduction of foundation overturning from Section 12.13.4 in ASCE 7 shall not be used. Where using these alternative basic *load* combinations for proportioning foundations for loadings, which include seismic *loads*, the vertical seismic *load effect*, E_v, in Equation 12.4-4 of ASCE 7 is permitted to be taken equal to zero. Where required by ASCE 7, Chapters 12, 13 and 15, the load combinations including overstrength of ASCE 7, Section 2.3.6 shall be used.

$D + L + (L_r \text{ or } S \text{ or } R)$ **(Equation 16-1)**

$D + L + 0.6W$ **(Equation 16-2)**

$D + L + 0.6W + S/2$ **(Equation 16-3)**

$D + L + S + 0.6W/2$ **(Equation 16-4)**

$D + L + S + E/1.4$ **(Equation 16-5)**

$0.9D + E/1.4$ **(Equation 16-6)**

Exceptions:

1. Crane hook *loads* need not be combined with *roof live loads* or with more than three-fourths of the snow load or one-half of the wind load.
2. Flat roof snow *loads* of 30 pounds per square foot (1.44 kN/m^2) or less and *roof live loads* of 30 pounds per square foot (1.44 kN/m^2) or less need not be combined with seismic loads. Where flat roof snow loads exceed 30 pounds per square foot (1.44 kN/m^2), 20 percent shall be combined with seismic loads.

SECTION 1606
DEAD LOADS

1606.1 General. *Dead loads* are those *loads* defined in Chapter 2 of this code. *Dead loads* shall be considered to be permanent loads.

1606.2 Weights of materials of construction. For purposes of design, the actual weights of materials of construction shall be used. In the absence of definite information, values used shall be subject to the approval of the *building official*.

1606.3 Weight of fixed service equipment. In determining dead loads for purposes of design, the weight of fixed service equipment, including the maximum weight of the contents of fixed service equipment, shall be included. The components of fixed service equipment that are variable, such as liquid contents and movable trays, shall not be used to counteract forces causing overturning, sliding, and uplift conditions in accordance with Section 1.3.6 of ASCE 7.

Exceptions:

1. Where force effects are the result of the presence of the variable components, the components are permitted to be used to counter those load effects. In such cases, the structure shall be designed for force effects with the variable components present and with them absent.
2. For the calculation of seismic force effects, the components of fixed service equipment that are variable, such as liquid contents and movable trays, need not exceed those expected during normal operation.

1606.4 Photovoltaic panel systems. The weight of *photovoltaic panel systems*, their support system, and ballast shall be considered as dead *load*.

1606.5 Vegetative and landscaped roofs. The weight of all landscaping and hardscaping materials for vegetative and landscaped roofs shall be considered as *dead load*. The weight shall be computed considering both fully saturated soil and drainage layer materials and fully dry soil and drainage layer materials to determine the most severe *load* effects on the structure.

SECTION 1607
LIVE LOADS

1607.1 General. *Live loads* are those loads defined in Chapter 2 of this code.

1607.2 Loads not specified. For occupancies or uses not designated in Section 1607, the *live load* shall be determined in accordance with a method *approved* by the *building official*.

1607.3 Uniform live loads. The *live loads* used in the design of buildings and *other structures* shall be the maximum loads expected by the intended use or occupancy but shall not be less than the minimum uniformly distributed *live loads* given in Table 1607.1.

1607.4 Concentrated live loads. Floors, roofs and other similar surfaces shall be designed to support the uniformly distributed *live loads* prescribed in Section 1607.3 or the concentrated *live loads*, given in Table 1607.1, whichever produces the greater *load effects*. Unless otherwise specified, the indicated concentration shall be assumed to be uniformly distributed over an area of $2^1/_2$ feet by $2^1/_2$ feet (762 mm by 762 mm) and shall be located so as to produce the maximum *load effects* in the structural members.

1607.5 Partition loads. In office buildings and in other buildings where partition locations are subject to change, provisions for partition weight shall be made, whether or not partitions are shown on the construction documents, unless the specified *live load* is 80 psf (3.83 kN/m^2) or greater. The partition *load* shall be not less than a uniformly distributed *live load* of 15 psf (0.72 kN/m^2).

1607.6 Helipads. Helipads shall be designed for the following *live loads*:

1. A uniform *live load*, *L*, as specified in Items 1.1 and 1.2. This *load* shall not be reduced.
   - 1.1. 40 psf (1.92 kN/m^2) where the design basis helicopter has a maximum take-off weight of 3,000 pounds (13.35 kN) or less.
   - 1.2. 60 psf (2.87 kN/m^2) where the design basis helicopter has a maximum take-off weight greater than 3,000 pounds (13.35 kN).
2. A single concentrated *live load*, *L*, of 3,000 pounds (13.35 kN) applied over an area of 4.5 inches by 4.5 inches (114 mm by 114 mm) and located so as to produce the maximum *load effects* on the structural elements under consideration. The concentrated *load* is not required to act concurrently with other uniform or concentrated *live loads*.
3. Two single concentrated *live loads*, *L*, 8 feet (2438 mm) apart applied on the landing pad (representing the helicopter's two main landing gear, whether skid type or wheeled type), each having a magnitude of 0.75 times the maximum take-off weight of the helicopter, and located so as to produce the maximum *load effects* on the structural elements under consideration. The concentrated loads shall be applied over an area of 8 inches by 8 inches (203 mm by 203 mm) and are not required to act concurrently with other uniform or concentrated *live loads*.

Landing areas designed for a design basis helicopter with maximum take-off weight of 3,000 pounds (13.35 kN) shall be identified with a 3,000-pound (13.34 kN) weight limitation. The landing area weight limitation shall be indicated by the numeral "3" (kips) located in the bottom right corner of the landing area as viewed from the primary approach path. The indication for the landing area weight limitation shall be a minimum 5 feet (1524 mm) in height.

1607.7 Passenger vehicle garages. Floors in garages or portions of a building used for the storage of motor vehicles shall be designed for the uniformly distributed *live loads* indicated in Table 1607.1 or the following concentrated *load*:

1. For garages restricted to passenger vehicles accommodating not more than nine passengers, 3,000 pounds (13.35 kN) acting on an area of 4.5 inches by 4.5 inches (114 mm by 114 mm).
2. For mechanical parking structures without slab or deck that are used for storing passenger vehicles only, 2,250 pounds (10 kN) per wheel.

1607.8 Heavy vehicle loads. Floors and other surfaces that are intended to support vehicle *loads* greater than a 10,000-pound (4536 kg) gross vehicle weight rating shall comply with Sections 1607.8.1 through 1607.8.5.

1607.8.1 Loads. Where any structure does not restrict access for vehicles that exceed a 10,000-pound (4536 kg) gross vehicle weight rating, those portions of the structure subject to such *loads* shall be designed using the vehicular *live loads*, including consideration of impact and fatigue, in accordance with the codes and specifications required by the jurisdiction having authority for the design and construction of the roadways and bridges in the same location of the structure.

1607.8.2 Fire truck and emergency vehicles. Where a structure or portions of a structure are accessed and loaded by fire department access vehicles and other similar emergency vehicles, the structure shall be designed for the greater of the following *loads*:

1. The actual operational *loads*, including outrigger reactions and contact areas of the vehicles as stipulated and *approved* by the *building official*.
2. The live loading specified in Section 1607.8.1.

TABLE 1607.1
MINIMUM UNIFORMLY DISTRIBUTED LIVE LOADS, L_o, AND MINIMUM CONCENTRATED LIVE LOADS

OCCUPANCY OR USE			UNIFORM (psf)	CONCENTRATED (pounds)	ALSO SEE SECTION
1.	Apartments (see residential)		—	—	—
2.	Access floor systems	Office use	50	2,000	—
		Computer use	100	2,000	—
3.	Armories and drill rooms		150[b]	—	—
4.	Assembly areas	Fixed seats (fastened to floor)	60[a]	—	—
		Follow spot, projections and control rooms	50		
		Lobbies	100[a]		
		Movable seats	100[a]		
		Stage floors	150[b]		
		Platforms (assembly)	100[a]		
		Bleachers, folding and telescopic seating and grandstands	100[a] (See Section 1607.19)		
		Stadiums and arenas with fixed seats (fastened to the floor)	60[a] (See Section 1607.19)		
		Other assembly areas	100[a]		
5.	Balconies and decks		1.5 times the live load for the area served, not required to exceed 100	—	—
6.	Catwalks for maintenance and service access		40	300	—
7.	Cornices		60	—	—
8.	Corridors	First floor	100	—	—
		Other floors	Same as occupancy served except as indicated		
9.	Dining rooms and restaurants		100[a]	—	—
10.	Dwellings (see residential)		—	—	—
11.	Elevator machine room and control room grating (on area of 2 inches by 2 inches)		—	300	—
12.	Finish light floor plate construction (on area of 1 inch by 1 inch)		—	200	—
13.	Fire escapes		100	—	—
		On single-family dwellings only	40		
14.	Fixed ladders		See Section 1607.17		—
15.	Garages	Passenger vehicles only	40[c]	See Section 1607.7	—
		Trucks and buses	See Section 1607.8		
16.	Handrails, guards and grab bars		See Section 1607.9		—
17.	Helipads		See Section 1607.6		—
18.	Hospitals	Corridors above first floor	80	1,000	—
		Operating rooms, laboratories	60	1,000	
		Patient rooms	40	1,000	
19.	Hotels (see residential)		—	—	—

(continued)

TABLE 1607.1—continued
MINIMUM UNIFORMLY DISTRIBUTED LIVE LOADS, L_o, AND MINIMUM CONCENTRATED LIVE LOADS

OCCUPANCY OR USE			UNIFORM (psf)	CONCENTRATED (pounds)	ALSO SEE SECTION
20.	Libraries	Corridors above first floor	80	1,000	—
		Reading rooms	60	1,000	—
		Stack rooms	150[b]	1,000	Section 1607.18
21.	Manufacturing	Heavy	250[b]	3,000	—
		Light	125[b]	2,000	
22.	Marquees, except one- and two-family dwellings		75	—	—
23.	Office buildings	Corridors above first floor	80	2,000	—
		File and computer rooms shall be designed for heavier loads based on anticipated occupancy	—	—	
		Lobbies and first-floor corridors	100	2,000	
		Offices	50	2,000	
24.	Penal institutions	Cell blocks	40	—	—
		Corridors	100		
25.	Recreational uses	Bowling alleys, poolrooms and similar uses	75[a]	—	—
		Dance halls and ballrooms	100[a]		
		Gymnasiums	100[a]		
		Ice skating rinks	250[b]		
		Roller skating rinks	100[a]		
26.	Residential	One- and two-family dwellings:		—	Section 1607.22
		Uninhabitable attics without storage	10		
		Uninhabitable attics with storage	20		
		Habitable attics and sleeping areas	30		
		Canopies, including marquees	20		
		All other areas	40		
		Hotels and multifamily dwellings:			
		Private rooms and corridors serving them	40		
		Public rooms[a] and corridors serving them	100		

(continued)

TABLE 1607.1—continued
MINIMUM UNIFORMLY DISTRIBUTED LIVE LOADS, L_0, AND MINIMUM CONCENTRATED LIVE LOADS

OCCUPANCY OR USE			UNIFORM (psf)	CONCENTRATED (pounds)	ALSO SEE SECTION
27.	Roofs	Ordinary flat, pitched, and curved roofs (that are not occupiable)	20	—	Section 1607.15.2
		Roof areas used for assembly purposes	100[a]	—	
		Roof areas used for occupancies other than assembly	Same as occupancy served	—	
		Vegetative and landscaped roofs:		—	
		Roof areas not intended for occupancy	20	—	
		Roof areas used for assembly purposes	100[a]	—	
		Roof areas used for other occupancies	Same as occupancy served	—	
		Awnings and canopies:		—	
		Fabric construction supported by a skeleton structure	5[a]	—	
		All other construction, except one- and two-family dwellings	20	—	
		Primary roof members exposed to a work floor:			Section 1607.15.2
		Single panel point of lower chord of roof trusses or any point along primary structural members supporting roofs over manufacturing, storage warehouses, and repair garages	—	2,000	
		All other primary roof members	—	300	
		All roof surfaces subject to maintenance workers	—	300	
28.	Schools	Classrooms	40	1,000	—
		Corridors above first floor	80	1,000	
		First-floor corridors	100	1,000	
29.	Scuttles, skylight ribs and accessible ceilings		—	200	—
30.	Sidewalks, vehicular driveways and yards, subject to trucking		250[b]	8,000	Section 1607.20
31.	Stairs and exits	One- and two-family dwellings	40	300	Section 1607.21
		All other	100	300	Section 1607.21
32.	Storage areas above ceilings		20	—	—
33.	Storage warehouses (shall be designed for heavier loads if required for anticipated storage)	Heavy	250[b]	—	—
		Light	125[b]		
34.	Stores	Retail:			—
		First floor	100	1,000	
		Upper floors	75	1,000	
		Wholesale, all floors	125[b]	1,000	
35.	Vehicle barriers		See Section 1607.10		—
36.	Walkways and elevated platforms (other than exitways)		60	—	—
37.	Yards and terraces, pedestrian		100[a]	—	—

For SI: 1 inch = 25.4 mm, 1 square inch = 645.16 mm^2, 1 square foot = 0.0929 m^2, 1 pound per square foot = 0.0479 kN/m^2, 1 pound = 0.004448 kN, 1 pound per cubic foot = 16 kg/m^3.

a. Live load reduction is not permitted.

b. Live load reduction is only permitted in accordance with Section 1607.12.1.2 or Item 1 of Section 1607.12.2.

c. Live load reduction is only permitted in accordance with Section 1607.12.1.3 or Item 2 of Section 1607.12.2.

1607.8.3 Heavy vehicle garages. Garages designed to accommodate vehicles that exceed a 10,000-pound (4536 kg) gross vehicle weight rating, shall be designed using the live loading specified by Section 1607.8.1. For garages the design for impact and fatigue is not required.

Exception: The vehicular *live loads* and *load* placement are allowed to be determined using the actual vehicle weights for the vehicles allowed onto the garage floors, provided that such *loads* and placement are based on rational engineering principles and are approved by the building official, but shall be not less than 50 psf (2.9 kN/m^2). This *live load* shall not be reduced.

1607.8.4 Forklifts and movable equipment. Where a structure is intended to have forklifts or other movable equipment present, the structure shall be designed for the total vehicle or equipment *load* and the individual wheel *loads* for the anticipated vehicles as specified by the owner of the facility. These *loads* shall be posted in accordance with Section 1607.8.5.

1607.8.4.1 Impact and fatigue. *Impact loads* and fatigue loading shall be considered in the design of the supporting structure. For the purposes of design, the vehicle and wheel *loads* shall be increased by 30 percent to account for impact.

1607.8.5 Posting. The maximum weight of vehicles allowed into or on a garage or other structure shall be posted by the owner or the owner's authorized agent in accordance with Section 106.1.

1607.9 Loads on handrails, guards, grab bars and seats. *Handrails* and *guards* shall be designed and constructed for the structural loading conditions set forth in Section 1607.9.1. Grab bars, shower seats and accessible benches shall be designed and constructed for the structural loading conditions set forth in Section 1607.9.2.

1607.9.1 Handrails and guards. *Handrails* and *guards* shall be designed to resist a linear *load* of 50 pounds per linear foot (plf) (0.73 kN/m) in accordance with Section 4.5.1.1 of ASCE 7. Glass *handrail* assemblies and *guards* shall comply with Section 2407.

Exceptions:

1. For one- and two-family dwellings, only the single concentrated *load* required by Section 1607.9.1.1 shall be applied.
2. In Group I-3, F, H and S occupancies, for areas that are not accessible to the general public and that have an *occupant load* less than 50, the minimum *load* shall be 20 pounds per foot (0.29 kN/m).

1607.9.1.1 Concentrated load. *Handrails* and *guards* shall be designed to resist a concentrated *load* of 200 pounds (0.89 kN) in accordance with Section 4.5.1 of ASCE 7.

1607.9.1.2 Guard component loads. Balusters, panel fillers and guard infill components, including all rails except the handrail and the top rail, shall be designed to resist a concentrated load of 50 pounds (0.22 kN) in accordance with Section 4.5.1.2 of ASCE 7.

1607.9.2 Grab bars, shower seats and accessible benches. Grab bars, shower seats and accessible benches shall be designed to resist a single concentrated *load* of 250 pounds (1.11 kN) applied in any direction at any point on the grab bar, shower seat, or seat of the accessible bench so as to produce the maximum *load effects*.

1607.10 Vehicle barriers. *Vehicle barriers* for passenger vehicles shall be designed to resist a concentrated *load* of 6,000 pounds (26.70 kN) in accordance with Section 4.5.3 of ASCE 7. Garages accommodating trucks and buses shall be designed in accordance with an *approved* method that contains provisions for traffic railings.

1607.11 Impact loads. The *live loads* specified in Sections 1607.3 through 1607.10 shall be assumed to include adequate allowance for ordinary impact conditions. Provisions shall be made in the structural design for uses and loads that involve unusual vibration and impact forces.

1607.11.1 Elevators. Members, elements and components subject to dynamic *loads* from elevators shall be designed for *impact loads* and deflection limits prescribed by ASME A17.1/CSA B44.

1607.11.2 Machinery. For the purpose of design, the weight of machinery and moving *loads* shall be increased as follows to allow for impact:

1. Light machinery, shaft- or motor-driven, 20 percent.
2. Reciprocating machinery or power-driven units, 50 percent.

Percentages shall be increased where specified by the manufacturer.

1607.11.3 Elements supporting hoists for façade access and building maintenance equipment. In addition to any other applicable *live loads*, structural elements that support hoists for façade access and building maintenance equipment shall be designed for a *live load* of 2.5 times the rated *load* of the hoist or the stall *load* of the hoist, whichever is larger.

1607.11.4 Fall arrest, lifeline, and rope descent system anchorages. In addition to any other applicable *live loads*, fall arrest, lifeline, and rope descent system anchorages and structural elements that support these anchorages shall be designed for a *live load* of not less than 3,100 pounds (13.8 kN) for each attached line, in any direction that the *load* can be applied.

Anchorages of horizontal lifelines and the structural elements that support these anchorages shall be designed for the maximum tension that develops in the horizontal lifeline from these *live loads*.

1607.12 Reduction in uniform live loads. Except for uniform *live loads* at roofs, all other minimum uniformly distributed *live loads*, L_o, in Table 1607.1 are permitted to be reduced in accordance with Section 1607.12.1 or 1607.12.2. Uniform *live loads* at roofs are permitted to be reduced in accordance with Section 1607.14.2.

1607.12.1 Basic uniform live load reduction. Subject to the limitations of Sections 1607.12.1.1 through 1607.12.1.3 and Table 1607.1, members for which a value of $K_{LL}A_T$ is 400 square feet (37.16 m^2) or more are permitted to be designed for a reduced uniformly distributed *live load*, L, in accordance with the following equation:

$$L = L_o\left(0.25 + \frac{15}{\sqrt{K_{LL}A_T}}\right) \quad \textbf{(Equation 16-7)}$$

For SI: $L = L_o\left(0.25 + \frac{4.57}{\sqrt{K_{LL}A_T}}\right)$

where:

L = Reduced design *live load* per square foot (m^2) of area supported by the member.

L_o = Unreduced design *live load* per square foot (m^2) of area supported by the member (see Table 1607.1).

K_{LL} = *Live load* element factor (see Table 1607.12.1).

A_T = Tributary area, in square feet (m^2).

L shall be not less than $0.50L_o$ for members supporting one floor and L shall be not less than $0.40L_o$ for members supporting two or more floors.

TABLE 1607.12.1
LIVE LOAD ELEMENT FACTOR, K_{LL}

ELEMENT	K_{LL}
Interior columns	4
Exterior columns without cantilever slabs	4
Edge columns with cantilever slabs	3
Corner columns with cantilever slabs	2
Edge beams without cantilever slabs	2
Interior beams	2
Members not previously identified including:	1
Edge beams with cantilever slabs	
Cantilever beams	
One-way slabs	
Two-way slabs	
Members without provisions for continuous shear transfer normal to their span	

1607.12.1.1 One-way slabs. The tributary area, A_T, for use in Equation 16-7 for one-way slabs shall not exceed an area defined by the slab span times a width normal to the span of 1.5 times the slab span.

1607.12.1.2 Heavy live loads. *Live loads* that exceed 100 psf (4.79 kN/m^2) shall not be reduced.

Exceptions:

1. The *live loads* for members supporting two or more floors are permitted to be reduced by not greater than 20 percent, but the *live load* shall be not less than L as calculated in Section 1607.12.1.
2. For uses other than storage, where *approved*, additional *live load* reductions shall be permitted where shown by the *registered design professional* that a rational approach has been used and that such reductions are warranted.

1607.12.1.3 Passenger vehicle garages. The *live loads* shall not be reduced in passenger vehicle garages.

Exception: The *live loads* for members supporting two or more floors are permitted to be reduced by not greater than 20 percent, but the *live load* be shall be not less than L as calculated in Section 1607.12.1.

1607.12.2 Alternative uniform live load reduction. As an alternative to Section 1607.12.1 and subject to the limitations of Table 1607.1, uniformly distributed *live loads* are permitted to be reduced in accordance with the following provisions. Such reductions shall apply to slab systems, beams, girders, columns, piers, walls and foundations.

1. A reduction shall not be permitted where the *live load* exceeds 100 psf (4.79 kN/m^2) except that the design *live load* for members supporting two or more floors is permitted to be reduced by not greater than 20 percent.

   Exception: For uses other than storage, where *approved*, additional *live load* reductions shall be permitted where shown by the *registered design professional* that a rational approach has been used and that such reductions are warranted.

2. A reduction shall not be permitted in passenger vehicle parking garages except that the *live loads* for members supporting two or more floors are permitted to be reduced by not greater than 20 percent.
3. For *live loads* not exceeding 100 psf (4.79 kN/m^2), the design *live load* for any structural member supporting 150 square feet (13.94 m^2) or more is permitted to be reduced in accordance with Equation 16-8
4. For one-way slabs, the area, A, for use in Equation 16-8 shall not exceed the product of the slab span and a width normal to the span of 0.5 times the slab span.

$$R = 0.08(A - 150) \quad \textbf{(Equation 16-8)}$$

For SI: $R = 0.861(A - 13.94)$

Such reduction shall not exceed the smallest of:

1. 40 percent for members supporting one floor.
2. 60 percent for members supporting two or more floors.
3. R as determined by the following equation:

$$R = 23.1(1 + D/L_o) \quad \textbf{(Equation 16-9)}$$

where:

A = Area of floor supported by the member, square feet (m^2).

D = *Dead load* per square foot (m^2) of area supported.

L_o = Unreduced *live load* per square foot (m^2) of area supported.

R = Reduction in percent.

1607.13 Distribution of floor loads. Where uniform floor *live loads* are involved in the design of structural members arranged so as to create continuity, the minimum applied loads shall be the full *dead loads* on all spans in combination with the floor *live loads* on spans selected to produce the greatest *load effect* at each location under consideration. Floor *live loads* are permitted to be reduced in accordance with Section 1607.12 .

1607.14 Roof loads. The structural supports of roofs and *marquees* shall be designed to resist wind and, where applicable, snow and earthquake *loads*, in addition to the *dead load* of construction and the appropriate *live loads* as prescribed in this section, or as set forth in Table 1607.1. The *live loads* acting on a sloping surface shall be assumed to act vertically on the horizontal projection of that surface.

1607.14.1 Distribution of roof loads. Where uniform roof live *loads* are reduced to less than 20 psf (0.96 kN/m^2) in accordance with Section 1607.14.2.1 and are applied to the design of structural members arranged so as to create continuity, the reduced roof *live load* shall be applied to adjacent spans or to alternate spans, whichever produces the most unfavorable *load effect*. See Section 1607.14.2 for reductions in minimum roof *live loads* and Section 7.5 of ASCE 7 for partial snow loading.

1607.14.2 Reduction in uniform roof live loads. The minimum uniformly distributed *live loads* of roofs and *marquees*, L_o, in Table 1607.1 are permitted to be reduced in accordance with Section 1607.14.2.1.

1607.14.2.1 Ordinary roofs, awnings and canopies. Ordinary flat, pitched and curved roofs, and *awnings* and canopies other than of fabric construction supported by a skeleton structure, are permitted to be designed for a reduced uniformly distributed *roof live load*, L_r, as specified in the following equations or other controlling combinations of *loads* as specified in Section 1605, whichever produces the greater *load effect*.

In structures such as *greenhouses*, where special scaffolding is used as a work surface for workers and materials during maintenance and repair operations, a lower roof *load* than specified in the following equations shall not be used unless *approved* by the *building official*. Such structures shall be designed for a minimum roof live *load* of 12 psf (0.58 kN/m^2).

$L_r = L_o R_1 R_2$ **(Equation 16-10)**

where: $12 \le L_r \le 20$

For SI: $L_r = L_o R_1 R_2$

where: $0.58 \le L_r \le 0.96$

L_o = Unreduced *roof live load* per square foot (m^2) of horizontal projection supported by the member (see Table 1607.1).

L_r = Reduced *roof live load* per square foot (m^2) of horizontal projection supported by the member.

The reduction factors R_1 and R_2 shall be determined as follows:

$R_1 = 1$ for $A_t \le 200$ square feet (18.58 m^2) **(Equation 16-11)**

$R_1 = 1.2 - 0.001A_t$ for 200 square feet $< A_t <$ 600 square feet **(Equation 16-12)**

For SI: $1.2 - 0.011A_t$ for 18.58 square meters $< A_t <$ 55.74 square meters

$R_1 = 0.6$ for $A_t \ge 600$ square feet (55.74 m^2) **(Equation 16-13)**

where:

A_t = Tributary area (span length multiplied by effective width) in square feet (m^2) supported by the member, and

$R_2 = 1$ for $F \le 4$ **(Equation 16-14)**

$R_2 = 1.2 - 0.05\,F$ for $4 < F < 12$ **(Equation 16-15)**

$R_2 = 0.6$ for $F \ge 12$ **(Equation 16-16)**

where:

F = For a sloped roof, the number of inches of rise per foot (for SI: $F = 0.12 \times$ slope, with slope expressed as a percentage), or for an arch or dome, the rise-to-span ratio multiplied by 32.

1607.14.2.2 Occupiable roofs. Areas of roofs that are occupiable, such as *vegetative roofs*, landscaped roofs or for assembly or other similar purposes, and *marquees* are permitted to have their uniformly distributed *live loads* reduced in accordance with Section 1607.12.

1607.14.3 Awnings and canopies. *Awnings* and canopies shall be designed for uniform *live loads* as required in Table 1607.1 as well as for snow *loads* and wind *loads* as specified in Sections 1608 and 1609.

1607.14.4 Photovoltaic panel systems. Roof structures that provide support for *photovoltaic panel systems* shall be designed in accordance with Sections 1607.14.4.1 through 1607.14.4.5 , as applicable.

1607.14.4.1 Roof live load. Roof structures that support *photovoltaic panel systems* shall be designed to resist each of the following conditions:

1. Applicable uniform and concentrated roof *loads* with the *photovoltaic panel system dead loads*.

   Exception: *Roof live loads* need not be applied to the area covered by *photovoltaic panels* where the clear space between the panels and the roof surface is 24 inches (610 mm) or less.

2. Applicable uniform and concentrated roof loads without the *photovoltaic panel system* present.

1607.14.4.2 Photovoltaic panels or modules. The structure of a roof that supports solar *photovoltaic panels* or modules shall be designed to accommodate the full solar *photovoltaic panels* or modules and ballast *dead load*, including concentrated *loads* from support frames in combination with the *loads* from Section 1607.14.4.1 and other applicable *loads*. Where applicable, snow drift *loads* created by the *photovoltaic panels* or modules shall be included.

1607.14.4.3 Photovoltaic panels installed on open grid roof structures. Structures with open grid framing and without a *roof deck* or sheathing supporting *photovoltaic panel systems* shall be designed to support the uniform and concentrated *roof live loads* specified in Section 1607.14.4.1, except that the uniform *roof live load* shall be permitted to be reduced to 12 psf (0.57 kN/m^2).

1607.14.4.4 Ground-mounted photovoltaic (PV) panel systems. Ground-mounted photovoltaic (PV) panel systems that are independent structures and do not have accessible/occupied space underneath are not required to accommodate a roof photovoltaic *live load*. Other *loads* and combinations in accordance with Section 1605 shall be accommodated.

1607.14.4.5 Ballasted photovoltaic panel systems. Roof structures that provide support for ballasted *photovoltaic panel systems* shall be designed, or analyzed, in accordance with Section 1604.4; checked in accordance with Section 1604.3.6 for deflections; and checked in accordance with Section 1611 for ponding.

1607.15 Crane loads. The crane *live load* shall be the rated capacity of the crane. Design *loads* for the runway beams, including connections and support brackets, of moving bridge cranes and monorail cranes shall include the maximum wheel *loads* of the crane and the vertical impact, lateral and longitudinal forces induced by the moving crane.

1607.15.1 Maximum wheel load. The maximum wheel *loads* shall be the wheel *loads* produced by the weight of the bridge, as applicable, plus the sum of the rated capacity and the weight of the trolley with the trolley positioned on its runway at the location where the resulting *load effect* is maximum.

1607.15.2 Vertical impact force. The maximum wheel *loads* of the crane shall be increased by the following percentages to account for the effects of vertical impact or vibration:

Monorail cranes (powered)	25 percent
Cab-operated or remotely operated bridge cranes (powered)	25 percent
Pendant-operated bridge cranes (powered)	10 percent
Bridge cranes or monorail cranes with hand-geared bridge, trolley and hoist	0 percent

1607.15.3 Lateral force. The lateral force on crane runway beams with electrically powered trolleys shall be calculated as 20 percent of the sum of the rated capacity of the crane and the weight of the hoist and trolley. The lateral force shall be assumed to act horizontally at the traction surface of a runway beam, in either direction perpendicular to the beam, and shall be distributed with due regard to the lateral stiffness of the runway beam and supporting structure.

1607.15.4 Longitudinal force. The longitudinal force on crane runway beams, except for bridge cranes with hand-geared bridges, shall be calculated as 10 percent of the maximum wheel *loads* of the crane. The longitudinal force shall be assumed to act horizontally at the traction surface of a runway beam, in either direction parallel to the beam.

1607.16 Interior walls and partitions. Interior walls and partitions that exceed 6 feet (1829 mm) in height, including their finish materials, shall have adequate strength and stiffness to resist the *loads* to which they are subjected but not less than a horizontal *load* of 5 psf (0.240 kN/m^2).

1607.16.1 Fabric partitions. *Fabric partitions* that exceed 6 feet (1829 mm) in height, including their finish materials, shall have adequate strength and stiffness to resist the following *load* conditions:

1. The horizontal distributed *load* need only be applied to the partition framing. The total area used to determine the distributed *load* shall be the area of the fabric face between the framing members to which the fabric is attached. The total distributed *load* shall be uniformly applied to such framing members in proportion to the length of each member.
2. A concentrated *load* of 40 pounds (0.176 kN) applied to an 8-inch-diameter (203 mm) area [50.3 square inches (32 452 mm^2)] of the fabric face at a height of 54 inches (1372 mm) above the floor.

1607.16.2 Fire walls. In order to meet the structural stability requirements of Section 706.2 where the structure on either side of the wall has collapsed, *fire walls* and their supports shall be designed to withstand a minimum horizontal allowable stress *load* of 5 psf (0.240 kN/m^2).

1607.17 Fixed ladders. Fixed ladders with rungs shall be designed to resist a single concentrated *load* of 300 pounds (1.33 kN) in accordance with Section 4.5.4 of ASCE 7. Where rails of fixed ladders extend above a floor or platform at the top of the ladder, each side rail extension shall be designed to resist a single concentrated *load* of 100 pounds (0.445 kN) in accordance with Section 4.5.4 of ASCE 7. Ship's ladders shall be designed to resist the *stair loads* given in Table 1607.1.

1607.18 Library stack rooms. The live loading indicated in Table 1607.1 for library stack rooms applies to stack room floors that support nonmobile, double-faced library book stacks, subject to the following limitations:

1. The nominal book stack unit height shall not exceed 90 inches (2290 mm).

2. The nominal shelf depth shall not exceed 12 inches (305 mm) for each face.
3. Parallel rows of double-faced book stacks shall be separated by aisles not less than 36 inches (914 mm) in width.

1607.19 Seating for assembly uses. *Bleachers*, *folding and telescopic seating* and *grandstands* shall be designed for the *loads* specified in ICC 300. Stadiums and arenas with fixed seats shall be designed for the horizontal sway *loads* in Section 1607.19.1.

1607.19.1 Horizontal sway loads. The design of stadiums and arenas with fixed seats shall include horizontal swaying forces applied to each row of seats as follows:

1. 24 pounds per linear foot (0.35 kN/m) of seat applied in a direction parallel to each row of seats.
2. 10 pounds per linear foot (0.15 kN/m) of seat applied in a direction perpendicular to each row of seats.

The parallel and perpendicular horizontal swaying forces are not required to be applied simultaneously.

1607.20 Sidewalks, vehicular driveways, and yards subject to trucking. The live loading indicated in Table 1607.1 for sidewalks, vehicular driveways, and yards subject to trucking shall comply with the requirements of this section.

1607.20.1 Uniform loads. In addition to the *loads* indicated in Table 1607.1, other uniform *loads* in accordance with an approved method that contains provisions for truck loading shall be considered where appropriate.

1607.20.2 Concentrated loads. The concentrated wheel *load* indicated in Table 1607.1 shall be applied on an area of $4^1/_2$ inches by $4^1/_2$ inches (114 mm by 114 mm).

1607.21 Stair treads. The concentrated *load* indicated in Table 1607.1 for *stair* treads shall be applied on an area of 2 inches by 2 inches (51 mm by 51 mm). This *load* need not be assumed to act concurrently with the uniform *load*.

1607.22 Residential attics. The *live loads* indicated in Table 1607.1 for *attics* in residential occupancies shall comply with the requirements of this section.

1607.22.1 Uninhabitable attics without storage. In residential occupancies, uninhabitable *attic* areas without storage are those where the maximum clear height between the joists and rafters is less than 42 inches (1067 mm), or where there are not two or more adjacent trusses with web configurations capable of accommodating an assumed rectangle 42 inches (1067 mm) in height by 24 inches (610 mm) in width, or greater, within the plane of the trusses. The *live load* in Table 1607.1 need not be assumed to act concurrently with any other *live load* requirement.

1607.22.2 Uninhabitable attics with storage. In residential occupancies, uninhabitable attic areas with storage are those where the maximum clear height between the joist and rafter is 42 inches (1067 mm) or greater, or where there are two or more adjacent trusses with web configurations capable of accommodating an assumed rectangle 42 inches (1067 mm) in height by 24 inches (610 mm) in width, or greater, within the plane of the trusses. The live load in Table 1607.1 need only be applied to those portions of the joists or truss bottom chords where both of the following conditions are met:

1. The attic area is accessed from an opening not less than 20 inches (508 mm) in width by 30 inches (762 mm) in length that is located where the clear height in the attic is not less than 30 inches (762 mm).
2. The slope of the joists or truss bottom chords is not greater than 2 units vertical in 12 units horizontal.

The remaining portions of the joists or truss bottom chords shall be designed for a uniformly distributed concurrent live load of not less than 10 pounds per square foot (0.48 kN/m^2).

1607.22.3 Attics served by stairs. *Attic* spaces served by *stairways* other than the pull-down type shall be designed to support the minimum *live load* specified for habitable *attics* and sleeping rooms.

SECTION 1608
SNOW LOADS

1608.1 General. Design snow *loads* shall be determined in accordance with Chapter 7 of ASCE 7, but the design roof *load* shall be not less than that determined by Section 1607.

1608.2 Ground snow loads. The ground snow *loads* to be used in determining the design snow *loads* for roofs shall be determined in accordance with ASCE 7 or Figures 1608.2(1) and 1608.2(2) for the contiguous United States and Table 1608.2 for Alaska. Site-specific case studies shall be made in areas designated "CS" in Figures 1608.2(1) and 1608.2(2). Ground snow *loads* for sites at elevations above the limits indicated in Figures 1608.2(1) and 1608.2(2) and for all sites within the CS areas shall be *approved*. Ground snow *load* determination for such sites shall be based on an extreme value statistical analysis of data available in the vicinity of the site using a value with a 2-percent annual probability of being exceeded (50-year mean recurrence interval). Snow loads are zero for Hawaii, except in mountainous regions as *approved* by the *building official*.

1608.3 Ponding instability. *Susceptible bays* of roofs shall be evaluated for ponding instability in accordance with Chapters 7 and 8 of ASCE 7.

SECTION 1609
WIND LOADS

1609.1 Applications. Buildings, structures and parts thereof shall be designed to withstand the minimum wind *loads* prescribed herein. Decreases in wind *loads* shall not be made for the effect of shielding by *other structures*.

TABLE 1608.2
GROUND SNOW LOADS, p_g, FOR ALASKAN LOCATIONS

LOCATION	POUNDS PER SQUARE FOOT
Adak	30
Anchorage	50
Angoon	70
Barrow	25
Barter Island	35
Bethel	40
Big Delta	50
Cold Bay	25
Cordova	100
Fairbanks	60
Fort Yukon	60
Galena	60
Gulkana	70
Homer	40
Juneau	60
Kenai	70
Kodiak	30
Kotzebue	60
McGrath	70
Nenana	80
Nome	70
Palmer	50
Petersburg	150
St. Paul Islands	40
Seward	50
Shemya	25
Sitka	50
Talkeetna	120
Unalakleet	50
Valdez	160
Whittier	300
Wrangell	60
Yakutat	150

For SI:1 pound per square foot = 0.0479 kN/m^2.

1609.1.1 Determination of wind loads. Wind *loads* on every building or structure shall be determined in accordance with Chapters 26 to 30 of ASCE 7. The type of opening protection required, the basic design *wind speed*, *V*, and the exposure category for a site is permitted to be determined in accordance with Section 1609 or ASCE 7. Wind shall be assumed to come from any horizontal direction and wind pressures shall be assumed to act normal to the surface considered.

Exceptions:

1. Subject to the limitations of Section 1609.1.1.1, the provisions of ICC 600 shall be permitted for applicable Group R-2 and R-3 buildings.
2. Subject to the limitations of Section 1609.1.1.1, residential structures using the provisions of AWC WFCM.
3. Subject to the limitations of Section 1609.1.1.1, residential structures using the provisions of AISI S230.
4. Designs using NAAMM FP 1001.
5. Designs using TIA-222 for antenna-supporting structures and antennas, provided that the horizontal extent of Topographic Category 2 escarpments in Section 2.6.6.2 of TIA-222 shall be 16 times the height of the escarpment.
6. Wind tunnel tests in accordance with ASCE 49 and Sections 31.4 and 31.5 of ASCE 7.

The wind speeds in Figures 1609.3(1) through 1609.3(12) are basic design wind speeds, *V*, and shall be converted in accordance with Section 1609.3.1 to allowable stress design wind speeds, V_{asd}, when the provisions of the standards referenced in Exceptions 4 and 5 are used.

1609.1.1.1 Applicability. The provisions of ICC 600 are applicable only to buildings located within Exposure B or C as defined in Section 1609.4. The provisions of ICC 600, AWC WFCM and AISI S230 shall not apply to buildings sited on the upper half of an isolated hill, ridge or escarpment meeting all of the following conditions:

1. The hill, ridge or escarpment is 60 feet (18 288 mm) or higher if located in Exposure B or 30 feet (9144 mm) or higher if located in Exposure C.
2. The maximum average slope of the hill exceeds 10 percent.
3. The hill, ridge or escarpment is unobstructed upwind by other such topographic features for a distance from the high point of 50 times the height of the hill or 2 miles (3.22 km), whichever is greater.

1609.2 Protection of openings. In *windborne debris regions*, glazing in buildings shall be impact resistant or protected with an impact-resistant covering meeting the requirements of an *approved* impact-resistant standard or ASTM E1996 referenced herein as follows:

1. Glazed openings located within 30 feet (9144 mm) of grade shall meet the requirements of the large missile test of ASTM E1996.
2. Glazed openings located more than 30 feet (9144 mm) above grade shall meet the provisions of the small missile test of ASTM E1996.

Exceptions:

1. *Wood structural panels* with a minimum thickness of $^{7}/_{16}$ inch (11.1 mm) and maximum panel span of 8 feet (2438 mm) shall be permitted for opening protection in buildings with a mean roof height of 33 feet (10 058 mm) or less that are classified as a Group R-3 or R-4 occupancy. Panels shall be

precut so that they shall be attached to the framing surrounding the opening containing the product with the glazed opening. Panels shall be predrilled as required for the anchorage method and shall be secured with the attachment hardware provided. Attachments shall be designed to resist the components and cladding *loads* determined in accordance with the provisions of ASCE 7, with corrosion-resistant attachment hardware provided and anchors permanently installed on the building. Attachment in accordance with Table 1609.2 with corrosion-resistant attachment hardware provided and anchors permanently installed on the building is permitted for buildings with a mean roof height of 45 feet (13 716 mm) or less where V_{asd} determined in accordance with Section 1609.3.1 does not exceed 140 mph (63 m/s).

2. Glazing in *Risk Category* I buildings, including *greenhouses* that are occupied for growing plants on a production or research basis, without public access shall be permitted to be unprotected.
3. Glazing in *Risk Category* II, III or IV buildings located over 60 feet (18 288 mm) above the ground and over 30 feet (9144 mm) above *aggregate* surface roofs located within 1,500 feet (458 m) of the building shall be permitted to be unprotected.

1609.2.1 Louvers. Louvers protecting intake and exhaust ventilation ducts not assumed to be open that are located within 30 feet (9144 mm) of grade shall meet the requirements of AMCA 540.

1609.2.2 Application of ASTM E1996. The text of Section 6.2.2 of ASTM E1996 shall be substituted as follows:

6.2.2 Unless otherwise specified, select the wind zone based on the basic design *wind speed, V*, as follows:

6.2.2.1 *Wind Zone 1*—130 mph ≤ basic design *wind speed, V* < 140 mph.

6.2.2.2 *Wind Zone 2*—140 mph ≤ basic design *wind speed, V* < 150 mph at greater than one mile (1.6 km) from the coastline. The coastline shall be measured from the mean high water mark.

6.2.2.3 *Wind Zone 3*—150 mph (67 m/s) ≤ basic design *wind speed, V* ≤ 160 mph (72 m/s), or 140 mph (63 m/s) ≤ basic design *wind speed, V* ≤ 160 mph (72 m/s) and within one mile (1.6 km) of the coastline. The coastline shall be measured from the mean high water mark.

6.2.2.4 *Wind Zone 4*— basic design *wind speed, V* > 160 mph (72 m/s).

1609.2.3 Garage doors. Garage door glazed opening protection for windborne debris shall meet the requirements of an *approved* impact-resisting standard or ANSI/DASMA 115.

1609.3 Basic design wind speed. The basic design *wind speed, V*, in mph, for the determination of the wind *loads* shall be determined by Figures 1609.3(1) through 1609.3(12). The basic design *wind speed, V*, for use in the design of *Risk Category* II buildings and structures shall be obtained from Figures 1609.3(1), 1609.3(5) and 1609.3(6). The basic design *wind speed, V*, for use in the design of *Risk Category* III buildings and structures shall be obtained from Figures 1609.3(2), 1609.3(7) and 1609.3(8). The basic design *wind speed, V*, for use in the design of *Risk Category* IV buildings and structures shall be obtained from Figures 1609.3(3), 1609.3(9) and 1609.3(10). The basic design *wind speed, V*, for use in the design of *Risk Category* I buildings and structures shall be obtained from Figures 1609.3(4), 1609.3(11) and 1609.3(12). The basic design *wind speed, V*, for the special wind regions indicated near mountainous terrain and near gorges shall be in accordance with local jurisdiction requirements. The basic design wind speeds, *V*, determined by the local jurisdiction shall be in accordance with Chapter 26 of ASCE 7.

In nonhurricane-prone regions, when the basic design *wind speed, V*, is estimated from regional climatic data, the basic design *wind speed, V*, shall be determined in accordance with Chapter 26 of ASCE 7.

TABLE 1609.2
WINDBORNE DEBRIS PROTECTION FASTENING SCHEDULE FOR WOOD STRUCTURAL PANELS[a, b, c, d]

FASTENER TYPE	FASTENER SPACING (inches)		
	Panel Span ≤ 4 feet	4 feet < Panel Span ≤ 6 feet	6 feet < Panel Span ≤ 8 feet
No. 8 wood-screw-based anchor with 2-inch embedment length	16	10	8
No. 10 wood-screw-based anchor with 2-inch embedment length	16	12	9
$^1/_4$-inch diameter lag-screw-based anchor with 2-inch embedment length	16	16	16

For SI: 1 inch = 25.4 mm, 1 foot = 304.8 mm, 1 pound = 4.448 N, 1 mile per hour = 0.447 m/s.

a. This table is based on 140 mph wind speeds and a 45-foot mean roof height.

b. Fasteners shall be installed at opposing ends of the wood structural panel. Fasteners shall be located not less than 1 inch from the edge of the panel.

c. Anchors shall penetrate through the exterior wall covering with an embedment length of 2 inches minimum into the building frame. Fasteners shall be located not less than $2^1/_2$ inches from the edge of concrete block or concrete.

d. Where panels are attached to masonry or masonry/stucco, they shall be attached using vibration-resistant anchors having a minimum ultimate withdrawal capacity of 1,500 pounds.

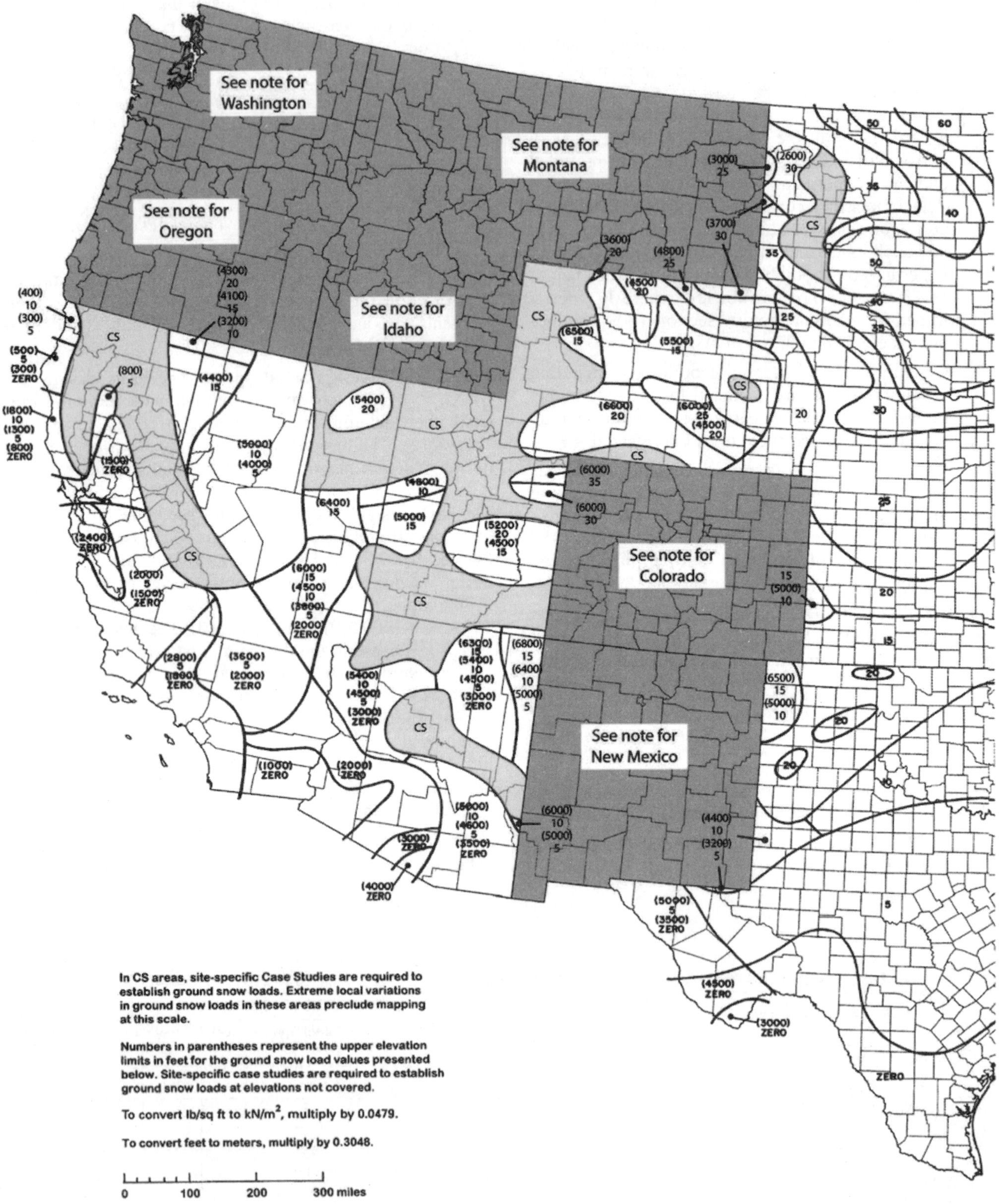

NOTE: See ASCE 7 Table 7.2-2 for Colorado, Table 7.2-3 for Idaho, Table 7.2-4 for Montana, Table 7.2-5 for Washington, Table 7.2-6 for New Mexico and Table 7.2-7 for Oregon.

FIGURE 1608.2(1)
GROUND SNOW LOADS, p_g, FOR THE UNITED STATES (psf)

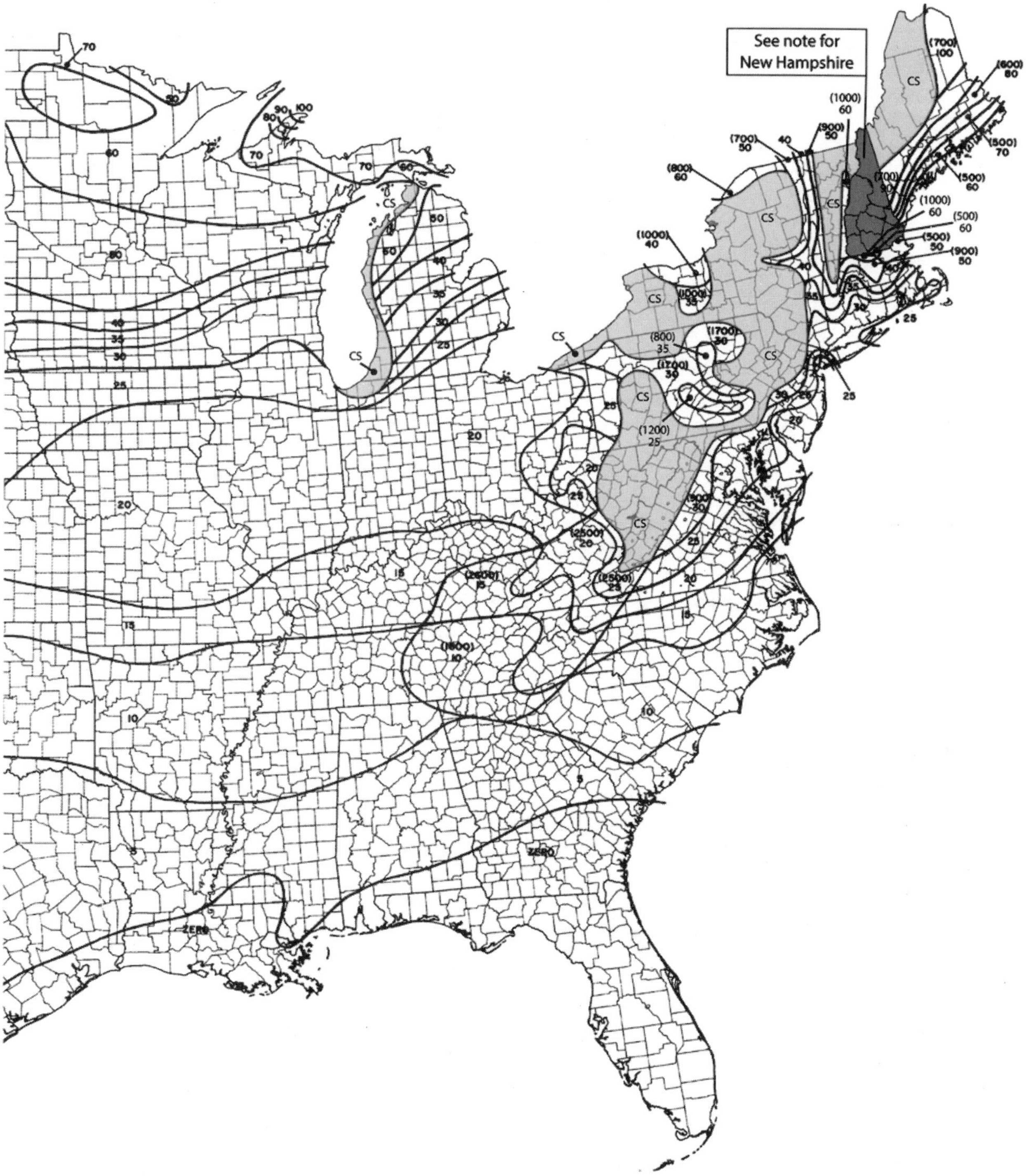

NOTE: See ASCE 7 Table 7.2-8 for New Hampshire.

FIGURE 1608.2(2)
GROUND SNOW LOADS, p_g, FOR THE UNITED STATES (psf)

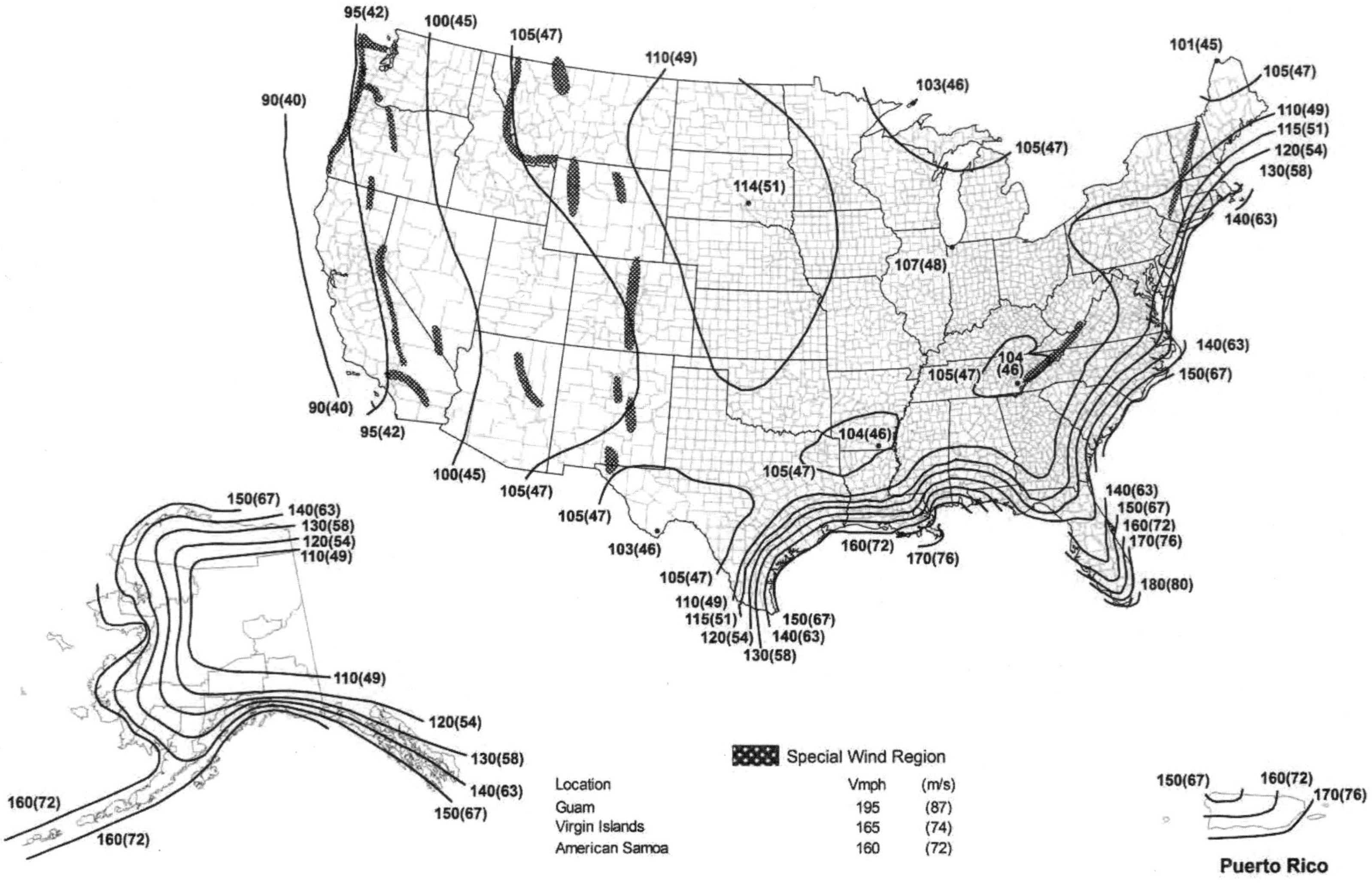

Location	Vmph	(m/s)
Guam	195	(87)
Virgin Islands	165	(74)
American Samoa	160	(72)

Notes:

1. Values are nominal design 3-second gust wind speeds in miles per hour (m/s) at 33 feet (10 m) above ground for Exposure C Category.
2. Linear interpolation between contours. Point values are provided to aid with interpolation.
3. Islands, coastal areas, and land boundaries outside the last contour shall use the last wind speed contour.
4. Mountainous terrain, gorges, ocean promontories, and special wind regions shall be examined for unusual wind conditions.
5. Wind speeds correspond to approximately a 7% probability of exceedance in 50 years (Annual Exceedance Probability = 0.00143, MRI = 700 Years).
6. Location-specific basic wind speeds shall be permitted to be determined using www.atcouncil.org/windspeed

FIGURE 1609.3(1)
BASIC DESIGN WIND SPEEDS, *V*, FOR RISK CATEGORY II BUILDINGS AND OTHER STRUCTURES

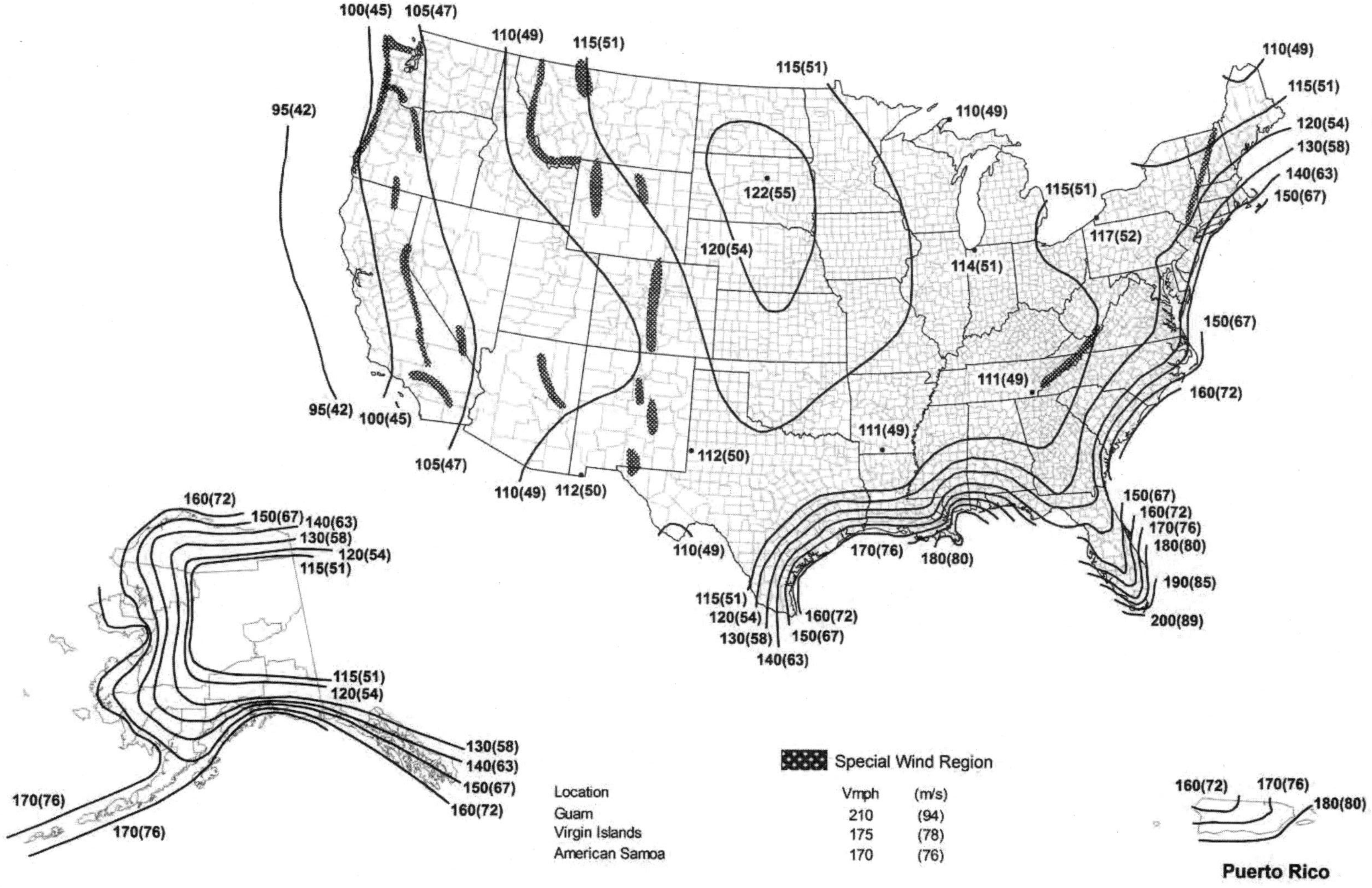

Location	Vmph	(m/s)
Guam	210	(94)
Virgin Islands	175	(78)
American Samoa	170	(76)

Notes:

1. Values are nominal design 3-second gust wind speeds in miles per hour (m/s) at 33 feet (10 m) above ground for Exposure C Category.
2. Linear interpolation between contours. Point values are provided to aid with interpolation.
3. Islands, coastal areas, and land boundaries outside the last contour shall use the last wind speed contour.
4. Mountainous terrain, gorges, ocean promontories, and special wind regions shall be examined for unusual wind conditions.
5. Wind speeds correspond to approximately a 3% probability of exceedance in 50 years (Annual Exceedance Probability = 0.00143, MRI = 700 Years).
6. Location-specific basic wind speeds shall be permitted to be determined using www.atcouncil.org/windspeed

FIGURE 1609.3(2)
BASIC DESIGN WIND SPEEDS, *V*, FOR RISK CATEGORY III BUILDINGS AND OTHER STRUCTURES

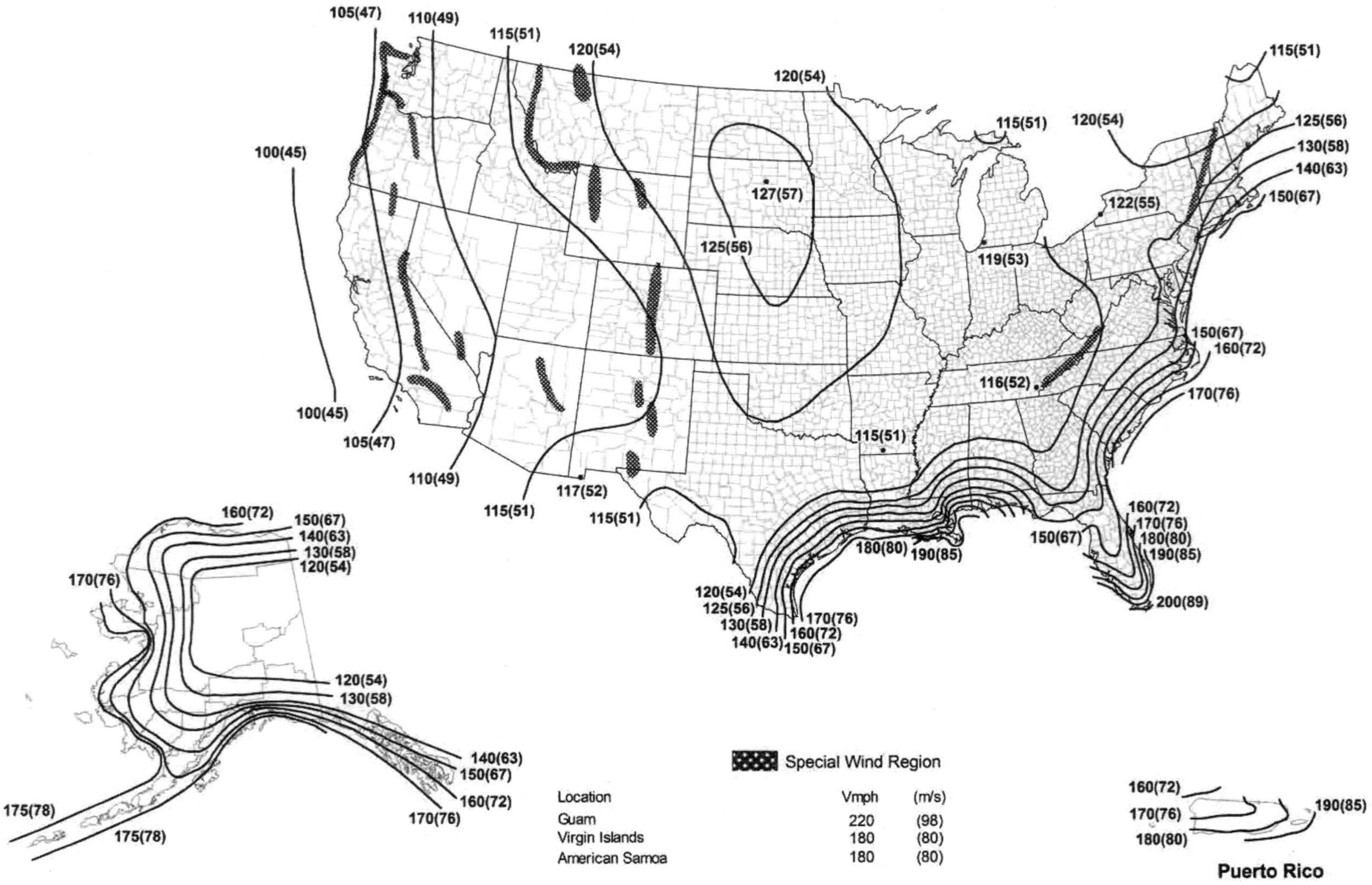

Notes:

1. Values are nominal design 3-second gust wind speeds in miles per hour (m/s) at 33 feet (10 m) above ground for Exposure C Category.
2. Linear interpolation between contours. Point values are provided to aid with interpolation.
3. Islands, coastal areas, and land boundaries outside the last contour shall use the last wind speed contour.
4. Mountainous terrain, gorges, ocean promontories, and special wind regions shall be examined for unusual wind conditions.
5. Wind speeds correspond to approximately a 1.6% probability of exceedance in 50 years (Annual Exceedance Probability = 0.00033, MRI = 3000 Years).
6. Location-specific basic wind speeds shall be permitted to be determined using www.atcouncil.org/windspeed

FIGURE 1609.3(3)
BASIC DESIGN WIND SPEEDS, *V*, FOR RISK CATEGORY IV BUILDINGS AND OTHER STRUCTURES

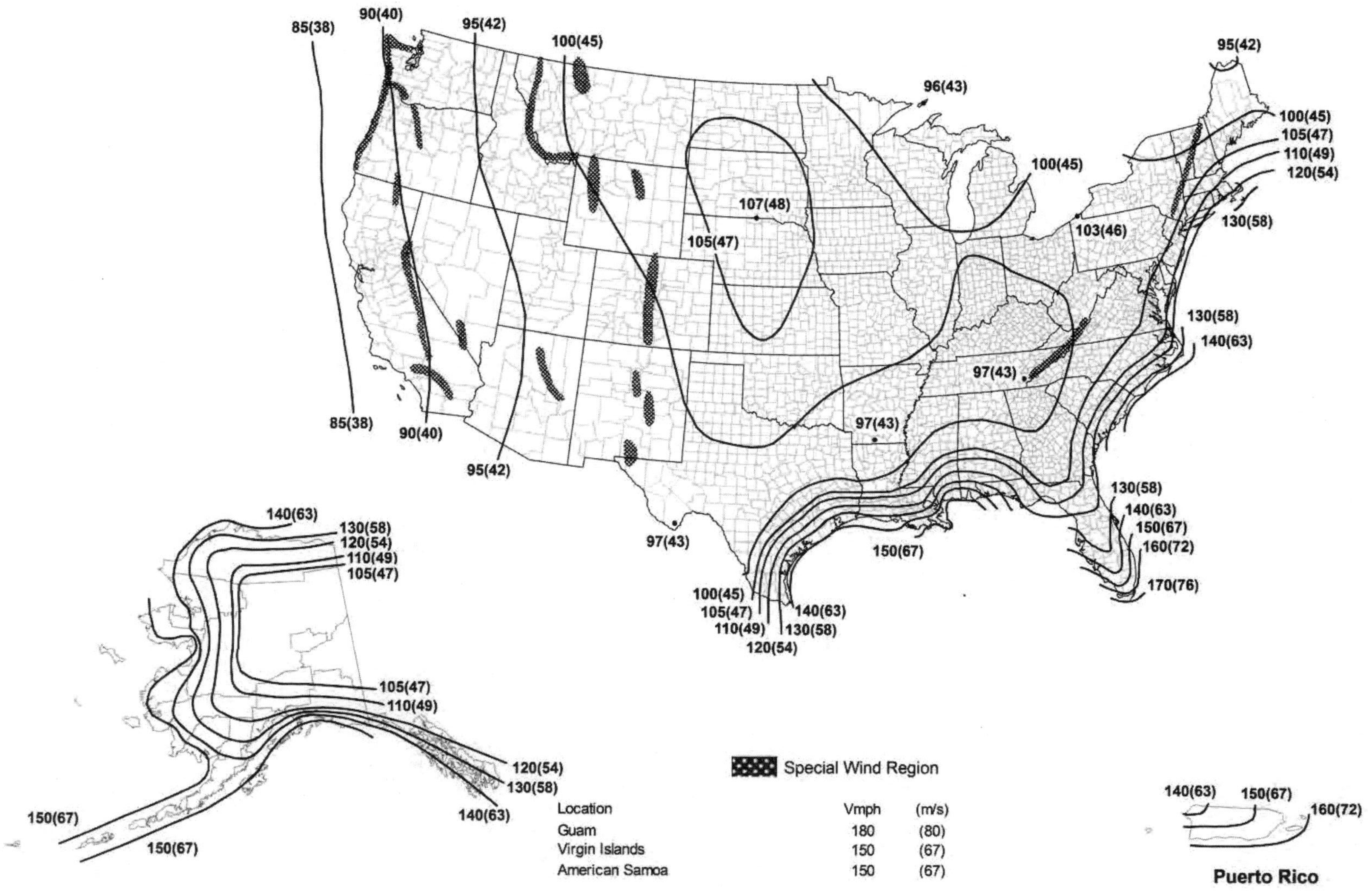

Location	Vmph	(m/s)
Guam	180	(80)
Virgin Islands	150	(67)
American Samoa	150	(67)

Notes:

1. Values are nominal design 3-second gust wind speeds in miles per hour (m/s) at 33 feet (10 m) above ground for Exposure C Category.
2. Linear interpolation between contours. Point values are provided to aid with interpolation.
3. Islands, coastal areas, and land boundaries outside the last contour shall use the last wind speed contour.
4. Mountainous terrain, gorges, ocean promontories, and special wind regions shall be examined for unusual wind conditions.
5. Wind speeds correspond to approximately a 15% probability of exceedance in 50 years (Annual Exceedance Probability = 0.00333, MRI = 300 Years).
6. Location-specific basic wind speeds shall be permitted to be determined using www.atcouncil.org/windspeed.

FIGURE 1609.3(4)
BASIC DESIGN WIND SPEEDS, *V*, FOR RISK CATEGORY I BUILDINGS AND OTHER STRUCTURES

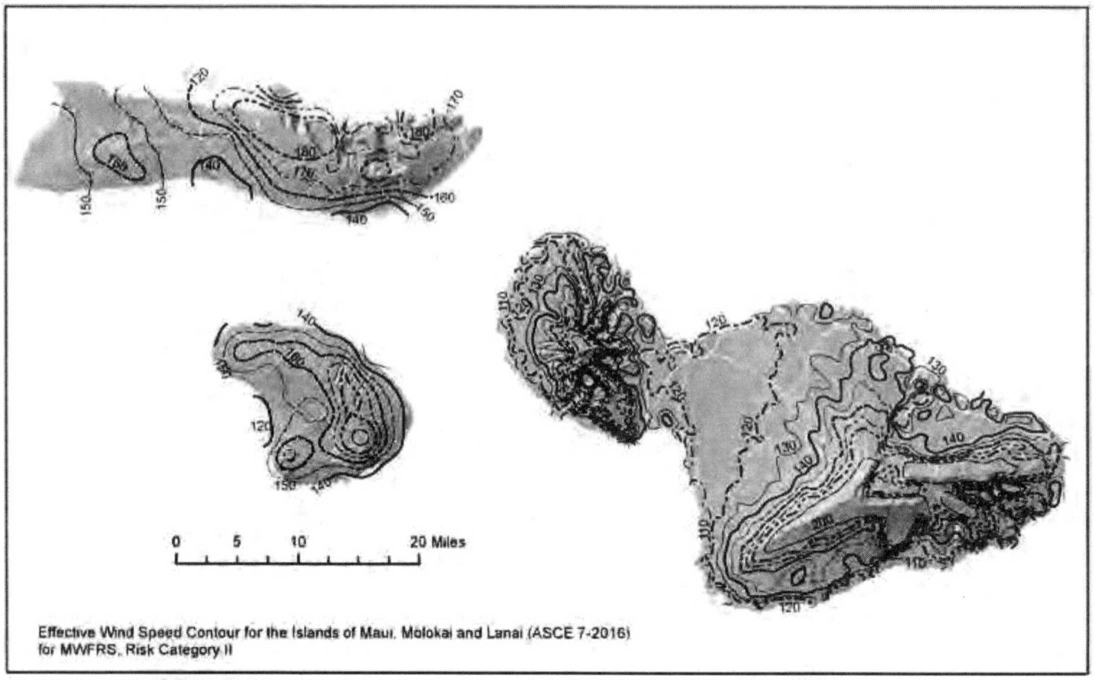

Effective Wind Speed Contour for the Islands of Maui, Molokai and Lanai (ASCE 7-2016) for MWFRS, Risk Category II

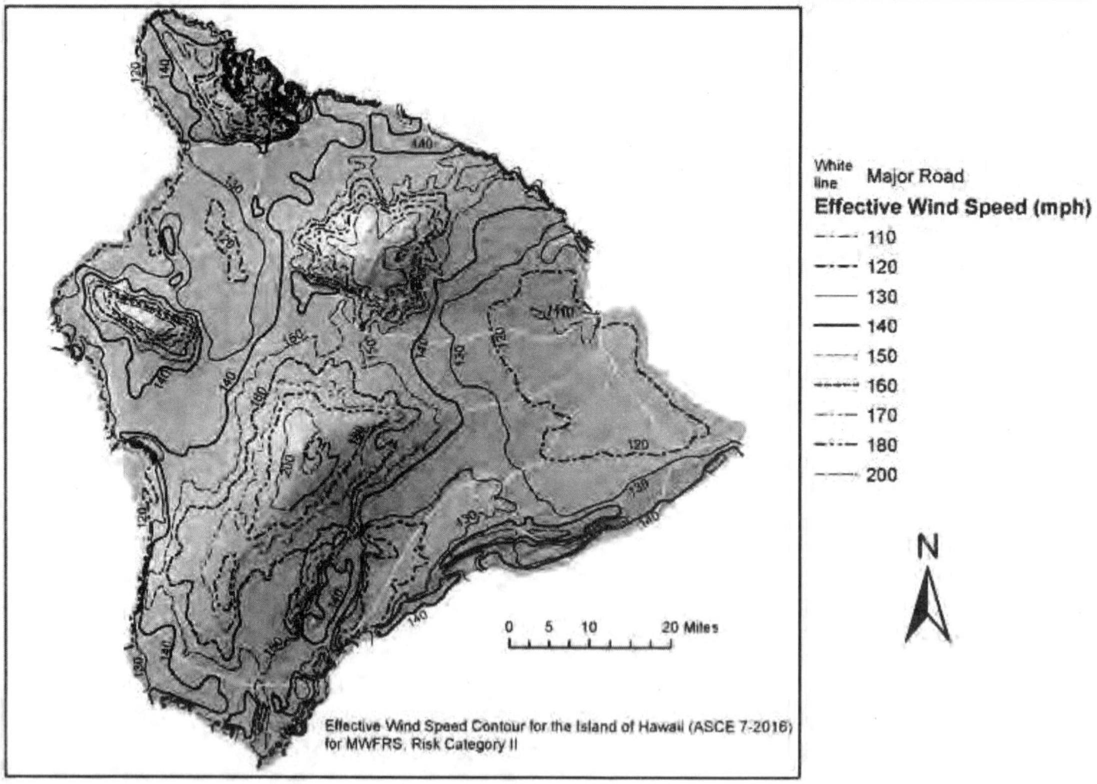

Effective Wind Speed Contour for the Island of Hawaii (ASCE 7-2016) for MWFRS, Risk Category II

Notes:

1. Values are nominal design 3-second gust wind speeds in miles per hour at 33 feet (10 m) above ground for Exposure C Category.
2. Linear interpolation between contours is permitted.
3. Islands and coastal areas outside the last contour shall use the las wind speed contour of the coastal area.
4. It is permitted to use the standard values of K_{zt} of 1.0 and K_d as given in Table 26.6-1 of ASCE 7.
5. Ocean promontories and local escarpments shall be examined for unusual wind conditions.
6. Wind speeds correspond to approximately a 7% probability of exceedance in 50 years (Annual Exceedance Probability = 0.00143, MRI = 700 Years).

FIGURE 1609.3(5)
BASIC DESIGN WIND SPEEDS, *V*, FOR RISK CATEGORY II BUILDINGS AND OTHER STRUCTURES IN HAWAI

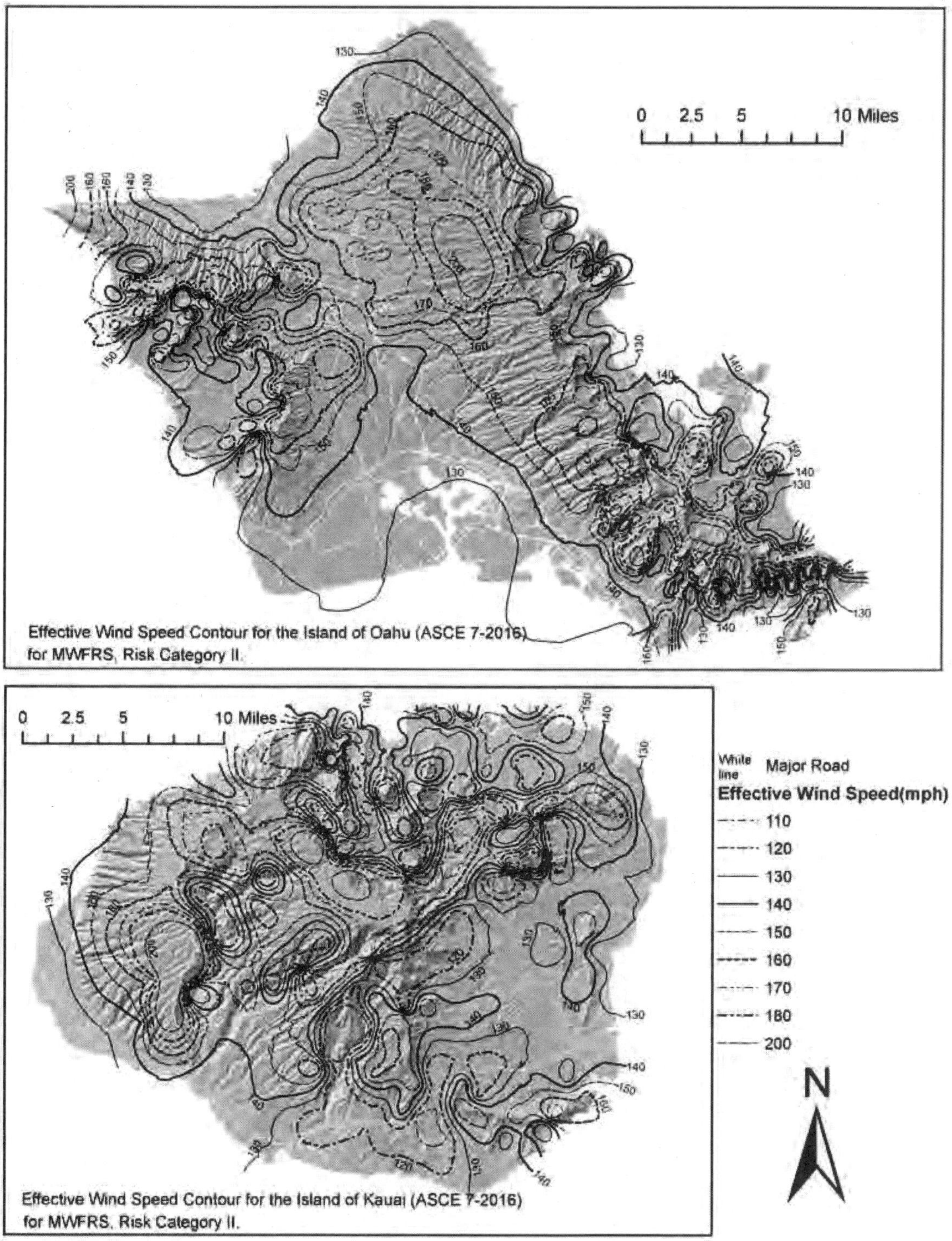

Effective Wind Speed Contour for the Island of Oahu (ASCE 7-2016) for MWFRS, Risk Category II.

Effective Wind Speed Contour for the Island of Kauai (ASCE 7-2016) for MWFRS, Risk Category II.

Notes:

1. Values are nominal design 3-second gust wind speeds in miles per hour at 33 feet (10 m) above ground for Exposure C Category.
2. Linear interpolation between contours is permitted.
3. Islands and coastal areas outside the last contour shall use the las wind speed contour of the coastal area.
4. It is permitted to use the standard values of K_{zt} of 1.0 and K_d as given in Table 26.6-1 of ASCE 7.
5. Ocean promontories and local escarpments shall be examined for unusual wind conditions.
6. Wind speeds correspond to approximately a 7% probability of exceedance in 50 years (Annual Exceedance Probability = 0.00143, MRI = 700 Years).

FIGURE 1609.3(6)
BASIC DESIGN WIND SPEEDS, *V*, FOR RISK CATEGORY II BUILDINGS AND OTHER STRUCTURES IN HAWAII (OAHU, KAUAI)

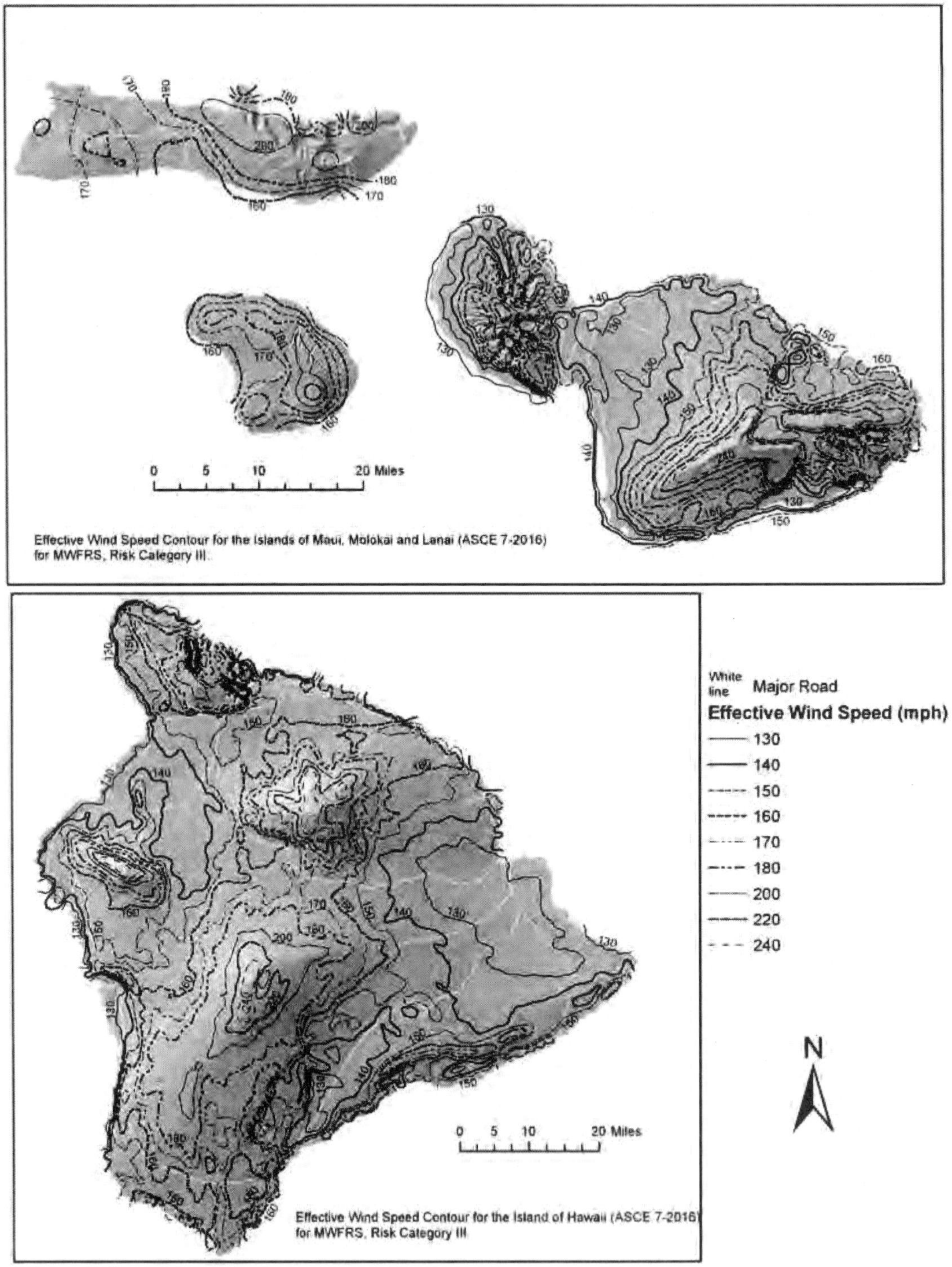

Notes:

1. Values are nominal design 3-second gust wind speeds in miles per hour at 33 feet (10 m) above ground for Exposure C Category.
2. Linear interpolation between contours is permitted.
3. Islands and coastal areas outside the last contour shall use the las wind speed contour of the coastal area.
4. It is permitted to use the standard values of K_{zt} of 1.0 and K_d as given in Table 26.6-1 of ASCE 7.
5. Ocean promontories and local escarpments shall be examined for unusual wind conditions.
6. Wind speeds correspond to approximately a 3% probability of exceedance in 50 years (Annual Exceedance Probability = 0.00143, MRI = 700 Years).

FIGURE 1609.3(7)
BASIC DESIGN WIND SPEEDS, *V*, FOR RISK CATEGORY III BUILDINGS AND OTHER STRUCTURES IN HAWAII

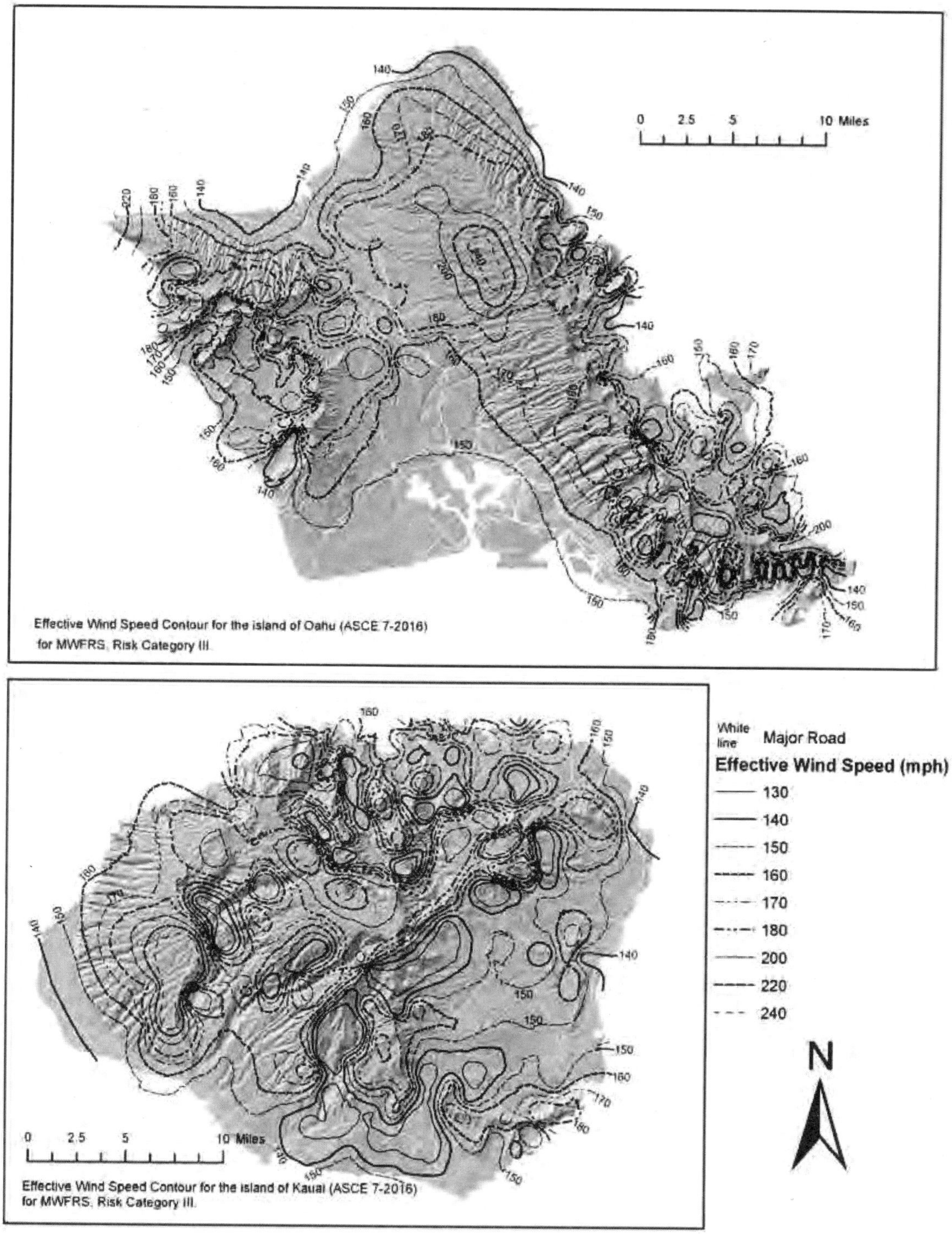

Notes:

1. Values are nominal design 3-second gust wind speeds in miles per hour at 33 feet (10 m) above ground for Exposure C Category.
2. Linear interpolation between contours is permitted.
3. Islands and coastal areas outside the last contour shall use the las wind speed contour of the coastal area.
4. It is permitted to use the standard values of K_{zt} of 1.0 and K_d as given in Table 26.6-1 of ASCE 7.
5. Ocean promontories and local escarpments shall be examined for unusual wind conditions.
6. Wind speeds correspond to approximately a 3% probability of exceedance in 50 years (Annual Exceedance Probability = 0.00143, MRI = 700 Years).

FIGURE 1609.3(8)
BASIC DESIGN WIND SPEEDS, *V*, FOR RISK CATEGORY III BUILDINGS AND OTHER STRUCTURES IN HAWAII (OAHU, KAUAI)

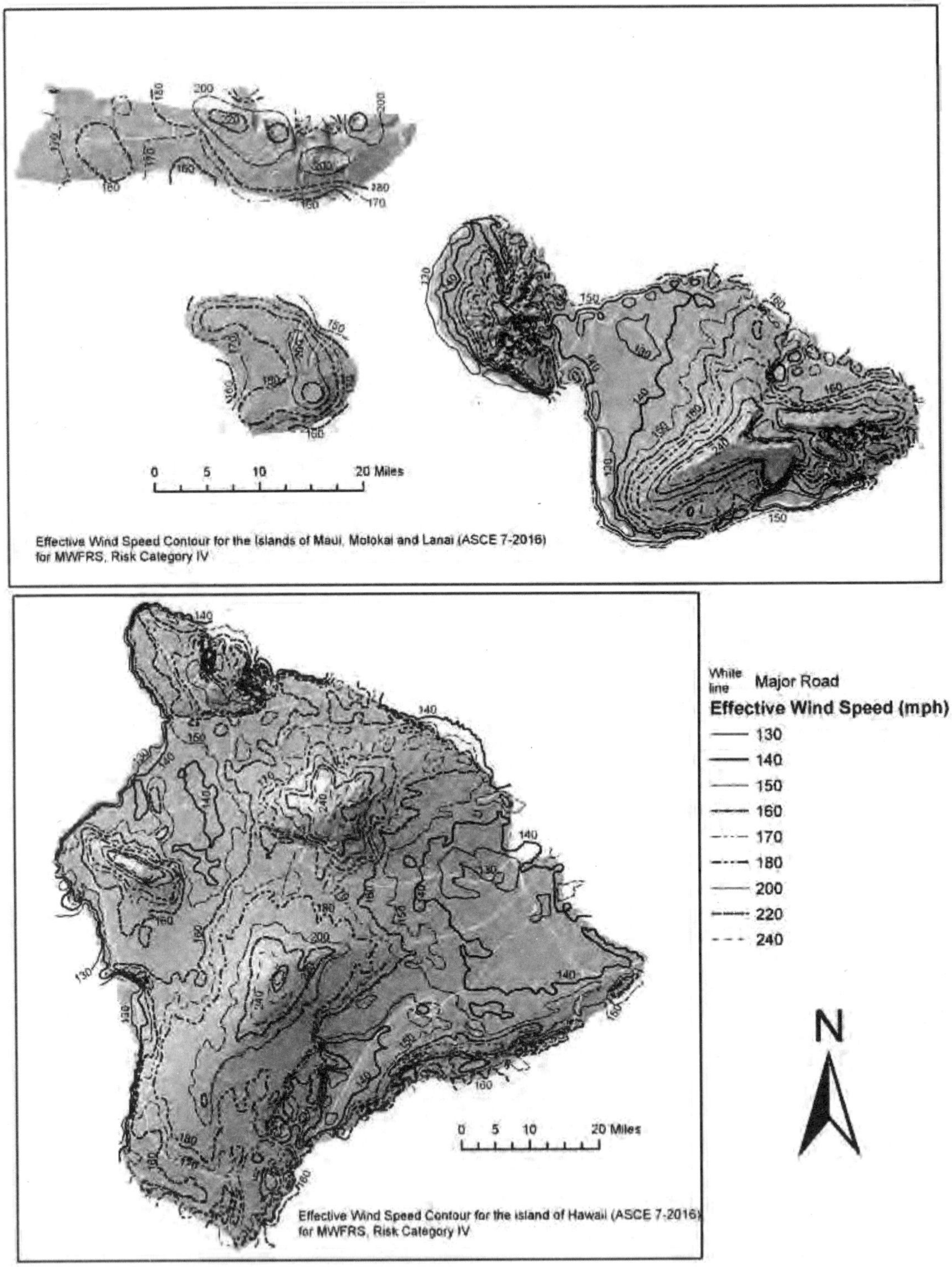

Notes:

1. Values are nominal design 3-second gust wind speeds in miles per hour at 33 feet (10 m) above ground for Exposure C category.
2. Linear interpolation between contours is permitted.
3. Islands and coastal areas outside the last contour shall use the last wind speed contour of the coastal area.
4. It is permitted to use the standard values of K_{zt} of 1.0 and K_d as given in Table 26.6-1 of ASCE 7.
5. Ocean promontories and local escarpments shall be examined for unusual wind conditions.
6. Wind speeds correspond to approximately a 1.7% probability of exceedance in 50 years (Annual Exceedance Probability = 0.00143, MRI = 700 Years).

FIGURE 1609.3(9)
BASIC DESIGN WIND SPEEDS, *V*, FOR RISK CATEGORY IV BUILDINGS AND OTHER STRUCTURES IN HAWAII

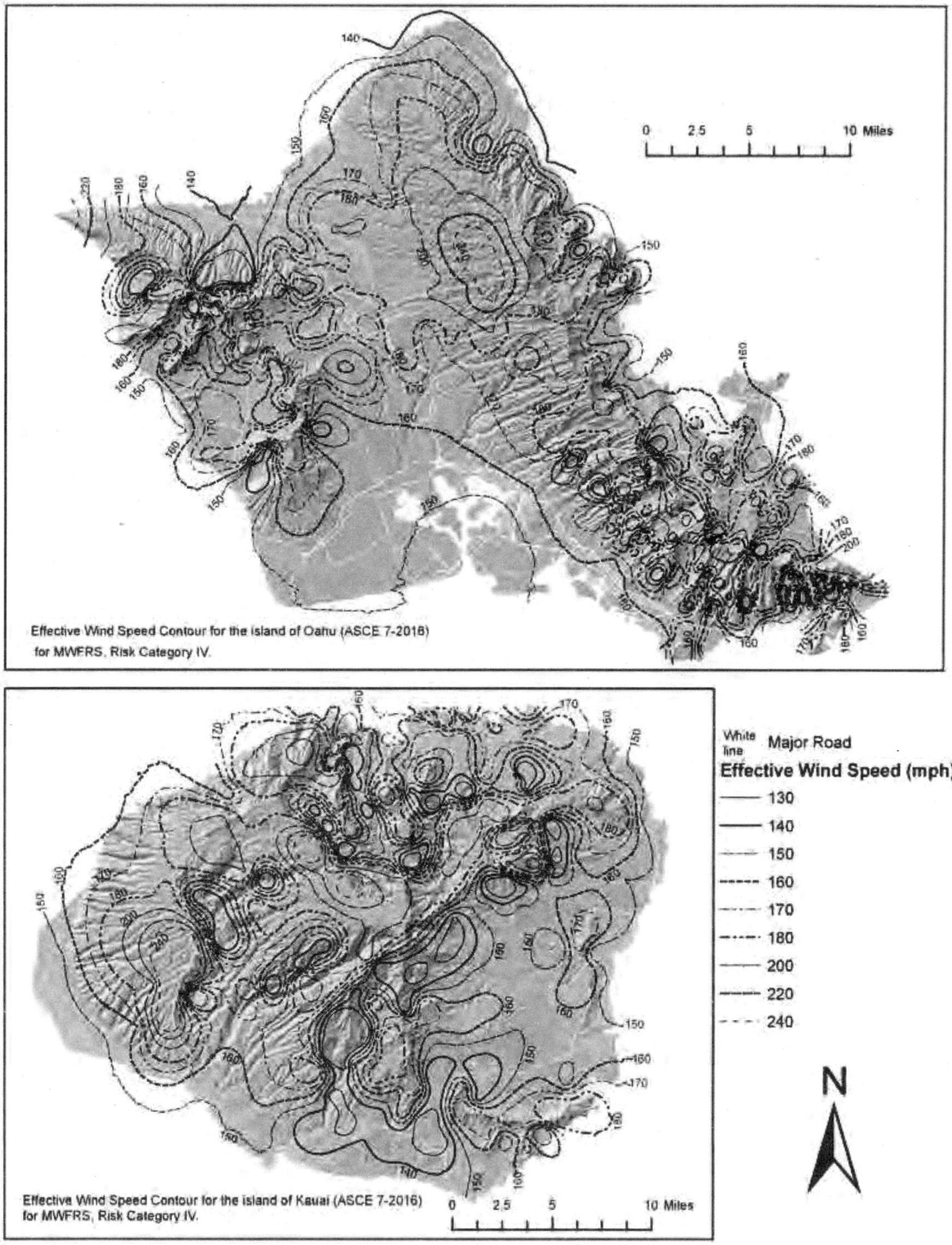

Notes:

1. Values are nominal design 3-second gust wind speeds in miles per hour at 33 feet (10 m) above ground for Exposure C category.
2. Linear interpolation between contours is permitted.
3. Islands and coastal areas outside the last contour shall use the last wind speed contour of the coastal area.
4. It is permitted to use the standard values of K_{zt} of 1.0 and K_d as given in Table 26.6-1 of ASCE 7.
5. Ocean promontories and local escarpments shall be examined for unusual wind conditions.
6. Wind speeds correspond to approximately a 1.7% probability of exceedance in 50 years (Annual Exceedance Probability = 0.00143, MRI = 700 Years).

FIGURE 1609.3(10)
BASIC DESIGN WIND SPEEDS, *V*, FOR RISK CATEGORY IV BUILDINGS AND OTHER STRUCTURES IN HAWAII (OAHU, KAUAI)

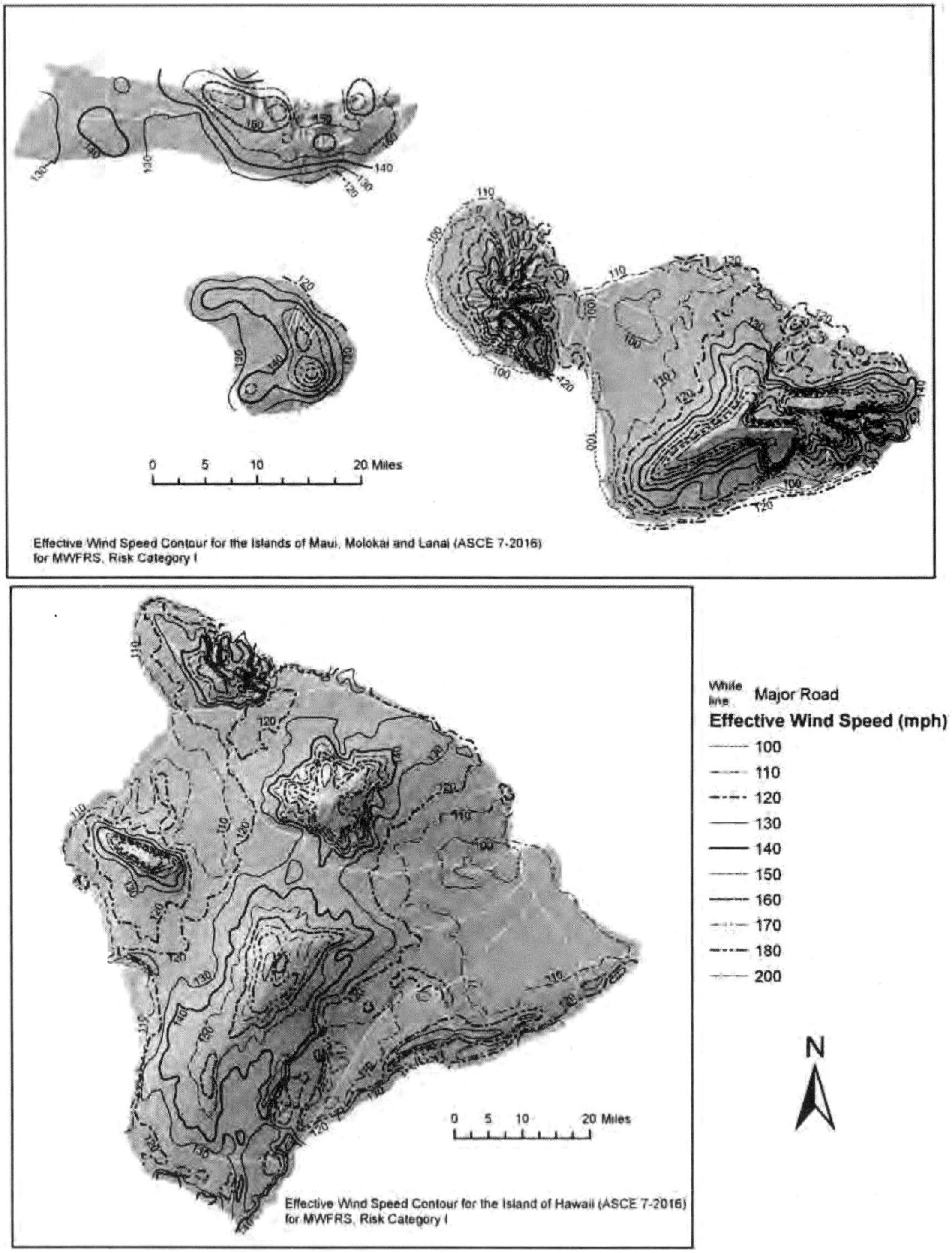

Notes:

1. Values are nominal design 3-second gust wind speeds in miles per hour at 33 feet (10 m) above ground for Exposure C category.
2. Linear interpolation between contours is permitted.
3. Islands and coastal areas outside the last contour shall use the last wind speed contour of the coastal area.
4. It is permitted to use the standard values of K_{zt} of 1.0 and K_d as given in Table 26.6-1 of ASCE 7.
5. Ocean promontories and local escarpments shall be examined for unusual wind conditions.
6. Wind speeds correspond to approximately a 15% probability of exceedance in 50 years (Annual Exceedance Probability = 0.00143, MRI = 700 Years).

FIGURE 1609.3(11)
BASIC DESIGN WIND SPEEDS, *V*, FOR RISK CATEGORY I BUILDINGS AND OTHER STRUCTURES IN HAWAII

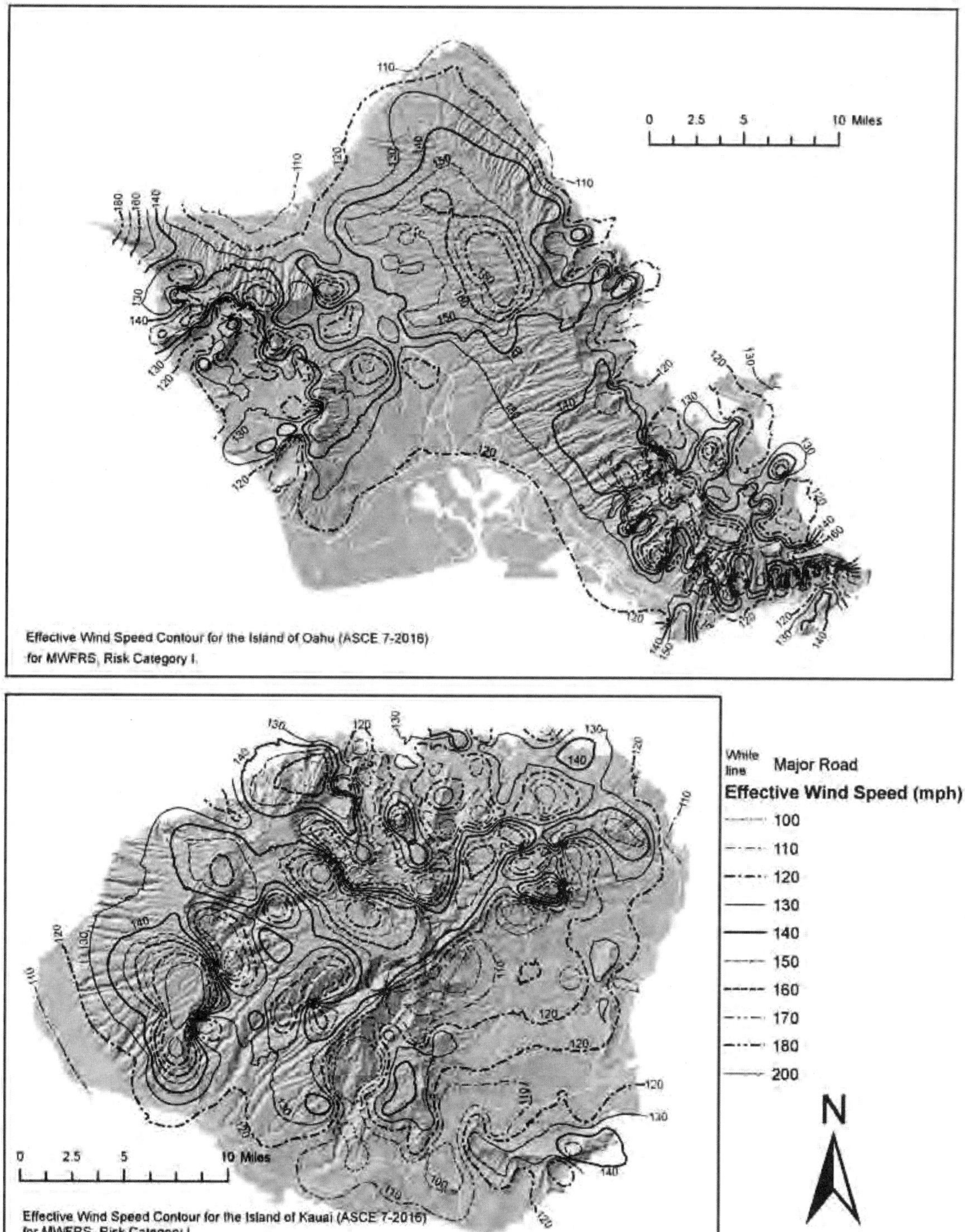

Notes:

1. Values are nominal design 3-second gust wind speeds in miles per hour at 33 feet (10 m) above ground for Exposure C category.
2. Linear interpolation between contours is permitted.
3. Islands and coastal areas outside the last contour shall use the last wind speed contour of the coastal area.
4. It is permitted to use the standard values of Kzt of 1.0 and Kd as given in Table 26.6-1 of ASCE 7.
5. Ocean promontories and local escarpments shall be examined for unusual wind conditions.
6. Wind speeds correspond to approximately a 15% probability of exceedance in 50 years (Annual Exceedance Probability = 0.00143, MRI = 700 Years).

FIGURE 1609.3(12)
BASIC DESIGN WIND SPEEDS, *V*, FOR RISK CATEGORY I BUILDINGS AND OTHER STRUCTURES IN HAWAII (OAHU, KAUAI)

1609.3.1 Wind speed conversion. Where required, the basic design wind speeds of Figures 1609.3(1) through 1609.3(12) shall be converted to *allowable stress design* wind speeds, V_{asd}, using Table 1609.3.1 or Equation 16-17.

$$V_{asd} = V\sqrt{0.6} \quad \textbf{(Equation 16-17)}$$

where:

V_{asd} = *Allowable stress design* wind speed applicable to methods specified in Exceptions 4 and 5 of Section 1609.1.1.

V = Basic design wind speeds determined from Figures 1609.3(1) through 1609.3(12).

1609.4 Exposure category. For each wind direction considered, an exposure category that adequately reflects the characteristics of ground surface irregularities shall be determined for the site at which the building or structure is to be constructed. Account shall be taken of variations in ground surface roughness that arise from natural topography and vegetation as well as from constructed features.

1609.4.1 Wind directions and sectors. For each selected wind direction at which the wind *loads* are to be evaluated, the exposure of the building or structure shall be determined for the two upwind sectors extending 45 degrees (0.79 rad) either side of the selected wind direction. The exposures in these two sectors shall be determined in accordance with Sections 1609.4.2 and 1609.4.3 and the exposure resulting in the highest wind *loads* shall be used to represent winds from that direction.

1609.4.2 Surface roughness categories. A ground surface roughness within each 45-degree (0.79 rad) sector shall be determined for a distance upwind of the site as defined in Section 1609.4.3 from the following categories, for the purpose of assigning an exposure category as defined in Section 1609.4.3.

Surface Roughness B. Urban and suburban areas, wooded areas or other terrain with numerous closely spaced obstructions having the size of single-family dwellings or larger.

Surface Roughness C. Open terrain with scattered obstructions having heights generally less than 30 feet (9144 mm). This category includes flat open country, and grasslands.

Surface Roughness D. Flat, unobstructed areas and water surfaces. This category includes smooth mud flats, salt flats and unbroken ice.

1609.4.3 Exposure categories. An exposure category shall be determined in accordance with the following:

Exposure B. For buildings with a mean roof height of less than or equal to 30 feet (9144 mm), Exposure B shall apply where the ground surface roughness, as defined by Surface Roughness B, prevails in the upwind direction for a distance of not less than 1,500 feet (457 m). For buildings with a mean roof height greater than 30 feet (9144 mm), Exposure B shall apply where Surface Roughness B prevails in the upwind direction for a distance of not less than 2,600 feet (792 m) or 20 times the height of the building, whichever is greater.

Exposure C. Exposure C shall apply for all cases where Exposure B or D does not apply.

Exposure D. Exposure D shall apply where the ground surface roughness, as defined by Surface Roughness D, prevails in the upwind direction for a distance of not less than 5,000 feet (1524 m) or 20 times the height of the building, whichever is greater. Exposure D shall apply where the ground surface roughness immediately upwind of the site is B or C, and the site is within a distance of 600 feet (183 m) or 20 times the *building height*, whichever is greater, from an Exposure D condition as defined in the previous sentence.

1609.5 Roof systems. Roof systems shall be designed and constructed in accordance with Sections 1609.5.1 through 1609.5.3, as applicable.

1609.5.1 Roof deck. The *roof deck* shall be designed to withstand the wind pressures determined in accordance with ASCE 7.

1609.5.2 Roof coverings. *Roof coverings* shall comply with Section 1609.5.1.

Exception: Rigid tile *roof coverings* that are air permeable and installed over a *roof deck* complying with Section 1609.5.1 are permitted to be designed in accordance with Section 1609.5.3.

Asphalt shingles installed over a *roof deck* complying with Section 1609.5.1 shall comply with the wind-resistance requirements of Section 1504.2.

TABLE 1609.3.1
WIND SPEED CONVERSIONS[a, b, c]

V	100	110	120	130	140	150	160	170	180	190	200
V_{asd}	78	85	93	101	108	116	124	132	139	147	155

For SI: 1 mile per hour = 0.44 m/s.

a. Linear interpolation is permitted.

b. V_{asd} = allowable stress design wind speed applicable to methods specified in Exceptions 1 through 5 of Section 1609.1.1.

c. V = basic design wind speeds determined from Figures 1609.3(1) through 1609.3(12).

1609.5.3 Rigid tile. Wind *loads* on rigid tile *roof coverings* shall be determined in accordance with the following equation:

$M_a = q_h C_L b L L_a [1.0 - GC_p]$ **(Equation 16-18)**

For SI:

$$M_a = \frac{q_h C_L b L L_a [1.0 - GC_p]}{1{,}000}$$

where:

b = Exposed width, feet (mm) of the roof tile.

C_L = Lift coefficient. The lift coefficient for concrete and clay tile shall be 0.2 or shall be determined by test in accordance with Section 1504.3.1.

GC_p = Roof pressure coefficient for each applicable roof zone determined from Chapter 30 of ASCE 7. Roof coefficients shall not be adjusted for internal pressure.

L = Length, feet (mm) of the roof tile.

L_a = Moment arm, feet (mm) from the axis of rotation to the point of uplift on the roof tile. The point of uplift shall be taken at 0.76L from the head of the tile and the middle of the exposed width. For roof tiles with nails or screws (with or without a tail clip), the axis of rotation shall be taken as the head of the tile for direct deck application or as the top edge of the batten for battened applications. For roof tiles fastened only by a nail or screw along the side of the tile, the axis of rotation shall be determined by testing. For roof tiles installed with battens and fastened only by a clip near the tail of the tile, the moment arm shall be determined about the top edge of the batten with consideration given for the point of rotation of the tiles based on straight bond or broken bond and the tile profile.

M_a = Aerodynamic uplift moment, feet-pounds (N-mm) acting to raise the tail of the tile.

q_h = Wind velocity pressure, psf (kN/m^2) determined from Section 26.10.2 of ASCE 7.

Concrete and clay roof tiles complying with the following limitations shall be designed to withstand the aerodynamic uplift moment as determined by this section.

1. The roof tiles shall be either loose laid on battens, mechanically fastened, *mortar* set or adhesive set.
2. The roof tiles shall be installed on solid sheathing that has been designed as components and cladding.
3. An *underlayment* shall be installed in accordance with Chapter 15.
4. The tile shall be single lapped interlocking with a minimum head lap of not less than 2 inches (51 mm).
5. The length of the tile shall be between 1.0 and 1.75 feet (305 mm and 533 mm).
6. The exposed width of the tile shall be between 0.67 and 1.25 feet (204 mm and 381 mm).
7. The maximum thickness of the tail of the tile shall not exceed 1.3 inches (33 mm).
8. Roof tiles using *mortar* set or adhesive set systems shall have not less than two-thirds of the tile's area free of *mortar* or adhesive contact.

SECTION 1610
SOIL LOADS AND HYDROSTATIC PRESSURE

1610.1 Lateral pressures. Foundation walls and retaining walls shall be designed to resist lateral soil *loads* from adjacent soil. Soil *loads* specified in Table 1610.1 shall be used as the minimum design lateral soil *loads* unless determined otherwise by a geotechnical investigation in accordance with Section 1803. Foundation walls and other walls in which horizontal movement is restricted at the top shall be designed for at-rest pressure. Retaining walls free to move and rotate at the top shall be permitted to be designed for active pressure. Lateral pressure from surcharge *loads* shall be added to the lateral soil *load*. Lateral pressure shall be increased if expansive soils are present at the site. Foundation walls shall be designed to support the weight of the full hydrostatic pressure of undrained backfill unless a drainage system is installed in accordance with Sections 1805.4.2 and 1805.4.3.

Exception: Foundation walls extending not more than 8 feet (2438 mm) below grade and laterally supported at the top by flexible *diaphragms* shall be permitted to be designed for active pressure.

1610.2 Uplift loads on floor and foundations. Basement floors, slabs on ground, foundations, and similar approximately horizontal elements below grade shall be designed to resist uplift *loads* where applicable. The upward pressure of water shall be taken as the full hydrostatic pressure applied over the entire area. The hydrostatic *load* shall be measured from the underside of the element being evaluated. The design for upward *loads* caused by expansive soils shall comply with Section 1808.6.

SECTION 1611
RAIN LOADS

1611.1 Design rain loads. Each portion of a roof shall be designed to sustain the *load* of rainwater as per the requirements of Chapter 8 of ASCE 7. The design rainfall shall be based on the 100-year 15-minute duration event, or on other rainfall rates determined from approved local weather data. Alternatively, a design rainfall of twice the 100-year hourly rainfall rate indicated in Figures 1611.1(1) through 1611.1(5) shall be permitted.

$R = 5.2(d_s + d_h)$ **(Equation 16-19)**

For SI: $R = 0.0098(d_s + d_h)$

where:

d_h = Additional depth of water on the undeflected roof above the inlet of secondary drainage system at its design flow (in other words, the hydraulic head), in inches (mm).

d_s = Depth of water on the undeflected roof up to the inlet of secondary drainage system when the primary drainage system is blocked (in other words, the static head), in inches (mm).

R = Rain load on the undeflected roof, in psf (kN/m^2). Where the phrase "undeflected roof" is used, deflections from *loads* (including *dead loads*) shall not be considered when determining the amount of rain on the roof.

1611.2 Ponding instability. *Susceptible bays* of roofs shall be evaluated for ponding instability in accordance with Chapters 7 and 8 of ASCE 7.

1611.3 Controlled drainage. Roofs equipped with hardware to control the rate of drainage shall be equipped with a secondary drainage system at a higher elevation that limits accumulation of water on the roof above that elevation. Such roofs shall be designed to sustain the *load* of rainwater that will accumulate on them to the elevation of the secondary drainage system plus the uniform *load* caused by water that rises above the inlet of the secondary drainage system at its design flow determined from Section 1611.1. Such roofs shall be checked for ponding instability in accordance with Section 1611.2.

SECTION 1612
FLOOD LOADS

1612.1 General. Within *flood hazard areas* as established in Section 1612.3, all new construction of buildings, structures and portions of buildings and structures, including *substantial improvement* and restoration of *substantial damage* to buildings and structures, shall be designed and constructed to resist the effects of flood hazards and *flood loads*. For buildings that are located in more than one *flood hazard area*, the provisions associated with the most restrictive *flood hazard area* shall apply.

1612.2 Design and construction. The design and construction of buildings and structures located in *flood hazard areas*, including *coastal high hazard areas* and *coastal A zones*, shall be in accordance with Chapter 5 of ASCE 7 and ASCE 24.

1612.3 Establishment of flood hazard areas. To establish *flood hazard areas*, the applicable governing authority shall adopt a flood hazard map and supporting data. The flood hazard map shall include, at a minimum, areas of special flood hazard as identified by the Federal Emergency Management Agency in an engineering report entitled "The *Flood Insurance Study* for **[INSERT NAME OF JURISDICTION]**," dated **[INSERT DATE OF ISSUANCE]**, as amended or revised with the accompanying *Flood Insurance Rate Map* (FIRM) and Flood Boundary and *Floodway* Map (FBFM) and related supporting data along with any revisions thereto. The adopted flood hazard map and supporting data are hereby adopted by reference and declared to be part of this section.

TABLE 1610.1
LATERAL SOIL LOAD

DESCRIPTION OF BACKFILL MATERIAL[c]	UNIFIED SOIL CLASSIFICATION	DESIGN LATERAL SOIL LOAD[a] (pound per square foot per foot of depth)	
		Active pressure	At-rest pressure
Well-graded, clean gravels; gravel-sand mixes	GW	30	60
Poorly graded clean gravels; gravel-sand mixes	GP	30	60
Silty gravels, poorly graded gravel-sand mixes	GM	40	60
Clayey gravels, poorly graded gravel-and-clay mixes	GC	45	60
Well-graded, clean sands; gravelly sand mixes	SW	30	60
Poorly graded clean sands; sand-gravel mixes	SP	30	60
Silty sands, poorly graded sand-silt mixes	SM	45	60
Sand-silt clay mix with plastic fines	SM-SC	45	100
Clayey sands, poorly graded sand-clay mixes	SC	60	100
Inorganic silts and clayey silts	ML	45	100
Mixture of inorganic silt and clay	ML-CL	60	100
Inorganic clays of low to medium plasticity	CL	60	100
Organic silts and silt clays, low plasticity	OL	Note b	Note b
Inorganic clayey silts, elastic silts	MH	Note b	Note b
Inorganic clays of high plasticity	CH	Note b	Note b
Organic clays and silty clays	OH	Note b	Note b

For SI: 1 pound per square foot per foot of depth = 0.157 kPa/m, 1 foot = 304.8 mm.

a. Design lateral soil loads are given for moist conditions for the specified soils at their optimum densities. Actual field conditions shall govern. Submerged or saturated soil pressures shall include the weight of the buoyant soil plus the hydrostatic loads.

b. Unsuitable as backfill material.

c. The definition and classification of soil materials shall be in accordance with ASTM D2487.

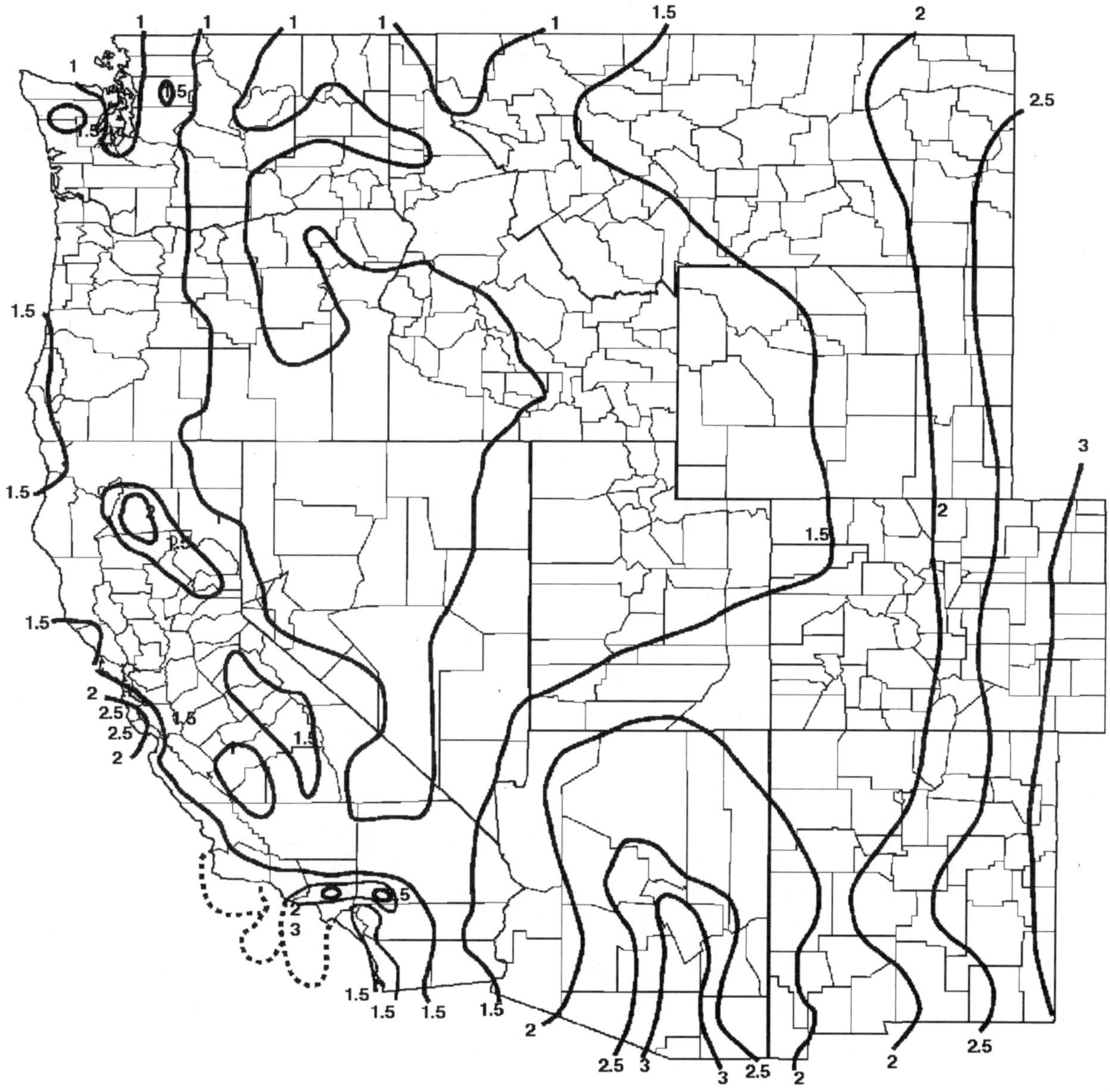

For SI: 1 inch = 25.4 mm.

Source: National Weather Service, National Oceanic and Atmospheric Administration, Washington, DC.

FIGURE 1611.1(1)
100-YEAR, 1-HOUR RAINFALL (INCHES) WESTERN UNITED STATES

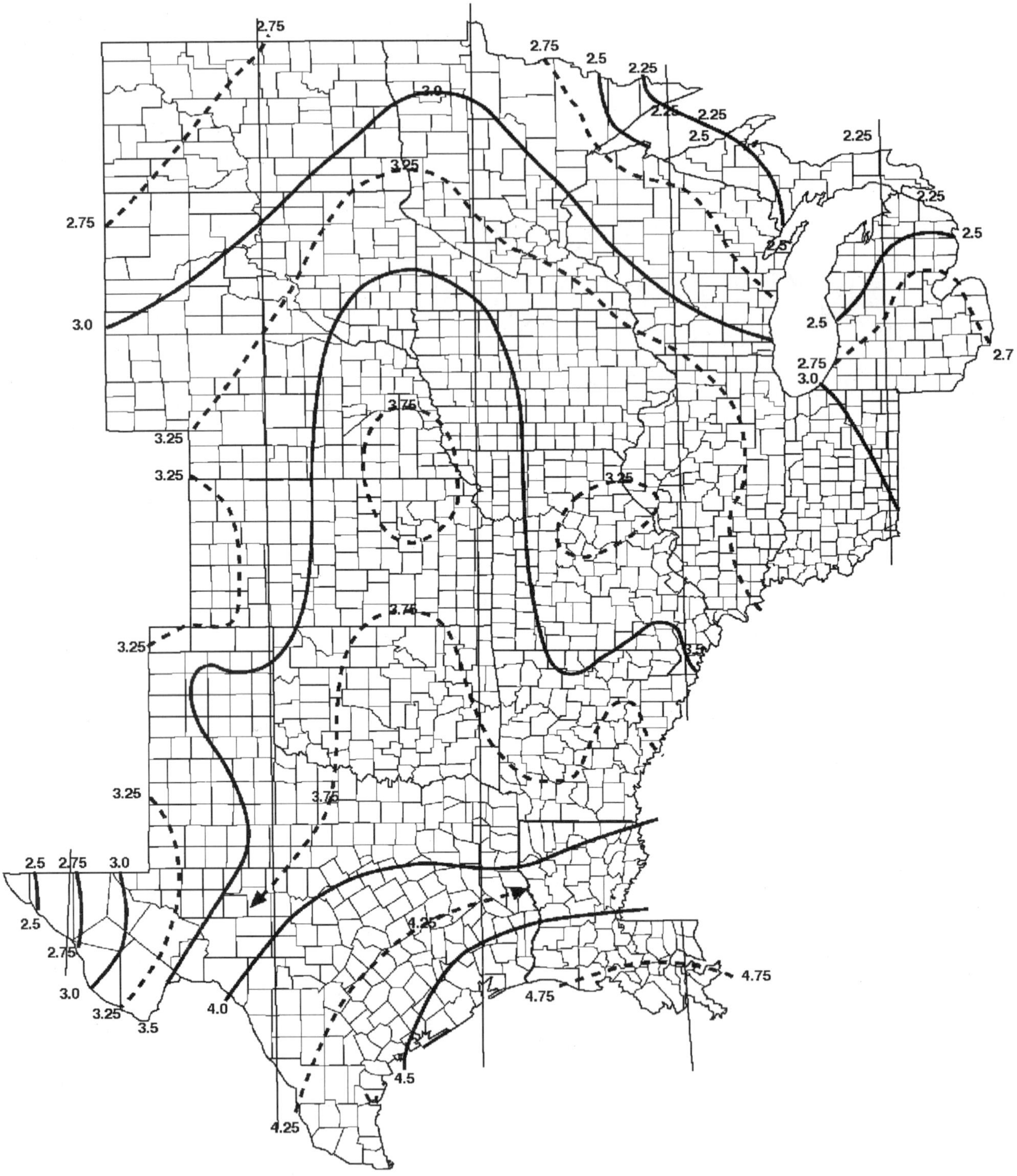

For SI:1 inch = 25.4 mm.

Source: National Weather Service, National Oceanic and Atmospheric Administration, Washington, DC.

FIGURE 1611.1(2)
100-YEAR, 1-HOUR RAINFALL (INCHES) CENTRAL UNITED STATES

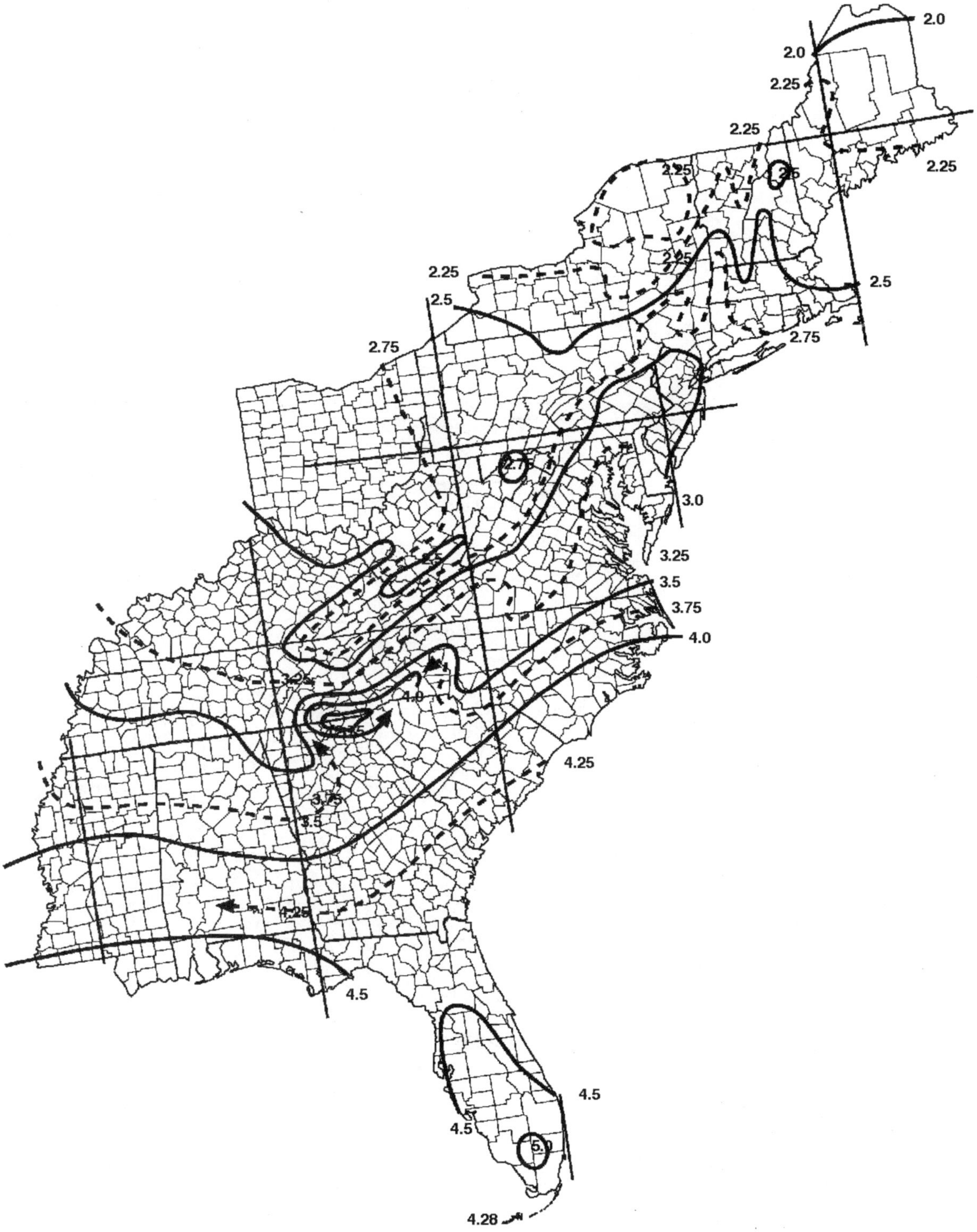

For SI:1 inch = 25.4 mm.
Source: National Weather Service, National Oceanic and Atmospheric Administration, Washington, DC.

FIGURE 1611.1(3)
100-YEAR, 1-HOUR RAINFALL (INCHES) EASTERN UNITED STATES

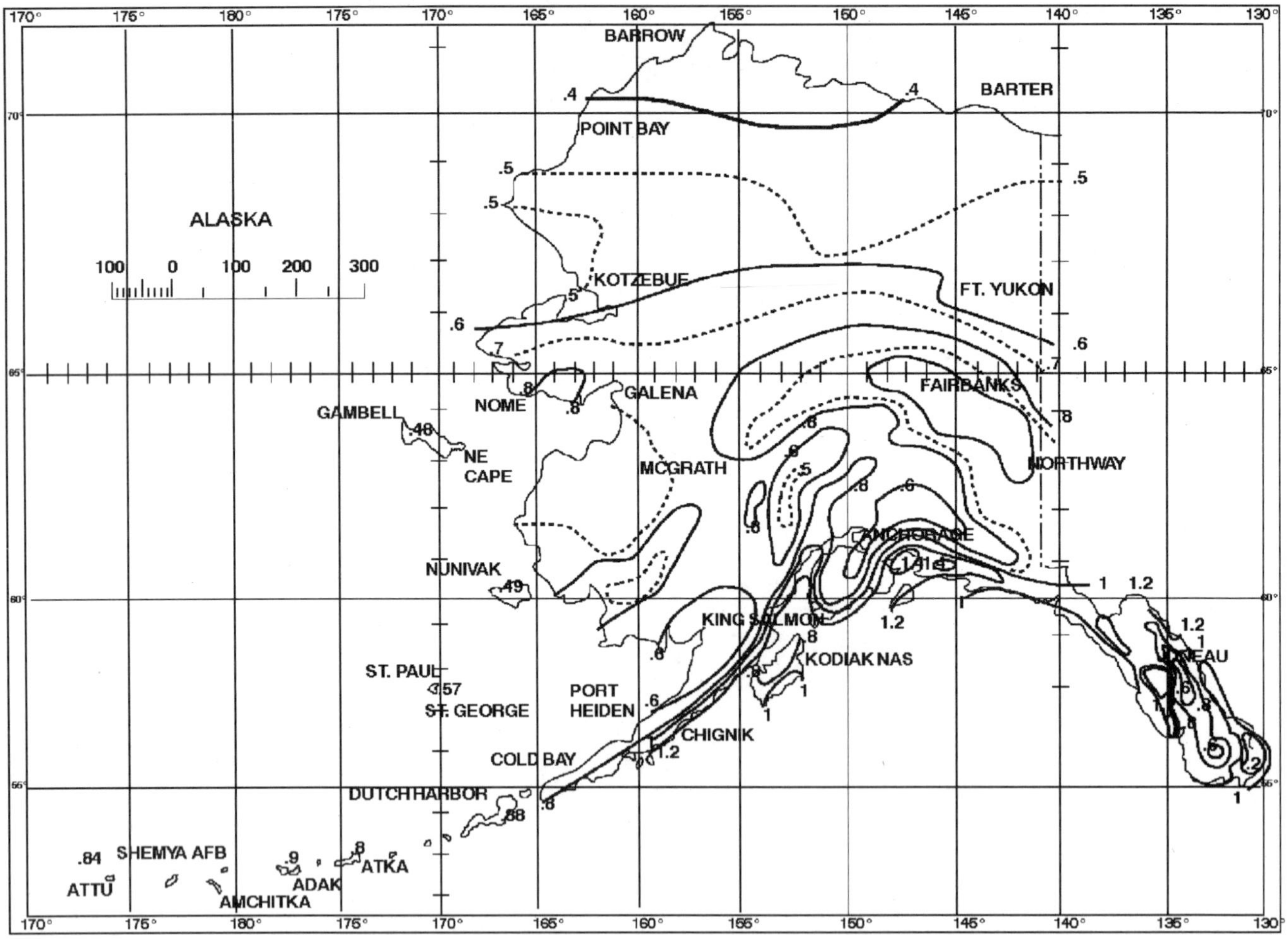

For SI:1 inch = 25.4 mm.
Source: National Weather Service, National Oceanic and Atmospheric Administration, Washington, DC.

FIGURE 1611.1(4)
100-YEAR, 1-HOUR RAINFALL (INCHES) ALASKA

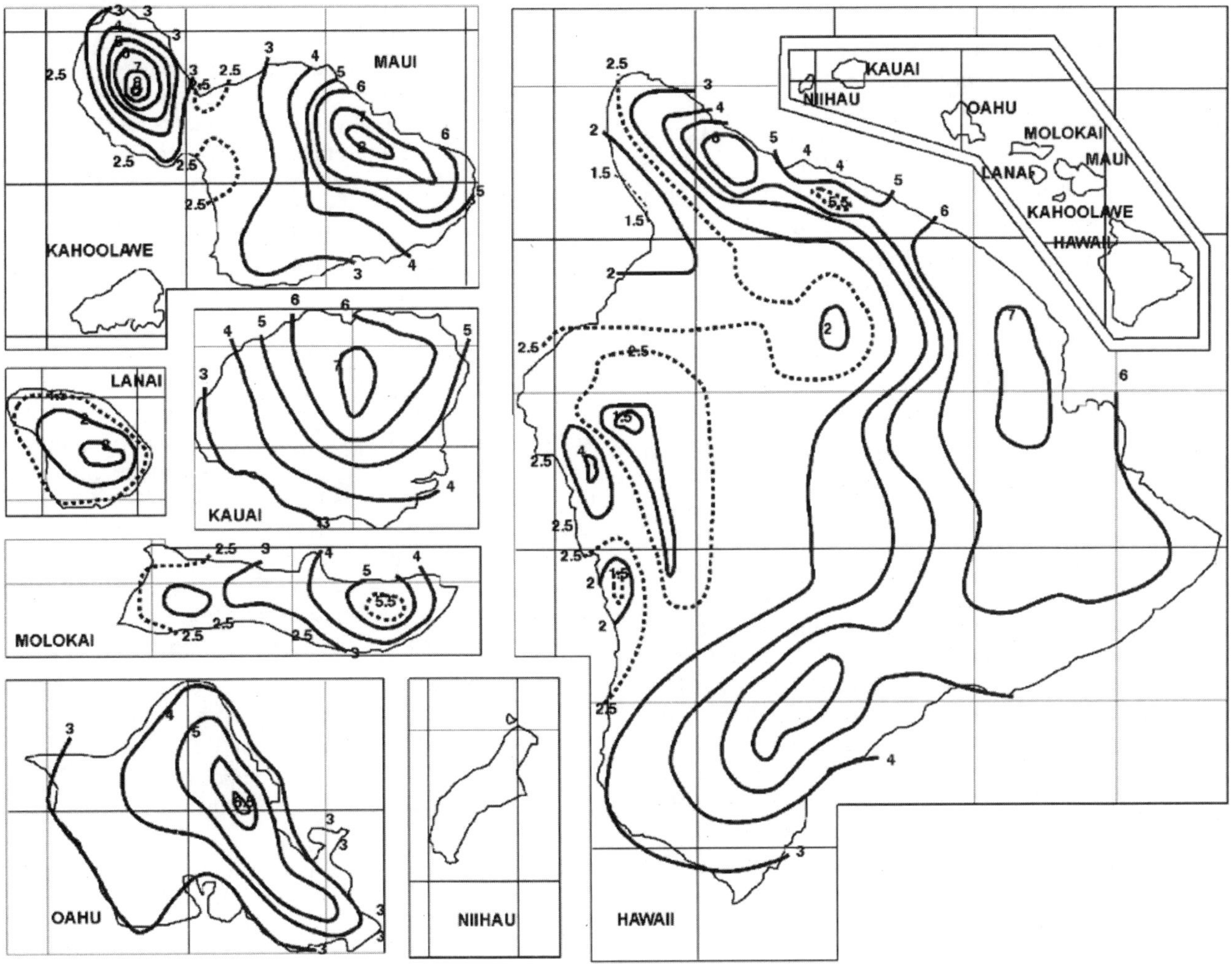

For SI:1 inch = 25.4 mm.
Source: National Weather Service, National Oceanic and Atmospheric Administration, Washington, DC.

FIGURE 1611.1(5)
100-YEAR, 1-HOUR RAINFALL (INCHES) HAWAII

1612.3.1 Design flood elevations. Where *design flood elevations* are not included in the *flood hazard areas* established in Section 1612.3, or where *floodways* are not designated, the *building official* is authorized to require the applicant to do one of the following:

1. Obtain and reasonably utilize any *design flood elevation* and *floodway* data available from a federal, state or other source.
2. Determine the *design flood elevation* or *floodway* in accordance with accepted hydrologic and hydraulic engineering practices used to define special *flood hazard areas*. Determinations shall be undertaken by a *registered design professional* who shall document that the technical methods used reflect currently accepted engineering practice.

1612.3.2 Determination of impacts. In riverine *flood hazard areas* where *design flood elevations* are specified but *floodways* have not been designated, the applicant shall provide a *floodway* analysis that demonstrates that the proposed work will not increase the *design flood elevation* more than 1 foot (305 mm) at any point within the jurisdiction of the applicable governing authority.

1612.4 Flood hazard documentation. The following documentation shall be prepared and sealed by a *registered design professional* and submitted to the *building official*:

1. For construction in *flood hazard areas* other than *coastal high hazard areas* or *coastal A zones*:
   - 1.1. The elevation of the *lowest floor*, including the basement, as required by the lowest floor elevation inspection in Section 110.3.3 and for the final inspection in Section 110.3.12.1.
   - 1.2. For fully enclosed areas below the *design flood elevation* where provisions to allow for the automatic entry and exit of floodwaters do not meet the minimum requirements in Section 2.7.2.1 of ASCE 24, *construction documents* shall include a statement that the design will provide for equalization of hydrostatic flood forces in accordance with Section 2.7.2.2 of ASCE 24.
   - 1.3. For *dry floodproofed* nonresidential buildings, *construction documents* shall include a statement that the *dry floodproofing* is designed in accordance with ASCE 24 and shall include the flood emergency plan specified in Chapter 6 of ASCE 24.
2. For construction in *coastal high hazard areas* and *coastal A zones*:
   - 2.1. The elevation of the bottom of the lowest horizontal structural member as required by the *lowest floor* elevation inspection in Section 110.3.3 and for the final inspection in Section 110.3.12.1.
   - 2.2. *Construction documents* shall include a statement that the building is designed in accordance with ASCE 24, including that the pile or column foundation and building or structure to be attached thereto is designed to be anchored to resist flotation, collapse and lateral movement due to the effects of wind and *flood loads* acting simultaneously on all building components, and other *load* requirements of Chapter 16.
   - 2.3. For breakaway walls designed to have a resistance of more than 20 psf (0.96 kN/m^2) determined using *allowable stress design*, *construction documents* shall include a statement that the breakaway wall is designed in accordance with ASCE 24.
   - 2.4 For breakaway walls where provisions to allow for the automatic entry and exit of floodwaters do not meet the minimum requirements in Section 2.7.2.1 of ASCE 24, construction documents shall include a statement that the design will provide for equalization of hydrostatic flood forces in accordance with Section 2.7.2.2 of ASCE 24.

SECTION 1613
EARTHQUAKE LOADS

1613.1 Scope. Every structure, and portion thereof, including nonstructural components that are permanently attached to structures and their supports and attachments, shall be designed and constructed to resist the effects of earthquake motions in accordance with Chapters 11, 12, 13, 15, 17 and 18 of ASCE 7, as applicable. The *seismic design category* for a structure is permitted to be determined in accordance with Section 1613 or ASCE 7.

Exceptions:

1. Detached one- and two-family dwellings, assigned to *Seismic Design Category* A, B or C, or located where the mapped short-period spectral response acceleration, S_S, is less than 0.4 g.
2. The *seismic force-resisting system* of wood-frame buildings that conform to the provisions of Section 2308 are not required to be analyzed as specified in this section.
3. Agricultural storage structures intended only for incidental human occupancy.
4. Structures that require special consideration of their response characteristics and environment that are not addressed by this code or ASCE 7 and for which other regulations provide seismic criteria, such as vehicular bridges, electrical transmission towers, hydraulic structures, buried utility lines and their appurtenances and nuclear reactors.
5. References within ASCE 7 to Chapter 14 shall not apply, except as specifically required herein.

1613.2 Seismic ground motion values. Seismic ground motion values shall be determined in accordance with this section.

1613.2.1 Mapped acceleration parameters. The parameters S_S and S_1 shall be determined from the 0.2 and 1-second spectral response accelerations shown on Figures 1613.2.1(1) through 1613.2.1(10). Where S_1 is less than or equal to 0.04 and S_S is less than or equal to 0.15, the structure is permitted to be assigned *Seismic Design Category* A.

1613.2.2 Site class definitions. Based on the site soil properties, the site shall be classified as *Site Class* A, B, C, D, E or F in accordance with Chapter 20 of ASCE 7.

Where the soil properties are not known in sufficient detail to determine the site class, *Site Class* D, subjected to the requirements of Section 1613.2.3, shall be used unless the *building official* or geotechnical data determines that *Site Class* E or F soils are present at the site.

Where site investigations that are performed in accordance with Chapter 20 of ASCE 7 reveal rock conditions consistent with *Site Class* B, but site-specific velocity measurements are not made, the *site coefficients* F_a and F_v shall be taken at unity (1.0).

1613.2.3 Site coefficients and adjusted maximum considered earthquake spectral response acceleration parameters. The maximum considered earthquake spectral response acceleration for short periods, S_{MS}, and at 1-second period, S_{M1}, adjusted for site class effects shall be determined by Equations 16-20 and 16-21, respectively:

$$S_{MS} = F_a S_S \quad \textbf{(Equation 16-20)}$$

$$S_{M1} = F_v S_1 \quad \textbf{(Equation 16-21)}$$

but S_{MS} shall not be taken less than S_{M1} except when determining the seismic design category in accordance with Section 1613.2.5.

where:

F_a = Site coefficient defined in Table 1613.2.3(1).

F_v = Site coefficient defined in Table 1613.2.3(2).

S_S = The mapped spectral accelerations for short periods as determined in Section 1613.2.1.

S_1 = The mapped spectral accelerations for a 1-second period as determined in Section 1613.2.1.

Where *Site Class* D is selected as the default site class per Section 1613.2.2, the value of F_a shall be not less than 1.2. Where the simplified design procedure of ASCE 7 Section 12.14 is used, the value of F_a shall be determined in accordance with ASCE 7 Section 12.14.8.1, and the values of F_v, S_{MS} and S_{M1} need not be determined.

1613.2.4 Design spectral response acceleration parameters. Five-percent damped design spectral response acceleration at short periods, S_{DS}, and at 1-second period, S_{D1}, shall be determined from Equations 16-22 and Equation 16-23, respectively:

$$S_{DS} = \frac{2}{3} S_{MS} \quad \textbf{(Equation 16-22)}$$

$$S_{D1} = \frac{2}{3} S_{M1} \quad \textbf{(Equation 16-23)}$$

where:

S_{MS} = The maximum considered earthquake spectral response accelerations for short period as determined in Section 1613.2.3.

S_{M1} = The maximum considered earthquake spectral response accelerations for 1-second period as determined in Section 1613.2.3.

1613.2.5 Determination of seismic design category. Structures classified as *Risk Category* I, II or III that are located where the mapped spectral response acceleration parameter at 1-second period, S_1, is greater than or equal to 0.75 shall be assigned to *Seismic Design Category* E. Structures classified as *Risk Category* IV that are located where the mapped spectral response acceleration parameter at 1-second period, S_1, is greater than or equal to 0.75 shall be assigned to *Seismic Design Category* F. Other structures shall be assigned to a *seismic design category* based on their *risk category* and the design spectral response acceleration parameters, S_{DS} and S_{D1}, determined in accordance with Section 1613.2.4 or the site-specific procedures of ASCE 7. Each building and structure shall be assigned to the more severe *seismic design category* in accordance with Table 1613.2.5(1) or 1613.2.5(2), irrespective of the fundamental period of vibration of the structure, *T*.

1613.2.5.1 Alternative seismic design category determination. Where S_1 is less than 0.75, the *seismic design category* is permitted to be determined from Table 1613.2.5(1) alone where all of the following apply:

1. In each of the two *orthogonal* directions, the approximate fundamental period of the structure, T_a, in each of the two *orthogonal* directions determined in accordance with Section 12.8.2.1 of ASCE 7, is less than 0.8 T_s determined in accordance with Section 11.8.6 of ASCE 7.
2. In each of the two *orthogonal* directions, the fundamental period of the structure used to calculate the *story* drift is less than T_s.
3. Equation 12.8-2 of ASCE 7 is used to determine the seismic response coefficient, C_s.
4. The *diaphragms* are rigid or are permitted to be idealized as rigid in accordance with Section 12.3.1 of ASCE 7 or, for *diaphragms* permitted to be idealized as flexible in accordance with Section 12.3.1 of ASCE 7, the distances between vertical elements of the *seismic force-resisting system* do not exceed 40 feet (12 192 mm).

1613.2.5.2 Simplified design procedure. Where the alternate simplified design procedure of ASCE 7 is used, the *seismic design category* shall be determined in accordance with ASCE 7.

TABLE 1613.2.3(1)
VALUES OF SITE COEFFICIENT F_a[a]

SITE CLASS	MAPPED RISK TARGETED MAXIMUM CONSIDERED EARTHQUAKE (MCE_R) SPECTRAL RESPONSE ACCELERATION PARAMETER AT SHORT PERIOD					
	$S_s \le 0.25$	$S_s = 0.50$	$S_s = 0.75$	$S_s = 1.00$	$S_s = 1.25$	$S_s \ge 1.5$
A	0.8	0.8	0.8	0.8	0.8	0.8
B	0.9	0.9	0.9	0.9	0.9	0.9
C	1.3	1.3	1.2	1.2	1.2	1.2
D	1.6	1.4	1.2	1.1	1.0	1.0
E	2.4	1.7	1.3	Note b	Note b	Note b
F	Note b	Note b	Note b	Note b	Note b	Note b

a. Use straight-line interpolation for intermediate values of mapped spectral response acceleration at short period, S_s.
b. Values shall be determined in accordance with Section 11.4.8 of ASCE 7.

TABLE 1613.2.3(2)
VALUES OF SITE COEFFICIENT F_V[a]

SITE CLASS	MAPPED RISK TARGETED MAXIMUM CONSIDERED EARTHQUAKE (MCE_R) SPECTRAL RESPONSE ACCELERATION PARAMETER AT 1-SECOND PERIOD					
	$S_1 \le 0.1$	$S_1 = 0.2$	$S_1 = 0.3$	$S_1 = 0.4$	$S_1 = 0.5$	$S_1 \ge 0.6$
A	0.8	0.8	0.8	0.8	0.8	0.8
B	0.8	0.8	0.8	0.8	0.8	0.8
C	1.5	1.5	1.5	1.5	1.5	1.4
D	2.4	2.2[c]	2.0[c]	1.9[c]	1.8[c]	1.7[c]
E	4.2	3.3[c]	2.8[c]	2.4[c]	2.2[c]	2.0[c]
F	Note b	Note b	Note b	Note b	Note b	Note b

a. Use straight-line interpolation for intermediate values of mapped spectral response acceleration at 1-second period, S_1.
b. Values shall be determined in accordance with Section 11.4.8 of ASCE 7.
c. See requirements for site-specific ground motions in Section 11.4.8 of ASCE 7.

TABLE 1613.2.5(1)
SEISMIC DESIGN CATEGORY BASED ON SHORT-PERIOD (0.2 second) RESPONSE ACCELERATION

VALUE OF S_{DS}	RISK CATEGORY		
	I or II	III	IV
$S_{DS} < 0.167g$	A	A	A
$0.167g \le S_{DS} < 0.33g$	B	B	C
$0.33g \le S_{DS} < 0.50g$	C	C	D
$0.50g \le S_{DS}$	D	D	D

TABLE 1613.2.5(2)
SEISMIC DESIGN CATEGORY BASED ON 1-SECOND PERIOD RESPONSE ACCELERATION

VALUE OF S_{D1}	RISK CATEGORY		
	I or II	III	IV
$S_{D1} < 0.067g$	A	A	A
$0.067g \le S_{D1} < 0.133g$	B	B	C
$0.133g \le S_{D1} < 0.20g$	C	C	D
$0.20g \le S_{D1}$	D	D	D

1613.3 Ballasted photovoltaic panel systems. Ballasted, roof-mounted *photovoltaic panel systems* need not be rigidly attached to the roof or supporting structure. Ballasted nonpenetrating systems shall be designed and installed only on roofs with slopes not more than one unit vertical in 12 units horizontal. Ballasted nonpenetrating systems shall be designed to resist sliding and uplift resulting from lateral and vertical forces as required by Section 1605, using a coefficient of friction determined by acceptable engineering principles. In structures assigned to *Seismic Design Category* C, D, E or F, ballasted nonpenetrating systems shall be designed to accommodate seismic displacement determined by nonlinear response-hi *story* or other *approved* analysis or shake-table testing, using input motions consistent with ASCE 7 lateral and vertical seismic forces for nonstructural components on roofs.

SECTION 1614 ATMOSPHERIC ICE LOADS

1614.1 General. *Ice-sensitive structures* shall be designed for atmospheric ice *loads* in accordance with Chapter 10 of ASCE 7.

SECTION 1615 TSUNAMI LOADS

1615.1 General. The design and construction of *Risk Category* III and IV buildings and structures located in the *Tsunami Design Zones* defined in the *Tsunami Design Geodatabase* shall be in accordance with Chapter 6 of ASCE 7, except as modified by this code.

SECTION 1616 STRUCTURAL INTEGRITY

1616.1 General. *High-rise buildings* that are assigned to *Risk Category* III or IV shall comply with the requirements of Section 1616.2 if they are frame structures, or Section 1616.3 if they are *bearing wall structures*.

1616.2 Frame structures. *Frame structures* shall comply with the requirements of this section.

1616.2.1 Concrete frame structures. *Frame structures* constructed primarily of reinforced or prestressed concrete, either cast-in-place or precast, or a combination of these, shall conform to the requirements of Section 4.10 of ACI 318. Where ACI 318 requires that nonprestressed reinforcing or prestressing steel pass through the region bounded by the longitudinal column reinforcement, that reinforcing or prestressing steel shall have a minimum nominal tensile strength equal to two-thirds of the required one-way vertical strength of the connection of the floor or roof system to the column in each direction of beam or slab reinforcement passing through the column.

Exception: Where concrete slabs with continuous reinforcement having an area not less than 0.0015 times the concrete area in each of two *orthogonal* directions are present and are either monolithic with or equivalently bonded to beams, girders or columns, the longitudinal reinforcing or prestressing steel passing through the column reinforcement shall have a nominal tensile strength of one-third of the required one-way vertical strength of the connection of the floor or roof system to the column in each direction of beam or slab reinforcement passing through the column.

1616.2.2 Structural steel, open web steel joist or joist girder, or composite steel and concrete *frame structures*. *Frame structures* constructed with a structural steel frame or a frame composed of open web *steel joists*, joist girders with or without other *structural steel elements* or a frame composed of composite steel or composite *steel joists* and reinforced concrete elements shall conform to the requirements of this section.

1616.2.2.1 Columns. Each column splice shall have the minimum *design strength* in tension to transfer the design dead and *live load* tributary to the column between the splice and the splice or base immediately below.

1616.2.2.2 Beams. End connections of all beams and girders shall have a minimum nominal axial tensile strength equal to the required vertical shear strength for *allowable stress design* (ASD) or two-thirds of the required shear strength for *load and resistance factor design* (*LRFD*) but not less than 10 kips (45 kN). For the purpose of this section, the shear force and the axial tensile force need not be considered to act simultaneously.

Exception: Where beams, girders, open web joist and joist girders support a concrete slab or concrete slab on metal deck that is attached to the beam or girder with not less than 3/8-inch-diameter (9.5 mm) headed shear studs, at a spacing of not more than 12 inches (305 mm) on center, averaged over the length of the member, or other attachment having equivalent shear strength, and the slab contains continuous distributed reinforcement in each of two *orthogonal* directions with an area not less than 0.0015 times the concrete area, the nominal axial tension strength of the end connection shall be permitted to be taken as half the required vertical shear strength for ASD or one-third of the required shear strength for *LRFD*, but not less than 10 kips (45 kN).

1616.3 Bearing wall structures. *Bearing wall structures* shall have vertical ties in all *load-bearing walls* and longitudinal ties, transverse ties and perimeter ties at each floor level in accordance with this section and as shown in Figure 1616.3.

1616.3.1 Concrete wall structures. Precast *bearing wall structures* constructed solely of reinforced or prestressed concrete, or combinations of these shall conform to the requirements of Sections 16.2.4 and 16.2.5 of ACI 318.

1616.3.2 Other bearing wall structures. Ties in *bearing wall structures* other than those covered in Section 1616.3.1 shall conform to this section.

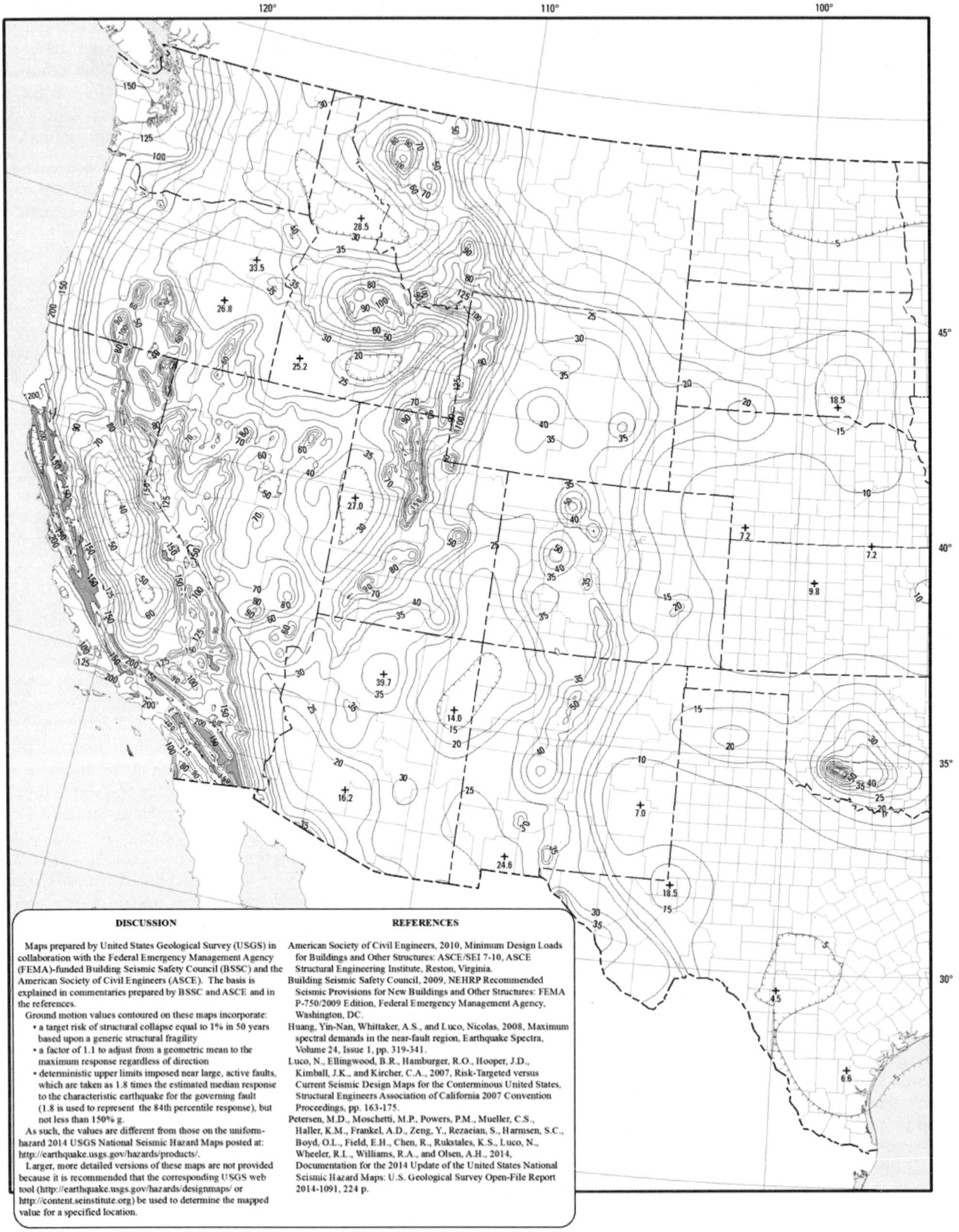

DISCUSSION

Maps prepared by United States Geological Survey (USGS) in collaboration with the Federal Emergency Management Agency (FEMA)-funded Building Seismic Safety Council (BSSC) and the American Society of Civil Engineers (ASCE). The basis is explained in commentaries prepared by BSSC and ASCE and in the references.

Ground motion values contoured on these maps incorporate:

- a target risk of structural collapse equal to 1% in 50 years based upon a generic structural fragility
- a factor of 1.1 to adjust from a geometric mean to the maximum response regardless of direction
- deterministic upper limits imposed near large, active faults, which are taken as 1.8 times the estimated median response to the characteristic earthquake for the governing fault (1.8 is used to represent the 84th percentile response), but not less than 150% g.

As such, the values are different from those on the uniform-hazard 2014 USGS National Seismic Hazard Maps posted at: http://earthquake.usgs.gov/hazards/products/.

Larger, more detailed versions of these maps are not provided because it is recommended that the corresponding USGS web tool (http://earthquake.usgs.gov/hazards/designmaps/ or http://content.seinstitute.org) be used to determine the mapped value for a specified location.

REFERENCES

American Society of Civil Engineers, 2010, Minimum Design Loads for Buildings and Other Structures: ASCE/SEI 7-10, ASCE Structural Engineering Institute, Reston, Virginia.

Building Seismic Safety Council, 2009, NEHRP Recommended Seismic Provisions for New Buildings and Other Structures: FEMA P-750/2009 Edition, Federal Emergency Management Agency, Washington, DC.

Huang, Yin-Nan, Whittaker, A.S., and Luco, Nicolas, 2008, Maximum spectral demands in the near-fault region, Earthquake Spectra, Volume 24, Issue 1, pp. 319-341.

Luco, N., Ellingwood, B.R., Hamburger, R.O., Hooper, J.D., Kimball, J.K., and Kircher, C.A., 2007, Risk-Targeted versus Current Seismic Design Maps for the Conterminous United States, Structural Engineers Association of California 2007 Convention Proceedings, pp. 163-175.

Petersen, M.D., Moschetti, M.P., Powers, P.M., Mueller, C.S., Haller, K.M., Frankel, A.D., Zeng, Y., Rezaeian, S., Harmsen, S.C., Boyd, O.L., Field, E.H., Chen, R., Rukstales, K.S., Luco, N., Wheeler, R.L., Williams, R.A., and Olsen, A.H., 2014, Documentation for the 2014 Update of the United States National Seismic Hazard Maps: U.S. Geological Survey Open-File Report 2014-1091, 224 p.

FIGURE 1613.2.1(1)
RISK-TARGETED MAXIMUM CONSIDERED EARTHQUAKE (MCE_R) GROUND MOTION RESPONSE ACCELERATIONS FOR THE CONTERMINOUS UNITED STATES OF 0.2-SECOND SPECTRAL RESPONSE ACCELERATION (5% OF CRITICAL DAMPING)

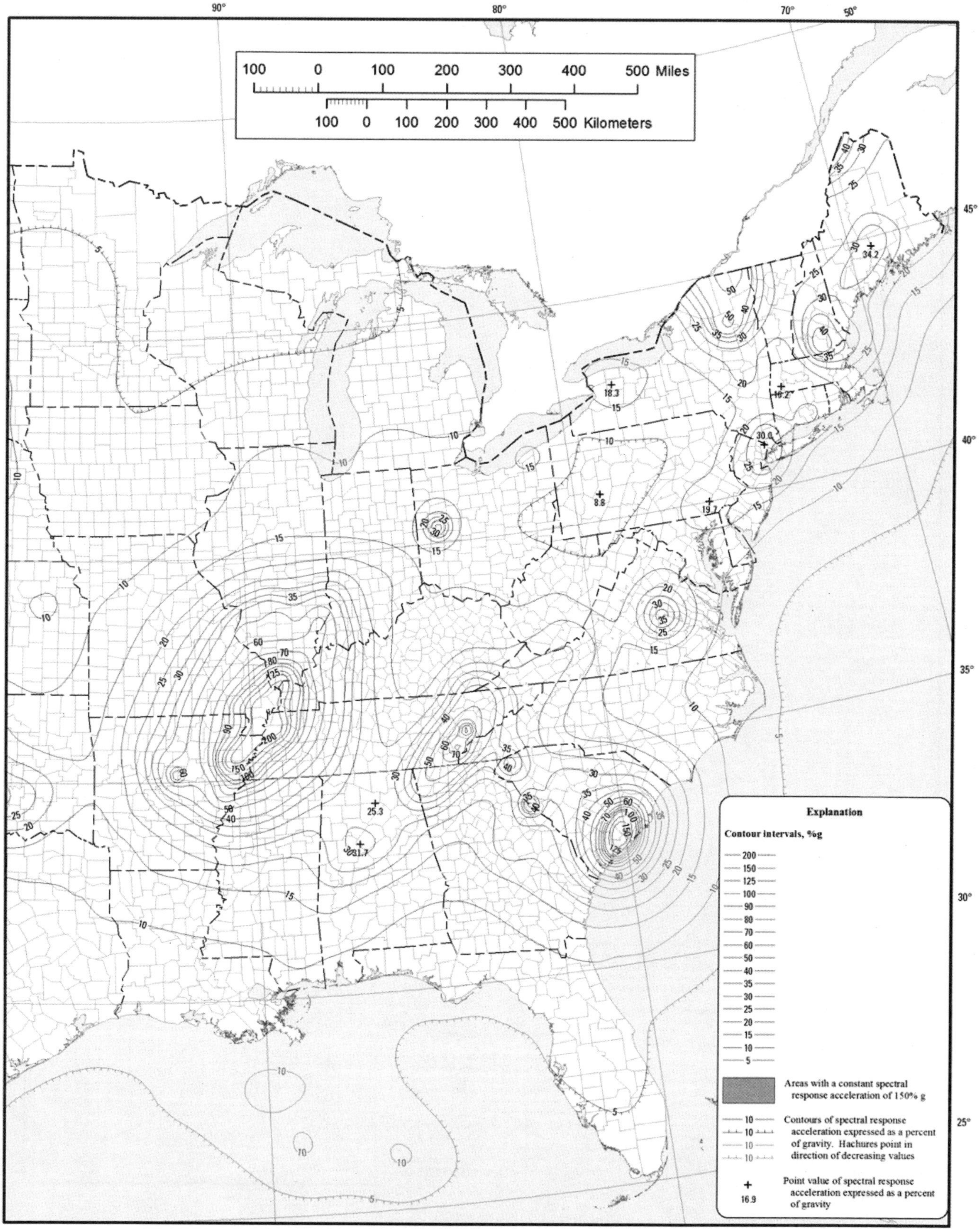

FIGURE 1613.2.1(2)
RISK-TARGETED MAXIMUM CONSIDERED EARTHQUAKE (MCE_R) GROUND MOTION RESPONSE ACCELERATIONS FOR THE CONTERMINOUS UNITED STATES OF 0.2-SECOND SPECTRAL RESPONSE ACCELERATION (5% OF CRITICAL DAMPING)

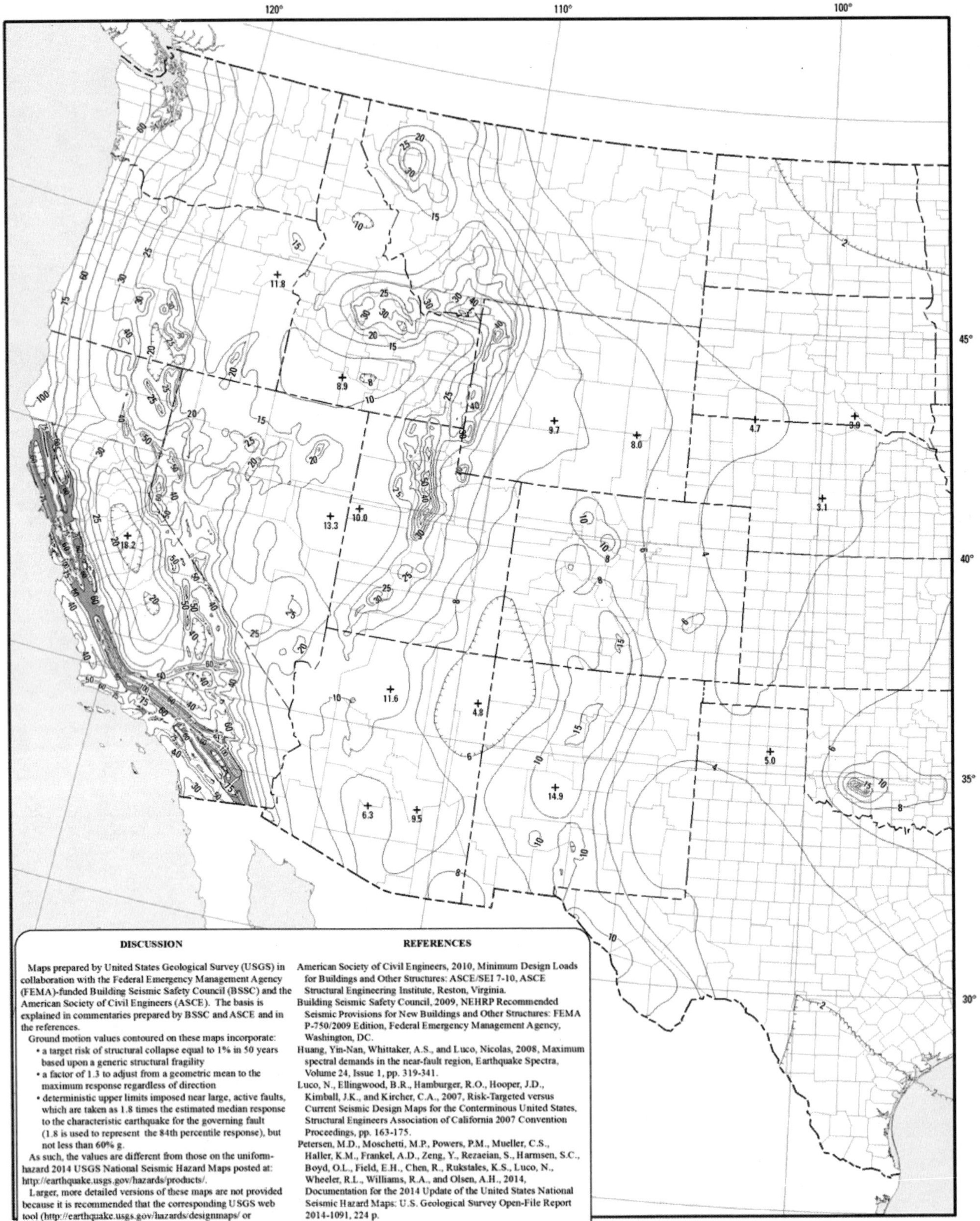

DISCUSSION

Maps prepared by United States Geological Survey (USGS) in collaboration with the Federal Emergency Management Agency (FEMA)-funded Building Seismic Safety Council (BSSC) and the American Society of Civil Engineers (ASCE). The basis is explained in commentaries prepared by BSSC and ASCE and in the references.

Ground motion values contoured on these maps incorporate:

- a target risk of structural collapse equal to 1% in 50 years based upon a generic structural fragility
- a factor of 1.3 to adjust from a geometric mean to the maximum response regardless of direction
- deterministic upper limits imposed near large, active faults, which are taken as 1.8 times the estimated median response to the characteristic earthquake for the governing fault (1.8 is used to represent the 84th percentile response), but not less than 60% g.

As such, the values are different from those on the uniform-hazard 2014 USGS National Seismic Hazard Maps posted at: http://earthquake.usgs.gov/hazards/products/.

Larger, more detailed versions of these maps are not provided because it is recommended that the corresponding USGS web tool (http://earthquake.usgs.gov/hazards/designmaps/ or http://content.seinstitute.org) be used to determine the mapped value for a specified location.

REFERENCES

American Society of Civil Engineers, 2010, Minimum Design Loads for Buildings and Other Structures: ASCE/SEI 7-10, ASCE Structural Engineering Institute, Reston, Virginia.

Building Seismic Safety Council, 2009, NEHRP Recommended Seismic Provisions for New Buildings and Other Structures: FEMA P-750/2009 Edition, Federal Emergency Management Agency, Washington, DC.

Huang, Yin-Nan, Whittaker, A.S., and Luco, Nicolas, 2008, Maximum spectral demands in the near-fault region, Earthquake Spectra, Volume 24, Issue 1, pp. 319-341.

Luco, N., Ellingwood, B.R., Hamburger, R.O., Hooper, J.D., Kimball, J.K., and Kircher, C.A., 2007, Risk-Targeted versus Current Seismic Design Maps for the Conterminous United States, Structural Engineers Association of California 2007 Convention Proceedings, pp. 163-175.

Petersen, M.D., Moschetti, M.P., Powers, P.M., Mueller, C.S., Haller, K.M., Frankel, A.D., Zeng, Y., Rezaeian, S., Harmsen, S.C., Boyd, O.L., Field, E.H., Chen, R., Rukstales, K.S., Luco, N., Wheeler, R.L., Williams, R.A., and Olsen, A.H., 2014, Documentation for the 2014 Update of the United States National Seismic Hazard Maps: U.S. Geological Survey Open-File Report 2014-1091, 224 p.

FIGURE 1613.2.1(3)
RISK-TARGETED MAXIMUM CONSIDERED EARTHQUAKE (MCE_R) GROUND MOTION RESPONSE ACCELERATIONS FOR THE CONTERMINOUS UNITED STATES OF 1-SECOND SPECTRAL RESPONSE ACCELERATION (5% OF CRITICAL DAMPING)

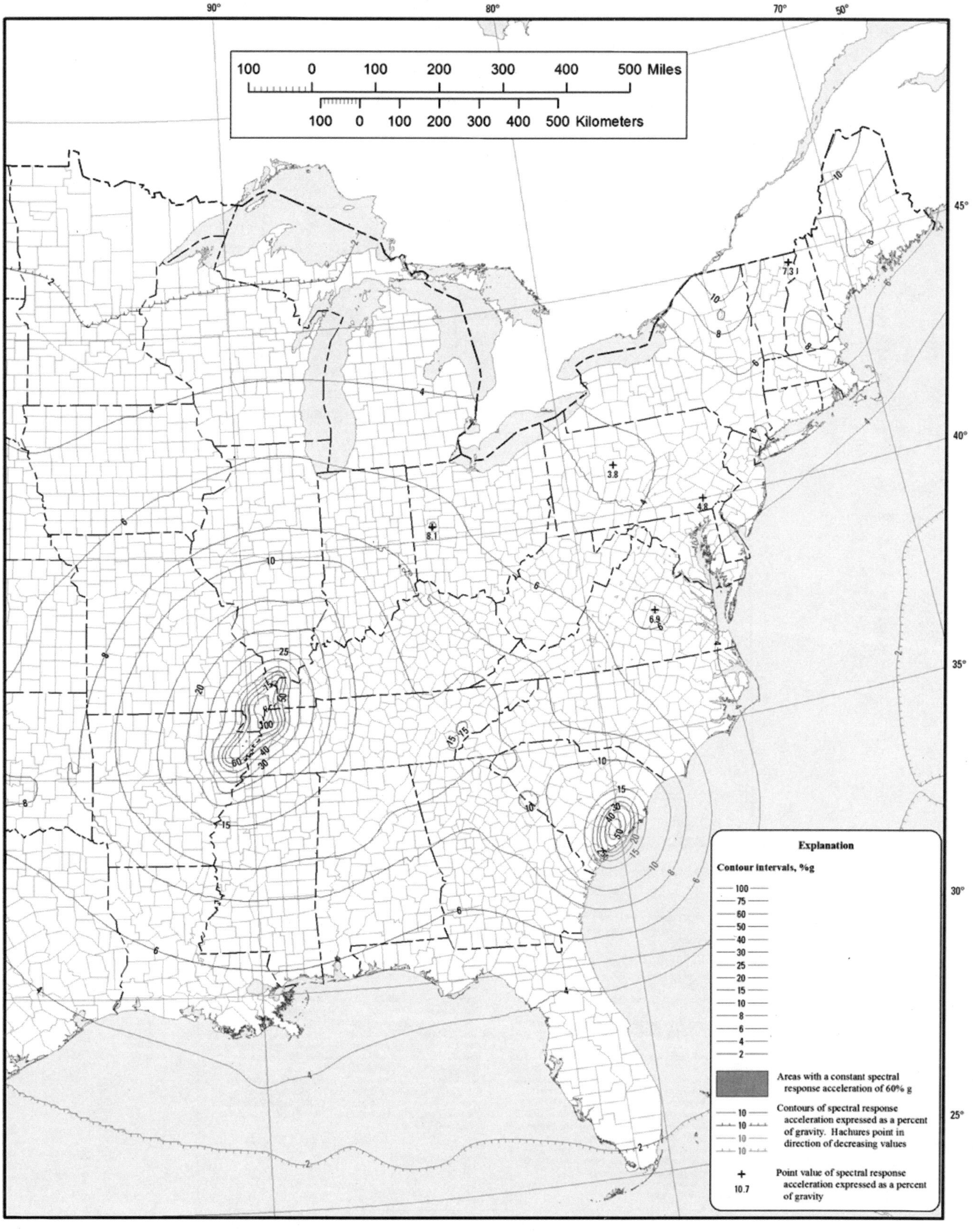

FIGURE 1613.2.1(4)
RISK-TARGETED MAXIMUM CONSIDERED EARTHQUAKE (MCE_R) GROUND MOTION RESPONSE ACCELERATIONS FOR THE COTERMINOUS UNITED STATES OF 1-SECOND SPECTRAL RESPONSE ACCELERATION (5% OF CRITICAL DAMPING)

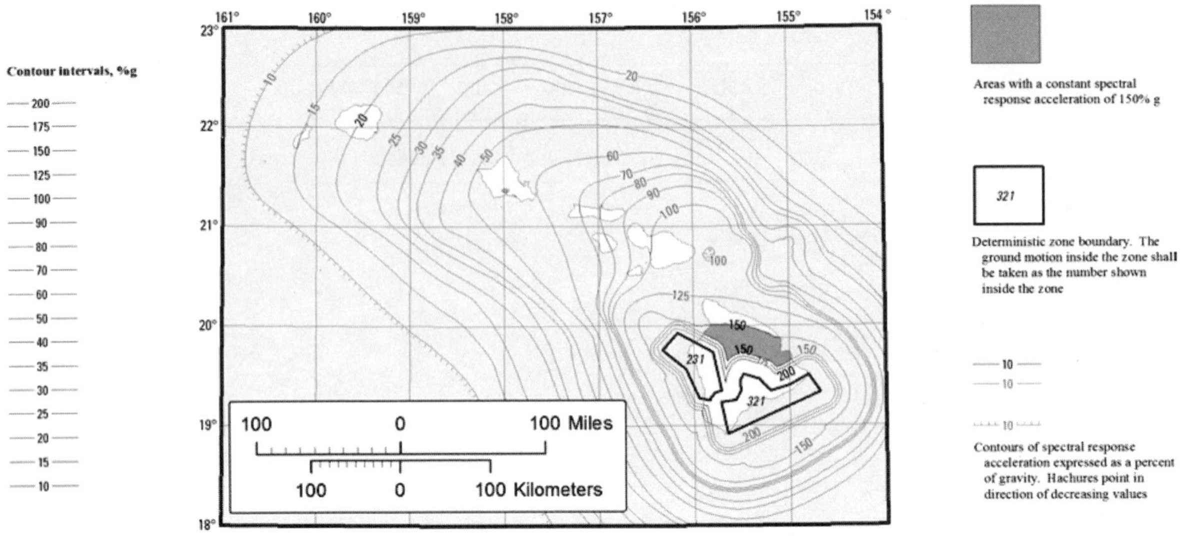

0.2 Second Spectral Response Acceleration (5% of Critical Damping)

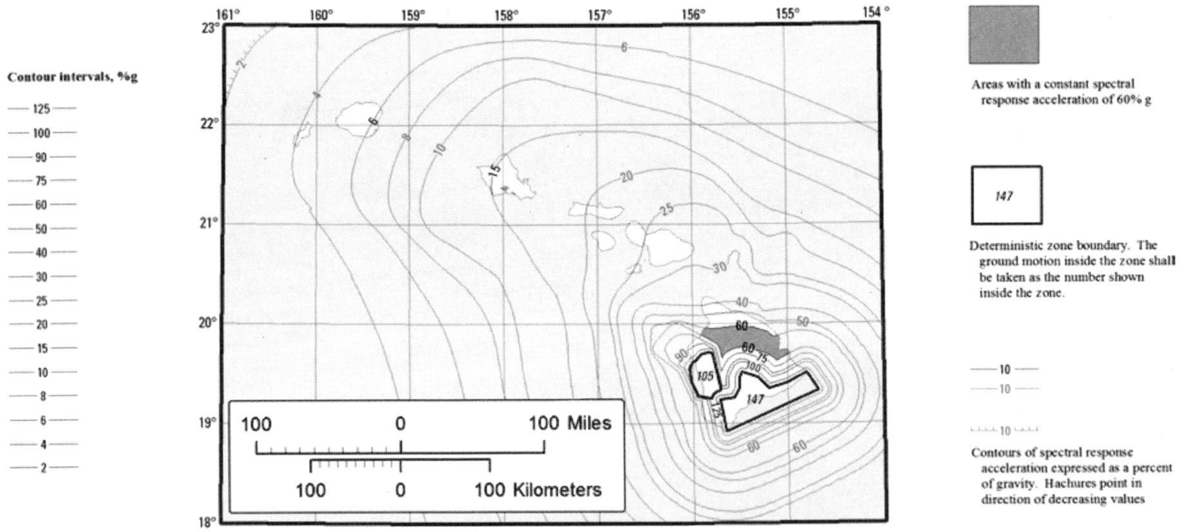

1.0 Second Spectral Response Acceleration (5% of Critical Damping)

DISCUSSION

Maps prepared by United States Geological Survey (USGS) in collaboration with the Federal Emergency Management Agency (FEMA)-funded Building Seismic Safety Council (BSSC) and the American Society of Civil Engineers (ASCE). The basis is explained in commentaries prepared by BSSC and ASCE and in the references.

Ground motion values contoured on these maps incorporate:

- a target risk of structural collapse equal to 1% in 50 years based upon a generic structural fragility
- deterministic upper limits imposed near large, active faults, which are taken as 1.8 times the estimated median response to the characteristic earthquake for the fault (1.8 is used to represent the 84th percentile response), but not less than 150% and 60% g for 0.2 and 1.0 sec, respectively.

As such, the values are different from those on the uniform-hazard 1998 USGS National Seismic Hazard Maps for Hawaii posted at http://earthquake.usgs.gov/hazmaps.

Larger, more detailed versions of these maps are not provided because it is recommended that the corresponding USGS web tool (http://earthquake.usgs.gov/designmaps or http://content.seinstitute.org) be used to determine the mapped value for a specified location.

REFERENCES

Building Seismic Safety Council, 2009, NEHRP Recommended Seismic Provisions for New Buildings and Other Structures: FEMA P-750/2009 Edition, Federal Emergency Management Agency, Washington, DC.

Huang, Yin-Nan, Whittaker, A.S., and Luco, Nicolas, 2008, Maximum spectral demands in the near-fault region, Earthquake Spectra, Volume 24, Issue 1, pp. 319-341.

Klein, F., Frankel, A.D., Mueller, C.S., Wesson, R.L., and Okubo, P., 2001, Seismic hazard in Hawaii: high rate of large earthquakes and probabilistic ground-motion maps, Bulletin of the Seismological Society of America, Volume 91, pp. 479-498.

Luco, Nicolas, Ellingwood, B.R., Hamburger, R.O., Hooper, J.D., Kimball, J.K., and Kircher, C.A., 2007, Risk-Targeted versus Current Seismic Design Maps for the Conterminous United States, Structural Engineers Association of California 2007 Convention Proceedings, pp. 163-175.

FIGURE 1613.2.1(5)
RISK-TARGETED MAXIMUM CONSIDERED EARTHQUAKE (MCE_R) GROUND MOTION RESPONSE ACCELERATIONS FOR HAWAII OF 0.2- AND 1-SECOND SPECTRAL RESPONSE ACCELERATION (5% OF CRITICAL DAMPING)

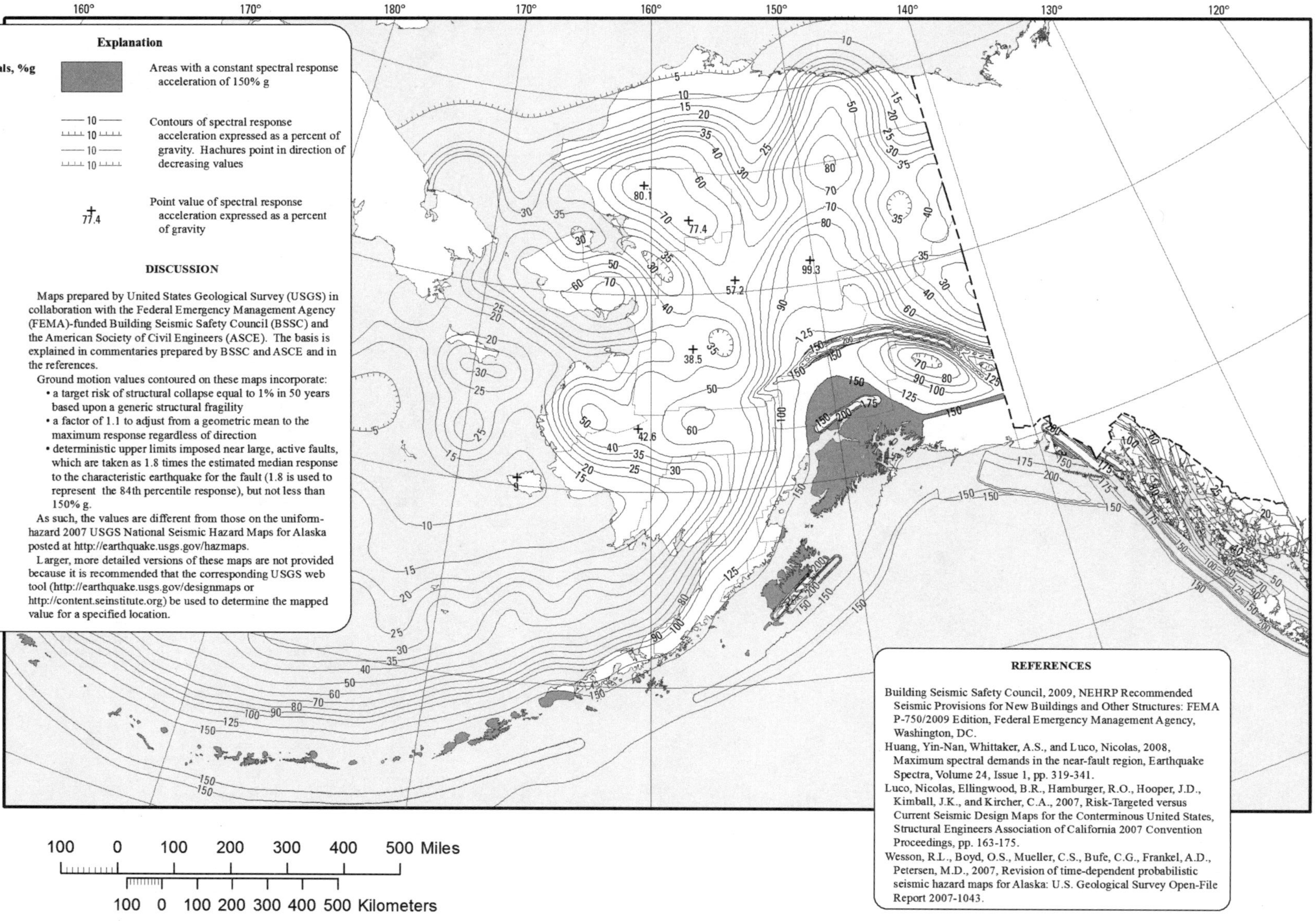

Explanation

Contour intervals, %g

200, 175, 150, 125, 100, 90, 80, 70, 60, 50, 40, 35, 30, 25, 20, 15, 10, 5

Areas with a constant spectral response acceleration of 150% g

Contours of spectral response acceleration expressed as a percent of gravity. Hachures point in direction of decreasing values

Point value of spectral response acceleration expressed as a percent of gravity (77.4)

DISCUSSION

Maps prepared by United States Geological Survey (USGS) in collaboration with the Federal Emergency Management Agency (FEMA)-funded Building Seismic Safety Council (BSSC) and the American Society of Civil Engineers (ASCE). The basis is explained in commentaries prepared by BSSC and ASCE and in the references.

Ground motion values contoured on these maps incorporate:

- a target risk of structural collapse equal to 1% in 50 years based upon a generic structural fragility
- a factor of 1.1 to adjust from a geometric mean to the maximum response regardless of direction
- deterministic upper limits imposed near large, active faults, which are taken as 1.8 times the estimated median response to the characteristic earthquake for the fault (1.8 is used to represent the 84th percentile response), but not less than 150% g.

As such, the values are different from those on the uniform-hazard 2007 USGS National Seismic Hazard Maps for Alaska posted at http://earthquake.usgs.gov/hazmaps.

Larger, more detailed versions of these maps are not provided because it is recommended that the corresponding USGS web tool (http://earthquake.usgs.gov/designmaps or http://content.seinstitute.org) be used to determine the mapped value for a specified location.

REFERENCES

Building Seismic Safety Council, 2009, NEHRP Recommended Seismic Provisions for New Buildings and Other Structures: FEMA P-750/2009 Edition, Federal Emergency Management Agency, Washington, DC.

Huang, Yin-Nan, Whittaker, A.S., and Luco, Nicolas, 2008, Maximum spectral demands in the near-fault region, Earthquake Spectra, Volume 24, Issue 1, pp. 319-341.

Luco, Nicolas, Ellingwood, B.R., Hamburger, R.O., Hooper, J.D., Kimball, J.K., and Kircher, C.A., 2007, Risk-Targeted versus Current Seismic Design Maps for the Conterminous United States, Structural Engineers Association of California 2007 Convention Proceedings, pp. 163-175.

Wesson, R.L., Boyd, O.S., Mueller, C.S., Bufe, C.G., Frankel, A.D., Petersen, M.D., 2007, Revision of time-dependent probabilistic seismic hazard maps for Alaska: U.S. Geological Survey Open-File Report 2007-1043.

FIGURE 1613.2.1(6)
RISK-TARGETED MAXIMUM CONSIDERED EARTHQUAKE (MCE_R) GROUND MOTION RESPONSE ACCELERATIONS FOR ALASKA OF 0.2-SECOND SPECTRAL RESPONSE ACCELERATION (5% OF CRITICAL DAMPING)

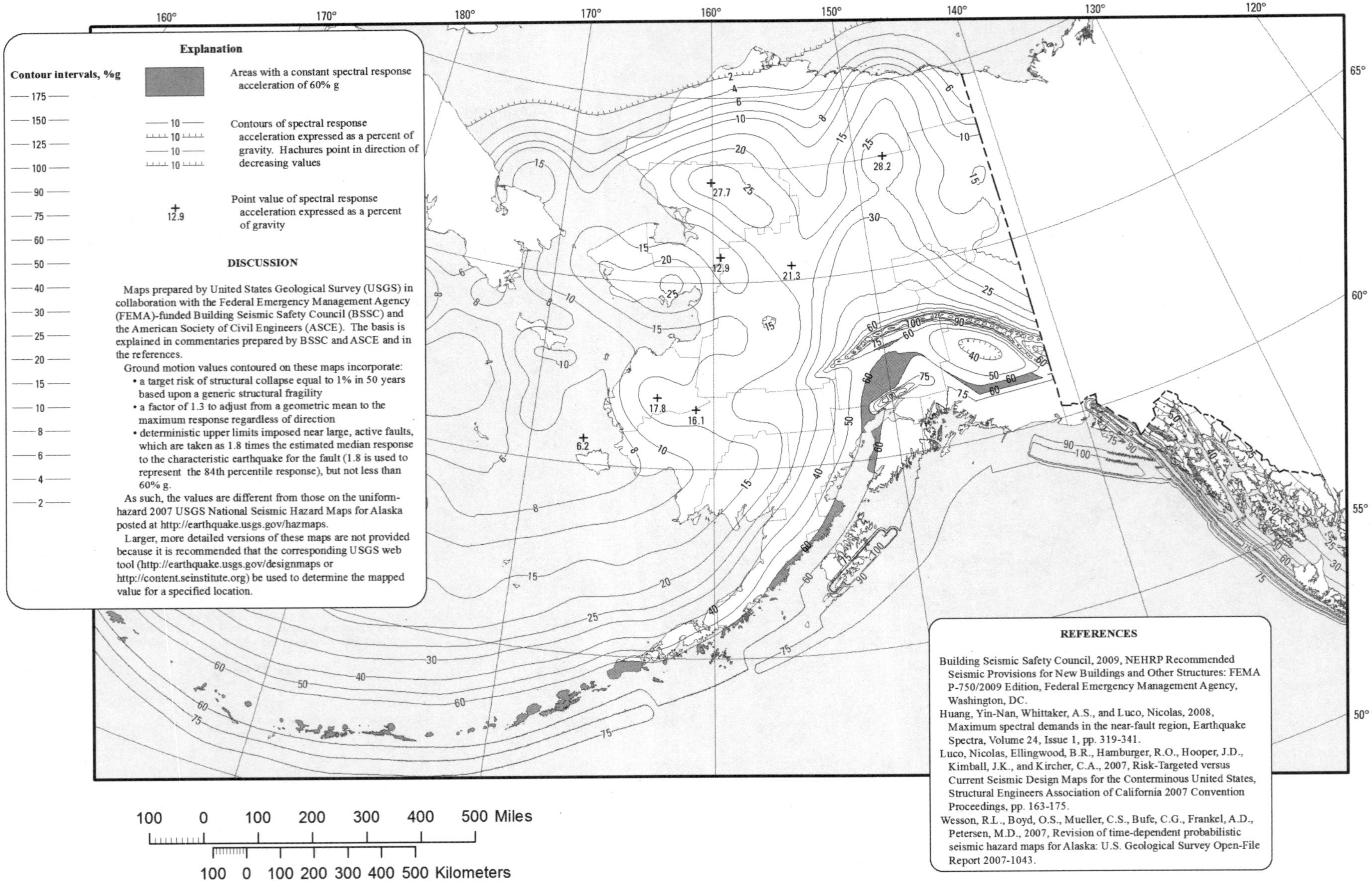

FIGURE 1613.2.1(7)
RISK-TARGETED MAXIMUM CONSIDERED EARTHQUAKE (MCE_R) GROUND MOTION RESPONSE ACCELERATIONS FOR ALASKA OF 1.0-SECOND SPECTRAL RESPONSE ACCELERATION (5% OF CRITICAL DAMPING)

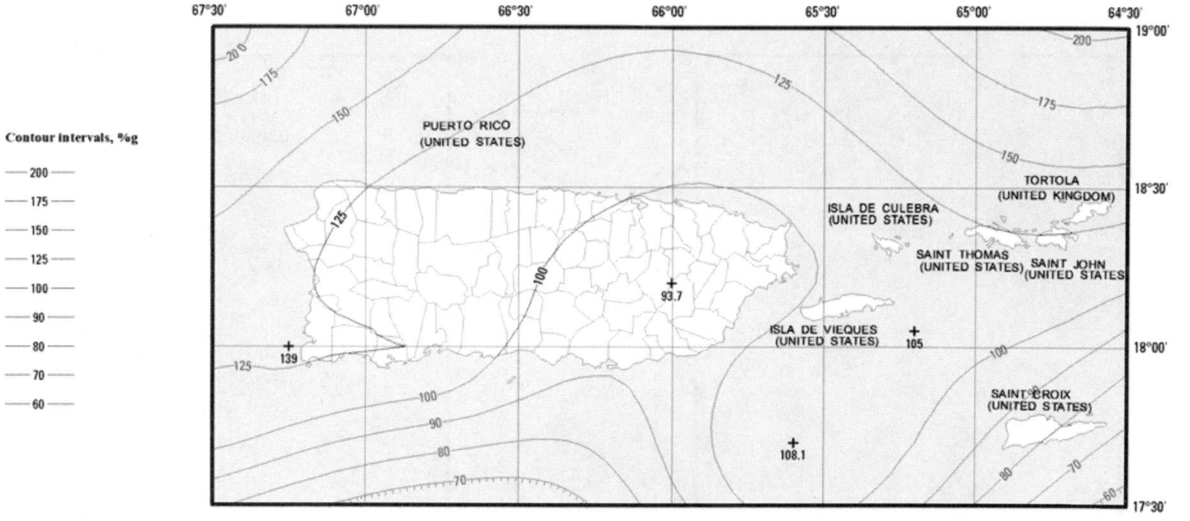

0.2 Second Spectral Response Acceleration (5% of Critical Damping)

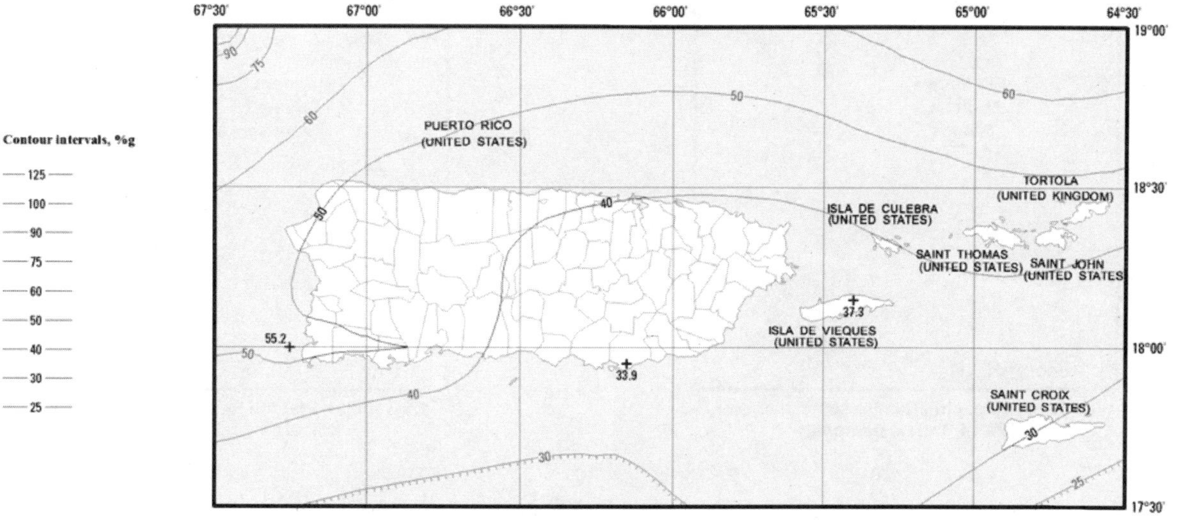

1.0 Second Spectral Response Acceleration (5% of Critical Damping)

Explanation

10
10
10

Contours of spectral response acceleration expressed as a percent of gravity. Hachures point in direction of decreasing values

+
93.7

Point value of spectral response acceleration expressed as a percent of gravity

DISCUSSION

Maps prepared by United States Geological Survey (USGS) in collaboration with the Federal Emergency Management Agency (FEMA)-funded Building Seismic Safety Council (BSSC) and the American Society of Civil Engineers (ASCE). The basis is explained in commentaries prepared by BSSC and ASCE and in the references.

Ground motion values contoured on these maps incorporate:

- a target risk of structural collapse equal to 1% in 50 years based upon a generic structural fragility
- a factor of 1.1 and 1.3 for 0.2 and 1.0 sec, respectively, to adjust from a geometric mean to the maximum response regardless of direction
- deterministic upper limits imposed near large, active faults, which are taken as 1.8 times the estimated median response to the characteristic earthquake for the fault (1.8 is used to represent the 84th percentile response), but not less than 150% and 60% g for 0.2 and 1.0 sec, respectively.

As such, the values are different from those on the uniform-hazard 2003 USGS National Seismic Hazard Maps for Puerto Rico and the U.S. Virgin Islands posted at http://earthquake.usgs.gov/hazmaps.

Larger, more detailed versions of these maps are not provided because it is recommended that the corresponding USGS web tool (http://earthquake.usgs.gov/designmaps or http://content.seinstitute.org) be used to determine the mapped value for a specified location.

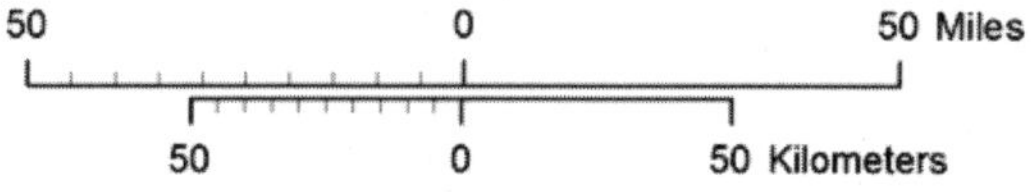

REFERENCES

Building Seismic Safety Council, 2009, NEHRP Recommended Seismic Provisions for New Buildings and Other Structures: FEMA P-750/2009 Edition, Federal Emergency Management Agency, Washington, DC.

Huang, Yin-Nan, Whittaker, A.S., and Luco, Nicolas, 2008, Maximum spectral demands in the near-fault region, Earthquake Spectra, Volume 24, Issue 1, pp. 319-341.

Luco, Nicolas, Ellingwood, B.R., Hamburger, R.O., Hooper, J.D., Kimball, J.K., and Kircher, C.A., 2007, Risk-Targeted versus Current Seismic Design Maps for the Conterminous United States, Structural Engineers Association of California 2007 Convention Proceedings, pp. 163-175.

Mueller, C.S., Frankel, A.D., Petersen, M.D., and Leyendecker, E.V., 2003, Documentation for the 2003 USGS Seismic Hazard Maps for Puerto Rico and the U.S. Virgin Islands: U.S. Geological Survey Open-File Report 03-379.

FIGURE 1613.2.1(8)
RISK-TARGETED MAXIMUM CONSIDERED EARTHQUAKE (MCE_R) GROUND MOTION RESPONSE ACCELERATIONS FOR PUERTO RICO AND THE UNITED STATES VIRGIN ISLANDS OF 0.2- AND 1-SECOND SPECTRAL RESPONSE ACCELERATION (5% OF CRITICAL DAMPING)

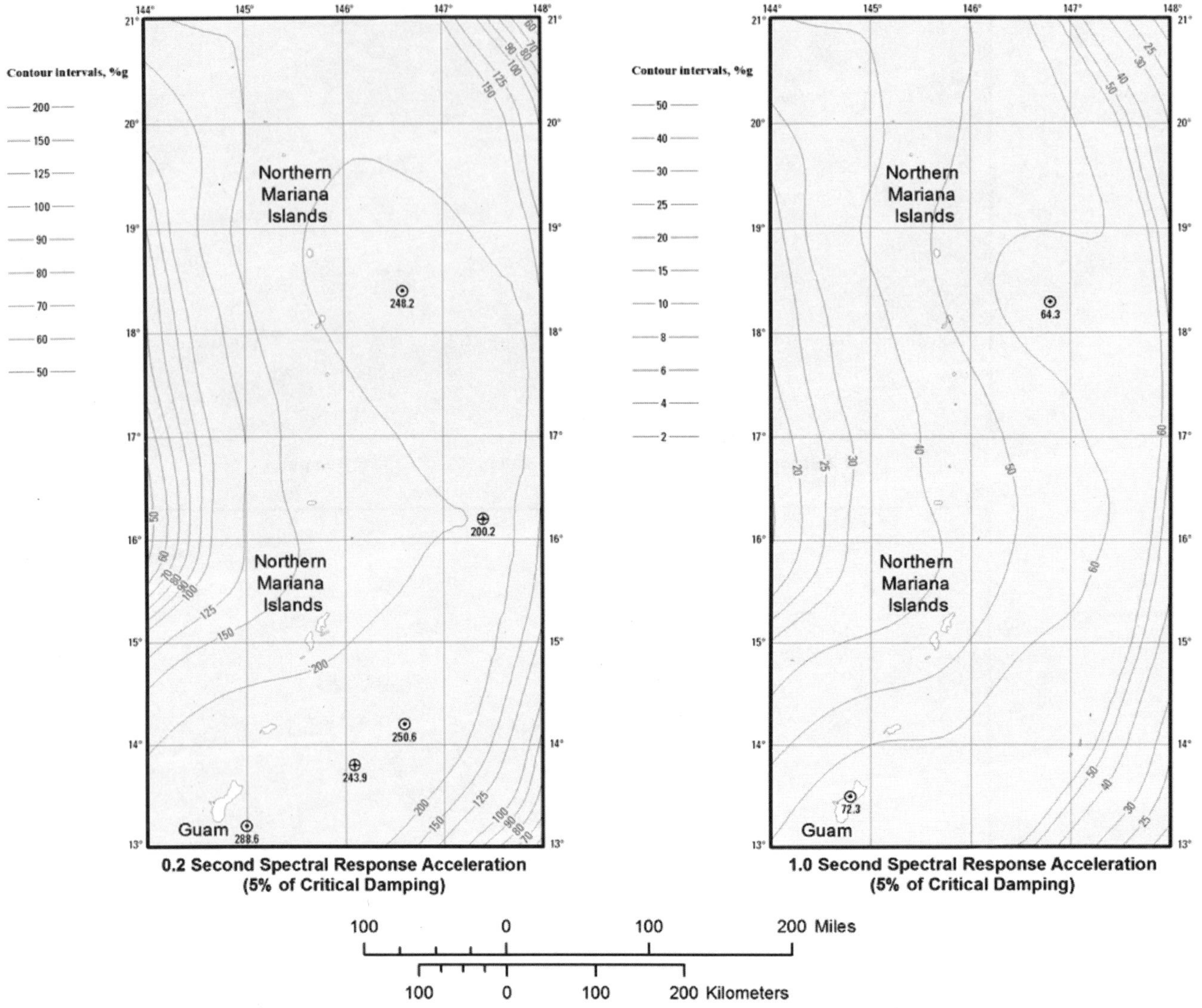

Explanation

Contours of spectral response acceleration expressed as a percent of gravity.

— 10 —

— 10 —

Point values of spectral response acceleration expressed as a percent of gravity.

⊕ 200.2 Local minimum

⊙ 250.6 Local maximum

⊕ 243.9 Saddle point

DISCUSSION

Maps prepared by United States Geological Survey (USGS) in collaboration with the Federal Emergency Management Agency (FEMA)-funded Building Seismic Safety Council (BSSC). The basis is explained in commentary prepared by BSSC and in the references.

Ground motion values contoured on these maps incorporate:

- a target risk of structural collapse equal to 1% in 50 years based upon a generic structural fragility
- a factor of 1.1 and 1.3 for 0.2 and 1.0 sec, respectively, to adjust from a geometric mean to the maximum response regardless of direction
- deterministic upper limits imposed near large, active faults, which are taken as 1.8 times the estimated median response to the characteristic earthquake for the fault (1.8 is used to represent the 84th percentile response), but not less than 150% and 60% g for 0.2 and 1.0 sec, respectively.

As such, the values are different from those on the uniform-hazard 2012 USGS National Seismic Hazard Maps for Guam and the Northern Mariana Islands posted at http://earthquake.usgs.gov/hazmaps.

Larger, more detailed versions of these maps are not provided because it is recommended that the corresponding USGS web tool (http://earthquake.usgs.gov/designmaps) be used to determine the mapped value for a specified location.

REFERENCES

Building Seismic Safety Council, 2009, NEHRP Recommended Seismic Provisions for New Buildings and Other Structures: FEMA P-750/2009 Edition, Federal Emergency Management Agency, Washington, DC.

Huang, Yin-Nan, Whittaker, A.S., and Luco, Nicolas, 2008, Maximum spectral demands in the near-fault region, Earthquake Spectra, Volume 24, Issue 1, pp. 319-341.

Luco, Nicolas, Ellingwood, B.R., Hamburger, R.O., Hooper, J.D., Kimball, J.K., and Kircher, C.A., 2007, Risk-Targeted versus Current Seisimic Design Maps for the Conterminous United States, Structural Engineers Association of California 2007 Convention Proceedings, pp. 163-175.

Mueller, C.S., Haller, K.M., Luco, Nicolas, Petersen, M.D., and Frankel, A.D., 2012, Seismic Hazard Assessment for Guam and the Northern Mariana Islands: U.S. Geological Survey Open-File Report 2012–1015.

FIGURE 1613.2.1(9)
RISK-TARGETED MAXIMUM CONSIDERED EARTHQUAKE (MCE_R) GROUND MOTION RESPONSE ACCELERATIONS FOR GUAM AND THE NORTHERN MARIANA ISLANDS OF 0.2- AND 1-SECOND SPECTRAL RESPONSE ACCELERATION (5% OF CRITICAL DAMPING)

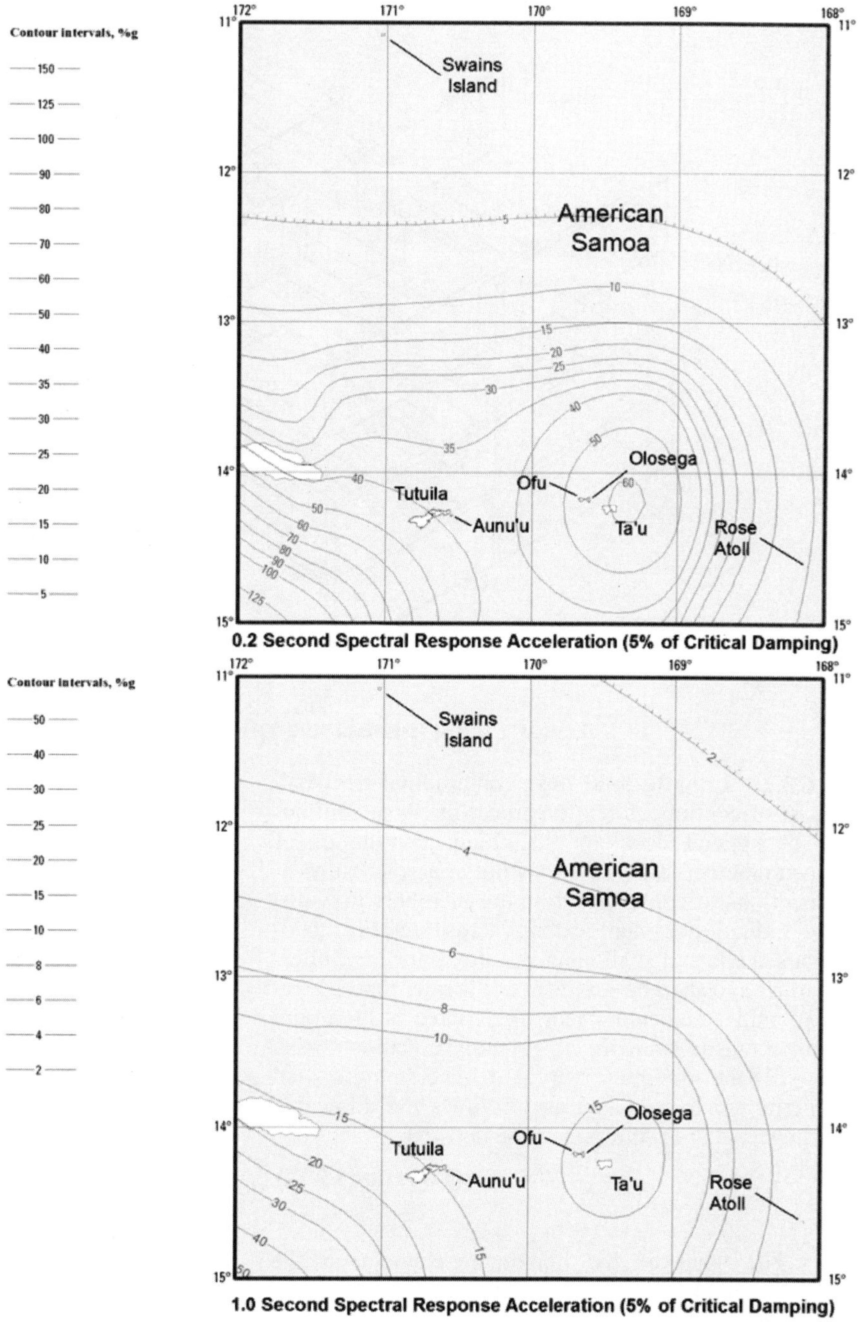

Explanation

10

10

10

Contours of spectral response acceleration expressed as a percent of gravity. Hachures point in direction of decreasing values

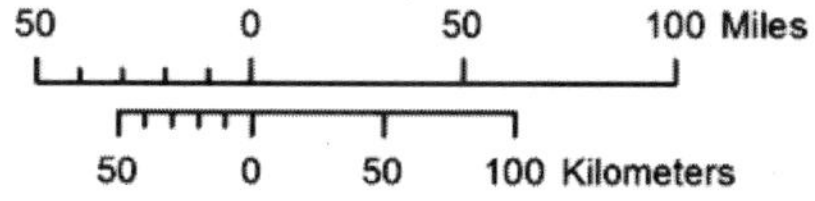

DISCUSSION

Maps prepared by United States Geological Survey (USGS) in collaboration with the Federal Emergency Management Agency (FEMA)-funded Building Seismic Safety Council (BSSC). The basis is explained in commentary prepared by BSSC and in the references.

Ground motion values contoured on these maps incorporate:

- a target risk of structural collapse equal to 1% in 50 years based upon a generic structural fragility
- a factor of 1.1 and 1.3 for 0.2 and 1.0 sec, respectively, to adjust from a geometric mean to the maximum response regardless of direction
- deterministic upper limits imposed near large, active faults, which are taken as 1.8 times the estimated median response to the characteristic earthquake for the fault (1.8 is used to represent the 84th percentile response), but not less than 150% and 60% g for 0.2 and 1.0 sec, respectively.

As such, the values are different from those on the uniform-hazard 2012 USGS National Seismic Hazard Maps for American Samoa posted at http://earthquake.usgs.gov/hazmaps.

Larger, more detailed versions of these maps are not provided because it is recommended that the corresponding USGS web tool (http://earthquake.usgs.gov/designmaps) be used to determine the mapped value for a specified location.

REFERENCES

Building Seismic Safety Council, 2009, NEHRP Recommended Seismic Provisions for New Buildings and Other Structures: FEMA P-750/2009 Edition, Federal Emergency Management Agency, Washington, DC.

Huang, Yin-Nan, Whittaker, A.S., and Luco, Nicolas, 2008, Maximum spectral demands in the near-fault region, Earthquake Spectra, Volume 24, Issue 1, pp. 319-341.

Luco, Nicolas, Ellingwood, B.R., Hamburger, R.O., Hooper, J.D., Kimball, J.K., and Kircher, C.A., 2007, Risk-Targeted versus Current Seismic Design Maps for the Conterminous United States, Structural Engineers Association of California 2007 Convention Proceedings, pp. 163-175.

Petersen, M.D., Harmsen, S.C., Rukstales, K.S., Mueller, C.S., McNamara, D.E., Luco, Nicolas, and Walling, Melanie, 2012, Seismic Hazard of American Samoa and Neighboring South Pacific Islands: Data, Methods, Parameters, and Results: U.S. Geological Survey Open-File Report 2012–1087.

FIGURE 1613.2.1(10)
RISK-TARGETED MAXIMUM CONSIDERED EARTHQUAKE (MCE_R) GROUND MOTION RESPONSE ACCELERATIONS FOR AMERICAN SAMOA OF 0.2- AND 1-SECOND SPECTRAL RESPONSE ACCELERATION (5% OF CRITICAL DAMPING)

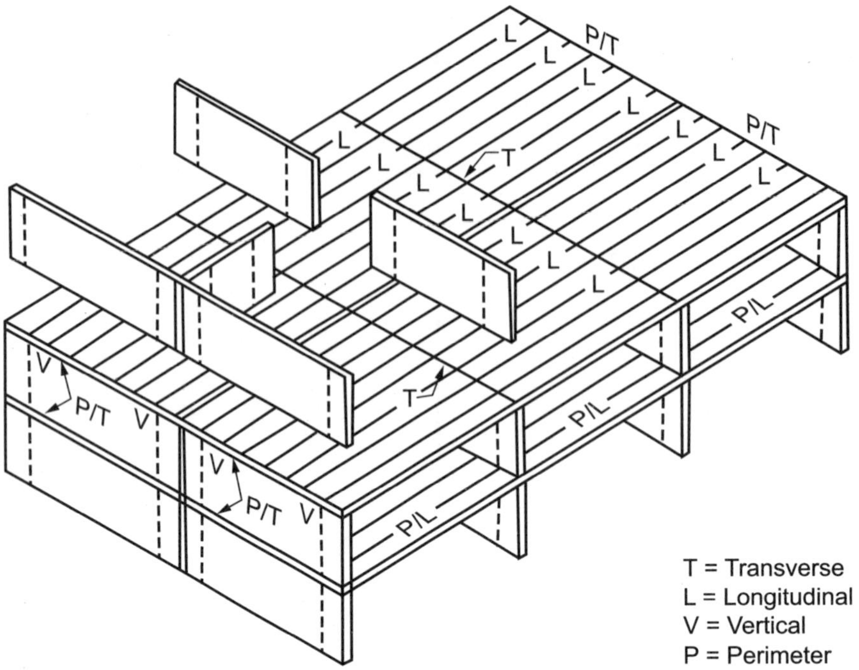

FIGURE 1616.3
LONGITUDINAL, PERIMETER, TRANSVERSE AND VERTICAL TIES

1616.3.2.1 Longitudinal ties. Longitudinal ties shall consist of continuous reinforcement in slabs; continuous or spliced decks or sheathing; continuous or spliced members framing to, within or across walls; or connections of continuous framing members to walls. Longitudinal ties shall extend across interior load-bearing walls and shall connect to exterior *load-bearing walls* and shall be spaced at not greater than 10 feet (3038 mm) on center. Ties shall have a minimum nominal tensile strength, T_T, given by Equation 16-24. For ASD the minimum nominal tensile strength shall be permitted to be taken as 1.5 times the allowable tensile stress times the area of the tie.

$$T_T = w\,LS \leq \alpha_T S \qquad \textbf{(Equation 16-24)}$$

where:

L = The span of the horizontal element in the direction of the tie, between bearing walls, feet (m).

w = The weight per unit area of the floor or roof in the span being tied to or across the wall, psf (N/m^2).

S = The spacing between ties, feet (m).

α_T = A coefficient with a value of 1,500 pounds per foot (2.25 kN/m) for masonry *bearing wall structures* and a value of 375 pounds per foot (0.6 kN/m) for structures with bearing walls of cold-formed steel *light-frame construction*.

1616.3.2.2 Transverse ties. Transverse ties shall consist of continuous reinforcement in slabs; continuous or spliced decks or sheathing; continuous or spliced members framing to, within or across walls; or connections of continuous framing members to walls. Transverse ties shall be placed not farther apart than the spacing of *load-bearing walls*. Transverse ties shall have minimum nominal tensile strength T_T, given by Equation 16-24. For ASD the minimum nominal tensile strength shall be permitted to be taken as 1.5 times the allowable tensile stress times the area of the tie.

1616.3.2.3 Perimeter ties. Perimeter ties shall consist of continuous reinforcement in slabs; continuous or spliced decks or sheathing; continuous or spliced members framing to, within or across walls; or connections of continuous framing members to walls. Ties around the perimeter of each floor and roof shall be located within 4 feet (1219 mm) of the edge and shall provide a nominal strength in tension not less than T_p, given by Equation 16-25. For ASD the minimum nominal tensile strength shall be permitted to be taken as 1.5 times the allowable tensile stress times the area of the tie.

$$T_p = 200w \leq \beta_T \qquad \textbf{(Equation 16-25)}$$

For SI: $T_p = 90.7w \leq \beta_T$

where:

w = As defined in Section 1616.3.2.1.

β_T = A coefficient with a value of 16,000 pounds (7200 kN) for structures with masonry bearing walls and a value of 4,000 pounds (1300 kN) for structures with bearing walls of cold-formed steel *light-frame construction*.

1616.3.2.4 Vertical ties. Vertical ties shall consist of continuous or spliced reinforcing, continuous or spliced members, wall sheathing or other engineered systems. Vertical tension ties shall be provided in bearing walls and shall be continuous over the height of the building. The minimum nominal tensile strength for vertical ties within a bearing wall shall be equal to the weight of the wall within that *story* plus the weight of the *diaphragm* tributary to the wall in the *story* below. Not fewer than two ties shall be provided for each wall. The strength of each tie need not exceed 3,000 pounds per foot (450 kN/m) of wall tributary to the tie for walls of masonry construction or 750 pounds per foot (140 kN/m) of wall tributary to the tie for walls of cold-formed steel *light-frame construction*.

CHAPTER 17

SPECIAL INSPECTIONS AND TESTS

User notes:

About this chapter: *Chapter 17 provides a variety of procedures and criteria for testing materials and assemblies, and labeling materials and assemblies. Its key purposes are to establish where additional inspections/observations and testing must be provided, and the submittals and verifications that must be provided to the building official. This chapter expands on the inspections of Chapter 1 by requiring special inspection by a qualified individual where indicated and, in some cases, structural observation by a registered design professional. Quality assurance measures that verify proper assembly of structural components and the suitability of the installed materials are intended to provide a building that, once constructed, complies with the minimum structural and fire-resistance code requirements as well as the approved design. To determine this compliance often requires frequent inspections and testing at specific stages of construction.*

Code development reminder: *Code change proposals to sections preceded by the designation [BF] will be considered by the IBC—Fire Safety Code Development Committee during the 2021 (Group A) Code Development Cycle. Sections preceded by the designation [F] will be considered by the International Fire Code Development Committee during the 2021 (Group A) Code Development Cycle. All other code change proposals will be considered by the IBC—Structural Code Development Committee during the Group B cycle.*

SECTION 1701
GENERAL

1701.1 Scope. The provisions of this chapter shall govern the quality, workmanship and requirements for materials covered. Materials of construction and tests shall conform to the applicable standards listed in this code.

SECTION 1702
NEW MATERIALS

1702.1 General. New building materials, equipment, appliances, systems or methods of construction not provided for in this code, and any material of questioned suitability proposed for use in the construction of a building or structure, shall be subjected to the tests prescribed in this chapter and in the approved rules to determine character, quality and limitations of use.

SECTION 1703
APPROVALS

1703.1 Approved agency. An approved agency shall provide all information as necessary for the *building official* to determine that the agency meets the applicable requirements specified in Sections 1703.1.1 through 1703.1.3.

1703.1.1 Independence. An *approved agency* shall be objective, competent and independent from the contractor responsible for the work being inspected. The agency shall disclose to the *building official* and the *registered design professional in responsible charge* possible conflicts of interest so that objectivity can be confirmed.

1703.1.2 Equipment. An *approved agency* shall have adequate equipment to perform required tests. The equipment shall be periodically calibrated.

1703.1.3 Personnel. An *approved agency* shall employ experienced personnel educated in conducting, supervising and evaluating tests and *special inspections*.

1703.2 Written approval. Any material, appliance, equipment, system or method of construction meeting the requirements of this code shall be *approved* in writing after satisfactory completion of the required tests and submission of required test reports.

1703.3 Record of approval. For any material, appliance, equipment, system or method of construction that has been *approved*, a record of such approval, including the conditions and limitations of the approval, shall be kept on file in the *building official*'s office and shall be available for public review at appropriate times.

1703.4 Performance. Specific information consisting of test reports conducted by an *approved agency* in accordance with the appropriate referenced standards, or other such information as necessary, shall be provided for the *building official* to determine that the product, material or assembly meets the applicable code requirements.

1703.4.1 Research and investigation. Sufficient technical data shall be submitted to the *building official* to substantiate the proposed use of any product, material or assembly. If it is determined that the evidence submitted is satisfactory proof of performance for the use intended, the *building official* shall approve the use of the product, material or assembly subject to the requirements of this code. The costs, reports and investigations required under these provisions shall be paid by the owner or the owner's authorized agent.

1703.4.2 Research reports. Supporting data, where necessary to assist in the approval of products, materials or assemblies not specifically provided for in this code, shall consist of valid research reports from *approved* sources.

1703.5 Labeling. Products, materials or assemblies required to be *labeled* shall be *labeled* in accordance with the procedures set forth in Sections 1703.5.1 through 1703.5.4.

1703.5.1 Testing. An *approved agency* shall test a representative sample of the product, material or assembly being *labeled* to the relevant standard or standards. The *approved agency* shall maintain a record of the tests performed. The record shall provide sufficient detail to verify compliance with the test standard.

1703.5.2 Inspection and identification. The *approved agency* shall periodically perform an inspection, which shall be in-plant if necessary, of the product or material that is to be *labeled*. The inspection shall verify that the labeled product, material or assembly is representative of the product, material or assembly tested.

1703.5.3 Label information. The *label* shall contain the manufacturer's identification, model number, serial number or definitive information describing the performance characteristics of the product, material or assembly and the *approved agency*'s identification.

1703.5.4 Method of labeling. Information required to be permanently identified on the product, material or assembly shall be acid etched, sand blasted, ceramic fired, laser etched, embossed or of a type that, once applied, cannot be removed without being destroyed.

1703.6 Evaluation and follow-up inspection services. Where structural components or other items regulated by this code are not visible for inspection after completion of a prefabricated assembly, the owner or the owner's authorized agent shall submit a report of each prefabricated assembly. The report shall indicate the complete details of the assembly, including a description of the assembly and its components, the basis upon which the assembly is being evaluated, test results and similar information and other data as necessary for the *building official* to determine conformance to this code. Such a report shall be *approved* by the *building official*.

1703.6.1 Follow-up inspection. The owner or the owner's authorized agent shall provide for *special inspections* of *fabricated items* in accordance with Section 1704.2.5.

1703.6.2 Test and inspection records. Copies of necessary test and *special inspection* records shall be filed with the *building official*.

SECTION 1704 SPECIAL INSPECTIONS AND TESTS, CONTRACTOR RESPONSIBILITY AND STRUCTURAL OBSERVATION

1704.1 General. *Special inspections* and tests, statements of *special inspections*, responsibilities of contractors, submittals to the *building official* and *structural observations* shall meet the applicable requirements of this section.

1704.2 Special inspections and tests. Where application is made to the *building official* for construction as specified in Section 105, the owner or the owner's authorized agent, other than the contractor, shall employ one or more *approved agencies* to provide *special inspections* and tests during construction on the types of work specified in Section 1705 and identify the *approved agencies* to the *building official*. These *special inspections* and tests are in addition to the inspections by the *building official* that are identified in Section 110.

Exceptions:

1. *Special inspections* and tests are not required for construction of a minor nature or as warranted by conditions in the jurisdiction as *approved* by the *building official*.
2. Unless otherwise required by the *building official*, *special inspections* and tests are not required for Group U occupancies that are accessory to a residential occupancy including, but not limited to, those listed in Section 312.1.
3. *Special inspections* and tests are not required for portions of structures designed and constructed in accordance with the cold-formed steel *light-frame construction* provisions of Section 2211.1.2 or the *conventional light-frame construction* provisions of Section 2308.
4. The contractor is permitted to employ the *approved agencies* where the contractor is also the owner.

1704.2.1 Special inspector qualifications. Prior to the start of the construction, the *approved agencies* shall provide written documentation to the *building official* demonstrating the competence and relevant experience or training of the *special inspectors* who will perform the *special inspections* and tests during construction. Experience or training shall be considered to be relevant where the documented experience or training is related in complexity to the same type of *special inspection* or testing activities for projects of similar complexity and material qualities. These qualifications are in addition to qualifications specified in other sections of this code.

The *registered design professional in responsible charge* and engineers of record involved in the design of the project are permitted to act as an *approved agency* and their personnel are permitted to act as *special inspectors* for the work designed by them, provided they qualify as *special inspectors*.

1704.2.2 Access for special inspection. The construction or work for which *special inspection* or testing is required shall remain accessible and exposed for *special inspection* or testing purposes until completion of the required *special inspections* or tests.

1704.2.3 Statement of special inspections. The applicant shall submit a statement of *special inspections* in accordance with Section 107.1 as a condition for permit issuance. This statement shall be in accordance with Section 1704.3.

Exception: A statement of *special inspections* is not required for portions of structures designed and constructed in accordance with the cold-formed steel *light-frame construction* provisions of Section 2211.1.2 or the *conventional light-frame construction* provisions of Section 2308.

1704.2.4 Report requirement. *Approved agencies* shall keep records of *special inspections* and tests. The *approved agency* shall submit reports of *special inspections* and tests to the *building official* and to the *registered design professional in responsible charge*. Reports shall indicate that work inspected or tested was or was not completed in conformance to *approved construction documents*. Discrepancies shall be brought to the immediate attention of the contractor for correction. If they are not corrected, the discrepancies shall be brought to the attention of the *building official* and to the *registered design professional in responsible charge* prior to the completion of that phase of the work. A final report documenting required *special inspections* and tests, and correction of any discrepancies noted in the inspections or tests, shall be submitted at a point in time agreed upon prior to the start of work by the owner or the owner's authorized agent to the *building official*.

1704.2.5 Special inspection of fabricated items. Where fabrication of structural, load-bearing or lateral load-resisting members or assemblies is being conducted on the premises of a fabricator's shop, *special inspections* of the *fabricated items* shall be performed during fabrication, except where the fabricator has been *approved* to perform work without *special inspections* in accordance with Section 1704.2.5.1.

1704.2.5.1 Fabricator approval. *Special inspections* during fabrication are not required where the work is done on the premises of a fabricator *approved* to perform such work without *special inspection*. Approval shall be based on review of the fabricator's written fabrication procedures and quality control manuals that provide a basis for control of materials and workmanship, with periodic auditing of fabrication and quality control practices by an *approved agency* or the *building official*. At completion of fabrication, the *approved fabricator* shall submit a *certificate of compliance* to the owner or the owner's authorized agent for submittal to the *building official* as specified in Section 1704.5 stating that the work was performed in accordance with the *approved construction documents*.

1704.3 Statement of special inspections. Where *special inspections* or tests are required by Section 1705, the *registered design professional in responsible charge* shall prepare a statement of *special inspections* in accordance with Section 1704.3.1 for submittal by the applicant in accordance with Section 1704.2.3.

Exception: The statement of *special inspections* is permitted to be prepared by a qualified person *approved* by the *building official* for construction not designed by a *registered design professional*.

1704.3.1 Content of statement of special inspections. The statement of *special inspections* shall identify the following:

1. The materials, systems, components and work required to have *special inspections* or tests by the *building official* or by the *registered design professional* responsible for each portion of the work.
2. The type and extent of each *special inspection*.
3. The type and extent of each test.
4. Additional requirements for *special inspections* or tests for seismic or wind resistance as specified in Sections 1705.12, 1705.13 and 1705.14 .
5. For each type of *special inspection*, identification as to whether it will be continuous *special inspection*, periodic *special inspection* or performed in accordance with the notation used in the referenced standard where the inspections are defined.

1704.3.2 Seismic requirements in the statement of special inspections. Where Section 1705.13 or 1705.14 specifies *special inspections* or tests for seismic resistance, the statement of *special inspections* shall identify the *designated seismic systems* and *seismic force-resisting systems* that are subject to the *special inspections* or tests.

1704.3.3 Wind requirements in the statement of special inspections. Where Section 1705.12 specifies *special inspection* for wind resistance, the statement of *special inspections* shall identify the *main windforce-resisting systems* and wind-resisting components that are subject to *special inspections*.

1704.4 Contractor responsibility. Each contractor responsible for the construction of a main wind- or *seismic force-resisting system*, *designated seismic system* or a wind- or seismic force-resisting component listed in the statement of *special inspections* shall submit a written statement of responsibility to the *building official* and the owner or the owner's authorized agent prior to the commencement of work on the system or component. The contractor's statement of responsibility shall contain acknowledgement of awareness of the special requirements contained in the statement of *special inspections*.

1704.5 Submittals to the building official. In addition to the submittal of reports of *special inspections* and tests in accordance with Section 1704.2.4, reports and certificates shall be submitted by the owner or the owner's authorized agent to the *building official* for each of the following:

1. *Certificates of compliance* for the fabrication of structural, load-bearing or lateral load-resisting members or assemblies on the premises of an *approved fabricator* in accordance with Section 1704.2.5.1.
2. *Certificates of compliance* for the seismic qualification of nonstructural components, supports and attachments in accordance with Section 1705.14.2.
3. *Certificates of compliance* for *designated seismic systems* in accordance with Section 1705.14.3.
4. Reports of preconstruction tests for shotcrete in accordance with ACI 318.
5. *Certificates of compliance* for open web *steel joists* and joist girders in accordance with Section 2207.5.

6. Reports of material properties verifying compliance with the requirements of AWS D1.4 for weldability as specified in Section 26.6.4 of ACI 318 for reinforcing bars in concrete complying with a standard other than ASTM A706 that are to be welded.
7. Reports of mill tests in accordance with Section 20.2.2.5 of ACI 318 for reinforcing bars complying with ASTM A615 and used to resist earthquake-induced flexural or axial forces in the special moment frames, special structural walls or coupling beams connecting special structural walls of *seismic force-resisting systems* in structures assigned to *Seismic Design Category* B, C, D, E or F.

1704.6 Structural observations. Where required by the provisions of Section 1704.6.1, the owner or the owner's authorized agent shall employ a *registered design professional* to perform *structural observations*. The structural observer shall visually observe representative locations of structural systems, details and load paths for general conformance to the approved construction documents. *Structural observation* does not include or waive the responsibility for the inspections in Section 110 or the *special inspections* in Section 1705 or other sections of this code. Prior to the commencement of observations, the structural observer shall submit to the *building official* a written statement identifying the frequency and extent of *structural observations*. At the conclusion of the work included in the permit, the structural observer shall submit to the *building official* a written statement that the site visits have been made and identify any reported deficiencies that, to the best of the structural observer's knowledge, have not been resolved.

1704.6.1 Structural observations for structures. *Structural observations* shall be provided for those structures where one or more of the following conditions exist:

1. The structure is classified as *Risk Category III or* IV.
2. The structure is a *high-rise building*.
3. The structure is assigned to *Seismic Design Category* E, and is greater than two stories above the grade plane.
4. Such observation is required by the *registered design professional* responsible for the structural design.
5. Such observation is specifically required by the *building official*.

SECTION 1705
REQUIRED SPECIAL INSPECTIONS AND TESTS

1705.1 General. *Special inspections* and tests of elements and nonstructural components of buildings and structures shall meet the applicable requirements of this section.

1705.1.1 Special cases. *Special inspections* and tests shall be required for proposed work that is, in the opinion of the *building official*, unusual in its nature, such as, but not limited to, the following examples:

1. Construction materials and systems that are alternatives to materials and systems prescribed by this code.
2. Unusual design applications of materials described in this code.
3. Materials and systems required to be installed in accordance with additional manufacturer's instructions that prescribe requirements not contained in this code or in standards referenced by this code.

1705.2 Steel construction. The *special inspections* and nondestructive testing of steel construction in buildings, structures, and portions thereof shall be in accordance with this section.

Exception: *Special inspections* of the steel fabrication process shall not be required where the fabrication process for the entire building or structure does not include any welding, thermal cutting or heating operation of any kind. In such cases, the fabricator shall be required to submit a detailed procedure for material control that demonstrates the fabricator's ability to maintain suitable records and procedures such that, at any time during the fabrication process, the material specification and grade for the main stress-carrying elements are capable of being determined. Mill test reports shall be identifiable to the main stress-carrying elements where required by the *approved construction documents*.

1705.2.1 Structural steel. *Special inspections* and nondestructive testing of *structural steel elements* in buildings, structures and portions thereof shall be in accordance with the quality assurance inspection requirements of AISC 360.

Exception: *Special inspection* of railing systems composed of *structural steel elements* shall be limited to welding inspection of welds at the base of cantilevered rail posts.

1705.2.2 Cold-formed steel deck. *Special inspections* and qualification of welding *special inspectors* for cold-formed steel floor and *roof deck* shall be in accordance with the quality assurance inspection requirements of SDI QA/QC.

1705.2.3 Open-web steel joists and joist girders. *Special inspections* of open-web *steel joists* and joist girders in buildings, structures and portions thereof shall be in accordance with Table 1705.2.3.

1705.2.4 Cold-formed steel trusses spanning 60 feet or greater. Where a cold-formed steel truss clear span is 60 feet (18 288 mm) or greater, the *special inspector* shall verify that the temporary installation restraint/bracing and the permanent *individual truss member* restraint/bracing are installed in accordance with the *approved* truss submittal package.

TABLE 1705.2.3
REQUIRED SPECIAL INSPECTIONS OF OPEN-WEB STEEL JOISTS AND JOIST GIRDERS

TYPE	CONTINUOUS SPECIAL INSPECTION	PERIODIC SPECIAL INSPECTION	REFERENCED STANDARD[a]
1. Installation of open-web steel joists and joist girders.			
a. End connections – welding or bolted.	—	X	SJI specifications listed in Section 2207.1.
b. Bridging – horizontal or diagonal.	—	—	—
1. Standard bridging.	—	X	SJI specifications listed in Section 2207.1.
2. Bridging that differs from the SJI specifications listed in Section 2207.1.	—	X	—

For SI: 1 inch = 25.4 mm.
a. Where applicable, see Section 1705.13.

1705.3 Concrete construction. *Special inspections* and tests of concrete construction shall be performed in accordance with this section and Table 1705.3.

Exceptions: *Special inspections* and tests shall not be required for:

1. Isolated spread concrete footings of buildings three stories or less above *grade plane* that are fully supported on earth or rock.
2. Continuous concrete footings supporting walls of buildings three stories or less above *grade plane* that are fully supported on earth or rock where:
   2.1. The footings support walls of *light-frame construction.*
   2.2. The footings are designed in accordance with Table 1809.7.
   2.3. The structural design of the footing is based on a specified compressive strength, f'_c, not more than 2,500 pounds per square inch (psi) (17.2 MPa), regardless of the compressive strength specified in the *approved construction documents* or used in the footing construction.
3. *Nonstructural concrete* slabs supported directly on the ground, including prestressed slabs on grade, where the effective prestress in the concrete is less than 150 psi (1.03 MPa).
4. Concrete foundation walls constructed in accordance with Table 1807.1.6.2.
5. Concrete patios, driveways and sidewalks, on grade.

1705.3.1 Welding of reinforcing bars. *Special inspections* of welding and qualifications of *special inspectors* for reinforcing bars shall be in accordance with the requirements of AWS D1.4 for *special inspection* and of AWS D1.4 for *special inspector* qualification.

1705.3.2 Material tests. In the absence of sufficient data or documentation providing evidence of conformance to quality standards for materials in Chapters 19 and 20 of ACI 318, the *building official* shall require testing of materials in accordance with the appropriate standards and criteria for the material in Chapters 19 and 20 of ACI 318.

1705.4 Masonry construction. *Special inspections* and tests of masonry construction shall be performed in accordance with the quality assurance program requirements of TMS 402 and TMS 602.

Exception: *Special inspections* and tests shall not be required for:

1. Empirically designed masonry, *glass unit masonry* or masonry *veneer* designed in accordance with Section 2109, Section 2110 or Chapter 14, respectively, where they are part of a structure classified as *Risk Category* I, II or III.
2. Masonry foundation walls constructed in accordance with Table 1807.1.6.3(1), 1807.1.6.3(2), 1807.1.6.3(3) or 1807.1.6.3(4).
3. Masonry fireplaces, masonry heaters or masonry chimneys installed or constructed in accordance with Section 2111, 2112 or 2113, respectively.

1705.4.1 Glass unit masonry and masonry veneer in Risk Category IV. *Special inspections* and tests for glass unit masonry or masonry *veneer* designed in accordance with Section 2110 or Chapter 14, respectively, where they are part of a structure classified as *Risk Category* IV shall be performed in accordance with TMS 602 Level 2.

1705.4.2 Vertical masonry foundation elements. *Special inspections* and tests of vertical masonry foundation elements shall be performed in accordance with Section 1705.4.

1705.5 Wood construction. *Special inspections* of prefabricated wood structural elements and assemblies shall be in accordance with Section 1704.2.5. *Special inspections* of site-built assemblies shall be in accordance with this section.

1705.5.1 High-load diaphragms. High-load *diaphragms* designed in accordance with Section 2306.2 shall be installed with *special inspections* as indicated in Section 1704.2. The *special inspector* shall inspect the *wood structural panel* sheathing to ascertain whether it is of the grade and thickness shown on the *approved* construction documents. Additionally, the *special inspector* must verify the *nominal size* of framing members at adjoining panel edges, the nail or staple diameter and length, the number of fastener lines and that the spacing between fasteners in each line and at edge margins agrees with the *approved construction documents*.

TABLE 1705.3
REQUIRED SPECIAL INSPECTIONS AND TESTS OF CONCRETE CONSTRUCTION

TYPE	CONTINUOUS SPECIAL INSPECTION	PERIODIC SPECIAL INSPECTION	REFERENCED STANDARD[a]	IBC REFERENCE
1. Inspect reinforcement, including prestressing tendons, and verify placement.	—	X	ACI 318: Ch. 20, 25.2, 25.3, 26.6.1-26.6.3	—
2. Reinforcing bar welding: a. Verify weldability of reinforcing bars other than ASTM A706; b. Inspect single-pass fillet welds, maximum $^5/_{16}$″; and c. Inspect all other welds.	 — — X	 X X —	AWS D1.4 ACI 318: 26.6.4	—
3. Inspect anchors cast in concrete.	—	X	ACI 318: 17.8.2	—
4. Inspect anchors post-installed in hardened concrete members.[b] a. Adhesive anchors installed in horizontally or upwardly inclined orientations to resist sustained tension loads. b. Mechanical anchors and adhesive anchors not defined in 4.a.	 X —	 — X	 ACI 318: 17.8.2.4 ACI 318: 17.8.2	—
5. Verify use of required design mix.	—	X	ACI 318: Ch. 19, 26.4.3, 26.4.4	1904.1, 1904.2
6. Prior to concrete placement, fabricate specimens for strength tests, perform slump and air content tests, and determine the temperature of the concrete.	X	—	ASTM C31 ASTM C172 ACI 318: 26.5, 26.12	—
7. Inspect concrete and shotcrete placement for proper application techniques.	X	—	ACI 318: 26.5	—
8. Verify maintenance of specified curing temperature and techniques.	—	X	ACI 318: 26.5.3-26.5.5	—
9. Inspect prestressed concrete for: a. Application of prestressing forces; and b. Grouting of bonded prestressing tendons.	 X X	 — —	ACI 318: 26.10	—
10. Inspect erection of precast concrete members.	—	X	ACI 318: 26.9	—
11. For precast concrete diaphragm connections or reinforcement at joints classified as moderate or high deformability elements (MDE or HDE) in structures assigned to Seismic Design Category C, D, E or F, inspect such connections and reinforcement in the field for: a. Installation of the embedded parts b. Completion of the continuity of reinforcement across joints. c. Completion of connections in the field.	 X X X	 — — —	ACI 318: 26.13.1.3 ACI 550.5	—
12. Inspect installation tolerances of precast concrete diaphragm connections for compliance with ACI 550.5.	—	X	ACI 318: 26.13.1.3	—
13. Verify in-situ concrete strength, prior to stressing of tendons in post-tensioned concrete and prior to removal of shores and forms from beams and structural slabs.	—	X	ACI 318: 26.11.2	—
14. Inspect formwork for shape, location and dimensions of the concrete member being formed.	—	X	ACI 318: 26.11.1.2(b)	—

For SI: 1 inch = 25.4 mm.
a. Where applicable, see Section 1705.13.
b. Specific requirements for special inspection shall be included in the research report for the anchor issued by an approved source in accordance with 17.8.2 in ACI 318, or other qualification procedures. Where specific requirements are not provided, special inspection requirements shall be specified by the registered design professional and shall be approved by the building official prior to the commencement of the work.

1705.5.2 Metal-plate-connected wood trusses spanning 60 feet or greater. Where a truss clear span is 60 feet (18 288 mm) or greater, the *special inspector* shall verify that the temporary installation restraint/bracing and the permanent individual truss member restraint/bracing are installed in accordance with the approved truss submittal package.

1705.5.3 Mass timber construction. *Special inspections* of *mass timber* elements in Types IV-A, IV-B and IV-C construction shall be in accordance with Table 1705.5.3.

1705.6 Soils. *Special inspections* and tests of existing site soil conditions, fill placement and load-bearing requirements shall be performed in accordance with this section and Table 1705.6. The *approved* geotechnical report and the *construction documents* prepared by the *registered design professionals* shall be used to determine compliance.

Exception: Where Section 1803 does not require reporting of materials and procedures for fill placement, the special inspector shall verify that the in-place dry density of the compacted fill is not less than 90 percent of the maximum dry density at optimum moisture content determined in accordance with ASTM D1557.

1705.7 Driven deep foundations. *Special inspections* and tests shall be performed during installation of driven *deep foundation* elements as specified in Table 1705.7. The approved geotechnical report and the construction documents prepared by the *registered design professionals* shall be used to determine compliance.

1705.8 Cast-in-place deep foundations. *Special inspections* and tests shall be performed during installation of cast-in-place *deep foundation* elements as specified in Table 1705.8. The *approved* geotechnical report and the *construction documents* prepared by the *registered design professionals* shall be used to determine compliance.

1705.9 Helical pile foundations. *Continuous special inspections* shall be performed during installation of *helical pile* foundations. The information recorded shall include installation equipment used, pile dimensions, tip elevations, final depth, final installation torque and other pertinent installation data as required by the *registered design professional* in responsible charge. The *approved* geotechnical report and the *construction documents* prepared by the *registered design professional* shall be used to determine compliance.

1705.10 Structural integrity of deep foundation elements. Whenever there is a reasonable doubt as to the structural integrity of a *deep foundation* element, an engineering assessment shall be required. The engineering assessment shall include tests for defects performed in accordance with

TABLE 1705.5.3
REQUIRED SPECIAL INSPECTIONS OF MASS TIMBER CONSTRUCTION

TYPE			CONTINUOUS SPECIAL INSPECTION	PERIODIC SPECIAL INSPECTION
1.	Inspection of anchorage and connections of mass timber construction to timber deep foundation systems.		—	X
2.	Inspect erection of mass timber construction.		—	X
3.	Inspection of connections where installation methods are required to meet design loads.			
	Threaded fasteners	Verify use of proper installation equipment.	—	X
		Verify use of pre-drilled holes where required.	—	X
		Inspect screws, including diameter, length, head type, spacing, installation angle and depth.	—	X
	Adhesive anchors installed in horizontal or upwardly inclined orientation to resist sustained tension loads.		X	—
	Adhesive anchors not defined in preceding cell.		—	X
	Bolted connections.		—	X
	Concealed connections.		—	X

TABLE 1705.6
REQUIRED SPECIAL INSPECTIONS AND TESTS OF SOILS

TYPE	CONTINUOUS SPECIAL INSPECTION	PERIODIC SPECIAL INSPECTION
1. Verify materials below *shallow foundations* are adequate to achieve the design bearing capacity.	—	X
2. Verify excavations are extended to proper depth and have reached proper material.	—	X
3. Perform classification and testing of compacted fill materials.	—	X
4. During fill placement, verify use of proper materials and procedures in accordance with the provisions of the approved geotechnical report. Verify densities and lift thicknesses during placement and compaction of compacted fill.	X	—
5. Prior to placement of compacted fill, inspect subgrade and verify that site has been prepared properly.	—	X

ASTM D4945, ASTM D5882, ASTM D6760 or ASTM D7949, or other *approved method*.

1705.11 Fabricated items. *Special inspections* of *fabricated items* shall be performed in accordance with Section 1704.2.5.

1705.12 Special inspections for wind resistance. *Special inspections* for wind resistance specified in Sections 1705.12.1 through 1705.12.3, unless exempted by the exceptions to Section 1704.2, are required for buildings and structures constructed in the following areas:

1. In wind Exposure Category B, where *V* is 150 miles per hour (67 m/sec) or greater.
2. In wind Exposure Category C or D, where *V* is 140 mph (62.6 m/sec) or greater.

1705.12.1 Structural wood. *Continuous special inspection* is required during field gluing operations of elements of the *main windforce-resisting system*. *Periodic special inspection* is required for nailing, bolting, anchoring and other fastening of elements of the *main windforce-resisting system*, including wood *shear walls*, wood *diaphragms*, *drag struts*, braces and *hold-downs*.

Exception: *Special inspections* are not required for wood *shear walls*, shear panels and *diaphragms*, including nailing, bolting, anchoring and other fastening to other elements of the *main windforce-resisting system*, where the lateral resistance is provided by structural sheathing and the specified fastener spacing at panel edges is more than 4 inches (102 mm) on center.

1705.12.2 Cold-formed steel light-frame construction. *Periodic special inspection* is required for welding operations of elements of the *main windforce-resisting system*. *Periodic special inspection* is required for screw attachment, bolting, anchoring and other fastening of elements of the *main windforce-resisting system*, including shear walls, braces, *diaphragms*, *collectors* (*drag struts*) and *hold-downs*.

Exception: *Special inspections* are not required for cold-formed steel light-frame shear walls and *diaphragms*, including screwing, bolting, anchoring and other fastening to components of the windforce-resisting system, where either of the following applies:

1. The sheathing is *gypsum board* or *fiberboard*.
2. The sheathing is *wood structural panel* or steel sheets on only one side of the *shear wall*, shear panel or *diaphragm* assembly and the specified fastener spacing at the panel or sheet edges is more than 4 inches (102 mm) on center (o.c.).

1705.12.3 Wind-resisting components. *Periodic special inspection* is required for fastening of the following systems and components:

1. *Roof covering*, *roof deck* and roof framing connections.
2. Exterior wall covering and wall connections to roof and floor *diaphragms* and framing.

1705.13 Special inspections for seismic resistance. *Special inspections* for seismic resistance shall be required as speci-

TABLE 1705.7
REQUIRED SPECIAL INSPECTIONS AND TESTS OF DRIVEN DEEP FOUNDATION ELEMENTS

TYPE	CONTINUOUS SPECIAL INSPECTION	PERIODIC SPECIAL INSPECTION
1. Verify element materials, sizes and lengths comply with the requirements.	X	—
2. Determine capacities of test elements and conduct additional load tests, as required.	X	—
3. Inspect driving operations and maintain complete and accurate records for each element.	X	—
4. Verify placement locations and plumbness, confirm type and size of hammer, record number of blows per foot of penetration, determine required penetrations to achieve design capacity, record tip and butt elevations and document any damage to foundation element.	X	—
5. For steel elements, perform additional special inspections in accordance with Section 1705.2.	In accordance with Section 1705.2	
6. For concrete elements and concrete-filled elements, perform tests and additional special inspections in accordance with Section 1705.3.	In accordance with Section 1705.3	
7. For specialty elements, perform additional inspections as determined by the registered design professional in responsible charge.	In accordance with Statement of Special Inspections	

TABLE 1705.8
REQUIRED SPECIAL INSPECTIONS AND TESTS OF CAST-IN-PLACE DEEP FOUNDATION ELEMENTS

TYPE	CONTINUOUS SPECIAL INSPECTION	PERIODIC SPECIAL INSPECTION
1. Inspect drilling operations and maintain complete and accurate records for each element.	X	—
2. Verify placement locations and plumbness, confirm element diameters, bell diameters (if applicable), lengths, embedment into bedrock (if applicable) and adequate end-bearing strata capacity. Record concrete or grout volumes.	X	—
3. For concrete elements, perform tests and additional *special inspections* in accordance with Section 1705.3.	In accordance with Section 1705.3	

fied in Sections 1705.13.1 through 1705.13.9, unless exempted by the exceptions of Section 1704.2.

Exception: The *special inspections* specified in Sections 1705.13.1 through 1705.13.9 are not required for structures designed and constructed in accordance with one of the following:

1. The structure consists of *light-frame construction*; the design spectral response acceleration at short periods, S_{DS}, as determined in Section 1613.2.4, does not exceed 0.5; and the *building height* of the structure does not exceed 35 feet (10 668 mm).
2. The *seismic force-resisting system* of the structure consists of *reinforced masonry* or reinforced concrete; the design spectral response acceleration at short periods, S_{DS}, as determined in Section 1613.2.4, does not exceed 0.5; and the *building height* of the structure does not exceed 25 feet (7620 mm).
3. The structure is a detached one- or two-family dwelling not exceeding two *stories above grade plane* and does not have any of the following horizontal or vertical irregularities in accordance with Section 12.3 of ASCE 7:
   - 3.1. Torsional or extreme torsional irregularity.
   - 3.2. Nonparallel systems irregularity.
   - 3.3. Stiffness-soft story or stiffness-extreme soft story irregularity.
   - 3.4. Discontinuity in lateral strength-weak story irregularity.

1705.13.1 Structural steel. *Special inspections* for seismic resistance shall be in accordance with Section 1705.13.1.1 or 1705.13.1.2, as applicable.

1705.13.1.1 Seismic force-resisting systems. *Special inspections* of structural steel in the *seismic force-resisting systems* in buildings and structures assigned to *Seismic Design Category* B, C, D, E or F shall be performed in accordance with the quality assurance requirements of AISC 341.

Exceptions:

1. In buildings and structures assigned to *Seismic Design Category* B or C, *special inspections* are not required for structural steel *seismic force-resisting systems* where the response modification coefficient, R, designated for "Steel systems not specifically detailed for seismic resistance, excluding cantilever column systems" in ASCE 7, Table 12.2-1, has been used for design and detailing.
2. In structures assigned to *Seismic Design Category* D, E, or F, *special inspections* are not required for structural steel *seismic force-resisting systems* where design and detailing in accordance with AISC 360 is permitted by ASCE 7, Table 15.4-1.

1705.13.1.2 Structural steel elements. *Special inspections* of *structural steel elements* in the *seismic force-resisting systems* of buildings and structures assigned to *Seismic Design Category* B, C, D, E or F other than those covered in Section 1705.13.1.1, including struts, *collectors*, chords and foundation elements, shall be performed in accordance with the quality assurance requirements of AISC 341.

Exceptions:

1. In buildings and structures assigned to *Seismic Design Category* B or C, *special inspections* of *structural steel elements* are not required for *seismic force-resisting systems* with a response modification coefficient, R, of 3 or less.
2. In structures assigned to *Seismic Design Category* D, E, or F, *special inspections* of *structural steel elements* are not required for *seismic force-resisting systems* where design and detailing other than AISC 341 is permitted by ASCE 7, Table 15.4-1. *Special inspection* shall be in accordance with the applicable referenced standard listed in ASCE 7, Table 15.4-1.

1705.13.2 Structural wood. For the *seismic force-resisting systems* of structures assigned to *Seismic Design Category* C, D, E or F:

1. *Continuous special inspection* shall be required during field gluing operations of elements of the *seismic force-resisting system*.
2. *Periodic special inspection* shall be required for nailing, bolting, anchoring and other fastening of elements of the *seismic force-resisting system*, including wood *shear walls*, wood *diaphragms*, *drag struts*, braces, shear panels and *hold-downs*.

Exception: *Special inspections* are not required for wood *shear walls*, shear panels and *diaphragms*, including nailing, bolting, anchoring and other fastening to other elements of the *seismic force-resisting system*, where the lateral resistance is provided by structural sheathing, and the specified fastener spacing at the panel edges is more than 4 inches (102 mm) on center.

1705.13.3 Cold-formed steel light-frame construction. For the *seismic force-resisting systems* of structures assigned to *Seismic Design Category* C, D, E or F, *periodic special inspection* shall be required for both:

1. Welding operations of elements of the *seismic force-resisting system*.
2. Screw attachment, bolting, anchoring and other fastening of elements of the *seismic force-resisting system*, including shear walls, braces, *diaphragms*, *collectors* (*drag struts*) and *hold-downs*.

Exception: *Special inspections* are not required for cold-formed steel light-frame shear walls and *diaphragms*, including screw installation, bolting, anchoring and other fastening to components of the *seismic force-resisting system*, where either of the following applies:

1. The sheathing is gypsum board or *fiberboard*.
2. The sheathing is *wood structural panel* or steel sheets on only one side of the *shear wall*, shear panel or *diaphragm* assembly and the specified fastener spacing at the panel or sheet edge is more than 4 inches (102 mm) on center.

1705.13.4 Designated seismic systems. For structures assigned to *Seismic Design Category* C, D, E or F, the *special inspector* shall examine *designated seismic systems* requiring seismic qualification in accordance with Section 13.2.2 of ASCE 7 and verify that the label, anchorage and mounting conform to the *certificate of compliance*.

1705.13.5 Architectural components. *Periodic special inspection* is required for the erection and fastening of exterior cladding, interior and exterior nonbearing walls and interior and exterior *veneer* in structures assigned to *Seismic Design Category* D, E or F.

Exception: *Periodic special inspection* is not required for the following:

1. Exterior cladding, interior and exterior nonbearing walls and interior and exterior *veneer* 30 feet (9144 mm) or less in height above grade or walking surface.
2. Exterior cladding and interior and exterior *veneer* weighing 5 psf (0.24 kN/m^2) or less.
3. Interior nonbearing walls weighing 15 psf (0.72 kN/m^2) or less.

1705.13.5.1 Access floors. Periodic *special inspection* is required for the anchorage of access floors in structures assigned to *Seismic Design Category* D, E or F.

1705.13.6 Plumbing, mechanical and electrical components. *Periodic special inspection* of plumbing, mechanical and electrical components shall be required for the following:

1. Anchorage of electrical equipment for emergency and standby power systems in structures assigned to *Seismic Design Category* C, D, E or F.
2. Anchorage of other electrical equipment in structures assigned to *Seismic Design Category* E or F.
3. Installation and anchorage of piping systems designed to carry hazardous materials and their associated mechanical units in structures assigned to *Seismic Design Category* C, D, E or F.
4. Installation and anchorage of ductwork designed to carry hazardous materials in structures assigned to *Seismic Design Category* C, D, E or F.
5. Installation and anchorage of vibration isolation systems in structures assigned to *Seismic Design Category* C, D, E or F where the *approved construction documents* require a nominal clearance of $^1/_4$ inch (6.4 mm) or less between the equipment support frame and restraint.
6. Installation of mechanical and electrical equipment, including duct work, piping systems and their structural supports, where automatic sprinkler systems are installed in structures assigned to *Seismic Design Category* C, D, E or F to verify one of the following:
   - 6.1. Minimum clearances have been provided as required by Section 13.2.3 ASCE/SEI 7.
   - 6.2. A nominal clearance of not less than 3 inches (76 mm) has been be provided between automatic sprinkler system drops and sprigs and: structural members not used collectively or independently to support the sprinklers; equipment attached to the building structure; and other systems' piping.

Where flexible sprinkler hose fittings are used, *special inspection* of minimum clearances is not required.

1705.13.7 Storage racks. Steel storage racks and steel cantilevered storage racks that are 8 feet (2438 mm) in height or greater and assigned to *Seismic Design Category* D, E or F shall be provided with periodic special inspection as required by Table 1705.13.7.

1705.13.8 Seismic isolation systems. *Periodic special inspection* shall be provided for seismic isolation systems in seismically isolated structures assigned to *Seismic Design Category* B, C, D, E or F during the fabrication and installation of isolator units and energy dissipation devices.

TABLE 1705.13.7
REQUIRED INSPECTIONS OF STORAGE RACK SYSTEMS

TYPE	CONTINUOUS INSPECTION	PERIODIC INSPECTION	REFERENCED STANDARD	IBC REFERENCE
1. Materials used, to verify compliance with one or more of the material test reports in accordance with the approved construction documents.	—	X	—	—
2. Fabricated storage rack elements.	—	X	—	Section 1704.2.5
3. Storage rack anchorage installation.	—	X	ANSI/MH16.1 Section 7.3.2	—
4. Completed storage rack system, to indicate compliance with the approved construction documents.	—	X	—	—

1705.13.9 Cold-formed steel special bolted moment frames. *Periodic special inspection* shall be provided for the installation of cold-formed steel special bolted moment frames in the *seismic force-resisting systems* of structures assigned to *Seismic Design Category* D, E or F.

1705.14 Testing for seismic resistance. Testing for seismic resistance shall be required as specified in Sections 1705.14.1 through 1705.14.4, unless exempted from *special inspections* by the exceptions of Section 1704.2.

1705.14.1 Structural steel. Nondestructive testing for seismic resistance shall be in accordance with Section 1705.14.1.1 or 1705.14.1.2, as applicable.

1705.14.1.1 Seismic force-resisting systems. Nondestructive testing of structural steel in the *seismic force-resisting systems* in buildings and structures assigned to *Seismic Design Category* B, C, D, E or F shall be performed in accordance with the quality assurance requirements of AISC 341.

Exceptions:

1. In buildings and structures assigned to *Seismic Design Category* B or C, nondestructive testing is not required for structural steel *seismic force-resisting systems* where the response modification coefficient, R, designated for "Steel systems not specifically detailed for seismic resistance, excluding cantilever column systems" in ASCE 7, Table 12.2-1, has been used for design and detailing.
2. In structures assigned to *Seismic Design Category* D, E, or F, nondestructive testing is not required for structural steel *seismic force-resisting systems* where design and detailing in accordance with AISC 360 is permitted by ASCE 7, Table 15.4-1.

1705.14.1.2 Structural steel elements. Nondestructive testing of *structural steel elements* in the *seismic force-resisting systems* of buildings and structures assigned to *Seismic Design Category* B, C, D, E or F other than those covered in Section 1705.14.1.1, including struts, *collectors*, chords and foundation elements, shall be performed in accordance with the quality assurance requirements of AISC 341.

Exceptions:

1. In buildings and structures assigned to *Seismic Design Category* B or C, nondestructive testing of *structural steel elements* is not required for *seismic force-resisting systems* with a response modification coefficient, R, of 3 or less.
2. In structures assigned to *Seismic Design Category* D, E or F, nondestructive testing of *structural steel elements* is not required for *seismic force-resisting systems* where design and detailing other than AISC 341 is permitted by ASCE 7, Table 15.4-1. Nondestructive testing of *structural steel elements* shall be in accordance with the applicable referenced standard listed in ASCE 7, Table 15.4-1.

1705.14.2 Nonstructural components. For structures assigned to *Seismic Design Category* B, C, D, E or F, where the requirements of Section 13.2.1 of ASCE 7 for nonstructural components, supports or attachments are met by seismic qualification as specified in Item 2 therein, the *registered design professional* shall specify on the *approved construction documents* the requirements for seismic qualification by analysis, testing or experience data. *Certificates of compliance* for the seismic qualification shall be submitted to the *building official* as specified in Section 1704.5.

1705.14.3 Designated seismic systems. For structures assigned to *Seismic Design Category* C, D, E or F and with *designated seismic systems* that are subject to the requirements of Section 13.2.2 of ASCE 7 for certification, the *registered design professional* shall specify on the *approved construction documents* the requirements to be met by analysis, testing or experience data as specified therein. *Certificates of compliance* documenting that the requirements are met shall be submitted to the *building official* as specified in Section 1704.5.

1705.14.4 Seismic isolation systems. Seismic isolation systems in seismically isolated structures assigned to *Seismic Design Category* B, C, D, E or F shall be tested in accordance with Section 17.8 of ASCE 7.

[BF] 1705.15 Sprayed fire-resistant materials. *Special inspections* and tests of sprayed fire-resistant materials applied to floor, roof and wall assemblies and structural members shall be performed in accordance with Sections 1705.15.1 through 1705.15.6. *Special inspections* shall be based on the fire-resistance design as designated in the *approved construction documents*. The tests set forth in this section shall be based on samplings from specific floor, roof and wall assemblies and structural members. *Special inspections* and tests shall be performed during construction with an additional visual inspection after the rough installation of electrical, automatic sprinkler, mechanical and plumbing systems and suspension systems for ceilings, and before concealment where applicable. The required sample size shall not exceed 110 percent of that specified by the referenced standards in Sections 1705.15.4.1 through 1705.15.4.9.

[BF] 1705.15.1 Physical and visual tests. The *special inspections* and tests shall include the following to demonstrate compliance with the listing and the *fire-resistance rating*:

1. Condition of substrates.
2. Thickness of application.
3. Density in pounds per cubic foot (kg/m^3).
4. Bond strength adhesion/cohesion.
5. Condition of finished application.

[BF] 1705.15.2 Structural member surface conditions. The surfaces shall be prepared in accordance with the *approved* fire-resistance design and the written instructions of *approved* manufacturers. The prepared surface of structural members to be sprayed shall be inspected by the *special inspector* before the application of the sprayed fire-resistant material.

[BF] 1705.15.3 Application. The substrate shall have a minimum ambient temperature before and after application as specified in the written instructions of *approved* manufacturers. The area for application shall be ventilated during and after application as required by the written instructions of *approved* manufacturers.

[BF] 1705.15.4 Thickness. Not more than 10 percent of the thickness measurements of the sprayed fire-resistant materials applied to floor, roof and wall assemblies and structural members shall be less than the thickness required by the *approved* fire-resistance design, and none shall be less than the minimum allowable thickness required by Section 1705.15.4.1.

[BF] 1705.15.4.1 Minimum allowable thickness. For design thicknesses 1 inch (25 mm) or greater, the minimum allowable individual thickness shall be the design thickness minus 1/4 inch (6.4 mm). For design thicknesses less than 1 inch (25 mm), the minimum allowable individual thickness shall be the design thickness minus 25 percent. Thickness shall be determined in accordance with ASTM E605. Samples of the sprayed fire-resistant materials shall be selected in accordance with Sections 1705.15.4.2 and 1705.15.4.3.

[BF] 1705.15.4.2 Floor, roof and wall assemblies. The thickness of the sprayed fire-resistant material applied to floor, roof and wall assemblies shall be determined in accordance with ASTM E605, making not less than four measurements for each 1,000 square feet (93 m^2) of the sprayed area, or portion thereof, in each story.

[BF] 1705.15.4.3 Cellular decks. Thickness measurements shall be selected from a square area, 12 inches by 12 inches (305 mm by 305 mm) in size. Not fewer than four measurements shall be made, located symmetrically within the square area.

[BF] 1705.15.4.4 Fluted decks. Thickness measurements shall be selected from a square area, 12 inches by 12 inches (305 mm by 305 mm) in size. Not fewer than four measurements shall be made, located symmetrically within the square area, including one each of the following: valley, crest and sides. The average of the measurements shall be reported.

[BF] 1705.15.4.5 Structural members. The thickness of the sprayed fire-resistant material applied to structural members shall be determined in accordance with ASTM E605. Thickness testing shall be performed on not less than 25 percent of the structural members on each floor.

[BF] 1705.15.4.6 Beams and girders. At beams and girders thickness measurements shall be made at nine locations around the beam or girder at each end of a 12-inch (305 mm) length.

[BF] 1705.15.4.7 Joists and trusses. At joists and trusses, thickness measurements shall be made at seven locations around the joist or truss at each end of a 12-inch (305 mm) length.

[BF] 1705.15.4.8 Wide-flanged columns. At wide-flanged columns, thickness measurements shall be made at 12 locations around the column at each end of a 12-inch (305 mm) length.

[BF] 1705.15.4.9 Hollow structural section and pipe columns. At hollow structural section and pipe columns, thickness measurements shall be made at not fewer than four locations around the column at each end of a 12-inch (305 mm) length.

[BF] 1705.15.5 Density. The density of the sprayed fire-resistant material shall be not less than the density specified in the *approved* fire-resistance design. Density of the sprayed fire-resistant material shall be determined in accordance with ASTM E605. The test samples for determining the density of the sprayed fire-resistant materials shall be selected as follows:

1. From each floor, roof and wall assembly at the rate of not less than one sample for every 2,500 square feet (232 m^2) or portion thereof of the sprayed area in each story.
2. From beams, girders, trusses and columns at the rate of not less than one sample for each type of structural member for each 2,500 square feet (232 m^2) of floor area or portion thereof in each *story*.

[BF] 1705.15.6 Bond strength. The cohesive/adhesive bond strength of the cured sprayed fire-resistant material applied to floor, roof and wall assemblies and structural members shall be not less than 150 pounds per square foot (psf) (7.18 kN/m^2). The cohesive/adhesive bond strength shall be determined in accordance with the field test specified in ASTM E736 by testing in-place samples of the sprayed fire-resistant material selected in accordance with Sections 1705.15.6.1 through 1705.15.6.3.

[BF] 1705.15.6.1 Floor, roof and wall assemblies. The test samples for determining the cohesive/adhesive bond strength of the sprayed fire-resistant materials shall be selected from each floor, roof and wall assembly at the rate of not less than one sample for every 2,500 square feet (232 m^2) of the sprayed area, or portion thereof, in each *story*.

[BF] 1705.15.6.2 Structural members. The test samples for determining the cohesive/adhesive bond strength of the sprayed fire-resistant materials shall be selected from beams, girders, trusses, columns and other structural members at the rate of not less than one sample for each type of structural member for each 2,500 square feet (232 m^2) of floor area or portion thereof in each *story*.

[BF] 1705.15.6.3 Primer, paint and encapsulant bond tests. Bond tests to qualify a primer, paint or encapsulant shall be conducted where the sprayed fire-resistant material is applied to a primed, painted or encapsulated surface for which acceptable bond-strength performance between these coatings and the fire-resistant material has not been determined. A bonding agent *approved* by the SFRM manufacturer shall be applied to a primed, painted or encapsulated surface where the bond strengths are found to be less than required values.

[BF] 1705.16 Mastic and intumescent fire-resistant coatings. *Special inspections* and tests for mastic and *intumescent fire-resistant coatings* applied to structural elements and decks shall be performed in accordance with AWCI 12-B. *Special inspections* and tests shall be based on the fire-resistance design as designated in the *approved construction documents*. *Special inspections* and tests shall be performed during construction. Additional visual inspection shall be performed after the rough installation and, where applicable, prior to the concealment of electrical, automatic sprinkler, mechanical and plumbing systems.

1705.17 Exterior insulation and finish systems (EIFS). *Special inspections* shall be required for all EIFS applications.

Exceptions:

1. *Special inspections* shall not be required for EIFS applications installed over a *water-resistive barrier* with a means of draining moisture to the exterior.
2. *Special inspections* shall not be required for EIFS applications installed over masonry or concrete walls.

1705.17.1 Water-resistive barrier coating. A *water-resistive barrier* coating complying with ASTM E2570 requires *special inspection* of the *water-resistive barrier* coating where installed over a sheathing substrate.

[BF] 1705.18 Fire-resistant penetrations and joints. In high-rise buildings, in buildings assigned to *Risk Category* III or IV, or in *fire areas* containing Group R occupancies with an *occupant load* greater than 250, *special inspections* for *through-penetrations*, *membrane penetration* firestops, *fire-resistant joint systems* and perimeter fire containment systems that are tested and *listed* in accordance with Sections 714.4.1.2, 714.5.1.2, 715.3.1 and 715.4 shall be in accordance with Section 1705.18.1 or 1705.18.2.

[BF] 1705.18.1 Penetration firestops. Inspections of *penetration firestop* systems that are tested and *listed* in accordance with Sections 714.4.1.2 and 714.5.1.2 shall be conducted by an *approved agency* in accordance with ASTM E2174.

[BF] 1705.18.2 Fire-resistant joint systems. Inspection of *fire-resistant joint systems* that are tested and *listed* in accordance with Sections 715.3.1 and 715.4 shall be conducted by an *approved agency* in accordance with ASTM E2393.

[F] 1705.19 Testing for smoke control. Smoke control systems shall be tested by a *special inspector*.

[F] 1705.19.1 Testing scope. The test scope shall be as follows:

1. During erection of ductwork and prior to concealment for the purposes of leakage testing and recording of device location.
2. Prior to occupancy and after sufficient completion for the purposes of pressure difference testing, flow measurements and detection and control verification.

[F] 1705.19.2 Qualifications. *Approved* agencies for smoke control testing shall have expertise in fire protection engineering, mechanical engineering and certification as air balancers.

1705.20 Sealing of mass timber. Periodic *special inspections* of sealants or adhesives shall be conducted where sealant or adhesive required by Section 703.7 is applied to *mass timber building elements* as designated in the *approved* construction documents.

SECTION 1706
DESIGN STRENGTHS OF MATERIALS

1706.1 Conformance to standards. The *design strengths* and permissible stresses of any structural material that are identified by a manufacturer's designation as to manufacture and grade by mill tests, or the strength and stress grade is otherwise confirmed to the satisfaction of the *building official*, shall conform to the specifications and methods of design of accepted engineering practice or the *approved* rules in the absence of applicable standards.

1706.2 New materials. For materials that are not specifically provided for in this code, the *design strengths* and permissible stresses shall be established by tests as provided for in Section 1707.

SECTION 1707
ALTERNATIVE TEST PROCEDURE

1707.1 General. In the absence of *approved* rules or other *approved* standards, the *building official* shall make, or cause to be made, the necessary tests and investigations; or the *building official* shall accept duly authenticated reports from *approved agencies* in respect to the quality and manner of use of new materials or assemblies as provided for in Section 104.11. The cost of all tests and other investigations required under the provisions of this code shall be borne by the owner or the owner's authorized agent.

SECTION 1708
IN-SITU LOAD TESTS

1708.1 General. Whenever there is a reasonable doubt as to the stability or load-bearing capacity of a completed building, structure or portion thereof for the expected *loads*, an engineering assessment shall be required. The engineering assessment shall involve either a structural analysis or an in-situ load test, or both. The structural analysis shall be based on actual material properties and other as-built conditions that affect stability or load-bearing capacity, and shall be conducted in accordance with the applicable design standard. The in-situ load tests shall be conducted in accordance with Section 1708.2. If the building, structure or portion thereof is found to have inadequate stability or load-bearing capacity for the expected *loads*, modifications to ensure structural adequacy or the removal of the inadequate construction shall be required.

1708.2 In-situ load tests. In-situ load tests shall be conducted in accordance with Section 1708.2.1 or 1708.2.2 and shall be supervised by a *registered design professional*. The test shall simulate the applicable loading conditions specified in Chapter 16 as necessary to address the concerns regarding structural stability of the building, structure or portion thereof.

1708.2.1 Load test procedure specified. Where a referenced material standard contains an applicable load test procedure and acceptance criteria, the test procedure and acceptance criteria in the standard shall apply. In the absence of specific *load factors* or acceptance criteria, the *load factors* and acceptance criteria in Section 1708.2.2 shall apply.

1708.2.2 Load test procedure not specified. In the absence of applicable load test procedures contained within a material standard referenced by this code or acceptance criteria for a specific material or method of construction, such *existing structure* shall be subjected to an approved test procedure developed by a *registered design professional* that simulates applicable loading and deformation conditions. For components that are not a part of the *seismic force-resisting system*, at a minimum the test load shall be equal to the specified factored design *loads*. For materials such as wood that have strengths that are dependent on load duration, the test load shall be adjusted to account for the difference in load duration of the test compared to the expected duration of the design *loads* being considered. For statically loaded components, the test load shall be left in place for a period of 24 hours. For components that carry dynamic *loads* (for example, machine supports or fall arrest anchors), the load shall be left in place for a period consistent with the component's actual function. The structure shall be considered to have successfully met the test requirements where the following criteria are satisfied:

1. Under the design *load*, the deflection shall not exceed the limitations specified in Section 1604.3.
2. Within 24 hours after removal of the test load, the structure shall have recovered not less than 75 percent of the maximum deflection.
3. During and immediately after the test, the structure shall not show evidence of failure.

SECTION 1709
PRECONSTRUCTION LOAD TESTS

1709.1 General. Where proposed construction is not capable of being designed by *approved* engineering analysis, or where proposed construction design method does not comply with the applicable material design standard, the system of construction or the structural unit and the connections shall be subjected to the tests prescribed in Section 1709. The *building official* shall accept certified reports of such tests conducted by an *approved* testing agency, provided that such tests meet the requirements of this code and *approved* procedures.

1709.2 Load test procedures specified. Where specific load test procedures, *load factors* and acceptance criteria are included in the applicable referenced standards, such test procedures, *load factors* and acceptance criteria shall apply. In the absence of specific test procedures, *load factors* or acceptance criteria, the corresponding provisions in Section 1709.3 shall apply.

1709.3 Load test procedures not specified. Where load test procedures are not specified in the applicable referenced standards, the load-bearing and deformation capacity of structural components and assemblies shall be determined on the basis of a test procedure developed by a *registered design professional* that simulates applicable loading and deformation conditions. For components and assemblies that are not a part of the *seismic force-resisting system*, the test shall be as specified in Section 1709.3.1. Load tests shall simulate the applicable loading conditions specified in Chapter 16.

1709.3.1 Test procedure. The test assembly shall be subjected to an increasing superimposed load equal to not less than two times the superimposed design load. The test load shall be left in place for a period of 24 hours. The tested assembly shall be considered to have successfully met the test requirements if the assembly recovers not less than 75 percent of the maximum deflection within 24 hours after the removal of the test load. The test assembly shall then be reloaded and subjected to an increasing superimposed load until either structural failure occurs or the superimposed load is equal to two and one-half times the load at which the deflection limitations specified in Section 1709.3.2 were reached, or the load is equal to two and one-half times the superimposed design load. In the case of structural components and assemblies for which deflection limitations are not specified in Section 1709.3.2, the test specimen shall be subjected to an increasing superimposed load until structural failure occurs or the load is equal to two and one-half times the desired superimposed design load. The allowable superimposed design load shall be taken as the least of:

1. The load at the deflection limitation given in Section 1709.3.2.
2. The failure load divided by 2.5.
3. The maximum load applied divided by 2.5.

1709.3.2 Deflection. The deflection of structural members under the design *load* shall not exceed the limitations in Section 1604.3.

1709.4 Wall and partition assemblies. *Load-bearing wall* and partition assemblies shall sustain the test load both with and without window framing. The test load shall include all design load components. Wall and partition assemblies shall be tested both with and without door and window framing.

1709.5 Exterior window and door assemblies. The design pressure rating of exterior windows and doors in buildings shall be determined in accordance with Section 1709.5.1 or 1709.5.2. For exterior windows and doors tested in accordance with Section 1709.5.1 or 1709.5.2, required design wind pressures determined from ASCE 7 shall be permitted to be converted to allowable stress design by multiplying by 0.6.

Exception: Structural wind load design pressures for window or door assemblies other than the size tested in accordance with Section 1709.5.1 or 1709.5.2 shall be permitted to be different than the design value of the tested assembly, provided that such pressures are determined by accepted engineering analysis or validated by an additional test of the window or door assembly to the alternative allowable design pressure in accordance with Section 1709.5.2. Components of the alternate size assembly shall be the same as the tested or labeled assembly. Where engineering analysis is used, it shall be performed in accordance with the analysis procedures of AAMA 2502.

1709.5.1 Exterior windows and doors. Exterior windows and sliding doors shall be tested and *labeled* as conforming to AAMA/WDMA/CSA101/I.S.2/A440. The *label* shall state the name of the manufacturer, the *approved* labeling agency and the product designation as specified in AAMA/WDMA/CSA101/I.S.2/A440. Exterior side-hinged doors shall be tested and *labeled* as conforming to AAMA/WDMA/CSA101/I.S.2/A440 or comply with Section 1709.5.2. Products tested and *labeled* as conforming to AAMA/WDMA/CSA 101/I.S.2/A440 shall not be subject to the requirements of Sections 2403.2 and 2403.3.

1709.5.2 Exterior windows and door assemblies not provided for in Section 1709.5.1. Exterior window and door assemblies shall be tested in accordance with ASTM E330. Exterior window and door assemblies containing glass shall comply with Section 2403. The design pressure for testing shall be calculated in accordance with Chapter 16. Each assembly shall be tested for 10 seconds at a load equal to 1.5 times the design pressure.

1709.5.2.1 Garage doors and rolling doors. Garage doors and rolling doors shall be tested in accordance with either ASTM E330 or ANSI/DASMA 108, and shall meet the pass/fail criteria of ANSI/DASMA 108. Garage doors and rolling doors shall be labeled with a permanent label identifying the door manufacturer, the door model/series number, the positive and negative design wind pressure rating, the installation instruction drawing reference number, and the applicable test standard.

1709.5.3 Windborne debris protection. Protection of exterior glazed openings in buildings located in *windborne debris regions* shall be in accordance with Section 1609.2.

1709.5.3.1 Impact protective systems testing and labeling. *Impact protective systems* shall be tested for impact resistance by an approved independent laboratory for compliance with ASTM E1886 and ASTM E1996 and for design wind pressure for compliance with ASTM E330. Required design wind pressures shall be determined in accordance with ASCE 7, and for the purposes of this section, multiplied by 0.6 to convert to *allowable stress design*.

Impact protective systems shall have a permanent label applied in accordance with Section 1703.5.4, identifying the manufacturer, product designation, performance characteristics, and approved inspection agency.

1709.6 Skylights and sloped glazing. *Skylights and sloped glazing* shall comply with the requirements of Chapter 24.

1709.7 Test specimens. Test specimens and construction shall be representative of the materials, workmanship and details normally used in practice. The properties of the materials used to construct the test assembly shall be determined on the basis of tests on samples taken from the load assembly or on representative samples of the materials used to construct the load test assembly. Required tests shall be conducted or witnessed by an *approved agency*.

CHAPTER 18

SOILS AND FOUNDATIONS

User notes:

About this chapter: *Chapter 18 provides criteria for geotechnical and structural considerations in the selection, design and installation of foundation systems to support the loads imposed by the structure above. This chapter includes requirements for soils investigation and site preparation for receiving a foundation, including the load-bearing values for soils and protection for the foundation from frost and water intrusion. Section 1808 addresses the basic requirements for all foundation types while subsequent sections address foundation requirements that are specific to shallow foundations and deep foundations.*

Code development reminder: *Code change proposals to this chapter will be considered by the IBC–Structural Code Development Committee during the 2022 (Group B) Code Development Cycle.*

SECTION 1801 GENERAL

1801.1 Scope. The provisions of this chapter shall apply to building and foundation systems.

SECTION 1802 DESIGN BASIS

1802.1 General. Allowable bearing pressures, allowable stresses and design formulas provided in this chapter shall be used with the *allowable stress design* load combinations specified in ASCE 7, Section 2.4 or the alternative allowable stress design load combinations of Section 1605.2. The quality and design of materials used structurally in excavations and foundations shall comply with the requirements specified in Chapters 16, 19, 21, 22 and 23. Excavations and fills shall comply with Chapter 33.

SECTION 1803 GEOTECHNICAL INVESTIGATIONS

1803.1 General. Geotechnical investigations shall be conducted in accordance with Section 1803.2 and reported in accordance with Section 1803.6. Where required by the *building official* or where geotechnical investigations involve in-situ testing, laboratory testing or engineering calculations, such investigations shall be conducted by a *registered design professional*.

1803.2 Investigations required. Geotechnical investigations shall be conducted in accordance with Sections 1803.3 through 1803.5.

Exception: The *building official* shall be permitted to waive the requirement for a geotechnical investigation where satisfactory data from adjacent areas is available that demonstrates an investigation is not necessary for any of the conditions in Sections 1803.5.1 through 1803.5.6 and Sections 1803.5.10 and 1803.5.11.

1803.3 Basis of investigation. Soil classification shall be based on observation and any necessary tests of the materials disclosed by borings, test pits or other subsurface exploration made in appropriate locations. Additional studies shall be made as necessary to evaluate slope stability, soil strength, position and adequacy of load-bearing soils, the effect of moisture variation on soil-bearing capacity, compressibility, liquefaction and expansiveness.

1803.3.1 Scope of investigation. The scope of the geotechnical investigation including the number and types of borings or soundings, the equipment used to drill or sample, the in-situ testing equipment and the laboratory testing program shall be determined by a *registered design professional.*

1803.4 Qualified representative. The investigation procedure and apparatus shall be in accordance with generally accepted engineering practice. The *registered design professional* shall have a fully qualified representative on site during all boring or sampling operations.

1803.5 Investigated conditions. Geotechnical investigations shall be conducted as indicated in Sections 1803.5.1 through 1803.5.12.

1803.5.1 Classification. Soil materials shall be classified in accordance with ASTM D2487.

1803.5.2 Questionable soil. Where the classification, strength or compressibility of the soil is in doubt or where a load-bearing value superior to that specified in this code is claimed, the *building official* shall be permitted to require that a geotechnical investigation be conducted.

1803.5.3 Expansive soil. In areas likely to have expansive soil, the *building official* shall require soil tests to determine where such soils do exist.

Soils meeting all four of the following provisions shall be considered to be expansive, except that tests to show compliance with Items 1, 2 and 3 shall not be required if the test prescribed in Item 4 is conducted:

1. Plasticity index (PI) of 15 or greater, determined in accordance with ASTM D4318.
2. More than 10 percent of the soil particles pass a No.200 sieve (75 μm), determined in accordance with ASTM D422.
3. More than 10 percent of the soil particles are less than 5 micrometers in size, determined in accordance with ASTM D422.

4. Expansion index greater than 20, determined in accordance with ASTM D4829.

1803.5.4 Ground-water table. A subsurface soil investigation shall be performed to determine whether the existing ground-water table is above or within 5 feet (1524 mm) below the elevation of the *lowest floor* level where such floor is located below the finished ground level adjacent to the foundation.

Exception: A subsurface soil investigation to determine the location of the ground-water table shall not be required where waterproofing is provided in accordance with Section 1805.

1803.5.5 Deep foundations. Where *deep foundations* will be used, a geotechnical investigation shall be conducted and shall include all of the following, unless sufficient data on which to base the design and installation is otherwise available:

1. Recommended *deep foundation* types and installed capacities.
2. Recommended center-to-center spacing of *deep foundation* elements.
3. Driving criteria.
4. Installation procedures.
5. Field inspection and reporting procedures (to include procedures for verification of the installed bearing capacity where required).
6. Load test requirements.
7. Suitability of *deep foundation* materials for the intended environment.
8. Designation of bearing stratum or strata.
9. Reductions for group action, where necessary.

1803.5.6 Rock strata. Where subsurface explorations at the project site indicate variations in the structure of rock on which foundations are to be constructed, a sufficient number of borings shall be drilled to sufficient depths to assess the competency of the rock and its load-bearing capacity.

1803.5.7 Excavation near foundations. Where excavation will reduce support from any foundation, a *registered design professional* shall prepare an assessment of the structure as determined from examination of the structure, available design documents, available subsurface data, and, if necessary, excavation of test pits. The *registered design professional* shall determine the requirements for support and protection of any existing foundation and prepare site-specific plans, details and sequence of work for submission. Such support shall be provided by underpinning, bracing, excavation retention systems, or by other means acceptable to the *building official*.

1803.5.8 Compacted fill material. Where *shallow foundations* will bear on compacted fill material more than 12 inches (305 mm) in depth, a geotechnical investigation shall be conducted and shall include all of the following:

1. Specifications for the preparation of the site prior to placement of compacted fill material.
2. Specifications for material to be used as compacted fill.
3. Test methods to be used to determine the maximum dry density and optimum moisture content of the material to be used as compacted fill.
4. Maximum allowable thickness of each lift of compacted fill material.
5. Field test method for determining the in-place dry density of the compacted fill.
6. Minimum acceptable in-place dry density expressed as a percentage of the maximum dry density determined in accordance with Item 3.
7. Number and frequency of field tests required to determine compliance with Item 6.

1803.5.9 Controlled low-strength material (CLSM). Where *shallow foundations* will bear on *controlled low-strength material* (*CLSM*), a geotechnical investigation shall be conducted and shall include all of the following:

1. Specifications for the preparation of the site prior to placement of the *CLSM*.
2. Specifications for the *CLSM*.
3. Laboratory or field test method(s) to be used to determine the compressive strength or bearing capacity of the *CLSM*.
4. Test methods for determining the acceptance of the *CLSM* in the field.
5. Number and frequency of field tests required to determine compliance with Item 4.

1803.5.10 Alternate setback and clearance. Where setbacks or clearances other than those required in Section 1808.7 are desired, the *building official* shall be permitted to require a geotechnical investigation by a *registered design professional* to demonstrate that the intent of Section 1808.7 would be satisfied. Such an investigation shall include consideration of material, height of slope, slope gradient, *load* intensity and erosion characteristics of slope material.

1803.5.11 Seismic Design Categories C through F. For structures assigned to *Seismic Design Category* C, D, E or F, a geotechnical investigation shall be conducted, and shall include an evaluation of all of the following potential geologic and seismic hazards:

1. Slope instability.
2. Liquefaction.
3. Total and differential settlement.
4. Surface displacement due to faulting or seismically induced lateral spreading or lateral flow.

1803.5.12 Seismic Design Categories D through F. For structures assigned to *Seismic Design Category* D, E or F, the geotechnical investigation required by Section 1803.5.11 shall include all of the following as applicable:

1. The determination of dynamic seismic lateral earth pressures on foundation walls and retaining walls supporting more than 6 feet (1.83 m) of

backfill height due to *design earthquake ground motions*.

2. The potential for liquefaction and soil strength loss evaluated for site peak ground acceleration, earthquake magnitude and source characteristics consistent with the maximum considered earthquake ground motions. Peak ground acceleration shall be determined based on one of the following:
   2.1. A site-specific study in accordance with Chapter 21 of ASCE 7.
   2.2. In accordance with Section 11.8.3 of ASCE 7.
3. An assessment of potential consequences of liquefaction and soil strength loss including, but not limited to, the following:
   3.1. Estimation of total and differential settlement.
   3.2. Lateral soil movement.
   3.3. Lateral soil *loads* on foundations.
   3.4. Reduction in foundation soil-bearing capacity and lateral soil reaction.
   3.5. Soil downdrag and reduction in axial and lateral soil reaction for pile foundations.
   3.6. Increases in soil lateral pressures on retaining walls.
   3.7. Flotation of buried structures.
4. Discussion of mitigation measures such as, but not limited to, the following:
   4.1. Selection of appropriate foundation type and depths.
   4.2. Selection of appropriate structural systems to accommodate anticipated displacements and forces.
   4.3. Ground stabilization.
   4.4. Any combination of these measures and how they shall be considered in the design of the structure.

1803.6 Reporting. Where geotechnical investigations are required, a written report of the investigations shall be submitted to the *building official* by the permit applicant at the time of permit application. This geotechnical report shall include, but need not be limited to, the following information:

1. A plot showing the location of the soil investigations.
2. A complete record of the soil boring and penetration test logs and soil samples.
3. A record of the soil profile.
4. Elevation of the water table, if encountered.
5. Recommendations for foundation type and design criteria, including but not limited to: bearing capacity of natural or compacted soil; provisions to mitigate the effects of expansive soils; mitigation of the effects of liquefaction, differential settlement and varying soil strength; and the effects of adjacent *loads*.
6. Expected total and differential settlement.
7. *Deep foundation* information in accordance with Section 1803.5.5.
8. Special design and construction provisions for foundations of structures founded on expansive soils, as necessary.
9. Compacted fill material properties and testing in accordance with Section 1803.5.8.
10. *Controlled low-strength material* properties and testing in accordance with Section 1803.5.9.

SECTION 1804
EXCAVATION, GRADING AND FILL

1804.1 Excavation near foundations. Excavation for any purpose shall not reduce vertical or lateral support for any foundation or adjacent foundation without first *underpinning* or protecting the foundation against detrimental lateral or vertical movement, or both, in accordance with Section 1803.5.7.

1804.2 Underpinning. Where *underpinning* is chosen to provide the protection or support of adjacent structures, the *underpinning* system shall be designed and installed in accordance with provisions of this chapter and Chapter 33.

1804.2.1 Underpinning sequencing. *Underpinning* shall be installed in a sequential manner that protects the neighboring structure and the working construction site. The sequence of installation shall be identified in the *approved construction documents*.

1804.3 Placement of backfill. The excavation outside the foundation shall be backfilled with soil that is free of organic material, construction debris, cobbles and boulders or with a *controlled low-strength material* (*CLSM*). The backfill shall be placed in lifts and compacted in a manner that does not damage the foundation or the waterproofing or dampproofing material.

Exception: *CLSM* need not be compacted.

1804.4 Site grading. The ground immediately adjacent to the foundation shall be sloped away from the building at a slope of not less than 1 unit vertical in 20 units horizontal (5-percent slope) for a minimum distance of 10 feet (3048 mm) measured perpendicular to the face of the wall. If physical obstructions or lot lines prohibit 10 feet (3048 mm) of horizontal distance, a 5-percent slope shall be provided to an *approved* alternative method of diverting water away from the foundation. Swales used for this purpose shall be sloped not less than 2 percent where located within 10 feet (3048 mm) of the building foundation. Impervious surfaces within 10 feet (3048 mm) of the building foundation shall be sloped not less than 2 percent away from the building.

Exceptions:

1. Where climatic or soil conditions warrant, the slope of the ground away from the building foun-

dation shall be permitted to be reduced to not less than 1 unit vertical in 48 units horizontal (2-percent slope).

2. Impervious surfaces shall be permitted to be sloped less than 2 percent where the surface is a door landing or *ramp* that is required to comply with Section 1010.1.4, 1012.3 or 1012.6.1.

The procedure used to establish the final ground level adjacent to the foundation shall account for additional settlement of the backfill.

1804.5 Grading and fill in flood hazard areas. In *flood hazard areas* established in Section 1612.3, grading, fill, or both, shall not be *approved*:

1. Unless such fill is placed, compacted and sloped to minimize shifting, slumping and erosion during the rise and fall of *flood* water and, as applicable, wave action.
2. In *floodways*, unless it has been demonstrated through hydrologic and hydraulic analyses performed by a *registered design professional* in accordance with standard engineering practice that the proposed grading or fill, or both, will not result in any increase in *flood* levels during the occurrence of the *design flood*.
3. In *coastal high hazard areas,* unless such fill is conducted or placed to avoid diversion of water and waves toward any building or structure.
4. Where *design flood elevations* are specified but *floodways* have not been designated, unless it has been demonstrated that the cumulative effect of the proposed *flood hazard area* encroachment, when combined with all other existing and anticipated *flood hazard area* encroachment, will not increase the *design flood elevation* more than 1 foot (305 mm) at any point.

1804.6 Compacted fill material. Where *shallow foundations* will bear on compacted fill material, the compacted fill shall comply with the provisions of an *approved* geotechnical report, as set forth in Section 1803.

Exception: Compacted fill material 12 inches (305 mm) in depth or less need not comply with an *approved* report, provided that the in-place dry density is not less than 90 percent of the maximum dry density at optimum moisture content determined in accordance with ASTM D1557. The compaction shall be verified by *special inspection* in accordance with Section 1705.6.

1804.7 Controlled low-strength material (CLSM). Where *shallow foundations* will bear on *controlled low-strength material* (*CLSM*), the *CLSM* shall comply with the provisions of an *approved* geotechnical report, as set forth in Section 1803.

SECTION 1805
DAMPPROOFING AND WATERPROOFING

1805.1 General. Walls or portions thereof that retain earth and enclose interior spaces and floors below grade shall be waterproofed and dampproofed in accordance with this section, with the exception of those spaces containing groups other than residential and institutional where such omission is not detrimental to the building or occupancy.

Ventilation for crawl spaces shall comply with Section 1202.4.

1805.1.1 Story above grade plane. Where a *basement* is considered a *story above grade plane* and the finished ground level adjacent to the basement wall is below the basement floor elevation for 25 percent or more of the perimeter, the floor and walls shall be dampproofed in accordance with Section 1805.2 and a foundation drain shall be installed in accordance with Section 1805.4.2. The foundation drain shall be installed around the portion of the perimeter where the basement floor is below ground level. The provisions of Sections 1803.5.4, 1805.3 and 1805.4.1 shall not apply in this case.

1805.1.2 Under-floor space. The finished ground level of an under-floor space such as a crawl space shall not be located below the bottom of the footings. Where there is evidence that the ground-water table rises to within 6 inches (152 mm) of the ground level at the outside building perimeter, or that the surface water does not readily drain from the building site, the ground level of the under-floor space shall be as high as the outside finished ground level, unless an *approved* drainage system is provided. The provisions of Sections 1803.5.4, 1805.2, 1805.3 and 1805.4 shall not apply in this case.

1805.1.2.1 Flood hazard areas. For buildings and structures in *flood hazard areas* as established in Section 1612.3, the finished ground level of an under-floor space such as a crawl space shall be equal to or higher than the outside finished ground level on one side or more.

Exception: Under-floor spaces of Group R-3 buildings that meet the requirements of FEMA TB 11.

1805.1.3 Ground-water control. Where the ground-water table is lowered and maintained at an elevation not less than 6 inches (152 mm) below the bottom of the *lowest floor*, the floor and walls shall be dampproofed in accordance with Section 1805.2. The design of the system to lower the ground-water table shall be based on accepted principles of engineering that shall consider, but not necessarily be limited to, permeability of the soil, rate at which water enters the drainage system, rated capacity of pumps, head against which pumps are to operate and the rated capacity of the disposal area of the system.

1805.2 Dampproofing. Where hydrostatic pressure will not occur as determined by Section 1803.5.4, floors and walls for other than wood foundation systems shall be dampproofed in accordance with this section. Wood foundation systems shall be constructed in accordance with AWC PWF.

1805.2.1 Floors. Dampproofing materials for floors shall be installed between the floor and the base course required by Section 1805.4.1, except where a separate floor is provided above a concrete slab.

Where installed beneath the slab, dampproofing shall consist of not less than 6-mil (0.006 inch; 0.152 mm) polyethylene with joints lapped not less than 6 inches (152 mm), or other *approved* methods or materials. Where permitted to be installed on top of the slab, dampproofing shall consist of mopped-on bitumen, not less than 4-mil (0.004 inch; 0.102 mm) polyethylene, or other *approved* methods or materials. Joints in the membrane shall be lapped and sealed in accordance with the manufacturer's installation instructions.

1805.2.2 Walls. Dampproofing materials for walls shall be installed on the exterior surface of the wall, and shall extend from the top of the footing to above ground level.

Dampproofing shall consist of a bituminous material, 3 pounds per square *yard* (16 N/m^2) of acrylic modified cement, $^1/_8$ inch (3.2 mm) coat of *surface-bonding mortar* complying with ASTM C887, any of the materials permitted for waterproofing by Section 1805.3.2 or other *approved* methods or materials.

1805.2.2.1 Surface preparation of walls. Prior to application of dampproofing materials on concrete walls, holes and recesses resulting from the removal of form ties shall be sealed with a bituminous material or other *approved* methods or materials. Unit masonry walls shall be parged on the exterior surface below ground level with not less than $^3/_8$ inch (9.5 mm) of Portland cement *mortar*. The parging shall be coved at the footing.

Exception: Parging of unit masonry walls is not required where a material is *approved* for direct application to the masonry.

1805.3 Waterproofing. Where the ground-water investigation required by Section 1803.5.4 indicates that a hydrostatic pressure condition exists, and the design does not include a ground-water control system as described in Section 1805.1.3, walls and floors shall be waterproofed in accordance with this section.

1805.3.1 Floors. Floors required to be waterproofed shall be of concrete and designed and constructed to withstand the hydrostatic pressures to which the floors will be subjected.

Waterproofing shall be accomplished by placing a membrane of rubberized asphalt, butyl rubber, fully adhered/fully bonded HDPE or polyolefin composite membrane or not less than 6-mil [0.006 inch (0.152 mm)] polyvinyl chloride with joints lapped not less than 6 inches (152 mm) or other *approved* materials under the slab. Joints in the membrane shall be lapped and sealed in accordance with the manufacturer's installation instructions.

1805.3.2 Walls. Walls required to be waterproofed shall be of concrete or masonry and shall be designed and constructed to withstand the hydrostatic pressures and other lateral *loads* to which the walls will be subjected.

Waterproofing shall be applied from the bottom of the wall to not less than 12 inches (305 mm) above the maximum elevation of the ground-water table. The remainder of the wall shall be dampproofed in accordance with Section 1805.2.2. Waterproofing shall consist of two-ply hot-mopped felts, not less than 6-mil (0.006 inch; 0.152 mm) polyvinyl chloride, 40-mil (0.040 inch; 1.02 mm) polymer-modified asphalt, 6-mil (0.006 inch; 0.152 mm) polyethylene or other *approved* methods or materials capable of bridging nonstructural cracks. Joints in the membrane shall be lapped and sealed in accordance with the manufacturer's installation instructions.

1805.3.2.1 Surface preparation of walls. Prior to the application of waterproofing materials on concrete or masonry walls, the walls shall be prepared in accordance with Section 1805.2.2.1.

1805.3.3 Joints and penetrations. *Joints* in walls and floors, *joints* between the wall and floor and penetrations of the wall and floor shall be made watertight utilizing *approved* methods and materials.

1805.4 Subsoil drainage system. Where a hydrostatic pressure condition does not exist, dampproofing shall be provided and a base shall be installed under the floor and a drain installed around the foundation perimeter. A subsoil drainage system designed and constructed in accordance with Section 1805.1.3 shall be deemed adequate for lowering the ground-water table.

1805.4.1 Floor base course. Floors of basements, except as provided for in Section 1805.1.1, shall be placed over a floor base course not less than 4 inches (102 mm) in thickness that consists of gravel or crushed stone containing not more than 10 percent of material that passes through a No. 4 (4.75 mm) sieve.

Exception: Where a site is located in well-drained gravel or sand/gravel mixture soils, a floor base course is not required.

1805.4.2 Foundation drain. A drain shall be placed around the perimeter of a foundation that consists of gravel or crushed stone containing not more than 10-percent material that passes through a No. 4 (4.75 mm) sieve. The drain shall extend not less than 12 inches (305 mm) beyond the outside edge of the footing. The thickness shall be such that the bottom of the drain is not higher than the bottom of the base under the floor, and that the top of the drain is not less than 6 inches (152 mm) above the top of the footing. The top of the drain shall be covered with an *approved* filter membrane material. Where a drain tile or perforated pipe is used, the invert of the pipe or tile shall not be higher than the floor elevation. The top of joints or the top of perforations shall be protected with an *approved* filter membrane material. The pipe or tile shall be placed on not less than 2 inches (51 mm) of gravel or crushed stone complying with Section 1805.4.1, and shall be covered with not less than 6 inches (152 mm) of the same material.

1805.4.3 Drainage discharge. The floor base and foundation perimeter drain shall discharge by gravity or mechanical means into an *approved* drainage system that complies with the *International Plumbing Code*.

Exception: Where a site is located in well-drained gravel or sand/gravel mixture soils, a dedicated drainage system is not required.

SECTION 1806 PRESUMPTIVE LOAD-BEARING VALUES OF SOILS

1806.1 Load combinations. The presumptive load-bearing values provided in Table 1806.2 shall be used with the *allowable stress design* load combinations specified in ASCE 7, Section 2.4 or the alternative allowable stress design load combinations of Section 1605.2. The values of vertical foundation pressure and lateral bearing pressure given in Table 1806.2 shall be permitted to be increased by one-third where used with the alternative allowable stress design load combinations of Section 1605.2 that include wind or earthquake *loads*.

1806.2 Presumptive load-bearing values. The load-bearing values used in design for supporting soils near the surface shall not exceed the values specified in Table 1806.2 unless data to substantiate the use of higher values are submitted and *approved*. Where the *building official* has reason to doubt the classification, strength or compressibility of the soil, the requirements of Section 1803.5.2 shall be satisfied.

Presumptive load-bearing values shall apply to materials with similar physical characteristics and dispositions. Mud, organic silt, organic clays, peat or unprepared fill shall not be assumed to have a presumptive load-bearing capacity unless data to substantiate the use of such a value are submitted.

Exception: A presumptive load-bearing capacity shall be permitted to be used where the *building official* deems the load-bearing capacity of mud, organic silt or unprepared fill is adequate for the support of lightweight or temporary structures.

1806.3 Lateral load resistance. Where the presumptive values of Table 1806.2 are used to determine resistance to lateral *loads*, the calculations shall be in accordance with Sections 1806.3.1 through 1806.3.4.

1806.3.1 Combined resistance. The total resistance to lateral *loads* shall be permitted to be determined by combining the values derived from the lateral bearing pressure and the lateral sliding resistance specified in Table 1806.2.

1806.3.2 Lateral sliding resistance limit. For clay, sandy clay, silty clay, clayey silt, silt and sandy silt, the lateral sliding resistance shall not exceed one-half the *dead load*.

1806.3.3 Increase for depth. The lateral bearing pressures specified in Table 1806.2 shall be permitted to be increased by the tabular value for each additional foot (305 mm) of depth to a value that is not greater than 15 times the tabular value.

1806.3.4 Increase for poles. Isolated poles for uses such as flagpoles or signs and poles used to support buildings that are not adversely affected by a $^1/_2$-inch (12.7 mm) motion at the ground surface due to short-term lateral *loads* shall be permitted to be designed using lateral bearing pressures equal to two times the tabular values.

SECTION 1807 FOUNDATION WALLS, RETAINING WALLS AND EMBEDDED POSTS AND POLES

1807.1 Foundation walls. Foundation walls shall be designed and constructed in accordance with Sections 1807.1.1 through 1807.1.6. Foundation walls shall be supported by foundations designed in accordance with Section 1808.

1807.1.1 Design lateral soil loads. Foundation walls shall be designed for the lateral soil *loads* set forth in Section 1610.

1807.1.2 Unbalanced backfill height. Unbalanced backfill height is the difference in height between the exterior

TABLE 1806.2
PRESUMPTIVE LOAD-BEARING VALUES

CLASS OF MATERIALS	VERTICAL FOUNDATION PRESSURE (psf)	LATERAL BEARING PRESSURE (psf/ft below natural grade)	LATERAL SLIDING RESISTANCE	
			Coefficient of friction[a]	Cohesion (psf)[b]
1. Crystalline bedrock	12,000	1,200	0.70	—
2. Sedimentary and foliated rock	4,000	400	0.35	—
3. Sandy gravel and gravel (GW and GP)	3,000	200	0.35	—
4. Sand, silty sand, clayey sand, silty gravel and clayey gravel (SW, SP, SM, SC, GM and GC)	2,000	150	0.25	—
5. Clay, sandy clay, silty clay, clayey silt, silt and sandy silt (CL, ML, MH and CH)	1,500	100	—	130

For SI: 1 pound per square foot = 0.0479kPa, 1 pound per square foot per foot = 0.157 kPa/m.

a. Coefficient to be multiplied by the dead load.

b. Cohesion value to be multiplied by the contact area, as limited by Section 1806.3.2.

finish ground level and the lower of the top of the concrete footing that supports the foundation wall or the *interior finish* ground level. Where an interior concrete slab on grade is provided and is in contact with the interior surface of the foundation wall, the unbalanced backfill height shall be permitted to be measured from the exterior finish ground level to the top of the interior concrete slab.

1807.1.3 Rubble stone foundation walls. Foundation walls of rough or random rubble stone shall be not less than 16 inches (406 mm) thick. Rubble stone shall not be used for foundation walls of structures assigned to *Seismic Design Category* C, D, E or F.

1807.1.4 Permanent wood foundation systems. Permanent wood foundation systems shall be designed and installed in accordance with AWC PWF. Lumber and plywood shall be preservative treated in accordance with AWPA U1 (Commodity Specification A, Special Requirement 4.2) and shall be identified in accordance with Section 2303.1.9.1.

1807.1.5 Concrete and masonry foundation walls. Concrete and masonry foundation walls shall be designed in accordance with Chapter 19 or 21, as applicable.

Exception: Concrete and masonry foundation walls shall be permitted to be designed and constructed in accordance with Section 1807.1.6.

1807.1.6 Prescriptive design of concrete and masonry foundation walls. Concrete and masonry foundation walls that are laterally supported at the top and bottom shall be permitted to be designed and constructed in accordance with this section.

1807.1.6.1 Foundation wall thickness. The thickness of prescriptively designed foundation walls shall be not less than the thickness of the wall supported, except that foundation walls of not less than 8-inch (203 mm) nominal width shall be permitted to support brick-veneered frame walls and 10-inch-wide (254 mm) *cavity walls* provided that the requirements of Section 1807.1.6.2 or 1807.1.6.3 are met.

1807.1.6.2 Concrete foundation walls. Concrete foundation walls shall comply with the following:

1. The thickness shall comply with the requirements of Table 1807.1.6.2.
2. The size and spacing of vertical reinforcement shown in Table 1807.1.6.2 are based on the use of reinforcement with a minimum yield strength of 60,000 pounds per square inch (psi) (414 MPa). Vertical reinforcement with a minimum yield strength of 40,000 psi (276 MPa) or 50,000 psi (345 MPa) shall be permitted, provided that the same size bar is used and the spacing shown in the table is reduced by multiplying the spacing by 0.67 or 0.83, respectively.
3. Vertical reinforcement, where required, shall be placed nearest the inside face of the wall a distance, *d*, from the outside face (soil face) of the wall. The distance, *d*, is equal to the wall thickness, *t*, minus 1.25 inches (32 mm) plus one-half the bar diameter, d_b, $[d = t - (1.25 + d_b / 2)]$. The reinforcement shall be placed within a tolerance of ± $^3/_8$ inch (9.5 mm) where *d* is less than or equal to 8 inches (203 mm) or ± $^1/_2$ inch (12.7 mm) where d is greater than 8 inches (203 mm).
4. In lieu of the reinforcement shown in Table 1807.1.6.2, smaller reinforcing bar sizes with closer spacings that provide an equivalent cross-sectional area of reinforcement per unit length shall be permitted.
5. Concrete cover for reinforcement measured from the inside face of the wall shall be not less than $^3/_4$ inch (19.1 mm). Concrete cover for reinforcement measured from the outside face of the wall shall be not less than $1^1/_2$ inches (38 mm) for No. 5 bars and smaller, and not less than 2 inches (51 mm) for larger bars.
6. Concrete shall have a specified compressive strength, f'_c, of not less than 2,500 psi (17.2 MPa).
7. The unfactored axial *load* per linear foot of wall shall not exceed 1.2 $t f'_c$ where t is the specified wall thickness in inches.

1807.1.6.2.1 Seismic requirements. Based on the *seismic design category* assigned to the structure in accordance with Section 1613, concrete foundation walls designed using Table 1807.1.6.2 shall be subject to the following limitations:

1. *Seismic Design Categories* A and B. Not less than one No. 5 bar shall be provided around window, door and similar sized openings. The bar shall be anchored to develop f_y in tension at the corners of openings.
2. *Seismic Design Categories* C, D, E and F. Tables shall not be used except as allowed for plain concrete members in Section 1905.1.7.

1807.1.6.3 Masonry foundation walls. Masonry foundation walls shall comply with the following:

1. The thickness shall comply with the requirements of Table 1807.1.6.3(1) for *plain masonry* walls or Table 1807.1.6.3(2), 1807.1.6.3(3) or 1807.1.6.3(4) for masonry walls with reinforcement.
2. Vertical reinforcement shall have a minimum yield strength of 60,000 psi (414 MPa).
3. The specified location of the reinforcement shall equal or exceed the effective depth distance, *d*, noted in Tables 1807.1.6.3(2), 1807.1.6.3(3) and 1807.1.6.3(4) and shall be measured from the face of the exterior (soil) side of the wall to the center of the vertical reinforcement. The reinforcement shall be placed within the tolerances specified in TMS 602, Article 3.4.B.11, of the specified location.
4. Grout shall comply with Section 2103.3.

5. Concrete *masonry units* shall comply with ASTM C90.
6. Clay *masonry units* shall comply with ASTM C652 for hollow brick, except compliance with ASTM C62 or ASTM C216 shall be permitted where solid *masonry units* are installed in accordance with Table 1807.1.6.3(1) for *plain masonry*.
7. *Masonry units* shall be laid in *running bond* and installed with Type M or S *mortar* in accordance with Section 2103.2.1.
8. The unfactored axial *load* per linear foot of wall shall not exceed 1.2 $t f'_m$ where t is the specified wall thickness in inches and f'_m is the *specified compressive strength of masonry* in pounds per square inch.
9. Not less than 4 inches (102 mm) of *solid masonry* shall be provided at girder supports at the top of hollow *masonry unit* foundation walls.
10. Corbeling of masonry shall be in accordance with Section 2104.1. Where an 8-inch (203 mm) wall is corbeled, the top corbel shall not extend higher than the bottom of the floor framing and shall be a full course of headers not less than 6 inches (152 mm) in length or the top course *bed joint* shall be tied to the vertical wall projection. The tie shall be W2.8 (4.8 mm) and spaced at a maximum horizontal distance of 36 inches (914 mm). The hollow space behind the corbelled masonry shall be filled with *mortar* or grout.

TABLE 1807.1.6.2
CONCRETE FOUNDATION WALLS[b, c]

MAXIMUM WALL HEIGHT (feet)	MAXIMUM UNBALANCED BACKFILL HEIGHT[e] (feet)	MINIMUM VERTICAL REINFORCEMENT-BAR SIZE AND SPACING (inches)								
		Design lateral soil load[a] (psf per foot of depth)								
		30[d]			45[d]			60		
		Minimum wall thickness (inches)								
		7.5	9.5	11.5	7.5	9.5	11.5	7.5	9.5	11.5
5	4	PC	PC	PC	PC	PC	PC	PC	PC	PC
	5	PC	PC	PC	PC	PC	PC	PC	PC	PC
6	4	PC	PC	PC	PC	PC	PC	PC	PC	PC
	5	PC	PC	PC	PC	PC	PC	PC	PC	PC
	6	PC	PC	PC	PC	PC	PC	PC	PC	PC
7	4	PC	PC	PC	PC	PC	PC	PC	PC	PC
	5	PC	PC	PC	PC	PC	PC	PC	PC	PC
	6	PC	PC	PC	PC	PC	PC	#5 at 48	PC	PC
	7	PC	PC	PC	#5 at 46	PC	PC	#6 at 48	PC	PC
8	4	PC	PC	PC	PC	PC	PC	PC	PC	PC
	5	PC	PC	PC	PC	PC	PC	PC	PC	PC
	6	PC	PC	PC	PC	PC	PC	#5 at 43	PC	PC
	7	PC	PC	PC	#5 at 41	PC	PC	#6 at 43	PC	PC
	8	#5 at 47	PC	PC	#6 at 43	PC	PC	#6 at 32	#6 at 44	PC
9	4	PC	PC	PC	PC	PC	PC	PC	PC	PC
	5	PC	PC	PC	PC	PC	PC	PC	PC	PC
	6	PC	PC	PC	PC	PC	PC	#5 at 39	PC	PC
	7	PC	PC	PC	#5 at 37	PC	PC	#6 at 38	#5 at 37	PC
	8	#5 at 41	PC	PC	#6 at 38	#5 at 37	PC	#7 at 39	#6 at 39	#4 at 48
	9[d]	#6 at 46	PC	PC	#7 at 41	#6 at 41	PC	#7 at 31	#7 at 41	#6 at 39
10	4	PC	PC	PC	PC	PC	PC	PC	PC	PC
	5	PC	PC	PC	PC	PC	PC	PC	PC	PC
	6	PC	PC	PC	PC	PC	PC	#5 at 37	PC	PC
	7	PC	PC	PC	#6 at 48	PC	PC	#6 at 35	#6 at 48	PC
	8	#5 at 38	PC	PC	#7 at 47	#6 at 47	PC	#7 at 35	#7 at 47	#6 at 45
	9[d]	#6 at 41	#4 at 48	PC	#7 at 37	#7 at 48	#4 at 48	#6 at 22	#7 at 37	#7 at 47
	10[d]	#7 at 45	#6 at 45	PC	#7 at 31	#7 at 40	#6 at 38	#6 at 22	#7 at 30	#7 at 38

For SI: 1 inch = 25.4 mm, 1 foot = 304.8 mm, 1 pound per square foot per foot = 0.157 kPa/m.

a. For design lateral soil loads, see Section 1610.
b. Provisions for this table are based on design and construction requirements specified in Section 1807.1.6.2.
c. PC = Plain Concrete.
d. Where unbalanced backfill height exceeds 8 feet and design lateral soil loads from Table 1610.1 are used, the requirements for 30 and 45 psf per foot of depth are not applicable (see Section 1610).
e. For height of unbalanced backfill, see Section 1807.1.2.

1807.1.6.3.1 Alternative foundation wall reinforcement. In lieu of the reinforcement provisions for masonry foundation walls in Table 1807.1.6.3(2), 1807.1.6.3(3) or 1807.1.6.3(4), alternative reinforcing bar sizes and spacings having an equivalent cross-sectional area of reinforcement per linear foot (mm) of wall shall be permitted to be used, provided that the spacing of reinforcement does not exceed 72 inches (1829 mm) and reinforcing bar sizes do not exceed No. 11.

1807.1.6.3.2 Seismic requirements. Based on the *seismic design category* assigned to the structure in accordance with Section 1613, masonry foundation walls designed using Tables 1807.1.6.3(1) through 1807.1.6.3(4) shall be subject to the following limitations:

1. *Seismic Design Categories* A and B. No additional seismic requirements.
2. *Seismic Design Category* C. A design using Tables 1807.1.6.3(1) through 1807.1.6.3(4) is subject to the seismic requirements of Section 7.4.3 of TMS 402.
3. *Seismic Design Category* D. A design using Tables 1807.1.6.3(2) through 1807.1.6.3(4) is subject to the seismic requirements of Section 7.4.4 of TMS 402.
4. *Seismic Design Categories* E and F. A design using Tables 1807.1.6.3(2) through 1807.1.6.3(4) is subject to the seismic requirements of Section 7.4.5 of TMS 402.

1807.2 Retaining walls. Retaining walls shall be designed in accordance with Sections 1807.2.1 through 1807.2.4.

1807.2.1 General. Retaining walls shall be designed to ensure stability against overturning, sliding, excessive foundation pressure and water uplift.

1807.2.2 Design lateral soil loads. Retaining walls shall be designed for the lateral soil *loads* set forth in Section 1610. For structures assigned to *Seismic Design Category* D, E, or F, the design of retaining walls supporting more than 6 feet (1829 mm) of backfill height shall incorporate the additional seismic lateral earth pressure in accordance with the geotechnical investigation where required in Section 1803.2.

1807.2.3 Safety factor. Retaining walls shall be designed to resist the lateral action of soil to produce sliding and overturning with a minimum safety factor of 1.5 in each case. The load combinations of Section 1605 shall not apply to this requirement. Instead, design shall be based on 0.7 times nominal earthquake *loads*, 1.0 times other *nominal loads*, and investigation with one or more of the variable *loads* set to zero. The safety factor against lateral sliding shall be taken as the available soil resistance at the base of the retaining wall foundation divided by the net lateral force applied to the retaining wall.

Exception: Where earthquake loads are included, the minimum safety factor for retaining wall sliding and overturning shall be 1.1.

1807.2.4 Segmental retaining walls. Dry-cast concrete units used in the construction of segmental retaining walls shall comply with ASTM C1372.

TABLE 1807.1.6.3(1)
PLAIN MASONRY FOUNDATION WALLS[a, b, c]

MAXIMUM WALL HEIGHT (feet)	MAXIMUM UNBALANCED BACKFILL HEIGHT[e] (feet)	MINIMUM NOMINAL WALL THICKNESS (inches)		
		Design lateral soil loada (psf per foot of depth)		
		30[f]	45[f]	60
7	4 (or less)	8	8	8
	5	8	10	10
	6	10	12	10 (solid[c])
	7	12	10 (solid[c])	10 (solid[c])
8	4 (or less)	8	8	8
	5	8	10	12
	6	10	12	12 (solid[c])
	7	12	12 (solid[c])	Note d
	8	10 (solid[c])	12 (solid[c])	Note d
9	4 (or less)	8	8	8
	5	8	10	12
	6	12	12	12 (solid[c])
	7	12(solid[c])	12 (solid[c])	Note d
	8	12(solid[c])	Note d	Note d
	9[f]	Note d	Note d	Note d

For SI: 1 inch = 25.4 mm, 1 foot = 304.8 mm, 1 pound per square foot per foot = 0.157 kPa/m.
a. For design lateral soil loads, see Section 1610.
b. Provisions for this table are based on design and construction requirements specified in Section 1807.1.6.3.
c. Solid grouted hollow units or solid masonry units.
d. A design in compliance with Chapter 21 or reinforcement in accordance with Table 1807.1.6.3(2) is required.
e. For height of unbalanced backfill, see Section 1807.1.2.
f. Where unbalanced backfill height exceeds 8 feet and design lateral soil loads from Table 1610.1 are used, the requirements for 30 and 45 psf per foot of depth are not applicable (see Section 1610).

1807.3 Embedded posts and poles. Designs to resist both axial and lateral *loads* employing posts or poles as columns embedded in earth or in concrete footings in earth shall be in accordance with Sections 1807.3.1 through 1807.3.3.

1807.3.1 Limitations. The design procedures outlined in this section are subject to the following limitations:

1. The frictional resistance for structural walls and slabs on silts and clays shall be limited to one-half of the normal force imposed on the soil by the weight of the footing or slab.
2. Posts embedded in earth shall not be used to provide lateral support for structural or nonstructural materials such as plaster, masonry or concrete unless bracing is provided that develops the limited deflection required.

Wood poles shall be treated in accordance with AWPA U1 for sawn timber posts (Commodity Specification A, Use Category 4B) and for round timber posts (Commodity Specification B, Use Category 4B).

1807.3.2 Design criteria. The depth to resist lateral *loads* shall be determined using the design criteria established in Sections 1807.3.2.1 through 1807.3.2.3, or by other methods *approved* by the *building official*.

1807.3.2.1 Nonconstrained. The following formula shall be used in determining the depth of embedment required to resist lateral *loads* where lateral constraint is not provided at the ground surface, such as by a rigid floor or rigid ground surface pavement, and where lateral constraint is not provided above the ground surface, such as by a structural *diaphragm*.

$$d = 0.5A\{1 + [1 + (4.36h/A)]^{1/2}\}$$ **(Equation 18-1)**

where:

$A = 2.34P/(S_1 b)$.

TABLE 1807.1.6.3(2)
8-INCH MASONRY FOUNDATION WALLS WITH REINFORCEMENT WHERE d ≥ 5 INCHES[a, b, c]

MAXIMUM WALL HEIGHT (feet-inches)	MAXIMUM UNBALANCED BACKFILL HEIGHT[d] (feet-inches)	MINIMUM VERTICAL REINFORCEMENT-BAR SIZE AND SPACING (inches)		
		Design lateral soil load[a] (psf per foot of depth)		
		30[e]	45[e]	60
7-4	4-0 (or less)	#4 at 48	#4 at 48	#4 at 48
	5-0	#4 at 48	#4 at 48	#4 at 48
	6-0	#4 at 48	#5 at 48	#5 at 48
	7-4	#5 at 48	#6 at 48	#7 at 48
8-0	4-0 (or less)	#4 at 48	#4 at 48	#4 at 48
	5-0	#4 at 48	#4 at 48	#4 at 48
	6-0	#4 at 48	#5 at 48	#5 at 48
	7-0	#5 at 48	#6 at 48	#7 at 48
	8-0	#5 at 48	#6 at 48	#7 at 48
8-8	4-0 (or less)	#4 at 48	#4 at 48	#4 at 48
	5-0	#4 at 48	#4 at 48	#5 at 48
	6-0	#4 at 48	#5 at 48	#6 at 48
	7-0	#5 at 48	#6 at 48	#7 at 48
	8-8[e]	#6 at 48	#7 at 48	#8 at 48
9-4	4-0 (or less)	#4 at 48	#4 at 48	#4 at 48
	5-0	#4 at 48	#4 at 48	#5 at 48
	6-0	#4 at 48	#5 at 48	#6 at 48
	7-0	#5 at 48	#6 at 48	#7 at 48
	8-0	#6 at 48	#7 at 48	#8 at 48
	9-4[e]	#7 at 48	#8 at 48	#9 at 48
10-0	4-0 (or less)	#4 at 48	#4 at 48	#4 at 48
	5-0	#4 at 48	#4 at 48	#5 at 48
	6-0	#4 at 48	#5 at 48	#6 at 48
	7-0	#5 at 48	#6 at 48	#7 at 48
	8-0	#6 at 48	#7 at 48	#8 at 48
	9-0[e]	#7 at 48	#8 at 48	#9 at 48
	10-10[e]	#7 at 48	#9 at 48	#9 at 48

For SI: 1 inch = 25.4 mm, 1 foot = 304.8 mm, 1 pound per square foot per foot = 0.157 kPa/m.

a. For design lateral soil loads, see Section 1610.
b. Provisions for this table are based on design and construction requirements specified in Section 1807.1.6.3.
c. For alternative reinforcement, see Section 1807.1.6.3.1.
d. For height of unbalanced backfill, see Section 1807.1.2.
e. Where unbalanced backfill height exceeds 8 feet and design lateral soil loads from Table 1610.1 are used, the requirements for 30 and 45 psf per foot of depth are not applicable. See Section 1610.

b = Diameter of round post or footing or diagonal dimension of square post or footing, feet (m).

d = Depth of embedment in earth in feet (m) but not over 12 feet (3658 mm) for purpose of computing lateral pressure.

h = Distance in feet (m) from ground surface to point of application of "P."

P = Applied lateral force in pounds (kN).

S_1 = Allowable lateral soil-bearing pressure as set forth in Section 1806.2 based on a depth of one-third the depth of embedment in pounds per square foot (psf) (kPa).

1807.3.2.2 Constrained. The following formula shall be used to determine the depth of embedment required to resist lateral *loads* where lateral constraint is provided at the ground surface, such as by a rigid floor or pavement.

$$d = \sqrt{\frac{4.25Ph}{S_3 b}} \quad \textbf{(Equation 18-2)}$$

or alternatively

$$d = \sqrt{\frac{4.25M_g}{S_3 b}} \quad \textbf{(Equation 18-3)}$$

where:

M_g = Moment in the post at grade, in foot-pounds (kN-m).

S_3 = Allowable lateral soil-bearing pressure as set forth in Section 1806.2 based on a depth equal to the depth of embedment in pounds per square foot (kPa).

1807.3.2.3 Vertical load. The resistance to vertical *loads* shall be determined using the vertical foundation pressure set forth in Table 1806.2.

TABLE 1807.1.6.3(3)
10-INCH MASONRY FOUNDATION WALLS WITH REINFORCEMENT WHERE d ≥ 6.75 INCHES[a, b, c]

MAXIMUM WALL HEIGHT (feet-inches)	MAXIMUM UNBALANCED BACKFILL HEIGHT[d] (feet-inches)	MINIMUM VERTICAL REINFORCEMENT-BAR SIZE AND SPACING (inches)		
		Design lateral soil load[a] (psf per foot of depth)		
		30[e]	45[e]	60
7-4	4-0 (or less)	#4 at 56	#4 at 56	#4 at 56
	5-0	#4 at 56	#4 at 56	#4 at 56
	6-0	#4 at 56	#4 at 56	#5 at 56
	7-4	#4 at 56	#5 at 56	#6 at 56
8-0	4-0 (or less)	#4 at 56	#4 at 56	#4 at 56
	5-0	#4 at 56	#4 at 56	#4 at 56
	6-0	#4 at 56	#4 at 56	#5 at 56
	7-0	#4 at 56	#5 at 56	#6 at 56
	8-0	#5 at 56	#6 at 56	#7 at 56
8-8	4-0 (or less)	#4 at 56	#4 at 56	#4 at 56
	5-0	#4 at 56	#4 at 56	#4 at 56
	6-0	#4 at 56	#4 at 56	#5 at 56
	7-0	#4 at 56	#5 at 56	#6 at 56
	8-8[e]	#5 at 56	#7 at 56	#8 at 56
9-4	4-0 (or less)	#4 at 56	#4 at 56	#4 at 56
	5-0	#4 at 56	#4 at 56	#4 at 56
	6-0	#4 at 56	#5 at 56	#5 at 56
	7-0	#4 at 56	#5 at 56	#6 at 56
	8-0	#5 at 56	#6 at 56	#7 at 56
	9-4[e]	#6 at 56	#7 at 56	#7 at 56
10-0	4-0 (or less)	#4 at 56	#4 at 56	#4 at 56
	5-0	#4 at 56	#4 at 56	#4 at 56
	6-0	#4 at 56	#5 at 56	#5 at 56
	7-0	#5 at 56	#6 at 56	#7 at 56
	8-0	#5 at 56	#7 at 56	#8 at 56
	9-0[e]	#6 at 56	#7 at 56	#9 at 56
	10-0[e]	#7 at 56	#8 at 56	#9 at 56

For SI: 1 inch = 25.4 mm, 1 foot = 304.8 mm, 1 pound per square foot per foot = 1.157 kPa/m.
a. For design lateral soil loads, see Section 1610.
b. Provisions for this table are based on design and construction requirements specified in Section 1807.1.6.3.
c. For alternative reinforcement, see Section 1807.1.6.3.1.
d. For height of unbalanced backfill, see Section 1807.1.2.
e. Where unbalanced backfill height exceeds 8 feet and design lateral soil loads from Table 1610.1 are used, the requirements for 30 and 45 psf per foot of depth are not applicable. See Section 1610.

1807.3.3 Backfill. The backfill in the *annular space* around columns not embedded in poured footings shall be by one of the following methods:

1. Backfill shall be of concrete with a specified compressive strength of not less than 2,000 psi (13.8 MPa). The hole shall be not less than 4 inches (102 mm) larger than the diameter of the column at its bottom or 4 inches (102 mm) larger than the diagonal dimension of a square or rectangular column.
2. Backfill shall be of clean sand. The sand shall be thoroughly compacted by tamping in layers not more than 8 inches (203 mm) in depth.
3. Backfill shall be of *controlled low-strength material* (*CLSM*).

SECTION 1808 FOUNDATIONS

1808.1 General. Foundations shall be designed and constructed in accordance with Sections 1808.2 through 1808.9. *Shallow foundations* shall satisfy the requirements of Section 1809. *Deep foundations* shall satisfy the requirements of Section 1810.

1808.2 Design for capacity and settlement. Foundations shall be so designed that the allowable bearing capacity of the soil is not exceeded, and that differential settlement is minimized. Foundations in areas with expansive soils shall be designed in accordance with the provisions of Section 1808.6.

1808.3 Design loads. Foundations shall be designed for the most unfavorable effects due to the combinations of *loads* specified in Section 2.3 or 2.4 of ASCE 7 or the alternative

TABLE 1807.1.6.3(4)
12-INCH MASONRY FOUNDATION WALLS WITH REINFORCEMENT WHERE d ≥ 8.75 INCHES[a, b, c]

MAXIMUM WALL HEIGHT (feet-inches)	MAXIMUM UNBALANCED BACKFILL HEIGHT[d] (feet-inches)	MINIMUM VERTICAL REINFORCEMENT-BAR SIZE AND SPACING (inches) Design lateral soil load[a] (psf per foot of depth)		
		30[e]	45[e]	60
7-4	4 (or less)	#4 at 72	#4 at 72	#4 at 72
	5-0	#4 at 72	#4 at 72	#4 at 72
	6-0	#4 at 72	#4 at 72	#5 at 72
	7-4	#4 at 72	#5 at 72	#6 at 72
8-0	4 (or less)	#4 at 72	#4 at 72	#4 at 72
	5-0	#4 at 72	#4 at 72	#4 at 72
	6-0	#4 at 72	#4 at 72	#5 at 72
	7-0	#4 at 72	#5 at 72	#6 at 72
	8-0	#5 at 72	#6 at 72	#8 at 72
8-8	4 (or less)	#4 at 72	#4 at 72	#4 at 72
	5-0	#4 at 72	#4 at 72	#4 at 72
	6-0	#4 at 72	#4 at 72	#5 at 72
	7-0	#4 at 72	#5 at 72	#6 at 72
	8-8[e]	#5 at 72	#7 at 72	#8 at 72
9-4	4 (or less)	#4 at 72	#4 at 72	#4 at 72
	5-0	#4 at 72	#4 at 72	#4 at 72
	6-0	#4 at 72	#5 at 72	#5 at 72
	7-0	#4 at 72	#5 at 72	#6 at 72
	8-0	#5 at 72	#6 at 72	#7 at 72
	9-4[e]	#6 at 72	#7 at 72	#8 at 72
10-0	4 (or less)	#4 at 72	#4 at 72	#4 at 72
	5-0	#4 at 72	#4 at 72	#4 at 72
	6-0	#4 at 72	#5 at 72	#5 at 72
	7-0	#4 at 72	#6 at 72	#6 at 72
	8-0	#5 at 72	#6 at 72	#7 at 72
	9-0[e]	#6 at 72	#7 at 72	#8 at 72
	10-0[e]	#7 at 72	#8 at 72	#9 at 72

For SI: 1 inch = 25.4 mm, 1 foot = 304.8 mm, 1 pound per square foot per foot = 0.157 kPa/m.
a. For design lateral soil loads, see Section 1610.
b. Provisions for this table are based on design and construction requirements specified in Section 1807.1.6.3.
c. For alternative reinforcement, see Section 1807.1.6.3.1.
d. For height of unbalanced backfill, see Section 1807.1.2.
e. Where unbalanced backfill height exceeds 8 feet and design lateral soil loads from Table 1610.1 are used, the requirements for 30 and 45 psf per foot of depth are not applicable. See Section 1610.

allowable stress design load combinations of Section 1605.2. The *dead load* is permitted to include the weight of foundations and overlying fill. Reduced *live loads*, as specified in Sections 1607.12 and 1607.14, shall be permitted to be used in the design of foundations.

1808.3.1 Seismic overturning. Where foundations are proportioned using the load combinations of Section 2.3 or 2.4 of ASCE 7 and the computation of seismic overturning effects is by equivalent lateral force analysis or modal analysis, the proportioning shall be in accordance with Section 12.13.4 of ASCE 7.

1808.3.2 Surcharge. Fill or other surcharge *loads* shall not be placed adjacent to any building or structure unless such building or structure is capable of withstanding the additional *loads* caused by the fill or the surcharge. Existing footings or foundations that will be affected by any excavation shall be underpinned or otherwise protected against settlement and shall be protected against detrimental lateral or vertical movement or both.

Exception: Minor grading for landscaping purposes shall be permitted where done with walk-behind equipment, where the grade is not increased more than 1 foot (305 mm) from original design grade or where *approved* by the *building official*.

1808.4 Vibratory loads. Where machinery operations or other vibrations are transmitted through the foundation, consideration shall be given in the foundation design to prevent detrimental disturbances of the soil.

1808.5 Shifting or moving soils. Where it is known that the shallow subsoils are of a shifting or moving character, foundations shall be carried to a sufficient depth to ensure stability.

1808.6 Design for expansive soils. Foundations for buildings and structures founded on expansive soils shall be designed in accordance with Section 1808.6.1 or 1808.6.2.

Exceptions: Foundation design need not comply with Section 1808.6.1 or 1808.6.2 where one of the following conditions is satisfied:

1. The soil is removed in accordance with Section 1808.6.3.
2. The *building official* approves stabilization of the soil in accordance with Section 1808.6.4.

1808.6.1 Foundations. Foundations placed on or within the active zone of expansive soils shall be designed to resist differential volume changes and to prevent structural damage to the supported structure. Deflection and racking of the supported structure shall be limited to that which will not interfere with the usability and serviceability of the structure.

Foundations placed below where volume change occurs or below expansive soil shall comply with the following provisions:

1. Foundations extending into or penetrating expansive soils shall be designed to prevent uplift of the supported structure.
2. Foundations penetrating expansive soils shall be designed to resist forces exerted on the foundation due to soil volume changes or shall be isolated from the expansive soil.

1808.6.2 Slab-on-ground foundations. Moments, shears and deflections for use in designing slab-on-ground, mat or raft foundations on expansive soils shall be determined in accordance with WRI/CRSI *Design of Slab-on-Ground Foundations* or PTI DC 10.5. Using the moments, shears and deflections determined above, nonprestressed slabs-on-ground, mat or raft foundations on expansive soils shall be designed in accordance with WRI/CRSI *Design of Slab-on-Ground Foundations* and post-tensioned slab-on-ground, mat or raft foundations on expansive soils shall be designed in accordance with PTI DC 10.5. It shall be permitted to analyze and design such slabs by other methods that account for soil-structure interaction, the deformed shape of the soil support, the plate or stiffened plate action of the slab as well as both center lift and edge lift conditions. Such alternative methods shall be rational and the basis for all aspects and parameters of the method shall be available for peer review.

1808.6.3 Removal of expansive soil. Where expansive soil is removed in lieu of designing foundations in accordance with Section 1808.6.1 or 1808.6.2, the soil shall be removed to a depth sufficient to ensure a constant moisture content in the remaining soil. Fill material shall not contain expansive soils and shall comply with Section 1804.5 or 1804.6.

Exception: Expansive soil need not be removed to the depth of constant moisture, provided that the confining pressure in the expansive soil created by the fill and supported structure exceeds the swell pressure.

1808.6.4 Stabilization. Where the active zone of expansive soils is stabilized in lieu of designing foundations in accordance with Section 1808.6.1 or 1808.6.2, the soil shall be stabilized by chemical, dewatering, presaturation or equivalent techniques.

1808.7 Foundations on or adjacent to slopes. The placement of buildings and structures on or adjacent to slopes steeper than one unit vertical in three units horizontal (33.3-percent slope) shall comply with Sections 1808.7.1 through 1808.7.5.

1808.7.1 Building clearance from ascending slopes. In general, buildings below slopes shall be set a sufficient distance from the slope to provide protection from slope drainage, erosion and shallow failures. Except as provided in Section 1808.7.5 and Figure 1808.7.1, the following criteria will be assumed to provide this protection. Where the existing slope is steeper than one unit vertical in one unit horizontal (100-percent slope), the toe of the slope shall be assumed to be at the intersection of a horizontal plane drawn from the top of the foundation and a plane drawn tangent to the slope at an angle of 45 degrees (0.79 rad) to the horizontal. Where a retaining wall is constructed at the toe of the slope, the height of the slope shall be measured from the top of the wall to the top of the slope.

1808.7.2 Foundation setback from descending slope surface. Foundations on or adjacent to slope surfaces shall be founded in firm material with an embedment and set back from the slope surface sufficient to provide vertical and lateral support for the foundation without detrimental settlement. Except as provided for in Section 1808.7.5 and Figure 1808.7.1, the following setback is deemed adequate to meet the criteria. Where the slope is steeper than 1 unit vertical in 1 unit horizontal (100-percent slope), the required setback shall be measured from an imaginary plane 45 degrees (0.79 rad) to the horizontal, projected upward from the toe of the slope.

1808.7.3 Pools. The setback between pools regulated by this code and slopes shall be equal to one-half the building footing setback distance required by this section. That portion of the pool wall within a horizontal distance of 7 feet (2134 mm) from the top of the slope shall be capable of supporting the water in the pool without soil support.

1808.7.4 Foundation elevation. On graded sites, the top of any exterior foundation shall extend above the elevation of the street gutter at point of discharge or the inlet of an *approved* drainage device not less than 12 inches (305 mm) plus 2 percent. Alternate elevations are permitted subject to the approval of the *building official*, provided that it can be demonstrated that required drainage to the point of discharge and away from the structure is provided at all locations on the site.

1808.7.5 Alternate setback and clearance. Alternate setbacks and clearances are permitted, subject to the approval of the *building official*. The *building official* shall be permitted to require a geotechnical investigation as set forth in Section 1803.5.10.

1808.8 Concrete foundations. The design, materials and construction of concrete foundations shall comply with Sections 1808.8.1 through 1808.8.6 and the provisions of Chapter 19.

Exception: Where concrete footings supporting walls of *light-frame construction* are designed in accordance with Table 1809.7, a specific design in accordance with Chapter 19 is not required.

1808.8.1 Concrete or grout strength and mix proportioning. Concrete or grout in foundations shall have a specified compressive strength (f'_c) not less than the largest applicable value indicated in Table 1808.8.1.

Where concrete or grout is to be pumped, the mix design including slump shall be adjusted to produce a pumpable mixture.

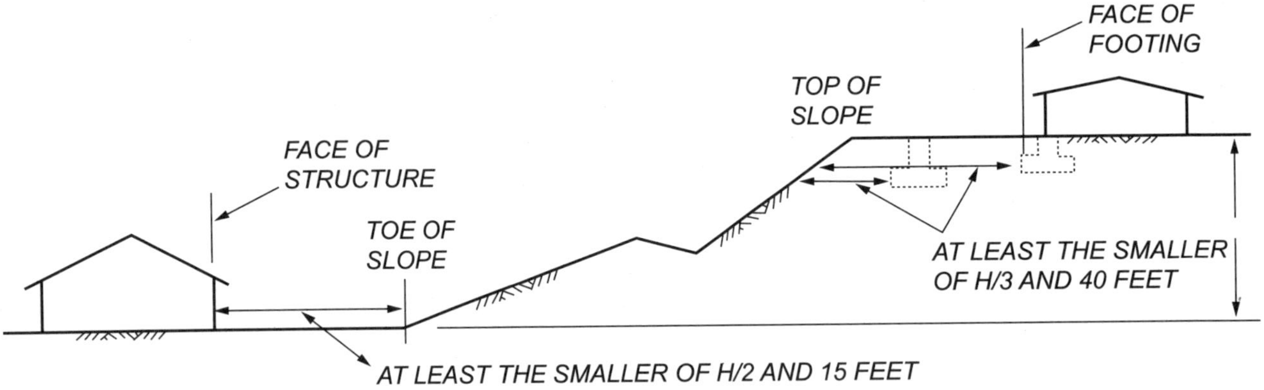

For SI: 1 foot = 304.8 mm.

FIGURE 1808.7.1
FOUNDATION CLEARANCES FROM SLOPES

TABLE 1808.8.1
MINIMUM SPECIFIED COMPRESSIVE STRENGTH f'_c OF CONCRETE OR GROUT

FOUNDATION ELEMENT OR CONDITION	SPECIFIED COMPRESSIVE STRENGTH, f'_c
1. Foundations for structures assigned to Seismic Design Category A, B or C	2,500 psi
2a. Foundations for Group R or U occupancies of light-frame construction, two stories or less in height, assigned to Seismic Design Category D, E or F	2,500 psi
2b. Foundations for other structures assigned to Seismic Design Category D, E or F	3,000 psi
3. Precast nonprestressed driven piles	4,000 psi
4. Socketed drilled shafts	4,000 psi
5. Micropiles	4,000 psi
6. Precast prestressed driven piles	5,000 psi

For SI: 1 pound per square inch = 0.00689 MPa.

1808.8.2 Concrete cover. The concrete cover provided for prestressed and nonprestressed reinforcement in foundations shall be not less than the largest applicable value specified in Table 1808.8.2. Longitudinal bars spaced less than $1^{1}/_{2}$ inches (38 mm) clear distance apart shall be considered to be bundled bars for which the concrete cover provided shall be not less than that required by Section 20.6.1.3.4 of ACI 318. Concrete cover shall be measured from the concrete surface to the outermost surface of the steel to which the cover requirement applies. Where concrete is placed in a temporary or permanent casing or a mandrel, the inside face of the casing or mandrel shall be considered to be the concrete surface.

1808.8.3 Placement of concrete. Concrete shall be placed in such a manner as to ensure the exclusion of any foreign matter and to secure a full-size foundation. Concrete shall not be placed through water unless a tremie or other method *approved* by the *building official* is used. Where placed under or in the presence of water, the concrete shall be deposited by *approved* means to ensure minimum segregation of the mix and negligible turbulence of the water. Where depositing concrete from the top of a *deep foundation* element, the concrete shall be chuted directly into smooth-sided pipes or tubes or placed in a rapid and continuous operation through a funnel hopper centered at the top of the element.

1808.8.4 Protection of concrete. Concrete foundations shall be protected from freezing during depositing and for a period of not less than 5 days thereafter. Water shall not be allowed to flow through the deposited concrete.

1808.8.5 Forming of concrete. Concrete foundations are permitted to be cast against the earth where, in the opinion of the *building official*, soil conditions do not require formwork. Where formwork is required, it shall be in accordance with Section 26.11 of ACI 318.

1808.8.6 Seismic requirements. See Section 1905 for additional requirements for foundations of structures assigned to *Seismic Design Category* C, D, E or F.

For structures assigned to *Seismic Design Category* D, E or F, provisions of Section 18.13 of ACI 318 shall apply where not in conflict with the provisions of Sections 1808 through 1810.

Exceptions:

1. Detached one- and two-family dwellings of *light-frame construction* and two stories or less above *grade plane* are not required to comply with the provisions of Section 18.13 of ACI 318.
2. Section 18.13.4.3(a) of ACI 318 shall not apply.

1808.9 Vertical masonry foundation elements. Vertical masonry foundation elements that are not *foundation piers* as defined in Section 202 shall be designed as piers, walls or columns, as applicable, in accordance with TMS 402.

SECTION 1809
SHALLOW FOUNDATIONS

1809.1 General. *Shallow foundations* shall be designed and constructed in accordance with Sections 1809.2 through 1809.13.

1809.2 Supporting soils. *Shallow foundations* shall be built on undisturbed soil, compacted fill material or *controlled low-strength material* (*CLSM*). Compacted fill material shall be placed in accordance with Section 1804.5. *CLSM* shall be placed in accordance with Section 1804.6.

1809.3 Stepped footings. The top surface of footings shall be level. The bottom surface of footings shall be permitted to have a slope not exceeding 1 unit vertical in 10 units horizontal (10-percent slope). Footings shall be stepped where it is necessary to change the elevation of the top surface of the footing or where the surface of the ground slopes more than 1 unit vertical in 10 units horizontal (10-percent slope).

1809.4 Depth and width of footings. The minimum depth of footings below the undisturbed ground surface shall be 12 inches (305 mm). Where applicable, the requirements of

TABLE 1808.8.2
MINIMUM CONCRETE COVER

FOUNDATION ELEMENT OR CONDITION	MINIMUM COVER
1. Shallow foundations	In accordance with Section 20.6 of ACI 318
2. Precast nonprestressed deep foundation elements Exposed to seawater Not manufactured under plant conditions Manufactured under plant control conditions	 3 inches 2 inches In accordance with Section 20.6.1.3.3 of ACI 318
3. Precast prestressed deep foundation elements Exposed to seawater Other	 2.5 inches In accordance with Section 20.6.1.3.3 of ACI 318
4.Cast-in-place deep foundation elements not enclosed by a steel pipe, tube or permanent casing	2.5 inches
5. Cast-in-place deep foundation elements enclosed by a steel pipe, tube or permanent casing	1 inch
6. Structural steel core within a steel pipe, tube or permanent casing	2 inches
7. Cast-in-place drilled shafts enclosed by a stable rock socket	1.5 inches

For SI:1 inch = 25.4 mm.

Section 1809.5 shall be satisfied. The minimum width of footings shall be 12 inches (305 mm).

1809.5 Frost protection. Except where otherwise protected from frost, foundations and other permanent supports of buildings and structures shall be protected from frost by one or more of the following methods:

1. Extending below the frost line of the locality.
2. Constructing in accordance with ASCE 32.
3. Erecting on solid rock.

Exception: Free-standing buildings meeting all of the following conditions shall not be required to be protected:

1. Assigned to *Risk Category* I.
2. Area of 600 square feet (56 m^2) or less for *light-frame construction* or 400 square feet (37 m^2) or less for other than *light-frame construction.*
3. Eave height of 10 feet (3048 mm) or less.

Shallow foundations shall not bear on frozen soil unless such frozen condition is of a permanent character.

1809.5.1 Frost protection at required exits. Frost protection shall be provided at exterior landings for all required exits with outward-swinging doors. Frost protection shall only be required to the extent necessary to ensure the unobstructed opening of the required *exit* doors.

1809.6 Location of footings. Footings on granular soil shall be so located that the line drawn between the lower edges of adjoining footings shall not have a slope steeper than 30 degrees (0.52 rad) with the horizontal, unless the material supporting the higher footing is braced or retained or otherwise laterally supported in an *approved* manner or a greater slope has been properly established by engineering analysis.

1809.7 Prescriptive footings for light-frame construction. Where a specific design is not provided, concrete or masonry-unit footings supporting walls of *light-frame construction* shall be permitted to be designed in accordance with Table 1809.7.

1809.8 Plain concrete footings. The edge thickness of plain concrete footings supporting walls of other than *light-frame construction* shall be not less than 8 inches (203 mm) where placed on soil or rock.

Exception: For plain concrete footings supporting Group R-3 occupancies, the edge thickness is permitted to be 6 inches (152 mm), provided that the footing does not extend beyond a distance greater than the thickness of the footing on either side of the supported wall.

1809.9 Masonry-unit footings. The design, materials and construction of masonry-unit footings shall comply with Sections 1809.9.1 and 1809.9.2, and the provisions of Chapter 21.

Exception: Where a specific design is not provided, masonry-unit footings supporting walls of *light-frame construction* shall be permitted to be designed in accordance with Table 1809.7.

1809.9.1 Dimensions. Masonry-unit footings shall be laid in Type M or S *mortar* complying with Section 2103.2.1 and the depth shall be not less than twice the projection beyond the wall, pier or column. The width shall be not less than 8 inches (203 mm) wider than the wall supported thereon.

1809.9.2 Offsets. The maximum offset of each course in brick foundation walls stepped up from the footings shall be $1^1/_2$ inches (38 mm) where laid in single courses, and 3 inches (76 mm) where laid in double courses.

1809.10 Pier and curtain wall foundations. Except in *Seismic Design Categories* D, E and F, pier and curtain wall foundations shall be permitted to be used to support *light-frame construction* not more than two *stories above grade plane*, provided that the following requirements are met:

1. All *load-bearing walls* shall be placed on continuous concrete footings bonded integrally with the *exterior wall* footings.
2. The minimum actual thickness of a load-bearing masonry wall shall be not less than 4 inches (102 mm) nominal or $3^5/_8$ inches (92 mm) actual thickness, and shall be bonded integrally with piers spaced 6 feet (1829 mm) on center (o.c.).
3. Piers shall be constructed in accordance with Chapter 21 and the following:
   3.1. The unsupported height of the masonry piers shall not exceed 10 times their least dimension.
   3.2. Where *structural clay tile* or hollow concrete *masonry units* are used for piers supporting beams and girders, the cellular

TABLE 1809.7
PRESCRIPTIVE FOOTINGS SUPPORTING WALLS OF LIGHT-FRAME CONSTRUCTION[a, b, c, d, e]

NUMBER OF FLOORS SUPPORTED BY THE FOOTING[f]	WIDTH OF FOOTING (inches)	THICKNESS OF FOOTING (inches)
1	12	6
2	15	6
3	18	8[g]

For SI: 1 inch = 25.4 mm, 1 foot = 304.8 mm.

a. Depth of footings shall be in accordance with Section 1809.4.

b. The ground under the floor shall be permitted to be excavated to the elevation of the top of the footing.

c. Interior stud-bearing walls shall be permitted to be supported by isolated footings. The footing width and length shall be twice the width shown in this table, and footings shall be spaced not more than 6 feet on center.

d. See Section 1905 for additional requirements for concrete footings of structures assigned to Seismic Design Category C, D, E or F.

e. For thickness of foundation walls, see Section 1807.1.6.

f. Footings shall be permitted to support a roof in addition to the stipulated number of floors. Footings supporting roof only shall be as required for supporting one floor.

g. Plain concrete footings for Group R-3 occupancies shall be permitted to be 6 inches thick.

spaces shall be filled solidly with concrete or Type M or S *mortar*.

Exception: Unfilled hollow piers shall be permitted where the unsupported height of the pier is not more than four times its least dimension.

3.3. Hollow piers shall be capped with 4 inches (102 mm) of *solid masonry* or concrete or the cavities of the top course shall be filled with concrete or grout.

4. The maximum height of a 4-inch (102 mm) load-bearing masonry foundation wall supporting wood frame walls and floors shall not be more than 4 feet (1219 mm) in height.
5. The unbalanced fill for 4-inch (102 mm) foundation walls shall not exceed 24 inches (610 mm) for *solid masonry*, nor 12 inches (305 mm) for hollow masonry.

1809.11 Steel grillage footings. Grillage footings of *structural steel elements* shall be separated with *approved* steel spacers and be entirely encased in concrete with not less than 6 inches (152 mm) on the bottom and not less than 4 inches (102 mm) at all other points. The spaces between the shapes shall be completely filled with concrete or cement grout.

1809.12 Timber footings. Timber footings shall be permitted for buildings of Type V construction and as otherwise *approved* by the *building official*. Such footings shall be treated in accordance with AWPA U1 (Commodity Specification A, Use Category 4B). Treated timbers are not required where placed entirely below permanent water level, or where used as capping for wood piles that project above the water level over submerged or marsh lands. The compressive stresses perpendicular to grain in untreated timber footings supported on treated piles shall not exceed 70 percent of the allowable stresses for the species and grade of timber as specified in the ANSI/AWC NDS.

1809.13 Footing seismic ties. Where a structure is assigned to *Seismic Design Category* D, E or F, individual spread footings founded on soil defined in Chapter 20 of ASCE 7 as *Site Class* E or F shall be interconnected by ties. Unless it is demonstrated that equivalent restraint is provided by reinforced concrete beams within slabs on grade or reinforced concrete slabs on grade, ties shall be capable of carrying, in tension or compression, a force equal to the lesser of the product of the larger footing design gravity *load* times the seismic coefficient, S_{DS}, divided by 10 and 25 percent of the smaller footing design gravity load.

SECTION 1810
DEEP FOUNDATIONS

1810.1 General. *Deep foundations* shall be analyzed, designed, detailed and installed in accordance with Sections 1810.1 through 1810.4.

1810.1.1 Geotechnical investigation. *Deep foundations* shall be designed and installed on the basis of a geotechnical investigation as set forth in Section 1803.

1810.1.2 Use of existing deep foundation elements. *Deep foundation* elements left in place where a structure has been demolished shall not be used for the support of new construction unless satisfactory evidence is submitted to the *building official*, which indicates that the elements are sound and meet the requirements of this code. Such elements shall be load tested or redriven to verify their capacities. The design *load* applied to such elements shall be the lowest allowable *load* as determined by tests or redriving data.

1810.1.3 Deep foundation elements classified as columns. *Deep foundation* elements standing unbraced in air, water or fluid soils shall be classified as columns and designed as such in accordance with the provisions of this code from their top down to the point where adequate lateral support is provided in accordance with Section 1810.2.1.

Exception: Where the unsupported height to least horizontal dimension of a cast-in-place *deep foundation* element does not exceed three, it shall be permitted to design and construct such an element as a pedestal in accordance with ACI 318.

1810.1.4 Special types of deep foundations. The use of types of *deep foundation* elements not specifically mentioned herein is permitted, subject to the approval of the *building official*, upon the submission of acceptable test data, calculations and other information relating to the structural properties and load capacity of such elements. The allowable stresses for materials shall not in any case exceed the limitations specified herein.

1810.2 Analysis. The analysis of *deep foundations* for design shall be in accordance with Sections 1810.2.1 through 1810.2.5.

1810.2.1 Lateral support. Any soil other than fluid soil shall be deemed to afford sufficient lateral support to prevent buckling of *deep foundation* elements and to permit the design of the elements in accordance with accepted engineering practice and the applicable provisions of this code.

Where *deep foundation* elements stand unbraced in air, water or fluid soils, it shall be permitted to consider them laterally supported at a point 5 feet (1524 mm) into stiff soil or 10 feet (3048 mm) into soft soil unless otherwise *approved* by the *building official* on the basis of a geotechnical investigation by a *registered design professional*.

1810.2.2 Stability. *Deep foundation* elements shall be braced to provide lateral stability in all directions. Three or more elements connected by a rigid cap shall be considered to be braced, provided that the elements are located in radial directions from the centroid of the group not less than 60 degrees (1 rad) apart. A two-element group in a rigid cap shall be considered to be braced along the axis connecting the two elements. Methods used to brace *deep foundation* elements shall be subject to the approval of the *building official*.

Deep foundation elements supporting walls shall be placed alternately in lines spaced not less than 1 foot (305 mm) apart and located symmetrically under the center of gravity of the wall load carried, unless effective measures are taken to provide for eccentricity and lateral forces, or the foundation elements are adequately braced to provide for lateral stability.

Exceptions:

1. Isolated cast-in-place *deep foundation* elements without lateral bracing shall be permitted where the least horizontal dimension is not less than 2 feet (610 mm), adequate lateral support in accordance with Section 1810.2.1 is provided for the entire height and the height does not exceed 12 times the least horizontal dimension.
2. A single row of *deep foundation* elements without lateral bracing is permitted for one- and two-family dwellings and lightweight construction not exceeding two *stories above grade plane* or 35 feet (10 668 mm) in *building height*, provided that the centers of the elements are located within the width of the supported wall.

1810.2.3 Settlement. The settlement of a single *deep foundation* element or group thereof shall be estimated based on *approved* methods of analysis. The predicted settlement shall cause neither harmful distortion of, nor instability in, the structure, nor cause any element to be loaded beyond its capacity.

1810.2.4 Lateral loads. The moments, shears and lateral deflections used for design of *deep foundation* elements shall be established considering the nonlinear interaction of the shaft and soil, as determined by a *registered design professional*. Where the ratio of the depth of embedment of the element to its least horizontal dimension is less than or equal to six, it shall be permitted to assume the element is rigid.

1810.2.4.1 Seismic Design Categories D through F. For structures assigned to *Seismic Design Category* D, E or F, *deep foundation* elements on *Site Class* E or F sites, as determined in Section 1613.2.2, shall be designed and constructed to withstand maximum imposed curvatures from earthquake ground motions and structure response. Curvatures shall include free-field soil strains modified for soil-foundation-structure interaction coupled with foundation element deformations associated with earthquake *loads* imparted to the foundation by the structure.

Exception: *Deep foundation* elements that satisfy the following additional detailing requirements shall be deemed to comply with the curvature capacity requirements of this section.

1. Precast prestressed concrete piles detailed in accordance with Section 1810.3.8.
2. Cast-in-place *deep foundation* elements with a minimum longitudinal reinforcement ratio of 0.005 extending the full length of the element and detailed in accordance with Sections 18.7.5.2, 18.7.5.3 and 18.7.5.4 of ACI 318 as required by Section 1810.3.9.4.2.2.

1810.2.5 Group effects. The analysis shall include group effects on lateral behavior where the center-to-center spacing of *deep foundation* elements in the direction of lateral force is less than eight times the least horizontal dimension of an element. The analysis shall include group effects on axial behavior where the center-to-center spacing of *deep foundation* elements is less than three times the least horizontal dimension of an element. Group effects shall be evaluated using a generally accepted method of analysis; the analysis for uplift of grouped elements with center-to-center spacing less than three times the least horizontal dimension of an element shall be evaluated in accordance with Section 1810.3.3.1.6.

1810.3 Design and detailing. *Deep foundations* shall be designed and detailed in accordance with Sections 1810.3.1 through 1810.3.13.

1810.3.1 Design conditions. Design of *deep foundations* shall include the design conditions specified in Sections 1810.3.1.1 through 1810.3.1.6, as applicable.

1810.3.1.1 Design methods for concrete elements. Where concrete *deep foundations* are laterally supported in accordance with Section 1810.2.1 for the entire height and applied forces cause bending moments not greater than those resulting from accidental eccentricities, structural design of the element using the allowable stress design load combinations specified in ASCE 7, Section 2.4 or the alternative allowable stress design load combinations of Section 1605.2 and the allowable stresses specified in this chapter shall be permitted. Otherwise, the structural design of concrete *deep foundation* elements shall use the strength load combinations specified in ASCE 7, Section 2.3 and *approved* strength design methods.

1810.3.1.2 Composite elements. Where a single *deep foundation* element comprises two or more sections of different materials or different types spliced together, each section of the composite assembly shall satisfy the applicable requirements of this code, and the maximum allowable *load* in each section shall be limited by the structural capacity of that section.

1810.3.1.3 Mislocation. The foundation or superstructure shall be designed to resist the effects of the mislocation of any *deep foundation* element by not less than 3 inches (76 mm). To resist the effects of mislocation, compressive overload of *deep foundation* elements to 110 percent of the allowable design load shall be permitted.

1810.3.1.4 Driven piles. Driven piles shall be designed and manufactured in accordance with accepted engineering practice to resist all stresses induced by handling, driving and service loads.

1810.3.1.5 Helical piles. *Helical piles* shall be designed and manufactured in accordance with accepted engineering practice to resist all stresses induced by installation into the ground and service loads.

1810.3.1.6 Casings. Temporary and permanent casings shall be of steel and shall be sufficiently strong to resist collapse and sufficiently watertight to exclude any foreign materials during the placing of concrete. Where a permanent casing is considered reinforcing steel, the steel shall be protected under the conditions specified in Section 1810.3.2.5. Horizontal joints in the casing shall be spliced in accordance with Section 1810.3.6.

1810.3.2 Materials. The materials used in *deep foundation* elements shall satisfy the requirements of Sections 1810.3.2.1 through 1810.3.2.8, as applicable.

1810.3.2.1 Concrete. Where concrete is cast in a steel pipe or where an enlarged base is formed by compacting concrete, the maximum size for coarse aggregate shall be $^3/_4$ inch (19.1 mm). Concrete to be compacted shall have a zero slump.

1810.3.2.1.1 Seismic hooks. For structures assigned to *Seismic Design Category* C, D, E or F, the ends of hoops, spirals and ties used in concrete *deep foundation* elements shall be terminated with seismic hooks, as defined in ACI 318, and shall be turned into the confined concrete core.

1810.3.2.1.2 ACI 318 Equation (25.7.3.3). Where this chapter requires detailing of concrete *deep foundation* elements in accordance with Section 18.7.5.4 of ACI 318, compliance with Equation (25.7.3.3) of ACI 318 shall not be required.

1810.3.2.2 Prestressing steel. Prestressing steel shall conform to ASTM A416.

1810.3.2.3 Steel. Structural steel H-piles and structural steel sheet piling shall conform to the material requirements in ASTM A6. Steel pipe piles shall conform to the material requirements in ASTM A252. Fully welded steel piles shall be fabricated from plates that conform to the material requirements in ASTM A36, ASTM A283, ASTM A572, ASTM A588 or ASTM A690.

1810.3.2.4 Timber. Timber *deep foundation* elements shall be designed as piles or poles in accordance with ANSI/AWC NDS. Round timber elements shall conform to ASTM D25. Sawn timber elements shall conform to DOC PS-20.

1810.3.2.4.1 Preservative treatment. Timber *deep foundation* elements used to support permanent structures shall be treated in accordance with this section unless it is established that the tops of the untreated timber elements will be below the lowest ground-water level assumed to exist during the life of the structure. Preservative and minimum final retention shall be in accordance with AWPA U1 (Commodity Specification E, Use Category 4C) for round timber elements and AWPA U1 (Commodity Specification A, Use Category 4B) for sawn timber elements. Preservative-treated timber elements shall be subject to a quality control program administered by an *approved* agency. Element cutoffs shall be treated in accordance with AWPA M4.

1810.3.2.5 Protection of materials. Where boring records or site conditions indicate possible deleterious action on the materials used in deep foundation elements because of soil constituents, changing water levels or other factors, the elements shall be adequately protected by materials, methods or processes *approved* by the *building official*. Protective materials shall be applied to the elements so as not to be rendered ineffective by installation. The effectiveness of such protective measures for the particular purpose shall have been thoroughly established by satisfactory service records or other evidence.

1810.3.2.6 Allowable stresses. The allowable stresses for materials used in deep foundation elements shall not exceed those specified in Table 1810.3.2.6.

1810.3.2.7 Increased allowable compressive stress for cased mandrell-driven cast-in-place elements. The allowable compressive stress in the concrete shall be permitted to be increased as specified in Table 1810.3.2.6 for those portions of permanently cased cast-in-place elements that satisfy all of the following conditions:

1. The design shall not use the casing to resist any portion of the axial load imposed.
2. The casing shall have a sealed tip and be mandrel driven.
3. The thickness of the casing shall be not less than manufacturer's standard gage No.14 (0.068 inch) (1.75 mm).
4. The casing shall be seamless or provided with seams of strength equal to the basic material and be of a configuration that will provide confinement to the cast-in-place concrete.
5. The ratio of steel yield strength (F_y) to specified compressive strength (f'_c) shall be not less than six.
6. The nominal diameter of the element shall not be greater than 16 inches (406 mm).

1810.3.2.8 Justification of higher allowable stresses. Use of allowable stresses greater than those specified in Section 1810.3.2.6 shall be permitted where supporting data justifying such higher stresses is filed with the *building official*. Such substantiating data shall include the following:

1. A geotechnical investigation in accordance with Section 1803.
2. Load tests in accordance with Section 1810.3.3.1.2, regardless of the load supported by the element.

The design and installation of the deep foundation elements shall be under the direct supervision of a *registered design professional* knowledgeable in the field of soil mechanics and deep foundations who shall submit a report to the *building official* stating that the elements as installed satisfy the design criteria.

1810.3.3 Determination of allowable loads. The allowable axial and lateral loads on deep foundation elements shall be determined by an *approved* formula, load tests or method of analysis.

1810.3.3.1 Allowable axial load. The allowable axial load on a deep foundation element shall be determined in accordance with Sections 1810.3.3.1.1 through 1810.3.3.1.9.

Exception: Where approved by the *building official,* load testing is not required.

1810.3.3.1.1 Driving criteria. The allowable compressive load on any driven deep foundation element where determined by the application of an *approved* driving formula shall not exceed 40 tons (356 kN). For allowable loads above 40 tons (356 kN), the wave equation method of analysis shall be used to estimate driveability for both driving stresses and net displacement per blow at the ultimate load. Allowable loads shall be verified by load tests in accordance with Section 1810.3.3.1.2. The formula or wave equation load shall be determined for gravity-drop or power-actuated hammers and the hammer energy used shall be the maximum consistent with the size, strength and weight of the driven elements. The use of a follower is permitted only with the approval of the *building official.* The introduction of fresh hammer cushion or pile cushion material just prior to final penetration is not permitted.

1810.3.3.1.2 Load tests. Where design compressive *loads* are greater than those determined using the allowable stresses specified in Section 1810.3.2.6, where the design *load* for any deep foundation element is in doubt, or where cast-in-place deep foundation elements have an enlarged base formed either by compacting concrete or by driving a precast base, control test elements shall be tested in accordance with ASTM D1143 or ASTM D4945. One element or more shall be load tested in each area of uniform subsoil conditions. Where required by the *building official*, additional elements shall be load tested where necessary to establish the safe design capacity. The resulting allowable *loads* shall

TABLE 1810.3.2.6
ALLOWABLE STRESSES FOR MATERIALS USED IN DEEP FOUNDATION ELEMENTS

MATERIAL TYPE AND CONDITION	MAXIMUM ALLOWABLE STRESS[a]
1. Concrete or grout in compression[b]	
Cast-in-place with a permanent casing in accordance with Section 1810.3.2.7 or Section 1810.3.5.3.4	$0.4\ f'_c$
Cast-in-place in other permanent casing or rock	$0.33\ f'_c$
Cast-in-place without a permanent casing	$0.3\ f'_c$
Precast nonprestressed	$0.33\ f'_c$
Precast prestressed	$0.33\ f'_c - 0.27\ f_{pc}$
2. Nonprestressed reinforcement in compression	$0.4\ f_y \leq 30{,}000$ psi
3. Steel in compression	
Cores within concrete-filled pipes or tubes	$0.5\ F_y \leq 32{,}000$ psi
Pipes, tubes or H-piles, where justified in accordance with Section 1810.3.2.8	$0.5\ F_y \leq 32{,}000$ psi
Pipes or tubes for micropiles	$0.4\ F_y \leq 32{,}000$ psi
Other pipes, tubes or H-piles	$0.35\ F_y \leq 24{,}000$ psi
Helical piles	$0.6\ F_y \leq 0.5\ F_u$
4. Nonprestressed reinforcement in tension	
Within micropiles	$0.6\ f_y$
Other conditions	
For load combinations that do not include wind or seismic loads	$0.5\ f_y \leq 30{,}000$ psi
For load combinations that include wind or seismic loads	$0.5\ f_y \leq 40{,}000$ psi
5. Steel in tension	
Pipes, tubes or H-piles, where justified in accordance with Section 1810.3.2.8	$0.5\ F_y \leq 32{,}000$ psi
Other pipes, tubes or H-piles	$0.35\ F_y \leq 24{,}000$ psi
Helical piles	$0.6\ F_y \leq 0.5\ F_u$
6. Timber	In accordance with the ANSI/AWC NDS

a. f'_c is the specified compressive strength of the concrete or grout; f_{pc} is the compressive stress on the gross concrete section due to effective prestress forces only; f_y is the specified yield strength of reinforcement; F_y is the specified minimum yield stress of steel; F_u is the specified minimum tensile stress of structural steel.

b. The stresses specified apply to the gross cross-sectional area of the concrete for precast prestressed piles and to the net cross-sectional area for all other piles. Where a temporary or permanent casing is used, the inside face of the casing shall be considered the outer edge of the concrete cross-section.

not be more than one-half of the ultimate axial load capacity of the test element as assessed by one of the published methods listed in Section 1810.3.3.1.3 with consideration for the test type, duration and subsoil. The ultimate axial load capacity shall be determined by a *registered design professional* with consideration given to tolerable total and differential settlements at design *load* in accordance with Section 1810.2.3. In subsequent installation of the balance of deep foundation elements, all elements shall be deemed to have a supporting capacity equal to that of the control element where such elements are of the same type, size and relative length as the test element; are installed using the same or comparable methods and equipment as the test element; are installed in similar subsoil conditions as the test element; and, for driven elements, where the rate of penetration (for example, net displacement per blow) of such elements is equal to or less than that of the test element driven with the same hammer through a comparable driving distance.

1810.3.3.1.3 Load test evaluation methods. It shall be permitted to evaluate load tests of *deep foundation* elements using any of the following methods:

1. Davisson Offset Limit.
2. Brinch-Hansen 90-percent Criterion.
3. Butler-Hoy Criterion.
4. Other methods *approved* by the *building official*.

1810.3.3.1.4 Allowable shaft resistance. The assumed shaft resistance developed by any uncased cast-in-place *deep foundation* element shall not exceed one-sixth of the bearing value of the soil material at minimum depth as set forth in Table 1806.2, up to 500 psf (24 kPa), unless a greater value is allowed by the *building official* on the basis of a geotechnical investigation as specified in Section 1803 or a greater value is substantiated by a load test in accordance with Section 1810.3.3.1.2. Shaft resistance and end-bearing resistance shall not be assumed to act simultaneously unless determined by a geotechnical investigation in accordance with Section 1803.

1810.3.3.1.5 Uplift capacity of a single *deep foundation* element. Where required by the design, the uplift capacity of a single deep foundation element shall be determined by an *approved* method of analysis based on a minimum factor of safety of three or by load tests conducted in accordance with ASTM D3689. The maximum allowable uplift *load* shall not exceed the ultimate load capacity as determined in Section 1810.3.3.1.2, using the results of load tests conducted in accordance with ASTM D3689, divided by a factor of safety of two.

Exception: Where uplift is due to wind or seismic loading, the minimum factor of safety shall be two where capacity is determined by an analysis and one and one-half where capacity is determined by load tests.

1810.3.3.1.6 Allowable uplift load of grouped *deep foundation* elements. For grouped *deep foundation* elements subjected to uplift, the allowable uplift *load* for the group shall be calculated by a generally accepted method of analysis. Where the *deep foundation* elements in the group are placed at a center-to-center spacing less than three times the least horizontal dimension of the largest single element, the allowable uplift *load* for the group is permitted to be calculated as the lesser of:

1. The proposed individual allowable uplift *load* times the number of elements in the group.
2. Two-thirds of the effective weight of the group and the soil contained within a block defined by the perimeter of the group and the length of the element, plus two-thirds of the ultimate shear resistance along the soil block.

1810.3.3.1.7 Load-bearing capacity. *Deep foundation* elements shall develop ultimate load capacities of not less than twice the design working *loads* in the designated load-bearing layers. Analysis shall show that soil layers underlying the designated load-bearing layers do not cause the load-bearing capacity safety factor to be less than two.

1810.3.3.1.8 Bent deep foundation elements. The load-bearing capacity of *deep foundation* elements discovered to have a sharp or sweeping bend shall be determined by an *approved* method of analysis or by load testing a representative element.

1810.3.3.1.9 Helical piles. The allowable axial design *load*, P_a, of *helical piles* shall be determined as follows:

$$P_a = 0.5\ P_u \qquad \textbf{(Equation 18-4)}$$

where P_u is the least value of:

1. Base capacity plus shaft resistance of the *helical pile*. The base capacity is equal to the sum of the areas of the helical bearing plates times the ultimate bearing capacity of the soil or rock comprising the bearing stratum. The shaft resistance is equal to the area of the shaft above the uppermost helical bearing plate times the ultimate skin resistance.
2. Ultimate capacity determined from well-documented correlations with installation torque.
3. Ultimate capacity determined from load tests where required by Section 1810.3.3.1.2.
4. Ultimate axial capacity of pile shaft.

5. Ultimate axial capacity of pile shaft couplings.
6. Sum of the ultimate axial capacity of helical bearing plates affixed to pile.

1810.3.3.2 Allowable lateral load. Where required by the design, the lateral load capacity of a single *deep foundation* element or a group thereof shall be determined by an *approved* method of analysis or by lateral load tests to not less than twice the proposed design working *load*. The resulting allowable *load* shall not be more than one-half of the *load* that produces a gross lateral movement of 1 inch (25 mm) at the lower of the top of foundation element and the ground surface, unless it can be shown that the predicted lateral movement shall cause neither harmful distortion of, nor instability in, the structure, nor cause any element to be loaded beyond its capacity.

1810.3.4 Subsiding soils or strata. Where *deep foundation* elements are installed through subsiding soils or other subsiding strata and derive support from underlying firmer materials, consideration shall be given to the downward frictional forces potentially imposed on the elements by the subsiding upper strata.

Where the influence of subsiding soils or strata is considered as imposing *loads* on the element, the allowable stresses specified in this chapter shall be permitted to be increased where satisfactory substantiating data are submitted.

1810.3.5 Dimensions of deep foundation elements. The dimensions of *deep foundation* elements shall be in accordance with Sections 1810.3.5.1 through 1810.3.5.3, as applicable.

1810.3.5.1 Precast. The minimum lateral dimension of precast concrete *deep foundation* elements shall be 8 inches (203 mm). Corners of square elements shall be chamfered.

1810.3.5.2 Cast-in-place or grouted-in-place. Cast-in-place and grouted-in-place *deep foundation* elements shall satisfy the requirements of this section.

1810.3.5.2.1 Cased. Cast-in-place or grouted-in-place *deep foundation* elements with a permanent casing shall have a nominal outside diameter of not less than 8 inches (203 mm).

1810.3.5.2.2 Uncased. Cast-in-place or grouted-in-place *deep foundation* elements without a permanent casing shall have a specified diameter of not less than 12 inches (305 mm). The element length shall not exceed 30 times the specified diameter.

Exception: The length of the element is permitted to exceed 30 times the specified diameter, provided that the design and installation of the deep foundations are under the direct supervision of a *registered design professional* knowledgeable in the field of soil mechanics and deep foundations. The *registered design professional* shall submit a report to the *building official* stating that the elements were installed in compliance with the *approved construction documents*.

1810.3.5.2.3 Micropiles. Micropiles shall have a nominal diameter of 12 inches (305 mm) or less. The minimum diameter set forth elsewhere in Section 1810.3.5 shall not apply to micropiles.

1810.3.5.3 Steel. Steel *deep foundation* elements shall satisfy the requirements of this section.

1810.3.5.3.1 Structural steel H-piles. Sections of structural steel H-piles shall comply with the requirements for HP shapes in ASTM A6, or the following:

1. The flange projections shall not exceed 14 times the minimum thickness of metal in either the flange or the web and the flange widths shall be not less than 80 percent of the depth of the section.
2. The nominal depth in the direction of the web shall be not less than 8 inches (203 mm).
3. Flanges and web shall have a minimum nominal thickness of $^3/_8$ inch (9.5 mm).

For structures assigned to *Seismic Design Category* D, E or F, design and detailing of H-piles shall also conform to the requirements of AISC 341.

1810.3.5.3.2 Fully welded steel piles fabricated from plates. Sections of fully welded steel piles fabricated from plates shall comply with the following:

1. The flange projections shall not exceed 14 times the minimum thickness of metal in either the flange or the web and the flange widths shall be not less than 80 percent of the depth of the section.
2. The nominal depth in the direction of the web shall be not less than 8 inches (203 mm).
3. Flanges and web shall have a minimum nominal thickness of $^3/_8$ inch (9.5 mm).

1810.3.5.3.3 Structural steel sheet piling. Individual sections of structural steel sheet piling shall conform to the profile indicated by the manufacturer, and shall conform to the general requirements specified by ASTM A6.

1810.3.5.3.4 Steel pipes and tubes. Steel pipes and tubes used as *deep foundation* elements shall have a nominal outside diameter of not less than 8 inches (203 mm). Where steel pipes or tubes are driven open ended, they shall have not less than 0.34 square inch (219 mm^2) of steel in cross section to resist each 1,000 foot-pounds (1356 Nm) of pile hammer energy, or shall have the equivalent strength for steels having a yield strength greater than 35,000 psi (241 MPa) or the wave equation analysis shall be permitted to be used to assess compression stresses induced by driving to evaluate if the pile section is appropriate for the selected hammer. Where a pipe or tube with wall thickness less than 0.179 inch (4.6 mm) is driven open ended, a suitable cutting shoe shall be provided. Concrete-filled steel pipes or tubes in structures assigned to *Seismic Design Category* C, D, E or F shall have a wall thickness of not less than $^{3}/_{16}$ inch (5 mm). The pipe or tube casing for socketed *drilled shafts* shall have a nominal outside diameter of not less than 18 inches (457 mm), a wall thickness of not less than $^{3}/_{8}$ inch (9.5 mm) and a suitable steel driving shoe welded to the bottom; the diameter of the rock socket shall be approximately equal to the inside diameter of the casing.

Exceptions:

1. There is no minimum diameter for steel pipes or tubes used in micropiles.
2. For mandrel-driven pipes or tubes, the minimum wall thickness shall be $^{1}/_{10}$ inch (2.5 mm).

1810.3.5.3.5 Helical piles. Dimensions of the central shaft and the number, size and thickness of helical bearing plates shall be sufficient to support the design loads.

1810.3.6 Splices. Splices shall be constructed so as to provide and maintain true alignment and position of the component parts of the *deep foundation* element during installation and subsequent thereto and shall be designed to resist the axial and shear forces and moments occurring at the location of the splice during driving and for design load combinations. Where *deep foundation* elements of the same type are being spliced, splices shall develop not less than 50 percent of the bending strength of the weaker section. Where *deep foundation* elements of different materials or different types are being spliced, splices shall develop the full compressive strength and not less than 50 percent of the tension and bending strength of the weaker section. Where structural steel cores are to be spliced, the ends shall be milled or ground to provide full contact and shall be full-depth welded.

Exception: For buildings assigned to *Seismic Design Category* A or B, splices need not comply with the 50-percent tension and bending strength requirements where justified by supporting data.

Splices occurring in the upper 10 feet (3048 mm) of the embedded portion of an element shall be designed to resist at allowable stresses the moment and shear that would result from an assumed eccentricity of the axial *load* of 3 inches (76 mm), or the element shall be braced in accordance with Section 1810.2.2 to other deep foundation elements that do not have splices in the upper 10 feet (3048 mm) of embedment.

1810.3.6.1 Seismic Design Categories C through F. For structures assigned to *Seismic Design Category* C, D, E or F splices of *deep foundation* elements shall develop the lesser of the following:

1. The nominal strength of the *deep foundation* element.
2. The axial and shear forces and moments from the seismic *load effects* including overstrength factor in accordance with Section 2.3.6 or 2.4.5 of ASCE 7.

1810.3.7 Top of element detailing at cutoffs. Where a minimum length for reinforcement or the extent of closely spaced confinement reinforcement is specified at the top of a *deep foundation* element, provisions shall be made so that those specified lengths or extents are maintained after cutoff.

1810.3.8 Precast concrete piles. Precast concrete piles shall be designed and detailed in accordance with ACI 318.

Exceptions:

1. For precast prestressed piles in *Seismic Design Category* C, the minimum volumetric ratio of spirals or circular hoops required by Section 18.13.5.10.4 of ACI 318 shall not apply in cases where the design includes full consideration of load combinations specified in ASCE 7, Section 2.3.6 or Section 2.4.5 and the applicable overstrength factor, Ω_0. In such cases, minimum transverse reinforcement index shall be as specified in Section 13.4.5.6 of ACI 318.
2. For precast prestressed piles in *Seismic Design Categories* D through F, the minimum volumetric ratio of spirals or circular hoops required by Section 18.13.5.10.5(c) of ACI 318 shall not apply in cases where the design includes full consideration of load combinations specified in ASCE 7, Section 2.3.6 or Section 2.4.5 and the applicable overstrength factor, Ω_0. In such cases, minimum transverse reinforcement shall be as specified in Section 13.4.5.6 of ACI 318.

1810.3.9 Cast-in-place deep foundations. Cast-in-place *deep foundation* elements shall be designed and detailed in accordance with Sections 1810.3.9.1 through 1810.3.9.6.

1810.3.9.1 Design cracking moment. The design cracking moment (ϕM_n) for a cast-in-place *deep foundation* element not enclosed by a structural steel pipe or tube shall be determined using the following equation:

$$\phi M_n = 3\sqrt{f'_c}S_m \qquad \textbf{(Equation 18-5)}$$

For SI: $\phi M_n = 0.25\sqrt{f'_c}S_m$

where:

f'_c = Specified compressive strength of concrete or grout, psi (MPa).

S_m = Elastic section modulus, neglecting reinforcement and casing, cubic inches (mm^3).

1810.3.9.2 Required reinforcement. Where subject to uplift or where the required moment strength determined using the load combinations of ASCE 7, Section 2.3 exceeds the design cracking moment determined in accordance with Section 1810.3.9.1 , cast-in-place deep foundations not enclosed by a structural steel pipe or tube shall be reinforced.

1810.3.9.3 Placement of reinforcement. Reinforcement where required shall be assembled and tied together and shall be placed in the *deep foundation* element as a unit before the reinforced portion of the element is filled with concrete.

Exceptions:

1. Steel dowels embedded 5 feet (1524 mm) or less shall be permitted to be placed after concreting, while the concrete is still in a semifluid state.
2. For *deep foundation* elements installed with a hollow-stem auger, tied reinforcement shall be placed after elements are concreted, while the concrete is still in a semifluid state. Longitudinal reinforcement without lateral ties shall be placed either through the hollow stem of the auger prior to concreting or after concreting, while the concrete is still in a semifluid state.
3. For Group R-3 and U occupancies not exceeding two stories of *light-frame construction*, reinforcement is permitted to be placed after concreting, while the concrete is still in a semifluid state, and the concrete cover requirement is permitted to be reduced to 2 inches (51 mm), provided that the construction method can be demonstrated to the satisfaction of the *building official*.

1810.3.9.4 Seismic reinforcement. Where a structure is assigned to *Seismic Design Category* C, reinforcement shall be provided in accordance with Section 1810.3.9.4.1. Where a structure is assigned to *Seismic Design Category* D, E or F, reinforcement shall be provided in accordance with Section 1810.3.9.4.2.

Exceptions:

1. Isolated *deep foundation* elements supporting posts of Group R-3 and U occupancies not exceeding two stories of *light-frame construction* shall be permitted to be reinforced as required by rational analysis but with not less than one No. 4 bar, without ties or spirals, where detailed so the element is not subject to lateral loads and the soil provides adequate lateral support in accordance with Section 1810.2.1 .
2. Isolated *deep foundation* elements supporting posts and bracing from decks and patios appurtenant to Group R-3 and U occupancies not exceeding two stories of *light-frame construction* shall be permitted to be reinforced as required by rational analysis but with not less than one No. 4 bar, without ties or spirals, where the lateral *load*, E, to the top of the element does not exceed 200 pounds (890 N) and the soil provides adequate lateral support in accordance with Section 1810.2.1.
3. *Deep foundation* elements supporting the concrete foundation wall of Group R-3 and U occupancies not exceeding two stories of light-frame construction shall be permitted to be reinforced as required by rational analysis but with not less than two No. 4 bars, without ties or spirals, where the design cracking moment determined in accordance with Section 1810.3.9.1 exceeds the required moment strength determined using the load combinations with overstrength factor in Section 2.3.6 or 2.4.5 of ASCE 7 and the soil provides adequate lateral support in accordance with Section 1810.2.1.
4. Closed ties or spirals where required by Section 1810.3.9.4.2 shall be permitted to be limited to the top 3 feet (914 mm) of *deep foundation* elements 10 feet (3048 mm) or less in depth supporting Group R-3 and U occupancies of *Seismic Design Category* D, not exceeding two stories of *light-frame construction*.

1810.3.9.4.1 Seismic reinforcement in Seismic Design Category C. For structures assigned to *Seismic Design Category* C, cast-in-place *deep foundation* elements shall be reinforced as specified in this section. Reinforcement shall be provided where required by analysis.

Not fewer than four longitudinal bars, with a minimum longitudinal reinforcement ratio of 0.0025, shall be provided throughout the minimum reinforced length of the element as defined in this section starting at the top of the element. The minimum reinforced length of the element shall be taken as the greatest of the following:

1. One-third of the element length.
2. A distance of 10 feet (3048 mm).
3. Three times the least element dimension.
4. The distance from the top of the element to the point where the design cracking moment determined in accordance with Section 1810.3.9.1 exceeds the required moment strength determined using the load combinations of ASCE 7, Section 2.3.

Transverse reinforcement shall consist of closed ties or spirals with a minimum $^{3}/_{8}$ inch (9.5 mm) diameter. Spacing of transverse reinforcement shall not exceed the smaller of 6 inches (152 mm) or 8-longitudinal-bar diameters, within a distance of three times the least element dimension from the bottom of the pile cap. Spacing of transverse reinforcement shall not exceed 16 longitudinal bar diameters throughout the remainder of the reinforced length.

Exceptions:

1. The requirements of this section shall not apply to concrete cast in structural steel pipes or tubes.
2. A spiral-welded metal casing of a thickness not less than the manufacturer's standard No. 14 gage (0.068 inch) is permitted to provide concrete confinement in lieu of the closed ties or spirals. Where used as such, the metal casing shall be protected against possible deleterious action due to soil constituents, changing water levels or other factors indicated by boring records of site conditions.

1810.3.9.4.2 Seismic reinforcement in Seismic Design Categories D through F. For structures assigned to *Seismic Design Category* D, E or F, cast-in-place *deep foundation* elements shall be reinforced as specified in this section. Reinforcement shall be provided where required by analysis.

Not fewer than four longitudinal bars, with a minimum longitudinal reinforcement ratio of 0.005, shall be provided throughout the minimum reinforced length of the element as defined in this section starting at the top of the element. The minimum reinforced length of the element shall be taken as the greatest of the following:

1. One-half of the element length.
2. A distance of 10 feet (3048 mm).
3. Three times the least element dimension.
4. The distance from the top of the element to the point where the design cracking moment determined in accordance with Section 1810.3.9.1 exceeds the required moment strength determined using the load combinations of ASCE 7, Section 2.3.

Transverse reinforcement shall consist of closed ties or spirals not smaller than No. 3 bars for elements with a least dimension up to 20 inches (508 mm), and No. 4 bars for larger elements. Throughout the remainder of the reinforced length outside the regions with transverse confinement reinforcement, as specified in Section 1810.3.9.4.2.1 or 1810.3.9.4.2.2, the spacing of transverse reinforcement shall not exceed the least of the following:

1. 12 longitudinal bar diameters.
2. One-half the least dimension of the element.
3. 12 inches (305 mm).

Exceptions:

1. The requirements of this section shall not apply to concrete cast in structural steel pipes or tubes.
2. A spiral-welded metal casing of a thickness not less than manufacturer's standard No. 14 gage (0.068 inch) is permitted to provide concrete confinement in lieu of the closed ties or spirals. Where used as such, the metal casing shall be protected against possible deleterious action due to soil constituents, changing water levels or other factors indicated by boring records of site conditions.

1810.3.9.4.2.1 Site Classes A through D. For *Site Class* A, B, C or D sites, transverse confinement reinforcement shall be provided in the element in accordance with Sections 18.7.5.2, 18.7.5.3 and 18.7.5.4 of ACI 318 within three times the least element dimension of the bottom of the pile cap. A transverse spiral reinforcement ratio of not less than one-half of that required in Table 18.10.6.4(g) of ACI 318 shall be permitted.

1810.3.9.4.2.2 Site Classes E and F. For *Site Class* E or F sites, transverse confinement reinforcement shall be provided in the element in accordance with Sections 18.7.5.2, 18.7.5.3 and 18.7.5.4 of ACI 318 within seven times the least element dimension of the pile cap and within seven times the least element dimension of the interfaces of strata that are hard or stiff and strata that are liquefiable or are composed of soft- to medium-stiff clay.

1810.3.9.5 Belled drilled shafts. Where *drilled shafts* are belled at the bottom, the edge thickness of the bell shall be not less than that required for the edge of footings. Where the sides of the bell slope at an angle less than 60 degrees (1 rad) from the horizontal, the effects of vertical shear shall be considered.

1810.3.9.6 Socketed drilled shafts. Socketed *drilled shafts* shall have a permanent pipe or tube casing that extends down to bedrock and an uncased socket drilled into the bedrock, both filled with concrete. Socketed *drilled shafts* shall have reinforcement or a structural steel core for the length as indicated by an *approved* method of analysis.

The depth of the rock socket shall be sufficient to develop the full load-bearing capacity of the element with a minimum safety factor of two, but the depth shall be not less than the outside diameter of the pipe or tube casing. The design of the rock socket is permitted to be predicated on the sum of the allowable load-bearing pressure on the bottom of the socket plus bond along the sides of the socket.

Where a structural steel core is used, the gross cross-sectional area of the core shall not exceed 25 percent of the gross area of the *drilled shaft*.

1810.3.10 Micropiles. Micropiles shall be designed and detailed in accordance with Sections 1810.3.10.1 through 1810.3.10.4.

1810.3.10.1 Construction. Micropiles shall develop their load-carrying capacity by means of a bond zone in soil, bedrock or a combination of soil and bedrock. Micropiles shall be grouted and have either a steel pipe or tube or steel reinforcement at every section along the length. It shall be permitted to transition from deformed reinforcing bars to steel pipe or tube reinforcement by extending the bars into the pipe or tube section by not less than their development length in tension in accordance with ACI 318.

1810.3.10.2 Materials. Reinforcement shall consist of deformed reinforcing bars in accordance with ASTM A615 Grade 60 or 75 or ASTM A722 Grade 150.

The steel pipe or tube shall have a minimum wall thickness of $^{3}/_{16}$ inch (4.8 mm). Splices shall comply with Section 1810.3.6. The steel pipe or tube shall have a minimum yield strength of 45,000 psi (310 MPa) and a minimum elongation of 15 percent as shown by mill certifications or two coupon test samples per 40,000 pounds (18 160 kg) of pipe or tube.

1810.3.10.3 Reinforcement. For micropiles or portions thereof grouted inside a temporary or permanent casing or inside a hole drilled into bedrock or a hole drilled with grout, the steel pipe or tube or steel reinforcement shall be designed to carry not less than 40 percent of the design compression load. Micropiles or portions thereof grouted in an open hole in soil without temporary or permanent casing and without suitable means of verifying the hole diameter during grouting shall be designed to carry the entire compression *load* in the reinforcing steel. Where a steel pipe or tube is used for reinforcement, the portion of the grout enclosed within the pipe is permitted to be included in the determination of the allowable stress in the grout.

1810.3.10.4 Seismic reinforcement. For structures assigned to *Seismic Design Category* C, a permanent steel casing shall be provided from the top of the *micropile* down to the point of zero curvature. For structures assigned to *Seismic Design Category* D, E or F, the *micropile* shall be considered as an alternative system in accordance with Section 104.11. The alternative system design, supporting documentation and test data shall be submitted to the *building official* for review and approval.

1810.3.11 Pile caps. Pile caps shall conform with ACI 318 and this section. Pile caps shall be of reinforced concrete, and shall include all elements to which vertical *deep foundation* elements are connected, including grade beams and mats. The soil immediately below the pile cap shall not be considered as carrying any vertical *load*, with the exception of a *combined pile raft*. The tops of vertical *deep foundation* elements shall be embedded not less than 3 inches (76 mm) into pile caps and the caps shall extend not less than 4 inches (102 mm) beyond the edges of the elements. The tops of elements shall be cut or chipped back to sound material before capping.

1810.3.11.1 Seismic Design Categories C through F. For structures assigned to *Seismic Design Category* C, D, E or F, concrete *deep foundation* elements shall be connected to the pile cap in accordance with ACI 318.

For resistance to uplift forces, anchorage of steel pipes, tubes or H-piles to the pile cap shall be made by means other than concrete bond to the bare steel section. Concrete-filled steel pipes or tubes shall have reinforcement of not less than 0.01 times the cross-sectional area of the concrete fill developed into the cap and extending into the fill a length equal to two times the required cap embedment, but not less than the development length in tension of the reinforcement.

1810.3.11.2 Seismic Design Categories D through F. For structures assigned to *Seismic Design Category* D, E or F, *deep foundation* element resistance to uplift forces or rotational restraint shall be provided by anchorage into the pile cap, designed considering the combined effect of axial forces due to uplift and bending moments due to fixity to the pile cap. Anchorage shall develop not less than 25 percent of the strength of the element in tension. Anchorage into the pile cap shall comply with the following:

1. In the case of uplift, the anchorage shall be capable of developing the least of the following:
   - 1.1. The nominal tensile strength of the longitudinal reinforcement in a concrete element.
   - 1.2. The nominal tensile strength of a steel element.
   - 1.3. The frictional force developed between the element and the soil multiplied by 1.3.

   Exception: The anchorage is permitted to be designed to resist the axial tension force resulting from the seismic *load effects* including overstrength factor in accordance with Section 2.3.6 or 2.4.5 of ASCE 7.
2. In the case of rotational restraint, the anchorage shall be designed to resist the axial and shear forces, and moments resulting from the seismic *load effects* including overstrength factor in accordance with Section 2.3.6 or 2.4.5 of ASCE 7 or the anchorage shall be capable of

developing the full axial, bending and shear nominal strength of the element.

3. The connection between the pile cap and the steel H-piles or unfilled steel pipe piles in structures assigned to *Seismic Design Category* D, E or F shall be designed for a tensile force of not less than 10 percent of the pile compression capacity.

Exceptions:

1. Connection tensile capacity need not exceed the strength required to resist seismic load effects including overstrength of ASCE 7 Section 12.4.3 or 12.14.3.2.
2. Connections need not be provided where the foundation or supported structure does not rely on the tensile capacity of the piles for stability under the design seismic force.

Where the vertical lateral-force-resisting elements are columns, the pile cap flexural strengths shall exceed the column flexural strength. The connection between batter piles and pile caps shall be designed to resist the nominal strength of the pile acting as a short column. Batter piles and their connection shall be designed to resist forces and moments that result from the application of seismic *load effects* including overstrength factor in accordance with Section 2.3.6 or 2.4.5 of ASCE 7.

1810.3.12 Grade beams. Grade beams shall comply with the provisions of ACI 318.

Exception: Grade beams designed to resist the seismic load effects including overstrength factor in accordance with Section 2.3.6 or 2.4.5 of ASCE 7.

1810.3.13 Seismic ties. Seismic ties shall comply with the provisions of ACI 318.

Exception: In Group R-3 and U occupancies of *light-frame construction*, *deep foundation* elements supporting foundation walls, isolated interior posts detailed so the element is not subject to lateral *loads* or exterior decks and patios are not subject to interconnection where the soils are of adequate stiffness, subject to the approval of the *building official*.

1810.4 Installation. Deep foundations shall be installed in accordance with Section 1810.4. Where a single *deep foundation* element comprises two or more sections of different materials or different types spliced together, each section shall satisfy the applicable conditions of installation.

1810.4.1 Structural integrity. *Deep foundation* elements shall be installed in such a manner and sequence as to prevent distortion or damage that would adversely affect the structural integrity of adjacent structures or of foundation elements being installed or already in place and as to avoid compacting the surrounding soil to the extent that other foundation elements cannot be installed properly.

1810.4.1.1 Compressive strength of precast concrete piles. A precast concrete pile shall not be driven before the concrete has attained a compressive strength of not less than 75 percent of the specified compressive strength (f'_c), but not less than the strength sufficient to withstand handling and driving forces.

1810.4.1.2 Shafts in unstable soils. Where cast-in-place *deep foundation* elements are formed through unstable soils, the open hole shall be stabilized by a casing, slurry, or other approved method prior to placing the concrete. Where the casing is withdrawn during concreting, the level of concrete shall be maintained above the bottom of the casing at a sufficient height to offset any hydrostatic or lateral soil pressure. Driven casings shall be mandrel driven their full length in contact with the surrounding soil.

1810.4.1.3 Driving near uncased concrete. Deep foundation elements shall not be driven within six element diameters center to center in granular soils or within one-half the element length in cohesive soils of an uncased element filled with concrete less than 48 hours old unless *approved* by the *building official*. If driving near uncased concrete elements causes the concrete surface in any completed element to rise or drop significantly or bleed additional water, the completed element shall be replaced.

1810.4.1.4 Driving near cased concrete. *Deep foundation* elements shall not be driven within four and one-half average diameters of a cased element filled with concrete less than 24 hours old unless *approved* by the *building official*. Concrete shall not be placed in casings within heave range of driving.

1810.4.1.5 Defective timber piles. Any substantial sudden change in rate of penetration of a timber pile shall be investigated for possible damage. If the sudden change in rate of penetration cannot be correlated to soil strata, the pile shall be removed for inspection or rejected.

1810.4.2 Identification. *Deep foundation* materials shall be identified for conformity to the specified grade with this identity maintained continuously from the point of manufacture to the point of installation or shall be tested by an *approved agency* to determine conformity to the specified grade. The *approved agency* shall furnish an affidavit of compliance to the *building official*.

1810.4.3 Location plan. A plan showing the location and designation of *deep foundation* elements by an identification system shall be filed with the *building official* prior to installation of such elements. Detailed records for elements shall bear an identification corresponding to that shown on the plan.

1810.4.4 Preexcavation. The use of jetting, augering or other methods of preexcavation shall be subject to the approval of the *building official*. Where permitted, preexcavation shall be carried out in the same manner as used for *deep foundation* elements subject to load tests and in such a manner that will not impair the carrying capacity of the elements already in place or damage adjacent structures. Element tips shall be advanced below the preexcavated depth until the required resistance or penetration is obtained.

1810.4.5 Vibratory driving. Vibratory drivers shall only be used to install *deep foundation* elements where the element load capacity is verified by load tests in accordance with Section 1810.3.3.1.2. The installation of production elements shall be controlled according to power consumption, rate of penetration or other *approved* means that ensure element capacities equal or exceed those of the test elements.

Exceptions:

1. The pile installation is completed by driving with an impact hammer in accordance with Section 1810.3.3.1.1.
2. The pile is to be used only for lateral resistance.

1810.4.6 Heaved elements. *Deep foundation* elements that have heaved during the driving of adjacent elements shall be redriven as necessary to develop the required capacity and penetration, or the capacity of the element shall be verified by load tests in accordance with Section 1810.3.3.1.2.

1810.4.7 Enlarged base cast-in-place elements. Enlarged bases for cast-in-place *deep foundation* elements formed by compacting concrete or by driving a precast base shall be formed in or driven into granular soils. Such elements shall be constructed in the same manner as successful prototype test elements driven for the project. Shafts extending through peat or other organic soil shall be encased in a permanent steel casing. Where a cased shaft is used, the shaft shall be adequately reinforced to resist column action or the *annular space* around the shaft shall be filled sufficiently to reestablish lateral support by the soil. Where heave occurs, the element shall be replaced unless it is demonstrated that the element is undamaged and capable of carrying twice its design *load*.

1810.4.8 Hollow-stem augered, cast-in-place elements. Where concrete or grout is placed by pumping through a hollow-stem auger, the auger shall be permitted to rotate in a clockwise direction during withdrawal. As the auger is withdrawn at a steady rate or in increments not to exceed 1 foot (305 mm), concreting or grouting pumping pressures shall be measured and maintained high enough at all times to offset hydrostatic and lateral earth pressures. Concrete or grout volumes shall be measured to ensure that the volume of concrete or grout placed in each element is equal to or greater than the theoretical volume of the hole created by the auger. Where the installation process of any element is interrupted or a loss of concreting or grouting pressure occurs, the element shall be redrilled to 5 feet (1524 mm) below the elevation of the tip of the auger when the installation was interrupted or concrete or grout pressure was lost and reformed. Augered cast-in-place elements shall not be installed within six diameters center to center of an element filled with concrete or grout less than 12 hours old, unless *approved* by the *building official*. If the concrete or grout level in any completed element drops due to installation of an adjacent element, the element shall be replaced.

1810.4.9 Socketed drilled shafts. The rock socket and pipe or tube casing of socketed *drilled shafts* shall be thoroughly cleaned of foreign materials before filling with concrete. Steel cores shall be bedded in cement grout at the base of the rock socket.

1810.4.10 Micropiles. *Micropile deep foundation* elements shall be permitted to be formed in holes advanced by rotary or percussive drilling methods, with or without casing. The elements shall be grouted with a fluid cement grout. The grout shall be pumped through a tremie pipe extending to the bottom of the element until grout of suitable quality returns at the top of the element. The following requirements apply to specific installation methods:

1. For micropiles grouted inside a temporary casing, the reinforcing bars shall be inserted prior to withdrawal of the casing. The casing shall be withdrawn in a controlled manner with the grout level maintained at the top of the element to ensure that the grout completely fills the drill hole. During withdrawal of the casing, the grout level inside the casing shall be monitored to verify that the flow of grout inside the casing is not obstructed.
2. For a *micropile* or portion thereof grouted in an open drill hole in soil without temporary casing, the minimum design diameter of the drill hole shall be verified by a suitable device during grouting.
3. For micropiles designed for end bearing, a suitable means shall be employed to verify that the bearing surface is properly cleaned prior to grouting.
4. Subsequent micropiles shall not be drilled near elements that have been grouted until the grout has had sufficient time to harden.
5. Micropiles shall be grouted as soon as possible after drilling is completed.
6. For micropiles designed with a full-length casing, the casing shall be pulled back to the top of the bond zone and reinserted or some other suitable means employed to ensure grout coverage outside the casing.

1810.4.11 Helical piles. *Helical piles* shall be installed to specified embedment depth and torsional resistance criteria as determined by a *registered design professional*. The torque applied during installation shall not exceed the manufacturer's rated maximum installation torque resistance of the *helical pile*.

1810.4.12 Special inspection. *Special inspections* in accordance with Sections 1705.7 and 1705.8 shall be provided for driven and cast-in-place *deep foundation* elements, respectively. *Special inspections* in accordance with Section 1705.9 shall be provided for *helical piles*.

CHAPTER 19
CONCRETE

Italics are used for text within Sections 1903 through 1905 of this code to indicate provisions that differ from ACI 318.

User notes:

About this chapter: *Chapter 19 provides minimum accepted practices for the design and construction of buildings and structural components using concrete—both plain and reinforced. Chapter 19 relies primarily on the reference to American Concrete Institute (ACI) 318, Building Code Requirements for Structural Concrete. Structural concrete must be designed and constructed to comply with this code and all listed standards. There are also specific provisions addressing concrete slabs and shotcrete.*

Code development reminder: *Code change proposals to this chapter will be considered by the IBC—Structural Code Development Committee during the 2022 (Group B) Code Development Cycle.*

SECTION 1901
GENERAL

1901.1 Scope. The provisions of this chapter shall govern the materials, quality control, design and construction of concrete used in structures.

1901.2 Plain and reinforced concrete. Structural concrete shall be designed and constructed in accordance with the requirements of this chapter and ACI 318 as amended in Section 1905 of this code. Except for the provisions of Sections 1904 and 1907, the design and construction of slabs on grade shall not be governed by this chapter unless they transmit vertical *loads* or lateral forces from other parts of the structure to the soil.

1901.3 Anchoring to concrete. Anchoring to concrete shall be in accordance with ACI 318 as amended in Section 1905, and applies to cast-in (headed bolts, headed studs and hooked J- or L-bolts), post-installed expansion (torque-controlled and displacement-controlled), undercut, screw, and adhesive anchors.

1901.4 Composite structural steel and concrete structures. Systems of structural steel acting compositely with reinforced concrete shall be designed in accordance with Section 2206 of this code.

1901.5 Construction documents. The *construction documents* for structural concrete construction shall include:

1. The specified compressive strength of concrete at the stated ages or stages of construction for which each concrete element is designed.
2. The specified strength or grade of reinforcement.
3. The size and location of structural elements, reinforcement and anchors.
4. Provision for dimensional changes resulting from creep, shrinkage and temperature.
5. The magnitude and location of prestressing forces.
6. Anchorage length of reinforcement and location and length of lap splices.
7. Type and location of mechanical and welded splices of reinforcement.
8. Details and location of contraction or isolation *joints* specified for plain concrete.
9. Minimum concrete compressive strength at time of posttensioning.
10. Stressing sequence for posttensioning tendons.
11. For structures assigned to *Seismic Design Category* D, E or F, a statement if slab on grade is designed as a structural *diaphragm*.

1901.6 Special inspections and tests. *Special inspections* and tests of concrete elements of buildings and structures and concreting operations shall be as required by Chapter 17.

1901.7 Tolerances for structural concrete. Where not indicated in *construction documents*, structural tolerances for concrete structural elements shall be in accordance with this section.

1901.7.1 Cast-in-place concrete tolerances. Structural tolerances for cast-in-place concrete structural elements shall be in accordance with ACI 117.

Exceptions:

1. Group R-3 detached one- or two-family dwellings are not required to comply with this section.
2. Shotcrete is not required to comply with this section.

1901.7.2 Precast concrete tolerances. Structural tolerances for precast concrete structural elements shall be in accordance with ACI ITG-7.

Exception: Group R-3 detached one- or two-family dwellings are not required to comply with this section.

SECTION 1902
COORDINATION OF TERMINOLOGY

1902.1 General. Coordination of terminology used in ACI 318 and ASCE 7 shall be in accordance with Sections 1902.1.1 and 1902.1.2.

1902.1.1 Design displacement. Design displacement at each level shall be the total lateral deflection at the level calculated for the design earthquake using the procedures defined in Section 12.8.6 of ASCE 7.

1902.1.2 Special structural wall. Special structural walls made of cast-in-place or precast concrete shall comply with the requirements of Sections 18.2.4 through 18.2.8, 18.10 and 18.11 of ACI 318, as applicable, in addition to the requirements for *ordinary reinforced concrete structural walls* or *ordinary precast structural walls*, as applicable. Where ASCE 7 refers to a "special reinforced concrete shear wall," it shall be deemed to mean a "special structural wall."

SECTION 1903 SPECIFICATIONS FOR TESTS AND MATERIALS

1903.1 General. Materials used to produce concrete, concrete itself and testing thereof shall comply with the applicable standards listed in ACI 318.

1903.2 Special inspections. *Where required, special inspections and tests shall be in accordance with Chapter 17.*

1903.3 Glass fiber-reinforced concrete. *Glass fiber-reinforced concrete (GFRC) and the materials used in such concrete shall be in accordance with the PCI MNL 128 standard.*

1903.4 Flat wall insulating concrete form (ICF) systems. *Insulating concrete form material used for forming flat concrete walls shall conform to ASTM E2634.*

SECTION 1904 DURABILITY REQUIREMENTS

1904.1 Structural concrete. Structural concrete shall conform to the durability requirements of ACI 318.

Exception: *For Group R-2 and R-3 occupancies not more than three stories above grade plane, the specified compressive strength, f′c, for concrete in basement walls, foundation walls, exterior walls and other vertical surfaces exposed to the weather shall be not less than 3,000 psi (20.7 MPa).*

1904.2 Nonstructural concrete. *The registered design professional shall assign nonstructural concrete a freeze-thaw exposure class, as defined in ACI 318, based on the anticipated exposure of nonstructural concrete. Nonstructural concrete shall have a minimum specified compressive strength, f′c, of 2,500 psi (17.2 MPa) for Class F0; 3,000 psi (20.7 MPa) for Class F1; and 3,500 psi (24.1 MPa) for Classes F2 and F3. Nonstructural concrete shall be air entrained in accordance with ACI 318.*

SECTION 1905 MODIFICATIONS TO ACI 318

1905.1 General. The text of ACI 318 shall be modified as indicated in Sections 1905.1.1 through 1905.1.8.

1905.1.1 ACI 318, Section 2.3. Modify existing definitions and add the following definitions to ACI 318, Section 2.3.

DETAILED PLAIN CONCRETE STRUCTURAL WALL. A wall complying with the requirements of Chapter 14, including 14.6.2.

ORDINARY PRECAST STRUCTURAL WALL. A precast wall complying with the requirements of Chapters 1 through 13, 15, 16 and 19 through 26.

ORDINARY REINFORCED CONCRETE STRUCTURAL WALL. A *cast-in-place* wall complying with the requirements of Chapters 1 through 13, 15, 16 and 19 through 26.

ORDINARY STRUCTURAL PLAIN CONCRETE WALL. A wall complying with the requirements of Chapter 14, *excluding 14.6.2.*

1905.1.2 ACI 318, Section 18.2.1. Modify ACI 318 Sections 18.2.1.2 and 18.2.1.6 to read as follows:

- *18.2.1.2 – Structures assigned to Seismic Design Category* A shall satisfy requirements of Chapters 1 through 17 and 19 through 26; *Chapter 18 does not apply.* Structures assigned to *Seismic Design Category* B, C, D, E or F shall satisfy 18.2.1.3 through 18.2.1.7, as applicable. *Except for structural elements of plain concrete complying with Section 1905.1.7 of the International Building Code, structural elements of plain concrete are prohibited in structures assigned to Seismic Design Category C, D, E or F.*
- 18.2.1.6 – Structural systems designated as part of the *seismic force-resisting system* shall be restricted to those *permitted by ASCE 7.* Except for *Seismic Design Category* A, for which Chapter 18 does not apply, the following provisions shall be satisfied for each structural system designated as part of the *seismic force-resisting system*, regardless of the *seismic design category*:
  - (a) Ordinary moment frames shall satisfy 18.3.
  - (b) *Ordinary reinforced concrete structural walls and ordinary precast structural walls* need not satisfy any provisions in Chapter 18.
  - (c) Intermediate moment frames shall satisfy 18.4.
  - (d) Intermediate precast *structural* walls shall satisfy 18.5.
  - (e) Special moment frames shall satisfy 18.6 through 18.9.
  - (f) Special structural walls shall satisfy 18.10.
  - (g) Special structural walls constructed using precast concrete shall satisfy 18.11.

Special moment frames and special structural walls shall also satisfy 18.2.4 through 18.2.8.

1905.1.3 ACI 318, Section 18.5. Modify ACI 318, Section 18.5 by adding new Section 18.5.2.2 and renumbering existing Sections 18.5.2.2 and 18.5.2.3 to become 18.5.2.3 and 18.5.2.4, respectively.

18.5.2.2 – Connections that are designed to yield shall be capable of maintaining 80 percent of their design strength at the deformation induced by the design displacement or shall use Type 2 mechanical splices.

18.5.2.3 – Elements of the connection that are not designed to yield shall develop at least 1.5 S_y.

18.5.2.4 – In structures assigned to SDC D, E or F, wall piers shall be designed in accordance with 18.10.8 or 18.14 in ACI 318.

1905.1.4 ACI 318, Section 18.11. Modify ACI 318, Section 18.11.2.1 to read as follows:

18.11.2.1 – Special structural walls constructed using precast concrete shall satisfy all the requirements of 18.10 *for cast-in-place special structural walls* in addition to 18.5.2.

1905.1.5 ACI 318, Section 18.13.1.1. Modify ACI 318, Section 18.13.1.1 to read as follows:

18.13.1.1 – Foundations resisting earthquake-induced forces or transferring earthquake-induced forces between a structure and ground shall comply with the requirements of 18.13 and other applicable provisions of ACI 318 *unless modified by Chapter 18 of the International Building Code.*

1905.1.6 ACI 318, Section 14.6. Modify ACI 318, Section 14.6 by adding new Section 14.6.2 to read as follows:

14.6.2 – Detailed plain concrete structural walls.

14.6.2.1 – Detailed plain concrete structural walls are walls conforming to the requirements of ordinary structural plain concrete walls and 14.6.2.2.

14.6.2.2 – Reinforcement shall be provided as follows:

- Vertical reinforcement of at least 0.20 square inch (129 mm^2) in cross-sectional area shall be provided continuously from support to support at each corner, at each side of each opening and at the ends of walls. The continuous vertical bar required beside an opening is permitted to substitute for one of the two No. 5 bars required by 14.6.1.
- *Horizontal reinforcement at least 0.20 square inch (129 mm^2) in cross-sectional area shall be provided:*
  1. *Continuously at structurally connected roof and floor levels and at the top of walls.*
  2. *At the bottom of load-bearing walls or in the top of foundations where doweled to the wall.*
  3. *At a maximum spacing of 120 inches (3048 mm).*

Reinforcement at the top and bottom of openings, where used in determining the maximum spacing specified in Item 3 above, shall be continuous in the wall.

1905.1.7 ACI 318, Section 14.1.4. Delete ACI 318, Section 14.1.4 and replace with the following:

14.1.4 – Plain concrete in structures assigned to Seismic Design Category C, D, E or F.

14.1.4.1 – Structures assigned to Seismic Design Category C, D, E or F shall not have elements of structural plain concrete, except as follows:

- *Structural plain concrete basement, foundation or other walls below the base as defined in ASCE 7 are permitted in detached one- and two-family dwellings three stories or less in height constructed with stud-bearing walls. In dwellings assigned to Seismic Design Category D or E, the height of the wall shall not exceed 8 feet (2438 mm), the thickness shall be not less than $7^1/_2$ inches (190 mm), and the wall shall retain no more than 4 feet (1219 mm) of unbalanced fill. Walls shall have reinforcement in accordance with 14.6.1.*
- *Isolated footings of plain concrete supporting pedestals or columns are permitted, provided the projection of the footing beyond the face of the supported member does not exceed the footing thickness.*

  ***Exception:** In detached one- and two-family dwellings three stories or less in height, the projection of the footing beyond the face of the supported member is permitted to exceed the footing thickness.*
- *Plain concrete footings supporting walls are permitted, provided the footings have at least two continuous longitudinal reinforcing bars. Bars shall not be smaller than No. 4 and shall have a total area of not less than 0.002 times the gross cross-sectional area of the footing. For footings that exceed 8 inches (203 mm) in thickness, a minimum of one bar shall be provided at the top and bottom of the footing. Continuity of reinforcement shall be provided at corners and intersections.*

  Exceptions:

  1. *In Seismic Design Categories A, B and C, detached one- and two-family dwellings three stories or less in height constructed with stud-bearing walls are permitted to have plain concrete footings without longitudinal reinforcement.*
  2. *For foundation systems consisting of a plain concrete footing and a*

plain concrete stemwall, a minimum of one bar shall be provided at the top of the stemwall and at the bottom of the footing.

3. *Where a slab on ground is cast monolithically with the footing, one No. 5 bar is permitted to be located at either the top of the slab or bottom of the footing.*

1905.1.8 ACI 318, Section 17.2.3. Modify ACI 318 Sections 17.2.3.4.2, 17.2.3.4.3(d) and 17.2.3.5.2 to read as follows:

- 17.2.3.4.2 – Where the tensile component of the strength-level earthquake force applied to anchors exceeds 20 percent of the total factored anchor tensile force associated with the same load combination, anchors and their attachments shall be designed in accordance with 17.2.3.4.3. The anchor design tensile strength shall be determined in accordance with 17.2.3.4.4.

  Exception: Anchors designed to resist wall out-of-plane forces with *design strengths* equal to or greater than the force determined in accordance with ASCE 7 Equation 12.11-1 or 12.14-10 shall be deemed to satisfy Section 17.2.3.4.3(d).

- 17.2.3.4.3(d) – The anchor or group of anchors shall be designed for the maximum tension obtained from design load combinations that include E, with E increased by Ω_0. The anchor design tensile strength shall *be calculated from 17.2.3.4.4.*

- 17.2.3.5.2 – Where the shear component of the strength-level earthquake force applied to anchors exceeds 20 percent of the total factored anchor shear force associated with the same load combination, anchors and their attachments shall be designed in accordance with 17.2.3.5.3. The anchor design shear strength for resisting earthquake forces shall be determined in accordance with 17.5.

  Exceptions:

  1. *For the calculation of the in-plane shear strength of anchor bolts attaching wood sill plates of bearing or nonbearing walls of light-frame wood structures to foundations or foundation stem walls, the in-plane shear strength in accordance with 17.5.2 and 17.5.3 need not be computed and 17.2.3.5.3 shall be deemed to be satisfied provided all of the following are met:*

     1.1. *The allowable in-plane shear strength of the anchor is determined in accordance with ANSI/AWC NDS Table 12E for lateral design values parallel to grain.*

     1.2. *The maximum anchor nominal diameter is $^5/_8$ inch (16 mm).*

     1.3. *Anchor bolts are embedded into concrete a minimum of 7 inches (178 mm).*

     1.4. *Anchor bolts are located a minimum of $1^3/_4$ inches (45 mm) from the edge of the concrete parallel to the length of the wood sill plate.*

     1.5. *Anchor bolts are located a minimum of 15 anchor diameters from the edge of the concrete perpendicular to the length of the wood sill plate.*

     1.6. *The sill plate is 2-inch (51 mm) or 3-inch (76 mm) nominal thickness.*

  2. *For the calculation of the in-plane shear strength of anchor bolts attaching cold-formed steel track of bearing or nonbearing walls of light-frame construction to foundations or foundation stem walls, the in-plane shear strength in accordance with 17.5.2 and 17.5.3 need not be computed and 17.2.3.5.3 shall be deemed to be satisfied provided all of the following are met:*

     2.1. The maximum anchor nominal diameter is $^5/_8$ inch (16 mm).

     2.2. Anchors are embedded into concrete a minimum of 7 inches (178 mm).

     2.3. Anchors are located a minimum of $1^3/_4$ inches (45 mm) from the edge of the concrete parallel to the length of the track.

     2.4. Anchors are located a minimum of 15 anchor diameters from the edge of the concrete perpendicular to the length of the track.

     2.5. The track is 33 to 68 mil (0.84 mm to 1.73 mm) designation thickness.

  Allowable in-plane shear strength of exempt anchors, parallel to the edge of concrete, shall be permitted to be determined in accordance with AISI S100 Section E3.3.1.

3. *In light-frame construction bearing or nonbearing walls, shear strength of concrete anchors less than or equal to 1 inch [25 mm] in diameter attaching sill plate or track to foundation or foundation stem wall need not satisfy 17.2.3.5.3(a) through (c) when the design strength of the anchors is determined in accordance with 17.5.2.1(c).*

SECTION 1906
FOOTINGS FOR LIGHT-FRAME CONSTRUCTION

1906.1 Plain concrete footings. For Group R-3 occupancies and buildings of other occupancies less than two stories above grade plane of *light-frame construction*, the required thickness of plain concrete footings is permitted to be 6 inches (152 mm), provided that the footing does not extend more than 4 inches (102 mm) on either side of the supported wall.

SECTION 1907
MINIMUM SLAB PROVISIONS

1907.1 General. The thickness of concrete floor slabs supported directly on the ground shall be not less than $3^1/_2$ inches (89 mm). A 6-mil (0.006 inch; 0.15 mm) polyethylene vapor retarder with joints lapped not less than 6 inches (152 mm) shall be placed between the base course or subgrade and the concrete floor slab, or other *approved* equivalent methods or materials shall be used to retard vapor transmission through the floor slab.

Exception: A vapor retarder is not required:

1. For detached structures accessory to occupancies in Group R-3, such as garages, utility buildings or other unheated facilities.
2. For unheated storage rooms having an area of less than 70 square feet (6.5 m^2) and carports attached to occupancies in Group R-3.
3. For buildings of other occupancies where migration of moisture through the slab from below will not be detrimental to the intended occupancy of the building.
4. For driveways, walks, patios and other flatwork that will not be enclosed at a later date.
5. Where *approved* based on local site conditions.

SECTION 1908
SHOTCRETE

1908.1 General. Shotcrete shall be in accordance with the requirements of ACI 318.

CHAPTER 20
ALUMINUM

User notes:

About this chapter: *Chapter 20 contains standards for the use of aluminum in building construction. Only the structural applications of aluminum are addressed so it would not apply to the use of aluminum in specialty products such as storefront or window framing or architectural hardware. The use of aluminum in heating, ventilating or air-conditioning systems is addressed in the* International Mechanical Code®. *This chapter references national standards from the Aluminum Association for use of aluminum in building construction, AA ASM 35, Aluminum Sheet Metal Work in Building Construction, and AA ADM, Aluminum Design Manual.*

Code development reminder: *Code change proposals to this chapter will be considered by the IBC—Structural Code Development Committee during the 2022 (Group B) Code Development Cycle.*

SECTION 2001
GENERAL

2001.1 Scope. This chapter shall govern the quality, design, fabrication and erection of aluminum.

SECTION 2002
MATERIALS

2002.1 General. Aluminum used for structural purposes in buildings and structures shall comply with AA ASM 35 and AA ADM. The *nominal loads* shall be the minimum design *loads* required by Chapter 16.

CHAPTER 21
MASONRY

User notes:

About this chapter: *Chapter 21 establishes minimum requirements for masonry construction. The provisions address: material specifications and test methods; types of wall construction; criteria for engineered and empirical designs; and required details of construction, including the execution of construction. The provisions provide a framework for applying applicable standards to the design and construction of masonry structures. Masonry design methodologies including allowable stress design, strength design and empirical design are covered by the provisions of this chapter. Also addressed are masonry fireplaces and chimneys, masonry heaters and glass unit masonry.*

Code development reminder: *Code change proposals to this chapter will be considered by the IBC—Structural Code Development Committee during the 2022 (Group B) Code Development Cycle.*

SECTION 2101
GENERAL

2101.1 Scope. This chapter shall govern the materials, design, construction and quality of masonry.

2101.2 Design methods. Masonry shall comply with the provisions of TMS 402, TMS 403 or TMS 404 as well as applicable requirements of this chapter.

2101.2.1 Masonry veneer. Masonry *veneer* shall comply with the provisions of Chapter 14.

2101.3 Special inspection. The *special inspection* of masonry shall be as defined in Chapter 17, or an itemized testing and inspection program shall be provided that meets or exceeds the requirements of Chapter 17.

SECTION 2102
NOTATIONS

2102.1 General. The following notations are used in the chapter:

NOTATIONS.

d_b = Diameter of reinforcement, inches (mm).

F_s = Allowable tensile or compressive stress in reinforcement, psi (MPa).

f_r = Modulus of rupture, psi (MPa).

f'_{AAC} = Specified compressive strength of AAC masonry, the minimum compressive strength for a class of AAC masonry as specified in TMS 602, psi (MPa).

f'_m = Specified compressive strength of masonry at age of 28 days, psi (MPa).

f'_{mi} = Specified compressive strength of masonry at the time of prestress transfer, psi (MPa).

K = The lesser of the masonry cover, clear spacing between adjacent reinforcement, or five times db, inches (mm).

L_s = Distance between supports, inches (mm).

l_d = Required development length or lap length of reinforcement, inches (mm).

P = The applied load at failure, pounds (N).

S_t = Thickness of the test specimen measured parallel to the direction of load, inches (mm).

S_w = Width of the test specimen measured parallel to the loading cylinder, inches (mm).

SECTION 2103
MASONRY CONSTRUCTION MATERIALS

2103.1 Masonry units. Concrete *masonry units*, clay or shale *masonry units*, stone *masonry units*, *glass unit masonry* and *AAC masonry* units shall comply with Article 2.3 of TMS 602. Architectural *cast stone* shall conform to ASTM C1364 and TMS 504. Adhered manufactured stone masonry *veneer* units shall conform to ASTM C1670.

Exception: *Structural clay tile* for nonstructural use in fireproofing of structural members and in wall furring shall not be required to meet the compressive strength specifications. The *fire-resistance rating* shall be determined in accordance with ASTM E119 or UL 263 and shall comply with the requirements of Table 705.5.

2103.1.1 Second-hand units. Second-hand *masonry units* shall not be reused unless they conform to the requirements of new units. The units shall be of whole, sound materials and free from cracks and other defects that will interfere with proper laying or use. Old *mortar* shall be cleaned from the unit before reuse.

2103.2 Mortar. *Mortar* for masonry construction shall comply with Section 2103.2.1, 2103.2.2, 2103.2.3 or 2103.2.4.

2103.2.1 Masonry mortar. *Mortar* for use in masonry construction shall conform to Articles 2.1 and 2.6 A of TMS 602.

2103.2.2 Surface-bonding mortar. *Surface-bonding mortar* shall comply with ASTM C887. Surface bonding of concrete *masonry units* shall comply with ASTM C946.

2103.2.3 Mortars for ceramic wall and floor tile. Portland cement *mortars* for installing ceramic wall and floor tile shall comply with ANSI A108.1A and ANSI A108.1B and be of the compositions indicated in Table 2103.2.3.

TABLE 2103.2.3
CERAMIC TILE MORTAR COMPOSITIONS

LOCATION	MORTAR	COMPOSITION
Walls	Scratchcoat	1 cement; $^1/_5$ hydrated lime; 4 dry or 5 damp sand
	Setting bed and leveling coat	1 cement; $^1/_2$ hydrated lime; 5 damp sand to 1 cement; 1 hydrated lime, 7 damp sand
Floors	Setting bed	1 cement; $^1/_{10}$ hydrated lime; 5 dry or 6 damp sand; or 1 cement; 5 dry or 6 damp sand
Ceilings	Scratchcoat and sand bed	1 cement; $^1/_2$ hydrated lime; $2^1/_2$ dry sand or 3 damp sand

2103.2.3.1 Dry-set Portland cement mortars. Premixed prepared Portland cement *mortars*, which require only the addition of water and are used in the installation of ceramic tile, shall comply with ANSI A118.1. The shear bond strength for tile set in such *mortar* shall be as required in accordance with ANSI A118.1. Tile set in dry-set Portland cement *mortar* shall be installed in accordance with ANSI A108.5.

2103.2.3.2 Latex-modified Portland cement mortar. Latex-modified Portland cement thin-set *mortars* in which latex is added to dry-set *mortar* as a replacement for all or part of the gauging water that are used for the installation of ceramic tile shall comply with ANSI A118.4. Tile set in latex-modified Portland cement shall be installed in accordance with ANSI A108.5.

2103.2.3.3 Epoxy mortar. Ceramic tile set and grouted with chemical-resistant epoxy shall comply with ANSI A118.3. Tile set and grouted with epoxy shall be installed in accordance with ANSI A108.6.

2103.2.3.4 Furan mortar and grout. Chemical-resistant furan *mortar* and grout that are used to install ceramic tile shall comply with ANSI A118.5. Tile set and grouted with furan shall be installed in accordance with ANSI A108.8.

2103.2.3.5 Modified epoxy-emulsion mortar and grout. Modified epoxy-emulsion *mortar* and grout that are used to install ceramic tile shall comply with ANSI A118.8. Tile set and grouted with modified epoxy-emulsion *mortar* and grout shall be installed in accordance with ANSI A108.9.

2103.2.3.6 Organic adhesives. Water-resistant organic adhesives used for the installation of ceramic tile shall comply with ANSI A136.1. The shear bond strength after water immersion shall be not less than 40 psi (275 kPa) for Type I adhesive and not less than 20 psi (138 kPa) for Type II adhesive when tested in accordance with ANSI A136.1. Tile set in organic adhesives shall be installed in accordance with ANSI A108.4.

2103.2.3.7 Portland cement grouts. Portland cement grouts used for the installation of ceramic tile shall comply with ANSI A118.6. Portland cement grouts for tile work shall be installed in accordance with ANSI A108.10.

2103.2.4 Mortar for adhered masonry veneer. *Mortar* for use with *adhered masonry veneer* shall conform to ASTM C270 for Type N or S, or shall comply with ANSI A118.4 for latex-modified Portland cement *mortar*.

2103.3 Grout. Grout shall comply with Article 2.2 of TMS 602.

2103.4 Metal reinforcement and accessories. Metal reinforcement and accessories shall conform to Article 2.4 of TMS 602. Where unidentified reinforcement is *approved* for use, not less than three tension and three bending tests shall be made on representative specimens of the reinforcement from each shipment and grade of reinforcing steel proposed for use in the work.

SECTION 2104 CONSTRUCTION

2104.1 Masonry construction. Masonry construction shall comply with the requirements of Sections 2104.1.1 and 2104.1.2 and with the requirements of either TMS 602 or TMS 604.

2104.1.1 Support on wood. Masonry shall not be supported on wood girders or other forms of wood construction except as permitted in Section 2304.12.

2104.1.2 Molded cornices. Unless structural support and anchorage are provided to resist the overturning moment, the center of gravity of projecting masonry or molded *cornices* shall lie within the middle one-third of the supporting wall. Terra cotta and metal *cornices* shall be provided with a structural frame of *approved* noncombustible material anchored in an *approved* manner.

SECTION 2105 QUALITY ASSURANCE

2105.1 General. A quality assurance program shall be used to ensure that the constructed masonry is in compliance with the *approved construction documents*.

The quality assurance program shall comply with the inspection and testing requirements of Chapter 17 and TMS 602.

SECTION 2106 SEISMIC DESIGN

2106.1 Seismic design requirements for masonry. Masonry structures and components shall comply with the requirements in Chapter 7 of TMS 402 depending on the structure's *seismic design category*.

SECTION 2107 ALLOWABLE STRESS DESIGN

2107.1 General. The design of masonry structures using *allowable stress design* shall comply with Section 2106 and the requirements of Chapters 1 through 8 of TMS 402 except as modified by Sections 2107.2 through 2107.3.

2107.2 TMS 402, Section 6.1.6.1.1, lap splices. As an alternative to Section 6.1.6.1.1, it shall be permitted to design lap splices in accordance with Section 2107.2.1.

2107.2.1 Lap splices. The minimum length of lap splices for reinforcing bars in tension or compression, l_d, shall be:

$$l_d = 0.002 d_b f_s \quad \textbf{(Equation 21-1)}$$

For SI: $l_d = 0.29 d_b f_s$

but not less than 12 inches (305 mm). The length of the lapped splice shall be not less than 40 bar diameters. where:

d_b = Diameter of reinforcement, inches (mm)

f_s = Computed stress in reinforcement due to design loads, psi (MPa).

In regions of moment where the design tensile stresses in the reinforcement are greater than 80 percent of the allowable steel tension stress, F_s, the lap length of splices shall be increased not less than 50 percent of the minimum required length, but need not be greater than 72 d_b. Other equivalent means of stress transfer to accomplish the same 50 percent increase shall be permitted. Where epoxy coated bars are used, lap length shall be increased by 50 percent.

2107.3 TMS 402, Section 6.1.6.1, splices of reinforcement. Modify Section 6.1.6.1 as follows:

6.1.6.1 – Splices of reinforcement. Lap splices, welded splices or mechanical splices are permitted in accordance with the provisions of this section. Welding shall conform to AWS D1.4. Welded splices shall be of ASTM A706 steel reinforcement. Reinforcement larger than No. 9 (M #29) shall be spliced using mechanical connections in accordance with Section 6.1.6.1.3.

SECTION 2108
STRENGTH DESIGN OF MASONRY

2108.1 General. The design of masonry structures using strength design shall comply with Section 2106 and the requirements of Chapters 1 through 7 and Chapter 9 of TMS 402, except as modified by Sections 2108.2 through 2108.3.

Exception: *AAC masonry* shall comply with the requirements of Chapters 1 through 7 and Chapter 11 of TMS 402.

2108.2 TMS 402, Section 6.1.5.1.1, development. Modify the second paragraph of Section 6.1.5.1.1 as follows:

The required development length of reinforcement shall be determined by Equation (6-1), but shall be not less than 12 inches (305 mm) and need not be greater than 72 d_b.

2108.3 TMS 402, Section 6.1.6.1.1, splices. Modify Sections 6.1.6.1.2 and 6.1.6.1.3 as follows:

6.1.6.1.2 – A welded splice shall have the bars butted and welded to develop not less than 125 percent of the yield strength, f_y, of the bar in tension or compression, as required. Welded splices shall be of ASTM A706 steel reinforcement. Welded splices shall not be permitted in plastic hinge zones of intermediate or special reinforced walls.

6.1.6.1.3 – Mechanical splices shall be classified as Type 1 or 2 in accordance with Section 18.2.7.1 of ACI 318. Type 1 mechanical splices shall not be used within a plastic hinge zone or within a beam-column joint of intermediate or special *reinforced masonry* shear walls. Type 2 mechanical splices are permitted in any location within a member.

SECTION 2109
EMPIRICAL DESIGN OF ADOBE MASONRY

2109.1 General. Empirically designed adobe masonry shall conform to the requirements of Appendix A of TMS 402, except where otherwise noted in this section.

2109.1.1 Limitations. The use of empirical design of adobe masonry shall be limited as noted in Section A.1.2 of TMS 402. In buildings that exceed one or more of the limitations of Section A.1.2 of TMS 402, masonry shall be designed in accordance with the engineered design provisions of Section 2101.2 or the foundation wall provisions of Section 1807.1.5.

Section A.1.2.2 of TMS 402 shall be modified as follows:

A.1.2.2 – *Wind*. Empirical requirements shall not apply to the design or construction of masonry for buildings, parts of buildings, or other structures to be located in areas where V_{asd} as determined in accordance with Section 1609.3.1 of the *International Building Code* exceeds 110 mph.

2109.2 Adobe construction. *Adobe construction* shall comply with this section and shall be subject to the requirements of this code for Type V construction, Appendix A of TMS 402, and this section.

2109.2.1 Unstabilized adobe. *Unstabilized adobe* shall comply with Sections 2109.2.1.1 through 2109.2.1.4.

2109.2.1.1 Compressive strength. Adobe units shall have an average compressive strength of 300 psi (2068 kPa) when tested in accordance with ASTM C67. Five samples shall be tested and individual units are not permitted to have a compressive strength of less than 250 psi (1724 kPa).

2109.2.1.2 Modulus of rupture. Adobe units shall have an average modulus of rupture of 50 psi (345 kPa) when tested in accordance with the following procedure. Five samples shall be tested and individual units shall not have a modulus of rupture of less than 35 psi (241 kPa).

2109.2.1.2.1 Support conditions. A cured unit shall be simply supported by 2-inch-diameter (51 mm) cylindrical supports located 2 inches (51 mm) in from each end and extending the full width of the unit.

2109.2.1.2.2 Loading conditions. A 2-inch-diameter (51 mm) cylinder shall be placed at midspan parallel to the supports.

2109.2.1.2.3 Testing procedure. A vertical *load* shall be applied to the cylinder at the rate of 500 pounds per minute (37 N/s) until failure occurs.

2109.2.1.2.4 Modulus of rupture determination. The modulus of rupture shall be determined by the equation:

$$f_r = 3\,PL_s/[2\,S_w(S_t^2)] \qquad \textbf{(Equation 21-2)}$$

where, for the purposes of this section only:

S_w = Width of the test specimen measured parallel to the loading cylinder, inches (mm).

f_r = Modulus of rupture, psi (MPa).

L_s = Distance between supports, inches (mm).

S_t = Thickness of the test specimen measured parallel to the direction of load, inches (mm).

P = The applied *load* at failure, pounds (N).

2109.2.1.3 Moisture content requirements. Adobe units shall have a moisture content not exceeding 4 percent by weight.

2109.2.1.4 Shrinkage cracks. Adobe units shall not contain more than three shrinkage cracks and any single shrinkage crack shall not exceed 3 inches (76 mm) in length or $^1/_8$ inch (3.2 mm) in width.

2109.2.2 Stabilized adobe. *Stabilized adobe* shall comply with Section 2109.2.1 for *unstabilized adobe* in addition to Sections 2109.2.2.1 and 2109.2.2.2.

2109.2.2.1 Soil requirements. Soil used for *stabilized adobe* units shall be chemically compatible with the stabilizing material.

2109.2.2.2 Absorption requirements. A 4-inch (102 mm) cube, cut from a *stabilized adobe* unit dried to a constant weight in a ventilated oven at 212°F to 239°F (100°C to 115°C), shall not absorb more than $2^1/_2$ percent moisture by weight when placed on a constantly water-saturated, porous surface for seven days. Not fewer than five specimens shall be tested and each specimen shall be cut from a separate unit.

2109.2.3 Allowable stress. The allowable compressive stress based on gross cross-sectional area of adobe shall not exceed 30 psi (207 kPa).

2109.2.3.1 Bolts. Bolt values shall not exceed those set forth in Table 2109.2.3.1.

2109.2.4 Detailed requirements. *Adobe construction* shall comply with Sections 2109.2.4.1 through 2109.2.4.9.

2109.2.4.1 Number of stories. *Adobe construction* shall be limited to buildings not exceeding one *story*, except that two-*story* construction is allowed where designed by a *registered design professional.*

2109.2.4.2 Mortar. *Mortar* for *adobe construction* shall comply with Sections 2109.2.4.2.1 and 2109.2.4.2.2.

2109.2.4.2.1 General. *Mortar* for adobe units shall be in accordance with Section 2103.2.1, or be composed of adobe soil of the same composition and stabilization as the adobe brick units. *Unstabilized adobe* soil *mortar* is permitted in conjunction with *unstabilized adobe* brick units.

2109.2.4.2.2 Mortar joints. Adobe units shall be laid with full head and *bed joints* and in full *running bond.*

2109.2.4.3 Parapet walls. *Parapet walls* constructed of adobe units shall be waterproofed.

2109.2.4.4 Wall thickness. The minimum thickness of *exterior walls* in one-*story* buildings shall be 10 inches (254 mm). The walls shall be laterally supported at intervals not exceeding 24 feet (7315 mm). The minimum thickness of interior *load-bearing walls* shall be 8 inches (203 mm). The unsupported height of any wall constructed of adobe units shall not exceed 10 times the thickness of such wall.

2109.2.4.5 Foundations. Foundations for *adobe construction* shall be in accordance with Sections 2109.2.4.5.1 and 2109.2.4.5.2.

2109.2.4.5.1 Foundation support. Walls and partitions constructed of adobe units shall be supported by foundations or footings that extend not less than 6 inches (152 mm) above adjacent ground surfaces and are constructed of *solid masonry* (excluding adobe) or concrete. Footings and foundations shall comply with Chapter 18.

2109.2.4.5.2 Lower course requirements. *Stabilized adobe* units shall be used in adobe walls for the first 4 inches (102 mm) above the finished first-floor elevation.

2109.2.4.6 Isolated piers or columns. Adobe units shall not be used for isolated piers or columns in a load-bearing capacity. Walls less than 24 inches (610 mm) in length shall be considered to be isolated piers or columns.

2109.2.4.7 Tie beams. *Exterior walls* and interior *load-bearing walls* constructed of adobe units shall have a continuous tie beam at the level of the floor or roof bearing and meeting the following requirements.

TABLE 2109.2.3.1
ALLOWABLE SHEAR ON BOLTS IN ADOBE MASONRY

DIAMETER OF BOLTS (inches)	MINIMUM EMBEDMENT (inches)	SHEAR (pounds)
$^1/_2$	—	—
$^5/_8$	12	200
$^3/_4$	15	300
$^7/_8$	18	400
1	21	500
$1^1/_8$	24	600

For SI: 1 inch = 25.4 mm, 1 pound = 4.448 N.

2109.2.4.7.1 Concrete tie beams. Concrete tie beams shall be 6 inches (152 mm) or more in depth and 10 inches (254 mm) or more in width. Concrete tie beams shall be continuously reinforced with not fewer than two No. 4 reinforcing bars. The specified compressive strength of concrete shall be not less than 2,500 psi (17.2 MPa).

2109.2.4.7.2 Wood tie beams. Wood tie beams shall be solid or built up of lumber having a nominal thickness of not less than 1 inch (25 mm), and shall have a depth of not less than 6 inches (152 mm) and a width of not less than 10 inches (254 mm). Joints in wood tie beams shall be spliced not less than 6 inches (152 mm). Splices shall not be allowed within 12 inches (305 mm) of an opening. Wood used in tie beams shall be *approved* naturally decay-resistant or *preservative-treated wood.*

2109.2.4.8 Exterior finish. Exterior finishes applied to adobe masonry walls shall be of any type permitted by this section or Chapter 14, except where stated otherwise in this section.

2109.2.4.8.1 Where required. *Unstabilized adobe* masonry walls shall receive a weather protective exterior finish in accordance with Section 2109.2.4.8.

2109.2.4.8.2 Vapor permeance. Plaster and finish assemblies shall have a vapor permeance of not less than 5 perms.

Exception: Insulation products applied to the exterior of *stabilized adobe* masonry walls in Climate Zones 2B, 3B, 4B and 5B shall not have a vapor permeance requirement.

2109.2.4.8.3 Plaster thickness and coats. Plaster applied to adobe masonry shall be not less than $^7/_8$ inch (22 mm) and not greater than 2 inches (51 mm) thick. Plaster shall be applied in not less than two coats.

2109.2.4.8.4 Plaster application. Where plaster is applied directly to adobe masonry walls, no intermediate membrane shall be used.

2109.2.4.8.5 Lath for plaster. Lath shall be provided for all plasters, except where not required elsewhere in Section 2109.2.4.8. Fasteners shall be corrosion resistant and spaced at a maximum of 16 inches (406 mm) on center with a minimum $1^1/_2$-inch (38 mm) penetration into the adobe wall. Metal lath shall comply with ASTM C1063, as modified by this section, and shall be corrosion resistant. Plastic lath shall comply with ASTM C1788, as modified by this section. Wood substrates shall be protected with No. 15 asphalt felt, an approved wood preservative or other protective coating prior to lath application.

2109.2.4.8.6 Cement plaster. *Cement plaster* shall conform to ASTM C926 and shall comply with Chapter 25, except that the proportion of lime in plaster coats shall be not less than 1 part lime to 4 parts cement. The combined thickness of *cement plaster* coats shall not exceed 1 inch (25 mm).

2109.2.4.8.7 Lime plaster. Lime plaster is any plaster with a binder composed of calcium hydroxide, including Type N or S hydrated lime, hydraulic lime, natural hydraulic lime, or slaked quicklime. Hydrated lime shall comply with ASTM C206. Hydraulic lime shall comply with ASTM C1707. Natural hydraulic lime shall comply with ASTM C141 and EN 459. Quicklime shall comply with ASTM C5.

2109.2.4.8.8 Cement-lime plaster. Cement-lime plaster shall be any plaster mix type CL, F or FL, as described in ASTM C926.

2109.2.4.8.9 Clay plaster. Clay plaster shall comply with this section.

2109.2.4.8.9.1 General. Clay plaster shall be any plaster having a clay or clay subsoil binder. Such plaster shall contain sufficient clay to fully bind the aggregate and shall be permitted to contain reinforcing fibers. Acceptable reinforcing fibers include chopped straw, sisal, and animal hair.

2109.2.4.8.9.2 Clay subsoil requirements. The suitability of clay subsoil shall be determined in accordance with the Figure 2 Ribbon Test and the Figure 3 Ball Test in the appendix of ASTM E2392/E2392M.

2109.2.4.8.9.3 Weather-exposed locations. Clay plaster exposed to water from direct or wind-driven rain or snow shall be finished with an approved erosion-resistant finish. The use of clay plasters shall not be permitted on weather-exposed parapets.

2109.2.4.8.9.4 Prohibited finish coat. Plaster containing Portland cement shall not be permitted as a finish over clay plaster.

2109.2.4.8.9.5 Conditions where lathing is not required. For *unstabilized adobe* walls finished with unstabilized clay plaster, lathing shall not be required.

2109.2.4.9 Lintels. Lintels shall be considered to be structural members and shall be designed in accordance with the applicable provisions of Chapter 16.

SECTION 2110
GLASS UNIT MASONRY

2110.1 General. *Glass unit masonry* construction shall comply with Chapter 13 of TMS 402 and this section.

2110.1.1 Limitations. Solid or hollow *approved* glass block shall not be used in *fire walls*, party walls, fire barriers, *fire partitions* or *smoke barriers*, or for load-bearing construction. Such blocks shall be erected with *mortar* and reinforcement in metal channel-type frames, structural frames, masonry or concrete recesses, embedded panel anchors as provided for both exterior and interior walls or other *approved* joint materials. Wood strip framing shall not be used in walls required to have a *fire-resistance rating* by other provisions of this code.

Exceptions:

1. Glass-block assemblies having a *fire protection rating* of not less than $^3/_4$ hour shall be permitted as opening protectives in accordance with Section 716 in *fire barriers*, *fire partitions* and *smoke barriers* that have a required *fire-resistance rating* of 1 hour or less and do not enclose exit *stairways* and *ramps* or exit passageways.
2. Glass-block assemblies as permitted in Section 404.6, Exception 2.

SECTION 2111 MASONRY FIREPLACES

2111.1 General. The construction of masonry fireplaces, consisting of concrete or masonry, shall be in accordance with this section.

2111.2 Fireplace drawings. The *construction documents* shall describe in sufficient detail the location, size and construction of masonry fireplaces. The thickness and characteristics of materials and the clearances from walls, partitions and ceilings shall be indicated.

2111.3 Footings and foundations. Footings for masonry fireplaces and their chimneys shall be constructed of concrete or *solid masonry* not less than 12 inches (305 mm) thick and shall extend not less than 6 inches (153 mm) beyond the face of the fireplace or foundation wall on all sides. Footings shall be founded on natural undisturbed earth or engineered fill below frost depth. In areas not subjected to freezing, footings shall be not less than 12 inches (305 mm) below finished grade.

2111.3.1 Ash dump cleanout. Cleanout openings, located within foundation walls below fireboxes, where provided, shall be equipped with ferrous metal or masonry doors and frames constructed to remain tightly closed, except when in use. Cleanouts shall be accessible and located so that ash removal will not create a hazard to combustible materials.

2111.4 Seismic reinforcement. In structures assigned to *Seismic Design Category* A or B, seismic reinforcement is not required. In structures assigned to *Seismic Design Category* C or D, masonry fireplaces shall be reinforced and anchored in accordance with Sections 2111.4.1, 2111.4.2 and 2111.5. In structures assigned to *Seismic Design Category* E or F, masonry fireplaces shall be reinforced in accordance with the requirements of Sections 2101 through 2108.

2111.4.1 Vertical reinforcing. For fireplaces with chimneys up to 40 inches (1016 mm) wide, four No. 4 continuous vertical bars, anchored in the foundation, shall be placed in the concrete between *wythes* of *solid masonry* or within the *cells* of hollow unit masonry and grouted in accordance with Section 2103.3. For fireplaces with chimneys greater than 40 inches (1016 mm) wide, two additional No. 4 vertical bars shall be provided for each additional 40 inches (1016 mm) in width or fraction thereof.

2111.4.2 Horizontal reinforcing. Vertical reinforcement shall be placed enclosed within $^1/_4$-inch (6.4 mm) ties or other reinforcing of equivalent net cross-sectional area, spaced not to exceed 18 inches (457 mm) on center in concrete; or placed in the *bed joints* of unit masonry at not less than every 18 inches (457 mm) of vertical height. Two such ties shall be provided at each bend in the vertical bars.

2111.5 Seismic anchorage. Masonry fireplaces and foundations shall be anchored at each floor, ceiling or roof line more than 6 feet (1829 mm) above grade with two $^3/_{16}$-inch by 1-inch (4.8 mm by 25 mm) straps embedded not less than 12 inches (305 mm) into the chimney. Straps shall be hooked around the outer bars and extend 6 inches (152 mm) beyond the bend. Each strap shall be fastened to not fewer than four floor joists with two $^1/_2$-inch (12.7 mm) bolts.

Exception: Seismic anchorage is not required for the following:

1. In structures assigned to *Seismic Design Category* A or B.
2. Where the masonry fireplace is constructed completely within the *exterior walls*.

2111.6 Firebox walls. Masonry fireboxes shall be constructed of solid *masonry units*, hollow *masonry units* grouted solid, stone or concrete. Where a lining of firebrick not less than 2 inches (51 mm) in thickness or other *approved* lining is provided, the minimum thickness of back and sidewalls shall each be 8 inches (203 mm) of *solid masonry*, including the lining. The width of joints between firebricks shall be not greater than $^1/_4$ inch (6.4 mm). Where a lining is not provided, the total minimum thickness of back and sidewalls shall be 10 inches (254 mm) of *solid masonry*. Firebrick shall conform to ASTM C27 or ASTM C1261 and shall be laid with medium-duty refractory *mortar* conforming to ASTM C199.

2111.6.1 Steel fireplace units. Steel fireplace units are permitted to be installed with *solid masonry* to form a masonry fireplace provided that they are installed according to either the requirements of their listing or the requirements of this section. Steel fireplace units incorporating a steel firebox lining shall be constructed with steel not less than $^1/_4$ inch (6.4 mm) in thickness, and an air-circulating chamber that is ducted to the interior of the building. The firebox lining shall be encased with *solid masonry* to provide a total thickness at the back and sides of not less than 8 inches (203 mm), of which not less than 4 inches (102 mm) shall be of *solid masonry* or concrete. Circulating air ducts employed with steel fireplace units shall be constructed of metal or masonry.

2111.7 Firebox dimensions. The firebox of a concrete or masonry fireplace shall have a minimum depth of 20 inches (508 mm). The throat shall be not less than 8 inches (203 mm) above the fireplace opening. The throat opening shall be not less than 4 inches (102 mm) in depth. The cross-sectional area of the passageway above the firebox, including the throat, damper and smoke chamber, shall be not less than the cross-sectional area of the flue.

Exception: Rumford fireplaces shall be permitted provided that the depth of the fireplace is not less than 12 inches (305 mm) and not less than one-third of the width of the fireplace opening, and the throat is not less than 12 inches (305 mm) above the lintel, and not less than $^1/_{20}$ the cross-sectional area of the fireplace opening.

2111.8 Lintel and throat. Masonry over a fireplace opening shall be supported by a lintel of noncombustible material. The minimum required bearing length on each end of the fireplace opening shall be 4 inches (102 mm). The *fireplace throat* or damper shall be located not less than 8 inches (203 mm) above the top of the fireplace opening.

2111.8.1 Damper. Masonry fireplaces shall be equipped with a ferrous metal damper located not less than 8 inches (203 mm) above the top of the fireplace opening. Dampers shall be installed in the fireplace or at the top of the flue venting the fireplace, and shall be operable from the room containing the fireplace. Damper controls shall be permitted to be located in the fireplace.

2111.9 Smoke chamber walls. Smoke chamber walls shall be constructed of solid *masonry units*, hollow *masonry units* grouted solid, stone or concrete. The total minimum thickness of front, back and sidewalls shall be 8 inches (203 mm) of *solid masonry*. The inside surface shall be parged smooth with refractory mortar conforming to ASTM C199. Where a lining of firebrick not less than 2 inches (51 mm) thick, or a lining of vitrified clay not less than $^5/_8$ inch (15.9 mm) thick, is provided, the total minimum thickness of front, back and sidewalls shall be 6 inches (152 mm) of *solid masonry*, including the lining. Firebrick shall conform to ASTM C1261 and shall be laid with refractory *mortar* conforming to ASTM C199. Vitrified clay linings shall conform to ASTM C315.

2111.9.1 Smoke chamber dimensions. The inside height of the smoke chamber from the *fireplace throat* to the beginning of the flue shall be not greater than the inside width of the fireplace opening. The inside surface of the smoke chamber shall not be inclined more than 45 degrees (0.76 rad) from vertical where prefabricated smoke chamber linings are used or where the smoke chamber walls are rolled or sloped rather than corbeled. Where the inside surface of the smoke chamber is formed by corbeled masonry, the walls shall not be corbeled more than 30 degrees (0.52 rad) from vertical.

2111.10 Hearth and hearth extension. Masonry fireplace hearths and hearth extensions shall be constructed of concrete or masonry, supported by noncombustible materials, and reinforced to carry their own weight and all imposed *loads*. Combustible material shall not remain against the underside of hearths or hearth extensions after construction.

2111.10.1 Hearth thickness. The minimum thickness of fireplace hearths shall be 4 inches (102 mm).

2111.10.2 Hearth extension thickness. The minimum thickness of hearth extensions shall be 2 inches (51 mm).

Exception: Where the bottom of the firebox opening is raised not less than 8 inches (203 mm) above the top of the hearth extension, a hearth extension of not less than $^3/_8$-inch-thick (9.5 mm) brick, concrete, stone, tile or other *approved* noncombustible material is permitted.

2111.11 Hearth extension dimensions. Hearth extensions shall extend not less than 16 inches (406 mm) in front of, and not less than 8 inches (203 mm) beyond, each side of the fireplace opening. Where the fireplace opening is 6 square feet (0.557 m^2) or larger, the hearth extension shall extend not less than 20 inches (508 mm) in front of, and not less than 12 inches (305 mm) beyond, each side of the fireplace opening.

2111.12 Fireplace clearance. Any portion of a masonry fireplace located in the interior of a building or within the *exterior wall* of a building shall have a clearance to combustibles of not less than 2 inches (51 mm) from the front faces and sides of masonry fireplaces and not less than 4 inches (102 mm) from the back faces of masonry fireplaces. The airspace shall not be filled, except to provide *fireblocking* in accordance with Section 2111.13.

Exceptions:

1. Masonry fireplaces *listed* and *labeled* for use in contact with combustibles in accordance with UL 127 and installed in accordance with the manufacturer's instructions are permitted to have combustible material in contact with their exterior surfaces.
2. Where masonry fireplaces are constructed as part of masonry or concrete walls, combustible materials shall not be in contact with the masonry or concrete walls less than 12 inches (306 mm) from the inside surface of the nearest firebox lining.
3. Exposed combustible *trim* and the edges of sheathing materials, such as wood siding, flooring and drywall, are permitted to abut the masonry fireplace sidewalls and hearth extension, in accordance with Figure 2111.12, provided that such combustible *trim* or sheathing is not less than 12 inches (306 mm) from the inside surface of the nearest firebox lining.
4. Exposed combustible mantels or *trim* is permitted to be placed directly on the masonry fireplace front surrounding the fireplace opening, provided that such combustible materials shall not be placed within 6 inches (153 mm) of a fireplace opening. Combustible material directly above and within 12 inches (305 mm) of the fireplace opening shall not project more than $^1/_8$ inch (3.2 mm) for each 1-inch (25 mm) distance from such opening. Combustible materials located along the sides of the fireplace opening that project more than $1^1/_2$ inches (38 mm) from the face of the fireplace shall have an additional clearance equal to the projection.

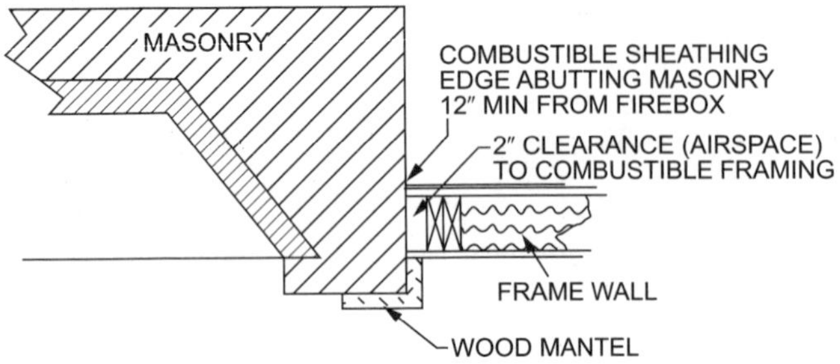

Note: 1 inch = 25.4 mm

FIGURE 2111.12
ILLUSTRATION OF EXCEPTION TO FIREPLACE CLEARANCE PROVISION

2111.13 Fireplace fireblocking. All spaces between fireplaces and floors and ceilings through which fireplaces pass shall be fireblocked with noncombustible material securely fastened in place. The *fireblocking* of spaces between wood joists, beams or headers shall be to a depth of 1 inch (25 mm) and shall only be placed on strips of metal or metal lath laid across the spaces between combustible material and the chimney.

2111.14 Exterior air. Factory-built or masonry fireplaces covered in this section shall be equipped with an exterior air supply to ensure proper fuel combustion unless the room is mechanically ventilated and controlled so that the indoor pressure is neutral or positive.

2111.14.1 Factory-built fireplaces. Exterior combustion air ducts for factory-built fireplaces shall be *listed* components of the fireplace, and installed according to the fireplace manufacturer's instructions.

2111.14.2 Masonry fireplaces. *Listed* combustion air ducts for masonry fireplaces shall be installed according to the terms of their listing and manufacturer's instructions.

2111.14.3 Exterior air intake. The exterior air intake shall be capable of providing all combustion air from the exterior of the *dwelling*. The exterior air intake shall not be located within a garage, *attic*, *basement* or crawl space of the *dwelling* nor shall the air intake be located at an elevation higher than the firebox. The exterior air intake shall be covered with a corrosion-resistant screen of $^1/_4$-inch (6.4 mm) mesh.

2111.14.4 Clearance. Unlisted combustion air ducts shall be installed with a minimum 1-inch (25 mm) clearance to combustibles for all parts of the duct within 5 feet (1524 mm) of the duct outlet.

2111.14.5 Passageway. The combustion air passageway shall be not less than 6 square inches (3870 mm^2) and not more than 55 square inches (0.035 m^2), except that combustion air systems for *listed* fireplaces or for fireplaces tested for emissions shall be constructed according to the fireplace manufacturer's instructions.

2111.14.6 Outlet. The exterior air outlet is permitted to be located in the back or sides of the firebox chamber or within 24 inches (610 mm) of the firebox opening on or near the floor. The outlet shall be closable and designed to prevent burning material from dropping into concealed combustible spaces.

SECTION 2112
MASONRY HEATERS

2112.1 Definition. A masonry heater is a heating appliance constructed of concrete or *solid masonry*, hereinafter referred to as "masonry," which is designed to absorb and store heat from a solid fuel fire built in the firebox by routing the exhaust gases through internal heat exchange channels in which the flow path downstream of the firebox includes flow in either a horizontal or downward direction before entering the chimney and which delivers heat by radiation from the masonry surface of the heater.

2112.2 Installation. Masonry heaters shall be installed in accordance with this section and comply with one of the following:

1. Masonry heaters shall comply with the requirements of ASTM E1602.
2. Masonry heaters shall be *listed* and *labeled* in accordance with UL 1482 or EN 15250 and installed in accordance with the manufacturer's instructions.

2112.3 Footings and foundation. The firebox floor of a masonry heater shall be a minimum thickness of 4 inches (102 mm) of noncombustible material and be supported on a noncombustible footing and foundation in accordance with Section 2113.2.

2112.4 Seismic reinforcing. In structures assigned to *Seismic Design Category* D, E or F, masonry heaters shall be anchored to the masonry foundation in accordance with Section 2113.3. Seismic reinforcing shall not be required within the body of a masonry heater with a height that is equal to or less than 3.5 times its body width and where the masonry chimney serving the heater is not supported by the body of the heater. Where the masonry chimney shares a common wall with the facing of the masonry heater, the chimney portion of the structure shall be reinforced in accordance with Section 2113.

2112.5 Masonry heater clearance. Combustible materials shall not be placed within 36 inches (914 mm) or the distance of the allowed reduction method from the outside surface of a masonry heater in accordance with NFPA 211, Section 12.6, and the required space between the heater and combustible material shall be fully vented to permit the free flow of air around all heater surfaces.

Exceptions:

1. Where the masonry heater wall thickness is not less than 8 inches (203 mm) of *solid masonry* and the wall thickness of the heat exchange channels is not less than 5 inches (127 mm) of *solid masonry*,

combustible materials shall not be placed within 4 inches (102 mm) of the outside surface of a masonry heater. A clearance of not less than 8 inches (203 mm) shall be provided between the gas-tight capping slab of the heater and a combustible ceiling.

2. Masonry heaters *listed* and *labeled* in accordance with UL 1482 or EN 15250 and installed in accordance with the manufacturer's instructions.

SECTION 2113
MASONRY CHIMNEYS

2113.1 General. The construction of masonry chimneys consisting of solid *masonry units*, hollow *masonry units* grouted solid, stone or concrete shall be in accordance with this section.

2113.2 Footings and foundations. Footings for masonry chimneys shall be constructed of concrete or *solid masonry* not less than 12 inches (305 mm) thick and shall extend not less than 6 inches (152 mm) beyond the face of the foundation or support wall on all sides. Footings shall be founded on natural undisturbed earth or engineered fill below frost depth. In areas not subjected to freezing, footings shall be not less than 12 inches (305 mm) below finished grade.

2113.3 Seismic reinforcement. In structures assigned to *Seismic Design Category* A or B, seismic reinforcement is not required. In structures assigned to *Seismic Design Category* C or D, masonry chimneys shall be reinforced and anchored in accordance with Sections 2113.3.1, 2113.3.2 and 2113.4. In structures assigned to *Seismic Design Category* E or F, masonry chimneys shall be reinforced in accordance with the requirements of Sections 2101 through 2108 and anchored in accordance with Section 2113.4.

2113.3.1 Vertical reinforcement. For chimneys up to 40 inches (1016 mm) wide, four No. 4 continuous vertical bars anchored in the foundation shall be placed in the concrete between *wythes* of *solid masonry* or within the *cells* of hollow unit masonry and grouted in accordance with Section 2103.3. Grout shall be prevented from bonding with the flue liner so that the flue liner is free to move with thermal expansion. For chimneys greater than 40 inches (1016 mm) wide, two additional No. 4 vertical bars shall be provided for each additional 40 inches (1016 mm) in width or fraction thereof.

2113.3.2 Horizontal reinforcement. Vertical reinforcement shall be placed enclosed within $^1/_4$-inch (6.4 mm) ties, or other reinforcing of equivalent net cross-sectional area, spaced not to exceed 18 inches (457 mm) on center in concrete, or placed in the *bed joints* of unit masonry, at not less than every 18 inches (457 mm) of vertical height. Two such ties shall be provided at each bend in the vertical bars.

2113.4 Seismic anchorage. Masonry chimneys and foundations shall be anchored at each floor, ceiling or roof line more than 6 feet (1829 mm) above grade with two $^3/_{16}$-inch by 1-inch (4.8 mm by 25 mm) straps embedded not less than 12 inches (305 mm) into the chimney. Straps shall be hooked around the outer bars and extend 6 inches (152 mm) beyond the bend. Each strap shall be fastened to not less than four floor joists with two $^1/_2$-inch (12.7 mm) bolts.

Exception: Seismic anchorage is not required for the following:

1. In structures assigned to *Seismic Design Category* A or B.
2. Where the masonry fireplace is constructed completely within the *exterior walls*.

2113.5 Corbeling. Masonry chimneys shall not be corbeled more than half of the chimney's wall thickness from a wall or foundation, nor shall a chimney be corbeled from a wall or foundation that is less than 12 inches (305 mm) in thickness unless it projects equally on each side of the wall, except that on the second *story* of a two-story *dwelling*, corbeling of chimneys on the exterior of the enclosing walls is permitted to equal the wall thickness. The projection of a single course shall not exceed one-half the unit height or one-third of the unit bed depth, whichever is less.

2113.6 Changes in dimension. The chimney wall or chimney flue lining shall not change in size or shape within 6 inches (152 mm) above or below where the chimney passes through floor components, ceiling components or roof components.

2113.7 Offsets. Where a masonry chimney is constructed with a fireclay flue liner surrounded by one *wythe* of masonry, the maximum offset shall be such that the centerline of the flue above the offset does not extend beyond the center of the chimney wall below the offset. Where the chimney offset is supported by masonry below the offset in an *approved* manner, the maximum offset limitations shall not apply. Each individual corbeled masonry course of the offset shall not exceed the projection limitations specified in Section 2113.5.

2113.8 Additional load. Chimneys shall not support *loads* other than their own weight unless they are designed and constructed to support the additional *load*. Masonry chimneys are permitted to be constructed as part of the masonry walls or concrete walls of the building.

2113.9 Termination. Chimneys shall extend not less than 2 feet (610 mm) higher than any portion of the building within 10 feet (3048 mm), but shall be not less than 3 feet (914 mm) above the highest point where the chimney passes through the roof.

2113.9.1 Chimney caps. Masonry chimneys shall have a concrete, metal or stone cap, sloped to shed water, a drip edge and a caulked bond break around any flue liners in accordance with ASTM C1283.

2113.9.2 Spark arrestors. Where a spark arrestor is installed on a masonry chimney, the spark arrestor shall meet all of the following requirements:

1. The net free area of the arrestor shall be not less than four times the net free area of the outlet of the chimney flue it serves.

2. The arrestor screen shall have heat and *corrosion resistance* equivalent to 19-gage galvanized steel or 24-gage stainless steel.
3. Openings shall not permit the passage of spheres having a diameter greater than $^1/_2$ inch (12.7 mm) nor block the passage of spheres having a diameter less than $^3/_8$ inch (9.5 mm).
4. The spark arrestor shall be accessible for cleaning and the screen or chimney cap shall be removable to allow for cleaning of the chimney flue.

2113.9.3 Rain caps. Where a masonry or metal rain cap is installed on a masonry chimney, the net free area under the cap shall be not less than four times the net free area of the outlet of the chimney flue it serves.

2113.10 Wall thickness. Masonry chimney walls shall be constructed of concrete, solid *masonry units* or hollow *masonry units* grouted solid with not less than 4 inches (102 mm) nominal thickness.

2113.10.1 Masonry veneer chimneys. Where masonry is used as *veneer* for a framed chimney, through flashing and weep holes shall be provided as required by Chapter 14.

2113.11 Flue lining (material). Masonry chimneys shall be lined. The lining material shall be appropriate for the type of appliance connected, according to the terms of the appliance listing and the manufacturer's instructions.

2113.11.1 Residential-type appliances (general). Flue lining systems shall comply with one of the following:

1. Clay flue lining complying with the requirements of ASTM C315.
2. *Listed* chimney lining systems complying with UL 1777.
3. Factory-built chimneys or chimney units *listed* for installation within masonry chimneys.
4. Other *approved* materials that will resist corrosion, erosion, softening or cracking from flue gases and condensate at temperatures up to 1,800°F (982°C).

2113.11.1.1 Flue linings for specific appliances. Flue linings other than those covered in Section 2113.11.1 intended for use with specific appliances shall comply with Sections 2113.11.1.2 through 2113.11.1.4, 2113.11.2 and 2113.11.3.

2113.11.1.2 Gas appliances. Flue lining systems for gas appliances shall be in accordance with the *International Fuel Gas Code*.

2113.11.1.3 Pellet fuel-burning appliances. Flue lining and vent systems for use in masonry chimneys with pellet fuel-burning appliances shall be limited to flue lining systems complying with Section 2113.11.1 and pellet vents *listed* for installation within masonry chimneys (see Section 2113.11.1.5 for marking).

2113.11.1.4 Oil-fired appliances approved for use with L-vent. Flue lining and vent systems for use in masonry chimneys with oil-fired appliances *approved* for use with Type L vent shall be limited to flue lining systems complying with Section 2113.11.1 and *listed* chimney liners complying with UL 641 (see Section 2113.11.1.5 for marking).

2113.11.1.5 Notice of usage. When a flue is relined with a material not complying with Section 2113.11.1, the chimney shall be plainly and permanently identified by a *label* attached to a wall, ceiling or other conspicuous location adjacent to where the connector enters the chimney. The *label* shall include the following message or equivalent language: "This chimney is for use only with (type or category of appliance) that burns (type of fuel). Do not connect other types of appliances."

2113.11.2 Concrete and masonry chimneys for medium-heat appliances. Concrete and masonry chimneys for medium-heat appliances shall comply with Sections 2113.11.2.1 through 2113.11.2.5.

2113.11.2.1 Construction. Chimneys for medium-heat appliances shall be constructed of solid *masonry units* or of concrete with walls not less than 8 inches (203 mm) thick, or with stone masonry not less than 12 inches (305 mm) thick.

2113.11.2.2 Lining. Concrete and masonry chimneys shall be lined with an *approved* medium-duty refractory brick not less than 4$^1/_2$ inches (114 mm) thick laid on the 4$^1/_2$-inch bed (114 mm) in an *approved* medium-duty refractory *mortar*. The lining shall start 2 feet (610 mm) or more below the lowest chimney connector entrance. Chimneys terminating 25 feet (7620 mm) or less above a chimney connector entrance shall be lined to the top.

2113.11.2.3 Multiple passageway. Concrete and masonry chimneys containing more than one passageway shall have the liners separated by a minimum 4-inch-thick (102 mm) concrete or *solid masonry* wall.

2113.11.2.4 Termination height. Concrete and masonry chimneys for medium-heat appliances shall extend not less than 10 feet (3048 mm) higher than any portion of any building within 25 feet (7620 mm).

2113.11.2.5 Clearance. A minimum clearance of 4 inches (102 mm) shall be provided between the exterior surfaces of a concrete or masonry chimney for medium-heat appliances and combustible material.

2113.11.3 Concrete and masonry chimneys for high-heat appliances. Concrete and masonry chimneys for high-heat appliances shall comply with 2113.11.3.1 through 2113.11.3.4.

2113.11.3.1 Construction. Chimneys for high-heat appliances shall be constructed with double walls of solid *masonry units* or of concrete, each wall to be not less than 8 inches (203 mm) thick with a minimum airspace of 2 inches (51 mm) between the walls.

2113.11.3.2 Lining. The inside of the interior wall shall be lined with an *approved* high-duty refractory brick, not less than $4^1/_2$ inches (114 mm) thick laid on the $4^1/_2$-inch bed (114 mm) in an *approved* high-duty refractory *mortar*. The lining shall start at the base of the chimney and extend continuously to the top.

2113.11.3.3 Termination height. Concrete and masonry chimneys for high-heat appliances shall extend not less than 20 feet (6096 mm) higher than any portion of any building within 50 feet (15 240 mm).

2113.11.3.4 Clearance. Concrete and masonry chimneys for high-heat appliances shall have *approved* clearance from buildings and structures to prevent overheating combustible materials, permit inspection and maintenance operations on the chimney and prevent danger of burns to persons.

2113.12 Clay flue lining (installation). Clay flue liners shall be installed in accordance with ASTM C1283 and extend from a point not less than 8 inches (203 mm) below the lowest inlet or, in the case of fireplaces, from the top of the smoke chamber to a point above the enclosing walls. The lining shall be carried up vertically, with a maximum slope not greater than 30 degrees (0.52 rad) from the vertical.

Clay flue liners shall be laid in medium-duty nonwater-soluble refractory *mortar* conforming to ASTM C199 with tight *mortar* joints left smooth on the inside and installed to maintain an airspace or insulation not to exceed the thickness of the flue liner separating the flue liners from the interior face of the chimney masonry walls. Flue lining shall be supported on all sides. Only enough *mortar* shall be placed to make the joint and hold the liners in position.

2113.13 Additional requirements.

2113.13.1 Listed materials. *Listed* materials used as flue linings shall be installed in accordance with the terms of their listings and the manufacturer's instructions.

2113.13.2 Space around lining. The space surrounding a chimney lining system or vent installed within a masonry chimney shall not be used to vent any other appliance.

Exception: This shall not prevent the installation of a separate flue lining in accordance with the manufacturer's instructions.

2113.14 Multiple flues. Where two or more flues are located in the same chimney, masonry *wythes* shall be built between adjacent flue linings. The masonry *wythes* shall be not less than 4 inches (102 mm) thick and bonded into the walls of the chimney.

Exception: Where venting only one appliance, two flues are permitted to adjoin each other in the same chimney with only the flue lining separation between them. The joints of the adjacent flue linings shall be staggered not less than 4 inches (102 mm).

2113.15 Flue area (appliance). Chimney flues shall not be smaller in area than the area of the connector from the appliance. Chimney flues connected to more than one appliance shall be not less than the area of the largest connector plus 50 percent of the areas of additional chimney connectors.

Exceptions:

1. Chimney flues serving oil-fired appliances sized in accordance with NFPA 31.
2. Chimney flues serving gas-fired appliances sized in accordance with the *International Fuel Gas Code*.

2113.16 Flue area (masonry fireplace). Flue sizing for chimneys serving fireplaces shall be in accordance with Section 2113.16.1 or 2113.16.2.

2113.16.1 Minimum area. Round chimney flues shall have a minimum net cross-sectional area of not less than $^1/_{12}$ of the fireplace opening. Square chimney flues shall have a minimum net cross-sectional area of not less than $^1/_{10}$ of the fireplace opening. Rectangular chimney flues with an aspect ratio less than 2 to 1 shall have a minimum net cross-sectional area of not less than $^1/_{10}$ of the fireplace opening. Rectangular chimney flues with an aspect ratio of 2 to 1 or more shall have a minimum net cross-sectional area of not less than $^1/_8$ of the fireplace opening.

2113.16.2 Determination of minimum area. The minimum net cross-sectional area of the flue shall be determined in accordance with Figure 2113.16. A flue size providing not less than the equivalent net cross-sectional area shall be used. Cross-sectional areas of clay flue linings are as provided in Tables 2113.16(1) and 2113.16(2) or as provided by the manufacturer or as measured in the field. The height of the chimney shall be measured from the firebox floor to the top of the chimney flue.

2113.17 Inlet. Inlets to masonry chimneys shall enter from the side. Inlets shall have a thimble of fireclay, rigid refractory material or metal that will prevent the connector from pulling out of the inlet or from extending beyond the wall of the liner.

2113.18 Masonry chimney cleanout openings. Cleanout openings shall be provided within 6 inches (152 mm) of the base of each flue within every masonry chimney. The upper edge of the cleanout shall be located not less than 6 inches (152 mm) below the lowest chimney inlet opening. The height of the opening shall be not less than 6 inches (152 mm). The cleanout shall be provided with a noncombustible cover.

Exception: Chimney flues serving masonry fireplaces, where cleaning is possible through the fireplace opening.

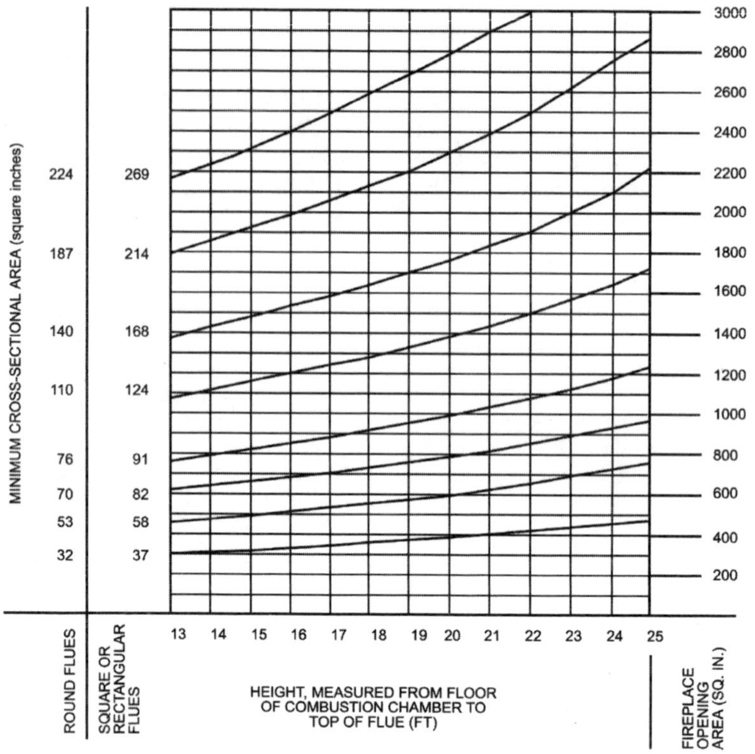

Note: 1 inch = 25.4 mm, 1 square inch = 645 mm2.

FIGURE 2113.16
FLUE SIZES FOR MASONRY CHIMNEYS

TABLE 2113.16(1)
NET CROSS-SECTIONAL AREA OF ROUND FLUE SIZES[a]

FLUE SIZE, INSIDE DIAMETER (inches)	CROSS-SECTIONAL AREA (square inches)
6	28
7	38
8	50
10	78
$10^3/_4$	90
12	113
15	176
18	254

For SI: 1 inch = 25.4 mm, 1 square inch = 645.16 mm^2.
a. Flue sizes are based on ASTM C315.

TABLE 2113.16(2)
NET CROSS-SECTIONAL AREA OF SQUARE AND RECTANGULAR FLUE SIZES

FLUE SIZE, OUTSIDE NOMINAL DIMENSIONS (inches)	CROSS-SECTIONAL AREA (square inches)
4.5 × 8.5	23
4.5 × 13	34
8 × 8	42
8.5 × 8.5	49
8 × 12	67
8.5 × 13	76
12 × 12	102
8.5 × 18	101
13 ×13	127
12 × 16	131
13 × 18	173
16 × 16	181
16 × 20	222
18 × 18	233
20 × 20	298
20 × 24	335
24 × 24	431

For SI: 1 inch = 25.4 mm, 1 square inch = 645.16 mm^2.

2113.19 Chimney clearances. Any portion of a masonry chimney located in the interior of the building or within the *exterior wall* of the building shall have a minimum airspace clearance to combustibles of 2 inches (51 mm). Chimneys located entirely outside the *exterior walls* of the building, including chimneys that pass through the soffit or *cornice*, shall have a minimum airspace clearance of 1 inch (25 mm). The airspace shall not be filled, except to provide *fireblocking* in accordance with Section 2113.20.

Exceptions:

1. Masonry chimneys equipped with a chimney lining system *listed* and *labeled* for use in chimneys in contact with combustibles in accordance with UL 1777, and installed in accordance with the manufacturer's instructions, are permitted to have combustible material in contact with their exterior surfaces.
2. Where masonry chimneys are constructed as part of masonry or concrete walls, combustible materials shall not be in contact with the masonry or concrete wall less than 12 inches (305 mm) from the inside surface of the nearest flue lining.
3. Exposed combustible *trim* and the edges of sheathing materials, such as wood siding, are permitted to abut the masonry chimney sidewalls, in accordance with Figure 2113.19, provided that such combustible *trim* or sheathing is not less than 12 inches (305 mm) from the inside surface of the nearest flue lining. Combustible material and *trim* shall not overlap the corners of the chimney by more than 1 inch (25 mm).

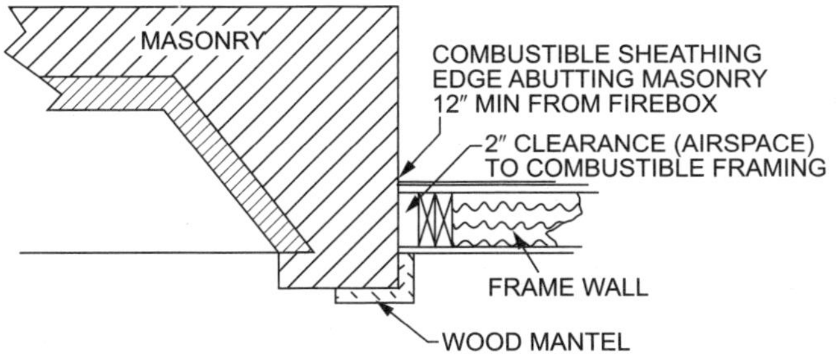

FIGURE 2113.19
ILLUSTRATION OF EXCEPTION THREE
CHIMNEY CLEARANCE PROVISION

2113.20 Chimney fireblocking. All spaces between chimneys and floors and ceilings through which chimneys pass shall be fireblocked with noncombustible material securely fastened in place. The *fireblocking* of spaces between wood joists, beams or headers shall be self-supporting or be placed on strips of metal or metal lath laid across the spaces between combustible material and the chimney.

SECTION 2114
DRY-STACK MASONRY

2114.1 General. The design of dry-stack masonry structures shall comply with the requirements of Chapters 1 through 8 of TMS 402 except as modified by Sections 2114.2 through 2114.5.

2114.2 Limitations. Dry-stack masonry shall be prohibited in *Risk Category* IV structures.

2114.3 Materials. Concrete *masonry units* complying with ASTM C90 shall be used.

2114.4 Strength. Dry-stack masonry shall be of adequate strength and proportions to support all superimposed *loads* without exceeding the allowable stresses listed in Table 2114.4. Allowable stresses not specified in Table 2114.4 shall comply with the requirements of Chapter 8 of TMS 402.

TABLE 2114.4
GROSS CROSS-SECTIONAL AREA
ALLOWABLE STRESS FOR DRY-STACK MASONRY

DESCRIPTION	MAXIMUM ALLOWABLE STRESS (psi)
Compression	45
Flexural tension	
Horizontal span	30
Vertical span	18
Shear	10

For SI: 1 pound per square inch = 0.006895 MPa.

2114.5 Construction. Construction of dry-stack masonry shall comply with ASTM C946.

CHAPTER 22

STEEL

User notes:

About this chapter: *Chapter 22 provides the minimum requirements for the design and construction of structural steel (including composite construction), cold-formed steel, steel joists, steel cable structures and steel storage racks. This chapter specifies appropriate design and construction standards for these types of structures. It also provides a road map of the applicable technical requirements for steel structures. Chapter 22 requires that the design and use of steel structures and components be in accordance with the applicable specifications and standards of the American Institute of Steel Construction, the American Iron and Steel Institute, the Steel Joist Institute and the American Society of Civil Engineers.*

Code development reminder: *Code change proposals to this chapter will be considered by the IBC—Structural Code Development Committee during the 2022 (Group B) Code Development Cycle.*

SECTION 2201 GENERAL

2201.1 Scope. The provisions of this chapter govern the quality, design, fabrication and erection of steel construction.

SECTION 2202 IDENTIFICATION OF STEEL FOR STRUCTURAL PURPOSES

2202.1 General. Identification of *structural steel elements* shall be in accordance with AISC 360. Identification of cold-formed steel members shall be in accordance with AISI S100. Identification of cold-formed steel *light-frame construction* shall also comply with the requirements contained in AISI S240 or AISI S220, as applicable. Other steel furnished for structural load-carrying purposes shall be properly identified for conformity to the ordered grade in accordance with the specified ASTM standard or other specification and the provisions of this chapter. Where the steel grade is not readily identifiable from marking and test records, the steel shall be tested to verify conformity to such standards.

SECTION 2203 PROTECTION OF STEEL FOR STRUCTURAL PURPOSES

2203.1 General. Painting of *structural steel elements* shall be in accordance with AISC 360. Painting of open-web steel joists and joist girders shall be in accordance with SJI 100 and SJI 200. Individual structural members and assembled panels of *cold-formed steel construction* shall be protected against corrosion in accordance with the requirements contained in AISI S100. Protection of cold-formed steel *light-frame construction* shall be in accordance with AISI S240 or AISI S220, as applicable.

SECTION 2204 CONNECTIONS

2204.1 Welding. The details of design, workmanship and technique for welding and qualification of welding personnel shall be in accordance with the specifications listed in Sections 2205, 2206, 2207, 2208, 2210 and 2211. For *special inspection* of welding, see Section 1705.2.

2204.2 Bolting. The design, installation and inspection of bolts shall be in accordance with the requirements of Sections 2205, 2206, 2207, 2210 and 2211. For *special inspection* of the installation of high-strength bolts, see Section 1705.2.

2204.3 Anchor rods. Anchor rods shall be set in accordance with the *approved construction documents*. The protrusion of the threaded ends through the connected material shall fully engage the threads of the nuts but shall not be greater than the length of the threads on the bolts.

SECTION 2205 STRUCTURAL STEEL

2205.1 General. The design, fabrication and erection of *structural steel elements* in buildings, structures and portions thereof shall be in accordance with AISC 360.

2205.2 Seismic design. Where required, the seismic design, fabrication and erection of buildings, structures and portions thereof shall be in accordance with Section 2205.2.1 or 2205.2.2, as applicable.

2205.2.1 Structural steel seismic force-resisting systems. The design, detailing, fabrication and erection of structural steel *seismic force-resisting systems* shall be in accordance with the provisions of Section 2205.2.1.1 or 2205.2.1.2, as applicable.

2205.2.1.1 Seismic Design Category B or C. Structures assigned to *Seismic Design Category* B or C shall be of any construction permitted in Section 2205.

Where a response modification coefficient, *R*, in accordance with ASCE 7, Table 12.2-1, is used for the design of structures assigned to *Seismic Design Category* B or C, the structures shall be designed and detailed in accordance with the requirements of AISC 341. Beam-to-column moment connections in special moment frames and intermediate moment frames shall be prequalified in accordance with AISC 341, Section K1, qualified by testing in accordance with AISC 341, Section K2, or shall be prequalified in accordance with AISC 358.

Exception: The response modification coefficient, *R*, designated for "Steel systems not specifically detailed for seismic resistance, excluding cantilever column systems" in ASCE 7, Table 12.2-1, shall be permitted for systems designed and detailed in accordance with AISC 360, and need not be designed and detailed in accordance with AISC 341.

2205.2.1.2 Seismic Design Category D, E or F. Structures assigned to *Seismic Design Category* D, E or F shall be designed and detailed in accordance with AISC 341, except as permitted in ASCE 7, Table 15.4-1. Beam-to-column moment connections in special moment frames and intermediate moment frames shall be prequalified in accordance with AISC 341, Section K1, qualified by testing in accordance with AISC 341, Section K2, or shall be prequalified in accordance with AISC 358.

2205.2.2 Structural steel elements. The design, detailing, fabrication and erection of *structural steel elements* in *seismic force-resisting systems* other than those covered in Section 2205.2.1, including struts, *collectors*, chords and foundation elements, shall be in accordance with AISC 341 where either of the following applies:

1. The structure is assigned to *Seismic Design Category* D, E or F, except as permitted in ASCE 7, Table 15.4-1.
2. A response modification coefficient, R, greater than 3 in accordance with ASCE 7, Table 12.2-1, is used for the design of the structure assigned to *Seismic Design Category* B or C.

SECTION 2206
COMPOSITE STRUCTURAL STEEL AND CONCRETE STRUCTURES

2206.1 General. Systems of *structural steel elements* acting compositely with reinforced concrete shall be designed in accordance with AISC 360 and ACI 318, excluding ACI 318 Chapter 14.

2206.2 Seismic design. Where required, the seismic design, fabrication and erection of composite steel and concrete systems shall be in accordance with Section 2206.2.1.

2206.2.1 Seismic requirements for composite structural steel and concrete construction. Where a response modification coefficient, *R*, in accordance with ASCE 7, Table 12.2-1, is used for the design of systems of structural steel acting compositely with reinforced concrete, the structures shall be designed and detailed in accordance with the requirements of AISC 341.

SECTION 2207
STEEL JOISTS

2207.1 General. The design, manufacture and use of open-web *steel joists* and joist girders shall be in accordance with either SJI 100 or SJI 200, as applicable.

2207.1.1 Seismic design. Where required, the seismic design of buildings shall be in accordance with the additional provisions of Section 2205.2 or 2211.1.1.

2207.2 Design. The *registered design professional* shall indicate on the *construction documents* the *steel joist* and *steel joist* girder designations from the specifications listed in Section 2207.1; and shall indicate the requirements for joist and joist girder design, layout, end supports, anchorage, bridging design that differs from the SJI specifications listed in Section 2207.1, bridging termination connections and bearing connection design to resist uplift and lateral *loads*. These documents shall indicate special requirements as follows:

1. Special *loads* including:
   - 1.1. Concentrated *loads*.
   - 1.2. Nonuniform *loads*.
   - 1.3. Net uplift *loads*.
   - 1.4. Axial *loads*.
   - 1.5. End moments.
   - 1.6. Connection forces.
2. Special considerations including:
   - 2.1. Profiles for joist and joist girder configurations that differ from those defined by the SJI specifications listed in Section 2207.1.
   - 2.2. Oversized or other nonstandard web openings.
   - 2.3. Extended ends.
3. Live and total *load* deflection criteria for joists and joist girder configurations that differ from those defined by the SJI specifications listed in Section 2207.1.

2207.3 Calculations. The *steel joist* and joist girder manufacturer shall design the *steel joists* and *steel joist* girders in accordance with the SJI specifications listed in Section 2207.1 to support the *load* requirements of Section 2207.2. The *registered design professional* shall be permitted to require submission of the *steel joist* and joist girder calculations as prepared by a *registered design professional* responsible for the product design. Where requested by the *registered design professional*, the *steel joist* manufacturer shall submit design calculations with a cover letter bearing the seal and signature of the joist manufacturer's *registered design professional*. In addition to the design calculations submitted under seal and signature, the following shall be included:

1. Bridging design that differs from the SJI specifications listed in Section 2207.1, such as cantilevered conditions and net uplift.
2. Connection design for:
   - 2.1. Connections that differ from the SJI specifications listed in Section 2207.1, such as flush-framed or framed connections.
   - 2.2. Field splices.
   - 2.3. Joist headers.

2207.4 Steel joist drawings. *Steel joist* placement plans shall be provided to show the *steel joist* products as specified on the *approved construction documents* and are to be utilized for field installation in accordance with specific project requirements as stated in Section 2207.2. *Steel joist* placement plans shall include, at a minimum, the following:

1. Listing of applicable *loads* as stated in Section 2207.2 and used in the design of the *steel joists* and joist girders as specified in the *approved construction documents*.
2. Profiles for joist and joist girder configurations that differ from those defined by the SJI specifications listed in Section 2207.1.
3. Connection requirements for:
   - 3.1. Joist supports.
   - 3.2. Joist girder supports.
   - 3.3. Field splices.
   - 3.4. Bridging attachments.
4. Live and total *load* deflection criteria for joists and joist girder configurations that differ from those defined by the SJI specifications listed in Section 2207.1.
5. Size, location and connections for bridging.
6. Joist headers.

Steel joist placement plans do not require the seal and signature of the joist manufacturer's *registered design professional*.

2207.5 Certification. At completion of manufacture, the *steel joist* manufacturer shall submit a *certificate of compliance* to the owner or the owner's authorized agent for submittal to the *building official* as specified in Section 1704.5 stating that work was performed in accordance with *approved construction documents* and with SJI specifications listed in Section 2207.1.

SECTION 2208
STEEL CABLE STRUCTURES

2208.1 General. The design, fabrication and erection including related connections, and protective coatings of steel cables for buildings shall be in accordance with ASCE 19.

SECTION 2209
STEEL STORAGE RACKS

2209.1 Steel storage racks. The design, testing and utilization of steel *storage racks* made of cold-formed or hot-rolled steel structural members shall be in accordance with RMI ANSI/MH 16.1. Where required by ASCE 7, the seismic design of steel *storage racks* shall be in accordance with Section 15.5.3 of ASCE 7.

2209.2 Steel cantilevered storage racks. The design, testing and utilization of steel cantilevered *storage racks* made of cold-formed or hot-rolled steel structural members shall be in accordance with RMI ANSI/MH 16.3. Where required by ASCE 7, the seismic design of steel cantilevered *storage racks* shall be in accordance with Section 15.5.3 of ASCE 7.

2209.3 Certification. For rack storage structures that are 8 feet (2438 mm) in height or greater to the top *load* level and assigned to *Seismic Design Category* D, E, or F at completion of the *storage rack* installation, a *certificate of compliance* shall be submitted to the owner or the owner's authorized agent stating that the work was performed in accordance with approved construction documents.

SECTION 2210
COLD-FORMED STEEL

2210.1 General. The design of cold-formed carbon and low-alloy steel structural members shall be in accordance with AISI S100. The design of cold-formed stainless-steel structural members shall be in accordance with ASCE 8. Cold-formed steel *light-frame construction* shall comply with Section 2211. Where required, the seismic design of cold-formed steel structures shall be in accordance with the additional provisions of Section 2210.2.

2210.1.1 Steel decks. The design and construction of cold-formed steel decks shall be in accordance with this section.

2210.1.1.1 Noncomposite steel floor decks. Noncomposite steel floor decks shall be permitted to be designed and constructed in accordance with ANSI/SDI-NC1.0.

2210.1.1.2 Steel roof deck. Steel *roof decks* shall be permitted to be designed and constructed in accordance with ANSI/SDI-RD1.0.

2210.1.1.3 Composite slabs on steel decks. Composite slabs of concrete and steel deck shall be permitted to be designed and constructed in accordance with SDI-C.

2210.2 Seismic requirements for cold-formed steel structures. Where a response modification coefficient, R, in accordance with ASCE 7, Table 12.2-1, is used for the design of cold-formed steel structures, the structures shall be designed and detailed in accordance with the requirements of AISI S100, ASCE 8, or, for cold-formed steel special-bolted moment frames, AISI S400.

SECTION 2211
COLD-FORMED STEEL
LIGHT-FRAME CONSTRUCTION

2211.1 Structural framing. For cold-formed steel *light-frame construction*, the design and installation of the following structural framing systems, including their members and connections, shall be in accordance with AISI S240, and Sections 2211.1.1 through 2211.1.3, as applicable:

1. Floor and roof systems.
2. Structural walls.
3. Shear walls, strap-braced walls and diaphragms that resist in-plane lateral loads.
4. Trusses.

2211.1.1 Seismic requirements for cold-formed steel structural systems. The design of cold-formed steel *light-frame construction* to resist seismic forces shall be in accordance with the provisions of Section 2211.1.1.1 or 2211.1.1.2, as applicable.

2211.1.1.1 Seismic Design Categories B and C. Where a response modification coefficient, *R*, in accordance with ASCE 7, Table 12.2-1 is used for the design of cold-formed steel *light-frame construction* assigned to *Seismic Design Category* B or C, the *seismic force-resisting system* shall be designed and detailed in accordance with the requirements of AISI S400.

Exception: The response modification coefficient, *R*, designated for "Steel systems not specifically detailed for seismic resistance, excluding cantilever column systems" in ASCE 7, Table 12.2-1, shall be permitted for systems designed and detailed in accordance with AISI S240 and need not be designed and detailed in accordance with AISI S400

2211.1.1.2 Seismic Design Categories D through F. In cold-formed steel *light-frame construction* assigned to *Seismic Design Category* D, E or F, the *seismic force-resisting system* shall be designed and detailed in accordance with AISI S400.

2211.1.2 Prescriptive framing. Detached one- and two-family *dwellings* and *townhouses*, less than or equal to three *stories above grade plane*, shall be permitted to be constructed in accordance with AISI S230 subject to the limitations therein.

2211.1.3 Truss design. Cold-formed steel trusses shall comply with the additional provisions of Sections 2211.1.3.1. through 2211.1.3.3.

2211.1.3.1 Truss design drawings. The truss design drawings shall conform to the requirements of Section I1 of AISI S202 and shall be provided with the shipment of trusses delivered to the job site. The truss design drawings shall include the details of permanent *individual truss member* restraint/bracing in accordance with Section I1.6 of AISI S202 where these methods are utilized to provide restraint/bracing.

2211.1.3.2 Trusses spanning 60 feet or greater. The owner or the owner's authorized agent shall contract with a *registered design professional* for the design of the temporary installation restraint/bracing and the permanent *individual truss member* restraint/bracing for trusses with clear spans 60 feet (18 288 mm) or greater. *Special inspection* of trusses over 60 feet (18 288 mm) in length shall be in accordance with Section 1705.2.

2211.1.3.3 Truss quality assurance. Trusses not part of a manufacturing process that provides requirements for quality control done under the supervision of a third-party quality control agency in accordance with AISI S240 Chapter D shall be fabricated in compliance with Sections 1704.2.5 and 1705.2, as applicable.

2211.2 Nonstructural members. For cold-formed steel *light-frame construction*, the design and installation of nonstructural members and connections shall be in accordance with AISI S220.

CHAPTER 23

WOOD

User notes:

About this chapter: *Chapter 23 provides minimum requirements for the design of buildings and structures that use wood and wood-based products. The chapter is organized around three design methodologies: allowable stress design (ASD), load and resistance factor design (LRFD) and conventional light-frame construction. In addition it allows the use of the American Wood Council Wood Frame Construction Manual for a limited range of structures. Included in the chapter are references to design and manufacturing standards for various wood and wood-based products; general construction requirements; design criteria for lateral force-resisting systems and specific requirements for the application of the three design methods.*

Code development reminder: *Code change proposals to this chapter will be considered by the IBC—Structural Code Development Committee during the 2022 (Group B) Code Development Cycle.*

SECTION 2301
GENERAL

2301.1 Scope. The provisions of this chapter shall govern the materials, design, construction and quality of wood members and their fasteners.

2301.2 Nominal sizes. For the purposes of this chapter, where dimensions of lumber are specified, they shall be deemed to be nominal dimensions unless specifically designated as actual dimensions (see Section 2304.2).

SECTION 2302
DESIGN REQUIREMENTS

2302.1 General. The design of structural elements or systems, constructed partially or wholly of wood or wood-based products, shall be in accordance with one of the following methods:

1. *Allowable stress design* in accordance with Sections 2304, 2305 and 2306.
2. *Load and resistance factor design* in accordance with Sections 2304, 2305 and 2307.
3. *Conventional light-frame construction* in accordance with Sections 2304 and 2308.
4. AWC WFCM in accordance with Section 2309.
5. The design and construction of log structures in accordance with the provisions of ICC 400.

SECTION 2303
MINIMUM STANDARDS AND QUALITY

2303.1 General. Structural sawn lumber; end-jointed lumber; *prefabricated wood I-joists*; *structural glued-laminated timber*; *wood structural panels*; fiberboard sheathing (where used structurally); *hardboard* siding (where used structurally); *particleboard*; *preservative-treated wood*; structural log members; *structural composite lumber*; round timber poles and piles; *fire-retardant-treated wood*; hardwood plywood; wood trusses; joist hangers; nails; and staples shall conform to the applicable provisions of this section.

2303.1.1 Sawn lumber. Sawn lumber used for load-supporting purposes, including end-jointed or edge-glued lumber, machine stress-rated or machine-evaluated lumber, shall be identified by the grade *mark* of a lumber grading or inspection agency that has been approved by an *accreditation body* that complies with DOC PS 20 or equivalent. Grading practices and identification shall comply with rules published by an agency approved in accordance with the procedures of DOC PS 20 or equivalent procedures.

2303.1.1.1 Certificate of inspection. In lieu of a grade *mark* on the material, a certificate of inspection as to species and grade issued by a lumber grading or inspection agency meeting the requirements of this section is permitted to be accepted for precut, remanufactured or rough-sawn lumber and for sizes larger than 3 inches (76 mm) nominal thickness.

2303.1.1.2 End-jointed lumber. *Approved* end-jointed lumber is permitted to be used interchangeably with solid-sawn members of the same species and grade. End-jointed lumber used in an assembly required to have a *fire-resistance rating* shall have the designation "Heat Resistant Adhesive" or "HRA" included in its grade *mark*.

2303.1.2 Prefabricated wood I-joists. Structural capacities and design provisions for *prefabricated wood I-joists* shall be established and monitored in accordance with ASTM D5055.

2303.1.3 Structural glued-laminated timber. Glued-laminated timbers shall be manufactured and identified as required in ANSI/APA 190.1 and ASTM D3737.

2303.1.4 Structural glued cross-laminated timber. Cross-laminated timbers shall be manufactured and identified in accordance with ANSI/APA PRG 320.

2303.1.5 Wood structural panels. *Wood structural panels*, where used structurally (including those used for siding, roof and wall sheathing, subflooring, *diaphragms* and built-up members), shall conform to the requirements for their type in DOC PS 1, DOC PS 2 or ANSI/APA PRP 210. Each panel or member shall be identified for grade, bond classification, and *Performance Category* by the trademarks of an *approved* testing and grading agency. The *Performance Category* value shall be used as the "nominal panel thickness" or "panel thickness" whenever referenced in this code. *Wood structural panel* components shall be designed and fabricated in accordance with the applicable standards listed in Section 2306.1 and identified by the trademarks of an *approved* testing and inspection agency indicating conformance to the applicable standard. In addition, *wood structural panels* where permanently exposed in outdoor applications shall be of exterior type, except that *wood structural panel* roof sheathing exposed to the outdoors on the underside is permitted to be Exposure 1 type.

2303.1.6 Fiberboard. *Fiberboard* for its various uses shall conform to ASTM C208. *Fiberboard* sheathing, where used structurally, shall be identified by an *approved* agency as conforming to ASTM C208.

2303.1.6.1 Jointing. To ensure tight-fitting assemblies, edges shall be manufactured with square, shiplapped, beveled, tongue-and-groove or U-shaped joints.

2303.1.6.2 Roof insulation. Where used as roof insulation in all types of construction, *fiberboard* shall be protected with an *approved roof covering.*

2303.1.6.3 Wall insulation. Where installed and fireblocked to comply with Chapter 7, *fiberboards* are permitted as wall insulation in all types of construction. In *fire walls* and *fire barriers*, unless treated to comply with Section 803.1 for Class A materials, the boards shall be cemented directly to the concrete, masonry or other noncombustible base and shall be protected with an *approved* noncombustible *veneer* anchored to the base without intervening airspaces.

2303.1.6.3.1 Protection. *Fiberboard* wall insulation applied on the exterior of foundation walls shall be protected below ground level with a bituminous coating.

2303.1.7 Hardboard. *Hardboard* siding shall conform to the requirements of ANSI A135.6 and, where used structurally, shall be identified by the *label* of an *approved agency*. *Hardboard* underlayment shall meet the strength requirements of $^{7}/_{32}$-inch (5.6 mm) or $^{1}/_{4}$-inch (6.4 mm) service class *hardboard* planed or sanded on one side to a uniform thickness of not less than 0.200 inch (5.1 mm). Prefinished *hardboard* paneling shall meet the requirements of ANSI A135.5. Other basic *hardboard* products shall meet the requirements of ANSI A135.4. *Hardboard* products shall be installed in accordance with manufacture's recommendations.

2303.1.8 Particleboard. *Particleboard* shall conform to ANSI A208.1. *Particleboard* shall be identified by the grade *mark* or certificate of inspection issued by an *approved agency*. *Particleboard* shall not be utilized for applications other than indicated in this section unless the *particleboard* complies with the provisions of Section 2306.3.

2303.1.8.1 Floor underlayment. *Particleboard* floor underlayment shall conform to Type PBU of ANSI A208.1. Type PBU underlayment shall be not less than $^{1}/_{4}$-inch (6.4 mm) thick and shall be installed in accordance with the instructions of the Composite Panel Association.

2303.1.9 Preservative-treated wood. Lumber, timber, plywood, piles and poles supporting permanent structures required by Section 2304.12 to be preservative treated shall conform to AWPA U1 and M4. Lumber and plywood used in permanent wood foundation systems shall conform to Chapter 18.

2303.1.9.1 Identification. Wood required by Section 2304.12 to be preservative treated shall bear the quality *mark* of an inspection agency that maintains continuing supervision, testing and inspection over the quality of the *preservative-treated wood*. Inspection agencies for *preservative-treated wood* shall be *listed* by an *accreditation body* that complies with the requirements of the American Lumber Standards Treated Wood Program, or equivalent. The quality *mark* shall be on a stamp or *label* affixed to the *preservative-treated wood*, and shall include the following information:

1. Identification of treating manufacturer.
2. Type of preservative used.
3. Minimum preservative retention (pcf).
4. End use for which the product is treated.
5. AWPA standard to which the product was treated.
6. Identity of the accredited inspection agency.

2303.1.9.2 Moisture content. Where *preservative-treated wood* is used in enclosed locations where drying in service cannot readily occur, such wood shall be at a moisture content of 19 percent or less before being covered with insulation, interior wall finish, floor covering or other materials.

2303.1.10 Structural composite lumber. Structural capacities for *structural composite lumber* shall be established and monitored in accordance with ASTM D5456.

2303.1.11 Structural log members. Stress grading of structural log members of nonrectangular shape, as typically used in log buildings, shall be in accordance with ASTM D3957. Such structural log members shall be identified by the grade *mark* of an *approved* lumber grading or inspection agency. In lieu of a grade *mark* on the material, a certificate of inspection as to species and grade issued by a lumber grading or inspection agency meeting the requirements of this section shall be permitted.

2303.1.12 Round timber poles and piles. Round timber poles and piles shall comply with ASTM D3200 and ASTM D25, respectively.

2303.1.13 Engineered wood rim board. *Engineered wood rim boards* shall conform to ANSI/APA PRR 410 or shall be evaluated in accordance with ASTM D7672.

Structural capacities shall be in accordance with ANSI/APA PRR 410 or established in accordance with ASTM D7672. Rim boards conforming to ANSI/APA PRR 410 shall be marked in accordance with that standard.

2303.2 Fire-retardant-treated wood. *Fire-retardant-treated wood* is any wood product that, when impregnated with chemicals by a pressure process or other means during manufacture, shall have, when tested in accordance with ASTM E84 or UL 723, a *listed flame spread index* of 25 or less. Additionally, the ASTM E84 or UL 723 test shall be continued for a 20-minute period and the flame front shall not progress more than $10^1/_2$ feet (3200 mm) beyond the centerline of the burners at any time during the test.

2303.2.1 Pressure process. For wood products impregnated with chemicals by a pressure process, the process shall be performed in closed vessels under pressures not less than 50 pounds per square inch gauge (psig) (345 kPa).

2303.2.2 Other means during manufacture. For wood products impregnated with chemicals by other means during manufacture, the treatment shall be an integral part of the manufacturing process of the wood product. The treatment shall provide permanent protection to all surfaces of the wood product. The use of paints, coating, stains or other surface treatments is not an approved method of protection as required in this section.

2303.2.3 Fire testing of wood structural panels. Wood structural panels shall be tested with a ripped or cut longitudinal gap of $^1/_8$ inch (3.2 mm).

2303.2.4 Labeling. In addition to the labels required in Section 2303.1.1 for sawn lumber and Section 2303.1.5 for *wood structural panels*, each piece of fire-retardant-treated lumber and *wood structural panels* shall be labeled. The *label* shall contain the following items:

1. The identification *mark* of an *approved agency* in accordance with Section 1703.5.
2. Identification of the treating manufacturer.
3. The name of the fire-retardant treatment.
4. The species of wood treated.
5. *Flame spread* and *smoke-developed index*.
6. Method of drying after treatment.
7. Conformance with appropriate standards in accordance with Sections 2303.2.5 through 2303.2.8.
8. For *fire-retardant-treated wood* exposed to weather, damp or wet locations, include the words "No increase in the *listed* classification when subjected to the Standard Rain Test" (ASTM D2898).

2303.2.5 Strength adjustments. Design values for untreated lumber and *wood structural panels*, as specified in Section 2303.1, shall be adjusted for *fire-retardant-treated wood.* Adjustments to design values shall be based on an *approved* method of investigation that takes into consideration the effects of the anticipated temperature and humidity to which the *fire-retardant-treated wood* will be subjected, the type of treatment and redrying procedures.

2303.2.5.1 Wood structural panels. The effect of treatment and the method of redrying after treatment, and exposure to high temperatures and high humidities on the flexure properties of fire-retardant-treated softwood plywood shall be determined in accordance with ASTM D5516. The test data developed by ASTM D5516 shall be used to develop adjustment factors, maximum *loads* and spans, or both, for untreated plywood design values in accordance with ASTM D6305. Each manufacturer shall publish the allowable maximum *loads* and spans for service as floor and roof sheathing for its treatment.

2303.2.5.2 Lumber. For each species of wood that is treated, the effects of the treatment, the method of redrying after treatment and exposure to high temperatures and high humidities on the allowable design properties of fire-retardant-treated lumber shall be determined in accordance with ASTM D5664. The test data developed by ASTM D5664 shall be used to develop modification factors for use at or near room temperature and at elevated temperatures and humidity in accordance with ASTM D6841. Each manufacturer shall publish the modification factors for service at temperatures of not less than 80°F (27°C) and for roof framing. The roof framing modification factors shall take into consideration the climatological location.

2303.2.6 Exposure to weather, damp or wet locations. Where *fire-retardant-treated wood* is exposed to weather, or damp or wet locations, it shall be identified as "Exterior" to indicate there is no increase in the *listed flame spread index* as defined in Section 2303.2 when subjected to ASTM D2898.

2303.2.7 Interior applications. Interior *fire-retardant-treated wood* shall have moisture content of not over 28 percent when tested in accordance with ASTM D3201 procedures at 92-percent relative humidity. Interior *fire-retardant-treated wood* shall be tested in accordance with Section 2303.2.5.1 or 2303.2.5.2. Interior *fire-retardant-treated wood* designated as Type A shall be tested in accordance with the provisions of this section.

2303.2.8 Moisture content. *Fire-retardant-treated wood* shall be dried to a moisture content of 19 percent or less for lumber and 15 percent or less for *wood structural panels* before use. For wood kiln-dried after treatment (KDAT), the kiln temperatures shall not exceed those used in kiln drying the lumber and plywood submitted for the tests described in Section 2303.2.5.1 for plywood and 2303.2.5.2 for lumber.

2303.2.9 Types I and II construction applications. See Section 603.1 for limitations on the use of *fire-retardant-treated wood* in buildings of Type I or II construction.

2303.3 Hardwood and plywood. Hardwood and decorative plywood shall be manufactured and identified as required in HPVA HP-1.

2303.4 Trusses. Wood trusses shall comply with Sections 2303.4.1 through 2303.4.7.

2303.4.1 Design. Wood trusses shall be designed in accordance with the provisions of this code and accepted engineering practice. Members are permitted to be joined by nails, glue, bolts, timber connectors, metal connector plates or other *approved* framing devices.

2303.4.1.1 Truss design drawings. The written, graphic and pictorial depiction of each individual truss shall be provided to the *building official* for approval prior to installation. Truss design drawings shall be provided with the shipment of trusses delivered to the job site. Truss design drawings shall include, at a minimum, the following information:

1. Slope or depth, span and spacing.
2. Location of all joints and support locations.
3. Number of plies if greater than one.
4. Required bearing widths.
5. Design *loads* as applicable, including:
   - 5.1. Top chord *live load.*
   - 5.2. Top chord *dead load.*
   - 5.3. Bottom chord *live load.*
   - 5.4. Bottom chord *dead load.*
   - 5.5. Additional *loads* and locations.
   - 5.6. Environmental design criteria and *loads* (such as wind, rain, snow, seismic).
6. Other lateral *loads*, including *drag strut loads*.
7. Adjustments to wood member and metal connector plate design value for conditions of use.
8. Maximum reaction force and direction, including maximum uplift reaction forces where applicable.
9. Joint connection type and description, such as size and thickness or gage, and the dimensioned location of each joint connector except where symmetrically located relative to the joint interface.
10. Size, species and grade for each wood member.
11. Truss-to-truss connections and truss field assembly requirements.
12. Calculated span-to-deflection ratio and maximum vertical and horizontal deflection for live and total *load* as applicable.
13. Maximum axial tension and compression forces in the truss members.
14. Required permanent *individual truss member* restraint location and the method and details of restraint and diagonal bracing to be used in accordance with Section 2303.4.1.2.

2303.4.1.2 Permanent individual truss member restraint (PITMR) and permanent individual truss member diagonal bracing (PITMDB). Where the truss design drawings designate the need for *permanent individual truss member restraint*, it shall be accomplished by one of the following methods:

1. *PITMR* and *PITMDB* installed using standard industry lateral restraint and diagonal bracing details in accordance with TPI 1, Section 2.3.3.1.1, accepted engineering practice, or Figures 2303.4.1.2(1), (3), and (5).
2. *Individual truss member* reinforcement in place of the specified lateral restraints (i.e., buckling reinforcement such as T-reinforcement, L-reinforcement, proprietary reinforcement, etc.) such that the buckling of any individual truss member is resisted internally by the individual truss. The buckling reinforcement of individual truss members shall be installed as shown on the truss design drawing, on supplemental truss member buckling reinforcement details provided by the truss designer or in accordance with Figures 2303.4.1.2 (2) and (4).
3. A project-specific *PITMR* and *PITMDB* design provided by any *registered design professional.*

2303.4.1.2.1 Trusses installed without a diaphragm. Trusses installed without a *diaphragm* on the top or bottom chord shall require a project specific *PITMR* and *PITMDB* design prepared by a *registered design professional*.

Exception: Group U occupancies.

2303.4.1.3 Trusses spanning 60 feet or greater. The owner or the owner's authorized agent shall contract with any qualified *registered design professional* for the design of the temporary installation restraint and diagonal bracing and the *PITMR* and *PITMDB* for all trusses with clear spans 60 feet (18 288 mm) or greater.

2303.4.1.4 Truss designer. The individual or organization responsible for the design of trusses.

2303.4.1.4.1 Truss design drawings. Where required by the *registered design professional*, the *building official* or the statutes of the jurisdiction in which the project is to be constructed, each individual truss design drawing shall bear the seal and signature of the truss designer.

Exceptions:

1. Where a cover sheet and truss index sheet are combined into a single sheet and attached to the set of truss design drawings, the single cover/truss index sheet is the only document required to be signed and sealed by the truss designer.
2. Where a cover sheet and a truss index sheet are separately provided and attached to the set of truss design drawings, the cover sheet and the truss index sheet are the only documents required to be signed and sealed by the truss designer.

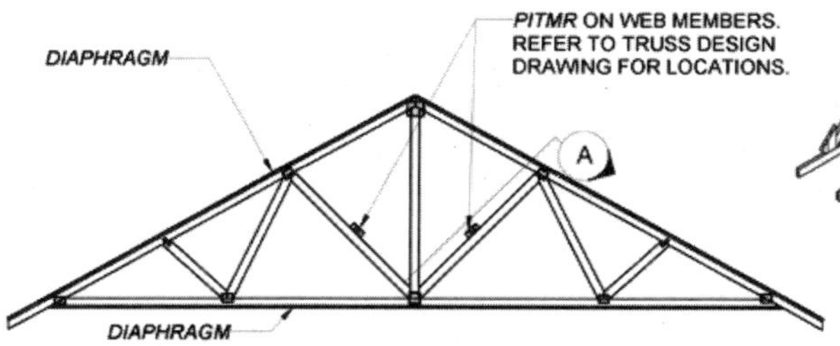

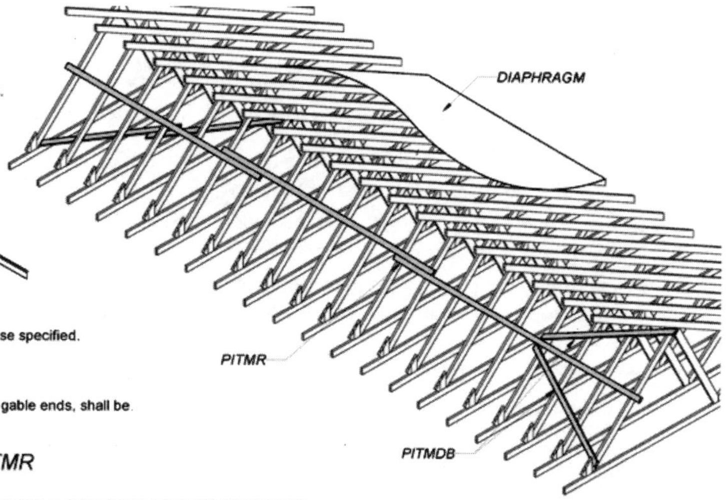

a.) Use minimum 2x4 stress - graded lumber for *PITMR* and *PITMDB* unless otherwise specified.

b.) Truss top chord and bottom chord members shall be restrained and braced.

c.) Bracing to resist forces applied perpendicular to the truss, such as wind bracing at gable ends, shall be specified by the building designer.

ELEVATION VIEW OF TRUSS WITH SINGLE ROW, *PITMR*

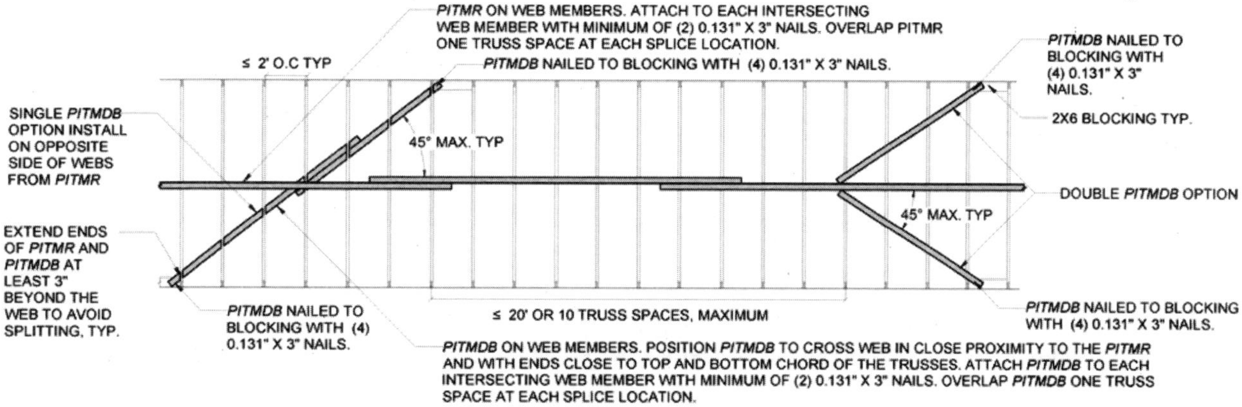

A SECTION (EXAMPLE OF SINGLE ROW OF *PITMR* WITH *PITMDB* ON WEB MEMBERS)

For SI: 1 inch = 25.4 mm, 1 foot = 304.8 mm.

FIGURE 2303.4.1.2 (1)
PITMR AND PITMDB FOR TRUSS WEB MEMBERS REQUIRING ONE ROW OF PITMR

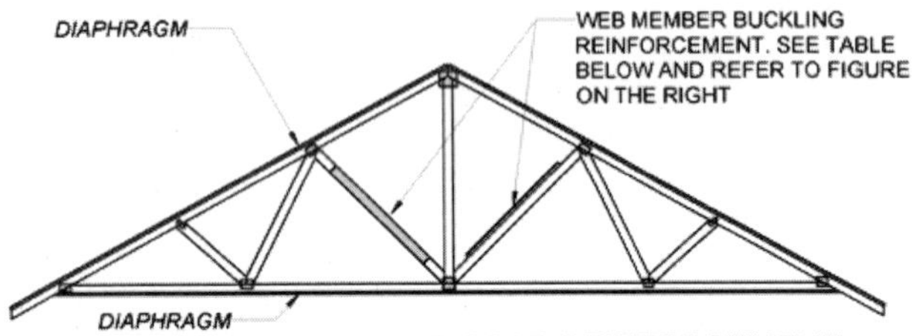

ELEVATION VIEW OF L, T OR SCAB REINFORCEMENT

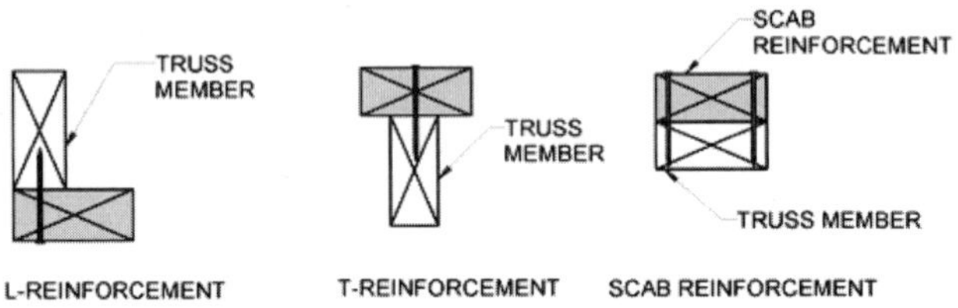

TYPES OF WEB MEMBER BUCKLING REINFORCEMENT

a.) Truss top chord and bottom chord members shall be restrained and braced.

b.) Bracing to resist forces applied perpendicular to the truss, such as wind bracing at gable ends, shall be specified by the building designer.

c.) Use the table below unless project specific web member reinforcement is provided on the truss design drawing or supplemental truss buckling reinforcement details are provided by the truss designer.

NUMBER OF ROWS OF PITMR SPECIFIED ON WEB MEMBER	SIZE OF TRUSS WEB	TYPE AND SIZE OF WEB REINFORCEMENT[1] FOR T, L OR SCAB[2]	GRADE OF WEB REINFORCEMENT	MINIMUM LENGTH OF WEB REINFORCEMENT	MINIMUM CONNECTION OF WEB REINFORCEMENT TO WEB
ONE	2x4	2x4	Same species and grade or better than web member	90% of web or extend to within 6" of end of web member, whichever is greater	(0.131" x 3") nails at 6" on-center[2]
	2x6	2x6			
	2x8	2x8			

[1]Maximum allowable web length is 14'

[2]Attach Scab reinforcement to web with two rows of minimum 0.131" x 3" nails at 6" on-center

For SI:1 inch - 25.4 mm, 1 foot = 304.8 mm.

FIGURE 2303.4.1.2(2)
ALTERNATIVE INSTALLATION USING BUCKLING REINFORCEMENT FOR TRUSS WEB MEMBERS IN LIEU OF ONE ROW OF PITMR

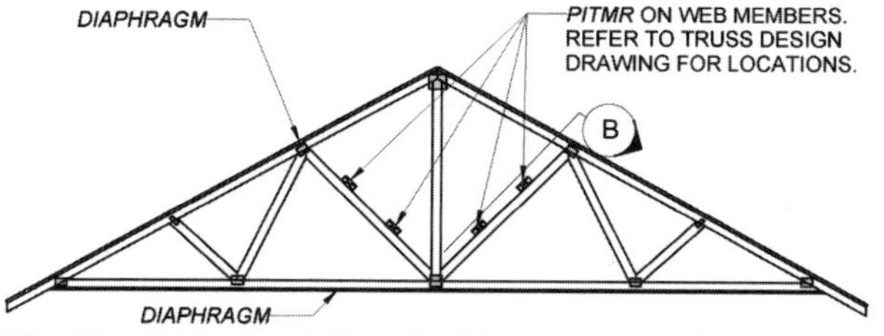

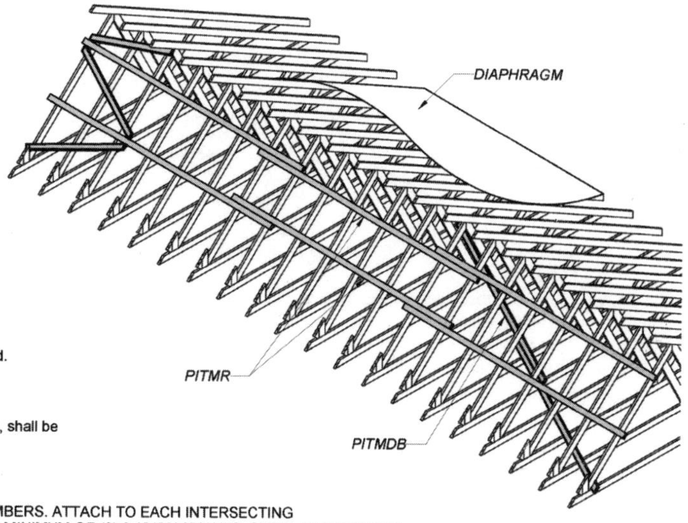

a.) Use minimum 2x4 stress - graded lumber for *PITMR* and *PITMDB* unless otherwise specified.

b.) Truss top chord and bottom chord members shall be restrained and braced.

c.) Bracing to resist forces applied perpendicular to the truss, such as wind bracing at gable ends, shall be specified by the building designer.

ELEVATION VIEW OF TRUSS WITH DOUBLE ROW, *PITMR*

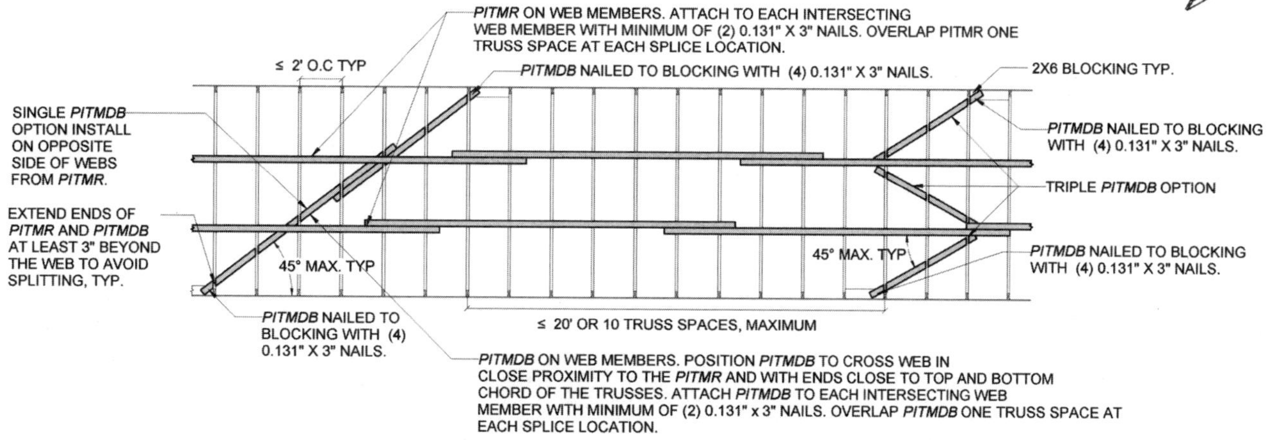

(B) SECTION (EXAMPLE OF DOUBLE ROW OF *PITMR* WITH *PITMDB* ON WEB MEMBERS)

For SI:1 inch = 25.4 mm, 1 foot = 304.8 mm.

FIGURE 2303.4.1.2(3)
PITMR AND PITMDB FOR TRUSS WEB MEMBERS REQUIRING TWO ROWS OF PITMR

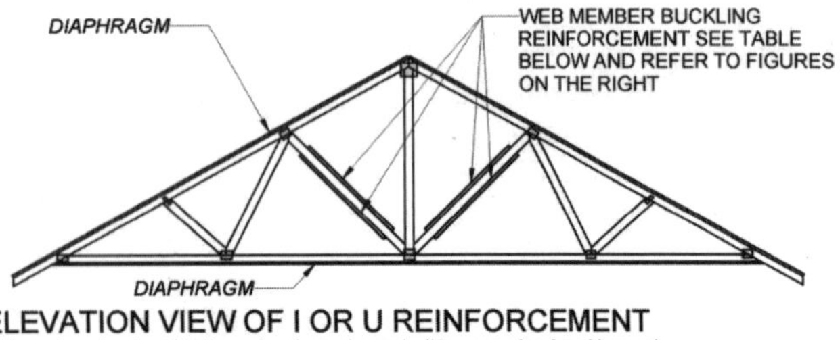

a.) Truss top chord and bottom chord members shall be restrained and braced.

b.) Bracing to resist forces applied perpendicular to the truss, such as wind bracing at gable ends, shall be specified by the building designer.

c.) Use the table below unless project specific web member reinforcement is provided on the truss design drawing or supplemental truss buckling reinforcement details are provided by the truss designer.

NUMBER OF ROWS OF PITMR SPECIFIED ON WEB MEMBER	SIZE OF TRUSS WEB	TYPE AND SIZE OF WEB FOR I OR U REINFORCEMENT[1]	GRADE OF WEB REINFORCEMENT	MINIMUM LENGTH OF WEB REINFORCEMENT	MINIMUM CONNECTION OF WEB REINFORCEMENT TO WEB
TWO	2x4	(2) -2x4	Same species and grade or better than web member	90% of web or extend to within 6" of end of web member, whichever is greater	(0.131" x 3") nails at 6" on-center
	2x6	(2) -2x6			
	2x8	(2) -2x8			

[1]Maximum allowable web length is 14'

For SI: 1 inch = 25.4 mm, 1 foot = 304.8 mm.

FIGURE 2303.4.1.2(4)
ALTERNATIVE INSTALLATION USING BUCKLING REINFORCEMENT FOR TRUSS WEB MEMBERS IN LIEU OF TWO ROWS OF PITMR

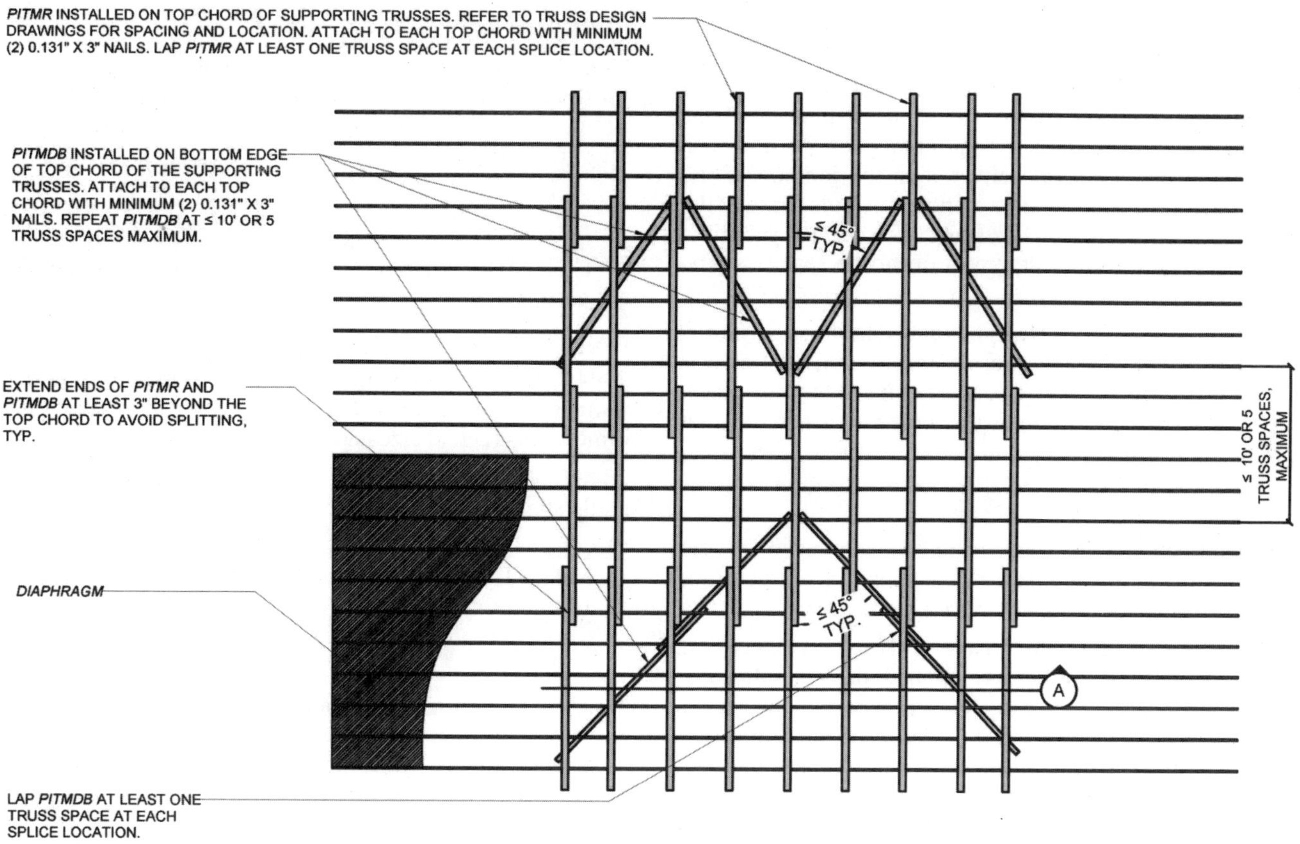

PLAN VIEW

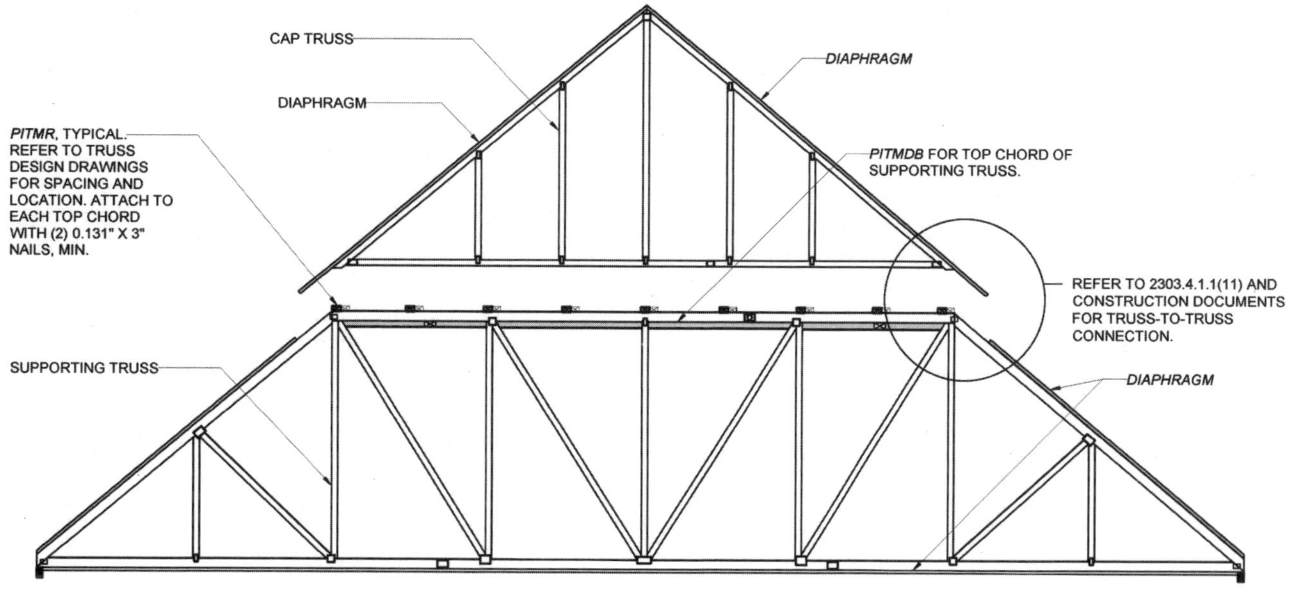

a.) Use minimum 2x4 stress - graded lumber for *PITMR* and *PITMDB* unless otherwise specified.
b.) Web *PITMR* and *PITMDB* not shown for clarity.
c.) Truss top chord and bottom chord members shall be restrained and braced.
d.) Bracing to resist forces applied perpendicular to the truss, such as wind bracing at gable ends, shall be specified by the building designer.

SECTION AT A

For SI:1 inch = 25.4 mm, 1 foot = 304.8 mm.

FIGURE 2303.4.1.2(5)
PITMR AND PITMDB FOR FLAT PORTION OF TOP CHORD IN A PIGGYBACK ASSEMBLY

2303.4.2 Truss placement diagram. The truss manufacturer shall provide a truss placement diagram that identifies the proposed location for each individually designated truss and references the corresponding truss design drawing. The truss placement diagram shall be provided as part of the truss submittal package, and with the shipment of trusses delivered to the job site. Truss placement diagrams that serve only as a guide for installation and do not deviate from the *permit* submittal drawings shall not be required to bear the seal or signature of the truss designer.

2303.4.3 Truss submittal package. The truss submittal package provided by the truss manufacturer shall consist of each individual truss design drawing, the truss placement diagram, the permanent *individual truss member* restraint/bracing method and details and any other structural details germane to the trusses; and, as applicable, the cover/truss index sheet.

2303.4.4 Anchorage. The design for the transfer of *loads* and anchorage of each truss to the supporting structure is the responsibility of the *registered design professional*.

2303.4.5 Alterations to trusses. Truss members and components shall not be cut, notched, drilled, spliced or otherwise altered in any way without written concurrence and approval of a *registered design professional*. Alterations resulting in the addition of *loads* to any member (for example, HVAC equipment, piping, additional roofing or insulation) shall not be permitted without verification that the truss is capable of supporting such additional loading.

2303.4.6 TPI 1 specifications. In addition to Sections 2303.4.1 through 2303.4.5, the design, manufacture and quality assurance of metal-plate-connected wood trusses shall be in accordance with TPI 1. Job-site inspections shall be in compliance with Section 110.4, as applicable.

2303.4.7 Truss quality assurance. Trusses not part of a manufacturing process in accordance with either Section 2303.4.6 or a referenced standard, which provides requirements for quality control done under the supervision of a third-party quality control agency, shall be manufactured in compliance with Sections 1704.2.5 and 1705.5, as applicable.

2303.5 Test standard for joist hangers. Joist hangers shall be in accordance with ASTM D7147.

2303.6 Nails and staples. Nails and staples shall conform to requirements of ASTM F1667, including Supplement 1. Nails used for framing and sheathing connections shall have minimum average bending yield strengths as follows: 80 kips per square inch (ksi) (551 MPa) for shank diameters larger than 0.177 inch (4.50 mm) but not larger than 0.254 inch (6.45 mm), 90 ksi (620 MPa) for shank diameters larger than 0.142 inch (3.61 mm) but not larger than 0.177 inch (4.50 mm) and 100 ksi (689 MPa) for shank diameters of not less than 0.099 inch (2.51 mm) but not larger than 0.142 inch (3.61 mm). Staples used for framing and sheathing connections shall have minimum average bending moments as follows: 3.6 in.-lbs (0.41 N-m) for No. 16 gage staples, 4.0 in.-lbs (0.45 N-m) for No. 15 gage staples, and 4.3 in.-lbs (0.49 N-m) for No. 14 gage staples.

2303.7 Shrinkage. Consideration shall be given in design for the effects of wood cross-grain dimensional changes that occur as a result of changes in the wood moisture content after installation.

SECTION 2304
GENERAL CONSTRUCTION REQUIREMENTS

2304.1 General. The provisions of this section apply to design methods specified in Section 2302.1.

2304.2 Size of structural members. Computations to determine the required sizes of members shall be based on the net dimensions (actual sizes) and not nominal sizes.

2304.3 Wall framing. The framing of exterior and interior walls shall be in accordance with the provisions specified in Section 2308 unless a specific design is furnished.

2304.3.1 Bottom plates. Studs shall have full bearing on a 2-inch-thick (actual $1^1/_2$-inch, 38 mm) or larger plate or sill having a width not less than equal to the width of the studs.

2304.3.2 Framing over openings. Headers, double joists, trusses or other *approved* assemblies that are of adequate size to transfer *loads* to the vertical members shall be provided over window and door openings in *load-bearing walls* and partitions.

2304.3.3 Shrinkage. Wood walls and bearing partitions shall not support more than two floors and a roof unless an analysis satisfactory to the *building official* shows that shrinkage of the wood framing will not have adverse effects on the structure or any plumbing, electrical or mechanical systems or other equipment installed therein due to excessive shrinkage or differential movements caused by shrinkage. The analysis shall show that the roof drainage system and the foregoing systems or equipment will not be adversely affected or, as an alternate, such systems shall be designed to accommodate the differential shrinkage or movements.

2304.4 Floor and roof framing. The framing of wood-joisted floors and wood-framed roofs shall be in accordance with the provisions specified in Section 2308 unless a specific design is furnished.

2304.5 Framing around flues and chimneys. Combustible framing shall be not less than 2 inches (51 mm), but shall be not less than the distance specified in Sections 2111 and 2113 and the *International Mechanical Code*, from flues, chimneys and fireplaces, and 6 inches (152 mm) away from flue openings.

2304.6 Exterior wall sheathing. Wall sheathing on the outside of *exterior walls*, including *gables*, and the connection of the sheathing to framing shall be designed in accordance with the general provisions of this code and shall be capable of resisting wind pressures in accordance with Section 1609.

2304.6.1 Wood structural panel sheathing. Where *wood structural panel* sheathing is used as the exposed finish on the outside of *exterior walls*, it shall have an exterior exposure durability classification. Where *wood structural panel* sheathing is used elsewhere, but not as the exposed finish, it shall be of a type manufactured with exterior glue (Exposure 1 or Exterior). *Wood structural panel* sheathing, connections and framing spacing shall be in accordance with Table 2304.6.1 for the applicable wind speed and exposure category where used in enclosed buildings with a mean roof height not greater than 30 feet (9144 mm) and a topographic factor (K_{zt}) of 1.0.

2304.7 Interior paneling. Softwood *wood structural panels* used for interior paneling shall conform to the provisions of Chapter 8 and shall be installed in accordance with Table 2304.10.2. Panels shall comply with DOC PS 1, DOC PS 2 or ANSI/APA PRP 210. Prefinished *hardboard* paneling shall meet the requirements of ANSI A135.5. Hardwood plywood shall conform to HPVA HP-1.

2304.8 Floor and roof sheathing. Structural floor sheathing and structural roof sheathing shall comply with Sections 2304.8.1 and 2304.8.2, respectively.

2304.8.1 Structural floor sheathing. Structural floor sheathing shall be designed in accordance with the general provisions of this code.

Floor sheathing conforming to the provisions of Table 2304.8(1), 2304.8(2), 2304.8(3) or 2304.8(4) shall be deemed to meet the requirements of this section.

2304.8.2 Structural roof sheathing. Structural roof sheathing shall be designed in accordance with the general provisions of this code and the special provisions in this section.

Roof sheathing conforming to the provisions of Table 2304.8(1), 2304.8(2), 2304.8(3) or 2304.8(5) shall be deemed to meet the requirements of this section. *Wood structural panel* roof sheathing shall be of a type manufactured with exterior glue (Exposure 1 or Exterior).

TABLE 2304.6.1
MAXIMUM ALLOWABLE STRESS DESIGN WIND SPEED, V_{asd} PERMITTED FOR WOOD STRUCTURAL PANEL WALL SHEATHING USED TO RESIST WIND PRESSURES[a, b, c]

MINIMUM NAIL		MINIMUM WOOD STRUCTURAL PANEL SPAN RATING	MINIMUM NOMINAL PANEL THICKNESS (inches)	MAXIMUM WALL STUD SPACING (inches)	PANEL NAIL SPACING		MAXIMUM ALLOWABLE STRESS DESIGN WIND SPEED, V_{asd}[d] (MPH)		
Size	Penetration (inches)				Edges (inches o.c.)	Field (inches o.c.)	Wind exposure category		
							B	C	D
6d common (2.0" × 0.113")	1.5	24/0	$^{3}/_{8}$	16	6	12	110	90	85
		24/16	$^{7}/_{16}$	16	6	12	110	100	90
						6	150	125	110
8d common (2.5" × 0.131")	1.75	24/16	$^{7}/_{16}$	16	6	12	130	110	105
						6	150	125	110
				24	6	12	110	90	85
						6	110	90	85

For SI: 1 inch = 25.4 mm, 1 mile per hour = 0.447 m/s.

a. Panel strength axis shall be parallel or perpendicular to supports. Three-ply plywood sheathing with studs spaced more than 16 inches on center shall be applied with panel strength axis perpendicular to supports.

b. The table is based on wind pressures acting toward and away from building surfaces in accordance with Section 30.7 of ASCE 7. Lateral requirements shall be in accordance with Section 2305 or 2308.

c. Wood structural panels with span ratings of wall-16 or wall-24 shall be permitted as an alternative to panels with a 24/0 span rating. Plywood siding rated 16 on center or 24 on center shall be permitted as an alternative to panels with a 24/16 span rating. Wall-16 and plywood siding 16 on center shall be used with studs spaced not more than 16 inches on center.

d. V_{asd} shall be determined in accordance with Section 1609.3.1.

TABLE 2304.8(1)
ALLOWABLE SPANS FOR LUMBER FLOOR AND ROOF SHEATHING

SPAN (inches)	MINIMUM NET THICKNESS (inches) OF LUMBER PLACED			
	Perpendicular to supports		Diagonally to supports	
	Surfaced dry[a]	Surfaced unseasoned	Surfaced dry[a]	Surfaced unseasoned
Floors				
24	$^{3}/_{4}$	$^{25}/_{32}$	$^{3}/_{4}$	$^{25}/_{32}$
16	$^{5}/_{8}$	$^{11}/_{16}$	$^{5}/_{8}$	$^{11}/_{16}$
Roofs				
24	$^{5}/_{8}$	$^{11}/_{16}$	$^{3}/_{4}$	$^{25}/_{32}$

For SI: 1 inch = 25.4 mm.

a. Maximum 19-percent moisture content.

TABLE 2304.8(2)
SHEATHING LUMBER, MINIMUM GRADE REQUIREMENTS: BOARD GRADE

SOLID FLOOR OR ROOF SHEATHING	SPACED ROOF SHEATHING	GRADING RULES
Utility	Standard	NLGA, WCLIB, WWPA
4 common or utility	3 common or standard	NLGA, WCLIB, WWPA, NSLB or NELMA
No. 3	No. 2	SPIB
Merchantable	Construction common	RIS

TABLE 2304.8(3)
ALLOWABLE SPANS AND LOADS FOR WOOD STRUCTURAL PANEL SHEATHING AND SINGLE-FLOOR GRADES CONTINUOUS OVER TWO OR MORE SPANS WITH STRENGTH AXIS PERPENDICULAR TO SUPPORTS[a]

SHEATHING GRADES		ROOF[b]				FLOOR[c]
Panel span rating roof/floor span	Panel thickness (inches)	Maximum span (inches)		Load[d] (psf)		Maximum span (inches)
		With edge support[e]	Without edge support	Total load	Live load	
16/0	$^3/_8$	16	16	40	30	0
20/0	$^3/_8$	20	20	40	30	0
24/0	$^3/_8$, $^7/_{16}$, $^1/_2$	24	20[f]	40	30	0
24/16	$^7/_{16}$, $^1/_2$	24	24	50	40	16
32/16	$^{15}/_{32}$, $^1/_2$, $^5/_8$	32	28	40	30	16[g]
40/20	$^{19}/_{32}$, $^5/_8$, $^3/_4$, $^7/_8$	40	32	40	30	20[g,h]
48/24	$^{23}/_{32}$, $^3/_4$, $^7/_8$	48	36	45	35	24
54/32	$^7/_8$, 1	54	40	45	35	32
60/32	$^7/_8$, $1^1/_8$	60	48	45	35	32
SINGLE FLOOR GRADES		**ROOF[b]**				**FLOOR[c]**
Panel span rating	**Panel thickness (inches)**	**Maximum span (inches)**		**Load[e] (psf)**		**Maximum span (inches)**
		With edge support[e]	**Without edge support**	**Total load**	**Live load**	
16 o.c.	$^1/_2$, $^{19}/_{32}$, $^5/_8$	24	24	50	40	16[g]
20 o.c.	$^{19}/_{32}$, $^5/_8$, $^3/_4$	32	32	40	30	20[g,h]
24 o.c.	$^{23}/_{32}$, $^3/_4$	48	36	35	25	24
32 o.c.	$^7/_8$, 1	48	40	50	40	32
48 o.c.	$1^3/_{32}$, $1^1/_8$	60	48	50	40	48

For SI: 1 inch = 25.4 mm, 1 pound per square foot = 0.0479 kN/m^2.

a. Applies to panels 24 inches or wider.

b. Uniform load deflection limitations $^1/_{180}$ of span under live load plus dead load, $^1/_{240}$ under live load only.

c. Panel edges shall have approved tongue-and-groove joints or shall be supported with blocking unless $^1/_4$-inch minimum thickness underlayment or $1^1/_2$ inches of approved cellular or lightweight concrete is placed over the subfloor, or finish floor is $^3/_4$-inch wood strip. Allowable uniform load based on deflection of $^1/_{360}$ of span is 100 pounds per square foot except the span rating of 48 inches on center is based on a total load of 65 pounds per square foot.

d. Allowable load at maximum span.

e. Tongue-and-groove edges, panel edge clips (one midway between each support, except two equally spaced between supports 48 inches on center), lumber blocking or other. Only lumber blocking shall satisfy blocked diaphragm requirements.

f. For $^1/_2$-inch panel, maximum span shall be 24 inches.

g. Span is permitted to be 24 inches on center where $^3/_4$-inch wood strip flooring is installed at right angles to joist.

h. Span is permitted to be 24 inches on center for floors where $1^1/_2$ inches of cellular or lightweight concrete is applied over the panels.

TABLE 2304.8(4)
ALLOWABLE SPAN FOR WOOD STRUCTURAL PANEL COMBINATION SUBFLOOR-UNDERLAYMENT (SINGLE FLOOR)[a]
(Panels Continuous Over Two or More Spans and Strength Axis Perpendicular to Supports)

IDENTIFICATION	MAXIMUM SPACING OF JOISTS (inches)				
	16	20	24	32	48
Species group[b]	Thickness (inches)				
1	$^1/_2$	$^5/_8$	$^3/_4$	—	—
2, 3	$^5/_8$	$^3/_4$	$^7/_8$	—	—
4	$^3/_4$	$^7/_8$	1	—	—
Single floor span rating[c]	16 o.c.	20 o.c.	24 o.c.	32 o.c.	48 o.c.

For SI: 1 inch = 25.4 mm, 1 pound per square foot = 0.0479 kN/m^2.

a. Spans limited to value shown because of possible effects of concentrated loads. Allowable uniform loads based on deflection of $^1/_{360}$ of span is 100 pounds per square foot except allowable total uniform load for 1$^1/_8$-inch wood structural panels over joists spaced 48 inches on center is 65 pounds per square foot. Panel edges shall have approved tongue-and-groove joints or shall be supported with blocking, unless $^1/_4$-inch minimum thickness underlayment or 1$^1/_2$ inches of approved cellular or lightweight concrete is placed over the subfloor, or finish floor is $^3/_4$-inch wood strip.

b. Applicable to all grades of sanded exterior-type plywood. See DOC PS 1 for plywood species groups.

c. Applicable to Underlayment grade, C-C (Plugged) plywood, and Single Floor grade wood structural panels.

TABLE 2304.8(5)
ALLOWABLE LOAD (PSF) FOR WOOD STRUCTURAL PANEL ROOF SHEATHING CONTINUOUS OVER TWO OR MORE SPANS AND STRENGTH AXIS PARALLEL TO SUPPORTS
(Plywood structural panels are five-ply, five-layer unless otherwise noted)[a]

PANEL GRADE	THICKNESS (inch)	MAXIMUM SPAN (inches)	LOAD AT MAXIMUM SPAN (psf)	
			Live	Total
Structural I sheathing	$^7/_{16}$	24	20	30
	$^{15}/_{32}$	24	35[b]	45[b]
	$^1/_2$	24	40[b]	50[b]
	$^{19}/_{32}$, $^5/_8$	24	70	80
	$^{23}/_{32}$, $^3/_4$	24	90	100
Sheathing, other grades covered in DOC PS 1 or DOC PS 2	$^7/_{16}$	16	40	50
	$^{15}/_{32}$	24	20	25
	$^1/_2$	24	25	30
	$^{19}/_{32}$	24	40[b]	50[b]
	$^5/_8$	24	45[b]	55[b]
	$^{23}/_{32}$, $^3/_4$	24	60[b]	65[b]

For SI: 1 inch = 25.4 mm, 1 pound per square foot = 0.0479 kN/m^2.

a. Uniform load deflection limitations $^1/_{180}$ of span under live load plus dead load, $^1/_{240}$ under live load only. Edges shall be blocked with lumber or other approved type of edge supports.

b. For composite and four-ply plywood structural panel, load shall be reduced by 15 pounds per square foot.

2304.9 Lumber decking. Lumber decking shall be designed and installed in accordance with the general provisions of this code and Sections 2304.9.1 through 2304.9.5.3. Other lumber decking patterns and connection designs shall be substantiated through engineering analysis.

2304.9.1 General. Each piece of lumber decking shall be square-end trimmed. Where random lengths are furnished, each piece shall be square end trimmed across the face so that not less than 90 percent of the pieces are within 0.5 degrees (0.00873 rad) of square. The ends of the pieces shall be permitted to be beveled up to 2 degrees (0.0349 rad) from the vertical with the exposed face of the piece slightly longer than the opposite face of the piece. Tongue-and-groove decking shall be installed with the tongues up on sloped or pitched roofs with pattern faces down.

2304.9.2 Layup patterns. Lumber decking is permitted to be laid up following one of five standard patterns as defined in Sections 2304.9.2.1 through 2304.9.2.5.

2304.9.2.1 Simple span pattern. All pieces shall be supported on their ends (in other words, by two supports).

2304.9.2.2 Two-span continuous pattern. All pieces shall be supported by three supports, and all end joints shall occur in line on alternating supports. Supporting members shall be designed to accommodate the *load* redistribution caused by this pattern.

2304.9.2.3 Combination simple and two-span continuous pattern. Courses in end spans shall be alternating simple-span pattern and two-span continuous pattern. End joints shall be staggered in adjacent courses and shall bear on supports.

2304.9.2.4 Cantilevered pieces intermixed pattern. The decking shall extend across not fewer than three spans. Pieces in each starter course and every third course shall be simple span pattern. Pieces in other courses shall be cantilevered over the supports with end joints at alternating quarter or third points of the spans. Each piece shall bear on one support or more.

2304.9.2.5 Controlled random pattern. The decking shall extend across not fewer than three spans. End joints of pieces within 6 inches (152 mm) of the end joints of the adjacent pieces in either direction shall be separated by not fewer than two intervening courses. In the end bays, each piece shall bear on one support or more. Where an end joint occurs in an end bay, the next piece in the same course shall continue over the first inner support for not less than 24 inches (610 mm). The details of the controlled random pattern shall be as specified for each decking material in Section 2304.9.3.3, 2304.9.4.3 or 2304.9.5.3.

Decking that cantilevers beyond a support for a horizontal distance greater than 18 inches (457 mm), 24 inches (610 mm) or 36 inches (914 mm) for 2-inch (51 mm), 3-inch (76 mm) and 4-inch (102 mm) nominal thickness decking, respectively, shall comply with the following:

1. The maximum cantilevered length shall be 30 percent of the length of the first adjacent interior span.
2. A structural fascia shall be fastened to each decking piece to maintain a continuous, straight line.
3. End joints shall not be in the decking between the cantilevered end of the decking and the centerline of the first adjacent interior span.

2304.9.3 Mechanically laminated decking. Mechanically laminated decking shall comply with Sections 2304.9.3.1 through 2304.9.3.3.

2304.9.3.1 General. Mechanically laminated decking consists of square-edged dimension lumber laminations set on edge and nailed to the adjacent pieces and to the supports.

2304.9.3.2 Nailing. The length of nails connecting laminations shall be not less than two and one-half times the net thickness of each lamination. Where decking supports are 48 inches (1219 mm) on center or less, side nails shall be installed not more than 30 inches (762 mm) on center alternating between top and bottom edges, and staggered one-third of the spacing in adjacent laminations. Where supports are spaced more than 48 inches (1219 mm) on center, side nails shall be installed not more than 18 inches (457 mm) on center alternating between top and bottom edges and staggered one-third of the spacing in adjacent laminations. For mechanically laminated decking constructed with laminations of 2-inch (51 mm) nominal thickness, nailing in accordance with Table 2304.9.3.2 shall be permitted. Two side nails shall be installed at each end of butt-jointed pieces.

Laminations shall be toenailed to supports with 20d or larger common nails. Where the supports are 48 inches (1219 mm) on center or less, alternate laminations shall be toenailed to alternate supports; where supports are spaced more than 48 inches (1219 mm) on center, alternate laminations shall be toenailed to every support. For mechanically laminated decking constructed with laminations of 2-inch (51 mm) nominal thickness, toenailing in accordance with Table 2304.9.3.2 shall be permitted.

TABLE 2304.9.3.2
FASTENING SCHEDULE FOR MECHANICALLY LAMINATED DECKING USING LAMINATIONS OF 2-INCH NOMINAL THICKNESS

MINIMUM NAIL SIZE (Length x Diameter) (inches)	MAXIMUM SPACING BETWEEN FACE NAILS [a, b] (inches)		NUMBER OF TOENAILS INTO SUPPORTS[c]
	Decking Supports ≤ 48 inches o.c.	Decking Supports > 48 inches o.c.	
4 × 0.192	30	18	1
4 × 0.162	24	14	2
4 × 0.148	22	13	2
$3^1/_2$ × 0.162	20	12	2
$3^1/_2$ × 0.148	19	11	2
$3^1/_2$ × 0.135	17	10	2
3 × 0.148	11	7	2
3 × 0.128	9	5	2
$2^3/_4$ × 0.148	10	6	2
$2^3/_4$ × 0.131	9	6	3
$2^3/_4$ × 0.120	8	5	3

For SI: 1 inch = 25.4 mm

a. Nails shall be driven perpendicular to the lamination face, alternating between top and bottom edges.

b. Where nails penetrate through two laminations and into the third, they shall be staggered one-third of the spacing in adjacent laminations. Otherwise, nails shall be staggered one-half of the spacing in adjacent laminations.

c. Where supports are 48 inches on center or less, alternate laminations shall be toenailed to alternate supports; where supports are spaced more than 48 inches on center, alternate laminations shall be toenailed to every support.

2304.9.3.3 Controlled random pattern. There shall be a minimum distance of 24 inches (610 mm) between end joints in adjacent courses. The pieces in the first and second courses shall bear on not fewer than two supports with end joints in these two courses occurring on alternate supports. Not more than seven intervening courses shall be permitted before this pattern is repeated.

2304.9.4 Two-inch sawn tongue-and-groove decking. Two-inch (51 mm) sawn tongue-and-groove decking shall comply with Sections 2304.9.4.1 through 2304.9.4.3.

2304.9.4.1 General. Two-inch (51 mm) decking shall have a maximum moisture content of 15 percent. Decking shall be machined with a single tongue-and-groove pattern. Each decking piece shall be nailed to each support.

2304.9.4.2 Nailing. Each piece of decking shall be toenailed at each support with one 16d common nail through the tongue and face-nailed with one 16d common nail.

2304.9.4.3 Controlled random pattern. There shall be a minimum distance of 24 inches (610 mm) between end joints in adjacent courses. The pieces in the first and second courses shall bear on not fewer than two supports with end joints in these two courses occurring on alternate supports. Not more than seven intervening courses shall be permitted before this pattern is repeated.

2304.9.5 Three- and four-inch sawn tongue-and-groove decking. Three- and four-inch (76 mm and 102 mm) sawn tongue-and-groove decking shall comply with Sections 2304.9.5.1 through 2304.9.5.3.

2304.9.5.1 General. Three-inch (76 mm) and four-inch (102 mm) decking shall have a maximum moisture content of 19 percent. Decking shall be machined with a double tongue-and-groove pattern. Decking pieces shall be interconnected and nailed to the supports.

2304.9.5.2 Nailing. Each piece shall be toenailed at each support with one 40d common nail and face-nailed with one 60d common nail. Courses shall be spiked to each other with 8-inch (203 mm) spikes at maximum intervals of 30 inches (762 mm) through predrilled edge holes penetrating to a depth of approximately 4 inches (102 mm). One spike shall be installed at a distance not exceeding 10 inches (254 mm) from the end of each piece.

2304.9.5.3 Controlled random pattern. There shall be a minimum distance of 48 inches (1219 mm) between end joints in adjacent courses. Pieces not bearing on a support are permitted to be located in interior bays provided that the adjacent pieces in the same course continue over the support for not less than 24 inches (610 mm). This condition shall not occur more than once in every six courses in each interior bay.

2304.10 Connectors and fasteners. Connectors and fasteners shall comply with the applicable provisions of Sections 2304.10.1 through 2304.10.8 .

2304.10.1 Connection fire-resistance rating. *Fire-resistance* ratings for connections in Type IV-A, IV-B, or IV-C construction shall be determined by one of the following:

1. Testing in accordance with Section 703.2 where the connection is part of the *fire resistance* test.
2. Engineering analysis that demonstrates that the temperature rise at any portion of the connection is limited to an average temperature rise of 250°F (139°C), and a maximum temperature rise of 325°F (181°C), for a time corresponding to the required *fire-resistance* rating of the structural element being connected. For the purposes of this analysis, the connection includes connectors, fasteners, and portions of wood members included in the structural design of the connection.

2304.10.2 Fastener requirements. Connections for wood members shall be designed in accordance with the appropriate methodology in Section 2302.1. The number and size of fasteners connecting wood members shall be not less than that set forth in Table 2304.10.2.

2304.10.3 Sheathing fasteners. Sheathing nails or other *approved* sheathing connectors shall be driven so that their head or crown is flush with the surface of the sheathing.

2304.10.4 Joist hangers and framing anchors. Connections depending on joist hangers or framing anchors, ties and other mechanical fastenings not otherwise covered are permitted where *approved*. The vertical load-bearing capacity, torsional moment capacity and deflection characteristics of joist hangers shall be determined in accordance with ASTM D7147.

2304.10.5 Other fasteners. Clips, staples, glues and other *approved* methods of fastening are permitted where *approved*.

2304.10.6 Fasteners and connectors in contact with preservative-treated and fire-retardant-treated wood. Fasteners, including nuts and washers, and connectors in contact with *preservative-treated* and *fire-retardant-treated wood* shall be in accordance with Sections 2304.10.6.1 through 2304.10.6.4. The coating weights for zinc-coated fasteners shall be in accordance with ASTM A153. Stainless steel driven fasteners shall be in accordance with the material requirements of ASTM F1667.

TABLE 2304.10.2
FASTENING SCHEDULE

DESCRIPTION OF BUILDING ELEMENTS	NUMBER AND TYPE OF FASTENER[g]	SPACING AND LOCATION
Roof		
1. Blocking between ceiling joists, rafters or trusses to top plate or other framing below	4-8d box (2$^{1}/_{2}$″ x 0.113″); or 3-8d common (2$^{1}/_{2}$″ × 0.131″); or 3-10d box (3″ × 0.128″); or 3-3″ × 0.131″ nails; or 3-3″14 gage staples, $^{7}/_{16}$″ crown	Each end, toenail
Blocking between rafters or truss not at the wall top plate, to rafter or truss	2-8d common (2$^{1}/_{2}$″ × 0.131″) 2-3″ × 0.131″ nails 2-3″ 14 gage staples	Each end, toenail
	2-16 d common (3$^{1}/_{2}$″ × 0.162″) 3-3″ × 0.131″ nails 3-3″ 14 gage staples	End nail
Flat blocking to truss and web filler	16d common (3$^{1}/_{2}$″ × 0.162″) @ 6″ o.c. 3″ × 0.131″ nails @ 6″ o.c. 3″ × 14 gage staples @ 6″ o.c	Face nail
2. Ceiling joists to top plate	4-8d box (2$^{1}/_{2}$″ x 0.113″); or 3-8d common (2$^{1}/_{2}$″ × 0.131″); or 3-10d box (3″ × 0.128″); or 3-3″ × 0.131″ nails; or 3-3″ 14 gage staples, $^{7}/_{16}$″ crown	Each joist, toenail
3. Ceiling joist not attached to parallel rafter, laps over partitions (no thrust) (see Section 2308.7.3.1, Table 2308.7.3.1)	3-16d common (3$^{1}/_{2}$″ × 0.162″); or 4-10d box (3″ × 0.128″); or 4-3″ × 0.131″ nails; or 4-3″ 14 gage staples, $^{7}/_{16}$″ crown	Face nail
4. Ceiling joist attached to parallel rafter (heel joint) (see Section 2308.7.3.1, Table 2308.7.3.1)	Per Table 2308.7.3.1	Face nail
5. Collar tie to rafter	3-10d common (3″ × 0.148″); or 4-10d box (3″ × 0.128″); or 4-3″ × 0.131″ nails; or 4-3″ 14 gage staples, $^{7}/_{16}$″ crown	Face nail
6. Rafter or roof truss to top plate (See Section 2308.7.5, Table 2308.7.5)	3-10 common (3″ × 0.148″); or 3-16d box (3$^{1}/_{2}$″ × 0.135″); or 4-10d box (3″ × 0.128″); or 4-3″ × 0.131 nails; or 4-3″ 14 gage staples, $^{7}/_{16}$″ crown	2 toenails on one side and 1 toenail on opposite side of rafter or truss[c]
7. Roof rafters to ridge valley or hip rafters; or roof rafter to 2-inch ridge beam	2-16d common (3$^{1}/_{2}$″ × 0.162″); or 3-16d box (3$^{1}/_{2}$″ × 0.135″); or 3-10d box (3″ × 0.128″); or 3-3″ × 0.131″ nails; or 3-3″ 14 gage staples, $^{7}/_{16}$″ crown	End nail
	3-10d common (3$^{1}/_{2}$″ × 0.148″); or 4-16d box (3$^{1}/_{2}$″ × 0.135″); or 4-10d box (3″ × 0.128″); or 4-3″ × 0.131″ nails; or 4-3″ 14 gage staples, $^{7}/_{16}$″ crown	Toenail
Wall		
8. Stud to stud (not at braced wall panels)	16d common (3$^{1}/_{2}$″ × 0.162″);	24″ o.c. face nail
	10d box (3″ × 0.128″); or 3″ × 0.131″ nails; or 3-3″ 14 gage staples, $^{7}/_{16}$″ crown	16″ o.c. face nail

(continued)

TABLE 2304.10.2—continued
FASTENING SCHEDULE

DESCRIPTION OF BUILDING ELEMENTS	NUMBER AND TYPE OF FASTENER[g]	SPACING AND LOCATION
Wall		
9. Stud to stud and abutting studs at intersecting wall corners (at braced wall panels)	16d common ($3^1/_2$″ × 0.162″)	16″ o.c. face nail
	16d box ($3^1/_2$″ × 0.135″); or 3″ × 0.131″ nails; or 3-3″ 14 gage staples, $^7/_{16}$″ crown	12″ o.c. face nail
10. Built-up header (2″ to 2″ header)	16d common ($3^1/_2$″ × 0.162″)	16″ o.c. each edge, face nail
	16d box ($3^1/_2$″ × 0.135″)	12″ o.c. each edge, face nail
11. Continuous header to stud	4-8d common ($2^1/_2$″ × 0.131″); or 4-10d box (3″ × 0.128″); or 5-8d box ($2^1/_2$″ x 0.113″)	Toenail
12. Top plate to top plate	16d common ($3^1/_2$″ × 0.162″)	16″ o.c. face nail
	10d box (3″ × 0.128″); or 3″ × 0.131″ nails; or 3″ 14 gage staples, $^7/_{16}$″ crown	12″ o.c. face nail
13. Top plate to top plate, at end joints	8-16d common ($3^1/_2$″ × 0.162″); or 12-16d box ($3^1/_2$″ x 0.135″); or 12-10d box (3″ × 0.128″); or 12-3″ × 0.131″ nails; or 12-3″ 14 gage staples, $^7/_{16}$″ crown	Each side of end joint, face nail (minimum 24" lap splice length each side of end joint)
14. Bottom plate to joist, rim joist, band joist or blocking (not at braced wall panels)	16d common ($3^1/_2$″ × 0.162″)	16″ o.c. face nail
	16d box ($3^1/_2$″ × 0.135″); or 3″ × 0.131″ nails; or 3″ 14 gage staples, $^7/_{16}$″ crown	12″ o.c. face nail
15. Bottom plate to joist, rim joist, band joist or blocking at braced wall panels	2-16d common ($3^1/_2$″ × 0.162″); or 3-16d box ($3^1/_2$″ × 0.135″); or 4-3″ × 0.131″ nails; or 4-3″ 14 gage staples, $^7/_{16}$″ crown	16″ o.c. face nail
16. Stud to top or bottom plate	3-16d box ($3^1/_2$″ x 0.135″); or 4-8d common ($2^1/_2$″ × 0.131″); or 4-10d box (3″ × 0.128″); or 4-3″ × 0.131″ nails; or 4-8d box ($2^1/_2$″ x 0.113″); or 4-3″ 14 gage staples, $^7/_{16}$″ crown	Toenail
	2-16d common ($3^1/_2$″ × 0.162″); or 3-16d box ($3^1/_2$″ x 0.135″); or 3-10d box (3″ × 0.128″); or 3-3″ × 0.131″ nails; or 3-3″ 14 gage staples, $^7/_{16}$″ crown	End nail
17. Top plates, laps at corners and intersections	2-16d common ($3^1/_2$″ × 0.162″); or 3-10d box (3″ × 0.128″); or 3-3″ × 0.131″ nails; or 3-3″ 14 gage staples, $^7/_{16}$″ crown	Face nail
18. 1″ brace to each stud and plate	3-8d box ($2^1/_2$″ x 0.113″); or 2-8d common ($2^1/_2$″ × 0.131″); or 2-10d box (3″ × 0.128″); or 2-3″ × 0.131″ nails; or 2-3″ 14 gage staples, $^7/_{16}$″ crown	Face nail
19. 1″ × 6″ sheathing to each bearing	3-8d box ($2^1/_2$″ x 0.113″); or 2-8d common ($2^1/_2$″ × 0.131″); or 2-10d box (3″ × 0.128″); or 2-$1^3/_4$″ 16 gage staples, 1″ crown	Face nail

(continued)

TABLE 2304.10.2—continued
FASTENING SCHEDULE

DESCRIPTION OF BUILDING ELEMENTS	NUMBER AND TYPE OF FASTENER[g]	SPACING AND LOCATION
Wall		
20. 1″ × 8″ and wider sheathing to each bearing	3-8d common (2$^1/_2$″ × 0.131″); or 3-8d box (2$^1/_2$″ x 0.113″); or 3-10d box (3″ × 0.128″); or 3-1$^3/_4$″ 16 gage staples, 1″ crown	Face nail
	Wider than 1″ × 8″ 3-8d common (2$^1/_2$″ x 0.131″); or 4-8d box (2$^1/_2$″ × 0.113″); or 3-10d box (3″ × 0.128″); or 4-1$^3/_4$″ 16 gage staples, 1″ crown	
Floor		
21. Joist to sill, top plate, or girder	4-8d box (2$^1/_2$″ × 0.113″); or 3-8d common (2$^1/_2$″ × 0.131″); or floor 3-10d box (3″ × 0.128″); or 3-3″ × 0.131″ nails; or 3-3″ 14 gage staples, $^7/_{16}$″ crown	Toenail
22. Rim joist, band joist, or blocking to top plate, sill or other framing below	8d box (2$^1/_2$″ × 0.113″)	4″ o.c., toenail
	8d common (2$^1/_2$″ × 0.131″); or 10d box (3″ × 0.128″); or 3″ × 0.131″ nails; or 3″ 14 gage staples, $^7/_{16}$″ crown	6″ o.c., toenail
23. 1″ × 6″ subfloor or less to each joist	3-8d box (2$^1/_2$″ × 0.113″); or 2-8d common (2$^1/_2$″ × 0.131″); or 3-10d box (3″ × 0.128″); or 2-1$^3/_4$″ 16 gage staples, 1″ crown	Face nail
24. 2 subfloor to joist or girder	3-16d box (3$^1/_2$″ × 0.135″); or 2-16d common (3$^1/_2$″ × 0.162″)	Blind and face nail
25. 2″ planks (plank & beam – floor & roof)	3-16d box (3$^1/_2$″ × 0.135″); or 2-16d common (3$^1/_2$″ × 0.162″)	Each bearing, face nail
26. Built-up girders and beams, 2″ lumber layers	20d common (4″ × 0.192″)	32″ o.c., face nail at top and bottom staggered on opposite sides
	10d box (3″ × 0.128″); or 3″ × 0.131″ nails; or 3″ 14 gage staples, $^7/_{16}$″ crown	24″ o.c. face nail at top and bottom staggered on opposite sides
	And: 2-20d common (4″ × 0.192″); or 3-10d box (3″ × 0.128″); or 3-3″ × 0.131″ nails; or 3-3″ 14 gage staples, $^7/_{16}$″ crown	Ends and at each splice, face nail
27. Ledger strip supporting joists or rafters	3-16d common (3$^1/_2$″ × 0.162″); or 4-16d box (3$^1/_2$″ × 0.135″); or 4-10d box (3″ × 0.128″); or 4-3″ × 0.131″ nails; or 4-3″ 14 gage staples, $^7/_{16}$″ crown	Each joist or rafter, face nail
28. Joist to band joist or rim joist	3-16d common (3$^1/_2$″ × 0.162″); or 4-10d box (3″ × 0.128″); or 4-3″ × 0.131″ nails; or 4-3″ 14 gage staples, $^7/_{16}$″ crown	End nail
29. Bridging or blocking to joist, rafter or truss	2-8d common (2$^1/_2$″ × 0.131″); or 2-10d box (3″ × 0.128″); or 2-3″ × 0.131″ nails; or 2-3″ 14 gage staples, $^7/_{16}$″ crown	Each end, toenail

(continued)

TABLE 2304.10.2—continued
FASTENING SCHEDULE

DESCRIPTION OF BUILDING ELEMENTS	NUMBER AND TYPE OF FASTENER[g]	SPACING AND LOCATION	
Wood structural panels (WSP), subfloor, roof and interior wall sheathing to framing and particleboard wall sheathing to framing[a]			
		Edges (inches)	**Intermediate supports (inches)**
30. 3/8″ – 1/2″	6d common or deformed (2″ × 0.113″); or 2 3/8″ × 0.113″ nail (subfloor and wall)	6	12
	8d common or deformed (2 1/2″ × 0.131″× 0.281″ head) (roof) or RSRS-01 (2 3/8″ × 0.113″) nail (roof)[d]	6[e]	6[e]
	1 3/4″ 16 gage staple, 7/16″ crown (subfloor and wall)	4	8
	2 3/8″ × 0.113″× 0.266″ head nail (roof)	3[f]	3[f]
	1 3/4″ 16 gage staple, 7/16″ crown (roof)	3[f]	3[f]
31. 19/32″ – 3/4″	8d common (2 1/2″ × 0.131″); or deformed (2″ × 0.113″) (subfloor and wall)	6	12
	8d common or deformed (2 1/2″ × 0.131″ × 0.281″ head) (roof) or RSRS-01 (2 3/8″ × 0.113″) nail (roof)[d]	6[e]	6[e]
	2 3/8″ × 0.113″× 0.266″ head nail; or 2″ 16 gage staple, 7/16″ crown	4	8
32. 7/8″ – 1 1/4″	10d common (3″ × 0.148″); or deformed (2 1/2″ × 0.131″ × 0.281″ head)	6	12
Other exterior wall sheathing			
33. 1/2″ fiberboard sheathing[b]	1 1/2″ × 0.120″, galvanized roofing nail (7/16″ head diameter); or 1 1/4″ 16 gage staple with 7/16″ or 1″ crown	3	6
34. 25/32″ fiberboard sheathing[b]	1 3/4″ × 0.120″ galvanized roofing nail (7/16″ diameter head); or 1 1/2″ 16 gage staple with 7/16″ or 1″ crown	3	6
Wood structural panels, combination subfloor underlayment to framing			
35. 3/4″ and less	8d common (2 1/2″ × 0.131″); or deformed (2″ × 0.113″); or deformed (2″ × 0.120″)	6	12
36. 7/8″ – 1″	8d common (2 1/2″ × 0.131″); or deformed (2 1/2″ × 0.131″); or deformed (2 1/2″ × 0.120″)	6	12
37. 1 1/8″ – 1 1/4″	10d common (3″ × 0.148″); or deformed (2 1/2″ × 0.131″); or deformed (2 1/2″ × 0.120″)	6	12
Panel siding to framing			
38. 1/2″ or less	6d corrosion-resistant siding (1 7/8″ × 0.106″); or 6d corrosion-resistant casing (2″ × 0.099″)	6	12
39. 5/8″	8d corrosion-resistant siding (2 3/8″ × 0.128″); or 8d corrosion-resistant casing (2 1/2″ × 0.113″)	6	12

(continued)

TABLE 2304.10.2—continued
FASTENING SCHEDULE

DESCRIPTION OF BUILDING ELEMENTS	NUMBER AND TYPE OF FASTENER[g]	SPACING AND LOCATION	
Wood structural panels (WSP), subfloor, roof and interior wall sheathing to framing and particleboard wall sheathing to framing[a]			
		Edges (inches)	Intermediate supports (inches)
Interior paneling			
40. $^1/_4''$	4d casing ($1^1/_2'' \times 0.080''$); or 4d finish ($1^1/_2'' \times 0.072''$)	6	12
41. $^3/_8''$	6d casing ($2'' \times 0.099''$); or 6d finish ($2'' \times 0.092''$) (Panel supports at 24 inches)	6	12

For SI: 1 inch = 25.4 mm.

a. Nails spaced at 6 inches at intermediate supports where spans are 48 inches or more. For nailing of wood structural panel and particleboard diaphragms and shear walls, refer to Section 2305. Nails for wall sheathing are permitted to be common, box or casing.

b. Spacing shall be 6 inches on center on the edges and 12 inches on center at intermediate supports for nonstructural applications. Panel supports at 16 inches (20 inches if strength axis in the long direction of the panel, unless otherwise marked).

c. Where a rafter is fastened to an adjacent parallel ceiling joist in accordance with this schedule and the ceiling joist is fastened to the top plate in accordance with this schedule, the number of toenails in the rafter shall be permitted to be reduced by one nail.

d. RSRS-01 is a Roof Sheathing Ring Shank nail meeting the specifications in ASTM F1667.

e. Tabulated fastener requirements apply where the ultimate design wind speed is less than 140 mph. For wood structural panel roof sheathing attached to gable-end roof framing and to intermediate supports within 48 inches of roof edges and ridges, nails shall be spaced at 4 inches on center where the ultimate design wind speed is greater than 130 mph in Exposure B or greater than 110 mph in Exposure C. Spacing exceeding 6 inches on center at intermediate supports shall be permitted where the fastening is designed per the AWC NDS.

f. Fastening is only permitted where the ultimate design wind speed is less than or equal to 110 mph.

g. Nails and staples are carbon steel meeting the specifications of ASTM F1667. Connections using nails and staples of other materials, such as stainless steel, shall be designed by acceptable engineering practice or approved under Section 104.11.

2304.10.6.1 Fasteners and connectors for preservative-treated wood. Fasteners, including nuts and washers, in contact with *preservative-treated wood* shall be of hot-dipped zinc-coated galvanized steel, stainless steel, silicon bronze or copper. Staples shall be of stainless steel. Fasteners other than nails, staples, timber rivets, wood screws and lag screws shall be permitted to be of mechanically deposited zinc-coated steel with coating weights in accordance with ASTM B695, Class 55 minimum. Connectors that are used in exterior applications and in contact with *preservative-treated wood* shall have coating types and weights in accordance with the treated wood or connector manufacturer's recommendations. In the absence of manufacturer's recommendations, not less than ASTM A653, Type G185 zinc-coated galvanized steel, or equivalent, shall be used.

Exception: Plain carbon steel fasteners, including nuts and washers, in SBX/DOT and zinc borate *preservative-treated wood* in an interior, dry environment shall be permitted.

2304.10.6.2 Fastenings for wood foundations. Fastenings, including nuts and washers, for wood foundations shall be as required in AWC PWF.

2304.10.6.3 Fasteners for fire-retardant-treated wood used in exterior applications or wet or damp locations. Fasteners, including nuts and washers, for *fire-retardant-treated wood* used in exterior applications or wet or damp locations shall be of hot-dipped zinc-coated galvanized steel, stainless steel, silicon bronze or copper. Staples shall be of stainless steel. Fasteners other than nails, staples, timber rivets, wood screws and lag screws shall be permitted to be of mechanically deposited zinc-coated steel with coating weights in accordance with ASTM B695, Class 55 minimum.

2304.10.6.4 Fasteners for fire-retardant-treated wood used in interior applications. Fasteners, including nuts and washers, for *fire-retardant-treated wood* used in interior locations shall be in accordance with the manufacturer's recommendations. In the absence of manufacturer's recommendations, Section 2304.10.6.3 shall apply.

2304.10.7 Load path. Where wall framing members are not continuous from the foundation sill to the roof, the members shall be secured to ensure a continuous *load* path. Where required, sheet metal clamps, ties or clips shall be formed of galvanized steel or other *approved* corrosion-resistant material not less than 0.0329-inch (0.836 mm) base metal thickness.

2304.10.8 Framing requirements. Wood columns and posts shall be framed to provide full end bearing. Alternatively, column-and-post end connections shall be designed to resist the full compressive *loads*, neglecting end-bearing capacity. Column-and-post end connections shall be fastened to resist lateral and net induced uplift forces.

2304.11 Heavy timber construction. Where a structure, portion thereof or individual structural elements are required by provisions of this code to be of heavy timber, the *building elements* therein shall comply with the applicable provisions of Sections 2304.11.1 through 2304.11.4. Minimum dimensions of heavy timber shall comply with the applicable requirements in Table 2304.11 based on roofs or floors supported and the configuration of each structural element, or in Sections 2304.11.2 through 2304.11.4. Lumber decking shall be in accordance with Section 2304.9.

2304.11.1 Details of heavy timber structural members. Heavy timber structural members shall be detailed and constructed in accordance with Sections 2304.11.1 through 2304.11.1.3.

2304.11.1.1 Columns. Minimum dimensions of columns shall be in accordance with Table 2304.11. Columns shall be continuous or superimposed throughout all stories and connected in an *approved* manner. Girders and beams at column connections shall be closely fitted around columns and adjoining ends shall be cross tied to each other, or intertied by caps or ties, to transfer horizontal *loads* across joints. Wood bolsters shall not be placed on tops of columns unless the columns support roof *loads* only. Where traditional heavy timber detailing is used, connections shall be by means of reinforced concrete or metal caps with brackets, by properly designed steel or iron caps, with pintles and base plates, by timber splice plates affixed to the columns by metal connectors housed within the contact faces, or by other *approved* methods.

2304.11.1.2 Floor framing. Minimum dimensions of floor framing shall be in accordance with Table 2304.11. *Approved* wall plate boxes or hangers shall be provided where wood beams, girders or trusses rest on masonry or concrete walls. Where intermediate beams are used to support a floor, they shall rest on top of girders, or shall be supported by an *approved* metal hanger into which the ends of the beams shall be closely fitted. Where traditional heavy timber detailing is used, these connections shall be permitted to be supported by ledgers or blocks securely fastened to the sides of the girders.

2304.11.1.3 Roof framing. Minimum dimensions of roof framing shall be in accordance with Table 2304.11. Every roof girder and not less than every alternate roof beam shall be anchored to its supporting member to resist forces as required in Chapter 16.

2304.11.2 Partitions and walls. Partitions and walls shall comply with Section 2304.11.2.1 or 2304.11.2.2.

2304.11.2.1 Exterior walls. *Exterior walls* shall be permitted to be *cross-laminated timber* not less than 4 inches (102 mm) in thickness meeting the requirements of Section 2303.1.4.

2304.11.2.2 Interior walls and partitions. Interior walls and partitions shall be of solid wood construction formed by not less than two layers of 1-inch (25 mm) matched boards or laminated construction 4 inches (102 mm) thick, or of 1-hour fire-resistance-rated construction.

2304.11.3 Floors. Floors shall be without concealed spaces or with concealed spaces complying with Section 602.4.4.3. Wood floors shall be constructed in accordance with Section 2304.11.3.1 or 2304.11.3.2.

2304.11.3.1 Cross-laminated timber floors. *Cross-laminated timber* shall be not less than 4 inches (102 mm) in actual thickness. *Cross-laminated timber* shall be continuous from support to support and mechanically fastened to one another. *Cross-laminated timber*

TABLE 2304.11
MINIMUM DIMENSIONS OF HEAVY TIMBER STRUCTURAL MEMBERS

		MINIMUM NOMINAL SOLID SAWN SIZE		MINIMUM GLUED-LAMINATED NET SIZE		MINIMUM STRUCTURAL COMPOSITE LUMBER NET SIZE	
SUPPORTING	HEAVY TIMBER STRUCTURAL ELEMENTS	Width, inch	Depth, inch	Width, inch	Depth, inch	Width, inch	Depth, inch
Floor loads only or combined floor and roof loads	Columns; Framed sawn or glued-laminated timber arches that spring from the floor line; Framed timber trusses	8	8	$6^{3}/_{4}$	$8^{1}/_{4}$	7	$7^{1}/_{2}$
	Wood beams and girders	6	10	5	$10^{1}/_{2}$	$5^{1}/_{4}$	$9^{1}/_{2}$
	Columns (roof and ceiling loads); Lower half of: wood-frame or glued-laminated arches that spring from the floor line or from grade	6	8	5	$8^{1}/_{4}$	$5^{1}/_{4}$	$7^{1}/_{2}$
Roof loads only	Upper half of: wood-frame or glued-laminated arches that spring from the floor line or from grade	6	6	5	6	$5^{1}/_{4}$	$5^{1}/_{2}$
	Framed timber trusses and other roof framing;[a] Framed or glued-laminated arches that spring from the top of walls or wall abutments	4[b]	6	3[b]	$6^{7}/_{8}$	$3^{1}/_{2}$[b]	$5^{1}/_{2}$

For SI: 1 inch = 25.4 mm.

a. Spaced members shall be permitted to be composed of two or more pieces not less than 3 inches nominal in thickness where blocked solidly throughout their intervening spaces or where spaces are tightly closed by a continuous wood cover plate of not less than 2 inches nominal in thickness secured to the underside of the members. Splice plates shall be not less than 3 inches nominal in thickness.

b. Where protected by approved automatic sprinklers under the roof deck, framing members shall be not less than 3 inches nominal in width.

shall be permitted to be connected to walls without a shrinkage gap providing swelling or shrinking is considered in the design. Corbelling of masonry walls under the floor shall be permitted to be used.

2304.11.3.2 Sawn or glued-laminated plank floors. Sawn or glued-laminated plank floors shall be one of the following:

1. Sawn or glued-laminated planks, splined or tongue-and-groove, of not less than 3 inches (76 mm) nominal in thickness covered with 1-inch (25 mm) nominal dimension tongue-and-groove flooring, laid crosswise or diagonally, $^{15}/_{32}$-inch (12 mm) *wood structural panel* or $^{1}/_{2}$-inch (12.7 mm) *particleboard.*
2. Planks not less than 4 inches (102 mm) nominal in width set on edge close together and well spiked and covered with 1-inch (25 mm) nominal dimension flooring or $^{15}/_{32}$-inch (12 mm) *wood structural panel* or $^{1}/_{2}$-inch (12.7 mm) *particleboard.*

The lumber shall be laid so that continuous lines of joints will occur only at points of support. Floors shall not extend closer than $^{1}/_{2}$ inch (12.7 mm) to walls. Such $^{1}/_{2}$-inch (12.7 mm) space shall be covered by a molding fastened to the wall and so arranged that it will not obstruct the swelling or shrinkage movements of the floor. Corbelling of masonry walls under the floor shall be permitted to be used in place of molding.

2304.11.4 Roof decks. Roofs shall be without concealed spaces or with concealed spaces complying with Section 602.4.4.3. Roof decks shall be constructed in accordance with Section 2304.11.4.1 or 2304.11.4.2. Other types of decking shall be an alternative that provides equivalent *fire resistance* and structural properties. Where supported by a wall, *roof decks* shall be anchored to walls to resist forces determined in accordance with Chapter 16. Such anchors shall consist of steel bolts, lags, screws or *approved* hardware of sufficient strength to resist prescribed forces.

2304.11.4.1 Cross-laminated timber roofs. *Cross-laminated timber* roofs shall be not less than 3 inches (76 mm) nominal in thickness and shall be continuous from support to support and mechanically fastened to one another.

2304.11.4.2 Sawn, wood structural panel, or glued-laminated plank roofs. Sawn, *wood structural panel,* or glued-laminated plank roofs shall be one of the following:

1. Sawn or glued laminated, splined or tongue-and-groove plank, not less than 2 inches (51 mm) nominal in thickness.
2. $1^{1}/_{8}$-inch-thick (32 mm) *wood structural panel* (exterior glue).
3. Planks not less than 3 inches (76 mm) nominal in width, set on edge close together and laid as required for floors.

2304.12 Protection against decay and termites. Wood shall be protected from decay and termites in accordance with the applicable provisions of Sections 2304.12.1 through 2304.12.4.

2304.12.1 Locations requiring waterborne preservatives or naturally durable wood. Wood used above ground in the locations specified in Sections 2304.12.1.1 through 2304.12.1.5 shall be *naturally durable wood* or *preservative-treated wood* using waterborne preservatives, in accordance with AWPA U1 for above-ground use.

2304.12.1.1 Joists, girders and subfloor. Wood joists or wood structural floors that are closer than 18 inches (457 mm) or wood girders that are closer than 12 inches (305 mm) to the exposed ground in crawl spaces or unexcavated areas located within the perimeter of the building foundation shall be of naturally durable or *preservative-treated wood.*

2304.12.1.2 Wood supported by exterior foundation walls. Wood framing members, including wood sheathing, that are in contact with exterior foundation walls and are less than 8 inches (203 mm) from exposed earth shall be of naturally durable or *preservative-treated wood.*

2304.12.1.3 Exterior walls below grade. Wood framing members and furring strips in direct contact with the interior of exterior masonry or concrete walls below grade shall be of naturally durable or *preservative-treated wood.*

2304.12.1.4 Sleepers and sills. Sleepers and sills on a concrete or masonry slab that is in direct contact with earth shall be of naturally durable or *preservative-treated wood.*

2304.12.1.5 Wood siding. Clearance between wood siding and earth on the exterior of a building shall be not less than 6 inches (152 mm) or less than 2 inches (51 mm) vertical from concrete steps, porch slabs, patio slabs and similar horizontal surfaces exposed to the weather except where siding, sheathing and wall framing are of naturally durable or *preservative-treated wood.*

2304.12.2 Other locations. Wood used in the locations specified in Sections 2304.12.2.1 through 2304.12.2.8 shall be *naturally durable wood* or *preservative-treated* wood in accordance with AWPA U1. *Preservative-treated* wood used in interior locations shall be protected with two coats of urethane, shellac, latex epoxy or varnish unless waterborne preservatives are used. Prior to application of the protective finish, the wood shall be dried in accordance with the manufacturer's recommendations.

2304.12.2.1 Girder ends. The ends of wood girders entering exterior masonry or concrete walls shall be provided with a $^{1}/_{2}$-inch (12.7 mm) airspace on top, sides and end, unless naturally durable or *preservative-treated wood* is used.

2304.12.2.2 Posts or columns. Posts or columns supporting permanent structures and supported by a concrete or masonry slab or footing that is in direct contact with the earth shall be of naturally durable or *preservative-treated wood.*

Exception: Posts or columns that meet all of the following:

1. Are not exposed to the weather, or are protected by a roof, eave, overhang, or other covering if exposed to the weather.
2. Are supported by concrete piers or metal pedestals projected not less than 1 inch (25 mm) above the slab or deck and are separated from the concrete pier by an impervious moisture barrier.
3. Are located not less than 8 inches (203 mm) above exposed earth.

2304.12.2.3 Supporting member for permanent appurtenances. Naturally durable or *preservative-treated wood* shall be utilized for those portions of wood members that form the structural supports of buildings, balconies, porches or similar permanent building appurtenances where such members are exposed to the weather without adequate protection from a roof, eave, overhang or other covering to prevent moisture or water accumulation on the surface or at joints between members.

Exception: Sawn lumber in buildings located in a geographical region where experience has demonstrated that climatic conditions preclude the need to use durable materials where the structure is exposed to the weather.

2304.12.2.4 Supporting members for permeable floors and roofs. Wood structural members that support moisture-permeable floors or roofs that are exposed to the weather, such as concrete or masonry slabs, shall be of naturally durable or *preservative-treated wood* unless separated from such floors or roofs by an impervious moisture barrier. The impervious moisture barrier system protecting the structure supporting floors shall provide positive drainage of water that infiltrates the moisture-permeable floor topping.

2304.12.2.5 Ventilation beneath balcony or elevated walking surfaces. Enclosed framing in exterior balconies and elevated walking surfaces that have *weather-exposed surfaces* shall be provided with openings that provide a net free cross-ventilation area not less than $^{1}/_{150}$ of the area of each separate space.

2304.12.2.6 Wood in contact with the ground or fresh water. Wood used in contact with exposed earth shall be naturally durable for both decay and termite resistance or preservative treated in accordance with AWPA U1 for soil or fresh water use.

Exception: Untreated wood is permitted where such wood is continuously and entirely below the groundwater level or submerged in fresh water.

2304.12.2.6.1 Posts or columns. Posts and columns that are supporting permanent structures and embedded in concrete that is exposed to the weather or in direct contact with the earth shall be of *preservative-treated wood.*

2304.12.2.7 Termite protection. In geographical areas where hazard of termite damage is known to be very heavy, wood floor framing in the locations specified in Section 2304.12.1.1 and exposed framing of exterior decks or balconies shall be of *naturally durable species* (*termite resistant*) or preservative treated in accordance with AWPA U1 for the species, product preservative and end use or provided with *approved* methods of termite protection.

2304.12.2.8 Wood used in retaining walls and cribs. Wood installed in retaining or crib walls shall be preservative treated in accordance with AWPA U1 for soil and fresh water use.

2304.12.3 Attic ventilation. For *attic* ventilation, see Section 1202.2.2.

2304.12.4 Under-floor ventilation (crawl space). For under-floor ventilation (crawl space), see Section 1202.4.

2304.13 Long-term loading. Wood members supporting concrete, masonry or similar materials shall be checked for the effects of long-term loading using the provisions of the ANSI/AWC NDS. The total deflection, including the effects of long-term loading, shall be limited in accordance with Section 1604.3.1 for these supported materials.

Exception: Horizontal wood members supporting masonry or concrete nonstructural floor or roof surfacing not more than 4 inches (102 mm) thick need not be checked for long-term loading.

SECTION 2305 GENERAL DESIGN REQUIREMENTS FOR LATERAL FORCE-RESISTING SYSTEMS

2305.1 General. Structures using wood-frame *shear walls* or wood-frame *diaphragms* to resist wind, seismic or other lateral *loads* shall be designed and constructed in accordance with AWC SDPWS and the applicable provisions of Sections 2305, 2306 and 2307.

2305.1.1 Openings in shear panels. Openings in shear panels that materially affect their strength shall be detailed on the plans and shall have their edges adequately reinforced to transfer all shearing stresses.

2305.2 Diaphragm deflection. The deflection of wood-frame *diaphragms* shall be determined in accordance with AWC SDPWS. The deflection (Δ_{dia}) of a blocked *wood structural panel* diaphragm uniformly fastened throughout with staples is permitted to be calculated in accordance with Equation 23-1. If not uniformly fastened, the constant 0.188 (For SI: 1/1627) in the third term shall be modified by an approved method.

$$\Delta_{dia} = 5vL^3/8EAW + vL/4Gt + 0.188Le_n + \Sigma(x\Delta_c)/2W$$

(Equation 23-1)

For SI: $\Delta_{dia} = 0.052vL^3/EAW + vL/4Gt + Le_n/1627 + \Sigma(x\Delta_c)/2W$

where:

A = Area of chord cross section, in square inches (mm^2).

E = Modulus of elasticity of *diaphragm* chords, in pounds per square inch (N/mm2).

e_n = Staple slip, in inches (mm) [see Table 2305.2(1)].

Gt = Panel rigidity through the thickness, in pounds per inch (N/mm) of panel width or depth [see Table 2305.2(2)].

L = *Diaphragm* length (dimension perpendicular to the direction of the applied *load*), in feet (mm).

v = Induced unit shear in pounds per linear foot (plf) (N/mm).

W = *Diaphragm* width [in the direction of applied force, in feet (mm)].

x = Distance from chord splice to nearest support, in feet (mm).

Δ_c = Diaphragm chord splice slip at the induced unit shear, in inches (mm).

Δ_{dia} = Maximum mid-span *diaphragm* deflection determined by elastic analysis, in inches (mm).

TABLE 2305.2(1)
e_n VALUES (inches) FOR USE IN CALCULATING DIAPHRAGM AND SHEAR WALL DEFLECTION DUE TO FASTENER SLIP (Structural I)[a, c]

LOAD PER FASTENER[b] (pounds)	FASTENER DESIGNATIONS
	14-Ga staple × 2 inches long
60	0.011
80	0.018
100	0.028
120	0.04
140	0.053
160	0.068

For SI: 1 inch = 25.4 mm, 1 foot = 304.8 mm, 1 pound = 4.448 N.

a. Increase e_n values 20 percent for plywood grades other than Structural I.

b. Load per fastener = maximum shear per foot divided by the number of fasteners per foot at interior panel edges.

c. Decrease e_n values 50 percent for seasoned lumber (moisture content < 19 percent).

TABLE 2305.2(2)
VALUES OF *Gt* FOR USE IN CALCULATING DEFLECTION OF WOOD STRUCTURAL PANEL SHEAR WALLS AND DIAPHRAGMS

PANEL TYPE	SPAN RATING	VALUES OF *Gt* (lb/in. panel depth or width)							
		Structural Sheathing				Structural I			
		Plywood			OSB	Plywood			OSB
		3-ply	4-ply	5-ply[a]		3-ply	4-ply	5-ply[a]	
Sheathing	24/0	25,000	32,500	37,500	77,500	32,500	42,500	41,500	77,500
	24/16	27,000	35,000	40,500	83,500	35,000	45,500	44,500	83,500
	32/16	27,000	35,000	40,500	83,500	35,000	45,500	44,500	83,500
	40/20	28,500	37,000	43,000	88,500	37,000	48,000	47,500	88,500
	48/24	31,000	40,500	46,500	96,000	40,500	52,500	51,000	96,000
Single Floor	16 o.c.	27,000	35,000	40,500	83,500	35,000	45,500	44,500	83,500
	20 o.c.	28,000	36,500	42,000	87,000	36,500	47,500	46,000	87,000
	24 o.c.	30,000	39,000	45,000	93,000	39,000	50,500	49,500	93,000
	32 o.c.	36,000	47,000	54,000	110,000	47,000	61,000	59,500	110,000
	48 o.c.	50,500	65,500	76,000	155,000	65,500	85,000	83,500	155,000

		Structural Sheathing			Structural I		
	Thickness (in.)	A-A, A-C	Marine	All Other Grades	A-A, A-C	Marine	All Other Grades
Sanded Plywood	1/4	24,000	31,000	24,000	31,000	31,000	31,000
	11/32	25,500	33,000	25,500	33,000	33,000	33,000
	3/8	26,000	34,000	26,000	34,000	34,000	34,000
	15/32	38,000	49,500	38,000	49,500	49,500	49,500
	1/2	38,500	50,000	38,500	50,000	50,000	50,000
	19/32	49,000	63,500	49,000	63,500	63,500	63,500
	5/8	49,500	64,500	49,500	64,500	64,500	64,500
	23/32	50,500	65,500	50,500	65,500	65,500	65,500
	3/4	51,000	66,500	51,000	66,500	66,500	66,500
	7/8	52,500	68,500	52,500	68,500	68,500	68,500
	1	73,500	95,500	73,500	95,500	95,500	95,500
	1 1/8	75,000	97,500	75,000	97,500	97,500	97,500

For SI: 1 inch = 25.4 mm, 1 pound/inch = 0.1751 N/mm.

a. 5-ply applies to plywood with five or more layers. For 5-ply plywood with three layers, use values for 4-ply panels.

2305.3 Shear wall deflection. The deflection of wood-frame *shear walls* shall be determined in accordance with AWC SDPWS. The deflection (Δ_{sw}) of a blocked *wood structural panel* shear wall uniformly fastened throughout with staples is permitted to be calculated in accordance with Equation 23-2.

$$\Delta_{sw} = 8vh^3/EAb + vh/4Gt + 0.75he_n + d_a h/b \quad \textbf{(Equation 23-2)}$$

For SI: $$\Delta_{sw} = vh^3/3EAb + vh/Gt + \frac{he_n}{407.6} + d_a h/b$$

where:

A = Area of end-post cross section in square inches (mm^2).

b = Shear wall length, in feet (mm).

d_a = Total vertical elongation of wall anchorage system (such as fastener slip, device elongation, rod elongation) in inches (mm), at the induced unit shear in the shear wall (v).

E = Modulus of elasticity of end posts, in pounds per square inch (N/mm^2).

e_n = Staple slip, in inches (mm) [see Table 2305.2(1)].

Gt = Panel rigidity through the thickness, in pounds per inch (N/mm) of panel width or depth [see Table 2305.2(2)].

h = *Shear wall* height, in feet (mm).

v = Induced unit shear, in pounds per linear foot (N/mm).

Δ_{sw} = Maximum *shear wall* deflection determined by elastic analysis, in inches (mm).

SECTION 2306 ALLOWABLE STRESS DESIGN

2306.1 Allowable stress design. The design and construction of wood elements in structures using *allowable stress design* shall be in accordance with the following applicable standards:

2306.1.1 Joists and rafters. The design of rafter spans is permitted to be in accordance with the AWC STJR.

2306.1.2 Plank and beam flooring. The design of plank and beam flooring is permitted to be in accordance with the AWC Wood Construction Data No. 4.

2306.1.3 Treated wood stress adjustments. The allowable unit stresses for *preservative-treated wood* need not be adjusted for treatment, but are subject to other adjustments.

The allowable unit stresses for *fire-retardant-treated wood*, including fastener values, shall be developed from an *approved* method of investigation that considers the effects of anticipated temperature and humidity to which the *fire-retardant-treated wood* will be subjected, the type of treatment and the redrying process. Other adjustments are applicable except that the *impact load* duration shall not apply.

2306.1.4 Lumber decking. The capacity of lumber decking arranged according to the patterns described in Section 2304.9.2 shall be the lesser of the capacities determined for moment and deflection according to the formulas in Table 2306.1.4.

**TABLE 2306.1
STANDARDS FOR DESIGN AND CONSTRUCTION OF WOOD ELEMENTS IN STRUCTURES USING ALLOWABLE STRESS DESIGN**

STANDARDS PROMULGATOR	STANDARD	TITLE
American Wood Council		
	ANSI/AWC NDS	National Design Specification for Wood Construction
	SDPWS	Special Design Provisions for Wind and Seismic
American Society of Agricultural and Biological Engineers		
	ASABE EP 484.3	Diaphragm Design of Metal-clad, Wood-Frame Rectangular Buildings
	ASABE EP 486.3	Shallow Post and Pier Foundation Design
	ASABE EP 559.1	Design Requirements and Bending Properties for Mechanically Laminated Wood Assemblies
APA—The Engineered Wood Association		
	ANSI 117	Standard Specifications for Structural Glued Laminated Timber of Softwood Species
	ANSI A190.1	Structural Glued Laminated Timber
		Panel Design Specification
		Plywood Design Specification Supplement 1—Design & Fabrication of Plywood Curved Panel
		Plywood Design Specification Supplement 2—Design & Fabrication of Glued Plywood-lumber Beams
		Plywood Design Specification Supplement 3—Design & Fabrication of Plywood Stressed-skin Panels
		Plywood Design Specification Supplement 4—Design & Fabrication of Plywood Sandwich Panels
		Plywood Design Specification Supplement 5—Design & Fabrication of All-plywood Beams
	APA T300	Glulam Connection Details
	APA S560	Field Notching and Drilling of Glued Laminated Timber Beams
	APA S475	Glued Laminated Beam Design Tables
	APA X450	Glulam in Residential Construction
	APA X440	Product and Application Guide: Glulam
	APA R540	Builders Tips: Proper Storage and Handling of Glulam Beams
Truss Plate Institute, Inc.		
	TPI 1	National Design Standard for Metal Plate Connected Wood Truss Construction
West Coast Lumber Inspection Bureau		
	AITC 104	Typical Construction Details
	AITC 110	Standard Appearance Grades for Structural Glued Laminated Timber
	AITC 113	Standard for Dimensions of Structural Glued Laminated Timber
	AITC 119	Standard Specifications for Structural Glued Laminated Timber of Hardwood Species
	AITC 200	Inspection Manual

TABLE 2306.1.4
ALLOWABLE LOADS FOR LUMBER DECKING

PATTERN	ALLOWABLE AREA LOAD[a]	
	Moment	Deflection
Simple span	$w_b = \frac{8F'_b d^2}{l^2 6}$	$w_\Delta = \frac{384\Delta E' d^3}{5l^4 \; 12}$
Two-span continuous	$w_b = \frac{8F'_b d^2}{l^2 6}$	$w_\Delta = \frac{185\Delta E' d^3}{l^4 \; 12}$
Combination simple- and two-span continuous	$w_b = \frac{8F'_b d^2}{l^2 6}$	$w_\Delta = \frac{131\Delta E' d^3}{l^4 \; 12}$
Cantilevered pieces intermixed	$w_b = \frac{20F'_b d^2}{3l^2 6}$	$w_\Delta = \frac{105\Delta E' d^3}{l^4 \; 12}$
Controlled random layup		
Mechanically laminated decking	$w_b = \frac{20F'_b d^2}{3l^2 6}$	$w_\Delta = \frac{100\Delta E' d^3}{l^4 \; 12}$
2-inch decking	$w_b = \frac{20F'_b d^2}{3l^2 6}$	$w_\Delta = \frac{100\Delta E' d^3}{l^4 \; 12}$
3-inch and 4-inch decking	$w_b = \frac{8F'_b d^2}{l^2 6}$	$w_\Delta = \frac{116\Delta E' d^3}{l^4 \; 12}$

For SI: 1 inch = 25.4 mm.

a. w_b = Allowable total uniform load limited by moment.
w_Δ = Allowable total uniform load limited by deflection.
d = Actual decking thickness.
l = Span of decking.
F_b' = Allowable bending stress adjusted by applicable factors.
E' = Modulus of elasticity adjusted by applicable factors.

2306.2 Wood-frame diaphragms. Wood-frame *diaphragms* shall be designed and constructed in accordance with AWC SDPWS. Where panels are fastened to framing members with staples, requirements and limitations of AWC SDPWS shall be met and the allowable shear values set forth in Table 2306.2(1) or 2306.2(2) shall be permitted. The allowable shear values in Tables 2306.2(1) and 2306.2(2) are permitted to be increased 40 percent for wind design.

2306.2.1 Gypsum board diaphragm ceilings. *Gypsum board diaphragm* ceilings shall be in accordance with Section 2508.6.

2306.3 Wood-frame shear walls. Wood-frame *shear walls* shall be designed and constructed in accordance with AWC SDPWS. Where panels are fastened to framing members with staples, requirements and limitations of AWC SDPWS shall be met and the allowable shear values set forth in Table 2306.3(1), 2306.3(2) or 2306.3(3) shall be permitted. The allowable shear values in Tables 2306.3(1) and 2306.3(2) are permitted to be increased 40 percent for wind design. Panels complying with ANSI/APA PRP-210 shall be permitted to use design values for Plywood Siding in the AWC SDPWS.

SECTION 2307
LOAD AND RESISTANCE FACTOR DESIGN

2307.1 Load and resistance factor design. The design and construction of wood elements and structures using *load and resistance factor design* shall be in accordance with ANSI/AWC NDS and AWC SDPWS.

SECTION 2308
CONVENTIONAL LIGHT-FRAME CONSTRUCTION

2308.1 General. The requirements of this section are intended for *conventional light-frame construction*. Other construction methods are permitted to be used, provided that a satisfactory design is submitted showing compliance with other provisions of this code. Interior nonload-bearing partitions, ceilings and curtain walls of *conventional light-frame construction* are not subject to the limitations of Section 2308.2. Detached one- and two-family dwellings and townhouses not more than three *stories above grade plane* in height with a separate *means of egress* and their accessory structures shall comply with the *International Residential Code*.

2308.1.1 Portions exceeding limitations of conventional light-frame construction. Where portions of a building of otherwise *conventional light-frame construction* exceed the limits of Section 2308.2, those portions and the supporting load path shall be designed in accordance with accepted engineering practice and the provisions of this code. For the purposes of this section, the term "portions" shall mean parts of buildings containing volume and area such as a room or a series of rooms. The extent of such design need only demonstrate compliance of the nonconventional light-framed elements with other applicable provisions of this code and shall be compatible with the performance of the conventional light-framed system.

2308.1.2 Connections and fasteners. Connectors and fasteners used in conventional construction shall comply with the requirements of Section 2304.10.

2308.2 Limitations. Buildings are permitted to be constructed in accordance with the provisions of *conventional light-frame construction*, subject to the limitations in Sections 2308.2.1 through 2308.2.6.

TABLE 2306.2(1)
ALLOWABLE SHEAR VALUES (POUNDS PER FOOT) FOR WOOD STRUCTURAL PANEL DIAPHRAGMS UTILIZING STAPLES WITH FRAMING OF DOUGLAS FIR-LARCH, OR SOUTHERN PINE[a] FOR WIND OR SEISMIC LOADING[f]

PANEL GRADE	STAPLE LENGTH AND GAGE[d]	MINIMUM FASTENER PENETRATION IN FRAMING (inches)	MINIMUM NOMINAL PANEL THICKNESS (inch)	MINIMUM NOMINAL WIDTH OF FRAMING MEMBERS AT ADJOINING PANEL EDGES AND BOUNDARIES[e] (inches)	BLOCKED DIAPHRAGMS				UNBLOCKED DIAPHRAGMS	
					Fastener spacing (inches) at diaphragm boundaries (all cases) at continuous panel edges parallel to load (Cases 3, 4), and at all panel edges (Cases 5, 6)[b]				Fasteners spaced 6 inches max. at supported edges[b]	
					6	4	$2^1/_2$[c]	2[c]	Case 1 (No unblocked edges or continuous joints parallel to load)	All other configurations (Cases 2, 3, 4, 5 and 6)[g]
					Fastener spacing (inches) at other panel edges (Cases 1, 2, 3 and 4)[b]					
					6	6	4	3		
Structural I grades	$1^1/_2$ 16 gage	1	$^3/_8$	2	175	235	350	400	155	115
				3	200	265	395	450	175	130
			$^{15}/_{32}$	2	175	235	350	400	155	120
				3	200	265	395	450	175	130
Sheathing, single floor and other grades covered in DOC PS 1 and PS 2	$1^1/_2$ 16 gage	1	$^3/_8$	2	160	210	315	360	140	105
				3	180	235	355	400	160	120
			$^7/_{16}$	2	165	225	335	380	150	110
				3	190	250	375	425	165	125
			$^{15}/_{32}$	2	160	210	315	360	140	105
				3	180	235	355	405	160	120
			$^{19}/_{32}$	2	175	235	350	400	155	115
				3	200	265	395	450	175	130

For SI: 1 inch = 25.4 mm, 1 pound per foot = 14.5939 N/m.

a. For framing of other species: (1) Find specific gravity for species of lumber in ANSI/AWC NDS. (2) For staples find shear value from table for Structural I panels (regardless of actual grade) and multiply value by 0.82 for species with specific gravity of 0.42 or greater, or 0.65 for all other species.

b. Space fasteners maximum 12 inches on center along intermediate framing members (6 inches on center where supports are spaced 48 inches on center).

c. Framing at adjoining panel edges shall be 3 inches nominal or wider.

d. Staples shall have a minimum crown width of $^7/_{16}$ inch and shall be installed with their crowns parallel to the long dimension of the framing members.

e. The minimum nominal width of framing members not located at boundaries or adjoining panel edges shall be 2 inches.

f. For shear loads of normal or permanent load duration as defined by the ANSI/AWC NDS, the values in the table shall be multiplied by 0.63 or 0.56, respectively.

g. For Case 1 through 6 descriptions see Figure 2306.2(1).

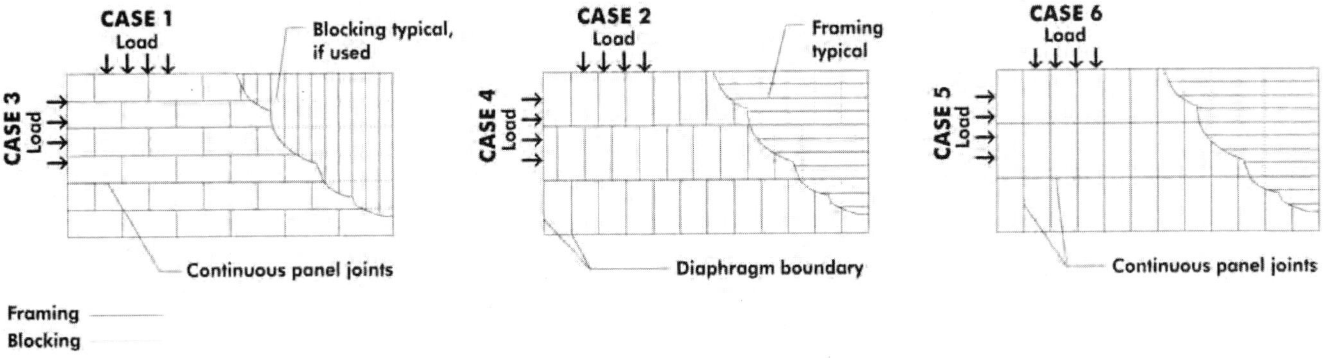

FIGURE 2306.2(1)
CASES 1 THROUGH 6 FOR USE WITH TABLE 2306.2(1)

TABLE 2306.2(2)
ALLOWABLE SHEAR VALUES (POUNDS PER FOOT) FOR WOOD STRUCTURAL PANEL BLOCKED DIAPHRAGMS UTILIZING MULTIPLE ROWS OF STAPLES (HIGH-LOAD DIAPHRAGMS) WITH FRAMING OF DOUGLAS FIR-LARCH OR SOUTHERN PINE[a] FOR WIND OR SEISMIC LOADING[b, g, h]

PANEL GRADE[c]	STAPLE GAGE[f]	MINIMUM FASTENER PENETRATION IN FRAMING (inches)	MINIMUM NOMINAL PANEL THICKNESS (inch)	MINIMUM NOMINAL WIDTH OF FRAMING MEMBER AT ADJOINING PANEL EDGES AND BOUNDARIES[e]	LINES OF FASTENERS	BLOCKED DIAPHRAGMS					
						Cases 1 and 2[d]					
						Fastener Spacing Per Line at Boundaries (inches)[i]					
						4		$2^1/_2$		2	
						Fastener Spacing Per Line at Other Panel Edges (inches)[i]					
						6	4	4	3	3	2
Structural I grades	14 gage staples	2	$^{15}/_{32}$	3	2	600	600	860	960	1,060	1,200
				4	3	860	900	1,160	1,295	1,295	1,400
			$^{19}/_{32}$	3	2	600	600	875	960	1,075	1,200
				4	3	875	900	1,175	1,440	1,475	1,795
Sheathing single floor and other grades covered in DOC PS 1 and PS 2	14 gage staples	2	$^{15}/_{32}$	3	2	540	540	735	865	915	1,080
				4	3	735	810	1,005	1,105	1,105	1,195
			$^{19}/_{32}$	3	2	600	600	865	960	1,065	1,200
				4	3	865	900	1,130	1,430	1,370	1,485
			$^{23}/_{32}$	4	3	865	900	1,130	1,490	1,430	1,545

For SI: 1 inch = 25.4 mm, 1 pound per foot = 14.5939 N/m.

a. For framing of other species: (1) Find specific gravity for species of framing lumber in ANSI/AWC NDS. (2) For staples, find shear value from table for Structural I panels (regardless of actual grade) and multiply value by 0.82 for species with specific gravity of 0.42 or greater, or 0.65 for all other species.

b. Fastening along intermediate framing members: Space fasteners not greater than 12 inches on center, except 6 inches on center for spans greater than 32 inches.

c. Panels conforming to DOC PS 1 or PS 2.

d. This table gives shear values for Cases 1 and 2 as shown in Table 2306.2(1). The values shown are applicable to Cases 3, 4, 5 and 6 as shown in Table 2306.2(1), providing fasteners at all continuous panel edges are spaced in accordance with the boundary fastener spacing.

e. The minimum nominal depth of framing members shall be 3 inches nominal. The minimum nominal width of framing members not located at boundaries or adjoining panel edges shall be 2 inches.

f. Staples shall have a minimum crown width of $^7/_{16}$ inch, and shall be installed with their crowns parallel to the long dimension of the framing members.

g. High-load diaphragms shall be subject to special inspection in accordance with Section 1705.5.1.

h. For shear loads of normal or permanent load duration as defined by the ANSI/AWC NDS, the values in the table shall be multiplied by 0.63 or 0.56, respectively.

i. For fastener spacing diagrams see Figure 2306.2(2).

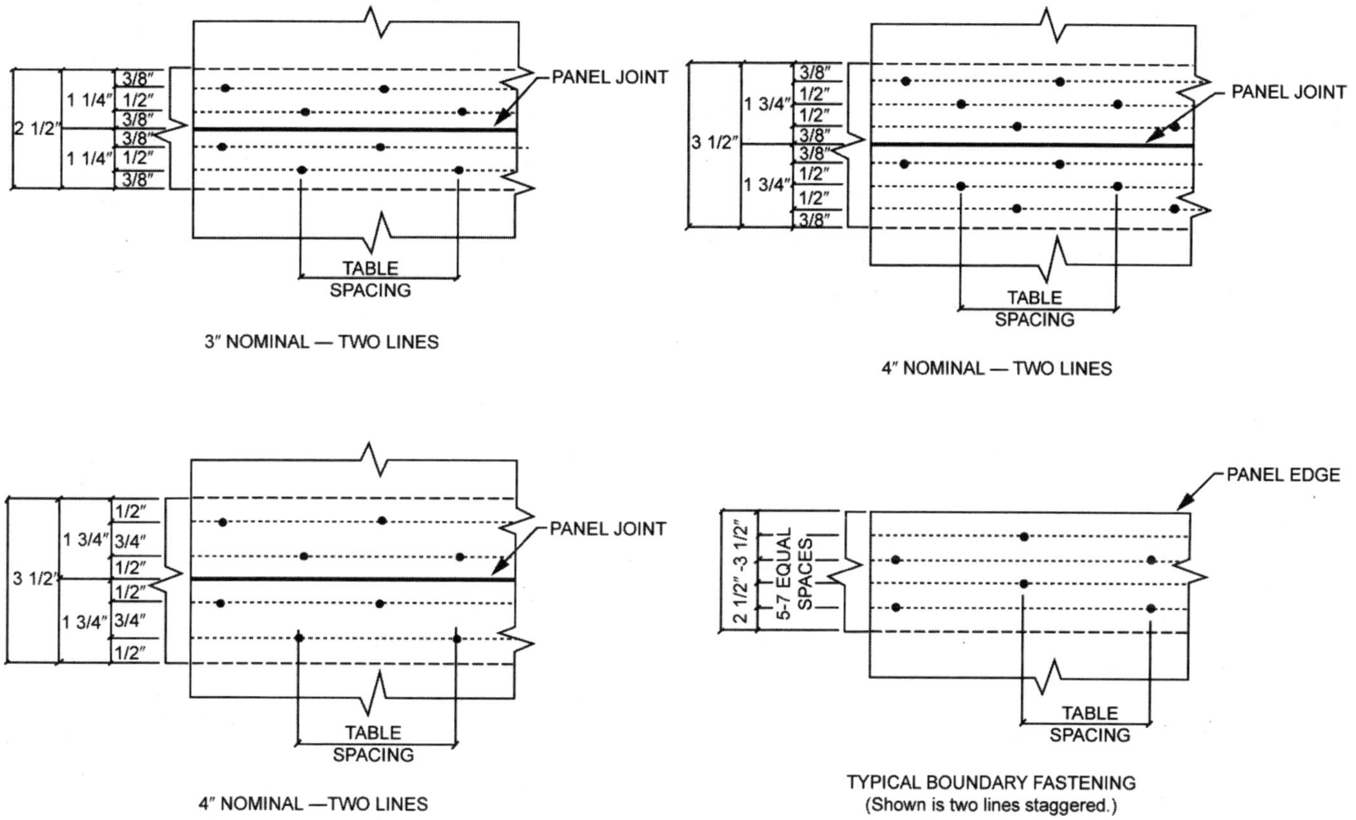

NOTE: SPACE PANEL END AND EDGE JOINT 1/8-INCH. REDUCE SPACING BETWEEN LINES OF NAILS AS NECESSARY TO MAINTAIN MINIMUM 3/8=INCH FASTENER EDGE MARGINS, MINIMUM SPACING BETWEEN LINES IS 3/8-INCH

For SI: 1 inch = 25.4 mm.

FIGURE 2306.2(2)
FASTENER SPACING DIAGRAMS FOR USE WITH TABLE 2306.2(2)

TABLE 2306.3(1)
ALLOWABLE SHEAR VALUES (POUNDS PER FOOT) FOR WOOD STRUCTURAL PANEL SHEAR WALLS UTILIZING STAPLES WITH FRAMING OF DOUGLAS FIR-LARCH OR SOUTHERN PINE[a] FOR WIND OR SEISMIC LOADING[b, f, g, i]

PANEL GRADE	MINIMUM NOMINAL PANEL THICKNESS (inch)	MINIMUM FASTENER PENETRATION IN FRAMING (inches)	PANELS APPLIED DIRECT TO FRAMING					PANELS APPLIED OVER $^1/_2$" OR $^5/_8$" GYPSUM SHEATHING				
			Staple length and gage[h] (inches)	Fastener spacing at panel edges (inches)				Staple length and gage[h] (inches)	Fastener spacing at panel edges (inches)			
				6	4	3	2[d]		6	4	3	2[d]
Structural I sheathing	$^3/_8$	1	$1^1/_2$ 16 Gage	155	235	315	400	2 16 Gage	155	235	310	400
	$^7/_{16}$			170	260	345	440		155	235	310	400
	$^{15}/_{32}$			185	280	375	475		155	235	300	400
Sheathing, plywood siding[e] except Group 5 Species, ANSI/APA PRP 210siding[e]	$^5/_{16}$[c] or $^1/_4$[c]	1	$1^1/_2$ 16 Gage	145	220	295	375	2 16 Gage	110	165	220	285
	$^3/_8$			140	210	280	360		140	210	280	360
	$^7/_{16}$			155	230	310	395		140	210	280	360
	$^{15}/_{32}$			170	255	335	430		140	210	280	360
	$^{19}/_{32}$		$1^3/_4$ 16 Gage	185	280	375	475	—	—	—	—	—

For SI: 1 inch = 25.4 mm, 1 pound per foot = 14.5939 N/m.

a. For framing of other species: (1) Find specific gravity for species of lumber in ANSI/AWC NDS. (2) For staples find shear value from table for Structural I panels (regardless of actual grade) and multiply value by 0.82 for species with specific gravity of 0.42 or greater, or 0.65 for all other species.

b. Panel edges backed with 2-inch nominal or wider framing. Install panels either horizontally or vertically. Space fasteners maximum 6 inches on center along intermediate framing members for $^3/_8$-inch and $^7/_{16}$-inch panels installed on studs spaced 24 inches on center. For other conditions and panel thickness, space fasteners maximum 12 inches on center on intermediate supports.

c. $^3/_8$-inch panel thickness or siding with a span rating of 16 inches on center is the minimum recommended where applied directly to framing as exterior siding. For grooved panel siding, the nominal panel thickness is the thickness of the panel measured at the point of fastening.

d. Framing at adjoining panel edges shall be 3 inches nominal or wider.

e. Values apply to all-veneer plywood. Thickness at point of fastening on panel edges governs shear values.

f. Where panels are applied on both faces of a wall and fastener spacing is less than 6 inches on center on either side, panel joints shall be offset to fall on different framing members, or framing shall be 3 inches nominal or thicker at adjoining panel edges.

g. In Seismic Design Category D, E or F, where shear design values exceed 350 pounds per linear foot, all framing members receiving edge fastening from abutting panels shall be not less than a single 3-inch nominal member, or two 2-inch nominal members fastened together in accordance with Section 2306.1 to transfer the design shear value between framing members. Wood structural panel joint and sill plate nailing shall be staggered at all panel edges. See AWC SDPWS for sill plate size and anchorage requirements.

h. Staples shall have a minimum crown width of $^7/_{16}$ inch and shall be installed with their crowns parallel to the long dimension of the framing members.

i. For shear loads of normal or permanent load duration as defined by the ANSI/AWC NDS, the values in the table shall be multiplied by 0.63 or 0.56, respectively.

TABLE 2306.3(2)
ALLOWABLE SHEAR VALUES (plf) FOR WIND OR SEISMIC LOADING ON SHEAR WALLS OF FIBERBOARD SHEATHING BOARD CONSTRUCTION UTILIZING STAPLES FOR TYPE V CONSTRUCTION ONLY[a, b, c, d, e]

THICKNESS AND GRADE (inches)	STAPLE GAGE AND DIMENSIONS	ALLOWABLE SHEAR VALUE (pounds per linear foot) STAPLE SPACING AT PANEL EDGES (inches)[a]		
		4	3	2
$^1/_2$ or $^{25}/_{32}$ Structural	No. 16 gage galvanized staple, $^7/_{16}$″ crown $1^3/_4$ inches long	150	200	225
	No. 16 gage galvanized staple, 1″ crown $1^3/_4$ inches long	220	290	325

For SI: 1 inch = 25.4 mm, 1 pound per foot = 14.5939 N/m.

a. Fiberboard sheathing shall not be used to brace concrete or masonry walls.

b. Panel edges shall be backed with 2-inch or wider framing of Douglas Fir-larch or Southern pine. For framing of other species: (1) Find specific gravity for species of framing lumber in ANSI/AWC NDS. (2) For staples, multiply the shear value from the table by 0.82 for species with specific gravity of 0.42 or greater, or 0.65 for all other species.

c. Values shown are for fiberboard sheathing on one side only with long panel dimension either parallel or perpendicular to studs.

d. Fastener shall be spaced 6 inches on center along intermediate framing members.

e. Values are not permitted in Seismic Design Category D, E or F.

TABLE 2306.3(3)
ALLOWABLE SHEAR VALUES FOR WIND OR SEISMIC FORCES FOR SHEAR WALLS OF LATH AND PLASTER OR GYPSUM BOARD WOOD FRAMED WALL ASSEMBLIES UTILIZING STAPLES

TYPE OF MATERIAL	THICKNESS OF MATERIAL	WALL CONSTRUCTION	STAPLE SPACING[b] MAXIMUM (inches)	SHEAR VALUE[a, c] (plf)	MINIMUM STAPLE SIZE [f, g]
1. Expanded metal or woven wire lath and Portland cement plaster	7/8″	Unblocked	6	180	No. 16 gage galv. staple, 7/8″ legs
2. Gypsum lath, plain or perforated	3/8″ lath and 1/2″ plaster	Unblocked	5	100	No. 16 gage galv. staple, 1 1/8″ long
3. Gypsum sheathing	1/2″ × 2′ × 8′	Unblocked	4	75	No. 16 gage galv. staple, 1 3/4″ long
	1/2″ ×4′	Blocked[d]	4	175	
		Unblocked	7	100	
4. Gypsum board, gypsum veneer base or water-resistant gypsum backing board	1/2″	Unblocked[d]	7	75	No. 16 gage galv. staple, 1 1/2″ long
		Unblocked[d]	4	110	
		Unblocked	7	100	
		Unblocked	4	125	
		Blocked[e]	7	125	
		Blocked[e]	4	150	
	5/8″	Unblocked[d]	7	115	No. 16 gage galv. staple, 1 5/8″ long
			4	145	
		Blocked[e]	7	145	
			4	175	
		Blocked[e] Two-ply	Base ply: 9 Face ply: 7	250	No. 16 gage galv. staple 1 5/8″ long No. 15 gage galv. staple, 2 1/4″ long

For SI: 1 inch = 25.4 mm, 1 foot = 304.8 mm, 1 pound per foot = 14.5939 N/m.

a. These shear walls shall not be used to resist loads imposed by masonry or concrete walls (see AWC SDPWS). Values shown are for short-term loading due to wind or seismic loading. Walls resisting seismic loads shall be subject to the limitations in Section 12.2.1 of ASCE 7. Values shown shall be reduced 25 percent for normal loading.

b. Applies to fastening at studs, top and bottom plates and blocking.

c. Except as noted, shear values are based on a maximum framing spacing of 16 inches on center.

d. Maximum framing spacing of 24 inches on center.

e. All edges are blocked, and edge fastening is provided at all supports and all panel edges.

f. Staples shall have a minimum crown width of 7/16 inch, measured outside the legs, and shall be installed with their crowns parallel to the long dimension of the framing members.

g. Staples for the attachment of gypsum lath and woven-wire lath shall have a minimum crown width of 3/4 inch, measured outside the legs.

2308.2.1 Stories. Structures of *conventional light-frame construction* shall be limited in *story* height in accordance with Table 2308.2.1.

TABLE 2308.2.1
ALLOWABLE STORY HEIGHT

SEISMIC DESIGN CATEGORY	ALLOWABLE STORY ABOVE GRADE PLANE
A and B	Three stories
C	Two stories
D and E[a]	One story

For SI:1 inch = 25.4 mm.

a. For the purposes of this section, for buildings assigned to Seismic Design Category D or E, cripple walls shall be considered to be a story unless cripple walls are solid blocked and do not exceed 14 inches in height.

2308.2.2 Allowable floor-to-floor height. Maximum floor-to-floor height shall not exceed 11 feet, 7 inches (3531 mm). Exterior bearing wall and interior braced wall heights shall not exceed a stud height of 10 feet (3048 mm).

2308.2.3 Allowable loads. *Loads* shall be in accordance with Chapter 16 and shall not exceed the following:

1. Average *dead loads* shall not exceed 15 psf (718 N/m^2) for combined roof and ceiling, *exterior walls*, floors and partitions.

   Exceptions:

   1. Subject to the limitations of Section 2308.6.10, stone or masonry *veneer* up to the less of 5 inches (127 mm) thick or 50 pounds per square foot (2395 N/m^2) and installed in accordance with Chapter 14 is permitted to a height of 30 feet (9144 mm) above a noncombustible foundation, with an additional 8 feet (2439) permitted for *gable* ends.

2. Concrete or masonry fireplaces, heaters and chimneys shall be permitted in accordance with the provisions of this code.

2. *Live loads* shall not exceed 40 psf (1916 N/m^2) for floors.

Exception: *Live loads* for concrete slab-on-ground floors in *Risk Categories* I and II shall be not more than 125 psf.

3. Ground snow *loads* shall not exceed 50 psf (2395 N/m^2).

2308.2.4 Basic wind speed. *V* shall not exceed 130 miles per hour (57 m/s) (3-second gust).

Exceptions:

1. *V* shall not exceed 140 mph (63 m/s) (3-second gust) for buildings in Exposure Category B that are not located in a *hurricane-prone region*.
2. Where *V* exceeds 130 mph (3-second gust), the provisions of either AWC WFCM or ICC 600 are permitted to be used.

2308.2.5 Allowable roof span. Ceiling joist and rafter framing constructed in accordance with Section 2308.7 and trusses shall not span more than 40 feet (12 192 mm) between points of vertical support. A ridge board in accordance with Section 2308.7 or 2308.7.3.1 shall not be considered a vertical support.

2308.2.6 Risk category limitation. The use of the provisions for *conventional light-frame construction* in this section shall not be permitted for *Risk Category* IV buildings assigned to *Seismic Design Category* B, C, D or F.

2308.3 Foundations and footings. Foundations and footings shall be designed and constructed in accordance with Chapter 18. Connections to foundations and footings shall comply with this section.

2308.3.1 Foundation plates or sills. Foundation plates or sills resting on concrete or masonry foundations shall comply with Section 2304.3.1. Foundation plates or sills shall be bolted or anchored to the foundation with not less than $^1/_2$-inch-diameter (12.7 mm) steel bolts or *approved* anchors spaced to provide equivalent anchorage as the steel bolts. Bolts shall be embedded not less than 7 inches (178 mm) into concrete or masonry. The bolts shall be located in the middle third of the width of the plate. Bolts shall be spaced not more than 6 feet (1829 mm) on center and there shall be not less than two bolts or anchor straps per piece with one bolt or anchor strap located not more than 12 inches (305 mm) or less than 4 inches (102 mm) from each end of each piece. Bolts in sill plates of *braced wall lines* in structures over two stories above grade shall be spaced not more than 4 feet (1219 mm) on center. A properly sized nut and washer shall be tightened on each bolt to the plate.

2308.3.1.1 Braced wall line sill plate anchorage in Seismic Design Category D. Sill plates along *braced wall lines* in buildings assigned to *Seismic Design Category* D shall be anchored with not less than $^1/_2$-inch (12.7 mm) diameter anchor bolts with steel plate washers between the foundation sill plate and the nut, or approved anchor straps load-rated in accordance with Section 2304.10.4 and spaced to provide equivalent anchorage. Plate washers shall be not less than 0.229 inch by 3 inches by 3 inches (5.82 mm by 76 mm by 76 mm) in size. The hole in the plate washer is permitted to be diagonally slotted with a width of up to $^3/_{16}$ inch (4.76 mm) larger than the bolt diameter and a slot length not to exceed 1$^3/_4$ inches (44 mm), provided that a standard cut washer is placed between the plate washer and the nut.

2308.3.1.2 Braced wall line sill plate anchorage in Seismic Design Category E. Sill plates along *braced wall lines* in buildings assigned to *Seismic Design Category* E shall be anchored with not less than $^5/_8$-inch diameter (15.9 mm) anchor bolts with steel plate washers between the foundation sill plate and the nut, or approved anchor straps load-rated in accordance with Section 2304.10.4 and spaced to provide equivalent anchorage. Plate washers shall be not less than 0.229 inch by 3 inches by 3 inches (5.82 mm by 76 mm by 76 mm) in size. The hole in the plate washer is permitted to be diagonally slotted with a width of up to $^3/_{16}$ inch (4.76 mm) larger than the bolt diameter and a slot length not to exceed 1$^3/_4$ inches (44 mm), provided that a standard cut washer is placed between the plate washer and the nut.

2308.4 Floor framing. Floor framing shall comply with this section.

2308.4.1 Girders. Girders for single-story construction or girders supporting *loads* from a single floor shall be not less than 4 inches by 6 inches (102 mm by 152 mm) for spans 6 feet (1829 mm) or less, provided that girders are spaced not more than 8 feet (2438 mm) on center. Other girders shall be designed to support the *loads* specified in this code. Girder end joints shall occur over supports.

Where a girder is spliced over a support, an adequate tie shall be provided. The ends of beams or girders supported on masonry or concrete shall not have less than 3 inches (76 mm) of bearing.

2308.4.1.1 Allowable girder spans. The allowable spans of girders that are fabricated of dimension lumber shall not exceed the values set forth in Table 2308.4.1.1(1) or 2308.4.1.1(2).

2308.4.2 Floor joists. Floor joists shall comply with this section.

2308.4.2.1 Span. Spans for floor joists shall be in accordance with Table 2308.4.2.1(1), Table 2308.4.2.1(2) or the AWC STJR.

2308.4.2.2 Bearing. The ends of each joist shall have not less than 1$^1/_2$ inches (38 mm) of bearing on wood or metal, or not less than 3 inches (76 mm) on masonry, except where supported on a 1-inch by 4-inch (25 mm by 102 mm) ribbon strip and nailed to the adjoining stud.

TABLE 2308.4.1.1(1)
HEADER AND GIRDER SPANS[a, b] FOR EXTERIOR BEARING WALLS
(Maximum spans for Douglas fir-larch, hem-fir, Southern pine and spruce-pine-fir and required number of jack studs)

GIRDERS AND HEADERS SUPPORTING	SIZE	GROUND SNOW LOAD (psf)[e]																	
		30						50						70					
		Building width[c] (feet)																	
		12		24		36		12		24		36		12		24		36	
		Span[f]	NJ[d]	Span[f]	NJ[d]	Span[f]	NJ[d]	Span[f]	NJ[d]	Span[f]	NJ[d]	Span[f]	NJ[d]	Span[f]	NJ[d]	Span[f]	NJ[d]	Span[f]	NJ[d]
Roof and ceiling	1-2 × 6	4-0	1	3-1	2	2-7	2	3-5	1	2-8	2	2-3	2	3-0	2	2-4	2	2-0	2
	1-2 × 8	5-1	2	3-11	2	3-3	2	4-4	2	3-4	2	2-10	2	3-10	2	3-0	2	2-6	3
	1-2 × 10	6-0	2	4-8	2	3-11	2	5-2	2	4-0	2	3-4	3	4-7	2	3-6	3	3-0	3
	1-2 × 12	7-1	2	5-5	2	4-7	3	6-1	2	4-8	3	3-11	3	5-5	2	4-2	3	3-6	3
	2-2 × 4	4-0	1	3-1	1	2-7	1	3-5	1	2-7	1	2-2	1	3-0	1	2-4	1	2-0	1
	2-2 × 6	6-0	1	4-7	1	3-10	1	5-1	1	3-11	1	3-3	2	4-6	1	3-6	2	2-11	2
	2-2 × 8	7-7	1	5-9	1	4-10	2	6-5	1	5-0	2	4-2	2	5-9	1	4-5	2	3-9	2
	2-2 × 10	9-0	1	6-10	2	5-9	2	7-8	2	5-11	2	4-11	2	6-9	2	5-3	2	4-5	2
	2-2 × 12	10-7	2	8-1	2	6-10	2	9-0	2	6-11	2	5-10	2	8-0	2	6-2	2	5-2	3
	3-2 × 8	9-5	1	7-3	1	6-1	1	8-1	1	6-3	1	5-3	2	7-2	1	5-6	2	4-8	2
	3-2 × 10	11-3	1	8-7	1	7-3	2	9-7	1	7-4	2	6-2	2	8-6	1	6-7	2	5-6	2
	3-2 × 12	13-2	1	10-1	2	8-6	2	11-3	2	8-8	2	7-4	2	10-0	2	7-9	2	6-6	2
	4-2 × 8	10-11	1	8-4	1	7-0	1	9-4	1	7-2	1	6-0	1	8-3	1	6-4	1	5-4	2
	4-2 × 10	12-11	1	9-11	1	8-4	1	11-1	1	8-6	1	7-2	2	9-10	1	7-7	2	6-4	2
	4-2 × 12	15-3	1	11-8	1	9-10	2	13-0	1	10-0	2	8-5	2	11-7	1	8-11	2	7-6	2
Roof, ceiling and one center-bearing floor	1-2 × 6	3-3	1	2-7	2	2-2	2	3-0	2	2-4	2	2-0	2	2-9	2	2-2	2	1-10	2
	1-2 × 8	4-1	2	3-3	2	2-9	2	3-9	2	3-0	2	2-6	3	3-6	2	2-9	2	2-4	3
	1-2 × 10	4-11	2	3-10	2	3-3	3	4-6	2	3-6	3	3-0	3	4-1	2	3-3	3	2-9	3
	1-2 × 12	5-9	2	4-6	3	3-10	3	5-3	2	4-2	3	3-6	3	4-10	3	3-10	3	3-3	4
	2-2 × 4	3-3	1	2-6	1	2-2	1	3-0	1	2-4	1	2-0	1	2-8	1	2-2	1	1-10	1
	2-2 × 6	4-10	1	3-9	1	3-3	2	4-5	1	3-6	2	3-0	2	4-1	1	3-3	2	2-9	2
	2-2 × 8	6-1	1	4-10	2	4-1	2	5-7	2	4-5	2	3-9	2	5-2	2	4-1	2	3-6	2
	2-2 × 10	7-3	2	5-8	2	4-10	2	6-8	2	5-3	2	4-5	2	6-1	2	4-10	2	4-1	2
	2-2 × 12	8-6	2	6-8	2	5-8	2	7-10	2	6-2	2	5-3	3	7-2	2	5-8	2	4-10	3
	3-2 × 8	7-8	1	6-0	1	5-1	2	7-0	1	5-6	2	4-8	2	6-5	1	5-1	2	4-4	2
	3-2 × 10	9-1	1	7-2	2	6-1	2	8-4	1	6-7	2	5-7	2	7-8	2	6-1	2	5-2	2
	3-2 × 12	10-8	2	8-5	2	7-2	2	9-10	2	7-8	2	6-7	2	9-0	2	7-1	2	6-1	2
	4-2 × 8	8-10	1	6-11	1	5-11	1	8-1	1	6-4	1	5-5	2	7-5	1	5-11	1	5-0	2
	4-2 × 10	10-6	1	8-3	2	7-0	2	9-8	1	7-7	2	6-5	2	8-10	1	7-0	2	6-0	2
	4-2 × 12	12-4	1	9-8	2	8-3	2	11-4	2	8-11	2	7-7	2	10-4	2	8-3	2	7-0	2
Roof, ceiling and one clear span floor	1-2 × 6	2-11	2	2-3	2	1-11	2	2-9	2	2-1	2	1-9	2	2-7	2	2-0	2	1-8	2
	1-2 × 8	3-9	2	2-10	2	2-5	3	3-6	2	2-8	2	2-3	3	3-3	2	2-6	3	2-2	3
	1-2 × 10	4-5	2	3-5	3	2-10	3	4-2	2	3-2	3	2-8	3	3-11	2	3-0	3	2-6	3
	1-2 × 12	5-2	2	4-0	3	3-4	3	4-10	3	3-9	3	3-2	4	4-7	3	3-6	3	3-0	4
	2-2 × 4	2-11	1	2-3	1	1-10	1	2-9	1	2-1	1	1-9	1	2-7	1	2-0	1	1-8	1
	2-2 × 6	4-4	1	3-4	2	2-10	2	4-1	1	3-2	2	2-8	2	3-10	1	3-0	2	2-6	2
	2-2 × 8	5-6	2	4-3	2	3-7	2	5-2	2	4-0	2	3-4	2	4-10	2	3-9	2	3-2	2
	2-2 × 10	6-7	2	5-0	2	4-2	2	6-1	2	4-9	2	4-0	2	5-9	2	4-5	2	3-9	3
	2-2 × 12	7-9	2	5-11	2	4-11	3	7-2	2	5-7	2	4-8	3	6-9	2	5-3	3	4-5	3
	3-2 × 8	6-11	1	5-3	2	4-5	2	6-5	1	5-0	2	4-2	2	6-1	1	4-8	2	4-0	2
	3-2 × 10	8-3	2	6-3	2	5-3	2	7-8	2	5-11	2	5-0	2	7-3	2	5-7	2	4-8	2
	3-2 × 12	9-8	2	7-5	2	6-2	2	9-0	2	7-0	2	5-10	2	8-6	2	6-7	2	5-6	3
	4-2 × 8	8-0	1	6-1	1	5-1	2	7-5	1	5-9	2	4-10	2	7-0	1	5-5	2	4-7	2
	4-2 × 10	9-6	1	7-3	2	6-1	2	8-10	1	6-10	2	5-9	2	8-4	1	6-5	2	5-5	2
	4-2 × 12	11-2	2	8-6	2	7-2	2	10-5	2	8-0	2	6-9	2	9-10	2	7-7	2	6-5	2

(continued)

TABLE 2308.4.1.1(1)—continued
HEADER AND GIRDER SPANS[a, b] FOR EXTERIOR BEARING WALLS
(Maximum spans for Douglas fir-larch, hem-fir, Southern pine and spruce-pine-fir and required number of jack studs)

GIRDERS AND HEADERS SUPPORTING	SIZE	GROUND SNOW LOAD (psf)[e]																	
		30						50						70					
		Building width[c] (feet)																	
		12		24		36		12		24		36		12		24		36	
		Span[f]	NJ[d]	Span[f]	NJ[d]	Span[f]	NJ[d]	Span[f]	NJ[d]	Span[f]	NJ[d]	Span[f]	NJ[d]	Span[f]	NJ[d]	Span[f]	NJ[d]	Span[f]	NJ[d]
Roof, ceiling and two center-bearing floors	1-2 × 6	2-8	2	2-1	2	1-10	2	2-7	2	2-0	2	1-9	2	2-5	2	1-11	2	1-8	2
	1-2 × 8	3-5	2	2-8	2	2-4	3	3-3	2	2-7	2	2-2	3	3-1	2	2-5	3	2-1	3
	1-2 × 10	4-0	2	3-2	3	2-9	3	3-10	2	3-1	3	2-7	3	3-8	2	2-11	3	2-5	3
	1-2 × 12	4-9	3	3-9	3	3-2	4	4-6	3	3-7	3	3-1	4	4-3	3	3-5	3	2-11	4
	2-2 × 4	2-8	1	2-1	1	1-9	1	2-6	1	2-0	1	1-8	1	2-5	1	1-11	1	1-7	1
	2-2 × 6	4-0	1	3-2	2	2-8	2	3-9	1	3-0	2	2-7	2	3-7	1	2-10	2	2-5	2
	2-2 × 8	5-0	2	4-0	2	3-5	2	4-10	2	3-10	2	3-3	2	4-7	2	3-7	2	3-1	2
	2-2 × 10	6-0	2	4-9	2	4-0	2	5-8	2	4-6	2	3-10	3	5-5	2	4-3	2	3-8	3
	2-2 × 12	7-0	2	5-7	2	4-9	3	6-8	2	5-4	3	4-6	3	6-4	2	5-0	3	4-3	3
	3-2 × 8	6-4	1	5-0	2	4-3	2	6-0	1	4-9	2	4-1	2	5-8	2	4-6	2	3-10	2
	3-2 × 10	7-6	2	5-11	2	5-1	2	7-1	2	5-8	2	4-10	2	6-9	2	5-4	2	4-7	2
	3-2 × 12	8-10	2	7-0	2	5-11	2	8-5	2	6-8	2	5-8	3	8-0	2	6-4	2	5-4	3
	4-2 × 8	7-3	1	5-9	1	4-11	2	6-11	1	5-6	2	4-8	2	6-7	1	5-2	2	4-5	2
	4-2 × 10	8-8	1	6-10	2	5-10	2	8-3	2	6-6	2	5-7	2	7-10	2	6-2	2	5-3	2
	4-2 × 12	10-2	2	8-1	2	6-10	2	9-8	2	7-8	2	6-7	2	9-2	2	7-3	2	6-2	2
Roof, ceiling and two clear span floors	1-2 × 6	2-3	2	1-9	2	1-5	2	2-3	2	1-9	2	1-5	3	2-2	2	1-8	2	1-5	3
	1-2 × 8	2-10	2	2-2	3	1-10	3	2-10	2	2-2	3	1-10	3	2-9	2	2-1	3	1-10	3
	1-2 × 10	3-4	2	2-7	3	2-2	3	3-4	3	2-7	3	2-2	4	3-3	3	2-6	3	2-2	4
	1-2 × 12	4-0	3	3-0	3	2-7	4	4-0	3	3-0	4	2-7	4	3-10	3	3-0	4	2-6	4
	2-2 × 4	2-3	1	1-8	1	1-4	1	2-3	1	1-8	1	1-4	1	2-2	1	1-8	1	1-4	2
	2-2 × 6	3-4	1	2-6	2	2-2	2	3-4	2	2-6	2	2-2	2	3-3	2	2-6	2	2-1	2
	2-2 × 8	4-3	2	3-3	2	2-8	2	4-3	2	3-3	2	2-8	2	4-1	2	3-2	2	2-8	3
	2-2 × 10	5-0	2	3-10	2	3-2	3	5-0	2	3-10	2	3-2	3	4-10	2	3-9	3	3-2	3
	2-2 × 12	5-11	2	4-6	3	3-9	3	5-11	2	4-6	3	3-9	3	5-8	2	4-5	3	3-9	3
	3-2 × 8	5-3	1	4-0	2	3-5	2	5-3	2	4-0	2	3-5	2	5-1	2	3-11	2	3-4	2
	3-2 × 10	6-3	2	4-9	2	4-0	2	6-3	2	4-9	2	4-0	2	6-1	2	4-8	2	4-0	3
	3-2 × 12	7-5	2	5-8	2	4-9	3	7-5	2	5-8	2	4-9	3	7-2	2	5-6	3	4-8	3
	4-2 × 8	6-1	1	4-8	2	3-11	2	6-1	1	4-8	2	3-11	2	5-11	1	4-7	2	3-10	2
	4-2 × 10	7-3	2	5-6	2	4-8	2	7-3	2	5-6	2	4-8	2	7-0	2	5-5	2	4-7	2
	4-2 × 12	8-6	2	6-6	2	5-6	2	8-6	2	6-6	2	5-6	2	8-3	2	6-4	2	5-4	3

For SI: 1 inch = 25.4 mm, 1 pound per square foot = 0.0479 kPa.

a. Spans are given in feet and inches.

b. Spans are based on minimum design properties for No. 2 grade lumber of Douglas fir-larch, hem-fir, Southern pine and spruce-pine fir.

c. Building width is measured perpendicular to the ridge. For widths between those shown, spans are permitted to be interpolated.

d. NJ = Number of jack studs required to support each end. Where the number of required jack studs equals one, the header is permitted to be supported by an approved framing anchor attached to the full-height wall stud and to the header.

e. Use 30 psf ground snow load for cases in which ground snow load is less than 30 psf and the roof live load is equal to or less than 20 psf.

f. Spans are calculated assuming the top of the header or girder is laterally braced by perpendicular framing. Where the top of the header or girder is not laterally braced (for example, cripple studs bearing on the header), tabulated spans for headers consisting of 2 × 8, 2 × 10, or 2 × 12 sizes shall be multiplied by 0.70 or the header or girder shall be designed.

TABLE 2308.4.1.1(2)
HEADER AND GIRDER SPANS[a, b] FOR INTERIOR BEARING WALLS
(Maximum spans for Douglas fir-larch, hem-fir, Southern pine and spruce-pine-fir and required number of jack studs)

HEADERS AND GIRDERS SUPPORTING	SIZE	BUILDING WIDTH[c] (feet)					
		12		24		36	
		Span[e]	NJ[d]	Span[e]	NJ[d]	Span[e]	NJ[d]
One floor only	2-2 × 4	4-1	1	2-10	1	2-4	1
	2-2 × 6	6-1	1	4-4	1	3-6	1
	2-2 × 8	7-9	1	5-5	1	4-5	2
	2-2 × 10	9-2	1	6-6	2	5-3	2
	2-2 × 12	10-9	1	7-7	2	6-3	2
	3-2 × 8	9-8	1	6-10	1	5-7	1
	3-2 × 10	11-5	1	8-1	1	6-7	2
	3-2 × 12	13-6	1	9-6	2	7-9	2
	4-2 × 8	11-2	1	7-11	1	6-5	1
	4-2 × 10	13-3	1	9-4	1	7-8	1
	4-2 × 12	15-7	1	11-0	1	9-0	2
Two floors	2-2 × 4	2-7	1	1-11	1	1-7	1
	2-2 × 6	3-11	1	2-11	2	2-5	2
	2-2 × 8	5-0	1	3-8	2	3-1	2
	2-2 × 10	5-11	2	4-4	2	3-7	2
	2-2 × 12	6-11	2	5-2	2	4-3	3
	3-2 × 8	6-3	1	4-7	2	3-10	2
	3-2 × 10	7-5	1	5-6	2	4-6	2
	3-2 × 12	8-8	2	6-5	2	5-4	2
	4-2 × 8	7-2	1	5-4	1	4-5	2
	4-2 × 10	8-6	1	6-4	2	5-3	2
	4-2 × 12	10-1	1	7-5	2	6-2	2

For SI: 1 inch = 25.4 mm, 1 foot = 304.8 mm.

a. Spans are given in feet and inches.

b. Spans are based on minimum design properties for No. 2 grade lumber of Douglas fir-larch, hem-fir, Southern pine and spruce-pine fir.

c. Building width is measured perpendicular to the ridge. For widths between those shown, spans are permitted to be interpolated.

d. NJ = Number of jack studs required to support each end. Where the number of required jack studs equals one, the header is permitted to be supported by an approved framing anchor attached to the full-height wall stud and to the header.

e. Spans are calculated assuming the top of the header or girder is laterally braced by perpendicular framing. Where the top of the header or girder is not laterally braced (for example, cripple studs bearing on the header), tabulated spans for headers consisting of 2 × 8, 2 × 10, or 2 × 12 sizes shall be multiplied by 0.70 or the header or girder shall be designed.

TABLE 2308.4.2.1(1)
FLOOR JOIST SPANS FOR COMMON LUMBER SPECIES
(Residential sleeping areas, live load = 30 psf, L/Δ = 360)

JOIST SPACING (inches)	SPECIES AND GRADE		DEAD LOAD = 10 psf				DEAD LOAD = 20 psf			
			2 × 6	2 × 8	2 × 10	2 × 12	2 × 6	2 × 8	2 × 10	2 × 12
			Maximum floor joist spans							
			(ft. - in.)	(ft. - in.)	(ft. - in.)	(ft. - in.)	(ft. - in.)	(ft. - in.)	(ft. - in.)	(ft. - in.)
12	Douglas Fir-Larch	SS	12-6	16-6	21-0	25-7	12-6	16-6	21-0	25-7
	Douglas Fir-Larch	#1	12-0	15-10	20-3	24-8	12-0	15-7	19-0	22-0
	Douglas Fir-Larch	#2	11-10	15-7	19-10	23-0	11-6	14-7	17-9	20-7
	Douglas Fir-Larch	#3	9-8	12-4	15-0	17-5	8-8	11-0	13-5	15-7
	Hem-Fir	SS	11-10	15-7	19-10	24-2	11-10	15-7	19-10	24-2
	Hem-Fir	#1	11-7	15-3	19-5	23-7	11-7	15-2	18-6	21-6
	Hem-Fir	#2	11-0	14-6	18-6	22-6	11-0	14-4	17-6	20-4
	Hem-Fir	#3	9-8	12-4	15-0	17-5	8-8	11-0	13-5	15-7
	Southern Pine	SS	12-3	16-2	20-8	25-1	12-3	16-2	20-8	25-1
	Southern Pine	#1	11-10	15-7	19-10	24-2	11-10	15-7	18-7	22-0
	Southern Pine	#2	11-3	14-11	18-1	21-4	10-9	13-8	16-2	19-1
	Southern Pine	#3	9-2	11-6	14-0	16-6	8-2	10-3	12-6	14-9
	Spruce-Pine-Fir	SS	11-7	15-3	19-5	23-7	11-7	15-3	19-5	23-7
	Spruce-Pine-Fir	#1	11-3	14-11	19-0	23-0	11-3	14-7	17-9	20-7
	Spruce-Pine-Fir	#2	11-3	14-11	19-0	23-0	11-3	14-7	17-9	20-7
	Spruce-Pine-Fir	#3	9-8	12-4	15-0	17-5	8-8	11-0	13-5	15-7
16	Douglas Fir-Larch	SS	11-4	15-0	19-1	23-3	11-4	15-0	19-1	23-0
	Douglas Fir-Larch	#1	10-11	14-5	18-5	21-4	10-8	13-6	16-5	19-1
	Douglas Fir-Larch	#2	10-9	14-1	17-2	19-11	9-11	12-7	15-5	17-10
	Douglas Fir-Larch	#3	8-5	10-8	13-0	15-1	7-6	9-6	11-8	13-6
	Hem-Fir	SS	10-9	14-2	18-0	21-11	10-9	14-2	18-0	21-11
	Hem-Fir	#1	10-6	13-10	17-8	20-9	10-4	13-1	16-0	18-7
	Hem-Fir	#2	10-0	13-2	16-10	19-8	9-10	12-5	15-2	17-7
	Hem-Fir	#3	8-5	10-8	13-0	15-1	7-6	9-6	11-8	13-6
	Southern Pine	SS	11-2	14-8	18-9	22-10	11-2	14-8	18-9	22-10
	Southern Pine	#1	10-9	14-2	18-0	21-4	10-9	13-9	16-1	19-1
	Southern Pine	#2	10-3	13-3	15-8	18-6	9-4	11-10	14-0	16-6
	Southern Pine	#3	7-11	10-10	12-1	14-4	7-1	8-11	10-10	12-10
	Spruce-Pine-Fir	SS	10-6	13-10	17-8	21-6	10-6	13-10	17-8	21-4
	Spruce-Pine-Fir	#1	10-3	13-6	17-2	19-11	9-11	12-7	15-5	17-10
	Spruce-Pine-Fir	#2	10-3	13-6	17-2	19-11	9-11	12-7	15-5	17-10
	Spruce-Pine-Fir	#3	8-5	10-8	13-0	15-1	7-6	9-6	11-8	13-6

(continued)

TABLE 2308.4.2.1(1)—continued
FLOOR JOIST SPANS FOR COMMON LUMBER SPECIES
(Residential sleeping areas, live load = 30 psf, L/Δ = 360)

JOIST SPACING (inches)	SPECIES AND GRADE		DEAD LOAD = 10 psf				DEAD LOAD = 20 psf			
			2 × 6	2 × 8	2 × 10	2 × 12	2 × 6	2 × 8	2 × 10	2 × 12
			Maximum floor joist spans							
			(ft. - in.)	(ft. - in.)	(ft. - in.)	(ft. - in.)	(ft. - in.)	(ft. - in.)	(ft. - in.)	(ft. - in.)
19.2	Douglas Fir-Larch	SS	10-8	14-1	18-0	21-10	10-8	14-1	18-0	21-0
	Douglas Fir-Larch	#1	10-4	13-7	16-9	19-6	9-8	12-4	15-0	17-5
	Douglas Fir-Larch	#2	10-1	12-10	15-8	18-3	9-1	11-6	14-1	16-3
	Douglas Fir-Larch	#3	7-8	9-9	11-10	13-9	6-10	8-8	10-7	12-4
	Hem-Fir	SS	10-1	13-4	17-0	20-8	10-1	13-4	17-0	20-7
	Hem-Fir	#1	9-10	13-0	16-4	19-0	9-6	12-0	14-8	17-0
	Hem-Fir	#2	9-5	12-5	15-6	17-1	8-11	11-4	13-10	16-1
	Hem-Fir	#3	7-8	9-9	11-10	13-9	6-10	8-8	10-7	12-4
	Southern Pine	SS	10-6	13-10	17-8	21-6	10-6	13-10	17-8	21-6
	Southern Pine	#1	10-1	13-4	16-5	19-6	9-11	12-7	14-8	17-5
	Southern Pine	#2	9-6	12-1	14-4	16-10	8-6	10-10	12-10	15-1
	Southern Pine	#3	7-3	9-1	11-0	13-1	6-5	8-2	9-10	11-8
	Spruce-Pine-Fir	SS	9-10	13-0	16-7	20-2	9-10	13-0	16-7	19-6
	Spruce-Pine-Fir	#1	9-8	12-9	15-8	18-3	9-1	11-6	14-1	16-3
	Spruce-Pine-Fir	#2	9-8	12-9	15-8	18-3	9-1	11-6	14-1	16-3
	Spruce-Pine-Fir	#3	7-8	9-9	11-10	13-9	6-10	8-8	10-7	12-4
24	Douglas Fir-Larch	SS	9-11	13-1	16-8	20-3	9-11	13-1	16-2	18-9
	Douglas Fir-Larch	#1	9-7	12-4	15-0	17-5	8-8	11-0	13-5	15-7
	Douglas Fir-Larch	#2	9-1	11-6	14-1	16-3	8-1	10-3	12-7	14-7
	Douglas Fir-Larch	#3	6-10	8-8	10-7	12-4	6-2	7-9	9-6	11-0
	Hem-Fir	SS	9-4	12-4	15-9	19-2	9-4	12-4	15-9	18-5
	Hem-Fir	#1	9-2	12-0	14-8	17-0	8-6	10-9	13-1	15-2
	Hem-Fir	#2	8-9	11-4	13-10	16-1	8-0	10-2	12-5	14-4
	Hem-Fir	#3	6-10	8-8	10-7	12-4	6-2	7-9	9-6	11-0
	Southern Pine	SS	9-9	12-10	16-5	19-11	9-9	12-10	16-5	19-8
	Southern Pine	#1	9-4	12-4	14-8	17-5	8-10	11-3	13-1	15-7
	Southern Pine	#2	8-6	10-10	12-10	15-1	7-7	9-8	11-5	13-6
	Southern Pine	#3	6-5	8-2	9-10	11-8	5-9	7-3	8-10	10-5
	Spruce-Pine-Fir	SS	9-2	12-1	15-5	18-9	9-2	12-1	15-0	17-5
	Spruce-Pine-Fir	#1	8-11	11-6	14-1	16-3	8-1	10-3	12-7	14-7
	Spruce-Pine-Fir	#2	8-11	11-6	14-1	16-3	8-1	10-3	12-7	14-7
	Spruce-Pine-Fir	#3	6-10	8-8	10-7	12-4	6-2	7-9	9-6	11-0

For SI: 1 inch = 25.4 mm, 1 foot = 304.8 mm, 1 pound per square foot = 0.0479 kPa.
Note: Check sources for availability of lumber in lengths greater than 20 feet.

TABLE 2308.4.2.1(2)
FLOOR JOIST SPANS FOR COMMON LUMBER SPECIES
(Residential living areas, live load = 40 psf, L/Δ = 360)

JOIST SPACING (inches)	SPECIES AND GRADE		DEAD LOAD = 10 psf				DEAD LOAD = 20 psf			
			2 × 6	2 × 8	2 × 10	2 × 12	2 × 6	2 × 8	2 × 10	2 × 12
			Maximum floor joist spans							
			(ft. - in.)	(ft. - in.)	(ft. - in.)	(ft. - in.)	(ft. - in.)	(ft. - in.)	(ft. - in.)	(ft. - in.)
12	Douglas Fir-Larch	SS	11-4	15-0	19-1	23-3	11-4	15-0	19-1	23-3
	Douglas Fir-Larch	#1	10-11	14-5	18-5	22-0	10-11	14-2	17-4	20-1
	Douglas Fir-Larch	#2	10-9	14-2	17-9	20-7	10-6	13-3	16-3	18-10
	Douglas Fir-Larch	#3	8-8	11-0	13-5	15-7	7-11	10-0	12-3	14-3
	Hem-Fir	SS	10-9	14-2	18-0	21-11	10-9	14-2	18-0	21-11
	Hem-Fir	#1	10-6	13-10	17-8	21-6	10-6	13-10	16-11	19-7
	Hem-Fir	#2	10-0	13-2	16-10	20-4	10-0	13-1	16-0	18-6
	Hem-Fir	#3	8-8	11-0	13-5	15-7	7-11	10-0	12-3	14-3
	Southern Pine	SS	11-2	14-8	18-9	22-10	11-2	14-8	18-9	22-10
	Southern Pine	#1	10-9	14-2	18-0	21-11	10-9	14-2	16-11	20-1
	Southern Pine	#2	10-3	13-6	16-2	19-1	9-10	12-6	14-9	17-5
	Southern Pine	#3	8-2	10-3	12-6	14-9	7-5	9-5	11-5	13-6
	Spruce-Pine-Fir	SS	10-6	13-10	17-8	21-6	10-6	13-10	17-8	21-6
	Spruce-Pine-Fir	#1	10-3	13-6	17-3	20-7	10-3	13-3	16-3	18-10
	Spruce-Pine-Fir	#2	10-3	13-6	17-3	20-7	10-3	13-3	16-3	18-10
	Spruce-Pine-Fir	#3	8-8	11-0	13-5	15-7	7-11	10-0	12-3	14-3
16	Douglas Fir-Larch	SS	10-4	13-7	17-4	21-1	10-4	13-7	17-4	21-0
	Douglas Fir-Larch	#1	9-11	13-1	16-5	19-1	9-8	12-4	15-0	17-5
	Douglas Fir-Larch	#2	9-9	12-7	15-5	17-10	9-1	11-6	14-1	16-3
	Douglas Fir-Larch	#3	7-6	9-6	11-8	13-6	6-10	8-8	10-7	12-4
	Hem-Fir	SS	9-9	12-10	16-5	19-11	9-9	12-10	16-5	19-11
	Hem-Fir	#1	9-6	12-7	16-0	18-7	9-6	12-0	14-8	17-0
	Hem-Fir	#2	9-1	12-0	15-2	17-7	8-11	11-4	13-10	16-1
	Hem-Fir	#3	7-6	9-6	11-8	13-6	6-10	8-8	10-7	12-4
	Southern Pine	SS	10-2	13-4	17-0	20-9	10-2	13-4	17-0	20-9
	Southern Pine	#1	9-9	12-10	16-1	19-1	9-9	12-7	14-8	17-5
	Southern Pine	#2	9-4	11-10	14-0	16-6	8-6	10-10	12-10	15-1
	Southern Pine	#3	7-1	8-11	10-10	12-10	6-5	8-2	9-10	11-8
	Spruce-Pine-Fir	SS	9-6	12-7	16-0	19-6	9-6	12-7	16-0	19-6
	Spruce-Pine-Fir	#1	9-4	12-3	15-5	17-10	9-1	11-6	14-1	16-3
	Spruce-Pine-Fir	#2	9-4	12-3	15-5	17-10	9-1	11-6	14-1	16-3
	Spruce-Pine-Fir	#3	7-6	9-6	11-8	13-6	6-10	8-8	10-7	12-4

(continued)

TABLE 2308.4.2.1(2)—continued
FLOOR JOIST SPANS FOR COMMON LUMBER SPECIES
(Residential living areas, live load = 40 psf, L/Δ = 360)

JOIST SPACING (inches)	SPECIES AND GRADE		DEAD LOAD = 10 psf				DEAD LOAD = 20 psf			
			2 × 6	2 × 8	2 × 10	2 × 12	2 × 6	2 × 8	2 × 10	2 × 12
			Maximum floor joist spans							
			(ft. - in.)	(ft. - in.)	(ft. - in.)	(ft. - in.)	(ft. - in.)	(ft. - in.)	(ft. - in.)	(ft. - in.)
19.2	Douglas Fir-Larch	SS	9-8	12-10	16-4	19-10	9-8	12-10	16-4	19-2
	Douglas Fir-Larch	#1	9-4	12-4	15-0	17-5	8-10	11-3	13-8	15-11
	Douglas Fir-Larch	#2	9-1	11-6	14-1	16-3	8-3	10-6	12-10	14-10
	Douglas Fir-Larch	#3	6-10	8-8	10-7	12-4	6-3	7-11	9-8	11-3
	Hem-Fir	SS	9-2	12-1	15-5	18-9	9-2	12-1	15-5	18-9
	Hem-Fir	#1	9-0	11-10	14-8	17-0	8-8	10-11	13-4	15-6
	Hem-Fir	#2	8-7	11-3	13-10	16-1	8-2	10-4	12-8	14-8
	Hem-Fir	#3	6-10	8-8	10-7	12-4	6-3	7-11	9-8	11-3
	Southern Pine	SS	9-6	12-7	16-0	19-6	9-6	12-7	16-0	19-6
	Southern Pine	#1	9-2	12-1	14-8	17-5	9-0	11-5	13-5	15-11
	Southern Pine	#2	8-6	10-10	12-10	15-1	7-9	9-10	11-8	13-9
	Southern Pine	#3	6-5	8-2	9-10	11-8	5-11	7-5	9-0	10-8
	Spruce-Pine-Fir	SS	9-0	11-10	15-1	18-4	9-0	11-10	15-1	17-9
	Spruce-Pine-Fir	#	8-9	11-6	14-1	16-3	8-3	10-6	12-10	14-10
	Spruce-Pine-Fir	#2	8-9	11-6	14-1	16-3	8-3	10-6	12-10	14-10
	Spruce-Pine-Fir	#3	6-10	8-8	10-7	12-4	6-3	7-11	9-8	11-3
24	Douglas Fir-Larch	SS	9-0	11-11	15-2	18-5	9-0	11-11	14-9	17-1
	Douglas Fir-Larch	#1	8-8	11-0	13-5	15-7	7-11	10-0	12-3	14-3
	Douglas Fir-Larch	#2	8-1	10-3	12-7	14-7	7-5	9-5	11-6	13-4
	Douglas Fir-Larch	#3	6-2	7-9	9-6	11-0	5-7	7-1	8-8	10-1
	Hem-Fir	SS	8-6	11-3	14-4	17-5	8-6	11-3	14-4	16-10[a]
	Hem-Fir	#1	8-4	10-9	13-1	15-2	7-9	9-9	11-11	13-10
	Hem-Fir	#2	7-11	10-2	12-5	14-4	7-4	9-3	11-4	13-1
	Hem-Fir	#3	6-2	7-9	9-6	11-0	5-7	7-1	8-8	10-1
	Southern Pine	SS	8-10	11-8	14-11	18-1	8-10	11-8	14-11	18-0
	Southern Pine	#1	8-6	11-3	13-1	15-7	8-1	10-3	12-0	14-3
	Southern Pine	#2	7-7	9-8	11-5	13-6	7-0	8-10	10-5	12-4
	Southern Pine	#3	5-9	7-3	8-10	10-5	5-3	6-8	8-1	9-6
	Spruce-Pine-Fir	SS	8-4	11-0	14-0	17-0	8-4	11-0	13-8	15-11
	Spruce-Pine-Fir	#1	8-1	10-3	12-7	14-7	7-5	9-5	11-6	13-4
	Spruce-Pine-Fir	#2	8-1	10-3	12-7	14-7	7-5	9-5	11-6	13-4
	Spruce-Pine-Fir	#3	6-2	7-9	9-6	11-0	5-7	7-1	8-8	10-1

For SI: 1 inch = 25.4 mm, 1 foot = 304.8 mm, 1 pound per square foot = 0.0479 kPa.
Note: Check sources for availability of lumber in lengths greater than 20 feet.
a. End bearing length shall be increased to 2 inches.

2308.4.2.3 Framing details. Joists shall be supported laterally at the ends and at each support by solid blocking except where the ends of the joists are nailed to a header, band or rim joist or to an adjoining stud or by other means. Solid blocking shall be not less than 2 inches (51 mm) in thickness and the full depth of the joist. Joist framing from opposite sides of a beam, girder or partition shall be lapped not less than 3 inches (76 mm) or the opposing joists shall be tied together in an approved manner. Joists framing into the side of a wood girder shall be supported by framing anchors or on ledger strips not less than 2 inches by 2 inches (51 mm by 51 mm).

2308.4.2.4 Notches and holes. Notches on the ends of joists shall not exceed one-fourth the joist depth. Notches in the top or bottom of joists shall not exceed one-sixth the depth and shall not be located in the middle third of the span. Holes bored in joists shall not be within 2 inches (51 mm) of the top or bottom of the joist and the diameter of any such hole shall not exceed one-third the depth of the joist.

2308.4.3 Engineered wood products. Engineered wood products shall be installed in accordance with manufacturer's recommendations. Cuts, notches and holes bored in trusses, *structural composite lumber*, structural glued-laminated members or I-joists are not permitted except where permitted by the manufacturer's recommendations or where the effects of such alterations are specifically considered in the design of the member by a *registered design professional.*

2308.4.4 Framing around openings. Trimmer and header joists shall be doubled, or of lumber of equivalent cross section, where the span of the header exceeds 4 feet (1219 mm). The ends of header joists more than 6 feet (1829 mm) in length shall be supported by framing anchors or joist hangers unless bearing on a beam, partition or wall. Tail joists over 12 feet (3658 mm) in length shall be supported at the header by framing anchors or on ledger strips not less than 2 inches by 2 inches (51 mm by 51 mm).

2308.4.4.1 Openings in floor diaphragms in Seismic Design Categories B, C, D and E. Openings in horizontal *diaphragms* in *Seismic Design Categories* B, C, D and E with a dimension that is greater than 4 feet (1219 mm) shall be constructed with metal ties and blocking in accordance with this section and Figure 2308.4.4.1(1). Metal ties shall be not less than 0.058 inch [1.47 mm (16 galvanized gage)] in thickness by $1^1/_2$ inches (38 mm) in width and shall have a yield stress not less than 33,000 psi (227 Mpa). Blocking shall extend not less than the dimension of the opening in the direction of the tie and blocking. Ties shall be attached to blocking in accordance with the manufacturer's instructions but with not less than eight 16d common nails on each side of the header-joist intersection.

Openings in floor *diaphragms* in *Seismic Design Categories* D and E shall not have any dimension exceeding 50 percent of the distance between *braced wall lines* or an area greater than 25 percent of the area between *orthogonal* pairs of *braced wall lines* [see Figure 2308.4.4.1(2)]; or the portion of the structure containing the opening shall be designed in accordance with accepted engineering practice to resist the forces specified in Chapter 16, to the extent such irregular opening affects the performance of the conventional framing system.

2308.4.4.2 Vertical offsets in floor diaphragms in Seismic Design Categories D and E. In *Seismic Design Categories* D and E, portions of a floor level shall not be vertically offset such that the framing members on either side of the offset cannot be lapped or tied together in an *approved* manner in accordance with Figure 2308.4.4.2 unless the portion of the structure containing the irregular offset is designed in accordance with accepted engineering practice.

Exception: Framing supported directly by foundations need not be lapped or tied directly together.

2308.4.5 Joists supporting bearing partitions. Bearing partitions parallel to joists shall be supported on beams, girders, doubled joists, walls or other bearing partitions. Bearing partitions perpendicular to joists shall not be offset from supporting girders, walls or partitions more than the joist depth unless such joists are of sufficient size to carry the additional *load.*

2308.4.6 Lateral support. Floor and ceiling framing with a nominal depth-to-thickness ratio not less than 5 to 1 shall have one edge held in line for the entire span. Where the nominal depth-to-thickness ratio of the framing member exceeds 6 to 1, there shall be one line of bridging for each 8 feet (2438 mm) of span, unless both edges of the member are held in line. The bridging shall consist of not less than 1-inch by 3-inch (25 mm by 76 mm) lumber, double nailed at each end, or equivalent metal bracing of equal rigidity, full-depth solid blocking or other *approved* means. A line of bridging shall be required at supports where equivalent lateral support is not otherwise provided.

2308.4.7 Structural floor sheathing. Structural floor sheathing shall comply with the provisions of Section 2304.8.1.

2308.4.8 Under-floor ventilation. For under-floor ventilation, see Section 1202.4.

2308.4.9 Floor framing supporting braced wall panels. Where *braced wall panels* are supported by cantilevered floors or are set back from the floor joist support, the floor framing shall comply with Section 2308.6.7.

2308.4.10 Anchorage of exterior *means of egress* components in Seismic Design Categories D and E. Exterior egress balconies, exterior *stairways* and *ramps* and similar *means of egress* components in structures assigned to *Seismic Design Category* D or E shall be positively anchored to the primary structure at not more than 8 feet (2438 mm) on center or shall be designed for lateral forces. Such attachment shall not be accomplished by use of toenails or nails subject to withdrawal.

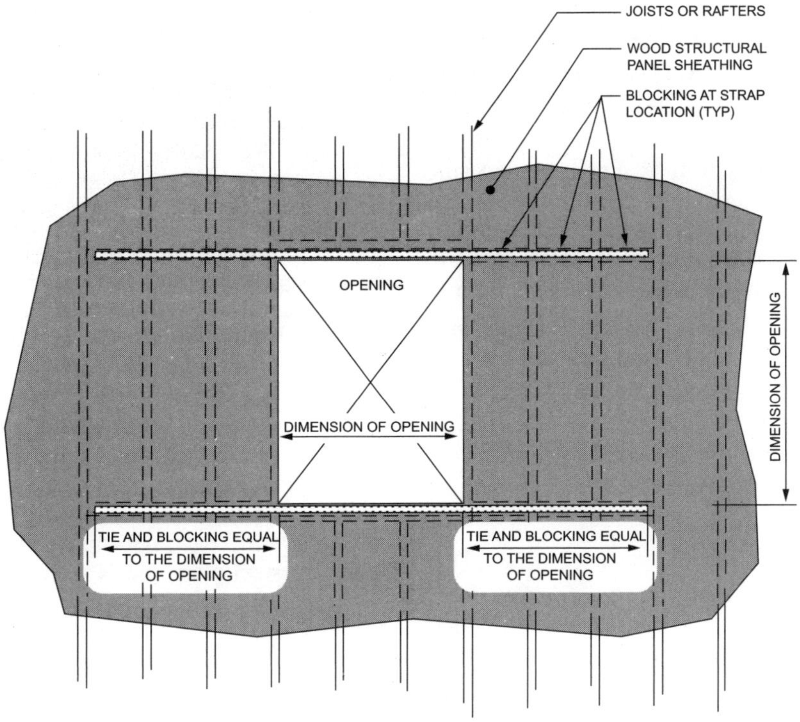

FIGURE 2308.4.4.1(1)
OPENINGS IN FLOOR AND ROOF DIAPHRAGMS

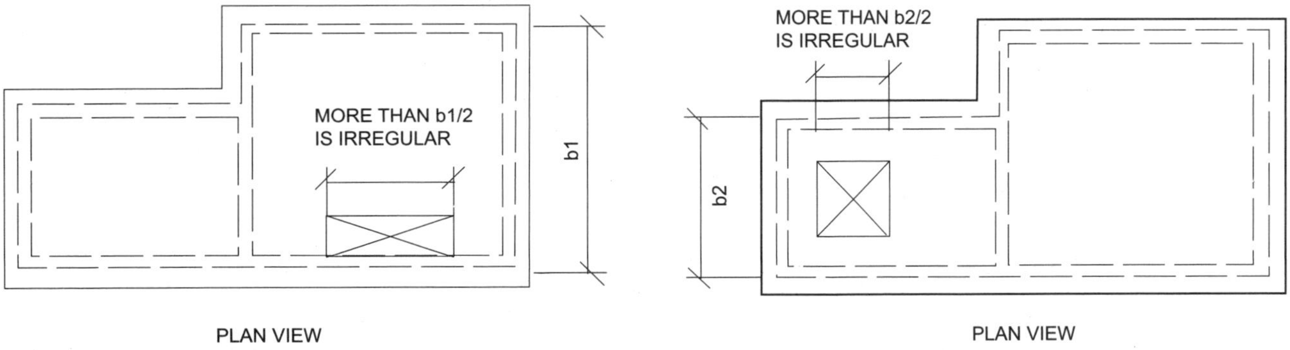

FIGURE 2308.4.4.1(2)
OPENING LIMITATIONS FOR FLOOR AND ROOF DIAPHRAGMS

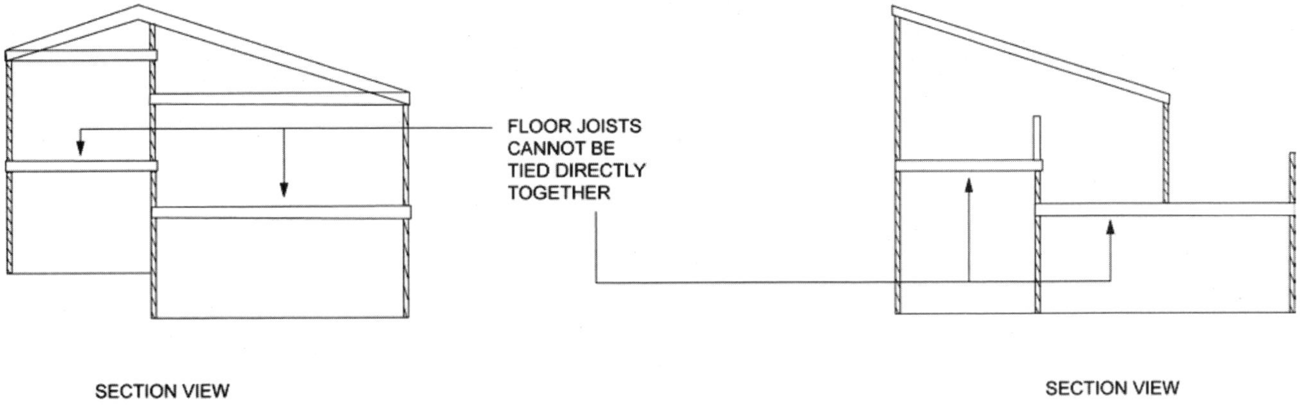

FIGURE 2308.4.4.2
PORTIONS OF FLOOR LEVEL OFFSET VERTICALLY

2308.5 Wall construction. Walls of *conventional light-frame construction* shall be in accordance with this section.

2308.5.1 Stud size, height and spacing. The size, height and spacing of studs shall be in accordance with Table 2308.5.1.

Studs shall be continuous from a support at the sole plate to a support at the top plate to resist *loads* perpendicular to the wall. The support shall be a foundation or floor, ceiling or roof *diaphragm* or shall be designed in accordance with accepted engineering practice.

Exception: Jack studs, trimmer studs and cripple studs at openings in walls that comply with Table 2308.4.1.1(1) or 2308.4.1.1(2).

2308.5.2 Framing details. Studs shall be placed with their wide dimension perpendicular to the wall. Not less than three studs shall be installed at each corner of an *exterior wall*.

Exceptions:

1. In interior nonbearing walls and partitions, studs are permitted to be set with the long dimension parallel to the wall.
2. At corners, two studs are permitted, provided that wood spacers or backup cleats of $^3/_8$-inch-thick (9.5 mm) *wood structural panel*, $^3/_8$-inch (9.5 mm) Type M "Exterior Glue" *particleboard*, 1-inch-thick (25 mm) lumber or other approved devices that will serve as an adequate backing for the attachment of facing materials are used. Where *fire-resistance ratings* or shear values are involved, wood spacers, backup cleats or other devices shall not be used unless specifically *approved* for such use.

2308.5.3 Plates and sills. Studs shall have plates and sills in accordance with this section.

2308.5.3.1 Bottom plate or sill. Studs shall have full bearing on a plate or sill. Plates or sills shall be not less than 2 inches (51 mm) nominal in thickness and have a width not less than the width of the wall studs.

2308.5.3.2 Top plates. Bearing and exterior wall studs shall be capped with double top plates installed to provide overlapping at corners and at intersections with other partitions. End joints in double top plates shall be offset not less than 48 inches (1219 mm), and shall be nailed in accordance with Table 2304.10.2. Plates shall be a nominal 2 inches (51 mm) in depth and have a width not less than the width of the studs.

Exception: A single top plate is permitted, provided that the plate is adequately tied at corners and intersecting walls by not less than the equivalent of 3-inch by 6-inch (76 mm by 152 mm) by 0.036-inch-thick (0.914 mm) galvanized steel plate that is nailed to each wall or segment of wall by six 8d [$2^1/_2$" × 0.113" (64-mm by 2.87 mm)] box nails or equivalent on each side of the joint. For the butt-joint splice between adjacent single top plates, not less than the equivalent of a 3-inch by 12-inch (76 mm by 304 mm) by 0.036-inch-thick (0.914 mm) galvanized steel plate that is nailed to each wall or segment of wall by 12 8d [$2^1/_2$-inch × 0.113-inch (64 mm by 2.87 mm)] box nails on each side of the joint shall be required, provided that the rafters, joists or trusses are centered over the studs with a tolerance of not more than 1 inch (25 mm). The top plate shall not be required over headers that are in the same plane and in line with the upper surface of the adjacent top plates and are tied to adjacent wall sections as required for the butt joint splice between adjacent single top plates.

Where bearing studs are spaced at 24-inch (610 mm) intervals, top plates are less than two 2-inch by 6-inch (51 mm by 152 mm) or two 3-inch by 4-inch (76 mm by 102 mm) members and the floor joists, floor trusses or roof trusses that they support are spaced at more than 16-inch (406 mm) intervals, such joists or trusses shall bear within 5 inches (127 mm) of the studs beneath or a third plate shall be installed.

2308.5.4 Nonload-bearing walls and partitions. In *nonload-bearing walls* and partitions, that are not part of a *braced wall panel*, studs shall be spaced not more than

TABLE 2308.5.1
SIZE, HEIGHT AND SPACING OF WOOD STUDS[c]

STUD SIZE (inches)	BEARING WALLS				NONBEARING WALLS	
	Laterally unsupported stud height[a] (feet)	Supporting roof and ceiling only	Supporting one floor, roof and ceiling	Supporting two floors, roof and ceiling	Laterally unsupported stud height[a] (feet)	Spacing (inches)
		Spacing (inches)				
2 × 3[b]	—	—	—	—	10	16
2 × 4	10	24	16	—	14	24
3 × 4	10	24	24	16	14	24
2 × 5	10	24	24	—	16	24
2 × 6	10	24	24	16	20	24

For SI: 1 inch = 25.4 mm, 1 foot = 304.8 mm.

a. Listed heights are distances between points of lateral support placed perpendicular to the plane of the wall. Increases in unsupported height are permitted where justified by an analysis.

b. Shall not be used in exterior walls.

c. Utility-grade studs shall not be spaced more than 16 inches on center or support more than a roof and ceiling, or exceed 8 feet in height for exterior walls and load-bearing walls or 10 feet for interior nonload-bearing walls.

24 inches (610 mm) on center. In interior *nonload-bearing walls* and partitions, studs are permitted to be set with the long dimension parallel to the wall. Where studs are set with the long dimensions parallel to the wall, use of utility *grade lumber* or studs exceeding 10 feet (3048 mm) is not permitted. Interior nonload-bearing partitions shall be capped with not less than a single top plate installed to provide overlapping at corners and at intersections with other walls and partitions. The plate shall be continuously tied at joints by solid blocking not less than 16 inches (406 mm) in length and equal in size to the plate or by $^1/_2$-inch by $1^1/_2$-inch (12.7 mm by 38 mm) metal ties with spliced sections fastened with two 16d nails on each side of the joint.

2308.5.5 Openings in walls and partitions. Openings in exterior and interior walls and partitions shall comply with Sections 2308.5.5.1 through 2308.5.5.3.

2308.5.5.1 Openings in exterior bearing walls. Headers shall be provided over each opening in exterior bearing walls. The size and spans in Table 2308.4.1.1(1) are permitted to be used for one- and two-family *dwellings*. Headers for other buildings shall be designed in accordance with Section 2302.1, Item 1 or 2. Headers of two or more pieces of nominal 2-inch (51 mm) framing lumber set on edge shall be permitted in accordance with Table 2308.4.1.1(1) and nailed together in accordance with Table 2304.10.2 or of solid lumber of equivalent size.

Single-member headers of nominal 2-inch (51 mm) thickness shall be framed with a single flat 2-inch-nominal (51 mm) member or wall plate not less in width than the wall studs on the top and bottom of the header in accordance with Figures 2308.5.5.1(1) and 2308.5.5.1(2) and face nailed to the top and bottom of the header with 10d box nails [3 inches × 0.128 inches (76 mm × 3.3 mm)] spaced 12 inches (305 mm) on center.

Wall studs shall support the ends of the header in accordance with Table 2308.4.1.1(1). Each end of a lintel or header shall have a bearing length of not less than $1^1/_2$ inches (38 mm) for the full width of the lintel.

2308.5.5.2 Openings in interior bearing partitions. Headers shall be provided over each opening in interior bearing partitions as required in Section 2308.5.5.1. The spans in Table 2308.4.1.1(2) are permitted to be used. Wall studs shall support the ends of the header in accordance with Table 2308.4.1.1(1) or 2308.4.1.1(2), as applicable.

2308.5.5.3 Openings in interior nonbearing partitions. Openings in nonbearing partitions are permitted to be framed with single studs and headers. Each end of a lintel or header shall have a bearing length of not less than $1^1/_2$ inches (38 mm) for the full width of the lintel.

2308.5.6 Cripple walls. Foundation *cripple walls* shall be framed of studs that are not less than the size of the studding above. Exterior *cripple wall* studs shall be not less than 14 inches (356 mm) in length, or shall be framed of solid blocking. Where exceeding 4 feet (1219 mm) in height, such walls shall be framed of studs having the size required for an additional *story*. See Section 2308.6.6 for *cripple wall* bracing.

2308.5.7 Bridging. Unless covered by interior or *exterior wall coverings* or sheathing meeting the minimum requirements of this code, stud partitions or walls with studs having a height-to-least-thickness ratio exceeding 50 shall have bridging that is not less than 2 inches (51 mm) in thickness and of the same width as the studs fitted snugly and nailed thereto to provide adequate lateral support. Bridging shall be placed in every stud cavity and at a frequency such that studs so braced shall not have a height-to-least-thickness ratio exceeding 50 with the height of the stud measured between horizontal framing and bridging or between bridging, whichever is greater.

2308.5.8 Pipes in walls. Stud partitions containing plumbing, heating or other pipes shall be framed and the joists underneath spaced to provide proper clearance for the piping. Where a partition containing piping runs parallel to the floor joists, the joists underneath such partitions shall be doubled and spaced to permit the passage of pipes and shall be bridged. Where plumbing, heating or other pipes are placed in, or partly in, a partition, necessitating the cutting of the soles or plates, a metal tie not less

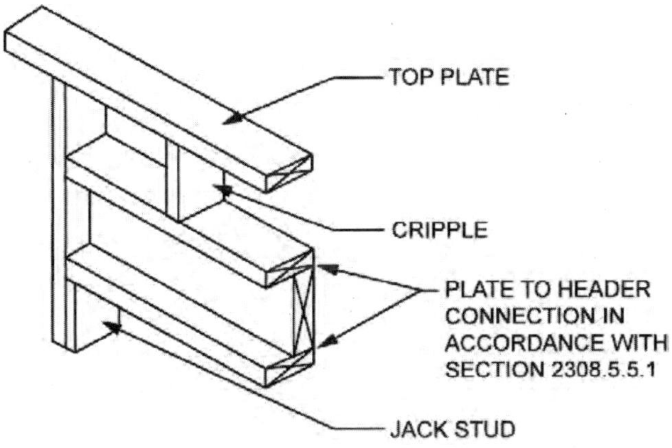

FIGURE 2308.5.5.1(1)
SINGLE-MEMBER HEADER IN EXTERIOR BEARING WALL

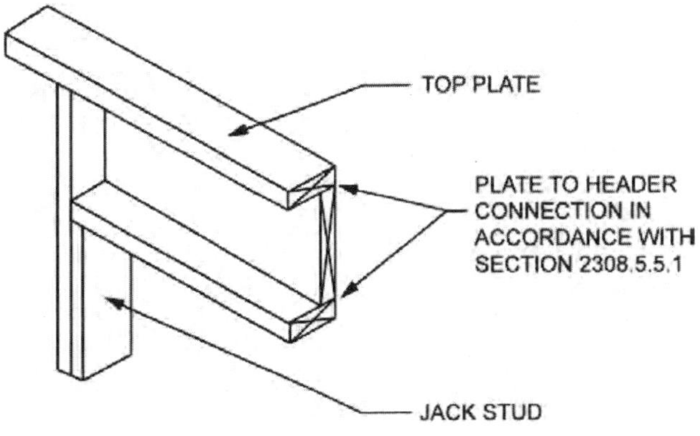

FIGURE 2308.5.5.1(2)
ALTERNATIVE SINGLE-MEMBER HEADER WITHOUT CRIPPLE

than 0.058 inch (1.47 mm) (16 galvanized gage) and $1^1/_2$ inches (38 mm) in width shall be fastened to each plate across and to each side of the opening with not less than six 16d nails.

2308.5.9 Cutting and notching. In *exterior walls* and bearing partitions, a wood stud shall not be cut or notched in excess of 25 percent of its depth. In nonbearing partitions that do not support *loads* other than the weight of the partition, a stud shall not be cut or notched in excess of 40 percent of its depth.

2308.5.10 Bored holes. The diameter of bored holes in wood studs shall not exceed 40 percent of the stud depth. The diameter of bored holes in wood studs shall not exceed 60 percent of the stud depth in nonbearing partitions. The diameter of bored holes in wood studs shall not exceed 60 percent of the stud depth in any wall where each stud is doubled, provided that not more than two such successive doubled studs are so bored. The edge of the bored hole shall not be closer than $^5/_8$ inch (15.9 mm) to the edge of the stud. Bored holes shall not be located at the same section of stud as a cut or notch.

2308.5.11 Exterior wall sheathing. Except where stucco construction that complies with Section 2510 is installed, the outside of *exterior walls*, including *gables*, of enclosed buildings shall be sheathed with one of the materials of the nominal thickness specified in Table 2308.5.11 with fasteners in accordance with the requirements of Section 2304.10 or fasteners designed in accordance with accepted engineering practice. Alternatively, sheathing materials and fasteners complying with Section 2304.6 shall be permitted.

2308.6 Wall bracing. Buildings shall be provided with exterior and interior *braced wall lines* as described in Sections 2308.6.1 through 2308.6.10.2.

2308.6.1 Braced wall lines. For the purpose of determining the amount and location of bracing required along each story level of a building, *braced wall lines* shall be designated as straight lines through the building plan in both the longitudinal and transverse direction and placed in accordance with Table 2308.6.1 and Figure 2308.6.1. *Braced wall line* spacing shall not exceed the distance specified in Table 2308.6.1. In structures assigned to *Seismic Design Category* D or E, *braced wall lines* shall intersect perpendicularly to each other.

2308.6.2 Braced wall panels. *Braced wall panels* shall be placed along *braced wall lines* in accordance with Table 2308.6.1 and Figure 2308.6.1 and as specified in Table 2308.6.3(1). A *braced wall panel* shall be located at each end of the *braced wall line* and at the corners of intersecting *braced wall lines* or shall begin within the maximum distance from the end of the *braced wall line* in accordance with Table 2308.6.1. *Braced wall panels* in a *braced wall line* shall not be offset from each other by more than 4 feet (1219 mm). *Braced wall panels* shall be clearly indicated on the plans.

2308.6.3 Braced wall panel methods. Construction of *braced wall panels* shall be by one or a combination of the methods in Table 2308.6.3(1). *Braced wall panel* length shall be in accordance with Section 2308.6.4 or 2308.6.5.

2308.6.4 Braced wall panel construction. For Methods DWB, WSP, SFB, PBS, PCP and HPS, each panel must be not less than 48 inches (1219 mm) in length, covering three stud spaces where studs are spaced 16 inches (406 mm) on center and covering two stud spaces where studs are spaced 24 inches (610 mm) on center. *Braced wall panels* less than 48 inches (1219 mm) in length shall not contribute toward the amount of required bracing. *Braced wall panels* that are longer than the required length shall be credited for their actual length. For Method GB, each panel must be not less than 96 inches (2438 mm) in length where applied to one side of the studs or 48 inches (1219 mm) in length where applied to both sides.

Vertical joints of panel sheathing shall occur over studs and adjacent panel joints shall be nailed to common framing members. Horizontal joints shall occur over blocking or other framing equal in size to the studding except where waived by the installation requirements for the specific sheathing materials. Sole plates shall be nailed to the floor framing in accordance with Section 2308.6.7 and top plates shall be connected to the framing above in accordance with Section 2308.6.7.2. Where joists are perpendicular to *braced wall lines* above, blocking shall be provided under and in line with the braced *wall panels*.

TABLE 2308.5.11
MINIMUM THICKNESS OF WALL SHEATHING

SHEATHING TYPE	MINIMUM THICKNESS	MAXIMUM WALL STUD SPACING
Diagonal wood boards	$^5/_8$ inch	24 inches on center
Structural fiberboard	$^1/_2$ inch	16 inches on center
Wood structural panel	In accordance with Tables 2308.6.3(2) and 2308.6.3(3)	—
M-S "Exterior Glue" and M-2 "Exterior Glue" particleboard	In accordance with Section 2306.3 and Table 2308.6.3(4)	—
Gypsum sheathing	$^1/_2$ inch	16 inches on center
Reinforced cement mortar	1 inch	24 inches on center
Hardboard panel siding	In accordance with Table 2308.6.3(5)	—

For SI: 1 inch = 25.4 mm.

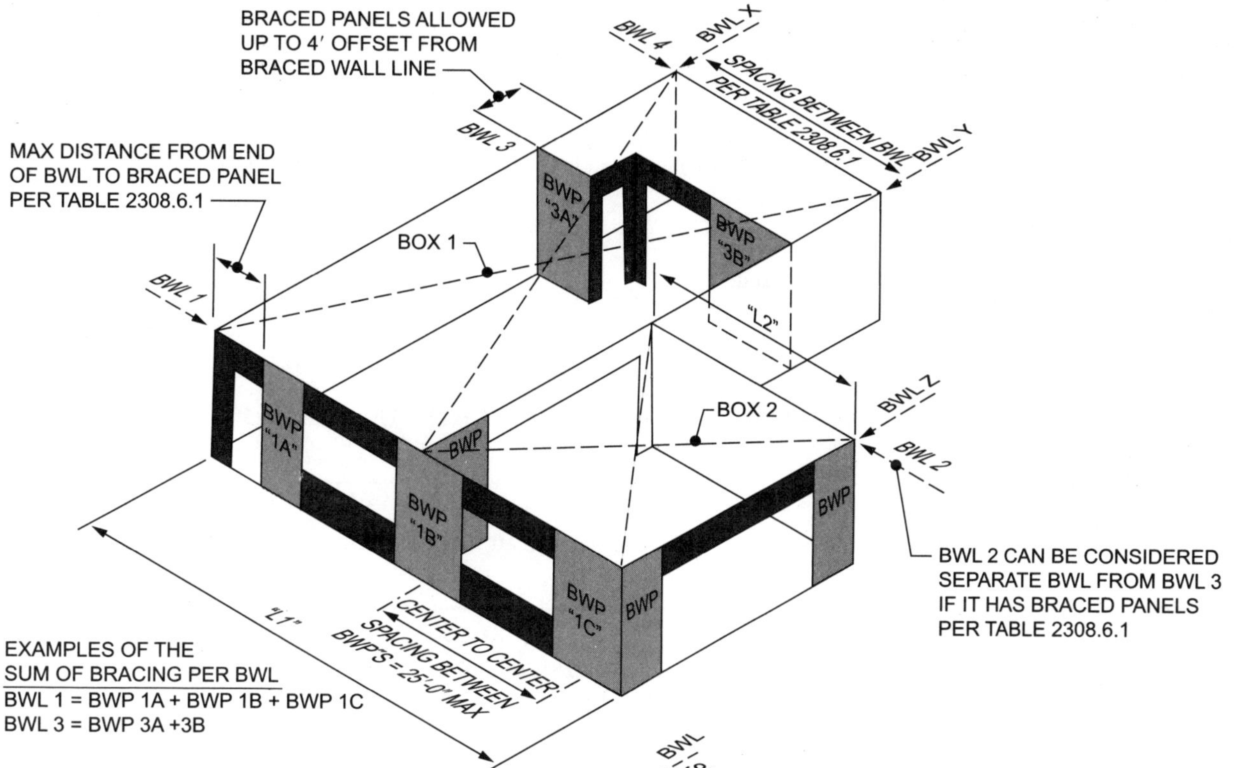

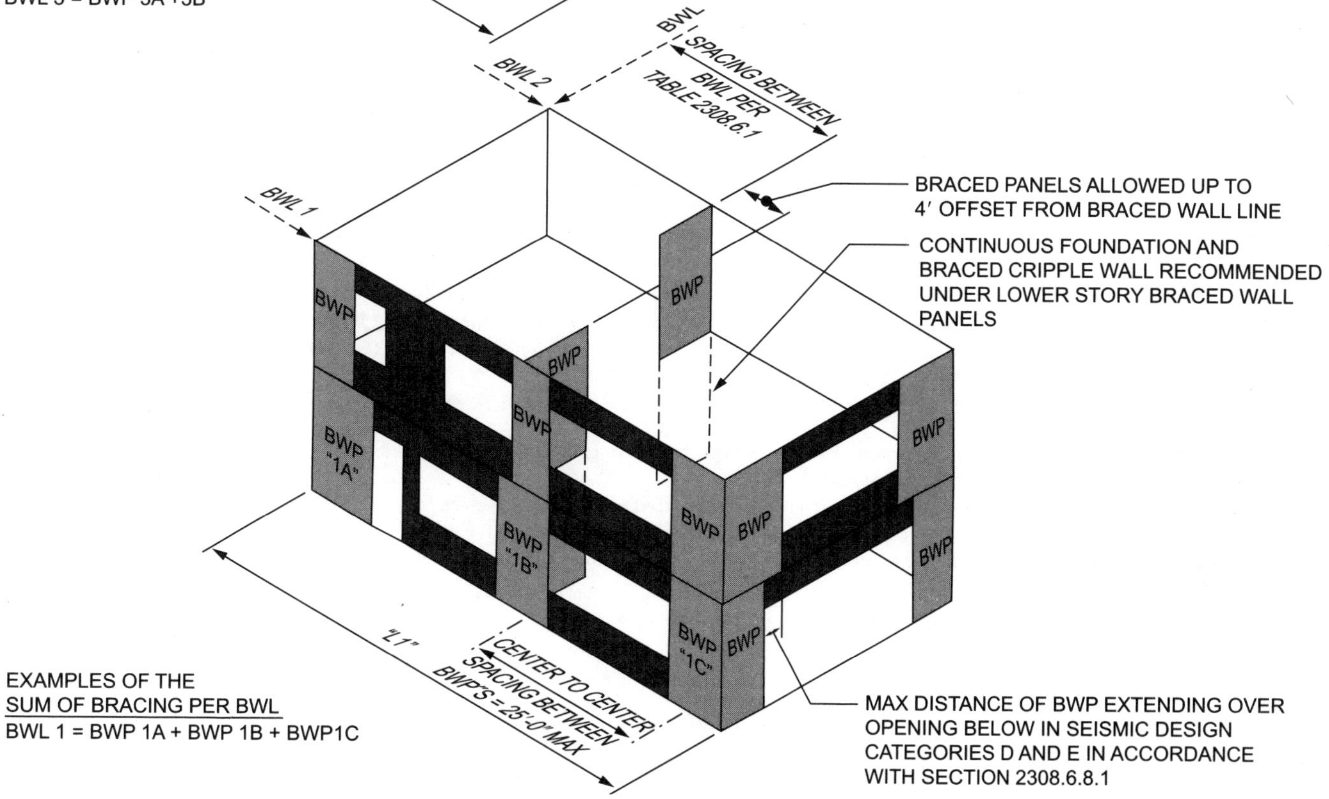

For SI: 1 foot = 304.8 mm.

FIGURE 2308.6.1
BASIC COMPONENTS OF THE LATERAL BRACING SYSTEM

TABLE 2308.6.1
WALL BRACING REQUIREMENTS[a]

SEISMIC DESIGN CATEGORY	STORY CONDITION (SEE SECTION 2308.2)	MAXIMUM SPACING OF BRACED WALL LINES	BRACED PANEL LOCATION, SPACING (O.C.) AND MINIMUM PERCENTAGE (X)			MAXIMUM DISTANCE OF BRACED WALL PANELS FROM EACH END OF BRACED WALL LINE
			Bracing method[b]			
			LIB	DWB, WSP	SFB, PBS, PCP, HPS, GB[c, d]	
A and B	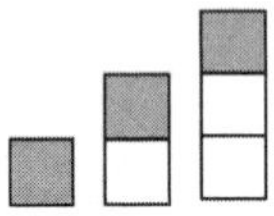	35′ - 0″	Each end and ≤ 25′- 0″ o.c.	Each end and ≤ 25′- 0″ o.c.	Each end and ≤ 25′- 0″ o.c.	12′- 6″
	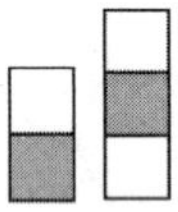	35′- 0″	Each end and ≤ 25′- 0″ o.c.	Each end and ≤ 25′- 0″ o.c.	Each end and ≤ 25′- 0″ o.c.	12′- 6″
	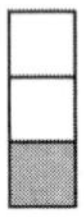	35′- 0″	NP	Each end and ≤ 25′- 0″ o.c.	Each end and ≤ 25′- 0″ o.c.	12′- 6″
C		35′- 0″	NP	Each end and ≤ 25′- 0″ o.c.	Each end and ≤ 25′- 0″ o.c.	12′- 6″
	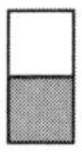	35′- 0″	NP	Each end and ≤ 25′- 0″ o.c. (minimum 25% of wall length)[e]	Each end and ≤ 25′- 0″ o.c. (minimum 25% of wall length)[e]	12′- 6″
D and E		25′- 0″	NP	$S_{DS} < 0.50$: Each end and ≤ 25′- 0″ o.c. (minimum 21% of wall length)[e]	$S_{DS} < 0.50$: Each end and ≤ 25′- 0″ o.c. (minimum 43% of wall length)[e]	8′- 0″
				$0.5 \leq S_{DS} < 0.75$: Each end and ≤ 25′- 0″ o.c. (minimum 32% of wall length)[e]	$0.5 \leq S_{DS} < 0.75$: Each end and ≤ 25′- 0″ o.c. (minimum 59% of wall length)[e]	
				$0.75 \leq S_{DS} \leq 1.00$: Each end and ≤ 25′- 0″ o.c. (minimum 37% of wall length)[e]	$0.75 \leq S_{DS} \leq 1.00$: Each end and ≤ 25′- 0″ o.c. (minimum 75% of wall length)	
				$S_{DS} > 1.00$: Each end and ≤ 25′- 0″ o.c. (minimum 48% of wall length)[e]	$S_{DS} > 1.00$: Each end and ≤ 25′- 0″ o.c. (minimum 100% of wall length)[e]	

For SI: 1 inch = 25.4 mm, 1 foot = 304.8 mm.

NP = Not Permitted.

a. This table specifies minimum requirements for braced wall panels along interior or exterior braced wall lines.

b. See Section 2308.6.3 for full description of bracing methods.

c. For Method GB, gypsum wallboard applied to framing supports that are spaced at 16 inches on center.

d. The required lengths shall be doubled for gypsum board applied to only one face of a braced wall panel.

e. Percentage shown represents the minimum amount of bracing required along the building length (or wall length if the structure has an irregular shape).

TABLE 2308.6.3(1)
BRACING METHODS

METHODS, MATERIAL	MINIMUM THICKNESS	FIGURE	CONNECTION CRITERIA[a]	
			Fasteners	Spacing
LIB[a] Let-in-bracing	1″ × 4″ wood or approved metal straps attached at 45° to 60° angles to studs at maximum of 16″ o.c.	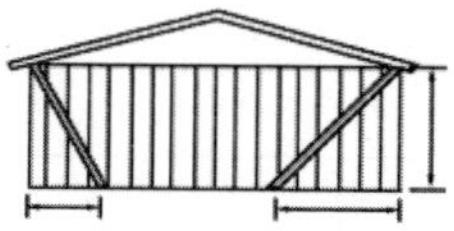	Table 2304.10.2	Wood: per stud plus top and bottom plates
			Metal strap: installed in accordance with manufacturer's recommendations	Metal strap: installed in accordance with manufacturer's recommendations
DWB Diagonal wood boards	$^3/_4$″ thick (1″ nominal) × 6″ minimum width to studs at maximum of 24″ o.c.	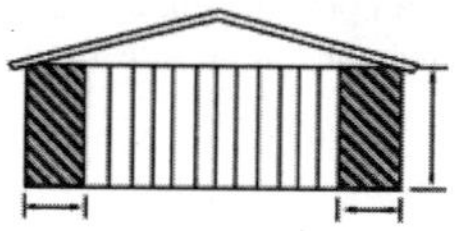	Table 2304.10.2	Per stud
WSP Wood structural panel	$^3/_8$″ in accordance with Table 2308.6.3(2) or 2308.6.3(3)	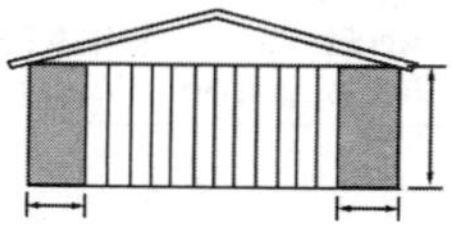	Table 2304.10.2	6″ edges 12″ field
SFB Structural fiberboard sheathing	$^1/_2$″ in accordance with Table 2304.10.2 to studs at maximum 16″ o.c.	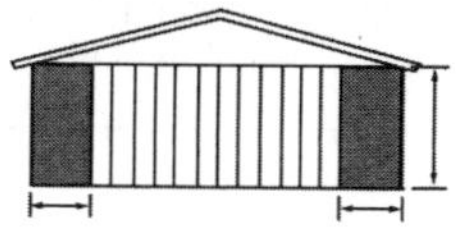	Table 2304.10.2	3″ edges 6″ field
GB Gypsum board (Double sided)	$^1/_2$″ or $^5/_8$″ by not less than 4′ wide to studs at maximum of 24" o.c.	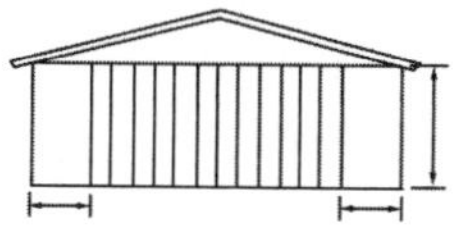	Section 2506.2 for exterior and interior sheathing: 5d annual ringed cooler nails (1$^5/_8$″ × 0.086″) or 1$^1/_4$″ screws (Type W or S) for $^1/_2$″ gypsum board or 1$^5/_8$″ screws (Type W or S) for $^5/_8$″ gypsum board	For all braced wall panel locations: 7″ o.c. along panel edges (including top and bottom plates) and 7" o.c. in the field
PBS Particleboard sheathing	$^3/_8$″ or $^1/_2$″ in accordance with Table 2308.6.3(4) to studs at maximum of 16″ o.c.	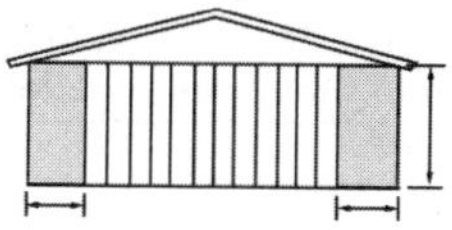	6d common (2″ long × 0.113″ dia.) nails for $^3/_8$″ thick sheathing or 8d common (2$^1/_2$″ long × 0.131″ dia.) nails for $^1/_2$″ thick sheathing	3″ edges 6″ field
PCP Portland cement plaster	Section 2510 to studs at maximum of 16″ o.c.	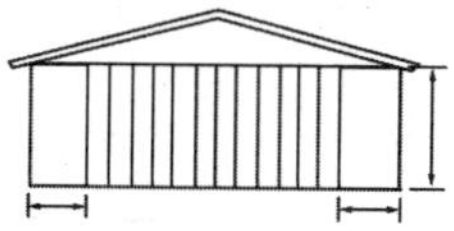	1$^1/_2$″ long, 11 gage, 0.120" dia., $^7/_{16}$″ dia. head nails or $^7/_8$″ long, 16 gage staples	6″ o.c. on all framing members
HPS Hardboard panel siding	$^7/_{16}$″ in accordance with Table 2308.6.3(5)	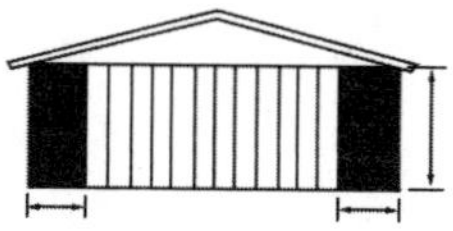	Table 2304.10.2	4″ edges 8″ field
ABW Alternate braced wall	$^3/_8$″	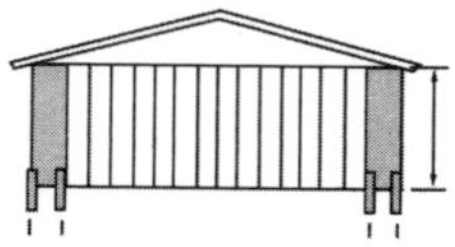	Figure 2308.6.5.1 and Section 2308.6.5.1	Figure 2308.6.5.1
PFH Portal frame with hold-downs	$^3/_8$″	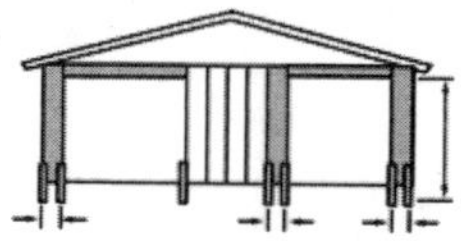	Figure 2308.6.5.2 and Section 2308.6.5.2	Figure 2308.6.5.2

For SI: 1 foot = 304.8 mm, 1 degree = 0.01745 rad.

a. Method LIB shall have gypsum board fastened to one or more side(s) with nails or screws

TABLE 2308.6.3(2)
EXPOSED PLYWOOD PANEL SIDING

MINIMUM THICKNESS[a] (inch)	MINIMUM NUMBER OF PLIES	STUD SPACING (inches) Plywood siding applied directly to studs or over sheathing
$^{3}/_{8}$	3	16[b]
$^{1}/_{2}$	4	24

For SI: 1 inch = 25.4 mm.

a. Thickness of grooved panels is measured at bottom of grooves.

b. Spans are permitted to be 24 inches if plywood siding applied with face grain perpendicular to studs or over one of the following: 1-inch board sheathing; $^{7}/_{16}$ - inch wood structural panel sheathing; or $^{3}/_{8}$-inch wood structural panel sheathing with strength axis (which is the long direction of the panel unless otherwise marked) of sheathing perpendicular to studs.

TABLE 2308.6.3(3)
WOOD STRUCTURAL PANEL WALL SHEATHING[b]
(Not exposed to the weather, strength axis parallel or perpendicular to studs except as indicated)

MINIMUM THICKNESS (inch)	PANEL SPAN RATING	STUD SPACING (inches)		
		Siding nailed to studs	Nailable sheathing	
			Sheathing parallel to studs	Sheathing perpendicular to studs
$^{3}/_{8}$, $^{15}/_{32}$, $^{1}/_{2}$	16/0, 20/0, 24/0, 32/16 Wall—24″ o.c.	24	16	24
$^{7}/_{16}$, $^{15}/_{32}$, $^{1}/_{2}$	24/0, 24/16, 32/16 Wall—24″ o.c.	24	24[a]	24

For SI: 1 inch = 25.4 mm.

a. Plywood shall consist of four or more plies.

b. Blocking of horizontal joints shall not be required except as specified in Section 2308.6.4.

TABLE 2308.6.3(4)
ALLOWABLE SPANS FOR PARTICLEBOARD WALL SHEATHING
(Not exposed to the weather, long dimension of the panel parallel or perpendicular to studs)

GRADE	THICKNESS (inch)	STUD SPACING (inches)	
		Siding nailed to studs	Sheathing under coverings specified in Section 2308.6.3 parallel or perpendicular to studs
M-S “Exterior Glue” and M-2 “Exterior Glue”	$^{3}/_{8}$	16	—
	$^{1}/_{2}$	16	16

For SI: 1 inch = 25.4 mm.

TABLE 2308.6.3(5)
HARDBOARD SIDING

SIDING	MINIMUM NOMINAL THICKNESS (inch)	2 × 4 FRAMING MAXIMUM SPACING	NAIL SIZE[a, b, d]	NAIL SPACING	
				General	Bracing panels[c]
1. Lap siding					
Direct to studs	$^3/_8$	16″ o.c.	8d	16″ o.c.	Not applicable
Over sheathing	$^3/_8$	16″ o.c.	10d	16″ o.c.	Not applicable
2. Square edge panel siding					
Direct to studs	$^3/_8$	24″ o.c.	6d	6″ o.c. edges; 12″ o.c. at intermediate supports	4″ o.c. edges; 8″ o.c. at intermediate supports
Over sheathing	$^3/_8$	24″ o.c.	8d	6″ o.c. edges; 12″ o.c. at intermediate supports	4″ o.c. edges; 8″ o.c. at intermediate supports
3. Shiplap edge panel siding					
Direct to studs	$^3/_8$	16″ o.c.	6d	6″ o.c. edges; 12″ o.c. at intermediate supports	4″ o.c. edges; 8″ o.c. at intermediate supports
Over sheathing	$^3/_8$	16″ o.c.	8d	6″ o.c. edges; 12″ o.c. at intermediate supports	4″ o.c. edges; 8″ o.c. at intermediate supports

For SI: 1 inch = 25.4 mm.

a. Nails shall be corrosion resistant.

b. Minimum acceptable nail dimensions:

	Panel Siding (inch)	Lap Siding (inch)
Shank diameter	0.092	0.099
Head diameter	0.225	0.240

c. Where used to comply with Section 2308.6.

d. Nail length must accommodate the sheathing and penetrate framing $1^1/_2$ inches.

2308.6.5 Alternative bracing. An alternate braced wall (ABW) or a portal frame with *hold-downs* (PFH) described in this section is permitted to substitute for a 48-inch (1219 mm) *braced wall panel* of Method DWB, WSP, SFB, PBS, PCP or HPS. For Method GB, each 96-inch (2438 mm) section (applied to one face) or 48-inch (1219 mm) section (applied to both faces) or portion thereof required by Table 2308.6.1 is permitted to be replaced by one panel constructed in accordance with Method ABW or PFH.

2308.6.5.1 Alternate braced wall (ABW). An ABW shall be constructed in accordance with this section and Figure 2308.6.5.1. In one-story buildings, each panel shall have a length of not less than 2 feet 8 inches (813 mm) and a height of not more than 10 feet (3048 mm). Each panel shall be sheathed on one face

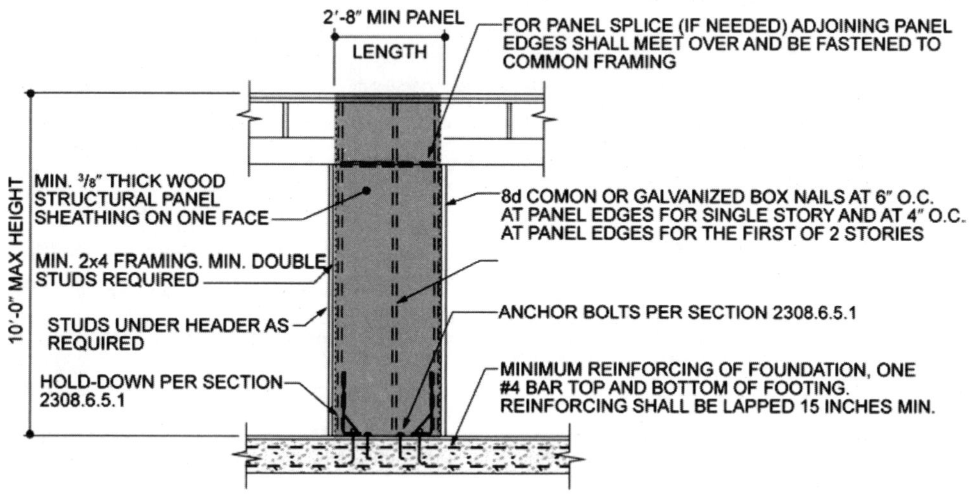

For SI:1 inch = 25.4 mm, 1 foot = 304.8 mm.

FIGURE 2308.6.5.1
ALTERNATE BRACED WALL PANEL (ABW)

with $^3/_8$-inch (3.2 mm) minimum-thickness *wood structural panel* sheathing nailed with 8d common or galvanized box nails in accordance with Table 2304.10.2 and blocked at *wood structural panel* edges. Two anchor bolts installed in accordance with Section 2308.3.1 shall be provided in each panel. Anchor bolts shall be placed at each panel outside quarter points. Each panel end stud shall have a *hold-down* device fastened to the foundation, capable of providing an *approved* uplift capacity of not less than 1,800 pounds (8006 N). The *hold-down* device shall be installed in accordance with the manufacturer's recommendations. The ABW shall be supported directly on a foundation or on floor framing supported directly on a foundation that is continuous across the entire length of the *braced wall line*. This foundation shall be reinforced with not less than one No. 4 bar top and bottom. Where the continuous foundation is required to have a depth greater than 12 inches (305 mm), a minimum 12-inch by 12-inch (305 mm by 305 mm) continuous footing or turned-down slab edge is permitted at door openings in the *braced wall line*. This continuous footing or turned-down slab edge shall be reinforced with not less than one No. 4 bar top and bottom. This reinforcement shall be lapped 15 inches (381 mm) with the reinforcement required in the continuous foundation located directly under the *braced wall line*.

Where the ABW is installed at the first *story* of two-story buildings, the *wood structural panel* sheathing shall be provided on both faces, three anchor bolts shall be placed at one-quarter points and *tie-down* device uplift capacity shall be not less than 3,000 pounds (13 344 N).

2308.6.5.2 Portal frame with hold-downs (PFH). A PFH shall be constructed in accordance with this section and Figure 2308.6.5.2. The adjacent door or window opening shall have a full-length header.

In one-story buildings, each panel shall have a length of not less than 16 inches (406 mm) and a height of not more than 10 feet (3048 mm). Each panel shall be sheathed on one face with a single layer of $^3/_8$-inch (9.5 mm) minimum-thickness *wood structural panel* sheathing nailed with 8d common or galvanized box nails in accordance with Figure 2308.6.5.2. The *wood structural panel* sheathing shall extend up over the solid sawn or glued-laminated header and shall be nailed in accordance with Figure 2308.6.5.2. A built-up header consisting of not fewer than two 2-inch by 12-inch (51 mm by 305 mm) boards, fastened in accordance with Item 24 of Table 2304.10.2 shall be permitted to be used. A spacer, if used, shall be placed on the side of the built-up beam opposite the *wood structural panel* sheathing. The header shall extend between the inside faces of the first full-length outer studs of each panel. The clear span of the header between the inner studs of each panel shall be not less than 6 feet (1829 mm) and not more than 18 feet (5486 mm) in length. A strap with an uplift capacity of not

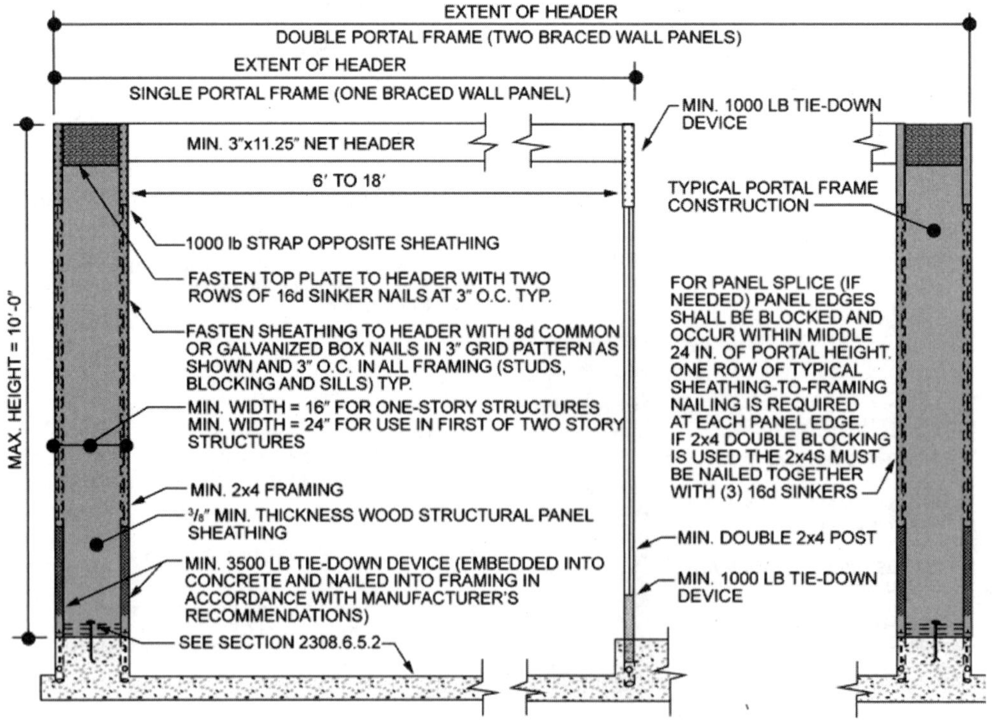

For SI:1 inch = 25.4 mm, 1 foot = 304.8 mm, 1 pound = 4.448 N.

FIGURE 2308.6.5.2
PORTAL FRAME WITH HOLD-DOWNS (PFH)

less than 1,000 pounds (4,400 N) shall fasten the header to the inner studs opposite the sheathing. One anchor bolt not less than $^5/_8$ inch (15.9 mm) diameter and installed in accordance with Section 2308.3.1 shall be provided in the center of each sill plate. The studs at each end of the panel shall have a *hold-down* device fastened to the foundation with an uplift capacity of not less than 3,500 pounds (15 570 N).

Where a panel is located on one side of the opening, the header shall extend between the inside face of the first full-length stud of the panel and the bearing studs at the other end of the opening. A strap with an uplift capacity of not less than 1,000 pounds (4400 N) shall fasten the header to the bearing studs. The bearing studs shall have a *hold-down* device fastened to the foundation with an uplift capacity of not less than 1,000 pounds (4400 N). The *hold-down* devices shall be an embedded strap type, installed in accordance with the manufacturer's recommendations. The PFH panels shall be supported directly on a foundation that is continuous across the entire length of the *braced wall line*. This foundation shall be reinforced with not less than one No. 4 bar top and bottom. Where the continuous foundation is required to have a depth greater than 12 inches (305 mm), a minimum 12-inch by 12-inch (305 mm by 305 mm) continuous footing or turned-down slab edge is permitted at door openings in the *braced wall line*. This continuous footing or turned-down slab edge shall be reinforced with not less than one No. 4 bar top and bottom. This reinforcement shall be lapped not less than 15 inches (381 mm) with the reinforcement required in the continuous foundation located directly under the *braced wall line*.

Where a PFH is installed at the first *story* of two-story buildings, each panel shall have a length of not less than 24 inches (610 mm).

2308.6.6 Cripple wall bracing. *Cripple walls* shall be braced in accordance with Section 2308.6.6.1 or 2308.6.6.2.

2308.6.6.1 Cripple wall bracing in Seismic Design Categories A, B and C. For the purposes of this section, *cripple walls* in *Seismic Design Categories* A, B and C having a stud height exceeding 14 inches (356 mm) shall be considered to be a *story* and shall be braced in accordance with Table 2308.6.1. Spacing of *edge nailing* for required *cripple wall* bracing shall not exceed 6 inches (152 mm) on center along the foundation plate and the top plate of the *cripple wall*. Nail size, nail spacing for *field nailing* and more restrictive *boundary nailing* requirements shall be as required elsewhere in the code for the specific bracing material used.

2308.6.6.2 Cripple wall bracing in Seismic Design Categories D and E. For the purposes of this section, *cripple walls* in *Seismic Design Categories* D and E shall not have a stud height exceeding 14 inches (356 mm), and studs shall be solid blocked in accordance with Section 2308.5.6 for the full dwelling perimeter and for the full length of interior braced walls lines supported on foundations, excepting ventilation and access openings.

2308.6.7 Connections of braced wall panels. *Braced wall panel* joints shall occur over studs or blocking. *Braced wall panels* shall be fastened to studs, top and bottom plates and at panel edges. *Braced wall panels* shall be applied to nominal 2-inch-wide [actual $1^1/_2$-inch (38 mm)] or larger stud framing.

2308.6.7.1 Bottom plate connection. *Braced wall line* bottom plates shall be connected to joists or full-depth blocking below in accordance with Table 2304.10.2, or to foundations in accordance with Section 2308.6.7.3.

2308.6.7.2 Top plate connection. Where joists or rafters are used, *braced wall line* top plates shall be fastened over the full length of the *braced wall line* to joists, rafters, rim boards or full-depth blocking above in accordance with Table 2304.10.2, as applicable, based on the orientation of the joists or rafters to the *braced wall line*. Blocking shall be not less than 2 inches (51 mm) in nominal thickness and shall be fastened to the *braced wall line* top plate as specified in Table 2304.10.2. Notching or drilling of holes in blocking in accordance with the requirements of Section 2308.4.2.4 or 2308.7.4 shall be permitted.

At exterior *gable* end walls, *braced wall panel* sheathing in the top story shall be extended and fastened to the roof framing where the spacing between parallel exterior braced wall lines is greater than 50 feet (15 240 mm).

Where roof trusses are used and are installed perpendicular to an exterior *braced wall line*, lateral forces shall be transferred from the roof *diaphragm* to the braced wall over the full length of the *braced wall line* by blocking of the ends of the trusses or by other *approved* methods providing equivalent lateral force transfer. Blocking shall be not less than 2 inches (51 mm) in nominal thickness and equal to the depth of the truss at the wall line and shall be fastened to the *braced wall line* top plate as specified in Table 2304.10.2. Notching or drilling of holes in blocking in accordance with the requirements of Section 2308.4.2.4 or 2308.7.4 shall be permitted.

Exception: Where the roof sheathing is greater than $9^1/_4$ inches (235 mm) above the top plate, solid blocking is not required where the framing members are connected using one of the following methods:

1. In accordance with Figure 2308.6.7.2(1).
2. In accordance with Figure 2308.6.7.2(2).
3. Full-height engineered blocking panels designed for values listed in AWC WFCM.
4. A design in accordance with accepted engineering methods.

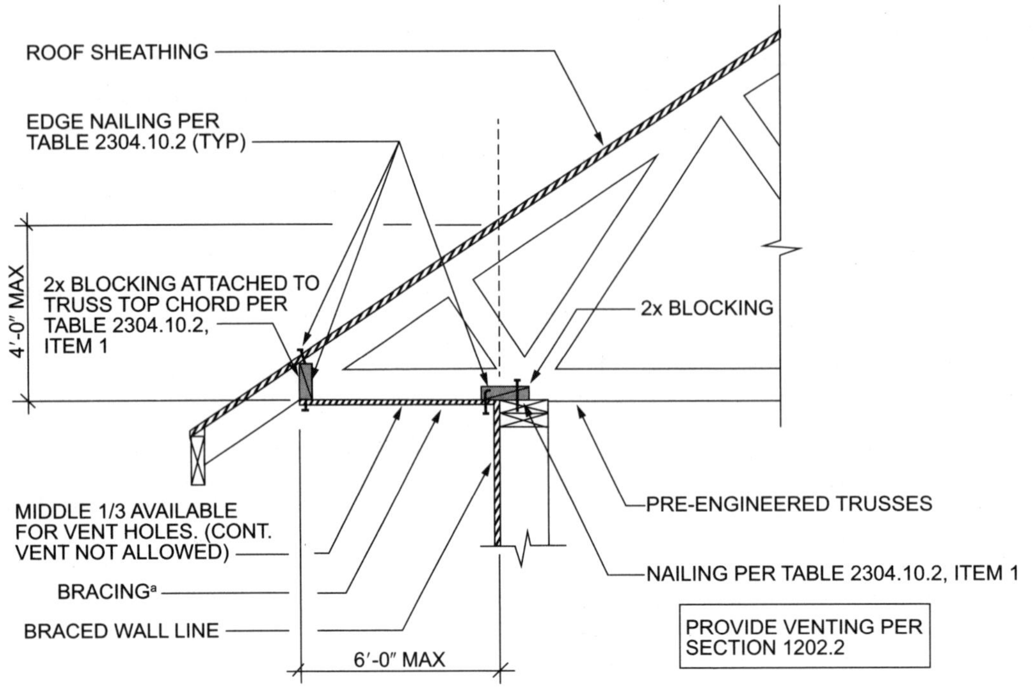

For SI:1 foot = 304.8 mm.

a. Methods of bracing shall be as described in Table 2308.6.3(1) DWB, WSP, SFB, GB, PBS, PCP or HPS.

FIGURE 2308.6.7.2(1)
BRACED WALL LINE TOP PLATE CONNECTION

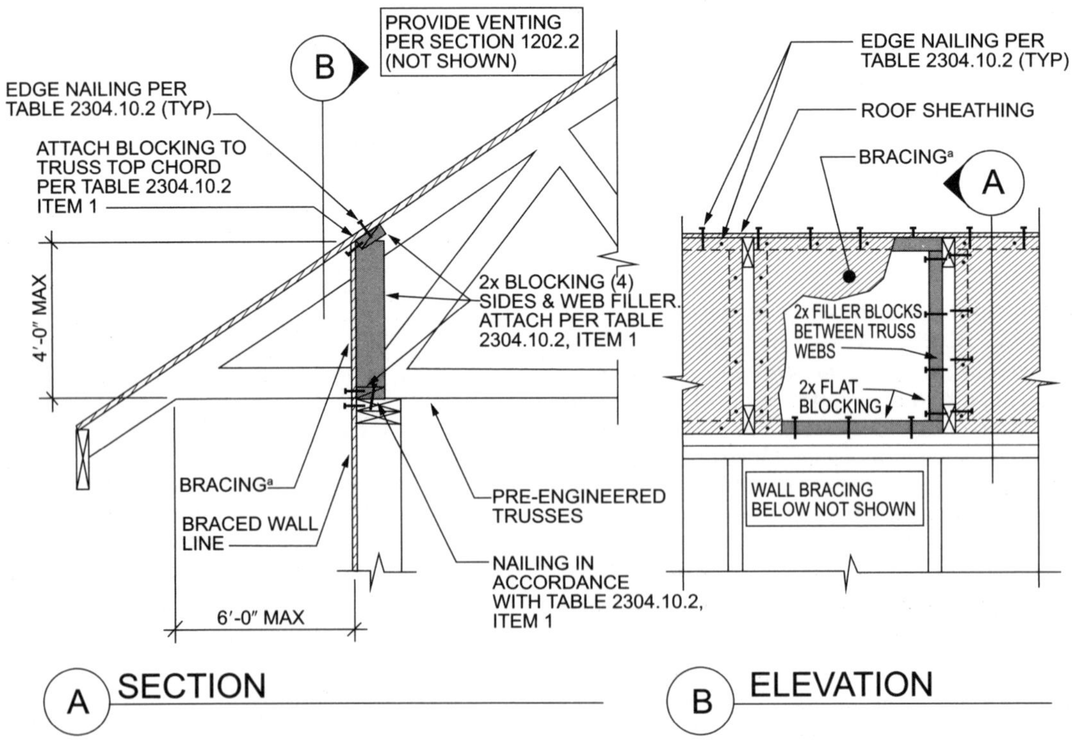

For SI:1 foot = 304.8 mm.

a. Methods of bracing shall be as described in Table 2308.6.3(1) DWB, WSP, SFB, GB, PBS, PCP or HPS.

FIGURE 2308.6.7.2(2)
BRACED WALL PANEL TOP PLATE CONNECTION

2308.6.7.3 Sill anchorage. Where foundations are required by Section 2308.6.8, *braced wall line* sills shall be anchored to concrete or masonry foundations. Such anchorage shall conform to the requirements of Section 2308.3. The anchors shall be distributed along the length of the *braced wall line*. Other anchorage devices having equivalent capacity are permitted.

2308.6.7.4 Anchorage to all-wood foundations. Where all-wood foundations are used, the force transfer from the *braced wall lines* shall be determined based on calculation and shall have a capacity that is not less than the connections required by Section 2308.3.

2308.6.8 Braced wall line and diaphragm support. *Braced wall lines* and floor and roof *diaphragms* shall be supported in accordance with this section.

2308.6.8.1 Foundation requirements. *Braced wall lines* shall be supported by continuous foundations.

Exception: For structures with a maximum plan dimension not more than 50 feet (15 240 mm), continuous foundations are required at *exterior walls* only.

For structures in *Seismic Design Categories* D and E, exterior *braced wall panels* shall be in the same plane vertically with the foundation or the portion of the structure containing the offset shall be designed in accordance with accepted engineering practice and Section 2308.1.1.

Exceptions:

1. Exterior *braced wall panels* shall be permitted to be located not more than 4 feet (1219 mm) from the foundation below where supported by a floor constructed in accordance with all of the following:
   - 1.1. Cantilevers or setbacks shall not exceed four times the nominal depth of the floor joists.
   - 1.2. Floor joists shall be 2 inches by 10 inches (51 mm by 254 mm) or larger and spaced not more than 16 inches (406 mm) on center.
   - 1.3. The ratio of the back span to the cantilever shall be not less than 2 to 1.
   - 1.4. Floor joists at ends of *braced wall panels* shall be doubled.
   - 1.5. A continuous rim joist shall be connected to the ends of cantilevered joists. The rim joist is permitted to be spliced using a metal tie not less than 0.058 inch (1.47 mm) (16 galvanized gage) and $1^1/_2$ inches (38 mm) in width fastened with six 16d common nails on each side. The metal tie shall have a yield stress not less than 33,000 psi (227 MPa).
   - 1.6. Joists at setbacks or the end of cantilevered joists shall not carry gravity *loads* from more than a single *story* having uniform wall and roof *loads* nor carry the reactions from headers having a span of 8 feet (2438 mm) or more.
2. The end of a required *braced wall panel* shall be allowed to extend not more than 1 foot (305 mm) over an opening in the wall below. This requirement is applicable to *braced wall panels* offset in plane and *braced wall panels* offset out of plane as permitted by Exception 1. *Braced wall panels* are permitted to extend over an opening not more than 8 feet (2438 mm) in width where the header is a 4-inch by 12-inch (102 mm by 305 mm) or larger member.

2308.6.8.2 Floor and roof diaphragm support in Seismic Design Categories D and E. In structures assigned to *Seismic Design Categories* D or E, floor and roof *diaphragms* shall be laterally supported by *braced wall lines* on all edges and connected in accordance with Section 2308.6.7 [see Figure 2308.6.8.2(1)].

Exception: Portions of roofs or floors that do not support *braced wall panels* above are permitted to extend up to 6 feet (1829 mm) beyond a *braced wall line* [see Figure 2308.6.8.2(2)] provided that the framing members are connected to the *braced wall line* below in accordance with Section 2308.6.7.

2308.6.8.3 Stepped footings in Seismic Design Categories B, C, D and E. In *Seismic Design Categories* B, C, D and E, where the height of a required *braced wall panel* extending from foundation to floor above varies more than 4 feet (1219 mm), the following construction shall be used:

1. Where the bottom of the footing is stepped and the *lowest floor* framing rests directly on a sill bolted to the footings, the sill shall be anchored as required in Section 2308.3.
2. Where the *lowest floor* framing rests directly on a sill bolted to a footing not less than 8 feet (2438 mm) in length along a line of bracing, the line shall be considered to be braced. The double plate of the cripple stud wall beyond the segment of footing extending to the lowest framed floor shall be spliced to the sill plate with metal ties, one on each side of the sill and plate. The metal ties shall be not less than 0.058 inch [1.47 mm (16 galvanized gage)] by $1^1/_2$ inches (38 mm) in width by 48 inches (1219 mm) with eight 16d common nails on each side of the splice location (see Figure

2308.6.8.3). The metal tie shall have a yield stress not less than 33,000 pounds per square inch (psi) (227 MPa).

3. Where *cripple walls* occur between the top of the footing and the *lowest floor* framing, the bracing requirements for a *story* shall apply.

2308.6.9 Attachment of sheathing. Fastening of *braced wall panel* sheathing shall be not less than that prescribed in Tables 2308.6.1 and 2304.10.2. Wall sheathing shall not be attached to framing members by adhesives.

2308.6.10 Limitations of concrete or masonry veneer. Concrete or masonry *veneer* shall comply with Chapter 14 and this section.

2308.6.10.1 Limitations of concrete or masonry veneer in Seismic Design Category B or C. In *Seismic Design Categories* B and C, concrete or masonry walls and stone or masonry *veneer* shall not extend above a basement.

Exceptions:

1. In structures assigned to *Seismic Design Category* B, stone and masonry *veneer* is permitted to be used in the first two *stories above grade plane* or the first three *stories above grade plane* where the lowest *story* has concrete or masonry walls, provided that *wood structural panel* wall bracing is used and the length of bracing provided is

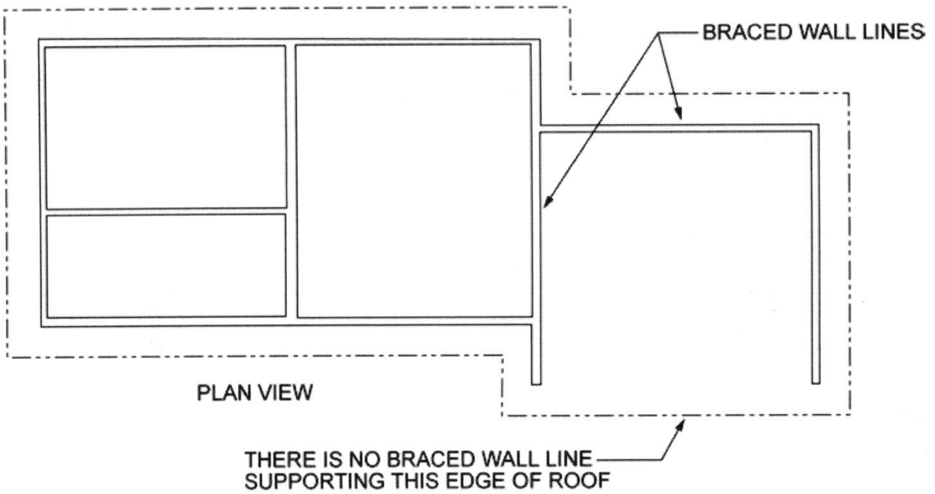

FIGURE 2308.6.8.2(1)
ROOF IN SDC D OR E NOT SUPPORTED ON ALL EDGES

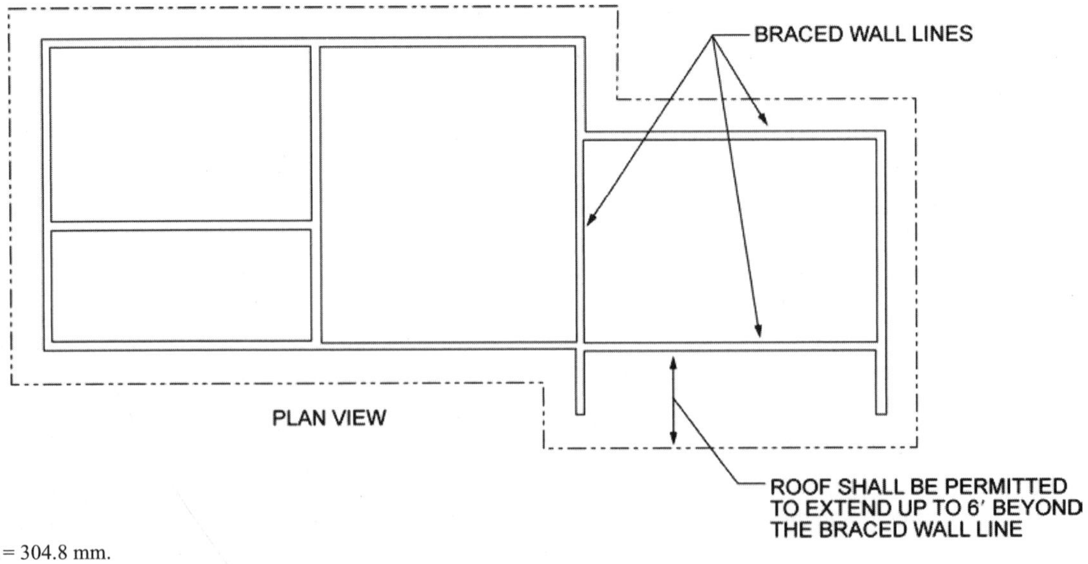

For SI: 1 foot = 304.8 mm.

FIGURE 2308.6.8.2(2)
ROOF EXTENSION IN SDC D OR E BEYOND BRACED WALL LINE

one and one-half times the required length specified in Table 2308.6.1.

2. Stone and masonry *veneer* is permitted to be used in the first *story above grade plane* or the first two *stories above grade plane* where the lowest *story* has concrete or masonry walls.

3. Stone and masonry *veneer* is permitted to be used in both *stories* of buildings with two *stories above grade plane*, provided that the following criteria are met:

   3.1. Type of brace in accordance with Section 2308.6.1 shall be WSP and the allowable shear capacity in accordance with Section 2306.3 shall be not less than 350 plf (5108 N/m).

   3.2. *Braced wall panels* in the second *story* shall be located in accordance with Section 2308.6.1 and not more than 25 feet (7620 mm) on center, and the total length of *braced wall panels* shall be not less than 25 percent of the *braced wall line* length. *Braced wall panels* in the first *story* shall be located in accordance with Section 2308.6.1 and not more than 25 feet (7620 mm) on center, and the total length of *braced wall panels* shall be not less than 45 percent of the *braced wall line* length.

   3.3. *Hold-down* connectors with an allowable capacity of 2,000 pounds (8896 N) shall be provided at the ends of each *braced wall panel* for the second *story* to the first *story* connection. *Hold-down* connectors with an allowable capacity of 3,900 pounds (17 347 N) shall be provided at the ends of each *braced wall panel* for the first *story* to the foundation connection. In all cases, the *hold-down* connector force shall be transferred to the foundation.

   3.4. *Cripple walls* shall not be permitted.

2308.6.10.2 Limitations of concrete or masonry in Seismic Design Categories D and E. In *Seismic Design Categories* D and E, concrete or masonry walls and stone or masonry *veneer* shall not extend above a basement.

Exception: In structures assigned to *Seismic Design Category* D, stone and masonry *veneer* is permitted to be used in the first *story above grade plane*, provided that the following criteria are met:

1. Type of brace in accordance with Section 2308.6.1 shall be WSP and the allowable shear capacity in accordance with Section 2306.3 shall be not less than 350 plf (5108 N/m).

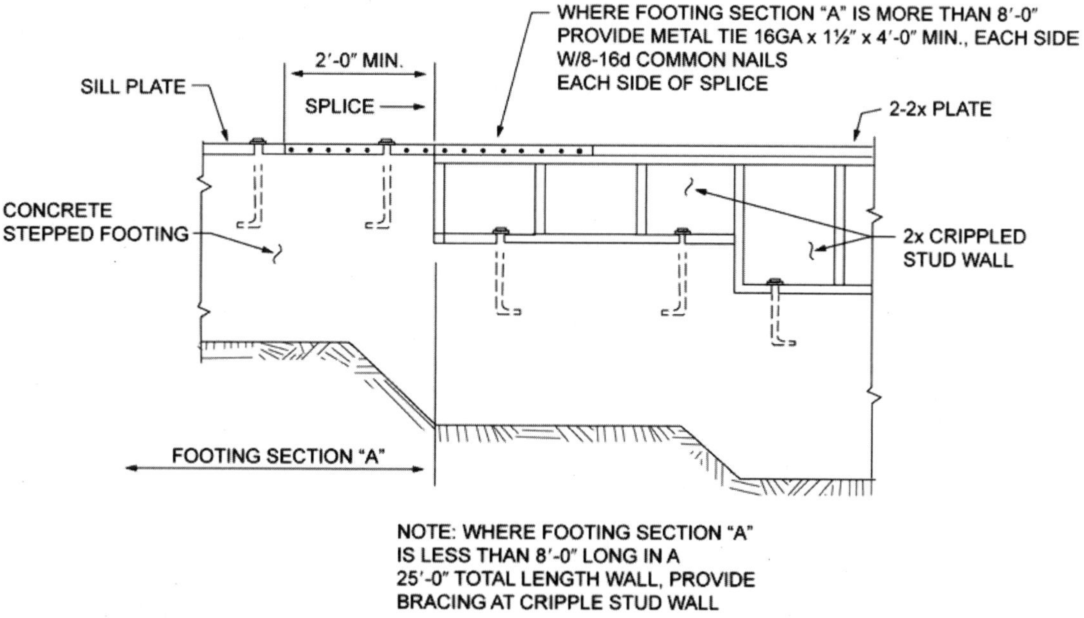

For SI: 1 inch = 25.4 mm, 1 foot = 304.8 mm.

FIGURE 2308.6.8.3
STEPPED FOOTING CONNECTION DETAILS

2. The *braced wall panels* in the first *story* shall be located at each end of the *braced wall line* and not more than 25 feet (7620 mm) on center, and the total length of *braced wall panels* shall be not less than 45 percent of the *braced wall line* length.
3. *Hold-down* connectors shall be provided at the ends of braced walls for the first floor to foundation with an allowable capacity of 2,100 pounds (9341 N).
4. *Cripple walls* shall not be permitted.

2308.7 Roof and ceiling framing. The framing details required in this section apply to roofs having a slope of not less than three units vertical in 12 units horizontal (25-percent slope). Where the roof slope is less than three units vertical in 12 units horizontal (25-percent slope), members supporting rafters and ceiling joists such as ridge board, hips and valleys shall be designed as beams.

2308.7.1 Ceiling joist spans. Spans for ceiling joists shall be in accordance with Table 2308.7.1(1) or 2308.7.1(2). For other grades and species, and other loading conditions, refer to the AWC STJR.

2308.7.2 Rafter spans. Spans for rafters shall be in accordance with Table 2308.7.2(1), 2308.7.2(2), 2308.7.2(3), 2308.7.2(4), 2308.7.2(5) or 2308.7.2(6). For other grades and species and other loading conditions, refer to the AWC STJR. The span of each rafter shall be measured along the horizontal projection of the rafter.

2308.7.3 Ceiling joist and rafter framing. Rafters shall be framed directly opposite each other at the ridge. There shall be a ridge board not less than 1-inch (25 mm) nominal thickness at ridges and not less in depth than the cut end of the rafter. At valleys and hips, there shall be a single valley or hip rafter not less than 2-inch (51 mm) nominal thickness and not less in depth than the cut end of the rafter.

2308.7.3.1 Ceiling joist and rafter connections. Ceiling joists and rafters shall be nailed to each other and the assembly shall be nailed to the top wall plate in accordance with Tables 2304.10.2 and 2308.7.5. Ceiling joists shall be continuous or securely joined where they meet over interior partitions and be fastened to adjacent rafters in accordance with Tables 2304.10.2 and 2308.7.3.1 to provide a continuous rafter tie across the building where such joists are parallel to the rafters. Ceiling joists shall have a bearing surface of not less than $1^1/_2$ inches (38 mm) on the top plate at each end.

Where ceiling joists are not parallel to rafters, an equivalent rafter tie shall be installed in a manner to provide a continuous tie across the building, at a spacing of not more than 4 feet (1219 mm) on center. The connections shall be in accordance with Tables 2308.7.3.1 and 2304.10.2, or connections of equivalent capacities shall be provided. Where ceiling joists or rafter ties are not provided at the top of the rafter support walls, the ridge formed by these rafters shall be supported by a girder conforming to Section 2308.8. Rafter ties shall be spaced not more than 4 feet (1219 mm) on center.

Rafter tie connections shall be based on the equivalent rafter spacing in Table 2308.7.3.1. Rafter-to-ceiling joist connections and rafter tie connections shall be of sufficient size and number to prevent splitting from nailing.

Roof framing member connection to *braced wall lines* shall be in accordance with Section 2308.6.7.2.

2308.7.4 Notches and holes. Notching at the ends of rafters or ceiling joists shall not exceed one-fourth the depth. Notches in the top or bottom of the rafter or ceiling joist shall not exceed one-sixth the depth and shall not be located in the middle one-third of the span, except that a notch not more than one-third of the depth is permitted in the top of the rafter or ceiling joist not further from the face of the support than the depth of the member. Holes bored in rafters or ceiling joists shall not be within 2 inches (51 mm) of the top and bottom and their diameter shall not exceed one-third the depth of the member.

2308.7.5 Wind uplift. The roof construction shall have rafter and truss ties to the wall below. Resultant uplift *loads* shall be transferred to the foundation using a continuous *load* path. The rafter or truss to wall connection shall comply with Tables 2304.10.2 and 2308.7.5.

2308.7.6 Framing around openings. Trimmer and header rafters shall be doubled, or of lumber of equivalent cross section, where the span of the header exceeds 4 feet (1219 mm). The ends of header rafters that are more than 6 feet (1829 mm) in length shall be supported by framing anchors or rafter hangers unless bearing on a beam, partition or wall.

2308.7.6.1 Openings in roof diaphragms in Seismic Design Categories B, C, D and E. In buildings classified as *Seismic Design Category* B, C, D or E. openings in horizontal *diaphragms* with a dimension that is greater than 4 feet (1219 mm) shall be constructed with metal ties and blocking in accordance with this section and Figure 2308.4.4.1(1). Metal ties shall be not less than 0.058 inch [1.47 mm (16 galvanized gage)] in thickness by $1^1/_2$ inches (38 mm) in width and shall have a yield stress not less than 33,000 psi (227 Mpa). Blocking shall extend not less than the dimension of the opening in the direction of the tie and blocking. Ties shall be attached to blocking in accordance with the manufacturer's instructions but with not less than eight 16d common nails on each side of the header-joist intersection.

2308.7.7 Purlins. Purlins to support roof *loads* are permitted to be installed to reduce the span of rafters within allowable limits and shall be supported by struts to bearing walls. The maximum span of 2-inch by 4-inch (51 mm by 102 mm) purlins shall be 4 feet (1219 mm). The maximum span of the 2-inch by 6-inch (51 mm by 152 mm) purlin shall be 6 feet (1829 mm), but the purlin shall not be smaller than the supported rafter. Struts shall be not less than 2-inch by 4-inch (51 mm by 102 mm) members. The unbraced length of struts shall not exceed 8 feet (2438 mm) and the slope of the struts shall be not less than 45 degrees (0.79 rad) from the horizontal.

2308.7.8 Blocking. Roof rafters and ceiling joists shall be supported laterally to prevent rotation and lateral displacement in accordance with Section 2308.4.6 and connected to *braced wall lines* in accordance with Section 2308.6.7.2.

2308.7.9 Engineered wood products. *Prefabricated wood I-joists*, *structural glued-laminated timber* and *structural composite lumber* shall not be notched or drilled except where permitted by the manufacturer's recommendations or where the effects of such alterations are specifically considered in the design of the member by a *registered design professional*.

2308.7.10 Roof sheathing. Roof sheathing shall be in accordance with Tables 2304.8(3) and 2304.8(5) for *wood structural panels*, and Tables 2304.8(1) and 2304.8(2) for lumber and shall comply with Section 2304.8.2.

2308.7.11 Joints. Joints in lumber sheathing shall occur over supports unless *approved* end-matched lumber is used, in which case each piece shall bear on not fewer than two supports.

2308.7.12 Roof planking. Planking shall be designed in accordance with the general provisions of this code.

In lieu of such design, 2-inch (51 mm) tongue-and groove planking is permitted in accordance with Table 2308.7.12. Joints in such planking are permitted to be randomly spaced, provided that the system is applied to not less than three continuous spans, planks are center matched and end matched or splined, each plank bears on one support or more, and joints are separated by not less than 24 inches (610 mm) in adjacent pieces.

2308.7.13 Wood trusses. Wood trusses shall be designed in accordance with Section 2303.4. Connection to *braced wall lines* shall be in accordance with Section 2308.6.7.2.

2308.7.14 Attic ventilation. For *attic* ventilation, see Section 1202.2.1.

2308.8 Design of elements. Combining of engineered elements or systems and conventionally specified elements or systems shall be permitted subject to the limits of Sections 2308.8.1 and 2308.8.2.

2308.8.1 Elements exceeding limitations of conventional construction. Where a building of otherwise conventional construction contains structural elements exceeding the limits of Section 2308.2, these elements and the supporting *load* path shall be designed in accordance with accepted engineering practice and the provisions of this code.

TABLE 2308.7.5
REQUIRED RATING OF APPROVED UPLIFT CONNECTORS (pounds)[a, b, c, e, f, g, h]

NOMINAL DESIGN WIND SPEED, V_{asd}[i]	ROOF SPAN (feet)							OVERHANGS (pounds/feet)[d]
	12	20	24	28	32	36	40	
85	-72	-120	-145	-169	-193	-217	-241	-38.55
90	-91	-151	-181	-212	-242	-272	-302	-43.22
100	-131	-281	-262	-305	-349	-393	-436	-53.36
110	-175	-292	-351	-409	-467	-526	-584	-64.56

For SI: 1 inch = 25.4 mm, 1 foot = 304.8 mm, 1 mile per hour = 1.61 km/hr, 1 pound = 0.454 Kg, 1 pound/foot = 14.5939 N/m.

a. The uplift connection requirements are based on a 30-foot mean roof height located in Exposure B. For Exposure C or D and for other mean roof heights, multiply the loads by the following adjustment coefficients:

	Mean Roof Height (feet)									
EXPOSURE	15	20	25	30	35	40	45	50	55	60
B	1.00	1.00	1.00	1.00	1.05	1.09	1.12	1.16	1.19	1.22
C	1.21	1.29	1.35	1.40	1.45	1.49	1.53	1.56	1.59	1.62
D	1.47	1.55	1.61	1.66	1.70	1.74	1.78	1.81	1.84	1.87

b. The uplift connection requirements are based on the framing being spaced 24 inches on center. Multiply by 0.67 for framing spaced 16 inches on center and multiply by 0.5 for framing spaced 12 inches on center.

c. The uplift connection requirements include an allowance for 10 pounds of dead load.

d. The uplift connection requirements do not account for the effects of overhangs. The magnitude of the loads shall be increased by adding the overhang loads found in the table. The overhang loads are based on framing spaced 24 inches on center. The overhang loads given shall be multiplied by the overhang projection and added to the roof uplift value in the table.

e. The uplift connection requirements are based on wind loading on end zones as defined in Figure 28.5-1 of ASCE 7. Connection loads for connections located a distance of 20 percent of the least horizontal dimension of the building from the corner of the building are permitted to be reduced by multiplying the table connection value by 0.7 and multiplying the overhang load by 0.8.

f. For wall-to-wall and wall-to-foundation connections, the capacity of the uplift connector is permitted to be reduced by 100 pounds for each full wall above. (For example, if a 500-pound rated connector is used on the roof framing, a 400-pound rated connector is permitted at the next floor level down).

g. Interpolation is permitted for intermediate values of V_{asd} and roof spans.

h. The rated capacity of approved tie-down devices is permitted to include up to a 60-percent increase for wind effects where allowed by material specifications.

i. V_{asd} shall be determined in accordance with Section 1609.3.1.

2308.8.2 Structural elements or systems not described herein. Where a building of otherwise conventional construction contains structural elements or systems not described in Section 2308, these elements or systems shall be designed in accordance with accepted engineering practice and the provisions of this code. The extent of such design need only demonstrate compliance of the nonconventional elements with other applicable provisions of this code and shall be compatible with the performance of the conventionally framed system.

SECTION 2309
WOOD FRAME CONSTRUCTION MANUAL

2309.1 Wood Frame Construction Manual. Structural design in accordance with the AWC WFCM shall be permitted for buildings assigned to *Risk Category* I or II subject to the limitations of Section 1.1.3 of the AWC WFCM and the *load* assumptions contained therein. Structural elements beyond these limitations shall be designed in accordance with accepted engineering practice.

TABLE 2308.7.1(1)
CEILING JOIST SPANS FOR COMMON LUMBER SPECIES
(Uninhabitable attics without storage, live load = 10 psf, L/Δ = 240)

CEILING JOIST SPACING (inches)	SPECIES AND GRADE		DEAD LOAD = 5 psf			
			2 × 4	2 × 6	2 × 8	2 × 10
			Maximum ceiling joist spans			
			(ft. - in.)	(ft. - in.)	(ft. - in.)	(ft. - in.)
12	Douglas Fir-Larch	SS	13-2	20-8	Note a	Note a
	Douglas Fir-Larch	#1	12-8	19-11	Note a	Note a
	Douglas Fir-Larch	#2	12-5	19-6	25-8	Note a
	Douglas Fir-Larch	#3	10-10	15-10	20-1	24-6
	Hem-Fir	SS	12-5	19-6	25-8	Note a
	Hem-Fir	#1	12-2	19-1	25-2	Note a
	Hem-Fir	#2	11-7	18-2	24-0	Note a
	Hem-Fir	#3	10-10	15-10	20-1	24-6
	Southern Pine	SS	12-11	20-3	Note a	Note a
	Southern Pine	#1	12-5	19-6	25-8	Note a
	Southern Pine	#2	11-10	18-8	24-7	Note a
	Southern Pine	#3	10-1	14-11	18-9	22-9
	Spruce-Pine-Fir	SS	12-2	19-1	25-2	Note a
	Spruce-Pine-Fir	#1	11-10	18-8	24-7	Note a
	Spruce-Pine-Fir	#2	11-10	18-8	24-7	Note a
	Spruce-Pine-Fir	#3	10-10	15-10	20-1	24-6
16	Douglas Fir-Larch	SS	11-11	18-9	24-8	Note a
	Douglas Fir-Larch	#1	11-6	18-1	23-10	Note a
	Douglas Fir-Larch	#2	11-3	17-8	23-0	Note a
	Douglas Fir-Larch	#3	9-5	13-9	17-5	21-3
	Hem-Fir	SS	11-3	17-8	23-4	Note a
	Hem-Fir	#1	11-0	17-4	22-10	Note a
	Hem-Fir	#2	10-6	16-6	21-9	Note a
	Hem-Fir	#3	9-5	13-9	17-5	21-3
	Southern Pine	SS	11-9	18-5	24-3	Note a
	Southern Pine	#1	11-3	17-8	23-4	Note a
	Southern Pine	#2	10-9	16-11	21-7	25-7
	Southern Pine	#3	8-9	12-11	16-3	19-9
	Spruce-Pine-Fir	SS	11-0	17-4	22-10	Note a
	Spruce-Pine-Fir	#1	10-9	16-11	22-4	Note a
	Spruce-Pine-Fir	#2	10-9	16-11	22-4	Note a
	Spruce-Pine-Fir	#3	9-5	13-9	17-5	21-3

(continued)

TABLE 2308.7.1(1)—continued
CEILING JOIST SPANS FOR COMMON LUMBER SPECIES
(Uninhabitable attics without storage, live load = 10 psf, L/Δ = 240)

CEILING JOIST SPACING (inches)	SPECIES AND GRADE		DEAD LOAD = 5 psf			
			2 × 4	2 × 6	2 × 8	2 × 10
			Maximum ceiling joist spans			
			(ft. - in.)	(ft. - in.)	(ft. - in.)	(ft. - in.)
19.2	Douglas Fir-Larch	SS	11-3	17-8	23-3	Note a
	Douglas Fir-Larch	#1	10-10	17-0	22-5	Note a
	Douglas Fir-Larch	#2	10-7	16-7	21-0	25-8
	Douglas Fir-Larch	#3	8-7	12-6	15-10	19-5
	Hem-Fir	SS	10-7	16-8	21-11	Note a
	Hem-Fir	#1	10-4	16-4	21-6	Note a
	Hem-Fir	#2	9-11	15-7	20-6	25-3
	Hem-Fir	#3	8-7	12-6	15-10	19-5
	Southern Pine	SS	11-0	17-4	22-10	Note a
	Southern Pine	#1	10-7	16-8	22-0	Note a
	Southern Pine	#2	10-2	15-7	19-8	23-5
	Southern Pine	#3	8-0	11-9	14-10	18-0
	Spruce-Pine-Fir	SS	10-4	16-4	21-6	Note a
	Spruce-Pine-Fir	#1	10-2	15-11	21-0	25-8
	Spruce-Pine-Fir	#2	10-2	15-11	21-0	25-8
	Spruce-Pine-Fir	#3	8-7	12-6	15-10	19-5
24	Douglas Fir-Larch	SS	10-5	16-4	21-7	Note a
	Douglas Fir-Larch	#1	10-0	15-9	20-1	24-6
	Douglas Fir-Larch	#2	9-10	14-10	18-9	22-11
	Douglas Fir-Larch	#3	7-8	11-2	14-2	17-4
	Hem-Fir	SS	9-10	15-6	20-5	Note a
	Hem-Fir	#1	9-8	15-2	19-7	23-11
	Hem-Fir	#2	9-2	14-5	18-6	22-7
	Hem-Fir	#3	7-8	11-2	14-2	17-4
	Southern Pine	SS	10-3	16-1	21-2	Note a
	Southern Pine	#1	9-10	15-6	20-5	24-0
	Southern Pine	#2	9-3	13-11	17-7	20-11
	Southern Pine	#3	7-2	10-6	13-3	16-1
	Spruce-Pine-Fir	SS	9-8	15-2	19-11	25-5
	Spruce-Pine-Fir	#1	9-5	14-9	18-9	22-11
	Spruce-Pine-Fir	#2	9-5	14-9	18-9	22-11
	Spruce-Pine-Fir	#3	7-8	11-2	14-2	17-4

Check sources for availability of lumber in lengths greater than 20 feet.

For SI: 1 inch = 25.4 mm, 1 foot = 304.8 mm, 1 pound per square foot = 0.0479 kPa.

a. Span exceeds 26 feet in length.

TABLE 2308.7.1(2)
CEILING JOIST SPANS FOR COMMON LUMBER SPECIES
(Uninhabitable attics with limited storage, live load = 20 psf, L/Δ = 240)

CEILING JOIST SPACING (inches)	SPECIES AND GRADE		DEAD LOAD = 10 psf			
			2 × 4	2 × 6	2 × 8	2 × 10
			Maximum ceiling joist spans			
			(ft. - in.)	(ft. - in.)	(ft. - in.)	(ft. - in.)
12	Douglas Fir-Larch	SS	10-5	16-4	21-7	Note a
	Douglas Fir-Larch	#1	10-0	15-9	20-1	24-6
	Douglas Fir-Larch	#2	9-10	14-10	18-9	22-11
	Douglas Fir-Larch	#3	7-8	11-2	14-2	17-4
	Hem-Fir	SS	9-10	15-6	20-5	Note a
	Hem-Fir	#1	9-8	15-2	19-7	23-11
	Hem-Fir	#2	9-2	14-5	18-6	22-7
	Hem-Fir	#3	7-8	11-2	14-2	17-4
	Southern Pine	SS	10-3	16-1	21-2	Note a
	Southern Pine	#1	9-10	15-6	20-5	24-0
	Southern Pine	#2	9-3	13-11	17-7	20-11
	Southern Pine	#3	7-2	10-6	13-3	16-1
	Spruce-Pine-Fir	SS	9-8	15-2	19-11	25-5
	Spruce-Pine-Fir	#1	9-5	14-9	18-9	22-11
	Spruce-Pine-Fir	#2	9-5	14-9	18-9	22-11
	Spruce-Pine-Fir	#3	7-8	11-2	14-2	17-4
16	Douglas Fir-Larch	SS	9-6	14-11	19-7	25-0
	Douglas Fir-Larch	#1	9-1	13-9	17-5	21-3
	Douglas Fir-Larch	#2	8-9	12-10	16-3	19-10
	Douglas Fir-Larch	#3	6-8	9-8	12-4	15-0
	Hem-Fir	SS	8-11	14-1	18-6	23-8
	Hem-Fir	#1	8-9	13-5	16-10	20-8
	Hem-Fir	#2	8-4	12-8	16-0	19-7
	Hem-Fir	#3	6-8	9-8	12-4	15-0
	Southern Pine	SS	9-4	14-7	19-3	24-7
	Southern Pine	#1	8-11	14-0	17-9	20-9
	Southern Pine	#2	8-0	12-0	15-3	18-1
	Southern Pine	#3	6-2	9-2	11-6	14-0
	Spruce-Pine-Fir	SS	8-9	13-9	18-1	23-1
	Spruce-Pine-Fir	#1	8-7	12-10	16-3	19-10
	Spruce-Pine-Fir	#2	8-7	12-10	16-3	19-10
	Spruce-Pine-Fir	#3	6-8	9-8	12-4	15-0

(continued)

TABLE 2308.7.1(2)—continued
CEILING JOIST SPANS FOR COMMON LUMBER SPECIES
(Uninhabitable attics with limited storage, live load = 20 psf, L/Δ = 240)

CEILING JOIST SPACING (inches)	SPECIES AND GRADE		DEAD LOAD = 10 psf			
			2 × 4	2 × 6	2 × 8	2 × 10
			Maximum ceiling joist spans			
			(ft. - in.)	(ft. - in.)	(ft. - in.)	(ft. - in.)
19.2	Douglas Fir-Larch	SS	8-11	14-0	18-5	23-4
	Douglas Fir-Larch	#1	8-7	12-6	15-10	19-5
	Douglas Fir-Larch	#2	8-0	11-9	14-10	18-2
	Douglas Fir-Larch	#3	6-1	8-10	11-3	13-8
	Hem-Fir	SS	8-5	13-3	17-5	22-3
	Hem-Fir	#1	8-3	12-3	15-6	18-11
	Hem-Fir	#2	7-10	11-7	14-8	17-10
	Hem-Fir	#3	6-1	8-10	11-3	13-8
	Southern Pine	SS	8-9	13-9	18-2	23-1
	Southern Pine	#1	8-5	12-9	16-2	18-11
	Southern Pine	#2	7-4	11-0	13-11	16-6
	Southern Pine	#3	5-8	8-4	10-6	12-9
	Spruce-Pine-Fir	SS	8-3	12-11	17-1	21-8
	Spruce-Pine-Fir	#1	8-0	11-9	14-10	18-2
	Spruce-Pine-Fir	#2	8-0	11-9	14-10	18-2
	Spruce-Pine-Fir	#3	6-1	8-10	11-3	13-8
24	Douglas Fir-Larch	SS	8-3	13-0	17-1	20-11
	Douglas Fir-Larch	#1	7-8	11-2	14-2	17-4
	Douglas Fir-Larch	#2	7-2	10-6	13-3	16-3
	Douglas Fir-Larch	#3	5-5	7-11	10-0	12-3
	Hem-Fir	SS	7-10	12-3	16-2	20-6
	Hem-Fir	#1	7-6	10-11	13-10	16-11
	Hem-Fir	#2	7-1	10-4	13-1	16-0
	Hem-Fir	#3	5-5	7-11	10-0	12-3
	Southern Pine	SS	8-1	12-9	16-10	21-6
	Southern Pine	#1	7-8	11-5	14-6	16-11
	Southern Pine	#2	6-7	9-10	12-6	14-9
	Southern Pine	#3	5-1	7-5	9-5	11-5
	Spruce-Pine-Fir	SS	7-8	12-0	15-10	19-5
	Spruce-Pine-Fir	#1	7-2	10-6	13-3	16-3
	Spruce-Pine-Fir	#2	7-2	10-6	13-3	16-3
	Spruce-Pine-Fir	#3	5-5	7-11	10-0	12-3

Check sources for availability of lumber in lengths greater than 20 feet.

For SI: 1 inch = 25.4 mm, 1 foot = 304.8 mm, 1 pound per square foot = 0.0479 kPa.

a. Span exceeds 26 feet in length.

TABLE 2308.7.2(1)
RAFTER SPANS FOR COMMON LUMBER SPECIES
(Roof live load = 20 psf, ceiling not attached to rafters, L/Δ = 180)

RAFTER SPACING (inches)	SPECIES AND GRADE		DEAD LOAD = 10 psf					DEAD LOAD = 20 psf				
			2 × 4	2 × 6	2 × 8	2 × 10	2 × 12	2 × 4	2 × 6	2 × 8	2 × 10	2 × 12
			Maximum rafter spans[a]									
			(ft. - in.)	(ft. - in.)	(ft. - in.)	(ft. - in.)	(ft. - in.)	(ft. - in.)	(ft. - in.)	(ft. - in.)	(ft. - in.)	(ft. - in.)
12	Douglas Fir-Larch	SS	11-6	18-0	23-9	Note b	Note b	11-6	18-0	23-5	Note b	Note b
	Douglas Fir-Larch	#1	11-1	17-4	22-5	Note b	Note b	10-6	15-4	19-5	23-9	Note b
	Douglas Fir-Larch	#2	10-10	16-7	21-0	25-8	Note b	9-10	14-4	18-2	22-3	25-9
	Douglas Fir-Larch	#3	8-7	12-6	15-10	19-5	22-6	7-5	10-10	13-9	16-9	19-6
	Hem-Fir	SS	10-10	17-0	22-5	Note b	Note b	10-10	17-0	22-5	Note b	Note b
	Hem-Fir	#1	10 -7	16-8	21-10	Note b	Note b	10-3	14-11	18-11	23-2	Note b
	Hem-Fir	#2	10-1	15-11	20-8	25-3	Note b	9-8	14-2	17-11	21-11	25-5
	Hem-Fir	#3	8-7	12-6	15-10	19-5	22-6	7-5	10-10	13-9	16-9	19-6
	Southern Pine	SS	11-3	17-8	23-4	Note b	Note b	11-3	17-8	23-4	Note b	Note b
	Southern Pine	#1	10-10	17-0	22-5	26-0	26-0	10-6	15-8	19-10	23-2	Note b
	Southern Pine	#2	10-4	15-7	19-8	23-5	26-0	9-0	13-6	17-1	20-3	23-10
	Southern Pine	#3	8-0	11-9	14-10	18-0	21-4	6-11	10-2	12-10	15-7	18-6
	Spruce-Pine-Fir	SS	10-7	16-8	21-11	Note b	Note b	10-7	16-8	21-9	Note b	Note b
	Spruce-Pine-Fir	#1	10-4	16-3	21-0	25-8	Note b	9-10	14-4	18-2	22-3	25-9
	Spruce-Pine-Fir	#2	10-4	16-3	21-0	25-8	Note b	9-10	14-4	18-2	22-3	25-9
	Spruce-Pine-Fir	#3	8-7	12-6	15-10	19-5	22-6	7-5	10-10	13-9	16-9	19-6
16	Douglas Fir-Larch	SS	10-5	16-4	21-7	Note b	Note b	10-5	16-0	20-3	24-9	Note b
	Douglas Fir-Larch	#1	10-0	15-4	19-5	23-9	Note b	9-1	13-3	16-10	20-7	23-10
	Douglas Fir-Larch	#2	9-10	14-4	18-2	22-3	25-9	8-6	12-5	15-9	19-3	22-4
	Douglas Fir-Larch	#3	7-5	10-10	13-9	16-9	19-6	6-5	9-5	11-11	14-6	16-10
	Hem-Fir	SS	9-10	15-6	20-5	Note b	Note b	9-10	15-6	19-11	24-4	Note b
	Hem-Fir	#1	9-8	14-11	18-11	23-2	Note b	8-10	12-11	16-5	20-0	23-3
	Hem-Fir	#2	9-2	14-2	17-11	21-11	25-5	8-5	12-3	15-6	18-11	22-0
	Hem-Fir	#3	7-5	10-10	13-9	16-9	19-6	6-5	9-5	11-11	14-6	16-10
	Southern Pine	SS	10-3	16-1	21-2	Note b	Note b	10-3	16-1	21-2	25-7	Note b
	Southern Pine	#1	9-10	15-6	19-10	23-2	26-0	9-1	13-7	17-2	20-1	23-10
	Southern Pine	#2	9-0	13-6	17-1	20-3	23-10	7-9	11-8	14-9	17-6	20-8
	Southern Pine	#3	6-11	10-2	12-10	15-7	18-6	6-0	8-10	11-2	13-6	16-0
	Spruce-Pine-Fir	SS	9-8	15-2	19-11	25-5	Note b	9-8	14-10	18-10	23-0	Note b
	Spruce-Pine-Fir	#1	9-5	14-4	18-2	22-3	25-9	8-6	12-5	15-9	19-3	22-4
	Spruce-Pine-Fir	#2	9-5	14-4	18-2	22-3	25-9	8-6	12-5	15-9	19-3	22-4
	Spruce-Pine-Fir	#3	7-5	10-10	13-9	16-9	19-6	6-5	9-5	11-11	14-6	16-10

(continued)

TABLE 2308.7.2(1)—continued
RAFTER SPANS FOR COMMON LUMBER SPECIES
(Roof live load = 20 psf, ceiling not attached to rafters, L/Δ = 180)

RAFTER SPACING (inches)	SPECIES AND GRADE		DEAD LOAD = 10 psf					DEAD LOAD = 20 psf				
			2 × 4	2 × 6	2 × 8	2 × 10	2 × 12	2 × 4	2 × 6	2 × 8	2 × 10	2 × 12
			Maximum rafter spans[a]									
			(ft. - in.)	(ft. - in.)	(ft. - in.)	(ft. - in.)	(ft. - in.)	(ft. - in.)	(ft. - in.)	(ft. - in.)	(ft. - in.)	(ft. - in.)
19.2	Douglas Fir-Larch	SS	9-10	15-5	20-4	25-11	Note b	9-10	14-7	18-6	22-7	Note b
	Douglas Fir-Larch	#1	9-5	14-0	17-9	21-8	25-2	8-4	12-2	15-4	18-9	21-9
	Douglas Fir-Larch	#2	8-11	13-1	16-7	20-3	23-6	7-9	11-4	14-4	17-7	20-4
	Douglas Fir-Larch	#3	6-9	9-11	12-7	15-4	17-9	5-10	8-7	10-10	13-3	15-5
	Hem-Fir	SS	9-3	14-7	19-2	24-6	Note b	9-3	14-4	18-2	22-3	25-9
	Hem-Fir	#1	9-1	13-8	17-4	21-1	24-6	8-1	11-10	15-0	18-4	21-3
	Hem-Fir	#2	8-8	12-11	16-4	20-0	23-2	7-8	11-2	14-2	17-4	20-1
	Hem-Fir	#3	6-9	9-11	12-7	15-4	17-9	5-10	8-7	10-10	13-3	15-5
	Southern Pine	SS	9-8	15-2	19-11	25-5	Note b	9-8	15-2	19-7	23-4	Note b
	Southern Pine	#1	9-3	14-3	18-1	21-2	25-2	8-4	12-4	15-8	18-4	21-9
	Southern Pine	#2	8-2	12-3	15-7	18-6	21-9	7-1	10-8	13-6	16-0	18-10
	Southern Pine	#3	6-4	9-4	11-9	14-3	16-10	5-6	8-1	10-2	12-4	14-7
	Spruce-Pine-Fir	SS	9-1	14-3	18-9	23-11	Note b	9-1	13-7	17-2	21-0	24-4
	Spruce-Pine-Fir	#1	8-10	13-1	16-7	20-3	23-6	7-9	11-4	14-4	17-7	20-4
	Spruce-Pine-Fir	#2	8-10	13-1	16-7	20-3	23-6	7-9	11-4	14-4	17-7	20-4
	Spruce-Pine-Fir	#3	6-9	9-11	12-7	15-4	17-9	5-10	8-7	10-10	13-3	15-5
24	Douglas Fir-Larch	SS	9-1	14-4	18-10	23-4	Note b	8-11	13-1	16-7	20-3	23-5
	Douglas Fir-Larch	#1	8-7	12-6	15-10	19-5	22-6	7-5	10-10	13-9	16-9	19-6
	Douglas Fir-Larch	#2	8-0	11-9	14-10	18-2	21-0	6-11	10-2	12-10	15-8	18-3
	Douglas Fir-Larch	#3	6-1	8-10	11-3	13-8	15-11	5-3	7-8	9-9	11-10	13-9
	Hem-Fir	SS	8-7	13-6	17-10	22-9	Note b	8-7	12-10	16-3	19-10	23-0
	Hem-Fir	#1	8-4	12-3	15-6	18-11	21-11	7-3	10-7	13-5	16-4	19-0
	Hem-Fir	#2	7-11	11-7	14-8	17-10	20-9	6-10	10-0	12-8	15-6	17-11
	Hem-Fir	#3	6-1	8-10	11-3	13-8	15-11	5-3	7-8	9-9	11-10	13-9
	Southern Pine	SS	8-11	14-1	18-6	23-8	Note b	8-11	13-10	17-6	20-10	24-8
	Southern Pine	#1	8-7	12-9	16-2	18-11	22-6	7-5	11-1	14-0	16-5	19-6
	Southern Pine	#2	7-4	11-0	13-11	16-6	19-6	6-4	9-6	12-1	14-4	16-10
	Southern Pine	#3	5-8	8-4	10-6	12-9	15-1	4-11	7-3	9-1	11-0	13-1
	Spruce-Pine-Fir	SS	8-5	13-3	17-5	21-8	25-2	8-4	12-2	15-4	18-9	21-9
	Spruce-Pine-Fir	#1	8-0	11-9	14-10	18-2	21-0	6-11	10-2	12-10	15-8	18-3
	Spruce-Pine-Fir	#2	8-0	11-9	14-10	18-2	21-0	6-11	10-2	12-10	15-8	18-3
	Spruce-Pine-Fir	#3	6-1	8-10	11-3	13-8	15-11	5-3	7-8	9-9	11-10	13-9

Check sources for availability of lumber in lengths greater than 20 feet.

For SI: 1 inch = 25.4 mm, 1 foot = 304.8 mm, 1 pound per square foot = 0.0479 kPa.

a. The tabulated rafter spans assume that ceiling joists are located at the bottom of the attic space or that some other method of resisting the outward push of the rafters on the bearing walls, such as rafter ties, is provided at that location. Where ceiling joists or rafter ties are located higher in the attic space, the rafter spans shall be multiplied by the adjustment factors in Table 2308.7.2(7).

b. Span exceeds 26 feet in length.

TABLE 2308.7.2(2)
RAFTER SPANS FOR COMMON LUMBER SPECIES
(Roof live load = 20 psf, ceiling attached to rafters, L/Δ = 240)

RAFTER SPACING (inches)	SPECIES AND GRADE		DEAD LOAD = 10 psf					DEAD LOAD = 20 psf				
			2 × 4	2 × 6	2 × 8	2 × 10	2 × 12	2 × 4	2 × 6	2 × 8	2 × 10	2 × 12
			Maximum rafter spans[a]									
			(ft. - in.)	(ft. - in.)	(ft. - in.)	(ft. - in.)	(ft. - in.)	(ft. - in.)	(ft. - in.)	(ft. - in.)	(ft. - in.)	(ft. - in.)
12	Douglas Fir-Larch	SS	10-5	16-4	21-7	Note b	Note b	10-5	16-4	21-7	Note b	Note b
	Douglas Fir-Larch	#1	10-0	15-9	20-10	Note b	Note b	10-0	15-4	19-5	23-9	Note b
	Douglas Fir-Larch	#2	9-10	15-6	20-5	25-8	Note b	9-10	14-4	18-2	22-3	25-9
	Douglas Fir-Larch	#3	8-7	12-6	15-10	19-5	22-6	7-5	10-10	13-9	16-9	19-6
	Hem-Fir	SS	9-10	15-6	20-5	Note b	Note b	9-10	15-6	20-5	Note b	Note b
	Hem-Fir	#1	9-8	15-2	19-11	25-5	Note b	9-8	14-11	18-11	23-2	Note b
	Hem-Fir	#2	9-2	14-5	19-0	24-3	Note b	9-2	14-2	17-11	21-11	25-5
	Hem-Fir	#3	8-7	12-6	15-10	19-5	22-6	7-5	10-10	13-9	16-9	19-6
	Southern Pine	SS	10-3	16-1	21-2	Note b	Note b	10-3	16-1	21-2	Note b	Note b
	Southern Pine	#1	9-10	15-6	20-5	26-0	26-0	9-10	15-6	19-10	23-2	26-0
	Southern Pine	#2	9-5	14-9	19-6	23-5	26-0	9-0	13-6	17-1	20-3	23-10
	Southern Pine	#3	8-0	11-9	14-10	18-0	21-4	6-11	10-2	12-10	15-7	18-6
	Spruce-Pine-Fir	SS	9-8	15-2	19-11	25-5	Note b	9-8	15-2	19-11	25-5	Note b
	Spruce-Pine-Fir	#1	9-5	14-9	19-6	24-10	Note b	9-5	14-4	18-2	22-3	25-9
	Spruce-Pine-Fir	#2	9-5	14-9	19-6	24-10	Note b	9-5	14-4	18-2	22-3	25-9
	Spruce-Pine-Fir	#3	8-7	12-6	15-10	19-5	22-6	7-5	10-10	13-9	16-9	19-6
16	Douglas Fir-Larch	SS	9-6	14-11	19-7	25-0	Note b	9-6	14-11	19-7	24-9	Note b
	Douglas Fir-Larch	#1	9-1	14-4	18-11	23-9	Note b	9-1	13-3	16-10	20-7	23-10
	Douglas Fir-Larch	#2	8-11	14-1	18-2	22-3	25-9	8-6	12-5	15-9	19-3	22-4
	Douglas Fir-Larch	#3	7-5	10-10	13-9	16-9	19-6	6-5	9-5	11-11	14-6	16-10
	Hem-Fir	SS	8-11	14-1	18-6	23-8	Note b	8-11	14-1	18-6	23-8	Note b
	Hem-Fir	#1	8-9	13-9	18-1	23-1	Note b	8-9	12-11	16-5	20-0	23-3
	Hem-Fir	#2	8-4	13-1	17-3	21-11	25-5	8-4	12-3	15-6	18-11	22-0
	Hem-Fir	#3	7-5	10-10	13-9	16-9	19-6	6-5	9-5	11-11	14-6	16-10
	Southern Pine	SS	9-4	14-7	19-3	24-7	Note b	9-4	14-7	19-3	24-7	Note b
	Southern Pine	#1	8-11	14-1	18-6	23-2	26-0	8-11	13-7	17-2	20-1	23-10
	Southern Pine	#2	8-7	13-5	17-1	20-3	23-10	7-9	11-8	14-9	17-6	20-8
	Southern Pine	#3	6-11	10-2	12-10	15-7	18-6	6-0	8-10	11-2	13-6	16-0
	Spruce-Pine-Fir	SS	8-9	13-9	18-1	23-1	Note b	8-9	13-9	18-1	23-0	Note b
	Spruce-Pine-Fir	#1	8-7	13-5	17-9	22-3	25-9	8-6	12-5	15-9	19-3	22-4
	Spruce-Pine-Fir	#2	8-7	13-5	17-9	22-3	25-9	8-6	12-5	15-9	19-3	22-4
	Spruce-Pine-Fir	#3	7-5	10-10	13-9	16-9	19-6	6-5	9-5	11-11	14-6	16-10

(continued)

TABLE 2308.7.2(2)—continued
RAFTER SPANS FOR COMMON LUMBER SPECIES
(Roof live load = 20 psf, ceiling attached to rafters, L/Δ = 240)

RAFTER SPACING (inches)	SPECIES AND GRADE		DEAD LOAD = 10 psf					DEAD LOAD = 20 psf				
			2 × 4	2 × 6	2 × 8	2 × 10	2 × 12	2 × 4	2 × 6	2 × 8	2 × 10	2 × 12
			Maximum rafter spans[a]									
			(ft. - in.)	(ft. - in.)	(ft. - in.)	(ft. - in.)	(ft. - in.)	(ft. - in.)	(ft. - in.)	(ft. - in.)	(ft. - in.)	(ft. - in.)
19.2	Douglas Fir-Larch	SS	8-11	14-0	18-5	23-7	Note b	8-11	14-0	18-5	22-7	Note b
	Douglas Fir-Larch	#1	8-7	13-6	17-9	21-8	25-2	8-4	12-2	15-4	18-9	21-9
	Douglas Fir-Larch	#2	8-5	13-1	16-7	20-3	23-6	7-9	11-4	14-4	17-7	20-4
	Douglas Fir-Larch	#3	6-9	9-11	12-7	15-4	17-9	5-10	8-7	10-10	13-3	15-5
	Hem-Fir	SS	8-5	13-3	17-5	22-3	Note b	8-5	13-3	17-5	22-3	25-9
	Hem-Fir	#1	8-3	12-11	17-1	21-1	24-6	8-1	11-10	15-0	18-4	21-3
	Hem-Fir	#2	7-10	12-4	16-3	20-0	23-2	7-8	11-2	14-2	17-4	20-1
	Hem-Fir	#3	6-9	9-11	12-7	15-4	17-9	5-10	8-7	10-10	13-3	15-5
	Southern Pine	SS	8-9	13-9	18-2	23-1	Note b	8-9	13-9	18-2	23-1	Note b
	Southern Pine	#1	8-5	13-3	17-5	21-2	25-2	8-4	12-4	15-8	18-4	21-9
	Southern Pine	#2	8-1	12-3	15-7	18-6	21-9	7-1	10-8	13-6	16-0	18-10
	Southern Pine	#3	6-4	9-4	11-9	14-3	16-10	5-6	8-1	10-2	12-4	14-7
	Spruce-Pine-Fir	SS	8-3	12-11	17-1	21-9	Note b	8-3	12-11	17-1	21-0	24-4
	Spruce-Pine-Fir	#1	8-1	12-8	16-7	20-3	23-6	7-9	11-4	14-4	17-7	20-4
	Spruce-Pine-Fir	#2	8-1	12-8	16-7	20-3	23-6	7-9	11-4	14-4	17-7	20-4
	Spruce-Pine-Fir	#3	6-9	9-11	12-7	15-4	17-9	5-10	8-7	10-10	13-3	15-5
24	Douglas Fir-Larch	SS	8-3	13-0	17-2	21-10	Note b	8-3	13-0	16-7	20-3	23-5
	Douglas Fir-Larch	#1	8-0	12-6	15-10	19-5	22-6	7-5	10-10	13-9	16-9	19-6
	Douglas Fir-Larch	#2	7-10	11-9	14-10	18-2	21-0	6-11	10-2	12-10	15-8	18-3
	Douglas Fir-Larch	#3	6-1	8-10	11-3	13-8	15-11	5-3	7-8	9-9	11-10	13-9
	Hem-Fir	SS	7-10	12-3	16-2	20-8	25-1	7-10	12-3	16-2	19-10	23-0
	Hem-Fir	#1	7-8	12-0	15-6	18-11	21-11	7-3	10-7	13-5	16-4	19-0
	Hem-Fir	#2	7-3	11-5	14-8	17-10	20-9	6-10	10-0	12-8	15-6	17-11
	Hem-Fir	#3	6-1	8-10	11-3	13-8	15-11	5-3	7-8	9-9	11-10	13-9
	Southern Pine	SS	8-1	12-9	16-10	21-6	Note b	8-1	12-9	16-10	20-10	24-8
	Southern Pine	#1	7-10	12-3	16-2	18-11	22-6	7-5	11-1	14-0	16-5	19-6
	Southern Pine	#2	7-4	11-0	13-11	16-6	19-6	6-4	9-6	12-1	14-4	16-10
	Southern Pine	#3	5-8	8-4	10-6	12-9	15-1	4-11	7-3	9-1	11-0	13-1
	Spruce-Pine-Fir	SS	7-8	12-0	15-10	20-2	24-7	7-8	12-0	15-4	18-9	21-9
	Spruce-Pine-Fir	#1	7-6	11-9	14-10	18-2	21-0	6-11	10-2	12-10	15-8	18-3
	Spruce-Pine-Fir	#2	7-6	11-9	14-10	18-2	21-0	6-11	10-2	12-10	15-8	18-3
	Spruce-Pine-Fir	#3	6-1	8-10	11-3	13-8	15-11	5-3	7-8	9-9	11-10	13-9

Check sources for availability of lumber in lengths greater than 20 feet.

For SI: 1 inch = 25.4 mm, 1 foot = 304.8 mm, 1 pound per square foot = 0.0479 kPa.

a. The tabulated rafter spans assume that ceiling joists are located at the bottom of the attic space or that some other method of resisting the outward push of the rafters on the bearing walls, such as rafter ties, is provided at that location. Where ceiling joists or rafter ties are located higher in the attic space, the rafter spans shall be multiplied by the adjustment factors in Table 2308.7.2(7).

b. Span exceeds 26 feet in length.

TABLE 2308.7.2(3)
RAFTER SPANS FOR COMMON LUMBER SPECIES
(Ground snow load = 30 psf, ceiling not attached to rafters, L/Δ = 180)

RAFTER SPACING (inches)	SPECIES AND GRADE		DEAD LOAD = 10 psf					DEAD LOAD = 20 psf				
			2 × 4	2 × 6	2 × 8	2 × 10	2 × 12	2 × 4	2 × 6	2 × 8	2 × 10	2 × 12
			Maximum rafter spans[a]									
			(ft. - in.)	(ft. - in.)	(ft. - in.)	(ft. - in.)	(ft. - in.)	(ft. - in.)	(ft. - in.)	(ft. - in.)	(ft. - in.)	(ft. - in.)
12	Douglas Fir-Larch	SS	10-0	15-9	20-9	Note b	Note b	10-0	15-9	20-1	24-6	Note b
	Douglas Fir-Larch	#1	9-8	14-9	18-8	22-9	Note b	9-0	13-2	16-8	20-4	23-7
	Douglas Fir-Larch	#2	9-5	13-9	17-5	21-4	24-8	8-5	12-4	15-7	19-1	22-1
	Douglas Fir-Larch	#3	7-1	10-5	13-2	16-1	18-8	6-4	9-4	11-9	14-5	16-8
	Hem-Fir	SS	9-6	14-10	19-7	25-0	Note b	9-6	14-10	19-7	24-1	Note b
	Hem-Fir	#1	9-3	14-4	18-2	22-2	25-9	8-9	12-10	16-3	19-10	23-0
	Hem-Fir	#2	8-10	13-7	17-2	21-0	24-4	8-4	12-2	15-4	18-9	21-9
	Hem-Fir	#3	7-1	10-5	13-2	16-1	18-8	6-4	9-4	11-9	14-5	16-8
	Southern Pine	SS	9-10	15-6	20-5	Note b	Note b	9-10	15-6	20-5	25-4	Note b
	Southern Pine	#1	9-6	14-10	19-0	22-3	26-0	9-0	13-5	17-0	19-11	23-7
	Southern Pine	#2	8-7	12-11	16-4	19-5	22-10	7-8	11-7	14-8	17-4	20-5
	Southern Pine	#3	6-7	9-9	12-4	15-0	17-9	5-11	8-9	11-0	13-5	15-10
	Spruce-Pine-Fir	SS	9-3	14-7	19-2	24-6	Note b	9-3	14-7	18-8	22-9	Note b
	Spruce-Pine-Fir	#1	9-1	13-9	17-5	21-4	24-8	8-5	12-4	15-7	19-1	22-1
	Spruce-Pine-Fir	#2	9-1	13-9	17-5	21-4	24-8	8-5	12-4	15-7	19-1	22-1
	Spruce-Pine-Fir	#3	7-1	10-5	13-2	16-1	18-8	6-4	9-4	11-9	14-5	16-8
16	Douglas Fir-Larch	SS	9-1	14-4	18-10	23-9	Note b	9-1	13-9	17-5	21-3	24-8
	Douglas Fir-Larch	#1	8-9	12-9	16-2	19-9	22-10	7-10	11-5	14-5	17-8	20-5
	Douglas Fir-Larch	#2	8-2	11-11	15-1	18-5	21-5	7-3	10-8	13-6	16-6	19-2
	Douglas Fir-Larch	#3	6-2	9-0	11-5	13-11	16-2	5-6	8-1	10-3	12-6	14-6
	Hem-Fir	SS	8-7	13-6	17-10	22-9	Note b	8-7	13-6	17-1	20-10	24-2
	Hem-Fir	#1	8-5	12-5	15-9	19-3	22-3	7-7	11-1	14-1	17-2	19-11
	Hem-Fir	#2	8-0	11-9	14-11	18-2	21-1	7-2	10-6	13-4	16-3	18-10
	Hem-Fir	#3	6-2	9-0	11-5	13-11	16-2	5-6	8-1	10-3	12-6	14-6
	Southern Pine	SS	8-11	14-1	18-6	23-8	Note b	8-11	14-1	18-5	21-11	25-11
	Southern Pine	#1	8-7	13-0	16-6	19-3	22-10	7-10	11-7	14-9	17-3	20-5
	Southern Pine	#2	7-6	11-2	14-2	16-10	19-10	6-8	10-0	12-8	15-1	17-9
	Southern Pine	#3	5-9	8-6	10-8	13-0	15-4	5-2	7-7	9-7	11-7	13-9
	Spruce-Pine-Fir	SS	8-5	13-3	17-5	22-1	25-7	8-5	12-9	16-2	19-9	22-10
	Spruce-Pine-Fir	#1	8-2	11-11	15-1	18-5	21-5	7-3	10-8	13-6	16-6	19-2
	Spruce-Pine-Fir	#2	8-2	11-11	15-1	18-5	21-5	7-3	10-8	13-6	16-6	19-2
	Spruce-Pine-Fir	#3	6-2	9-0	11-5	13-11	16-2	5-6	8-1	10-3	12-6	14-6

(continued)

TABLE 2308.7.2(3)—continued
RAFTER SPANS FOR COMMON LUMBER SPECIES
(Ground snow load = 30 psf, ceiling not attached to rafters, L/Δ = 180)

RAFTER SPACING (inches)	SPECIES AND GRADE		DEAD LOAD = 10 psf					DEAD LOAD = 20 psf				
			2 × 4	2 × 6	2 × 8	2 × 10	2 × 12	2 × 4	2 × 6	2 × 8	2 × 10	2 × 12
			Maximum rafter spans[a]									
			(ft. - in.)	(ft. - in.)	(ft. - in.)	(ft. - in.)	(ft. - in.)	(ft. - in.)	(ft. - in.)	(ft. - in.)	(ft. - in.)	(ft. - in.)
19.2	Douglas Fir-Larch	SS	8-7	13-6	17-9	21-8	25-2	8-7	12-6	15-10	19-5	22-6
	Douglas Fir-Larch	#1	7-11	11-8	14-9	18-0	20-11	7-1	10-5	13-2	16-1	18-8
	Douglas Fir-Larch	#2	7-5	10-11	13-9	16-10	19-6	6-8	9-9	12-4	15-1	17-6
	Douglas Fir-Larch	#3	5-7	8-3	10-5	12-9	14-9	5-0	7-4	9-4	11-5	13-2
	Hem-Fir	SS	8-1	12-9	16-9	21-4	24-8	8-1	12-4	15-7	19-1	22-1
	Hem-Fir	#1	7-9	11-4	14-4	17-7	20-4	6-11	10-2	12-10	15-8	18-2
	Hem-Fir	#2	7-4	10-9	13-7	16-7	19-3	6-7	9-7	12-2	14-10	17-3
	Hem-Fir	#3	5-7	8-3	10-5	12-9	14-9	5-0	7-4	9-4	11-5	13-2
	Southern Pine	SS	8-5	13-3	17-5	22-3	Note b	8-5	13-3	16-10	20-0	23-7
	Southern Pine	#1	8-0	11-10	15-1	17-7	20-11	7-1	10-7	13-5	15-9	18-8
	Southern Pine	#2	6-10	10-2	12-11	15-4	18-1	6-1	9-2	11-7	13-9	16-2
	Southern Pine	#3	5-3	7-9	9-9	11-10	14-0	4-8	6-11	8-9	10-7	12-6
	Spruce-Pine-Fir	SS	7-11	12-5	16-5	20-2	23-4	7-11	11-8	14-9	18-0	20-11
	Spruce-Pine-Fir	#1	7-5	10-11	13-9	16-10	19-6	6-8	9-9	12-4	15-1	17-6
	Spruce-Pine-Fir	#2	7-5	10-11	13-9	16-10	19-6	6-8	9-9	12-4	15-1	17-6
	Spruce-Pine-Fir	#3	5-7	8-3	10-5	12-9	14-9	5-0	7-4	9-4	11-5	13-2
24	Douglas Fir-Larch	SS	7-11	12-6	15-10	19-5	22-6	7-8	11-3	14-2	17-4	20-1
	Douglas Fir-Larch	#1	7-1	10-5	13-2	16-1	18-8	6-4	9-4	11-9	14-5	16-8
	Douglas Fir-Larch	#2	6-8	9-9	12-4	15-1	17-6	5-11	8-8	11-0	13-6	15-7
	Douglas Fir-Larch	#3	5-0	7-4	9-4	11-5	13-2	4-6	6-7	8-4	10-2	11-10
	Hem-Fir	SS	7-6	11-10	15-7	19-1	22-1	7-6	11-0	13-11	17-0	19-9
	Hem-Fir	#1	6-11	10-2	12-10	15-8	18-2	6-2	9-1	11-6	14-0	16-3
	Hem-Fir	#2	6-7	9-7	12-2	14-10	17-3	5-10	8-7	10-10	13-3	15-5
	Hem-Fir	#3	5-0	7-4	9-4	11-5	13-2	4-6	6-7	8-4	10-2	11-10
	Southern Pine	SS	7-10	12-3	16-2	20-0	23-7	7-10	11-10	15-0	17-11	21-2
	Southern Pine	#1	7-1	10-7	13-5	15-9	18-8	6-4	9-6	12-0	14-1	16-8
	Southern Pine	#2	6-1	9-2	11-7	13-9	16-2	5-5	8-2	10-4	12-3	14-6
	Southern Pine	#3	4-8	6-11	8-9	10-7	12-6	4-2	6-2	7-10	9-6	11-2
	Spruce-Pine-Fir	SS	7-4	11-7	14-9	18-0	20-11	7-1	10-5	13-2	16-1	18-8
	Spruce-Pine-Fir	#1	6-8	9-9	12-4	15-1	17-6	5-11	8-8	11-0	13-6	15-7
	Spruce-Pine-Fir	#2	6-8	9-9	12-4	15-1	17-6	5-11	8-8	11-0	13-6	15-7
	Spruce-Pine-Fir	#3	5-0	7-4	9-4	11-5	13-2	4-6	6-7	8-4	10-2	11-10

Check sources for availability of lumber in lengths greater than 20 feet.

For SI: 1 inch = 25.4 mm, 1 foot = 304.8 mm, 1 pound per square foot = 0.0479 kPa.

a. The tabulated rafter spans assume that ceiling joists are located at the bottom of the attic space or that some other method of resisting the outward push of the rafters on the bearing walls, such as rafter ties, is provided at that location. Where ceiling joists or rafter ties are located higher in the attic space, the rafter spans shall be multiplied by the adjustment factors in Table 2308.7.2(7).

b. Span exceeds 26 feet in length.

TABLE 2308.7.2(4)
RAFTER SPANS FOR COMMON LUMBER SPECIES
(Ground snow load = 50 psf, ceiling not attached to rafters, L/Δ = 180)

RAFTER SPACING (inches)	SPECIES AND GRADE		DEAD LOAD = 10 psf					DEAD LOAD = 20 psf				
			2 × 4	2 × 6	2 × 8	2 × 10	2 × 12	2 × 4	2 × 6	2 × 8	2 × 10	2 × 12
			Maximum rafter spans[a]									
			(ft. - in.)	(ft. - in.)	(ft. - in.)	(ft. - in.)	(ft. - in.)	(ft. - in.)	(ft. - in.)	(ft. - in.)	(ft. - in.)	(ft. - in.)
12	Douglas Fir-Larch	SS	8-5	13-3	17-6	22-4	26-0	8-5	13-3	17-0	20-9	24-0
	Douglas Fir-larch	#1	8-2	12-0	15-3	18-7	21-7	7-7	11-2	14-1	17-3	20-0
	Douglas Fir-larch	#2	7-8	11-3	14-3	17-5	20-2	7-1	10-5	13-2	16-1	18-8
	Douglas Fir-larch	#3	5-10	8-6	10-9	13-2	15-3	5-5	7-10	10-0	12-2	14-1
	Hem-Fir	SS	8-0	12-6	16-6	21-1	25-6	8-0	12-6	16-6	20-4	23-7
	Hem-Fir	#1	7-10	11-9	14-10	18-1	21-0	7-5	10-10	13-9	16-9	19-5
	Hem-Fir	#2	7-5	11-1	14-0	17-2	19-11	7-0	10-3	13-0	15-10	18-5
	Hem-Fir	#3	5-10	8-6	10-9	13-2	15-3	5-5	7-10	10-0	12-2	14-1
	Southern Pine	SS	8-4	13-1	17-2	21-11	Note b	8-4	13-1	17-2	21-5	25-3
	Southern Pine	#1	8-0	12-3	15-6	18-2	21-7	7-7	11-4	14-5	16-10	20-0
	Southern Pine	#2	7-0	10-6	13-4	15-10	18-8	6-6	9-9	12-4	14-8	17-3
	Southern Pine	#3	5-5	8-0	10-1	12-3	14-6	5-0	7-5	9-4	11-4	13-5
	Spruce-Pine-Fir	SS	7-10	12-3	16-2	20-8	24-1	7-10	12-3	15-9	19-3	22-4
	Spruce-Pine-Fir	#1	7-8	11-3	14-3	17-5	20-2	7-1	10-5	13-2	16-1	18-8
	Spruce-Pine-Fir	#2	7-8	11-3	14-3	17-5	20-2	7-1	10-5	13-2	16-1	18-8
	Spruce-Pine-Fir	#3	5-10	8-6	10-9	13-2	15-3	5-5	7-10	10-0	12-2	14-1
16	Douglas Fir-Larch	SS	7-8	12-1	15-10	19-5	22-6	7-8	11-7	14-8	17-11	20-10
	Douglas Fir-Larch	#1	7-1	10-5	13-2	16-1	18-8	6-7	9-8	12-2	14-11	17-3
	Douglas Fir-Larch	#2	6-8	9-9	12-4	15-1	17-6	6-2	9-0	11-5	13-11	16-2
	Douglas Fir-Larch	#3	5-0	7-4	9-4	11-5	13-2	4-8	6-10	8-8	10-6	12-3
	Hem-Fir	SS	7-3	11-5	15-0	19-1	22-1	7-3	11-5	14-5	17-8	20-5
	Hem-Fir	#1	6-11	10-2	12-10	15-8	18-2	6-5	9-5	11-11	14-6	16-10
	Hem-Fir	#2	6-7	9-7	12-2	14-10	17-3	6-1	8-11	11-3	13-9	15-11
	Hem-Fir	#3	5-0	7-4	9-4	11-5	13-2	4-8	6-10	8-8	10-6	12-3
	Southern Pine	SS	7-6	11-10	15-7	19-11	23-7	7-6	11-10	15-7	18-6	21-10
	Southern Pine	#1	7-1	10-7	13-5	15-9	18-8	6-7	9-10	12-5	14-7	17-3
	Southern Pine	#2	6-1	9-2	11-7	13-9	16-2	5-8	8-5	10-9	12-9	15-0
	Southern Pine	#3	4-8	6-11	8-9	10-7	12-6	4-4	6-5	8-1	9-10	11-7
	Spruce-Pine-Fir	SS	7-1	11-2	14-8	18-0	20-11	7-1	10-9	13-8	15-11	19-4
	Spruce-Pine-Fir	#1	6-8	9-9	12-4	15-1	17-6	6-2	9-0	11-5	13-11	16-2
	Spruce-Pine-Fir	#2	6-8	9-9	12-4	15-1	17-6	6-2	9-0	11-5	13-11	16-2
	Spruce-Pine-Fir	#3	5-0	7-4	9-4	11-5	13-2	4-8	6-10	8-8	10-6	12-3

(continued)

TABLE 2308.7.2(4)—continued
RAFTER SPANS FOR COMMON LUMBER SPECIES
(Ground snow load = 50 psf, ceiling not attached to rafters, L/Δ = 180)

RAFTER SPACING (inches)	SPECIES AND GRADE		DEAD LOAD = 10 psf					DEAD LOAD = 20 psf				
			2 × 4	2 × 6	2 × 8	2 × 10	2 × 12	2 × 4	2 × 6	2 × 8	2 × 10	2 × 12
			Maximum rafter spans[a]									
			(ft. - in.)	(ft. - in.)	(ft. - in.)	(ft. - in.)	(ft. - in.)	(ft. - in.)	(ft. - in.)	(ft. - in.)	(ft. - in.)	(ft. - in.)
19.2	Douglas Fir-Larch	SS	7-3	11-4	14-6	17-8	20-6	7-3	10-7	13-5	16-5	19-0
	Douglas Fir-Larch	#1	6-6	9-6	12-0	14-8	17-1	6-0	8-10	11-2	13-7	15-9
	Douglas Fir-Larch	#2	6-1	8-11	11-3	13-9	15-11	5-7	8-3	10-5	12-9	14-9
	Douglas Fir-Larch	#3	4-7	6-9	8-6	10-5	12-1	4-3	6-3	7-11	9-7	11-2
	Hem-Fir	SS	6-10	10-9	14-2	17-5	20-2	6-10	10-5	13-2	16-1	18-8
	Hem-Fir	#1	6-4	9-3	11-9	14-4	16-7	5-10	8-7	10-10	13-3	15-5
	Hem-Fir	#2	6-0	8-9	11-1	13-7	15-9	5-7	8-1	10-3	12-7	14-7
	Hem-Fir	#3	4-7	6-9	8-6	10-5	12-1	4-3	6-3	7-11	9-7	11-2
	Southern Pine	SS	7-1	11-2	14-8	18-3	21-7	7-1	11-2	14-2	16-11	20-0
	Southern Pine	#1	6-6	9-8	12-3	14-4	17-1	6-0	9-0	11-4	13-4	15-9
	Southern Pine	#2	5-7	8-4	10-7	12-6	14-9	5-2	7-9	9-9	11-7	13-8
	Southern Pine	#3	4-3	6-4	8-0	9-8	11-5	4-0	5-10	7-4	8-11	10-7
	Spruce-Pine-Fir	SS	6-8	10-6	13-5	16-5	19-1	6-8	9-10	12-5	15-3	17-8
	Spruce-Pine-Fir	#1	6-1	8-11	11-3	13-9	15-11	5-7	8-3	10-5	12-9	14-9
	Spruce-Pine-Fir	#2	6-1	8-11	11-3	13-9	15-11	5-7	8-3	10-5	12-9	14-9
	Spruce-Pine-Fir	#3	4-7	6-9	8-6	10-5	12-1	4-3	6-3	7-11	9-7	11-2
24	Douglas Fir-Larch	SS	6-8	10-3	13-0	15-10	18-4	6-6	9-6	12-0	14-8	17-0
	Douglas Fir-Larch	#1	5-10	8-6	10-9	13-2	15-3	5-5	7-10	10-0	12-2	14-1
	Douglas Fir-Larch	#2	5-5	7-11	10-1	12-4	14-3	5-0	7-4	9-4	11-5	13-2
	Douglas Fir-Larch	#3	4-1	6-0	7-7	9-4	10-9	3-10	5-7	7-1	8-7	10-0
	Hem-Fir	SS	6-4	9-11	12-9	15-7	18-0	6-4	9-4	11-9	14-5	16-8
	Hem-Fir	#1	5-8	8-3	10-6	12-10	14-10	5-3	7-8	9-9	11-10	13-9
	Hem-Fir	#2	5-4	7-10	9-11	12-1	14-1	4-11	7-3	9-2	11-3	13-0
	Hem-Fir	#3	4-1	6-0	7-7	9-4	10-9	3-10	5-7	7-1	8-7	10-0
	Southern Pine	SS	6-7	10-4	13-8	16-4	19-3	6-7	10-0	12-8	15-2	17-10
	Southern Pine	#1	5-10	8-8	11-0	12-10	15-3	5-5	8-0	10-2	11-11	14-1
	Southern Pine	#2	5-0	7-5	9-5	11-3	13-2	4-7	6-11	8-9	10-5	12-3
	Southern Pine	#3	3-10	5-8	7-1	8-8	10-3	3-6	5-3	6-7	8-0	9-6
	Spruce-Pine-Fir	SS	6-2	9-6	12-0	14-8	17-1	6-0	8-10	11-2	13-7	15-9
	Spruce-Pine-Fir	#1	5-5	7-11	10-1	12-4	14-3	5-0	7-4	9-4	11-5	13-2
	Spruce-Pine-Fir	#2	5-5	7-11	10-1	12-4	14-3	5-0	7-4	9-4	11-5	13-2
	Spruce-Pine-Fir	#3	4-1	6-0	7-7	9-4	10-9	3-10	5-7	7-1	8-7	10-0

Check sources for availability of lumber in lengths greater than 20 feet.

For SI: 1 inch = 25.4 mm, 1 foot = 304.8 mm, 1 pound per square foot = 0.0479 kPa.

a. The tabulated rafter spans assume that ceiling joists are located at the bottom of the attic space or that some other method of resisting the outward push of the rafters on the bearing walls, such as rafter ties, is provided at that location. Where ceiling joists or rafter ties are located higher in the attic space, the rafter spans shall be multiplied by the adjustment factors in Table 2308.7.2(7).

b. Span exceeds 26 feet in length.

TABLE 2308.7.2(5)
RAFTER SPANS FOR COMMON LUMBER SPECIES
(Ground snow load = 30 psf, ceiling attached to rafters, L/Δ = 240)

RAFTER SPACING (inches)	SPECIES AND GRADE		DEAD LOAD = 10 psf					DEAD LOAD = 20 psf				
			2 × 4	2 × 6	2 × 8	2 × 10	2 × 12	2 × 4	2 × 6	2 × 8	2 × 10	2 × 12
			Maximum rafter spans[a]									
			(ft. - in.)	(ft. - in.)	(ft. - in.)	(ft. - in.)	(ft. - in.)	(ft. - in.)	(ft. - in.)	(ft. - in.)	(ft. - in.)	(ft. - in.)
12	Douglas Fir-Larch	SS	9-1	14-4	18-10	24-1	Note b	9-1	14-4	18-10	24-1	Note b
	Douglas Fir-Larch	#1	8-9	13-9	18-2	22-9	Note b	8-9	13-2	16-8	20-4	23-7
	Douglas Fir-Larch	#2	8-7	13-6	17-5	21-4	24-8	8-5	12-4	15-7	19-1	22-1
	Douglas Fir-Larch	#3	7-1	10-5	13-2	16-1	18-8	6-4	9-4	11-9	14-5	16-8
	Hem-Fir	SS	8-7	13-6	17-10	22-9	Note b	8-7	13-6	17-10	22-9	Note b
	Hem-Fir	#1	8-5	13-3	17-5	22-2	25-9	8-5	12-10	16-3	19-10	23-0
	Hem-Fir	#2	8-0	12-7	16-7	21-0	24-4	8-0	12-2	15-4	18-9	21-9
	Hem-Fir	#3	7-1	10-5	13-2	16-1	18-8	6-4	9-4	11-9	14-5	16-8
	Southern Pine	SS	8-11	14-1	18-6	23-8	Note b	8-11	14-1	18-6	23-8	Note b
	Southern Pine	#1	8-7	13-6	17-10	22-3	Note b	8-7	13-5	17-0	19-11	23-7
	Southern Pine	#2	8-3	12-11	16-4	19-5	22-10	7-8	11-7	14-8	17-4	20-5
	Southern Pine	#3	6-7	9-9	12-4	15-0	17-9	5-11	8-9	11-0	13-5	15-10
	Spruce-Pine-Fir	SS	8-5	13-3	17-5	22-3	Note b	8-5	13-3	17-5	22-3	Note b
	Spruce-Pine-Fir	#1	8-3	12-11	17-0	21-4	24-8	8-3	12-4	15-7	19-1	22-1
	Spruce-Pine-Fir	#2	8-3	12-11	17-0	21-4	24-8	8-3	12-4	15-7	19-1	22-1
	Spruce-Pine-Fir	#3	7-1	10-5	13-2	16-1	18-8	6-4	9-4	11-9	14-5	16-8
16	Douglas Fir-Larch	SS	8-3	13-0	17-2	21-10	Note b	8-3	13-0	17-2	21-3	24-8
	Douglas Fir-Larch	#1	8-0	12-6	16-2	19-9	22-10	7-10	11-5	14-5	17-8	20-5
	Douglas Fir-Larch	#2	7-10	11-11	15-1	18-5	21-5	7-3	10-8	13-6	16-6	19-2
	Douglas Fir-Larch	#3	6-2	9-0	11-5	13-11	16-2	5-6	8-1	10-3	12-6	14-6
	Hem-Fir	SS	7-10	12-3	16-2	20-8	25-1	7-10	12-3	16-2	20-8	24-2
	Hem-Fir	#1	7-8	12-0	15-9	19-3	22-3	7-7	11-1	14-1	17-2	19-11
	Hem-Fir	#2	7-3	11-5	14-11	18-2	21-1	7-2	10-6	13-4	16-3	18-10
	Hem-Fir	#3	6-2	9-0	11-5	13-11	16-2	5-6	8-1	10-3	12-6	14-6
	Southern Pine	SS	8-1	12-9	16-10	21-6	Note b	8-1	12-9	16-10	21-6	25-11
	Southern Pine	#1	7-10	12-3	16-2	19-3	22-10	7-10	11-7	14-9	17-3	20-5
	Southern Pine	#2	7-6	11-2	14-2	16-10	19-10	6-8	10-0	12-8	15-1	17-9
	Southern Pine	#3	5-9	8-6	10-8	13-0	15-4	5-2	7-7	9-7	11-7	13-9
	Spruce-Pine-Fir	SS	7-8	12-0	15-10	20-2	24-7	7-8	12-0	15-10	19-9	22-10
	Spruce-Pine-Fir	#1	7-6	11-9	15-1	18-5	21-5	7-3	10-8	13-6	16-6	19-2
	Spruce-Pine-Fir	#2	7-6	11-9	15-1	18-5	21-5	7-3	10-8	13-6	16-6	19-2
	Spruce-Pine-Fir	#3	6-2	9-0	11-5	13-11	16-2	5-6	8-1	10-3	12-6	14-6

(continued)

TABLE 2308.7.2(5)—continued
RAFTER SPANS FOR COMMON LUMBER SPECIES
(Ground snow load = 30 psf, ceiling attached to rafters, L/Δ = 240)

RAFTER SPACING (inches)	SPECIES AND GRADE		DEAD LOAD = 10 psf					DEAD LOAD = 20 psf				
			2 × 4	2 × 6	2 × 8	2 × 10	2 × 12	2 × 4	2 × 6	2 × 8	2 × 10	2 × 12
			Maximum rafter spans[a]									
			(ft. - in.)	(ft. - in.)	(ft. - in.)	(ft. - in.)	(ft. - in.)	(ft. - in.)	(ft. - in.)	(ft. - in.)	(ft. - in.)	(ft. - in.)
19.2	Douglas Fir-Larch	SS	7-9	12-3	16-1	20-7	25-0	7-9	12-3	15-10	19-5	22-6
	Douglas Fir-Larch	#1	7-6	11-8	14-9	18-0	20-11	7-1	10-5	13-2	16-1	18-8
	Douglas Fir-Larch	#2	7-4	10-11	13-9	16-10	19-6	6-8	9-9	12-4	15-1	17-6
	Douglas Fir-Larch	#3	5-7	8-3	10-5	12-9	14-9	5-0	7-4	9-4	11-5	13-2
	Hem-Fir	SS	7-4	11-7	15-3	19-5	23-7	7-4	11-7	15-3	19-1	22-1
	Hem-Fir	#1	7-2	11-4	14-4	17-7	20-4	6-11	10-2	12-10	15-8	18-2
	Hem-Fir	#2	6-10	10-9	13-7	16-7	19-3	6-7	9-7	12-2	14-10	17-3
	Hem-Fir	#3	5-7	8-3	10-5	12-9	14-9	5-0	7-4	9-4	11-5	13-2
	Southern Pine	SS	7-8	12-0	15-10	20-2	24-7	7-8	12-0	15-10	20-0	23-7
	Southern Pine	#1	7-4	11-7	15-1	17-7	20-11	7-1	10-7	13-5	15-9	18-8
	Southern Pine	#2	6-10	10-2	12-11	15-4	18-1	6-1	9-2	11-7	13-9	16-2
	Southern Pine	#3	5-3	7-9	9-9	11-10	14-0	4-8	6-11	8-9	10-7	12-6
	Spruce-Pine-Fir	SS	7-2	11-4	14-11	19-0	23-1	7-2	11-4	14-9	18-0	20-11
	Spruce-Pine-Fir	#1	7-0	10-11	13-9	16-10	19-6	6-8	9-9	12-4	15-1	17-6
	Spruce-Pine-Fir	#2	7-0	10-11	13-9	16-10	19-6	6-8	9-9	12-4	15-1	17-6
	Spruce-Pine-Fir	#3	5-7	8-3	10-5	12-9	14-9	5-0	7-4	9-4	11-5	13-2
24	Douglas Fir-Larch	SS	7-3	11-4	15-0	19-1	22-6	7-3	11-3	14-2	17-4	20-1
	Douglas Fir-Larch	#1	7-0	10-5	13-2	16-1	18-8	6-4	9-4	11-9	14-5	16-8
	Douglas Fir-Larch	#2	6-8	9-9	12-4	15-1	17-6	5-11	8-8	11-0	13-6	15-7
	Douglas Fir-Larch	#3	5-0	7-4	9-4	11-5	13-2	4-6	6-7	8-4	10-2	11-10
	Hem-Fir	SS	6-10	10-9	14-2	18-0	21-11	6-10	10-9	13-11	17-0	19-9
	Hem-Fir	#1	6-8	10-2	12-10	15-8	18-2	6-2	9-1	11-6	14-0	16-3
	Hem-Fir	#2	6-4	9-7	12-2	14-10	17-3	5-10	8-7	10-10	13-3	15-5
	Hem-Fir	#3	5-0	7-4	9-4	11-5	13-2	4-6	6-7	8-4	10-2	11-10
	Southern Pine	SS	7-1	11-2	14-8	18-9	22-10	7-1	11-2	14-8	17-11	21-2
	Southern Pine	#1	6-10	10-7	13-5	15-9	18-8	6-4	9-6	12-0	14-1	16-8
	Southern Pine	#2	6-1	9-2	11-7	13-9	16-2	5-5	8-2	10-4	12-3	14-6
	Southern Pine	#3	4-8	6-11	8-9	10-7	12-6	4-2	6-2	7-10	9-6	11-2
	Spruce-Pine-Fir	SS	6-8	10-6	13-10	17-8	20-11	6-8	10-5	13-2	16-1	18-8
	Spruce-Pine-Fir	#1	6-6	9-9	12-4	15-1	17-6	5-11	8-8	11-0	13-6	15-7
	Spruce-Pine-Fir	#2	6-6	9-9	12-4	15-1	17-6	5-11	8-8	11-0	13-6	15-7
	Spruce-Pine-Fir	#3	5-0	7-4	9-4	11-5	13-2	4-6	6-7	8-4	10-2	11-10

Check sources for availability of lumber in lengths greater than 20 feet.

For SI: 1 inch = 25.4 mm, 1 foot = 304.8 mm, 1 pound per square foot = 0.0479 kPa.

a. The tabulated rafter spans assume that ceiling joists are located at the bottom of the attic space or that some other method of resisting the outward push of the rafters on the bearing walls, such as rafter ties, is provided at that location. Where ceiling joists or rafter ties are located higher in the attic space, the rafter spans shall be multiplied by the adjustment factors in Table 2308.7.2(7).

b. Span exceeds 26 feet in length.

TABLE 2308.7.2(6)
RAFTER SPANS FOR COMMON LUMBER SPECIES
(Ground snow load = 50 psf, ceiling attached to rafters, L/Δ = 240)

RAFTER SPACING (inches)	SPECIES AND GRADE		DEAD LOAD = 10 psf					DEAD LOAD = 20 psf				
			2 × 4	2 × 6	2 × 8	2 × 10	2 × 12	2 × 4	2 × 6	2 × 8	2 × 10	2 × 12
			Maximum rafter spans[a]									
			(ft. - in.)	(ft. - in.)	(ft. - in.)	(ft. - in.)	(ft. - in.)	(ft. - in.)	(ft. - in.)	(ft. - in.)	(ft. - in.)	(ft. - in.)
12	Douglas Fir-Larch	SS	7-8	12-1	15-11	20-3	24-8	7-8	12-1	15-11	20-3	24-0
	Douglas Fir-Larch	#1	7-5	11-7	15-3	18-7	21-7	7-5	11-2	14-1	17-3	20-0
	Douglas Fir-Larch	#2	7-3	11-3	14-3	17-5	20-2	7-1	10-5	13-2	16-1	18-8
	Douglas Fir-Larch	#3	5-10	8-6	10-9	13-2	15-3	5-5	7-10	10-0	12-2	14-1
	Hem-Fir	SS	7-3	11-5	15-0	19-2	23-4	7-3	11-5	15-0	19-2	23-4
	Hem-Fir	#1	7-1	11-2	14-8	18-1	21-0	7-1	10-10	13-9	16-9	19-5
	Hem-Fir	#2	6-9	10-8	14-0	17-2	19-11	6-9	10-3	13-0	15-10	18-5
	Hem-Fir	#3	5-10	8-6	10-9	13-2	15-3	5-5	7-10	10-0	12-2	14-1
	Southern Pine	SS	7-6	11-10	15-7	19-11	24-3	7-6	11-10	15-7	19-11	24-3
	Southern Pine	#1	7-3	11-5	15-0	18-2	21-7	7-3	11-4	14-5	16-10	20-0
	Southern Pine	#2	6-11	10-6	13-4	15-10	18-8	6-6	9-9	12-4	14-8	17-3
	Southern Pine	#3	5-5	8-0	10-1	12-3	14-6	5-0	7-5	9-4	11-4	13-5
	Spruce-Pine-Fir	SS	7-1	11-2	14-8	18-9	22-10	7-1	11-2	14-8	18-9	22-4
	Spruce-Pine-Fir	#1	6-11	10-11	14-3	17-5	20-2	6-11	10-5	13-2	16-1	18-8
	Spruce-Pine-Fir	#2	6-11	10-11	14-3	17-5	20-2	6-11	10-5	13-2	16-1	18-8
	Spruce-Pine-Fir	#3	5-10	8-6	10-9	13-2	15-3	5-5	7-10	10-0	12-2	14-1
16	Douglas Fir-Larch	SS	7-0	11-0	14-5	18-5	22-5	7-0	11-0	14-5	17-11	20-10
	Douglas Fir-Larch	#1	6-9	10-5	13-2	16-1	18-8	6-7	9-8	12-2	14-11	17-3
	Douglas Fir-Larch	#2	6-7	9-9	12-4	15-1	17-6	6-2	9-0	11-5	13-11	16-2
	Douglas Fir-Larch	#3	5-0	7-4	9-4	11-5	13-2	4-8	6-10	8-8	10-6	12-3
	Hem-Fir	SS	6-7	10-4	13-8	17-5	21-2	6-7	10-4	13-8	17-5	20-5
	Hem-Fir	#1	6-5	10-2	12-10	15-8	18-2	6-5	9-5	11-11	14-6	16-10
	Hem-Fir	#2	6-2	9-7	12-2	14-10	17-3	6-1	8-11	11-3	13-9	15-11
	Hem-Fir	#3	5-0	7-4	9-4	11-5	13-2	4-8	6-10	8-8	10-6	12-3
	Southern Pine	SS	6-10	10-9	14-2	18-1	22-0	6-10	10-9	14-2	18-1	21-10
	Southern Pine	#1	6-7	10-4	13-5	15-9	18-8	6-7	9-10	12-5	14-7	17-3
	Southern Pine	#2	6-1	9-2	11-7	13-9	16-2	5-8	8-5	10-9	12-9	15-0
	Southern Pine	#3	4-8	6-11	8-9	10-7	12-6	4-4	6-5	8-1	9-10	11-7
	Spruce-Pine-Fir	SS	6-5	10-2	13-4	17-0	20-9	6-5	10-2	13-4	16-8	19-4
	Spruce-Pine-Fir	#1	6-4	9-9	12-4	15-1	17-6	6-2	9-0	11-5	13-11	16-2
	Spruce-Pine-Fir	#2	6-4	9-9	12-4	15-1	17-6	6-2	9-0	11-5	13-11	16-2
	Spruce-Pine-Fir	#3	5-0	7-4	9-4	11-5	13-2	4-8	6-10	8-8	10-6	12-3

(continued)

TABLE 2308.7.2(6)—continued
RAFTER SPANS FOR COMMON LUMBER SPECIES
(Ground snow load = 50 psf, ceiling attached to rafters, L/Δ = 240)

RAFTER SPACING (inches)	SPECIES AND GRADE		DEAD LOAD = 10 psf					DEAD LOAD = 20 psf				
			2 × 4	2 × 6	2 × 8	2 × 10	2 × 12	2 × 4	2 × 6	2 × 8	2 × 10	2 × 12
			Maximum rafter spans[a]									
			(ft. - in.)	(ft. - in.)	(ft. - in.)	(ft. - in.)	(ft. - in.)	(ft. - in.)	(ft. - in.)	(ft. - in.)	(ft. - in.)	(ft. - in.)
19.2	Douglas Fir-Larch	SS	6-7	10-4	13-7	17-4	20-6	6-7	10-4	13-5	16-5	19-0
	Douglas Fir-Larch	#1	6-4	9-6	12-0	14-8	17-1	6-0	8-10	11-2	13-7	15-9
	Douglas Fir-Larch	#2	6-1	8-11	11-3	13-9	15-11	5-7	8-3	10-5	12-9	14-9
	Douglas Fir-Larch	#3	4-7	6-9	8-6	10-5	12-1	4-3	6-3	7-11	9-7	11-2
	Hem-Fir	SS	6-2	9-9	12-10	16-5	19-11	6-2	9-9	12-10	16-1	18-8
	Hem-Fir	#1	6-1	9-3	11-9	14-4	16-7	5-10	8-7	10-10	13-3	15-5
	Hem-Fir	#2	5-9	8-9	11-1	13-7	15-9	5-7	8-1	10-3	12-7	14-7
	Hem-Fir	#3	4-7	6-9	8-6	10-5	12-1	4-3	6-3	7-11	9-7	11-2
	Southern Pine	SS	6-5	10-2	13-4	17-0	20-9	6-5	10-2	13-4	16-11	20-0
	Southern Pine	#1	6-2	9-8	12-3	14-4	17-1	6-0	9-0	11-4	13-4	15-9
	Southern Pine	#2	5-7	8-4	10-7	12-6	14-9	5-2	7-9	9-9	11-7	13-8
	Southern Pine	#3	4-3	6-4	8-0	9-8	11-5	4-0	5-10	7-4	8-11	10-7
	Spruce-Pine-Fir	SS	6-1	9-6	12-7	16-0	19-1	6-1	9-6	12-5	15-3	17-8
	Spruce-Pine-Fir	#1	5-11	8-11	11-3	13-9	15-11	5-7	8-3	10-5	12-9	14-9
	Spruce-Pine-Fir	#2	5-11	8-11	11-3	13-9	15-11	5-7	8-3	10-5	12-9	14-9
	Spruce-Pine-Fir	#3	4-7	6-9	8-6	10-5	12-1	4-3	6-3	7-11	9-7	11-2
24	Douglas Fir-Larch	SS	6-1	9-7	12-7	15-10	18-4	6-1	9-6	12-0	14-8	17-0
	Douglas Fir-Larch	#1	5-10	8-6	10-9	13-2	15-3	5-5	7-10	10-0	12-2	14-1
	Douglas Fir-Larch	#2	5-5	7-11	10-1	12-4	14-3	5-0	7-4	9-4	11-5	13-2
	Douglas Fir-Larch	#3	4-1	6-0	7-7	9-4	10-9	3-10	5-7	7-1	8-7	10-0
	Hem-Fir	SS	5-9	9-1	11-11	15-2	18-0	5-9	9-1	11-9	14-5	15-11
	Hem-Fir	#1	5-8	8-3	10-6	12-10	14-10	5-3	7-8	9-9	11-10	13-9
	Hem-Fir	#2	5-4	7-10	9-11	12-1	14-1	4-11	7-3	9-2	11-3	13-0
	Hem-Fir	#3	4-1	6-0	7-7	9-4	10-9	3-10	5-7	7-1	8-7	10-0
	Southern Pine	SS	6-0	9-5	12-5	15-10	19-3	6-0	9-5	12-5	15-2	17-10
	Southern Pine	#1	5-9	8-8	11-0	12-10	15-3	5-5	8-0	10-2	11-11	14-1
	Southern Pine	#2	5-0	7-5	9-5	11-3	13-2	4-7	6-11	8-9	10-5	12-3
	Southern Pine	#3	3-10	5-8	7-1	8-8	10-3	3-6	5-3	6-7	8-0	9-6
	Spruce-Pine-Fir	SS	5-8	8-10	11-8	14-8	17-1	5-8	8-10	11-2	13-7	15-9
	Spruce-Pine-Fir	#1	5-5	7-11	10-1	12-4	14-3	5-0	7-4	9-4	11-5	13-2
	Spruce-Pine-Fir	#2	5-5	7-11	10-1	12-4	14-3	5-0	7-4	9-4	11-5	13-2
	Spruce-Pine-Fir	#3	4-1	6-0	7-7	9-4	10-9	3-10	5-7	7-1	8-7	10-0

Check sources for availability of lumber in lengths greater than 20 feet.

For SI: 1 inch = 25.4 mm, 1 foot = 304.8 mm, 1 pound per square foot = 0.0479 kPa.

a. The tabulated rafter spans assume that ceiling joists are located at the bottom of the attic space or that some other method of resisting the outward push of the rafters on the bearing walls, such as rafter ties, is provided at that location. Where ceiling joists or rafter ties are located higher in the attic space, the rafter spans shall be multiplied by the adjustment factors in Table 2308.7.2(7).

TABLE 2308.7.2(7)
RAFTER SPAN ADJUSTMENT FACTOR

H_C/H_R[a]	RAFTER SPAN ADJUSTMENT FACTOR
1/3	0.67
1/4	0.76
1/5	0.83
1/6	0.90
1/7.5 or less	1.00

a. H_C = Height of ceiling joists or rafter ties measured vertically above the top of the rafter support walls;
H_R = Height of roof ridge measured vertically above the top of the rafter support walls.

TABLE 2308.7.3.1 RAFTER TIE CONNECTIONS[i]

RAFTER SLOPE	TIE SPACING (inches)	LIVE LOAD ONLY[g]			GROUND SNOW LOAD (pounds per square foot)					
					30 pounds per square foot			50 pounds per square foot		
		Roof span (feet)								
		12	24	36	12	24	36	12	24	36
		Required number of 16d common ($3^1/_2$″ x 0.162″) nails per connection[a, b, c, d, e, f, h]								
3:12	12	3	5	8	3	6	9	5	9	13
	16	4	7	10	4	8	12	6	12	17
	19.2	4	8	12	5	10	14	7	14	21
	24	5	10	15	6	12	18	9	17	26
	32	7	13	20	8	16	24	12	23	34
	48	10	20	29	12	24	35	17	34	51
4:12	12	3	4	6	3	5	7	4	7	10
	16	3	5	8	3	6	9	5	9	13
	19.2	3	6	9	4	7	11	6	11	16
	24	4	8	11	5	9	13	7	13	19
	32	5	10	15	6	12	18	9	17	26
	48	8	15	22	9	18	26	13	26	38
5:12	12	3	3	5	3	4	6	3	6	8
	16	3	4	6	3	5	7	4	7	11
	19.2	3	5	7	3	6	9	5	9	13
	24	3	6	9	4	7	11	6	11	16
	32	4	8	12	5	10	14	7	14	21
	48	6	12	18	7	14	21	11	21	31
7:12	12	3	3	4	3	3	4	3	4	6
	16	3	3	5	3	4	5	3	5	8
	19.2	3	4	5	3	4	6	3	6	9
	24	3	5	7	3	5	8	4	8	11
	32	3	6	9	4	7	10	5	10	15
	48	5	9	13	5	10	15	8	15	22
9:12	12	3	3	3	3	3	3	3	3	5
	16	3	3	4	3	3	4	3	4	6
	19.2	3	3	4	3	4	5	3	5	7
	24	3	4	5	3	4	6	3	6	9
	32	3	5	7	3	6	8	4	8	12
	48	4	7	10	4	8	12	6	12	17
12:12	12	3	3	3	3	3	3	3	3	4
	16	3	3	3	3	3	3	3	3	5
	19.2	3	3	3	3	3	4	3	4	6
	24	3	3	4	3	3	5	3	5	7
	32	3	4	5	3	4	6	3	6	9
	48	3	5	8	3	6	9	5	9	13

For SI: 1 inch = 25.4 mm, 1 foot = 304.8 mm, 1 pound per square foot = 47.8 N/m^2.

a. 10d common (3″ × 0.148″) nails shall be permitted to be substituted for 16d common ($3^1/_2$″ × 0.162″) nails where the required number of nails is taken as 1.2 times the required number of 16d common nails, rounded up to the next full nail.

b. Rafter tie heel joint connections are not required where the ridge is supported by a load-bearing wall, header or ridge beam.

c. Where intermediate support of the rafter is provided by vertical struts or purlins to a load-bearing wall, the tabulated heel joint connection requirements are permitted to be reduced proportionally to the reduction in span.

d. Equivalent nailing patterns are required for ceiling joist to ceiling joist lap splices.

e. Connected members shall be of sufficient size to prevent splitting due to nailing.

f. For snow loads less than 30 pounds per square foot, the required number of nails is permitted to be reduced by multiplying by the ratio of actual snow load plus 10 divided by 40, but not less than the number required for no snow load.

g. Applies to roof live load of 20 psf or less.

h. Tabulated heel joint connection requirements assume that ceiling joists or rafter ties are located at the bottom of the attic space. Where ceiling joists or rafter ties are located higher in the attic, heel joint connection requirements shall be increased by the adjustment factors in Table 2308.7.3.1(1).

i. Tabulated requirements are based on 10 psf roof dead load in combination with the specified roof snow load and roof live load.

TABLE 2308.7.3.1(1)
HEEL JOINT CONNECTION ADJUSTMENT FACTORS

H_C/H_R [a, b]	HEEL JOINT CONNECTION ADJUSTMENT FACTOR
1/3	1.5
1/4	1.33
1/5	1.25
1/6	1.2
1/10 or less	1.11

a. H_C = Height of ceiling joists or rafter ties measured vertically from the top of the rafter support walls to the bottom of the ceiling joists or rafter ties; H_R = Height of roof ridge measured vertically from the top of the rafter support walls to the bottom of the roof ridge.
b. Where H_C/H_R exceeds 1/3, connections shall be designed in accordance with accepted engineering practice.

TABLE 2308.7.12
ALLOWABLE SPANS FOR 2-INCH TONGUE-AND-GROOVE DECKING

SPAN[a] (feet)	LIVE LOAD (pounds per square foot)	DEFLECTION LIMIT	BENDING STRESS (f) (pounds per square inch)	MODULUS OF ELASTICITY (E) (pounds per square inch)
		Roofs		
4	20	1/240	160	170,000
		1/360		256,000
	30	1/240	210	256,000
		1/360		384,000
	40	1/240	270	340,000
		1/360		512,000
4.5	20	1/240	200	242,000
		1/360		305,000
	30	1/240	270	363,000
		1/360		405,000
	40	1/240	350	484,000
		1/360		725,000
5.0	20	1/240	250	332,000
		1/360		500,000
	30	1/240	330	495,000
		1/360		742,000
	40	1/240	420	660,000
		1/360		1,000,000
5.5	20	1/240	300	442,000
		1/360		660,000
	30	1/240	400	662,000
		1/360		998,000
	40	1/240	500	884,000
		1/360		1,330,000
6.0	20	1/240	360	575,000
		1/360		862,000
	30	1/240	480	862,000
		1/360		1,295,000
	40	1/240	600	1,150,000
		1/360		1,730,000

(continued)

TABLE 2308.7.12—continued
ALLOWABLE SPANS FOR 2-INCH TONGUE-AND-GROOVE DECKING

SPAN[a] (feet)	LIVE LOAD (pounds per square foot)	DEFLECTION LIMIT	BENDING STRESS (f) (pounds per square inch)	MODULUS OF ELASTICITY (E) (pounds per square inch)
Roofs				
6.5	20	1/240 1/360	420	595,000 892,000
	30	1/240 1/360	560	892,000 1,340,000
	40	1/240 1/360	700	1,190,000 1,730,000
7.0	20	1/240 1/360	490	910,000 1,360,000
	30	1/240 1/360	650	1,370,000 2,000,000
	40	1/240 1/360	810	1,820,000 2,725,000
7.5	20	1/240 1/360	560	1,125,000 1,685,000
	30	1/240 1/360	750	1,685,000 2,530,000
	40	1/240 1/360	930	2,250,000 3,380,000
8.0	20	1/240 1/360	640	1,360,000 2,040,000
	30	1/240 1/360	850	2,040,000 3,060,000
Floors				
4 4.5 5.0	40	1/360	840 950 1,060	1,000,000 1,300,000 1,600,000

For SI: 1 inch = 25.4 mm, 1 foot = 304.8 mm, 1 pound per square foot = 0.0479 kN/m^2, 1 pound per square inch = 0.00689 N/mm^2.

a. Spans are based on simple beam action with 10 pounds per square foot dead load and provisions for a 300-pound concentrated load on a 12-inch width of decking. Random layup is permitted in accordance with the provisions of Section 2308.7.12. Lumber thickness is 1$^1/_2$ inches nominal.

CHAPTER 24
GLASS AND GLAZING

User notes:

About this chapter: *Chapter 24 establishes regulations for glass and glazing used in buildings and structures. Engineering and design requirements are included in the chapter for glazing that is subjected to wind and snow loads. Another concern of this chapter is glass and glazing used in areas where it is likely to be impacted by the occupants. Section 2406 identifies hazardous locations where glazing must either be safety glazing or protected to prevent impacts by occupants. Safety glazing must meet stringent standards and be appropriately marked or identified. Additional requirements are provided for glass and glazing in guards, handrails, elevator hoistways and elevator cars, as well as in athletic facilities.*

Code development reminder: *Code change proposals to this chapter will be considered by the IBC—Structural Code Development Committee during the 2022 (Group B) Code Development Cycle.*

SECTION 2401
GENERAL

2401.1 Scope. The provisions of this chapter shall govern the materials, design, construction and quality of glass, light-transmitting ceramic and light-transmitting plastic panels for exterior and interior use in both vertical and sloped applications in buildings and structures. Light-transmitting plastic glazing shall also meet the applicable requirements of Chapter 26.

SECTION 2402
GLAZING REPLACEMENT

2402.1 General. The installation of replacement glass shall be as required for new installations.

SECTION 2403
GENERAL REQUIREMENTS FOR GLASS

2403.1 Identification. Each pane shall bear the manufacturer's *mark* designating the type and thickness of the glass or glazing material. The identification shall not be omitted unless *approved* and an affidavit is furnished by the glazing contractor certifying that each light is glazed in accordance with *approved construction documents* that comply with the provisions of this chapter. Safety glazing shall be identified in accordance with Section 2406.3.

Each pane of tempered glass, except tempered spandrel glass, shall be permanently identified by the manufacturer. The identification *mark* shall be acid etched, sand blasted, ceramic fired, laser etched, embossed or of a type that, once applied, cannot be removed without being destroyed.

Tempered spandrel glass shall be provided with a removable paper marking by the manufacturer.

2403.2 Glass supports. Where one or more sides of any pane of glass are not firmly supported, or are subjected to unusual *load* conditions, detailed *construction documents*, detailed shop drawings and analysis or test data ensuring safe performance for the specific installation shall be prepared by a *registered design professional*.

2403.3 Glass framing. To be considered firmly supported, the framing members for each individual pane of glass shall be designed so that the deflection of the edge of the glass perpendicular to the glass pane does not exceed $^1/_{175}$ of the glass edge length where the glass edge length is not more than 13 feet 6 inches (4115 mm), or $^1/_{240}$ of the glass edge length + $^1/_4$ inch (6.4 mm) where the glass edge length is greater than 13 feet 6 inches (4115 mm), when subjected to the larger of the positive or negative load where loads are combined as specified in Section 1605.

2403.4 Interior glazed areas. Where interior glazing is installed adjacent to a walking surface, the differential deflection of two adjacent unsupported edges shall be not greater than the thickness of the panels when a force of 50 pounds per linear foot (plf) (730 N/m) is applied horizontally to one panel at any point up to 42 inches (1067 mm) above the walking surface.

2403.5 Louvered windows or jalousies. Float, wired and patterned glass in louvered windows and jalousies shall be not thinner than nominal $^3/_{16}$ inch (4.8 mm) and not longer than 48 inches (1219 mm). Exposed glass edges shall be smooth.

Wired glass with wire exposed on longitudinal edges shall not be used in louvered windows or jalousies.

Where other glass types are used, the design shall be submitted to the *building official* for approval.

SECTION 2404
WIND, SNOW, SEISMIC AND DEAD LOADS ON GLASS

2404.1 Vertical glass. Glass sloped 15 degrees (0.26 rad) or less from vertical in windows, curtain and window walls, doors and other exterior applications shall be designed to resist the wind *loads* due to basic design *wind speed*, *V*, in Section 1609 for components and cladding. Glass in glazed curtain walls, glazed storefronts and glazed partitions shall meet the seismic requirements of ASCE 7, Section 13.5.9. The load resistance of glass under uniform *load* shall be determined in accordance with ASTM E1300.

The design of vertical glazing shall be based on Equation 24-1.

$$0.6F_{gw} \leq F_{ga} \quad \textbf{(Equation 24-1)}$$

where:

F_{gw} = Wind *load* on the glass due to basic design *wind speed*, *V*, computed in accordance with Section 1609.

F_{ga} = Short duration *load* on the glass as determined in accordance with ASTM E1300.

2404.2 Sloped glass. Glass sloped more than 15 degrees (0.26 rad) from vertical in skylights, *sunrooms*, sloped roofs and other exterior applications shall be designed to resist the most critical combinations of loads determined by Equations 24-2, 24-3 and 24-4.

$$F_g = 0.6W_o - D \quad \textbf{(Equation 24-2)}$$

$$F_g = 0.6W_i + D + 0.5\,S \quad \textbf{(Equation 24-3)}$$

$$F_g = 0.3\,W_i + D + S \quad \textbf{(Equation 24-4)}$$

where:

D = Glass *dead load* psf (kN/m^2).

For glass sloped 30 degrees (0.52 rad) or less from horizontal,

= $13\,t_g$ (For SI: $0.0245\,t_g$).

For glass sloped more than 30 degrees (0.52 rad) from horizontal,

= $13\,t_g \cos\theta$ (For SI: $0.0245\,t_g \cos\theta$).

F_g = Total *load*, psf (kN/m^2) on glass.

S = Snow *load*, psf (kN/m^2) as determined in Section 1608.

t_g = Total glass thickness, inches (mm) of glass panes and plies.

W_i = Inward wind force, psf (kN/m^2) due to basic design *wind speed*, *V*, as calculated in Section 1609.

W_o = Outward wind force, psf (kN/m^2) due to basic design wind speed, *V*, as calculated in Section 1609.

θ = Angle of slope from horizontal.

Exception: The performance grade rating of *unit skylights* and *tubular daylighting devices* shall be determined in accordance with Section 2405.5.

The design of sloped glazing shall be based on Equation 24-5.

$$F_g \leq F_{ga} \quad \textbf{(Equation 24-5)}$$

where:

F_g = Total *load* on the glass as determined by Equations 24-2, 24-3 and 24-4.

F_{ga} = Short duration *load* resistance of the glass as determined in accordance with ASTM E1300 for Equations 24-2 and 24-3; or the long duration *load* resistance of the glass as determined in accordance with ASTM E1300 for Equation 24-4.

2404.3 Wired, patterned and sandblasted glass.

2404.3.1 Vertical wired glass. Wired glass sloped 15 degrees (0.26 rad) or less from vertical in windows, curtain and window walls, doors and other exterior applications shall be designed to resist the wind *loads* in Section 1609 for components and cladding according to the following equation:

$$0.6F_{gw} < 0.5\,F_{ge} \quad \textbf{(Equation 24-6)}$$

where:

F_{gw} = Wind *load* on the glass due to basic design *wind speed*, *V*, computed in accordance with Section 1609.

F_{ge} = Nonfactored *load* from ASTM E1300 using a thickness designation for monolithic glass that is not greater than the thickness of wired glass.

2404.3.2 Sloped wired glass. Wired glass sloped more than 15 degrees (0.26 rad) from vertical in skylights, sunspaces, sloped roofs and other exterior applications shall be designed to resist the most critical of the combinations of *loads* from Section 2404.2.

For Equations 24-2 and 24-3:

$$F_g < 0.5\,F_{ge} \quad \textbf{(Equation 24-7)}$$

For Equation 24-4:

$$F_g < 0.3\,F_{ge} \quad \textbf{(Equation 24-8)}$$

where:

F_g = Total *load* on the glass as determined by Equations 24-2, 24-3 and 24-4.

F_{ge} = Nonfactored *load* in accordance with ASTM E1300.

2404.3.3 Vertical patterned glass. Patterned glass sloped 15 degrees (0.26 rad) or less from vertical in windows, curtain and window walls, doors and other exterior applications shall be designed to resist the wind *loads* in Section 1609 for components and cladding according to Equation 24-9.

$$F_{gw} < 1.0\,F_{ge} \quad \textbf{(Equation 24-9)}$$

where:

F_{gw} = Wind *load* on the glass due to basic design *wind speed*, *V*, computed in accordance with Section 1609.

F_{ge} = Nonfactored *load* in accordance with ASTM E1300. The value for patterned glass shall be based on the thinnest part of the glass. Interpolation between nonfactored *load* charts in ASTM E1300 shall be permitted.

2404.3.4 Sloped patterned glass. Patterned glass sloped more than 15 degrees (0.26 rad) from vertical in skylights, sunspaces, sloped roofs and other exterior applications shall be designed to resist the most critical of the combinations of loads from Section 2404.2.

For Equations 24-2 and 24-3:

$$F_g < 1.0 F_{ge} \qquad \textbf{(Equation 24-10)}$$

For Equation 24-4:

$$F_g < 0.6F_{ge} \qquad \textbf{(Equation 24-11)}$$

where:

F_g = Total *load* on the glass as determined by Equations 24-2, 24-3 and 24-4.

F_{ge} = Nonfactored *load* in accordance with ASTM E1300. The value for patterned glass shall be based on the thinnest part of the glass. Interpolation between the nonfactored *load* charts in ASTM E1300 shall be permitted.

2404.3.5 Vertical sandblasted glass. Sandblasted glass sloped 15 degrees (0.26 rad) or less from vertical in windows, curtain and window walls, doors, and other exterior applications shall be designed to resist the wind *loads* in Section 1609 for components and cladding according to Equation 24-12.

$$0.6F_{gw} < 0.5 F_{ge} \qquad \textbf{(Equation 24-12)}$$

where:

F_g = Wind *load* on the glass due to basic design *wind speed*, *V*, computed in accordance with Section 1609.

F_{ge} = Nonfactored *load* in accordance with ASTM E1300. The value for sandblasted glass is for moderate levels of sandblasting.

2404.4 Other designs. For designs outside the scope of this section, an analysis or test data for the specific installation shall be prepared by a *registered design professional*.

SECTION 2405
SLOPED GLAZING AND SKYLIGHTS

2405.1 Scope. This section applies to the installation of glass and other transparent, translucent or opaque glazing material installed at a slope of more than 15 degrees (0.26 rad) from the vertical plane, including glazing materials in skylights, roofs and sloped walls.

2405.2 Allowable glazing materials and limitations. Sloped glazing shall be any of the following materials, subject to the listed limitations.

1. For monolithic glazing systems, the glazing material of the single light or layer shall be laminated glass with a minimum 30-mil (0.76 mm) polyvinyl butyral (or equivalent) interlayer, wired glass, light-transmitting plastic materials meeting the requirements of Section 2607, heat-strengthened glass or fully tempered glass.
2. For multiple-layer glazing systems, each light or layer shall consist of any of the glazing materials specified in Item 1.

Annealed glass is permitted to be used as specified in Exceptions 2 and 3 of Section 2405.3.

Laminated glass and plastic materials described in Items 1 and 2 shall not require the screening or height restrictions provided in Section 2405.3.

For additional requirements for plastic skylights, see Section 2610. Glass-block construction shall conform to the requirements of Section 2110.1.

2405.3 Screening. Where used in monolithic glazing systems, annealed, heat-strengthened, fully tempered and wired glass shall have broken glass retention screens installed below the glazing material. The screens and their fastenings shall be: capable of supporting twice the weight of the glazing; firmly and substantially fastened to the framing members; and installed within 4 inches (102 mm) of the glass. The screens shall be constructed of a noncombustible material not thinner than No. 12 B&S gage (0.0808 inch) with mesh not larger than 1 inch by 1 inch (25 mm by 25 mm). In a corrosive atmosphere, structurally equivalent noncorrosive screen materials shall be used. Annealed, heat-strengthened, fully tempered and wired glass, where used in multiple-layer glazing systems as the bottom glass layer over the walking surface, shall be equipped with screening that conforms to the requirements for monolithic glazing systems.

Exception: In monolithic and multiple-layer sloped glazing systems, the following applies:

1. Fully tempered glass installed without protective screens where glazed between intervening floors at a slope of 30 degrees (0.52 rad) or less from the vertical plane shall have the highest point of the glass 10 feet (3048 mm) or less above the walking surface.
2. Screens are not required below any glazing material, including annealed glass, where the walking surface below the glazing material is permanently protected from the risk of falling glass or the area below the glazing material is not a walking surface.
3. Any glazing material, including annealed glass, is permitted to be installed without screens in the sloped glazing systems of commercial or detached noncombustible *greenhouses* used exclusively for growing plants and not open to the public, provided that the height of the *greenhouse* at the ridge does not exceed 30 feet (9144 mm) above grade.
4. Screens shall not be required in individual *dwelling units* in Groups R-2, R-3 and R-4 where fully tempered glass is used as single glazing or as both panes in an insulating glass unit, and the following conditions are met:
    - 4.1. Each pane of the glass is 16 square feet (1.5 m^2) or less in area.
    - 4.2. The highest point of the glass is 12 feet (3658 mm) or less above any walking surface or other accessible area.
    - 4.3. The glass thickness is $^3/_{16}$ inch (4.8 mm) or less.

5. Screens shall not be required for laminated glass with a 15-mil (0.38 mm) polyvinyl butyral (or equivalent) interlayer used in individual *dwelling units* in Groups R-2, R-3 and R-4 within the following limits:
   5.1. Each pane of glass is 16 square feet (1.5 m^2) or less in area.
   5.2. The highest point of the glass is 12 feet (3658 mm) or less above a walking surface or other accessible area.

2405.4 Framing. In Types I and II construction, sloped glazing and skylight frames shall be constructed of noncombustible materials. In structures where acid fumes deleterious to metal are incidental to the use of the buildings, *approved* pressure-treated wood or other *approved* noncorrosive materials are permitted to be used for sash and frames. Framing supporting sloped glazing and skylights shall be designed to resist the tributary roof *loads* in Chapter 16. Skylights set at an angle of less than 45 degrees (0.79 rad) from the horizontal plane shall be mounted not less than 4 inches (102 mm) above the plane of the roof on a curb constructed as required for the frame. Skylights shall not be installed in the plane of the roof where the roof pitch is less than 45 degrees (0.79 rad) from the horizontal.

Exception: Installation of a skylight without a curb shall be permitted on roofs with a minimum slope of 14 degrees (three units vertical in 12 units horizontal) in Group R-3 occupancies. *Unit skylights* installed in a roof with a pitch flatter than 14 degrees (0.25 rad) shall be mounted not less than 4 inches (102 mm) above the plane of the roof on a curb constructed as required for the frame unless otherwise specified in the manufacturer's installation instructions.

2405.5 Unit skylights and tubular daylighting devices. *Unit skylights* and *tubular daylighting devices* shall be tested and labeled as complying with AAMA/WDMA/CSA 101/I.S./A440. The *label* shall state the name of the manufacturer, the *approved* labeling agency, the product designation and the performance grade rating as specified in AAMA/WDMA/CSA 101/I.S.2/A440. Where the product manufacturer has chosen to have the performance grade of the skylight rated separately for positive and negative design pressure, then the *label* shall state both performance grade ratings as specified in AAMA/WDMA/CSA 101/I.S.2/A440 and the skylight shall comply with Section 2405.5.2. Where the skylight is not rated separately for positive and negative pressure, then the performance grade rating shown on the *label* shall be the performance grade rating determined in accordance with AAMA/WDMA/CSA 101/I.S.2/A440 for both positive and negative design pressure and the skylight shall conform to Section 2405.5.1.

2405.5.1 Skylights rated for the same performance grade for both positive and negative design pressure. The design of skylights shall be based on Equation 24-13.

$$F_g \leq PG \qquad \textbf{(Equation 24-13)}$$

where:

F_g = Maximum *load* on the skylight determined from Equations 24-2 through 24-4 in Section 2404.2.

PG = Performance grade rating of the skylight.

2405.5.2 Skylights rated for separate performance grades for positive and negative design pressure. The design of skylights rated for performance grade for both positive and negative design pressures shall be based on Equations 24-14 and 24-15.

$$F_{gi} \leq PG_{Pos} \qquad \textbf{(Equation 24-14)}$$

$$F_{go} \leq PG_{Neg} \qquad \textbf{(Equation 24-15)}$$

where:

PG_{Pos} = Performance grade rating of the skylight under positive design pressure;

PG_{Neg} = Performance grade rating of the skylight under negative design pressure; and

F_{gi} and F_{go} are determined in accordance with the following:

For $0.6W_o \geq D$,

where:

W_o = Outward wind force, psf (kN/m^2) due to basic design wind speed, V, as calculated in Section 1609.

D = The dead weight of the glazing, psf (kN/m^2) as determined in Section 2404.2 for glass, or by the weight of the plastic, psf (kN/m^2) for plastic glazing.

F_{gi} = Maximum load on the skylight determined from Equations 24-3 and 24-4 in Section 2404.2.

F_{go} = Maximum load on the skylight determined from Equation 24-2.

For $0.6W_o < D$,

where:

W_o = The outward wind force, psf (kN/m^2) due to basic design wind speed, V, as calculated in Section 1609.

D = The dead weight of the glazing, psf (kN/m^2) as determined in Section 2404.2 for glass, or by the weight of the plastic for plastic glazing.

F_{gi} = Maximum load on the skylight determined from Equations 24-2 through 24-4 in Section 2404.2.

F_{go} = 0.

SECTION 2406
SAFETY GLAZING

2406.1 Human impact loads. Individual glazed areas, including glass mirrors, in hazardous locations as defined in Section 2406.4 shall comply with Sections 2406.1.1 through 2406.1.4.

Exception: Mirrors and other glass panels mounted or hung on a surface that provides a continuous backing support.

2406.1.1 Impact test. Except as provided in Sections 2406.1.2 through 2406.1.4, all glazing shall pass the impact test requirements of Section 2406.2.

2406.1.2 Plastic glazing. Plastic glazing shall meet the weathering requirements of ANSI Z97.1.

2406.1.3 Glass block. Glass-block walls shall comply with Section 2110.

2406.1.4 Louvered windows and jalousies. Louvered windows and jalousies shall comply with Section 2403.5.

2406.2 Impact test. Where required by other sections of this code, glazing shall be tested in accordance with CPSC 16 CFR Part 1201. Glazing shall comply with the test criteria for Category II, unless otherwise indicated in Table 2406.2(1).

Exception: Glazing not in doors or enclosures for hot tubs, whirlpools, saunas, steam rooms, bathtubs and showers shall be permitted to be tested in accordance with ANSI Z97.1. Glazing shall comply with the test criteria for Class A, unless otherwise indicated in Table 2406.2(2).

2406.3 Identification of safety glazing. Except as indicated in Section 2406.3.1, each pane of safety glazing installed in hazardous locations shall be identified by a manufacturer's designation specifying who applied the designation, the manufacturer or installer and the safety glazing standard with which it complies, as well as the information specified in Section 2403.1. The designation shall be acid etched, sand blasted, ceramic fired, laser etched, embossed or of a type that once applied, cannot be removed without being destroyed. A *label* meeting the requirements of this section shall be permitted in lieu of the manufacturer's designation.

Exceptions:

1. For other than tempered glass, manufacturer's designations are not required, provided that the *building official* approves the use of a certificate, affidavit or other evidence confirming compliance with this code.
2. Tempered spandrel glass is permitted to be identified by the manufacturer with a removable paper designation.

2406.3.1 Multipane assemblies. Multipane glazed assemblies having individual panes not exceeding 1 square foot (0.09 m^2) in exposed areas shall have one pane or more in the assembly marked as indicated in Section 2406.3. Other panes in the assembly shall be marked "CPSC 16 CFR Part 1201" or "ANSI Z97.1," as appropriate.

2406.4 Hazardous locations. The locations specified in Sections 2406.4.1 through 2406.4.7 shall be considered to be specific hazardous locations requiring safety glazing materials.

2406.4.1 Glazing in doors. Glazing in all fixed and operable panels of swinging, sliding and bifold doors shall be considered to be a hazardous location.

Exceptions:

1. Glazed openings of a size through which a 3-inch-diameter (76 mm) sphere is unable to pass.
2. Decorative glazing.
3. Glazing materials used as curved glazed panels in revolving doors.
4. Commercial refrigerated cabinet glazed doors.

2406.4.2 Glazing adjacent to doors. Glazing in an individual fixed or operable panel adjacent to a door where the nearest vertical edge of the glazing is within a 24-inch (610 mm) arc of either vertical edge of the door in a closed position and where the bottom exposed edge of the glazing is less than 60 inches (1524 mm) above the walking surface shall be considered to be a hazardous location.

Exceptions:

1. Decorative glazing.
2. Where there is an intervening wall or other permanent barrier between the door and glazing.
3. Where access through the door is to a closet or storage area 3 feet (914 mm) or less in depth.

TABLE 2406.2(1)
MINIMUM CATEGORY CLASSIFICATION OF GLAZING USING CPSC 16 CFR PART 1201

EXPOSED SURFACE AREA OF ONE SIDE OF ONE LITE	GLAZING IN STORM OR COMBINATION DOORS (Category class)	GLAZING IN DOORS (Category class)	GLAZED PANELS REGULATED BY SECTION 2406.4.3 (Category class)	GLAZED PANELS REGULATED BY SECTION 2406.4.2 (Category class)	DOORS AND ENCLOSURES REGULATED BY SECTION 2406.4.5 (Category class)	SLIDING GLASS DOORS PATIO TYPE (Category class)
9 square feet or less	I	I	No requirement	I	II	II
More than 9 square feet	II	II	II	II	II	II

For SI: 1 square foot = 0.0929 m^2.

TABLE 2406.2(2)
MINIMUM CATEGORY CLASSIFICATION OF GLAZING USING ANSI Z97.1

EXPOSED SURFACE AREA OF ONE SIDE OF ONE LITE	GLAZED PANELS REGULATED BY SECTION 2406.4.3 (Category class)	GLAZED PANELS REGULATED BY SECTION 2406.4.2 (Category class)	DOORS AND ENCLOSURES REGULATED BY SECTION 2406.4.5[a] (Category class)
9 square feet or less	No requirement	B	A
More than 9 square feet	A	A	A

For SI: square foot = 0.0929 m^2.

a. Use is only permitted by the exception to Section 2406.2.

Glazing in this application shall comply with Section 2406.4.3.

4. Glazing in walls on the latch side of and perpendicular to the plane of the door in a closed position in one- and two-family dwellings or within dwelling units in Group R-2.

2406.4.3 Glazing in windows. Glazing in an individual fixed or operable panel that meets all of the following conditions shall be considered to be a hazardous location:

1. The exposed area of an individual pane is greater than 9 square feet (0.84 m^2).
2. The bottom edge of the glazing is less than 18 inches (457 mm) above the floor.
3. The top edge of the glazing is greater than 36 inches (914 mm) above the floor.
4. One or more walking surface(s) are within 36 inches (914 mm), measured horizontally and in a straight line, of the plane of the glazing.

Exceptions:

1. Decorative glazing.
2. Where a horizontal rail is installed on the accessible side(s) of the glazing 34 to 38 inches (864 to 965 mm) above the walking surface. The rail shall be capable of withstanding a horizontal *load* of 50 pounds per linear foot (730 N/m) without contacting the glass and be not less than $1^1/_2$ inches (38 mm) in cross-sectional height.
3. Outboard panes in insulating glass units or multiple glazing where the bottom exposed edge of the glass is 25 feet (7620 mm) or more above any grade, roof, walking surface or other horizontal or sloped (within 45 degrees of horizontal) (0.79 rad) surface adjacent to the glass exterior.

2406.4.4 Glazing in guards and railings. Glazing in *guards* and railings, including structural baluster panels and nonstructural in-fill panels, regardless of area or height above a walking surface shall be considered to be a hazardous location.

2406.4.5 Glazing and wet surfaces. Glazing in walls, enclosures or fences containing or facing hot tubs, spas, whirlpools, saunas, steam rooms, bathtubs, showers and indoor or outdoor *swimming pools* where the bottom exposed edge of the glazing is less than 60 inches (1524 mm) measured vertically above any standing or walking surface shall be considered to be a hazardous location. This shall apply to single glazing and all panes in multiple glazing.

Exception: Glazing that is more than 60 inches (1524 mm), measured horizontally and in a straight line, from the water's edge of a bathtub, hot tub, spa, whirlpool or *swimming pool*.

2406.4.6 Glazing adjacent to stairways and *ramps*. Glazing where the bottom exposed edge of the glazing is less than 60 inches (1524 mm) above the plane of the adjacent walking surface of *stairways*, landings between flights of *stairs* and ramps shall be considered to be a hazardous location.

Exceptions:

1. The side of a *stairway*, landing or *ramp* that has a *guard* complying with the provisions of Sections 1015 and 1607.9, and the plane of the glass is greater than 18 inches (457 mm) from the railing.
2. Glazing 36 inches (914 mm) or more measured horizontally from the walking surface.

2406.4.7 Glazing adjacent to the bottom stairway landing. Glazing adjacent to the landing at the bottom of a *stairway* where the glazing is less than 60 inches (1524 mm) above the landing and within a 60-inch (1524 mm) horizontal arc that is less than 180 degrees (3.14 rad) from the bottom tread *nosing* shall be considered to be a hazardous location.

Exception: Glazing that is protected by a *guard* complying with Sections 1015 and 1607.9 where the plane of the glass is greater than 18 inches (457 mm) from the *guard*.

2406.5 Fire department access panels. Fire department glass access panels shall be of tempered glass. For insulating glass units, all panes shall be tempered glass.

SECTION 2407
GLASS IN HANDRAILS AND GUARDS

2407.1 Materials. Glass used in a *handrail* or a *guard* shall be laminated glass constructed of fully tempered or heat-strengthened glass and shall comply with Category II of CPSC 16 CFR Part 1201 or Class A of ANSI Z97.1. Glazing in a *handrail* or a *guard* shall be of an *approved* safety glazing material that conforms to the provisions of Section 2406.1.1. For all glazing types, the minimum nominal thickness shall be $^1/_4$ inch (6.4 mm).

Exception: Single fully tempered glass complying with Category II of CPSC 16 CFR Part 1201 or Class A of ANSI Z97.1 shall be permitted to be used in *handrails* and guards where there is no walking surface beneath them or the walking surface is permanently protected from the risk of falling glass.

2407.1.1 Loads. Glass *handrails* and guards and their support systems shall be designed to withstand the *loads* specified in Section 1607.9. Glass *handrails* and *guards* shall be designed using a factor of safety of four.

2407.1.2 Guards with structural glass balusters. *Guards* with structural glass balusters, whether vertical posts, columns or panels, shall be installed with an attached top rail or *handrail*. The top rail or *handrail* shall be supported by not fewer than three glass balusters, or shall be otherwise supported to remain in place should one glass baluster fail.

Exception: An attached top rail or *handrail* is not required where the glass baluster panels are laminated glass with two or more glass plies of equal thickness and of the same glass type. The balusters shall be tested to remain in place as a barrier following impact or glass breakage in accordance with ASTM E2353.

2407.1.3 Parking garages. Glazing materials shall not be installed in *handrails* or *guards* in parking garages except for pedestrian areas not exposed to impact from vehicles.

2407.1.4 Glazing in windborne debris regions. Glazing installed in exterior *handrails* or *guards* in *windborne debris regions* shall be laminated glass complying with Category II of CPSC 16 CFR 1201 or Class A of ANSI Z97.1. Where the top rail is supported by glass, the assembly shall be tested according to the impact requirements of Section 1609.2 and the top rail shall remain in place after impact.

SECTION 2408
GLAZING IN ATHLETIC FACILITIES

2408.1 General. Glazing in athletic facilities and similar uses subject to impact loads, which forms whole or partial wall sections or which is used as a door or part of a door, shall comply with this section.

2408.2 Racquetball and squash courts.

2408.2.1 Testing. Test methods and loads for individual glazed areas in racquetball and squash courts subject to impact *loads* shall conform to those of CPSC 16 CFR Part 1201 or ANSI Z97.1 with impacts being applied at a height of 59 inches (1499 mm) above the playing surface to an actual or simulated glass wall installation with fixtures, fittings and methods of assembly identical to those used in practice.

Glass walls shall comply with the following conditions:

1. A glass wall in a racquetball or squash court, or similar use subject to impact loads, shall remain intact following a test impact.
2. The deflection of such walls shall be not greater than $1^1/_2$ inches (38 mm) at the point of impact for a drop height of 48 inches (1219 mm).

Glass doors shall comply with the following conditions:

1. Glass doors shall remain intact following a test impact at the prescribed height in the center of the door.
2. The relative deflection between the edge of a glass door and the adjacent wall shall not exceed the thickness of the wall plus $^1/_2$ inch (12.7 mm) for a drop height of 48 inches (1219 mm).

2408.3 Gymnasiums and basketball courts. Glazing in multipurpose gymnasiums, basketball courts and similar athletic facilities subject to human impact loads shall comply with Category II of CPSC 16 CFR Part 1201 or Class A of ANSI Z97.1.

SECTION 2409
GLASS IN WALKWAYS, ELEVATOR HOISTWAYS AND ELEVATOR CARS

2409.1 Glass walkways. Glass installed as a part of a floor/ceiling assembly as a walking surface and constructed with laminated glass shall comply with ASTM E2751 or with the *load* requirements specified in Chapter 16. Such assemblies shall comply with the *fire-resistance rating* and marking requirements of this code where applicable.

2409.2 Glass in elevator hoistway enclosures. Glass in elevator hoistway enclosures and hoistway doors shall be laminated glass conforming to ANSI Z97.1 or CPSC 16 CFR Part 1201.

2409.2.1 Fire-resistance-rated hoistways. Glass installed in hoistways and hoistway doors where the hoistway is required to have a *fire-resistance rating* shall comply with Section 716.

2409.2.2 Glass hoistway doors. The glass in glass hoistway doors shall be not less than 60 percent of the total visible door panel surface area as seen from the landing side.

2409.3 Visions panels in elevator hoistway doors. Glass in vision panels in elevator hoistway doors shall be permitted to be any transparent glazing material not less than $^1/_4$ inch (6.4 mm) in thickness conforming to Class A in accordance with ANSI Z97.1 or Category II in accordance with CPSC 16 CFR Part 1201. The area of any single vision panel shall be not less than 24 square inches (15 484 mm^2) and the total area of one or more vision panels in any hoistway door shall be not more than 85 square inches (54 839 mm^2).

2409.4 Glass in elevator cars. Glass in elevator cars shall be in accordance with this section.

2409.4.1 Glass types. Glass in elevator car enclosures, glass elevator car doors and glass used for lining walls and ceilings of elevator cars shall be laminated glass conforming to Class A in accordance with ANSI Z97.1 or Category II in accordance with CPSC 16 CFR Part 1201.

Exception: Tempered glass shall be permitted to be used for lining walls and ceilings of elevator cars provided that:

1. The glass is bonded to a nonpolymeric coating, sheeting or film backing having a physical integrity to hold the fragments when the glass breaks.
2. The glass is not subjected to further treatment such as sandblasting; etching; heat treatment or painting that could alter the original properties of the glass.
3. The glass is tested to the acceptance criteria for laminated glass as specified for Class A in accordance with ANSI Z97.1 or Category II in accordance with CPSC 16 CFR Part 1201.

2409.4.2 Surface area. The glass in glass elevator car doors shall be not less than 60 percent of the total visible door panel surface area as seen from the car side of the doors.

CHAPTER 25

GYPSUM BOARD, GYPSUM PANEL PRODUCTS AND PLASTER

User notes:

About this chapter: *Chapter 25 contains the provisions and referenced standards that regulate the design, construction and quality of gypsum board, gypsum panel products and plaster and, in addition, addresses reinforced gypsum concrete. These materials are some of the most commonly used interior and exterior finish materials in the building industry. This chapter primarily addresses quality-control-related issues with regard to material specifications and installation requirements. Most products are manufactured in accordance with industry standards. The building official or inspector needs to verify that the appropriate product is used and properly installed for the intended use and location. Proper design and installation of these materials are necessary to provide weather resistance and required fire protection for both structural and nonstructural building components.*

Code development reminder: *Code change proposals to this chapter will be considered by the IBC—Structural Code Development Committee during the 2022 (Group B) Code Development Cycle.*

SECTION 2501 GENERAL

2501.1 Scope. Provisions of this chapter shall govern the materials, design, construction and quality of *gypsum board*, *gypsum panel products*, lath, *gypsum plaster*, *cement plaster* and reinforced gypsum concrete.

2501.2 Other materials. Other *approved* wall or ceiling coverings shall be permitted to be installed in accordance with the recommendations of the manufacturer and the conditions of approval.

SECTION 2502 PERFORMANCE

2502.1 General. Lathing, plastering and *gypsum board* and *gypsum panel product* construction shall be done in the manner and with the materials specified in this chapter and, where required for fire protection, shall comply with the provisions of Chapter 7.

SECTION 2503 INSPECTION

2503.1 Inspection. Lath, *gypsum board* and *gypsum panel products* shall be inspected in accordance with Section 110.3.6.

SECTION 2504 VERTICAL AND HORIZONTAL ASSEMBLIES

2504.1 Scope. The following requirements shall be met where construction involves *gypsum board*, *gypsum panel products* or lath and plaster in vertical and *horizontal assemblies*.

2504.1.1 Wood framing. Wood supports for lath, *gypsum board* or *gypsum panel products*, as well as wood stripping or furring, shall be not less than 2 inches (51 mm) nominal thickness in the least dimension.

Exception: The minimum nominal dimension of wood furring strips installed over solid backing shall be not less than 1 inch by 2 inches (25 mm by 51 mm).

2504.1.2 Studless partitions. The minimum thickness of vertically erected studless solid plaster partitions of $^3/_8$-inch (9.5 mm) and $^3/_4$-inch (19.1 mm) rib metal lath, $^1/_2$-inch-thick (12.7 mm) gypsum lath, *gypsum board* or *gypsum panel product* shall be 2 inches (51 mm).

SECTION 2505 SHEAR WALL CONSTRUCTION

2505.1 Resistance to shear (wood framing). Wood-frame *shear walls* sheathed with *gypsum board*, *gypsum panel products* or lath and plaster shall be designed and constructed in accordance with Section 2306.3 and are permitted to resist wind and seismic *loads*. Walls resisting seismic *loads* shall be subject to the limitations in Section 12.2.1 of ASCE 7.

2505.2 Resistance to shear (steel framing). Cold-formed steel-frame shear walls sheathed with *gypsum board* or *gypsum panel products* and constructed in accordance with the materials and provisions of Section 2211.1.1 are permitted to resist wind and seismic *loads*. Walls resisting seismic *loads* shall be subject to the limitations in Section 12.2.1 of ASCE 7.

SECTION 2506 GYPSUM BOARD AND GYPSUM PANEL PRODUCT MATERIALS

2506.1 General. *Gypsum board*, *gypsum panel products* and accessories shall be identified by the manufacturer's designation to indicate compliance with the appropriate standards referenced in this section and stored to protect such materials from the weather.

2506.2 Standards. *Gypsum board* and *gypsum panel products* shall conform to the appropriate standards listed in Table 2506.2 and Chapter 35 and, where required for fire protection, shall conform to the provisions of Chapter 7.

2506.2.1 Other materials. Metal suspension systems for acoustical and lay-in panel ceilings shall comply with ASTM C635 listed in Chapter 35 and Section 13.5.6 of ASCE 7 for installation in high seismic areas.

SECTION 2507 LATHING AND PLASTERING

2507.1 General. Lathing and plastering materials and accessories shall be marked by the manufacturer's designation to indicate compliance with the appropriate standards referenced in this section and stored in such a manner to protect them from the weather.

2507.2 Standards. Lathing and plastering materials shall conform to the standards listed in Table 2507.2 and Chapter 35 and, where required for fire protection, shall conform to the provisions of Chapter 7.

SECTION 2508 GYPSUM CONSTRUCTION

2508.1 General. *Gypsum board*, *gypsum panel products* and *gypsum plaster* construction shall be of the materials listed in Tables 2506.2 and 2507.2. These materials shall be assembled and installed in compliance with the appropriate standards listed in Tables 2508.1 and 2511.1.1 and Chapter 35.

2508.2 Limitations. *Gypsum wallboard* or *gypsum plaster* shall not be used in any exterior surface where such gypsum construction will be exposed directly to the weather. *Gypsum wallboard* shall not be used where there will be direct exposure to water or continuous high humidity conditions. *Gypsum sheathing* shall be installed on exterior surfaces in accordance with ASTM C1280.

2508.2.1 Weather protection. *Gypsum wallboard*, gypsum lath or *gypsum plaster* shall not be installed until weather protection for the installation is provided.

2508.3 Single-ply application. Edges and ends of *gypsum board* and *gypsum panel products* shall occur on the framing members, except those edges and ends that are perpendicular to the framing members. Edges and ends of *gypsum board* and *gypsum panel products* shall be in moderate contact except in concealed spaces where fire-resistance-rated construction, shear resistance or *diaphragm* action is not required.

2508.3.1 Floating angles. Fasteners at the top and bottom plates of vertical assemblies, or the edges and ends of *horizontal assemblies* perpendicular to supports, and at the wall line are permitted to be omitted except on shear resisting elements or fire-resistance-rated assemblies. Fasteners shall be applied in such a manner as not to fracture the face paper with the fastener head.

TABLE 2508.1
INSTALLATION OF GYPSUM CONSTRUCTION

MATERIAL	STANDARD
Gypsum board and gypsum panel products	GA 216; ASTM C840
Gypsum sheathing and gypsum panel products	ASTM C1280
Gypsum veneer base	ASTM C844
Interior lathing and furring	ASTM C841
Steel framing for gypsum board and gypsum panel products	ASTM C754; C1007

TABLE 2506.2
GYPSUM BOARD AND GYPSUM PANEL PRODUCTS MATERIALS AND ACCESSORIES

MATERIAL	STANDARD
Accessories for gypsum board	ASTM C1047
Adhesives for fastening gypsum board	ASTM C557
Cold-formed steel studs and track, structural	AISI S240
Cold-formed steel studs and track, nonstructural	AISI S220
Elastomeric joint sealants	ASTM C920
Expandable foam adhesives for fastening gypsum wallboard	ASTM D6464
Factory-laminated gypsum panel products	ASTM C1766
Fiber-reinforced gypsum panels	ASTM C1278
Glass mat gypsum backing panel	ASTM C1178
Glass mat gypsum panel 5	ASTM C1658
Glass mat gypsum substrate	ASTM C1177
Joint reinforcing tape and compound	ASTM C474; C475
Nails for gypsum boards	ASTM C514, F547, F1667
Steel screws	ASTM C954; C1002
Standard specification for gypsum board	ASTM C1396
Testing gypsum and gypsum products	ASTM C22; C472; C473

2508.4 Adhesives. *Gypsum board* and *gypsum panel products* secured to framing with adhesives in ceiling assemblies shall be attached using an approved fastening schedule. Expandable foam adhesives for fastening *gypsum wallboard* shall conform to ASTM D6464. Other adhesives for the installation of *gypsum wallboard* shall conform to ASTM C557.

2508.5 Joint treatment. *Gypsum board* and *gypsum panel product* fire-resistance-rated assemblies shall have joints and fasteners treated.

Exception: Joint and fastener treatment need not be provided where any of the following conditions occur:

1. Where the *gypsum board* or the *gypsum panel product* is to receive a decorative finish such as wood paneling, battens, acoustical finishes or any similar application that would be equivalent to joint treatment.
2. On single-layer systems where joints occur over wood framing members.
3. Square edge or tongue-and-groove edge *gypsum board* (V-edge), *gypsum panel products*, gypsum backing board or *gypsum sheathing*.
4. On multilayer systems where the joints of adjacent layers are offset.
5. Assemblies tested without joint treatment.

2508.6 Horizontal gypsum board or gypsum panel product diaphragm ceilings. *Gypsum board* or *gypsum panel products* shall be permitted to be used on wood joists to create a horizontal *diaphragm* ceiling in accordance with Table 2508.6.

TABLE 2507.2
LATH, PLASTERING MATERIALS AND ACCESSORIES

MATERIAL	STANDARD
Accessories for gypsum veneer base	ASTM C1047
Blended cement	ASTM C595
Cold-formed steel studs and track, structural	AISI S240
Cold-formed steel studs and track, nonstructural	AISI S220
Exterior plaster bonding compounds	ASTM C932
Hydraulic cement	ASTM C1157; C1600
Gypsum casting and molding plaster	ASTM C59
Gypsum Keene's cement	ASTM C61
Gypsum plaster	ASTM C28
Gypsum veneer plaster	ASTM C587
Interior bonding compounds, gypsum	ASTM C631
Lime plasters	ASTM C5; C206
Masonry cement	ASTM C91
Metal lath	ASTM C847
Plaster aggregates	
Sand	ASTM C35; C897
Perlite	ASTM C35
Vermiculite	ASTM C35
Plastic cement	ASTM C1328
Portland cement	ASTM C150
Steel screws	ASTM C1002; C954
Welded wire lath	ASTM C933
Woven wire plaster base	ASTM C1032

TABLE 2508.6
SHEAR CAPACITY FOR HORIZONTAL WOOD-FRAME GYPSUM BOARD DIAPHRAGM CEILING ASSEMBLIES

MATERIAL	THICKNESS OF MATERIAL (MINIMUM) (inches)	SPACING OF FRAMING MEMBERS (inches)	SHEAR VALUE[a, b] (PLF OF CEILING)	MIMIMUM FASTENER SIZE
Gypsum board or gypsum panel product	$^1/_2$	16 o.c.	90	5d cooler or wallboard nail; $1^5/_8$-inch long; 0.086-inch shank; $^{15}/_{64}$-inch head[c]
Gypsum board or gypsum panel product	$^1/_2$	24 o.c.	70	5d cooler or wallboard nail; $1^5/_8$-inch long; 0.086-inch shank; $^{15}/_{64}$-inch head[c]

For SI: 1 inch = 25.4 mm, 1 pound per foot = 14.59 N/m.

a. Values are not cumulative with other horizontal diaphragm values and are for short-term wind or seismic loading. Values shall be reduced 25 percent for normal loading.

b. Values shall be reduced 50 percent in Seismic Design Categories D, E and F.

c. $1^1/_4$-inch, No. 6 Type S or W screws are permitted to be substituted for the listed nails.

2508.6.1 Diaphragm proportions. The maximum allowable *diaphragm* proportions shall be $1^1/_2$:1 between shear resisting elements. Rotation or cantilever conditions shall not be permitted.

2508.6.2 Installation. *Gypsum board* or *gypsum panel products* used in a horizontal *diaphragm* ceiling shall be installed perpendicular to ceiling framing members. End joints of adjacent courses of *gypsum board* shall not occur on the same joist.

2508.6.3 Blocking of perimeter edges. Perimeter edges shall be blocked using a wood member not less than 2-inch by 6-inch (51 mm by 152 mm) nominal dimension. Blocking material shall be installed flat over the top plate of the wall to provide a nailing surface not less than 2 inches (51 mm) in width for the attachment of the *gypsum board* or *gypsum panel product*.

2508.6.4 Fasteners. Fasteners used for the attachment of *gypsum board* or *gypsum panel products* to a horizontal *diaphragm* ceiling shall be as defined in Table 2508.6. Fasteners shall be spaced not more than 7 inches (178 mm) on center at all supports, including perimeter blocking, and not more than $^3/_8$ inch (9.5 mm) from the edges and ends of the *gypsum board* or *gypsum panel product*.

2508.6.5 Lateral force restrictions. *Gypsum board* or *gypsum panel products* shall not be used in *diaphragm* ceilings to resist lateral forces imposed by masonry or concrete construction.

SECTION 2509
SHOWERS AND WATER CLOSETS

2509.1 Wet areas. Showers and public toilet walls shall conform to Section 1210.2.

2509.2 Base for tile. Materials used as a base for wall tile in tub and shower areas and wall and ceiling panels in shower areas shall be of materials listed in Table 2509.2 and installed in accordance with the manufacturer's recommendations. Water-resistant gypsum backing board shall be used as a base for tile in water closet compartment walls when installed in accordance with GA 216 or ASTM C840 and the manufacturer's recommendations. Regular *gypsum wallboard* is permitted under tile or wall panels in other wall and ceiling areas when installed in accordance with GA 216 or ASTM C840.

TABLE 2509.2
BACKERBOARD MATERIALS

MATERIAL	STANDARD
Glass mat gypsum backing panel	ASTM C1178
Nonasbestos fiber-cement backer board	ASTM C1288 or ISO 8336, Category C
Nonasbestos fiber-mat reinforced cementitious backer unit	ASTM C1325

2509.3 Limitations. Water-resistant gypsum backing board shall not be used in the following locations:

1. Over a vapor retarder in shower or bathtub compartments.
2. Where there will be direct exposure to water or in areas subject to continuous high humidity.

SECTION 2510
LATHING AND FURRING FOR CEMENT PLASTER (STUCCO)

2510.1 General. Exterior and interior *cement plaster* and lathing shall be done with the appropriate materials listed in Table 2507.2 and Chapter 35.

2510.2 Weather protection. Materials shall be stored in such a manner as to protect them from the weather.

2510.3 Installation. Installation of these materials shall be in compliance with ASTM C926 and ASTM C1063.

2510.4 Corrosion resistance. Metal lath and lath attachments shall be of corrosion-resistant material.

2510.5 Backing. Backing or a lath shall provide sufficient rigidity to permit plaster applications.

2510.5.1 Support of lath. Where lath on vertical surfaces extends between rafters or other similar projecting members, solid backing shall be installed to provide support for lath and attachments.

2510.5.2 Use of gypsum backing board. Gypsum backing for cement plaster shall be in accordance with Section 2510.5.2.1 or 2510.5.2.2.

2510.5.2.1 Gypsum board as a backing board. Gypsum lath or *gypsum wallboard* shall not be used as a backing for *cement plaster*.

Exception: Gypsum lath or *gypsum wallboard* is permitted, with a *water-resistive barrier*, as a backing for self-furred metal lath or self-furred wire fabric lath and *cement plaster* where either of the following conditions occur:

1. On horizontal supports of ceilings or roof soffits.
2. On interior walls.

2510.5.2.2 Gypsum sheathing backing. *Gypsum sheathing* is permitted as a backing for metal or wire fabric lath and *cement plaster* on walls. A *water-resistive barrier* shall be provided in accordance with Section 2510.6.

2510.5.3 Backing not required. *Wire backing* is not required under expanded metal lath or paperbacked wire fabric lath.

2510.6 Water-resistive barriers. *Water-resistive barriers* shall be installed as required in Section 1403.2 and, where applied over wood-based sheathing, shall comply with Section 2510.6.1 or 2510.6.2.

2510.6.1 Dry climates. One of the following shall apply for dry (B) climate zones:

1. The water-resistive barrier shall be two layers of 10-minute Grade D paper or have a water resistance equal to or greater than two layers of water-resistive barrier complying with ASTM E2556,

Type I. The individual layers shall be installed independently such that each layer provides a separate continuous plane and any flashing, installed in accordance with Section 1404.4 and intended to drain to the water-resistive barrier, is directed between the layers.

2. The water-resistive barrier shall be 60-minute Grade D paper or have a water resistance equal to or greater than one layer of water-resistive barrier complying with ASTM E2556, Type II. The water-resistive barrier shall be separated from the stucco by a layer of foam plastic insulating sheathing or other nonwater absorbing layer, or a drainage space.

2510.6.2 Moist or marine climates. In moist (A) or marine (C) climate zones, *water-resistive barrier* shall comply with one of the following:

1. In addition to complying with Item 1 or 2 of Section 2510.6.1, a space or drainage material not less than $^{3}/_{16}$ inch (4.8 mm) in depth shall be applied to the exterior side of the *water-resistive barrier*.
2. In addition to complying with Item 2 of Section 2510.6.1, drainage on the exterior side of the *water-resistive barrier* shall have a minimum drainage efficiency of 90 percent as measured in accordance with ASTM E2273 or Annex A2 of ASTM E2925.

2510.7 Preparation of masonry and concrete. Surfaces shall be clean, free from efflorescence, sufficiently damp and rough for proper bond. If the surface is insufficiently rough, *approved* bonding agents or a Portland cement dash bond coat mixed in proportions of not more than two parts volume of sand to one part volume of Portland cement or plastic cement shall be applied. The dash bond coat shall be left undisturbed and shall be moist cured not less than 24 hours.

SECTION 2511
INTERIOR PLASTER

2511.1 General. Plastering gypsum plaster or *cement plaster* shall be not less than three coats where applied over metal lath or wire fabric lath and not less than two coats where applied over other bases permitted by this chapter.

Exception: *Gypsum veneer plaster* and *cement plaster* specifically designed and *approved* for one-coat applications.

2511.1.1 Installation. Installation of lathing and plaster materials shall conform to Table 2511.1.1 and Section 2507.

2511.2 Limitations. Plaster shall not be applied directly to fiber insulation board. *Cement plaster* shall not be applied directly to gypsum lath or *gypsum plaster* except as specified in Sections 2510.5.1 and 2510.5.2.

2511.3 Grounds. Where installed, grounds shall ensure the minimum thickness of plaster as set forth in ASTM C842 and ASTM C926. Plaster thickness shall be measured from the face of lath and other bases.

2511.4 Interior masonry or concrete. Condition of surfaces shall be as specified in Section 2510.7. *Approved* specially prepared *gypsum plaster* designed for application to concrete surfaces or *approved* acoustical plaster is permitted. The total thickness of base coat plaster applied to concrete ceilings shall be as set forth in ASTM C842 or ASTM C926. Should ceiling surfaces require more than the maximum thickness permitted in ASTM C842 or ASTM C926, metal lath or wire fabric lath shall be installed on such surfaces before plastering.

2511.5 Wet areas. Showers and public toilet walls shall conform to Sections 1210.2 and 1210.3. Where wood frame walls and partitions are covered on the interior with *cement plaster* or tile of similar material and are subject to water splash, the framing shall be protected with an *approved* moisture barrier.

SECTION 2512
EXTERIOR PLASTER

2512.1 General. Plastering with *cement plaster* shall be not less than three coats where applied over metal lath or wire fabric lath or *gypsum board* backing as specified in Section 2510.5 and shall be not less than two coats where applied over masonry or concrete. If the plaster surface is to be completely covered by *veneer* or other facing material, or is completely concealed by another wall, plaster application need only be two coats, provided that the total thickness is as set forth in ASTM C926.

2512.1.1 On-grade floor slab. On wood frame or steel stud construction with an on-grade concrete floor slab system, exterior plaster shall be applied in such a manner as to cover, but not to extend below, the lath and paper. The application of lath, paper and flashing or drip screeds shall comply with ASTM C1063.

2512.1.2 Weep screeds. A minimum 0.019-inch (0.48 mm) (No. 26 galvanized sheet gage), corrosion-resistant weep screed with a minimum vertical attachment flange of $3^{1}/_{2}$ inches (89 mm) shall be provided at or below the foundation plate line on exterior stud walls in accordance with ASTM C926. The weep screed shall be placed not less than 4 inches (102 mm) above the earth or 2 inches

TABLE 2511.1.1
INSTALLATION OF PLASTER CONSTRUCTION

MATERIAL	STANDARD
Cement plaster	ASTM C926
Gypsum plaster	ASTM C842
Gypsum veneer plaster	ASTM C843
Interior lathing and furring (gypsum plaster)	ASTM C841
Lathing and furring (cement plaster)	ASTM C1063
Steel framing	ASTM C754; C1007

(51 mm) above paved areas and be of a type that will allow trapped water to drain to the exterior of the building. The *water-resistive barrier* shall lap the attachment flange. The exterior lath shall cover and terminate on the attachment flange of the weep screed.

2512.2 Plasticity agents. Only *approved* plasticity agents and *approved* amounts thereof shall be added to Portland cement or blended cements. Where plastic cement or masonry cement is used, additional lime or plasticizers shall not be added. Hydrated lime or the equivalent amount of lime putty used as a plasticizer is permitted to be added to *cement plaster* or cement and lime plaster in an amount not to exceed that set forth in ASTM C926.

2512.3 Limitations. *Gypsum plaster* shall not be used on exterior surfaces.

2512.4 Cement plaster. Plaster coats shall be protected from freezing for a period of not less than 24 hours after set has occurred. Plaster shall be applied when the ambient temperature is higher than 40°F (4°C), unless provisions are made to keep *cement plaster* work above 40°F (4°C) during application and 48 hours thereafter.

2512.5 Second-coat application. The second coat shall be brought out to proper thickness, rodded and floated sufficiently rough to provide adequate bond for the finish coat. The second coat shall not have variations greater than $^1/_4$ inch (6.4 mm) in any direction under a 5-foot (1524 mm) straight edge.

2512.6 Curing and interval. First and second coats of *cement plaster* shall be applied and moist cured as set forth in ASTM C926 and Table 2512.6.

**TABLE 2512.6
CEMENT PLASTERS**

COAT	MINIMUM PERIOD MOIST CURING	MINIMUM INTERVAL BETWEEN COATS
First	48 hours[a]	48 hours[b]
Second	48 hours	7 days[c]
Finish	—	Note c

a. The first two coats shall be as required for the first coats of exterior plaster, except that the moist-curing time period between the first and second coats shall be not less than 24 hours. Moist curing shall not be required where job and weather conditions are favorable to the retention of moisture in the cement plaster for the required time period.

b. Twenty-four-hour minimum interval between coats of interior cement plaster. For alternative method of application, see Section 2512.8.

c. Finish coat plaster is permitted to be applied to interior cement plaster base coats after a 48-hour period.

2512.7 Application to solid backings. Where applied over gypsum backing as specified in Section 2510.5 or directly to unit masonry surfaces, the second coat is permitted to be applied as soon as the first coat has attained sufficient hardness.

2512.8 Alternate method of application. The second coat is permitted to be applied as soon as the first coat has attained sufficient rigidity to receive the second coat.

2512.8.1 Admixtures. Where using this method of application, calcium aluminate cement up to 15 percent of the weight of the Portland cement is permitted to be added to the mix.

2512.8.2 Curing. Curing of the first coat is permitted to be omitted and the second coat shall be cured as set forth in ASTM C926 and Table 2512.6.

2512.9 Finish coats. *Cement plaster* finish coats shall be applied over base coats that have been in place for the time periods set forth in ASTM C926. The third or finish coat shall be applied with sufficient material and pressure to bond and to cover the brown coat and shall be of sufficient thickness to conceal the brown coat.

SECTION 2513
EXPOSED AGGREGATE PLASTER

2513.1 General. Exposed natural or integrally colored aggregate is permitted to be partially embedded in a natural or colored bedding coat of *cement plaster* or *gypsum plaster*, subject to the provisions of this section.

2513.2 Aggregate. The aggregate shall be applied manually or mechanically and shall consist of marble chips, pebbles or similar durable, moderately hard (three or more on the Mohs hardness scale), nonreactive materials.

2513.3 Bedding coat proportions. The bedding coat for interior or exterior surfaces shall be composed of one part Portland cement and one part Type S lime; or one part blended cement and one part Type S lime; or masonry cement; or plastic cement and not more than three parts of graded white or natural sand by volume. The bedding coat for *interior surfaces* shall be composed of 100 pounds (45.4 kg) of neat *gypsum plaster* and not more than 200 pounds (90.8 kg) of graded white sand. A factory-prepared bedding coat for interior or exterior use is permitted. The bedding coat for exterior surfaces shall have a minimum compressive strength of 1,000 pounds per square inch (6895 kPa).

2513.4 Application. The bedding coat is permitted to be applied directly over the first (scratch) coat of plaster, provided that the ultimate overall thickness is not less than $^7/_8$ inch (22 mm), including lath. Over concrete or masonry surfaces, the overall thickness shall be not less than $^1/_2$ inch (12.7 mm).

2513.5 Bases. Exposed aggregate plaster is permitted to be applied over concrete, masonry, *cement plaster* base coats or *gypsum plaster* base coats installed in accordance with Section 2511 or 2512.

2513.6 Preparation of masonry and concrete. Masonry and concrete surfaces shall be prepared in accordance with the provisions of Section 2510.7.

2513.7 Curing of base coats. *Cement plaster* base coats shall be cured in accordance with ASTM C926. Cement plaster bedding coats shall retain sufficient moisture for hydration (hardening) for 24 hours minimum or, where necessary, shall be kept damp for 24 hours by light water spraying.

SECTION 2514
REINFORCED GYPSUM CONCRETE

2514.1 General. Reinforced gypsum concrete shall comply with the requirements of ASTM C317 and ASTM C956.

2514.2 Minimum thickness. The minimum thickness of reinforced gypsum concrete shall be 2 inches (51 mm) except the minimum required thickness shall be reduced to $1^1/_2$ inches (38 mm), provided that the following conditions are satisfied:

1. The overall thickness, including the formboard, is not less than 2 inches (51 mm).
2. The clear span of the gypsum concrete between supports does not exceed 33 inches (838 mm).
3. *Diaphragm* action is not required.
4. The design *live load* does not exceed 40 pounds per square foot (psf) (1915 Pa).

CHAPTER 26
PLASTIC

User note:

About this chapter: *The use of plastics in building construction and components is addressed in Chapter 26. This chapter provides standards addressing foam plastic insulation, foam plastics used as interior finish and trim, and other plastic veneers used on the inside or outside of a building. This chapter addresses the use of light-transmitting plastics in various configurations such as walls, roof panels, skylights, signs and glazing. Requirements for the use of fiber-reinforced polymers, fiberglass-reinforced polymers and reflective plastic core insulation are also contained in this chapter. Additionally, requirements specific to the use of wood-plastic composites and plastic lumber are contained in this chapter.*

SECTION 2601
GENERAL

2601.1 Scope. These provisions shall govern the materials, design, application, construction and installation of foam plastic, foam plastic insulation, plastic *veneer*, interior plastic finish and *trim*, light-transmitting plastics and plastic composites, including plastic lumber.

SECTION 2602
FINISH AND TRIM

2602.1 Exterior wall covering and architectural trim. See Chapter 14 for requirements for *exterior wall* covering and architectural trim.

2602.2 Interior finish and trim. See Section 2604 for requirements for *interior finish* and *trim*.

SECTION 2603
FOAM PLASTIC INSULATION

2603.1 General. The provisions of this section shall govern the requirements and uses of foam plastic insulation in buildings and structures.

2603.1.1 Spray-applied foam plastic. Single- and multiple-component spray-applied foam plastic insulation shall comply with the provisions of Section 2603 and ICC 1100-2018.

2603.2 Labeling and identification. Packages and containers of foam plastic insulation and foam plastic insulation components delivered to the job site shall bear the *label* of an *approved agency* showing the manufacturer's name, product listing, product identification and information sufficient to determine that the end use will comply with the code requirements.

2603.3 Surface-burning characteristics. Unless otherwise indicated in this section, foam plastic insulation and foam plastic cores of manufactured assemblies shall have a *flame spread index* of not more than 75 and a *smoke-developed index* of not more than 450 where tested in the maximum thickness intended for use in accordance with ASTM E84 or UL 723. Loose fill-type foam plastic insulation shall be tested as board stock for the *flame spread* and smoke-developed indices.

Exceptions:

1. *Smoke-developed index* for interior *trim* as provided for in Section 2604.2.
2. In cold storage buildings, ice plants, food plants, food processing rooms and similar areas, foam plastic insulation where tested in a thickness of 4 inches (102 mm) shall be permitted in a thickness up to 10 inches (254 mm) where the building is equipped throughout with an automatic sprinkler system in accordance with Section 903.3.1.1. The approved *automatic sprinkler system* shall be provided in both the room and that part of the building in which the room is located.
3. Foam plastic insulation that is a part of a Class A, B or C roof-covering assembly provided that the assembly with the foam plastic insulation satisfactorily passes NFPA 276 or UL 1256. The *smoke-developed index* shall not be limited for roof applications.
4. Foam plastic insulation greater than 4 inches (102 mm) in thickness shall have a maximum *flame spread index* of 75 and a *smoke-developed index* of 450 where tested at a minimum thickness of 4 inches (102 mm), provided that the end use is approved in accordance with Section 2603.9 using the maximum thickness and density intended for use.
5. *Flame spread* and smoke-developed indices for foam plastic interior signs in *covered and open mall buildings* provided that the signs comply with Section 402.6.4.

2603.4 Thermal barrier. Except as provided for in Sections 2603.4.1 and 2603.9, foam plastic shall be separated from the interior of a building by an approved thermal barrier of $^1/_2$-inch (12.7 mm) *gypsum wallboard*, heavy timber in accordance with Section 602.4 or a material that is tested in accordance with and meets the acceptance criteria of both the Temperature Transmission Fire Test and the Integrity Fire Test of NFPA 275. Combustible concealed spaces shall comply with Section 718.

2603.4.1 Thermal barrier not required. The thermal barrier specified in Section 2603.4 is not required under the conditions set forth in Sections 2603.4.1.1 through 2603.4.1.14.

2603.4.1.1 Masonry or concrete construction. A thermal barrier is not required for foam plastic installed in a masonry or concrete wall, floor or roof system where the foam plastic insulation is covered on each face by not less than 1-inch (25 mm) thickness of masonry or concrete.

2603.4.1.2 Cooler and freezer walls. Foam plastic installed in a maximum thickness of 10 inches (254 mm) in cooler and freezer walls shall:

1. Have a *flame spread index* of 25 or less and a *smoke-developed index* of not more than 450, where tested in a minimum 4-inch (102 mm) thickness.
2. Have flash ignition and self-ignition temperatures of not less than 600°F and 800°F (316°C and 427°C), respectively.
3. Have a covering of not less than 0.032-inch (0.8 mm) aluminum or corrosion-resistant steel having a base metal thickness not less than 0.0160 inch (0.4 mm) at any point.
4. Be protected by an *automatic sprinkler system* in accordance with Section 903.3.1.1. Where the cooler or freezer is within a building, both the cooler or freezer and that part of the building in which it is located shall be sprinklered.

2603.4.1.3 Walk-in coolers. In nonsprinklered buildings, foam plastic having a thickness that does not exceed 4 inches (102 mm) and a maximum *flame spread index* of 75 is permitted in walk-in coolers or freezer units where the aggregate floor area does not exceed 400 square feet (37 m^2) and the foam plastic is covered by a metal facing not less than 0.032-inch-thick (0.81 mm) aluminum or corrosion-resistant steel having a minimum base metal thickness of 0.016 inch (0.41 mm). A thickness of up to 10 inches (254 mm) is permitted where protected by a thermal barrier.

2603.4.1.4 Exterior walls, one-story buildings. For one-*story* buildings, foam plastic having a *flame spread index* of 25 or less, and a *smoke-developed index* of not more than 450, shall be permitted without thermal barriers in or on *exterior walls* in a thickness not more than 4 inches (102 mm) where the foam plastic is covered by a thickness of not less than 0.032-inch-thick (0.81 mm) aluminum or corrosion-resistant steel having a base metal thickness of 0.0160 inch (0.41 mm) and the building is equipped throughout with an *automatic sprinkler system* in accordance with Section 903.3.1.1.

2603.4.1.5 Roofing. A thermal barrier is not required for foam plastic insulation that is a part of a Class A, B or C roof-covering assembly that is installed in accordance with the code and the manufacturer's instructions and is either constructed as described in Item 1 or tested as described in Item 2.

1. The roof assembly is separated from the interior of the building by *wood structural panel* sheathing not less than 0.47 inch (11.9 mm) in thickness bonded with exterior glue, with edges supported by blocking, tongue-and-groove joints, other approved type of edge support or an equivalent material.
2. The assembly with the foam plastic insulation satisfactorily passes NFPA 276 or UL 1256.

2603.4.1.6 Attics and crawl spaces. Within an *attic* or crawl space where entry is made only for service of utilities, foam plastic insulation shall be protected against ignition by $1^1/_2$-inch-thick (38 mm) mineral fiber insulation; $^1/_4$-inch-thick (6.4 mm) *wood structural panel*, *particleboard* or *hardboard*; $^3/_8$-inch (9.5 mm) *gypsum wallboard*, corrosion-resistant steel having a base metal thickness of 0.016 inch (0.4 mm); $1^1/_2$-inch-thick (38 mm) self-supported spray-applied cellulose insulation in *attic* spaces only or other approved material installed in such a manner that the foam plastic insulation is not exposed. The protective covering shall be consistent with the requirements for the type of construction.

2603.4.1.7 Doors not required to have a fire protection rating. Where pivoted or side-hinged doors are permitted without a *fire protection rating*, foam plastic insulation, having a *flame spread index* of 75 or less and a *smoke-developed index* of not more than 450, shall be permitted as a core material where the door facing is of aluminum not less than 0.032 inch (0.8 mm) in thickness or steel having a base metal thickness of not less than 0.016 inch (0.4 mm) at any point.

2603.4.1.8 Exterior doors in buildings of Group R-2 or R-3. In occupancies classified as Group R-2 or R-3, foam-filled exterior entrance doors to individual *dwelling units* that do not require a *fire-resistance rating* shall be faced with aluminum, steel, fiberglass, wood or other approved materials.

2603.4.1.9 Garage doors. Where garage doors are permitted without a *fire-resistance rating* and foam plastic is used as a core material, the door facing shall be metal having a minimum thickness of 0.032-inch (0.8 mm) aluminum or 0.010-inch (0.25 mm) steel or the facing shall be minimum 0.125-inch-thick (3.2 mm) wood. Garage doors having facings other than those described in this section shall be tested in accordance with, and meet the acceptance criteria of, DASMA 107.

> **Exception:** Garage doors using foam plastic insulation complying with Section 2603.3 in detached and attached garages associated with one- and two-family dwellings need not be provided with a thermal barrier.

2603.4.1.10 Siding backer board. Foam plastic insulation of not more than 2,000 British thermal units per square feet (Btu/sq. ft.) (22.7 mJ/m^2) as determined by NFPA 259 shall be permitted as a siding backer board with a maximum thickness of $^1/_2$ inch (12.7 mm), provided that it is separated from the interior of the building by not less than 2 inches (51 mm) of mineral fiber insulation or equivalent or where applied as insulation with re-siding over existing wall construction.

2603.4.1.11 Interior trim. Foam plastic used as interior *trim* in accordance with Section 2604 shall be permitted without a thermal barrier.

2603.4.1.12 Interior signs. Foam plastic used for interior signs in *covered mall buildings* in accordance with Section 402.6.4 shall be permitted without a thermal barrier. Foam plastic signs that are not affixed to interior building surfaces shall comply with Chapter 8 of the *International Fire Code*.

2603.4.1.13 Type V construction. Foam plastic spray applied to a sill plate, joist header and rim joist in Type V construction is subject to all of the following:

1. The maximum thickness of the foam plastic shall be 3$^1/_4$ inches (82.6 mm).
2. The density of the foam plastic shall be in the range of 1.5 to 2.0 pcf (24 to 32 kg/m^3).
3. The foam plastic shall have a *flame spread index* of 25 or less and an accompanying *smoke-developed index* of 450 or less when tested in accordance with ASTM E84 or UL 723.

2603.4.1.14 Floors. The thermal barrier specified in Section 2603.4 is not required to be installed on the walking surface of a structural floor system that contains foam plastic insulation where the foam plastic is covered by a minimum nominal $^1/_2$-inch-thick (12.7 mm) *wood structural panel* or approved equivalent. The thermal barrier specified in Section 2603.4 is required on the underside of the structural floor system that contains foam plastic insulation where the underside of the structural floor system is exposed to the interior of the building.

Exception: Foam plastic used as part of an *interior floor finish*.

2603.5 Exterior walls of buildings of any height. *Exterior walls* of buildings of Type I, II, III or IV construction of any height shall comply with Sections 2603.5.1 through 2603.5.7. *Exterior walls* of cold storage buildings required to be constructed of noncombustible materials, where the building is more than one *story* in height, shall comply with the provisions of Sections 2603.5.1 through 2603.5.7. *Exterior walls* of buildings of Type V construction shall comply with Sections 2603.2, 2603.3 and 2603.4. *Fireblocking* shall be in accordance with Section 718.2.

2603.5.1 Fire-resistance-rated walls. Where the wall is required to have a *fire-resistance rating*, data based on tests conducted in accordance with ASTM E119 or UL 263 shall be provided to substantiate that the *fire-resistance rating* is maintained.

2603.5.2 Thermal barrier. Any foam plastic insulation shall be separated from the building interior by a thermal barrier meeting the provisions of Section 2603.4, unless special approval is obtained on the basis of Section 2603.9.

Exception: One-*story* buildings complying with Section 2603.4.1.4.

2603.5.3 Potential heat. The potential heat of foam plastic insulation in any portion of the wall or panel shall not exceed the potential heat expressed in Btu per square feet (mJ/m^2) of the foam plastic insulation contained in the wall assembly tested in accordance with Section 2603.5.5. The potential heat of the foam plastic insulation shall be determined by tests conducted in accordance with NFPA 259 and the results shall be expressed in Btu per square feet (mJ/m^2).

Exception: One-*story* buildings complying with Section 2603.4.1.4.

2603.5.4 *Flame spread* and smoke-developed indices. Foam plastic insulation, exterior coatings and facings shall be tested separately in the thickness intended for use, but not to exceed 4 inches (102 mm), and shall each have a *flame spread index* of 25 or less and a *smoke-developed index* of 450 or less as determined in accordance with ASTM E84 or UL 723.

Exception: Prefabricated or factory-manufactured panels having minimum 0.020-inch (0.51 mm) aluminum facings and a total thickness of $^1/_4$ inch (6.4 mm) or less are permitted to be tested as an assembly where the foam plastic core is not exposed in the course of construction.

2603.5.5 Vertical and lateral fire propagation. The *exterior wall* assembly shall be tested in accordance with and comply with the acceptance criteria of NFPA 285.

Exceptions:

1. One-*story* buildings complying with Section 2603.4.1.4.
2. Wall assemblies where the foam plastic insulation is covered on each face by not less than 1-inch (25 mm) thickness of masonry or concrete and meeting one of the following:
   2.1. There is no airspace between the insulation and the concrete or masonry.
   2.2. The insulation has a *flame spread index* of not more than 25 as determined in accordance with ASTM E84 or UL 723 and the maximum airspace between the insulation and the concrete or masonry is not more than 1 inch (25 mm).

2603.5.6 Label required. The edge or face of each piece, package or container of foam plastic insulation shall bear the *label* of an *approved agency*. The *label* shall contain the manufacturer's or distributor's identification, model number, serial number or definitive information describing the product or materials' performance characteristics and *approved agency*'s identification.

2603.5.7 Ignition. *Exterior walls* shall not exhibit sustained flaming where tested in accordance with NFPA 268. Where a material is intended to be installed in more than one thickness, tests of the minimum and maximum thickness intended for use shall be performed.

Exception: Assemblies protected on the outside with one of the following:

1. A thermal barrier complying with Section 2603.4.
2. A minimum 1-inch (25 mm) thickness of concrete or masonry.
3. Glass-fiber-reinforced concrete panels of a minimum thickness of $^3/_8$ inch (9.5 mm).
4. Metal-faced panels having minimum 0.019-inch-thick (0.48 mm) aluminum or 0.016-inch-thick (0.41 mm) corrosion-resistant steel outer facings.
5. A minimum $^7/_8$-inch (22.2 mm) thickness of stucco complying with Section 2510.
6. A minimum $^1/_4$-inch (6.4 mm) thickness of *fiber-cement* lap, panel or shingle siding complying with Section 1404.16 and Section 1404.16.1 or 1404.16.2.

2603.6 Roofing. Foam plastic insulation meeting the requirements of Sections 2603.2, 2603.3 and 2603.4 shall be permitted as part of a roof-covering assembly, provided that the assembly with the foam plastic insulation is a Class A, B or C roofing assembly where tested in accordance with ASTM E108 or UL 790.

2603.7 Foam plastic in plenums as interior finish or interior trim. Foam plastic in plenums used as interior wall or ceiling finish, or interior *trim*, shall exhibit a *flame spread index* of 25 or less and a *smoke-developed index* of 50 or less when tested in accordance with ASTM E84 or UL 723 at the maximum thickness and density intended for use, and shall be tested in accordance with NFPA 286 and meet the acceptance criteria of Section 803.1.1. As an alternative to testing to NFPA 286, the foam plastic shall be approved based on tests conducted in accordance with Section 2603.9.

Exceptions:

1. Foam plastic in plenums used as interior wall or ceiling finish, or interior *trim*, shall exhibit a *flame spread index* of 75 or less and a *smoke-developed index* of 450 or less when tested in accordance with ASTM E84 or UL 723 at the maximum thickness and density intended for use, where it is separated from the airflow in the plenum by a thermal barrier complying with Section 2603.4.
2. Foam plastic in plenums used as interior wall or ceiling finish, or interior *trim*, shall exhibit a *flame spread index* of 75 or less and a *smoke-developed index* of 450 or less when tested in accordance with ASTM E84 or UL 723 at the maximum thickness and density intended for use, where it is separated from the airflow in the plenum by corrosion-resistant steel having a base metal thickness of not less than 0.0160 inch (0.4 mm).
3. Foam plastic in plenums used as interior wall or ceiling finish, or interior *trim*, shall exhibit a *flame spread index* of 75 or less and a *smoke-developed index* of 450 or less when tested in accordance with ASTM E84 or UL 723 at the maximum thickness and density intended for use, where it is separated from the airflow in the plenum by not less than a 1-inch (25 mm) thickness of masonry or concrete.

2603.8 Protection against termites. In areas where the probability of termite infestation is very heavy in accordance with Figure 2603.8, extruded and expanded polystyrene, polyisocyanurate and other foam plastics shall not be installed on the exterior face or under interior or exterior foundation walls or slab foundations located below grade. The clearance between foam plastics installed above grade and exposed earth shall be not less than 6 inches (152 mm).

Exceptions:

1. Buildings where the structural members of walls, floors, ceilings and roofs are entirely of noncombustible materials or *preservative-treated wood*.
2. An approved method of protecting the foam plastic and structure from subterranean termite damage is provided.
3. On the interior side of basement walls.

2603.9 Special approval. Foam plastic shall not be required to comply with the requirements of Section 2603.4 or those of Section 2603.6 where specifically approved based on large-scale tests such as, but not limited to, NFPA 286 (with the acceptance criteria of Section 803.1.1.1), FM 4880, UL 1040 or UL 1715. Such testing shall be related to the actual end-use configuration and be performed on the finished manufactured foam plastic assembly in the maximum thickness intended for use. Foam plastics that are used as *interior finish* on the basis of special tests shall conform to the *flame spread* and smoke-developed requirements of Chapter 8. Assemblies tested shall include seams, joints and other typical details used in the installation of the assembly and shall be tested in the manner intended for use.

[BS] 2603.10 Wind resistance. Foam plastic insulation complying with ASTM C578 and ASTM C1289 and used as *exterior wall* sheathing on framed wall assemblies shall comply with ANSI/FS 100 for wind pressure resistance.

[BS] 2603.11 Cladding attachment over foam sheathing to masonry or concrete wall construction. Cladding shall be specified and installed in accordance with Chapter 14 and the cladding manufacturer's installation instructions or an approved design. Foam sheathing shall be attached to masonry or concrete construction in accordance with the insulation manufacturer's installation instructions or an approved design. Furring and furring attachments through foam sheathing shall be designed to resist design *loads* determined in accordance with Chapter 16, including support of cladding weight as applicable. Fasteners used to attach cladding or furring through foam sheathing to masonry or concrete substrates shall be approved for application into masonry or concrete material and shall be installed in accordance with the fastener manufacturer's installation instructions.

Exceptions:

1. Where the cladding manufacturer has provided approved installation instructions for application over foam sheathing and connection to a masonry or concrete substrate, those requirements shall apply.
2. For *exterior insulation and finish systems*, refer to Section 1407.
3. For anchored masonry or stone *veneer* installed over foam sheathing, refer to Section 1404.

[BS] 2603.12 Cladding attachment over foam sheathing to cold-formed steel framing. Cladding shall be specified and installed in accordance with Chapter 14 and the cladding manufacturer's approved installation instructions, including any limitations for use over foam plastic sheathing, or an approved design. Where used, furring and furring attachments shall be designed to resist design *loads* determined in accordance with Chapter 16. In addition, the cladding or furring attachments through foam sheathing to cold-formed steel framing shall meet or exceed the minimum fastening requirements of Sections 2603.12.1 and 2603.12.2, or an approved design for support of cladding weight.

Exceptions:

1. Where the cladding manufacturer has provided approved installation instructions for application over foam sheathing, those requirements shall apply.
2. For *exterior insulation and finish systems*, refer to Section 1407.
3. For anchored masonry or stone *veneer* installed over foam sheathing, refer to Section 1404.

[BS] 2603.12.1 Direct attachment. Where cladding is installed directly over foam sheathing without the use of furring, cladding minimum fastening requirements to support the cladding weight shall be as specified in Table 2603.12.1.

[BS] 2603.12.2 Furred cladding attachment. Where steel or wood furring is used to attach cladding over foam sheathing, furring minimum fastening requirements to support the cladding weight shall be as specified in Table 2603.12.2. Where placed horizontally, wood furring shall be *preservative-treated wood* in accordance with Section 2303.1.9 or *naturally durable wood* and fasteners shall be corrosion resistant in accordance Section 2304.10.6. Steel furring shall have a minimum G60 galvanized coating.

[BS] 2603.13 Cladding attachment over foam sheathing to wood framing. Cladding shall be specified and installed in accordance with Chapter 14 and the cladding manufacturer's installation instructions. Where used, furring and furring attachments shall be designed to resist design *loads* determined in accordance with Chapter 16. In addition, the cladding or furring attachments through foam sheathing to framing shall meet or exceed the minimum fastening requirements of Section 2603.13.1 or 2603.13.2, or an approved design for support of cladding weight.

Exceptions:

1. Where the cladding manufacturer has provided approved installation instructions for application over foam sheathing, those requirements shall apply.

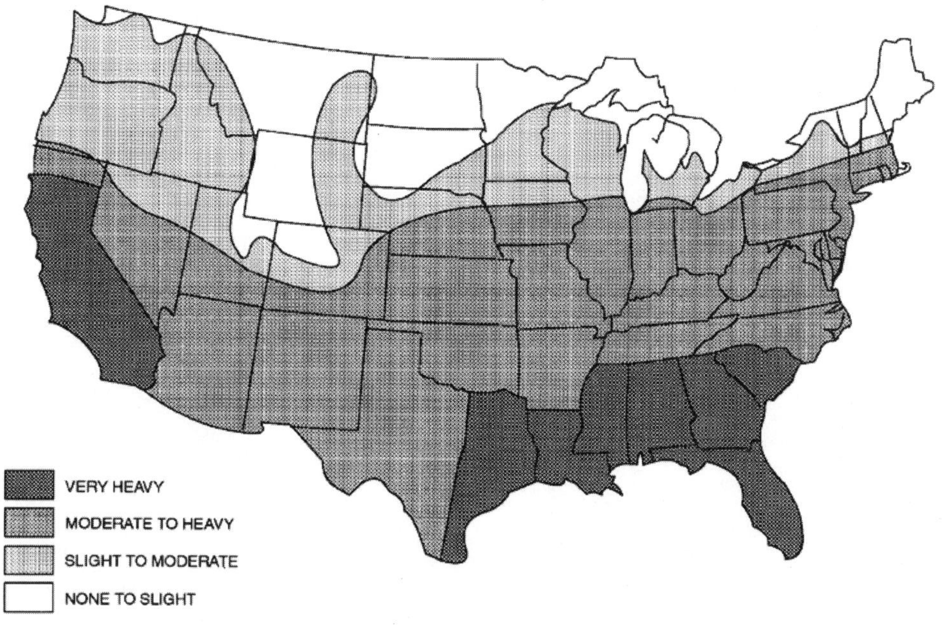

FIGURE 2603.8
TERMITE INFESTATION PROBABILITY MAP

[BS]TABLE 2603.12.1
CLADDING MINIMUM FASTENING REQUIREMENTS FOR DIRECT ATTACHMENT OVER FOAM PLASTIC SHEATHING TO SUPPORT CLADDING WEIGHT[a]

CLADDING FASTENER THROUGH FOAM SHEATHING INTO:	CLADDING FASTENER TYPE AND MINIMUM SIZE[b]	CLADDING FASTENER VERTICAL SPACING (inches)	MAXIMUM THICKNESS OF FOAM SHEATHING[c] (inches)							
			16" o.c. fastener horizontal spacing				24" o.c. fastener horizontal spacing			
			Cladding weight				Cladding weight			
			3 psf	11 psf	18 psf	25 psf	3 psf	11 psf	18 psf	25 psf
Cold-formed steel framing (minimum penetration of steel thickness plus 3 threads)	#8 screw into 33 mil steel or thicker	6	3.00	2.95	2.20	1.45	3.00	2.35	1.25	DR
		8	3.00	2.55	1.60	0.60	3.00	1.80	DR	DR
		12	3.00	1.80	DR	DR	3.00	0.65	DR	DR
	#10 screw into 33 mil steel	6	4.00	3.50	2.70	1.95	4.00	2.90	1.70	0.55
		8	4.00	3.10	2.05	1.00	4.00	2.25	0.70	DR
		12	4.00	2.25	0.70	DR	3.70	1.05	DR	DR
	#10 screw into 43 mil steel or thicker	6	4.00	4.00	4.00	3.60	4.00	4.00	3.45	2.70
		8	4.00	4.00	3.70	3.00	4.00	3.85	2.80	1.80
		12	4.00	3.85	2.80	1.80	4.00	3.05	1.50	DR

For SI: 1 inch = 25.4 mm, 1 pound per square foot (psf) = 0.0479 kPa, 1 pound per square inch = 0.00689 MPa.
DR = design required, o.c. = on center.
a. Cold-formed steel framing shall be minimum 33 ksi steel for 33 mil and 43 mil steel and 50 ksi steel for 54 mil steel or thicker.
b. Screws shall comply with the requirements of AISI S240.
c. Foam sheathing shall have a minimum compressive strength of 15 pounds per square inch in accordance with ASTM C578 or ASTM C1289.

[BS]TABLE 2603.12.2
FURRING MINIMUM FASTENING REQUIREMENTS FOR APPLICATION OVER FOAM PLASTIC SHEATHING TO SUPPORT CLADDING WEIGHT[a]

FURRING MATERIAL	FRAMING MEMBER	FASTENER TYPE AND MINIMUM SIZE[b]	MINIMUM PENETRATION INTO WALL FRAMING (inches)	FASTENER SPACING IN FURRING (inches)	MAXIMUM THICKNESS OF FOAM SHEATHING[d] (inches)							
					16" o.c. furring[e]				24" o.c. furring[e]			
					Cladding weight				Cladding weight			
					3 psf	11 psf	18 psf	25 psf	3 psf	11 psf	18 psf	25 psf
Minimum 33 mil steel furring or minimum 1x wood furring[c]	33 mil cold-formed steel stud	#8 screw	Steel thickness plus 3 threads	12	3.00	1.80	DR	DR	3.00	0.65	DR	DR
				16	3.00	1.00	DR	DR	2.85	DR	DR	DR
				24	2.85	DR	DR	DR	2.20	DR	DR	DR
		#10 screw	Steel thickness plus 3 threads	12	4.00	2.25	0.70	DR	3.70	1.05	DR	DR
				16	3.85	1.45	DR	DR	3.40	DR	DR	DR
				24	3.40	DR	DR	DR	2.70	DR	DR	DR
	43 mil or thicker cold-formed steel stud	#8 Screw	Steel thickness plus 3 threads	12	3.00	1.80	DR	DR	3.00	0.65	DR	DR
				16	3.00	1.00	DR	DR	2.85	DR	DR	DR
				24	2.85	DR	DR	DR	2.20	DR	DR	DR
		#10 screw	Steel thickness plus 3 threads	12	4.00	3.85	2.80	1.80	4.00	3.05	1.50	DR
				16	4.00	3.30	1.95	0.60	4.00	2.25	DR	DR
				24	4.00	2.25	DR	DR	4.00	0.65	DR	DR

For SI: 1 inch = 25.4 mm, 1 pound per square foot (psf) = 0.0479 kPa, 1 pound per square inch = 0.00689 MPa.
DR = Design Required, o.c. = on center.
a. Wood furring shall be spruce-pine-fir or any softwood species with a specific gravity of 0.42 or greater. Steel furring shall be minimum 33 ksi steel. Cold-formed steel studs shall be minimum 33 ksi steel for 33 mil and 43 mil thickness and 50 ksi steel for 54 mil steel or thicker.
b. Screws shall comply with the requirements of AISI S240.
c. Where the required cladding fastener penetration into wood material exceeds 3/4 inch and is not more than 11/2 inches, a minimum 2-inch nominal wood furring or an approved design shall be used.
d. Foam sheathing shall have a minimum compressive strength of 15 pounds per square inch in accordance with ASTM C578 or ASTM C1289.
e. Furring shall be spaced not more than 24 inches on center, in a vertical or horizontal orientation. In a vertical orientation, furring shall be located over wall studs and attached with the required fastener spacing. In a horizontal orientation, the indicated 8-inch and 12-inch fastener spacing in furring shall be achieved by use of two fasteners into studs at 16 inches and 24 inches on center, respectively.

2. For *exterior insulation and finish systems*, refer to Section 1407.
3. For anchored masonry or stone *veneer* installed over foam sheathing, refer to Section 1404.

[BS] 2603.13.1 Direct attachment. Where cladding is installed directly over foam sheathing without the use of furring, minimum fastening requirements to support the cladding weight shall be as specified in Table 2603.13.1.

[BS] 2603.13.2 Furred cladding attachment. Where wood furring is used to attach cladding over foam sheathing, furring minimum fastening requirements to support the cladding weight shall be as specified in Table 2603.13.2. Where placed horizontally, wood furring shall be *preservative-treated wood* in accordance with Section 2303.1.9 or *naturally durable wood* and fasteners shall be corrosion resistant in accordance with Section 2304.10.6.

SECTION 2604 INTERIOR FINISH AND TRIM

2604.1 General. Plastic materials installed as *interior finish* or *trim* shall comply with Chapter 8. Foam plastics shall only be installed as *interior finish* where approved in accordance with the special provisions of Section 2603.9. Foam plastics that are used as *interior finish* shall meet the *flame spread* and *smoke-developed index* requirements for *interior finish* in accordance with Chapter 8. Foam plastics installed as interior *trim* shall comply with Section 2604.2.

2604.1.1 Plenums. Foam plastics installed in plenums as interior wall or ceiling finish shall comply with Section 2603.7. Foam plastics installed in plenums as interior *trim* shall comply with Sections 2604.2 and 2603.7.

[F] 2604.2 Interior trim. Foam plastic used as interior *trim* shall comply with Sections 2604.2.1 through 2604.2.4.

[F] 2604.2.1 Density. The minimum density of the interior *trim* shall be 20 pcf (320 kg/m^3).

[F] 2604.2.2 Thickness. The maximum thickness of the interior *trim* shall be $^1/_2$ inch (12.7 mm) and the maximum width shall be 8 inches (204 mm).

[F] 2604.2.3 Area limitation. The interior *trim* shall not constitute more than 10 percent of the specific wall or ceiling areas to which it is attached.

[F] 2604.2.4 Flame spread. The *flame spread index* shall not exceed 75 where tested in accordance with ASTM E84 or UL 723. The *smoke-developed index* shall not be limited.

Exception: Where the interior *trim* material has been tested as an *interior finish* in accordance with NFPA 286 and complies with the acceptance criteria in Section 803.1.1.1, it shall not be required to be tested for *flame spread index* in accordance with ASTM E84 or UL 723.

SECTION 2605 PLASTIC VENEER

2605.1 Interior use. Where used within a building, plastic *veneer* shall comply with the *interior finish* requirements of Chapter 8.

[BS]TABLE 2603.13.1
CLADDING MINIMUM FASTENING REQUIREMENTS FOR DIRECT ATTACHMENT OVER FOAM PLASTIC SHEATHING TO SUPPORT CLADDING WEIGHT[a]

CLADDING FASTENER THROUGH FOAM SHEATHING INTO:	CLADDING FASTENER TYPE AND MINIMUM SIZE[b]	CLADDING FASTENER VERTICAL SPACING (INCHES)	MAXIMUM THICKNESS OF FOAM SHEATHING[c] (INCHES)							
			16" o.c. fastener horizontal spacing				24" o.c. fastener horizontal spacing			
			Cladding weight:				Cladding weight:			
			3 psf	11 psf	18 psf	25 psf	3 psf	11 psf	18 psf	25 psf
Wood Framing (minimum 1$^1/_4$- inch penetration)	0.113" diameter nail	6	2.00	1.45	0.75	DR	2.00	0.85	DR	DR
		8	2.00	1.00	DR	DR	2.00	0.55	DR	DR
		12	2.00	0.55	DR	DR	1.85	DR	DR	DR
	0.120" diameter nail	6	3.00	1.70	0.90	0.55	3.00	1.05	0.50	DR
		8	3.00	1.20	0.60	DR	3.00	0.70	DR	DR
		12	3.00	0.70	DR	DR	2.15	DR	DR	DR
	0.131" diameter nail	6	4.00	2.15	1.20	0.75	4.00	1.35	0.70	DR
		8	4.00	1.55	0.80	DR	4.00	0.90	DR	DR
		12	4.00	0.90	DR	DR	2.70	0.50	DR	DR
	0.162" diameter nail	6	4.00	3.55	2.05	1.40	4.00	2.25	1.25	0.80
		8	4.00	2.55	1.45	0.95	4.00	1.60	0.85	0.50
		12	4.00	1.60	0.85	0.50	4.00	0.95	DR	DR

For SI: 1 inch = 25.4 mm, 1 pound per square foot (psf) = 0.0479 kPa.
DR = Design Required, o.c. = on center.
a. Wood framing shall be spruce-pine-fir or any wood species with a specific gravity of 0.42 or greater in accordance with ANSI/AWC NDS.
b. Nail fasteners shall comply with ASTM F1667, except nail length shall be permitted to exceed ASTM F1667 standard lengths.
c. Foam sheathing shall have a minimum compressive strength of 15 psi in accordance with ASTM C578 or ASTM C1289.

[BS]TABLE 2603.13.2
FURRING MINIMUM FASTENING REQUIREMENTS FOR APPLICATION OVER FOAM PLASTIC SHEATHING TO SUPPORT CLADDING WEIGHT[a, b]

FURRING MATERIAL	FRAMING MEMBER	FASTENER TYPE AND MINIMUM SIZE	MINIMUM PENETRATION INTO WALL FRAMING (INCHES)	FASTENER SPACING IN FURRING (INCHES)	MAXIMUM THICKNESS OF FOAM SHEATHING[d] (INCHES)							
					16" o.c. furring[e]				24" o.c. furring[e]			
					Siding weight:				Siding weight:			
					3 psf	11 psf	18 psf	25 psf	3 psf	11 psf	18 psf	25 psf
Minimum 1x Wood Furring[c]	Minimum 2x Wood Stud	0.131" diameter nail	$1^1/_4$	8	4.00	2.45	1.45	0.95	4.00	1.60	0.85	DR
				12	4.00	1.60	0.85	DR	4.00	0.95	DR	DR
				16	4.00	1.10	DR	DR	3.05	0.60	DR	DR
		0.162" diameter nail	$1^1/_4$	8	4.00	4.00	2.45	1.60	4.00	2.75	1.45	0.85
				12	4.00	2.75	1.45	0.85	4.00	1.65	0.75	DR
				16	4.00	1.90	0.95	DR	4.00	1.05	DR	DR
		No. 10 wood screw	1	12	4.00	2.30	1.20	0.70	4.00	1.40	0.60	DR
				16	4.00	1.65	0.75	DR	4.00	0.90	DR	DR
				24	4.00	0.90	DR	DR	2.85	DR	DR	DR
		$^1/_4$" lag screw	$1^1/_2$	12	4.00	2.65	1.50	0.90	4.00	1.65	0.80	DR
				16	4.00	1.95	0.95	0.50	4.00	1.10	DR	DR
				24	4.00	1.10	DR	DR	3.25	0.50	DR	DR

For SI: 1 inch = 25.4 mm, 1 pound per square foot (psf) = 0.0479 kPa, 1 pound per square inch = 0.00689 MPa.
DR = Design Required, o.c. = on center.

a. Wood framing and furring shall be spruce-pine-fir or any wood species with a specific gravity of 0.42 or greater in accordance with ANSI/AWC NDS.
b. Nail fasteners shall comply with ASTM F1667, except nail length shall be permitted to exceed ASTM F1667 standard lengths.
c. Where the required cladding fastener penetration into wood material exceeds 3/4 inch and is not more than 11/2 inches, a minimum 2-inch nominal wood furring or an approved design shall be used.
d. Foam sheathing shall have a minimum compressive strength of 15 psi in accordance with ASTM C578 or ASTM C1289.
e. Furring shall be spaced not greater than 24 inches on center in a vertical or horizontal orientation. In a vertical orientation, furring shall be located over wall studs and attached with the required fastener spacing. In a horizontal orientation, the indicated 8-inch and 12-inch fastener spacing in furring shall be achieved by use of two fasteners into studs at 16 inches and 24 inches on center, respectively.

2605.2 Exterior use. Exterior plastic *veneer*, other than plastic siding, shall be permitted to be installed on the *exterior walls* of buildings of any type of construction in accordance with all of the following requirements:

1. Plastic *veneer* shall comply with Section 2606.4.
2. Plastic *veneer* shall not be attached to any *exterior wall* to a height greater than 50 feet (15 240 mm) above grade.
3. Sections of plastic *veneer* shall not exceed 300 square feet (27.9 m^2) in area and shall be separated by not less than 4 feet (1219 mm) vertically.

Exception: The area and separation requirements and the smoke-density limitation are not applicable to plastic *veneer* applied to buildings constructed of Type VB construction, provided that the walls are not required to have a *fire-resistance rating*.

2605.3 Plastic siding. Plastic siding shall comply with the requirements of Sections 1403 and 1404.

SECTION 2606
LIGHT-TRANSMITTING PLASTICS

2606.1 General. The provisions of this section and Sections 2607 through 2611 shall govern the quality and methods of application of light-transmitting plastics for use as light-transmitting materials in buildings and structures. Foam plastics shall comply with Section 2603. Light-transmitting plastic materials that meet the other code requirements for walls and roofs shall be permitted to be used in accordance with the other applicable chapters of the code.

2606.2 Approval for use. Sufficient technical data shall be submitted to substantiate the proposed use of any light-transmitting material, as approved by the *building official* and subject to the requirements of this section.

2606.3 Identification. Each unit or package of light-transmitting plastic shall be identified with a *mark* or decal satisfactory to the *building official*, which includes identification as to the material classification.

2606.4 Specifications. Light-transmitting plastics, including thermoplastic, thermosetting or reinforced thermosetting plastic material, shall have a self-ignition temperature of 650°F (343°C) or greater where tested in accordance with ASTM D1929; a *smoke-developed index* not greater than 450 where tested in the manner intended for use in accordance with ASTM E84 or UL 723, or a maximum average smoke density rating not greater than 75 where tested in the thickness intended for use in accordance with ASTM D2843 and shall conform to one of the following combustibility classifications:

Class CC1: Plastic materials that have a burning extent of 1 inch (25 mm) or less where tested at a nominal thickness of 0.060 inch (1.5 mm), or in the thickness intended for use, in accordance with ASTM D635.

Class CC2: Plastic materials that have a burning rate of $2^1/_2$ inches per minute (1.06 mm/s) or less where tested at a nominal thickness of 0.060 inch (1.5 mm), or in the thickness intended for use, in accordance with ASTM D635.

[BS] 2606.5 Structural requirements. Light-transmitting plastic materials in their assembly shall be of adequate strength and durability to withstand the *loads* indicated in Chapter 16. Technical data shall be submitted to establish stresses, maximum unsupported spans and such other information for the various thicknesses and forms used as deemed necessary by the *building official*.

[BS] 2606.6 Fastening. Fastening shall be adequate to withstand the *loads* in Chapter 16. Proper allowance shall be made for expansion and contraction of light-transmitting plastic materials in accordance with accepted data on the coefficient of expansion of the material and other material in conjunction with which it is employed.

2606.7 Light-diffusing systems. Unless the building is equipped throughout with an *automatic sprinkler system* in accordance with Section 903.3.1.1, *light-diffusing systems* shall not be installed in the following occupancies and locations:

1. Group A with an *occupant load* of 1,000 or more.
2. Theaters with a *stage* and proscenium opening and an *occupant load* of 700 or more.
3. Group I-2.
4. Group I-3.
5. *Interior exit stairways* and *ramps* and exit passageways.

2606.7.1 Support. Light-transmitting plastic diffusers shall be supported directly or indirectly from ceiling or roof construction by use of noncombustible hangers. Hangers shall be not less than No. 12 steel-wire gage (0.106 inch) galvanized wire or equivalent.

2606.7.2 Installation. Light-transmitting plastic diffusers shall comply with Chapter 8 unless the light-transmitting plastic diffusers will fall from the mountings before igniting, at an ambient temperature of not less than 200°F (111°C) below the ignition temperature of the panels. The panels shall remain in place at an ambient room temperature of 175°F (79°C) for a period of not less than 15 minutes.

2606.7.3 Size limitations. Individual panels or units shall not exceed 10 feet (3048 mm) in length nor 30 square feet (2.79 m^2) in area.

2606.7.4 Automatic sprinkler system. In buildings that are equipped throughout with an *automatic sprinkler system* in accordance with Section 903.3.1.1, plastic *light-diffusing systems* shall be protected both above and below unless the sprinkler system has been specifically approved for installation only above the *light-diffusing system*, or the *light-diffusing system* is listed and labeled in accordance with UL 723S. Areas of *light-diffusing systems* that are protected in accordance with this section shall not be limited.

2606.7.5 Electrical luminaires. Light-transmitting plastic panels and light-diffuser panels that are installed in approved electrical luminaires shall comply with the requirements of Chapter 8 unless the light-transmitting plastic panels conform to the requirements of Section 2606.7.2. The area of approved light-transmitting plastic materials that is used in required *exits* or *corridors* shall not exceed 30 percent of the aggregate area of the ceiling in which such panels are installed, unless the building is equipped throughout with an *automatic sprinkler system* in accordance with Section 903.3.1.1.

2606.8 Partitions. Light-transmitting plastics used in or as partitions shall comply with the requirements of Chapters 6 and 8.

2606.9 Bathroom accessories. Light-transmitting plastics shall be permitted as glazing in shower stalls, shower doors, bathtub enclosures and similar accessory units. Safety glazing shall be provided in accordance with Chapter 24.

2606.10 Awnings, patio covers and similar structures. *Awnings* constructed of light-transmitting plastics shall be constructed in accordance with the provisions specified in Section 3105 and Chapter 32 for projections. Patio covers constructed of light-transmitting plastics shall comply with Section 2606. Light-transmitting plastics used in canopies at motor fuel-dispensing facilities shall comply with Section 2606, except as modified by Section 406.7.2.

2606.11 Greenhouses. Light-transmitting plastics shall be permitted in lieu of glass in *greenhouses*.

2606.12 Solar collectors. Light-transmitting plastic covers on solar collectors having noncombustible sides and bottoms shall be permitted on buildings not over three *stories above grade plane* or 9,000 square feet (836.1 m^2) in total floor area, provided that the light-transmitting plastic cover does not exceed 33.33 percent of the roof area for CC1 materials or 25 percent of the roof area for CC2 materials.

Exception: Light-transmitting plastic covers having a thickness of 0.010 inch (0.3 mm) or less shall be permitted to be of any plastic material provided that the area of the solar collectors does not exceed 33.33 percent of the roof area.

SECTION 2607
LIGHT-TRANSMITTING PLASTIC WALL PANELS

2607.1 General. Light-transmitting plastics shall not be used as wall panels in *exterior walls* in occupancies in Groups A-1, A-2, H, I-2 and I-3. In other groups, light-transmitting plastics shall be permitted to be used as wall panels in *exterior walls*, provided that the walls are not required to have a *fire-resistance rating* and the installation conforms to the requirements of this section. Such panels shall be erected and anchored on a foundation, waterproofed or otherwise protected from moisture absorption and sealed with a coat of mastic or other approved waterproof coating. Light-transmitting plastic wall panels shall comply with Section 2606.

2607.2 Installation. *Exterior wall* panels installed as provided for herein shall not alter the type of construction classification of the building.

2607.3 Height limitation. Light-transmitting plastics shall not be installed more than 75 feet (22 860 mm) above *grade plane*.

2607.4 Area limitation and separation. The maximum area of a single wall panel and minimum vertical and horizontal separation requirements for exterior light-transmitting plastic wall panels shall be as provided for in Table 2607.4. The maximum percentage of wall area of any *story* in light-transmitting plastic wall panels shall not exceed that indicated in Table 2607.4 or the percentage of unprotected openings permitted by Section 705.8, whichever is smaller.

Exceptions:

1. In structures provided with approved flame barriers extending 30 inches (760 mm) beyond the *exterior wall* in the plane of the floor, a vertical separation is not required at the floor except that provided by the vertical thickness of the flame barrier projection.
2. *Veneers* of approved weather-resistant light-transmitting plastics used as exterior siding in buildings of Type V construction in compliance with Section 1405.
3. The area of light-transmitting plastic wall panels in *exterior walls* of *greenhouses* shall be exempt from the area limitations of Table 2607.4 but shall be limited as required for unprotected openings in accordance with Section 705.8.

2607.5 Automatic sprinkler system. Where the building is equipped throughout with an *automatic sprinkler system* in accordance with Section 903.3.1.1, the maximum percentage area of *exterior wall* in any *story* in light-transmitting plastic wall panels and the maximum square footage of a single area given in Table 2607.4 shall be increased 100 percent, but the area of light-transmitting plastic wall panels shall not exceed 50 percent of the wall area in any *story*, or the area permitted by Section 705.8 for unprotected openings, whichever is smaller. These installations shall not be installed more than 75 feet (22 860 mm) above *grade plane*.

2607.6 Combinations of glazing and wall panels. Combinations of light-transmitting plastic glazing and light-transmitting plastic wall panels shall be subject to the area, height and percentage limitations and the separation requirements applicable to the class of light-transmitting plastic as prescribed for light-transmitting plastic wall panel installations.

SECTION 2608
LIGHT-TRANSMITTING PLASTIC GLAZING

2608.1 Buildings of Type VB construction. Openings in the *exterior walls* of buildings of Type VB construction, where not required to be protected by Section 705, shall be permitted to be glazed or equipped with light-transmitting plastic. Light-transmitting plastic glazing shall comply with Section 2606.

2608.2 Buildings of other types of construction. Openings in the *exterior walls* of buildings of types of construction other than Type VB, where not required to be protected by Section 705, shall be permitted to be glazed or equipped with light-transmitting plastic in accordance with Section 2606 and all of the following:

1. The aggregate area of light-transmitting plastic glazing shall not exceed 25 percent of the area of any wall face of the *story* in which it is installed. The area of a single pane of glazing installed above the first *story above grade plane* shall not exceed 16 square feet (1.5 m^2) and the vertical dimension of a single pane shall not exceed 4 feet (1219 mm).

   Exception: Where an *automatic sprinkler system* is provided throughout in accordance with Section 903.3.1.1, the area of allowable glazing shall be increased to not more than 50 percent of the wall face of the *story* in which it is installed with no limit on the maximum dimension or area of a single pane of glazing.

2. Approved flame barriers extending 30 inches (762 mm) beyond the *exterior wall* in the plane of the floor, or vertical panels not less than 4 feet (1219

TABLE 2607.4
AREA LIMITATION AND SEPARATION REQUIREMENTS FOR LIGHT-TRANSMITTING PLASTIC WALL PANELS[a]

FIRE SEPARATION DISTANCE (feet)	CLASS OF PLASTIC	MAXIMUM PERCENTAGE AREA OF EXTERIOR WALL IN PLASTIC WALL PANELS	MAXIMUM SINGLE AREA OF PLASTIC WALL PANELS (square feet)	MINIMUM SEPARATION OF PLASTIC WALL PANELS (feet)	
				Vertical	Horizontal
Less than 6	—	Not Permitted	Not Permitted	—	—
6 or more but less than 11	CC1	10	50	8	4
	CC2	Not Permitted	Not Permitted	—	—
11 or more but less than or equal to 30	CC1	25	90	6	4
	CC2	15	70	8	4
Over 30	CC1	50	Not Limited	3[b]	0
	CC2	50	100	6[b]	3

For SI: 1 foot = 304.8 mm, 1 square foot = 0.0929 m2.

a. For combinations of plastic glazing and plastic wall panel areas permitted, see Section 2607.6.

b. For reductions in vertical separation allowed, see Section 2607.4.

mm) in height, shall be installed between glazed units located in adjacent stories.

Exception: Buildings equipped throughout with an *automatic sprinkler system* in accordance with Section 903.3.1.1.

3. Light-transmitting plastics shall not be installed more than 75 feet (22 860 mm) above grade level.

Exception: Buildings equipped throughout with an *automatic sprinkler system* in accordance with Section 903.3.1.1.

SECTION 2609 LIGHT-TRANSMITTING PLASTIC ROOF PANELS

2609.1 General. Light-transmitting plastic roof panels shall comply with this section and Section 2606. Light-transmitting plastic roof panels shall not be installed in Groups H, I-2 and I-3. In all other groups, light-transmitting plastic roof panels shall comply with any one of the following conditions:

1. The building is equipped throughout with an *automatic sprinkler system* in accordance with Section 903.3.1.1.
2. The roof construction is not required to have a *fire-resistance rating* by Table 601.
3. The roof panels meet the requirements for roof coverings in accordance with Chapter 15.

2609.2 Separation. Individual roof panels shall be separated from each other by a distance of not less than 4 feet (1219 mm) measured in a horizontal plane.

Exceptions:

1. The separation between roof panels is not required in a building equipped throughout with an *automatic sprinkler system* in accordance with Section 903.3.1.1.
2. The separation between roof panels is not required in low-hazard occupancy buildings complying with the conditions of Section 2609.4, Exception 2 or 3.

2609.3 Location. Where *exterior wall* openings are required to be protected by Section 705.8, a roof panel shall not be installed within 6 feet (1829 mm) of such *exterior wall.*

2609.4 Area limitations. Roof panels shall be limited in area and the aggregate area of panels shall be limited by a percentage of the floor area of the room or space sheltered in accordance with Table 2609.4.

Exceptions:

1. The area limitations of Table 2609.4 shall be permitted to be increased by 100 percent in buildings equipped throughout with an *automatic sprinkler system* in accordance with Section 903.3.1.1.
2. Low-hazard occupancy buildings, such as swimming pool shelters, shall be exempt from the area limitations of Table 2609.4, provided that the buildings do not exceed 5,000 square feet (465 m^2) in area and have a minimum *fire separation distance* of 10 feet (3048 mm).
3. *Greenhouses* that are occupied for growing or maintaining plants, without public access, shall be exempt from the area limitations of Table 2609.4 provided that they have a minimum *fire separation distance* of 4 feet (1220 mm).
4. Roof coverings over terraces and patios in occupancies in Group R-3 shall be exempt from the area limitations of Table 2609.4 and shall be permitted with light-transmitting plastics.

TABLE 2609.4 AREA LIMITATIONS FOR LIGHT-TRANSMITTING PLASTIC ROOF PANELS

CLASS OF PLASTIC	MAXIMUM AREA OF INDIVIDUAL ROOF PANELS (square feet)	MAXIMUM AGGREGATE AREA OF ROOF PANELS (percent of floor area)
CC1	300	30
CC2	100	25

For SI: 1 square foot = 0.0929 m^2.

SECTION 2610 LIGHT-TRANSMITTING PLASTIC SKYLIGHT GLAZING

2610.1 Light-transmitting plastic glazing of skylight assemblies. Skylight assemblies glazed with light-transmitting plastic shall conform to the provisions of this section and Section 2606.

Exception: Skylights in which the light-transmitting plastic conforms to the required roof-covering class in accordance with Section 1505.

2610.1.1 Unit skylights. *Unit skylights* glazed with light-transmitting plastic shall comply with Section 2405.5.

2610.2 Mounting. The light-transmitting plastic shall be mounted above the plane of the roof on a curb constructed in accordance with the requirements for the type of construction classification, but not less than 4 inches (102 mm) above the plane of the roof. Edges of the light-transmitting plastic skylights or domes shall be protected by metal or other approved noncombustible material, or the light transmitting plastic dome or skylight shall be shown to be able to resist ignition where exposed at the edge to a flame from a Class B brand as described in ASTM E108 or UL 790. The Class B brand test shall be conducted on a skylight that is elevated to a height as specified in the manufacturer's installation instructions, but not less than 4 inches (102 mm).

Exceptions:

1. Curbs shall not be required for skylights used on roofs having a minimum slope of three units vertical in 12 units horizontal (25-percent slope) in occupancies in Group R-3 and on buildings with a nonclassified *roof covering*.

2. The metal or noncombustible edge material is not required where nonclassified roof coverings are permitted.

2610.3 Slope. Flat or corrugated light-transmitting plastic skylights shall slope not less than four units vertical in 12 units horizontal (4:12). Dome-shaped skylights shall rise above the mounting flange a minimum distance equal to 10 percent of the maximum width of the dome but not less than 3 inches (76 mm).

Exception: Skylights that pass the Class B Burning Brand Test specified in ASTM E108 or UL 790.

2610.4 Maximum area of skylights. Each skylight shall have a maximum area within the curb of 100 square feet (9.3 m^2).

Exception: The area limitation shall not apply where the building is equipped throughout with an *automatic sprinkler system* in accordance with Section 903.3.1.1 or the building is equipped with smoke and heat vents in accordance with Section 910.

2610.5 Aggregate area of skylights. The aggregate area of skylights shall not exceed $33^1/_3$ percent of the floor area of the room or space sheltered by the roof in which such skylights are installed where Class CC1 materials are utilized, and 25 percent where Class CC2 materials are utilized.

Exception: The aggregate area limitations of light-transmitting plastic skylights shall be increased 100 percent beyond the limitations set forth in this section where the building is equipped throughout with an *automatic sprinkler system* in accordance with Section 903.3.1.1 or the building is equipped with smoke and heat vents in accordance with Section 910.

2610.6 Separation. Skylights shall be separated from each other by a distance of not less than 4 feet (1219 mm) measured in a horizontal plane.

Exceptions:

1. Buildings equipped throughout with an *automatic sprinkler system* in accordance with Section 903.3.1.1.
2. In Group R-3, multiple skylights located above the same room or space with a combined area not exceeding the limits set forth in Section 2610.4.

2610.7 Location. Where *exterior wall* openings are required to be protected in accordance with Section 705, a skylight shall not be installed within 6 feet (1829 mm) of such *exterior wall.*

2610.8 Combinations of roof panels and skylights. Combinations of light-transmitting plastic roof panels and skylights shall be subject to the area and percentage limitations and separation requirements applicable to roof panel installations.

SECTION 2611
LIGHT-TRANSMITTING PLASTIC INTERIOR SIGNS

2611.1 General. Light-transmitting plastic interior signs shall be limited as specified in Sections 2606 and 2611.2 through 2611.4.

Exception: Light-transmitting plastic interior wall signs in *covered and open mall buildings* shall comply with Section 402.6.4.

2611.2 Maximum area. The aggregate area of all light-transmitting plastics shall not exceed 24 square feet (2.23 m^2).

Exception: In buildings equipped throughout with an automatic sprinkler system in accordance with Section 903.3.1.1, the aggregate area of light-transmitting plastics shall not exceed 100 square feet (9.29 m^2), provided that all plastics are Class CC1 in accordance with Section 2606.4.

2611.3 Separation. Signs exceeding the aggregate area of Section 2611.2 shall be separated from each other by not less than 4 feet (1219 mm) horizontally and 8 feet (2438 mm) vertically.

2611.4 Encasement. Backs of wall-mounted signs and non-illuminated portions of all signs regulated by this section shall be fully encased in metal.

SECTION 2612
PLASTIC COMPOSITES

[BS] 2612.1 General. Plastic composites shall consist of either wood/plastic composites or plastic lumber. Plastic composites shall comply with the provisions of this code and with the additional requirements of Section 2612.

[BS] 2612.2 Labeling. Plastic composite deck boards and *stair* treads, or their packaging, shall bear a *label* that indicates compliance with ASTM D7032 and includes the allowable load and maximum allowable span determined in accordance with ASTM D7032. Plastic composite *handrails* and *guards*, or their packaging, shall bear a label that indicates compliance with ASTM D7032 and includes the maximum allowable span determined in accordance with ASTM D7032.

2612.3 Flame spread index. Plastic composite deck boards, *stair* treads, *handrails* and *guards* shall exhibit a *flame spread index* not exceeding 200 when tested in accordance with ASTM E84 or UL 723 with the test specimen remaining in place during the test.

Exception: Materials determined to be noncombustible in accordance with Section 703.3.

[BS] 2612.4 Termite and decay resistance. Where required by Section 2304.12, plastic composite deck boards, *stair* treads, *handrails* and *guards* containing wood, cellulosic or any other biodegradable materials shall be termite and *decay resistant* as determined in accordance with ASTM D7032.

[BS] 2612.5 Construction requirements. Plastic composites meeting the requirements of Section 2612 shall be permitted to be used as exterior deck boards, *stair* treads, *handrails* and *guards* where combustible construction is permitted.

[BS] 2612.5.1 Span rating. Plastic composites used as exterior deck boards shall have a span rating determined in accordance with ASTM D7032.

[BS] 2612.6 Plastic composite deck boards, stair treads, handrails and guards. Plastic composite deck boards, *stair* treads, *handrails* and *guards* shall be installed in accordance with this code and the manufacturer's instructions.

SECTION 2613 FIBER-REINFORCED POLYMER

2613.1 General. The provisions of this section shall govern the requirements and uses of *fiber-reinforced polymer* in and on buildings and structures.

2613.2 Labeling and identification. Packages and containers of *fiber-reinforced polymer* and their components delivered to the job site shall bear the *label* of an *approved agency* showing the manufacturer's name, product listing, product identification and information sufficient to determine that the end use will comply with the code requirements.

2613.3 Interior finishes. *Fiber-reinforced polymer* used as *interior finishes, decorative materials or trim* shall comply with Chapter 8.

2613.3.1 Foam plastic cores. *Fiber-reinforced polymer* used as *interior finish* and that contains foam plastic cores shall comply with Chapter 8 and this chapter.

2613.4 Light-transmitting materials. *Fiber-reinforced polymer* used as light-transmitting materials shall comply with Sections 2606 through 2611 as required for the specific application.

2613.5 Exterior use. *Fiber-reinforced polymer* shall be permitted to be installed on the *exterior walls* of buildings of any type of construction where such polymers meet the requirements of Section 2603.5. *Fireblocking* shall be installed in accordance with Section 718.

Exceptions:

1. Compliance with Section 2603.5 is not required where all of the following conditions are met:
   - 1.1. The *fiber-reinforced polymer* shall not exceed an aggregate total of 20 percent of the area of the specific wall to which it is attached, and single architectural elements shall not exceed 10 percent of the area of the specific wall to which it is attached, and no contiguous sets of architectural elements shall not exceed 10 percent of the area of the specific wall to which they are attached.
   - 1.2. The *fiber-reinforced polymer* shall have a *flame spread index* of 25 or less. The *flame spread index* requirement shall not be required for coatings or paints having a thickness of less than 0.036 inch (0.9 mm) that are applied directly to the surface of the *fiber-reinforced polymer*.
   - 1.3. *Fireblocking* complying with Section 718.2.6 shall be installed.
   - 1.4. The *fiber-reinforced polymer* shall be installed directly to a noncombustible substrate or be separated from the *exterior wall* by one of the following materials: corrosion-resistant steel having a minimum base metal thickness of 0.016 inch (0.41 mm) at any point, aluminum having a minimum thickness of 0.019 inch (0.5 mm) or other approved noncombustible material.
2. Compliance with Section 2603.5 is not required where the *fiber-reinforced polymer* is installed on buildings that are 40 feet (12 190 mm) or less above grade and the following conditions are met:
   - 2.1. The *fiber-reinforced polymer* shall meet the requirements of Section 1405.1.
   - 2.2. Where the *fire separation distance* is 5 feet (1524 mm) or less, the area of the fiber-reinforced polymer shall not exceed 10 percent of the wall area. Where the *fire separation distance* is greater than 5 feet (1524 mm), the area of the *exterior wall* coverage using *fiber-reinforced polymer* shall not be limited.
   - 2.3. The *fiber-reinforced polymer* shall have a *flame spread index* of 200 or less. The *flame spread index* requirements do not apply to coatings or paints having a thickness of less than 0.036 inch (0.9 mm) that are applied directly to the surface of the *fiber-reinforced polymer*.
   - 2.4. *Fireblocking* complying with Section 718.2.6 shall be installed.

SECTION 2614 REFLECTIVE PLASTIC CORE INSULATION

2614.1 General. The provisions of this section shall govern the requirements and uses of reflective plastic core insulation in buildings and structures. Reflective plastic core insulation shall comply with the requirements of Section 2614 and of Section 2614.3 or 2614.4.

2614.2 Identification. Packages and containers of reflective plastic core insulation delivered to the job site shall show the manufacturer's or supplier's name, product identification and information sufficient to determine that the end use will comply with the code requirements.

2614.3 Surface-burning characteristics. Reflective plastic core insulation shall have a *flame spread index* of not more than 25 and a *smoke-developed index* of not more than 450 when tested in accordance with ASTM E84 or UL 723. The reflective plastic core insulation shall be tested at the maximum thickness intended for use. Test specimen preparation and mounting shall be in accordance with ASTM E2599.

2614.4 Room corner test heat release. Reflective plastic core insulation shall comply with the acceptance criteria of Section 803.1.1.1 when tested in accordance with NFPA 286 or UL 1715 in the manner intended for use and at the maximum thickness intended for use.

CHAPTER 27
ELECTRICAL

User note:

About this chapter: *Electrical systems and components are integral to most structures; therefore, it is necessary for the code to address their installation and protection. Structures depend on electricity for the operation of many life safety systems including fire alarm, smoke control and exhaust, fire suppression, fire command and communication systems. Since power supply to these systems is essential, Chapter 27 addresses where standby and emergency power must be provided.*

SECTION 2701
GENERAL

2701.1 Scope. The provisions of this chapter and NFPA 70 shall govern the design, construction, erection and installation of the electrical components, appliances, equipment and systems used in buildings and structures covered by this code. The *International Fire Code*, the *International Property Maintenance Code* and NFPA 70 shall govern the use and maintenance of electrical components, appliances, equipment and systems. The *International Existing Building Code* and NFPA 70 shall govern the alteration, repair, relocation, replacement and addition of electrical components, appliances, or equipment and systems.

SECTION 2702
EMERGENCY AND STANDBY POWER SYSTEMS

[F] 2702.1 General. Emergency power systems and standby power systems shall comply with Sections 2702.1.1 through 2702.1.8.

[F] 2702.1.1 Stationary generators. Stationary emergency and standby power generators required by this code shall be listed in accordance with UL 2200.

[F] 2702.1.2 Fuel-line piping protection. Fuel lines supplying a generator set inside a *high-rise building* shall be separated from areas of the *building* other than the room the generator is located in by one of the following methods:

1. A fire-resistant pipe-protection system that has been tested in accordance with UL 1489. The system shall be installed as tested and in accordance with the manufacturer's installation instructions, and shall have a rating of not less than 2 hours. Where the *building* is protected throughout with an automatic sprinkler system installed in accordance with Section 903.3.1.1, the required rating shall be reduced to 1 hour.
2. An assembly that has a fire-resistance rating of not less than 2 hours. Where the building is protected throughout with an automatic sprinkler system installed in accordance with Section 903.3.1.1, the required fire-resistance rating shall be reduced to 1 hour.
3. Other approved methods.

[F] 2702.1.3 Installation. Emergency power systems and standby power systems required by this code or the *International Fire Code* shall be installed in accordance with the *International Fire Code*, NFPA 70, NFPA 110 and NFPA 111.

[F] 2702.1.4 Load transfer. Emergency power systems shall automatically provide secondary power within 10 seconds after primary power is lost, unless specified otherwise in this code. Standby power systems shall automatically provide secondary power within 60 seconds after primary power is lost, unless specified otherwise in this code.

[F] 2702.1.5 Load duration. Emergency power systems and standby power systems shall be designed to provide the required power for a minimum duration of 2 hours without being refueled or recharged, unless specified otherwise in this code.

[F] 2702.1.6 Uninterruptable power source. An uninterrupted source of power shall be provided for equipment where required by the manufacturer's instructions, the listing, this code or applicable referenced standards.

[F] 2702.1.7 Interchangeability. Emergency power systems shall be an acceptable alternative for installations that require standby power systems.

[F] 2702.1.8 Group I-2 occupancies. In Group I-2 occupancies located in *flood hazard areas* established in Section 1612.3, where new essential electrical systems are installed, and where new essential electrical system generators are installed, the systems and generators shall be located and installed in accordance with ASCE 24. Where connections for hookup of temporary generators are provided, the connections shall be located at or above the elevation required in ASCE 24.

[F] 2702.2 Where required. Emergency and standby power systems shall be provided where required by Sections 2702.2.1 through 2702.2.19.

[F] 2702.2.1 Ambulatory care facilities. Essential electrical systems for *ambulatory care facilities* shall comply with Section 422.6.

[F] 2702.2.2 Elevators and platform lifts. Standby power shall be provided for elevators and platform lifts as required in Sections 1009.4.1, 1009.5, 3003.1, 3007.8 and 3008.8.

[F] 2702.2.3 Emergency responder communication coverage systems. Standby power shall be provided for in-building 2-way emergency responder communication coverage systems required in Section 918 and the *International Fire Code*. The standby power supply shall be capable of operating the in-building 2-way emergency responder communication coverage system at 100-percent system operation capacity for a duration of not less than 12 hours.

[F] 2702.2.4 Emergency voice/alarm communication systems. Standby power shall be provided for emergency voice/alarm communication systems in accordance with NFPA 72.

[F] 2702.2.5 Exhaust systems. Standby power shall be provided for common exhaust systems for domestic kitchens located in multistory structures as required in Section 505.5 of the *International Mechanical Code*. Standby power shall be provided for common exhaust systems for clothes dryers located in multistory structures as required in Section 504.11 of the *International Mechanical Code* and Section 614.11 of the *International Fuel Gas Code*.

[F] 2702.2.6 Exit signs. Emergency power shall be provided for exit signs as required in Section 1013.6.3. The system shall be capable of powering the required load for a duration of not less than 90 minutes.

[F] 2702.2.7 Gas detection system. Emergency or standby power shall be provided for gas detection systems in accordance with the *International Fire Code*.

[F] 2702.2.8 Group I-2 occupancies. Essential electrical systems for Group I-2 occupancies shall be in accordance with Section 407.11.

[F] 2702.2.9 Group I-3 occupancies. Emergency power shall be provided for *power-operated* doors and locks in Group I-3 occupancies as required in Section 408.4.2.

[F] 2702.2.10 Hazardous materials. Emergency or standby power shall be provided in occupancies with hazardous materials where required by the *International Fire Code*.

[F] 2702.2.11 High-rise buildings. Emergency and standby power shall be provided in high-rise buildings as required in Section 403.4.8.

[F] 2702.2.12 Hydrogen fuel gas rooms. Standby power shall be provided for hydrogen fuel gas rooms as required by the *International Fire Code*.

[F] 2702.2.13 Laboratory suites. Standby or emergency power shall be provided in accordance with Section 5004.7 of the *International Fire Code* where *laboratory suites* are located above the sixth *story above grade plane* or located in a story below grade plane.

[F] 2702.2.14 Means of egress illumination. Emergency power shall be provided for means of egress illumination as required in Section 1008.3. The system shall be capable of powering the required load for a duration of not less than 90 minutes.

[F] 2702.2.15 Membrane structures. Standby power shall be provided for auxiliary inflation systems in permanent membrane structures as required in Section 3102.8.2. Standby power shall be provided for a duration of not less than 4 hours. Auxiliary inflation systems in temporary *air-supported* and air-inflated membrane structures shall be provided in accordance with Section 3103.10.4 of the *International Fire Code*.

[F] 2702.2.16 Semiconductor fabrication facilities. Emergency power shall be provided for semiconductor fabrication facilities as required in Section 415.11.11.

[F] 2702.2.17 Smoke control systems. Standby power shall be provided for smoke control systems as required in Sections 404.7, 909.11, 909.20.7.2 and 909.21.5.

[F] 2702.2.18 Special purpose horizontal sliding, accordion or folding doors. Standby power shall be provided for special purpose horizontal sliding, accordion or folding doors as required in Section 1010.3.3. The standby power supply shall have a capacity to operate not fewer than 50 closing cycles of the door.

[F] 2702.2.19 Underground buildings. Emergency and standby power shall be provided in underground buildings as required in Section 405.

[F] 2702.3 Critical circuits. Required critical circuits shall be protected using one of the following methods:

1. Cables, used for survivability of required critical circuits, that are listed in accordance with UL 2196 and have a *fire-resistance rating* of not less than 1 hour.
2. *Electrical circuit protective systems* having a *fire-resistance rating* of not less than 1 hour. *Electrical circuit protective systems* are installed in accordance with their listing requirements.
3. Construction having a *fire-resistance rating* of not less than 1 hour.

[F] 2702.4 Maintenance. Emergency and standby power systems shall be maintained and tested in accordance with the *International Fire Code*.

CHAPTER 28

MECHANICAL SYSTEMS

User note:

About this chapter: *Mechanical systems are a key element of any building. Chapter 28 regulates such systems by linking to the* International Mechanical Code® and International Fuel Gas Code®, *where details of mechanical system requirements are provided.*

SECTION 2801 GENERAL

[M] 2801.1 Scope. The provisions of this chapter, the *International Mechanical Code* and the *International Fuel Gas Code* shall govern the design, construction, erection and installation of mechanical appliances, equipment and systems used in buildings and structures covered by this code. Masonry chimneys, fireplaces and barbecues shall comply with the *International Mechanical Code* and Chapter 21 of this code. The *International Fire Code*, the *International Property Maintenance Code*, the *International Mechanical Code* and the *International Fuel Gas Code* shall govern the use and maintenance of mechanical components, appliances, equipment and systems. The *International Existing Building Code*, the *International Mechanical Code* and the *International Fuel Gas Code* shall govern the alteration, repair, relocation, replacement and addition of mechanical components, appliances, equipment and systems.

CHAPTER 29
PLUMBING SYSTEMS

User note:

About this chapter: *Plumbing systems are another key element of any building. Chapter 29 provides the necessary number of plumbing fixtures, including water closets, lavatories, bathtubs and showers. The quality and design of each fixture must be in accordance with the* International Plumbing Code®.

SECTION 2901
GENERAL

[P] 2901.1 Scope. The provisions of this chapter and the *International Plumbing Code* shall govern the design, construction, erection and installation of plumbing components, appliances, equipment and systems used in *buildings* and structures covered by this code. Toilet and bathing rooms shall be constructed in accordance with Section 1210. Private sewage disposal systems shall conform to the *International Private Sewage Disposal Code*. The *International Fire Code*, the *International Property Maintenance Code* and the *International Plumbing Code* shall govern the use and maintenance of plumbing components, appliances, equipment and systems. The *International Existing Building Code* and the *International Plumbing Code* shall govern the alteration, *repair*, relocation, replacement and *addition* of plumbing components, *appliances*, *equipment* and systems.

SECTION 2902
MINIMUM PLUMBING FACILITIES

[P] 2902.1 Minimum number of fixtures. Plumbing fixtures shall be provided in the minimum number as shown in Table 2902.1 based on the actual use of the building or space. Uses not shown in Table 2902.1 shall be considered individually by the code official. The number of occupants shall be determined by this code.

[P] 2902.1.1 Fixture calculations. To determine the *occupant load* of each sex, the total *occupant load* shall be divided in half. To determine the required number of fixtures, the fixture ratio or ratios for each fixture type shall be applied to the *occupant load* of each sex in accordance with Table 2902.1. Fractional numbers resulting from applying the fixture ratios of Table 2902.1 shall be rounded up to the next whole number. For calculations involving multiple occupancies, such fractional numbers for each occupancy shall first be summed and then rounded up to the next whole number.

Exceptions:

1. The total *occupant load* shall not be required to be divided in half where approved statistical data indicates a distribution of the sexes of other than 50 percent of each sex.
2. Where multiple-user facilities are designed to serve all genders, the minimum fixture count shall be calculated 100 percent, based on total *occupant load*. In such multiple-user user facilities, each fixture type shall be in accordance with ICC A117.1 and each urinal that is provided shall be located in a stall.
3. Distribution of the sexes is not required where single-user water closets and bathing room fixtures are provided in accordance with Section 2902.1.2.

[P] 2902.1.2 Single-user toilet and bathing room fixtures. The plumbing fixtures located in single-user toilet and bathing rooms, including family or assisted-use toilet and bathing rooms that are required by Section 1110.2.1, shall contribute toward the total number of required plumbing fixtures for a building or tenant space. Single-user toilet and bathing rooms, and family or assisted-use toilet rooms and bathing rooms shall be identified as being available for use by all persons regardless of their sex.

The total number of fixtures shall be permitted to be based on the required number of separate facilities or based on the aggregate of any combination of single-user or separate facilities.

[P] TABLE 2902.1
MINIMUM NUMBER OF REQUIRED PLUMBING FIXTURES[a] (See Sections 2902.1.1 and 2902.2)

No.	CLASSIFICATION	DESCRIPTION	WATER CLOSETS (URINALS SEE SECTION 424.2 OF THE *INTERNATIONAL PLUMBING CODE*)		LAVATORIES		BATHTUBS/ SHOWERS	DRINKING FOUNTAINS (SEE SECTION 410 OF THE *INTERNATIONAL PLUMBING CODE*)	OTHER
			Male	Female	Male	Female			
1	Assembly	Theaters and other buildings for the performing arts and motion pictures[d]	1 per 125	1 per 65	1 per 200		—	1 per 500	1 service sink
		Nightclubs, bars, taverns, dance halls and buildings for similar purposes[d]	1 per 40	1 per 40	1 per 75		—	1 per 500	1 service sink
		Restaurants, banquet halls and food courts[d]	1 per 75	1 per 75	1 per 200		—	1 per 500	1 service sink
		Casino gaming areas	1 per 100 for the first 400 and 1 per 250 for the remainder exceeding 400	1 per 50 for the first 400 and 1 per 150 for the remainder exceeding 400	1 per 250 for the first 750 and 1 per 500 for the remainder exceeding 750		—	1 per 1,000	1 service sink
		Auditoriums without permanent seating, art galleries, exhibition halls, museums, lecture halls, libraries, arcades and gymnasiums[d]	1 per 125	1 per 65	1 per 200		—	1 per 500	1 service sink
		Passenger terminals and transportation facilities[d]	1 per 500	1 per 500	1 per 750		—	1 per 1,000	1 service sink
		Places of worship and other religious services[d]	1 per 150	1 per 75	1 per 200		—	1 per 1,000	1 service sink
		Coliseums, arenas, skating rinks, pools and tennis courts for indoor sporting events and activities	1 per 75 for the first 1,500 and 1 per 120 for the remainder exceeding 1,500	1 per 40 for the first 1,520 and 1 per 60 for the remainder exceeding 1,520	1 per 200	1 per 150	—	1 per 1,000	1 service sink
		Stadiums, amusement parks, bleachers and grandstands for outdoor sporting events and activities[f]	1 per 75 for the first 1,500 and 1 per 120 for the remainder exceeding 1,500	1 per 40 for the first 1,520 and 1 per 60 for the remainder exceeding 1,520	1 per 200	1 per 150	—	1 per 1,000	1 service sink

(continued)

[P] TABLE 2902.1—continued
MINIMUM NUMBER OF REQUIRED PLUMBING FIXTURES[a] (See Sections 2902.1.1 and 2902.2)

No.	CLASSIFICATION	DESCRIPTION	WATER CLOSETS (URINALS SEE SECTION 424.2 OF THE *INTERNATIONAL PLUMBING CODE*)		LAVATORIES		BATHTUBS/ SHOWERS	DRINKING FOUNTAINS (SEE SECTION 410 OF THE *INTERNATIONAL PLUMBING CODE*)	OTHER
			Male	Female	Male	Female			
2	Business	Buildings for the transaction of business, professional services, other services involving merchandise, office buildings, banks, light industrial, ambulatory care and similar uses	1 per 25 for the first 50 and 1 per 50 for the remainder exceeding 50		1 per 40 for the first 80 and 1 per 80 for the remainder exceeding 80		—	1 per 100	1 service sink[e]
3	Educational	Educational facilities	1 per 50		1 per 50		—	1 per 100	1 service sink
4	Factory and industrial	Structures in which occupants are engaged in work fabricating, assembly or processing of products or materials	1 per 100		1 per 100		—	1 per 400	1 service sink
5	Institutional	Custodial care facilities	1 per 10		1 per 10		1 per 8	1 per 100	1 service sink
		Medical care recipients in hospitals and nursing homes[b]	1 per room[c]		1 per room[c]		1 per 15	1 per 100	1 service sink
		Employees in hospitals and nursing homes[b]	1 per 25		1 per 35		—	1 per 100	—
		Visitors in hospitals and nursing homes	1 per 75		1 per 100		—	1 per 500	—
		Prisons[b]	1 per cell		1 per cell		1 per 15	1 per 100	1 service sink
		Reformatories, detention centers and correctional centers[b]	1 per 15		1 per 15		1 per 15	1 per 100	1 service sink
		Employees in reformitories, detention centers and correctional centers[b]	1 per 25		1 per 35		—	1 per 100	—
		Adult day care and child day care	1 per 15		1 per 15		1	1 per 100	1 service sink
6	Mercantile	Retail stores, service stations, shops, salesrooms, markets and shopping centers	1 per 500		1 per 750		—	1 per 1,000	1 service sink[e]

(continued)

[P] TABLE 2902.1—continued
MINIMUM NUMBER OF REQUIRED PLUMBING FIXTURES[a] (See Sections 2902.1.1 and 2902.2)

No.	CLASSIFICATION	DESCRIPTION	WATER CLOSETS (URINALS SEE SECTION 424.2 OF THE *INTERNATIONAL PLUMBING CODE*)		LAVATORIES		BATHTUBS/ SHOWERS	DRINKING FOUNTAINS (SEE SECTION 410 OF THE *INTERNATIONAL PLUMBING CODE*)	OTHER
			Male	Female	Male	Female			
7	Residential	Hotels, motels, boarding houses (transient)	1 per sleeping unit		1 per sleeping unit		1 per sleeping unit	—	1 service sink
		Dormitories, fraternities, sororities and boarding houses (not transient)	1 per 10		1 per 10		1 per 8	1 per 100	1 service sink
		Apartment house	1 per dwelling unit		1 per dwelling unit		1 per dwelling unit	—	1 kitchen sink per dwelling unit; 1 automatic clothes washer connection per 20 dwelling units
		One- and two-family dwellings and lodging houses with five or fewer guestrooms	1 per dwelling unit		1 per 10		1 per dwelling unit	—	1 kitchen sink per dwelling unit; 1 automatic clothes washer connection per dwelling unit
		Congregate living facilities with 16 or fewer persons	1 per 10		1 per 10		1 per 8	1 per 100	1 service sink
8	Storage	Structures for the storage of goods, warehouses, storehouses and freight depots, low and moderate hazard	1 per 100		1 per 100		—	1 per 1,000	1 service sink

a. The fixtures shown are based on one fixture being the minimum required for the number of persons indicated or any fraction of the number of persons indicated. The number of occupants shall be determined by this code.
b. Toilet facilities for employees shall be separate from facilities for inmates or care recipients.
c. A single-occupant toilet room with one water closet and one lavatory serving not more than two adjacent patient sleeping units shall be permitted, provided that each patient sleeping unit has direct access to the toilet room and provisions for privacy for the toilet room user are provided.
d. The occupant load for seasonal outdoor seating and entertainment areas shall be included when determining the minimum number of facilities required.
e. For business and mercantile classifications with an occupant load of 15 or fewer, a service sink shall not be required.
f. The required number and type of plumbing fixtures for outdoor swimming pools shall be in accordance with Section 609 of the *International Swimming Pool and Spa Code*.

[P] 2902.1.3 Lavatory distribution. Where two or more toilet rooms are provided for each sex, the required number of lavatories shall be distributed proportionately to the required number of water closets.

[P] 2902.2 Separate facilities. Where plumbing fixtures are required, separate facilities shall be provided for each sex.

Exceptions:

1. Separate facilities shall not be required for *dwelling units* and *sleeping units*.
2. Separate facilities shall not be required in structures or tenant spaces with a total *occupant load*, including both employees and customers, of 15 or fewer.
3. Separate facilities shall not be required in mercantile occupancies in which the maximum *occupant load* is 100 or fewer.
4. Separate facilities shall not be required in business occupancies in which the maximum *occupant load* is 25 or fewer.
5. Separate facilities shall not be required to be designated by sex where single-user toilets rooms are provided in accordance with Section 2902.1.2.
6. Separate facilities shall not be required where rooms having both water closets and lavatory fixtures are designed for use by both sexes and privacy for water closets are installed in accordance with Section 405.3.4 of the *International Plumbing Code*. Urinals shall be located in an area visually separated from the remainder of the facility or each urinal that is provided shall be located in a stall.

[P] 2902.2.1 Family or assisted-use toilet facilities serving as separate facilities. Where a building or tenant space requires a separate toilet facility for each sex and each toilet facility is required to have only one water closet, two family or assisted-use toilet facilities shall be permitted to serve as the required separate facilities. Family or assisted-use toilet facilities shall not be required to be identified for exclusive use by either sex as required by Section 2902.4.

[P] 2902.3 Employee and public toilet facilities. For structures and tenant spaces intended for public utilization, customers, patrons and visitors shall be provided with public toilet facilities. Employees associated with structures and tenant spaces shall be provided with toilet facilities. The number of plumbing fixtures located within the required toilet facilities shall be provided in accordance with Section 2902 for all users. Employee toilet facilities shall be either separate or combined employee and public toilet facilities.

Exception: Public toilet facilities shall not be required for:

1. Parking garages where operated without parking attendants.
2. Structures and tenant spaces intended for quick transactions, including takeout, pickup and drop-off, having a public access area less than or equal to 300 square feet (28 m^2).

[P] 2902.3.1 Access. The route to the public toilet facilities required by Section 2902.3 shall not pass through kitchens, storage rooms or closets. Access to the required facilities shall be from within the building or from the exterior of the building. The public shall have access to the required toilet facilities at all times that the building is occupied.

[P] 2902.3.2 Prohibited toilet room location. Toilet rooms shall not open directly into a room used for the preparation of food for service to the public.

[P] 2902.3.3 Location of toilet facilities in occupancies other than malls. In occupancies other than covered and *open mall buildings*, the required public and employee toilet facilities shall be located not more than one *story* above or below the space required to be provided with toilet facilities, and the path of travel to such facilities shall not exceed a distance of 500 feet (152 m).

Exceptions:

1. The location and maximum distances of travel to required employee facilities in factory and industrial *occupancies* shall be permitted to exceed that required by this section, provided that the location and maximum distances of travel are *approved*.
2. The location and maximum distances of travel to required public and employee facilities in Group S *occupancies* shall be permitted to exceed that required by this section, provided that the location and maximum distances of travel are *approved*.

[P] 2902.3.4 Location of toilet facilities in malls. In covered and *open mall buildings*, the required public and employee toilet facilities shall be located not more than one *story* above or below the space required to be provided with toilet facilities, and the path of travel to such facilities shall not exceed a distance of 300 feet (91 m). In mall buildings, the required facilities shall be based on total square footage (m^2) within a *covered mall building* or within the perimeter line of an *open mall building*, and facilities shall be installed in each individual store or in a central toilet area located in accordance with this section. The maximum distance of travel to central toilet facilities in mall buildings shall be measured from the main entrance of any store or tenant space. In mall buildings, where employees' toilet facilities are not provided in the individual store, the maximum distance of travel shall be measured from the employees' work area of the store or tenant space.

[P] 2902.3.5 Pay facilities. Where pay facilities are installed, such facilities shall be in excess of the required minimum facilities. Required facilities shall be free of charge.

[P] 2902.3.6 Door locking. Where a toilet room is provided for the use of multiple occupants, the egress door for the room shall not be lockable from the inside of the room. This section does not apply to family or assisted-use toilet rooms.

[P] 2902.4 Signage. Required public facilities shall be provided with signs that designate the sex as required by Section 2902.2. Signs shall be readily visible and located near the entrance to each toilet facility. Signs for *accessible* toilet facilities shall comply with Section 1112.

[P] 2902.4.1 Directional signage. Directional signage indicating the route to the required public toilet facilities shall be posted in a lobby, corridor, aisle or similar space, such that the sign can be readily seen from the main entrance to the building or tenant space.

[P] 2902.5 Drinking fountain location. Drinking fountains shall not be required to be located in individual tenant spaces provided that public drinking fountains are located within a distance of travel of 500 feet (152 m) of the most remote location in the tenant space and not more than one story above or below the tenant space. Where the tenant space is in a covered or open mall, such distance shall not exceed 300 feet (91 m).

[P] 2902.6 Small occupancies. Drinking fountains shall not be required for an *occupant load* of 15 or fewer.

[P] 2902.7 Service sink location. Service sinks shall not be required to be located in individual tenant spaces in a covered mall provided that service sinks are located within a distance of travel of 300 feet (91 m) of the most remote location in the tenant space and not more than one story above or below the tenant space. Service sinks shall be located on an *accessible route*.

[P] 2903 INSTALLATION OF FIXTURES

[P] 2903.1 Setting. Fixtures shall be set level and in proper alignment with reference to adjacent walls.

[P] 2903.1.1 Water closets, urinals, lavatories and bidets. A water closet, urinal, lavatory or bidet shall not be set closer than 15 inches (381 mm) from its center to any side wall, partition, vanity or other obstruction. Where partitions or other obstructions do not separate adjacent fixtures, fixtures shall not be set closer than 30 inches (762 mm) center to center between adjacent fixtures. There shall be not less than a 21-inch (533 mm) clearance in front of a water closet, urinal, lavatory or bidet to any wall, fixture or door. Water closet compartments shall be not less than 30 inches (762 mm) in width and not less than 60 inches (1524 mm) in depth for floor-mounted water closets and not less than 30 inches (762 mm) in width and 56 inches (1422 mm) in depth for wall-hung water closets.

Exception: An accessible children's water closet shall be set not closer than 12 inches (305 mm) from its center to the required partition or to the wall on one side.

[P] 2903.1.2 Public lavatories. In employee and public toilet rooms, the required lavatory shall be located in the same room as the required water closet.

[P] 2903.1.3 Location of fixtures and piping. Piping, fixtures or equipment shall not be located in such a manner as to interfere with the normal operation of windows, doors or other means of egress openings.

[P] 2903.1.4 Water closet compartment. Each water closet utilized by the public or employees shall occupy a separate compartment with walls or partitions and a door enclosing the fixtures to ensure privacy.

Exceptions:

1. Water closet compartments shall not be required in a single-occupant toilet room with a lockable door.
2. Toilet rooms located in child day care facilities and containing two or more water closets shall be permitted to have one water closet without an enclosing compartment.
3. This provision is not applicable to toilet areas located within Group I-3 housing areas.

[P] 2903.1.5 Urinal partitions. Each urinal utilized by the public or employees shall occupy a separate area with walls or partitions to provide privacy. The horizontal dimension between walls or partitions at each urinal shall be not less than 30 inches (762 mm). The walls or partitions shall begin at a height not greater than 12 inches (305 mm) from and extend not less than 60 inches (1524 mm) above the finished floor surface. The walls or partitions shall extend from the wall surface at each side of the urinal not less than 18 inches (457 mm) or to a point not less than 6inches (152 mm) beyond the outermost front lip of the urinal measured from the finished backwall surface, whichever is greater.

Exceptions:

1. Urinal partitions shall not be required in a single-occupant or family/assisted-use toilet room with a lockable door.
2. Toilet rooms located in child day care facilities and containing two or more urinals shall be permitted to have one urinal without partitions.

CHAPTER 30

ELEVATORS AND CONVEYING SYSTEMS

User note:

About this chapter: *Chapter 30 contains the provisions that regulate vertical and horizontal transportation and material-handling systems installed in buildings. This chapter also provides several elements that protect occupants and assist emergency responders during fires.*

SECTION 3001
GENERAL

3001.1 Scope. This chapter governs the design, construction, installation, *alteration* and repair of elevators and conveying systems and their components.

3001.2 Emergency elevator communication systems for the deaf, hard of hearing and speech impaired. An emergency two-way communication system shall be provided. The system shall provide visible text and audible modes that meet all of the following requirements:

1. When operating in each mode, include a live interactive system that allows back and forth conversation between the elevator occupants and emergency personnel.
2. Is operational when the elevator is operational.
3. Allows elevator occupants to select the text-based or audible mode depending on their communication needs to interact with emergency personnel.

3001.3 Referenced standards. Except as otherwise provided for in this code, the design, construction, installation, alteration, repair and maintenance of elevators and conveying systems and their components shall conform to the applicable standard specified in Table 3001.3 and ASCE 24 for construction in *flood hazard areas* established in Section 1612.3.

TABLE 3001.3
ELEVATORS AND CONVEYING SYSTEMS AND COMPONENTS

TYPE	STANDARD
Automotive lifts	ALI ALCTV
Belt manlifts	ASME A90.1
Conveyors and related equipment	ASME B20.1
Elevators, escalators, dumbwaiters, moving walks, material lifts	ASME A17.1/CSA B44, ASME A17.7/CSA B44.7
Industrial scissor lifts	ANSI MH29.1
Platform lifts, stairway chairlifts, wheelchair lifts	ASME A18.1

3001.4 Accessibility. Passenger elevators required to be accessible or to serve as part of an *accessible* means of egress shall comply with Sections 1009 and 1110.8.

3001.5 Change in use. A change in use of an elevator from freight to passenger, passenger to freight, or from one freight class to another freight class shall comply with Section 8.7 of ASME A17.1/CSA B44.

SECTION 3002
HOISTWAY ENCLOSURES

3002.1 Hoistway enclosure protection. Elevator, dumbwaiter and other hoistway enclosures shall be *shaft enclosures* complying with Sections 712 and 713.

3002.1.1 Opening protectives. Openings in hoistway enclosures shall be protected as required in Chapter 7.

Exception: The elevator car doors and the associated hoistway enclosure doors at the floor level designated for recall in accordance with Section 3003.2 shall be permitted to remain open during Phase I Emergency Recall Operation.

3002.1.2 Hardware. Hardware on opening protectives shall be of an *approved* type installed as tested, except that *approved* interlocks, mechanical locks and electric contacts, door and gate electric contacts and door-operating mechanisms shall be exempt from the fire test requirements.

3002.2 Number of elevator cars in a hoistway. Where four or more elevator cars serve all or the same portion of a building, the elevators shall be located in not fewer than two separate hoistways. Not more than four elevator cars shall be located in any single hoistway enclosure.

3002.3 Emergency signs. An *approved* pictorial sign of a standardized design shall be posted adjacent to each elevator call station on all floors instructing occupants to use the exit stairways and not to use the elevators in case of fire. The sign shall read: IN CASE OF FIRE, ELEVATORS ARE OUT OF SERVICE. USE EXIT STAIRS.

Exceptions:

1. The emergency sign shall not be required for elevators that are part of an *accessible* means of egress complying with Section 1009.4.
2. The emergency sign shall not be required for elevators that are used for occupant self-evacuation in accordance with Section 3008.

3002.4 Elevator car to accommodate ambulance stretcher. Where elevators are provided in buildings four or more *stories* above, or four or more *stories* below, *grade plane*, not fewer than one elevator shall be provided for fire department emergency access to all floors. The elevator car shall be of such a size and arrangement to accommodate an ambulance stretcher 24 inches by 84 inches (610 mm by 2134 mm) with not less than 5-inch (127 mm) radius corners, in the horizontal, open position and shall be identified by the international symbol for emergency medical services (star of life). The symbol shall be not less than 3 inches (76 mm) in height and shall be placed inside on both sides of the hoistway door frame.

3002.5 Emergency doors. Where an elevator is installed in a single blind hoistway or on the outside of a building, there shall be installed in the blind portion of the hoistway or blank face of the building, an emergency door in accordance with ASME A17.1/CSA B44.

3002.6 Prohibited doors. Doors, other than hoistway doors and the elevator car door, shall be prohibited at the point of access to an elevator car unless such doors are readily openable from the car side without a key, tool, special knowledge or effort.

3002.7 Common enclosure with *stairway*. Elevators shall not be in a common *shaft enclosure* with a *stairway*.

> **Exception:** Elevators within *open parking garages need not be separated from stairway enclosures.*

3002.8 Glass in elevator enclosures. Glass in elevator enclosures shall comply with Section 2409.2.

3002.9 Plumbing and mechanical systems. Plumbing and mechanical systems shall not be located in an elevator hoistway enclosure.

> **Exception:** Floor drains, sumps and sump pumps shall be permitted at the base of the hoistway enclosure provided that they are indirectly connected to the plumbing system.

SECTION 3003 EMERGENCY OPERATIONS

[F] 3003.1 Standby power. In buildings and structures where standby power is required or furnished to operate an elevator, the operation shall be in accordance with Sections 3003.1.1 through 3003.1.4.

[F] 3003.1.1 Manual transfer. Standby power shall be manually transferable to all elevators in each bank.

[F] 3003.1.2 One elevator. Where only one elevator is installed, the elevator shall automatically transfer to standby power within 60 seconds after failure of normal power.

[F] 3003.1.3 Two or more elevators. Where two or more elevators are controlled by a common operating system, all elevators shall automatically transfer to standby power within 60 seconds after failure of normal power where the standby power source is of sufficient capacity to operate all elevators at the same time. Where the standby power source is not of sufficient capacity to operate all elevators at the same time, all elevators shall transfer to standby power in sequence, return to the designated landing and disconnect from the standby power source. After all elevators have been returned to the designated level, not less than one elevator shall remain operable from the standby power source.

[F] 3003.1.4 Venting. Where standby power is connected to elevators, the machine room *ventilation* or air conditioning shall be connected to the standby power source.

[F] 3003.2 Fire fighters' emergency operation. Elevators shall be provided with Phase I emergency recall operation and Phase II emergency in-car operation in accordance with ASME A17.1/CSA B44.

[F] 3003.3 Standardized fire service elevator keys. All elevators shall be equipped to operate with a standardized fire service elevator key in accordance with the *International Fire Code*.

SECTION 3004 CONVEYING SYSTEMS

3004.1 General. Escalators, moving walks, conveyors, personnel hoists and material hoists shall comply with the provisions of Sections 3004.2 through 3004.4.

3004.2 Escalators and moving walks. Escalators and moving walks shall be constructed of *approved* noncombustible and fire-retardant materials. This requirement shall not apply to electrical equipment, wiring, wheels, handrails and the use of $^{1}/_{28}$-inch (0.9 mm) wood *veneers* on balustrades backed up with noncombustible materials.

3004.2.1 Enclosure. Escalator floor openings shall be enclosed with *shaft enclosures* complying with Section 713.

3004.2.2 Escalators. Where provided in below-grade transportation stations, escalators shall have a clear width of not less than 32 inches (815 mm).

3004.3 Conveyors. Conveyors and conveying systems shall comply with ASME B20.1.

3004.3.1 Enclosure. Conveyors and related equipment connecting successive floors or levels shall be enclosed with *shaft enclosures* complying with Section 713.

3004.3.2 Conveyor safeties. Power-operated conveyors, belts and other material-moving devices shall be equipped with automatic limit switches that will shut off the power in an emergency and automatically stop all operation of the device.

3004.4 Personnel and material hoists. Personnel and material hoists shall be designed utilizing an *approved* method that accounts for the conditions imposed during the intended operation of the hoist device. The design shall include, but is not limited to, anticipated loads, structural stability, impact, vibration, stresses and seismic restraint. The design shall account for the construction, installation, operation and inspection of the hoist tower, car, machinery and control equipment, guide members and hoisting mechanism. Additionally, the design of personnel hoists shall include provisions for field testing and maintenance that will demonstrate that the hoist device functions in accordance with the design. Field tests shall be conducted upon the completion of an installation or following a major *alteration* of a personnel hoist.

SECTION 3005
MACHINE ROOMS

3005.1 Access. An *approved* means of access shall be provided to elevator machine rooms, control rooms, control spaces and machinery spaces.

3005.2 Venting. Elevator machine rooms, machinery spaces that contain the driving machine, and control rooms or spaces that contain the operation or motion controller for elevator operation shall be provided with an independent *ventilation* or air-conditioning system to protect against the overheating of the electrical equipment. The system shall be capable of maintaining temperatures within the range established for the elevator equipment.

3005.3 Pressurization. The elevator machine room, control rooms or control space with openings into a pressurized elevator hoistway shall be pressurized upon activation of a heat or smoke detector located in the elevator machine room, control room or control space.

3005.4 Machine rooms, control rooms, machinery spaces, and control spaces. The following rooms and spaces shall be enclosed with *fire barriers* constructed in accordance with Section 707 or *horizontal assemblies* constructed in accordance with Section 711, or both:

1. Machine rooms
2. Control rooms
3. Control spaces
4. Machinery spaces outside of the hoistway enclosure

The *fire-resistance rating* shall be not less than the required rating of the hoistway enclosure served by the machinery. Openings in the *fire barriers* shall be protected with assemblies having a *fire protection rating* not less than that required for the hoistway enclosure doors.

Exceptions:

1. For other than fire service access elevators and occupant evacuation elevators, where machine rooms, machinery spaces, control rooms and control spaces do not abut and do not have openings to the hoistway enclosure they serve, the *fire barriers* constructed in accordance with Section 707 or *horizontal assemblies* constructed in accordance with Section 711, or both, shall be permitted to be reduced to a 1-hour *fire-resistance rating*.
2. For other than fire service access elevators and occupant evacuation elevators, in buildings four *stories* or less above *grade plane* where machine room, machinery spaces, control rooms and control spaces do not abut and do not have openings to the hoistway enclosure they serve, the machine room, machinery spaces, control rooms and control spaces are not required to be fire-resistance rated.

3005.5 Shunt trip. Where elevator hoistways, elevator machine rooms, control rooms and control spaces containing elevator control equipment are protected with automatic sprinklers, a means installed in accordance with Section 21.4 of NFPA 72 shall be provided to automatically disconnect the main line power supply to the affected elevator prior to the application of water. This means shall not be self-resetting. The activation of automatic sprinklers outside the hoistway, machine room, machinery space, control room or control space shall not disconnect the main line power supply.

3005.6 Plumbing systems. Plumbing systems shall not be located in elevator equipment rooms.

SECTION 3006
ELEVATOR LOBBIES AND HOISTWAY OPENING PROTECTION

3006.1 General. Elevator hoistway openings and enclosed elevator lobbies shall be provided in accordance with the following:

1. Where hoistway opening protection is required by Section 3006.2, such protection shall be in accordance with Section 3006.3.
2. Where enclosed elevator lobbies are required for underground buildings, such lobbies shall comply with Section 405.4.3.
3. Where an *area of refuge* is required and an enclosed elevator lobby is provided to serve as an *area of refuge*, the enclosed elevator lobby shall comply with Section 1009.6.
4. Where fire service access elevators are provided, enclosed elevator lobbies shall comply with Section 3007.6.
5. Where occupant evacuation elevators are provided, enclosed elevator lobbies shall comply with Section 3008.6.

3006.2 Hoistway opening protection required. Elevator hoistway door openings shall be protected in accordance with Section 3006.3 where an elevator hoistway connects more than three *stories*, is required to be enclosed within a *shaft enclosure* in accordance with Section 712.1.1 and any of the following conditions apply:

1. The building is not protected throughout with an *automatic sprinkler system* in accordance with Section 903.3.1.1 or 903.3.1.2.
2. The building contains a Group I-1, Condition 2 occupancy.
3. The building contains a Group I-2 occupancy.
4. The building contains a Group I-3 occupancy.
5. The building is a high rise and the elevator hoistway is more than 75 feet (22 860 mm) in height. The height of the hoistway shall be measured from the *lowest floor* to the highest floor of the floors served by the hoistway.

Exceptions:

1. Protection of elevator hoistway door openings is not required where the elevator serves only *open parking garages* in accordance with Section 406.5.

2. Protection of elevator hoistway door openings is not required at the level(s) of exit discharge, provided that the level(s) of exit discharge is equipped with an *automatic sprinkler system* in accordance with Section 903.3.1.1.
3. Enclosed elevator lobbies and protection of elevator hoistway door openings are not required on levels where the elevator hoistway opens to the exterior.

3006.2.1 Rated corridors. Where *corridors* are required to be fire-resistance rated in accordance with Section 1020.2, elevator hoistway openings shall be protected in accordance with Section 3006.3.

3006.3 Hoistway opening protection. Where Section 3006.2 requires protection of the elevator hoistway door opening, the protection shall be provided by one of the following:

1. An enclosed elevator lobby shall be provided at each floor to separate the elevator hoistway *shaft enclosure* doors from each floor by *fire partitions* in accordance with Section 708. In addition, doors protecting openings in the elevator lobby enclosure walls shall comply with Section 716.2.2.1 as required for *corridor* walls. Penetrations of the enclosed elevator lobby by ducts and air transfer openings shall be protected as required for *corridors* in accordance with Section 717.5.4.1.
2. An enclosed elevator lobby shall be provided at each floor to separate the elevator hoistway *shaft enclosure* doors from each floor by *smoke partitions* in accordance with Section 710 where the building is equipped throughout with an *automatic sprinkler system* installed in accordance with Section 903.3.1.1 or 903.3.1.2. In addition, doors protecting openings in the *smoke partitions* shall comply with Sections 710.5.2.2, 710.5.2.3 and 716.2.6.1. Penetrations of the enclosed elevator lobby by ducts and air transfer openings shall be protected as required for *corridors* in accordance with Section 717.5.4.1.
3. Additional doors shall be provided at each elevator hoistway door opening in accordance with Section 3002.6. Such door shall comply with the smoke and draft control door assembly requirements in Section 716.2.2.1.1 when tested in accordance with UL 1784 without an artificial bottom seal.
4. The elevator hoistway shall be pressurized in accordance with Section 909.21.

3006.4 Means of egress. Elevator lobbies shall be provided with not less than one *means of egress* complying with Chapter 10 and other provisions in this code. Egress through an enclosed elevator lobby shall be permitted in accordance with Item 1 of Section 1016.2.

SECTION 3007
FIRE SERVICE ACCESS ELEVATOR

3007.1 General. Where required by Section 403.6.1, every floor above and including the lowest level of fire department vehicle access of the building shall be served by fire service access elevators complying with Sections 3007.1 through 3007.9. Except as modified in this section, fire service access elevators shall be installed in accordance with this chapter and ASME A17.1/CSA B44.

Exceptions:

1. Elevators that only service an open or enclosed parking garage and the lobby of the building shall not be required to serve as fire service access elevators.
2. The elevator shall not be required to serve the top floor of a building where that floor is utilized only for equipment for building systems.

3007.2 Automatic sprinkler system. The building shall be equipped throughout with an *automatic sprinkler system* in accordance with Section 903.3.1.1, except as otherwise permitted by Section 903.3.1.1.1 and as prohibited by Section 3007.2.1.

3007.2.1 Prohibited locations. Automatic sprinklers shall not be installed in machine rooms, elevator machinery spaces, control rooms, control spaces and elevator hoistways of fire service access elevators.

3007.2.2 Sprinkler system monitoring. The sprinkler system shall have a sprinkler control valve supervisory switch and water-flow-initiating device provided for each floor that is monitored by the building's *fire alarm system*.

3007.3 Water protection. Water from the operation of an automatic sprinkler system outside the enclosed lobby shall be prevented from infiltrating into the hoistway enclosure in accordance with an approved method.

3007.4 Shunt trip. Means for elevator shutdown in accordance with Section 3005.5 shall not be installed on elevator systems used for fire service access elevators.

3007.5 Hoistway enclosures. The fire service access elevator hoistway shall be located in a *shaft enclosure* complying with Section 713.

3007.5.1 Structural integrity of hoistway enclosures. The fire service access elevator hoistway enclosure shall comply with Sections 403.2.2.1 through 403.2.2.4.

3007.5.2 Hoistway lighting. When fire-fighters' emergency operation is active, the entire height of the hoistway shall be illuminated at not less than 1 footcandle (11 lux) as measured from the top of the car of each fire service access elevator.

3007.6 Fire service access elevator lobby. The fire service access elevator shall open into an enclosed fire service access elevator lobby in accordance with Sections 3007.6.1 through 3007.6.5. Egress is permitted through the enclosed elevator lobby in accordance with Item 1 of Section 1016.2.

Exception: Where a fire service access elevator has two entrances onto a floor, the second entrance shall be permitted to be protected in accordance with Section 3006.3.

3007.6.1 Access to interior exit stairway or ramp. The enclosed fire service access elevator lobby shall have *direct access* from the enclosed elevator lobby to an enclosure for an *interior exit stairway* or *ramp*.

Exception: Access to an *interior exit stairway* or *ramp* shall be permitted to be through a protected path of travel that has a level of fire protection not less than the elevator lobby enclosure. The protected path shall be separated from the enclosed elevator lobby through an opening protected by a smoke and draft control assembly in accordance Section 716.2.2.1.

3007.6.2 Lobby enclosure. The fire service access elevator lobby shall be enclosed with a *smoke barrier* having a *fire-resistance rating* of not less than 1 hour, except that lobby doorways shall comply with Section 3007.6.3.

Exception: Enclosed fire service access elevator lobbies are not required at the *levels of exit discharge*.

3007.6.3 Lobby doorways. Other than doors to the hoistway, elevator control room or elevator control space, each doorway to an enclosed fire service access elevator lobby shall be provided with a $^{3}/_{4}$-hour *fire door assembly* complying with Section 716. The *fire door assembly* shall comply with the smoke and draft control door assembly requirements of Section 716.2.2.1.1 and be tested in accordance with UL 1784 without an artificial bottom seal.

3007.6.4 Lobby size. Regardless of the number of fire service access elevators served by the same elevator lobby, the enclosed fire service access elevator lobby shall be not less than 150 square feet (14 m^2) in an area with a dimension of not less than 8 feet (2440 mm).

3007.6.5 Fire service access elevator symbol. A pictorial symbol of a standardized design designating which elevators are fire service access elevators shall be installed on each side of the hoistway door frame on the portion of the frame at right angles to the fire service access elevator lobby. The fire service access elevator symbol shall be designed as shown in Figure 3007.6.5 and shall comply with the following:

1. The fire service access elevator symbol shall be not less than 3 inches (76 mm) in height.
2. The helmet shall contrast with the background, with either a light helmet on a dark background or a dark helmet on a light background.
3. The vertical center line of the fire service access elevator symbol shall be centered on the hoistway door frame. Each symbol shall be not less than 78 inches (1981 mm), and not more than 84 inches (2134 mm) above the finished floor at the threshold.

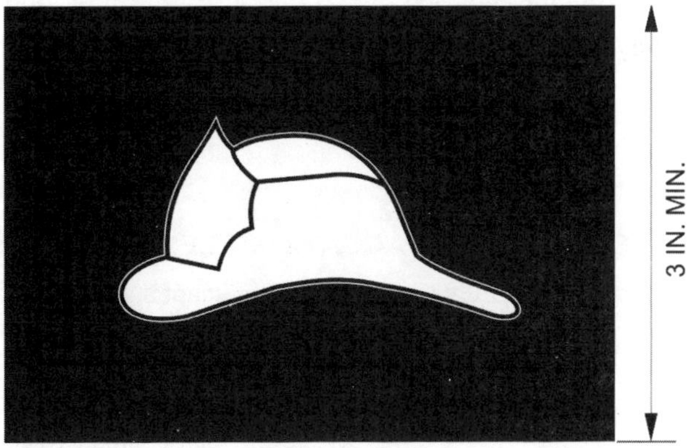

For S.I. 1 inch = 25.4 mm.

FIGURE 3007.6.5
FIRE SERVICE ACCESS ELEVATOR SYMBOL

3007.7 Elevator system monitoring. The fire service access elevator shall be continuously monitored at the *fire command center* by a standard emergency service interface system meeting the requirements of NFPA 72.

3007.8 Electrical power. The following features serving each fire service access elevator shall be supplied by both normal power and Type 60/Class 2/Level 1 standby power:

1. Elevator equipment.
2. Elevator hoistway lighting.
3. *Ventilation* and cooling equipment for elevator machine rooms, control rooms, machine spaces and control spaces.
4. Elevator car lighting.

3007.8.1 Protection of wiring or cables. Wires or cables that are located outside of the elevator hoistway and machine room and that provide normal or standby power, control signals, communication with the car, lighting, heating, air conditioning, *ventilation* and fire-detecting systems to fire service access elevators shall be protected using one of the following methods:

1. Cables used for survivability of required critical circuits shall be listed in accordance with UL 2196 and shall have a *fire-resistance rating* of not less than 2 hours.
2. *Electrical circuit protective systems* shall have a *fire-resistance rating* of not less than 2 hours. *Electrical circuit protective systems* shall be installed in accordance with their listing requirements.
3. Construction having a *fire-resistance rating* of not less than 2 hours.

Exception: Wiring and cables to control signals are not required to be protected provided that wiring and cables do not serve Phase II emergency in-car operations.

3007.9 Standpipe hose connection. A Class I standpipe hose connection in accordance with Section 905 shall be provided in the *interior exit stairway* and *ramp* having *direct access* from the enclosed fire service access elevator lobby.

3007.9.1 Access. The exit enclosure containing the standpipe shall have access to the floor without passing through the enclosed fire service access elevator lobby.

SECTION 3008
OCCUPANT EVACUATION ELEVATORS

3008.1 General. Elevators used for occupant self-evacuation during fires shall comply with Sections 3008.1 through 3008.10.

3008.1.1 Number of occupant evacuation elevators. The number of elevators available for occupant evacuation shall be determined based on an egress analysis that addresses one of the following scenarios:

1. Full-building evacuation where the analysis demonstrates that the number of elevators provided for evacuation results in an evacuation time less than 1 hour.
2. Evacuation of the five consecutive floors with the highest cumulative *occupant load* where the analysis demonstrates that the number of elevators provided for evacuation results in an evacuation time less than 15 minutes.

Not less than one elevator in each bank shall be designated for occupant evacuation. Not less than two shall be provided in each occupant evacuation elevator lobby where more than one elevator opens into the lobby. Signage shall be provided to denote which elevators are available for occupant evacuation.

3008.1.2 Additional exit stairway. Where an additional *means of egress* is required in accordance with Section 403.5.2, an additional *exit stairway* shall not be required to be installed in buildings provided with occupant evacuation elevators complying with Section 3008.1.

3008.1.3 Fire safety and evacuation plan. The building shall have an *approved* fire safety and evacuation plan in accordance with the applicable requirements of Section 404 of the *International Fire Code*. The fire safety and evacuation plan shall incorporate specific procedures for the occupants using evacuation elevators.

3008.1.4 Operation. The occupant evacuation elevators shall be used for occupant self-evacuation in accordance with the occupant evacuation operation requirements in ASME A17.1/CSA B44 and the building's fire safety and evacuation plan.

3008.2 Automatic sprinkler system. The building shall be equipped throughout with an *approved*, electrically supervised *automatic sprinkler system* in accordance with Section 903.3.1.1, except as otherwise permitted by Section 903.3.1.1.1 and as prohibited by Section 3008.2.1.

3008.2.1 Prohibited locations. Automatic sprinklers shall not be installed in elevator machine rooms, machinery spaces, control rooms, control spaces and elevator hoistways of occupant evacuation elevators.

3008.2.2 Sprinkler system monitoring. The *automatic sprinkler system* shall have a sprinkler control valve supervisory switch and water-flow-initiating device provided for each floor that is monitored by the building's *fire alarm system*.

3008.3 Water protection. Water from the operation of an *automatic sprinkler system* outside the enclosed lobby shall be prevented from infiltrating into the hoistway enclosure in accordance with an *approved* method.

3008.4 Shunt trip. Means for elevator shutdown in accordance with Section 3005.5 shall not be installed on elevator systems used for occupant evacuation elevators.

3008.5 Hoistway enclosure protection. Occupant evacuation elevator hoistways shall be located in *shaft enclosures* complying with Section 713.

3008.5.1 Structural integrity of hoistway enclosures. Occupant evacuation elevator hoistway enclosures shall comply with Sections 403.2.2.1 through 403.2.2.4.

3008.6 Occupant evacuation elevator lobby. Occupant evacuation elevators shall open into an enclosed elevator lobby in accordance with Sections 3008.6.1 through 3008.6.6. Egress is permitted through the elevator lobby in accordance with Item 1 of Section 1016.2.

3008.6.1 Access to interior exit stairway or ramp. The occupant evacuation elevator lobby shall have *direct access* from the enclosed elevator lobby to an *interior exit stairway or ramp*.

Exceptions:

1. Access to an *interior exit stairway or ramp* shall be permitted to be through a protected path of travel that has a level of fire protection not less than the elevator lobby enclosure. The protected path shall be separated from the enclosed elevator lobby through an opening protected by a smoke and draft control assembly in accordance Section 716.2.2.1.
2. Elevators that only service an *open parking garage* and the lobby of the building shall not be required to provide *direct access*.

3008.6.2 Lobby enclosure. The occupant evacuation elevator lobby shall be enclosed with a *smoke barrier* having a *fire-resistance rating* of not less than 1 hour, except that lobby doorways shall comply with Section 3008.6.3.

Exception: Enclosed occupant evacuation elevator lobbies are not required at the *levels of exit discharge*.

3008.6.3 Lobby doorways. Other than the doors to the hoistway, elevator machine rooms, machinery spaces, control rooms and control spaces within the lobby enclosure *smoke barrier*, each doorway to an occupant evacuation elevator lobby shall be provided with a $^{3}/_{4}$-hour *fire door assembly* complying with Section 716. The *fire door assembly* shall comply with the smoke and draft control assembly requirements of Section 716.2.2.1.1 and be tested in accordance with UL 1784 without an artificial bottom seal.

3008.6.3.1 Vision panel. A vision panel shall be installed in each *fire door assembly* protecting the lobby doorway. The vision panel shall consist of fire-protection-rated glazing, shall comply with the requirements of Section 716 and shall be located to furnish clear vision of the occupant evacuation elevator lobby.

3008.6.3.2 Door closing. Each *fire door assembly* protecting the lobby doorway shall be automatic-closing upon receipt of any fire alarm signal from the *emergency voice/alarm communication system* serving the building.

3008.6.4 Lobby size. Each occupant evacuation elevator lobby shall have minimum floor area as follows:

1. The occupant evacuation elevator lobby floor area shall accommodate, at 3 square feet (0.28 m^2) per person, not less than 25 percent of the *occupant load* of the floor area served by the lobby.
2. The occupant evacuation elevator lobby floor area shall accommodate one *wheelchair space* of 30 inches by 52 inches (760 mm by 1320 mm) for each 50 persons, or portion thereof, of the *occupant load* of the floor area served by the lobby.

Exception: The size of lobbies serving multiple banks of elevators shall have the minimum floor area *approved* on an individual basis and shall be consistent with the building's fire safety and evacuation plan.

3008.6.5 Signage. An *approved* sign indicating elevators are suitable for occupant self-evacuation shall be posted on all floors adjacent to each elevator call station serving occupant evacuation elevators.

3008.6.6 Two-way communication system. A two-way communication system shall be provided in each occupant evacuation elevator lobby for the purpose of initiating communication with the *fire command center* or an alternate location *approved* by the fire department. The two-way communication system shall be designed and installed in accordance with Sections 1009.8.1 and 1009.8.2.

3008.7 Elevator system monitoring. The occupant evacuation elevators shall be continuously monitored at the *fire command center* or a central control point *approved* by the fire department and arranged to display all of the following information:

1. Floor location of each elevator car.
2. Direction of travel of each elevator car.
3. Status of each elevator car with respect to whether it is occupied.
4. Status of normal power to the elevator equipment, elevator machinery and electrical apparatus cooling equipment where provided, elevator machine room, control room and control space *ventilation* and cooling equipment.
5. Status of standby or emergency power system that provides backup power to the elevator equipment, elevator machinery and electrical cooling equipment where provided, elevator machine room, control room and control space *ventilation* and cooling equipment.
6. Activation of any fire alarm initiating device in any elevator lobby, elevator machine room, machine space containing a motor controller or electric driving machine, control space, control room or elevator hoistway.

3008.7.1 Elevator recall. The *fire command center* or an alternate location *approved* by the fire department shall be provided with the means to manually initiate a Phase I Emergency Recall of the occupant evacuation elevators in accordance with ASME A17.1/CSA B44.

3008.8 Electrical power. The following features serving each occupant evacuation elevator shall be supplied by both normal power and Type 60/Class 2/Level 1 standby power:

1. Elevator equipment.
2. *Ventilation* and cooling equipment for elevator machine rooms, control rooms, machinery spaces and control spaces.
3. Elevator car lighting.

3008.8.1 Determination of standby power load. Standby power loads shall be based on the determination of the number of occupant evacuation elevators in Section 3008.1.1.

3008.8.2 Protection of wiring or cables. Wires or cables that are located outside of the elevator hoistway, machine room, control room and control space and that provide normal or standby power, control signals, communication with the car, lighting, heating, air conditioning, *ventilation* and fire-detecting systems to occupant evacuation elevators shall be protected using one of the following methods:

1. Cables used for survivability of required critical circuits shall be listed in accordance with UL 2196 and shall have a *fire-resistance rating* of not less than 2 hours.
2. *Electrical circuit protective systems* shall have a *fire-resistance rating* of not less than 2 hours. *Electrical circuit protective systems* shall be installed in accordance with their listing requirements.
3. Construction having a *fire-resistance rating* of not less than 2 hours.

Exception: Wiring and cables to control signals are not required to be protected provided that wiring and cables do not serve Phase II emergency in-car operation.

3008.9 Emergency voice/alarm communication system. The building shall be provided with an *emergency voice/alarm communication system*. The *emergency voice/alarm communication system* shall be accessible to the fire department. The system shall be provided in accordance with Section 907.5.2.2.

3008.9.1 Notification appliances. Not fewer than one audible and one visible notification appliance shall be installed within each occupant evacuation elevator lobby.

3008.10 Hazardous material areas. Building areas shall not contain hazardous materials exceeding the maximum allowable quantities per *control area* as addressed in Section 414.2.

CHAPTER 31
SPECIAL CONSTRUCTION

User notes:

About this chapter: *Chapter 31 provides regulations for unique buildings and building elements. Those include buildings such as membrane structures, greenhouses and relocatable buildings. Special elements include pedestrian walkways and tunnels, awnings, canopies and marquees, vehicular gates, solar energy systems, public use restrooms in flood hazard areas, and intermodal shipping containers.*

Code development reminder: *Code change proposals to sections preceded by the designation [BS] will be considered by the IBC—Structural Code Development Committee during the 2022 (Group B) Code Development Cycle.*

SECTION 3101
GENERAL

3101.1 Scope. The provisions of this chapter shall govern special building construction including *membrane structures*, temporary structures, *pedestrian walkways* and tunnels, automatic *vehicular gates*, *awnings* and *canopies*, *marquees*, signs, towers, antennas, relocatable buildings, swimming pool enclosures and safety devices, solar energy systems, public use restroom buildings on publicly owned lands in *flood hazard areas* and *intermodal shipping containers*.

SECTION 3102
MEMBRANE STRUCTURES

3102.1 General. The provisions of Sections 3102.1 through 3102.8 shall apply to *air-supported*, *air-inflated*, *membrane-covered cable*, *membrane-covered frame* and tensile *membrane structures*, collectively known as *membrane structures*, erected for a period of 180 days or longer. Those erected for a shorter period of time shall comply with the *International Fire Code*. Membrane structures covering water storage facilities, water clarifiers, water treatment plants, sewage treatment plants, *greenhouses* and similar facilities not used for human occupancy are required to meet only the requirements of Sections 3102.3.1 and 3102.7. Membrane structures erected on a building, balcony, deck or other structure for any period of time shall comply with this section.

3102.2 Tensile membrane structures and air-supported structures. *Tensile membrane structures* and *air-supported structures*, including permanent and temporary structures, shall be designed and constructed in accordance with ASCE 55. The provisions in Sections 3102.3 through 3102.6 shall apply.

3102.3 Type of construction. *Noncombustible membrane structures* shall be classified as Type IIB construction. Noncombustible frame or cable-supported structures covered by an *approved* membrane in accordance with Section 3102.3.1 shall be classified as Type IIB construction. Heavy timber frame-supported structures covered by an *approved* membrane in accordance with Section 3102.3.1 shall be classified as Type IV-HT construction. Other membrane structures shall be classified as Type V construction.

Exception: Plastic less than 30 feet (9144 mm) above any floor used in *greenhouses*, where occupancy by the general public is not authorized, and for aquaculture pond covers is not required to meet the fire propagation performance criteria of Test Method 1 or Test Method 2, as appropriate, of NFPA 701.

3102.3.1 Membrane and interior liner material. Membranes and interior liners shall be either noncombustible as set forth in Section 703.3 or meet the fire propagation performance criteria of Test Method 1 or Test Method 2, as appropriate, of NFPA 701 and the manufacturer's test protocol.

Exception: Plastic less than 20 mil (0.5 mm) in thickness used in *greenhouses*, where occupancy by the general public is not authorized, and for aquaculture pond covers is not required to meet the fire propagation performance criteria of Test Method 1 or Test Method 2, as appropriate, of NFPA 701.

3102.4 Allowable floor areas. The area of a membrane structure shall not exceed the limitations specified in Section 506.

3102.5 Maximum height. Membrane structures shall not exceed one *story* nor shall such structures exceed the height limitations in feet specified in Section 504.3.

Exception: *Noncombustible membrane structures* serving as roofs only.

3102.6 Mixed construction. Membrane structures shall be permitted to be utilized as specified in this section as a portion of buildings of other types of construction. Height and area limits shall be as specified for the type of construction and occupancy of the building.

3102.6.1 Noncombustible membrane. A noncombustible membrane shall be permitted for use as the roof or as a skylight of any building or *atrium* of a building of any type of construction provided that the membrane is not less than 20 feet (6096 mm) above any floor, balcony or gallery.

3102.6.1.1 Membrane. A membrane meeting the fire propagation performance criteria of Test Method 1 or Test Method 2, as appropriate, of NFPA 701 shall be permitted to be used as the roof or as a skylight on buildings of Type IIB, III, IV-HT and V construction, provided that the membrane is not less than 20 feet (6096 mm) above any floor, balcony or gallery.

3102.7 Engineering design. The structure shall be designed and constructed to sustain *dead loads*; *loads* due to tension or inflation; *live loads* including wind, snow or *flood* and seismic loads and in accordance with Chapter 16.

3102.7.1 Lateral restraint. For *membrane-covered frame structures*, the membrane shall not be considered to provide lateral restraint in the calculation of the capacities of the frame members.

3102.8 Inflation systems. *Air-supported* and *air-inflated structures* shall be provided with primary and auxiliary inflation systems to meet the minimum requirements of Sections 3102.8.1 through 3102.8.3.

3102.8.1 Equipment requirements. The inflation system shall consist of one or more blowers and shall include provisions for automatic control to maintain the required inflation pressures. The system shall be so designed as to prevent overpressurization of the system.

3102.8.1.1 Auxiliary inflation system. In addition to the primary inflation system, in buildings larger than 1,500 square feet (140 m^2) in area, an auxiliary inflation system shall be provided with sufficient capacity to maintain the inflation of the structure in case of primary system failure. The auxiliary inflation system shall operate automatically when there is a loss of internal pressure and when the primary blower system becomes inoperative.

3102.8.1.2 Blower equipment. Blower equipment shall meet all of the following requirements:

1. Blowers shall be powered by continuous-rated motors at the maximum power required for any flow condition as required by the structural design.
2. Blowers shall be provided with inlet screens, belt guards and other protective devices as required by the *building official* to provide protection from injury.
3. Blowers shall be housed within a weather-protecting structure.
4. Blowers shall be equipped with backdraft check dampers to minimize air loss when inoperative.
5. Blower inlets shall be located to provide protection from air contamination. The location of inlets shall be *approved*.

3102.8.2 Standby power. Wherever an auxiliary inflation system is required, an *approved* standby power-generating system shall be provided. The system shall be equipped with a suitable means for automatically starting the generator set upon failure of the normal electrical service and for automatic transfer and operation of all of the required electrical functions at full power within 60 seconds of such service failure. Standby power shall be capable of operating independently for not less than 4 hours.

3102.8.3 Support provisions. A system capable of supporting the membrane in the event of deflation shall be provided for in *air-supported* and *air-inflated structures* having an *occupant load* of 50 or more or where covering a swimming pool regardless of *occupant load*. The support system shall be capable of maintaining membrane structures used as a roof for Type I construction not less than 20 feet (6096 mm) above floor or seating areas. The support system shall be capable of maintaining other membranes not less than 7 feet (2134 mm) above the floor, seating area or surface of the water.

SECTION 3103
TEMPORARY STRUCTURES

3103.1 General. The provisions of Sections 3103.1 through 3103.4 shall apply to structures erected for a period of less than 180 days. *Special event structures*, tents, umbrella structures and other membrane structures erected for a period of less than 180 days shall also comply with the *International Fire Code*. Those erected for a longer period of time shall comply with applicable sections of this code.

3103.1.1 Conformance. Temporary structures and uses shall conform to the structural strength, fire safety, *means of egress*, accessibility, light, *ventilation* and sanitary requirements of this code as necessary to ensure public health, safety and general welfare.

3103.1.2 Permit required. Temporary structures that cover an area greater than 120 square feet (11.16 m^2), including connecting areas or spaces with a common *means of egress* or entrance that are used or intended to be used for the gathering together of 10 or more persons, shall not be erected, operated or maintained for any purpose without obtaining a *permit* from the *building official*.

3103.2 Construction documents. A *permit* application and *construction documents* shall be submitted for each installation of a temporary structure. The *construction documents*, shall include a site plan indicating the location of the temporary structure and information delineating the *means of egress* and the *occupant load*.

3103.3 Location. Temporary structures shall be located in accordance with the requirements of Table 705.5 based on the *fire-resistance rating* of the *exterior walls* for the proposed type of construction.

3103.4 Means of egress. Temporary structures shall conform to the *means of egress* requirements of Chapter 10 and shall have an *exit access* travel distance of 100 feet (30 480 mm) or less.

SECTION 3104
PEDESTRIAN WALKWAYS AND TUNNELS

3104.1 General. This section shall apply to connections between buildings such as *pedestrian walkways* or tunnels, located at, above or below grade level, that are used as a means of travel by persons. The *pedestrian walkway* shall not contribute to the *building area* or the number of *stories* or height of connected buildings.

3104.1.1 Application. *Pedestrian walkways* shall be designed and constructed in accordance with Sections 3104.2 through 3104.9. Tunnels shall be designed and constructed in accordance with Sections 3104.2 and 3104.10.

3104.2 Separate structures. Buildings connected by *pedestrian walkways* or tunnels shall be considered to be separate structures.

Exceptions:

1. Buildings that are on the same lot and considered as portions of a single building in accordance with Section 503.1.2.
2. For purposes of calculating the number of *Type B units* required by Chapter 11, structurally connected buildings and buildings with multiple wings shall be considered to be one structure.

3104.3 Construction. The *pedestrian walkway* shall be of noncombustible construction.

Exceptions:

1. Combustible construction shall be permitted where connected buildings are of combustible construction.
2. *Fire-retardant-treated wood*, in accordance with Section 603.1, Item 1.3, shall be permitted for the roof construction of the *pedestrian walkway* where connected buildings are not less than Type I or II construction.

3104.4 Contents. Only materials and decorations *approved* by the *building official* shall be located in the *pedestrian walkway*.

3104.5 Connections of pedestrian walkways to buildings. The connection of a *pedestrian walkway* to a building shall comply with Section 3104.5.1, 3104.5.2, 3104.5.3 or 3104.5.4.

Exception: Buildings that are on the same lot and considered as portions of a single building in accordance with Section 503.1.2.

3104.5.1 Fire barriers. *Pedestrian walkways* shall be separated from the interior of the building by not less than 2-hour *fire barriers* constructed in accordance with Section 707 and Sections 3104.5.1.1 through 3104.5.1.3.

3104.5.1.1 Exterior walls. *Exterior walls* of buildings connected to *pedestrian walkways* shall be 2-hour fire-resistance rated. This protection shall extend not less than 10 feet (3048 mm) in every direction surrounding the perimeter of the *pedestrian walkway*.

3104.5.1.2 Openings in exterior walls of connected buildings. Openings in *exterior walls* required to be fire-resistance rated in accordance with Section 3104.5.1.1 shall be equipped with opening protectives providing a not less than $^{3}/_{4}$-hour *fire protection rating* in accordance with Section 716.

3104.5.1.3 Supporting construction. The *fire barrier* shall be supported by construction as required by Section 707.5.1.

3104.5.2 Alternative separation. The wall separating the *pedestrian walkway* and the building shall comply with Section 3104.5.2.1 or 3104.5.2.2 where:

1. The distance between the connected buildings is more than 10 feet (3048 mm).
2. The *pedestrian walkway* and connected buildings are equipped throughout with an *automatic sprinkler system* in accordance with Section 903.3.1.1, and the roof of the walkway is not more than 55 feet (16 764 mm) above grade connecting to the fifth, or lower, *story above grade plane*, of each building.

Exception: *Open parking garages* need not be equipped with an *automatic sprinkler system.*

3104.5.2.1 Passage of smoke. The wall shall be capable of resisting the passage of smoke.

3104.5.2.2 Glass. The wall shall be constructed of a tempered, wired or laminated glass and doors separating the interior of the building from the *pedestrian walkway*. The glass shall be protected by an *automatic sprinkler system* in accordance with Section 903.3.1.1 that, when actuated, shall completely wet the entire surface of interior sides of the wall or glass. Obstructions shall not be installed between the sprinkler heads and the wall or glass. The glass shall be in a gasketed frame and installed in such a manner that the framing system will deflect without breaking (loading) the glass before the sprinkler operates.

3104.5.3 Open sides on walkway. Where the distance between the connected buildings is more than 10 feet (3048 mm), the walls at the intersection of the *pedestrian walkway* and each building need not be fire-resistance rated provided that both sidewalls of the *pedestrian walkway* are not less than 50 percent open with the open area uniformly distributed to prevent the accumulation of smoke and *toxic* gases. The roof of the walkway shall be located not more than 40 feet (12 160 mm) above *grade plane*, and the walkway shall only be permitted to connect to the third or lower *story* of each building.

Exception: Where the *pedestrian walkway* is protected with an automatic sprinkler system in accordance with Section 903.3.1.1, the roof of the walkway shall be located not more than 55 feet (16 764 mm) above *grade plane* and the walkway shall only be permitted to connect to the fifth or lower *story* of each building.

3104.5.4 Exterior walls greater than 2 hours. Where *exterior walls* of connected buildings are required by Section 705 to have a *fire-resistance rating* greater than 2 hours, the walls at the intersection of the *pedestrian walkway* and each building need not be fire-resistance rated provided:

1. The *pedestrian walkway* is equipped throughout with an *automatic sprinkler system* installed in accordance with Section 903.3.1.1.
2. The roof of the walkway is not located more than 55 feet (16 764 mm) above *grade plane* and the walkway connects to the fifth, or lower, *story above grade plane* of each building.

3104.6 Public way. *Pedestrian walkways* over a *public way* shall comply with Chapter 32.

3104.7 Egress. Access shall be provided at all times to a *pedestrian walkway* that serves as a required *exit.*

3104.8 Width. The unobstructed width of *pedestrian walkways* shall be not less than 36 inches (914 mm). The total width shall be not greater than 30 feet (9144 mm).

3104.9 Exit access travel. The length of *exit access* travel shall be 200 feet (60 960 mm) or less.

Exceptions:

1. *Exit access* travel distance on a *pedestrian walkway* equipped throughout with an *automatic sprinkler system* in accordance with Section 903.3.1.1 shall be 250 feet (76 200 mm) or less.
2. *Exit access* travel distance on a *pedestrian walkway* constructed with both sides not less than 50 percent open shall be 300 feet (91 440 mm) or less.
3. *Exit access* travel distance on a *pedestrian walkway* constructed with both sides not less than 50 percent open, and equipped throughout with an *automatic sprinkler system* in accordance with Section 903.3.1.1, shall be 400 feet (122 m) or less.

3104.10 Tunneled walkway. Separation between the tunneled walkway and the building to which it is connected shall be not less than 2-hour fire-resistant construction and openings therein shall be protected in accordance with Section 716.

SECTION 3105
AWNINGS AND CANOPIES

3105.1 General. *Awnings* and *canopies* shall comply with the requirements of Sections 3105.2 and 3105.3 and other applicable sections of this code.

3105.2 Design and construction. *Awnings* and *canopies* shall be designed and constructed to withstand wind or other lateral *loads* and *live loads* as required by Chapter 16 with due allowance for shape, open construction and similar features that relieve the pressures or loads. Structural members shall be protected to prevent deterioration. *Awnings* shall have frames of noncombustible material, *fire-retardant-treated wood*, heavy timber complying with Section 2304.11, or 1-hour construction with combustible or noncombustible covers and shall be either fixed, retractable, folding or collapsible.

3105.3 Awnings and canopy materials. *Awnings* and *canopies* shall be provided with an *approved* covering that complies with one of the following:

1. The fire propagation performance criteria of Test Method 1 or Test Method 2, as appropriate, of NFPA 701.
2. Has a *flame spread index* not greater than 25 when tested in accordance with ASTM E84 or UL 723.
3. Meets all of the following criteria when tested in accordance with NFPA 286:
   3.1. During the 40 kW exposure, flames shall not spread to the ceiling.
   3.2. Flashover, as defined in NFPA 286, shall not occur.
   3.3. The flame shall not spread to the outer extremity of the sample on any wall or ceiling.
   3.4. The peak heat release rate throughout the test shall not exceed 800 kW.

Exception: The fire propagation performance and *flame spread index* requirements shall not apply to awnings installed on detached one- and two-family dwellings.

SECTION 3106
MARQUEES

3106.1 General. *Marquees* shall comply with Sections 3106.2 through 3106.5 and other applicable sections of this code.

3106.2 Thickness. The height or thickness of a *marquee* measured vertically from its lowest to its highest point shall be not greater than 3 feet (914 mm) where the *marquee* projects more than two-thirds of the distance from the *lot line* to the curb line, and shall be not greater than 9 feet (2743 mm) where the *marquee* is less than two-thirds of the distance from the *lot line* to the curb line.

3106.3 Roof construction. Where the roof or any part thereof is a skylight, the skylight shall comply with the requirements of Chapter 24. Every roof and skylight of a *marquee* shall be sloped to downspouts that shall conduct any drainage from the *marquee* in such a manner so as not to spill over the sidewalk.

3106.4 Location prohibited. Every *marquee* shall be so located as not to interfere with the operation of any exterior standpipe, and such that the *marquee* does not obstruct the clear passage of *stairways* or *exit discharge* from the building or the installation or maintenance of street lighting.

3106.5 Construction. A *marquee* shall be supported entirely from the building and constructed of noncombustible materials. *Marquees* shall be designed as required in Chapter 16. Structural members shall be protected to prevent deterioration.

SECTION 3107 SIGNS

3107.1 General. Signs shall be designed, constructed and maintained in accordance with this code.

SECTION 3108 TELECOMMUNICATION AND BROADCAST TOWERS

[BS] 3108.1 General. Towers shall be designed and constructed in accordance with the provisions of TIA 222. Towers shall be designed for seismic *loads*; exceptions related to seismic design listed in Section 2.7.3 of TIA 222 shall not apply. In Section 2.6.6.2 of TIA 222, the horizontal extent of Topographic Category 2, escarpments, shall be 16 times the height of the escarpment.

Exception: Single free-standing poles used to support antennas not greater than 75 feet (22 860 mm), measured from the top of the pole to grade, shall not be required to be noncombustible.

[BS] 3108.2 Location and access. Towers shall be located such that guy wires and other accessories shall not cross or encroach on any street or other public space, or over above-ground electric utility lines, or encroach on any privately owned property without the written consent of the owner of the encroached-upon property, space or above-ground electric utility lines. Towers shall be equipped with climbing and working facilities in compliance with TIA 222. Access to the tower sites shall be limited as required by applicable OSHA, FCC and EPA regulations.

SECTION 3109 SWIMMING POOLS, SPAS AND HOT TUBS

3109.1 General. The design and construction of swimming pools, spas and hot tubs shall comply with the *International Swimming Pool and Spa Code.*

SECTION 3110 AUTOMATIC VEHICULAR GATES

3110.1 General. *Automatic vehicular gates* shall comply with the requirements of Sections 3110.2 and 3110.3 and other applicable sections of this code.

3110.2 Vehicular gates intended for automation. *Vehicular gates* intended for automation shall be designed, constructed and installed to comply with the requirements of ASTM F2200.

3110.3 Vehicular gate openers. *Vehicular gate* openers, where provided, shall be *listed* in accordance with UL 325.

SECTION 3111 SOLAR ENERGY SYSTEMS

3111.1 General. Solar energy systems shall comply with the requirements of this section.

3111.1.1 Wind resistance. Rooftop-mounted photovoltaic (PV) panel systems and solar thermal collectors shall be designed in accordance with Section 1609.

3111.1.2 Roof live load. Roof structures that provide support for solar energy systems shall be designed in accordance with Section 1607.14.4.

3111.2 Solar thermal systems. Solar thermal systems shall be designed and installed in accordance with this section, the *International Plumbing Code*, the *International Mechanical Code* and the *International Fire Code*. Where light-transmitting plastic covers are used, solar thermal collectors shall be designed in accordance with Section 2606.12.

3111.2.1 Equipment. Solar thermal systems and components shall be *listed* and *labeled* in accordance with ICC 900/SRCC 300 and ICC 901/SRCC 100.

3111.3 Photovoltaic solar energy systems. Photovoltaic solar energy systems shall be designed and installed in accordance with this section, the *International Fire Code*, NFPA 70 and the manufacturer's installation instructions.

3111.3.1 Equipment. *Photovoltaic panels* and modules shall be *listed* and *labeled* in accordance with UL 1703 or with both UL 61730-1 and UL 61730-2. Inverters shall be *listed* and *labeled* in accordance with UL 1741. Systems connected to the utility grid shall use inverters *listed* for utility interaction.

3111.3.2 Fire classification. Rooftop-mounted photovoltaic (PV) panel systems shall have a fire classification in accordance with Section 1505.9. Building-integrated photovoltaic (BIPV) systems installed as roof coverings shall have a fire classification in accordance with Section 1505.8.

3111.3.3 Building-integrated photovoltaic (BIPV) systems. BIPV systems installed as roof coverings shall be designed and installed in accordance with Section 1507.

3111.3.4 Access and pathways. Roof access, pathways and spacing requirements shall be provided in accordance with Section 1205 of the *International Fire Code.*

3111.3.5 Ground-mounted photovoltaic systems. Ground-mounted photovoltaic systems shall be designed and installed in accordance with Chapter 16 and the *International Fire Code.*

3111.3.5.1 Fire separation distances. Ground-mounted photovoltaic systems shall be subject to the *fire separation distance* requirements determined by the local jurisdiction.

SECTION 3112 GREENHOUSES

3112.1 General. The provisions of this section shall apply to *greenhouses* that are designed and used for the cultivation, maintenance, or protection of plants.

3112.2 Accessibility. *Greenhouses* shall be *accessible* in accordance with Chapter 11.

3112.3 Structural design. *Greenhouses* shall comply with the structural design requirements for *greenhouses* in Chapter 16.

3112.4 Glass and glazing. Glass and glazing used in *greenhouses* shall comply with Section 2405.

3112.5 Light-transmitting plastics. Light-transmitting plastics shall be permitted in lieu of plain glass in *greenhouses* and shall comply with Section 2606.

3112.6 Membrane structures. *Greenhouses* that are membrane structures shall comply with Section 3102.

3112.6.1 Plastic film. Plastic films used in *greenhouses* shall comply with Section 3102.3.

SECTION 3113 RELOCATABLE BUILDINGS

3113.1 General. The provisions of this section shall apply to relocatable buildings. Relocatable buildings manufactured after the effective date of this code shall comply with the applicable provisions of this code.

Exception: This section shall not apply to manufactured housing used as dwellings.

3113.1.1 Compliance. A newly constructed relocatable building shall comply with the requirements of this code for new construction. An existing relocatable building that is undergoing alteration, addition, change of occupancy or relocation shall comply with Chapter 14 of the *International Existing Building Code.*

3113.2 Supplemental information. Supplemental information specific to a relocatable building shall be submitted to the authority having jurisdiction. It shall, as a minimum, include the following in addition to the information required by Section 105:

1. Manufacturer's name and address.
2. Date of manufacture.
3. Serial number of module.
4. Manufacturer's design drawings.
5. Type of construction in accordance with Section 602.
6. Design *loads* including: *roof live load*, roof snow *load*, floor *live load*, wind *load* and seismic *site class*, use group and design category.
7. Additional building planning and structural design data.
8. Site-built structure or appurtenance attached to the relocatable building.

3113.3 Manufacturer's data plate. Each relocatable module shall have a data plate that is permanently attached on or adjacent to the electrical panel, and shall include the following information:

1. Occupancy group.
2. Manufacturer's name and address.
3. Date of manufacture.
4. Serial number of module.
5. Design *roof live load*, design floor *live load*, snow *load*, wind and seismic design.
6. *Approved* quality assurance agency or *approved* inspection agency.
7. Codes and standards of construction.
8. Envelope thermal resistance values.
9. Electrical service size.
10. Fuel-burning equipment and size.
11. Special limitations if any.

3113.4 Inspection agencies. The building official is authorized to accept reports of inspections conducted by *approved* inspection agencies during off-site construction of the relocatable building, and to satisfy the applicable requirements of Sections 110.3 through 110.3.12.1.

SECTION 3114 PUBLIC USE RESTROOM BUILDINGS IN FLOOD HAZARD AREAS

3114.1 General. For the purpose of this section, public restroom buildings are located on publicly owned lands in *flood hazard areas* and intended for public use. Public restroom buildings and portions of other buildings that contain public restrooms are limited to toilet rooms, bathrooms, showers and changing rooms. Public restroom buildings and portions of buildings that contain public restrooms shall comply with the requirements of this section. Public-use restrooms that are not elevated or *dry floodproofed* in accordance with Section 1612 shall comply with Section 3114.2. Portions of buildings that include uses other than public-use toilet rooms, bathrooms, showers and changing rooms shall comply with Section 1612.

3114.2 Flood resistance. Public-use restrooms on publicly owned lands in *flood hazard areas* shall comply with the requirements of ASCE 24, except for elevation requirements, and shall comply with all of the following criteria:

1. The building footprint is not more than 1,500 square feet (139 m^2).
2. Located, designed and constructed to resist the effects of flood hazards and flood loads to minimize flood damage from a combination of wind and water loads associated with the base flood.
3. Anchored to prevent flotation, collapse or lateral movement resulting from hydrodynamic and hydrostatic loads, including the effects of buoyancy during conditions of the base flood.
4. Constructed of flood-damage-resistant materials.
5. Where enclosed by walls, the walls have flood openings.
6. Mechanical and electrical systems are located above the base flood elevation.
7. Plumbing fixtures and plumbing connections are located above the base flood elevation.
8. An emergency plan, approved by the jurisdiction, is submitted to the building official and includes build-

ing design documents specifying implementation of protection measures prior to the onset of flooding conditions.

Exceptions:

1. Minimum necessary electric equipment required to address health, life safety and electric code requirements is permitted below the base flood elevation in accordance with ASCE 24 provisions for electric elements installed below the minimum elevations.
2. Plumbing fixtures and connections are permitted below the base flood elevation provided that the fixtures and connections are designed and installed to minimize or eliminate infiltration of floodwaters into the sanitary sewage system and discharges from sanitary sewage systems into floodwaters.

SECTION 3115
INTERMODAL SHIPPING CONTAINERS

3115.1 General. The provisions of Section 3115 and other applicable sections of this code shall apply to *intermodal shipping containers* that are repurposed for use as buildings or structures, or as a part of buildings or structures.

Exceptions:

1. Intermodal shipping containers previously approved as existing relocatable buildings complying with Chapter 14 of the International Existing Building Code.
2. Stationary storage battery arrays located in intermodal shipping containers complying with Chapter 12 of the International Fire Code.
3. Intermodal shipping containers that are listed as equipment complying with the standard for equipment, such as air chillers, engine generators, modular data centers, and other similar equipment.
4. Intermodal shipping containers housing or supporting experimental equipment are exempt from the requirements of Section 3115, provided that they comply with all of the following:
    4.1. Such units shall be single stand-alone units supported at grade level and used only for occupancies as specified under Risk Category I in Table 1604.5.
    4.2. Such units are located a minimum of 8 feet (2438 mm) from adjacent structures, and are not connected to a fuel gas system or fuel gas utility.
    4.3. In hurricane-prone regions and flood hazard areas, such units are designed in accordance with the applicable provisions of Chapter 16.

3115.2 Construction documents. The construction documents shall contain information to verify the dimensions and establish the physical properties of the steel components and wood floor components of the *intermodal shipping container,* in addition to the information required by Sections 107 and 1603.

3115.3 Intermodal shipping container information. Intermodal shipping containers shall bear an existing data plate containing the following information as required by ISO 6346 and verified by an approved agency. A report of the verification process and findings shall be provided to the building owner.

1. Manufacturer's name or identification number.
2. Date manufactured.
3. Safety approval number.
4. Identification number.
5. Maximum operating gross mass or weight (kg) (lbs).
6. Allowable stacking load for 1.8G (kg) (lbs).
7. Transverse racking test force (Newtons).
8. Valid maintenance examination date.

Where approved by the *building official*, the markings and existing data plate are permitted to be removed from the intermodal shipping containers before they are repurposed for use as buildings or structures or as a part of buildings or structures.

3115.4 Protection against decay and termites. Wood structural floors of *intermodal shipping containers* shall be protected from decay and termites in accordance with the applicable provisions of Section 2304.12.1.1.

3115.5 Under-floor ventilation. The space between the bottom of the floor joists and the earth under any *intermodal shipping container*, except spaces occupied by basements and cellars, shall be provided with ventilation in accordance with Section 1202.4.

3115.6 Roof assemblies. *Intermodal shipping container* roof assemblies shall comply with the applicable requirements of Chapter 15.

Exception: Single-unit, stand-alone intermodal shipping containers not attached to, or stacked vertically over, other intermodal shipping containers, buildings or structures.

3115.7 Joints and voids. Joints and voids that create concealed spaces between connected or stacked *intermodal shipping containers* at fire-resistance-rated walls, floor or floor/ceiling assemblies and roofs or roof/ceiling assemblies shall be protected by an approved *fire-resistant joint system* in accordance with Section 715.

3115.8 Structural. Intermodal shipping containers that conform to ISO 1496-1 and are repurposed for use as buildings or structures, or as a part of buildings or structures, shall be designed in accordance with Chapter 16 and this section.

3115.8.1 Foundations. *Intermodal shipping containers* repurposed for use as a permanent building or structure shall be supported on foundations or other supporting structures designed and constructed in accordance with Chapters 16 through 23.

3115.8.1.1 Anchorage. *Intermodal shipping containers* shall be anchored to foundations or other supporting structures as necessary to provide a continuous load path for all applicable design and environmental *loads* in accordance with Chapter 16.

3115.8.2 Welds. New welds and connections shall be equal to or greater than the original connections.

3115.8.3 Structural design. The structural design for the *intermodal shipping containers* repurposed for use as a building or structure, or as part of a building or structure, shall comply with Section 3115.8.4 or 3115.8.5.

3115.8.4 Detailed design procedure. A structural analysis meeting the requirements of this section shall be provided to the *building official* to demonstrate the structural adequacy of the intermodal shipping containers.

Exception: Intermodal shipping containers designed in accordance with Section 3115.8.5.

3115.8.4.1 Material properties. Structural material properties for existing intermodal shipping container steel components shall be established by material testing where the steel grade and composition cannot be identified by the manufacturer's designation as to manufacture and mill test.

3115.8.4.2 Seismic design parameters. The seismic force-resisting system shall be designed and detailed in accordance with one of the following:

1. Where all or portions of the corrugated steel container sides are considered to be the seismic force-resisting system, design and detailing shall be in accordance with the ASCE 7, Table 12.2-1 requirements for light-frame bearing-wall systems with shear panels of all other materials.
2. Where portions of the corrugated steel container sides are retained, but are not considered to be the seismic force-resisting system, an independent seismic force-resisting system shall be selected, designed and detailed in accordance with ASCE 7, Table 12.2-1.
3. Where portions of the corrugated steel container sides are retained and integrated into a seismic force-resisting system other than as permitted by Item 1, seismic design parameters shall be developed from testing and analysis in accordance with Section 104.11 and ASCE 7, Section 12.2.1.1 or 12.2.1.2.

3115.8.4.3 Allowable shear value. The allowable shear values for the *intermodal shipping container* corrugated steel sheet panel side walls and end walls shall be demonstrated by testing and analysis accordance with Section 104.11. Where penetrations are made in the side walls or end walls designated as part of the lateral force-resisting system, the penetrations shall be substantiated by rational analysis.

3115.8.5 Simplified structural design of single-unit containers. Single-unit *intermodal shipping containers* conforming to the limitations of Section 3115.8.5.1 shall be permitted to be designed in accordance with the simplified structural design provisions of Section 3115.8.5.2.

3115.8.5.1 Limitations. The use of Section 3115.8.5 is subject to the following limitations:

1. The *intermodal shipping container* shall be a single-unit, stand-alone unit supported on a foundation and shall not be in contact with or supporting any other shipping container or other structure.
2. The *intermodal shipping container* top and bottom rails, corner castings, and columns or any portion thereof shall not be notched, cut, or removed in any manner.
3. The *intermodal shipping container* shall be erected in a level and horizontal position with the floor located at the bottom.
4. The *intermodal shipping container* shall be located in Seismic Design Category A, B, C or D.

3115.8.5.2 Simplified structural design. Where permitted by Section 3115.8.5.1, single-unit, stand-alone intermodal shipping containers shall be designed using the following assumptions for the corrugated steel shear walls:

1. The appropriate detailing requirements contained in Chapters 16 through 23.
2. Response modification coefficient, R = 2.
3. Overstrength factor, $\Omega_0 = 2.5$.
4. Deflection amplification factor, $C_d = 2$.
5. Limits on structural height, $h_n = 9.5$ feet (2900 mm).

3115.8.5.3 Allowable shear. The allowable shear for the corrugated steel side walls (longitudinal) and end walls (transverse) for wind design and seismic design using the coefficients of Section 3115.8.5.2 shall be in accordance with Table 3115.8.5.3, provided that all of the following conditions are met:

1. The total linear length of all openings in any individual side wall or end wall shall be limited to not more than 50 percent of the length of that side wall or end wall, as shown in Figure 3115.8.5.3(1).
2. Any full-height wall length, or portion thereof, less than 4 feet (305 mm) shall not be considered as a portion of the lateral force-resisting system, as shown in Figure 3115.8.5.3(2).
3. All side walls or end walls used as part of the lateral force-resisting system shall have an existing or new boundary element on all sides to form a continuous load path, or paths, with adequate strength and stiffness to transfer all forces from the point of application to the final point of resistance, as shown in Figure 3115.8.5.3(3).

4. Where openings are made in container walls, floors or roofs, for doors, windows and other openings:
   - 4.1 The openings shall be framed with steel elements that are designed in accordance with Chapters 16 and 22.
   - 4.2 The cross section and material grade of any new steel element shall be equal to or greater than the steel element removed.
5. A maximum of one penetration not greater than 6 inches (152 mm) in diameter for conduits, pipes, tubes or vents, or not greater than 16 square inches (10 323 mm2) for electrical boxes, is permitted for each individual 8-foot (2438 mm) length of lateral force-resisting wall. Penetrations located in walls that are not part of the lateral force-resisting system shall not be limited in size or quantity. Existing intermodal shipping container vents shall not be considered a penetration, as shown in Figure 3115.8.5.3(4).
6. End wall doors designated as part of the lateral force-resisting system shall be welded closed.

TABLE 3115.8.5.3
ALLOWABLE SHEAR VALUES FOR INTERMODAL SHIPPING CONTAINER CORRUGATED STEEL WALLS FOR WIND OR SEISMIC LOADING

CONTAINER DESIGNATION[b]	CONTAINER DIMENSION (nominal length)	CONTAINER DIMENSION (nominal height)	ALLOWABLE SHEAR VALUES (PLF)[a, c]	
			Side Wall	End Wall
1EEE	45 feet	9.5 feet	75	843
1EE		8.5 feet		
1AAA	40 feet	9.5 feet	84	
1AA		8.5 feet		
1A		8.0 feet		
1AX		< 8.0 feet		
1BBB	30 feet	9.5 feet	112	
1BB		8.5 feet		
1B		8.0 feet		
1BX		< 8.0 feet		
1CC	20 feet	8.5 feet	168	
1C		8.0 feet		
1CX		< 8.0 feet		
1D	10 feet	8.0 feet	337	
1DX		< 8.0 feet		

For SI: 1 foot = 304.8 mm.

a. The allowable strength shear for the side walls and end walls of the intermodal shipping containers are derived from ISO 1496-1 and reduced by a factor of safety of 5.

b. Container designation type is derived from ISO 668.

c. Limitations of Section 3115.8.5.1 shall apply.

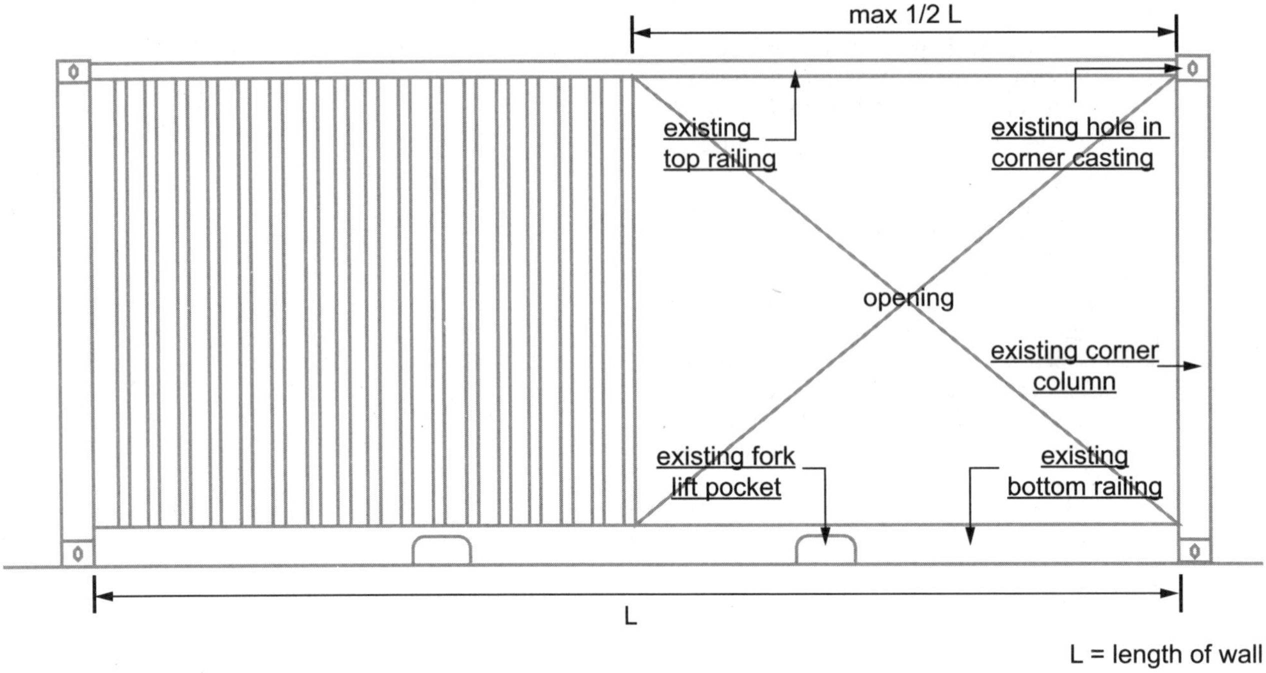

FIGURE 3115.8.5.3(1)
BRACING UNIT DISTRIBUTION—MAXIMUM LINEAR LENGTH

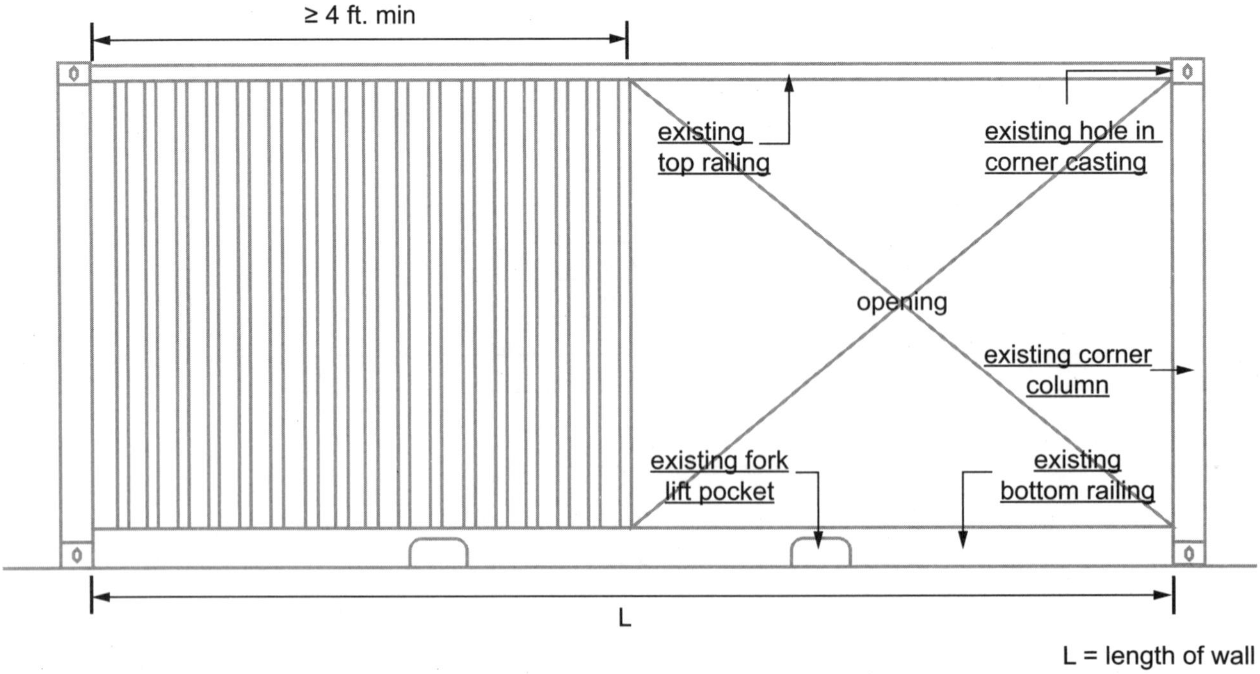

For SI: 1 foot = 304.8 mm.

FIGURE 3115.8.5.3(2)
BRACING UNIT DISTRIBUTION—MINIMUM LINEAR LENGTH

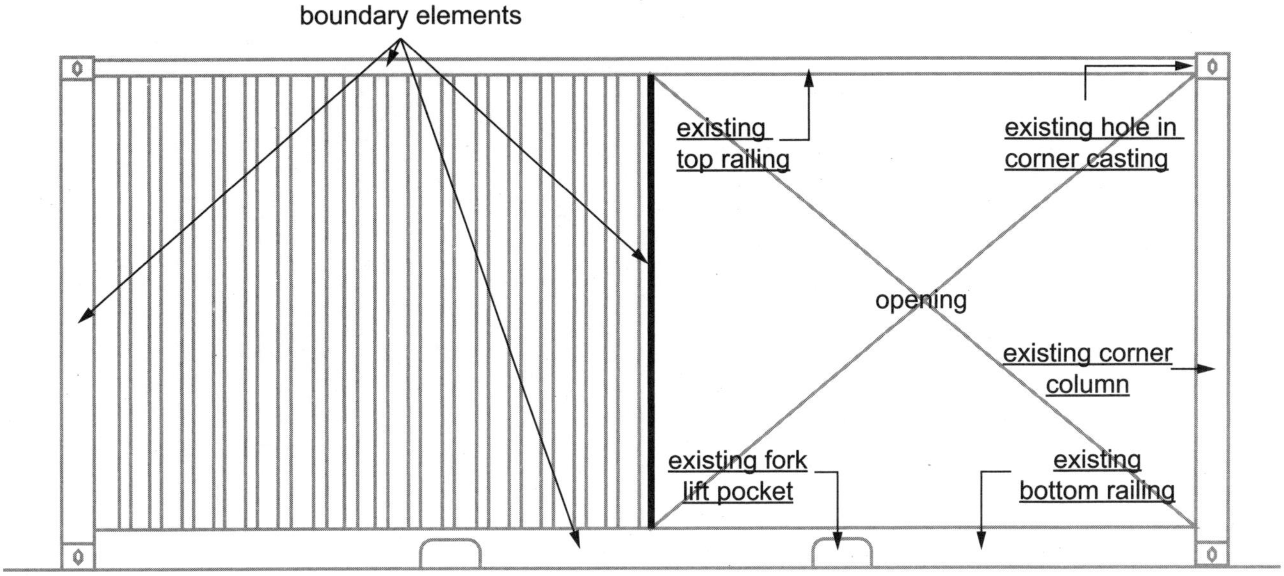

FIGURE 3115.8.5.3(3)
BRACING UNIT DISTRIBUTION—BOUNDARY ELEMENTS

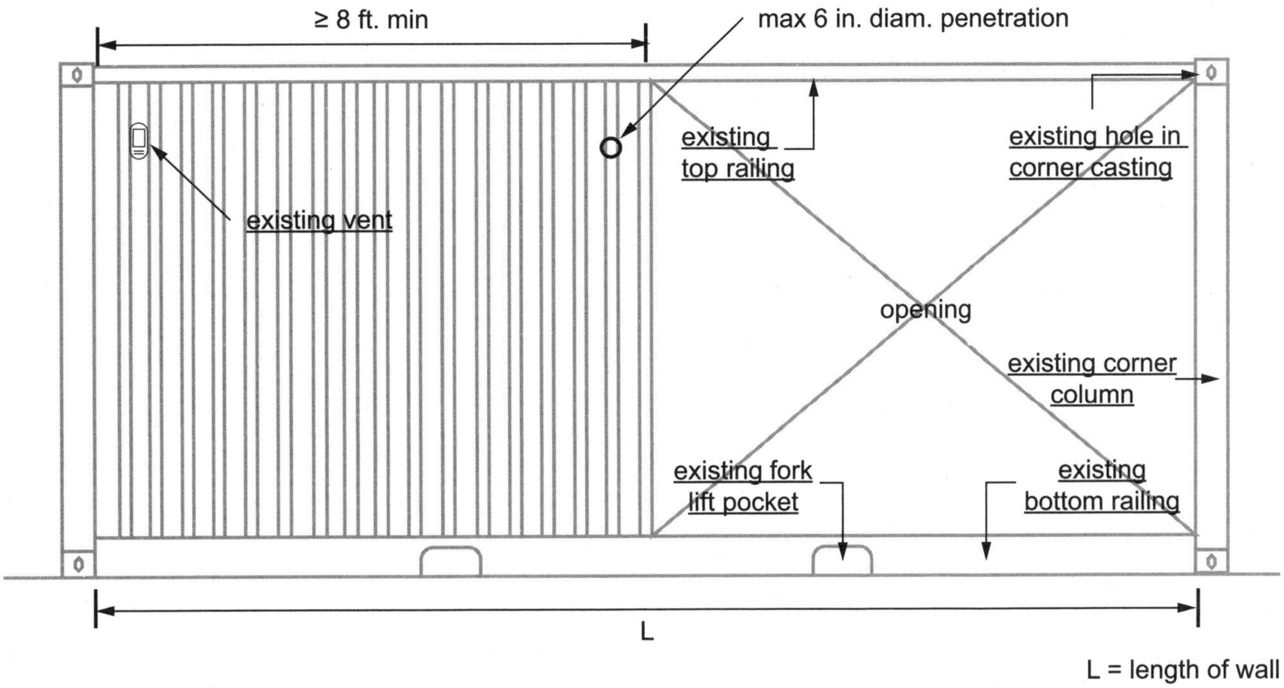

For SI: 1 inch = 25.4 mm, 1 foot = 304.8 mm.

FIGURE 3115.8.5.3(4)
BRACING UNIT DISTRIBUTION—PENETRATION LIMITATIONS

CHAPTER 32

ENCROACHMENTS INTO THE PUBLIC RIGHT-OF-WAY

User note:

About this chapter: *From time to time it is necessary or appropriate for a portion of a building to encroach onto an adjoining public right-of-way. Chapter 32 establishes parameters for such encroachments not only at grade but also above and below grade.*

SECTION 3201
GENERAL

3201.1 Scope. The provisions of this chapter shall govern the encroachment of structures into the public right-of-way.

3201.2 Measurement. The projection of any structure or portion thereof shall be the distance measured horizontally from the *lot line* to the outermost point of the projection.

3201.3 Other laws. The provisions of this chapter shall not be construed to permit the violation of other laws or ordinances regulating the use and occupancy of public property.

3201.4 Drainage. Drainage water collected from a roof, *awning*, *canopy* or *marquee*, and condensate from mechanical equipment shall not flow over a public walking surface.

SECTION 3202
ENCROACHMENTS

3202.1 Encroachments below grade. Encroachments below grade shall comply with Sections 3202.1.1 through 3202.1.3.

3202.1.1 Structural support. A part of a building erected below grade that is necessary for structural support of the building or structure shall not project beyond the *lot lines*, except that the footings of street walls or their supports that are located not less than 8 feet (2438 mm) below grade shall not project more than 12 inches (305 mm) beyond the street *lot line*.

3202.1.2 Vaults and other enclosed spaces. The construction and utilization of vaults and other enclosed spaces below grade shall be subject to the terms and conditions of the applicable governing authority.

3202.1.3 Areaways. *Areaways* shall be protected by grates, *guards* or other *approved* means.

3202.2 Encroachments above grade and below 8 feet in height. Encroachments into the public right-of-way above grade and below 8 feet (2438 mm) in height shall be prohibited except as provided for in Sections 3202.2.1 through 3202.2.3. Doors and windows shall not open or project into the public right-of-way.

3202.2.1 Steps. Steps shall not project more than 12 inches (305 mm) and shall be guarded by *approved* devices not less than 3 feet (914 mm) in height, or shall be located between columns or pilasters.

3202.2.2 Architectural features. Columns or pilasters, including bases and moldings, shall not project more than 12 inches (305 mm). Belt courses, lintels, sills, architraves, pediments and similar architectural features shall not project more than 4 inches (102 mm).

3202.2.3 Awnings. The vertical clearance from the public right-of-way to the lowest part of any *awning*, including valances, shall be not less than 7 feet (2134 mm).

3202.3 Encroachments 8 feet or more above grade. Encroachments 8 feet (2438 mm) or more above grade shall comply with Sections 3202.3.1 through 3202.3.4.

3202.3.1 Awnings, canopies, marquees and signs. *Awnings*, *canopies*, *marquees* and signs shall be constructed so as to support applicable *loads* as specified in Chapter 16. *Awnings*, *canopies*, *marquees* and signs with less than 15 feet (4572 mm) of clearance above the sidewalk shall not extend into or occupy more than two-thirds the width of the sidewalk measured from the building. Stanchions or columns that support *awnings*, *canopies*, *marquees* and signs shall be located not less than 2 feet (610 mm) in from the curb line.

3202.3.2 Windows, balconies, architectural features and mechanical equipment. Where the vertical clearance above grade to projecting windows, balconies, architectural features or mechanical equipment is more than 8 feet (2438 mm), 1 inch (25 mm) of encroachment is permitted for each additional 1 inch (25 mm) of clearance above 8 feet (2438 mm), but the maximum encroachment shall be 4 feet (1219 mm).

3202.3.3 Encroachments 15 feet or more above grade. Encroachments 15 feet (4572 mm) or more above grade shall not be limited.

3202.3.4 Pedestrian walkways. The installation of a *pedestrian walkway* over a public right-of-way shall be subject to the approval of the applicable governing authority. The vertical clearance from the public right-of-way to the lowest part of a *pedestrian walkway* shall be not less than 15 feet (4572 mm).

3202.4 Temporary encroachments. Where allowed by the applicable governing authority, vestibules and storm enclosures shall not be erected for a period of time exceeding 7 months in any 1 year and shall not encroach more than 3 feet (914 mm) nor more than one-fourth of the width of the sidewalk beyond the street *lot line*. Temporary entrance awnings shall be erected with a clearance of not less than 7 feet (2134 mm) to the lowest portion of the hood or awning where supported on removable steel or other *approved* noncombustible support.

CHAPTER 33

SAFEGUARDS DURING CONSTRUCTION

User notes:

About this chapter: *While the balance of the chapters in this code specify how a building is to be designed and constructed in order to be in compliance with the code, Chapter 33 looks to the actual construction process. Parameters are provided for demolition and for protecting adjacent property during demolition and construction. This chapter also addresses the need for a fire watch during nonworking hours for certain buildings once the construction has progressed significantly. Issues such as how to provide egress while the building is growing, the timing of standpipe and sprinkler installation, and protection of pedestrians are addressed.*

Code development reminder: *Code change proposals to sections preceded by the designation [BS] will be considered by the IBC—Structural Code Development Committee during the 2022 (Group B) Code Development Cycle.*

SECTION 3301 GENERAL

3301.1 Scope. The provisions of this chapter shall govern safety during construction and the protection of adjacent public and private properties.

3301.2 Storage and placement. Construction equipment and materials shall be stored and placed so as not to endanger the public, the workers or adjoining property for the duration of the construction project.

** **[BS] 3301.2.1 Structural and construction loads.** Structural roof components shall be capable of supporting the roof-covering system and the material and equipment *loads* that will be encountered during installation of the system.

SECTION 3302 CONSTRUCTION SAFEGUARDS

3302.1 Alterations, repairs and additions. Required *exits*, existing structural elements, fire protection devices and sanitary safeguards shall be maintained at all times during *alterations*, *repairs* or *additions* to any building or structure.

Exceptions:

1. Where such required elements or devices are being altered or repaired, adequate substitute provisions shall be made.
2. Maintenance of such elements and devices is not required where the existing building is not occupied.

3302.2 Manner of removal. Waste materials shall be removed in a manner that prevents injury or damage to persons, adjoining properties and public rights-of-way.

3302.3 Fire safety during construction. Fire safety during construction shall comply with the applicable requirements of this code and the applicable provisions of Chapter 33 of the *International Fire Code.*

SECTION 3303 DEMOLITION

3303.1 Construction documents. *Construction documents* and a schedule for demolition shall be submitted where required by the *building official*. Where such information is required, work shall not be done until such *construction documents* or schedule, or both, are *approved*.

3303.2 Pedestrian protection. The work of demolishing any building shall not be commenced until pedestrian protection is in place as required by this chapter.

3303.3 Means of egress. A *horizontal exit* shall not be destroyed unless and until a substitute *means of egress* has been provided and *approved*.

3303.4 Vacant lot. Where a structure has been demolished or removed, the vacant lot shall be filled and maintained to the existing grade or in accordance with the ordinances of the jurisdiction having authority.

3303.5 Water accumulation. Provision shall be made to prevent the accumulation of water or damage to any foundations on the premises or the adjoining property.

3303.6 Utility connections. Service utility connections shall be discontinued and capped in accordance with the *approved* rules and the requirements of the applicable governing authority.

3303.7 Fire safety during demolition. Fire safety during demolition shall comply with the applicable requirements of this code and the applicable provisions of *Chapter 33* of the *International Fire Code.*

SECTION 3304 SITE WORK

3304.1 Excavation and fill. Excavation and fill for buildings and structures shall be constructed or protected so as not to endanger life or property. Stumps and roots shall be removed from the soil to a depth of not less than 12 inches (305 mm) below the surface of the ground in the area to be occupied by the building. Wood forms that have been used in placing concrete, if within the ground or between foundation sills and the ground, shall be removed before a building is occupied or used for any purpose. Before completion, loose or casual wood shall be removed from direct contact with the ground under the building.

3304.1.1 Slope limits. Slopes for permanent fill shall be not steeper than one unit vertical in two units horizontal (50-percent slope). Cut slopes for permanent excavations shall be not steeper than one unit vertical in two units horizontal (50-percent slope). Deviation from the foregoing limitations for cut slopes shall be permitted only upon the presentation of a soil investigation report acceptable to the *building official.*

3304.1.2 Surcharge. Fill or other surcharge *loads* shall not be placed adjacent to any building or structure unless such building or structure is capable of withstanding the additional *loads* caused by the fill or surcharge. Existing footings or foundations that can be affected by any excavation shall be underpinned adequately or otherwise protected against settlement and shall be protected against lateral movement.

3304.1.3 Footings on adjacent slopes. For footings on adjacent slopes, see Chapter 18.

3304.1.4 Fill supporting foundations. Fill to be used to support the foundations of any building or structure shall comply with Section 1804.6. *Special inspections* of compacted fill shall be in accordance with Section 1705.6.

SECTION 3305 SANITARY

3305.1 Facilities required. Sanitary facilities shall be provided during construction, remodeling or demolition activities in accordance with the *International Plumbing Code.*

SECTION 3306 PROTECTION OF PEDESTRIANS

[BS] 3306.1 Protection required. Pedestrians shall be protected during construction, remodeling and demolition activities as required by this chapter and Table 3306.1. Signs shall be provided to direct pedestrian traffic.

[BS] 3306.2 Walkways. A walkway shall be provided for pedestrian travel in front of every construction and demolition site unless the applicable governing authority authorizes the sidewalk to be fenced or closed. A walkway shall be provided for pedestrian travel that leads from a *building* entrance or exit of an occupied structure to a public way. Walkways shall be of sufficient width to accommodate the pedestrian traffic, but shall be not less than 4 feet (1219 mm) in width. Walkways shall be provided with a durable walking surface. Walkways shall be *accessible* in accordance with Chapter 11 and shall be designed to support all imposed loads, and the design *live load* shall be not less than 150 pounds per square foot (psf) (7.2 kN/m^2).

[BS] 3306.3 Directional barricades. Pedestrian traffic shall be protected by a directional barricade where the walkway extends into the street. The directional barricade shall be of sufficient size and construction to direct vehicular traffic away from the pedestrian path.

[BS] 3306.4 Construction railings. Construction railings shall be not less than 42 inches (1067 mm) in height and shall be sufficient to direct pedestrians around construction areas.

[BS] 3306.5 Barriers. Barriers shall be not less than 8 feet (2438 mm) in height and shall be placed on the side of the walkway nearest the construction. Barriers shall extend the entire length of the construction site. Openings in such barriers shall be protected by doors that are normally kept closed.

[BS] 3306.6 Barrier design. Barriers shall be designed to resist *loads* required in Chapter 16 unless constructed as follows:

1. Barriers shall be provided with 2-inch by 4-inch (51 mm by 102 mm) top and bottom plates.
2. The barrier material shall be boards not less than $^{3}/_{4}$-inch (19.1 mm) thick or *wood structural panels* not less than $^{1}/_{4}$-inch (6.4 mm) thick.
3. Wood structural use panels shall be bonded with an adhesive identical to that for exterior wood structural use panels.
4. Wood structural use panels $^{1}/_{4}$ inch (6.4 mm) or $^{5}/_{16}$ inch (23.8 mm) in thickness shall have studs spaced not more than 2 feet (610 mm) on center.
5. Wood structural use panels $^{3}/_{8}$ inch (9.5 mm) or $^{1}/_{2}$ inch (12.7 mm) in thickness shall have studs spaced not more than 4 feet (1219 mm) on center provided

TABLE 3306.1
PROTECTION OF PEDESTRIANS

HEIGHT OF CONSTRUCTION	DISTANCE FROM CONSTRUCTION TO LOT LINE	TYPE OF PROTECTION REQUIRED
8 feet or less	Less than 5 feet	Construction railings
	5 feet or more	None
More than 8 feet	Less than 5 feet	Barrier and covered walkway
	5 feet or more, but not more than one-fourth the height of construction	Barrier and covered walkway
	5 feet or more, but between one-fourth and one-half the height of construction	Barrier
	5 feet or more, but exceeding one-half the height of construction	None

For SI: 1 foot = 304.8 mm.

that a 2-inch by 4-inch (51 mm by 102 mm) stiffener is placed horizontally at mid-height where the stud spacing is greater than 2 feet (610 mm) on center.

6. Wood structural use panels $^5/_8$ inch (15.9 mm) or thicker shall not span over 8 feet (2438 mm).

[BS] 3306.7 Covered walkways. Covered walkways shall have a clear height of not less than 8 feet (2438 mm) as measured from the floor surface to the canopy overhead. Adequate lighting shall be provided at all times. Covered walkways shall be designed to support all imposed *loads*. The design *live load* shall be not less than 150 psf (7.2 kN/m^2) for the entire structure.

Exception: Roofs and supporting structures of covered walkways for new, *light-frame construction* not exceeding two *stories* above grade plane are permitted to be designed for a *live load* of 75 psf (3.6kN/m^2) or the *loads* imposed on them, whichever is greater. In lieu of such designs, the roof and supporting structure of a covered walkway are permitted to be constructed as follows:

1. Footings shall be continuous 2-inch by 6-inch (51 mm by 152 mm) members.
2. Posts not less than 4 inches by 6 inches (102 mm by 152 mm) shall be provided on both sides of the roof and spaced not more than 12 feet (3658 mm) on center.
3. Stringers not less than 4 inches by 12 inches (102 mm by 305 mm) shall be placed on edge upon the posts.
4. Joists resting on the stringers shall be not less than 2 inches by 8 inches (51 mm by 203 mm) and shall be spaced not more than 2 feet (610 mm) on center.
5. The deck shall be planks not less than 2 inches (51 mm) thick or *wood structural panels* with an exterior exposure durability classification not less than $^{23}/_{32}$ inch (18.3 mm) thick nailed to the joists.
6. Each post shall be knee braced to joists and stringers by members not less than 2 inches by 4 inches (51 mm by 102 mm); 4 feet (1219 mm) in length.
7. A curb that is not less than 2 inches by 4 inches (51 mm by 102 mm) shall be set on edge along the outside edge of the deck.

[BS] 3306.8 Repair, maintenance and removal. Pedestrian protection required by this chapter shall be maintained in place and kept in good order for the entire length of time pedestrians are subject to being endangered. The *owner* or the *owner's* authorized agent, on completion of the construction activity, shall immediately remove walkways, debris and other obstructions and leave such public property in as good a condition as it was before such work was commenced.

[BS] 3306.9 Adjacent to excavations. Every excavation on a site located 5 feet (1524 mm) or less from the street *lot line* shall be enclosed with a barrier not less than 6 feet (1829 mm) in height. Where located more than 5 feet (1524 mm) from the street *lot line*, a barrier shall be erected where required by the *building official*. Barriers shall be of adequate strength to resist wind pressure as specified in Chapter 16.

SECTION 3307 PROTECTION OF ADJOINING PROPERTY

[BS] 3307.1 Protection required. Adjoining public and private property shall be protected from damage during construction, remodeling and demolition work. Protection shall be provided for footings, foundations, party walls, chimneys, skylights and roofs. Provisions shall be made to control water runoff and erosion during construction or demolition activities. The person making or causing an excavation to be made shall provide written notice to the *owners* of adjoining buildings advising them that the excavation is to be made and that the adjoining buildings should be protected. Said notification shall be delivered not less than 10 days prior to the scheduled starting date of the excavation.

[BS] 3307.2 Excavation retention systems. Where a retention system is used to provide support of an excavation for protection of adjacent *structures*, the system shall conform to the requirements in Sections 3307.2.1 through 3307.2.3.

[BS] 3307.2.1 Excavation retention system design. Excavation retention systems shall be designed by a *registered design professional* to provide vertical and lateral support.

[BS] 3307.2.2 Excavation retention system monitoring. The retention system design shall include requirements for monitoring of the system and adjacent structures for horizontal and vertical movement.

[BS] 3307.2.3 Retention system removal. Elements of the system shall only be removed or decommissioned where adequate replacement support is provided by backfill or by the new structure. Removal or decommissioning shall be performed in such a manner that protects the adjacent property.

SECTION 3308 TEMPORARY USE OF STREETS, ALLEYS AND PUBLIC PROPERTY

3308.1 Storage and handling of materials. The temporary use of streets or public property for the storage or handling of materials or of equipment required for construction or demolition, and the protection provided to the public shall comply with the provisions of the applicable governing authority and this chapter.

3308.1.1 Obstructions. Construction materials and equipment shall not be placed or stored so as to obstruct access to fire hydrants, standpipes, fire or police alarm boxes, catch basins or manholes, nor shall such material or equipment be located within 20 feet (6096 mm) of a street intersection, or placed so as to obstruct normal observations of traffic signals or to hinder the use of public transit loading platforms.

3308.2 Utility fixtures. Building materials, fences, sheds or any obstruction of any kind shall not be placed so as to obstruct free approach to any fire hydrant, fire department connection, utility pole, manhole, fire alarm box or catch basin, or so as to interfere with the passage of water in the gutter. Protection against damage shall be provided to such utility fixtures during the progress of the work, but sight of them shall not be obstructed.

SECTION 3309 FIRE EXTINGUISHERS

[F] 3309.1 Where required. Structures under construction, *alteration* or demolition shall be provided with not fewer than one *approved* portable fire extinguisher in accordance with Section 906 and sized for not less than ordinary hazard as follows:

1. At each *stairway* on all floor levels where combustible materials have accumulated.
2. In every storage and construction shed.
3. Additional portable fire extinguishers shall be provided where special hazards exist, such as the storage and use of flammable and combustible liquids.

[F] 3309.2 Fire hazards. The provisions of this code and the *International Fire Code* shall be strictly observed to safeguard against all fire hazards attendant upon construction operations.

SECTION 3310 MEANS OF EGRESS

3310.1 Stairways required. Where building construction exceeds 40 feet (12 192 mm) in height above the lowest level of fire department vehicle access, a temporary or permanent *stairway* shall be provided. As construction progresses, such *stairway* shall be extended to within one floor of the highest point of construction having secured decking or flooring.

[F] 3310.2 Maintenance of means of egress. *Means of egress* and required *accessible means of egress* shall be maintained at all times during construction, demolition, remodeling or *alterations* and *additions* to any building.

> **Exception:** Existing *means of egress* need not be maintained where *approved* temporary *means of egress* systems and facilities are provided.

SECTION 3311 STANDPIPES

[F] 3311.1 Where required. In buildings required to have standpipes by Section 905.3.1, not fewer than one standpipe shall be provided for use during construction. Such standpipes shall be installed prior to construction exceeding 40 feet (12 192 mm) in height above the lowest level of fire department vehicle access. Such standpipes shall be provided with fire department hose connections at locations adjacent to *stairways* complying with Section 3310.1. As construction progresses, such standpipes shall be extended to within one floor of the highest point of construction having secured decking or flooring.

[F] 3311.2 Buildings being demolished. Where a building is being demolished and a standpipe exists within such a building, such standpipe shall be maintained in an operable condition so as to be available for use by the fire department. Such standpipe shall be demolished with the building but shall not be demolished more than one floor below the floor being demolished.

[F] 3311.3 Detailed requirements. Standpipes shall be installed in accordance with the provisions of Chapter 9.

> **Exception:** Standpipes shall be either temporary or permanent in nature, and with or without a water supply, provided that such standpipes conform to the requirements of Section 905 as to capacity, outlets and materials.

SECTION 3312 AUTOMATIC SPRINKLER SYSTEM

[F] 3312.1 Completion before occupancy. In buildings where an *automatic sprinkler system* is required by this code, it shall be unlawful to occupy any portion of a building or structure until the *automatic sprinkler system* installation has been tested and *approved*, except as provided in Section 111.3.

[F] 3312.2 Operation of valves. Operation of sprinkler control valves shall be permitted only by properly authorized personnel and shall be accompanied by notification of duly designated parties. When the sprinkler protection is being regularly turned off and on to facilitate connection of newly completed segments, the sprinkler control valves shall be checked at the end of each work period to ascertain that protection is in service.

SECTION 3313 WATER SUPPLY FOR FIRE PROTECTION

[F] 3313.1 Where required. An *approved* water supply for fire protection, either temporary or permanent, shall be made available as soon as combustible building materials arrive on the site, on commencement of vertical combustible construction, and on installation of a standpipe system in buildings under construction, in accordance with Sections 3313.2 through 3313.5.

> **Exception**: The *fire code official* is authorized to reduce the fire-flow requirements for isolated buildings or a group of buildings in rural areas or small communities where the development of full fire-flow requirements is impractical.

[F] 3313.2 Combustible building materials. When combustible building materials of the building under construction are delivered to a site, a minimum fire flow of 500 gallons per minute (1893 L/m) shall be provided. The fire hydrant used to provide this fire flow supply shall be within 500 feet (152 m) of the combustible building materials, as measured along an approved fire apparatus access lane. Where the site configuration is such that one fire hydrant cannot be located within 500 feet (152 m) of all combustible building materials, additional fire hydrants shall be required to provide coverage in accordance with this section.

[F] 3313.3 Vertical construction of Types III, IV and V construction. Prior to commencement of vertical construction of Type III, IV or V buildings that utilize any combustible building materials, the fire flow required by Sections 3313.3.1 through 3313.3.3 shall be provided, accompanied by fire hydrants in sufficient quantity to deliver the required fire flow and proper coverage.

[F] 3313.3.1 Fire separation up to 30 feet. Where a building of Type III, IV or V construction has a *fire separation distance* of less than 30 feet (9144 mm) from property lot lines, and an adjacent property has an *existing structure* or otherwise can be built on, the water supply shall provide either a minimum of 500 gallons per minute (1893 L/m), or the entire fire flow required for the building when constructed, whichever is greater.

[F] 3313.3.2 Fire separation of 30 feet up to 60 feet. Where a building of Type III, IV or V construction has a *fire separation distance* of 30 feet (9144 mm) up to 60 feet (18 288 mm) from property lot lines, and an adjacent property has an *existing structure* or otherwise can be built on, the water supply shall provide a minimum of 500 gallons per minute (1893 L/m), or 50 percent of the fire flow required for the building when constructed, whichever is greater.

[F] 3313.3.3 Fire separation of 60 feet or greater. Where a building of Type III, IV or V construction has a fire separation of 60 feet (18 288 mm) or greater from a property lot line, a water supply of 500 gallons per minute (1893 L/m) shall be provided.

[F] 3313.4 Vertical construction, Types I and II construction. If combustible building materials are delivered to the construction site, water supply in accordance with Section 3313.2 shall be provided. Additional water supply for fire flow is not required prior to commencing vertical construction of Type I and II buildings.

[F] 3313.5 Standpipe supply. Regardless of the presence of combustible building materials, the construction type or the *fire separation distance*, where a standpipe is required in accordance with Section 3313, a water supply providing a minimum flow of 500 gallons per minute (1893 L/m) shall be provided. The fire hydrant used for this water supply shall be located within 100 feet (30 480 mm) of the fire department connection supplying the standpipe.

SECTION 3314
FIRE WATCH DURING CONSTRUCTON

[F] 3314.1 Fire watch during combustible construction. A fire watch shall be provided during nonworking hours for construction that exceeds 40 feet (12 192 mm) in height above the lowest adjacent grade at any point along the building perimeter, for new multistory construction with an aggregate area exceeding 50,000 square feet (4645 m^2) per story or as required by the fire code official.

CHAPTER 34

RESERVED

Action taken during the 2012 Code Development Process removed Chapter 34, Existing Structures, from the IBC. The provisions of this chapter are contained in the International Existing Building Code. See Section 101.4.7.

CHAPTER 35

REFERENCED STANDARDS

User note:

About this chapter: The International Building Code® *contains numerous references to standards promulgated by other organizations that are used to provide requirements for materials and methods of construction. This chapter contains a comprehensive list of all standards that are referenced in this code. These standards, in essence, are part of this code to the extent of the reference to the standard.*

This chapter lists the standards that are referenced in various sections of this document. The standards are listed herein by the promulgating agency of the standard, the standard identification, the effective date and title, and the section or sections of this document that reference the standard. The application of the referenced standards shall be as specified in Section 102.4.

AA

Aluminum Association
1400 Crystal Drive, Suite 430
Arlington, VA 22202

ADM—2020: Aluminum Design Manual
1604.3.5, 2002.1

ASM 35—00: Aluminum Sheet Metal Work in Building Construction (Fourth Edition)
2002.1

AAMA

American Architectural Manufacturers Association
1900 E Golf Road, Suite 1250
Schaumburg, IL 60173

711—20: Voluntary Specification for Self Adhering Flashing Used for Installation of Exterior Wall Fenestration Products
1404.4

714—20: Voluntary Specification for Liquid Applied Flashing Used to Create a Water-resistive Seal around Exterior Wall Openings in Buildings
1404.4

1402—09: Standard Specifications for Aluminum Siding, Soffit and Fascia
1403.5.1

2502—19: Comparative Analysis Procedure for Window and Door Products
1709.5

AAMA/WDMA/CSA 101/I.S.2/A440—17: North American Fenestration Standard/Specifications for Windows, Doors and Skylights
1709.5.1, 2405.5

ACI

American Concrete Institute
38800 Country Club Drive
Farmington Hills, MI 48331-3439

117—10: Specification for Tolerances for Concrete Construction and Materials
1901.7.1

216.1—14: Code Requirements for Determining Fire Resistance of Concrete and Masonry Construction Assemblies
Table 721.1(2), 722.1

318—19: Building Code Requirements for Structural Concrete
722.2.4.3, 1604.3.2, 1616.2.1, 1616.3.1, 1704.5, Table 1705.3, 1705.3.2, 1808.8.2, Table 1808.8.2, 1808.8.5, 1808.8.6, 1810.1.3, 1810.2.4.1, 1810.3.2.1.1, 1810.3.2.1.2, 1810.3.8, 1810.3.9.4.2.1, 1810.3.9.4.2.2, 1810.3.10.1, 1810.3.11, 1810.3.11.1, 1810.3.12, 1810.3.13, 1901.2, 1901.3, 1902.1, 1903.1, 1904.1, 1904.2, 1905.1, 1905.1.1, 1905.1.2, 1905.1.3, 1905.1.4, 1905.1.5, 1905.1.6, 1905.1.7, 1905.1.8, 1908.1, 2108.3, 2206.1

550.5—18: Code Requirements for the Design of Precast Concrete Diaphragms for Earthquake Motions
Table 1705.3

ITG—7-09: Specification for Tolerances for Precast Concrete
1901.7.2

AISC

American Institute of Steel
130 East Randolph Street, Suite 2000
Chicago, IL 60601-6219

ANSI/AISC 341—16: Seismic Provisions for Structural Steel Buildings
1705.13.1.1, 1705.13.1.2, 1705.14.1.1, 1705.14.1.2, 1810.3.5.3.1, 2205.2.1.1, 2205.2.1.2, 2205.2.2, 2206.2.1

ANSI/AISC 358—16/s1—18: Prequalified Connections for Special and Intermediate Steel Moment Frames for Seismic Applications, Including Supplement No. 1
2205.2.1.1, 2205.2.1.2

ANSI/AISC 360—16: Specification for Structural Steel Buildings
722.5.2.2.1, 1604.3.3, 1705.2.1, 2202.1, 2203.1, 2205.1, 2205.2.1.1, 2206.1

AISI

American Iron and Steel Institute
25 Massachusetts Avenue, NW Suite 800
Washington, DC 20001

AISI S100—16(2020) w/S2—20: North American Specification for the Design of Cold-Formed Steel Structural Members, 2016 Edition (Reaffirmed 2020), with Supplement 2, 2020 Edition
1604.3.3, 1905.1.8, 2202.1, 2203.1, 2210.1, 2210.2

AISI S202—20: Code of Standard Practice for Cold-formed Steel Framing, 2020
2211.1.3.1

AISI S220—20: North American Standard for Cold-Formed Steel Nonstructural Framing
2202.1, 2203.1, 2211.2, Table 2506.2, Table 2507.2

AISI S230—2019: Standard for Cold-formed Steel Framing—Prescriptive Method for One- and Two-family Dwellings, 2019
1609.1.1, 1609.1.1.1, 2211.1.2

AISI S240—20: North American Standard for Cold-Formed Steel Structuring Framing, 2020
2202.1, 2203.1, 2211.1, 2211.1.1.1, 2211.1.3.3, Table 2506.2, Table 2507.2, Table 2603.12.1, Table 2603.12.2

AISI S400—20: North American Standard for Seismic Design of Cold-formed Steel Structural Systems, 2020
2210.2, 2211.1.1.1, 2211.1.1.2

ALI

Automotive Lift Institute, Inc.
P.O. Box 85
Cortland, NY 13045

ALI ALCTV—2016: Standard for Automotive Lifts—Safety Requirements for Construction, Testing and Validation (ANSI)
Table 3001.3

AMCA

Air Movement and Control Association International
30 West University Drive
Arlington Heights, IL 60004

540—13: Test Method for Louvers Impacted by Wind Borne Debris
1609.2.1

ANSI

American National Standards Institute
25 West 43rd Street, Fourth Floor
New York, NY 10036

A13.1—2020: Scheme for the Identification of Piping Systems
415.11.7.5

A108.1A—17: Installation of Ceramic Tile in the Wet-set Method, with Portland Cement Mortar
2103.2.3

A108.1B—17: Installation of Ceramic Tile, Quarry Tile on a Cured Portland Cement Mortar Setting Bed with Dry-set or Latex-Portland Mortar
2103.2.3

ANSI—continued

A108.4—09: Installation of Ceramic Tile with Organic Adhesives or Water-cleanable Tile-setting Epoxy Adhesive
2103.2.3.6

A108.5—19: Installation of Ceramic Tile with Dry-set Portland Cement Mortar or Latex-Portland Cement Mortar
2103.2.3.1, 2103.2.3.2

A108.6—19: Installation of Ceramic Tile with Chemical-resistant, Water Cleanable Tile-setting and -grouting Epoxy
2103.2.3.3

A108.8—19: Installation of Ceramic Tile with Chemical-resistant Furan Resin Mortar and Grout
2103.2.3.4

A108.9—19: Installation of Ceramic Tile with Modified Epoxy Emulsion Mortar/Grout
2103.2.3.5

A108.10—17: Installation of Grout in Tilework
2103.2.3.7

A118.1—18: American National Standard Specifications for Dry-set Portland Cement Mortar
2103.2.3.1

A118.3—20: American National Standard Specifications for Chemical-resistant, Water-cleanable Tile-setting and -grouting Epoxy and Water Cleanable Tile-setting Epoxy Adhesive
2103.2.3.3

A118.4—18: American National Standard Specifications for Modified Dry-set Cement Mortar
2103.2.3.2, 2103.2.4

A118.5—99: American National Standard Specifications for Chemical Resistant Furan Mortar and Grouts for Tile Installation
2103.2.3.4

A118.6—19: American National Standard Specifications for Cement Grouts for Tile Installation
2103.2.3.7

A118.8—99: American National Standard Specifications for Modified Epoxy Emulsion Mortar/Grout
2103.2.3.5

A136.1—19: American National Standard Specifications for the Installation of Ceramic Tile
2103.2.3.6

A137.1—19: American National Standard Specifications for Ceramic Tile
202

A137.3—17: American National Standard Specifications for Gauged Porcelain Tiles and Gauged Porcelain Tile Panel/Slabs
202

Z 97.1—14: Safety Glazing Materials Used in Buildings—Safety Performance Specifications and Methods of Test
2406.1.2, 2406.2, Table 2406.2(2), 2406.3.1, 2407.1, 2407.1.4, 2408.2.1, 2408.3, 2409.2, 2409.3, 2409.4.1

APA

APA - Engineered Wood Association
7011 South 19th Street
Tacoma, WA 98466-7400

ANSI 117—2020: Standard Specification for Structural Glued Laminated Timber of Softwood Species
2306.1

ANSI/APA A190.1—2017: Structural Glued Laminated Timber
2303.1.3, 2306.1

ANSI/APA PRG 320—2019: Standard for Performance-rated Cross-laminated Timber
602.4, 2303.1.4

ANSI/APA PRP 210—2019: Standard for Performance-Rated Engineered Wood Siding
2303.1.5, 2304.7, 2306.3, Table 2306.3(1)

ANSI/APA PRR 410—16: Standard for Performance-Rated Engineered Wood Rim Boards
2303.1.13

APA PDS Supplement 1—12: Design and Fabrication of Plywood Curved Panels (revised 2013)
2306.1

APA PDS Supplement 2—12: Design and Fabrication of Plywood-lumber Beams (revised 2013)
2306.1

APA—continued

APA PDS Supplement 3—12: Design and Fabrication of Plywood Stressed-skin Panels (revised 2013)
2306.1

APA PDS Supplement 4—12: Design and Fabrication of Plywood Sandwich Panels (revised 2013)
2306.1

APA PDS Supplement 5—16: Design and Fabrication of All-plywood Beams (revised 2013)
2306.1

APA PDS—20: Panel Design Specification
2306.1

APA R540—19: Builder Tips: Proper Storage and Handling of Glulam Beams
2306.1

APA S475—20: Glued Laminated Beam Design Tables
2306.1

APA S560—20: Field Notching and Drilling of Glued Laminated Timber Beams
2306.1

APA T300—16: Glulam Connection Details
2306.1

APA X440—17: Product Guide: Glulam
2306.1

APA X450—18: Glulam in Residential Construction—Building—Construction Guide
2306.1

ASABE

American Society of Agricultural and Biological Engineers
2950 Niles Road
St. Joseph, MI 49085

EP 484.3 DEC2017: Diaphragm Design of Metal-clad, Wood-frame Rectangular Buildings
2306.1

EP 486.3 SEP2017: Shallow-post and Pier Foundation Design
2306.1

EP 559.1 W/Corr. AUG2010(R2014): Design Requirements and Bending Properties for Mechanically Laminated Wood Assemblies
2306.1

ASCE/SEI

American Society of Civil Engineers
Structural Engineering Institute
1801 Alexander Bell Drive
Reston, VA 20191

7—16 with Supplement 1: Minimum Design Loads and Associated Criteria for Buildings and Other Structures
202, Table 1504.2, 1602.1, Table 1604.3, 1604.5, Table 1604.5, 1604.8.2, 1604.9, 1605.1, 1605.1.1, 1605.2 , 1606.3, 1607.9.1, 1607.9.1.1, 1607.9.1.2, 1607.10, 1607.14.1, 1607.17, 1608.1, 1608.2, Figure 1608.2(1), 1608.3, 1609.1.1, 1609.2, 1609.3, Figure 1609.3(5), Figure 1609.3(6), Figure 1609.3(7), Figure 1609.3(8), Figure 1609.3(9), Figure 1609.3(10), Figure 1609.3(11), Figure 1609.3(12), 1609.5.1, 1609.5.3, 1611.1, 1611.2, 1612.2, 1613.1, 1613.2.2, 1613.2.3, Table 1613.2.3(1), Table 1613.2.3(2), 1613.2.5, 1613.2.5.1, 1613.2.5.2, 1613.3, 1614.1, 1615.1, 1705.13 , 1705.13.1.1, 1705.13.1.2, 1705.13.4, 1705.14.1.1, 1705.14.1.2, 1705.14.2, 1705.14.3, 1705.14.4, 1709.5, 1709.5.3.1, 1802.1, 1803.5.12, 1806.1, 1808.3, 1808.3.1, 1809.13, 1810.3.1.1, 1810.3.6.1, 1810.3.8, 1810.3.9.2, 1810.3.9.4, 1810.3.9.4.1, 1810.3.9.4.2, 1810.3.11.2, 1810.3.12, 1902.1, 1905.1.2, 1905.1.7, 1905.1.8, 2205.2.1.1, 2205.2.1.2, 2205.2.2, 2206.2.1, 2209.1, 2209.2, 2210.2, 2211.1.1.1, Table 2304.6.1, Table 2306.3(3), Table 2308.7.5, 2404.1, 2505.1, 2505.2, 2506.2.1

8—20: Standard Specification for the Design of Cold-formed Stainless Steel Structural Members
1604.3.3, 2210.1, 2210.2

19—16: Structural Applications of Steel Cables for Buildings
2208.1

24—14: Flood Resistant Design and Construction
1202.4.2, 1202.4.4, 1612.2, 1612.4, 2702.1.8, 3001.3

ASCE/SEI—continued

29—19: Standard Calculation Methods for Structural Fire Protection
722.1

32—01: Design and Construction of Frost Protected Shallow Foundations
1809.5

49—12: Wind Tunnel Testing for Buildings and Other Structures
1609.1.1

55—16: Tensile Membrane Structures
3102.2

ASHRAE

ASHRAE
1791 Tullie Circle NE
Atlanta, GA 30329USA

170—2017: Ventilation of Health Care Facilities
1020.6

ASME

American Society of Mechanical Engineers
Two Park Avenue
New York, NY 10016

A17.1—2019/CSA B44—19: Safety Code for Elevators and Escalators
907.3.3, 911.1.6, 1009.4.1, 1607.11.1, 3001.2, Table 3001.3, 3001.4, 3001.5, 3002.5, 3003.2, 3007.1, 3008.1.4, 3008.7.1

A17.7—2007/CSA B44—07(R2019): Performance-based Safety Code for Elevators and Escalators
Table 3001.3, 3001.5, 3002.5

A18.1—2020: Safety Standard for Platform Lifts and Stairway Chairlifts
1110.9, Table 3001.3

A90.1—2020: Safety Standard for Belt Manlifts
Table 3001.3

B16.18—2018: Cast Copper Alloy Solder Joint Pressure Fittings
909.13.1

B16.22—2018: Wrought Copper and Copper Alloy Solder Joint Pressure Fittings
909.13.1

B20.1—2021: Safety Standard for Conveyors and Related Equipment
Table 3001.3, 3004.3

B31.3—2020: Process Piping
415.11.7

ASSP

American Society of Safety Professionals
520 N. Northwest Highway
Park Ridge, IL 60068

ANSI/ASSE Z359.1—2019: The Fall Protection Code
1015.6, 1015.7

ASTM

ASTM International
100 Barr Harbor Drive, P.O. Box C700
West Conshohocken, PA 19428

A6/A6M—2017A: Standard Specification for General Requirements for Rolled Structural Steel Bars, Plates, Shapes and Sheet Piling
1810.3.2.3, 1810.3.5.3.1, 1810.3.5.3.3

A36/A36M—14: Specification for Carbon Structural Steel
1810.3.2.3

A153/A153M—2016A: Specification for Zinc Coating (Hot-dip) on Iron and Steel Hardware
2304.10.6

ASTM—continued

A240/A240M—17: Standard Specification for Chromium and Chromium-nickel Stainless Steel Plate, Sheet and Strip for Pressure Vessels and for General Applications
Table 1507.4.3(1)

A252—2010(2018): Specification for Welded and Seamless Steel Pipe Piles
1810.3.2.3

A283/A283M—2018: Specification for Low and Intermediate Tensile Strength Carbon Steel Plates
1810.3.2.3

A416/A416M—2017A: Specification for Steel Strand, Uncoated Seven-wire for Prestressed Concrete
1810.3.2.2

A463/A463M—15: Standard Specification for Steel Sheet, Aluminum-coated, by the Hot-dip Process
Table 1507.4.3(2)

A572/A572M—2018: Specification for High-strength Low-alloy Columbium-Vanadium Structural Steel
1810.3.2.3

A588/A588M—15: Specification for High-strength Low-alloy Structural Steel with 50 ksi (345 MPa) Minimum Yield Point with Atmospheric Corrosion Resistance
1810.3.2.3

A615/A615M—15ae1: Specification for Deformed and Plain Carbon-steel Bars for Concrete Reinforcement
1704.5, 1810.3.10.2

A653/A653M—2017: Specification for Steel Sheet, Zinc-coated Galvanized or Zinc-iron Alloy-coated Galvannealed by the Hot-dip Process
Table 1507.4.3(1), Table 1507.4.3(2), 2304.10.6.1

A690/A690M—13a(2018): Standard Specification for High-strength Low-alloy Nickel, Copper, Phosphorus Steel H-piles and Sheet Piling with Atmospheric Corrosion Resistance for Use in Marine Environments
1810.3.2.3

A706/A706M—2016: Specification for Low-alloy Steel Deformed and Plain Bars for Concrete Reinforcement
1704.5, Table 1705.3, 2107.3, 2108.3

A722/A722M—2018: Specification for High-strength Steel Bars for Prestressed Concrete
1810.3.10.2

A755/A755M—2016E1: Specification for Steel Sheet, Metallic-coated by the Hot-dip Process and Prepainted by the Coil-coating Process for Exterior Exposed Building Products
Table 1507.4.3(1), Table 1507.4.3(2)

A792/A792M—10(2015): Specification for Steel Sheet, 55% Aluminum-zinc Alloy-coated by the Hot-dip Process
Table 1507.4.3(1), Table 1507.4.3(2)

A875/A875M—13: Standard Specification for Steel Sheet, Zinc-5%, Aluminum Alloy-coated by the Hot-dip Process
Table 1507.4.3(2)

A924/A924M—2017A: Standard Specification for General Requirements for Steel Sheet, Metallic-coated by the Hot-dip Process
Table 1507.4.3(1)

B42—2015A: Specification for Seamless Copper Pipe, Standard Sizes
909.13.1

B43—15: Specification for Seamless Red Brass Pipe, Standard Sizes
909.13.1

B68/B68M—11: Specification for Seamless Copper Tube, Bright Annealed (Metric)
909.13.1

B88—2016: Specification for Seamless Copper Water Tube
909.13.1

B101—12: Specification for Lead-coated Copper Sheet and Strip for Building Construction
1403.5.3, Table 1507.2.8.2, Table 1507.4.3(1)

B209—14: Specification for Aluminum and Aluminum Alloy Steel and Plate
Table 1507.4.3(1)

B251—2017: Specification for General Requirements for Wrought Seamless Copper and Copper-alloy Tube
909.13.1

B280—2018: Specification for Seamless Copper Tube for Air Conditioning and Refrigeration Field Service
909.13.1

ASTM—continued

B370—12: Specification for Copper Sheet and Strip for Building Construction
1403.5.2, Table 1507.2.8.2, Table 1507.4.3(1)

B695—2004(2016): Standard Specification for Coatings of Zinc Mechanically Deposited on Iron and Steel Strip for Building Construction
2304.10.6.1, 2304.10.6.3

C5—2018: Specification for Quicklime for Structural Purposes
2109.2.4.8.7, Table 2507.2

C22/C22M—00(2015): Specification for Gypsum
Table 2506.2

C27—1998(2018): Specification for Classification of Fireclay and High-alumina Refractory Brick
2111.6

C28/C28M—10(2015): Specification for Gypsum Plasters
Table 2507.2

C31/C31M—2018B: Practice for Making and Curing Concrete Test Specimens in the Field
Table 1705.3

C33/C33M—2018: Specification for Concrete Aggregates
722.3.1.4, 722.4.1.1.3

C35/C35—01(2014): Specification for Inorganic Aggregates for Use in Gypsum Plaster
Table 2507.2

C55—2017: Specification for Concrete Building Brick
Table 722.3.2

C59/C59M—00(2015): Specification for Gypsum Casting Plaster and Molding Plaster
Table 2507.2

C61/C61M—00(2015): Specification for Gypsum Keene's Cement
Table 2507.2

C62—2017: Standard Specification for Building Brick (Solid Masonry Units Made from Clay or Shale)
1807.1.6.3

C67/C67M—2018: Test Methods of Sampling and Testing Brick and Structural Clay Tile
722.4.1.1.1, 2109.2.1.1

C73—2017: Specification for Calcium Silicate Brick (Sand-lime Brick)
Table 722.3.2

C90—2016A: Specification for Loadbearing Concrete Masonry Units
Table 722.3.2, 1807.1.6.3, 2114.3

C91/C91M—2018: Specification for Masonry Cement
Table 2507.2

C94/C94M—2017A: Specification for Ready-mixed Concrete
110.3.1

C140/C140M—2018: Test Method Sampling and Testing Concrete Masonry Units and Related Units
722.3.1.2

C141/C141M —14: Standard Specification for Hydrated Hydraulic Lime for Structural Purposes
2109.2.4.8.7

C150/C150M—2018: Specification for Portland Cement
Table 2507.2

C172/C172M—2017: Practice for Sampling Freshly Mixed Concrete
Table 1705.3

C199—1984(2016): Test Method for Pier Test for Refractory Mortars
2111.6, 2111.9, 2113.12

C206—14: Specification for Finishing Hydrated Lime
2109.2.4.8.7, Table 2507.2

C208—2012(2017)E1: Specification for Cellulosic Fiber Insulating Board
Table 1508.2, 2303.1.6

ASTM—continued

C216—2017A: Specification for Facing Brick (Solid Masonry Units Made from Clay or Shale)
Table 721.1(2), 1807.1.6.3

C270—14a: Specification for Mortar for Unit Masonry
2103.2.4

C315—2007(2016): Specification for Clay Flue Liners and Chimney Pots
2111.9, 2113.11.1, Table 2113.16(1)

C317/C317M—2000(2015): Specification for Gypsum Concrete
2514.1

C330/C330M—2017A: Specification for Lightweight Aggregates for Structural Concrete
202

C331/C331M—2017: Specification for Lightweight Aggregates for Concrete Masonry Units
722.3.1.4, 722.4.1.1.3

C406/C406M—15: Specification for Roofing Slate
1507.7.5

C472—99(2014): Standard Test Methods for Physical Testing of Gypsum, Gypsum Plasters and Gypsum Concrete
Table 2506.2

C473—2017: Test Methods for Physical Testing of Gypsum Panel Products
Table 2506.2

C474—15: Test Methods for Joint Treatment Materials for Gypsum Board Construction
Table 2506.2

C475/C475M—2017: Specification for Joint Compound and Joint Tape for Finishing Gypsum Board
Table 2506.2

C514—04(2014): Specification for Nails for the Application of Gypsum Board
Table 721.1(2), Table 721.1(3), Table 2506.2

C516—2008(2014)E1: Specifications for Vermiculite Loose Fill Thermal Insulation
722.3.1.4, 722.4.1.1.3

C547—2017: Specification for Mineral Fiber Pipe Insulation
Table 721.1(2), Table 721.1(3)

C549—06(2012): Specification for Perlite Loose Fill Insulation
722.3.1.4, 722.4.1.1.3

C552—2017E1: Standard Specification for Cellular Glass Thermal Insulation
Table 1508.2

C557—2003(2017): Specification for Adhesives for Fastening Gypsum Wallboard to Wood Framing
Table 2506.2, 2508.4

C578—2018: Standard Specification for Rigid, Cellular Polystyrene Thermal Insulation
Table 1508.2, 2603.10, Table 2603.12.1, Table 2603.12.2, Table 2603.13.1, Table 2603.13.2

C587—2004(2018): Specification for Gypsum Veneer Plaster
Table 2507.2

C595/C595M—2018: Specification for Blended Hydraulic Cements
Table 2507.2

C631—09(2014): Specification for Bonding Compounds for Interior Gypsum Plastering
Table 2507.2

C635/C635M—2017: Specification for the Manufacture, Performance and Testing of Metal Suspension Systems for Acoustical Tile and Lay-in Panel Ceilings
2506.2.1

C636/C636M—13: Practice for Installation of Metal Ceiling Suspension Systems for Acoustical Tile and Lay-in Panels
808.1.1.1

C652—2017A: Specification for Hollow Brick (Hollow Masonry Units Made from Clay or Shale)
1807.1.6.3

C726—2017: Standard Specification for Mineral Wool Roof Insulation Board
Table 1508.2

ASTM—continued

C728—2017A: Standard Specification for Perlite Thermal Insulation Board
Table 1508.2

C744—2016: Specification for Prefaced Concrete and Calcium Silicate Masonry Units
Table 722.3.2

C754—2018: Specification for Installation of Steel Framing Members to Receive Screw-attached Gypsum Panel Products
Table 2508.1, Table 2511.1.1

C836/C836M—2018: Specification for High-solids Content, Cold Liquid-applied Elastomeric Waterproofing Membrane for Use with Separate Wearing Course
1507.14.2

C840—2018A: Specification for Application and Finishing of Gypsum Board
Table 2508.1, 2509.2

C841—2003(2018): Specification for Installation of Interior Lathing and Furring
Table 2508.1, Table 2511.1.1

C842—05(2015): Specification for Application of Interior Gypsum Plaster
Table 2511.1.1, 2511.3, 2511.4

C843—2017: Specification for Application of Gypsum Veneer Plaster
Table 2511.1.1

C844—2015: Specification for Application of Gypsum Base to Receive Gypsum Veneer Plaster
Table 2508.1

C847—14a: Specification for Metal Lath
Table 2507.2

C887—13: Specification for Packaged, Dry Combined Materials for Surface Bonding Mortar
1805.2.2, 2103.2.2

C897—15: Specification for Aggregate for Job-mixed Portland Cement-based Plaster
Table 2507.2

C920—2018: Standard for Specification for Elastomeric Joint Sealants
703.7, Table 2506.2

C926—2018B: Specification for Application of Portland Cement-based Plaster
2109.2.4.8.6, 2109.2.4.8.8, 2510.3, Table 2511.1.1, 2511.3, 2511.4, 2512.1, 2512.1.2, 2512.2, 2512.6, 2512.8.2, 2512.9, 2513.7

C932—06(2013): Specification for Surface-applied Bonding Compounds for Exterior Plastering
Table 2507.2

C933—2018: Specification for Welded Wire Lath
Table 2507.2

C946—2018: Specification for Construction of Dry-stacked, Surface-bonded Walls
2103.2.2, 2114.5

C954—2018: Specification for Steel Drill Screws for the Application of Gypsum Panel Products or Metal Plaster Bases to Steel Studs from 0.033 inch (0.84 mm) to 0.112 inch (2.84 mm) in Thickness
Table 2506.2, Table 2507.2

C956—04(2015): Specification for Installation of Cast-in-place Reinforced Gypsum Concrete
2514.1

C957/C957M—2017: Specification for High-solids Content, Cold Liquid-applied Elastomeric Waterproofing Membrane with Integral Wearing Surface
1507.14.2

C1002—2018: Specification for Steel Self-piercing Tapping Screws for the Application of Gypsum Panel Products or Metal Plaster Bases to Wood Studs or Steel Studs
722.7.2.2, Table 2506.2, Table 2507.2

C1007—11a(2015): Specification for Installation of Load Bearing (Transverse and Axial) Steel Studs and Related Accessories
Table 2508.1, Table 2511.1.1

C1029—15: Specification for Spray-applied Rigid Cellular Polyurethane Thermal Insulation
1507.13.2

C1032—2018: Specification for Woven Wire Plaster Base
Table 2507.2

ASTM—continued

C1047—2018: Specification for Accessories for Gypsum Wallboard and Gypsum Veneer Base
Table 2506.2, Table 2507.2

C1063—2018B: Specification for Installation of Lathing and Furring to Receive Interior and Exterior Portland Cement-based Plaster
2109.2.4.8.5, 2510.3, Table 2511.1.1, 2512.1.1

C1088—2018: Specification for Thin Veneer Brick Units Made from Clay or Shale
Table 721.1(2)

C1157/C1157M—2017: Standard Performance Specification for Hydraulic Cement
Table 2507.2

C1167—2011(2017): Specification for Clay Roof Tiles
1507.3.4

C1177/C1177M—2017: Specification for Glass Mat Gypsum Substrate for Use as Sheathing
Table 1508.2, Table 2506.2

C1178/C1178M—2018: Specification for Coated Mat Water-resistant Gypsum Backing Panel
Table 2506.2, Table 2509.2

C1186—2008(2016): Specification for Flat Fiber Cement Sheets
1403.10, 1404.16.1, 1404.16.2

C1261—2013(2017)E1: Specification for Firebox Brick for Residential Fireplaces
2111.6, 2111.9

C1278/C1278M—2017: Specification for Fiber-reinforced Gypsum Panel
Table 1508.2, Table 2506.2

C1280—13a: Specification for Application of Exterior Gypsum Panel Products for Use as Sheathing
Table 2508.1, 2508.2

C1283—2015: Practice for Installing Clay Flue Lining
2113.9.1, 2113.12

C1288—2017: Standard Specification for Discrete Nonasbestos Fiber-cement Interior Substrate Sheets
Table 2509.2

C1289—2018: Standard Specification for Faced Rigid Cellular Polyisocyanurate Thermal Insulation Board
Table 1508.2, 2603.10, Table 2603.12.1, Table 2603.12.2

C1313/C1313M—13: Standard Specification for Sheet Radiant Barriers for Building Construction Applications
1510.4

C1325—2018: Standard Specification for Nonasbestos Fiber-mat Reinforced Cement Backer Units
Table 2509.2

C1328/C1328M—12: Specification for Plastic (Stucco Cement)
Table 2507.2

C1364—2017: Standard Specification for Architectural Cast Stone
2103.1

C1372—17: Standard Specification for Dry-Cast Segmental Retaining Wall Units
1807.2.4

C1396/C1396M—2017: Specification for Gypsum Board
Figure 722.5.1(2), Figure 722.5.1(3), Table 2506.2

C1492—2003(2016): Standard Specification for Concrete Roof Tile
1507.3.5

C1568—08(2013): Standard Test Method for Wind Resistance of Concrete and Clay Roof Tiles (Mechanical Uplift Resistance Method)
1504.3.1.1

C1569—03(2016): Standard Test Method for Wind Resistance of Concrete and Clay Roof Tiles (Wind Tunnel Method)
1504.3.1.2

C1570—03(2016): Standard Test Method for Wind Resistance of Concrete and Clay Roof Tiles (Air Permeability Method)
1504.3.1.3

C1600/C1600M—2017: Standard Specification for Rapid Hardening Hydraulic Cement
Table 2507.2

ASTM—continued

C1629/C1629M—2018A: Standard Classification for Abuse-resistant Nondecorated Interior Gypsum Panel Products and Fiber-reinforced Cement Panels
403.2.2.1, 403.2.2.2, 403.2.2.4

C1658/C1658M—2018: Standard Specification for Glass Mat Gypsum Panels
Table 2506.2

C1670/C1670M—2018: Standard Specification for Adhered Manufactured Stone Masonry Veneer Units
2103.1

C1707—18: Standard Specification for Pozzolanic Hydraulic Lime for Structural Purposes
2109.2.4.8.7

C1766—2015: Standard Specification for Factory-laminated Gypsum Panel Products
Table 2506.2

C1788—14: Standard Specification for Non Metallic Plaster Bases (Lath) Used with Portland Cement Based Plaster in Vertical Wall Applications
2109.2.4.8.5

D25—2012(2017): Specification for Round Timber Piles
1810.3.2.4, 2303.1.12

D41/D41M—2011(2016): Specification for Asphalt Primer Used in Roofing, Dampproofing and Waterproofing
Table 1507.10.2

D43/D43M—2000(2018): Specification for Coal Tar Primer Used in Roofing, Dampproofing and Waterproofing
Table 1507.10.2

D56—2016A: Test Method for Flash Point by Tag Closed Cup Tester
202

D86—2017: Test Method for Distillation of Petroleum Products and Liquid Fuels at Atmospheric Pressure
202

D92—12b: Test Method for Flash and Fire Points by Cleveland Open Cup Tester
202

D93—2018: Test Methods for Flash Point by Pensky-Martens Closed Cup Tester
202

D226/D226M—2017: Specification for Asphalt-saturated Organic Felt Used in Roofing and Waterproofing
1403.2, 1505.2, 1507.1.1, Table 1507.1.1(1), 1507.3.3, Table 1507.8, 1507.9.5, Table 1507.10.2, 1507.17.3, 1507.17.4.1

D227/D227M—2003(2018): Specification for Coal-tar-saturated Organic Felt Used in Roofing and Waterproofing
Table 1507.10.2

D312/D312M—2016M: Specification for Asphalt Used in Roofing
Table 1507.10.2

D422—63(2007)e2: Test Method for Particle-size Analysis of Soils
1803.5.3

D448—2012(2017): Standard Classification for Sizes of Aggregate for Road and Bridge Construction
1507.12.3

D450/D450M—2017(2018): Specification for Coal-tar Pitch Used in Roofing, Dampproofing and Waterproofing
Table 1507.10.2

D635—14: Test Method for Rate of Burning and/or Extent and Time of Burning of Plastics in a Horizontal Position
2606.4

D1143/D1143M—2007(2013)E1: Test Methods for Deep Foundations Under Static Axial Compressive Load
1810.3.3.1.2

D1227—13: Specification for Emulsified Asphalt Used as a Protective Coating for Roofing
Table 1507.10.2

D1557—12e1: Test Methods for Laboratory Compaction Characteristics of Soil Using Modified Effort [56,000 ft-lb/ft^3 (2,700 kN m/m^3)]
1705.6, 1804.6

D1863/D1863M—2005(2018): Specification for Mineral Aggregate Used on Built-up Roofs
Table 1504.9, Table 1507.10.2

D1929—16: Standard Test Method for Determining Ignition Temperature of Plastics
402.6.4.4, 406.7.2, 2606.4

ASTM—continued

D1970/D1970M—2017A: Specification for Self-adhering Polymer Modified Bituminous Sheet Materials Used as Steep Roof Underlayment for Ice Dam Protection
1507.1.1, 1507.3.9, 1507.5.7, 1507.8.8, 1507.9.9, 1507.11.2.1, 1507.17.4.1

D2178/D2178M—15A: Specification for Asphalt Glass Felt Used in Roofing and Waterproofing
Table 1507.10.2

D2487—2017: Practice for Classification of Soils for Engineering Purposes (Unified Soil Classification System)
Table 1610.1, 1803.5.1

D2626/D2626M—04(2012)e1: Specification for Asphalt Saturated and Coated Organic Felt Base Sheet Used in Roofing
Table 1507.1.1(1), 1507.3.3, Table 1507.10.2

D2822/D2822M—2005(2011): Specification for Asphalt Roof Cement, Asbestos Containing
Table 1507.10.2

D2823/D2823M—05(2011)e1: Specification for Asphalt Roof Coatings, Asbestos Containing
Table 1507.10.2

D2824/D2824M—2018: Standard Specification for Aluminum-pigmented Asphalt Roof Coatings, Nonfibered and Fibered without Asbestos
Table 1507.10.2

D2843—16: Standard Test Method for Density of Smoke from the Burning or Decomposition of Plastics
2606.4

D2859—2016: Standard Test Method for Ignition Characteristics of Finished Textile Floor Covering Materials
804.4.1, 804.4.2

D2898—2010(2017): Test Methods for Accelerated Weathering of Fire-retardant-treated Wood for Fire Testing
1505.1, 2303.2.4, 2303.2.6

D3019/D3019M—2017: Specification for Lap Cement Used with Asphalt Roll Roofing, Nonfibered, Asbestos Fibered and Nonasbestos Fibered
Table 1507.10.2

D3161/D3161M—2016A: Test Method for Wind Resistance of Steep Slope Roofing Products (Fan Induced Method)
1504.2 , Table 1504.2, 1504.4.3

D3200—1974(2017): Standard Specification and Test Method for Establishing Recommended Design Stresses for Round Timber Construction Poles
2303.1.12

D3201/D3201M—13: Test Method for Hygroscopic Properties of Fire-retardant-treated Wood and Wood-based Products
2303.2.7

D3278—96(2011): Test Methods for Flash Point of Liquids by Small Scale Closed-cup Apparatus
202

D3462/D3462M—2016: Specification for Asphalt Shingles Made from Glass Felt and Surfaced with Mineral Granules
1507.2.4

D3468/D3468M—99(2013)e1: Specification for Liquid-applied Neoprene and Chlorosulfonated Polyethylene Used in Roofing and Waterproofing
1507.14.2

D3498—03(2011): Standard Specification for Adhesives for Field-Gluing Plywood to Lumber Framing for Floor Systems
703.7

D3679—2017: Specification for Rigid Poly (Vinyl Chloride) (PVC) Siding
1403.9, 1404.14

D3689/D3698M—07(2013)e1: Test Methods for Deep Foundations under Static Axial Tensile Load
1810.3.3.1.5

D3737—2018E1: Practice for Establishing Allowable Properties for Structural Glued Laminated Timber (Glulam)
2303.1.3

D3746/D3746M—1985(2015)E1: Test Method for Impact Resistance of Bituminous Roofing Systems
1504.8

D3747—79(2007): Specification for Emulsified Asphalt Adhesive for Adhering Roof Insulation
Table 1507.10.2

D3909/D3909M—14: Specification for Asphalt Roll Roofing (Glass Felt) Surfaced with Mineral Granules
1507.2.8.2, 1507.6.5, Table 1507.10.2

ASTM—continued

D3957—2009(2015): Standard Practices for Establishing Stress Grades for Structural Members Used in Log Buildings
2303.1.11

D4022/D4022M—07(2012)e1: Specification for Coal Tar Roof Cement, Asbestos Containing
Table 1507.10.2

D4272—15: Test Method for Total Energy Impact of Plastic Films by Dart Drop
1504.8

D4318—10e1: Test Methods for Liquid Limit, Plastic Limit and Plasticity Index of Soils
1803.5.3

D4434/D4434M—2015: Specification for Poly (Vinyl Chloride) Sheet Roofing
Table 1507.12.2

D4479/D4479M—2007(2018): Specification for Asphalt Roof Coatings—Asbestos-free
Table 1507.10.2

D4586/D4586M—2007(2018): Specification for Asphalt Roof Cement—Asbestos-free
Table 1507.10.2

D4601/D4601M—04(2012)e1: Specification for Asphalt-coated Glass Fiber Base Sheet Used in Roofing
Table 1507.10.2, 1507.11.2.1

D4637/D4637M—2015: Specification for EPDM Sheet Used in Single-ply Roof Membrane
Table 1507.12.2

D4829—11: Test Method for Expansion Index of Soils
1803.5.3

D4869/D4869M—2016A: Specification for Asphalt-saturated (Organic Felt) Underlayment Used in Steep Slope Roofing
1507.1.1, Table 1507.1.1(1), 1507.17.3, 1507.17.4.1

D4897/D4897M—2016: Specification for Asphalt-coated Glass Fiber Venting Base Sheet Used in Roofing
Table 1507.10.2

D4945—2017: Test Method for High-strain Dynamic Testing of Deep Foundations
1705.10, 1810.3.3.1.2

D4990—97a(2013): Specification for Coal Tar Glass Felt Used in Roofing and Waterproofing
Table 1507.10.2

D5019—07a: Specification for Reinforced Nonvulcanized Polymeric Sheet Used in Roofing Membrane
Table 1507.12.2

D5055—2016: Specification for Establishing and Monitoring Structural Capacities of Prefabricated Wood I-joists
2303.1.2

D5456—2018: Specification for Evaluation of Structural Composite Lumber Products
2303.1.10

D5516—2018: Test Method of Evaluating the Flexural Properties of Fire-retardant Treated Softwood Plywood Exposed to Elevated Temperatures
2303.2.5.1

D5643/D5643M—2006(2018): Specification for Coal Tar Roof Cement, Asbestos-free
Table 1507.10.2

D5664—2017: Standard Test Method for Evaluating the Effects of Fire-retardant Treatment and Elevated Temperatures on Strength Properties of Fire-retardant Treated Lumber
2303.2.5.2

D5665/D5665M—99a(2014)e1: Specification for Thermoplastic Fabrics Used in Cold-applied Roofing and Waterproofing
Table 1507.10.2

D5726—98(2013): Specification for Thermoplastic Fabrics Used in Hot-applied Roofing and Waterproofing
Table 1507.10.2

D5882—16: Standard Test Method for Low Strain Impact Integrity Testing of Deep Foundations
1705.10

D6083/D6083M—2018: Specification for Liquid Applied Acrylic Coating Used in Roofing
Table 1507.10.2, Table 1507.13.3

D6162/D6162M—2016: Specification for Styrene-butadiene-styrene (SBS) Modified Bituminous Sheet Materials Using a Combination of Polyester and Glass Fiber Reinforcements
1507.11.2

ASTM—continued

D6163/D6163M—2016: Specification for Styrene-butadiene-styrene (SBS) Modified Bituminous Sheet Materials Using Glass Fiber Reinforcements
1507.11.2

D6164/D6164M—2016: Specification for Styrene-butadiene-styrene (SBS) Modified Bituminous Sheet Metal Materials Using Polyester Reinforcements
1507.11.2

D6222/D6222M—2016: Specification for Atactic Polypropylene (APP) Modified Bituminous Sheet Materials Using Polyester Reinforcements
1507.11.2

D6223/D6223M—2016: Specification for Atactic Polypropylene (APP) Modified Bituminous Sheet Materials Using a Combination of Polyester and Glass Fiber Reinforcements
1507.11.2

D6298—2016: Specification for Fiberglass Reinforced Styrene-butadiene-styrene (SBS) Modified Bituminous Sheets with a Factory Applied Metal Surface
1507.11.2

D6305—08(2015)e1: Practice for Calculating Bending Strength Design Adjustment Factors for Fire-retardant-treated Plywood Roof Sheathing
2303.2.5.1

D6380/D6380M—2003(2018): Standard Specification for Asphalt Roll Roofing (Organic) Felt
Table 1507.1.1(1), 1507.2.8.2, 1507.3.3, 1507.6.5

D6464—2003A(2017): Standard Specification for Expandable Foam Adhesives for Fastening Gypsum Wallboard to Wood Framing
Table 2506.2, 2508.4

D6509/D6509M—2016: Standard Specification for Atactic Polypropylene (APP) Modified Bituminous Base Sheet Materials Using Glass Fiber Reinforcements
1507.11.2

D6694/D6694M—08(2013)e1: Standard Specification for Liquid-applied Silicone Coating Used in Spray Polyurethane Foam Roofing Systems
Table 1507.13.3

D6754/D6754M—2015: Standard Specification for Ketone Ethylene Ester Based Sheet Roofing
Table 1507.12.2

D6757—2018: Specification for Underlayment Felt Containing Inorganic Fibers Used in Steep Slope Roofing
1507.1.1, Table 1507.1.1(1), 1507.17.3, 1507.17.4.1

D6760—16: Standard Test Method for Integrity Testing of Concrete Deep Foundations by Ultrasonic Crosshole Testing
1705.10

D6841—2016: Standard Practice for Calculating Design Value Treatment Adjustment Factors for Fire-retardant Treated Lumber
2303.2.5.2

D6878/D6878M—2017: Standard Specification for Thermoplastic Polyolefin Based Sheet Roofing
Table 1507.12.2

D6947/D6947M—2016: Standard Specification for Liquid Applied Moisture Cured Polyurethane Coating Used in Spray Polyurethane Foam Roofing System
Table 1507.13.3

D7032—2017: Standard Specification for Establishing Performance Ratings for Wood, Plastic Composite Deck Boards and Guardrail Systems (Guards or Rails)
705.2.3.1, 2612.2, 2612.4, 2612.5.1

D7147—2011(2018): Specification for Testing and Establishing Allowable Loads of Joist Hangers
2303.5, 2304.10.4

D7158/D7158M—2019: Standard Test Method for Wind Resistance of Asphalt Shingles (Uplift Force/Uplift Resistance Method)
1504.2 , Table 1504.2

D7254—2017: Standard Specification for Polypropylene (PP) Siding
1403.12

D7425/D7425M—13: Standard Specification for Spray Polyurethane Foam Used for Roofing Applications
1507.13.2

D7655/D7655M—2012(2017): Standard Classification for Size of Aggregate Used as Ballast for Roof Membrane Systems
1507.12.3

ASTM—continued

D7672—14E1: Standard Specification for Evaluating Structural Capacities of Rim Board Products and Assemblies
2303.1.13

D7949—14: Standard Test Methods for Thermal Integrity Profiling of Concrete Deep Foundations
1705.10

E84—2018B: Standard Test Methods for Surface Burning Characteristics of Building Materials
202, 402.6.4.4, 406.7.2, 424.2, 602.4.1.1, 602.4.2.1, 602.4.3.1, 720.1, 720.4, 803.1.2, 803.5.2, 803.10, 803.11, 803.12, 803.13, 806.7, 1402.5, 1403.12.1, 1406.9, 1406.10.1, 1408.9, 1408.10.1, 1511.6.2, 1511.6.3, 2303.2, 2603.3, 2603.4.1.13, 2603.5.4, 2603.5.5, 2603.7, 2604.2.4, 2606.4, 2612.3, 2614.3, 3105.3

E90—2009(2016): Test Method for Laboratory Measurement of Airborne Sound Transmission Loss of Building Partitions and Elements
1206.2, 1206.2.1

E96/E96M—2016: Standard Test Methods for Water Vapor Transmission of Materials
202, 1202.3, 1404.3.1

E108—2017: Standard Test Methods for Fire Tests of Roof Coverings
1505.1, 2603.6, 2610.2, 2610.3

E119—2018B: Standard Test Methods for Fire Tests of Building Construction and Materials
703.2.1.1, 703.2.1.3, 703.2.1.4, 703.2.1.5, 703.2.2, 703.4, 703.6, 704.12, 705.7, 705.8.5, 707.6, 712.1.13.2, 714.4.1, 714.5.1, 715.3, 715.4, 715.4.1, Table 716.1(1), Table 716.1(2), Table 716.1(3), 716.1.2.3, 716.2.5.1.1, 716.2.5.4, 716.3.2.1.1, 717.3.1, 717.5.2, 717.5.3, 717.6.1, 717.6.2, Table 721.1(1), 2103.1, 2603.5.1

E136—2019: Standard Test Method for Behavior of Materials in a Vertical Tube Furnace at 750°C
703.3.1

E283—04(2012): Standard Test Method for Determining Rate of Air Leakage through Exterior Windows, Curtain Walls and Doors Under Specified Pressure Differences across the Specimen
202

E330/E330M—14: Standard Test Method for Structural Performance of Exterior Windows, Doors, Skylights and Curtain Walls by Uniform Static Air Pressure Difference
1709.5.2, 1709.5.2.1, 1709.5.3.1

E331—2000(2016): Standard Test Method for Water Penetration of Exterior Windows, Skylights, Doors and Curtain Walls by Uniform Static Air Pressure Difference
1402.2, 1403.2

E336—17a: Standard Test Method for Measurement of Airborne Sound Attenuation between Rooms in Buildings
1206.2

E492—2009(2016)E1: Test Method for Laboratory Measurement of Impact Sound Transmission Through Floor-ceiling Assemblies Using the Tapping Machine
1206.3

E605/E605M—93(2015)e1: Test Method for Thickness and Density of Sprayed Fire-resistive Material (SFRM) Applied to Structural Members
1705.15.4.1, 1705.15.4.2, 1705.15.4.5, 1705.15.5

E648—2017A: Standard Test Method for Critical Radiant Flux of Floor-covering Systems Using a Radiant Heat Energy Source
406.2.4, 424.2, 804.2, 804.3

E681—09(2015): Test Methods for Concentration Limits of Flammability of Chemical Vapors and Gases
202

E736/E736M—2017: Test Method for Cohesion/Adhesion of Sprayed Fire-resistive Materials Applied to Structural Members
704.13.3.2, 1705.15.6

E814—2013A(2017): Test Method for Fire Tests of Penetration Firestop Systems
202, 714.4.1.2, 714.4.2, 714.5.1.2

E970—2017: Standard Test Method for Critical Radiant Flux of Exposed Attic Floor Insulation Using a Radiant Heat Energy Source
720.3.1

E1007—16: Test Method f or Field Measurement of Tapping Machine Impact Sound Transmission Through Floor-Ceiling Assemblies and Associated Support Structures
1206.3

E1300—2016: Practice for Determining Load Resistance of Glass in Buildings
2404.1, 2404.2, 2404.3.1, 2404.3.2, 2404.3.3, 2404.3.4, 2404.3.5

ASTM—continued

E1354—2017: Standard Test Method for Heat and Visible Smoke Release Rates for Materials and Products Using an Oxygen Consumption Calorimeter
424.2, 602.4.1.1, 602.4.2.1, 602.4.3.1, 1402.5

E1592—2005(2017): Test Method for Structural Performance of Sheet Metal Roof and Siding Systems by Uniform Static Air Pressure Difference
1504.4.2

E1602—2003(2017): Guide for Construction of Solid Fuel-burning Masonry Heaters
2112.2

E1886—2013A: Standard Test Method for Performance of Exterior Windows, Curtain Walls, Doors and Impact Protective Systems Impacted by Missile(s) and Exposed to Cyclic Pressure Differentials
1609.2, 1709.5.3.1

E1966—2017: Standard Test Method for Fire-resistive Joint Systems
202, 715.3.1, 1709.5.3.1

E1996—2017: Specification for Performance of Exterior Windows, Curtain Walls, Doors and Impact Protective Systems Impacted by Windborne Debris in Hurricanes
1609.2, 1609.2.2, 1709.5.3.1

E2072—14: Standard Specification for Photoluminescent (Phosphorescent) Safety Markings
1025.4

E2174—2018: Standard Practice for On-site Inspection of Installed Fire Stops
1705.18.1

E2178—13: Standard Test Method for Air Permeance of Building Materials
202

E2273—2018: Standard Test Method for Determining the Drainage Efficiency of Exterior Insulation and Finish Systems (EIFS) Clad Wall Assemblies
1407.4.1, 2510.6.2

E2307—15BE1: Standard Test Method for Determining Fire Resistance of Perimeter Fire Barriers Using the Intermediate-scale, Multistory Test Apparatus
715.4

E2353—2016: Standard Test Methods for Performance of Glazing in Permanent Railing Systems, Guards and Balustrades
2407.1.2

E2392 / E2392M—10(2016): Standard Guide for Design of Earthen Wall Building Systems
2109.2.4.8.9.2

E2393—10a(2015): Standard Practice for On-site Inspection of Installed Fire Resistive Joint Systems and Perimeter Fire Barriers
1705.18.2

E2404—2017: Practice for Specimen Preparation and Mounting of Textile, Paper or Polymeric (Including Vinyl) and Wood Wall or Ceiling Coverings, Facing and Veneers to Assess Surface Burning Characteristics
803.5.2, 803.12, 1402.5

E2556/E2556M—2010(2016): Standard Specification for Vapor Permeable Flexible Sheet Water-resistive Barriers Intended for Mechanical Attachment
1403.2, 2510.6.1

E2568—2017A: Standard Specification for PB Exterior Insulation and Finish Systems
1407.2

E2570/E2570M—07(2014)e1: Standard Test Method for Evaluating Water-resistive Barrier (WRB) Coatings Used under Exterior Insulation and Finish Systems (EIFS) for EIFS with Drainage
1407.4.1.1, 1705.17.1

E2573—2017: Standard Practice for Specimen Preparation and Mounting of Site-fabricated Stretch Systems to Assess Surface Burning Characteristics
803.10

E2579—2015: Standard Practice for Specimen Preparation and Mounting of Wood Products to Assess Surface Burning Characteristics
803.11

E2599—2018: Standard Practice for Specimen Preparation and Mounting of Reflective Insulation, Radiant Barrier and Vinyl Stretch Ceiling Materials for Building Applications to Assess Surface Burning Characteristics
2614.3

ASTM—continued

E2634—2018: Standard Specification for Flat Wall Insulating Concrete Form (ICF) Systems
1903.4

E2652—16: Standard Test Method for Behavior of Materials in a Tube Furnace with a Cone-shaped Airflow Stabilizer at 750°C
703.3.1

E2751/E2751M—2017A: Practice for Design and Performance of Supported Laminated Glass Walkways
2409.1

E2925—17: Standard Specification for Manufactured Polymeric Drainage and Ventilation Materials Used to Provide a Rainscreen Function
2510.6.2

F547—2017: Terminology of Nails for Use with Wood and Wood-base Materials
Table 2506.2

F1667—2018: Specification for Driven Fasteners: Nails, Spikes and Staples
Table 721.1(2), Table 721.1(3), 1507.2.5, 1507.16.5, 2303.6, Table 2304.10.2, 2304.10.6, Table 2506.2, Table 2603.13.1, Table 2603.13.2

F2006—17: Standard/Safety Specification for Window Fall Prevention Devices for Nonemergency Escape (Egress) and Rescue (Ingress) Windows
1015.8

F2090—17: Specification for Window Fall Prevention Devices with Emergency Escape (Egress) Release Mechanisms
1015.8, 1015.8.1

F2200—2017: Standard Specification for Automated Vehicular Gate Construction
3110.2

G152—13: Practice for Operating Open Flame Carbon Arc Light Apparatus for Exposure of Nonmetallic Materials
1504.7

G154—2016A: Practice for Operating Fluorescent Light Apparatus for UV Exposure of Nonmetallic Materials
1504.7

G155—13: Practice for Operating Xenon Arc Light Apparatus for Exposure of Nonmetallic Materials
1504.7

AWC

American Wood Council
222 Catoctin Circle SE, Suite 201
Leesburg, VA 20175

ANSI/AWC NDS—2018: National Design Specification (NDS) for Wood Construction—with 2018 NDS Supplement
202, 722.1, Table 1604.3, 1809.12, 1810.3.2.4, Table 1810.3.2.6, 1905.1.8, 2304.13, 2306.1, Table 2306.2(1), Table 2306.2(2), Table 2306.3(1), Table 2306.3(2), 2307.1, Table 2603.13.2

ANSI/AWC PWF—2021: Permanent Wood Foundation Design Specification
1805.2, 1807.1.4, 2304.10.6.2

ANSI/AWC SDPWS—2021: Special Design Provisions for Wind and Seismic
202, 2305.1, 2305.2, 2305.3, 2306.1, 2306.2, 2306.3, Table 2306.3(1), Table 2306.3(3), 2307.1

ANSI/AWC WFCM—2018: Wood Frame Construction Manual for One- and Two-Family Dwellings
1609.1.1, 1609.1.1.1, 2302.1, 2308.2.4, 2308.6.7.2, 2309.1

AWC STJR—2021: Span Tables for Joists and Rafters
2306.1.1, 2308.4.2.1, 2308.7.1, 2308.7.2

AWC WCD No. 4—2003: Wood Construction Data—Plank and Beam Framing for Residential Buildings
2306.1.2

AWCI

Association of the Wall and Ceiling Industry
513 West Broad Street, Suite 210
Falls Church, VA 22046

12-B—14: Technical Manual 12B, Third Edition; Standard Practice for the Testing and Inspection of Field Applied Thin Film Intumescent Fire-resistive Materials; an Annotated Guide
1705.16

AWPA

American Wood Protection Association
P.O. Box 361784
Birmingham, AL 35236-1784

C1—03: All Timber Products—Preservative Treatment by Pressure Processes
1505.6

M4—15: Standard for the Care of Preservative-treated Wood Products
1810.3.2.4.1, 2303.1.9

U1—20: USE CATEGORY SYSTEM: User Specification for Treated Wood Except Commodity Specification H
Table 1507.9.6, 1807.1.4, 1807.3.1, 1809.12, 1810.3.2.4.1, 2303.1.9, 2304.12.1, 2304.12.2, 2304.12.2.6, 2304.12.2.7 , 2304.12.2.8

AWS

American Welding Society
8669 NW 36 Street, #130
Miami, FL 33166-6672

D1.4/D1.4M—2018: Structural Welding Code—Steel Reinforcing Bars
1704.5, Table 1705.3, 1705.3.1, 2107.3

BHMA

Builders Hardware Manufacturers' Association
355 Lexington Avenue, 15th Floor
New York, NY 10017

A 156.10—2017: Power Operated Pedestrian Doors
1010.3.2

A 156.19—2020: Standard for Power Assist and Low Energy Power Operated Doors
1010.3.2

A 156.27—2019: Power and Manual Operated Revolving Pedestrian Doors
1010.3.1.1

A 156.38—2020: Low Energy Power Operated Sliding and Folding Doors
1010.3.2

CEN

European Committee for Standardization (CEN)
Rue de la Science 23
Brussels, Belgium 1000

BS EN 15250—2007: Slow Heat Release Appliances Fired by Solid Fuel Requirements and Test Methods
2112.2, 2112.5

EN 1081—98: Resilient Floor Coverings—Determination of the Electrical Resistance
406.7.1

CPA

Composite Panel Association
19465 Deerfield Avenue, Suite 306
Leesburg, VA 20176

ANSI A135.4—2012: Basic Hardboard
1403.3.1, 2303.1.7

ANSI A135.5—2012: Prefinished Hardboard Paneling
2303.1.7, 2304.7

ANSI A135.6—2012: Engineered Wood Siding
1403.3.2, 2303.1.7

ANSI A208.1—2016: Particleboard
2303.1.8, 2303.1.8.1

CPSC

Consumer Product Safety Commission
4330 East/West Highway
Bethesda, MD 20814

16 CFR Part 1201 (2002): Safety Standard for Architectural Glazing Material
2406.2, Table 2406.2(1), 2406.3.1, 2407.1, 2407.1.4, 2408.2.1, 2408.3, 2409.2, 2409.4.1

16 CFR Part 1209 (2002): Interim Safety Standard for Cellulose Insulation
720.6

16 CFR Part 1404 (2002): Cellulose Insulation
720.6

16 CFR Part 1500 (2009): Hazardous Substances and Articles; Administration and Enforcement Regulations
202

16 CFR Part 1500.44 (2009): Method for Determining Extremely Flammable and Flammable Solids
202

16 CFR Part 1507 (2002): Fireworks Devices
202

16 CFR Part 1630 (2007): Standard for the Surface Flammability of Carpets and Rugs
804.4.1

CSA

Canadian Standards Association
8501 East Pleasant Valley Road
Cleveland, OH 44131

AAMA/WDMA/CSA 101/I.S.2/A440—17: North American Fenestration Standard/Specifications for Windows, Doors and Unit Skylights
1709.5.1, 2405.5

ASME A17.1—2019/CSA B44—2019: Safety Code for Elevators and Escalators
907.3.3, 911.1.6, 1009.4.1, 1607.11.1, 3001.2, Table 3001.3, 3001.5, 3002.5, 3003.2, 3007.1, 3008.1.4, 3008.7.1

ASME A17.7—2007/CSA B44.7—07(R2017): Performance-based Safety Code for Elevators and Escalators
Table 3001.3, 3001.5, 3002.5

CSSB

Cedar Shake & Shingle Bureau
P. O. Box 1178
Sumas, WA 98295-1178

CSSB—97: Grading and Packing Rules for Western Red Cedar Shakes and Western Red Shingles of the Cedar Shake and Shingle Bureau
Table 1507.8.5, Table 1507.9.6

DASMA

Door & Access Systems Manufacturers Association International
1300 Sumner Avenue
Cleveland, OH 44115

ANSI/DASMA 107—2017: Room Fire Test Standard for Garage Doors Using Foam Plastic Insulation
2603.4.1.9

ANSI/DASMA 108—2017: Standard Method for Testing Sectional Garage Doors, Rolling Doors and Flexible Doors: Determination of Structural Performance Under Uniform Static Air Pressure Difference
1709.5.2.1

ANSI/DASMA 115—2017: Standard Method for Testing Sectional Garage Doors, Rolling Doors and Flexible Doors: Determination of Structural Performance Under Missile Impact and Cyclic Wind Pressure
1609.2.3

DHA

Decorative Hardwoods Association
42777 Trade West Dr
Sterling, VA 20166

ANSI/HPVA HP-1—2016: American National Standard for Hardwood and Decorative Plywood
2303.3, 2304.7

DOC

U.S. Department of Commerce
National Institute of Standards and Technology
100 Bureau Drive
Gaithersburg, MD 20899

PS 1—19: Structural Plywood
2303.1.5, 2304.7, Table 2304.8(4), Table 2304.8(5), Table 2306.2(1), Table 2306.2(2)

PS 2—18: Performance Standard for Wood-based Structural-use Panels
2303.1.5, 2304.7, Table 2304.8(5), Table 2306.2(1), Table 2306.2(2)

PS 20—05: American Softwood Lumber Standard
202, 1810.3.2.4, 2303.1.1

DOL

U.S. Department of Labor
Occupational Safety and Health Administration
c/o Superintendent of Documents
U.S. Government Printing Office
Washington, DC 20210

29 CFR Part 1910.1000 (2015): Air Contaminants
202

DOTn

U.S. Department of Transportation
Office of Hazardous Material Safety
1200 New Jersey Avenue, SE
East Building, 2nd Floor
Washington, DC 20590

49 CFR 173.192—2011: Packaging for Certain Toxic Gases in Hazard Zone A
Table 415.6.5

49 CFR Parts 100–185—2015: Hazardous Materials Regulations
202

49 CFR Parts 173–178—2015: Specification of Transportation of Explosive and Other Dangerous Articles, UN 0335, UN 0336 Shipping Containers
202

49 CFR Parts 173.137—(2009): Shippers—General Requirements for Shipments and Packaging—Class 8—Assignment of Packing Group
202

EN

European Committee for Standardization
Rue de la Science 23 B
Brussels, Belgium 1040 Belgium

EN 459-1—15: Building Lime. Definitions, Specifications and Conformity Criteria
2109.2.4.8.7

FEMA

Federal Emergency Management Agency
500 C Street S.W.
Washington, DC 20472

FEMA-TB-11—01: Crawlspace Construction for Buildings Located in Special Flood Hazard Areas
1805.1.2.1

FM

FM Approvals
Headquarters Office
1151 Boston-Providence Turnpike
P.O. Box 9102
Norwood, MA 02062

4430—2012: Approval Standard for Heat and Smoke Vents
910.3.1

4450—(1989): Approval Standard for Class 1 Insulated Steel Deck Roofs—with Supplements through July 1992
1510.2

4470—2016: Approval Standard for Single-ply Polymer-modified Bitumen Sheet, Built-up Roof (BUR) and Liquid Applied Roof Assemblies for Use in Class 1 and Noncombustible Roof Deck Construction
1504.8

4474—2011: American National Standard for Evaluating the Simulated Wind Uplift Resistance of Roof Assemblies Using Static Positive and/or Negative Differential Pressures
1504.4.1, 1504.4.2, 1504.4.3

4880—2017: American National Standard for Evaluating the Fire Performance Insulated Building Panel Assemblies and Interior Finish Materials
2603.4, 2603.9

GA

Gypsum Association
962 Wayne Avenue, Suite 620
Silver Spring, MD 20910

GA 216—2018: Application and Finishing of Gypsum Panel Products
Table 2508.1, 2509.2

GA 600—2018: Fire-resistance and Sound Control Design Manual, 22nd Edition
Table 721.1(1), Table 721.1(2), Table 721.1(3)

ICC

International Code Council, Inc.
500 New Jersey Ave NW 6th Floor
Washington, DC 20001

ICC 300—17: ICC Standard on Bleachers, Folding and Telescopic Seating and Grandstands
1030.1.1, 1030.7, 1607.19

ICC 400—17: Standard on Design and Construction of Log Structures
2302.1

ICC 500—2020: ICC/NSSA Standard for the Design and Construction of Storm Shelters
202, 423.1, 423.2, 423.3.1, 423.3.2, 423.4, 423.5, 1031.2, 1604.5.1, 1604.10

ICC 600—2020: Standard for Residential Construction in High-wind Regions
1609.1.1, 1609.1.1.1, 2308.2.4

ICC 900/SRCC 300—2020: Solar Thermal System Standard
3111.2.1

ICC 901/SRCC 100—2020: Solar Thermal Collector Standard
3111.2.1

ICC 1100—18: Standard for Spray-applied Foam Plastic Insulation
2603.1.1

ICC A117.1—17: Accessible and Usable Buildings and Facilities
202, 907.5.2.3.3, 1009.8.2, 1009.9, 1009.11, 1010.2.13.1, 1012.1, 1012.6.5, 1012.10, 1013.4, 1023.9, 1102.1, 1108.2, 1110.1, 1110.2, 1110.5.1, 1110.5.2, 1112.3, 1112.4, 1112.5, 1112.5.2, 1207.1

IEBC—21: International Existing Building Code®
101.4.7, 102.6, 116.5, 201.3, Table 504.3, Table 504.4, 2701.1, 2801.1, 3113.1.1

IECC—21: International Energy Conservation Code®
101.4.6, 201.3, 202, 1202.1, 1202.4.3.2, 1301.1.1

ICC—continued

IFC—21: International Fire Code®

101.4.5, 102.6, 102.6.2, 201.3, 202, 307.1, Table 307.1(1), Table 307.1(2), 307.1.1, 403.4.5, 404.2, 406.2.9, 406.6.4.4, 406.7, 406.8, 407.4, 410.2.6, 411.1, 412.1, 412.5.1, 413.1, 414.1.1, 414.1.2, 414.1.2.1, 414.2, Table 414.2.5(1), Table 414.2.5(2), 414.2.5.1, 414.2.5.2, 414.2.5.3, 414.3, 414.5, 414.5.1, Table 414.5.1, 414.5.2, 414.5.3, 414.6, 415.2, 415.6.1, 415.6.2, 415.6.3, 415.6.4, 415.6.4.1, 415.6.4.4, Table 415.6.5, 415.8.2, 415.9, 415.9.1, 415.9.1.3, 415.9.1.4, 415.9.1.6, 415.9.1.7, 415.9.1.8, 415.9.2, 415.9.3, 415.10, 415.11, 415.11.1.7, 415.11.5, 415.11.8.2, 415.11.10.3, 415.11.11.1, 416.1, 416.2.3, 416.4, 419.1, 422.3.1, 426.1, 426.1.4, 427.1, 427.2.3, 428.1, 428.2, 428.3, Table 504.3, Table 504.4, Table 506.2, 507.4, 507.8.1.1.1, 507.8.1.1.2, 507.8.1.1.3, Table 509.1, 705.8.1, 707.1, 707.4, Table 716.1(2), 716.2.5.4.1, 716.3.2.1.1.1, 806.3, 901.2, 901.3, 901.5, 901.6.3, 901.6.4, 903.2.7.1, Table 903.2.11.6, 903.2.12, 903.5, 904.2.2, 904.11.3, 905.1, 905.3.6, 905.3.7, 906.1, Table 906.1, 906.4, 907.1.1, 907.2.5, 907.2.6, 907.2.8, 907.2.13.2, 907.2.15, 907.2.16, 907.2.23, 907.5.2.2, 907.6.6, 907.6.6.3, 907.8, 909.6.3, 909.12.1, 909.19, 909.20, 910.2.2, 910.5, 912.4.3, 915.1, 915.6, 916.7, 916.11, 918.1, 1002.1, 1002.2, 1010.2.14, 1202.5.2, 1202.6, 1507.15, 1512.1, Table 1604.5, 2603.4.1.12, 2701.1, 2702.1.3, 2702.2.3, 2702.2.7, 2702.2.10, 2702.2.13, 2702.2.15, 2702.4, 3003.3, 3008.1.3, 3102.1, 3102.8.3, 3103.1, 3111.1.2, 3111.2, 3111.3, 3111.3.4, 3111.3.5, 3302.3, 3303.7, 3309.2

IFGC—21: International Fuel Gas Code®

101.4.1, 201.3, Table 307.1(1), 415.9.2, 2113.11.1.2, 2113.15, 2702.2.6, 2801.1

IMC—21: International Mechanical Code®

101.4.2, 201.3, Table 307.1(1), 406.2.9, 406.6.2, 406.8.1, 407.2.7, 409.3, 412.5.6, 414.1.2, 414.3, 415.9.1, 415.9.2, 415.9.3, 415.11.12, 416.2.3, 420.9, 420.10, 420.11.1, 421.4, 422.7, 426.1.4, 427.2.2, 427.2.3, 602.4.2.5, 602.4.3.5, 602.4.4.3, 603.1, 603.1.1, 603.1.2, 712.1.6, 715.4, 715.7, 717.2.2, 717.5.2, 717.5.3, 717.6.1, 717.6.2, 717.6.3, 718.5, 720.1, 720.7, 903.2.11.4, 904.2.2, 904.13, 907.3.1, 909.1, 909.10.2, 909.13.1, 910.4.7, 1006.2.2.3, 1011.16, 1020.6.1, 1202.1, 1202.2.2, 1202.4.3.2, 1202.5.2.1, 1202.6, 1209.3, 2702.2.5, 2801.1, 3111.2

IPC—21: International Plumbing Code®

101.4.3, 201.3, 415.9.3, 603.1.2, 718.5, 903.3.5, 1205.3.3, 1503.4, 1805.4.3, 2901.1, Table 2902.1, 3111.2, 3305.1

IPMC—21: International Property Maintenance Code®

101.4.4, 102.6, 102.6.2, 103.3, 2701.1, 2801.1

IPSDC—21: International Private Sewage Disposal Code®

101.4.3, 2901.1

IRC—21: International Residential Code®

101.2, 102.6.1, 104.2.1, 110.3.3, 305.2.3, 308.2.4, 308.3.2, 308.5.4, 310.1, 310.4.1, 310.4.2, 2308.1

ISPSC—21: International Swimming Pool and Spa Code®

3109.1

IWUIC—21: International Wildland-Urban Interface Code®

Table 1505.1

SBCCI SSTD 11—97: Test Standard for Determining Wind Resistance of Concrete or Clay Roof Tiles

1504.3.1.1, 1504.3.1.2, 1504.3.1.3

ISO

International Organization for Standardization
Chemin de Blandonnet 8
CP 401 1214 Vernier
Geneva, Switzerland

ISO 668—2013: Series 1 Freight Containers—Classifications, Dimensions and Ratings

Table 3115.8.5.3

ISO 1496-1—2013: Series 1 Freight Containers—Specification and Testing - Part 1: General Cargo Containers for General Purposes

3115.8, Table 3115.8.5.3

ISO 6346—1995: Freight Containers—Code, Identification and Marking with Amendment 3 - 2012

3115.3

ISO 8115—86: Cotton Bales—Dimensions and Density

Table 307.1(1), Table 415.11.1.1.1

ISO 8336—09: Fiber-cement Flat Sheets—Product Specification and Test Methods

1403.10, 1404.16.1, 1404.16.2, Table 2509.2

MHI

Material Handling Institute
8720 Red Oak Blvd. Suite 201
Charlotte, NC 28217

ANSI MH29.1—08: Safety Requirements for Industrial Scissors Lifts
Table 3001.3

ANSI/MH16.1—12: Specification for the Design, Testing and Utilization of Industrial Steel Storage Racks
Table 1705.13.7

NAAMM

National Association of Architectural Metal Manufacturers
800 Roosevelt Road, Bldg. C, Suite 312
Glen Ellyn, IL 60137

FP 1001—18: Guide Specifications for Design of Metal Flag Poles
1609.1.1

NCMA

National Concrete Masonry Association
13750 Sunrise Valley
Herndon, VA 20171

TEK 5—84(2005): Details for Concrete Masonry Fire Walls
Table 721.1(2)

NFPA

National Fire Protection Association
1 Batterymarch Park
Quincy, MA 02169-7471

04—21: Standard for Integrated Fire Protection and Life Safety System Testing
901.6.2.1 , 901.6.2.2

10—21: Standard for Portable Fire Extinguishers
906.2, Table 906.3(1), Table 906.3(2), 906.3.2, 906.3.4

11—16: Standard for Low-, Medium, and High Expansion Foam
904.7

12—15: Standard on Carbon Dioxide Extinguishing Systems
904.8, 904.13

12A—18: Standard on Halon 1301 Fire Extinguishing Systems
904.9

13—19: Standard for Installation of Sprinkler Systems
403.3.3, 712.1.3.1, 903.3.1.1, 903.3.2, 903.3.8.2, 903.3.8.5, 904.13, 905.3.4, 907.6.4, 1019.3

13D—19: Standard for the Installation of Sprinkler Systems in One- and Two-family Dwellings and Manufactured Homes
903.3.1.3

13R—19: Standard for the Installation of Sprinkler Systems in Low-rise Residential Occupancies
903.3.1.2, 903.3.5.2, 903.4

14—19: Standard for the Installation of Standpipe and Hose System
905.2, 905.3.4, 905.4.2, 905.6.2, 905.8

16—19: Standard for the Installation of Foam-water Sprinkler and Foam-water Spray Systems
904.7, 904.13

17—20: Standard for Dry Chemical Extinguishing Systems
904.6, 904.13

17A—20: Standard for Wet Chemical Extinguishing Systems
904.5, 904.13

20—19: Standard for the Installation of Stationary Pumps for Fire Protection
412.2.4.1, 913.1, 913.2, 913.2.1, 913.5

30—21: Flammable and Combustible Liquids Code
415.6.1, 415.6.2, 507.8.1.1.1, 507.8.1.1.2

NFPA—continued

30A—21: Code for Motor Fuel Dispensing Facilities and Repair Garages
406.2.9.2

31—20: Standard for the Installation of Oil-burning Equipment
2113.15

32—16: Standard for Dry Cleaning Facilities
415.9.3

40—19: Standard for the Storage and Handling of Cellulose Nitrate Film
409.1

45—19: Standard on Fire Protection Laboratories Using Chemicals (2015 Edition)
428.3.7

58—20: Liquefied Petroleum Gas Code
415.9.2

61—20: Standard for the Prevention of Fires and Dust Explosions in Agricultural and Food Product Facilities
426.1

70—20: National Electrical Code
108.3, 406.2.7, 406.2.9, 412.5.7, 415.11.1.8, Table 509.1, 904.3.1, 907.6.1, 909.12.2, 909.16.3, 910.4.6, 1204.4.1, 2701.1, 2702.1.3, 3111.3

72—19: National Fire Alarm and Signaling Code
407.4.4.5, 407.4.4.5.1, 901.6, 903.4.1, 904.3.5, 907.1.2, 907.2, 907.2.6, 907.2.9.3, 907.2.11, 907.2.13.2, 907.3, 907.3.3, 907.3.4, 907.5.2.1.2, 907.5.2.2, 907.5.2.2.5, 907.6, 907.6.1, 907.6.2, 907.6.6, 907.7, 907.7.1, 907.7.2, 911.1.6, 917.1, 2702.2.4, 3005.5, 3007.7

80—19: Standard for Fire Doors and Other Opening Protectives
410.2.5, 509.4.2, 716.1, 716.2.5.1, 716.2.6.4, 716.2.9, 716.3.4.1, 716.3.5, 716.4.3, 1010.3.3

82—19: Standard on Incinerators and Waste and Linen Handling Systems and Equipment
713.13

85—19: Boiler and Combustion System Hazards Code
426.1

92—18: Standard for Smoke Control Systems
909.7, 909.8

99—21: Health Care Facilities Code
407.11, 422.6, 425.1

101—21: Life Safety Code
1030.6.2

105—19: Standard for Smoke Door Assemblies and Other Opening Protectives
405.4.2, 710.5.2.2, 716.2.10, 909.20.4.1

110—19: Standard for Emergency and Standby Power Systems
2702.1.3

111—19: Standard on Stored Electrical Energy Emergency and Standby Power Systems
2702.1.3

120—20: Standard for Fire Prevention and Control in Coal Mines
426.1

170—18: Standard for Fire Safety and Emergency Symbols
1025.2.6.1

211—19: Standard for Chimneys, Fireplaces, Vents and Solid Fuel-burning Appliances
2112.5

221—21: Standard for High Challenge Fire Walls, Fire Walls and Fire Barrier Walls
706.2, Table 716.1(2)

252—17: Standard Methods of Fire Tests of Door Assemblies
Table 716.1(1), 716.1.1, 716.1.2.2.1, 716.2.1.1, 716.2.1.2, 716.2.2.1, 716.2.2.2, 716.2.2.3.1, 716.2.5.1.1

253—19: Standard Method of Test for Critical Radiant Flux of Floor Covering Systems Using a Radiant Heat Energy Source
406.2.4, 424.2, 804.2, 804.3

257—17: Standard for Fire Test for Window and Glass Block Assemblies
Table 716.1(1), 716.1.1, 716.1.2.2.2, 716.3.1.1, 716.3.1.2, 716.3.2.1.3, 716.3.4

NFPA—continued

259—18: Standard Test Method for Potential Heat of Building Materials
2603.4.1.10, 2603.5.3

265—19: Standard Methods of Fire Tests for Evaluating Room Fire Growth Contribution of Textile or Expanded Vinyl Wall Coverings on Full Height Panels and Walls
803.5.1, 803.5.1.1

268—19: Standard Test Method for Determining Ignitability of Exterior Wall Assemblies Using a Radiant Heat Energy Source
1405.1.1.1, 1405.1.1.1.1, 1405.1.1.1.2, 2603.5.7

275—17: Standard Method of Fire Tests for the Evaluation of Thermal Barriers
508.4.4.1, 509.4.1.1, 1406.10.2, 1408.10.2, 2603.4

276—19: Standard Method of Fire Tests for Determining the Heat Release Rate of Roofing Assemblies with Combustible Above-deck Roofing Components
1508.1, 2603.3, 2603.4.1.5

285—19: Standard Fire Test Method for the Evaluation of Fire Propagation Characteristics of Exterior Nonload-bearing Wall Assemblies Containing Combustible Components
718.2.6, 1402.5, 1406.10.3, 1408.10.4, 1511.6.2, 2603.5.5

286—15: Standard Methods of Fire Test for Evaluating Contribution of Wall and Ceiling Interior Finish to Room Fire Growth
402.6.4.4, 424.2, 803.1.1, 803.1.1.1, 803.11, 803.12, 803.13, 1406.10.2, 1408.10.3, 2603.7, 2603.9, 2604.2.4, 2614.4, 3105.3

288—17: Standard Methods of Fire Tests of Horizontal Fire Door Assemblies Installed in Horizontal in Fire-resistance-rated Floor Systems
712.1.13.1

289—19: Standard Method of Fire Test for Individual Fuel Packages
402.6.2, 402.6.4.5, 424.2, 806.4

409—16: Standard for Aircraft Hangars
412.3.6, Table 412.3.6, 412.3.6.1, 412.5.5

418—16: Standard for Heliports
412.7.4

484—19: Standard for Combustible Metals
426.1

652—19: Standard on the Fundamentals of Combustible Dust
426.1

654—20: Standard for the Prevention of Fire and Dust Explosions from the Manufacturing, Processing and Handling of Combustible Particulate Solids
426.1

655—17: Standard for the Prevention of Sulfur Fires and Explosions
426.1

664—20: Standard for the Prevention of Fires and Explosions in Wood Processing and Woodworking Facilities
426.1

701—19: Standard Methods of Fire Tests for Flame Propagation of Textiles and Films
410.2.6, 424.2, 806.4, 3102.3, 3102.3.1, 3102.6.1.1, 3105.3

704—17: Standard System for the Identification of the Hazards of Materials for Emergency Response
202, 415.5.2

720—15: Standard for the Installation of Carbon Monoxide (CO) Detection and Warning Equipment
915.5.1, 915.5.2

750—19: Standard on Water Mist Fire Protection Systems
202, 904.11.1.1, 904.13

1124—17: Code for the Manufacture, Transportation and Storage of Fireworks and Pyrotechnic Articles
415.6.4.1

2001—18: Standard on Clean Agent Fire Extinguishing Systems
904.10

2010—20: Standard for Fixed Aerosol Fire-extinguishing Systems
904.12

PCI

Precast Prestressed Concrete Institute
8770 West Bryn Mawr, Suite 1150
Chicago, IL 60631-3517

PCI 124—18: Specification for Fire Resistance of Precast Prestressed Concrete
722.1, 722.2.3.1

PCI 128—19: Specification for Glass Fiber Reinforced Concrete Panels
1903.3

PTI

Post-Tensioning Institute
38800 Country Club Drive
Farmington Hills, MI 48331

PTI DC—10.5-19: Standard Requirements for Design and Analysis of Shallow Post-Tensioned Concrete Foundations on Expansive and Stable Soils
1808.6.2

RMI

Rack Manufacturers Institute
8720 Red Oak Boulevard, Suite 201
Charlotte, NC 28217

ANSI/MH16.1—12: Specification for Design, Testing and Utilization of Industrial Steel Storage Racks
2209.1

ANSI/MH16.3—16: Specification for the Design, Testing and Utilization of Industrial Steel Cantilevered Storage Racks
2209.2

SBCA

Structural Building Components Association
6300 Enterprise Lane
Madison, WI 53719

ANSI/FS 100—12(R2018): Standard Requirements for Wind Pressure Resistance of Foam Plastic Insulating Sheathing Used in Exterior Wall Covering Assemblies
2603.10

SDI

Steel Deck Institute
2661 Clearview Road #3
Allison Park, PA 15101

SDI NC—2017: Standard for Noncomposite Steel Floor Deck
2210.1.1.1

SDI RD—2017: Standard for Steel Roof Deck
2210.1.1.2

SDI-C—2017: Standard for Composite Steel Floor Deck—Slabs
2210.1.1.3

SDI-QA/QC—2017: Standard for Quality Control and Quality Assurance for Installation of Steel Deck
1705.2.2

SJI

Steel Joist Institute
140 Evans Street, Suite 203
Florence, SC 29501

SJI 100—20: 45th Edition Standard Specifications, Load Tables and Weight Tables for K-Series, LH-Series, DLH-Series and Joist Girders
1604.3.3, 2203.1, 2207.1

SJI 200—15: 2nd Edition Standard Specifications, Weight Tables and Bridging Tables for CJ-Series Composite Steel Joists
1604.3.3, 2203.1, 2207.1

SPRI

Single-Ply Roofing Institute
465 Waverly Oaks Road, Suite 421
Waltham, MA 02452

ANSI/SPRI RP-4—19: Wind Design Guide for Ballasted Single-ply Roofing Systems
1504.5

ANSI/SPRI VF-1—17: External Fire Design Standard for Vegetative Roofs
1505.10

ANSI/SPRI/FM 4435-ES-1—17: Wind Test Design Standard for Edge Systems Used with Low Slope Roofing Systems
1504.6

ANSI/SPRI GT-1—2016: Test Standard for Gutter Systems
1504.6.1

SRCC

Solar Rating & Certification Corporation
400 High Point Drive, Suite 400
Cocoa, FL 32926

ICC 900/SRCC 300—2020: Solar Thermal System Standard
3111.2.1

ICC 901/SRCC 100—2020: Solar Thermal Collector Standard
3111.2.1

TIA

Telecommunications Industry Association
1320 N. Courthouse Road #200
Arlington, VA 22201

ANSI/TIA 222-H—2017: Structural Standard for Antenna Supporting Structures, Antennas and Small Wind Turbine Support Structures
1609.1.1, 3108.1, 3108.2

TMS

The Masonry Society
105 South Sunset Street, Suite Q
Longmont, CO 80501-6172

216—2013: Standard Method for Determining Fire Resistance of Concrete and Masonry Construction Assemblies
Table 721.1(2), 722.1

302—2018: Standard Method for Determining the Sound Transmission Class Rating for Masonry Walls
1208.2.1

402—2016: Building Code for Masonry Structures
1404.6, 1404.6.2, 1404.10, 1604.3.4, 1705.4, 1807.1.6.3.2, 1808.9, 2101.2, 2106.1, 2107.1, 2107.2, 2107.3, 2108.1, 2108.2, 2108.3, 2109.1, 2109.1.1, 2109.2, 2110.1, 2114.1, 2114.4

403—2017: Direct Design Handbook for Masonry Structures
2101.2

404—2016: Standard for the Design of Architectural Cast Stone
2101.2

504—2016: Standard for the Fabrication of Architectural Cast Stone
2103.1

602—2016: Specification for Masonry Structures
1404.6.1, 1705.4, 1705.4.1, 1807.1.6.3, 2103.1, 2103.2.1, 2103.3, 2103.4, 2104.1, 2105.1

604—2016: Standard for the Installation of Architectural Cast Stone
2104.1

TPI

Truss Plate Institute
2670 Crain Highway, Suite 203
Waldorf, MD 20601

TPI 1—2014: National Design Standard for Metal-plate-connected Wood Truss Construction
2303.4.6, 2306.1

UL

UL LLC
333 Pfingsten Road
Northbrook, IL 60062

9—2009: Fire Tests of Window Assemblies—with Revisions through February 2015
Table 716.1(1), 716.1.1, 716.1.2.2.2, 716.2.1.3, 716.3.1.1, 716.3.1.2, 716.3.2.1.3, 716.3.4, 1013.5

10A—2009: Tin Clad Fire Doors—with Revisions through July 2018
716.2.1

10B—2008: Fire Tests of Door Assemblies—with Revisions through February 2015
Table 716.1(1), 716.1.1, 716.1.2.2.1, 716.2.1.2, 716.2.2.2, 716.2.2.3.1, 716.2.5.1.1

10C—2016: Positive Pressure Fire Tests of Door Assemblies
Table 716.1(1), 716.1.1, 716.1.2.2.1, 716.2.1.1, 716.2.2.1, 716.2.2.2, 716.2.2.3.1, 716.2.5.1.1, 1010.2.9.3

10D—2017: Standard for Fire Tests of Fire Protective Curtain Assemblies
716.4

14B—2008: Sliding Hardware for Standard Horizontally Mounted Tin Clad Fire Doors—with Revisions through July 2017
716.2.1

14C—2006: Swinging Hardware for Standard Tin Clad Fire Doors Mounted Singly and in Pairs—with Revisions through July 2017
716.2.1

55A—2004: Materials for Built-up Roof Coverings
1507.10.2

103—2010: Factory-built Chimneys, for Residential Type and Building Heating Appliances—with Revisions through March 2017
718.2.5.1

127—2011: Factory-built Fireplaces—with Revisions through July 2016
718.2.5.1, 2111.12

199E—2004: Outline of Investigation for Fire Testing of Sprinklers and Water Spray Nozzles for Protection of Deep Fat Fryers
904.13.4.1

217—2015: Single and Multiple Station Smoke Alarms—with Revisions through November 2016
907.2.11

263—11: Fire Tests of Building Construction and Materials—with Revisions through March 2018
703.2, 703.2.1.3, 703.2.1.5, 703.2.2, 703.4, 703.4, 704.12, 705.7, 705.8.5, 707.6, 712.1.13.2, 714.4.1, 714.5.1, 715.3, Table 716.1(1), Table 716.1(3), 716.1.2.3, 716.2.5.1.1, 716.2.5.4, 716.3.2.1.1, 717.3.1, 717.5.2, 717.5.3, 717.6.1, 717.6.2, Table 721.1(1), 2103.1, 2603.5.1

268—2016: Smoke Detectors for Fire Alarm Systems-with revisions through July 2016
407.9, 907.2.6.2, 907.2.11.7

294—2018: Access Control System Units—with Revisions through October 2018
1010.2.11, 1010.2.12, 1010.2.13.1, 1010.2.14

300—2005(R2010): Fire Testing of Fire Extinguishing Systems for Protection of Commercial Cooking Equipment—with Revisions through December 2014
904.13

300A—2006: Outline of Investigation for Extinguishing System Units for Residential Range Top Cooking Surfaces
904.14.1.1

305—2012: Panic Hardware—with Revisions through March 2017
1010.2.9.3

325—2017: Door, Drapery, Gate, Louver and Window Operations and Systems
406.2.1, 3110.3

555—2006: Fire Dampers—with Revisions through October 2016
717.3.1

UL—continued

555C—2014: Ceiling Dampers—with Revisions through May 2017
717.3.1

555S—2014: Smoke Dampers—with Revisions through October 2016
717.3.1

580—2006: Test for Uplift Resistance of Roof Assemblies—with Revisions through October 2018
1504.4.1, 1504.4.2

641—2010: Type L Low-temperature Venting Systems—with Revisions through April 2018
2113.11.1.4

710B—2011: Recirculating Systems—with Revisions through August 2014
904.13

723—2018: Test for Surface Burning Characteristics of Building Materials
202, 402.6.4.4, 406.7.2, 720.1, 720.4, 803.1.2, 803.5.2, 803.10, 803.11, 803.12, 803.13, 806.7, 1402.5, 1403.12.1, 1406.9, 1406.10.1, 1408.9, 1408.10.1, 1511.6.2, 1511.6.3, 2303.2, 2603.3, 2603.4.1.13, 2603.5.4, 2603.5.5, 2603.7, 2604.2.4, 2606.4, 2612.3, 2614.3, 3105.3

723S—2006: Drop-Out Ceilings Installed Beneath Automatic Sprinklers
2606.7.4

790—2004: Standard Test Methods for Fire Tests of Roof Coverings—with Revisions through October 2018
1505.1, 2603.6, 2610.2, 2610.3

793—2008: Automatically Operated Roof Vents for Smoke and Heat—with Revisions through March 2017
910.3.1

864—2014: Control Units and Accessories for Fire Alarm Systems—with Revisions through March 2018
909.12

924—2016: Safety Emergency Lighting and Power Equipment—with Revisions through May 2018
1013.5

1040—1996: Fire Test of Insulated Wall Construction—with Revisions through April 2017
1406.10.2, 2603.9

1256—02: Fire Test of Roof Deck Construction—with Revisions through August 2018
1508.1, 2603.3, 2603.4.1.5

1479—2015: Fire Tests of Penetration Firestops
202, 714.4.1.2, 714.4.2, 714.5.1.2, 714.5.4

1482—2011: Solid-fuel Type Room Heaters—with Revisions through August 2015
2112.2, 2112.5

1489—2016: Fire Resistant Pipe Protection Systems Carrying Combustible Liquids
403.4.8.2

1703—2002: Flat-plate Photovoltaic Modules and Panels—with Revisions through September 2018
1507.17.5, 3111.3.1

1715—97: Fire Test of Interior Finish Material—with Revisions through April 2017
1406.10.2, 2603.9, 2614.4

1741—2010: Inverters, Converters, Controllers and Interconnection System Equipment for Use with Distributed Energy Resources—with Revisions through February 2018
3111.3.1

1777—2007: Chimney Liners—with Revisions through April 2014
2113.11.1, 2113.19

1784—2015: Air Leakage Tests of Door Assemblies
405.4.3, 710.5.2.2, 710.5.2.2.1, 716.2.1.4, 716.2.9.1, 716.2.9.3, 3006.3, 3007.6.3, 3008.6.3

1897—2015: Uplift Tests for Roof Covering Systems
1504.4.1, 1504.4.3

1975—06: Fire Tests for Foamed Plastics Used for Decorative Purposes
402.6.2, 402.6.4.5, 424.2

1994—2015: Luminous Egress Path Marking Systems
411.6, 1008.2.1, 1025.2.1, 1025.2.3, 1025.2.4, 1025.2.5, 1025.4

2034—2017: Single- and Multiple-station Carbon Monoxide Alarms—with Revisions through September 2018
915.4.2, 915.4.4

UL—continued

2075—2013: Standard for Gas and Vapor Detectors and Sensors—with Revisions through December 2017
915.5.1, 915.5.3

2079—2015: Tests for Fire Resistance of Building Joint Systems
202, 715.3.1, 715.8

2196—2017: Standard for Fire Test for Circuit Integrity of Fire-Resistive Power, Instrumentation, Control and Data Cables
909.20.7.1, 913.2.2, 2702.3, 3007.8.1, 3008.8.2

2200—2012: Stationary Engine Generator Assemblies—with Revisions through October 2015
2702.1.1

2202—2009: Electric Vehicle (EV) Charging System Equipment—with Revisions through February 2018
406.2.7

2594—2016: Electric Vehicle Supply Equipment
406.2.7

2703—2014: Mounting Systems, Mounting Devices, Clamping/Retention Devices and Ground Lugs for Use with Flat-plate Photovoltaic Modules and Panels—with Revisions through December 2019
1505.9

7103—19: Outline of Investigation for Building-Integrated Photovoltaic Roof Coverings
Table 1504.2, 1507.16.6

61730-1—2017: Photovoltaic (PV) Module Safety Qualification - Part 1: Requirements for Construction
1507.16.6, 1507.17.5, 3111.3.1

61730-2—2017: Photovoltaic (PV) Module Safety Qualification - Part 2: Requirements for Testing
1507.16.6, 1507.17.5, 3111.3.1

ULC

Underwriters Laboratories of Canada
13775 Commerce Parkway
Richmond, BC V6V 2V4

CAN/ULC S 102.2—2018: Standard Method of Test for Surface Burning Characteristics of Flooring, Floor Coverings and Miscellaneous Materials and Assemblies
720.2, 720.3, 720.4

USC

United States Code
732 North Capitol Street NW
Washington, DC 20401-0003

18 USC Part 1, Ch. 40: Importation, Manufacture, Distribution and Storage of Explosive Materials
202

WCLIB

West Coast Lumber Inspection Bureau
P.O. Box 23145
Portland, OR 97223

AITC 104—03: Typical Construction Details
2306.1

AITC 110—01: Standard Appearance Grades for Structural Glued Laminated Timber
2306.1

AITC 113—10: Standard for Dimensions of Structural Glued Laminated Timber
2306.1

AITC 119—96: Standard Specifications for Structural Glued Laminated Timber of Hardwood Species
2306.1

AITC 200—09: Manufacturing Quality Control Systems Manual for Structural Glued Laminated Timber
2306.1

WDMA

Window and Door Manufacturers Association
2025 M Street NW, Suite 800
Washington, DC 20006

AAMA/WDMA/CSA 101/I.S.2/A440—17: Specifications for Windows, Doors and Unit Skylights
1709.5.1, 2405.5

WRI

Wire Reinforcement Institute, Inc.
942 Main Street, Suite 300
Hartford, CT 06103

WRI/CRSI—81: Design of Slab-on-ground Foundations—with 1996 Update
1808.6.2

APPENDIX A
EMPLOYEE QUALIFICATIONS

The provisions contained in this appendix are not mandatory unless specifically referenced in the adopting ordinance.

User notes:

About this appendix: *Appendix A provides optional criteria for the qualifications for jurisdictions to consider when hiring personnel to enforce the building code. Criteria for the building official, plan reviewers and inspectors are provided.*

Code development reminder: *Code change proposals to this appendix will be considered by the Administrative Code Development Committee during the 2022 (Group B) Code Development Cycle.*

SECTION A101
BUILDING OFFICIAL QUALIFICATIONS

[A] A101.1 Building official. The *building official* shall have not fewer than 10 years' experience or equivalent as an architect, engineer, inspector, contractor or superintendent of construction, or any combination of these, 5 years of which shall have been supervisory experience. The *building official* should be certified as a *building official* through a recognized certification program. The building official shall be appointed or hired by the applicable governing authority.

[A] A101.2 Chief inspector. The *building official* can designate supervisors to administer the provisions of this code and the *International Mechanical*, *Plumbing* and *Fuel Gas Codes*. Each supervisor shall have not fewer than 10 years experience or equivalent as an architect, engineer, inspector, contractor or superintendent of construction, or any combination of these, 5 years of which shall have been in a supervisory capacity. They shall be certified through a recognized certification program for the appropriate trade.

[A] A101.3 Inspector and plans examiner. The *building official* shall appoint or hire such number of officers, inspectors, assistants and other employees as shall be authorized by the jurisdiction. A person who has fewer than 5 years of experience as a contractor, engineer, architect, or as a superintendent, foreman or competent mechanic in charge of construction shall not be appointed or hired as inspector of construction or plans examiner. The inspector or plans examiner shall be certified through a recognized certification program for the appropriate trade.

[A] A101.4 Termination of employment. Employees in the position of *building official*, chief inspector or inspector shall not be removed from office except for cause after full opportunity has been given to be heard on specific charges before such applicable governing authority.

SECTION A102
REFERENCED STANDARDS

[A] A102.1 General. See Table A102.1 for standards that are referenced in various sections of this appendix. Standards are listed by the standard identification with the effective date, standard title, and the section or sections of this appendix that reference the standard.

TABLE A102.1
REFERENCED STANDARDS

STANDARD ACRONYM	STANDARD NAME	SECTIONS HEREIN REFERENCED
IBC—21	*International Building Code*	A101.2
IMC—21	*International Mechanical Code*	A101.2
IPC—21	*International Plumbing Code*	A101.2
IFGC—21	*International Fuel Gas Code*	A101.2

APPENDIX B
BOARD OF APPEALS

The provisions contained in this appendix are not mandatory unless specifically referenced in the adopting ordinance.

User notes:

About this appendix: *Appendix B provides criteria for Board of Appeals members. Also provided are procedures by which the Board of Appeals should conduct its business.*

Code development reminder: *Code change proposals to this appendix will be considered by the Administrative Code Development Committee during the 2022 (Group B) Code Development Cycle.*

SECTION B101
GENERAL

[A] B101.1 Scope. A board of appeals shall be established within the jurisdiction for the purpose of hearing applications for modification of the requirements of this code pursuant to the provisions of Section 113. The board shall be established and operated in accordance with this section, and shall be authorized to hear evidence from appellants and the building official pertaining to the application and intent of this code for the purpose of issuing orders pursuant to these provisions.

[A] B101.2 Application for appeal. Any person shall have the right to appeal a decision of the building official to the board. An application for appeal shall be based on a claim that the intent of this code or the rules legally adopted hereunder have been incorrectly interpreted, the provisions of this code do not fully apply or an equally good or better form of construction is proposed. The application shall be filed on a form obtained from the building official within 20 days after the notice was served.

[A] B101.2.1 Limitation of authority. The board shall not have authority to waive requirements of this code or interpret the administration of this code.

[A] B101.2.2 Stays of enforcement. Appeals of notice and orders, other than Imminent Danger notices, shall stay the enforcement of the notice and order until the appeal is heard by the board.

[A] B101.3 Membership of board. The board shall consist of five voting members appointed by the chief appointing authority of the jurisdiction. Each member shall serve for **[INSERT NUMBER OF YEARS]** years or until a successor has been appointed. The board members' terms shall be staggered at intervals, so as to provide continuity. The building official shall be an ex officio member of said board but shall not vote on any matter before the board.

[A] B101.3.1 Qualifications. The board shall consist of five individuals, who are qualified by experience and training to pass on matters pertaining to building construction and are not employees of the jurisdiction.

[A] B101.3.2 Alternate members. The chief appointing authority is authorized to appoint two alternate members who shall be called by the board chairperson to hear appeals during the absence or disqualification of a member. Alternate members shall possess the qualifications required for board membership, and shall be appointed for the same term or until a successor has been appointed.

[A] B101.3.3 Vacancies. Vacancies shall be filled for an unexpired term in the same manner in which original appointments are required to be made.

[A] B101.3.4 Chairperson. The board shall annually select one of its members to serve as chairperson.

[A] B101.3.5 Secretary. The chief appointing authority shall designate a qualified clerk to serve as secretary to the board. The secretary shall file a detailed record of all proceedings, which shall set forth the reasons for the board's decision, the vote of each member, the absence of a member and any failure of a member to vote.

[A] B101.3.6 Conflict of interest. A member with any personal, professional or financial interest in a matter before the board shall declare such interest and refrain from participating in discussions, deliberations and voting on such matters.

[A] B101.3.7 Compensation of members. Compensation of members shall be determined by law.

[A] B101.3.8 Removal from the board. A member shall be removed from the board prior to the end of their term only for cause. Any member with continued absence from regular meeting of the board may be removed at the discretion of the chief appointing authority.

[A] B101.4 Rules and procedures. The board shall establish policies and procedures necessary to carry out its duties consistent with the provisions of this code and applicable state law. The procedures shall not require compliance with strict rules of evidence, but shall mandate that only relevant information be presented.

[A] B101.5 Notice of meeting. The board shall meet upon notice from the chairperson, within 10 days of the filing of an appeal or at stated periodic intervals.

[A] B101.5.1 Open hearing. All hearings before the board shall be open to the public. The appellant, the appellant's representative, the building official and any person whose interests are affected shall be given an opportunity to be heard.

B101.5.2 Quorum. Three members of the board shall constitute a quorum.

[A] B101.5.3 Postponed hearing. When five members are not present to hear an appeal, either the appellant or the appellant's representative shall have the right to request a postponement of the hearing.

[A] B101.6 Legal counsel. The jurisdiction shall furnish legal counsel to the board to provide members with general legal advice concerning matters before them for consideration. Members shall be represented by legal counsel at the jurisdiction's expense in all matters arising from service within the scope of their duties.

[A] B101.7 Board decision. The board shall only modify or reverse the decision of the building official by a concurring vote of three or more members.

[A] B101.7.1 Resolution. The decision of the board shall be by resolution. Every decision shall be promptly filed in writing in the office of the *building* official within three days and shall be open to the public for inspection. A certified copy shall be furnished to the appellant or the appellant's representative and to the *building official*.

[A] B101.7.2 Administration. The *building official* shall take immediate action in accordance with the decision of the board.

[A] B101.8 Court review. Any person, whether or not a previous party of the appeal, shall have the right to apply to the appropriate court for a writ of certiorari to correct errors of law. Application for review shall be made in the manner and time required by law following the filing of the decision in the office of the chief administrative officer.

APPENDIX C
GROUP U—AGRICULTURAL BUILDINGS

The provisions contained in this appendix are not mandatory unless specifically referenced in the adopting ordinance.

User note:

About this appendix: *Agricultural buildings are given special consideration in Appendix C. Often such buildings have unique uses and structural needs. Where an agricultural building is surrounded by 60 feet of open area on all sides, size limits are waived. Automatic sprinkler protection may be required.*

SECTION C101
GENERAL

C101.1 Scope. The provisions of this appendix shall apply exclusively to *agricultural buildings*. Such buildings shall be classified as Group U and shall include the following uses:

1. Livestock shelters or buildings, including shade structures and milking barns.
2. Poultry buildings or shelters.
3. Barns.
4. Storage of equipment and machinery used exclusively in agriculture.
5. Horticultural structures, including detached production *greenhouses* and crop protection shelters.
6. Sheds.
7. Grain silos.
8. Stables.

SECTION C102
ALLOWABLE HEIGHT AND AREA

C102.1 General. Buildings classified as Group U Agricultural shall not exceed the area or height limits specified in Table C102.1.

C102.2 One-story unlimited area. The area of a one-story Group U *agricultural building* shall not be limited if the building is surrounded and adjoined by *public ways* or *yards* not less than 60 feet (18 288 mm) in width.

C102.3 Two-story unlimited area. The area of a two-story Group U *agricultural building* shall not be limited if the building is surrounded and adjoined by *public ways* or *yards* not less than 60 feet (18 288 mm) in width and is provided with an *approved automatic sprinkler system* throughout in accordance with Section 903.3.1.1.

SECTION C103
MIXED OCCUPANCIES

C103.1 Mixed occupancies. Mixed occupancies shall be protected in accordance with Section 508.

SECTION C104
EXITS

C104.1 Exit facilities. Exits shall be provided in accordance with Chapters 10 and 11.

Exceptions:

1. The maximum travel distance from any point in the building to an approved *exit* shall not exceed 300 feet (91 440 mm).
2. One *exit* is required for each 15,000 square feet (1393.5 m^2) of area or fraction thereof.

TABLE C102.1
BASIC ALLOWABLE AREA FOR A GROUP U, ONE STORY IN HEIGHT AND MAXIMUM HEIGHT OF SUCH OCCUPANCY

I		II		III and IV		V	
A	B	A	B	III A and IV	III B	A	B
ALLOWABLE AREA (square feet)[a]							
Unlimited	60,000	27,100	18,000	27,100	18,000	21,100	12,000
MAXIMUM HEIGHT IN STORIES							
Unlimited	12	4	2	4	2	3	2
MAXIMUM HEIGHT IN FEET							
Unlimited	160	65	55	65	55	50	40

For SI: 1 square foot = 0.0929 m^2.

a. See Section C102 for unlimited area under certain conditions.

APPENDIX D
FIRE DISTRICTS

The provisions contained in this appendix are not mandatory unless specifically referenced in the adopting ordinance.

User note:

About this appendix: *Appendix D establishes a framework by which a jurisdiction can establish a portion of a jurisdiction as a fire district. Fire districts are often designated in a more densely developed portion of a city where limiting the potential spread of fire is a key consideration. Specific construction types and users are prohibited in a fire district.*

SECTION D101
GENERAL

D101.1 Scope. The fire district shall include such territory or portion as outlined in an ordinance or law entitled "An Ordinance (Resolution) Creating and Establishing a Fire District." Wherever, in such ordinance creating and establishing a fire district, reference is made to the fire district, it shall be construed to mean the fire district designated and referred to in this appendix.

D101.1.1 Mapping. The fire district complying with the provisions of Section D101.1 shall be shown on a map that shall be available to the public.

D101.2 Establishment of area. For the purpose of this code, the fire district shall include that territory or area as described in Sections D101.2.1 through D101.2.3.

D101.2.1 Adjoining blocks. Two or more adjoining blocks, exclusive of intervening streets, where not less than 50 percent of the ground area is built upon and more than 50 percent of the built-on area is devoted to hotels and motels of Group R-1; Group B occupancies; theaters, nightclubs, restaurants of Group A-1 and A-2 occupancies; garages, express and freight depots, warehouses and storage buildings used for the storage of finished products (not located with and forming a part of a manufactured or industrial plant); or Group S occupancy. Where the average height of a building is two and one-half *stories* or more, a block should be considered if the ground area built upon is not less than 40 percent.

D101.2.2 Buffer zone. Where four contiguous blocks or more comprise a fire district, there shall be a buffer zone of 200 feet (60 960 mm) around the perimeter of such district. Streets, rights-of-way and other open spaces not subject to building construction can be included in the 200- foot (60 960 mm) buffer zone.

D101.2.3 Developed blocks. Where blocks adjacent to the fire district have developed to the extent that not less than 25 percent of the ground area is built upon and 40 percent or more of the built-on area is devoted to the occupancies specified in Section D101.2.1, they can be considered for inclusion in the fire district, and can form all or a portion of the 200-foot (60 960 mm) buffer zone required in Section D101.2.2.

SECTION D102
BUILDING RESTRICTIONS

D102.1 Types of construction permitted. Within the fire district every building hereafter erected shall be Type I, II, III or IV, except as permitted in Section D104.

D102.2 Other specific requirements.

D102.2.1 Exterior walls. Exterior walls of buildings located in the fire district shall comply with the requirements in Table 601 except as required in Section D102.2.6.

D102.2.2 Group H prohibited. Group H occupancies shall be prohibited from location within the fire district.

D102.2.3 Construction type. Every building shall be constructed as required based on the type of construction indicated in Chapter 6.

D102.2.4 Roof covering. *Roof covering* in the fire district shall conform to the requirements of Class A or B roof coverings as defined in Section 1505.

D102.2.5 Structural fire rating. Walls, floors, roofs and their supporting structural members shall be not less than 1-hour fire-resistance-rated construction.

Exceptions:

1. Buildings of Type IV-HT construction.
2. Buildings equipped throughout with an *automatic sprinkler system* in accordance with Section 903.3.1.1.
3. Automobile parking structures.
4. Buildings surrounded on all sides by a permanently open space of not less than 30 feet (9144 mm).
5. Partitions complying with Section 603.1, Item 11.

D102.2.6 Exterior walls. Exterior *load-bearing walls* of Type II buildings shall have a *fire-resistance rating* of 2 hours or more where such walls are located within 30 feet (9144 mm) of a common property line or an assumed property line. Exterior *nonload-bearing walls* of Type II buildings located within 30 feet (9144 mm) of a common property line or an assumed property line shall have *fire-*

resistance ratings as required by Table 601, but not less than 1 hour. *Exterior walls* located more than 30 feet (9144 mm) from a common property line or an assumed property line shall comply with Table 601.

Exception: In the case of one-story buildings that are 2,000 square feet (186 m^2) or less in area, *exterior walls* located more than 15 feet (4572 mm) from a common property line or an assumed property line need only comply with Table 601.

D102.2.7 Architectural trim. Architectural trim on buildings located in the fire district shall be constructed of *approved* noncombustible materials or *fire-retardant-treated wood*.

D102.2.8 Permanent canopies. Permanent canopies are permitted to extend over adjacent open spaces provided that all of the following are met:

1. The canopy and its supports shall be of noncombustible material, *fire-retardant-treated wood*, Type IV construction or of 1-hour fire-resistance-rated construction.

   Exception: Any textile covering for the canopy shall be flame resistant as determined by tests conducted in accordance with NFPA 701 after both accelerated water leaching and accelerated weathering.

2. Any canopy covering, other than textiles, shall have a *flame spread index* not greater than 25 when tested in accordance with ASTM E84 or UL 723 in the form intended for use.
3. The canopy shall have one long side open.
4. The maximum horizontal width of the canopy shall be not greater than 15 feet (4572 mm).
5. The *fire resistance* of *exterior walls* shall not be reduced.

D102.2.9 Roof structures. Structures, except aerial supports 12 feet (3658 mm) high or less, flagpoles, water tanks and cooling towers, placed above the roof of any building within the fire district shall be of noncombustible material and shall be supported by construction of noncombustible material.

D102.2.10 Plastic signs. The use of plastics complying with Section 2611 for signs is permitted provided that the structure of the sign in which the plastic is mounted or installed is noncombustible.

D102.2.11 Plastic veneer. Exterior plastic *veneer* is not permitted in the fire district.

SECTION D103 CHANGES TO BUILDINGS

D103.1 Existing buildings within the fire district. An existing building shall not be increased in height or area unless it is of a type of construction permitted for new buildings within the fire district or is altered to comply with the requirements for such type of construction. Nor shall any existing building be extended on any side, nor square footage or floors added within the existing building unless such modifications are of a type of construction permitted for new buildings within the fire district.

D103.2 Other alterations. Nothing in Section D103.1 shall prohibit other alterations within the fire district provided that such *alterations* do not create a change of occupancy that is otherwise prohibited or increase the fire hazard.

D103.3 Moving buildings. Buildings shall not hereafter be moved into the fire district or to another lot in the fire district unless the building is of a type of construction permitted in the fire district.

SECTION D104 BUILDINGS LOCATED PARTIALLY IN THE FIRE DISTRICT

D104.1 General. Any building located partially in the fire district shall be of a type of construction required for the fire district, unless the major portion of such building lies outside of the fire district and all portions of it extend not more than 10 feet (3048 mm) inside the boundaries of the fire district.

SECTION D105 EXCEPTIONS TO RESTRICTIONS IN FIRE DISTRICT

D105.1 General. The preceding provisions of this appendix shall not apply in the following instances:

1. Temporary buildings used in connection with duly authorized construction.
2. A *private garage* used exclusively as such, not more than one *story* in height, nor more than 650 square feet (60 m^2) in area, located on the same lot with a *dwelling*.
3. Fences not over 8 feet (2438 mm) high.
4. Coal tipples, material bins and trestles of Type IV construction.
5. Water tanks and cooling towers conforming to Sections 1510.3 and 1510.4.
6. *Greenhouses* less than 15 feet (4572 mm) high.
7. Porches on dwellings not over one *story* in height, and not over 10 feet (3048 mm) wide from the face of the building, provided that such porch does not come within 5 feet (1524 mm) of any property line.
8. Sheds open on a long side not over 15 feet (4572 mm) high and 500 square feet (46 m^2) in area.
9. One- and two-family *dwellings* where of a type of construction not permitted in the fire district can be extended 25 percent of the floor area existing at the time of inclusion in the fire district by any type of construction permitted by this code.
10. Wood decks less than 600 square feet (56 m^2) where constructed of 2-inch (51 mm) nominal wood, pressure treated for exterior use.
11. Wood *veneers* on *exterior walls* conforming to Section 1404.5.

12. Exterior plastic *veneer* complying with Section 2605.2 where installed on *exterior walls* required to have a *fire-resistance rating* not less than 1 hour, provided that the exterior plastic *veneer* does not exhibit sustained flaming as defined in NFPA 268.

SECTION D106
REFERENCED STANDARDS

D106.1 General. See Table D106.1 for standards that are referenced in various sections of this appendix. Standards are listed by the standard identification with the effective date, standard title, and the section or sections of this appendix that reference the standard.

TABLE D106.1
REFERENCED STANDARDS

STANDARD ACRONYM	STANDARD NAME	SECTIONS HEREIN REFERENCED
ASTM E84—2018B	*Test Method for Surface Burning Characteristics of Building Materials*	D102.2.8
NFPA 268—17	*Test Method for Determining Ignitability of Exterior Wall Assemblies Using a Radiant Heat Energy Source*	D105.1
NFPA 701—19	*Methods of Fire Tests for Flame-Propagation of Textiles and Films*	D102.2.8
UL 723—2018	*Standard for Test for Surface Burning Characteristics of Building Materials*	D102.2.8

APPENDIX E
SUPPLEMENTARY ACCESSIBILITY REQUIREMENTS

The provisions contained in this appendix are not mandatory unless specifically referenced in the adopting ordinance.

User note:

About this appendix: *The Architectural and Transportation Barriers Compliance Board (U.S. Access Board) has revised and updated its accessibility guidelines for buildings and facilities covered by the Americans with Disabilities Act (ADA) and the Architectural Barriers Act (ABA). Appendix E includes scoping requirements contained in the 2010 ADA Standards for Accessible Design that are not in Chapter 11 and not otherwise mentioned or mainstreamed throughout the code. Items in this appendix address subjects not typically addressed in building codes (for example, beds, room signage, transportation facilities).*

SECTION E101 GENERAL

E101.1 Scope. The provisions of this appendix shall control the supplementary requirements for the design and construction of facilities for *accessibility* for individuals with disabilities.

E101.2 Design. Technical requirements for items herein shall comply with this code and ICC A117.1.

SECTION E102 DEFINITIONS

E102.1 General. The following words and terms shall, for the purposes of this appendix, have the meanings shown herein. Refer to Chapter 2 of this code for general definitions.

CLOSED-CIRCUIT TELEPHONE. A telephone with a dedicated line such as a house phone, courtesy phone or phone that must be used to gain entrance to a facility.

MAILBOXES. Receptacles for the receipt of documents, packages or other deliverable matter. *Mailboxes* include, but are not limited to, post office boxes and receptacles provided by commercial mail-receiving agencies, apartment houses and schools.

TRANSIENT LODGING. A building, facility or portion thereof, excluding inpatient *medical care* facilities and long-term care facilities, that contains one or more *dwelling units* or *sleeping units*. Examples of *transient lodging* include, but are not limited to, resorts, *group homes*, hotels, motels, *dormitories*, homeless shelters, halfway houses and social service lodging.

SECTION E103 ACCESSIBLE ROUTE

E103.1 Raised platforms. In banquet rooms or spaces where a head table or speaker's lectern is located on a raised platform, an *accessible* route shall be provided to the platform.

SECTION E104 SPECIAL OCCUPANCIES

E104.1 General. *Transient lodging* facilities shall be provided with accessible features in accordance with Section E104.2. Group I-3 occupancies shall be provided with accessible features in accordance with Section E104.2.

E104.2 Communication features. *Accessible* communication features shall be provided in accordance with Sections E104.2.1 through E104.2.4.

E104.2.1 Transient lodging. In *transient lodging* facilities, *sleeping units* with accessible communication features shall be provided in accordance with Table E104.2.1. Units required to comply with Table E104.2.1 shall be dispersed among the various classes of units.

E104.2.2 Group I-3. In Group I-3 occupancies at least 2 percent of the total number of general holding *cells* and general housing *cells* equipped with audible emergency notification systems, and not less than one cell, shall be provided with visual notification devices. Permanently installed telephones within the cell shall comply with Section E104.2.4.

E104.2.3 Dwelling units and sleeping units. Where *dwelling units* and *sleeping units* are altered or added, the requirements of Section E104.2 shall apply only to the units being altered or added until the number of units with accessible communication features complies with the minimum number required for new construction.

E104.2.4 Notification devices. Visual notification devices shall be provided to alert room occupants of incoming telephone calls and a door knock or bell. Notification devices shall not be connected to visual alarm signal appliances. Permanently installed telephones shall have volume controls and an electrical outlet complying with ICC A117.1 located within 48 inches (1219 mm) of the telephone to facilitate the use of a TTY.

SECTION E105
OTHER FEATURES AND FACILITIES

E105.1 Portable toilets and bathing rooms. Where multiple single-user portable toilet or bathing units are clustered at a single location, at least 5 percent, but not less than one toilet unit or bathing unit at each cluster, shall be *accessible*. Signs containing the International Symbol of Accessibility shall identify accessible portable toilets and bathing units.

Exception: Portable toilet units provided for use exclusively by construction personnel on a construction site.

E105.2 Laundry equipment. Where provided in spaces required to be *accessible*, washing machines and clothes dryers shall comply with this section.

E105.2.1 Washing machines. Where three or fewer washing machines are provided, one or more shall be *accessible*. Where more than three washing machines are provided, two or more shall be *accessible*.

E105.2.2 Clothes dryers. Where three or fewer clothes dryers are provided, one or more shall be *accessible*. Where more than three clothes dryers are provided, two or more shall be *accessible*.

E105.3 Gaming machines, depositories, vending machines, change machines and similar equipment. Not fewer than one of each type of depository, vending machine, change machine and similar equipment shall be *accessible*. Two percent of gaming machines shall be *accessible* and provided with a front approach. *Accessible* gaming machines shall be distributed throughout the different types of gaming machines provided.

Exception: Drive-up-only depositories are not required to comply with this section.

E105.4 Mailboxes. Where *mailboxes* are provided in an interior location, 5 percent of the total, but not less than one, of each type shall be *accessible*. In residential and institutional facilities, where *mailboxes* are provided for each *dwelling unit* or *sleeping unit*, *accessible mailboxes* shall be provided for each unit required to be an *Accessible unit*.

E105.5 Automatic teller machines and fare machines. Where automatic teller machines or self-service fare vending, collection or adjustment machines are provided, not fewer than one machine of each type at each location where such machines are provided shall be *accessible*. Where bins are provided for envelopes, wastepaper or other purposes, not fewer than one of each type shall be *accessible*.

E105.6 Two-way communication systems. Where two-way communication systems are provided to gain admittance to a building or facility or to restricted areas within a building or facility, the system shall be *accessible*.

SECTION E106
TELEPHONES

E106.1 General. Where coin-operated public pay telephones, coinless public pay telephones, public *closed-circuit telephones*, courtesy phones or other types of public telephones are provided, accessible public telephones shall be provided in accordance with Sections E106.2 through E106.5 for each type of public telephone provided. For purposes of this section, a bank of telephones shall be considered to consist of two or more adjacent telephones.

E106.2 Wheelchair-accessible telephones. Where public telephones are provided, wheelchair-accessible telephones shall be provided in accordance with Table E106.2.

Exception: Drive-up-only public telephones are not required to be *accessible*.

TABLE E106.2
WHEELCHAIR-ACCESSIBLE TELEPHONES

NUMBER OF TELEPHONES PROVIDED ON A FLOOR, LEVEL OR EXTERIOR SITE	MINIMUM REQUIRED NUMBER OF WHEELCHAIR-ACCESSIBLE TELEPHONES
1 or more single unit	1 per floor, level and exterior site
1 bank	1 per floor, level and exterior site
2 or more banks	1 per bank

TABLE E104.2.1
DWELLING OR SLEEPING UNITS WITH ACCESSIBLE COMMUNICATION FEATURES

TOTAL NUMBER OF DWELLING OR SLEEPING UNITS PROVIDED	MINIMUM REQUIRED NUMBER OF DWELLING OR SLEEPING UNITS WITH ACCESSIBLE COMMUNICATION FEATURES
1	1
2 to 25	2
26 to 50	4
51 to 75	7
76 to 100	9
101 to 150	12
151 to 200	14
201 to 300	17
301 to 400	20
401 to 500	22
501 to 1,000	5% of total
1,001 and over	50 plus 3 for each 100 over 1,000

E106.3 Volume controls. All public telephones provided shall have accessible volume control.

E106.4 TTYs. TTYs shall be provided in accordance with Sections E106.4.1 through E106.4.9.

E106.4.1 Bank requirement. Where four or more public pay telephones are provided at a bank of telephones, at least one public TTY shall be provided at that bank.

Exception: TTYs are not required at banks of telephones located within 200 feet (60 960 mm) of, and on the same floor as, a bank containing a public TTY.

E106.4.2 Floor requirement. Where four or more public pay telephones are provided on a floor of a privately owned building, one or more public TTY shall be provided on that floor. Where one public pay telephone or more, is provided on a floor of a publicly owned building, not fewer than one public TTY shall be provided on that floor.

E106.4.3 Building requirement. Where four or more public pay telephones are provided in a privately owned building, one or more public TTY shall be provided in the building. Where at least one public pay telephone is provided in a publicly owned building, one or more public TTY shall be provided in the building.

E106.4.4 Site requirement. Where four or more public pay telephones are provided on a site, one or more public TTY shall be provided on the site.

E106.4.5 Rest stops, emergency road stops and service plazas. Where a public pay telephone is provided at a public rest stop, emergency road stop or service plaza, at least one public TTY shall be provided.

E106.4.6 Hospitals. Where a public pay telephone is provided in or adjacent to a hospital emergency room, hospital recovery room or hospital waiting room, one or more public TTY shall be provided at each such location.

E106.4.7 Transportation facilities. Transportation facilities shall be provided with TTYs in accordance with Sections E109.2.5 and E110.2 in addition to the TTYs required by Sections E106.4.1 through E106.4.4.

E106.4.8 Detention and correctional facilities. In detention and correctional facilities, where a public pay telephone is provided in a secured area used only by detainees or inmates and security personnel, then not fewer than one TTY shall be provided in not fewer than one secured area.

E106.4.9 Signs. Public TTYs shall be identified by the International Symbol of TTY complying with ICC A117.1. Directional signs indicating the location of the nearest public TTY shall be provided at banks of public pay telephones not containing a public TTY. Additionally, where signs provide direction to public pay telephones, they shall provide direction to public TTYs. Such signs shall comply with visual signage requirements in ICC A117.1 and shall include the International Symbol of TTY.

E106.5 Shelves for portable TTYs. Where a bank of telephones in the interior of a building consists of three or more public pay telephones, not fewer than one public pay telephone at the bank shall be provided with a shelf and an electrical outlet.

Exceptions:

1. In secured areas of detention and correctional facilities, if shelves and outlets are prohibited for purposes of security or safety shelves and outlets for TTYs are not required to be provided.
2. The shelf and electrical outlet shall not be required at a bank of telephones with a TTY.

SECTION E107 SIGNAGE

E107.1 Signs. Required accessible portable toilets and bathing facilities shall be identified by the International Symbol of Accessibility.

E107.2 Designations. Interior and exterior signs identifying permanent rooms and spaces shall be visual characters, raised characters and braille complying with ICC A117.1. Where pictograms are provided as designations of interior rooms and spaces, the pictograms shall have visual characters, raised characters and braille complying with ICC A117.1.

Exceptions:

1. Exterior signs that are not located at the door to the space they serve are not required to comply.
2. Building directories, menus, seat and row designations in assembly areas, occupant names, building addresses and company names and logos are not required to comply.
3. Signs in parking facilities are not required to comply.
4. Temporary (seven days or less) signs are not required to comply.
5. In detention and correctional facilities, signs not located in public areas are not required to comply.

E107.3 Directional and informational signs. Signs that provide direction to, or information about, permanent interior spaces of the site and facilities shall contain visual characters complying with ICC A117.1.

Exception: Building directories, personnel names, company or occupant names and logos, menus and temporary (seven days or less) signs are not required to comply with ICC A117.1.

E107.4 Other signs. Signage indicating special accessibility provisions shall be provided as follows:

1. At bus stops and terminals, signage must be provided in accordance with Section E108.4.
2. At fixed facilities and stations, signage must be provided in accordance with Sections E109.2.2 through E109.2.2.3.
3. At airports, terminal information systems must be provided in accordance with Section E110.3.

SECTION E108
BUS STOPS

E108.1 General. Bus stops shall comply with Sections E108.2 through E108.5.

E108.2 Bus boarding and alighting areas. Bus boarding and alighting areas shall comply with Sections E108.2.1 through E108.2.4.

E108.2.1 Surface. Bus boarding and alighting areas shall have a firm, stable surface.

E108.2.2 Dimensions. Bus boarding and alighting areas shall have a clear length of 96 inches (2440 mm) minimum, measured perpendicular to the curb or vehicle roadway edge, and a clear width of 60 inches (1525 mm) minimum, measured parallel to the vehicle roadway.

E108.2.3 Connection. Bus boarding and alighting areas shall be connected to streets, sidewalks or pedestrian paths by an *accessible route* complying with Section 1104.

E108.2.4 Slope. Parallel to the roadway, the slope of the bus boarding and alighting area shall be the same as the roadway, to the maximum extent practicable. For water drainage, a maximum slope of 1:48 perpendicular to the roadway is allowed.

E108.3 Bus shelters. Where provided, new or replaced bus shelters shall provide a minimum clear floor or ground space complying with ICC A117.1, Section 305, entirely within the shelter. Such shelters shall be connected by an *accessible route* to the boarding area required by Section E108.2.

E108.4 Signs. New bus route identification signs shall have finish and contrast complying with ICC A117.1. Additionally, to the maximum extent practicable, new bus route identification signs shall provide visual characters complying with ICC A117.1.

Exception: Bus schedules, timetables and maps that are posted at the bus stop or bus bay are not required to meet this requirement.

E108.5 Bus stop siting. Bus stop sites shall be chosen such that, to the maximum extent practicable, the areas where lifts or *ramps* are to be deployed comply with Sections E108.2 and E108.3.

SECTION E109
TRANSPORTATION FACILITIES AND STATIONS

E109.1 General. Fixed transportation facilities and stations shall comply with the applicable provisions of Section E109.2.

E109.2 New construction. New stations in rapid rail, light rail, commuter rail, intercity rail, high speed rail and other fixed guideway systems shall comply with Sections E109.2.1 through E109.2.8.

E109.2.1 Station entrances. Where different entrances to a station serve different transportation fixed routes or groups of fixed routes, at least one entrance serving each group or route shall comply with Section 1104.

E109.2.2 Signs. Signage in fixed transportation facilities and stations shall comply with Sections E109.2.2.1 through E109.2.2.3.

E109.2.2.1 Raised character and braille signs. Where signs are provided at entrances to stations identifying the station or the entrance, or both, at least one sign at each entrance shall be raised characters and braille. A minimum of one raised character and braille sign identifying the specific station shall be provided on each platform or boarding area. Such signs shall be placed in uniform locations at entrances and on platforms or boarding areas within the transit system to the maximum extent practicable.

Exceptions:

1. Where the station does not have a defined entrance but signs are provided, the raised characters and braille signs shall be placed in a central location.
2. Signs are not required to be raised characters and braille where audible signs are remotely transmitted to hand-held receivers, or are user or proximity actuated.

E109.2.2.2 Identification signs. Stations covered by this section shall have identification signs containing visual characters complying with ICC A117.1. Signs shall be clearly visible and within the sightlines of a standing or sitting passenger from within the train on both sides when not obstructed by another train.

E109.2.2.3 Informational signs. Lists of stations, routes and destinations served by the station that are located on boarding areas, platforms or *mezzanines* shall provide visual characters complying with ICC A117.1. Signs covered by this provision shall, to the maximum extent practicable, be placed in uniform locations within the transit system.

E109.2.3 Fare machines. Self-service fare vending, collection and adjustment machines shall comply with ICC A117.1, Section 707. Where self-service fare vending, collection or adjustment machines are provided for the use of the general public, at least one accessible machine of each type provided shall be provided at each accessible point of entry and *exit*.

E109.2.4 Rail-to-platform height. Station platforms shall be positioned to coordinate with vehicles in accordance with the applicable provisions of 36 CFR, Part 1192. Low-level platforms shall be 8 inches (250 mm) minimum above top of rail.

Exception: Where vehicles are boarded from sidewalks or street level, low-level platforms shall be permitted to be less than 8 inches (250 mm).

E109.2.5 TTYs. Where a public pay telephone is provided in a transit facility (as defined by the Department of Transportation), at least one public TTY complying with ICC A117.1, Section 704.4, shall be provided in the station. In addition, where one or more public pay telephones serve a particular entrance to a transportation facility, at least one TTY telephone complying with ICC A117.1, Section 704.4, shall be provided to serve that entrance.

E109.2.6 Track crossings. Where a *circulation path* serving boarding platforms crosses tracks, an *accessible* route shall be provided.

Exception: Openings for wheel flanges shall be permitted to be $2^1/_2$ inches (64 mm) maximum.

E109.2.7 Public address systems. Where public address systems convey audible information to the public, the same or equivalent information shall be provided in a visual format.

E109.2.8 Clocks. Where clocks are provided for use by the general public, the clock face shall be uncluttered so that its elements are clearly visible. Hands, numerals and digits shall contrast with the background either light-on-dark or dark-on-light. Where clocks are mounted overhead, numerals and digits shall comply with visual character requirements.

SECTION E110 AIRPORTS

E110.1 New construction. New construction of airports shall comply with Sections E110.2 through E110.4.

E110.2 TTYs. Where public pay telephones are provided, at least one TTY shall be provided in compliance with ICC A117.1, Section 704.4. Additionally, if four or more public pay telephones are located in a main terminal outside the security areas, a concourse within the security areas or a baggage claim area in a terminal, at least one public TTY complying with ICC A117.1, Section 704.4, shall also be provided in each such location.

E110.3 Terminal information systems. Where terminal information systems convey audible information to the public, the same or equivalent information shall be provided in a visual format.

E110.4 Clocks. Where clocks are provided for use by the general public, the clock face shall be uncluttered so that its elements are clearly visible. Hands, numerals and digits shall contrast with the background either light-on-dark or dark-on-light. Where clocks are mounted overhead, numerals and digits shall comply with visual character requirements.

SECTION E111 REFERENCED STANDARDS

E111.1 General. See Table E111.1 for standards that are referenced in various sections of this appendix. Standards are listed by the standard identification with the effective date, standard title, and the section or sections of this appendix that reference the standard.

TABLE E111.1 REFERENCED STANDARDS

STANDARD ACRONYM	STANDARD NAME	SECTIONS HEREIN REFERENCED
DOJ 36 CFR Part 1192	*Americans with Disabilities Act (ADA) Accessibility Guidelines for Transportation Vehicles (ADAAG).* Washington, DC: Department of Justice, 1991	E109.2.4
IBC 2021	*International Building Code*	E102.1
ICC A117.1-2017	*Accessible and Usable Buildings and Facilities*	E101.2, E104.2.4, E106.4.9, E107.2, E107.3, E108.3, E108.4, E109.2.2.2, E109.2.2.3, E109.2.3, E109.2.5, E110.2

APPENDIX F
RODENTPROOFING

The provisions contained in this appendix are not mandatory unless specifically referenced in the adopting ordinance.

User notes:

About this appendix: *The provisions of Appendix F are minimum mechanical methods to prevent the entry of rodents into a building. These standards, when used in conjunction with cleanliness and maintenance programs, can significantly reduce the potential of rodents invading a building.*

Code development reminder: *Code change proposals to this appendix will be considered by the IBC—Structural Code Development Committee during the 2022 (Group B) Code Development Cycle.*

SECTION F101
GENERAL

F101.1 General. Buildings or structures and the walls enclosing habitable or occupiable rooms and spaces in which persons live, sleep or work, or in which feed, food or foodstuffs are stored, prepared, processed, served or sold, shall be constructed in accordance with the provisions of this section.

F101.2 Foundation wall ventilation openings. Foundation wall ventilation openings shall be covered for their height and width with perforated sheet metal plates not less than 0.070 inch (1.8 mm) thick, expanded sheet metal plates not less than 0.047 inch (1.2 mm) thick, cast-iron grills or grating, extruded aluminum load-bearing vents or with hardware cloth of 0.035 inch (0.89 mm) wire or heavier. The openings therein shall not exceed $^1/_4$ inch (6.4 mm).

F101.3 Foundation and exterior wall sealing. *Annular spaces* around pipes, electric cables, conduits or other openings in the walls shall be protected against the passage of rodents by closing such openings with cement *mortar*, concrete masonry or noncorrosive metal.

F101.4 Doors. Doors on which metal protection has been applied shall be hinged so as to be free swinging. When closed, the maximum clearance between any door, door jambs and sills shall be not greater than $^3/_8$ inch (9.5 mm).

F101.5 Windows and other openings. Windows and other openings for the purpose of light or ventilation located in *exterior walls* within 2 feet (610 mm) above the existing ground level immediately below such opening shall be covered for their entire height and width, including frame, with hardware cloth of not less than 0.035-inch (0.89 mm) wire or heavier.

F101.5.1 Rodent-accessible openings. Windows and other openings for the purpose of light and ventilation in the *exterior walls* not covered in this chapter, accessible to rodents by way of exposed pipes, wires, conduits and other appurtenances, shall be covered with wire cloth of at least 0.035-inch (0.89 mm) wire. In lieu of wire cloth covering, said pipes, wires, conduits and other appurtenances shall be blocked from rodent usage by installing solid sheet metal guards 0.024 inch (0.61 mm) thick or heavier. Guards shall be fitted around pipes, wires, conduits or other appurtenances. In addition, they shall be fastened securely to and shall extend perpendicularly from the *exterior wall* for not less than 12 inches (305 mm) beyond and on either side of pipes, wires, conduits or appurtenances.

F101.6 Pier and wood construction.

F101.6.1 Sill less than 12 inches above ground. Buildings not provided with a continuous foundation shall be provided with protection against rodents at grade by providing either an apron in accordance with Section F101.6.1.1 or a floor slab in accordance with Section F101.6.1.2.

F101.6.1.1 Apron. Where an apron is provided, the apron shall be not less than 8 inches (203 mm) above, nor less than 24 inches (610 mm) below, grade. The apron shall not terminate below the lower edge of the siding material. The apron shall be constructed of an approved nondecayable, water-resistant rodentproofing material of required strength and shall be installed around the entire perimeter of the building. Where constructed of masonry or concrete materials, the apron shall be not less than 4 inches (102 mm) in thickness.

F101.6.1.2 Grade floors. Where continuous concrete-grade floor slabs are provided, open spaces shall not be left between the slab and walls, and openings in the slab shall be protected.

F101.6.2 Sill at or above 12 inches above ground. Buildings not provided with a continuous foundation and that have sills 12 inches (305 mm) or more above ground level shall be provided with protection against rodents at grade in accordance with any of the following:

1. Section F101.6.1.1 or F101.6.1.2.
2. By installing solid sheet metal collars not less than 0.024 inch (0.6 mm) thick at the top of each pier or pile and around each pipe, cable, conduit, wire or other item that provides a continuous pathway from the ground to the floor.
3. By encasing the pipes, cables, conduits or wires in an enclosure constructed in accordance with Section F101.6.1.1.

APPENDIX G
FLOOD-RESISTANT CONSTRUCTION

The provisions contained in this appendix are not mandatory unless specifically referenced in the adopting ordinance.

User notes:

About this appendix: *Appendix G is intended to provide the additional flood-plain management and administrative requirements of the National Flood Insurance Program (NFIP) that are not included in the code. Communities that adopt the International Building Code® and Appendix G will meet the minimum requirements of NFIP as set forth in Title 44 of the Code of Federal Regulations.*

Code development reminder: *Code change proposals to this appendix will be considered by the IBC—Structural Code Development Committee during the 2022 (Group B) Code Development Cycle.*

SECTION G101
ADMINISTRATION

G101.1 Purpose. The purpose of this appendix is to promote the public health, safety and general welfare and to minimize public and private losses due to *flood* conditions in specific *flood hazard areas* through the establishment of comprehensive regulations for management of *flood hazard areas* designed to:

1. Prevent unnecessary disruption of commerce, access and public service during times of *flooding*.
2. Manage the alteration of natural flood plains, stream channels and shorelines.
3. Manage filling, grading, dredging and other development that may increase flood damage or erosion potential.
4. Prevent or regulate the construction of flood barriers that will divert floodwaters or that can increase flood hazards.
5. Contribute to improved construction techniques in the flood plain.

G101.2 Objectives. The objectives of this appendix are to protect human life, minimize the expenditure of public money for flood control projects, minimize the need for rescue and relief efforts associated with *flooding*, minimize prolonged business interruption, minimize damage to public facilities and utilities, help maintain a stable tax base by providing for the sound use and development of flood-prone areas, contribute to improved construction techniques in the flood plain and ensure that potential owners and occupants are notified that property is within *flood hazard areas*.

G101.3 Scope. The provisions of this appendix shall apply to all proposed development in a *flood hazard area* established in Section 1612 of this code, including certain building work exempt from permit under Section 105.2.

G101.4 Violations. Any violation of a provision of this appendix, or failure to comply with a *permit* or variance issued pursuant to this appendix or any requirement of this appendix, shall be handled in accordance with Section 114.

G101.5 Designation of floodplain administrator. The [INSERT JURISDICTION'S SELECTED POSITION TITLE] is designated as the floodplain administrator and is authorized and directed to enforce the provisions of this appendix. The floodplain administrator is authorized to delegate performance of certain duties to other employees of the jurisdiction. Such designation shall not alter any duties and powers of the building official.

SECTION G102
DEFINITIONS

G102.1 General. The following words and terms shall, for the purposes of this appendix, have the meanings shown herein. Refer to Chapter 2 of this code for general definitions.

DEVELOPMENT. Any man-made change to improved or unimproved real estate, including but not limited to, buildings or *other structures*, temporary structures, temporary or permanent storage of materials, mining, dredging, filling, grading, paving, excavations, operations and other land-disturbing activities.

FUNCTIONALLY DEPENDENT FACILITY. A facility that cannot perform its intended purpose unless it is located or carried out in close proximity to water. The term includes only docking facilities, port facilities necessary for the loading or unloading of cargo or passengers, and shipbuilding and ship repair facilities. The term does not include long-term storage, manufacture, sales or service facilities.

MANUFACTURED HOME. A structure that is transportable in one or more sections, built on a permanent chassis, designed for use with or without a permanent foundation when attached to the required utilities, and constructed to the Federal Manufactured Home Construction and Safety Standards and rules and regulations promulgated by the U.S. Department of Housing and Urban Development. The term also includes mobile homes, park trailers, travel trailers and similar transportable structures that are placed on a site for 180 consecutive days or longer.

MANUFACTURED HOME PARK OR SUBDIVISION. A parcel (or contiguous parcels) of land divided into two or more manufactured home lots for rent or sale.

RECREATIONAL VEHICLE. A vehicle that is built on a single chassis, 400 square feet (37.16 m^2) or less when measured at the largest horizontal projection, designed to be self-propelled or permanently towable by a light-duty truck, and designed primarily not for use as a permanent dwelling but as temporary living quarters for recreational, camping, travel or seasonal use. A recreational vehicle is ready for highway use if it is on its wheels or jacking system, is attached to the site only by quick disconnect-type utilities and security devices and has no permanently attached additions.

VARIANCE. A grant of relief from the requirements of this section that permits construction in a manner otherwise prohibited by this section where specific enforcement would result in unnecessary hardship.

VIOLATION. A development that is not fully compliant with this appendix or Section 1612, as applicable.

SECTION G103 APPLICABILITY

G103.1 General. This appendix, in conjunction with this code, provides minimum requirements for development located in *flood hazard areas*, including:

1. The subdivision of land.
2. Site improvements and installation of utilities.
3. Placement and replacement of manufactured homes.
4. Placement of recreational vehicles.
5. New construction and repair, reconstruction, rehabilitation or additions to new construction.
6. *Substantial improvement* of existing buildings and structures, including restoration after damage.
7. Installation of tanks.
8. Temporary structures.
9. Temporary or permanent storage, utility and miscellaneous Group U buildings and structures.
10. Certain building work exempt from permit under Section 105.2 and other buildings and development activities.

G103.2 Establishment of flood hazard areas. *Flood hazard areas* are established in Section 1612.3 of this code, adopted by the applicable governing authority on **[INSERT DATE]**.

SECTION G104 POWERS AND DUTIES

G104.1 Permit applications. All applications for permits shall comply with the following:

1. The floodplain administrator shall review all *permit* applications to determine whether proposed development is located in *flood hazard areas* established in Section G103.2.
2. Where a proposed development site is in a *flood hazard area*, all development to which this appendix is applicable as specified in Section G103.1 shall be designed and constructed with methods, practices and materials that minimize *flood* damage and that are in accordance with this code and ASCE 24.

G104.2 Other permits. It shall be the responsibility of the floodplain administrator to ensure that approval of a proposed development shall not be given until proof that necessary permits have been granted by federal or state agencies having jurisdiction over such development.

G104.3 Determination of design flood elevations. If *design flood elevations* are not specified, the floodplain administrator is authorized to require the applicant to meet one of the following:

1. Obtain, review and reasonably utilize data available from a federal, state or other source.
2. Determine the *design flood elevation* in accordance with accepted hydrologic and hydraulic engineering techniques. Such analyses shall be performed and sealed by a *registered design professional*. Studies, analyses and computations shall be submitted in sufficient detail to allow review and approval by the floodplain administrator. *The accuracy of data submitted for such determination shall be the responsibility of the applicant.*

G104.4 Activities in riverine flood hazard areas. In riverine *flood hazard areas* where *design flood elevations* are specified but *floodways* have not been designated, the floodplain administrator shall not permit any new construction, *substantial improvement* or other development, including fill, unless the applicant submits an engineering analysis prepared by a *registered design professional*, demonstrating that the cumulative effect of the proposed development, when combined with all other existing and anticipated *flood hazard area* encroachment, will not increase the *design flood elevation* more than 1 foot (305 mm) at any point within the community.

G104.5 Floodway encroachment. Prior to issuing a *permit* for any floodway encroachment, including fill, new construction, *substantial improvements* and other development or land-disturbing activity, the floodplain administrator shall require submission of a certification, prepared by a *registered design professional*, along with supporting technical data, demonstrating that such development will not cause any increase of the *base flood* level.

G104.5.1 Floodway revisions. A *floodway* encroachment that increases the level of the *base flood* is authorized if the applicant has applied for a conditional *Flood Insurance Rate Map* (FIRM) revision and has received the approval of the Federal Emergency Management Agency (FEMA).

G104.6 Watercourse alteration. Prior to issuing a *permit* for any alteration or relocation of any watercourse, the floodplain administrator shall require the applicant to provide notification of the proposal to the appropriate authorities of all adjacent government jurisdictions, as well as appropriate state agencies. A copy of the notification shall be maintained in the permit records and submitted to FEMA.

G104.6.1 Engineering analysis. The floodplain administrator shall require submission of an engineering analysis, prepared by a *registered design professional*, demonstrating that the flood-carrying capacity of the altered or relocated portion of the watercourse will not be decreased. Such watercourses shall be maintained in a manner that preserves the channel's flood-carrying capacity.

G104.7 Alterations in coastal areas. Prior to issuing a *permit* for any alteration of sand dunes and mangrove stands in *coastal high-hazard areas* and coastal A zones, the floodplain administrator shall require submission of an engineering analysis, prepared by a *registered design professional*, demonstrating that the proposed alteration will not increase the potential for flood damage.

G104.8 Records. The floodplain administrator shall maintain a permanent record of all *permits* issued in *flood hazard areas*, including supporting certifications and documentation required by this appendix and copies of inspection reports, design certifications and documentation of elevations required in Section 1612 of this code and Section R322 of the *International Residential Code*.

G104.9 Inspections. Development for which a *permit* under this appendix is required shall be subject to inspection. The floodplain administrator or the floodplain administrator's designee shall make, or cause to be made, inspections of all development in *flood hazard areas* authorized by issuance of a *permit* under this appendix.

G104.10 Use of changed technical data. The floodplain administrator and the applicant shall not use changed *flood hazard area* boundaries or base flood elevations for proposed buildings or developments unless the floodplain administrator or applicant has applied for a conditional *Flood Insurance Rate Map* (FIRM) revision and has received the approval of the Federal Emergency Management Agency (FEMA).

SECTION G105
PERMITS

G105.1 Required. Any person, owner or owner's authorized agent who intends to conduct any development in a flood hazard area shall first make application to the floodplain administrator and shall obtain the required *permit*.

G105.2 Application for permit. The applicant shall file an application in writing on a form furnished by the floodplain administrator. Such application shall:

1. Identify and describe the development to be covered by the *permit*.
2. Describe the land on which the proposed development is to be conducted by legal description, street address or similar description that will readily identify and definitely locate the site.
3. Include a site plan showing the delineation of *flood hazard areas*, *floodway* boundaries, *flood* zones, *design flood elevations*, ground elevations, proposed fill and excavation and drainage patterns and facilities.
4. Include in subdivision proposals and other proposed developments with more than 50 lots or larger than 5 acres (20 234 m^2), *base flood elevation* data in accordance with Section 1612.3.1 if such data are not identified for the *flood hazard areas* established in Section G103.2.
5. Indicate the use and occupancy for which the proposed development is intended.
6. Be accompanied by construction documents, grading and filling plans and other information deemed appropriate by the floodplain administrator.
7. State the valuation of the proposed work.
8. Be signed by the applicant or the applicant's authorized agent.

G105.3 Validity of permit. The issuance of a *permit* under this appendix shall not be construed to be a *permit* for, or approval of, any violation of this appendix or any other ordinance of the jurisdiction. The issuance of a *permit* based on submitted documents and information shall not prevent the floodplain administrator from requiring the correction of errors. The floodplain administrator is authorized to prevent occupancy or use of a structure or site that is in violation of this appendix or other ordinances of this jurisdiction.

G105.4 Expiration. A *permit* shall become invalid if the proposed development is not commenced within 180 days after its issuance, or if the work authorized is suspended or abandoned for a period of 180 days after the work commences. Extensions shall be requested in writing and justifiable cause demonstrated. The floodplain administrator is authorized to grant, in writing, one or more extensions of time, for periods not more than 180 days each.

G105.5 Suspension or revocation. The floodplain administrator is authorized to suspend or revoke a *permit* issued under this appendix wherever the *permit* is issued in error or on the basis of incorrect, inaccurate or incomplete information, or in violation of any ordinance or code of this jurisdiction.

SECTION G106
VARIANCES

G106.1 General. The *board of appeals* established pursuant to Section 113, or other established or designed board, shall hear and decide requests for variances. The board shall base its determination on technical justifications, and has the right to attach such conditions to variances as it deems necessary to further the purposes and objectives of this appendix and Section 1612.

G106.2 Records. The floodplain administrator shall maintain a permanent record of all variance actions, including justification for their issuance.

G106.3 Historic structures. A variance is authorized to be issued for the repair or rehabilitation of a historic structure upon a determination that the proposed repair or rehabilitation will not preclude the structure's continued designation as a historic structure, and the variance is the minimum necessary to preserve the historic character and design of the structure.

Exception: Within *flood hazard areas*, historic structures that do not meet one or more of the following designations:

1. Listed or preliminarily determined to be eligible for listing in the National Register of Historic Places.
2. Determined by the Secretary of the U.S. Department of Interior as contributing to the historical significance of a registered historic district or a district preliminarily determined to qualify as an historic district.
3. Designated as *historic* under a state or local historic preservation program that is approved by the Department of Interior.

G106.4 Functionally dependent facilities. A variance is authorized to be issued for the construction or *substantial improvement* of a functionally dependent facility provided that the criteria in Section 1612.1 are met and the variance is the minimum necessary to allow the construction or *substantial improvement*, and that all due consideration has been given to methods and materials that minimize *flood* damages during the *design flood* and do not create additional threats to public safety.

G106.5 Restrictions. The board shall not issue a variance for any proposed development in a *floodway* if any increase in flood levels would result during the *base flood* discharge.

G106.6 Considerations. In reviewing applications for variances, the board shall consider all technical evaluations, all relevant factors, all other portions of this appendix and the following:

1. The danger that materials and debris may be swept onto other lands resulting in further injury or damage.
2. The danger to life and property due to *flooding* or erosion damage.
3. The susceptibility of the proposed development, including contents, to *flood* damage and the effect of such damage on current and future owners.
4. The importance of the services provided by the proposed development to the community.
5. The availability of alternate locations for the proposed development that are not subject to *flooding* or erosion.
6. The compatibility of the proposed development with existing and anticipated development.
7. The relationship of the proposed development to the comprehensive plan and flood plain management program for that area.
8. The safety of access to the property in times of *flood* for ordinary and emergency vehicles.
9. The expected heights, velocity, duration, rate of rise and debris and sediment transport of the floodwaters and the effects of wave action, if applicable, expected at the site.
10. The costs of providing governmental services during and after *flood* conditions including maintenance and repair of public utilities and facilities such as sewer, gas, electrical and water systems, streets and bridges.

G106.7 Conditions for issuance. Variances shall only be issued by the board where all of the following criteria are met:

1. A technical showing of good and sufficient cause that the unique characteristics of the size, configuration or topography of the site renders the elevation standards inappropriate.
2. A determination that failure to grant the variance would result in exceptional hardship by rendering the lot undevelopable.
3. A determination that the granting of a variance will not result in increased *flood* heights, additional threats to public safety, extraordinary public expense, nor create nuisances, cause fraud on or victimization of the public or conflict with existing local laws or ordinances.
4. A determination that the variance is the minimum necessary, considering the *flood* hazard, to afford relief.
5. Notification to the applicant in writing over the signature of the floodplain administrator that the issuance of a variance to construct a structure below the *base flood* level will result in increased premium rates for flood insurance up to amounts as high as $25 for $100 of insurance coverage, and that such construction below the *base flood* level increases risks to life and property.

SECTION G107 SUBDIVISIONS

G107.1 General. Any subdivision proposal, including proposals for manufactured home parks and subdivisions, or other proposed new development in a *flood hazard area* shall be reviewed to verify all of the following:

1. Such proposals are consistent with the need to minimize *flood* damage.
2. Public utilities and facilities, such as sewer, gas, electric and water systems, are located and constructed to minimize or eliminate *flood* damage.
3. Adequate drainage is provided to reduce exposure to *flood* hazards.

G107.2 Subdivision requirements. The following requirements shall apply in the case of any proposed subdivision, including proposals for manufactured home parks and subdivisions, any portion of which lies within a *flood hazard area*:

1. The *flood hazard area*, including *floodways*, *coastal high-hazard areas* and coastal A zones, as appropriate, shall be delineated on tentative and final subdivision plats.

2. *Design flood elevations* shall be shown on tentative and final subdivision plats.
3. Residential building lots shall be provided with adequate buildable area outside the *floodway*.
4. The design criteria for utilities and facilities set forth in this appendix and appropriate International Codes shall be met.

SECTION G108
SITE IMPROVEMENT

G108.1 Development in floodways. Development or land-disturbing activity shall not be authorized in the *floodway* unless it has been demonstrated through hydrologic and hydraulic analyses performed in accordance with standard engineering practice, and prepared by a *registered design professional*, that the proposed encroachment will not result in any increase in the *base flood* level.

G108.2 Coastal high-hazard areas and coastal A zones. In *coastal high-hazard areas* and coastal A zones:

1. New buildings and buildings that are substantially improved shall only be authorized landward of the reach of mean high tide.
2. The use of fill for structural support of buildings is prohibited.

G108.3 Sewer facilities. All new or replaced sanitary sewer facilities, private sewage treatment plants (including all pumping stations and collector systems) and on-site waste disposal systems shall be designed in accordance with Chapter 7, ASCE 24, to minimize or eliminate infiltration of floodwaters into the facilities and discharge from the facilities into floodwaters, or impairment of the facilities and systems.

G108.4 Water facilities. All new or replacement water facilities shall be designed in accordance with the provisions of Chapter 7, ASCE 24, to minimize or eliminate infiltration of floodwaters into the systems.

G108.5 Storm drainage. Storm drainage shall be designed to convey the flow of surface waters to minimize or eliminate damage to persons or property.

G108.6 Streets and sidewalks. Streets and sidewalks shall be designed to minimize potential for increasing or aggravating *flood* levels.

SECTION G109
MANUFACTURED HOMES

G109.1 Elevation. All new and replacement manufactured homes to be placed or substantially improved in a *flood hazard area* shall be elevated such that the *lowest floor* of the manufactured home is elevated to or above the *design flood elevation*.

G109.2 Foundations. All new and replacement manufactured homes, including *substantial improvement* of existing manufactured homes, shall be placed on a permanent, reinforced foundation that is designed in accordance with Section R322 of the *International Residential Code*.

G109.3 Anchoring. All new and replacement manufactured homes to be placed or substantially improved in a *flood hazard area* shall be installed using methods and practices that minimize *flood* damage. Manufactured homes shall be securely anchored to an adequately anchored foundation system to resist flotation, collapse and lateral movement. Methods of anchoring are authorized to include, but are not limited to, use of over-the-top or frame ties to ground anchors. This requirement is in addition to applicable state and local anchoring requirements for resisting wind forces.

G109.4 Protection of mechanical equipment and outside appliances. Mechanical equipment and outside appliances shall be elevated to or above the *design flood elevation*.

Exception: Where such equipment and appliances are designed and installed to prevent water from entering or accumulating within their components and the systems are constructed to resist hydrostatic and hydrodynamic *loads* and stresses, including the effects of buoyancy, during the occurrence of *flooding* up to the elevation required by Section R322 of the *International Residential Code*, the systems and equipment shall be permitted to be located below the elevation required by Section R322 of the *International Residential Code*. Electrical wiring systems shall be permitted below the design *flood* elevation provided that they conform to the provisions of NFPA 70.

G109.5 Enclosures. Fully enclosed areas below elevated manufactured homes shall comply with the requirements of Section R322 of the *International Residential Code*.

SECTION G110
RECREATIONAL VEHICLES

G110.1 Placement prohibited. The placement of recreational vehicles shall not be authorized in *coastal high-hazard areas* and in *floodways*.

G110.2 Temporary placement. Recreational vehicles in *flood hazard areas* shall be fully licensed and ready for highway use, or shall be placed on a site for less than 180 consecutive days.

G110.3 Permanent placement. Recreational vehicles that are not fully licensed and ready for highway use, or that are to be placed on a site for more than 180 consecutive days, shall meet the requirements of Section G109 for manufactured homes.

SECTION G111
TANKS

G111.1 Tanks. Underground and above-ground tanks shall be designed, constructed, installed and anchored in accordance with ASCE 24.

SECTION G112 OTHER BUILDING WORK

G112.1 Garages and accessory structures. Garages and accessory structures shall be designed and constructed in accordance with ASCE 24.

G112.2 Fences. Fences in *floodways* that have the potential to block the passage of floodwaters, such as stockade fences and wire mesh fences, shall meet the requirement of Section G104.5.

G112.3 Oil derricks. Oil derricks located in *flood hazard areas* shall be designed in conformance with the *flood loads* in Sections 1603.1.7 and 1612.

G112.4 Retaining walls, sidewalks and driveways. Retaining walls, sidewalks and driveways shall meet the requirements of Section 1804.5.

G112.5 Swimming pools. Swimming pools shall be designed and constructed in accordance with ASCE 24. Above-ground swimming pools, on-ground swimming pools and in-ground swimming pools that involve placement of fill in *floodways* shall also meet the requirements of Section G104.5.

G112.6 Decks, porches, and patios. Decks, porches and patios shall be designed and constructed in accordance with ASCE 24.

G112.7 Nonstructural concrete slabs in coastal high-hazard areas and coastal A zones. In *coastal high-hazard areas* and coastal A zones, *nonstructural concrete* slabs used as parking pads, enclosure floors, landings, decks, walkways, patios and similar nonstructural uses are permitted beneath or adjacent to buildings and structures provided that the concrete slabs shall be constructed in accordance with ASCE 24.

G112.8 Roads and watercourse crossings in regulated floodways. Roads and watercourse crossings that encroach into regulated *floodways*, including roads, bridges, culverts, low-water crossings and similar means for vehicles or pedestrians to travel from one side of a watercourse to the other, shall meet the requirement of Section G104.5.

SECTION G113 TEMPORARY STRUCTURES AND TEMPORARY STORAGE

G113.1 Temporary structures. Temporary structures shall be erected for a period of less than 180 days. Temporary structures shall be anchored to prevent flotation, collapse or lateral movement resulting from hydrostatic *loads*, including the effects of buoyancy, during conditions of the *design flood*. Fully enclosed temporary structures shall have flood openings that are in accordance with ASCE 24 to allow for the automatic entry and exit of floodwaters.

G113.2 Temporary storage. Temporary storage includes storage of goods and materials for a period of less than 180 days. Stored materials shall not include hazardous materials.

G113.3 Floodway encroachment. Temporary structures and temporary storage in *floodways* shall meet the requirements of G104.5.

SECTION G114 UTILITY AND MISCELLANEOUS GROUP U

G114.1 Utility and miscellaneous Group U. Utility and miscellaneous Group U includes buildings that are accessory in character and miscellaneous structures not classified in any specific occupancy in this code, including, but not limited to, *agricultural buildings*, aircraft hangars (accessory to a one- or two-family residence), barns, carports, fences more than 6 feet (1829 mm) high, grain silos (accessory to a residential occupancy), greenhouses, livestock shelters, private garages, retaining walls, sheds, stables and towers.

G114.2 Flood loads. Utility and miscellaneous Group U buildings and structures, including *substantial improvement* of such buildings and structures, shall be anchored to prevent flotation, collapse or lateral movement resulting from *flood loads*, including the effects of buoyancy, during conditions of the *design flood*.

G114.3 Elevation. Utility and miscellaneous Group U buildings and structures, including *substantial improvement* of such buildings and structures, shall be elevated such that the *lowest floor*, including basement, is elevated to or above the *design flood elevation* in accordance with Section 1612 of this code.

G114.4 Enclosures below design flood elevation. Fully enclosed areas below the *design flood elevation* shall be constructed in accordance with ASCE 24.

G114.5 Flood-damage-resistant materials. Flood-damage-resistant materials shall be used below the *design flood elevation*.

G114.6 Protection of mechanical, plumbing and electrical systems. Mechanical, plumbing and electrical systems, including plumbing fixtures, shall be elevated to or above the *design flood elevation*.

> **Exception:** Electrical systems, equipment and components; heating, ventilating, air conditioning and plumbing appliances; plumbing fixtures, duct systems and other service equipment shall be permitted to be located below the *design flood elevation* provided that they are designed and installed to prevent water from entering or accumulating within the components and to resist hydrostatic and hydrodynamic *loads* and stresses, including the effects of buoyancy, during the occurrence of flooding to the *design flood elevation* in compliance with the flood-resistant construction requirements of this code. Electrical wiring systems shall be permitted to be located below the *design flood elevation* provided that they conform to the provisions of NFPA 70.

SECTION G115
REFERENCED STANDARDS

G115.1 General. See Table G115.1 for standards that are referenced in various sections of this appendix. Standards are listed by the standard identification with the effective date, standard title, and the section or sections of this appendix referenced in the standard.

TABLE G115.1
REFERENCED STANDARDS

STANDARD ACRONYM	STANDARD NAME	SECTIONS HEREIN REFERENCED
ASCE 24—14	*Flood Resistant Design and Construction*	G104.1, G108.3, G108.4, G111.1, G112.1, G112.5, G112.6, G112.7, G113.1, G114.4
HUD 24 CFR Part 3285 (2008)	*Manufactured Home Construction and Safety Standards*	G102
IBC—21	*International Building Code®*	G103.2, G114.1, G114.3
IRC—21	*International Residential Code®*	G109.2, G109.4, G109.5
NFPA 70—20	*National Electric Code®*	G109.4, G114.6

APPENDIX H

SIGNS

The provisions contained in this appendix are not mandatory unless specifically referenced in the adopting ordinance.

User notes:

About this appendix: *Appendix H gathers in one place the various standards that regulate the construction and protection of outdoor signs. Wherever possible, the appendix provides standards in performance language, thus allowing the widest possible application.*

Code development reminder: *Code change proposals to this appendix will be considered by the IBC—Structural Code Development Committee during the 2022 (Group B) Code Development Cycle.*

SECTION H101 GENERAL

H101.1 General. A sign shall not be erected in a manner that would confuse or obstruct the view of or interfere with exit signs required by Chapter 10 or with official traffic signs, signals or devices. Signs and sign support structures, together with their supports, braces, guys and anchors, shall be kept in repair and in proper state of preservation. The display surfaces of signs shall be kept neatly painted or posted at all times.

H101.2 Signs exempt from permits. The following signs are exempt from the requirements to obtain a *permit* before erection:

1. Painted nonilluminated signs.
2. Temporary signs announcing the sale or rent of property.
3. Signs erected by transportation authorities.
4. Projecting signs not exceeding 2.5 square feet (0.23 m^2).
5. The changing of moveable parts of an approved sign that is designed for such changes, or the repainting or repositioning of display matter shall not be deemed an alteration.

SECTION H102 DEFINITIONS

H102.1 General. The following words and terms shall, for the purposes of this appendix, have the meanings shown herein. Refer to Chapter 2 of this code for general definitions.

COMBINATION SIGN. A sign incorporating any combination of the features of pole, projecting and roof signs.

DISPLAY SIGN. The area made available by the sign structure for the purpose of displaying the advertising message.

ELECTRIC SIGN. A sign containing electrical wiring, but not including signs illuminated by an exterior light source.

GROUND SIGN. A billboard or similar type of sign that is supported by one or more uprights, poles or braces in or upon the ground other than a combination sign or pole sign, as defined by this code.

POLE SIGN. A sign wholly supported by a sign structure in the ground.

PORTABLE DISPLAY SURFACE. A display surface temporarily fixed to a standardized advertising structure that is regularly moved from structure to structure at periodic intervals.

PROJECTING SIGN. A sign other than a wall sign that projects from and is supported by a wall of a building or structure.

ROOF SIGN. A sign erected on or above a roof or parapet of a building or structure.

SIGN. Any letter, figure, character, mark, plane, point, marquee sign, design, poster, pictorial, picture, stroke, stripe, line, trademark, reading matter or illuminated service, which shall be constructed, placed, attached, painted, erected, fastened or manufactured in any manner whatsoever, so that the same shall be used for the attraction of the public to any place, subject, person, firm, corporation, public performance, article, machine or merchandise, whatsoever, which is displayed in any manner outdoors. Every sign shall be classified and conform to the requirements of that classification as set forth in this chapter.

SIGN STRUCTURE. Any structure that supports or is capable of supporting a sign as defined in this code. A sign structure is permitted to be a single pole and is not required to be an integral part of the building.

WALL SIGN. Any sign attached to or erected against the wall of a building or structure, with the exposed face of the sign in a plane parallel to the plane of said wall.

SECTION H103 LOCATION

H103.1 Location restrictions. Signs shall not be erected, constructed or maintained so as to obstruct any fire escape or any window or door or opening used as a *means of egress* or so as to prevent free passage from one part of a roof to any other part thereof. A sign shall not be attached in any form, shape or manner to a fire escape, nor be placed in such manner as to interfere with any opening required for ventilation.

SECTION H104 IDENTIFICATION

H104.1 Identification. Every outdoor advertising display sign hereafter erected, constructed or maintained, for which a permit is required, shall be plainly marked with the name of the person, firm or corporation erecting and maintaining such sign and shall have affixed on the front thereof the permit number issued for said sign or other method of identification *approved* by the *building official*.

SECTION H105 DESIGN AND CONSTRUCTION

H105.1 General requirements. Signs shall be designed and constructed to comply with the provisions of this code for use of materials, *loads* and stresses.

H105.2 Permits, drawings and specifications. Where a permit is required, as provided in Chapter 1, construction documents shall be required. These documents shall show the dimensions, material and required details of construction, including *loads*, stresses and anchors.

H105.3 Wind load. Signs shall be designed and constructed to withstand wind pressure as provided for in Chapter 16.

H105.4 Seismic load. Signs designed to withstand wind pressures shall be considered capable of withstanding earthquake *loads*, except as provided for in Chapter 16.

H105.5 Working stresses. In outdoor advertising display signs, the allowable working stresses shall conform to the requirements of Chapter 16. The working stresses of wire rope and its fastenings shall not exceed 25 percent of the ultimate strength of the rope or fasteners.

Exceptions:

1. The allowable working stresses for steel and wood shall be in accordance with the provisions of Chapters 22 and 23.
2. The working strength of chains, cables, guys or steel rods shall not exceed one-fifth of the ultimate strength of such chains, cables, guys or steel.

H105.6 Attachment. Signs attached to masonry, concrete or steel shall be safely and securely fastened by means of metal anchors, bolts or approved expansion screws of sufficient size and anchorage to safely support the *loads* applied.

SECTION H106 ELECTRICAL

H106.1 Illumination. A sign shall not be illuminated by other than electrical means, and electrical devices and wiring shall be installed in accordance with the requirements of NFPA 70. Any open spark or flame shall not be used for display purposes unless specifically *approved*.

H106.1.1 Internally illuminated signs. Except as provided for in Section 2611, where internally illuminated signs have facings of wood or of approved plastic complying with the requirements of Section 2606.4, the area of such facing section shall be not more than 120 square feet (11.16 m^2) and the wiring for electric lighting shall be entirely enclosed in the sign cabinet with a clearance of not less than 2 inches (51 mm) from the facing material. The dimensional limitation of 120 square feet (11.16 m^2) shall not apply to sign facing sections made from flame-resistant-coated fabric (ordinarily known as "flexible sign face plastic") that weighs less than 20 ounces per square yard (678 g/m^2) and that, when tested in accordance with NFPA 701, meets the fire propagation performance requirements of both Test 1 and Test 2 or that, when tested in accordance with an approved test method, exhibits an average burn time of 2 seconds or less and a burning extent of 5.9 inches (150 mm) or less for 10 specimens.

H106.2 Electrical service. Signs that require electrical service shall comply with NFPA 70.

SECTION H107 COMBUSTIBLE MATERIALS

H107.1 Use of combustibles. Wood, plastics complying with the requirements of Section H107.1.1 or plastic *veneer* panels as provided for in Chapter 26, or other materials of combustible characteristics similar to wood, used for moldings, cappings, nailing blocks, letters and latticing, shall comply with Section H109.1 and shall not be used for other ornamental features of signs, unless approved.

H107.1.1 Plastic materials. Notwithstanding any other provisions of this code, plastics that burn at a rate not faster than 2.5 inches per minute (64 mm/s) when tested in accordance with ASTM D635 shall be approved for use as the display surface material and for the letters, decorations and facings on signs and outdoor display structures.

H107.1.2 Electric sign faces. Individual plastic facings of electric signs shall not exceed 200 square feet (18.6 m^2) in area.

H107.1.3 Area limitation. If the area of a display surface exceeds 200 square feet (18.6 m^2), the area occupied or covered by plastics complying with the requirements of Section H107.1.1 shall be limited to 200 square feet (18.6 m^2) plus 50 percent of the difference between 200 square feet (18.6 m^2) and the area of display surface. The area of plastic on a display surface shall not in any case exceed 1,100 square feet (102 m^2).

H107.1.4 Plastic appurtenances. Letters and decorations mounted on a plastic facing or display surface can be made of plastics complying with the requirements of Section H107.1.1.

SECTION H108 ANIMATED DEVICES

H108.1 Fail-safe device. Signs that contain moving sections or ornaments shall have fail-safe provisions to prevent the section or ornament from releasing and falling or shifting its center of gravity more than 15 inches (381 mm). The fail-safe device shall be in addition to the mechanism and the mechanism's housing that operate the movable section or ornament. The fail-safe device shall be capable of supporting the full dead weight of the section or ornament when the moving mechanism releases.

SECTION H109 GROUND SIGNS

H109.1 Height restrictions. The structural frame of ground signs shall not be erected of combustible materials to a height of more than 35 feet (10 668 mm) above the ground. Ground signs constructed entirely of noncombustible material shall not be erected to a height of greater than 100 feet (30 480 mm) above the ground. Greater heights are permitted where approved and located so as not to create a hazard or danger to the public.

H109.2 Required clearance. The bottom coping of every ground sign shall be not less than 3 feet (914 mm) above the ground or street level, which space can be filled with platform decorative trim or light wooden construction.

H109.3 Wood anchors and supports. Where wood anchors or supports are embedded in the soil, the wood shall be pressure treated with an approved preservative.

SECTION H110 ROOF SIGNS

H110.1 General. Roof signs shall be constructed entirely of metal or other approved noncombustible material except as provided for in Sections H106.1.1 and H107.1. Provisions shall be made for electric grounding of metallic parts. Where combustible materials are permitted in letters or other ornamental features, wiring and tubing shall be kept free and insulated therefrom. Roof signs shall be so constructed as to leave a clear space of not less than 6 feet (1829 mm) between the roof level and the lowest part of the sign and shall have not less than 5 feet (1524 mm) clearance between the vertical supports thereof. Roof sign structures shall not project beyond an *exterior wall.*

> **Exception:** Signs on flat roofs with every part of the roof accessible.

H110.2 Bearing plates. The bearing plates of roof signs shall distribute the load directly to or on masonry walls, steel roof girders, columns or beams. The building shall be designed to avoid overstress of these members.

H110.3 Height of solid signs. A roof sign having a solid surface shall not exceed, at any point, a height of 24 feet (7315 mm) measured from the roof surface.

H110.4 Height of open signs. Open roof signs in which the uniform open area is not less than 40 percent of total gross area shall not exceed a height of 75 feet (22 860 mm) on buildings of Type 1 or Type 2 construction. On buildings of other construction types, the height shall not exceed 40 feet (12 192 mm). Such signs shall be thoroughly secured to the building on which they are installed, erected or constructed by iron, metal anchors, bolts, supports, chains, stranded cables, steel rods or braces and they shall be maintained in good condition.

H110.5 Height of closed signs. A closed roof sign shall not be erected to a height greater than 50 feet (15 240 mm) above the roof of buildings of Type 1 or 2 construction or more than 35 feet (10 668 mm) above the roof of buildings of Type 3, 4 or 5 construction.

SECTION H111 WALL SIGNS

H111.1 Materials. Wall signs that have an area exceeding 40 square feet (3.72 m^2) shall be constructed of metal or other approved noncombustible material, except for nailing rails and as provided for in Sections H106.1.1 and H107.1.

H111.2 Exterior wall mounting details. Wall signs attached to *exterior walls* of *solid masonry*, concrete or stone shall be safely and securely attached by means of metal anchors, bolts or expansion screws of not less than $^3/_8$ inch (9.5 mm) diameter and shall be embedded not less than 5 inches (127 mm). Wood blocks shall not be used for anchorage, except in the case of wall signs attached to buildings with walls of wood. A wall sign shall not be supported by anchorages secured to an unbraced *parapet wall.*

H111.3 Extension. Wall signs shall not extend above the top of the wall or beyond the ends of the wall to which the signs are attached unless such signs conform to the requirements for roof signs, projecting signs or ground signs.

SECTION H112 PROJECTING SIGNS

H112.1 General. Projecting signs shall be constructed entirely of metal or other noncombustible material and securely attached to a building or structure by metal supports such as bolts, anchors, supports, chains, guys or steel rods. Staples or nails shall not be used to secure any projecting sign to any building or structure. The *dead load* of projecting signs not parallel to the building or structure and the *load* due to wind pressure shall be supported with chains, guys or steel rods having net cross-sectional dimension of not less than $^3/_8$ inch (9.5 mm) diameter. Such supports shall be erected or maintained at an angle of not less than 45 percent (0.78 rad) with the horizontal to resist the *dead load* and at angle of 45 percent (0.78 rad) or more with the face of the sign to resist the specified wind pressure. If such projecting sign exceeds 30 square feet (2.8 m^2) in one facial area, there shall be provided not fewer than two such supports on each side not more than 8 feet (2438 mm) apart to resist the wind pressure.

H112.2 Attachment of supports. Supports shall be secured to a bolt or expansion screw that will develop the strength of the supporting chains, guys or steel rods, with a minimum $^5/_8$-inch (15.9 mm) bolt or lag screw, by an expansion shield. Turnbuckles shall be placed in chains, guys or steel rods supporting projecting signs.

H112.3 Wall mounting details. Chains, cables, guys or steel rods used to support the live or *dead load* of projecting signs are permitted to be fastened to *solid masonry* walls with expansion bolts or by machine screws in iron supports, but such supports shall not be attached to an unbraced *parapet wall*. Where the supports must be fastened to walls made of wood, the supporting anchor bolts must go through the wall and be plated or fastened on the inside in a secure manner.

H112.4 Height limitation. A projecting sign shall not be erected on the wall of any building so as to project above the roof or cornice wall or, on buildings without a cornice wall, above the roof level except that a sign erected at a right angle to the building, the horizontal width of which sign is perpendicular to such a wall and does not exceed 18 inches (457 mm), is permitted to be erected to a height not exceeding 2 feet (610 mm) above the roof or cornice wall or above the roof level where there is no cornice wall. A sign attached to a corner of a building and parallel to the vertical line of such corner shall be deemed to be erected at a right angle to the building wall.

H112.5 Additional loads. Projecting sign structures that will be used to support an individual on a ladder or other servicing device, whether or not specifically designed for the servicing device, shall be capable of supporting the anticipated additional *load*, but not less than a 100-pound (445 N) concentrated horizontal load and a 300-pound (1334 N) concentrated vertical *load* applied at the point of assumed or most eccentric loading. The building component to which the projecting sign is attached shall be designed to support the additional *loads*.

SECTION H113 MARQUEE SIGNS

H113.1 Materials. *Marquee* signs shall be constructed entirely of metal or other approved noncombustible material except as provided for in Sections H106.1.1 and H107.1.

H113.2 Attachment. *Marquee* signs shall be attached to approved *marquees* that are constructed in accordance with Section 3106.

H113.3 Dimensions. *Marquee* signs, whether on the front or side, shall not project beyond the perimeter of the *marquee*.

H113.4 Height limitation. *Marquee* signs shall not extend more than 6 feet (1829 mm) above, or 1 foot (305 mm) below such *marquee*. Signs shall not have a vertical dimension greater than 8 feet (2438 mm).

SECTION H114 PORTABLE SIGNS

H114.1 General. Portable signs shall conform to requirements for ground, roof, projecting, flat and temporary signs where such signs are used in a similar capacity. The requirements of this section shall not be construed to require portable signs to have connections to surfaces, *tie-downs* or foundations where provisions are made by temporary means or configuration of the structure to provide stability for the expected duration of the installation.

SECTION H115 THICKNESS OF SIGNS

H115.1 General. Tables H115.1(1) and H115.1(2) provide requirements for the size, thicknesses and types of glass panels and projection signs, respectively.

TABLE H115.1(1)
SIZE, THICKNESS AND TYPE OF GLASS PANELS IN SIGNS

MAXIMUM SIZE OF EXPOSED PANEL		MINIMUM THICKNESS OF GLASS (inches)	TYPE OF GLASS
Any dimension (inches)	Area (square inches)		
30	500	$^1/_8$	Plain, plate or wired
45	700	$^3/_{16}$	Plain, plate or wired
144	3,600	$^1/_4$	Plain, plate or wired
> 144	> 3,600	$^1/_4$	Wired glass

For SI: 1 inch = 25.4 mm, 1 square inch = 645.16 mm^2.

TABLE H115.1(2)
THICKNESS OF PROJECTION SIGN

PROJECTION (feet)	MAXIMUM THICKNESS (feet)
5	2
4	2.5
3	3
2	3.5
1	4

For SI: 1 foot = 304.8 mm.

SECTION H116 REFERENCED STANDARDS

H116.1 General. See Table H115.1 for standards that are referenced in various sections of this appendix. Standards are listed by the standard identification with the effective date, standard title, and the section or sections of this appendix that reference the standard.

TABLE H116.1
REFERENCED STANDARDS

STANDARD ACRONYM	STANDARD NAME	SECTIONS HEREIN REFERENCED
ASTM D635—14	*Test Method for Rate of Burning and/or Extent and Time of Burning of Plastics in a Horizontal Position*	H107.1.1
NFPA 70—20	*National Electrical Code*	H106.1, H106.2
NFPA 701—19	*Methods of Fire Test for Flame Propagation of Textiles and Films*	H106.1.1

APPENDIX I
PATIO COVERS

The provisions contained in this appendix are not mandatory unless specifically referenced in the adopting ordinance.

User notes:

About this appendix: *Appendix I provides standards applicable to the construction and use of patio covers. It is limited in application to patio covers accessory to dwelling units. Covers of patios and other outdoor areas associated with restaurants, mercantile buildings, offices, nursing homes or other nondwelling occupancies would be subject to standards in the main code and not this appendix.*

Code development reminder: *Code change proposals to this appendix will be considered by the IBC—Structural Code Development Committee during the 2022 (Group B) Code Development Cycle.*

SECTION I101
GENERAL

I101.1 General. Patio covers shall be permitted to be detached from or attached to *dwelling units*. Patio covers shall be used only for recreational, outdoor living purposes and not as carports, garages, storage rooms or habitable rooms.

SECTION I102
DEFINITION

I102.1 General. The following term shall, for the purposes of this appendix, have the meaning shown herein. Refer to Chapter 2 of this code for general definitions.

PATIO COVER. A structure with open or glazed walls that is used for recreational, outdoor living purposes associated with a *dwelling unit*.

SECTION I103
EXTERIOR WALLS AND OPENINGS

I103.1 Enclosure walls. Enclosure walls shall be permitted to be of any configuration, provided that the open or glazed area of the longer wall and one additional wall is equal to not less than 65 percent of the area below not less than 6 feet 8 inches (2032 mm) of each wall, measured from the floor. Openings shall be permitted to be enclosed with insect screening, translucent or transparent plastic conforming to the provisions of Sections 2606 through 2610, glass conforming to the provisions of Chapter 24 or any combination of the foregoing.

I103.2 Light, ventilation and emergency egress. Exterior openings of the *dwelling unit* required for light and ventilation shall be permitted to open into a patio structure. However, the patio structure shall be unenclosed if such openings are serving as emergency egress or rescue openings from sleeping rooms. Where such exterior openings serve as an exit from the dwelling unit, the patio structure, unless unenclosed, shall be provided with exits conforming to the provisions of Chapter 10.

SECTION I104
HEIGHT

I104.1 Height. Patio covers shall be limited to one-story structures not more than 12 feet (3657 mm) in height.

SECTION I105
STRUCTURAL PROVISIONS

I105.1 Design loads. Patio covers shall be designed and constructed to sustain, within the stress limits of this code, all *dead loads* plus a minimum vertical *live load* of 10 pounds per square foot (0.48 kN/m^2) except that snow *loads* shall be used where such snow *loads* exceed this minimum. Such patio covers shall be designed to resist the minimum wind and seismic loads set forth in this code.

I105.2 Footings. In areas with a frost depth of zero, a patio cover shall be permitted to be supported on a concrete slab on grade without footings, provided that the slab conforms to the provisions of Chapter 19 of this code and is not less than $3^1/_2$ inches (89 mm) thick, and the columns do not support *loads* in excess of 750 pounds (3.36 kN) per column.

APPENDIX J
GRADING

The provisions contained in this appendix are not mandatory unless specifically referenced in the adopting ordinance.

User notes:

About this appendix: *Appendix J provides standards for the grading of properties. The appendix also provides standards for the administration and enforcement of a grading program, including permit and inspection requirements. Appendix J was originally developed in the 1960s and used for many years in jurisdictions throughout the western United States. It is intended to provide consistent and uniform code requirements anywhere grading is considered an issue.*

Code development reminder: *Code change proposals to this appendix will be considered by the IBC—Structural Code Development Committee during the 2022 (Group B) Code Development Cycle.*

SECTION J101 GENERAL

J101.1 Scope. The provisions of this chapter apply to grading, excavation and earthwork construction, including fills and embankments. Where conflicts occur between the technical requirements of this chapter and the geotechnical report, the geotechnical report shall govern.

J101.2 Flood hazard areas. Unless the applicant has submitted an engineering analysis, prepared in accordance with standard engineering practice by a *registered design professional*, that demonstrates the proposed work will not result in any increase in the level of the *base flood*, grading, excavation and earthwork construction, including fills and embankments, shall not be permitted in *floodways* that are in *flood hazard areas* established in Section 1612.3 or in *flood hazard areas* where *design flood elevations* are specified but *floodways* have not been designated.

SECTION J102 DEFINITIONS

J102.1 Definitions. The following words and terms shall, for the purposes of this appendix, have the meanings shown herein. Refer to Chapter 2 of this code for general definitions.

BENCH. A relatively level step excavated into earth material on which fill is to be placed.

COMPACTION. The densification of a fill by mechanical means.

CUT. See "*Excavation.*"

DOWN DRAIN. A device for collecting water from a swale or ditch located on or above a slope, and safely delivering it to an approved drainage facility.

EROSION. The wearing away of the ground surface as a result of the movement of wind, water or ice.

EXCAVATION. The removal of earth material by artificial means, also referred to as a cut.

FILL. Deposition of earth materials by artificial means.

GRADE. The vertical location of the ground surface.

GRADE, EXISTING. The grade prior to grading.

GRADE, FINISHED. The grade of the site at the conclusion of all grading efforts.

GRADING. An excavation or fill or combination thereof.

KEY. A compacted fill placed in a trench excavated in earth material beneath the toe of a slope.

SLOPE. An inclined surface, the inclination of which is expressed as a ratio of horizontal distance to vertical distance.

TERRACE. A relatively level step constructed in the face of a graded slope for drainage and maintenance purposes.

SECTION J103 PERMITS REQUIRED

J103.1 Permits required. Except as exempted in Section J103.2, grading shall not be performed without first having obtained a *permit* therefor from the *building official*. A grading *permit* does not include the construction of retaining walls or other structures.

J103.2 Exemptions. A grading *permit* shall not be required for the following:

1. Grading in an isolated, self-contained area, provided that the public is not endangered and that such grading will not adversely affect adjoining properties.
2. Excavation for construction of a structure permitted under this code.
3. Cemetery graves.
4. Refuse disposal sites controlled by other regulations.
5. Excavations for wells, or trenches for utilities.
6. Mining, quarrying, excavating, processing or stockpiling rock, sand, gravel, aggregate or clay controlled by other regulations, provided that such operations do not affect the lateral support of, or significantly increase stresses in, soil on adjoining properties.
7. Exploratory excavations performed under the direction of a registered design professional.

Exemption from the *permit* requirements of this appendix shall not be deemed to grant authorization for any work to be done in any manner in violation of the provisions of this code or any other laws or ordinances of this jurisdiction.

SECTION J104 PERMIT APPLICATION AND SUBMITTALS

J104.1 Submittal requirements. In addition to the provisions of Section 105.3, the applicant shall state the estimated quantities of excavation and fill.

J104.2 Site plan requirements. In addition to the provisions of Section 107, a grading plan shall show the existing grade and finished grade in contour intervals of sufficient clarity to indicate the nature and extent of the work and show in detail that it complies with the requirements of this code. The plans shall show the existing grade on adjoining properties in sufficient detail to identify how grade changes will conform to the requirements of this code.

J104.3 Geotechnical report. A geotechnical report prepared by a *registered design professional* shall be provided. The report shall contain not less than the following:

1. The nature and distribution of existing soils.
2. Conclusions and recommendations for grading procedures.
3. Soil design criteria for any structures or embankments required to accomplish the proposed grading.
4. Where necessary, slope stability studies, and recommendations and conclusions regarding site geology.

Exception: A geotechnical report is not required where the *building official* determines that the nature of the work applied for is such that a report is not necessary.

J104.4 Liquefaction study. For sites with mapped maximum considered earthquake spectral response accelerations at short periods (S_s) greater than 0.5g as determined by Section 1613, a study of the liquefaction potential of the site shall be provided and the recommendations incorporated in the plans.

Exception: A liquefaction study is not required where the *building official* determines from established local data that the liquefaction potential is low.

SECTION J105 INSPECTIONS

J105.1 General. Inspections shall be governed by Section 110 of this code.

J105.2 Special inspections. The *special inspection* requirements of Section 1705.6 shall apply to work performed under a grading permit where required by the *building official*.

SECTION J106 EXCAVATIONS

J106.1 Maximum slope. The slope of cut surfaces shall be not steeper than is safe for the intended use, and shall be not more than one unit vertical in two units horizontal (50-percent slope) unless the owner or the owner's authorized agent furnishes a geotechnical report justifying a steeper slope.

Exceptions:

1. A cut surface shall be permitted to be at a slope of 1.5 units horizontal to 1 unit vertical (67-percent slope) provided that all of the following are met:
   1.1. It is not intended to support structures or surcharges.
   1.2. It is adequately protected against erosion.
   1.3. It is not more than 8 feet (2438 mm) in height.
   1.4. It is approved by the building code official.
   1.5. Ground water is not encountered.
2. A cut surface in bedrock shall be permitted to be at a slope of 1 unit horizontal to 1 unit vertical (100-percent slope).

SECTION J107 FILLS

J107.1 General. Unless otherwise recommended in the geotechnical report, fills shall comply with the provisions of this section.

J107.2 Surface preparation. The ground surface shall be prepared to receive fill by removing vegetation, topsoil and other unsuitable materials, and scarifying the ground to provide a bond with the fill material.

J107.3 Benching. Where existing grade is at a slope steeper than one unit vertical in five units horizontal (20-percent slope) and the depth of the fill exceeds 5 feet (1524 mm) benching shall be provided in accordance with Figure J107.3. A key shall be provided that is not less than 10 feet (3048 mm) in width and 2 feet (610 mm) in depth.

J107.4 Fill material. Fill material shall not include organic, frozen or other deleterious materials. Rock or similar irreducible material greater than 12 inches (305 mm) in any dimension shall not be included in fills.

J107.5 Compaction. All fill material shall be compacted to 90 percent of maximum density as determined by ASTM D1557, Modified Proctor, in lifts not exceeding 12 inches (305 mm) in depth.

J107.6 Maximum slope. The slope of fill surfaces shall be not steeper than is safe for the intended use. Fill slopes steeper than one unit vertical in two units horizontal (50-percent slope) shall be justified by a geotechnical report or engineering data.

SECTION J108 SETBACKS

J108.1 General. Cut and fill slopes shall be set back from the property lines in accordance with this section. Setback dimensions shall be measured perpendicular to the property line and shall be as shown in Figure J108.1, unless substantiating data is submitted justifying reduced setbacks.

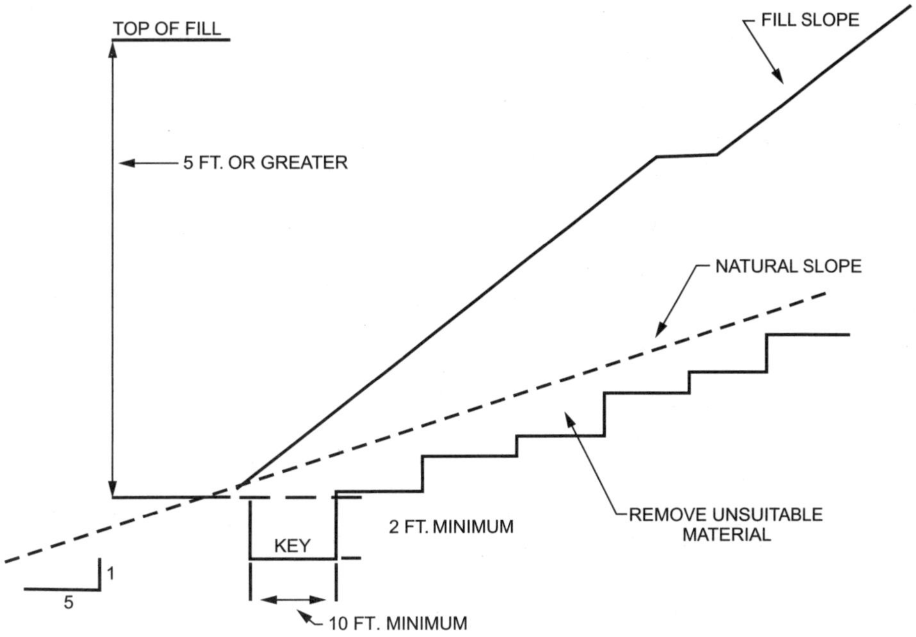

For SI: 1 foot = 304.8 mm.

FIGURE J107.3
BENCHING DETAILS

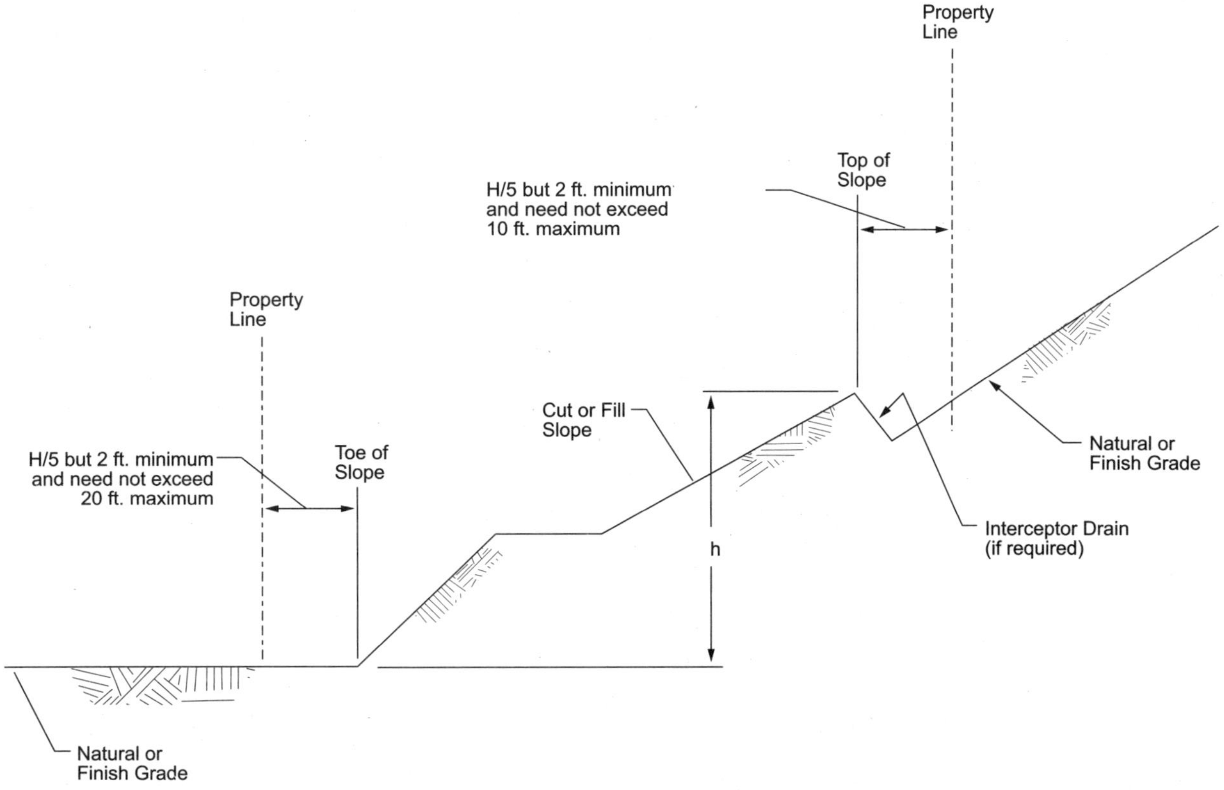

For SI: 1 foot = 304.8 mm.

FIGURE J108.1
DRAINAGE DIMENSIONS

J108.2 Top of slope. The setback at the top of a cut slope shall be not less than that shown in Figure J108.1, or than is required to accommodate any required interceptor drains, whichever is greater.

J108.3 Slope protection. Where required to protect adjacent properties at the toe of a slope from adverse effects of the grading, additional protection, *approved* by the *building official*, shall be included. Examples of such protection include but are not be limited to:

1. Setbacks greater than those required by Figure J108.1.
2. Provisions for retaining walls or similar construction.
3. Erosion protection of the fill slopes.
4. Provision for the control of surface waters.

SECTION J109 DRAINAGE AND TERRACING

J109.1 General. Unless otherwise recommended by a *registered design professional*, drainage facilities and terracing shall be provided in accordance with the requirements of this section.

Exception: Drainage facilities and terracing need not be provided where the ground slope is not steeper than one unit vertical in three units horizontal (33-percent slope).

J109.2 Terraces. Terraces not less than 6 feet (1829 mm) in width shall be established at not more than 30-foot (9144 mm) vertical intervals on all cut or fill slopes to control surface drainage and debris. Suitable access shall be provided to allow for cleaning and maintenance.

Where more than two terraces are required, one terrace, located at approximately mid-height, shall be not less than 12 feet (3658 mm) in width.

Swales or ditches shall be provided on terraces. They shall have a minimum gradient of one unit vertical in 20 units horizontal (5-percent slope) and shall be paved with concrete not less than 3 inches (76 mm) in thickness, or with other materials suitable to the application. They shall have a depth not less than 12 inches (305 mm) and a width not less than 5 feet (1524 mm).

A single run of swale or ditch shall not collect runoff from a tributary area exceeding 13,500 square feet (1256 m^2) (projected) without discharging into a down drain.

J109.3 Interceptor drains. Interceptor drains shall be installed along the top of cut slopes receiving drainage from a tributary width greater than 40 feet (12 192 mm), measured horizontally. They shall have a minimum depth of 1 foot (305 mm) and a minimum width of 3 feet (915 mm). The slope shall be *approved* by the *building official*, but shall be not less than one unit vertical in 50 units horizontal (2-percent slope). The drain shall be paved with concrete not less than 3 inches (76 mm) in thickness, or by other materials suitable to the application. Discharge from the drain shall be accomplished in a manner to prevent erosion and shall be approved by the *building official*.

J109.4 Drainage across property lines. Drainage across property lines shall not exceed that which existed prior to grading. Excess or concentrated drainage shall be contained on site or directed to an approved drainage facility. Erosion of the ground in the area of discharge shall be prevented by installation of nonerosive down drains or other devices.

SECTION J110 EROSION CONTROL

J110.1 General. The faces of cut and fill slopes shall be prepared and maintained to control erosion. This control shall be permitted to consist of effective planting.

Exception: Erosion control measures need not be provided on cut slopes not subject to erosion due to the erosion-resistant character of the materials.

Erosion control for the slopes shall be installed as soon as practicable and prior to calling for final inspection.

J110.2 Other devices. Where necessary, check dams, cribbing, riprap or other devices or methods shall be employed to control erosion and provide safety.

SECTION J111 REFERENCED STANDARDS

J111.1 General. See Table J111.1 for standards that are referenced in various sections of this appendix. Standards are listed by the standard identification with the effective date, standard title, and the section or sections of this appendix that reference the standard.

TABLE J111.1 REFERENCED STANDARDS

STANDARD ACRONYM	STANDARD NAME	SECTIONS HEREIN REFERENCED
ASTM D1557—12E1	*Test Method for Laboratory Compaction Characteristics of Soil Using Modified Effort* [56,000 ft-lb/ft^3 (2,700 kN-m/m^3)].	J107.5

APPENDIX K
ADMINISTRATIVE PROVISIONS

The provisions contained in this appendix are not mandatory unless specifically referenced in the adopting ordinance.

User note:

About this appendix: *Appendix K primarily provides the administrative mechanisms for the enforcement of NFPA 70, the* National Electrical Code. *While NFPA 70 includes an administrative annex, the provisions of Appendix K are designed to be compatible with the administrative provisions found in Chapter 1 of the* International Building Code® *and the other I-Codes.*

With the exception of Section K111, this appendix contains only administrative provisions that are intended to be used by a jurisdiction to implement and enforce NFPA 70, the National Electrical Code. Annex H of NFPA 70 also contains administrative and enforcement provisions, and these provisions may or may not be completely compatible with or consistent with Chapter 1 of the IBC, whereas the provisions in this appendix are compatible and consistent with Chapter 1 of the IBC and other I-Codes. Section K111 contains technical provisions that are unique to this appendix and are in addition to those of NFPA 70.

The provisions of Appendix K are specific to what might be designated as an Electrical Department of Inspection and Code Enforcement and could be implemented where other such provisions are not adopted.

SECTION K101
GENERAL

K101.1 Purpose. A purpose of this code is to establish minimum requirements to safeguard public health, safety and general welfare by regulating and controlling the design, construction, installation, quality of materials, location, operation and maintenance or use of electrical systems and equipment.

K101.2 Scope. This code applies to the design, construction, installation, *alteration*, repairs, relocation, replacement, *addition* to, use or maintenance of electrical systems and equipment.

SECTION K102
APPLICABILITY

K102.1 General. The provisions of this code apply to all matters affecting or relating to structures and premises, as set forth in Section K101.

K102.2 Existing installations. Except as otherwise provided for in this chapter, a provision in this code shall not require the removal, *alteration* or abandonment of, or prevent the continued utilization and maintenance of, existing electrical systems and equipment lawfully in existence at the time of the adoption of this code.

K102.3 Maintenance. Electrical systems, equipment, materials and appurtenances, both existing and new, and parts thereof shall be maintained in proper operating condition in accordance with the original design and in a safe, hazard-free condition. Devices or safeguards that are required by this code shall be maintained in compliance with the code edition under which installed. The *owner* or the *owner's* authorized agent shall be responsible for the maintenance of the electrical systems and equipment. To determine compliance with this provision, the *building official* shall have the authority to require that the electrical systems and equipment be reinspected.

K102.4 Additions, alterations and repairs. *Additions, alterations,* renovations and *repairs* to electrical systems and equipment shall conform to that required for new electrical systems and equipment without requiring that the existing electrical systems or equipment comply with all of the requirements of this code. *Additions, alterations* and *repairs* shall not cause existing electrical systems or equipment to become unsafe, hazardous or overloaded.

Minor *additions, alterations,* renovations and *repairs* to existing electrical systems and equipment shall meet the provisions for new construction, except where such work is performed in the same manner and arrangement as was in the existing system, is not hazardous and is *approved.*

K102.5 Subjects not regulated by this code. Where no applicable standards or requirements are set forth in this code, or are contained within other laws, codes, regulations, ordinances or bylaws adopted by the jurisdiction, compliance with applicable standards of nationally recognized standards as are *approved* shall be deemed as prima facie evidence of compliance with the intent of this code. Nothing herein shall derogate from the authority of the *building official* to determine compliance with codes or standards for those activities or installations within the building official's jurisdiction or responsibility.

SECTION K103
PERMITS

K103.1 Types of permits. An *owner*, authorized agent or contractor who desires to construct, enlarge, alter, *repair*, move, demolish or change the occupancy of a building or structure, or to erect, install, enlarge, alter, *repair*, remove, convert or replace electrical systems or equipment, the installation of which is regulated by this code, or to cause such work to be done, shall first make application to the *building official* and obtain the required *permit* for the work.

Exception: Where *repair* or replacement of electrical systems or equipment must be performed in an emergency situation, the *permit* application shall be submitted within the next working business day of the department of electrical inspection.

K103.2 Work exempt from permit. The following work shall be exempt from the requirement for a *permit*:

1. Listed cord- and plug-connected temporary decorative lighting.
2. Reinstallation of attachment plug receptacles, but not the outlets therefor.
3. Replacement of branch circuit overcurrent devices of the required capacity in the same location.
4. Temporary wiring for experimental purposes in suitable experimental laboratories.
5. Electrical wiring, devices, appliances, apparatus or equipment operating at less than 25 volts and not capable of supplying more than 50 watts of energy.

Exemption from the permit requirements of this code shall not be deemed to grant authorization for work to be done in violation of the provisions of this code or other laws or ordinances of this jurisdiction.

SECTION K104 CONSTRUCTION DOCUMENTS

K104.1 Information on construction documents. *Construction documents* shall be drawn to scale upon suitable material. Electronic media documents are permitted to be submitted where *approved* by the *building official*. *Construction documents* shall be of sufficient clarity to indicate the location, nature and extent of the work proposed and show in detail that such work will conform to the provisions of this code and relevant laws, ordinances, rules and regulations, as determined by the *building official*.

K104.2 Penetrations. *Construction documents* shall indicate where penetrations will be made for electrical systems and shall indicate the materials and methods for maintaining required structural safety, *fire-resistance rating* and *fireblocking*.

K104.3 Load calculations. Where an *addition* or *alteration* is made to an existing electrical system, an electrical load calculation shall be prepared to determine if the existing electrical service has the capacity to serve the added load.

SECTION K105 ALTERNATIVE ENGINEERED DESIGN

K105.1 General. The design, documentation, inspection, testing and approval of an alternative engineered design electrical system shall comply with this section.

K105.2 Design criteria. An alternative engineered design shall conform to the intent of the provisions of this code and shall provide an equivalent level of quality, strength, effectiveness, *fire resistance*, durability and safety. Materials, equipment or components shall be designed and installed in accordance with the manufacturer's instructions.

K105.3 Submittal. The *registered design professional* shall indicate on the *permit* application that the electrical system is an alternative engineered design. The *permit* and permanent *permit* records shall indicate that an alternative engineered design was part of the *approved* installation.

K105.4 Technical data. The *registered design professional* shall submit sufficient technical data to substantiate the proposed alternative engineered design and to prove that the performance meets the intent of this code.

K105.5 Construction documents. The *registered design professional* shall submit to the *building official* two complete sets of signed and sealed *construction documents* for the alternative engineered design. The *construction documents* shall include floor plans and a diagram of the work.

K105.6 Design approval. Where the *building official* determines that the alternative engineered design conforms to the intent of this code, the electrical system shall be *approved*. If the alternative engineered design is not *approved*, the *building official* shall notify the *registered design professional* in writing, stating the reasons therefor.

K105.7 Inspection and testing. The alternative engineered design shall be tested and inspected in accordance with the requirements of this code.

SECTION K106 REQUIRED INSPECTIONS

K106.1 General. The *building official*, upon notification, shall make the inspections set forth in this section.

K106.2 Underground. Underground inspection shall be made after trenches or ditches are excavated and bedded, piping and conductors installed, and before backfill is put in place. Where excavated soil contains rocks, broken concrete, frozen chunks and other rubble that would damage or break the raceway, cable or conductors, or where corrosive action will occur, protection shall be provided in the form of granular or selected material, *approved* running boards, sleeves or other means.

K106.3 Rough-in. Rough-in inspection shall be made after the roof, framing, *fireblocking* and bracing are in place and all wiring and other components to be concealed are complete, and prior to the installation of wall or ceiling membranes.

K106.4 Contractors' responsibilities. It shall be the responsibility of every contractor who enters into contracts for the installation or repair of electrical systems for which a *permit* is required to comply with adopted state and local rules and regulations concerning licensing.

SECTION K107 PREFABRICATED CONSTRUCTION

K107.1 Prefabricated construction. Prefabricated construction is subject to Sections K107.2 through K107.5.

K107.2 Evaluation and follow-up inspection services. Prior to the approval of a prefabricated construction assembly having concealed electrical work and the issuance of an electrical *permit*, the *building official* shall require the submittal of an evaluation report on each prefabricated construction assembly, indicating the complete details of the electrical system, including a description of the system and its components, the basis upon which the system is being evaluated, test results and similar information, and other data as necessary for the *building official* to determine conformance to this code.

K107.3 Evaluation service. The *building official* shall designate the evaluation service of an *approved* agency as the evaluation agency and review such agency's evaluation report for adequacy and conformance to this code.

K107.4 Follow-up inspection. Except where ready access is provided to electrical systems, service equipment and accessories for complete inspection at the site without disassembly or dismantling, the *building official* shall conduct the in-plant inspections as frequently as necessary to ensure conformance to the *approved* evaluation report or shall designate an independent, *approved* inspection agency to conduct such inspections. The inspection agency shall furnish the *building official* with the follow-up inspection manual and a report of inspections upon request, and the electrical system shall have an identifying label permanently affixed to the system indicating that factory inspections have been performed.

K107.5 Test and inspection records. Required test and inspection records shall be available to the *building official* at all times during the fabrication of the electrical system and the erection of the building; or such records as the *building official* designates shall be filed.

SECTION K108 TESTING

K108.1 Testing. Electrical work shall be tested as required in this code. Tests shall be performed by the *permit* holder and observed by the *building official*.

K108.1.1 Apparatus, material and labor for tests. Apparatus, material and labor required for testing an electrical system or part thereof shall be furnished by the *permit* holder.

K108.1.2 Reinspection and testing. Where any work or installation does not pass an initial test or inspection, the necessary corrections shall be made so as to achieve compliance with this code. The work or installation shall then be resubmitted to the *building official* for inspection and testing.

SECTION K109 RECONNECTION

K109.1 Connection after order to disconnect. A person shall not make utility service or energy source connections to systems regulated by this code, which have been disconnected or ordered to be disconnected by the *building official*, or the use of which has been ordered to be discontinued by the *building official* until the *building official* authorizes the reconnection and use of such systems.

SECTION K110 CONDEMNING ELECTRICAL SYSTEMS

K110.1 Authority to condemn electrical systems. Wherever the *building official* determines that any electrical system, or portion thereof, regulated by this code has become hazardous to life, health or property, the *building official* shall order in writing that such electrical systems either be removed or restored to a safe condition. A time limit for compliance with such order shall be specified in the written notice. A person shall not use or maintain a defective electrical system or equipment after receiving such notice.

Where such electrical system is to be disconnected, written notice as prescribed in this code shall be given. In cases of immediate danger to life or property, such disconnection shall be made immediately without such notice.

SECTION K111 ELECTRICAL PROVISIONS

K111.1 Adoption. Electrical systems and equipment shall be designed, constructed and installed in accordance with the *International Residential Code* or NFPA 70 as applicable, except as otherwise provided in this code.

[F] K111.2 Abatement of electrical hazards. All identified electrical hazards shall be abated. All identified hazardous electrical conditions in permanent wiring shall be brought to the attention of the *building official* responsible for enforcement of this code. Electrical wiring, devices, appliances and other equipment that is modified or damaged and constitutes an electrical shock or fire hazard shall not be used.

[F] K111.3 Appliance and fixture listing. Electrical appliances and fixtures shall be tested and listed in published reports of inspected electrical equipment by an *approved* agency and installed in accordance with all instructions included as part of such listing.

K111.4 Nonmetallic-sheathed cable. The use of Type NM, NMC and NMS (nonmetallic sheathed) cable wiring methods shall not be limited based on height, number of stories or construction type of the building or structure.

K111.5 Cutting, notching and boring. The cutting, notching and boring of wood and steel framing members, structural members and engineered wood products shall be in accordance with this code.

K111.6 Smoke alarm circuits. Single- and multiple-station smoke alarms required by this code and installed within *dwelling* units shall not be connected as the only load on a branch circuit. Such alarms shall be supplied by branch circuits having lighting loads consisting of lighting outlets in habitable spaces.

K111.7 Equipment and door labeling. Doors into electrical control panel rooms shall be marked with a plainly visible and legible sign stating "ELECTRICAL ROOM" or similar *approved* wording. The disconnecting means for each service, feeder or branch circuit originating on a switchboard or panelboard shall be legibly and durably marked to indicate its purpose unless such purpose is clearly evident.

APPENDIX L
EARTHQUAKE RECORDING INSTRUMENTATION

The provisions contained in this appendix are not mandatory unless specifically referenced in the adopting ordinance.

User notes:

About this appendix: *The purpose of Appendix L is to foster the collection of ground motion data, particularly from strong-motion earthquakes. When this ground motion data is synthesized, it may be useful in developing future improvements to the earthquake provisions of the code.*

Code development reminder: *Code change proposals to this appendix will be considered by the IBC—Structural Code Development Committee during the 2022 (Group B) Code Development Cycle.*

SECTION L101
GENERAL

L101.1 General. Every structure located where the 1-second spectral response acceleration, S_1, determined in accordance with Section 1613.2, is greater than 0.40 and either exceeds six stories in height with an aggregate floor area of 60,000 square feet (5574 m^2) or more, or exceeds 10 *stories* in height regardless of floor area, shall be equipped with not fewer than three approved recording accelerographs. The accelerographs shall be interconnected for common start and common timing.

L101.2 Location. As a minimum, instruments shall be located at the lowest level, mid-height, and near the top of the structure. Each instrument shall be located so that access is maintained at all times and is unobstructed by room contents. A sign stating "MAINTAIN CLEAR ACCESS TO THIS INSTRUMENT" in 1-inch (25 mm) block letters shall be posted in a conspicuous location.

L101.3 Maintenance. Maintenance and service of the instrumentation shall be provided by the owner of the structure. Data produced by the instrument shall be made available to the *building official* on request.

Maintenance and service of the instruments shall be performed annually by an *approved* testing agency. The owner shall file with the *building official* a written report from an approved testing agency certifying that each instrument has been serviced and is in proper working condition. This report shall be submitted when the instruments are installed and annually thereafter. Each instrument shall have affixed to it an externally visible tag specifying the date of the last maintenance or service and the printed name and address of the testing agency.

APPENDIX M
TSUNAMI-GENERATED FLOOD HAZARDS

The provisions contained in this appendix are not mandatory unless specifically referenced in the adopting ordinance.

User notes:

About this appendix: *Appendix M allows the adoption of guidelines for constructing vertical evacuation refuge structures within areas that are considered tsunami hazard zones.*

Code development reminder: *Code change proposals to this appendix will be considered by the IBC—Structural Code Development Committee during the 2022 (Group B) Code Development Cycle.*

SECTION M101
REFUGE STRUCTURES FOR VERTICAL EVACUATION FROM TSUNAMI-GENERATED FLOOD HAZARDS

M101.1 General. The purpose of this appendix is to provide tsunami vertical evacuation planning criteria for those coastal communities that have a tsunami hazard as shown in a *Tsunami Design Zone* Map.

M101.2 Definitions. The following term shall, for the purposes of this appendix, have the meaning shown herein. Refer to Chapter 2 of this code for general definitions.

TSUNAMI DESIGN ZONE MAP. A map that designates the extent of inundation by a Maximum Considered Tsunami, as defined by Chapter 6 of ASCE 7.

M101.3 Establishment of tsunami design zone. Where applicable, the *Tsunami Design Zone* Map shall meet or exceed the inundation limit given by the ASCE 7 *Tsunami Design Geodatabase*.

M101.4 Planning of tsunami vertical evacuation refuge structures within the tsunami design zone. Tsunami Vertical Evacuation Refuge Structures located within a tsunami hazard design zone shall be planned, sited, and developed in general accordance with the planning criteria of the FEMA P646 guidelines.

Exception: These criteria shall not be considered mandatory for evaluation of existing buildings for evacuation planning purposes.

SECTION M102
REFERENCED STANDARDS

M102.1 General. See Table M102.1 for standards that are referenced in various sections of this appendix. Standards are listed by the standard identification with the effective date, standard title, and the section or sections of this appendix that reference the standard.

TABLE M102.1
REFERENCED STANDARDS

STANDARD ACRONYM	STANDARD NAME	SECTIONS HEREIN REFERENCED
ASCE 7—16 with Supplement 1	*Minimum Design Load and Associated Criteria for Buildings and Other Structures*	M101.2, M101.3
FEMA P646—12	*Guidelines for Design of Structures for Vertical Evacuation from Tsunamis*	M101.4

APPENDIX N
REPLICABLE BUILDINGS

The provisions contained in this appendix are not mandatory unless specifically referenced in the adopting ordinance.

User notes:

About this appendix: *Appendix N provides jurisdictions with a means of incorporating guidelines for replicable buildings into their building code adoption process. The intent of these provisions is to give jurisdictions a means of streamlining their document review process while verifying code compliance.*

Code development reminder: *Code change proposals to this appendix will be considered by the IBC—Structural Code Development Committee during the 2022 (Group B) Code Development Cycle.*

SECTION N101 ADMINISTRATION

N101.1 Purpose. The purpose of this appendix is to provide a format and direction regarding the implementation of a replicable building program.

N101.2 Objectives. Such programs allow a jurisdiction to recover from a natural disaster faster and allow for consistent application of the codes for replicable building projects. It will result in faster turnaround for the end user, and a quicker turnaround through the plan review process.

SECTION N102 DEFINITIONS

N102.1 Definitions. The following words and terms shall, for the purposes of this appendix, have the meanings shown herein.

REPLICABLE BUILDING. A building or structure utilizing a replicable design.

REPLICABLE DESIGN. A prototypical design developed for application in multiple locations with minimal variation or modification.

SECTION N103 REPLICABLE DESIGN REQUIREMENTS

N103.1 Prototypical construction documents. A replicable design shall establish prototypical construction documents for application at multiple locations. The construction documents shall include details appropriate to each wind region, *seismic design category*, and climate zone for locations in which the replicable design is intended for application. Application of replicable design shall not vary with regard to the following, except for allowable variations in accordance with Section N106.

1. Use and occupancy classification.
2. Building heights and area limitations.
3. Type of construction classification.
4. *Fire-resistance ratings*.
5. *Interior finishes*.
6. Fire protection system.
7. *Means of egress*.
8. Accessibility.
9. Structural design criteria.
10. Energy efficiency.
11. Type of mechanical and electrical systems.
12. Type of plumbing system and number of fixtures.

SECTION N104 REPLICABLE DESIGN SUBMITTAL REQUIREMENTS

N104.1 General. A summary description of the replicable design and related construction documents shall be submitted to an *approved* agency. Where approval is requested for elements of the replicable design that is not within the scope of the *International Building Code*, the construction documents shall specifically designate the codes for which review is sought. Construction documents shall be signed, sealed and dated by a registered design professional.

N104.1.1 Architectural plans and specifications. Where approval of the architectural requirements of the replicable design is sought, the submittal documents shall include architectural plans and specifications as follows:

1. Description of uses and the proposed occupancy groups for all portions of the building.
2. Proposed type of construction of the building.
3. Fully dimensioned drawings to determine *building areas* and height.
4. Adequate details and dimensions to evaluate *means of egress*, including *occupant loads* for each floor, exit arrangement and sizes, corridors, doors and *stairs*.
5. Exit signs and means of egress lighting, including power supply.
6. Accessibility scoping provisions.
7. Description and details of proposed special occupancies such as a covered mall, high-rise, *mezzanine*, *atrium* and public garage.

8. Adequate details to evaluate fire-resistance-rated construction requirements, including data substantiating required ratings.
9. Details for plastics, insulation and safety glazing installation.
10. Details of required fire protection systems.
11. Material specifications demonstrating fire-resistance criteria.

N104.1.2 Structural plans, specifications and engineering details. Where approval of the structural requirements of the replicable design is sought, the submittal documents shall include details for each wind region, *seismic design category* and climate zone for which approval is sought; and shall include the following:

1. Signed and sealed structural design calculations that support the member sizes on the drawings.
2. Design *load* criteria, including: frost depth, *live loads*, snow *loads*, wind *loads*, earthquake design date, and other special *loads*
3. Details of foundations and superstructure.
4. Provisions for *special inspections*.

N104.1.3 Energy conservation details. Where approval of the energy conservation requirements of the replicable design is sought, the submittal documents shall include details for each climate zone for which approval is sought; and shall include the following:

1. Climate zones for which approval is sought.
2. Building envelope details.
3. Building mechanical system details.
4. Details of electrical power and lighting systems.
5. Provisions for system commissioning.

SECTION N105 REVIEW AND APPROVAL OF REPLICABLE DESIGN

N105.1 General. Proposed replicable designs shall be reviewed by an approved agency. The review shall be applicable only to the replicable design features submitted in accordance with Section N104. The review shall determine compliance with this code and additional codes specified in Section N104.1.

N105.2 Documentation. The results of the review shall be documented indicating compliance with the code requirements.

N105.3 Deficiencies. Where the review of the submitted construction documents identifies elements where the design is deficient and will not comply with the applicable code requirements, the approved agency shall notify the proponent of the replicable design, in writing, of the specific areas of noncompliance and request correction.

N105.4 Approval. Where the review of the submitted construction documents determines that the design is in compliance with the codes designated in Section N104.1, and where deficiencies identified in Section N105.3 have been corrected the approved agency shall issue a summary report of Approved Replicable Design. The summary report shall include any limitations on the approved replicable design including, but not limited to climate zones, wind regions and *seismic design categories*.

SECTION N106 SITE-SPECIFIC APPLICATION OF APPROVED REPLICABLE DESIGN

N106.1 General. Where site-specific application of a replicable design that has been approved under the provisions of Section N105 is sought, the construction documents submitted to the building official shall comply with this section.

N106.2 Submittal documents. A summary description of the replicable design and related construction document shall be submitted. Construction documents shall be signed, sealed and dated by the registered design professional. A statement, signed, sealed and dated by the registered design professional, that the replicable design submitted for local review is the same as the replicable design reviewed by the approved agency, shall be submitted.

N106.2.1 Architectural plans and specifications. Architectural plans and specifications shall include the following:

1. Construction documents for variations from the replicable design.
2. Construction for portions that are not part of the replicable design.
3. Documents for local requirements as identified by the building official.
4. Construction documents detailing the foundation system.

SECTION N107 SITE-SPECIFIC REVIEW AND APPROVAL OF REPLICABLE DESIGN

N107.1 General. Proposed site-specific application of replicable design shall be submitted to the building official in accordance with the provisions of Chapter 1 and Appendix N.

N107.2 Site-specific review and approval of replicable design. The building official shall verify that the replicable design submitted for site-specific application is the same as the approved replicable design reviewed by the approved agency. In addition, the building official shall review the following for code compliance.

1. Construction documents for variations from the replicable design.
2. Construction for portions of the building that are not part of the replicable design.
3. Documents for local requirements as identified by the building official.

[A] APPENDIX O

PERFORMANCE-BASED APPLICATION

The provisions contained in this appendix are not mandatory unless specifically referenced in the adopting ordinance.

User notes:

About this appendix: *Appendix O provides an optional design, review and approval framework for use by the building official. Typical uses would include cases of alternate methods in Chapter 1, select areas of the code that require a rational analysis such as Section 909 and elsewhere. It simply extracts the relevant administrative provisions from the ICC Performance Code into a more concise, usable appendix format for a jurisdiction confronted with such a need. Currently there are multiple, varying jurisdictional rules and procedures in many communities regarding procedure and none in even more. The building official is often left alone to reach decisions not just on the merits of a design, but must first also decide on the submittal and review process. As an appendix, the provisions herein are entirely optional to a jurisdiction. This appendix can be adopted, adopted with local modifications, or even used on a case-by-case basis as part of a Memorandum of Understanding or similar legal agreement between the jurisdiction and the owner/design team. It simply represents another tool for the jurisdiction to reach for in cases of need; it neither encourages nor creates any additional opportunity for performance-based design.*

Code development reminder: *Code change proposals to this appendix will be considered by the Administrative Code Development Committee during the 2022 (Group B) Code Development Cycle.*

[A] SECTION O101 GENERAL

O101.1 Introduction. The following administrative provisions are excerpted from the *ICC Performance Code for Buildings and Facilities* and can be used in conjunction with the Alternate Methods provisions in Chapter 1, or for a review of submittals requiring a rational analysis or performance-based design. These provisions provide an established framework for the building official in terms of the design expertise needed, the necessary submittals, a review framework and related items.

O101.2 Qualifications. Registered design professionals shall possess the knowledge, skills and abilities necessary to demonstrate compliance with this code.

O101.3 Construction document preparation. Construction documents required by this code shall be prepared in adequate detail and submitted for review and approval in accordance with Section 107.

O101.3.1 Review. Construction documents submitted in accordance with this code shall be reviewed for code compliance with the appropriate code provisions in accordance with Section 107.

O101.4 Construction. Construction shall comply with the approved construction documents submitted in accordance with this code, and shall be verified and approved to demonstrate compliance with this code.

O101.4.1 Facility operating policies and procedures. Policies, operations, training and procedures shall comply with approved documents submitted in accordance with this code, and shall be verified and approved to demonstrate compliance with this code.

O101.4.2 Maintenance. Maintenance of the performance-based design shall be ensured throughout the life of the building or portion thereof.

O101.4.3 Changes. The owner or the owner's authorized agent shall be responsible to ensure that any change to the facility, process, or system does not increase the hazard level beyond that originally designed without approval and that changes shall be documented in accordance with the code.

O101.5 Documentation. The registered design professional shall prepare appropriate documentation for the project, clearly detailing the approach and rationale for the design submittal, the construction and the future use of the building, facility or process.

O101.5.1 Reports and manuals. The design report shall document the steps taken in the design analysis, clearly identifying the criteria, parameters, inputs, assumptions, sensitivities and limitations involved in the analysis. The design report shall clearly identify bounding conditions, assumptions and sensitivities that clarify the expected uses and limitations of the performance analysis. This report shall verify that the design approach is in compliance with the applicable codes and acceptable methods and shall be submitted for concurrence by the building official prior to the construction documents being completed. The report shall document the design features to be incorporated based on the analysis.

The design report shall address the following:

1. Project scope.
2. Goals and objectives.
3. Performance criteria.
4. Hazard scenarios.
5. Design fire loads and hazards.
6. Final design.
7. Evaluation.
8. Bounding conditions and critical design assumptions.
9. Critical design features.
10. System design and operational requirements.
11. Operational and maintenance requirements.

12. Commissioning testing requirements and acceptance criteria.
13. Frequency of certificate renewal.
14. Supporting documents and references.
15. Preliminary site and floor plans.

O101.5.2 Design submittal. Applicable construction documents shall be submitted to the building official for review. The documents shall be submitted in accordance with the jurisdiction's procedures and in sufficient detail to obtain appropriate permits.

O101.6 Review. Construction documents submitted in accordance with this code shall be reviewed for code compliance with the appropriate code provisions.

O101.6.1 Peer review. The owner or the owner's authorized agent shall be responsible for retaining and furnishing the services of a registered design professional or recognized expert, who will perform as a peer reviewer, where required and approved by the building official.

O101.6.2 Costs. The costs of special services, including contract review, where required by the building official, shall be borne by the owner or the owner's authorized agent.

O101.7 Permits. Prior to the start of construction, appropriate permits shall be obtained in accordance with the jurisdiction's procedures and applicable codes.

O101.8 Verification of compliance. Upon completion of the project, documentation shall be prepared that verifies performance and prescriptive code provisions have been met. Where required by the building official, the registered design professional shall file a report that verifies bounding conditions are met.

O101.9 Extent of documentation. Approved construction documents, the operations and maintenance manual, inspection and testing records, and certificates of occupancy with conditions shall be included in the project documentation of the building official's records.

O101.10 Analysis of change. The registered design professional shall evaluate the existing building, facilities, premises, processes, and contents, and the applicable documentation of the proposed change as it affects portions of the building, facility, premises, processes and contents that were previously designed for compliance under a performance-based code. Prior to any change that was not documented in a previously approved design, the registered design professional shall examine the applicable design documents, bounding conditions, operation and maintenance manuals, and deed restrictions.

INDEX

A

B

C

D

E

F

G

H

I

M

N

O

P

R

S

T

U

V

W

Y

RESOURCE A

RECOMMENDED PRACTICES FOR REMOTE VIRTUAL INSPECTIONS (RVI)

User note:

About this resource: *The typical process of inspecting projects by inspectors driving to job sites and performing on-site inspections has certain challenges that impact timeliness and resource efficiencies both for building construction and safety industry and regulating jurisdictions. The time spent driving to job sites, particularly in larger cities with busy traffic patterns, takes up a substantial part of the day, reducing the number of inspections possible to complete and creating a backlog of requested inspections.*

To address some of the challenges, many jurisdictions have implemented remote virtual inspections (RVI) for more routine and simpler inspections such as water heater replacements or other, similar items. RVI is an alternative to on-site inspections using a video call with the inspector. With advances in technology and availability of sophisticated smart phones and tablets, RVI have become more common, and some jurisdictions plan to implement them for more complicated and larger inspection items or projects.

To assist the building construction industry and member jurisdictions in adoption of an RVI program, in May 2020, the International Code Council® (ICC®) published Recommended Practices for Remote Virtual Inspections (RVI). *This publication offers a comprehensive framework for both local jurisdictions and building industry professionals that desire to implement a remote inspection program.*

Recommended Practices for Remote Virtual Inspections (RVI)

ICC
INTERNATIONAL CODE COUNCIL®

Recommended Practices for Remote Virtual Inspections (RVI)

Second Printing: September 2020

ISBN: 978-1-952468-23-0

The information contained in this document is believed to be accurate; however, it is being provided for informational purposes only and is intended for use only as a reference for recommended practices. Publication of this document by the ICC should not be construed as the ICC engaging in or rendering engineering, legal or other professional services. Use of the information contained in this publication should not be considered by the user as a substitute for the advice of a registered professional engineer, attorney or other professional. If such advice is required, it should be sought through the services of a registered professional engineer, licensed attorney or other professional. This Recommended Practices document does not intend to overrule national building codes or local practices, but to provide a reference for minimum levels of quality to safeguard health, property and general welfare.

PRINTED IN THE USA

Table of Contents

Preface

Technological advances have created enormous possibilities in all aspects of life, including the building construction and safety industry. Digital and online tools for building design, construction and administrative functions, such as permit application, plan review, inspection and commissioning, have drastically increased the efficiency and accuracy of achieving safe and resilient communities. Local, state and national governments have taken advantage of advancing technologies and have incorporated various levels of digitization into their processes in order to save time and reduce costs. Examples of such efforts include online offering of permit applications, payment of permit fees, submittal of plans and digital plan review.

The speed of adoption and implementation of technology, however, varies by geographic region and depends on a number of factors, including the availability of financial resources and the infrastructure needed to support the technology. Many Authorities Having Jurisdiction (AHJs) have implemented technology at various levels with good success and have embraced greater reliance on digitization as time goes by.

The 2020 global coronavirus pandemic created an impetus in speeding the implementation of modern technologies and taking advantage of new ideas in a much shorter time frame. The spread of COVID-19 and the closing of most businesses and social activities in many parts of the world to create social distancing resulted in many sectors of the economy searching to find new solutions for conducting business.

Many AHJs needed to come up with solutions to perform all aspects of codes and standards administration from remote locations and/or home offices. One such solution using available technology is Remote Virtual Inspections (RVI).

RVI is a method of inspection that allows the needed inspections to proceed in a timely manner by the owner or contractor located on the jobsite and the inspector or inspection teams performing the inspection remotely. While this practice gained good acceptance and implementation during the weeks and months of COVID-19 social distancing, its advantages are so great that it will likely become a popular and routine tool for the foreseeable future.

The advantages and opportunities created by RVI locally, nationally and globally are enormous, allowing those with technical expertise in their specific subjects to offer their services across the globe. Building code specialists, inspectors and consultants will be able to provide services and consulting from far distances and to help building safety and resiliency anywhere needed at the local, national or global level.

Recommended Practices for Remote Virtual Inspections (RVI) was developed based on study, research, and discussions related to items that should be considered and addressed for an effective and consistent RVI program and to assist AHJs in implementing the readily available technologies in the adoption and implementation of their own RVI program.

ICC welcomes your comments and feedback to improve future editions of this Recommended Practices publication. Submit feedback at www.iccsafe.org/RVI.

About the International Code Council®

The International Code Council is a nonprofit association that provides a wide range of building safety solutions including product evaluation, accreditation, certification, codification and training. It develops model codes and standards used worldwide to construct safe, sustainable, affordable and resilient structures. The mission of the Code Council is to provide the highest quality codes, standards, products and services for all concerned with the safety and performance of the built environment. ICC Evaluation Service (ICC-ES) is the industry leader in performing technical evaluations for code compliance fostering safe and sustainable design and construction.

Washington DC Headquarters:
500 New Jersey Avenue, NW, 6th Floor
Washington, DC 20001-2070

Regional Offices:

- Eastern Regional Office (BIR)
- Central Regional Office (CH)
- Western Regional Office (LA)
- Distribution Center (Lenexa, KS)

888-ICC-SAFE (888-422-7233)
www.iccsafe.org

Family of Solutions:

1.0 Introduction

Hand-held devices such as smartphones and tablets have capabilities for real time, online communication of videos and photos. Use of advanced tools and technologies, combined with the power of such hand-held devices, has made it possible for anyone to observe the construction activities of a jobsite from any location, near or thousands of miles away. Using Remote Virtual Inspection (RVI) allows construction projects to continue without impediment and allows the Authority Having Jurisdiction (AHJ) to continue to provide the vital services needed for construction of safe buildings.

Purpose and Scope

The purpose and scope of these Recommended Practices is to provide guidance to the Authority Having Jurisdiction (AHJ) when implementing a Remote Virtual Inspection (RVI) program as well as to the construction industry user. This document specifically addresses implementation and administration of RVI. These procedures are organized in a fashion that can be readily implemented by the AHJ as part of their inspection procedures. This document also provides recommended practices to construction industry professionals submitting to an RVI.

Until recently, Remote Virtual Inspections have been conducted only by a few AHJs at varying levels. As a result, there has not been a standardized program that addresses how to prepare for, conduct and participate in these types of inspections.

2.0 Definitions and Acronyms

1. **RVI: Remote Virtual Inspection:** Remote Virtual Inspection, also known as RVI, is a form of visual inspection which uses visual or electronic aids to allow an inspector or team of inspectors to observe products and/or materials from a distance because the objects are inaccessible or are in dangerous environments, or whereby circumstances or conditions prevent an in-person inspection.
2. **AHJ: Authority Having Jurisdiction.**

3.0 Remote Virtual Inspection Process

Remote Virtual Inspections (RVI) may provide benefits to AHJs and customers alike. In certain circumstances, an RVI may provide a better quality inspection with an increase in efficiency and cost savings. It will increase the efficiency of the inspection process utilizing modern technology. Depending on the location and complexity of a project, some limitations may impact its use. In cases where an RVI is not suitable or technology fails to provide sufficient visual clarity (i.e., poor/no service or Wi-Fi, poor lighting, etc.), an onsite inspection may be required. Subject to local approval, the AHJ may choose to use an approved third-party inspection agency or utilize staff inspectors. Where Wi-Fi and/or cellular reception are poor or not available, some AHJs may consider allowing the contractor to provide an acceptable electronic documentation of the area that needs an inspection for review by the assigned inspector or team of inspectors.

A clear understanding of the RVI requirements and communication throughout the process by both parties is paramount to the completion of a successful inspection. The inspector will check all aspects of the permitted construction project to the adopted codes and other applicable laws and regulations no differently than if it were an onsite inspection. Identification of the project jobsite location, posted address and its location within the building will be a critical part of the process.

The applicable Codes and Standards to be used for RVI are the same as the adopted codes and referenced standards of the AHJ. The implementation of the RVI is intended to achieve the same results as the typical in-person site inspection by applying the provisions of adopted codes such as the IBC®, IRC®, IPC®, IFC® and other applicable and adopted International Codes.

Customer: Requesting a Remote Virtual Inspection

1. Ensure there is an active permit issued or certificate application filed or obtain the appropriate one prior to attempting to schedule an inspection for the project in question.
2. Electronically sign a notice indicating that the permit holder of record or representative:
   - 2.1. Consents to the use of the remote inspections.
   - 2.2. Is responsible for their own safety during the remote inspection.
   - 2.3. Allows the complete use of the videos and photos of the remote inspection by the AHJ.
   - 2.4. Certifies they are making available the site and inspection items truthfully and to the best of their ability.
   - 2.5. Is responsible for compliance with all codes and standards applicable to the project.
   - 2.6. Acknowledges that participation in the remote inspection program is voluntary (if not a mandatory program within the AHJ's jurisdiction).
   - 2.7. Acknowledges that the decision to perform an RVI is at the sole discretion of the AHJ.

3. Prior to contacting the department to schedule the inspection, confirm that the minimum criteria for a remote inspection are met. See Appendix A for examples of qualified inspection activities.
   3.1. Note that some types of inspections may be too complex or otherwise not compatible for remote inspections.
4. Call or go online to schedule an appointment for inspection with the AHJ.
5. Must be at least 18 years old or with an adult to perform the virtual inspection.
6. When scheduling the inspection, provide the address, permit number, and type and number of requested inspections.

AHJ: Scheduling Remote Virtual Inspection

1. Schedule Inspection Time.
   1.1. All remote inspections scheduled should be requested by the customer a minimum of one business day prior to the desired inspection date.
   1.2. Schedule inspection either online or by telephone.
   1.3. Schedule sufficient time for the type of inspection requested.
   1.4. AHJ to send an inspection confirmation email or text to the customer with the date, approximate time of RVI and name of inspector.
   1.5. Send notice of customer consent and acknowledgment for electronic signature. Must be returned by customer prior to inspection scheduled time.
2. Time slots for inspections.
   2.1. Anticipated length of inspections per type (i.e., water heater installation, HVAC replacement, etc.) needs to be established.
   2.2. Each customer will be given an approximate time window for inspection.
3. Post the earliest available time for remote inspections and the latest time of the day a remote inspection may be scheduled Monday through Friday or other days selected by the AHJ.
4. Schedule after-hours or emergency inspections on a case-by-case basis.
5. Determine the types of inspections allowed for remote inspections. See Appendix A for examples of qualified inspection activities.
   5.1. All inspections may qualify for an RVI, depending on the AHJ's resources and policies.
6. Determine which type of videotelephony is available for use and is compatible with the AHJ's permitting software and videotelephony equipment.
   6.1. Videotelephony platform examples: FaceTime, Google Duo, Zoom, WhatsApp, Skype, Tango, WebEx, Microsoft Teams, GoToMeeting, etc.

Customer: Prepare for Remote Virtual Inspection

1. Prior to the inspection, ensure that:
    1.1. The jobsite is safe at all times for the individual(s) using the device during the remote inspection including health safety.
    1.2. The device (smartphone, tablet, drone, etc.) is fully charged and has a suitably charged additional power supply (battery pack).
    1.3. The use of a noise-canceling headset is recommended.
    1.4. The jobsite has high-speed Wi-Fi connectivity or minimum 4G cellular service with a strong signal.
    1.5. The necessary tools based on type of inspection are readily available.
        1.5.1. For example, carry a flashlight, tape measure, level, step ladder (for close ups of ceiling), GFCI tester, etc. An extending pole for the video device, such as selfie pole, may be very helpful in taking the smartphone or other video device closer to the point of inspection in various places such as very high ceilings.
2. Have approved plans, permit card, and other necessary construction documents available onsite.
3. Make sure good lighting is available and clear the area of any unnecessary objects.
4. All features applicable to the required inspection must be visible at the time of the remote inspection. These features must be captured sufficiently and clearly for the inspector to evaluate.
5. If at any point the inspector believes that the remote inspection process is not allowing them to properly assess compliance, they may require that a site inspection be conducted at a future date or instruct the customer to make different arrangements.
    5.1. In areas within the jobsite where there is no Wi-Fi or cell service, at the sole discretion of the inspector, the contractor may be allowed to provide video and/or photographic documentation of the item(s) to be inspected for review by the authorized inspector at a later time.
6. The onsite inspection may be conducted by an approved third-party inspection agency or by the AHJ's inspection staff.

Customer: Prepare to Receive Remote Virtual Inspection Call

1. Ensure that the lens and screen of any device being used to capture images or video has been cleaned. Dust, grit, smudges, etc., might interfere with the image quality and distorting the inspector's view.
2. To minimize interruptions during the RVI and to ensure that the video feed will be uninterrupted, make sure that all notifications are turned off in the Settings of the mobile device used for the RVI. Should the video be interrupted, the inspection could be delayed or have to be rescheduled.
3. Be prepared to answer the inspector's call at any time during the scheduled timeframe. Be cooperative and closely follow the inspector's instructions.

4. As each site and inspection is different, allot the proper amount of time for the type of inspection and accessibility of the site.
5. Carefully follow the inspector's instructions for where to direct the device and for covering the site. Do not rush the inspector but allow him or her adequate time to conduct the RVI to his or her satisfaction.
6. As much as possible, minimize background noise as that can interfere with communication with the inspector.

What to Expect During the Inspection

1. Begin inspection at the street view looking at the structure with the address or other required jobsite identification in the video display.
   1.1. Inspector may also verify location through GPS/Geotagging where the service is available.
2. Follow the directions of the inspector with respect to the order and direction of inspection.
3. As the inspection progresses, write down any items that the inspector finds that need to be corrected. Be sure the notes are detailed and ask questions of or seek clarification from the inspector at the time of the RVI.
4. If provided a permit card, do not write on it. During the next in-person visit, the inspector should update it then.
5. In most cases, the inspector will relay the results of the inspection before the end of the RVI of passing, failing or not ready for inspection.
6. Do not cover any work needing corrections until corrections are verified by reinspection. Reinspection fees may apply in accordance with the AHJ's policies.
7. Note: At a minimum, there must be an adult of the required legal age on site who will represent the owner/representative during the entire duration of the RVI.
8. The owner/representative must be able to verbally communicate with the remote inspector at all times during the inspection.

Inspection Results

1. Results of the inspection will be entered into the AHJ's permit database as soon as practicable after the RVI is completed. It is important to note that the inspection was completed using the RVI process.
2. Where an approval tag for utility connections is required, the AHJ should work directly with the utility company.
3. Following the inspection:
   3.1. Inspection comments will be available on the AHJ's website, within the AHJ's normal timelines, indicating passing or failing with the list of corrections when applicable.
   3.2. In addition, the inspector may email the inspection information upon request to the customer as soon as inspection information is available.

   3.3. The inspector will determine whether additional fee(s) for reinspection is required.

4. Scheduling a reinspection or the next inspection needed is based on availability of time slots.
5. The authorized inspector may provide an option for the owner/representative to submit electronic documentation that a deficiency or deficiencies have been corrected.
6. It is incumbent on the owner/representative to provide the address and permit number on all submitted correspondence or communications.

Maintaining Records of Inspections

Required inspection records, including, but not limited to, correction notices, electronic media, recordings or photo documentation, shall be maintained in accordance with the AHJ's policy, laws, regulations, and applicable codes, and may be subject to disclosure.

4.0 Training and Communication

Training and effective communication of processes, procedures and requirements are essential and a critical part to the success of any program. This program is no different as it lends itself to new technology, new programs, and methods that are in many cases, new to the building construction and safety industry. Therefore, training of the AHJ's staff as well as the building industry on the various programs and procedures will save time and money and make the administrative and enforcement process a positive experience with minimal confusion. Training also leads to better communications between an AHJ and its customers.

Staff Training

1. Ensure all staff are trained in the appropriate areas of responsibility.
2. Permit Technicians:
   - 2.1. Review of approved permit applications relative to RVI requirements.
   - 2.2. Required departmental approvals are complete.
   - 2.3. Fee collection process.
   - 2.4. Required documents for the project (plans, calculations, etc.).
3. Remote Inspection Staff:
   - 3.1. Inspection software and hardware.
   - 3.2. Remote inspection procedures.
   - 3.3. Types of platforms used (Facetime, Skype, etc.).
   - 3.4. Reinspection fee procedures.
   - 3.5. Recording inspection results in permit tracking system.

Customer/Applicant

1. Ensure the owner and representative are trained in their areas of responsibility.
2. Permit applicant:
   2.1. Knowledge of the AHJ's departmental approvals required for the project.
   2.2. Knowledge of the AHJ's RVI protocol.
   2.3. Ensuring project meets RVI protocol.
   2.4. Ensure that the project is ready for the RVI at the scheduled time.
   2.5. Comply with the inspector's direction.
3. Owner/Contractor/Subcontractor:
   3.1. Requesting remote inspection process.
   3.2. Knowledge of remote inspections procedures.
   3.3. Platform required (Facetime, Skype, Google Duo, etc.).
   3.4. Jobsite communication requirements (Wi-Fi, 4G, etc.).
   3.5. Communication skills.

Additional Considerations

1. Adopt basic online security practices. Consult with your IT department for guidance.
2. Consult with your legal counsel to ensure compliance with all federal, state and local requirements related to your RVI program. For example, you may want to consult counsel to find out whether a homeowner's release is needed to conduct an RVI.
3. Ensure that all staff have access to the codes and standards that are applicable to what they are inspecting. The Code Council's Digital Codes Library (https://codes.iccsafe.org/) offers online access to all ICC model codes and standards and most state codes.
4. Document lessons learned to improve your RVI program and to support potential long-term establishment of virtual inspection processes.

5.0 Appendix A (Examples of Potential Activities)

The following are a few examples of construction activities that may be considered to be included in a RVI Program. This list is not all-inclusive. The determination of whether an inspection can be conducted remotely is at the sole discretion of the AHJ.

- Plumbing system repairs or fixture replacements.
- Construction trailer installations.
- Swimming pool excavations.
- Gas line repairs or gas utility clearance.
- Electric utility clearances.
- HVAC direct replacement or repair.
- Minor residential electrical.
- Miscellaneous repair/exterior repair or upgrades (stucco, windows, etc.).
- Re-roofing/roof covering replacement.
- Water heater or water softener direct replacement.
- New residential plumbing rough-in.
- New residential rough framing inspections.
- Residential rooftop-mounted photovoltaic panel systems.
- HUD manufactured home installation verification.
- Any other inspection approved by the AHJ.

ICC INTERNATIONAL CODE COUNCIL®

KEYS TO SUCCESS FOR REMOTE VIRTUAL INSPECTIONS

The Value of Communication

WHAT ARE REMOTE VIRTUAL INSPECTIONS (RVI)?

Remote Virtual Inspections, also known as RVI, are a form of inspections which use visual or electronic aids to allow an inspector or team of inspectors to observe certain types of construction, products and/or materials from a distance.

RVI are a solution to help inspectors observe construction and objects that might be inaccessible or in dangerous environments, or whereby circumstances or conditions prevent an in-person inspection.

RVI BENEFITS

- Construction projects can continue without impediment
- Building professionals can continue providing services with minimal health risk during pandemics such as COVID-19
- Authorities Having Jurisdiction (AHJs), testing agencies, manufacturers, laboratories, home builders and contractors are able to provide the vital services needed on all levels for the construction of safe buildings
- Inspectors can continue providing services remotely while saving time and money
- Safe and resilient construction projects can continue to grow and thrive anywhere needed at the local, national or global level

THE FOUNDATION OF SUCCESS: TRAINING AND COMMUNICATION

With the use of new technology, new programs, and methods, AHJs, testing agencies, manufacturers, laboratories, home builders and contractors can implement effective training of these programs to their staff to create a foundation for successful Remote Virtual Inspections.

A clear understanding of RVI requirements and communication throughout the process by all parties involved is paramount to the completion of successful inspections.

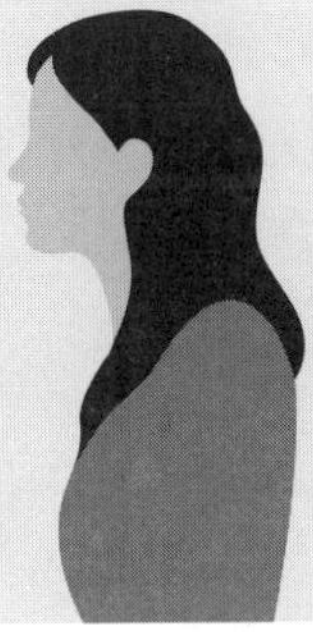

THE KEY STEPS TO A REMOTE VIRTUAL INSPECTION

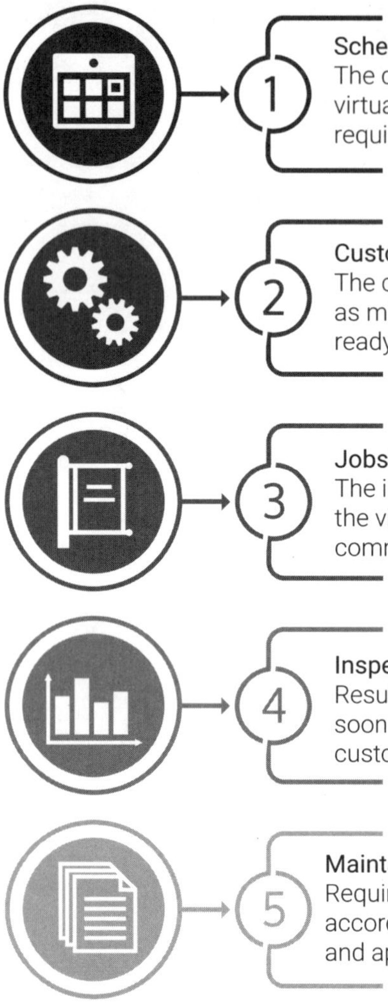

Scheduling
The customer works with the AHJ to schedule remote virtual inspections and ensure both parties have the required documents and equipment necessary

Customer Preparation
The customer prepares the jobsite and devices as well as minimizes disruptions for remote virtual inspection, ready to cooperate with the inspector when on the call

Jobsite Identification
The inspector makes sure the jobsite is identified on the video display and begins inspection, keeping close communication with the customer

Inspection Results
Results are entered into the AHJ's permit database as soon as RVI is completed and communicated with the customer

Maintaining Records of Inspections
Required inspection records are maintained in accordance with the AHJ's policies, laws, regulations, and applicable codes, and may be subject to disclosure

LEARN MORE

Read the full version of our Recommended Practices for Virtual Remote Inspections here.

Recommended Practices for Remote Virtual Inspections (RVI)

Recommended Practices for Remote Virtual Inspections (RVI) is the most complete source of information on remote inspections. RVI is an alternative to on-site inspections using a video call on a 4G or WiFi telephony (smartphone, tablet, etc.) in order to interact with the inspector. It is a comprehensive tool for local jurisdictions and the building industry alike that desire to implement a remote inspection program.

This publication covers the RVI process, inspection scheduling, preparation, what the owner/contractor should expect, training and communications, and recording and maintaining records. While all types of inspections may not be suitable for RVI, a list of potential construction activities suitable for remote inspections is provided.

RVI also lends itself to connect seamlessly as part of an overall online program that will allow jurisdictions to provide complete services to the public utilizing the latest technology. Online permitting and electronic plan review, together with remote virtual inspections, can provide a complete program that keeps the construction industry moving while providing a healthy environment for all participants.

ISBN 978-1-952468-23-0

9 781952 468230

Item No. 7072S1